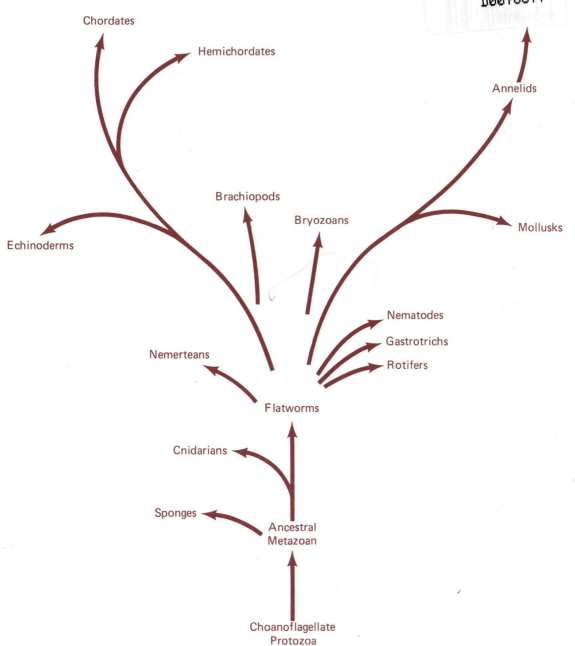

ENGELBRETA
BQ 215
Recommended
READINGS

PROTOZOANS 21-22, 26, 27, 29, 31 36-42 47-50 54-55 63-64
PORIFERA 73-80 85-88 102
CNIDARIA 103-104 108-110 113-118-122 127-131

CTENOPHORES 164-168
BILATERAL 175-174 181-187
PARASITISM 206-211
PLATYHELMINTHES 206-211-213 221-226 240-241 241-253
NEMATOMORPHA 304-310 316-316 323 325-328
PSEUDOMINTHES 281-283 287-288 290-292-255 300-303
NEMERTEANS 264-269
CRYPTOBIOSIS 280-291

KINORHYNCHS

INVERTEBRATE ZOOLOGY

SIXTH EDITION

Edward E. Ruppert

Clemson University, South Carolina

Robert D. Barnes

Gettysburg College, Pennsylvania

Saunders College Publishing
Harcourt Brace College Publishers
Fort Worth   Philadelphia   San Diego
New York   Orlando   San Antonio
Toronto   Montreal   London   Sydney   Tokyo

Text Typeface: Times Roman
Compositor: University Graphics, Inc.
Acquisitions Editor: Julie Levin Alexander
Developmental Editor: Christine Connelly
Managing Editor: Carol Field
Project Editor: Nancy Lubars
Copy Editor: Linda Davoli
Manager of Art and Design: Carol Bleistine
Art Director: Anne Muldrow, Susan Blaker
Text Designer: Julie Anderson
Cover Designer: Vincent Ceci
Text Artwork: John Norton/Vantage
Director of EDP: Tim Frelick
Production Manager: Joanne Cassetti
Cover Credit: Flip Nicklin/Minden Pictures

Printed in the United States of America

INVERTEBRATE ZOOLOGY, Sixth Edition

ISBN 0-03-026668-8

Library of Congress Catalog Card Number: 93-085930

3 4 5 6 016 9 8 7 6 5 4 3 2 1

For Robert D. Barnes (1927–1993)
who taught invertebrate zoology to generations of biologists

PREFACE

The study of invertebrate zoology is an opportunity to survey the spectacular array of animal forms and functions and to apprehend the often subtle concepts that unify animal diversity. With these general goals in mind, and with attention to living invertebrates, we have emphasized the correlations between structure and function and the development of conceptual themes.

Special Features of the Sixth Edition

The sixth edition adopts several new features to enhance the study of invertebrates. These include

- Conceptual preambles called **Principles and Emerging Patterns**
- New chapters covering an introduction to bilaterally symmetrical animals (5); the evolution of invertebrates (20); the protochordates, including the cephalochordates, and the chaetognaths (17)
- Extensive revision or reorganization of the chapters covering small-bodied or lesser-known invertebrates and new factual information added to all of the chapters
- Over 300 original new illustrations
- A new glossary

Principles and Emerging Patterns

In a systematic approach to the study of invertebrates, the new **Principles and Emerging Patterns** sections are designed to identify the evolutionary emergence of structural patterns, such as multicellularity, epithelial tissues, and extracellular compartmentation, and then to explain their functional significances. We have also used these sections to introduce functional principles from fields ranging from cell and developmental biology to ecology and evolution. Our guiding assumption is that principles and patterns are best learned when introduced sequentially and in conjunction with a growing knowledge of factual material. Thus, each chapter is preceded by at least one **Principles and Emerging Patterns** section, some by several, and the placement of a given section was dictated either by where the pattern first emerged or by a group of animals that illustrate a principle particularly well. In the interest of distributing these sections across all of the chapters and not encumbering any one chapter with too many sections, however, some sections that might have been developed earlier in the book have been placed later. For example, although sponges and cnidarians make wide use of chemical defenses, we postponed the general discussion of this topic to the chapter on insects, thus allowing for the development of more general themes in the early chapters. Nevertheless, the reader is free to use the **Principles and Emerging Patterns** sections selectively and in any order.

New Chapters

Chapters 5 and 20 are entire **Principles and Emerging Patterns** sections. Chapter 5, *Bilateral Animals,* identifies the structural patterns and functional principles that are common to all or most bilaterally symmetrical animals and is, therefore, a conceptual introduction to the remainder of the book. Chapter 20, *Invertebrate Evolution,* is an exercise in the construction and testing of evolutionary patterns that builds on the students' knowledge of facts and principles. It

v

provides a step-by-step analysis of patterns in invertebrate phylogeny with an emphasis on one of the central issues, the evolutionary origin of bilaterally symmetrical animals. The chapter also provides an up-to-date summary of discoveries resulting from the application of the techniques of electron microscopy and molecular biology to invertebrate systematics.

Chapter 17, *Protochordates and Chaetognaths,* includes a new section on cephalochordates and expanded coverage of the hemichordates and urochordates. The increased depth of coverage in this chapter was prompted by the the recent revival of interest in the evolutionary origin of vertebrates, an event for which the protochordates provide an ample number of facts, and the need to modernize the discussion of these taxa.

Chapter Revisions and Reorganizations

A number of chapters have been revised or reorganized in this edition to reflect new ideas on phylogeny or to underscore the absence of them. Thus, the molluscs once again precede the annelids and the molluscs are preceded by a reorganized chapter on non-segmented protostomes. Pogonophorans are discussed with the annelids and pentastomids with the arthropods. Mesozoans (Orthonectida and Rhombozoa) have been assigned to the chapter covering platyhelminths, and tardigrades to the chapter on aschelminths as a matter of pedagogical convenience, but the student is made aware of their likely phylogenetic placements. We have used a system of classification that is up to date and generally accepted, but in some cases, have allowed pedagogy to be the determining factor. For example, it is simpler to discuss Cubomedusae with the scyphozoans than as a separate class, and protobranchs, lamellibranchs, and septibranchs, though without formal taxonomic status, are still useful reference groups of bivalves. Nevertheless, the student is made aware of current thinking on the classification of these animals. The chapter on coral reefs, which stood alone in the Fifth Edition, has been shortened and incorporated into the chapter on cnidarians.

New Illustrations

The new artwork was developed in collaboration with artist and biologist, John Norton. Together, we spent many hours designing and executing new conceptual drawings as well as drawings that depict new structural information and patterns of movement and de-

velopment. Many of the new drawings are derived from our personal observations and those of our students.

Glossary

Most of the boldfaced terms in the text are defined in a new separate glossary at the end of the book.

Pedagogy

Although the terminology used in the Fifth Edition has not been significantly altered, a few changes have been made for the sake of consistency or in conformity with modern usage. The term septum, for example, is now used consistently throughout the coverage of anthozoans and replaces the alternate use of septum and mesentery for the same (homologous) structure in different organisms. Similarly, we employ the term gastrodermis not only for the epithelial lining of the coelenteron of cnidarians and ctenophores but also for the midgut lining of bilaterally symmetrical animals. An additional change has been made in reference to the common names of many gastropod and bivalve molluscs in an effort to promote conservation. Instead of referring to the living animals as "shells" (e.g., top shells, jingle shells, olive shells), we now refer to them using the terms snails and clams. These changes may at first seem awkward to those already acquainted with the molluscs, but they soon become familiar and help to shift the emphasis from shells as "collectors items" to snails and clams as living organisms.

Important common terms and terms that apply to characteristic structures and functions have been set in boldface type in this edition to highlight their importance and to facilitate reading and recollection. Most of these terms have been added to the glossary.

Systematic resumés follow the sections covering all but the smallest groups of invertebrates. Each resumé includes the formal classification of the taxonomic group in question and the definition of each taxon.

We continue to believe that this text should provide the student entry into the vast literature on invertebrates. The references at the end of each chapter therefore include most books and review articles on invertebrates, as well as many research papers. Some of these would be appropriate for further student reading if the instructor wished to designate them. The listings for all chapters have been subdivided by major taxonomic groups to make them easier to use.

Acknowledgements

We are grateful to the many people who, over the past seven years, have voluntarily provided corrections, information, and suggestions. They have contributed much to the improvement of this edition. We wish to especially aknowledge those who reviewed chapters or sections of chapters dealing with particular invertebrate groups or topics. They are not responsible for any errors or oversights that may remain. These reviewers were Mary Beverly-Burton, University of Guelph; Brian Bingham, Western Washington University, Shannon Point Marine Center; Kenneth Jay Boss, Harvard University; Steven Burian, Southern Connecticut State University; James M. Colacino, Clemson University; Ronald Dimock, Wake Forest University; Darryl Felder, University of Southwestern Louisiana; Thomas Fenchel, University of Copenhagen; Stephen L. Gardiner, Bryn Mawr College; Rick Grosberg, University of California, Davis; Frederick W. Harrison, Western Carolina University; David G. Heckel, Clemson University; Robert P. Higgins, Smithsonian Institution; Daniel Hornbach, Macalester College; Donald Kangas, Northeast Missouri State University; Roni Kingsley, University of Richmond; Charles Lambert, California State University-Fullerton; Roger Lloyd, Florida Community College-Jacksonville; Diane R. Nelson, East Tennessee State University; Gayle P. Noblet, Clemson University; Sidney Pierce, University of Maryland; Douglas Ruby, Drexel University; Eugene H. Schmitz, University of Arkansas; Erik Scully, Towson State University; George Shinn, Northeast Missouri State University; Stephen Shuster, Northern Arizona University; Stephen Stricker, University of New Mexico; James M. Turbeville, Indiana University; Seth Tyler, University of Maine; Elizabeth Waldorf, Mississippi Gulf Coast Community College-Jefferson Davis Campus; A. P. Wheeler, Clemson University; W. Herbert Wilson, Colby College; John P. Wourms, Clemson University.

We are grateful for the editorial assistance of Jennifer Frick (Clemson University, Western Carolina University) and Janice Diehl (Gettysburg College). We also wish to thank the staff at Saunders College Publishing and especially Julie Levin Alexander, Christine Connelly, and Nancy Lubars for their help in bringing the manuscript to final form. Parts of this revision were written at the Smithsonian Marine Station at Link Port (Mary E. Rice, Director) and the Darling Marine Center of the University of Maine (Kevin J. Eckelbarger, Director), and we thank the Directors of both institutions for their help.

One of us (EER) would like to thank his current and former graduate students for their contributions: Marianne Klauser Litvaitis, Susan M. Lester, C. Kathleen Heinsohn, Karen Carle Meyers, Elizabeth J. Balser, James M. Turbeville, Yun-Tao Ma, Edwin W. Edmondson, Deborah Y. Pinson, Christine A. Byrum, Peter R. Smith, Elizabeth Messenger Thomas, Penelope Travis, Karen McGlothlin, Kimberly Mathis Purcell, and Laura Corley.

Robert D. Barnes worked diligently on the revision leading up to this edition. Always an advocate of concepts that would help students to rationalize the diversity of invertebrates, he was enthusiastic about the **Principles and Emerging Patterns** approach, and he wrote many of the sections. He continued to revise the text and to plan for the next edition of *Invertebrate Zoology* up to his sudden and untimely death.

Edward E. Ruppert
December 1993

TABLE OF CONTENTS

INTRODUCTION

There are over a million described species of animals. Of this number about 5 percent possess a backbone and are known as vertebrates (see the figure on the inside front cover). All others, constituting the greater part of the Animal Kingdom, are invertebrates. These animals are the subject of this book.

Division of the Animal Kingdom into vertebrates and invertebrates is artificial and reflects a historical human bias in favor of humankind's own relatives. This one characteristic of a single subphylum of animals is used as the basis for separating the entire Animal Kingdom into two groups. One could just as logically divide animals into molluscs and nonmolluscs or arthropods and nonarthropods. The latter classification could be supported at least from the standpoint of numbers, because approximately 85 percent of all animals are arthropods (see the figure on the inside front cover).

The artificiality of the vertbrate/invertebrate concept is especially apparent when one considers the vast and heterogeneous assemblage of groups that are lumped together in this category. Invertebrates do not hold a single positive characteristic in common, aside from general animal features also shared with vertebrates. The range in size, in structural diversity, and in adaptations to different modes of existence is enormous. Some invertebrates have common phylogenetic origins; others are only remotely related. Some are much more closely related to the vertebrates than to other invertebrate groups.

Quite obviously, invertebrate zoology cannot be considered a special field of zoology, certainly not in the same sense as protozoology or entomology. A field that embraces all biological aspects—morphology, physiology, embryology, and ecology—of 95 percent of the Animal Kingdom represents no distinct area of zoology itself. For the same reason, no zoologist can truly be called an invertebrate zoologist. He or she is a protozoologist, a malacologist, or an acarologist; or studies some aspect of physiology, embryology, or ecology of one or more animal groups. Beyond such limited areas the number and diversity of invertebrates are too great to permit much more than a good general knowledge of the major groups.

In the following chapters there will be continual references to the many sorts of environments in which invertebrates are found. Because some may be unfamiliar, we briefly describe the more common ones here.

THE EARTH'S MAJOR AQUATIC ENVIRONMENTS

The Marine Environment

The Animal Kingdom is generally believed to have originated in Archeozoic oceans long before the first fossil record. Every major phylum of animals has at least some marine representatives; some groups, such as cnidarians and echinoderms, are largely or entirely marine. From the ancestral marine environment, different groups of animals have invaded fresh water; some have moved onto land.

Compared with fresh water and with land, the marine environment is relatively uniform. Oxygen is generally available, and the salinity of the open ocean is relatively constant, ranging from 34 to 36 parts per thousand (3.4–3.6 percent), depending on the latitude. Light and temperature vary greatly, however, largely as a consequence of depth. Thus, life is not uniformly distributed throughout the depth and breadth of the world's oceans, which cover approximately 71% of the earth's surface. The margins of the continents extend seaward in the form of underwater **shelves** to depths of 150 to 200 m and then slope more steeply to depths of 3000 m or more. Before reaching the ocean floor, the **continental slope** is interrupted by a terrace, or more gradual incline, formed by the **continental rise.** The floor of the ocean basins, called the **abyssal plain,** ranges from 3000 to 5000 m in depth and may be marked by such features as **sea mounts, ridges,** and **trenches.** Widths of the different continental shelves vary considerably. The edge of the western Atlantic shelf is some 75 mi from the shore, but along the Pacific coast of North America the continental shelf is very narrow.

Waters over the continental shelves constitute the **neritic zone,** and those beyond the shelf make up the **oceanic zone** (Fig. 1–1). The edge of the sea, which rises and falls with the tide, is the **intertidal (littoral) zone.** The region above is the **supratidal (suppralittoral),** and that below is the **subtidal (sublittoral).** The continental slopes form the **bathyal zone,** the abyssal plains form the **abyssal zone,** and the trenches form the **hadal zone.**

Vertical distribution of marine organisms is largely controlled by the depth of light penetration. Light sufficient for photosynthesis to exceed respiration penetrates to only a short distance below the sur-

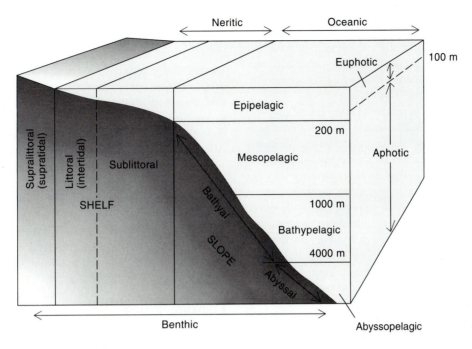

FIGURE 1–1 Marine environments.

face or to depths as great as 200 m, depending on the turbidity of the water. Below this upper **euphotic zone** is a transition zone where some photosynthesis can occur, but the production rate is less than the loss through respiration. From the transition zone down to the ocean floor, total darkness prevails. This region constitutes the **aphotic zone.** The animals that are permanent inhabitants of the aphotic and transition zones are carnivorous, suspension, or detritus feeders, and most depend indirectly on the photosynthetic activity of the microscopic algae in the upper, lighted regions.

The suspended or swimming animals of ocean waters constitute the **pelagic fauna,** and those that live on the bottom compose the **benthic fauna.** Bottom dwellers may live on the surface (**epifauna**) or beneath the surface (**infauna**) of the ocean floor and usually strikingly reflect the character of the substratum, that is, whether it is a hard bottom of coral or rock, or a soft bottom of sand or mud. Many animals are adapted for living in the spaces between sand grains and compose what is commonly referred to as the **interstitial fauna,** or **meiofauna.** This group includes representatives of virtually every major phylum of animals, and a number of previously unknown

groups of animals have been discovered here in recent years. Pelagic and benthic animals are found in all of the horizontal zones. For example, one can refer to neritic pelagic animals or to the infauna of the abyssal zone.

Many intertidal and shallow-water habitats contain a rich diversity of species. They also contain the best studied invertebrates, since these habitats are the most accessible to investigators. Surf beaches and rocky shorelines are certainly the most familiar shallow-water marine habitats, but there are other rich areas for collecting and study. In protected bays and sounds, there are usually shallow expanses of sand and mud that are exposed at low tide but covered at high water. Many marine animals live on such **sand** or **mud** flats, usually buried but often with evidence of occupancy on the surface above, such as burrow openings.

Sea grass beds of subtidal shallow water are also habitats for many invertebrates. Sea grasses, such as eel grass, turtle grass, manatee grass, and others of both temperate and tropical seas, are flowering plants adapted for a life of total submersion in sea water. The long leaves are flat or stringlike, and the plants are anchored to the bottom by roots.

Freshwater and Estuarine Environments

The lakes of the world also exhibit a horizontal and vertical zonation, but their smaller size, shallower depth, and freshwater content make them ecologically different in many ways from oceans.

Temperature is a primary factor controlling the environment of lakes. In contrast to salt water, which becomes increasingly dense at decreasing temperatures, fresh water reaches its greatest density at 4°C; thus, when lakes in temperate parts of the world are warmed during spring and summer, the warm water stays at the surface, and the heavier, colder water remains at the bottom. Little circulation occurs between the upper and lower levels, so that not only is the bottom zone dark, but it is also relatively stagnant from lack of oxygen and supports only a limited fauna. A general circulation occurs in fall and spring with temperature changes.

Tropical lakes either have a single winter turnover or exhibit a highly stable condition, with little vertical circulation. If deep and stable, the bottom layers of water are anaerobic.

The junction of freshwater rivers and streams with the sea is not abrupt. Rather, the two environments grade into one another, creating an **estuarine** environment, characterized by brackish water, that is, salinities are considerably below the 3.5% typical of the open sea. The estuarine environment embraces river mouths and surrounding deltas, coastal marshes, small embayments, and the finger-like extensions of the sea that probe the coast or margins of sounds. It is usually affected by tides, from which the word **estuary** (aestus, tide) is derived. The majority of animals living in the open ocean are osmoconformers and stenohaline and cannot survive greatly reduced salinities. The lower and fluctuating salinities of estuaries thus restrict the estuarine fauna to those euryhaline marine invaders and the few freshwater species that can tolerate these conditions. The fauna also contains some animals that have become especially adapted for estuarine conditions and are found nowhere else.

In temperate regions, a characteristic estuarine community is the **salt marsh,** composed principally of various grasses and sedges. Salt marshes differ from sea grass beds in being intertidal and emergent. Only the lower half of the plant is covered at high tide. Along the east coast of the United States, cord grass (*Spartina*) forms great expanses of salt marsh where the salinity is not too low.

In the tropics, the ecological counterpart of salt marshes is **mangroves.** Mangroves refers to species of small trees that can tolerate saline conditions. They occupy the intertidal zone and commonly possess prop roots or special aerial roots (pneumatophores) that project above the water's surface. The most highly developed mangrove communities are found in the Indo-Pacific, where numerous species form a number of zones extending seaward. Such mangroves may occupy vast coastal areas and are virtually impenetrable. Red mangrove, *Rhizophora mangle,* which possesses long prop roots extending straight downward from the limbs, is the common mangrove of tropical America (Fig. 1–2). Mangroves trap sediment, thereby contributing to land building. They create a habitat that is occupied by many animals and other plants.

Plankton, Primary Production, and Food Chains

Both oceans and freshwater lakes contain a large assemblage of microscopic organisms that are free swimming or suspended in the water. These organisms constitute the plankton and include both plants (**phytoplankton**) and animals (**zooplankton**). Although many planktonic organisms are capable of locomotion, they are too small to move independently of currents. Phytoplankton is composed of enormous numbers of diatoms and other microscopic algae. Marine zooplankton includes representatives from virtually every group of animals, either as adults or as developmental stages. Some species (**holoplankton**) spend their entire lives in the plankton; the larvae of others (**meroplankton**) enter and leave the plankton at different points in the course of their development. The animal constituents of freshwater plankton are more limited in number. Plankton, especially marine plankton, is of primary importance in the aquatic food chain. The photosynthetic phytoplankton—chiefly diatoms, dinoflagellates, minute flagellates, and cyanobacteria—form the primary trophic level and serve as food for larger animals. As would be expected, plankton attains its greatest density in the upper, lighted zone of waters with high nutrient levels (nitrates, phosphates, and so on). The inorganic nutrients are necessary in the synthesis of organic compounds by phytoplankton. In general, higher nutrient levels are found in shallow coastal waters, in areas of upwelling, and in the surface waters of cold and temper-

FIGURE 1–2   A mangrove at low tide. This is red mangrove, *Rhizophora mangle.* Note the bolsters of algae surrounding the prop roots. *(Photo courtesy of Betty M. Barnes.)*

ate seas, where mixing with deeper levels is not impeded.

Tropical and subtropical oceanic surface waters are generally impoverished because mixing with nutrient-rich deeper water is minimal. The surface water, which is warm, and thus, less dense, rides on top of the colder, and therefore, heavier water of deeper levels. Oceanic waters that have a low productivity, such as the Gulf Stream and the Sargasso Sea, are clear and blue. The low concentration of plankton allows light to penetrate to a considerable depth, and the blue wavelengths are reflected from the water molecules. Sea water that is rich in plankton is green or gray. Plankton and organic detritus reflect yellow wavelengths, which, combined with the blue wavelengths reflected by the water molecules, produce a green or gray color.

COPING WITH ANIMAL DIVERSITY

For the student who is attempting to study invertebrates in some depth for the first time, the task may seem overwhelming. Each group has certain structural peculiarities, in other words, a unique **groundplan,** or architectural design. Indeed, there are roughly some 30 different groundplans exhibited by multicel-

lular animals, and each has its own special anatomical terminology. Moreover, each of the 30 phyla of multicellular animals has a distinct classification, and some knowledge of that classification is necessary in order to discuss the diversity within the larger phyla. All of this tends to magnify the differences between groups and to obscure functional and structural similarities that result from similar modes of existence and similar environmental conditions, as well as to mask the homologies arising from close evolutionary relationships. An important way of coping with animal diversity is to understand underlying principles and patterns that are shared by numerous groups of animals, thereby enabling you to unite large assemblages of phyla and to make or even predict correlations between design, function, and environment.

To help you with these relationships, the section entitled **Principles and Emerging Patterns** which precedes each of the chapters in this text will discuss a few of the many structural, physiological, and developmental conditions that relate to the great diversity of animals. Some of these principles and patterns are encountered in many groups of animals; others are found in only a few. Some are **primary,** that is, they would have been found in the common ancestor of all of the groups in which the condition is present;

others are **secondary,** or **convergent,** meaning they have evolved independently in the groups in which they are present in response to similar environmental challenges. We place these topics before chapters for which the group to which the chapter is devoted provides especially good illustrations. However, to distribute the topics of these principles and emerging patterns across all of the chapters, they could not always be placed in chapters where they might first be illustrated. **Principles and Emerging Patterns** can greatly increase your understanding of the great diversity of animals with which most of each chapter deals. They can provide some important frameworks to which you can relate many aspects of invertebrate structure and function. At the same time, these opening sections are understandable if read independently of the rest of the chapters.

REFERENCES

The references at the end of each chapter are not intended to be a selection of titles recommended for further reading. The literature on invertebrates is enormous, as one would expect, considering the vast area of biology it covers. Most of this literature consists of research papers scattered through a great number of biological journals published throughout the world over the last 100 years. The references in this text may be placed in two categories. One category comprises research and review papers that focus on specific aspects of the biology of invertebrates, and the other category consists of larger reference works on the general biology or the systematics of the group of invertebrates with which the chapter deals. Those reference citations whose content is not clearly indicated by their titles have been provided with brief annotations. The papers, review articles, and books listed at the end of each chapter and the references that they provide in turn will lead the student to much of the literature available for any group of invertebrates.

The references below are composed of general works on invertebrates. Specific volumes especially relevant to certain groups are cited again in later chapters.

REFERENCES
Multivolume Works Covering Invertebrate Groups
Bronn, H. G. (Ed.): 1866– . Klassen und Ordnungen des Tierreichs. C. F. Winter, Leipzig and Heidelberg. (Many volumes; the series is still incomplete.)

Grassé, P. (Ed.): 1948– . Traité de Zoologie. Masson et Cie, Paris. (Covers the entire Animal Kingdom; still incomplete.)

Hyman, L. H., 1940–1967. The Invertebrates. McGraw-Hill Book Co., New York. (Six volumes. Volumes on annelids and arthropods were never completed.)

Kaestner, A., 1967–1970. Invertebrate Zoology. Interscience Publishers, New York. (Three volumes. Completed, although lophophorates and echinoderms are not included.)

Moore, R. C. (Ed.): 1952–   Treatise on Invertebrate Paleontology. Geological Society of America and University of Kansas Press, Kansas. (A detailed treatment of fossil invertebrates in many volumes.)

Parker, S. P. (Ed.): 1982. Synopsis and Classification of Living Organisms. McGraw-Hill Book Co., New York. (Vol. 1, 1166 pp.; Vol. 2, 1236 pp. A brief description of the families and all higher taxa of living organisms, including some information on the biology of the group. A very useful reference.)

Works on Morphology, Physiology, or Ecology of Invertebrates
Abbott, D. P., 1987. Observing Marine Invertebrates. In

Hilgard, G. H. (Ed.): Stanford University Press, Stanford, CA.

Alexander, R. M. 1982. Locomotion of Animals. Chapman and Hall, London. 192 pp.

Atema, J., Fay, R. R., Popper, A. N. and Tavolga, W. N. (Eds.): 1987. Sensory Biology of Aquatic Animals. Springer-Verlag, New York. 936 pp.

Autrum, H. (Ed.): 1980. Handbook of Sensory Physiology. Springer-Verlag, Berlin. (Eight volumes.)

Beklemishev, V. N. 1969. Principles of Comparative Anatomy of Invertebrates. University of Chicago Press, Chicago. (Two volumes.)

Bereiter-Hahn, J., Matoltsy, A. G., and Richards, K. S. (Eds.): 1984. Biology of the Integument. Springer-Verlag, Berlin. (Vol. 1, Invertebrates. 841 pp. The integument of each phylum of invertebrates is covered separately.)

Boardman, R. S., Cheetham, A. H., and Rowell, A. J. 1986. Fossil Invertebrates. Blackwell Scientific Publ., Boston. 713 pp.

Brusca, G. J., 1975. General Patterns of Invertebrate Development. Mad River Press, Eureka, CA. 134 pp.

Bullock, T. H., and Horridge, G. A. 1965. Structure and

Function of the Nervous System of Invertebrates. W. H. Freeman, San Francisco. (Two volumes.)

Carefoot, T. 1977. Pacific Seashores, A Guide to Intertidal Ecology. University of Washington Press, Seattle.

Chia, F., and Rice, M. E. (Eds.): 1978. Settlement and Metamorphosis of Marine Invertebrate Larvae. Elsevier North Holland, New York. 290 pp.

Clark, R. B. 1964. Dynamics in Metazoan Evolution: The Origin of the Coelom and Segments. Clarendon Press, Oxford. 313 pp.

Conway Morris, S., George, J. D., Gibson, R. et al (Eds.): 1985. The Origins and Relationships of Lower Invertebrates. Systematics Association Spec. Vol. No. 28. Clarendon Press, Oxford. 394 pp.

Crawford, C. S. 1981. Biology of Desert Invertebrates. Springer-Verlag, New York. 314 pp.

Daiber, F. C. 1982. Animals of the Tidal Marsh. Van Nostrand Reinhold, New York. 432 pp.

Dyer, J. C., and Schram, F. R. 1983. A Manual of Invertebrate Paleontology. Stipes, Champaign, IL.

Elder, H. Y., and Trueman, E. R. (Eds.): 1980. Aspects of Animal Movement. Cambridge University Press, New York. 250 pp.

Eltringham, S. K. 1971. Life in Mud and Sand. Crane, Russak, and Co., New York. 218 pp. An ecology of marine mud and sand habitats.

Fretter, V., and Graham, A. 1976. A Functional Anatomy of Invertebrates. Academic Press, London. 600 pp.

Gage, J. D., and Tyler, P.A. 1991. Deep-Sea Biology: A Natural History of Organisms at the Deep-Sea Floor. Cambridge University Press, Cambridge, U.K. 504 pp.

Giese, A. C., and Pearse, J. S. 1974–1991. Reproduction of Marine Invertebrates. Academic Press, New York. (Six volumes.)

Habermehl, G. G. 1981. Venomous Animals and Their Toxins. Springer-Verlag, Berlin. 195 pp.

Halstead, B. W. 1988. Poisonous and Venomous Marine Animals of the World. 3rd Edition. Darwin Press, Princeton. 1168 pp.

Hardy, A. C. 1956. The Open Sea. Houghton Mifflin, Boston.

Harris, V. A. 1990. Sessile Animals of the Seashore. Chapman and Hall, London. (379 pp. A biology of sessile animals.)

Harrison, F. W. (Ed.): 1991– . Microscopic Anatomy of Invertebrates. Wiley-Liss, New York. (Fifteen volumes, but all not yet completed.)

Harrison, F. W., and Cowden, R. R. (Eds.): 1982. Developmental Biology of Freshwater Invertebrates. Alan R. Liss, New York.

Highnam, K. C., and Hill, L. 1977. The Comparative Endocrinology of the Invertebrates. 2nd Edition. University Park Press, Baltimore. 357 pp.

House, M. R. (Ed.): 1979. The Origin of Major Invertebrate Groups. Systematics Association Spec. Vol. 12. Academic Press, London. 515 pp.

Huges, R. N. 1989. A Functional Biology of Clonal Animals. Chapman and Hall, London.

Kennedy, G. Y. 1979. Pigments of marine invertebrates. Adv. Mar. Biol., *16:*309–381.

Kerfoot, W. C. (Ed.): 1980. Evolution and Ecology of Zooplankton Communities. University Press of New England, Hanover, NH. 794 pp.

Kume, M., and Dan, K. 1968. Invertebrate Embryology. Clearinghouse for Federal Scientific and Technical Information, Springfield, VA.

Laufer, H., and Downer, R. G. H. (Eds.): 1988. Endocrinology of Selected Invertebrate Types. Invertebrate Endocrinology. Vol. 2. Alan R. Liss, New York. 500 pp.

Lincoln, R. J., and Sheals, J. G, 1979. Invertebrate Animals: Collection and Preservation. British Museum (Natural History) and Cambridge University Press, London and Cambridge, U.K.

Little, C. 1984. The Colonization of Land: Origins and Adaptations of Terrestrial Animals. Cambridge University Press, New York. 480 pp.

Little, C. 1990. The Terrestrial Invasion: An Ecophysiological Approach to the Origin of Land Animals. Cambridge University Press, Cambridge, U.K. 304 pp.

MacGinitie, G. E., and MacGinitie, N. 1968. Natural History of Marine Animals. 2nd Edition. McGraw-Hill, New York.

Mann, T. 1984. Spermatophores: Development, Structure, Biochemical Attributes and Role in the Transfer of Spermatozoa. Springer-Verlag, Berlin.

Newell, R. C. 1979. Biology of Intertidal Animals. 3rd Edition. Marine Ecological Surveys Ltd., Faversham, Kent, U.K. 560 pp.

Nicol, J. A. C. 1969. The Biology of Marine Animals. 2nd Edition. Wiley-Interscience, New York.

Pandian, T. J. and Vernberg, F. J. (Eds.): 1990. Animal Energetics. Academic Press, New York. (Two volumes. 1154 pp.)

Rankin, J. C., and Davenport, J. A. 1981. Animal Osmoregulation. John Wiley and Sons, New York. 202 pp.

Russell, F. E. 1984. Marine toxins and venomous and poisonous marine plants and animals. Adv. Mar. Biol. *21:*60–233.

Schaller, F. 1968. Soil Animals. University of Michigan Press, Ann Arbor. 114 pp.

Schmidt-Nielsen, K. 1990. Animal Physiology: Adaptation and Environment. 4th Edition. Cambridge University Press, Cambridge, U.K. 602 pp.

Smith, D. C., and Douglas, A. E. 1987. The Biology of Symbiosis. Edward Arnold Publ., London. 302 pp.

Smith, D. C., and Tiffon, Y. (Eds.): 1980. Nutrition in the Lower Metazoa. Pergamon Press, Elmsford, NY. (188 pp. Proceedings of a symposium.)

Stephenson, T. A., and Stephenson, A. 1972. Life Between Tidemarks on Rocky Shores. W. H. Freeman, San Francisco. 425 pp. (Ecology of the intertidal zone of rocky shores. Systematic coverage of specific regions of the world.)

Taylor, D. L. 1973. The cellular interactions of algal-invertebrate symbiosis. Adv. Mar. Biol. *11:*1–56.

Trueman, E. R. 1975. The Locomotion of Soft-Bodied Animals. American Elsevier Publishing Co., New York 200 pp.

Vernberg, F. J., and Vernberg, W. B. (Eds.): 1981. Functional Adaptions of Marine Organisms. Academic Press, New York. 347 pp.

Vogel, S. 1983. Life in Moving Fluids: The Physical Biology of Flow. Princeton University Press, Princeton. 368 pp.

Vogel, S. 1989. Life's Devices: The Physical World of Animals and Plants. Princeton University Press, Princeton. 350 pp.

Wainwright, S. A. 1988. Axis and Circumference: The Cylindrical Shape of Plants and Animals. Harvard University Press, Cambridge, MA. 176 pp.

Welsch, U., and Storch, V. 1976. Comparative Animal Cytology and Histology. University of Washington Press, Seattle. 343 pp.

General References for Collecting and Identification

Brusca, R. C. 1980. Common Intertidal Invertebrates of the Gulf of California. 2nd Edition. University of Arizona Press, Tucson. 513 pp.

Brusca, G. J., and Brusca, R. C. 1978. A Naturalist's Seashore Guide: Common Marine Life of the Northern California Coast and Adjacent Shores. Mad River Press, Eureka, CA. 215 pp.

Campbell, A. C. 1976. The Hamlyn Guide to the Seashore and Shallow Seas of Britain and Europe. The Hamlyn Publishing Group, London. 320 pp.

Colin, P. L. 1978. Caribbean Reef Invertebrates and Plants. T. F. H. Publishers, Neptune City, NJ. 478 pp.

Fielding A. 1982. Hawaiian Reefs and Tidepools. Oriental, Honolulu.

Fotheringham, N., and Brunenmeister, S. L. 1975. Common Marine Invertebrates of the Northwestern Gulf Coast. Gulf Publishing Co., Houston.

Gosner, K. L. 1979. A Field Guide to the Atlantic Seashore. The Peterson Field Guide Series. Houghton Mifflin Co., Boston. (329 pp. Covers an area from the Bay of Fundy to Cape Hatteras.)

Gunson, D. 1983. Collins Guide to the New Zealand Seashore. Collins, Auckland.

Hayward, P. J., and Ryland, J. S. (Eds.): 1991. The Marine Fauna of the British Isles and North-West Europe. Vol. 1: Introduction and Protozoans to Arthropods; Vol. 2: Molluscs to Chordates. Oxford University Press, Oxford. 996 pp.

Hurlbert, S. H., and Villalobos-Figueroa, A. (Eds.): 1982. Aquatic Biota of Mexico, Central America and the West Indies. Aquatic Biota-SDSU Foundation, San Diego State University, San Diego. 529 pp.

Kaplan, E. H. 1982. A Field Guide to Coral Reefs of the Caribbean and Florida. Peterson Field Guide Series. Houghton Mifflin Co., Boston. 289 pp.

Kermack, D. M., and Barnes, R. S. K. (Eds.): Synopses of the British Fauna. Academic Press, London. (The many volumes of this series [not yet complete] provide keys and descriptions for specific groups of invertebrates. These volumes are listed in the references at the end of various chapters of this text.)

Kerstitch, A. 1989. Sea of Cortez Marine Invertebrates. Sea Challengers, Monterey, CA.

Kozloff, E. N. 1983. Seashore Life of the Northern Pacific Coast: An Illustrated Guide to Northern California, Oregon, Washington and British Columbia. Revised Edition. University of Washington Press, Seattle. 370 pp.

Kozloff, E. N. 1988. Marine Invertebrates of the Pacific Northwest. University of Washington Press, Seattle. 511 pp.

Lincoln, R. J., and Sheals, J. G. 1979. Invertebrate Animals: Collection and Preservation. British Museum, Cambridge University Press, Cambridge, U.K. 150 pp.

Luther, W., and Fiedler, K. 1976. A Field Guide to the Mediterranean Sea Shore. Collins, London.

Meinkoth, N. A. 1981. The Audubon Society Field Guide to North American Seashore Creatures. Alfred Knopf, New York.799 pp.

Morris, R. H., Abbott, D. P., and Haderlie, E. C. 1980. Intertidal Invertebrates of California. Stanford University Press, Palo Alto, CA. (690 pp. This impressive work not only provides a guide to the intertidal invertebrates of California but also summarizes the information known about them and gives references to the literature.)

Morton, J., and Miller, M. 1973. The New Zealand Sea Shore. 2nd Edition. Collins, London. 653 pp.

Morton B., and Morton, J. 1983. The Sea Shore Ecology of Hong Kong. Hong Kong University Press, Hong Kong.

Newell, G. E., and Newell, R. C. 1973. Marine Plankton, A Practical Guide. Hutchinson Educational Ltd., London. 244 pp.

Peckarsky, B. L., Fraissinet, P. R., Penton, M. A., et al. 1990. Freshwater Macroinvertebrates of Northeastern North America. Comstock Pub. Co., Ithaca, NY. 442 pp.

Pennak, R. W. 1978. Freshwater Invertebrates of the United States. 2nd Edition. John Wiley and Sons, New York. 803 pp.

Pennak, R. W. 1989. Freshwater Invertebrates of the United States. Protozoa to Molluscs. 3rd Edition. John Wiley and Sons, New York. 768 pp.

Riedl, R. (Ed.): 1983. Fauna and Flora des Mittelmeeres. Verlag Paul Parey, Hamburg.

Ruppert, E., and Fox, R. 1988. Seashore Animals of the Southeast: A Guide to Common Shallow-water Invertebrates of the Southeastern Atlantic Coast. University of South Carolina Press, Columbia. 429 pp.

Sefton, N., and Webster, S. K. 1986. A Field Guide to Caribbean Reef Invertebrates. Sea Challengers, Monterey, CA.

Smith, D. L. 1977. A Guide to Marine Coastal Plankton and Marine Invertebrate Larvae. Kendall/Hunt Publishing Co., Dubuque, IA. 161 pp.

Smith, R. I. (Ed.): 1964. Keys to Marine Invertebrates of the Woods Hole Region. Contribution No. 11, Systematics-Ecology Program, Marine Biology Laboratory, Woods Hole, MA.

Sterrer, W. E. (Ed.): 1986. Marine Fauna and Flora of Bermuda. Wiley-Interscience, New York. (This work is also a valuable reference for Florida and the Caribbean.)

Wickstead, J. H. 1965. An Introduction to the Study of Tropical Plankton. Hutchinson and Co., London. 160 pp.

Zinn, D. J. 1985. Handbook for Beach Strollers from Maine to Cape Hatteras. Globe Pequot Press, Chester, CT.

Laboratory Guides

Dales, R. P. 1981. Practical Invertebrate Zoology. 2nd Edition. John Wiley and Sons, New York. 356 pp.

Freeman, W. H., and Bracegirdle, B. 1971. An Atlas of Invertebrate Structure. Heinemann Educational Books, London. 129 pp.

Pierce, S. K., and Maugel, T. K. 1989. Illustrated Invertebrate Anatomy. Oxford University Press, Oxford. 320 pp.

Sherman, I. W., and Sherman, V. G. 1976. The Invertebrates: Function and Form. 2nd Edition. Macmillan Co., New York. 334 pp.

Wallace, R. L., Taylor, W. K., and Litton, J. R. 1988. Invertebrate Zoology, 4th Edition. Macmillan Publishing Co., New York. 475 pp.

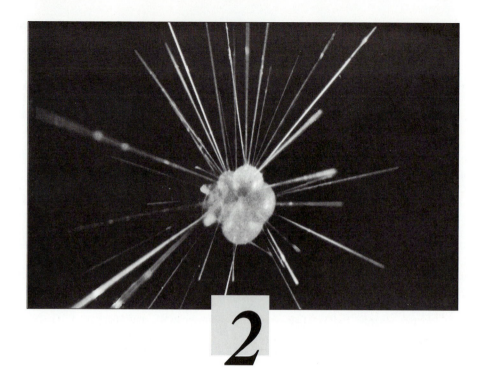

2

PROTOZOA

PRINCIPLES AND EMERGING PATTERNS

All animals must solve the same problems of existence—procurement of food and oxygen, maintenance of water and salt balance, removal of metabolic wastes, and perpetuation of the species. The body design necessary to meet these problems is, in large part, correlated with four factors: (1) the type of environment—marine, freshwater, or terrestrial—in which the animal lives; (2) the size of the animal; (3) the mode of existence of the animal, and (4) the constraints of the animal's genome. The last phrase refers to the limitations imposed by the ancestral design controlled by the animal's genetic makeup. For example, there is a marvelous diversity within the snail design, but the design itself imposes certain limitations on the snail's possibilities, for example, there are no flying snails.

EFFECTS OF ENVIRONMENTS

Let us briefly examine the first three factors. The three major types of environments—salt water, freshwater, and land—are markedly different in their physiological demands.

The Marine Environment

The marine environment is generally the most stable. Wave action, tides, and vertical and horizontal ocean currents produce a continual mixing of sea water and ensure a medium in which the concentration of dissolved gases and salts fluctuates relatively little. The buoyancy of sea water reduces the problem of support. It is therefore not surprising that the largest invertebrates have always been marine. Because sea water is more or less isosmotic with the body tissue fluids of most invertebrate marine animals, maintaining water and salt balance is relatively simple, and most marine animals are osmoconformers. The buoyancy and uniformity of sea water provide an ideal medium for animal reproduction. Eggs in sea water can be shed and fertilized and can undergo development as floating embryos with little danger of desiccation and salt imbalance or of being swept away by rapid currents into less favorable environments. Larvae are characteristic of many marine animals, and the general features of larvae will be discussed in Chapter 3's Principles and Emerging Patterns.

Freshwater Environments

Fresh water is a much less constant medium than sea water. Streams vary greatly in turbidity, velocity, and volume, not only along their course but also from time to time as a result of droughts or heavy rains. Small ponds and lakes fluctuate in oxygen content, turbidity, and water volume. In large lakes the environment changes radically with increasing depth.

Like salt water, fresh water is buoyant and aids in support. The low salt concentration, however, creates some difficulty in maintaining water and salt balance. Because the body of the animal contains a higher osmotic concentration than that of the external environment, water has a tendency to diffuse inward. The animal thus has the problem of getting rid of excess water. As a consequence, freshwater animals usually have some mechanism for pumping water out of the body while holding onto the salts; that is, they osmoregulate.

In general, the eggs of freshwater animals are either retained by the parent or attached to the bottom of the stream or lake, rather than being free floating (as is often true of marine animals). In addition, larval stages are usually absent. Floating eggs and free-swimming larvae are too easily swept away by currents. Because freshwater eggs develop directly into adults without an intermediate feeding larva, the eggs typically contain considerable amounts of yolk.

Nitrogenous wastes of aquatic animals, both marine and freshwater, are usually excreted as ammonia. Ammonia is very soluble and toxic and requires considerable water for its removal, but because there is no danger of water loss in aquatic animals, the excretion of ammonia presents no difficulty.

Terrestrial Environments

Terrestrial animals live in the harshest environment. The supporting buoyancy of water is absent. Most critical, however, is the problem of water loss by evaporation. The solution to this problem has been a primary factor in the evolution of many adaptations for life on land. The integument of terrestrial animals presents a better barrier between the inner and the outer environment than that of aquatic animals. There is more oxygen in a given volume of air than the same volume of water, but respiratory surfaces must be moist. They are, therefore, usually located in the interior of the body, reducing desiccation. Nitrogenous wastes are commonly excreted as urea or uric acid, which are less toxic and require less water for removal than does ammonia. Fertilization must be internal; the eggs are usually enclosed in a protective envelope or deposited in a moist environment. Except for insects and a few other arthropods, development is direct, and the eggs are usually endowed with large amounts of yolk. Terrestrial animals that are not well adapted to withstand desiccation are either nocturnal or restricted to humid or moist habitats.

EFFECTS OF BODY SIZE

The second factor correlated with the nature of animal structure is the size of the animal. As the body increases in size, the ratio of surface area to volume decreases, because volume increases by the cube, and surface area increases by the square of the linear dimension. In small animals the surface area is sufficiently great in comparison with the body volume that exchange of gases and waste can be carried out efficiently by diffusion through the general body surface. Also, internal transport can take place by diffusion alone. However, small animals that live on land are subject to greater water loss, which occurs primarily across the body surface.

As the body increases in size, distances become too great for internal transport to take place by diffusion alone, and mechanisms for bulk flow become necessary. In larger animals this has led to the development of coelomic and blood-vascular circulatory systems. Also, through folding and coiling, the surface area of internal organs and portions of the external surface may be increased to facilitate secretion, absorption, gas exchange, and other processes.

EFFECTS OF MODE OF EXISTENCE

The third factor related to the nature of animal structure is the mode of existence. Free-moving animals are generally bilaterally symmetrical. The nervous system and sense organs are usually concentrated at the anterior end of the body because this is the part that first comes in contact with the environment. We will discuss bilateral symmetry at greater length in Chapter 5.

Attached, or sessile, animals are radially symmetrical or tend toward a radial symmetry, in which the entire body (as in sea anemones) or some part of the body (as in barnacles) consists of a central axis around which similar parts are symmetrically arranged. Radial symmetry is an advantage for sessile existence because it allows the animal to meet the challenges of its environment from all directions. Skeletons, envelopes, or tubes are commonly present to support or protect sessile animals against motile predators and environmental extremes.

CELLULAR SPECIALIZATIONS

Protozoa are unicellular eukaryotic organisms. All of the multicellular organisms including the multicellular animals, evolved from various protozoan ancestors. The protozoa have remained at the unicellular level of organization but have evolved along numerous lines through the specialization of parts of the cytoplasm (organelles) or of skeletal structures. Thus, similarity and complexity in protozoa are expressed in the number and nature of their organelles and skeletons in the same way that simplicity and complexity in multicellular animals are seen in degrees of specialization of tissues and organ systems. Protozoan cells are more complex than metazoan cells because structural adaptations for survival can only occur at the cellular level of organization. Each protozoan cell, as an organism, must have structural specializations for all life-sustaining functions, such as locomotion, food acquisition, internal transport, and reproduction. In metazoans, on the other hand, these specializations, though ultimately a product of cells, are distributed over cells, tissues, organs, and organ systems. A single metazoan cell need only contribute an increment to the overall adaptive organization of the individual. As a result, metazoan cells are less complex but more specialized for particular functions than protozoan cells.

In addition to the basic metabolic machinery common to all cells, metazoans inherited many features of cell structure from their protozoan ancestors and have adapted them to a multicellular organization.

The Cell Membrane

The cell membrane separates the remainder of the cell from the exterior. In doing so, it regulates the cell interior for optimal functioning by maintaining more or less specific and constant internal conditions. The cell membrane controls what may enter and leave the cell, the responsiveness of the cell to external stimuli, the selectiveness with which the cell binds to other cells or to a substratum, and the maintenance of cell shape. The bilayered structure of the cell membrane results from the

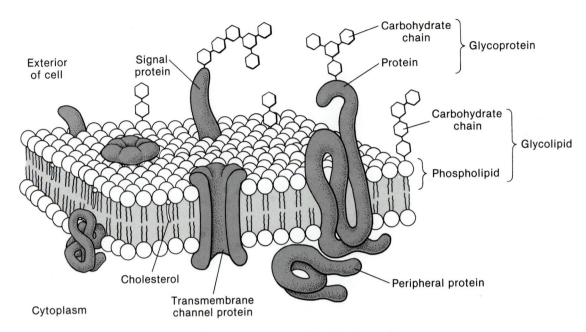

FIGURE 2-1 Ultrastructure of the cell membrane.

opposing phospholipids that compose it (Fig. 2–1). Proteins are also important membrane constituents and may span it or may be attached to the inner or outer surfaces. The exposed outer surfaces of membrane proteins and lipids may have carbohydrates attached to them like tails that radiate into the surrounding medium. Together these tails and especially their extracellular peripheral proteins form a **glycocalyx,** which is an important physiological barrier that forms a template on which exoskeletons are secreted, and regulates binding to signal molecules and to surfaces, such as other cells. Membrane proteins may receive and transmit signals to the interior of the cell and serve as points of anchorage for cytoskeletal fibers. The cell membrane itself can also play a skeletal role. If the membrane lipids are largely unsaturated, like some vegetable oils used in cooking, the membrane is relatively fluid and flexible. If, on the other hand, the lipids are mostly saturated, like lard, the membrane is less fluid and more stiff. Cholesterol, which is a common component of cell membranes, also reduces the fluidity of the membrane.

Flagella and Cilia

Flagella and cilia are characteristic of many protozoan and metazoan cells. In general, **flagella** are typically long and their motion is a complex whiplike undulation. **Cilia,** on the other hand, are short and their motion is stiff and oarlike. But the distinction is not sharp and their ultrastructure is identical. A single flagellum or cilium is constructed like an electric cable with an outer membranous sheath and an inner fibrous core (Fig. 2–2). The inner core is composed of microtubules and other proteins together called an **axoneme.** The axoneme consists of two central singlet microtubules encircled by nine doublet outer microtubules. One microtubule of each doublet bears two rows of projections, or arms (containing the enzyme dynein), directed toward the adjacent doublet. The axoneme is enclosed within a

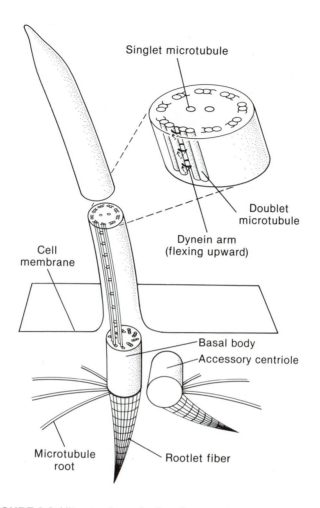

Singlet microtubule

Doublet microtubule

Dynein arm (flexing upward)

Cell membrane

Basal body

Accessory centriole

Microtubule root

Rootlet fiber

FIGURE 2-2 Ultrastructure of a flagellum or cilium.

sheath that is continuous with the cell membrane. Bending of the flagellum is caused by active sliding of adjacent doublets past each other. (Place your hands and fingers together, palm to palm, and simultaneously move your index fingers from side to side.) The dynein side arms on the doublets provide the sliding force. In the presence of adenosine triphosphate (ATP), the arm on one doublet attaches to an adjacent doublet and flexes, causing the doublets to slide past each other by one increment. Successive attachments and flexes cause the doublets to slide smoothly past one another over a distance sufficient to bend the flagellum. All flagella and cilia rise from and are anchored to a **basal body** that lies immediately below the cell surface (Fig. 2–2). When basal bodies are distributed to daughter cells during mitosis, they typically arrange themselves at each pole of the mitotic spindle and are then designated as **centrioles** (Fig. 2–3). A region around basal bodies and centrioles, called the **microtubule organizing center (MTOC),** controls the organized assembly of microtubules. A basal body forms the template on which developing axonemes are organized. Basal bodies, in fact, have an ultrastructure like that of an axoneme except that the central singlets are absent and the nine fibrils in the outer circle are in triplets, two of the three being continuous with the doublets of the flag-

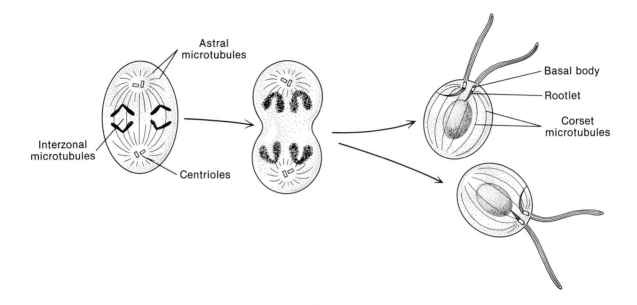

FIGURE 2-3 Relationship of centrioles and spindle fibers to basal bodies and cytoskeleton.

ellum. Dynein arms are absent on the triplets. A basal body (and its cilium or flagellum) is usually anchored in the cell, often to the nucleus and cell membrane, by one or more root structures (Fig. 2–2). These may be sprays of microtubules, tapered striated rootlets, or both. The proteinaceous rootlet fibers are contractile and can, on contraction, pull the flagellum into a shallow pocket or alter its orientation.

In most protozoan and metazoan cells, the flagellum propagates an undulatory wave from the cell to the flagellar tip that pushes the cell ahead or drives water away from a stationary cell (Fig. 2–4). (We will encounter some exceptions later in the chapter on protozoa.) As an undulatory wave moves along the flagellum, the advancing wavefront generates a longitudinal force to propel the cell similar to the

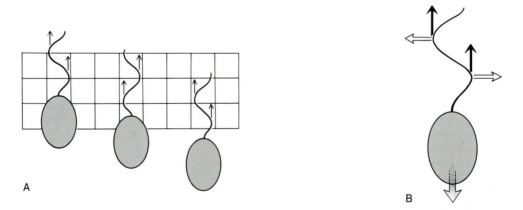

FIGURE 2-4 Flagellar propulsion. **A,** Base-to-tip wave propagation. **B,** Forces generated by base-to-tip wave propagation. Lateral forces (outlined arrows) cancel each other. Longitudinal forces (solid arrows) combine to produce forward thrust in the opposite direction.

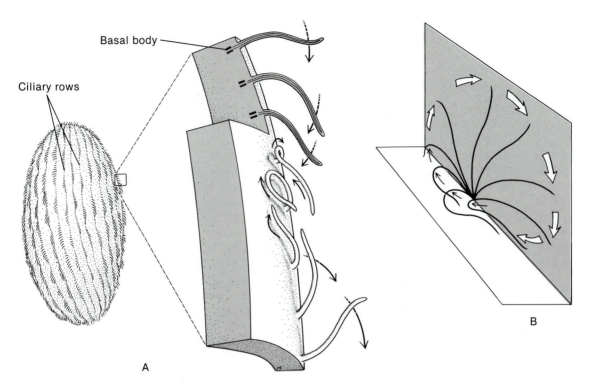

Basal body

Ciliary rows

A

B

FIGURE 2-5 A, Metachronal wave of ciliary beating. Along the length of a row the cilia are in sequential stages of the beat cycle. **B,** The effective (outlined arrows) and recovery (solid arrows) strokes in the beat cycle of a single cilium.

way in which a wavefront advancing on a beach produces a longitudinal force to move a swimmer onshore (Fig. 2–4B). Unlike a wave advancing on a beach, however, the sideways undulations of a flagellum also generate lateral forces. Because the lateral undulations are usually symmetrical, the left-directed forces cancel the right-directed forces, and only the longitudinal force remains to move the cell.

Cilia are short, commonly numerous, densely arranged, and especially well represented in the ciliate protozoans. Cilia do not beat simultaneously but rather sequentially in each longitudinal row (Fig. 2–5A). The sequential activation of cilia over the surface of the cell is seen as waves, called **metachronal waves.** In its beat each cilium performs an effective and a recovery stroke. During the effective stroke, the cilium is outstretched stiffly and moves, like an oar, from a forward to a backward position in a plane perpendicular to the body surface (Fig. 2–5B). In the recovery stroke the cilium flexes and sweeps forward, snakelike, close to and in a plane parallel to the body surface.

Movement in a Fluid Medium

Swimming animals and protozoa are subjected to two sources of water resistance, or drag—pressure drag and viscous drag. **Pressure drag** refers to the difference between the pressure at the front end (higher pressure) of a forward moving organism and that at the rear end (lower pressure). Streamlining reduces pressure drag. **Viscous,** or **adhesive, drag** results from the tendency of the water molecules, which are electrostatically polar, to stick to each other and to surfaces. A swimming organism is coated with a layer of nonmoving, adhering water molecules that, in turn, adhere to other water molecules. As the distance from the organism's surface increases, molecules steadily shear off until the full velocity of the current is attained. There is thus a coatlike boundary layer of nonmoving and slower moving water surrounding a flagellum or cilium, indeed around the entire protozoan cell or multicellular animal. This principal of aqueous viscosity is equally applicable to fluid moving within a tube and accounts for the peripheral resistance of blood circulating through blood vessels.

For very small organisms, especially protozoa and animals moving by means of flagella or cilia, or even stationary ciliated or flagellated cells in metazoans that are driving water, viscous drag is of much greater significance than pressure drag. The surface area over which the boundary layer develops is enormous compared with the very small cell volume. The viscous resistance is so great that it would be equivalent to a larger animal swimming in honey. It is not surprising that very small animals are not streamlined because streamlining increases the surface area and thus the viscous drag. In later chapters we will see how fluid viscosity influences the lives of various groups of invertebrate animals.

How Cells Take up Food and Other Substances

Substances enter protozoan cells and the cells of other eukaryotes in a variety of ways. The protein channels of the cell membrane provide for the passive diffusion of water, ions, and small molecules, such as sugars and amino acids. Some function as energy-requiring pumps, actively transporting certain molecules or moving ions in or out against their concentration gradient.

Some extracellular materials enter a cell in minute pits on the cell's membrane that later pinch off internally (Fig. 2–6)—a process called **endocytosis. Pinocytosis** is a nonspecific form of endocytosis in which the rate of uptake is in simple proportion to the external concentration of the material being absorbed (Fig. 2–6A).

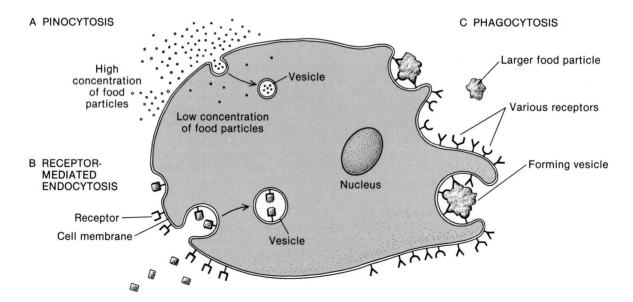

FIGURE 2-6 Modes of food uptake by vesicles. **A,** Pinocytosis. **B,** Receptor-mediated endocytosis. **C,** Phagocytosis.

Water, ions, and small molecules may be taken in by pinocytosis. **Receptor-mediated endocytosis** brings in proteins and other macromolecules at a rate greater than predicted by the concentration gradient. These substances bind to and are concentrated on specific membrane receptors before they are internalized in the vesicles (Fig. 2–6B). **Phagocytosis** is the engulfment of large particles, such as bacteria and protozoans, and their enclosure within relatively large vesicles (Fig. 2–6C). Phagocytosis requires binding of a particle to multiple membrane receptors and dynamic alteration not only of the cell membrane but of underlying cytoplasmic elements as well.

How Cells Digest Food

Most protozoans and some cells of many metazoans must take at least some nutrients in through the cell membrane and digest them before using the products in other cell activities. The organelles in which digestion occurs and the processes by which it occurs are virtually the same in all cells, whether protozoan or metazoan, although the terminology may vary from group to group.

Ingested food occupies a vesicle or **food vacuole,** which binds to and fuses with primary **lysosomes** already present in the cell (Fig. 2–7). As fusion occurs,

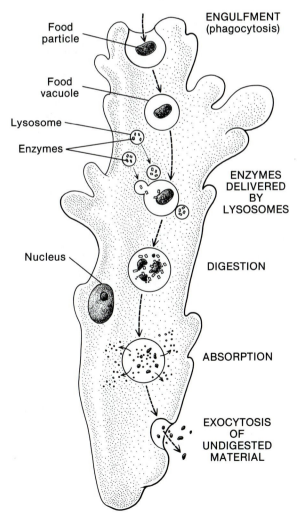

FIGURE 2-7 Digestion within a food vacuole.

lysosomal acids and hydrolytic enzymes are released into the vacuole. Products of intracellular digestion diffuse from the vacuole into the cytoplasm of the cell where they may be used in metabolism by other organelles or stored, after undergoing synthesis, in forms, such as glycogen and lipid. Indigestible material is released from the cell to the exterior by fusion of the residual vesicle with the cell membrane in a process called **exocytosis.**

Cell Secretions

Many cells synthesize various macromolecules, especially proteins and glycoproteins, that are passed to the exterior of the cell surface. Synthesis typically involves Golgi bodies and the endoplasmic reticulum, the latter forming the vesicle that transports the secretory product to the cell membrane. Here the membrane of the vesicle fuses with the cell membrane in a process of exocytosis, and the cell product is expelled to the exterior (Fig. 2–8). Many cells secretions, such as extracellular enzymes and pheromones, are exported, away from the producing cell. One of the most widespread animal secretions, **mucus,** is a mucopolysaccharide with a large carbohydrate and a smaller protein component. Mucus is utilized by animals in a variety of ways, as an adhesive, protective cover, or lubricant.

Some cell secretions remain associated with the external surface of the cell membrane to form extracellular skeletal materials (Fig. 2–8), of which **collagen** is a good example. The basic units (tropocollagen) of this common animal fibrous protein are discharged to the exterior of the cell by exocytosis, where they are assembled into collagen fibers. Although such fibers may become completely separated from the cell surface, they are often anchored to the cell membrane by means of special surface proteins (fibronectins) and glycoproteins (proteoglycans). It is by such cellular secretions that many multicellular animals, such as sponges, attach to surfaces or encase themselves within skeletal envelopes.

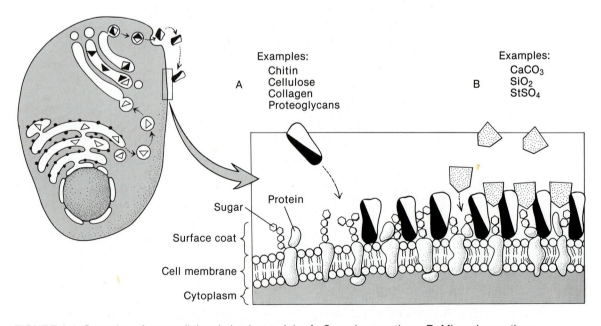

FIGURE 2-8 Secretion of extracellular skeletal materials. **A,** Organic secretions. **B,** Mineral secretions.

How Cells Receive Signals

Protozoan and metazoan cells respond to external chemical stimuli by means of receptors on the cell's surface. When a substance, such as a compound released by a prey organism, binds to a specific surface protein receptor, the combination molecule may open or close protein membrane channels or initiate reactions beneath the membrane.

Cellular Symbiosis

Symbiosis refers to the close physical interrelationship between two different species. When one member benefits and the other is harmed, the relationship is termed **parasitism;** when one benefits and the other neither benefits nor is harmed, the relationship is termed **commensalism;** when both benefit, the relationship is termed **mutualism.** A widespread mutualistic relationship in both protozoa and animals occurs with autotrophic organisms. The autotrophic symbionts may be cyanobacteria, diatoms, or green unicellular algae (**zoochlorellae**), but the most commonly occurring symbionts are a nonmotile stage of certain flagellated protozoa called dinoflagellates (**zooxanthellae**), which will be described later in this chapter (p. 31). The autotrophic member of the partnership is generally located within the host cell, although within a few metazoans, they are found between cells. The symbiosis probably arose through phagocytosis of the autotroph by the heterotrophic cells. Delayed digestion by the host cell resulted in some of the captured cells continuing to live and photosynthesize. Some of their excess photosynthate was shared with the host—a positive force selecting for factors that maintain the autotroph alive within its vesicular prison. Such a scenario must have occurred numerous times, considering the different sorts of autotrophs and their symbiotic partners. The benefits are probably basically the same where the symbiosis occurs. The autotroph provides excess organic carbon from photosynthesis to the host, and the host provides certain nutrients, such as nitrogen and phosphorus compounds, to the autotroph. Rarely does the protozoan or metazoan rely entirely on its autotrophs for nutrition; typically, the benefits of symbiosis supplement heterotrophic nutrition to varying degrees.

PROTOZOA

The protozoa is a diverse assemblage of some 80,000 single-cell organisms possessing typical (eukaryote) membrane-bound cellular organelles. Because many are animal-like, being motile and having heterotrophic nutrition, this assemblage was treated in the past as a single phylum within the Animal Kingdom—the phylum Protozoa. They are now placed within a number of different unicellular phyla, which together with most algal phyla constitute the kingdom Protista, also called Protoctista. Protozoa is now used as a convenient common name for unicellular, motile protists.

The unicellular level of organization is the only characteristic by which the protozoa as a whole can be described; in all other respects they display extreme diversity. Protozoa exhibit all types of symmetry, a great range of structural complexity, and adaptations for all types of environmental conditions.

Although most protozoa occur as solitary individuals, there are numerous colonial forms. Some of these, such as species of *Volvox,* attain such a degree of cellular interdependence that they approach a multicellular tissue level of structure. Individual independent protozoa range in size from a micrometer in the case of the planktonic *Micromonas* to a few millimeters in some dinoflagellates, amoebas, and ciliates.

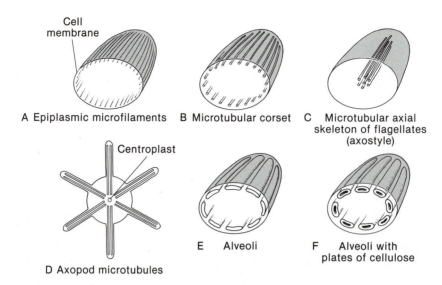

FIGURE 2-9 Microfilamentous, microtubular, and alveolar endoskeletons of protozoa.

Free living protozoa occur wherever moisture is present—in the sea, in all types of fresh water, and in the soil. There are also commensal, mutualistic, and many parasitic species.

PROTOZOAN ORGANELLES AND GENERAL PHYSIOLOGY

The protozoan body is usually bounded only by the cell membrane. The rigidity or flexibility of the body and its shape are largely dependent on the nature of the cytoskeleton. Cytoskeletons are typically but not always located immediately below the cell membrane, where, together with it and other organelles, they form the **pellicle,** a sort of protozoan "body wall." The cytoskeleton is often composed of slender filamentous proteins, microtubules or vesicles, or various combinations of all three. The filamentous proteins may form a dense supportive mesh called the **epiplasm** (Fig. 2–9A), as, for example, in *Euglena* and the ciliates. More conspicuous cytoskeletal organelles are the pellicular microtubules that occur in all flagellates, in spore-forming protozoa, and in ciliates. They can be arranged as a microtubular corset (Fig. 2–9B), or, as in some flagellates, the microtubules originate on the flagellar basal bodies and radiate rearward from them to the opposite extremity of the cell as a sort of axial skeleton (axostyle) (Fig. 2–9C). Such microtubules resemble the microtubules of a mitotic spindle, which radiate between centrioles

and form the mitotic apparatus (Fig. 2–3). In other protozoa, such as the radiolarians and heliozoans, bundles of microtubules radiate outward from a **centrosome** or similar structures (both MTOCs), each bundle extending into and supporting a raylike projection of the cell (axopod) (Fig. 2–9D). The centrosome and its microtubules resemble the starlike asters that form around centrioles at the poles of the mitotic spindle.

Vesicles occur immediately below the cell membrane in many protozoans, such as dinoflagellates, spore-forming protozoans, and ciliates. When these vesicles are flattened and form a more or less continuous layer beneath the cell membrane, they are referred to as **alveoli.** "Empty" alveoli, like those that occur in ciliates, may be turgid and help to support the cell (Fig. 2–9E). In dinoflagellates, plates of cellulose are secreted into the alveolar vesicles to form a rigid skeleton (Fig. 2–9F).

Protozoan skeletons, like those of metazoans, can also be exoskeletons. They are secreted onto the outer surface of the cell and are usually called tests.

The protozoan locomotor organelles may be **flagella, cilia,** or flowing extensions of the body called **pseudopodia.** These are important organelles for defining the protozoan groups and will be discussed extensively in later sections of the chapter.

All types of nutrition occur in protozoa. Some protozoa rely on photoautotrophy, others absorb soluble organic nutrients from their environment, and

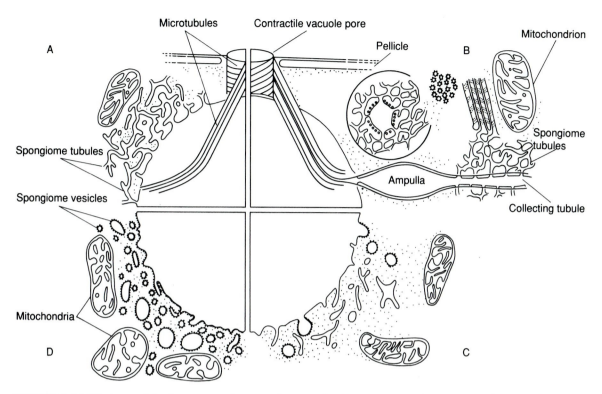

FIGURE 2-10 Diagram of the four types of contractile vacuoles found in protozoa. Types **A** and **B** are both characteristic of ciliates, where the spongiome is composed of irregular tubules and bands of microtubules support the pore and extend over the vacuole surface. **A,** The network of spongiome tubules empties directly into the vacuole. **B,** The network of irregular tubules of the spongiome first empties into straight collecting tubules, which may dilate as ampullae before discharging into the vacuole. **C,** Typical of amebas, the spongiome contains small vesicles and tubules. *(After Patterson, D. J. 1980. Contractile vacuoles and associated structures; their organization and function. Biol. Rev. 55:1–46. © Copyright Cambridge University Press; reprinted by permission.)*

many ingest food particles or prey and digest them intracellularly within food vacuoles. Like other organelles, the walls of food vacuoles are composed of lipid bilayer membranes, which can be synthesized or removed rapidly and recycled. Food reaches the vacuole by phagocytosis, often through a cell mouth or cytostome. Very fine particles in solution may enter by pinocytosis, which may occur over the entire surface of the cell. Intracellular digestion has been most studied in amoebas and ciliates, and, for the most part, follows the general pattern described on p. 19. Some specializations in ciliates will be described later (p. 60).

A few protozoa, especially those that live in the digestive tract of animals, are obligatory anaerobes. Those that live in water where there is active decomposition of organic matter may be facultative anaerobes, utilizing oxygen when present but also capable of anaerobic respiration. The changing availability of food and oxygen associated with decay typically re-

sults in a distinct succession of populations and protozoan species. Because of their short generation time, protozoa can be important in monitoring pollution. Their relatively rapid response to changes in the aquatic environment can be indicative of reduced oxygen.

Many protozoa osmoregulate to remove excess water (volume regulation) and to adjust the concentration and proportion of their internal ions (ionic regulation). Excess water enters protozoans by osmosis when their internal osmotic concentration exceeds that of the surrounding water. Additional water may enter with food in vacuoles and pinocytotic vesicles, for example, an amoeba that is experimentally fed on a protein solution imbibes by endocytosis a quantity of water equivalent to one third its body volume.

Osmoregulation is accomplished by active ion transport at the cell membrane and by a system of water- and ion-pumping organelles called the **contractile vacuole complex** (Fig. 2–10). The complex is composed of a large spherical vesicle—the **contractile vacuole** proper—and a surrounding system of small vesicles or tubules termed the **spongiome.** The spongiome provides for the collection of fluid, which is delivered to the contractile vacuole, which in turn expels the fluid to the outside of the organism through a temporary or permanent pore. In some protozoa (some amoebas and flagellates), the vacuole completely disappears following contraction and is reformed by fusion of small vesicles. In others (many ciliates), the vacuole collapses at discharge and is refilled by fluid from the surrounding tubules of the spongiome. The rate of discharge depends on the osmotic pressure of the external medium. *Paramecium caudatum,* which lives in fresh water, can complete a cycle of vacuole filling and discharge as rapidly as every 6 s and expel a volume equivalent to its entire body every 15 min. The basis for contraction may differ among groups of protozoans. In dinoflagellates, a flagellar rootlet branches to form a contractile sheath around the vacuole. In *Paramecium* actin filaments surround the vacuole.

In freshwater protozoa the vacuole contents are more dilute than the cytoplasm but more concentrated than the surrounding external environment. Although the mechanism of contractile vacuole function is not fully understood, it now seems likely that water enters the spongiome as the tubular membranes secrete ions from the cytoplasm into the tubule. Water and ions pass along the tubule to a water-impermeable region in which certain ions and perhaps other substances are selectively reabsorbed. The remaining water, ions, and perhaps secreted waste metabolites are then released from the cell as the vacuole contracts. The contractile vacuole system is of no particular significance in removing metabolic wastes, such as ammonia and CO_2; these simply diffuse to the outside of the organism.

Protozoa with cell walls, meaning those having a layer of cellulose or other material outside of the cell membrane, lack contractile vacuoles. As water enters the cell by osmosis, the rigid wall resists swelling and a hydrostatic pressure is established. When the rising hydrostatic pressure equals the osmotic pressure, no further water enters the cell.

Like animals, protozoa respond to chemical and physical cues in ways that enable them to avoid adverse conditions, locate food, and find mates. In this sense, each protozoan cell must be both receptor and effector. In their receptiveness to environmental stimuli, protozoans resemble the sensory nerve cells of animals.

Like animal receptors, protozoans often receive external stimuli as signal substances that bind to specific membrane molecules. Binding can cause a specific ion channel to open, allowing ions (often Na^+ and K^+) to flow down their concentration gradients (Na^+ in K^+ out). Because the resting cell membrane is polarized with respect to the distribution of these ions, opening of the ion channels depolarizes the membrane. (Depolarizations can be measured as changes in electrical potentials, or voltage, using electrodes and a voltmeter.) When the membrane is depolarized, Ca^{++} channels open, and calcium ions enter the cell. Entry of calcium into the cell triggers other changes, such as reversal of ciliary beat, which causes the cell to withdraw from the disturbance. *Paramecium,* for example, has at least nine different ion channels, some of which are localized at the front and others are at the rear of the cell. Such localized receptor fields differentiate "head" from "tail" and are thus analogous to the concentration of receptor organs of many animals. Intercellular chemical signalling (pheromones) in protozoans, in fact, often involves signal molecules, such as serotonin, β-endorphin, acetylcholine, and cyclic-AMP, which, in animals, function as neurotransmitters and internal messengers.

Reproduction and Life Cycles

Asexual reproduction by mitosis occurs in most protozoa and is the only known mode of reproduction in some species. Division of the organism into two or more progeny cells is called **fission.** When this process results in two similar progeny cells, it is termed **binary fission;** when one progeny cell is much smaller than the other, the process is called **budding.** In some protozoa, multiple fission, or schizogony, is the rule. In schizogony, after a varying number of nuclear divisions, the cell divides into a number of progeny cells. With few exceptions, asexual reproduction involves some replication of organelles following fission.

Sexual reproduction is not universal among protozoans and where it occurs, commonly does not lead directly to the formation of new individuals. There are many well-studied protozoans in which sexual reproduction has never been recorded. In some species this absence may be primitive; in others it may represent a secondary loss. The primitive protozoan life cycle may have been one in which haploid individuals reproduced solely by fission (Fig. 2–11A), for example, the living parasitic trypanosomatids. Sexual

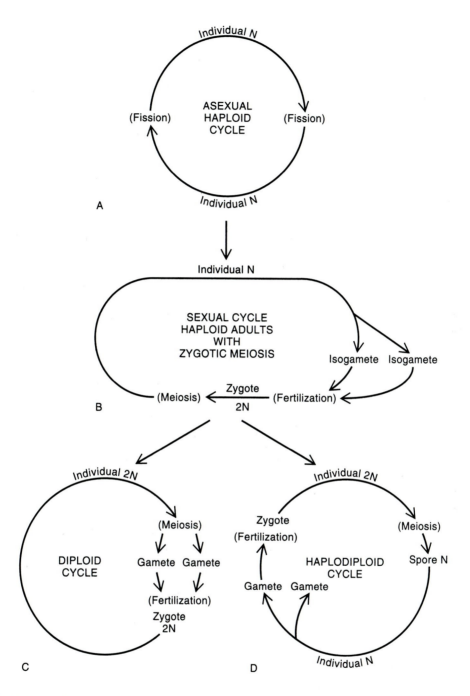

FIGURE 2-11 Protistan life cycles. **A,** A primitive eukaryote life cycle in which new individuals are produced sole-ly by fission. Illustrated among living species by the trypanosomatids. Fusion of two individuals could have led to life cycles **(B)** in which isogametes form a diploid zygote. The zygote then undergoes meiosis to form haploid in-dividuals. Protists with this type of life cycle include the volvocids, many dinoflagellates, hypermastigids, and sporozozoans. Extension of the diploid phase could have led to a diploid cycle and a haplodiploid cycle. In the diploid cycle **(C),** which characterizes opalinids, some hypermastigids, heliozoans, many green algae, diatoms and ciliates, the adults are diploid and meiosis occurs in the formation of gametes. Ciliates do not produce ga-metes but exchange haploid nuclei, which fuse. In the haplodiploid cycle **(D),** found in coccoliths, many forams, and many algae, the formation of spores and diploid individuals alternate with haploid individuals.

reproduction probably arose by the fusion of two similar haploid individuals producing a diploid zygote. Soon after fusion, the zygote may have divided by meiosis to restore the haploid chromosome number and to produce four new cells. This scenario is supported by the common occurrence of identical, flagellated gametes (isogametes) among protozoans, and there are many species in which the adults are haploid and the life cycle involves fusion (fertilization) of isogametes to form a diploid zygote (Fig. 2–11B). In these the zygote quickly undergoes meiosis to produce haploid adults. A delay in the meiosis of the zygote would prolong the diploid stage. In such a case diploidy could come to dominate the life cycle, and the haploid phase would be restricted to gametes, which would arise by meiosis and then fuse to form a diploid zygote (Fig. 2–11C). Such a diploid life cycle is characteristic of animals and is also found in a number of protistan groups, notably the ciliates. However, ciliates exchange haploid nuclei during conjugation rather than produce haploid gametes. Other species of protozoa contain haploid individuals that alternate in the life cycle with diploid individuals (Fig. 2–11D). In such a haplodiploid cycle meiosis does not occur in the formation of gametes but in the formation of haploid spores, from which the haploid individuals arise. The diploid individuals arise from the zygote. Multicellular plants also have this type of cycle.

Encystment is characteristic of the life cycle of many protozoa, including the majority of freshwater species. In forming a cyst, the protozoan secretes a thickened envelope about itself and becomes inactive. Depending on the species, the protective cyst is resistant to desiccation or low temperatures, and encystment enables the animal to pass through unfavorable environmental conditions. The simplest life cycle includes only two phases: an active phase and a protective, encysted phase. However, the more complex life cycles are often characterized by encysted zygotes or by formation of special reproductive cysts, in which fission, gametogenesis, or other reproductive processes take place.

Protozoa may be dispersed long distances in either the motile or encysted stages. Water currents, wind, and mud and debris on the bodies of water birds and other animals are common agents of dispersal.

SUMMARY

1. Protozoa are unicellular organisms that are animal-like, being commonly motile and heterotrophic. They are a very diverse assemblage, and the major groups are treated as separate phyla of eukaryote protists.

2. Most protozoa inhabit the sea or fresh water, but there are many parasitic, commensal, and mutualistic species.

3. Most protozoan phyla are distinguished, in part, by their type of locomotor organelles: flagella, pseudopodia, or cilia. The protozoan body (cell) may be supported by an exoskeleton (test) or an internal cytoskeleton. The latter may be composed of microfilaments, microtubules, or vesicles located beneath the cell membrane. The cell membrane and the underlying cytoskeleton form the pellicle.

4. Digestion occurs intracellularly within a food vacuole, and food reaches the vacuole through a cell mouth or by engulfment.

5. Water volume and ion regulation are usually accomplished by a contractile vacuole.

6. Reproduction by fission (mitosis) occurs at some time in the life history of almost all protozoa. Meiosis, gamete formation, and fertilization have been observed in many species. Production of isogametes is common. Depending on the group, meiosis occurs at the zygote stage, in the formation of gametes, or in the formation of haploid spores. Encystment is common.

PROTISTAN EVOLUTION

Protistan evolution is the history of the eukaryote cell, and living protists reflect different stages in that history. The earliest protists were probably amoeboid forms capturing food particles by phagocytosis. Most, but not all, protists possess mitochondria, which initially may have been symbiotic prokaryotic organisms acquired by engulfment. Flagella may also have been acquired early, both before and after the acquisition of mitochondria, because mitochondria are absent in some living flagellates. Many living flagellates possess chloroplasts, which are also believed to have arisen through a partnership between early flagellated phagocytic eukaryotes and prokaryotic symbionts. Some living flagellates retain this ancestral phagotrophy along with photosynthetic capabilities; others are solely photoautotrophs. Contributing to the confusing array of protistan lineages, some species have lost their chloroplasts and are secondarily heterotrophs, and some of these have lost their flagella and are secondarily amoeboid. Clearly then, amoebas, flagellates, heterotrophs, and autotrophs do not form neatly related groups. Some amoebas are probably more closely related to some flagellates than they are to other amoeboid groups.

For purposes of discussion we divide the protozoa into flagellates, amoebas and relatives, spore-producers, and ciliates, recognizing that only the ciliates are likely to be a natural group. A systematic résumé is provided at the end of the chapter. The system of classification published in 1980 by the Society of Protozoologists recognizes only six phyla. It is simple but certainly artificial. Corliss's system recognizes 45 phyla, which probably best reflects actual monophyletic assemblages, meaning that each has a common ancestor. We followed the system used by Sleigh, which attempts to convey the array of independent evolutionary lines but is a little more conservative than that of Corliss.

FLAGELLATED PROTOZOA

The flagellates, also called mastigophorans, include those protozoa that possess flagella. They include the largest number of species, some 6900. They are commonly divided into **phytoflagellates** and **zooflagellates,** a division of convenience, not one of evolutionary relationship. The phytoflagellates usually bear one or two flagella and typically possess chloroplasts. These organisms are thus plantlike, and many botanists treat species in this division as algae. The phytoflagellate division contains most of the free living members of the class and includes such common organisms as *Euglena, Chlamydomonas, Volvox,* and *Peranema.* The zooflagellates possess one to many flagella, lack chloroplasts, and are heterotrophs. Some are free living, but the majority are commensal, symbiotic, or parasitic in other animals, especially arthropods and vertebrates.

Locomotion

The presence of flagella is the distinguishing feature of flagellates, and most species possess two. They may be of equal or unequal length, and one may be leading and one trailing, as in *Peranema* (Fig. 2–13B) and the dinoflagellates.

As we described earlier (p. 16), flagellar propulsion follows essentially the same principle as that of a

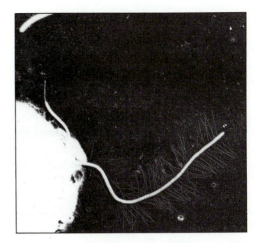

FIGURE 2-12 A phytoflagellate with a long flagellum bearing mastigonemes and a short smooth one. *(Photo courtesy of BioPhoto Associates.)*

propeller, the flagellum undergoes undulations that either push or pull. The undulatory waves pass from base to tip and drive the organism in the opposite direction (Fig. 2–4), or more rarely, the undulations pass from tip to base and pull the organism. Many flagellates have flagella with fine lateral branches (**mastigonemes**) (Fig. 2–12). The hydrodynamic effect of the branches is to cause the flagellum to pull rather than push even though the flagellar waves are passing from base to tip. Flagellates that have thin, flexible pellicles are often capable of amoeboid movement, but some chrysomonads and others may cast off their flagella and assume an amoeboid type of locomotion entirely.

Nutrition

Phytoflagellates are primarily autotrophic and contain chlorophyll. When the chlorophyll is not masked by other pigments, the flagellate appears green in color. If the xanthophylls dominate, the color is red, orange, yellow, or brown. Different phytoflagellate groups are characterized by different combinations of chlorophyll types and accessory pigments, and they store reserve foods as oils or fats or various forms of carbohydrates.

Strict heterotrophic nutrition occurs in zooflagellates as well as some other groups, and there are many parasitic species. The mechanisms of food capture and ingestion vary greatly, and the methods employed by some of the better known groups will be described in the following sections.

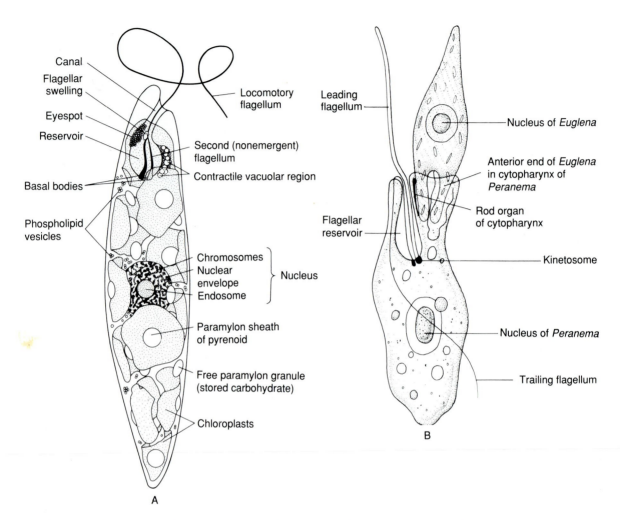

Canal
Flagellar swelling
Locomotory flagellum
Eyespot
Reservoir
Second (nonemergent) flagellum
Contractile vacuolar region
Basal bodies
Phospholipid vesicles
Chromosomes
Nuclear envelope
Endosome
Nucleus
Paramylon sheath of pyrenoid
Free paramylon granule (stored carbohydrate)
Chloroplasts

Leading flagellum
Nucleus of *Euglena*
Anterior end of *Euglena* in cytopharynx of *Peranema*
Rod organ of cytopharynx
Flagellar reservoir
Kinetosome
Nucleus of *Peranema*
Trailing flagellum

A

B

FIGURE 2-13 A, Structure of the phytoflagellate *Euglena gracilis.* **B,** *Peranema* swallowing a *Euglena.* *(A, From Leedale, G. F. 1967. Euglenoid Flagellates. Prentice-Hall, Inc., Englewood Cliffs, N.J. B, Modified after Chen.)*

Form and Structure

Flagellates differ so greatly in structure that a description of the assemblage as a whole is difficult. Most possess distinct anterior and posterior ends, although almost any plan of symmetry occurs. Most are free-swimming, but there are some sessile forms. There are also many colonial species. Space permits description of only a few of the many different groups.

Euglenophytes

The members of the marine and freshwater phylum Euglenophyta contain chlorophyll b as well as chlorophyll a. The genera *Peranema* and *Euglena* contain perhaps the most familiar flagellates. The body is elongated with an invagination (reservoir) at the anterior end. A contractile vacuole discharges into the pocket, and two "hairy" flagella arise from its wall. In *Euglena* one flagellum is very short and terminates at the base of the long flagellum (Fig. 2–13A). A pigment spot, or stigma, shades a swollen basal area of the long flagellum, which is thought to have a photoreceptive function. In the colorless *Peranema* both flagella are long, but one trails backward (Fig. 2–13B). The pellicle is commonly flexible, and helical striations or sculpturing can be seen with electron microscopy.

Green species, such as *Euglena,* are of course autotrophic, and carbohydrate is stored as granules of

paramylon. Colorless heterotrophic species may depend on organic compounds absorbed from the surrounding water or may be phagotrophs, like *Peranema*. In *Peranema* the anterior end of the body contains two parallel rods, making up a so-called rod organ, located adjacent to the reservoir (Fig. 2–13B). The anterior of the rod organ terminates at the cytostome, which is just below the outer opening of the canal leading from the reservoir. *Peranema* feeds on a wide variety of living organisms, including *Euglena,* and the cytostome can be greatly distended to permit engulfment of large prey. In feeding, the rod organ is protruded and used as an anchor to pull in prey, which is swallowed whole (Fig. 2–13B), or the rod organ can cut into the victim, whose contents are then sucked out. Following ingestion, the prey is digested within food vacuoles.

Chlorophytes: Order Volvocales

The Volvocales are part of the large phylum of marine and freshwater green algae (phylum Chlorophyta), which includes nonmotile filamentous and thalloid forms. The cells of the Volvocales are bounded by a secreted glycoprotein wall, and much of the interior is occupied by a large chloroplast. There are two to four smooth flagella. An eyespot and two contractile vacuoles also may be present. Among the flagellate species, some are solitary, such as *Chlamydomonas* (Fig. 2–14A), and others are colonial. The colonies have the form of flattened curved plates of from 4 to 64 cells (Fig. 2–14B) or hollow spheres. The largest colonies are the hollow spheres of *Volvox* and *Pleodorina,* which contain hundreds of cells interconnected by protoplasmic strands in addition to mucoid material (Fig. 2–14C).

Dinoflagellates

The phylum Dinophyta contains some 2000 marine and freshwater flagellates of considerable ecological importance in the sea, where they convert sunlight into photosynthate. The presence of xanthophyll pigments give them a brown or golden brown color. They lack chlorophyll b, although chlorophylls a and c are present. The nuclei contain permanently condensed chromosomes having relatively small amounts of protein. Each chromosome is attached to the nuclear membrane. Many biologists believe these nuclear features to be a very primitive eukaryote condition. The generalized dinoflagellate body is roughly

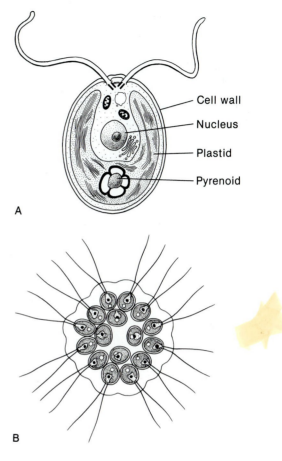

A

Cell wall
Nucleus
Plastid
Pyrenoid

B

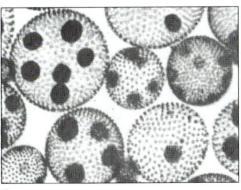

C

FIGURE 2-14 Volvocales. **A,** *Chlamydomonas reinhardtii,* a noncolonial species. **B,** *Gonium pectorale.* This genus of colonial species has the cells arranged as a flat, rather square plate, all embedded within a gelatinous envelope. **C,** *Volvox,* in which the colonies are hollow spheres. Note daughter colonies within parent colonies. *(A, From Sleigh, M. 1989: Protozoa and Other Protists. Edward Arnold, London. p. 140. B, C, Photomicrograph courtesy of General Biological Supply House, Inc.)*

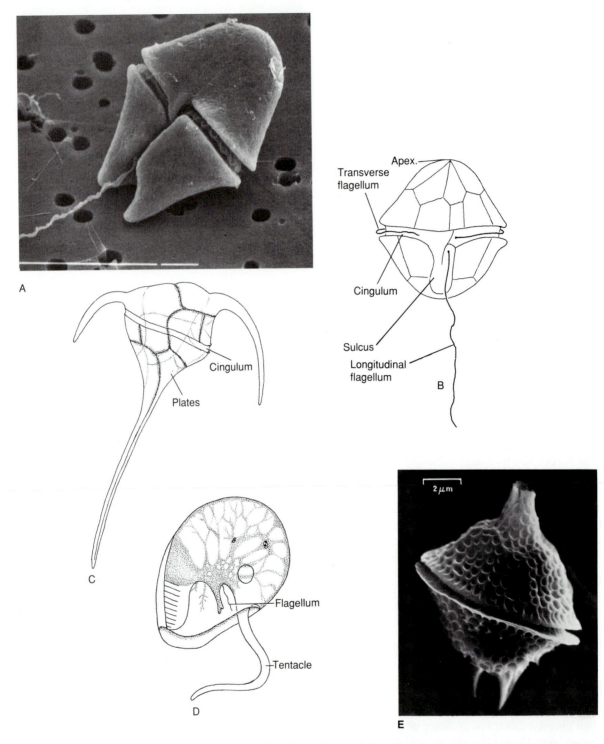

FIGURE 2-15 Dinoflagellates. **A,** A naked dinoflagellate, *Gymnodinium.* **B,** A freshwater armored dinoflagellate, *Glenodinium cinctum.* **C,** *Ceratium.* **D,** *Noctiluca.* An aberrant bioluminescent dinoflagellate. Only one of the small flagella is visible within the "oral" depression. **E,** Scanning electron micrograph of *Gonyaulax digitale,* a marine species that causes red tides. *(A, From Dodge, J. D., and Lee, J. J. 1985. Dinoflagellida. In Lee, J. J., et al. (Eds.): An Illustrated Guide To Protozoa. Society for Protozoology, Lawrence, KS. p. 29. B, After Pennak, R. W. 1978. Freshwater Invertebrate of the United States. 2nd Edition. John Wiley and Sons, New York. C, After Jorgenson. D, After Robin. E, Micrograph courtesy of Dodge, J. D.)*

ovoid but asymmetrical (Fig. 2–15A). Typical dinoflagellates possess two flagella. One is attached a short distance behind the middle of the body, is directed posteriorly, and lies in a longitudinal groove (**sulcus**). The other flagellum is transverse and located in a groove (**cingulum**) that either rings the body or forms a spiral of several turns. The transverse flagellum causes both rotation and forward movement; the longitudinal one drives water posteriorly and contributes to forward motion.

Dinoflagellates possess a relatively complex thickened pellicle, or theca, which contains deposits of skeletal cellulose within flattened intracellular vesicles, or alveoli. Where the theca is thin and flexible, as in the common freshwater and marine genus *Gymnodinium,* the dinoflagellate is said to be unarmored, or naked (Fig. 2–15A). Armored dinoflagellates have a highly developed theca composed of two valves or of plates (Fig. 2–15B). Frequently the armor is sculptured, and often long projections or winglike extensions protrude from the body, creating bizarre shapes (Fig. 2–15C). The large and aberrant *Noctiluca* (Fig. 2–15D) and many smaller species of some common genera are luminescent and are the principal contributors to planktonic bioluminescence (see p. 681). When present in large numbers, their greenish light in the wake of a boat or oars may produce a striking effect in a quiet sea at night.

Many dinoflagellates are autotrophic, but the colorless forms are heterotrophic, and some pigmented species exhibit both modes of nutrition. The prey is usually captured with pseudopodia and ingested through an oral opening associated with the longitudinal flagellar groove. There are parasitic and mutualistic dinoflagellates. The latter (*Symbiodinium microadriaticum*), symbionts of corals and other animals, are of great significance in reef ecology (pp. 157 and 160).

Dinoflagellates occur in countless numbers in marine plankton where they are important contributors to oceanic primary production, especially in the tropics. Marine species of the genera *Gymnodinium, Gonyaulax,* and others are responsible for outbreaks of the so-called red tides off the coasts of New England, Florida, California, Europe, and elsewhere (Fig. 2–15E). Under ideal environmental conditions and perhaps with the presence of a growth-promoting substance, populations of certain species increase to enormous numbers. Red tides are not always red. The water may be yellow, green, or brown, depending on the predominant pigments of the blooming organisms. Concentrations of certain toxic metabolic substances reach such high levels that other marine life may be killed. The 1972 red tides off the coasts of New England and Florida killed thousands of birds, fish, and other animals and wreaked havoc on the shellfish industry by infecting clams and oysters that fed on the dinoflagellates.

Ciguatera food poisoning is caused by a dinoflagellate that lives attached to multicellular algae. Ciguatoxin is acquired by grazing herbivorous fish that concentrate the toxin in their tissues and pass it up the food chain. The toxin can reach such high levels in the tissues of carnivorous fish that, when eaten by humans, it can produce serious poisoning and even death. In addition to gastrointestinal symptoms, such as diarrhea and nausea, there may be respiratory problems, muscle weakness, and long-lasting aberrant skin sensations.

Choanoflagellates

The members of this phylum of marine and freshwater zooflagellates are distinctive in having a cylindrical collar of microvilli around the base of the single flagellum (Fig. 2–16). Choanoflagellates may be solitary or colonial, attached or free swimming. Sessile species are attached by a stalk, part of a vaselike theca that is sometimes present (Fig. 2–16A). The individuals of colonial planktonic forms, such as species of *Proterospongia,* are united by a jelly-like matrix or by their collars (Fig. 2–16B and C). In the latter case the colony somewhat resembles a plate, with all of the collars and flagella located on the same side. The marine *Proterospongia choanojuncta* was found to include both a colonial planktonic stage and a solitary, aflagellate attached stage (Fig. 2–16C). Many zoologists now believe that the choanoflagellates are most closely related to the metazoan animals of any group of protozoans (p. 69).

Kinetoplastids

Zooflagellates of the phylum Kinetoplastida include a few common free-living species as well as a number of important parasites. All possess a conspicuous mass of DNA, called a **kinetoplast,** that is located within a single, large, elongated mitochondrion. The one or two flagella arise from a pit, and their basal bodies are located on or near the kinetoplast mitochondrion.

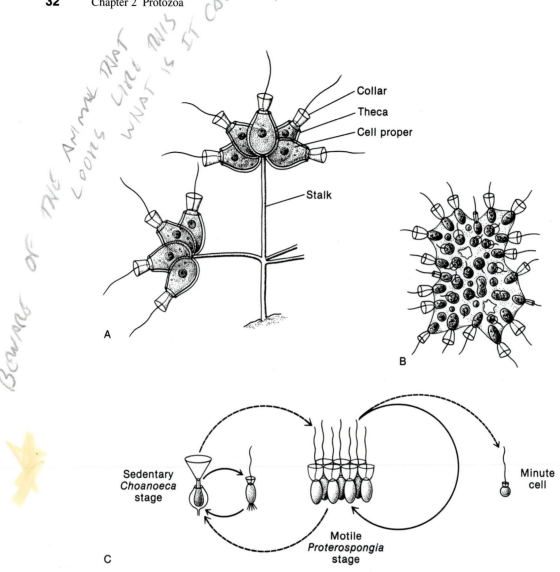

Beware of the animal that looks like this what is it colum 2,

FIGURE 2-16 Choanoflagellates, in which the flagellum is surrounded by a collar of microvilli. **A,** A stalked species. The stalk is an extension of a vaselike theca that surrounds the cell proper. **B,** *Proterospongia,* a colonial species, in which the individuals are connected by a jelly-like matrix. **C,** *Proterospongia choanojuncta* has both a sessile and planktonic stage, the latter being both colonial and solitary. *(A, From Farmer, J. N. 1980. the Protozoa. C. V. Mosby Co., St. Louis. B, From Leadbeater, B. S. C. 1983. Life-history and ultrastructure of a new marine species of Proterospongia. Jour. Mar. Biol. Assoc. U.K. 63:135–160.)*

Species of the free-living biflagellate *Bodo* (Fig. 2–17D) are commonly found in brackish and fresh water and in soil, where they feed on bacteria. The trypanosomatid kinetoplastids are gut parasites of insects and blood parasites of vertebrates. Only the anterior flagellum is present (Fig. 2–17C), the second flagellum being represented by a basal body. Commonly, the flagellum trails and is connected along the sides of the body by an undulating membrane.

Species of the trypanosomid genera *Leishmania* and *Trypanosoma* are agents of numerous diseases of humans and domesticated animals in subtropical and tropical regions of the world. Part of the life cycle is passed within or attached to gut cells of blood-sucking insects, mostly various kinds of flies, and another part of the cycle is spent in the blood or in white blood cells and lymphoid cells of the vertebrate host, although other tissues may be invaded. Intracellular stages are aflagellate, but during the life cycle there are motile, extracellular flagellate stages in the bloodstream or in the invertebrate host (Fig. 2–17B and C).

Leishmania is the agent of the widespread kala-azar and related diseases of Eurasia, Africa, and America and causes skin lesions (Fig. 2–17A) and interference with immune responses, among other effects. Sand flies are the blood-sucking insect host of this protist.

Chagas's disease of tropical America, which probably accounted for Darwin's chronic ill health following the voyage of the Beagle, is caused by *Trypanosoma cruzi* and is transmitted by blood-sucking hemipteran bugs. Extensive damage may be caused in the human host when the parasite leaves the circulatory system and invades the liver, spleen, and heart muscles. *Trypanosoma brucei rhodesiense* and *T. b. gambiense* are the causal agents of African sleeping sickness and are transmitted by the tsetse fly (Figs. 2–17B and C). The parasite invades the cerebrospinal fluid and brain, producing the lethargy, drowsiness, and mental deterioration that mark the terminal phase

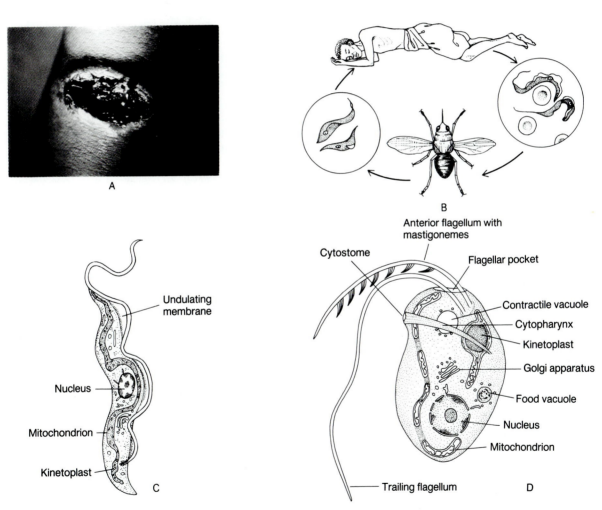

FIGURE 2-17 Kinetoplastids. **A,** Skin lesion on a boy's wrist, one of the symptoms of kala-azar disease caused by *Leishmania.* **B,** Life cycle of *Trypanosoma brucei.* **C,** Structure of *Trypanosoma brucei.* **D,** *Bodo saltans,* a free living member of the Kinetoplastida. Only sectioned parts of the long single looping mitochondrion are shown. *(A, Photo Courtesy of S. S. Hendrix. B, From Sleigh, M. A. 1973. The Biology of Protozoa. Edward Arnold, London. p. 141. C, Modified after Brooker from Farmer, J. N. 1980. The Protozoa: Introduction to Protozoology. C. V. Mosby Co., St. Louis. p. 214.)*

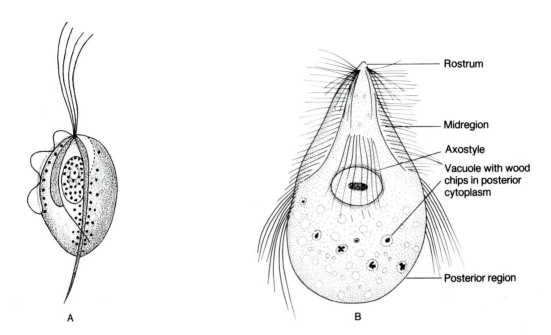

FIGURE 2-18 A, *Trichomonas vaginalis,* a trichomonad, parasitic in the human vagina and male reproductive tract. In addition to the four anterior flagella, there is a trailing flagellum bordering an undulating membrane. A supporting rodlike structure, the axostyle, extends posteriorly from the blepharoplast. **B,** *Trichonympha campanula,* a hypermastigote flagellate that lives in the gut of termites. *(A, After Wenrich. B, From Farmer, J. N. 1980. The Protozoa: Introduction to Protozoology. C. V. Mosby Co., St. Louis. p. 266.)*

of the disease. There are also various trypanosome diseases of horses, cattle, and sheep that are of considerable economic importance.

Multiflagellated Zooflagellates

Most of the remaining zooflagellates possess from four to many (sometimes thousands) of flagella. Groups of flagella are associated with several microtubular and fibrillar organelles, together forming a complex called a **mastigont system.** There are very few free-living species (*Hexamita*); most live anaerobically in the guts of vertebrates and insects, especially wood roaches and termites, and typically lack mitochondria. *Giardia intestinalis* occurs in the intestine of humans and generally produces no symptoms, but heavy infections can cause diarrhea. *Trichomonas vaginalis* is a small parasite with four anterior flagella (Fig. 2–18A) that inhabits the urogenital tract of humans and causes a widespread venereal disease. Living tissues can be invaded and the vagina of seriously infected women produces a greenish yellow discharge.

In the evolution of these flagellates great complexity has resulted from multiplication of the mastigont systems, leading to species with many flagella. The hypermastigote flagellates, which are gut symbionts of termites and wood roaches, are extremely complex. They are nearly all multiflagellate, with a saclike or elongated body, usually bearing an anterior rostrum and cap. The ultrastructure of *Trichonympha,* which possesses thousands of flagella, virtually defies description (Fig. 2–18B). However, there is also a hypermastigote (*Mixotricha*) in termites that has few flagella and is moved by attached bacteria (spirochetes). In many termites and wood-eating cockroaches, the host is dependent on its flagellate fauna for the digestion of wood. The wood consumed by the roach or termite is ingested by the flagellates, and the products of flagellate digestion are utilized by the insect. The ingestion of the wood particles occurs at the posterior end of the flagellate by pseudopodial engulfment. The termite host loses its fauna with each molt, but by licking other individuals, by rectal feeding, or by eating cysts passed in feces (in the case of roaches), a new fauna is obtained. In wood-eating cockroaches, the life cycles of the flagellates are closely correlated with the production of molting hormones by the late nymphal insect.

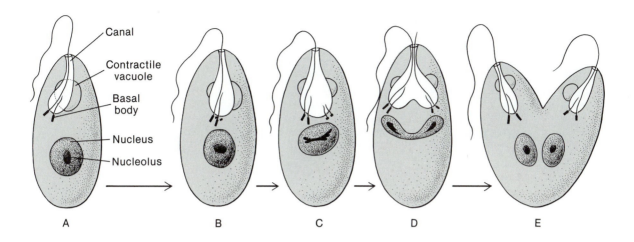

FIGURE 2-19 Asexual reproduction (symmetrogenic division) in *Euglena.* **A,** The centriole has already divided. **B,** Each centriole produces a new basal body and flagellum. The nucleus is in prophase, and the contractile vacuole is double. **C,** The old pair of flagellar roots separate and fuse with the new roots. **D,** Mitosis proceeds, and the gullet begins to divide. **E,** Anterior end dividing following duplication of organelles. *(Redrawn from Ratcliffe, 1927.)*

Reproduction and Life Cycles

In the majority of flagellates, asexual reproduction occurs by binary fission, and most commonly the organism divides longitudinally. Division is thus said to be **symmetrogenic,** that is, producing mirror-image daughter cells (Fig. 2–19). In multiflagellate species, the flagella are divided between the daughter cells. In those species with few flagella, the one or several flagella may duplicate prior to cell division, or they may be equally apportioned to each daughter cell, resorbed, and formed anew in each daughter cell, or even unequally apportioned. The same may apply to other organelles within the parent. Thus, division in many flagellates is not perfectly symmetrogenic.

In the armored dinoflagellates, the two fission products regenerate the missing plates or form a completely new armor, the old one having been shed before division. In *Volvox* any one of a certain few cells at the posterior of the colony may undergo fission to form a daughter colony within the parent (Fig. 2–14C). The daughter colonies usually escape by rupturing the wall of the parent colony.

Sexual reproduction is still poorly known for many flagellate groups, and in some, such as the euglenophytes, it apparently does not take place. In many dinoflagellates and metamonads the zygote is formed from the fusion of flagellated isogametes. The Volvocida displays all gradations of gamete differentiation, from isogamy to highly developed heterogamy, and meiosis takes place after the formation of the zygote, that is, it is postzygotic (Fig. 2–11B). In some species of *Chlamydomonas,* the cells act as gametes (isogametes). Other species show the beginnings of sex differentiation by having gametes that differ just slightly in size. In *Platydorina,* heterogamy is well developed, but the large macrogametes still retain flagella and are free swimming. Finally, in *Volvox,* true eggs and sperm develop from special reproductive cells at the posterior of the colony. The egg is stationary and is fertilized within the parent colony. Colonies may be either monoecious or dioecious.

Cysts are formed in many flagellate groups. Nonflagellated (**palmella**) stages are characteristic of most well-known phytoflagellates, such as dinoflagellates. In the palmella stage the organism loses its flagellum and becomes a ball-like, nonmotile, usually floating, relatively undifferentiated cell, located inside the original parental envelope when such an envelope is present. Fission often follows, so that the palmella consists of a cluster of cells. It is in the palmella stage that symbiotic dinoflagellates inhabit their host.

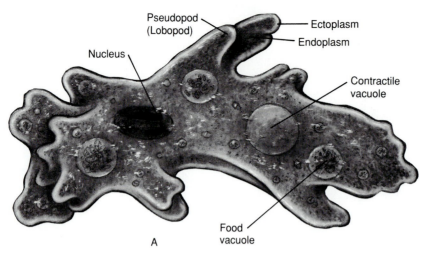

A

FIGURE 2-20 **A,** *Amoeba.* **B,** *Chaos carolinense,* a large ameba that moves by a single lobopodium. **C,** *Penardia mutabilis,* a freshwater ameba with branching, anastomosing filopods. *(B and C, From Bovee, E. C. 1985. Class lobosea and class filosea. In Lee, J. J., et al. (Eds.): Illus-trated Guide to the Protozoa. Society for Protozoology, Lawrence, KS. pp. 162 and 230.)*

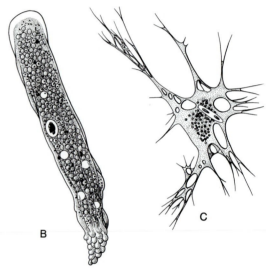

B

C

SUMMARY

1. Flagellates are distinguished by the presence of one or more flagella.

2. Flagellate assemblage include many autotrophic groups (phytoflagellates). They possess chlorophyll plus other pigments and store such food materials as oils, fats, and starches (other than glycogen). These groups are assigned to various algal phyla, some containing multicellular algae.

3. The remaining flagellates are a small, heterogeneous assemblage of heterotrophs (zooflagellates). A few are free living, but most are parasitic, commensal, or mutualistic in other animals.

4. Flagellar beat pushes or pulls the flagellate, and the path of movement depends on the point of flagellum attachment and the combined action when there is more than one flagellum.

5. Most flagellates divide longitudinally. Where sexual reproduction occurs, isogametes are common.

AMEBOID PROTOZOA

The ameboid protozoa, traditionally placed in an old subphylum, the Sarcodina, include those protozoa in which adults possess flowing extensions of the body called **pseudopodia.** Here belong the familiar amebas as well as many other marine, freshwater, and terrestrial taxa. Pseudopodia are used for capturing prey and in benthic groups for locomotion as well. The ameboid form of some species may be a retention of the ancestral protistan condition. However, in most species it has probably been secondarily derived after loss of flagella, for many groups possess flagellated gametes in their life cycle.

Ameboid protozoa are asymmetrical or have a radial symmetry. They possess relatively few organelles and in this respect are perhaps the simplest protozoa. However, skeletal structures, which are found in the majority of species, reach a complexity and beauty that is surpassed by few other organisms.

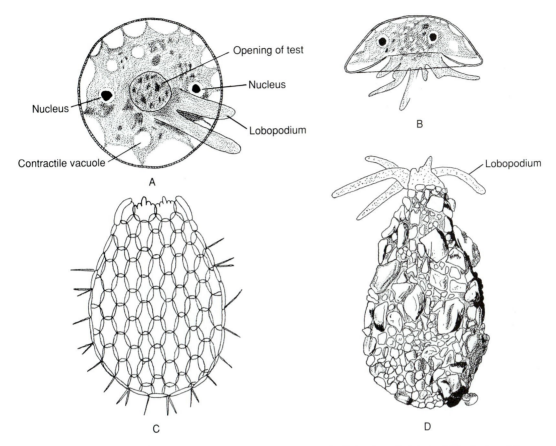

FIGURE 2-21 Shelled amebas. **A,** *Arcella vulgaris,* apical view. **B,** Side view. **C,** Test of *Euglypha strigosa,* composed of siliceous scales and spines. **D,** *Difflugia oblonga* with test of mineral particles. *(A, B, and D, After Deflandre, G. 1953. In Grassé, P. Traité de Zoologie, Masson and Co., Paris. Vol. I, pt. II. C, After Wailes.)*

Form and Structure

There are four principal groups of ameboid protozoa: the amebas, the foraminiferans, the heliozoans, and the radiolarians.

Amebas

Amebas may be naked or enclosed in a shell or test. The naked amebas, which include the genera *Amoeba* (Fig. 2–20A) and *Pelomyxa,* live in the sea, in fresh water, and in the water films around soil particles. The shape, although constantly changing, is characteristic of different species. Some giant forms reach several millimeters in length. The cytoplasm in amebas is divided into a stiff, clear, external ectoplasm and a more fluid internal endoplasm. The pseudopodia are one of two types. **Lobopodia,** which are typical of many amebas, are rather wide with rounded or blunt tips (Fig. 2–20A and B). They are commonly tubular and are composed of both ectoplasm and en-

doplasm. **Filopodia,** which occur in many small amebas, are narrow, clear and sometimes branched (Fig. 2–20C). Because lobopods, filopods, and the netlike pseudopodia of the next group (forams) are all modifications of the same pseudopodial type, the ameboid protozoa that possess them are called rhizopods, meaning "rootlike foot" (phylum Rhizopoda).

In shelled amebas, which are largely inhabitants of fresh water, damp soil, and mosses, either the radial or bilateral extracellular shell is secreted by the cytoplasm, in which case it is organic, siliceous, or both, or it is composed of foreign materials embedded in a cementing matrix. The ameba is attached by protoplasmic strands to the inner wall of the shell, and there is a large opening through which the pseudopodia or body can be protruded. Pseudopodia may be either lobopods or filopods. In *Arcella* (Figs. 2–21A and B), one of the most common freshwater amebas,

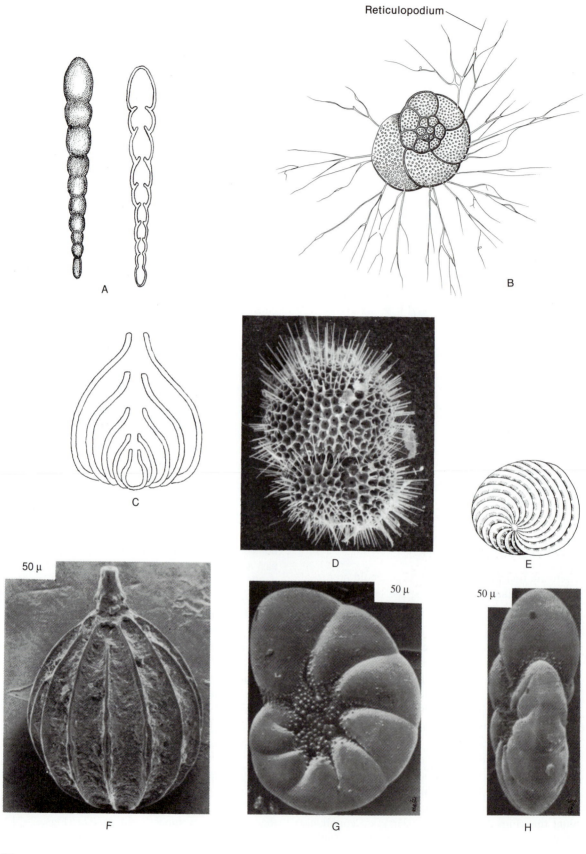

Reticulopodium

A

B

C

D

E

50 μ

F

50 μ

G

50 μ

H

38

the brown or straw-colored protein shell has the shape of a flattened dome with the aperture in the middle of the underside. In *Euglypha* the secreted shell is constructed of overlapping siliceous scales (Fig. 2–21C). *Difflugia* has a shell composed of mineral particles that are ingested by the animal and embedded in a secreted matrix (Fig. 2–21D).

Marine species commonly lack contractile vacuoles; freshwater species possess one to several, and at least in larger naked species they tend to form at the forward end of the organism and are expelled at the rear.

Foraminiferans

The large order Foraminiferida is primarily marine. The pseudopodia, called **reticulopodia,** form a threadlike branched mesh (Fig. 2–22B). Each reticulopodium contains axial microtubules and uses them to transport vesicles bidirectionally, like adjoining up and down escalators. Foraminiferans construct a shell of secreted organic material or of cemented foreign mineral particles or of secreted calcium carbonate. Calcareous shells are most common and well preserved in the fossil records; 28,000 of the 35,000 described forams are fossil species (see p. 597).

Some species of forams live within a single-chambered shell, but most forms have multichambered shells. Multichambered forams begin life in a single chamber, but as the organism increases in size, the cytoplasm overflows through a large opening in the first chamber and secretes another compartment. This process is continuous throughout the life of the foram and results in the formation of a series of many chambers, each of which may be larger than the preceding one. Because the addition of new chambers follows a symmetrical pattern, the shells have a distinctive shape and arrangement of chambers (Fig. 2–22).

The entire shell is filled with cytoplasm that is continuous from one chamber to the next. The outside of the foram shell is covered with a thin layer of cytoplasm that overflows from the large aperture. Pseudopodia may be restricted to the cytoplasm of the aperture, or they may arise from the layer over the shell (Figs. 2–22B). In some species they emerge through shell pores; however, there are forams with superficial pores that are sealed off from the interior.

Multichambered forams are not colonies but single cells. Many are visible to the naked eye and a few, such as the so-called mermaid's pennies of Australia, attain a size of a centimeter in diameter.

Most forams are benthic, but species of *Globigerina* and related genera are common planktonic forms. The chambers of these species are spherical but arranged in a somewhat spiral manner (Figs. 2–22B and D). Planktonic forams have more delicate shells than do benthic species, and the shells commonly bear spines. The spines are so long in some species that the foram is visible to the naked eye and can be collected undamaged with a jar by a scuba diver.

A few forams are sessile. *Homotrema* forms large, red, calcareous tubercles about the size of a wart on the underside of coral heads. The pink sands of the beaches of Bermuda result from the large number of pieces of *Homotrema* tests.

A number of forams harbor symbiotic autotrophic protists—such as chlorophytes, diatoms, or unicellular red algae, depending on the foram—an unusual diversity.

Heliozoans

The spherical protozoa called heliozoans occur in the sea and in still bodies of fresh water and may be floating or more commonly located on bottom debris. Some bottom-dwelling forms are stalked (Fig.

◀ **Figure 2-22** Foraminiferans. **A,** Shell of *Rheophax nodulosa,* entire, and in section. **B,** Living *Globigerina bulloides.* **C,** Shell of an ellipsoidinid foraminiferan, in section. **D,** Cleaned test of *Globigerinoides sacculifer,* a tropical planktonic foram with spines. **E,** *Archaias* sp., a common benthic foram of shallow tropical waters. **F–H,** Scanning electron micrographs of foram tests: **F,** *Lagena sulcata.* **G** and **H,** Surface and edge views of *Nonion depressulus. (A, After Brady. B, drawn from a photograph in Grell, K. G. 1973. Protozoology. Springer-Verlag, Berlin. p. 285. D, By Be, A. W. H. 1968. Science, 161:881–884. Copyright 1968 by the American Association for the Advancement of Science. F–H, From Murray, J. W. 1971. An Atlas of British Recent Foraminiferida. Heinemann Educational Books.)*

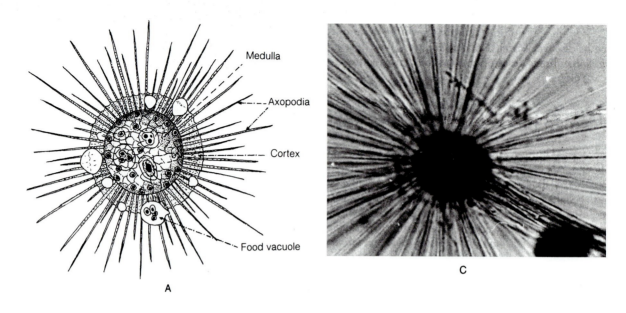

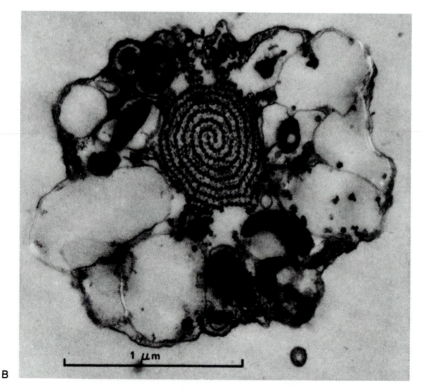

FIGURE 2-23 A multinucleate heliozoan, *Actinosphaerium eichorni*. **A,** Entire animal. **B,** Electron micrograph of a section through an axopod of *Actinosphaerium*. The axial rod is composed of a double spiral of microtubules. **C,** A sessile, stalked heliozoan. Stalk extends toward lower right corner. Medulla, cortex, and axopods can be clearly seen in photograph. *(A, After Doflein. B, Photomicrograph courtesy of Macdonald, A. C. In Sleigh, M. A., 1973. The Biology of Protozoa. Edward Arnold Publishers, London. p. 162. C, From Febre-Chevalier, C. 1985. Class Heliozoea. In Lee, J. J., et al. (Eds.): Illustrated Guide to the Protozoa. Society for Protozoology, Lawrence, KS. p. 307.)*

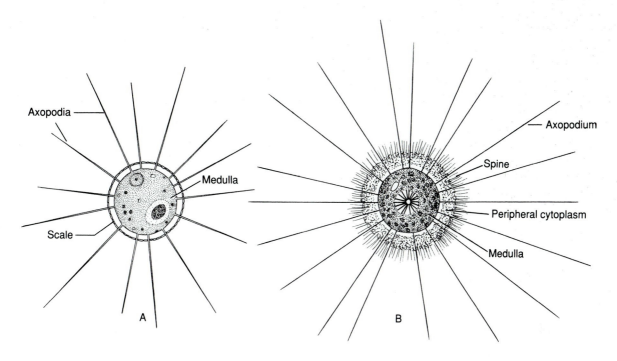

FIGURE 2-24 Heliozoans. **A,** *Pinaciophora fluviatilis* with skeleton of scales. **B,** *Heterophrys myriopoda* with skeleton in form of spines. *(A and B, From Penard from Hall.)*

2–23C). The fine, needle-like pseudopodia, called **axopodia,** radiate from the surface of the body (Fig. 2–23). Each axopod contains a central axial rod, which is covered with a moving, adhesive cytoplasm. Although the axial rod has a supporting function, it is not a permanent skeleton but a bundle of microtubules that can shorten or even "melt." Some species have long delicate filopodia in addition to the axopodia. Ameboid protozoa with axopods, which include the heliozoans and the next group, the radiolarians, are collectively called actinopods and compose the phylum Actinopoda.

The body of a heliozoan consists of an outer ectoplasmic sphere, or **cortex,** which is often greatly vacuolated, and an inner part of the body, or **medulla** (Fig. 2–24). The medulla is composed of dense endoplasm, containing one to many nuclei and the bases of the axial rods. Contractile vacuoles are present in freshwater species. There are also species that harbor symbiotic green algae (zoochlorellae).

Although heliozoans can be naked (Fig. 2–23), skeletons are present in some species and may be composed of organic or siliceous pieces secreted by the organism and embedded in an outer gelatinous covering. The siliceous pieces assume a great variety of forms, such as scales (Fig. 2–24A), tubes, spatulas, or needles. These siliceous pieces may be arranged tangentially to the body or, when the skeleton is composed of long needles (Fig. 2–24B), may radiate like the axopods. Regardless of the nature and arrangement of the skeleton, openings are present through which the axopods project.

Radiolarians

Among the most beautiful of the protozoa are members of three classes, collectively called radiolarians. They are entirely marine and primarily planktonic. Radiolarians are relatively large protozoa; some species are several millimeters in diameter, and some colonial species attain a length of up to 20 cm (*Collozoum*). Like heliozoans, the bodies of radiolarians are usually spherical and divided into inner and outer parts (Fig. 2–25). The inner region, which contains one to many nuclei, is bounded by a **central capsule** with a membranous wall, a distinctive feature of radiolarians. The capsule membrane is perforated by

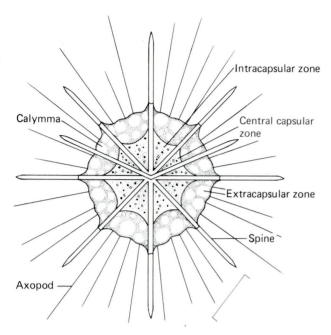

FIGURE 2-25 *Acanthometra,* a radiolarian with a skeleton of radiating strontium sulfate rods. *(From Farmer, J. N. 1980. The Protozoa. C. V. Mosby Co., St. Louis. p. 353.)*

openings (Fig. 2–26A), which allow the cytoplasm of the central capsule (or intracapsular cytoplasm) to be continuous with the cytoplasm of the outer division of the body. This extracapsular cytoplasm forms a broad vacuolated cortex, called the **calymma,** that surrounds the central capsule. In many species the calymma contains large numbers of symbiotic cryptomonads or more often dinoflagellates (zooxanthellae) in a nonflagellate (palmella) stage.

The pseudopodia are axopods and long delicate filopods that radiate from the surface of the body. Their axes of microtubules originate in the cytoplasm inside the central capsule and extend through the calymma.

A skeleton is almost always present in radiolarians and is usually siliceous, but in the related class Acantharea it is composed of strontium sulfate. Several types of skeletal arrangements occur. One type has a radiating structure, in which the skeleton is composed of long spines or needles that radiate from the center of the central capsule and extend beyond the outer surface of the body (Fig. 2–25). A second type of skeleton is constructed in the form of a spherical or bilateral lattice, which is often ornamented with barbs and spines (Fig. 2–26B).

The planktonic radiolarians display a distinct vertical stratification from the ocean surface down to 5000 m depths. The great numbers in which planktonic radiolarians as well as foraminiferans occur are indicated by the fact that their shells, sinking to the bottom at death, form a primary constituent of many ocean bottom sediments. Where they compose 30% or more of the sediment, it is called a **foram** or **radiolarian ooze.** At depths below 4000 m, however, the great pressure tends to dissolve foram shells. The strontium sulfate skeletons of acantharians do not seem to fossilize but dissolve rapidly with the death of the cell.

The ameboid protozoa are the only large group of protozoa that has an extensive fossil record. Fossil forms are, of course, restricted to those with skeletons or shells—the shelled amebas, the foraminiferans, and the radiolarians. The fossil record of the shelled amebas is relatively brief and recent. The group appears as fossils only in the Cenozoic era, and the fossils consist of forms that are virtually identical to those living today. However, the foraminiferans and radiolarians have a long and abundant fossil record. In fact, the radiolarians are among the oldest known fossils. The foraminiferans first appeared in the Cambrian period, and from the late Paleozoic era on, there is an abundant fossil record. Extensive accumulations of foram shells occurred during the Mesozoic and early Cenozoic eras and contributed to the formation of great limestone and chalk deposits in different parts of the world. The quarries from which the Egyptian pyramids were built are composed predominantly of foram shells.

The widespread occurrence of fossil forams and their long geologic history make them extremely useful. Index fossils for correlating sedimentary layers, that is, sediments of the same age contain the same forams, are important indicators for geologists involved in oil exploration. Certain marine planktonic forams spiral clockwise or counterclockwise depending on temperature, providing a record of past cold and warm periods. The varying ratios of oxygen isotopes in foram shells from deep sea sediments provide clues about glacial ice accumulation.

Locomotion

Flowing ameboid movement is limited to those rhizopods that possess lobopods or filopods and has been most studied in the large naked amebas with

FIGURE 2-26 A, Glass model of a colonial radiolarian, *Trypanosphaera transformata.* **B,** Spherical skeleton of *Hexacontium.* **C,** Bilateral skeleton of *Lamprocyclas. (A, Neg./Trans. No. 318863. Courtesy Department Library Services, American Museum of Natural History. B and C, From Cachon, J., and Cachon, M. 1985. Class Polycystinea. In Lee, J. J., et al. (Eds.): Illustrated Guide to the Protozoa. Society for Protozoology, Lawrence, KS. pp. 288 and 296.)*

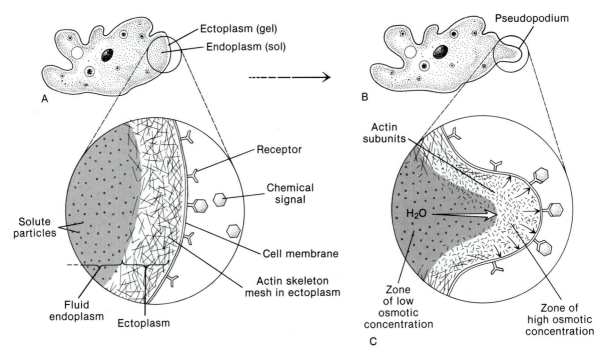

FIGURE 2-27 Osmotic theory of amoeboid flow. Actin mesh creates gel state of ectoplasm. **A,** Chemical signals bind to membrane receptors and initiate the depolymerization of the actin mesh. **B,** Depolymerized actin subunits raise osmotic concentration of this region of ectoplasm and water flows in. **C,** The influx of water causes the extension of the cell as a pseudopodium. Peripheral to the pseudopodial tip, actin subunits repolymerize as skeletal mesh, extending the ectoplasmic gel state forward like a progressing sleeve.

lobopods. In these animals locomotion may involve a single large, tubular lobopod or several small ones with caps of hyaline protoplasm at the tips. At points, a pseudopodium is anchored to the substratum.

The theories of ameboid movement accepted by most zoologists at present assume that cytoplasmic flow is related to the changes between the sol and gel states of the peripheral cytoplasm. The pseudopodial tip controls the change. As a result of some initial stimulus, the outer gelled ectoplasm becomes fluid at the site where the pseudopod will form, and internal pressure causes the inner fluid endoplasm to flow out at this point, forming a pseudopodium. In the interior of the pseudopodium, the endoplasm flows forward along the line of progression. Around the periphery, endoplasm is converted to ectoplasm, thus building up and extending the sides of the pseudopodium like a well-starched sleeve. In the conversion of endoplasm to ectoplasm actin subunits polymerize (become longer) and bond to each other at more or less right angles, creating a mesh of filaments. It is this mesh that accounts for the rigid gelatinous state of ectoplasm. The small mesh size excludes organelles and thus accounts for the hyaline appearance. At the pos-

terior end of the body, ectoplasm is converted to endoplasm by depolymerization. Cell membrane is also removed here, and new cell membrane is added at the pseudopodial tip.

The force for flow could be generated in one of two ways. Bonding of myosin with the actin mesh could convert the mesh into a contractile jacket, forcing the fluid interior endoplasm forward. However, myosin has been difficult to demonstrate in ectoplasm. Alternatively, the initial depolymerization of the actin mesh at the pseudopodial tip would increase the number of particles (actin subunits) (Figure 2–27). Particle increase would raise the osmotic concentration and water would flow from the endoplasm out into the tip.

The mechanism of flow in filopods and reticulopods is less clear, because cytoplasm appears to move along the outside of a more rigid axis. The axis is composed of microtubules, which together resemble an elaborate ciliary axoneme but which never have a 9+2 arrangement. A shearing bidirectional flow occurs in which the flow is toward the pseudopodial tip on one side and back on the other. Some filopods have been reported to fold up on retraction.

Creeping forams such as *Allogromia* extend a rigid reticulopod net for some distance from the body. The body is pulled or dragged along by the reticulopods, which may project several millimeters into the sand. Movement appears to involve lengthening and shortening of the axial microtubules.

The pseudopodia of most heliozoans and radiolarians are food-capturing rather than locomotor organelles. However, radiolarians are able to move vertically in the water by extending or contracting the calymma and axopods, by increasing or decreasing the vacuolated condition of the calymma, and by the presence of endoplasmic oil droplets.

Nutrition

Amoeboid protozoa are entirely heterotrophic. Their food consists of all types of small organisms: bacteria; algae; diatoms; protozoans; and even small multicellular animals, such as rotifers, copepod larvae, and nematodes. The prey is captured and engulfed by means of the pseudopodia.

In amebas, pseudopodia extend around the prey, eventually enveloping it completely with cytoplasm, or the body surface invaginates to form a food cup. The enclosing of the captured organism by cytoplasm (phagocytosis) results in the formation of a food vacuole within the ameba (Fig. 2–7B).

In foraminiferans, heliozoans, and radiolarians the numerous radiating pseudopodia (reticulopods or axopods) act primarily as traps in the capture of prey. Any organism that comes in contact with the pseudopodia becomes coated with an adhesive substance extruded to the pseudopod surface. This coating contains discharged lysosomes, and their proteolytic secretions paralyze the prey and initiate digestion even during capture. The long spines of many planktonic forams are also covered with such a film and are more important as a food-trapping surface than as a flotation mechanism. They are able to capture fairly large prey, such as small crustaceans. In all three groups food particles are enclosed in food vacuoles and drawn toward the interior of the body. The axial rods of heliozoans may contract, drawing the prey into the ectoplasmic cortex, or the axopods may liquefy and surround the food, forming a vacuole at the site of capture. The vacuole then moves inward. Digestion occurs in the cortex of heliozoans and the calymma of radiolarians. In foraminiferans food is initially digested outside the shell, and then digestion is completed inside small food vacuoles within the shell.

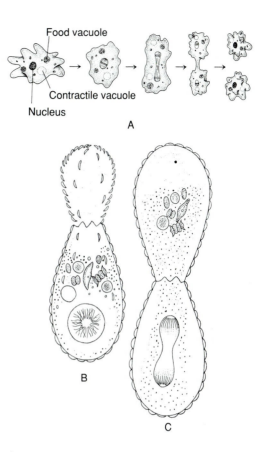

FIGURE 2-28 A, Fission in a naked ameba. **B** and **C,** Two stages in the division of *Euglypha,* a shelled ameba. **B,** Formation of skeletal plates on protoplasmic mass protruding from aperture. **C,** Division of nucleus. *(B and C after Sevajakov from Dogiel.)*

Egestion (exocytosis) can take place at any point on the surface of the body, and in the actively moving amebas, wastes are usually released at the rear, as the animal crawls forward.

Some naked amebas are parasitic. The majority are endoparasites in the digestive tracts of annelids, insects, and vertebrates. Several species occur in the human digestive tract, but of these only *Entamoeba histolytica,* which is responsible for amebic dysentery, is ordinarily pathogenic. The life cycle of these intestinal amebas is direct, and the parasites are usually transmitted from the digestive tract of one host to that of another by means of cysts that are passed in feces.

Reproduction and Life Cycle

Asexual reproduction in most amebas, heliozoans, and radiolarians is by binary fission (Fig. 2–28A). In

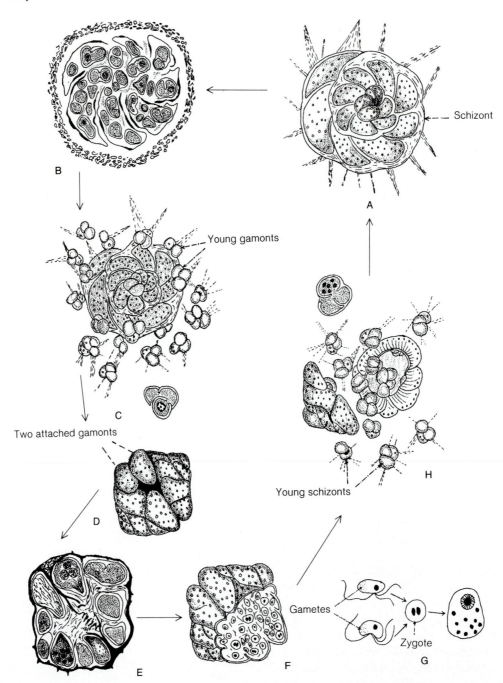

FIGURE 2-29 Life cycle of the foraminiferan, *Discorbis patelliformis*. **A,** Schizont. **B,** Asexual development of gamonts within chambers of schizont. **C,** Liberation of young gamonts from parent schizont. **D–G,** Formation and fusion of gametes within two attached gamonts. **H,** Separation of attached gamonts accompanied by liberation of young schizonts. *(A–H, After Myers from Calvez.)*

amebas with a soft shell, the shell divides into two parts, and each daughter cell forms a new half. When the shell is dense and continuous, as in *Arcella* and *Euglypha,* a mass of cytoplasm extrudes from the opening prior to division; this extruded mass secretes

a new shell (Fig. 2–28B). The double-shelled animal now divides. Multiple fission is common in multinucleated amebas and heliozoans.

Division in the radiolarians is somewhat similar to that in the shelled amebas. Either the skeleton itself divides and each progeny cell forms the lacking half, or one offspring receives the skeleton and the other secretes a new one.

Sexual reproduction has been observed infrequently in amebas. Among the heliozoans, sexual reproduction is known in some genera, such as *Actinosphaerium* and *Actinophrys* (diploid life cycle, Fig. 2–11), and a number of species produce flagellated stages. Flagellated stages have also been reported in radiolarians. All flagellated stages in ameboid protozoa, however, are not gametic; in some species they are asexual dispersive swarmers.

Reproduction in the foraminiferans commonly involves an alternation of asexual and sexual generations (haplodiploid cycle). An example of a life cycle in a multilocular foraminiferan is illustrated in Figure 2–29.

SUMMARY

1. Ameboid protozoa are distinguished by the presence of flowing extensions of cytoplasm called pseudopodia, which are used in feeding and, in some, for locomotion. The pseudopodia are given different names, depending on their shape and structure.

2. Although organelles have remained relatively simple, many species have evolved complex skeletons. The various groups of ameboid protozoa are distinguished by the nature of their skeletons and their pseudopodia.

3. The marine, freshwater, and parasitic naked amebas have no special skeletal structures and possess large, commonly tubular lobopodia or fine, strap-like filopodia, which are used for both feeding and locomotion.

4. Shelled amebas, which are found in the sea, fresh water, and soil, are covered by a shell composed of secreted organic material or of foreign mineral material cemented together. A large aperture permits the protrusion of lobopodia or filopodia.

5. Foraminiferans, which are largely benthic marine species, possess a calcareous test that is usually multichambered. A single large opening permits the protrusion of cytoplasm, which may cover the exterior of the test. Long, delicate, and often anastomosing reticulopodia extend from the protruded cytoplasm and are used in food trapping and locomotion.

6. Heliozoans are spherical, floating, and benthic ameboid protozoa found in the sea and fresh water. Long, radiating, needle-like pseudopodia (axopods) are used in trapping food. The axopods arise from the interior (medulla) and extend through an outer ectoplasmic cortex, which is commonly vacuo-

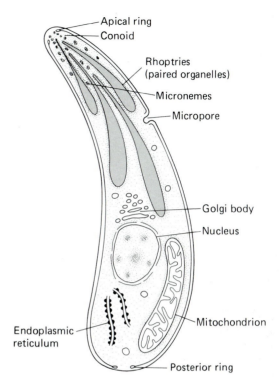

FIGURE 2-30 Lateral view of a generalized sporozoan (apicomplexan). *(From Farmer, J. N. 1980. The Protozoa. C. V. Mosby Co., St. Louis. p. 360.)*

lated. The cortex often contains a siliceous skeleton of plates, tubes, and needles.

7. Radiolarians are marine planktonic species with spherical bodies and radiating axopods. An organic capsule wall separates a central cortex from extracapsular cytoplasm. Radiolarians have complex skeletons of silicon dioxide or strontium sulfate within the extracapsular cytoplasm, organized in the form of lattice spheres or radiating spines or both.

Spore-Forming Protozoa

There are some 4000 species of endoparasitic protozoa that have sporelike stages in the life cycle. They lack cilia, flagella, or pseudopodia and live within or between cells of their invertebrate or vertebrate hosts. Most known species belong to the phylum Sporozoa, or Apicomplexa, so named because of a complex of ringlike, tubular, filamentous organelles at the apical end (Fig. 2–30), visible only with the electron microscope. The function of the apical complex is uncertain but may include entry into the host cell. One or more feeding pores are located on the side of the body.

Sporozoans are haploid except for the zygote. The life cycle typically involves an asexual and a sex-

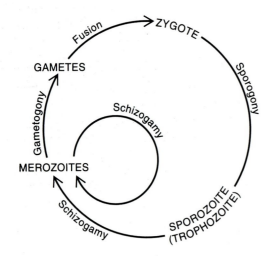

FIGURE 2-31 Life cycle of coccidean sporozoans. All stages are haploid except the zygote, which undergoes meiosis in the formation of spores (sporogony). The ability of merozoites to produce more merozoites (merogony) constitutes an asexual cycle within the sexual life cycle. *(From Levine, N. D. 1985. Phylum Apicomplexa. In Lee, J. J., et al. (Eds.): Illustrated Guide to the Protozoa. Society for Protozoology, Lawrence, KS. p. 325.)*

ual phase (Fig. 2–31). An infective sporelike stage, called a **sporozoite,** results from meiosis of the zygote. Subsequent multiple fission produces more sporozoites. The sporozoite invades the host and becomes a feeding **trophozoite.** In some sporozoans trophozoites undergo multiple fission (**schizogamy**) to produce individuals called **merozoites,** each of which may go through several additional cycles of schizogamy. Each merozoite becomes a feeding trophozoite. Eventually the trophozoite divides by multiple fission (**gamogony**) to form gametes that fuse to form a zygote. The zygote undergoes meiosis to form sporozoites.

The nature and life cycle of sporozoans can be illustrated by the coccidians, which include the parasites that cause malaria in humans. Malaria continues to be one of the worst scourges of humankind. About 300 million people are believed to be infected each year, and the annual death rate is about 1% of those infected. Left untreated, the disease can be long-lasting and terribly debilitating. Malaria has played a major and often unrecognized role in human history. The name means literally "bad air" because the disease was originally thought to be caused by the air of swamps and marshes. Although malaria had been recognized since ancient times, the causative agent was not discovered until 1880, when Louis Laveran, a physician with the French army in North Africa, identified the coccidian parasite *Plasmodium* in the blood cells of a malarial patient. In 1887, Ronald Ross, a physician in the British army in India, determined that the mosquito *(Anopheles)* was the vector.

Four species of *Plasmodium* infect humans. The introduction of the parasite into a human host is brought about by the bite of certain species of mosquitoes, which inject the sporozoites along with their salivary secretions into the capillaries of the skin (Fig. 2–32). The sporozoite is carried by the bloodstream to the liver, where it invades a liver cell and becomes a trophozoite. Here further development results in asexual reproduction through multiple fission (schizogony). The merozoite progeny cells reinvade host liver cells and continue to multiply by schizogony. After a week or so these merozoites leave the liver cells and invade red blood cells. Within the red cell the transformed merozoite or trophozoite parasite increases in size and undergoes another cycle of multiple fission. The individuals (merozoites) produced by fission within the red cells escape and invade other red cells. Liberation and reinvasion events do not occur continuously but rather in discrete episodes or pulses from all infected red blood cells. The time required to complete the developmental cycle within the host's cells determines the period between successive pulses. The periodic release of the merozoites, along with their metabolic products, causes chills and fever—the typical symptoms of malaria. More serious damage is caused by the blocking of capillaries by infected and less pliable red blood cells.

Eventually, some of the parasites invading red cells do not undergo fission but become transformed into gametocytes. The gametocyte remains within the red blood cell. If such a cell is ingested by a mosquito, the gametocyte is liberated within the new host's gut. After some further development, a male gametocyte (microgametocyte) fuses with a female gametocyte (macrogametocyte) to form a zygote. The zygote penetrates the stomach wall and gives rise to a large number of spore stages (sporozoites). It is these stages, which migrate to the salivary glands, that are introduced into the next human host by the bite of the mosquito.

The asexual stage of other coccidians occurs in blood cells or in gut cells. A number of diseases of domesticated animals are caused by coccidians. Species of the genus *Eimeria,* for example, affect chickens, turkeys, pigs, sheep, and cattle (Fig. 2–33).

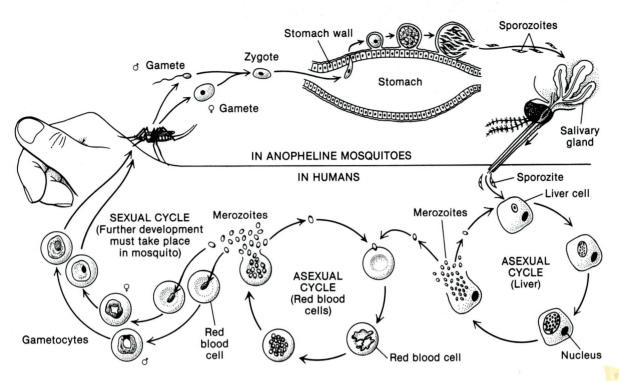

FIGURE 2-32 The life cycles of *Plasmodium* in a mosquito and in a human. Reinvasion of liver cells in the tissue cycle does not occur in *Plasmodium falciparum*. *(Redrawn and modified from Blacklock and Southwell.)*

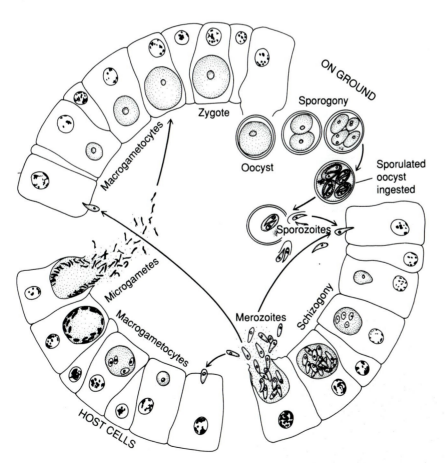

FIGURE 2-33 Life cycle of an eimeriid coccidian, a destructive intracellular parasite of the gut epithelium of many vertebrates, including domesticated birds and mammals. *(From Noble, E. R., and Noble, G. A. 1982: Parasitology. 5th Edition. Lea and Febiger, Philadelphia.)*

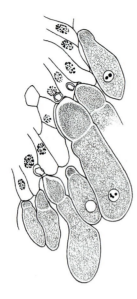

FIGURE 2-34 Trophozoites of the gregarine *Gregarina garnhami* attacking the midgut epithelium of a locust. *(After Canning. In Noble, E. R., and Noble, G. A. 1982. Parasitology. 5th Edition. Lea and Febiger, Philadelphia.)*

Another common group contains the gregarines, which attain the largest size among the sporozoans. They are extracellular parasites inhabiting the gut and other cavities of invertebrates, especially annelids and insects. Some reach 10 mm in length. The body of a gregarine trophozoite is elongate (Fig. 2–34), and the anterior part sometimes possesses hooks, a sucker or suckers, or a simple filament or knob for anchoring the parasite into the host's cells (Fig. 2–34). The host becomes infected through ingesting spores containing sporozoites of the parasites. Depending on the species, the liberated gregarine sporozoites either remain in the gut of the host or penetrate the gut wall to reach other areas of the body. The life cycle commonly lacks schizogony though gametocytes are formed as a result of meiosis followed by multiple fission.

The remaining groups of spore-forming protozoa contain relatively small numbers of species and are described in the Systematic Résumé on p. 66.

SUMMARY

1. Several phyla of parasitic protozoa possess sporelike infective stages and lack cilia, flagella, or pseudopodia as adults.

2. Members of the phylum Sporozoa, or Apicomplexa, possess a complex of distinctive organelles at the apical end. The phylum includes the gregarines, which are extracellular parasites of insects and annelids and other worms, and the coccidians, which are intracellular parasites of gut and blood cells of vertebrates and invertebrates. *Plasmodium,* the causal agent of malaria, is the best known and most familiar coccidian.

3. The complex life cycles of sporozoans usually involve fission (schizogony), sexual reproduction (gamogony), and spore formation (sporogony).

PHYLUM CILIOPHORA

The phylum Ciliophora is the largest and the most homogeneous of the principal protozoan phyla, and all evidence indicates that its members share a common evolutionary ancestry. Some 7200 species have been described. They are widely distributed in both fresh and marine waters and in the water films of soil. About one third of ciliate species are ecto- and endocommensals or parasites.

All possess cilia or compound ciliary structures as locomotor or food-acquiring organelles at some time in the life cycle. Also present is an **infraciliary system,** composed of ciliary basal bodies, or kinetosomes, below the level of the cell surface and associated with fibrils that run in various directions. This is a distinguishing characteristic of ciliates. Such an infraciliary system may be present at all stages in the life cycle even when the cilia themselves are reduced. Most ciliates possess a cell mouth, or **cytostome.** In contrast to the other protozoan classes, ciliates are characterized by the presence of two types of nuclei: one vegetative (the **macronucleus,** concerned with the synthesis of RNA as well as DNA) and the other reproductive (the **micronucleus,** concerned primarily with the synthesis of DNA). Fission is transverse, and sexual reproduction never involves the formation of free gametes.

Form and Structure

The body shape is usually constant and in general is asymmetrical; however, they are radial symmetrical with an anterior mouth (Fig. 2–35). Although the majority of ciliates are solitary and free-swimming, there are both sessile and colonial forms. The bodies of tintinnids and some heterotrichs, peritrichs, and suctorians are housed within a **lorica,** a girdle-like encasement, which is either secreted or composed of foreign material cemented together (Fig. 2–36).

The ciliate body is typically covered by a complex **pellicle,** usually containing a number of different organelles. The pellicular system has been studied in detail in numerous genera, including *Paramecium.*

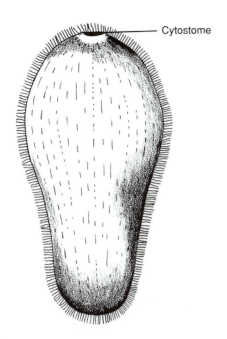

Cytostome

FIGURE 2-35 *Prorodon*, a primitive ciliate. *(After Fauré-Fremiet from Corliss.)*

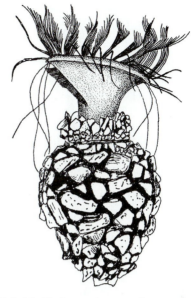

FIGURE 2-36 *Tintinnopsis,* a marine ciliate with lorica, or text, composed of foreign particles. Note conspicuous membranelles and tentacle-like organelles interspersed between them. *(After Fauré-Fremiet from Corliss.)*

There is an outer limiting plasma membrane, which is continuous with the membrane surrounding the cilia. Beneath the outer membrane is a single layer of closely packed vesicles, or **alveoli,** each of which is moderately to greatly flattened (Figs. 2–9E and 2–37). The outer and inner membranes bounding a flattened alveolus thus form a middle and inner membrane of the ciliate pellicle. Between adjacent alveoli emerge the cilia and mucigenic or other bodies (Fig. 2–37). Beneath the alveoli is located the infraciliary

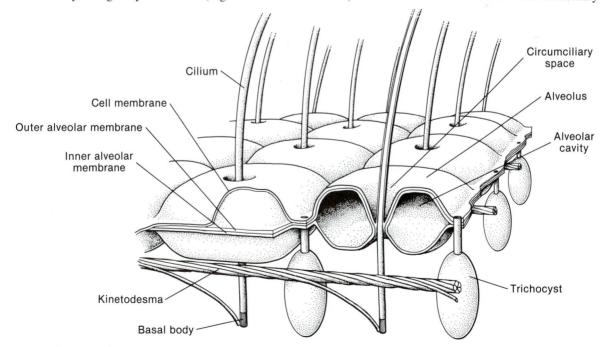

Cilium

Circumciliary space

Cell membrane

Alveolus

Outer alveolar membrane

Alveolar cavity

Inner alveolar membrane

Trichocyst

Kinetodesma

Basal body

FIGURE 2-37 Pellicular system in *Paramecium. (After Ehret and Powers from Corliss.)*

FIGURE 2-38 Electron micrograph of discharged trichocysts of *Paramecium.* Note golf-tee-shaped barb and long, striated shaft. *(By Jacus and Hall. 1946. Biol. Bull. 91:141.)*

system, that is, the kinetosomes and fibrils. The alveoli contribute to the stability of the pellicle and perhaps limit the permeability of the cell surface.

The pellicle of the familiar *Paramecium* has inflated donut-shaped alveoli (Fig. 2–37). The inflated condition and the shape of the alveoli produce a polygonal space about the one or two cilia that arise within them. Alternating with the alveoli are bottle-shaped organelles, the **trichocysts,** which form a second, deeper, compact layer of the pellicular system. The trichocyst is a peculiar rodlike organelle characteristic of some ciliates, which can be explosively discharged as a filament. In the undischarged state it is oriented at right angles to the body surface. At discharge the trichocyst produces a long, striated, thread-like shaft surmounted by a barb, which is shaped somewhat like a golf tee (Fig. 2–38). The shaft is not evident in the undischarged state and is probably polymerized in the process of discharge. Trichocysts appear to function in defense against predators.

Toxicysts are vesicular organelles found in the pellicle of gymnostomes (*Dileptus* and *Didinium*), which on discharge produce long threads with bulbous bases. Toxicysts are used for defense or for capturing prey by paralysis and cytolysis. They are commonly restricted to the parts of the ciliate body that contact prey, such as around the cytostome in *Didini-*

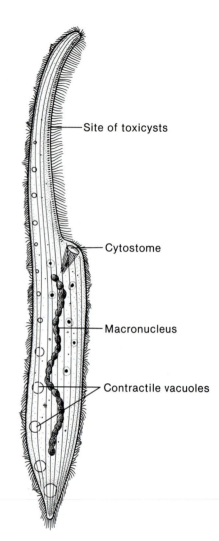

FIGURE 2-39 *Dileptus anser,* a carnivorous ciliate with a long row of toxicysts in the region in front of the cytostome. *(After Sleigh, M. 1989. Protozoa and Other Protists. Edward Arnold, London. p. 198.)*

Labels on figure: Site of toxicysts; Cytostome; Macronucleus; Contractile vacuoles

um or the anterior body region of *Dileptus* (Fig. 2–39).

Mucigenic bodies (**mucocysts**) are another group of pellicular organelles found in many ciliates. They are arranged in rows like trichocysts and discharge a mucoid material that may function in the formation of cysts or protective coverings.

The ciliature can be conveniently divided into the **body** (or **somatic**) **ciliature,** which occurs over the general body surface, and the **oral ciliature,** which is associated with the mouth region. Distribution of body cilia is quite variable. In some groups, cilia cover the entire cell and are arranged in longitudinal rows (Fig. 2–35), but in more specialized groups they have become limited to certain regions of the body (Fig. 2–40).

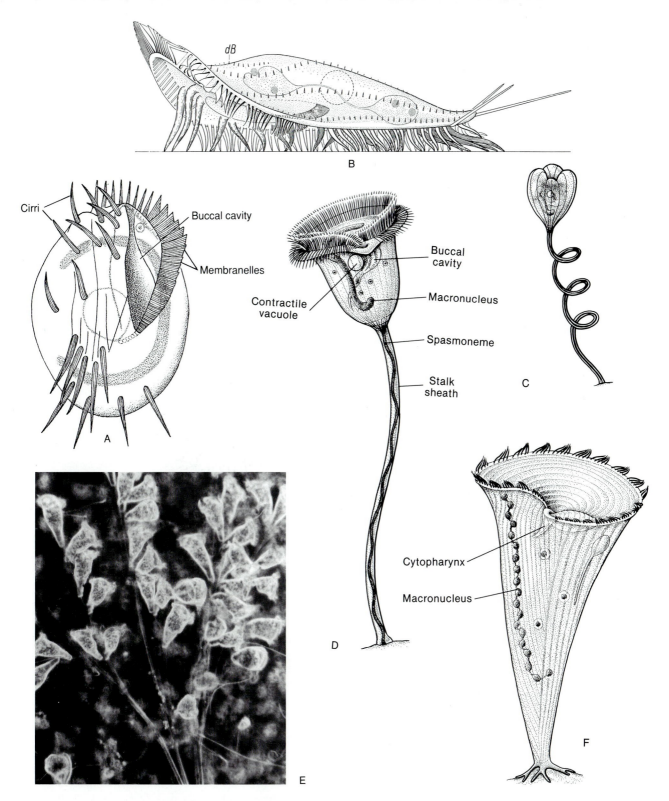

FIGURE 2-40 A, Ventral view of the hypotrich *Euplotes.* **B,** Lateral view of the hypotrich *Stylonychia mytilus.* The ventral view is rather similar to *Euplotes.* **C** and **D,** *Vorticella convallaria* in contracted state **(C);** in expanded state **(D). E,** *Carchesium polypinum,* a colonial peritrich, similar to *Vorticella.* **F,** *Stentor coeruleus. (A, After Pierson from Kudo. B, By Machemer, H. In Grell, K. G. 1973. Protozoology, Springer-Verlag, New York. p. 304. C, D, and F, from Sleigh, M. 1989. Protozoa and Other Protists. Edward Arnold, London. pp. 211 and 213. D, From Small, E. B., and Lynn, D. H. 1985. Phylum ciliophora. In Lee, J. J., et al (Eds.): Illustrated Guide to Protozoa. Society for Protozoology, Lawrence, KS. p. 555).*

As mentioned earlier, each cilium arises from a basal body or kinetosome located in the alveolar layer (Fig. 2–37). The kinetosomes that form a particular longitudinal row are connected by means of fine striated fibers, called a **kinetodesma.** A kinetodesma is actually a cable of still smaller kinetodesmal fibrils, each of which originates from a kinetosome. The cilia, kinetosomes, and kinetodesma together make up a **kinety.** The kinetodesma runs to the right side of the row of kinetosomes, as seen from the ciliate's perspective. At the kinetosome, the kinetodesmal fibrils are connected to certain of the kinetosome triplets.

A kinety system is characteristic of all ciliates, although there are variations in details of the pattern. Even groups, such as the Suctorida, which possess cilia only during developmental stages, retain part of the kinety system in the adult.

Locomotion

The ciliates are the fastest moving of the protozoa. The movements of adjacent cilia are coupled as a result of interference effects of the surrounding water layers. Thus, hydrodynamic forces impose a coordination on the cilia. The beat of individual cilia, rather than being random or synchronous, is part of the metachronal waves that sweep along the length of the body (Fig. 2–5A). There is no evidence that the infraciliature functions as a conducting system in coordination. It may serve primarily in ciliary anchorage and controlling cell shape.

In genera, such as *Paramecium,* the direction of the effective ciliary stroke is oblique to the long axis of the body (Fig. 2–41A). This causes the ciliate to swim in a spiral course and at the same time to rotate on its longitudinal axis. The ciliary beat can be reversed, and the ciliate can move backward. This backward movement is associated with the so-called avoiding reaction. In *Paramecium,* for example, when the ciliate comes in contact with some undesirable substance or object, the ciliary beat is reversed (Fig. 2–41B). The ciliate moves backward a short distance, turns slightly clockwise or counterclockwise, and moves forward again. If unfavorable conditions are still encountered, the avoiding reaction is repeated. External stimuli are probably detected through the cell membrane when it is mechanically modified by certain long, stiff cilia that play no role in movement. The direction and intensity of the beat are controlled by changing levels of Ca^{++} and K^+ ions.

The highly specialized hypotrichs, such as *Urostyla, Stylonychia,* and *Euplotes* (Figs. 2–40A and

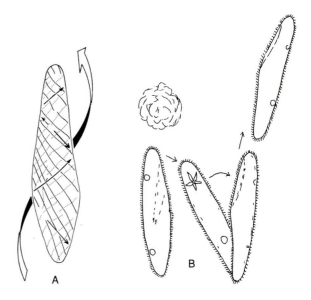

FIGURE 2-41 Ciliary locomotion in *Paramecium.* **A,** Metachronal waves in *Paramecium* during forward swimming. Wave crests are shown by lines and their direction by the small solid arrows. Dashed arrows indicate direction of effective stroke of cilia. Movement of ciliate is indicated by large arrow. **B,** The avoiding reaction of *Paramecium.* (A, From Machemer, H. 1974. Ciliary activity and metachronism in Protozoa. In Sleigh, M. A. (Ed.): Cilia and Flagella. Academic Press, London. p. 224. B, After Hyman, L. H. 1940. The Invertebrates, Vol. I. McGraw-Hill Book Co., New York.)

B), have greatly modified body cilia. The body has become differentiated into distinct dorsal and ventral surfaces, and cilia have largely disappeared except on certain areas of the ventral surface. Here the cilia occur as a number of tufts, called **cirri.** The cilia of a cirrus beat together, and coordination is believed to result from viscous forces between the closely associated cilia.

Some ciliates, especially sessile forms, can undergo contractile movements, either shortening the stalk by which the body is attached, as in *Vorticella* (Fig. 2–40D), or shortening the entire body, as in *Stentor* (Fig. 2–40F). Contraction in *Stentor* is brought about by bundles of contractile filaments, or **myonemes,** that lie in the pellicle. In *Vorticella* (Figs. 2–40C and D) and the colonial *Carchesium* (Fig. 2–40E), both of which have bell-shaped bodies attached by a long slender stalk, the contractile elements extend into the stalk as a single, large, spiral fiber. It functions very much like a coiled spring, producing the familiar popping movements that are so

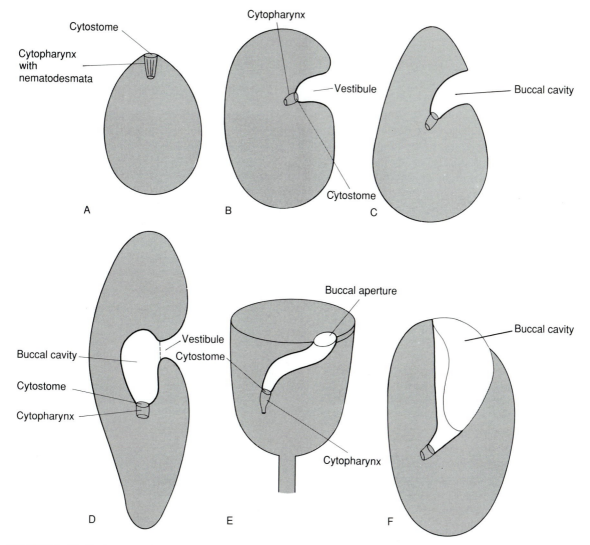

FIGURE 2-42 Oral areas of various ciliates. **A,** In rhabdophorine gymnostomes, such as *Coleps, Prorodon,* and *Didinium.* **B,** In a trichostome, such as *Colpoda,* with a vestibule that is displaced from anterior end. **C,** In a tetrahymenine hymenostome, such as *Tetrahymena.* **D,** In a peniculine hymenostome, such as *Paramecium.* **E,** In a peritrich, such as *Vorticella.* **F,** In a hypotrich, such as *Euplotes.* (A–F, Modified after Corliss, J. O. 1961. The Ciliated Protozoa. Pergamon Press, New York.)

characteristic of *Vorticella* and related genera. The fiber is not composed of the proteins actin and myosin, as in animal muscles, but rather another protein, called **spasmin,** which requires ATP for extension.

Nutrition

Feeding in ciliates parallels, on a microscopic level, feeding in multicellular animals. Typically a distinct mouth, or **cytostome,** is present, although it has been secondarily lost in some groups. In some groups the mouth is located anteriorly (Fig. 2–35), but in most ciliates it has been displaced posteriorly to varying degrees. The mouth opens into a canal or passageway

called the **cytopharynx,** which is separated from the endoplasm by a membrane. It is this membrane that enlarges and pinches off as a food vacuole. The wall of the cytopharynx is lined with microtubular rods (**nematodesmata**) arranged like the staves of a barrel. The microtubules support the walls of the pharynx and assist the inward transport of food vacuoles.

The ingestive organelles may consist only of the cytostome and cytopharynx (Figs. 2–35, 2–39, and 2–40F), but in the majority of ciliates the cytostome is preceded by a preoral chamber that aids in food capture and manipulation. The preoral chamber may take the form of a **vestibule,** which varies from a slight depression to a deep funnel, with the cytostome at its base (Fig. 2–42B). The vestibule is clothed with simple cilia derived from the somatic ciliature. In other ciliates the preoral chamber is typically a **buccal cavity,** which differs from a vestibule by containing compound ciliary organelles instead of simple cilia (Fig. 2–42C–F). There are two basic types of such ciliary organelles: the **undulating membrane** and the **membranelle.** An undulating membrane is a row of adhering cilia forming a sheet (Fig. 2–43A). A membranelle is derived from two or three short rows of cilia, all of which adhere to form a more or less triangular or fan-shaped plate. These membranelles typically occur in a series (Figs. 2–36, 2–40A, and 2–43A). Although there is no actual fusion of adjacent cilia in these compound organelles, their kinetosomes and bases are sufficiently close together to pro-

duce some sort of structural coupling that causes all of the cilia of a membranelle to beat together.

The term **peristome,** which is commonly encountered in the literature, is synonymous with **buccal cavity.** In members of a number of orders the buccal organelles project from the buccal cavity, or, as in the Hypotrichida (Fig. 2–40A), the buccal cavity is somewhat shallow so that the organelles occupy a flattened area around the oral region. Such an area is called the peristomial field. In forms like *Paramecium* there is a vestibule in front of the buccal cavity (Figs. 2–42D and 2–43B).

Free-swimming ciliates display various types of feeding habits. Some, often called raptorial, capture single organisms by direct interception. Prey consist

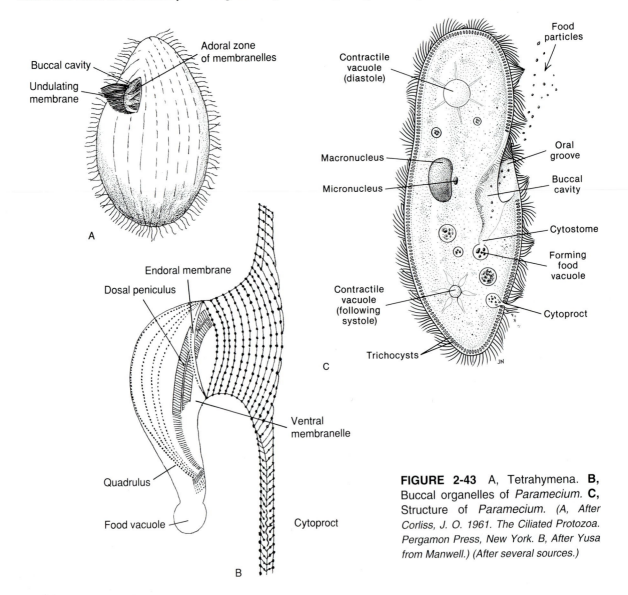

FIGURE 2-43 A, Tetrahymena. **B,** Buccal organelles of *Paramecium.* **C,** Structure of *Paramecium. (A, After Corliss, J. O. 1961. The Ciliated Protozoa. Pergamon Press, New York. B, After Yusa from Manwell.) (After several sources.)*

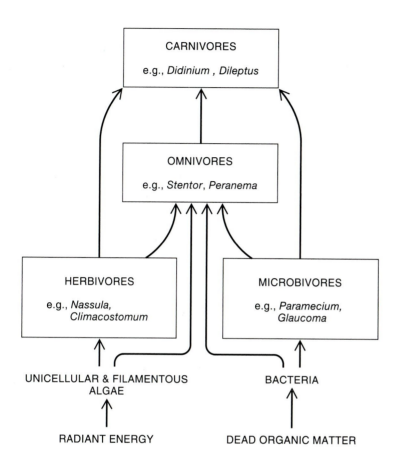

CARNIVORES

e.g., *Didinium , Dileptus*

OMNIVORES

e.g., *Stentor, Peranema*

HERBIVORES

e.g., *Nassula,
Climacostomum*

MICROBIVORES

e.g., *Paramecium,
Glaucoma*

UNICELLULAR & FILAMENTOUS
ALGAE

BACTERIA

RADIANT ENERGY

DEAD ORGANIC MATTER

FIGURE 2-44 The relationship of ciliates to each other as food sources (trophic levels) in a protozoan community. All genera are ciliates except *Peranema,* which is a flagellate. *(From Laybourn-Parry, J. 1984. A Functional Biology of Free-Living Protozoa. University of California Press, Berkeley, CA. p. 136.)*

of rotifers, gastrotrichs, and protozoa, including other ciliates. The oral apparatus of direct interceptors is typically limited to the cytostome and cytopharynx. Some feed on diatoms, and a few ciliates even ingest filamentous blue-green algae. Many ciliates are filter feeders, collecting organisms from ambient currents or from currents produced by the ciliate. These forms usually have a complex buccal ciliature. Food particles collected by direct interception in some small ciliates may be collected by filtering in larger species. Food may include bacteria and detritus, as well as other protozoa.

Given the range of feeding habits and diets of ciliates, their microbial world exhibits food webs and trophic energy relationships similar to those found among multicellular animals (Fig. 2–44). Moreover, a microhabitat supports a distinct ecological succession of species populations as trophic and other environmental conditions change over time (Fig. 2–45).

Didinium has perhaps been most studied of all the raptorial feeders. This little barrel-shaped ciliate feeds on other ciliates, particularly *Paramecium* (Fig. 2–46A). When *Didinium* attacks a *Paramecium,* it discharges toxicysts into the *Paramecium* and the proboscis-like anterior end attaches to the prey through the terminal mouth, which can open almost as wide as the diameter of the body.

The free-living members of the aberrant subclass Suctoria are all sessile and most are attached to marine and freshwater invertebrates (Figs. 2–46B and 2–47). Cilia are present only in the immature stages. The body bears tentacles, which may be knobbed at the tip or shaped like long spines (Fig. 2–46B). Each tentacle is supported by a cylinder of microtubules and carries special organelles, called **haptocysts** (Figs. 2–46C and D). Suctorians feed on other ciliates, and when prey strikes the tentacles, the haptocysts are discharged into the prey body, anchoring it to the tentacles (Fig. 2–47). The contents of the prey are then sucked through the tubular tentacle within a long food vacuole into the suctorian. "Suction" in suctorians is an example of rapid phagocytosis, which is accelerated by microtubles within the tentacle.

Typically characteristic of filter feeders is the buccal cavity. Food is brought to the body and into the buccal cavity by the compound ciliary organelles.

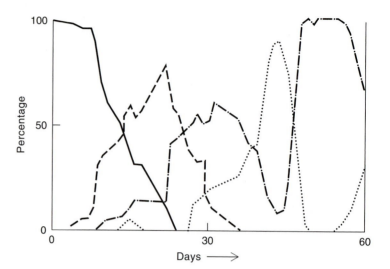

FIGURE 2-45 Succession of four protozoans, chiefly ciliates, in sewage sludge over 60 days. *(Modified from Curds, C. R. 1966. An ecological study of ciliated protozoa in activated-sludge. Oikos. 15:282–289.)*

From the buccal cavity the food particles are driven through the cytostome and into the cytopharynx. When the particles reach the cytopharynx, they collect within a food vacuole.

The order Hymenostomatida, meaning "membrane-mouthed," contains filter feeders like *Tetrahymena* (Fig. 2–43A). The cytostome is located a little behind the leading edge of the body. Just within the broad opening to the buccal cavity are four ciliary organelles—an undulating membrane on the right side of the chamber and three membranelles on the left. The three membranelles constitute an **adoral zone** of membranelles, which in many other groups of ciliates is much more developed and extensive.

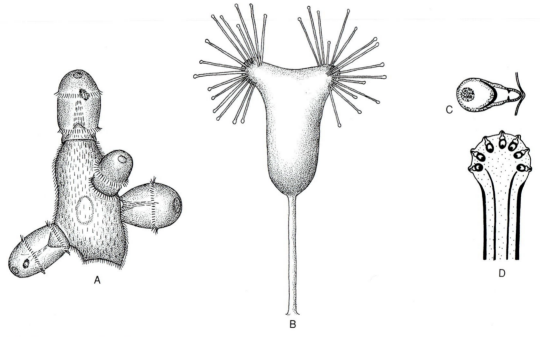

FIGURE 2-46 A, Four *Didinium*, raptorial ciliates, attacking one *Paramecium*. **B,** *Acineta*, a suctorian. **C** and **D,** Suctorian haptocysts and prey capture. Haptocyst isolated **(C)** and within tentacle tip **(D).** Attachment of tentacle to prey can be seen in Figure 2–47. *(A, After Mast from Dogiel. B, After Clakins from Hyman. C and D, from Sleigh, M. A. 1973. The Biology of Protozoa, Edward Arnold Publishers, London. p. 64. Based on micrographs of Rudzinska, Bardele, and Grell.)*

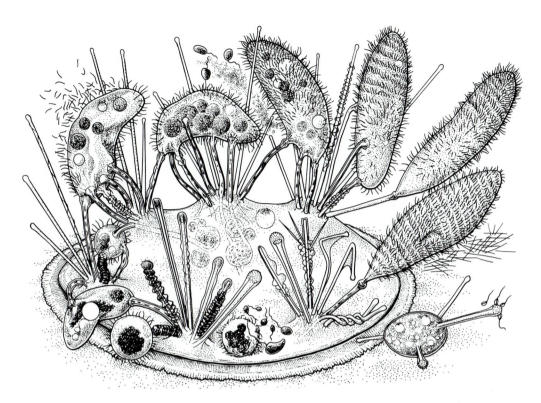

FIGURE 2-47 A colony of the suctorian *Heliophrya* feeding on *paramecia*. Paramecia on the right have just been captured. Those on the left have been ingested to various degrees. *(From Spoon, et. al. 1976. Observations on the behavior and feeding mechanisms of the suctorian Heliophrya erhardi preying on Paramecium. Trans. Am. Micros. Soc. 95:443–462.)*

In *Paramecium,* the most familiar genus of the order, an **oral groove** along the side of the body leads posteriorly to a vestibule, located about midway back from the anterior end (Fig. 2–43C). The vestibule, buccal cavity, and cytopharynx together form a large, curved funnel (Figs. 2–42D and 2–43B). The undulating membrane, here called the **endoral membrane,** runs transversely along the right wall and marks the junction of the vestibule and buccal cavity. The three membranelles are also modified. Two, called **peniculi,** are greatly lengthened and thus tend to be more similar to an undulating membrane in function than to the more typical membranelle.

In feeding, the cilia of the oral groove produce a current of water that sweeps in an arclike manner down the side of the body and over the oral region. The ciliature of the vestibule and buccal cavity pull in food particles and drive them into the forming food vacuole.

In the subclass Peritricha, whose members possess little or no somatic cilia, the buccal ciliary organelles are highly developed and form a large, disclike, peristomial field at the apical end of the animal. In the much-studied peritrich genus *Vorticella,* a peripheral shelflike projection can close over the disc when the animal is retracted (Figs. 2–40C and D and 2–42E). The ciliary organelles lie in a peristomial groove between the edge of the disc and the peripheral shelf and consist of two ciliary bands, which wind in a counterclockwise direction around the margin of the disc and then turn downward into the funnel-shaped buccal cavity (Fig. 2–40D). The inner ciliary band produces the water current, and the outer band acts as the filter. Suspended particles, mostly bacteria, are transported along with a stream of water between the two bands into the buccal cavity.

Ciliates of the subclass Spirotricha, which includes such familiar forms as *Stentor, Halteria, Spirostomum,* and *Euplotes,* are typically filter feeders. They usually possess a highly developed adoral zone of many membranelles (Figs. 2–40A and F). Within the cytopharynx of all ciliates, food particles

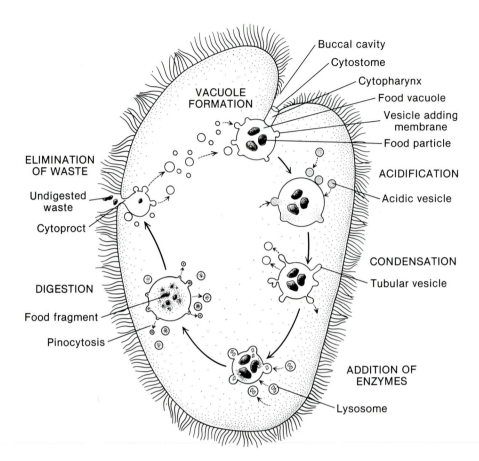

FIGURE 2-48 Intracellular digestion in ciliates. In the formation of the food vacuole, small discoidal vesicles add membrane to provide for its increase in size. Acidic vesicles (acidosomes) bring about a drop in pH. The vacuole becomes smaller by the removal of membrane. Lysosomes add digestive enzymes.

enter the food vacuole. When the food vacuole reaches a certain size, it breaks free from the cytopharynx, and a new vacuole forms in its place. Detached vacuoles then begin a more or less circulatory movement through the endoplasm.

Digestion follows the pattern described on p. 19 but is peculiar in developing a very low initial pH. In *Paramecium,* following the formation of the food vacuole (Fig. 2–48), acidic vesicles (acidosomes) fuse with the vacuole and some cell membrane is removed. As a result the vacuole becomes smaller and the pH drops to 3. Lysosomes now join the vacuole, but the contents are too acid for enzymatic efficiency. For reasons still unknown, the pH rises, and at pH 4.5 to 5 digestion occurs. This is the same pH characteristic of intracellular digestion in other organisms that have been studied. Following digestion, the waste-laden food vacuole moves to a fixed opening, or **cytoproct,** at the body surface and expels its contents.

Membrane material at the cytoproct is then recycled to the cytostome in the form of small vesicles.

About 15% of ciliates are parasitic, and there are many ecto- and endocommensals. Many suctorians are commensals, and a few are parasites. Hosts include fishes, mammals, various invertebrates, and other ciliates. *Endosphaera,* for example, is parasitic within the body of the peritrich *Telotrochidium.*

Other interesting commensal ciliates include *Kerona,* a little crawling hypotrich, and *Trichodina,* a mobile peritrich, which are ectocommensals on the surface of hydras. There are also some free-swimming peritrichs that occur on the body surfaces of freshwater planarians, tadpoles, sponges, and other animals. The genus *Balantidium* includes many species that are endocommensals or endoparasites in the guts of insects and many different vertebrates. *Balantidium coli* is an endocommensal in the intestines of pigs and is passed by means of cysts in the feces. This ciliate has occasionally been found in humans, where in conjunction with bacteria it erodes pits in the intestinal mucosa and produces pathogenic symptoms. The related highly specialized Entodiniomorphida live as harmless commensals in the digestive tracts of many different hoofed mammals. Like the flagellate symbionts of termites and roaches, some of them ingest and break down the cellulose of the vegetation eaten by their hosts. The products of digestion are utilized by the host.

Some ciliates display symbiotic relationships with algae. The most familiar of these is *Paramecium bursaria,* in which the endoplasm is filled with green zoochlorellae, but there are others, such as the mouthless marine *Mesodinium.* There are also ciliates that retain functional chloroplasts from their algal prey.

Water Balance

Contractile vacuoles are found in both marine and freshwater species, but in the latter they discharge more rapidly (see p. 23). In some species a single vacuole is located near the posterior, but many species possess more than one vacuole (Fig. 2–39). In *Paramecium* one vacuole is located at both the posterior and the anterior of the body (Fig. 2–43C). The vacuoles are always associated with the innermost region of the ectoplasm and empty through one or two permanent pores that penetrate the pellicle. The spongiome contains a network of irregular tubules, which may empty into the vacuole directly or by way of collecting tubules (Fig. 2–10).

Reproduction

Ciliates differ from almost all other organisms in possessing two distinct types of nuclei—a usually large **macronucleus** and one or more small **micronuclei.** The micronuclei are small, rounded bodies and vary in number from 1 to as many as 20, depending on the species. They are diploid, with little RNA. The micronucleus is a store of genetic material responsible for genetic exchange and nuclear reorganization and also gives rise to the macronuclei. The macronucleus is sometimes called the **vegetative nucleus,** because it is not essential in sexual reproduction. However, the macronucleus is necessary for normal metabolism, for mitotic division, and for the control of cellular differentiation, and it is responsible for the genetic control of the phenotype through protein synthesis. There is hundreds to thousands of times more DNA in the macronucleus than in the micronucleus because of

duplications following the micronuclear origin of the macronucleus. Furthermore, macronuclear DNA is not organized in chromosomes but, instead, in small subchromosomal or gene-size units. The macronuclei include numerous nucleoli in which ribosomal RNA is synthesized. The amplification of genes in the macronucleus probably increases the rate of synthesis of gene products to be used in the assembly of complex and numerous ciliate organelles.

One or more macronuclei are present, and they may assume a variety of shapes (Figs. 2–39, 2–40D and F). The large macronucleus of *Paramecium* is somewhat oval or bean-shaped and is located just anterior to the middle of the body. In *Stentor* and *Spirostomum* the macronuclei are long and arranged like a string of beads. Not infrequently the macronucleus is in the form of a long rod bent in different configurations, such as a C in *Euplotes* or a horseshoe in *Vorticella.* The unusual shape of many macronuclei may be an adaptation to reduce the distance between the center of the nucleus and the cytoplasm. Because diffusion of messenger RNA from the nucleus to the cytoplasm may be a rate-limiting step in protein synthesis and cell assembly, any reduction in diffusion distance accelerates the process.

Asexual Reproduction

Asexual reproduction is always by means of binary fission, which is typically **transverse,** the division plane cutting across the kineties—the longitudinal rows of cilia or basal bodies (Fig. 2–49A). This is in contrast to the symmetrogenic fission of flagellates (Fig. 2–19), in which the plane of division (longitudinal) cuts between the rows of basal bodies. Mitotic spindles are formed only in the division of the micronuclei. Division of the macronuclei is usually accomplished by constriction. When a number of macronuclei are present, they may first combine as a single body before dividing. The same is true of some forms with beaded or elongated macronuclei.

During cell division of multicellular animals, centrioles (basal bodies) occur at the poles of the mitotic spindle, probably to ensure that each daughter cell receives a template on which a cilium can later be produced. As a ciliate cell divides, however, each daughter cell receives cilia and basal bodies (kinetosomes) replicated earlier in the cell cortex. As a result of this cortical replication, centrioles are absent from the mitotic spindle of the ciliate micronucleus.

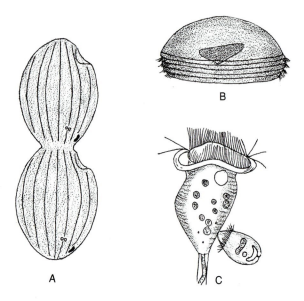

FIGURE 2-49 A, Homothetogenic type of fission, in which the plane of division cuts across the kineties. **B,** Detached bud of *Dendrocometes.* **C,** Conjugation in *Vorticella.* Note the small nonsessile microconjugant. *(A, After Corliss. B, After Pestel from Hyman. C, After Kent from Hyman.)*

Sexual Reproduction

Sexual reproduction involves an exchange of nuclear material by conjugation that usually precedes a series of transverse divisions. By contact in the course of swimming, often by the release of attractant substances, two sexually compatible members of a par-

ticular species adhere commonly in the oral or buccal region of the body. Following the initial attachment, there is degeneration of trichocysts and cilia (but not kinetosomes) and a fusion of membranes in the region of contact. Two such fused ciliates are called **conjugants;** attachment lasts for several hours. During this period a reorganization and exchange of nuclear material occurs (Fig. 2–50A–F). Only the micronuclei are involved in conjugation; the macronucleus breaks up and disappears either in the course of or following micronuclear exchange.

The steps leading to the exchange of micronuclear material between the two conjugants are fairly constant in all species. After two meiotic divisions of the micronuclei, all but one of them degenerate. This one then divides, producing two gametic micronuclei that are genetically identical. One is stationary; the other migrates into the opposite conjugant. The migrating nucleus in each conjugant moves through the region of fused protoplasm into the opposite member of the conjugating pair. There the gamete's nuclei fuse with one another to form a zygotic nucleus, or **synkaryon.** Shortly after nuclear fusion the two ciliates separate; each is now called an **exconjugant.** After separation, there follow in each exconjugant a varying number of nuclear divisions, leading to the reconstitution of the normal nuclear condition characteristic of the species. This reconstitution usually, but not always, involves a certain number of cytosomal divisions. For example, in some forms where there is but a single macronucleus and a single micronucleus

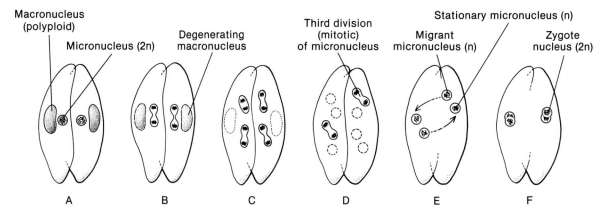

FIGURE 2-50 Sexual reproduction in *Paramecium caudatum,* in which there is one macronucleus and one micronucleus. **A,** Two individuals are united in conjugation. **B–D,** The micronucleus of each undergoes three divisions, the first two of which are meiotic. **E,** Migrant micronuclei are exchanged. **F,** They fuse with the stationary micronucleus of the opposite conjugant to form a synkaryon, or "zygote" nucleus.

in the adult, the synkaryon divides once. One of the nuclei forms a micronucleus; the other forms the macronucleus. Thus, the normal nuclear condition is restored without any cytosomal divisions.

However, in *Paramecium caudatum,* which also possesses a single nucleus of each type, the synkaryon divides three times, producing eight nuclei. Four become micronuclei, and four become macronuclei. Three of the micronuclei degenerate. The remaining micronucleus divides during each of the two subsequent cytosomal divisions, and each of the four resulting offspring cells receives one macronucleus and one micronucleus. In those species that have numerous nuclei of both types, there is no cytosomal division; the synkaryon merely divides a sufficient number of times to produce the requisite number of macronuclei and micronuclei.

In some of the more specialized ciliates, the conjugants are a little smaller than nonconjugating individuals, or the two members of a conjugating pair are of strikingly different sizes. Such "dioecious" macro- and microconjugants occur in *Vorticella* (Fig. 2–49C) and represent an adaptation for conjugation in sessile species. The macroconjugant remains attached while the small bell of the microconjugant breaks free from its stalk and swims about. On contact with an attached macroconjugant the two bells adhere. A synkaryon forms only in the macroconjugant from one gametic nucleus contributed by each conjugant. However, there is no separation after conjugation, and the little migratory conjugant degenerates. In the Suctoria conjugation takes place between two attached individuals that happen to be located side by side.

The frequency of conjugation is extremely variable. Members of some families have rarely been observed to undergo the sexual phenomenon of conjugation; others conjugate every few days or weeks. In some species a period of "immaturity," in which only fission occurs, precedes a period in which individuals are capable of conjugation. Numerous factors, such as temperature, light, and food supply, are known to induce or influence conjugation.

In some ciliates the nuclear reorganization following conjugation seems to have a rejuvenating effect and is necessary for continued asexual fission. For example, it has been shown that some species of *Paramecium* can pass through only approximately 350 continuous asexual generations. If nuclear reorganization does not occur, the asexual line (or clone) dies out, apparently because of decline in function of the macronucleus. Another type of nuclear reorganization called **autogamy** may take place and has the same effect on fission as does conjugation. Autogamy involves the same nuclear behavior as does conjugation, but there is no conjugation and no exchange of micronuclear material between two individuals. The macronucleus degenerates and the micronucleus divides a number of times to form eight or more nuclei. Two of these nuclei fuse to form a synkaryon; the others degenerate and disappear. The synkaryon then divides to form a new micronucleus and macronuclei, as occurs in conjugation.

Definite mating types have been shown to exist in species of *Paramecium, Tetrahymena, Euplotes, Stylonychia,* and some other ciliates. For example, a number of sibling species of what was formerly *Paramecium aurelia* have two or more mating types. Conjugation is always restricted to a member of the opposite mating type within the same syngen and does not occur between members of the same type, apparently owing to a failure of the surfaces to adhere. The mating types are hereditary.

Most ciliates are capable of forming resistant cysts in response to unfavorable conditions, such as lack of food or desiccation. **Encystment** is important in carrying the species through winter or dry periods and providing a condition in which the organism can be transported by wind or in mud on the feet of birds or other animals. In some forms, such as *Colpoda,* reproductive processes occur only during the encysted state. There are also some ciliates, including *Paramecium,* that are believed never to encyst.

SUMMARY

1. The ciliates constitute the largest phylum of protozoa (phylum Ciliophora). They are the most animal-like and exhibit a high level of organelle development.

2. They possess cilia for locomotion and, in many species, for suspension feeding. Associated with the kinetosomes is a complex anchorage system of fibrils, all of which make up the infraciliature.

3. The body wall of ciliates is a complex living pellicle, containing alveoli, trichocysts, and other organelles, in addition to the infraciliature.

4. The body surface may be covered with uniform cilia, which function in locomotion. There has been a tendency, however, for this somatic ciliature to become reduced or in some groups to become specialized as well (cirri).

5. The cilia around the mouth region (buccal ciliature) have become specialized as compound ciliary organelles called membranelles and undulating membranes in many ciliates that employ suspension feeding.

6. In addition to suspension feeding on bacteria and other particles, some ciliates feed on algae, and many are carnivorous on other protozoa and microscopic animals. The cytostome and cytopharynx open into a food vacuole. The undigestible remains are discharged through a fixed cell anus (cytoproct). The discharge position of the contractile vacuoles is also constant.

7. Ciliates reproduce asexually by transverse fission and sexually by conjugation. Conjugation involves an exchange of micronuclei, each of which fuses with a nonmigratory micronucleus to form a zygote nucleus. Conjugation is preceded by meiotic divisions of one micronucleus and is followed by reconstitution of the normal nuclear condition, which may involve fission.

SYSTEMATIC RÉSUMÉ OF THE PROTOZOA

Flagellated Protozoa

Phylum Dinophyta. Dinoflagellates. Phytoflagellates with an equatorial and a posterior longitudinal flagellum located in grooves. Body either naked or covered by cellulose plates or valves or by a cellulose membrane. Brown or yellow chromoplasts and stigma usually present, but there are many colorless species. Largely marine; some parasites. Includes the marine genera *Gonyaulax, Noctiluca, Histiophysis,* and *Ornithocercus,* and the marine and freshwater genera *Glenodinium, Gymnodinium, Ceratium, Oodinium,* and *Symbiodinium.*

Phylum Parabasalia. Zooflagellates with from four to many flagella. Without distinct mitochondria; Golgi apparatus plus filament associated with basal body composes a distinctive organelle, the parabasal body. Contains two groups. The trichomonads with four to six flagella per mastigont system, one flagellum trailing; found in the gut of vertebrates and insects: *Trichomonas* (parasitic in urogenital tract of vertebrates, including humans). The Hypermastigida with many flagella per mastigont system; found in the gut of termites and woodroaches: *Lophomonas, Trichonympha, Barbulanympha.*

Phylum Metamonada. Multiflagellate zooflagellates. One to several mastigont systems having between one and four flagella each. One flagellum in each system is turned backward.

Class Anaxostylea. Lacking a longitudinal microtubular rod as part of the mastigont system.

Order Retortamonadida. Gut parasites of insects or vertebrates, with two or four flagella. One flagellum associated with ventrally located cytostome. *Chilomastix.*

Order Diplomonadida. Bilaterally symmetrical flagellates, with two nuclei, each nucleus associated with between one and four flagella. Mitochondria absent. Mostly parasites. *Hexamita, Giardia.*

Class Axostylea. Having a longitudinal microtubular rod as part of the mastigont system.

Order Oxymonadida. Commensal or mutualistic flagellates in the guts of insects; a few in vertebrates. One to many nuclei, each nucleus associated with four flagella. *Oxymonas, Pyrsonympha.*

Phylum Kinetoplastida. One or two flagella emerging from a pit. A body containing DNA (kinetoplast) located within elongate mitochondrion and associated with flagellar basal bodies. Mostly parasitic. *Bodo, Leishmania, Trypanosoma.*

Phylum Euglenophyta. Elongated green or colorless flagellates with two flagella arising from an anterior recess. Stigma present in colored forms. Primarily freshwater. *Euglena, Phacus, Peranema, Rhabdomonas.*

Phylum Cryptophyta. Compressed, biflagellate phytoflagellates, with an anterior depression or reservoir. Two chromoplastids, usually yellow to brown or colorless. Marine and freshwater. *Chilomonas* is a common colorless genus in polluted water.

Phylum Opalinata. Body covered by longitudinal, oblique rows of cilia rising from anterior subterminal rows. Infraciliature characteristic of the true ciliates is lacking. Two or many monomorphic nuclei. Binary fission generally longitudinal. Sexual reproduction involves syngamy with flagellated gametes. Gut commensals of anurans; less commonly of fishes, salamanders, and reptiles. *Opalina, Zelleriella.*

Phylum Heterokonta. Having two dissimilar flagella, one with hairs and one smooth. Chloroplasts contain chlorophylls a and c. This large heterogeneous group of mostly autotrophic protistans includes the multicellular brown algae, some filamentous algae, and the diatoms. In these forms the flagellated condition appears only in reproductive stages.

The principal protozoan heterokonts are members of the class Chrysophyceae. These are small flagellates with chromoplast containing fucoxanthin. Most are naked but some have the body covered with siliceous scales. Largely inhabitants of fresh water. *Chromulina, Ochromonas, Synura.*

Phylum Chlorophyta. Autotrophic green protistans having chlorophylls a and b. Includes many multicellular species (green algae) with flagellated reproductive stages. Principal protozoan flagellates are members of the order Volvocales, unicellular and colonial species with two to four apical flagella and a single cup-shaped chloroplast per cell. Some colorless forms. Largely in fresh water. *Chlamydomonas, Polytomella, Haematococcus, Gonium, Pandorina, Platydorina, Eudorina, Pleodorina, Volvox.*

Phylum Haptophyta. Small phytoflagellates having a flagella-like organelle (haptonema) located between the two true flagella. Body covered by organic scales, although in the coccolithophorids, these scales are covered by calcium carbonate crystals. Largely marine. *Coccolithus.*

Phylum Choanoflagellida. Solitary and colonial zooflagellates with a single flagellum surrounded by a collar of microvilli. Some species sessile and stalked. Body may be naked or covered by a theca, which is some marine species is siliceous. Marine and freshwater. *Codonosiga, Proterospongia, Salpingoeca.*

Ameboid Protozoa. The slime molds have been omitted from this résumé as well as some small groups of ameboid forms.

Phylum Rhizopoda. Lobopodia, filipodia, or reticulopodia used for locomotion and feeding.

Class Lobosea. Pseudopodia, usually lobopods. Flagellated stages absent.

Subclass Gymnamoebia. Amebas without shells. Marine and freshwater; many parasites. *Amoeba, Chaos, Acanthamoeba, Entamoeba.*

Subclass Testacealobosia. Amebas with shells. Marine and freshwater. *Arcella, Difflugia, Centropyxis.*

Class Heteroblastea (Schizopyrenida). Naked amebas with flagellated stages. Marine, freshwater, and soil species. *Naegleria, Vahlkampfia.*

Class Karyoblastea (Pelobiontea). Naked, multinucleated amebas with one pseudopod and no flagellated stages. *Pelomyxa.*

Class Filosea. Amebas with filopods.

Subclass Aconchulinia. Naked amebas. Freshwater and parasites of algae. *Vampyrella.*

Subclass Testaceafilosia. Shelled amebas. Marine and freshwater; many species in mosses. *Gromia, Euglypha.*

Class Granuloreticulosea. Amoeboid protozoans with delicate granular reticulopodia.

Order Foraminiferida. Chiefly marine species with mostly multichambered shells. Shells may be organic, but most commonly are calcareous. *Globigerina, Orbulina, Discorbis, Spirillina, Numulites, Homotrema.*

The orders Athalamida and Monothalamida contain, respectively, a small number of naked species and a small number of species with shells of a single chamber that lack an alternation of generations in the life cycle.

Phylum Actinopoda. Primarily floating or sessile ameboid protozoa with actinopodia and delicate filopodia radiating from a spherical body.

Class Acantharea. Radiolarians with a radiating skeleton of strontium sulfate. Marine. *Acanthometra.*

Class Polycystinea. Radiolarians with a siliceous skeleton and a perforated capsular membrane. Marine. *Thassicola, Collozoum, Sphaerozoum.*

Class Phaeodarea. Radiolarians with a siliceous skeleton but a capsular membrane containing three large pores. Marine. *Aulacantha.*

Class Heliozoea. Without central capsule. Naked, or if skeleton present, of siliceous scales and spines. Marine and freshwater. *Actinophrys, Actinosphaerium, Camptonema.*

Spore-forming Protozoa

Phylum Sporozoa, or Apicomplexa. Spore-producing parasitic protozoa with apical complex at some stage. Spores lacking polar filaments.

Class Gregarinea. Mature trophozoites are large and occur in host's gut and body cavities. Parasites of annelids and arthropods. *Gregarina, Monocystis* (common parasite of earthworm's seminal receptacles).

Class Coccidea. Mature trophozoites are small and intracellular. *Eimeria, Isopora, Aggregata, Plasmodium, Toxoplasma.*

Class Piroplasmea. Parasites of vertebrate red blood cells transmitted by ticks. No spores. *Theileria, Babesia.*

Phylum Microspora. Parasitic protozoa having spores with a polar filament. *Nosema.*

Phylum Myxosporidia. Parasitic protozoa having spores with polar filament and encased in several valves. *Myxosoma.*

Ciliated Protozoa.

Phylum Ciliophora. Protozoans possessing an infraciliature and, at least at some time in their life, surface cilia. Nuclei are dimorphic.

Class Kinetofragminophora. Isolated kineties in oral region of body bearing cilia but not compound ciliary organelles.

Subclass Gymnostomata. Cytostome at or near surface of body and located at the anterior end or laterally. Somatic ciliation generally uniform. *Stephanopogon, Loxodes, Coleps, Prorodon, Actinobolina, Didinium, Dileptus, Lacrymaria, Litonotus,* and *Loxophyllum.*

Subclass Vestibulifera. Cytostome within a vestibulum (vestibule) bearing distinct ciliature. Free-living and symbiotic species. *Balantidium, Colpoda, Blepharocorys, Entodinium.*

Subclass Hypostomata. Body cylindrical or dorsoventrally flattened, with the mouth on the ventral side in either case. Somatic ciliature often reduced. Free-living and many symbiotic species. *Synhymenia, Nassula, Microthorax, Hypocoma, Trochilioides, Chilodochona, Lobochona, Spirochona, Stylochona, Ancistrocoma, Foettingeria, Chromidina, Ascophrys.*

Subclass Suctoria. Sessile, generally stalked, with tentacles at the free end. Cilia lacking in the adult but present in the free-swimming larval stage. Most are ectosymbionts on aquatic invertebrates. *Ephelota, Podophrya, Acineta.*

Class Oligohymenophora. Oral apparatus usually well developed and containing compound ciliary organelles.

Subclass Hymenostomata. Body ciliation commonly uniform and oral structures not conspicuous. *Colpidium, Glaucoma, Tetrahymena Paramecium, Pleuronema.*

Subclass Peritricha. Mostly sessile forms with reduced body ciliation. Oral ciliary band usually conspicuous. *Carchesium, Epistylis, Lagenophrys, Vorticella, Zoothamnium, Trichodina.*

Class Polyhymenophora. Oral region with conspicuous adoral zone of buccal membranelles. Some species with uniform body ciliation; others with compound organelles, such as cirri.

Subclass Spirotricha. With characteristics of class.

Order Heterotrichida. Mostly large ciliates with uniform body ciliation. *Blepharisma, Bursaria, Spirostomum, Stentor, Folliculina.*

Order Odontostomatida. Laterally compressed, wedge-shaped ciliates with reduced body ciliation. *Saprodinium.*

Order Oligotrichida. Ciliates with reduced somatic ciliature but with extensive projecting buccal ciliary organelles. *Halteria.* The suborder Tintinnina contains loricate species—*Codonella, Favella, Tintinnopsis, Tintinnus.*

Order Hypotrichida. Dorsoventrally flattened ciliates with cirri on the ventral side. *Urostyla, Euplotes, Uronychia, Stylonychia.*

REFERENCES

The general parasitology texts listed at the beginning of the references for the chapter on flatworms (p. 258) contain additional information on parasitic protozoa.

Anderson, O. R. 1983. Radiolaria. Springer-Verlag, Berlin. 355 pp.

Anderson, O. R. 1987. Comparative Protozoology: Ecology, Physiology, Life History. Springer-Verlag, Berlin. 482 pp.

Bannister, L. H. 1972. The structure of trichocyst in *Paramecium caudatum.* J. Cell Sci. *11*:899–929.

Be, A. W. H. 1982. Biology of planktonic Foraminifera. University of Tennessee, Stud. Geol. *6*:51–92.

Bick, H. 1972. Ciliate Protozoa. An Illustrated Guide to the Species as Biological Indicators in Freshwater Biology. American Public Health Association, Washington, DC. 198 pp.

Bick, H. 1973. Population dynamics of Protozoa associated with the decay of organic materials in fresh water. Am. Zool. *13*:149–160.

Borror, A. C. 1973. Protozoa: Ciliophora. Marine Flora and Fauna of the Northeastern United States. NOAA Tech. Report NMFS Circular 378. U. S. Printing Office. 62 pp.

Boynton, J. E., and Small, E. B. 1984. Ciliates by the slice. Sci. Teach. (Feb.): 35–38.

Bray, D., and White, J. G. 1988. Cortical flow in animal cells. Science. *239*:883–888.

Capriulo, G. M. (Ed.): 1990. Ecology of Marine Protozoa. Oxford University Press, 366 pp.

Corliss, J. O. 1979. The Ciliated Protozoa: Characterization, Classification, and Guide to the Literature. 2nd Edition. Pergamon Press, New York. 455 pp.

Corliss, J. O., and Esser, S. C. 1974. Comments on the role of the cyst in the life cycle and survival of free-living Protozoa. Trans. Am. Micros. Soc. *93(4)*:579–593.

Elliott, A. M. (Ed.): 1973. Biology of *Tetrahymena.* Dowden, Hutchinson, and Ross, Stroudsburg, PA. 508 pp.

Farmer, J. N. 1980. The Protozoa: Introduction to Protozoology. The C. V. Mosby Co., St. Louis. 732 pp.

Fenchel, T., 1987. Ecology of Protozoa: Biology of Free-Living Phagotrophic Protists. Science Tech Publishers, Madison, WI. 197 pp.

Fok, A. K., and Allen, R. D. 1990. The phagosome-lysosome membrane system and its regulation in *Paramecium.* Internat. Rev. Cytology. *123*:61–94.

Giese, A. C. 1973. *Blepharisma:* The Biology of a Light-Sensitive Protozoan. Stanford University Press, Stanford, CA. 366 pp.

Hammond, D. M., and Long, P. L. (Eds.): 1973. The Coccidia. University Park Press, Baltimore, MD. 482 pp.

Harrison, R. W., and Corliss, J. O. (Eds.): 1991. Microscopic Anatomy of Invertebrates. Vol. 1: Protozoa. Wiley-Liss, New York. 508 pp.

Harumoto, T., and Miyake, A. 1991. Defensive function of trichocysts in *Paramecium.* J. Exp. Zool. *260*:84.

Haynes, J. R. 1981. Foraminifera. John Wiley and Sons, New York. 434 pp.

Jeon, J. W. (Ed.): 1973. The Biology of Amoeba. Academic Press, New York.

Jones, A. R. 1974. The Ciliates. St. Martin's Press, New York. 207 pp. (A general biology of the ciliates.)

Laybourn-Parry, J. 1985. A Functional Biology of Free-Living Protozoa. University of California Press, Berkeley, CA.

Lee, J. J. 1985. The extent of algal and bacterial endosymbioses in protozoa. J. Protozool. *32(3)*:391–403.

Lee, J. J., Hutner, S. H., and Bovee, E. C. (Eds.): 1985. An Illustrated Guide to the Protozoa. Society of Protozoologists, Lawrence, KS. 629 pp. (An essential work for any exploration of the protozoa. Includes keys, descriptions, illustrations and much biological information on higher taxa.)

Levine, N. D., et al. 1980. A newly Revised Classification of the Protozoa. J. Protozool. *27(1)*:37–58.

Lynn, D. H. 1981. The organization and evolution of microtubular organelles in ciliated protozoa. Biol. Rev. *56*:243–292.

Margulis, L. 1974. Five-kingdom classification and the origin and evolution of cells. Evol. Biol. *7*:45–78.

Margulis, L., Corliss, J. O., Melkonian, M. et al. (Eds.): 1990. Handbook of Protoctista. Jones and Bartlett, Boston. 1024 pp.

Moore, R. C. (Ed.): 1964 and 1954. Treatise on Invertebrate Paleontology. Protista. Vols. C and D. Geological Society of America and University of Kansas Press, Lawrence.

Murray, J. W. 1973. Distribution and Ecology of Living Benthic Foraminiferids. Crane, Russak and Co., New York. 274 pp.

Murray J. W. 1979. British Nearshore Foraminiferids. Synopses of the British Fauna No. 16. Academic Press, London. 68 pp.

Ogden, C. G., and Hedley, R. H. 1980. An Atlas of Freshwater Testate Amoebae. British Museum, Oxford University Press, Oxford. 222 pp.

Patterson, D. J. 1980. Contractile vacuoles and associated structures: Their organization and function. Biol. Rev. *55*:1–46.

Sarjeant, W.A.S. 1974. Fossil and Living Dinoflagellates. Academic Press, London. 1002 pp.

Sleigh, M. A. 1989. Protozoa and Other Protists. Edward Arnold, London, 342 pp. (An excellent general account of the Protozoa.)

Sleigh, M. A. 1991. Mechanisms of flagellar propulsion. A biologist's view of the relation between structure, motion and fluid mechanics. Protoplasma. *164*:45–53.

Spoon, D. M., Chapman, G. B., Cheng, R. S., et al. 1976. Observations on the behavior and feeding mechanisms of the suctorian *Heliophyra erhardi* (Reider) Matthes preying on *Paramecium.* Trans. Am. Micros. Soc. *95*:443–462.

Steidinger, K. A., and Haddad, K. 1981. Biologic and hydrographic aspects of red tides. BioScienc. *31(11)*:814–819.

Stossel, T. P. 1990. How cells crawl. Am. Scien.

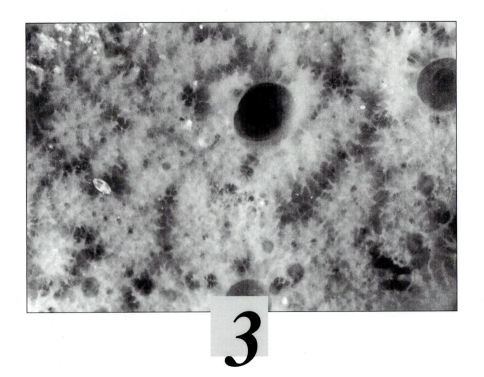

3

Sponges and Placozoans

PRINCIPLES AND EMERGING PATTERNS

Metazoans are multicellular, motile, heterotrophic organisms that pass through a blastula stage in the course of their early embryonic development. Although some, such as sponges and corals, have become sessile, they still retain motile larvae in their life cycle. Metazoans constitute almost all of what are generally considered to be animals. Their diversity is enormous, with many different ground plans, or body designs. There are some 29 phyla according to the most widely accepted groupings, and only 1, the Chordata, contains animals that are not invertebrates.

EVOLUTION OF METAZOANS

Most zoologists agree that metazoans have a common ancestry from some unicellular organisms. The **colonial theory,** in which the Metazoa are derived by way of a flagellated colony, is the classical and still most widely held hypothesis among contemporary zoologists. The colonial theory maintains that the flagellates are the ancestors of the metazoans; the following facts are cited as evidence in support of such an ancestry. Flagellated sperm cells occur throughout the Metazoa. Monociliated cells—cells with a single cilium—commonly occur among lower metazoans, particularly among sponges, hydras, sea anemones, and corals. Both eggs and sperm have evolved in some flagellates, such as the spherical, colonial *Volvox.* Although *Volvox* is frequently used as a model for the design of the flagellate colonial ancestor, these autotrophic organisms with plantlike cells are not likely ancestors of metazoans. Ultrastructural evidence points to the choanoflagellates (p. 31), a small group of animal-like, monoflagellated protozoa, as the best candidates. Some are solitary and some are colonial. Choanoflagellates have mitochondria and ciliary rootlets that are very similar to those in metazoan cells. Moreover, choanocytes—cells with a collar-like ring of microvilli around a flagellum as is typical of choanoflagellates—are found in a number of groups of metazoan animals, notably the sponges.

The colonial theory holds that the ancestral metazoan probably arose from a spherical, hollow, colonial flagellate (Fig. 3–1). The cells were monoflagellate on the outer surface; the colony possessed a distinct anterior/posterior axis, swimming with the anterior pole forward. The nonreproductive (somatic cells) were differenti-

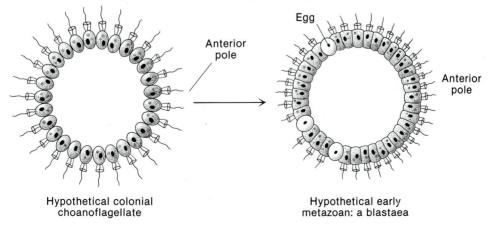

FIGURE 3–1 Origin of a simple hollow multicellular metazoan (blastaea) from a colonial choanoflagellate.

ated from reproductive ones. This hypothetical stage has been called a **blastaea** and is believed to reflect the blastula stage that occurs in the development of all animals. Further division of labor of the somatic cells would have led to increasing interdependence of cells until what was originally a colony of unicellular individuals became a single multicellular organism. The choanoflagellate colony might have been described as a superorganism because all of the individuals of the colony functioned collectively as a unit. This superorganism then provided the transition to a new level of organization, a multicellular one, in which specialization occurred not by organelles but by differentiation of cells.

EPITHELIAL AND CONNECTIVE TISSUES

Although differentiation of cells is a distinctive feature of metazoan organization, the different kinds of cells that compose the metazoan body are never distributed randomly. Similar cells are grouped together as **tissues.** Two types of tissues found in most animals are **epithelial** and **connective tissues.** Epithelial cells fit close together and form layers that bound spaces or form secretory masses (glands). Where epithelial cells serve primarily as a boundary layer, they are thin and flattened (**squamous epithelium).** Where they also have secretory, absorptive, or phagocytic functions, they tend to be **cuboidal** or **columnar** in shape (Fig. 3–2). Even sponges, which have a low level of cellular differentiation, have tissues that are epithelium-like.

  Epithelial cells are commonly ciliated. There may be one cilium or many per cell, but the monociliated condition is believed to be primitive, for it is found in all of the most primitive phyla of animals, as well as in a number of higher groups. Remember that cilia and flagella have identical ultrastructures, and where the term *flagellated* is applied to some animal cells, as in sponges, these cells could just as readily be called monociliated.

  Connective tissue is characterized by widely spaced cells and intervening extracellular matrix material that always contains at least some water and protein fibers (Fig. 3–2). A considerable part of the fibers is composed of collagen, a distinctive animal protein. Connective tissue can assume a skeletal function when a large number of fibers are present or when the matrix contains other components,

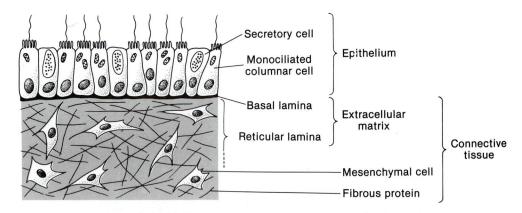

FIGURE 3–2 A layer of monciliated, columnar epithelium overlying connective tissue. Secretory cells are interspersed among the monociliated cells.

such as small pieces of silica or calcium carbonate. Such mineral **spicules** or **ossicles** impart rigidity to the connective tissue.

Connective tissue always lies between bounding layers of epithelium and thus constitutes a connective tissue compartment. The epithelium can regulate the passage of some substances passing into or out of the compartment. In a few animals, notably certain small jellyfish, the connective tissue (mesoglea) lacks cells, the matrix being derived entirely from the bounding epithelial layers.

METAZOAN LIFE CYCLES AND DEVELOPMENT

Metazoans are diploid organisms in which meiosis is restricted to the formation of haploid gametes. The diploid condition is restored in the zygote with fertilization. The zygote is a polarized cell with the polar axis running between **animal** and **vegetal hemispheres.** The first two division planes of the zygote are typically parallel to the polar axis, and the entire zygote is cleaved into two and four daughter cells, or **blastomeres** (Fig. 3–3). Such **complete cleavage** is also termed **holoblastic.** Eggs contain various amounts of yolk distributed in different ways. Because yolk tends to impede cleavage, both the amount and distribution of yolk profoundly affects the cleavage pattern. When there is a small amount of yolk rather evenly distributed in the egg (**homolecithal,** or **isolecithal** eggs), the blastomeres resulting from cleavage tend to be of equal size. Cleavage is said to be complete and **equal.** When a moderate amount of yolk is restricted to the vegetal hemisphere (**telolecithal eggs),** the blastomeres located in the animal hemisphere of the cleaving egg are smaller (**micromeres**) than those located in the yolky vegetal hemisphere (**macromeres**). Such cleavage is said to be complete but **unequal** (Fig. 3–3).

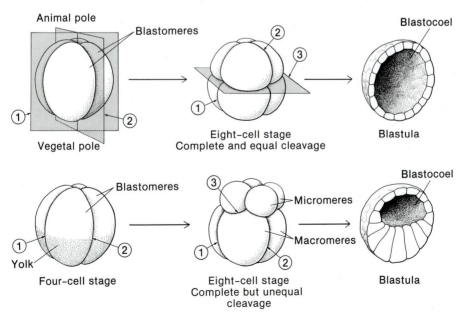

FIGURE 3–3 Early cleavage stages and blastulae from an egg with a small amount of yolk (top) and from an egg with a moderate amount of yolk concentrated in the vegetal hemisphere (bottom). The first two vertical cleavage planes would be the same in both. Numbers indicate furrows of cleavage planes.

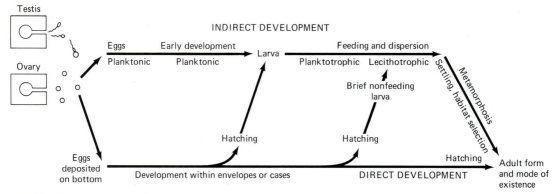

FIGURE 3–4 Patterns of development in animals.

Repeating cleavage divisions rapidly convert the zygote into a sphere of small blastomeres, called a **blastula** (Fig. 3–3). Where cleavage was complete and equal (eggs with little yolk), a single layer of blastomeres forms the blastula wall and encloses a large fluid-filled cavity, the **blastocoel.** Large amounts of yolk on the vegetal hemisphere can cause the blastocoel to be restricted to the animal hemisphere or to be obliterated completely so that the blastula is solid **(stereoblastula).**

In many animals, especially those that live in the sea, embryonic development leads to a motile, independent developmental stage called a **larva,** which typically is just barely visible to the naked eye and looks different from the adult. The larva may feed, attaining an external source of nutrition, and is carried by tidal and other currents, dispersing the species. After a planktonic life of hours, days, or even weeks, the larva settles to the bottom (if the species is benthic) and transforms **(metamorphoses)** into the adult form and assumes the adult life style. A life cycle with a larva is said to be **indirect** (Fig. 3–4).

An indirect life cycle with external fertilization and planktonic embryonic development is believed to be the primitive pattern in animals. In all groups of animals various reproductive adaptations that enhance the likelihood of fertilization and the survival of the embryo have evolved. Adaptations that enhance the likelihood of fertilization focus largely on increasing *synchrony* (gametes are produced and released at the same time) and *proximity* (gametes are released near each other). Synchronous production and release of gametes is largely under the control of environmental signals, such as temperature, light, and tides. Proximity has been achieved in a number of ways. **Hermaphroditism,** the presence of both male and female gonads in the same individual, is a common adaptation where population levels are low or where the adults live attached to some environmental structure, such as rock or coral. In these circumstances, any other adjacent individual is a potential mate. Note that most hermaphroditic animals utilize cross- rather than self-fertilization because the latter process reduces the possibility of genetic variation. However, there is an energy cost to hermaphroditism—the development of two types of gonads and associated systems as well as the production of both eggs and sperm (p. 209).

Many hermaphroditic animals are not **simultaneous hermaphrodites** but rather **protandric,** meaning that they develop testes first and ovaries later. That is, the sex of the individual shifts from male to female during the course of its reproductive life. The reverse sequence is uncommon.

We must caution that although there is a good correlation between hermaphroditism and sessility, they are some large groups of nonsessile animals, such as flatworms, in which the origin of their hermaphroditism is not so clear.

Another widely encountered adaptation for proximity is **internal fertilization,** which can lead to a wide range of modifications in the reproductive tract (these will be explored in later chapters). Where internal fertilization occurs, a smaller number of gametes are usually produced than when fertilization occurs externally in water.

The principal disadvantage of planktonic development is predation on the developing embryos and larvae. Encasing the eggs within protective envelopes attached to the bottom or brooding them within the parent removes the early embryonic stages from the plankton. Hatching or release from brooding at the larval stage still permits the advantages of planktonic feeding and dispersal. In many groups of animals with larvae, larval life is shortened by having sufficient yolk within the larval cells to take care of nutrition, and the nonfeeding larva has a dispersive function. In still others, the larva is completely dispensed with, and development is **direct**—the young hatching as miniature adults, or juveniles.

All of these life history modifications often can be encountered among different species within one class of animals. Direct development is commonly exhibited by freshwater animals, because of the hazards of currents and turbidity, but there are many exceptions.

SPONGES

Sponges, which constitute the phylum Porifera, are the most primitive of the multicellular animals. Sponges lack organs, but have well-developed connective tissue, in which cells perform a variety of functions. Compared with other metazoans, sponge cells show such a high degree of independence that the sponge body resembles a protozoan colony in some respects. Sponges are specialized in being sessile and in having an unusual body plan built around a system of water canals. Their sessility and lack of any conspicuous movement of body parts convinced Aristotle, Pliny, and other ancient naturalists that sponges were plants. In fact, it was not until 1765, when internal water currents were first observed, that the animal nature of sponges became clearly established.

Except for about 150 freshwater species, the approximately 5000 described species of sponges are marine animals. They abound in all seas, wherever rocks, shells, submerged timbers, or coral provide a suitable substratum. Some species even live on soft sand or mud bottoms. Most sponges prefer relatively shallow water, but some groups, including most glass sponges, live in deep water.

SPONGE STRUCTURE

Sponges vary greatly in size. Certain calcareous sponges are about the size of a grain of rice, but a large loggerhead sponge may exceed a meter in height and diameter. Some are radially symmetrical, but the majority are irregular and exhibit massive, erect, encrusting, or branching growth patterns (Figs. 3–5, 3–6, and 3–12). Most of the common species are brightly colored. Green, yellow, orange, red, and purple sponges are frequently encountered, but the significance of the coloration is uncertain. Protection

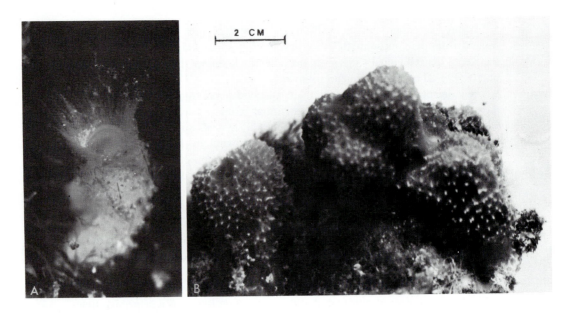

FIGURE 3–5 A, A small, calcareous syconoid sponge. The
vase-shaped body of this individual is no more than 5 mm in
length. Long spicules fringe the
large osculum. **B,** *Dysidea ethe-
ria,* a West Indian leuconoid
sponge, which is pale blue. One
osculum is visible at left. *(A and B
courtesy of Betty M. Barnes.)*

from solar radiation and warning coloration have been suggested for some species.

Sponge architecture is unique, being constructed around a system of water canals—an arrangement that is correlated with sponge sessility. This architecture is the key to understanding many aspects of sponge biology. The basic structure and histology of sponges is most easily understood by beginning with the simplest radial forms. Such sponges are called **asconoid sponges,** a structural term rather than a taxonomic one. The asconoid sponge is tubular and always small (Fig. 3–7). *Leucosolenia,* which is one of the few living genera of asconoid sponges, rarely exceeds 10 cm in height. Asconoid sponges are not usually solitary but are composed of clusters of tubes attached together along their long axes or at their bases.

The surface of an asconoid sponge is perforated by many small openings, called **ostia (or incurrent pores),** from which the name Porifera (pore-bearer) is derived. These pores open into the interior cavity, the **spongocoel (atrium),** which in turn opens to the outside through the **osculum,** a large opening at the top of the tube. A constant stream of water passes through the incurrent pores into the spongocoel and out through the osculum.

The body wall is relatively simple. The outer surface is covered by epithelial-like flattened cells, known as the **pinacocytes,** which together make up the pinacoderm. Unlike the epithelium of most other animals, however, a basal lamina and intercellular junctions are absent, and the margins of pinacocytes can be contracted or withdrawn so that the entire animal can decrease slightly in size. The basal pinacocytes secrete material that fixes the sponge to the substratum. Each pore is formed by a **porocyte,** a cell shaped like a ring that extends from the external surface to the spongocoel. The bore, or lumen, of the porocyte forms the incurrent pore, or ostium, and can be closed or opened by contraction.

Beneath the pinacoderm lies the **mesohyl** (sometimes referred to as mesenchyme), which consists of a gelatinous proteinaceous matrix containing skeletal

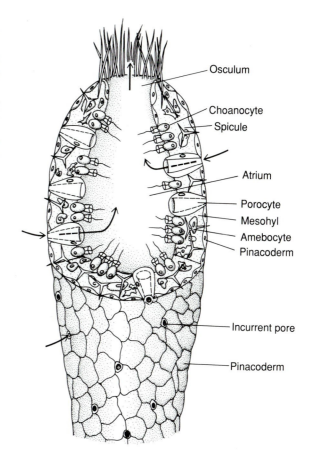

FIGURE 3–7 Diagram of a partially sectioned asconoid sponge. *(Based on a figure by Buchsbaum. B-E, After Minchin from Jones.)*

material and ameboid cells. The mesohyl is equivalent to the connective tissue of other metazoans.

The skeleton is relatively complex and provides a supporting framework for the living cells of the animal. (To avoid repetition, the discussion presented here on the sponge skeleton applies to the phylum in general, not just to the asconoid sponges.) The skeleton may be composed of calcareous spicules, siliceous spicules, protein spongin fibers, or a combination of the latter two. The spicules exist in a variety

◀**FIGURE 3–6** Relationship of sponge form to utilization of substratum. **A,** The two massive sponges at the right on top of the rock require an exposed surface, but their elevated form enables them to utilize water well above the substratum, and their attachment area is a relatively small part of the total body surface area. **B,** The encrusting sponges below the rock utilize much of their surface area for attachment, but their low encrusting form enables them to exploit the space of crevices and other confined areas. **C,** The sponge on the vertical surface at left utilizes space *within* the substratum. Small arrows indicate the movement of water into the sponge; large arrows indicate the exit of water from oscula.

of forms and are important in the identification and classification of species (Fig. 3–8). An extensive nomenclature has developed through the use of these structures in sponge taxonomy. The suffix, -axon, refers to the number of axes a spicule has, while -actine, refers to the number of rays or points. Monaxon spicules are shaped like needles or rods and may be curved or straight with pointed, knobbed, or hooked ends, while triaxons may have either three rays or six (hexactines). These terms apply to **megascleres,** the larger spicules forming the chief supporting elements in the skeleton. **Microscleres,** spicules that are are considerably smaller, have their own specialized terminology.

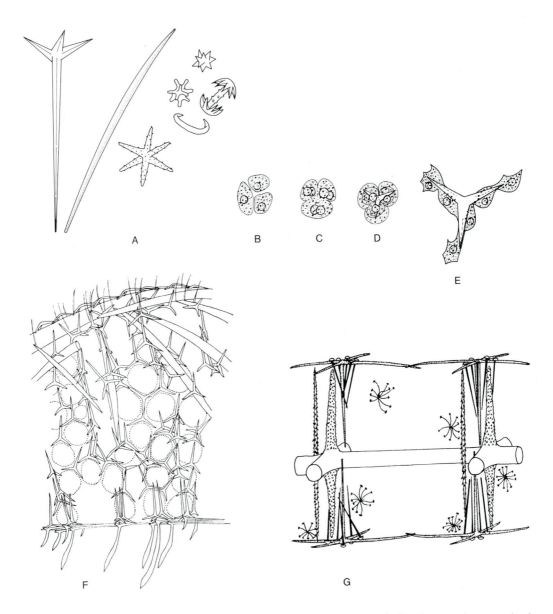

FIGURE 3–8 A, Sponge spicules, showing variations in shape and size (megascleres and microscleres). **B–E,** Secretion of a calcareous triradiate spicule. **F,** Section through a part of a leuconoid calcareous sponge, showing spicules in their natural position. **G,** Spicules of the hexactinellid sponge *Farrea sollasii* in their natural position. *(B–E, After Minchin from Jones. F after Borojevic. G after Schulze, both In Bergquist, P. R. 1978. Sponges. Hutchinson, London. pp. 147–151.)*

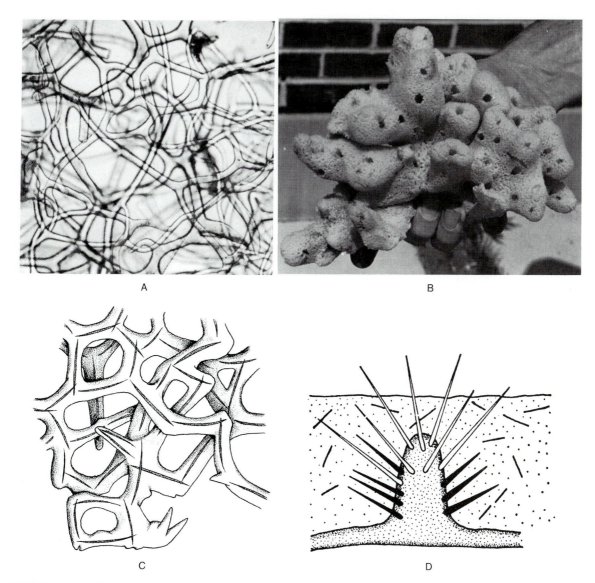

FIGURE 3–9 A, Photomicrograph of spongin fibers (they appear translucent). **B,** The spongin skeleton of a commercial sponge *(Spongia)* from the Mediterranean. The large openings are oscula. **C,** Spongin network of *Endectyon,* showing embedded spicules. **D,** Sponge with one type of spicule partially embedded within spongin. *(A, Courtesy of the General Biological Supply House, Inc. B, Courtesy of Betty M. Barnes. D, From Berquist, P. R. 1978. Sponges. Hutchinson, London. p.46.)*

The skeleton is located primarily in the mesohyl, but spicules frequently project through the pinacoderm. The arrangement of spicules is organized, with various types often combined in distinct groupings (Figs. 3–8F and G). They may interlock or be fused together. The organization in one part of the body may differ from that in another. Microscleres, for example, support the pinacoderm lining the water canals.

The mesohyl of all sponges contains dispersed collagen fibrils, but many sponges also possess a skeleton of coarse interconnecting fibers (Fig. 3–9). The fibers are composed of **spongin,** a fibrous protein similar to collagen. Some sponges contain so much spongin that they are tough and rubbery, and in many species siliceous spicules are embedded partially or completely in the fibers to help stiffen them (Figs. 3–9C and D).

Ameboid cells occur in the mesohyl and include a number of types. **Archeocytes,** large cells with large nuclei, are phagocytic and play a role in digestion. Archeocytes are said to be totipotent, that is, they are capable of transforming themselves into other types of cells needed within an animal. Fixed cells, called **collencytes,** are anchored by long, cytoplasmic strands and secrete the dispersed collagen fibers. Many sponges possess mobile cells, which also secrete these fibers.

The spicule skeleton is secreted by ameboid sclerocytes. One to several **sclerocytes** are usually involved in the secretion of a single spicule in the calcareous sponges, and the process is relatively complex. A three-pronged spicule, for example, originates in three sclerocytes derived from an amebocyte, called a scleroblast. The three sclerocytes partially fuse to form a trio of cells (Fig. 3–8B to E). Each member of the trio then divides, and between each pair of daughter cells one prong, or ray, of the spicule is secreted. The three prongs fuse at the base. Each of the three pairs of sclerocytes now moves outward along a ray, one cell secreting the end and one thickening the base of the spicule (Fig. 3–8E). The spongin skeleton is secreted by **spongocytes.**

On the inner side of the mesohyl, and lining the spongocoel, is a layer of cells, the **choanocytes,** which are very similar in structure to choanoflagellate protozoa (Fig. 2–16). The choanocyte is ovoid, with one end adjacent to the mesohyl. The opposite end of the choanocyte projects into the spongocoel and bears a flagellum surrounded by a collar of microvilli. The choanocytes are responsible for moving water through the sponge and for obtaining food. (Both of these processes are described in detail later.) It is important to understand that the spongocoel and its choanocyte lining are not homologous to the gut of other animals, and the choanocytes are not an endodermal derivative. Indeed, there is no endoderm in sponges. Sponges are gutless, and the gutless condition is primary.

The primitive asconoid structure imposes very definite size limitations. An increase in the volume of the spongocoel is not accompanied by a sufficient increase in surface area of the choanocyte layer to provide for water movement. Thus, asconoid sponges are always small.

The problems of water flow and surface area have been overcome during the evolution of sponges by the folding of the body wall and in many species by the reduction of the spongocoel. The folding increases the surface area of the choanocyte layer, and the reduction of the spongocoel lessens the volume of water that must be circulated. The net result of these changes is a greatly increased and more efficient water flow through the body. A greater size now becomes possible, although the primitive radial symmetry is commonly lost.

Sponges display various stages in the changes just described. Sponges that exhibit the first stages of body wall folding are called **syconoid sponges** and include the well-known genera *Grantia* and *Sycon* (=*Scypha*). In syconoid structure, the body wall has become "folded," forming external pockets extending inward from the outside, and evaginations, extending outward from the spongocoel (Fig. 3–10B and C). The many pockets produced by folding do not meet but bypass each other and are blind.

In this more advanced type of sponge, the choanocytes no longer line the spongocoel but are now confined to the evaginations, which are called **flagellated,** or **radial canals.** The corresponding invaginations from the pinacoderm side are known as **incurrent canals** and are lined by pinacocytes. The two canals are connected by openings called **prosopyles,** which are equivalent to the pores of asconoid sponges. Water now flows through the incurrent canals, the prosopyles, the flagellated canals, and the spongocoel and flows out the osculum.

A slightly more specialized stage of the syconoid structure develops when pinacocytes and mesohyl plug the open ends of the incurrent canals (Fig. 3–10D). Ostial openings remain to permit entrance of water into incurrent canals. Despite the folding of the body wall, syconoid sponges still retain a radial symmetry.

The highest degree of folding takes place in **leuconoid sponges** (Fig. 3–10E). The flagellated canals have transformed to form small, rounded, **flagellated chambers,** and the spongocoel has commonly been reduced to water canals leading to an osculum. Water enters the sponge through the ostia and passes into subdermal spaces leading into branching incurrent canals. These eventually open into the flagellated chambers through prosopyles. Water leaves the chamber through an **apopyle** and courses through excurrent canals, which become progressively larger as they are joined by other excurrent canals. A large canal eventually opens to the outside through the osculum. The canals are lined by pinacocytes. The num-

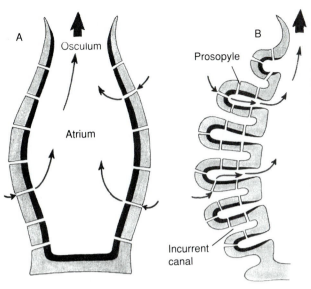

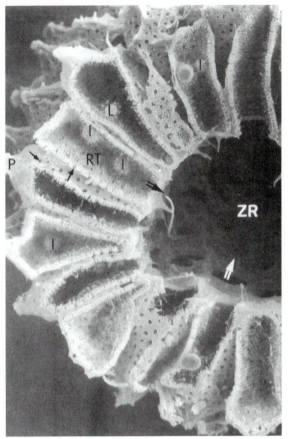

FIGURE 3–10 Morphological types of sponges (pinacoderm and mesohyl in pale gray; choanocyte layer, black). **A,** Asconoid type. **B,** Syconoid type. **C,** Scanning electron photomicrograph of a transverse section through a syconoid sponge *(Sycon)*. Pinacoderm (P) and ostia with porocytes (arrows) along incurrent canals. Openings into the central spongocoel (ZR) are shown with double arrows. L but not I is a section through a developing larva. **D,** More specialized syconoid type, in which entrance to incurrent canals has been partially filled with pinacoderm and mesohyl. **E,** Leuconoid type. *(A, B, D, and E Modified from Hyman L. H. 1940. The Invertebrates. Vol. I. McGraw-Hill Book Co., New York. C, From Weissenfels, N., and Schafer, D. 1985. Kombinierte phasenkontrast- und rasterelektronenmikrosopisch Histologie. Lietz-Mitt. Wiss. u. Techn. 8(8)244–246.)*

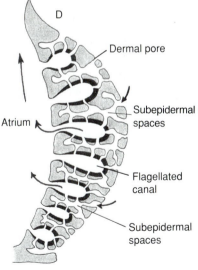

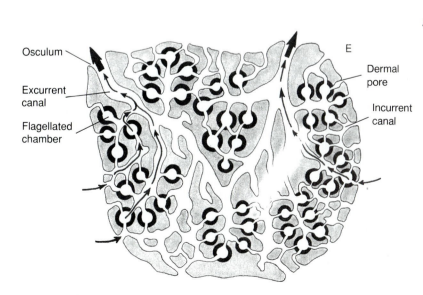

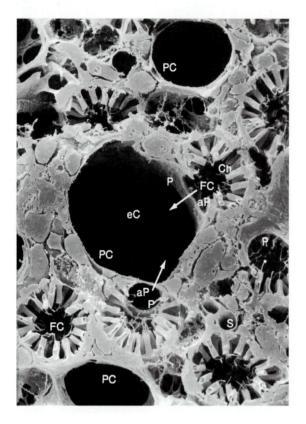

FIGURE 3–11 Scanning electron photomicrograph of a section taken from the freshwater sponge *Ephydatia fluviatilis*. In this section a number of flagellated chambers (FC) with choanocytes (Ch) surround a large excurrent canal (eC) with a wall formed by pinococytes (PC). The apopyles (aP) from the flagellated chambers pass through porocytes (P). Within the mesohyl (M) can be seen archeocytes (A), spicules (S), and spongin (Sp). *(From Weissenfels, N. 1982. Bau und Funktion des Süsswasserschwamms Ephydatia fluviatilis. IX. Rasterelektronenmikroskopische Histologie und Cytologie. Zoomorphology. 100:75–87.)*

ber of flagellated chambers may be enormous, for example, *Microciona prolifera* contains some 10,000 chambers per cubic millimeter, each 20 to 39 μm in diameter and containing about 57 choanocytes. The mesohyl is usually thicker than in asconoid sponges. Porocytes can form the ostia as well as prosopyles and apopyles in the endopinacoderm (Fig. 3–11).

Most sponges are built on the leuconoid plan, which is evidence of the efficiency of this type of structure. Leuconoid sponges can reach a large size because any addition to their mass increases the number of flagellated chambers necessary to propel water through the addition. Leuconoid sponges may be encrusting or erect with flattened or branching bodies, although many species are vase-shaped and tubular forms in which the excurrent canals empty into a large, central chamber (Figs. 3–12 and 3–14). Rather than a single osculum, there may be many oscula.

Most leuconoid sponges have one of two general internal designs, regardless of the external form. In one type the body is solid, and adjacent incurrent and excurrent canals commonly run parallel to each other, conducting water inward to flagellated chambers and outward to oscula, which are scattered over the surface (Fig. 3–13 A, B, and C). In the other type the body is hollow, with the oscula confined to the upper or distal parts of the body. Excurrent water canals do not return to the surface but open into the interior cavity, which leads to a distal osculum (Fig. 3–13D, E, F, and G).

The leuconoid plan clearly evolved more than once within sponges and, in some instances, may have involved a preceding syconoid stage.

PHYSIOLOGY

The physiology of a sponge largely depends on the current of water flowing through the body. The water brings in oxygen and food and removes waste. Even sperm are moved in and out and larvae released by the water currents. The volume of water pumped by a sponge is remarkable. A specimen of *Leuconia (Leucandra)*, a leuconoid sponge 10 cm in height and 1 cm in diameter, has roughly 2,250,000 flagellated chambers and pumps 22.5 l of water per day through its body. The flow velocity is greatest through the osculum and slowest through the flagellated chambers, because these two regions have, respectively, the smallest and the largest total cross-sectional areas of the various passageways (Fig. 3–15). By regulating the size of the osculum and closing the ostia, the animal can control the rate of flow and even stop it altogether. In some Demospongiae, control of the osculum is facilitated by a special type of mesohyl cell called a **myocyte,** which displays similarities to a smooth muscle cell in shape and contractility. However, unlike true muscle cells, the myocytes surrounding an osculum do not touch each other.

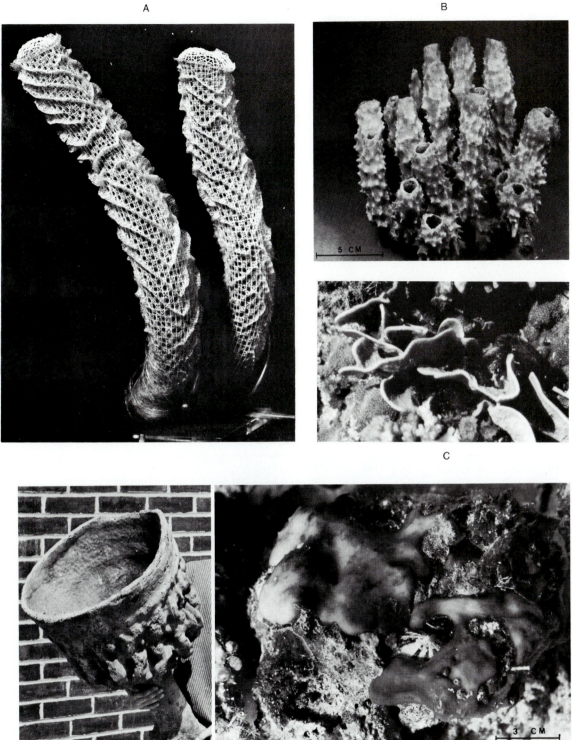

FIGURE 3–12 A, Venus's flower basket, *Euplectella,* a hexactinellid sponge in which the spicules are fused to form a lattice. **B,** *Callyspongia,* a tropical leuconoid sponge (Demospongiae) with a tubular body form. **C,** *Phyllospongia,* a leaflike sponge on a reef flat in Fiji. **D,** *Poterion,* a large, goblet-shaped leuconoid sponge (Demospongiae). **E,** "Chicken-liver," *Chondrilla,* a very common West Indian encrusting sponge, with a tough, almost cartilage-like spongin skeleton (Demospongiae). *(A, Courtesy of the American Museum of Natural History. B–E Courtesy of Betty M. Barnes)*

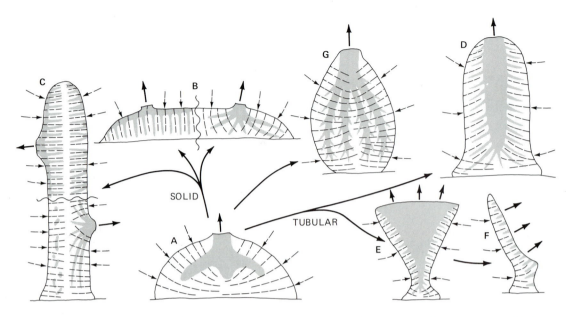

FIGURE 3–13 Diagram of the two types of sponge architecture, (**A, B,** and **C**) solid and (**D, E, F** and **G**) tubular, and their relationship to sponge form. The incurrent system is shown with dotted arrows and lines, and the excurrent system with heavy black arrows and lines. See text for further details. *(From Reiswig, H. M. 1975. The aquiferous systems of three marine Demospongiae. J. Morphol. 145(4):493–502.)*

FIGURE 3–14 A, A vaselike *Xestospongia muta* from the West Indies. A scleractinian coral is located to the left of the sponge. **B,** *Verongia,* a tubular West Indian sponge, which commonly reaches 50 cm or more in length. The surrounding plantlike animals are gorgonian corals. *(A and B, Courtesy of David Barnes.)*

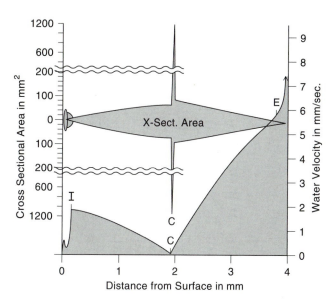

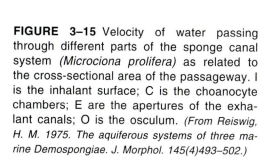

FIGURE 3–15 Velocity of water passing through different parts of the sponge canal system *(Microciona prolifera)* as related to the cross-sectional area of the passageway. I is the inhalant surface; C is the choanocyte chambers; E are the apertures of the exhalant canals; O is the osculum. *(From Reiswig, H. M. 1975. The aquiferous systems of three marine Demospongiae. J. Morphol. 145(4)493–502.)*

The water current is produced by the planar beating of the choanocyte flagella, but there is neither coordination nor synchrony of the flagella in a particular chamber. The choanocytes are oriented toward the apopyle, and each flagellum beats from base to tip, driving water from the flagellated chamber (Fig. 3–16A). As a result, water is sucked into the chamber from the incurrent canals through the small prosopyles located between the bases of the choanocytes. It is then driven to the center of the chamber and out the larger apopyle into an excurrent canal.

When a sponge is exposed to an external current, water flows passively through the body of a sponge, given certain structural conditions, such as elevated oscula. This hydrodynamic effect undoubtedly contributes to water passage in those sponges that live in strong to moderate currents. However, many sponges reach their greatest development in relatively quiet water or even in confined spaces.

Some sponges exhibit a diurnal rhythm in the propulsion of water through their bodies whereas others exhibit an erratic, endogenous water flow. External conditions, such as turbulent water caused by storms, may halt water flow regardless of internal conditions.

Sponges are filter feeders that depend on the stream of water passing through the body as a source of food. They feed on extremely fine particulate material. Studies on three species of Jamaican sponges have demonstrated that 80% of the organic matter consumed by these sponges is too small to be resolved with ordinary microscopy. The remaining 20% consists of bacteria, dinoflagellates, and other fine plankton.

In tropical waters, at least, there is about seven times more available carbon in the unresolved fraction than at the planktonic level. The sponges' ability to utilize this food source undoubtedly accounts for their success as sessile animals, especially in tropical waters.

Food particles are apparently selected largely on the basis of size and are screened in the course of their passage into the flagellated chambers. Only particles smaller than a certain size can enter the ostia or pass through the prosopyles. Screening is also provided by cytoplasmic strands stretched across the incurrent canals. The finest particles are removed by the choanocytes, perhaps by filtration across the choanocyte collar (Fig. 3–16B), but the mechanism is not fully understood. The microvilli are 0.03 to 0.10 μm apart and are interconnected by a fibril-like material.

All sponge cells can phagocytize particles. Large particles (5–50 μm) are phagocytized by pinacocytes lining the inhalant pathways or by archeocytes that move out of the mesohyl. Indeed, populations of archeocytes have been reported cleansing the outer pinacoderm surface. Particles of bacterial size and below (<1 μm) are removed and engulfed by the choanocytes on the surface of the cell and not on the microvillor collars. Both choanocytes and archeocytes can transfer their engulfed particles to cells, and

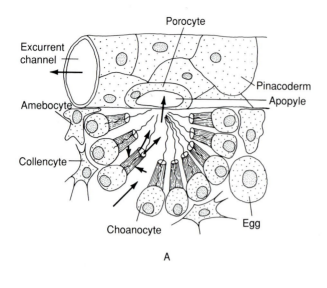

A

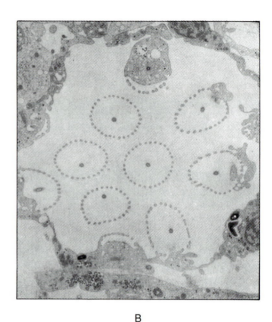

B

C

FIGURE 3–16 A, Section through flagellated chamber of freshwater sponge, *Ephydatia*. Arrows indicate direction of water currents. **B,** Cross section through a flagellated chamber of the boring sponge *Cliona lampa*. The rings are cross sections of the collars of choanocytes, showing the circles of microvilli and the central flagellum. **C,** Choanocyte chambers within the body wall of the hexactinellid *Euplectella* (Venus's flower basket). *(B, From Rutzler, K., and Rieger, G. 1973. Sponge burrowing: Fine structure of Cliona lampa penetrating calcareous substrata. Mar. Biol., 21(2):144–162. C, After Schulze, In Bergquist, P. R. 1978. Sponges. Hutchinson, London. pp. 59 and 26.)*

archeocytes rather than choanocytes appear to be the principal sites of digestion, which is within food vacuoles. The archeocytes probably also act as storage centers for food reserves.

Many marine sponges, both Demospongiae and Calcarea, are now known to harbor symbiotic photo-synthetic organisms. A few species contain nonmotile dinoflagellates (zooxanthellae), but the most common symbionts are cyanobacteria (blue-green algae), which live within the mesohyl or within specialized amebocytes. The zooxanthellae may give the sponge a yellowish hue and the cyanobacteria a green, violet, or brown color. The cyanobacterial symbionts of some keratose sponges, including *Verongia,* may make up more than 33% of the sponge. Such sponges live in shallow, well-lighted habitats and may have their symbiotic cyanobacteria restricted to the outer layers of the body. Excess photosynthate in the form of glycerol and a phosphorylated compound are utilized by the sponge host. Sponges studied on the Australian Great Barrier Reef obtain from 48 to 80% of their energy requirements from their cyanobacteria. Some sponges also contain intra- and extracellular bacteria, in addition to the cyanobacteria. However, the significance of such bacteria is still uncertain.

The great volume of water passing through the extensive canal system of a sponge means that most cells, even those deep within the body, are interacting directly with the external medium. Egested waste and nitrogenous waste (largely ammonia) leave the body in the water currents. Gas exchange occurs by simple diffusion between the flowing water and the cells in the sponge along the course of water flow.

The pinacocyte and choanocyte layers lack intercellular junctions and most other ultrastructural connections that make epithelial layers of other animals controlling barriers between internal and external environments. Sponges have sometimes been described as "leaky" animals. Their interstitial fluid (the fluid between cells) must be very similar to that of the environment even in freshwater species. Most cells of freshwater sponges possess contractile vacuoles, but the contractile vacuoles are osmoregulating for individual cells, not for extracellular compartments.

The presence of spicules increases the overall stiffness of sponge tissue by restricting the ability of organic components (polymers) to undergo molecular rearrangements in response to mechanical loads. Thus, for example, the spicule-filled sponge body resists deformation in strong water currents. Sponges that live in areas of high surge have a greater density of spicules, both in numbers and surface area, than individuals of the the same species living in quieter water. However, there is an upper limit to the increase in skeletal material, because greater density of spicules reduces the size of the intervening water canals. There is more resistance to water flow in smaller canals and thus a higher energy cost to pumping.

Sponges have no nervous system, and reactions are largely local. Coordination depends on transmission of messenger substances by diffusion within the mesohyl, by wandering ameboid cells, and along fixed cells in contact with each other. Electrical conduction, which does not involve action potentials, appears to take place by the latter route.

Many sponges produce metabolites that may prevent other organisms from settling on their surfaces or may deter some potential grazing predators. Nine out of 16 Antarctic sponges and 27 of 36 Caribbean species were found to be toxic to fish. However, compounds that repel fish do not necessarily deter other grazers. Turtles commonly feed on sponges, and 95% of their feces may consist of glass spicules. Some sponges utilize excreted metabolites in competing for space with other animals. For example, the Caribbean "chicken liver" sponge can kill adjacent stony corals and overgrow their skeletons. Some species have very distinctive odors (for example, one resembling garlic), and a few, like the red Caribbean fire sponge *Tedania ignis,* can cause a rash when handled. Various sponge biochemicals are being investigated for potential medical and commercial benefits.

Many sponges are hosts for other animals. Some large leuconoid sponges are veritable apartment houses for certain shrimps and brittle stars. One investigator collected over 16,000 snapping shrimps from within the water canals of one large loggerhead sponge. Some spider crabs put algae, sponges, and other sessile animals on their backs (p. 701). The collection attaches and grows on this mobile substratum, providing the crab with an effective camouflage. Certain other crabs (p. 701) cut out a cap of sponge that they fit over their back, and suberitid sponges growing on shells occupied by hermit crabs eventually overgrow and replace the shell.

THE CLASSES OF SPONGES

Approximately 5000 species of sponges have been described and are placed within four classes.

Class Calcarea, or Calcispongiae

Members of this class, known as calcareous sponges, are distinctive in having spicules composed of calcium carbonate. All the spicules are of the same general size and are monaxons or three- or four-pronged; they are usually separate. Spongin fibers are absent. All three grades of structure, asconoid, syconoid, and leuconoid, are encountered. Many Calcarea are drab, although brilliant yellow, red, and lavender species are known. They are not as large as species of other classes; most are less than 10 cm in height. Species of calcareous sponges exist throughout the oceans of the world, but most are restricted to relatively shallow coastal waters. Genera such as *Leucosolenia* and *Sycon* are commonly studied examples of asconoid and syconoid sponges.

The subclass Sphinctozoa contains a single recently discovered living representative *(Neocoelia)* from shaded recesses on Indo-Pacific reefs. Sphinctozoans were abundant from the late Paleozoic through the Mesozoic. They have no spicules, but a calcareous skeleton forms an outer perforated wall and the walls of interior chambers. Some fossil species have spicules embedded in the wall.

Class Hexactinellida, or Hyalospongiae

Representatives of this class are commonly known as glass sponges. The name Hexactinellida is derived from the fact that the spicules include a triaxon with six points (hexatine) (Fig. 3–8G). Furthermore, some of the spicules often are fused to form a skeleton that may be lattice-like and built of long, siliceous fibers that look like the loose fiberglass used in insulating. Thus, they are called glass sponges. The glass sponges, as a whole, are the most symmetrical and the most individualized of the sponges, that is, they show less tendency to form interconnecting clusters or large masses with many oscula. The shape is usually cup-, vase-, or urnlike, and they average 10 to 30 cm in height. The coloring in most of these sponges is pale. There is a well-developed spongocoel, and the single osculum is sometimes covered by a sieve plate—a gratelike covering formed from fused spicules. The filmy lattice-like skeletons composed of fused spicules in species, such as Venus's flower basket *(Euplectella),* retain the general body structure and symmetry of the living sponge and are very beautiful (Fig. 3–12). Basal tufts of spicule fibers implanted in sand or sediments adapt many species for living on soft bottoms.

The histology of hexactinellids is different from that of other sponges. All surfaces exposed to water are covered not by pinacoderm but by a framework of syncytial strands through which long spicules may project. Another syncytium, containing flagella with collars, lines the flagellated chambers. These collar bodies lose their nuclei after being formed. Archeocytes are one of the few discrete cell types. The flag-

ellated chambers are commonly thimble-shaped and oriented at right angles in parallel planes to the body wall and central spongocoel (Fig. 3–16C). Hexactinellids appear superficially to be somewhat syconoid in structure. However, there is nothing comparable to syconoid incurrent canals; water simply enters through the spaces within the meshlike outer trabecular syncytium.

In contrast to the Calcarea, the Hexactinellida are chiefly deepwater sponges. Most live between depths of 200 and 1000 m, but some have been dredged from the abyssal zone. Hexactinellids are found throughout the world, but in the Antarctic they are the dominant sponges.

Species of *Euplectella,* Venus's flower basket, display an interesting commensal relationship with certain species of shrimp *(Spongicola).* A young male and a young female shrimp enter the spongocoel where they grow and cannot escape through the sieve plate covering the osculum. Thus their entire life is spent within the sponge, where they feed on plankton brought in by the sponge's water currents. A spider crab *(Chorilla)* and an isopod *(Aega)* are also found as commensals with some species of *Euplectella.*

Class Demospongiae

This large class contains 90% of sponge species and includes most of the common and familiar forms. These sponges range in distribution from shallow water to great depths. Coloration is frequently brilliant because of pigment granules located in the amebocytes. Different species are characterized by different colors, and a complete array of hues is encountered. The skeleton of this class is variable. It may consist of siliceous spicules or spongin fibers or a combination of both. The genus *Oscarella* is unique in lacking both a spongin and a spicule skeleton. These Demospongiae with siliceous skeletons differ from the Hexactinellida in that their larger spicules are monoaxons or tetraxons, never triaxons (hexactines). When both spongin fibers and spicules are present, the spicules are usually connected to, or completely embedded in, the spongin fibers.

All Demospongiae are leuconoid, and the majority are irregular, but all types of growth patterns are displayed. Some are encrusting (Fig. 3–12E); some have an upright branching habit or form irregular mounds; others are stringlike or foliaceous (Fig. 3–12C). There are also species, such as *Poterion* (Fig.

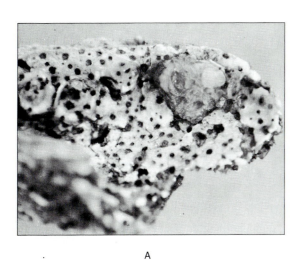

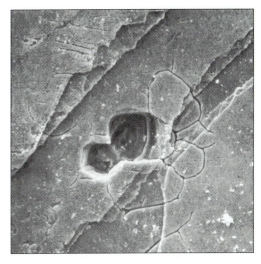

A   B

FIGURE 3–17 A, Remains of a clam shell that has been riddled with boring sponge. **B,** Calcareous surface from which two chips have been removed; four more are partially etched. *(A and B, From Rützler, K., and Rieger, G. 1973. Sponge burrowing: fine structure of Cliona lampa penetrating calcareous substrata. Mar. Biol. 21:144–162.)*

3–12D), that are goblet- or urn-shaped, and others, such as *Callyspongia* (Fig. 3–12B), that are tubular. The great variation in the shapes of members of the Demospongiae reflects, in part, adaptations to limitations of space, inclination of substrate, and current velocity. Large upright forms can exploit vertical space and use only a small part of their surface area for attachment. Encrusting forms, although they require more surface area for attachment, can utilize vertical surfaces and very confined habitats, such as crevices and spaces beneath stones (Fig. 3–6). The largest sponges are members of the Demospongiae; some of the tropical loggerhead sponges *(Spheciospongia)* form masses over a meter in height and diameter.

Several families of Demospongiae deserve mention. The boring sponges, composing the family Clionidae, are able to bore into calcareous structures, such as coral and mollusc shells (Fig. 3–6), forming channels that the body of the sponge then fills. At the surface the sponge body projects from the channel opening as small papillae. These papillae represent either clusters of ostia opening into an incurrent canal or an osculum. Excavation, which is begun by the larva, occurs when special amebocytes remove chips

of calcium carbonate. The amebocyte begins the process, etching the margins of the chip by digesting the organic framework material and dissolving the calcium carbonate (Fig. 3–17B). The chip is then undercut in the same manner, the amebocyte enveloping the chip in the process. Eventually, the chip is freed and is eliminated through the excurrent water canals. *Cliona celata,* a common boring sponge that lives in shallow waters along the Atlantic coast, inhabits old mollusc shells. The bright sulfur yellow of the sponge is visible where the bored channels reach the surface of the shell. *Cliona lampa* of the Caribbean is red, and it commonly overgrows the surface of the coral or coralline rock that it has penetrated as a thin encrusting sheet. Boring sponges are important agents in the decomposition of shell and coral (Fig. 3–17).

Members of two families of sponges occur in fresh water, but the family Spongillidae contains the majority of freshwater species. The Spongillidae are distributed worldwide and live in lakes, streams, and ponds where the water is not turbid. They have an encrusting growth pattern, and some are green because of the presence of symbiotic zoochlorellae in most of the sponge cells. The algae are brought in by water currents, taken up by pinacocytes and choanocytes, and transferred to other cells. The growth rate of sponges deprived of zoochlorellae is less than half the normal rate.

The family Spongiidae contains the marine species harvested as bath sponges. The skeleton is composed only of spongin fibers. *Spongia* and *Hippospongia,* the two genera of commercial value, are gathered from sponge-fishing grounds in the Gulf of Mexico, the Caribbean, and the Mediterranean. (There is no longer any large, commercial fishing for sponges in the United States.) The sponges are gathered by divers, and the living tissue is allowed to decompose in water. The remaining undecomposed skeleton of anastomosing spongin fibers is then washed (Fig. 3–9B). The colored block "sponges" seen on store counters are synthetic.

Class Sclerospongiae

A fourth class of sponges, the Sclerospongiae, contains a small number of species found in grottoes and tunnels within coral reefs. These leuconoid sponges have siliceous spicules and spongin fibers, but these elements and the surrounding living tissue rest on a solid basal skeleton of calcium carbonate or are enclosed within calcium carbonate chambers. Some specialists place the Sclerospongiae within the Demospongiae.

REGENERATION AND REPRODUCTION

Many sponges have remarkable powers of regeneration. Regeneration was employed in the propagation of commercial sponges in overfished areas off the Florida coast. Pieces of sponge called cuttings were attached to cement blocks and dumped into the water. Regeneration and several years' growth produced a sponge of marketable size. The classic experiment demonstrating the regenerative ability of sponges involves forcing living sponge tissue through a silk mesh. The separated cells quickly reorganize by progressive association of similar cells bounded by pinacocytes, forming themselves into several new sponges. Archeocytes are essential for reaggregation, as is a minimum number of cells. Calcium and magnesium ions plus some cell surface macromolecules are also necessary for reaggregation. Whether successful reaggregation will occur only with dissociated cells from the same species is still being debated. If

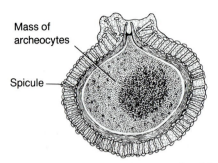

FIGURE 3–18 Section through gemmule of freshwater sponge. *(After Evans from Hyman.)*

an individual of certain sponges, for instance, *Tethya,* is sliced, and a piece from another sponge of the same species is inserted into the wound, host and graft grow together in a short time. In contrast, the host rejects a graft from a different species. There are also certain sponges, such as species of *Halichondria* and the freshwater *Spongilla,* in which developing individuals, following dense larval settlement, fuse and form sponges that are genetic mosaics. On the other hand, grafting experiments with the tropical *Callyspongia diffusa* indicate immunocompetence in some sponges, that is, the species distinguishes self from nonself: an individual accepts grafts that come from itself but rejects those from other members of the same species.

Asexual reproduction by the formation of buds that are liberated from the parent is not common in sponges, although it does occur in some species. Somewhat different from budding is the formation and release of packets of essential cells. Spongillid sponges, as well as some marine species, have such aggregates called **gemmules** (Fig. 3–18). In freshwater sponges, a mass of food-filled archeocytes becomes surrounded by other amebocytes (spongocytes) that deposit a hard covering composed of a material similar to spongin. Spicules are also incorporated, so that a thick resistant shell is formed. Gemmule formation takes place primarily in the fall when a large number of these bodies is formed by each sponge. With the onset of winter, the parent sponge disintegrates. The gemmules are able to withstand freezing and drying and thus are able to carry the species through the winter. In spring the interior cells undergo some initial development, and the primordium eventually emerges through an opening **(micropyle)** in the shell. The primordium continues development into an adult sponge and may attain a large biomass by the end of the summer.

Sexual reproduction and the development of sponges display a number of peculiar features. Both hermaphroditic and dioecious sponge species exist, although most are hermaphroditic, usually producing eggs and sperm at different times. The sperm arise from choanocytes. For example, the choanocytes of an entire flagellated chamber lose their collars and flagella and form spermatogonia, which then undergo meiosis. The cluster becomes surrounded by a cellular wall forming a spermatic cyst. Alternatively, a spermatic cyst may be derived from the division of a single sperm-mother cell.

Eggs arise from archeocytes or choanocytes. Eggs generally accumulate their food reserves by engulfing adjacent nurse cells and are usually located within a cluster of surrounding cells. Gamete production appears to be initiated by changes in water temperature, photoperiod, or cellular regression (p. 63), depending on the species.

Sperm leave the sponge by means of the exhalant water currents and are taken into other sponges in the inhalant stream. Certain tropical sponges have been observed to release their sperm suddenly in great milky clouds (Fig. 3–19), and sudden sperm release may be characteristic of most sponges.

After a sperm has reached a flagellated chamber, it is engulfed by a choanocyte, which transports the sperm to the egg. Both cells lose their flagella. After the carrier with its sperm has reached an egg (which would be close by in the surrounding mesohyl), the carrier either transfers the sperm nucleus or the carrier and sperm nucleus are engulfed by the egg. Fertilization thus occurs *in situ.* Only one species of sponge is known to liberate eggs that are then fertilized externally, outside of the sponge body. This is surprising given the wide occurrence of external fertilization in other marine animals.

In the majority of sponges, development to the larval stage takes place within the body of the parent. Among the Demospongiae, however, there are some species that liberate fertilized eggs, which develop in the sea water.

Cleavage is complete and generally radial. Development leads to a larval stage, which displays various degrees of differentiation. The larva is usually at the blastula stage of development. The majority of sponges possess a **parenchymella** larva, in which monociliated cells cover all of the outer surface, except, often, the posterior pole (Figs. 3–20A and B). Spicules are often present, and the interior of the larva commonly contains most of the cell types found in the adult, with the usual exception of choanocytes. The parenchymella larva breaks out of the mesohyl, exits through the parent's excurrent canal system, and has a brief free-swimming existence.

A few calcareous sponges, such as *Grantia, Sycon,* and *Leucosolenia,* and among the Demospongiae, *Oscarella,* have an **amphiblastula larva** (Fig. 3–20C). This larva is hollow, and one hemisphere is composed of small flagellated cells and the other of large nonflagellated macromeres.

Following settling and attachment by the anterior pole, the sponge larva undergoes an internal reorganization that is comparable to gastrulation in other ani-

FIGURE 3–19 Sperm release from a specimen of the tubular West Indian sponge *Verongia archeri.* Sponge is about 1.5 m long. *(From Reiswig, H. M. 1970. Science. 170:538–539.)*

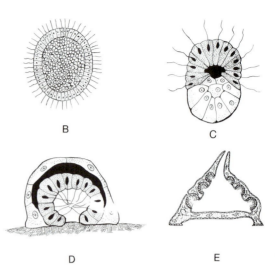

FIGURE 3–20 Sponge larvae and postlarval development. **A,** Scanning electron microphotograph of parenchymella larva of *Haliclona*. Bar = 50 μm. **B,** Parenchymella larvae in section. **C,** An amphiblastula larva. **D,** Gastrulation of an amphiblastula larva following settling. **E,** Postlarval rhagon stage. *(A, From Nielsen, C. 1987. Structure and function of metazoan ciliary bands and their phylogenetic significance. Acta Zoologica. 68(4):205–262. D, After Hammer. After Sollas.)*

A

mals. In the parenchymella, the external flagellated cells lose their flagella and move to the interior, where they regrow flagella and form choanocytes, and interior cells move to the periphery to form pinacocytes. The parenchymella larva of freshwater sponges and some marine species develops choanocytes before leaving the parent sponge. In these species, the external flagellated cells are sloughed off or move to the interior but are then phagocytized by amebocytes. In the hollow amphiblastula larva, reorganization following settling occurs either by epiboly or by invagination, or by both, but the macromeres overgrow the micromeres (Fig. 3–20D); in other metazoans the macromeres typically become internal. The macromeres in these sponges give rise to the pinacoderm and the micromeres to the choanocytes; both layers produce the amebocytes of the mesohyl. There is nothing equivalent to endoderm in sponges.

In most animals development proceeds from the establishment of gross form (morphogenesis) to the addition of more and more histological detail. However, given the absence of organs, cell differentiation in sponges precedes much of morphogenesis, that is,

the attainment of the definitive sponge form. Moreover, there is a great deal of cell mobility and reversal of cell differentiation.

In many of those calcareous sponges having a leuconoid structure, the final stages of development after attachment of the larva are preceded by stages resembling the asconoid and syconoid structures. In other leuconoid sponges, especially the Demospongiae, the leuconoid condition is attained more directly. The first stage is known as a **rhagon** (Fig. 3–20E). It resembles either the asconoid or the syconoid structure except that the walls are quite thick. The leuconoid plan develops directly from the rhagon stage by means of the formation of canals and flagellated chambers.

Some marine sponges live only one year; others live many years. Those in temperate regions are usually dormant in the winter. *Microciona* in Long Island Sound, for example, passes the winter in a reduced state, lacking flagellated chambers and other components of the water canal system. With an increase in water temperature, the sponge redevelops the adult functional condition. Freshwater sponges also overwinter in a regressed state or die, releasing gemmules.

A. L. Ayling at the marine laboratory of the University of Aukland in New Zealand studied the growth rates of 11 species of thin encrusting sponges living on a 12-m deep canyon wall. Ten clusters of sponge patches were marked with a masonry nail driven into the rock next to the sponges. The individual patches measured 1 to 20 mm thick, and the average patch surface area of the different species ranged from about 8 to 150 cm. The sponge patches were monitored at three-month intervals for two years. At each monitoring the sponge patches were photographed. The negatives were then projected at life size onto graph paper and the patch outlines traced for a permanent record, which could then be compared to others over the two-year period. These sponges were found to grow slowly. The most rapidly growing species would require 10 years to reach a diameter of 20 cm. Outward growth was not uniform around the margins, and the margins of some patches even regressed.

The Phylogenetic Position of Sponges

Sponges certainly arose prior to the Paleozoic era, and there have been a number of claims of pre-Cambrian fossils, although none has been clearly established. Beginning with the Cambrian period and extending to the present, however, the fossil record of sponges is abundant. Early Paleozoic reefs were composed of calcareous blue-green algae (stromatolites) and two now-extinct groups of calcareous spongelike animals (archaeocyathids and stromatoporoids*). There are also some flints composed entirely of fused sponge spicules. The phylum attained its greatest diversity and abundance during the Cretaceous.

The evolutionary origin of sponges poses a number of interesting problems. The absence of organs and the low level of cellular differentiation and interdependence in sponges certainly seem to be primitive characteristics. But a specialized body structure built around a water canal system and lacking distinct anterior and posterior ends is found in no other groups of animals. Moreover, the cellular differentiation is unlike that of other metazoans. All these features suggest that sponges are phylogenetically remote from other metazoans.

Some zoologists have suggested that the multi-

cellular condition of sponges evolved independently from that leading to the rest of the metazoan animals. However, most now believe that sponges had a common origin with other metazoans but diverged early in metazoan history. The sponge choanocyte is probably homologous to the flagellated collar cell of some choanoflagellate ancestor (p. 69). Many of the specialized sponge features, such as the water canal system, certainly developed in conjunction with a sessile existence; and flagellated, exterior cells of the motile ancestor probably moved into the interior to become choanocytes. Such movement occurs in the development of living sponges.

There can be little doubt that sponges diverged early from the main line of metazoan evolution and have given rise to no other members of the Animal Kingdom. Because of their isolated phylogenetic position, the sponges have often been placed in a separate subkingdom, the Parazoa, distinct from the other multicellular animals, the Eumetazoa.

PHYLUM PLACOZOA: THE MOST PRIMITIVE METAZOANS?

In 1883, a minute, animal-like, multicellular organism was discovered in a European seawater aquarium and named *Trichoplax adhaerens*. It has since been observed in various parts of the world and cultured numerous times. The flattened body, which reaches 2 to 3 mm in diameter, is composed of two outer layers of monociliated epithelial cells, which lack a basal lamina (Fig. 3–21). Between them lies an inner layer of loose, contractile, stellate cells—a form of connective tissue. Only four different types of somatic cells compose the tissues. The margins of the body are irregular and constantly change shape like an ameba (Fig. 3–21). The animal creeps over the substratum by means of its cilia, which are more numerous on the ventral surface, and feeds on protozoa and detritus, which are digested extracellularly.

Trichoplax undergoes asexual reproduction by fission and by budding. Eggs have been observed within the inner layer, but they may have arisen from the ventral epithelium. The DNA content is smaller than that determined for any other animal.

The phylum Placozoa has been created for *Trichoplax,* which like sponges is probably an early branch of the Animal Kingdom. *Trichoplax* is certainly the simplest known metazoan and may represent the most primitive. The flattened asymmetrical

*The stromatoporoids are now considered a subphylum of sponges similar to the sclerosponges.

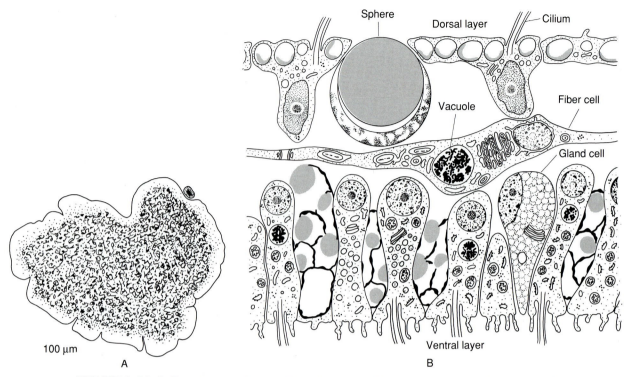

FIGURE 3–21 A, Dorsal view of *Trichoplax adhaerens.* The organism is greatly flattened, but note the irregular outline. **B,** Diagrammatic section through *Trichoplax* showing part of the ventral and dorsal layers and the cells that lie in the space between them. *(A, Drawn from a photograph by K. G. Grell. B, After Grell, K. G. 1981. Trichoplax adhaerens and the origin of Metazoa. International congress on the origin of the large phyla of metazoans. Accademia Nazionale dei Lincei. Atti dei Convegni Lincei. 49:113.)*

body, however, does not coincide with the radial symmetry postulated for the earliest metazoans, and this feature may be secondary, not primary.

SUMMARY

1. Sponges are sessile aquatic animals, mostly marine and largely inhabitants of hard substrata.

2. They are primitive in their lack of organs, including mouth and gut. There are different kinds of cells, but tissue differentiation, except for connective tissue, has not followed the common designs of other animals. These are neither neurons nor true muscle cells.

3. The bodies of sponges are organized around a system of water canals, a specialization correlated with sessility.

4. The small, vase-shaped asconoid body form, in which flagellated choanocytes line an interior spongocoel, is the primitive sponge form. The evolution of the common leuconoid form, in which the flagellated cells are distributed within a vast number of minute chambers, has permitted the attainment of much larger size and great diversity of shape, because each ad-

dition to the sponge body brings with it all of the units necessary to provide the required additional water flow.

5. The growth form of sponges is, in part, an adaptive response to the availability of space, the inclination of the substrate, and the current velocity.

6. Support is provided by a complex connective tissue, containing a skeleton of organic spongin fibers or siliceous or calcareous spicules, or a combination of spongin fibers and siliceous spicules.

7. Feeding, gas exchange, and waste removal depend on the flow of water through the body. The ability of the choanocyte collar to remove minute particles from the water stream has probably been an important factor in the long, successful history of sponges.

8. Most sponges are hermaphrodites. Sperm leave one sponge and enter another in the currents flowing through the water canals. Eggs in the mesohyl are fertilized *in situ.* They may then be released by way of the water canals or brooded until they reach the larval stage. In most sponges the flagellated larva is a blastula, and reorganization equivalent to gastrulation occurs following settling.

9. Sponges are probably an early evolutionary side branch that gave rise to no other groups of animals.

10. *Trichoplax adhaerens,* the only member of the phylum Placozoa, is a minute marine animal composed of ventral and dorsal epithelial layers enclosing loose mesenchyme-like cells.

REFERENCES

The literature included here is restricted to books and papers on sponges alone. The introductory references on pages 6–9 list many general works and field guides that contain sections on sponges.

Ayling, A. L. 1983. Growth and regeneration rates in thinly encrusting Demospongiae from temperate waters. Biol. Bull. *165*:343–352.

Bergquist, P. R. 1978. Sponges. Hutchinson and Co., London. 268 pp. (An excellent general account of sponges.)

Bergquist, P. R. 1985. Poriferan relationships. In Conway Morris, S., George, J. D., Gibson, R. et al. (Eds.): 1985. The Origins and Relationships of Lower Invertebrates. Systematics Association, Special Vol. No. 28. Clarendon Press, Oxford. 344 pp.

Brauer, E. B. 1975. Osmoregulation in the freshwater sponge, *Spongilla lacustris.* J. Exp. Zool. *192(2)*:181–192.

Brien, P., Levi, C., Sara, M. et al. 1973. Spongaires. Traite de Zoologie. Vol. III. Pt. 1. Masson et Cie. 716 pp.

Brill, B. 1973. Untersuchungen zur Ultrastruktur der Choanocyte von *Ephydatia fluviatilis.* L. Z. Zellforsch. *144*:231–245.

Cobb, W. R. 1969. Penetration of calcium carbonate substrates by the boring sponge, *Cliona.* Am. Zool. *9*:783–790.

De Vos, L. 1991. Atlas of Sponge Morphology. Smithsonian Institution Press, Washington, DC.

Elvin, D. W. 1976. Seasonal growth and reproduction of an intertidal sponge, *Haliclona permollis* (Bowerbank). Biol. Bull. *151*:108–125.

Fell, P. E., and Jacob, W. F. 1979. Reproduction and development of *Halichondria* sp. in the Mystic estuary, Connecticut. Biol. Bull. *156*:62–75.

Fransen, W. 1988. Oogenesis and larval development of *Scypha ciliata.* Zoomorphology. *107*:349–357.

Frost, T. M., Nagy, G. S., and Gilbert, J. J. 1982. Population dynamics and standing biomass of the freshwater sponge *Spongilla lacustris.* Ecology. *63(5)*:1203–1210.

Frost, T. M., and Williamson, C. E. 1980. *In situ* determination of the effect of symbiotic algae on the growth of the freshwater sponge *Spongilla lacustris.* Ecology. *61(6)*:1361–1370.

Fry, W. G. (Ed.): 1970. The Biology of Porifera. Academic Press, New York. 512 pp. (A collection of papers presented at a symposium of the Zoological Society of London.)

Green, G. 1977. Ecology of toxicity in marine sponges. Mar. Biol. *40*:207–215.

Harrison, F. W., and Cowden, R. R. 1976. Aspects of Sponge Biology. Academic Press, New York. 354 pp. (Papers presented at a symposium held in Albany, NY in 1975.)

Harrison, F. W., and De Vos, L. 1990. Porifera. *In* Harrison, F. W. (Ed.): Microscopic Anatomy of the Invertebrates. Vol. 2. Alan Liss, New York.

Hartman, W. D., and Goreau, T. F. 1970. Jamaican coralline sponges: Their morphology, ecology, and fossil relatives. *In* Fry, W. G. (Ed.): The Biology of Porifera. Academic Press, New York. pp. 205–240.

Hildemann, W. H., Johnson, I. S., and Jokiel, P. L. 1979. Immunocompetence in the lowest metazoan phylum: Transplantation immunity in sponges. Science. *204*:420–422.

Hohr, D. 1977. Differenzierungsvorgänge in der keimenden Gemmula von *Ephydatia fluviatilis.* Wilhelm Roux's Archives, *182*:329–346.

Jackson, J. B. C., Goreau, T. F., and Hartman, W. D. 1971. Recent brachiopod-coralline sponge communities and their paleoecological significance. Science. *173*:623–625.

Koehl, M. A. R. 1982. Mechanical design of spicule-reinforced connective tissue: Stiffness. Exp. Biol. *98*:239–267.

Langenbruch, P-F. 1988. Body structure of marine sponges. V. Structure of choanocyte chambers in some Mediterranean and Caribbean haplosclerid sponges. Zoomorphology. *108*:13–21.

Langenbruch, P.-F., and Scalera-Liaci, L. 1986. Body structure of marine sponges: IV. Aquiferous system and choanocyte chambers in *Haliclona elegans.* Zoomorphology, *106(4)*:205–211.

Langenbruch, P.-F., and Weissenfels, N. 1987. Canal systems and choanocyte chambers in freshwater sponges. Zoomorphology. *107*:11–16.

Lawn, I. D., Mackie, G. O., and Silver, G. 1981. Conduction system in a sponge. Science. *211*:1169–1171.

Lévi, C., and Boury-Esnault, N. (Eds.): 1979. Biologie des Spongiaires. Colloques International du Centre National de la Réchèrche Scientfique No. 291. (A collection of papers presented at a symposium.)

McClintock, J. B. 1987. Investigation of the relationship between invertebrate predation and biochemical composition, energy content, spicule armament and toxicity of benthic sponges at McMurdo Sound, Antarctica. Marine Biology. *94*:479–487.

Moore, R. D. (Ed.): 1955. Treatise on Invertebrate Paleontology. Archaeocyatha, Porifera. Vol. E. Geological Society of America and University of Kansas Press, Lawrence.

Nielsen, C. 1985. Animal phylogeny in the light of the trochaea theory. Biol. Linn. Soc. *25*:243–299.

Palumbi, S. R. 1986. How body plans limit acclimation: Responses of a demosponge to wave force. Ecology. *67(1)*:208–214.

Paulus, W. 1989. Ultrastructural investigation of spermatogenesis in *Spongilla lacustris* and *Ephydatia fluviatilis.* Zoomorphologie. *109*:123–130.

Pavans de Ceccatty, M. 1974. Coordination in sponges. The foundations of integration. Am. Zool. *14*:895–903.

Pomponi, S. A. 1979. Ultrastructure and cytochemistry of the etching area of boring sponges. *In* Lévi, C., and Boury-Esnault, N. (Eds.): Biologie des Spongiaires. Colloques Internationale du Centre National de la Réchèrche Scientifique No. 291. pp. 317–323.

Porter, J. W., and Targett, N. M. 1988. Allelochemical interactions between sponges and corals. Biol. Bull. *175*:230–239.

Reiswig, H. M. 1971a. *In situ* pumping activities of tropical Demospongiae. Mar. Biol. *9(1)*:38–50.

Reiswig, H. M. 1971b. Particle feeding in natural populations of three marine demosponges. Biol. Bull. *141(3)*:568–591.

Reiswig, H. M. 1975a. Bacteria as food for temperate-water marine sponges. Can. J. Zool. *53(5)*:582–589.

Reiswig, H. M. 1975b. The aquiferous systems of three marine Demospongiae. J. Morph. *145(4)*:493–502.

Rützler, K. 1990. New Perspectives in Sponge Biology. Smithsonian Institution Press, Washington, DC. 533 pp. (Papers from the 3rd International Conference on Sponge Biology.)

Rützler, K., and Rieger, G. 1973. Sponge burrowing: Fine structure of *Cliona lampa* penetrating calcareous substrata. Mar. Biol. *21*:144–162.

Saller, U. 1989. Microscopical aspects on symbiosis of *Spongilla lacustris* and green algae. Zoomorphology. *108*:291–296.

Schultz, B. A., and Bakus, G. J. 1992. Predation deterrence in marine sponges: Laboratory versus field studies. Bull. Mar. Sci. *50(1)*:205–211.

Simpson, T. L. 1968. The biology of the marine sponge *Microciona prolifera.* Temperature related, annual changes in functional and reproductive elements with a description of larval metamorphosis. J. Exp. Mar. Biol. Ecol. *2*:252–277.

Simpson, T. L. 1984. The Cell Biology of Sponges. Springer-Verlag, New York. 662 pp. (A detailed review of most aspects of sponge biology.)

Simpson, T. L., and Gilbert, J. J. 1973. Gemmulation, gemmule hatching, and sexual reproduction in freshwater sponges: The life cycle of *Spongilla lacustris* and *Tubella pennsylvanica.* Trans. Am. Micros. Soc. *92(3)*:422–433.

Stearn, C. W. 1975. The stromatoporoid animal. Lethaia. *8*:89–100.

Van de Vyver, G., and Willenz, P. 1975. An experimental study of the life-cycle of the fresh-water sponge *Ephydatia fluviatilis* in its natural surroundings. Wilhelm Roux' Archiv. *177*:41–52.

Vogel, S. 1974. Current induced flow through the sponge, *Halichondria.* Biol. Bull. *147(2)*:443–456.

Weissenfels, N. 1976. Bau und Funktion des Susswasserschwamms *Ephydatia fluviatilis.* III. Nahrungsaufnahme, Verdauung und Defakation. Zoomorphologie. *85*:73–88.

Weissenfels, N. 1980. Bau und Funktion des Susswasserschwamms *Ephydatia fluviatilis.* VII. Die Porocyten. Zoomorphologie. *95*:27–40.

Weissenfels, N. 1983. Bau und Funktion des Susswasserschwamms *Ephydatia fluviatilis.* X. Der Nachweis des offenen Mesenchyms durch Verfutterung von Backerhefe. Zoomorphologie. *103*:15–23.

Weissenfels, N., and Landschoff, H. W. 1977. Bau und Funktion des Susswasserschwamms *Ephydatia fluviatilis.* IV. Die Entwicklung der monaxialen SiO-Nadeln in Sandwich-Kulturen. Zool. Jb. Anat. *98*:355–371.

Wiedenmayer, F. 1977. Shallow-Water Sponges of the Western Bahamas. Birkhauser Verlag. Basel, Switzerland. 287 pp.

Wielsputz, C., and Saller, U. 1990. The metamorphosis of the parenchymula larva of *Ephydatia fluviatilis.* Zoomorphology. *109*:173–177.

Wilkinson, C. R. 1978. Microbial associations in sponges: III. Ultrastructure in the *in situ* associations in coral reef sponges. Mar. Biol. *49(2)*:177–185.

Wilkinson, C. R. 1979. Nutrient translocation from symbiotic cyanobacteria to coral reef sponges. *In* Levi, C., and Boury-Esnault, N. (Eds.):1979. Biologie des Spongiaires. Colloques International du Centre National de la Recherche Scientifique No. 291.

Wilkinson, C. R. 1983. Net primary productivity in coral reef sponges. Science. *219*:410–411.

Wilkinson, C. R. 1987. Productivity and abundance of large sponge populations on Flinders Reef flats, Coral Sea. Coral Reefs. *5*:183–188.

Willenz, P. 1980. Kinetic and morphological aspects of particle ingestion by the freshwater sponge *Ephydatia fluviatilis. In* Smith, D. C., and Tiffon, Y. (Eds.): Nutrition in the Lower Metazoa. Pergamon Press, Oxford. pp. 163–178.

Willenz, P., and Hartman, W. D. 1989. Micromorphology and ultrastructure of Caribbean sclerosponges. I. *Ceratoporella* and *Nicholsoni* and *Stromatospongia norae.* Mar. Biol. *103*:387–401.

4

CNIDARIANS AND CTENOPHORES

PRINCIPLES AND EMERGING PATTERNS

SIGNIFICANCE OF EPITHELIA

An epithelium is an organized layer of cells that generally separates body compartments of different chemical composition. Thus the evolution of epithelia was a precondition for the separation and physiological regulation of internal extracellular compartments in the metazoan body. Once regulated independently, each body compartment could then adopt one or more specialized functions, dividing the labor among several compartments.

An epithelium has four characteristics that are responsible for its function: it are organized as a continuous **sheet of cells,** the cells have **apical–basal polarity,** the cells rest on a **basal lamina,** and the cells are joined by **intercellular junctions.** For an epithelium to maintain a chemical difference between adjacent compartments, it must be in the form of an unbroken layer. Apical–basal polarity means that the epithelial apex, which faces one compartment, differs functionally from the base, which faces the other compartment. For example, the apical surface may absorb proteins, while the basal surface releases amino acids. The basal lamina is an extremely thin fibrous sheet that is secreted by the epithelial cells which rest on it. (Fig. 4–1). It functions to anchor the cells and may limit transport across the epithelium to ions and small molecules. Intercellular junctions bind the cell membrane of one epithelial cell to the next. Some, called **adhering junctions** (spot desmosomes and belt junctions), are mechanical in function and strengthen the epithelial pavement. **Sealing junctions** (septate junctions, tight junctions) increase control of the passage of ions and molecules across the epithelial layer, meaning that, molecules cannot easily slip between cells as appears to be true in sponges. **Gap junctions** allow intercellular communication.

The exposed outer epithelial surface of many animals has evolved the ability to secrete protective and/or supportive nonliving extracellular layers. Where the extracellular material is primarily organic, the layer is called a **cuticle.** Cuticles are composed of protein, which may be linked to complex polysaccharides, such as **chitin,** and may be thin and flexible or thick and rigid. In many animals, such as corals and molluscs, the principal component is inorganic **calcium carbonate,** which is laid down within a secreted organic framework.

SKELETONS 1

A **skeleton** is any structure that maintains shape, supports, or protects a body and allows for the transmission of forces. Examples encountered already include the internal cytoskeletal filaments and external tests of protozoans, and the connective tissue proteins (spongin) and mineral spicules of sponges. Among metazoans, a skeleton may be classified as an **endoskeleton** if it is inside the body, or an **exoskeleton** if it forms the outer covering of the body. Thus, insects have an exoskeletal cuticle and vertebrates, like ourselves, are provided with endoskeletal bones and cartilage.

An organism that uses water contained in a body compartment for support and for transmission of muscular forces is said to have a **hydrostatic skeleton.** Solid skeletal materials that are rubbery and elastic, such as ear cartilage, are **pliant skeletons;** materials that resist any change of shape, such as bones or shells, form **rigid skeletons.**

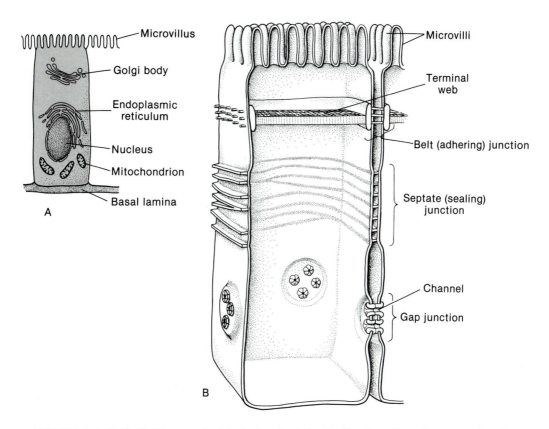

FIGURE 4–1 Epithelial tissues. **A,** A typical epithelial cell with microvilli on its exposed surface. **B,** types of junctions between epithelial cells. At sealing septate junctions, apposing cell membranes are bound together by surface proteins occluding the intercellular space. At belt junctions the cells are united by microfilaments that emerge from a beltlike thickening on the cytoplasmic side of each cell. At gap junctions the apposing cell membranes are close together, and the narrowed intercellular space is bridged by similar membrane channels.

Hydrostatic skeletons are enclosed by the body wall and are, by necessity, endoskeletons. An inflated sea anemone is a good example of an animal supported by a hydrostatic skeleton, although many other animals also use them. Among animals having a hydrostatic skeleton, or hydrostat, the body is supported by the slightly pressurized water within, much as a balloon maintains its shape when inflated with water or air. Because water is virtually incompressible, hydrostats are constant in volume and any localized increase in pressure as a result of muscular contraction will be transmitted equally throughout the hydrostat. Thus the force generated by the displacement of water in one region can be used to do work in another. (Using your hand, constrict the middle of a water-filled, cylindrical balloon and observe the effect on the two free ends.) Fluid movements are used to inflate and extend body parts, such as sea anemone tentacles, to anchor in crevices or sediments by swelling a part of the body, and in the production of coordinated peristaltic waves along the surface of the body for burrowing or burrow ventilation (p. 337).

Many animals with hydrostatic skeletons, such as sea anemones and worms, have more or less cylindrical bodies, and often the body wall is reinforced with a mesh of inelastic fibers, similar to the windings in the wall of a garden hose or the wire belts of a radial tire. The fibrous mesh toughens the body wall but also pre-

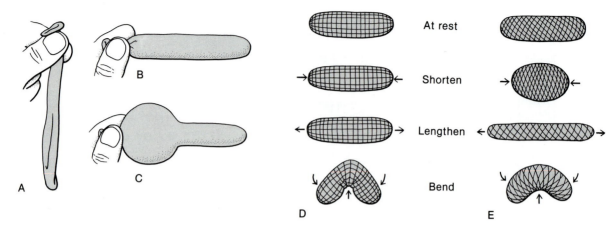

FIGURE 4–2 A balloon (**A**), when filled (**B**), approximates a hydrostatic skeleton. **C**, Addition of more air or water causes an aneurism as the balloon begins to inflate. Although both orthogonal (**D**) and helical (**E**) meshes protect against aneurisms, only the helical pattern prevents kinks and allows the body to elongate and shorten. (*Redrawn from Wainwright, S. A. 1988. Axis and Circumference. Harvard University Press. Cambridge, MA. 132 pp.*)

vents uncontrollable bulges, or aneurisms, from developing as the hydrostat is pressurized (Fig. 4–2A–C). To prevent aneurisms from developing, the meshwork fibers may be oriented in either an orthogonal pattern, having one set parallel to the long axis of the body and the other perpendicular to it, or the fibers may wind around the body in two helixes, one left handed and one right handed (crossed helical pattern) (Fig. 4–2D, E).

Although both orthogonal and helical patterns prevent aneurisms, only the helical pattern of fiber reinforcement is suitable for animal body walls. The orthogonal pattern (Fig. 4–2D) does not allow the body to increase in length or diameter, because the inelastic fibers must be stretched for this to occur. The orthogonal pattern also does not resist kinks as the body bends, and kinks, like aneurisms, are maladaptive for streamlined animals. The helical pattern, in contrast, resists kinks and permits changes in body length and diameter because the fibers are always at an angle to the primary axes of the body (Fig. 4–2E). The behavior of helical fibers in the body wall of an anemone is roughly similar to that of a coil spring; the spring is composed of inelastic steel wire, yet it freely can be extended and shortened, and does not kink when bent.

Helical reinforcing fibers are often proteins, such as collagen, that are embedded in invertebrate cuticles, for example in earthworms (annelids), or in the connective tissue (basement membrane) beneath the epidermis of the body, as in lancelets (cephalochordates) or ribbon worms (nemerteans), although they can also occur around specialized body structures, such as tentacles, blood vessels, and notochords. In humans, for example, the failure of the fibers to contain internal pressures may result in arterial aneurisms, varicose veins, ruptured ("slipped") vertebral discs, and hernias.

Pliant skeletons, like hydrostatic skeletons, deform when stressed, but unlike them, spring back to their original shape when the stress is removed because they are composed of elastic material. Among invertebrates, pliant skeletal material is used in the hinge of clams and in the supportive structures of horny corals, such as the sea fans. Pliant skeletons may be either endoskeletons or exoskeletons. Although there are a few specialized rubbery proteins, most pliant skeletons are composites, formed of a combination of proteins, polysaccharides, and water. They range in consistency from watery gels, as in the mesohyl of sponges and the connective tissue (mesoglea) of a few comb jellies, to the stiff, springy jelly (also mesoglea) of many jellyfishes. One obvious advantage of a pliant skeleton is that muscle power is required only for the initial change of shape, not to restore the body to its starting position. Pliant skeletons can be made stiffer by the addition of rigid pieces, such as mineral spicules, sand grains, or organic fibers, for example, collagen (spongin), chitin, or cellulose.

Rigid skeletons may form supportive platforms as in stony corals, a lattice arrangement as in glass sponges, or a framework of beams and levers, for example, in insects and vertebrates. Rigid skeletons may be endoskeletons or exoskeletons, but they always are composed of composite materials. Although there are exceptions, rigid skeletons tend to be common among terrestrial animals, which must support themselves in air, animals that move rapidly in water, such as crustaceans and fishes, and in exposed sessile and slow-moving benthic animals, for example barnacles and snails, for whom the skeleton is not only a support but also a protective retreat. Rigid skeletal frameworks, in conjunction with specialized muscles and ligaments, provide leverage for the precise application of large forces by certain appendages and the rapid but less forceful movements of others.

MOVEMENT

Movement is a distinguishing characteristic of animals, absent only in adult sponges. Although some animals like sea anemones, oysters, and barnacles live attached to rocks and other objects, they can still move parts of the body. Movement from place to place is achieved in very small aquatic animals or larval stages by the beating of surface cilia. Like ciliated protozoa, their surface-area-to-volume ratios are large and their sinking speeds are low. However, as animals become larger their volume increases by the cube and their surface area only by the square of the linear dimension. They have inadequate locomotor surface area, and their sinking speed exceeds their speed of swimming. Some gelatinous marine animals, such as ctenophores and salps, have solved this problem by reducing their density and thus their tendency to sink, that is, they are more buoyant. Most animals have abandoned ciliary swimming and are propelled by muscle contraction. Somewhat larger animals can use cilia for creeping over the bottom, provided there is a relatively large surface area in contact with the substratum.

Even if an animal transports itself by means of cilia, there is a need in all animals to move body parts. Elongation and shortening is the simplest form of movement and usually also involves changes in diameter. If elongation or shortening is asymmetrical (i.e., occurs on one side only), bending takes place. Such movements became possible with the evolution of contractile fibrils (myofibrils) located within the lower part of some epithelial cells. Such cells, called **myoepithelial cells,** are characteristic of many living cnidarians and are of scattered occurrence elsewhere in the Animal Kingdom. For example, human sweat glands, mammary glands, and the iris of the eye all contain myoepithelium. When the contractile fibers are all oriented parallel to the long axis of the body or part, they form a longitudinal muscular layer; when they are at right angles to the long axis of the body or part, they form a circular layer. True muscle cells—cells that have only a contractile function—probably evolved from myoepithelial cells through suppression (loss) of the epithelial portion of the cell and the assumption of a subsurface position.

NERVOUS SYSTEMS

Body movements through contraction of myofibrils must be coordinated and should occur rapidly in response to external tactile, chemical, or photic stimuli. **Neurons,** or nerve cells, evolved in conjunction with myoepithelium as a means of meeting that need. Neurons can monitor changes in external energy levels (stimuli) and transmit information in the form of a rapid wave of depolarization (nerve impulse) along the neuron membrane (axon) to the target cells (effectors). The evolution of neurons exploited the potential of all cells to regulate the ion concentrations on the inner and outer sides of the cell membrane, thereby creating a difference in the charge on the two sides. The nerve impulse is simply a progressive loss (depolarization) of this charge difference along the length of the membrane. Perhaps what is unique to neurons is their threadlike shape, making possible the conduction of signals over a long distance.

Primitively, neuron systems are closely associated with the outer epithelia of animals, usually located at the base of the epithelium. This position probably reflects the fact that nervous tissue and the outer epithelium are both derived from the same embryonic germ layer, as we shall see later in this chapter.

PHOTORECEPTORS AND SIMPLE EYES

The photoreceptor cells of most animals have evaginations on some part of the surface, providing greater surface area for photochemical reactions. There are two principal types of photoreceptors: ciliary and rhabdomeric. In **ciliary photoreceptors** the photosensitive surface is derived from the membrane of a cilium (Fig. 4–3A), and in **rhabdomeric photoreceptors** it is derived from the microvilli of the cell surface (Fig. 4–3B). Both types certainly evolved numerous times in the Animal Kingdom, and epidermal cells were probably the usual precursors; it is epidermal cells that usually bear cilia and microvilli, and outer epidermal cells would be the first cells through which light would pass on entering an animal's body. A third, less common type, called an epigenous photoreceptor, is derived from the dendritic surface of a sensory neuron and is best known in rotifers and nematodes.

Both rhabdomeric and ciliary photoreceptors have been found in larvae or adults of many of the animal groups. However, the rhabdomeric type predominates in the line of evolution embracing the flatworms, molluscs, annelids, and arthropods, and is the only type found among arthropods. Certain polyclad flatworms, which have rhabdomeric photoreceptors in the adult eye, have ciliary eyes in addition to rhabdomeric eyes in the larval stage.

Within the echinoderm/vertebrate line of evolution, vertebrates possess only the ciliary type, and echinoderms are rhabdomeric.

The photoreceptors of some animals, such as earthworms, are dispersed over the integument. Thus, earthworms can respond to light even through they have no eyes. However, in most animals the photoreceptors are concentrated. Such a concentration, called an **eye,** enables an animal to utilize other information provided by light besides general light intensity.

The simplest eye, called an **ocellus,** is organized as a pigment spot or pigment cup and is most commonly associated with the integument. A pigment spot ocellus

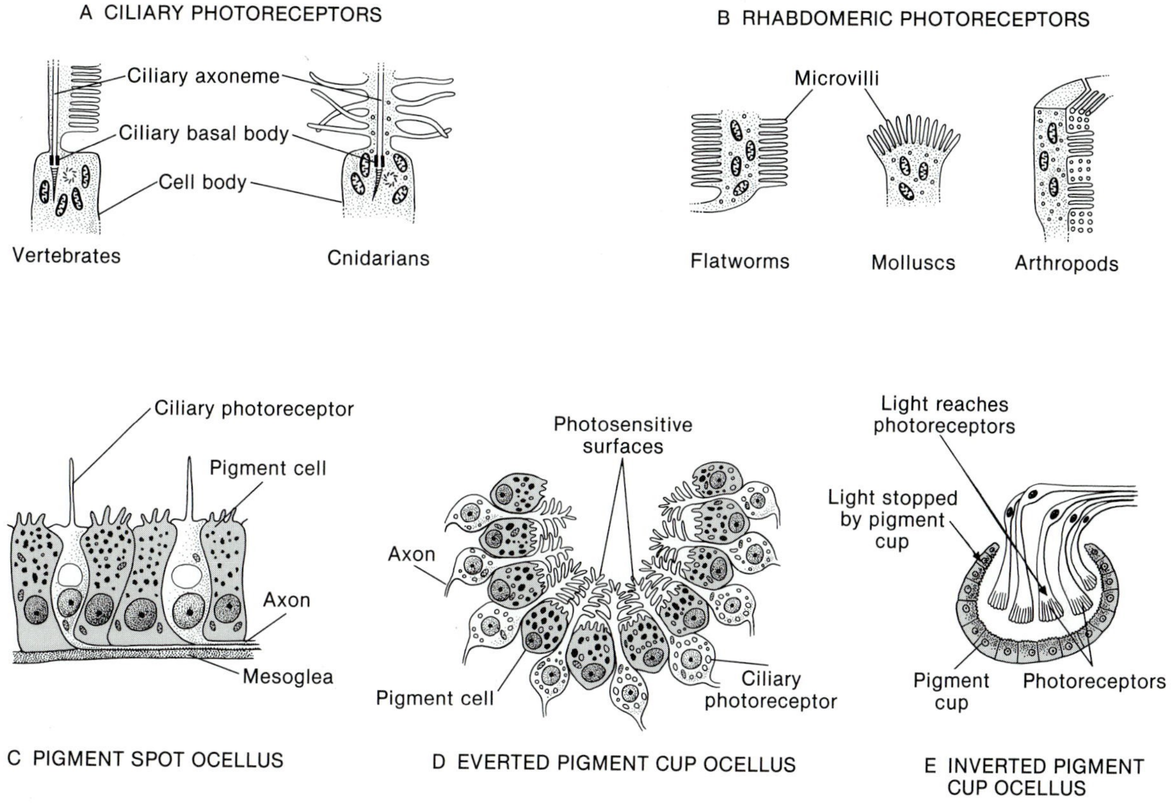

A CILIARY PHOTORECEPTORS

- Ciliary axoneme
- Ciliary basal body
- Cell body

Vertebrates   Cnidarians

B RHABDOMERIC PHOTORECEPTORS

Microvilli

Flatworms   Molluscs   Arthropods

- Ciliary photoreceptor
- Pigment cell
- Axon
- Mesoglea

C PIGMENT SPOT OCELLUS

Photosensitive surfaces

- Axon
- Pigment cell
- Ciliary photoreceptor

D EVERTED PIGMENT CUP OCELLUS

- Light reaches photoreceptors
- Light stopped by pigment cup
- Pigment cup
- Photoreceptors

E INVERTED PIGMENT CUP OCELLUS

FIGURE 4–3 Photoreceptors and ocelli. **A,** Ciliary photoreceptors. **B,** Rhabdomeric photoreceptors. **C,** A cnidarian pigment spot ocellus. **D,** An everted pigment cup ocellus. **E,** An inverted pigment cup ocellus. *(A and B, Greatly modified from Eakin, R. M. 1968. Evolution of photoreceptors. In Dobzhansky, T. et al (Eds.): Evolutionary Biology. Vol. 2. Appleton-Century-Crofts, New York. p. 206. C and D, Adapted from Singla, C. L. 1974. Ocelli of hydromedusae. Cell Tiss. Res. 49:413–429.)*

is a patch of photoreceptors interspersed with pigment cells (Fig. 4–3C). In a pigment cup ocellus the pigment cells form a cup into which the photoreceptor elements project (Fig. 4–3E). When the photoreceptors project between the pigment cells into the lumen of the cup, the ocellus is said to be everted (Fig. 4–3D); this arrangement is typical of integumental ocelli. In flatworms, the pigment cup lies below the epidermis and the photoreceptors project into the cup opening, not between pigment cells (Fig. 4–3E). An ocellus of this type is said to be inverted. The pigment cells shade the photoreceptors, enabling the animal, depending on the number of photoreceptors shaded, to determine the direction of the light source.

GUT CAVITIES

Aside from sponges, almost all animals possess a **gut cavity** that opens to the outside through a **mouth.** A gut cavity makes possible the utilization of large food masses and discontinuous feeding. Primitively, the gut cavity is lined with phagocytic cells, and digestion is in part intracellular, as is true of sponges and protozoa.

Extracellular digestion, however, is a feature of most metazoan animals. Certain of the gut epithelial cells secrete enzymes into the gut lumen, and within this confined space the concentration of enzymes can be high so that large food masses can be broken down extracellularly into small particles that are phagocytized or absorbed by other cells of the lining epithelium.

EMBRYONIC DEVELOPMENT

Sponge larvae are essentially blastulas. Indeed, one could argue that the adult sponge body is constructed within the embryonic framework of a blastula. In all other animals, embryonic development proceeds from a blastula through **gastrulation,** converting the spherical hollow or solid blastula into a two-layered embryo, or **gastrula,** that possesses the adult symmetry (Fig. 4–4). Gastrulation can take place in a number of ways and, like cleavage, is strongly influenced by the amount of yolk in the egg. Perhaps the easiest to visualize is gastrulation by **invagination,** whereby the cells of the vegetal hemisphere of the blastula fold inward into the interior (Fig. 4–4A). The new cavity of the infolding is called the **archenteron,** or **primitive gut.** It is surrounded by the old blastocoel and opens to the exterior by the

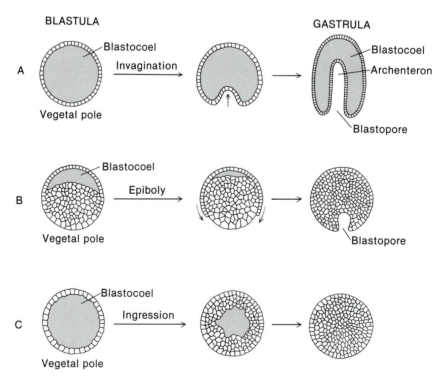

Figure 4–4 Modes of gastrulation. **A,** By invagination. Cells of the vegetal pole fold inward to form an archenteron. Opening to the exterior is a blastopore. **B,** By epiboly. Cells of the animal's pole, or blastula, grow down and over the yolk-laden cell of the vegetal hemisphere. **C,** By ingression. Cells cleaved from the inner side of the blastula wall fill the interior of the blastocoel. Interior mass of the cells later hollows out to form the cavity of the archenteron.

blastopore. Other modes of gastrulation include **epiboly** (Fig. 4–4B), in which cells in the animal half of the blastula overgrow those in the vegetal half; and **ingression** (Fig. 4–4C), in which cells of the blastula wall proliferate cells into the blastocoel.

With gastrulation the primary embryonic germ layers are defined, and from these layers all adult tissues are derived. Cells composing the outer wall of the gastrula constitute the embryonic **ectoderm** (Fig. 4–4A) and will form all of the surface epithelia and nervous system. Cells composing the archenteron wall constitute the **endoderm** (Fig. 4–4A), at least in part, and will form the gut epithelia and its derivatives. The third embryonic germ layer, the **mesoderm,** forms all of the intervening tissues. It has a more complicated origin and will be described in the next chapter. The embryos of most animals possess all three germ layers and are said to be **triploblastic.**

Cnidarians possess only ectoderm and endoderm and are therefore **diploblastic.** Moreover, the cnidarian gastrula retains a radial symmetry, which is the symmetry of the adult. The blastopore becomes the mouth, and the archenteron becomes the definitive coelenteron, which combines in one cavity the digestive and coelomic functions of higher animals.

COLONIAL ORGANIZATION 1

Colonial organization has evolved in many different animal groups, such as corals, bryozoans, and tunicates. The members, or **zooids,** of the **colony** are morphologically connected, they share resources and the colony arises by budding from the first individual derived from the larva or egg. Individuals that live in close proximity but lack a physical connection and do not share resources are **aggregations.** Some sea anemones, for example, reproduce asexually to form dense aggregations of separate but closely approximated individuals. Most colonial animals are sessile, but we will encounter some exceptions in later chapters. Although the colonies may be large, the colony members are generally smaller than solitary individuals of related species, probably because of the amount of substratum that would be necessary to support a colony composed of hundreds of large members.

In some colonial animals, such as stony corals, every member within the colony performs all functions and all have the same structure. However, in many colonial species, a division of labor has evolved in which some members perform only certain functions within the colony and have become structurally adapted for these tasks. Such a division of labor, whereby the members within a colony have become modified for different functions, is known as **morphological polymorphism.**

THE CNIDARIANS

The phylum Cnidaria, or Coelenterata, includes the familiar hydras, jellyfish, sea anemones, and corals, and the bright color of many species combined with their radial symmetry is often incredibly beautiful. Because radial symmetry is clearly primary, (primitive), it is commonly considered justification for placing the cnidarians within a division of the Animal Kingdom called the Radiata.

In contrast to sponges, cnidarians possess a gut cavity lined by endoderm, as do most other animals.

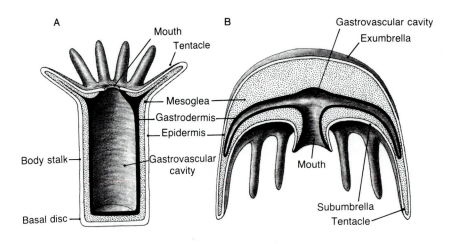

FIGURE 4–5 A, Polypoid body form. **B,** Medusoid body form.

In cnidarians, however, it is referred to as the **coelenteron,** or **gastrovascular cavity** (Fig. 4–5), because it functions in circulation in addition to digestion. The gastrovascular cavity lies along the polar axis of the animal and opens to the outside at one end to form a mouth. In cnidarians a circle of **tentacles,** representing evaginations of the body wall, surrounds the mouth to aid in the capture and ingestion of food.

The cnidarian body wall consists of three basic layers (Fig. 4–5): an outer epithelium, the **epidermis;** an inner epithelium, the **gastrodermis,** lining the gastrovascular cavity; and between these an extracellular layer called the **mesoglea.** The mesoglea ranges from a thin, noncellular basal lamina, as in hydras and many other hydrozoans, to a thick, fibrous, jelly-like, connective tissue with or without mesenchymal cells. Thus, cnidarians are **diploblastic,** that is, their bodies are constructed from only two germ layers, ectoderm and endoderm, from which the two adult epithelia arise.

Histologically, the cnidarians have remained rather primitive, although their structure anticipates some of the specializations that are found in higher metazoans. A considerable number of cell types composes the epidermis and gastrodermis, but there is only a limited degree of organ development.

Although all cnidarians are basically tentaculate and radially symmetrical, two basic forms are encountered within the phylum. One form, which is generally sessile, is the **polyp.** The other form is generally free-swimming and is called the **medusa.** Typically, the body of a polyp is tubular or cylindrical, with the oral end, bearing the mouth and tentacles, directed upward, while the opposite, or aboral, end is attached (Fig. 4–5A).

The medusoid body resembles a bell or umbrella, with the convex side upward and the mouth located in the center of the concave undersurface (Fig. 4–5B). The tentacles hang down from the margin of the bell. In contrast to the polypoid mesoglea, which is more or less thin, the medusoid mesoglea is extremely thick and constitutes the bulk of the animal. Because of this mass of jelly-like mesogleal material, these cnidarian forms are commonly known as **jellyfish.** Note that the medusoid and polypoid body forms are more or less inversions of each other. If a polyp were turned upside down and the aboral end rounded over, it would be similar to a medusa. Some cnidarians exhibit only the polypoid form, some only the medusoid form, and others pass through both in their life cycle. Colonial organization has evolved numerous times within the phylum, especially in polypoid forms.

Cnidarians are marine except for the hydras and a few other freshwater hydrozoans. Most are inhabitants of shallow water, and sessile forms abound on rocky coasts or on coral formations in tropical waters. The phylum is composed of approximately 9000 living species. Fossil cnidarians are known from the pre-

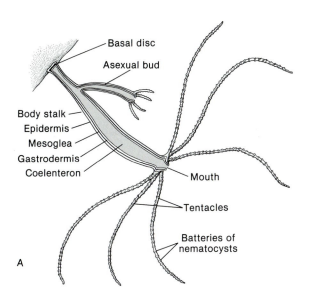

Basal disc
Asexual bud
Body stalk
Epidermis
Mesoglea
Gastrodermis
Coelenteron
Mouth
Tentacles
Batteries of
nematocysts

A

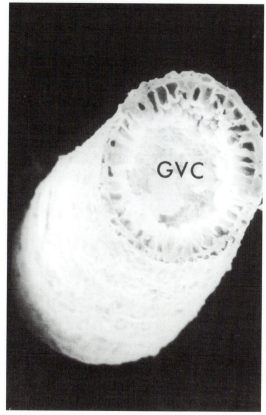

GVC

B

FIGURE 4–6. A, Section through a hydra with an asexual bud. **B,** Scanning electron photomicrographic view into the severed column of a hydra. *(A, After Hyman, L. H. 1940. The Invertebrates: Protozoa through Ctenophora. Vol. 1. McGraw-Hill Book Co., New York. B, From Thomas, M. B., and Edwards, N.C. 1990. Hydrozoa. In F. W. Harrison (Ed.): Microscopic Anatomy of the Invertebrates, Vol. 2. Wiley Liss, New York.)*

Cambrian, and a rich fossil record dates from the Cambrian.

HISTOLOGY AND PHYSIOLOGY: HYDRAS

For the sake of simplicity, the following introduction to the histology and physiology of cnidarians is primarily based on the familiar freshwater hydras, but the greater part of this discussion also applies to the other cnidarians; the more important exceptions are included here, as well as being described in the survey of the three classes later in the chapter.

Hydras are cylindrical, solitary polyps that range from a few millimeters to 1 cm or more in length (Fig. 4–6). However, the diameter seldom exceeds 1 mm. The aboral end of the cylindrical body stalk forms a basal disc, by which the animal attaches to

the substratum. The oral end contains a mound, or cone, called the **hypostome,** with the mouth at the top. Around the base of the cone is a circle of about six **tentacles.**

The Epidermis

The epidermis is composed of five principal types of cells.

Epitheliomuscular Cells

These myoepithelial cells (p. 99) are the most common type of cell in cnidarians. Called **epitheliomuscular cells,** they are somewhat columnar in shape, with the base resting against the mesoglea, and the

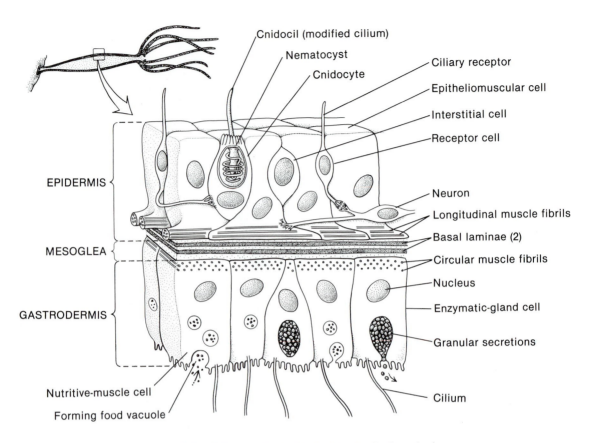

FIGURE 4–7 Body wall of a hydra (longitudinal section).

slightly expanded distal end forming most of the epidermal surface (Figs. 4–7 and 4–8A). However, unlike columnar epithelium, epitheliomuscular cells possess two, three, or more basal extensions, each containing a contractile myofibril. The basal extensions are oriented parallel to the axis of the body stalk and tentacles and interwoven with those of other epitheliomuscular cells. The ends of successive extensions are connected to each other. Collectively, they form a cylindrical, longitudinal, contractile layer. The epitheliomuscular cell as just described is widespread among cnidarians, but there are also many modifications resulting from suppression of either the epithelial or the contractile part of the cell. More-

FIGURE 4–8 **A,** Diagram of the cnidarian epidermis, showing epitheliomuscular cells, sensory cell, and nerve net. **B,** Diagram of a cnidocyte from the hydrozoan jellyfish *Gonionemus*. **C,** Puncturing of prey integument in the discharge of a penetrant (stenotele) nematocyst of hydra. See text for description. **D,** Open-tubed nematocyst of the hydroid *Laomedea*. (*A, From Mackie, G. O., and Passano, L. M. 1968. Epithelial conduction in hydromedusae. J. Gen. Physiol. 52:600. B, From Westfall, J. 1970: Z. Zellforsch., 110:457–470. C, From Tardent, P., and Holstein, T. 1982. Morphology and morphodynamics of the stenotele nematocyst of Hydra attenuata Pall. Cell Tissue Res. 224:269–290. D, From Ostman, C. 1982. Nematocysts and taxonomy in Laomedea, Gonothyraea and Obelia. Zoologica Scripta, 11(4):227–241.*)

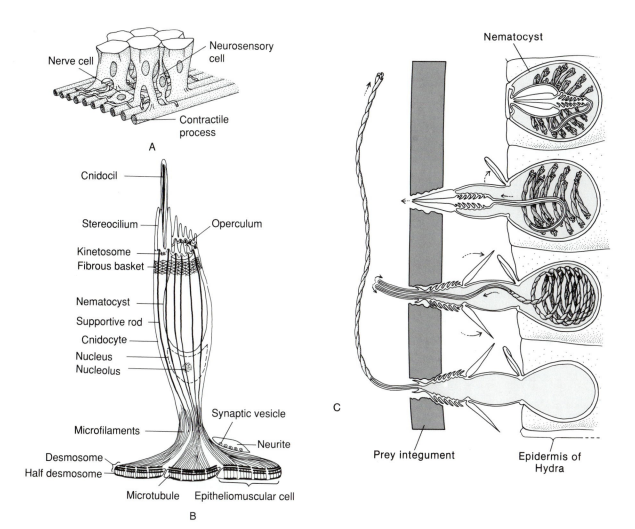

A — Nerve cell, Neurosensory cell, Contractile process

B — Cnidocil, Stereocilium, Operculum, Kinetosome, Fibrous basket, Nematocyst, Supportive rod, Cnidocyte, Nucleus, Nucleolus, Synaptic vesicle, Microfilaments, Neurite, Desmosome, Half desmosome, Microtubule, Epitheliomuscular cell

C — Nematocyst, Prey integument, Epidermis of Hydra

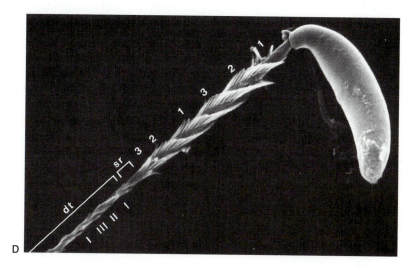

D — dt, sr

over, the cell may be flattened or squamous, rather than columnar (Fig. 4–9).

As is true of the epithelium of other animals, epitheliomuscular cells are bound together by intercellular junctions—belt junctions, septate sealing junctions, and gap junctions. The first two types run around the lateral margins of the apical end of the cell (See Fig. 4–1). The cells may also interdigitate into one another and be locked together by large foldlike couplings. Gap junctions, an important means of intercellular communication, are only found in the class Hydrozoa.

Interstitial Cells

Located beneath the epidermal surface and wedged between the epitheliomuscular cells are small, rounded cells with relatively large nuclei—**interstitial cells** (Fig. 4–7). The interstitial cells, acting as the germinal, or formative, cells of the animal, give rise to the sperm and eggs as well as to any other type of cells, that is, they are totipotent.

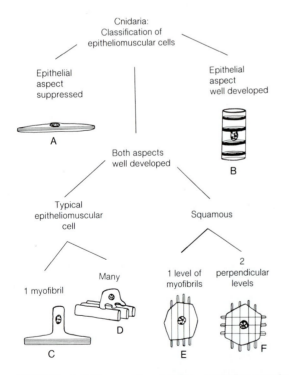

FIGURE 4–9 Types of cnidarian epitheliomuscular cells. *(From Chapman, D. M. 1974. Cnidarian histology. In Muscatine, L., and Lenhoff, H. M. (Eds.): Coelenterate Biology. Academic Press, New York. p. 45.)*

Cnidocytes

Cnidocytes are located throughout the epidermis and are lodged between or invaginated within epitheliomuscular cells. They are especially abundant on the tentacles. These specialized cells, which are unique to and characteristic of all cnidarians, contain organelles capable of eversion known as **cnidae.** The commonest type are the stinging structures called **nematocysts.** A cnidocyte is a rounded or an ovoid cell with a basal nucleus (Figs. 4–7 and 4–8B). In hydrozoans and scyphozoans, one end of the cell contains a short, stiff, bristle-like process called a **cnidocil,** which has an ultrastructure similar to that of a cilium and is exposed to the surface. In anthozoans, the cnidocils are not present, although a ciliary cone complex of similar function is associated with at least some of the types of cnidocytes found in this class (Figs. 4–10C and D). The interior of the cell is filled by a capsule containing a coiled, usually pleated tube (Fig. 4–8B), and the end of the capsule that is directed toward the outside is covered by an operculum or by lidlike flaps. The base is anchored to the lateral extensions of one or more epitheliomuscular cells and may also be associated with a neuron terminal.

The discharge mechanism apparently involves a rapid change in osmotic pressure within the capsule. Under the combined influence of mechanical and chemical stimuli, which are initially received and conducted by the cnidocyte surface structure, there is possibly a sudden release of calcium within the capsule. Water rushes into the capsule, and the intracapsular pressure rises. This pressure plus some intrinsic tension within the capsule structure itself causes the operculum or apical flaps of the cnida to open and the tube to evert (turns it inside out). In hydras the entire discharge process takes 3 ms. Studies on sea anemones have shown that chemical and mechanical stimuli may interact to discharge nematocysts.

Although cnidocytes usually fire as independent effectors, discharge can apparently be effected by nerve impulses from an associated neuron terminal, and neuronal connections may serve to coordinate firing by a large number of nematocysts.

A discharged nematocyst consists of a capsule and a threadlike tube of varying length (Figs. 4–8C and D). The tube or thread is commonly armed with spines, particularly around the base, and may be open or closed at the tip. Variations in the arrangement and length of the spines and the diameter of the tube give rise to some 30 kinds of cnidae, which are constant

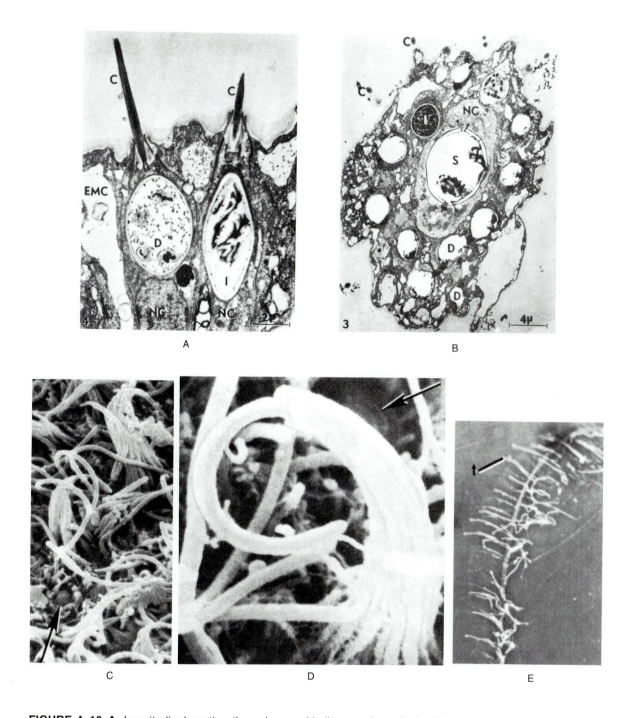

FIGURE 4–10 A, Longitudinal section through an epitheliomuscular cell showing two nematocyst capsules (D and I) with their projecting cnidocils (C). **B,** Cross section of an epitheliomuscular cell containing a central cnidocyte with a stenotele nematocyst (S) surrounded by cnidocytes with smaller nematocysts (D and I). Cnidocils of adjacent cnidocytes are labeled C, and NC is the central cnidocyte. **C,** Tentacle surface of the coral *Balanophyllia,* showing a number of ciliary cone complexes. Arrow indicates a probable spirocyst. **D,** A ciliary cone complex on the tentacle of the sea anemone *Corynactis californica.* A single central cilium is surrounded by shorter microvilli. **E,** Spirocyst thread from the coral *Paracyathus,* showing large numbers of unsolubilized tubules. *(B, From Westfall, J., Yamataka, S., and Enos, P. 1971. 20th Annual Proceedings of the Electron Microscopy Society of America. D, From Mariscal, R. N. 1974. Scanning electron microscopy of the sensory surface of the tentacles of sea anemones and corals. Z. Zellforsch. 147:149–156. E, From Mariscal, R. N., McLean, R. B., and Hand, C. 1977. The form and function of cnidarian spirocysts:3. Ultrastructure of the thread and the function of spirocysts. Cell Tissue Res. 178:427–433.)*

for the species and of great value to cnidarian taxonomists.

Nematocysts are the most important type of cnidae and are found in all cnidarians. The other two types, spirocysts and ptychocysts, are found only in anthozoans and will be described later. Nematocysts function in prey capture, and many can inject a toxin (Fig. 4–8C). The thread is generally open at the tip and frequently armed with spines. Upon discharge, the thread bores its way into the tissues of the prey and injects a protein toxin that has a paralyzing action. At least in some types, the spines aid in puncturing the integument of the prey. Under high-speed microcinematography the eversion of a barbed nematocyst (stenotele) of hydras is seen to occur in two phases. In the rapid first phase, the capsule lid opens, and the everting barbed region with three stylets directed forward punches a hole in the prey's integument (Fig. 4–8C). In the second phase the stylets flip back, and the thread everts into the body of the prey through the opening created by the stylets.

The toxic effect of the nematocysts of most cnidarians is not perceptible to humans. However, some marine species have nematocysts that can produce a painful burning sensation and irritation or even death. Other nematocysts, such as the desmonemes, do not possess any known toxic properties; instead, they function by adhesion or by wrapping and entangling small prey. The thread is closed at the end and may be unarmed and coiled or have a long spiny shaft. The spines appear to be an adaptation for adhesion to the prey surface.

Different species of cnidarians possess one to seven structural types of nematocysts, the number and types depending on the nature of the prey. *Hydra,* for example, possesses four types, which are arranged on the tentacles in groups of batteries (Fig. 4–6A). Each battery represents a large number of cnidocytes that have become invaginated within one epitheliomuscular cell (a battery cell) (Fig. 4–10B). The batteries appear as bumps or warts when the tentacles are extended.

Nematocysts and other cnidae are used but once; new cnidocytes are formed from nearby interstitial cells. About 25% of the nematocysts of *Hydra littoralis* are lost from the tentacles in the process of eating a brine shrimp. The discharged nematocysts are replaced within 48 h.

Mucus-Secreting Cells

Mucus-secreting gland cells are found in the epidermis. They are particularly abundant in the adhesive basal disc of *Hydra* and possess contractile extensions similar to the epitheliomuscular cells.

Receptor and Nerve Cells

The remaining types of cells are receptor cells and nerve cells. Receptor cells are elongated cells oriented at right angles to the epidermal surface (Figs. 4–7 and 4–8A). The base of each cell gives rise to a number of neuron processes, and the distal end terminates in a sphere or a sensory bristle (modified cilium). Receptor cells are particularly abundant on the tentacles and, like cnidocytes, may be invaginated within epitheliomuscular cells.

The nerve cells, which are superficially similar to the multipolar neurons of other animals, are located at the base of the epidermis next to the mesoglea (Fig. 4–7).

The Gastrodermis

The histology of the gastrodermis, or inner layer of the body wall, is somewhat similar to that of the epidermis (Fig. 4–7), but it is not strictly homologous to the gut lining of bilateral animals (p. 181). The cells corresponding to epitheliomuscular cells in the epidermis are called **nutritive-muscle cells** in the gastrodermis. The two types are similar in shape, but the nutritive-muscle cells are usually monociliated, and the basal contractile extensions, which develop to the highest degree in the hypostome and tentacle bases, are more delicate than those of the epidermis. Furthermore, the gastrodermal contractile fibers in hydras and other hydrozoans are oriented at right angles to the long axis of the body stalk and thus form a circular contractile layer.

Interspersed among the nutritive-muscle cells are **enzymatic-gland cells.** These are wedge-shaped, ciliated cells with their tapered ends directed toward the mesoglea. Enzymatic-gland cells do not possess the basal contractile processes.

Mucus-secreting gland cells are abundant around the mouth. Nerve cells also exist but in far fewer numbers than in the epidermis. Although nematocysts are lacking in the gastrodermis of hydras and other hydrozoans, they are present in restricted areas of this layer in the other classes of cnidarians.

In many cnidarians the gastrodermal cells contain symbiotic algae. Some species of *Hydra* harbor green **zoochlorellae,** as do certain freshwater sponges. However, the symbiotic algae of most marine cnidarians are **zooxanthellae** (p. 148). The green or yellow-brown color of these algae gives a similar color to the cnidarian host. Both zoochlorellae and zooxanthellae provide their hosts with excess photosynthate (p. 149).

MOVEMENT

The body stalk and tentacles of cnidarians can extend, contract, or bend to one side or the other. In hydras the gastrodermal fibers in most parts of the body are so poorly developed that movement is due almost entirely to the contractions of the longitudinal, epidermal fibers. Fluid within the gastrovascular cavity plays an important role as a hydraulic skeleton. By taking in water through the mouth as a result of the beating of the gastrodermal flagella, a relaxed hydra may stretch out to a length of 20 mm, whereas contraction of the epidermal fibers can reduce it to a mere 0.5 mm. Hydras can detach and shift locations by somersaulting or floating.

NUTRITION

Almost all cnidarians, including hydras, are carnivorous and feed mainly on small crustaceans. Contact with the tentacles brings about a discharge of nematocysts, which entangle and paralyze the prey. The tentacles then pull the captured animal toward the mouth, which opens to receive it. All of these feeding responses are initiated by various amino acids and peptides liberated from the prey, presumably through nematocyst puncture wounds. Mucus secretions aid in swallowing, and the mouth can be greatly distended. Oddly, there is no permanent mouth in hydras. When the animal is not feeding, the epidermal and gastrodermal cells around the mouth form sealing septate junctions with each other, and opening seems to be due to muscle contraction. Sealing the mouth with junctions that reduce transport between cells may isolate the coelenteron from the environment and allow its more effective use in digestion and, especially, osmoregulation.

Eventually the prey is pushed into the gastrovascular cavity, and enzymatic-gland cells discharge proteolytic enzymes, gradually reducing prey tissues to a soupy broth. The beating of the flagella of the gastrodermal cells ensures mixing.

After this initial extracellular phase, digestion continues intracellularly. The nutritive-muscle cells engulf small fragments of tissue. Continued digestion of proteins and the digestion of fats occur within food vacuoles in the nutritive-muscle cells, and the food vacuoles undergo the acid and alkaline phases characteristic of protozoa. Products of digestion are distributed by diffusion. Indigestible materials are ejected from the mouth when the body contracts.

GAS EXCHANGE AND EXCRETION

Gas exchange occurs across the general body surface. Nitrogenous wastes (ammonia) also diffuse through the general body surface. As in many other freshwater animals, there is a continual influx of water into the bodies of hydras through the body wall. Excess water, which is hypoosmotic to the tissue fluids, is removed periodically from the gastrovascular cavity via the mouth. The gastrovascular cavity thus acts like a giant contractile vacuole or a nephridium of a higher animal.

THE NERVOUS SYSTEM

The nerve cells are arranged in an irregular **nerve net,** or plexus, in the base of the epidermis and gastrodermis and are particularly concentrated around the mouth. Some synapses are symmetrical; that is, neuron terminals on both sides of the synapse secrete a transmitter substance, and an impulse can be initiated in either direction across the synapse. Neurons serving symmetrical synapses obviously transmit impulses in both directions and are thus directionally nonpolarized, in contrast to the neurons of higher animals.

The association of nerve cells to form conducting chains between receptor and effector shows all degrees of complexity. The neurons contain two, three, or more processes. These processes may terminate in muscle fibers, in sensory cells, or with the processes of other ganglion cells. Some neurons have even been shown to possess two branches, each serving a different kind of effector (cnidocyte and muscle fiber). The length of the processes varies, and they may give rise to motor and sensory branches. Further, some neurons have only motor processes, some have only sensory processes, and some act as interneurons.

A double nerve net system in the same body layer is common in cnidarians other than hydras. One nerve net acts as a diffuse, slow-conducting system of multipolar neurons; the other as a rapid, through-conducting system of bipolar neurons.

REPRODUCTION

In hydras, asexual **budding** is the usual means of reproduction during the warmer months of the year. A bud develops as a simple evagination of the body wall and contains an extension of the gastrovascular cavity. The mouth and tentacles form at the distal end, and eventually, the bud detaches from the parent to become an independent hydra. Budding occurs in many other cnidarians and is the means by which colonies form in colonial species.

Considering their facility for asexual reproduction, it is not surprising that hydras, like many cnidarians, have considerable powers of regeneration. A classical experiment is that of Abraham Trembley, who in 1744, inserted a knotted thread through the basal disc of a hydra and pulled it out through the mouth, turning the animal inside out. After a short period the gastrodermal cells reoriented themselves on the inside of the mesoglea, and the epidermal cells migrated to the outside. Although interstitial cells can contribute to regeneration, the process is not dependent on these cells; rather, dedifferentiated epidermal and gastrodermal cells are the principal source of regenerate material. In some cnidarians, entire animals have been produced from gastrodermal or epidermal cells alone.

A polarity, or gradient of dominance, exists from oral to aboral end. If the body stalk of a hydra is severed into several sections, each regenerates into a new individual. Furthermore, the original polarity is retained, so the tentacles always form on the end that was closest to the oral end of the intact animal, and a basal disc forms at the other end. A piece of the oral end grafted onto the middle of the body stalk induces the formation of another oral end. The oral/aboral gradient is also reflected in the rate of regeneration, because an aboral piece regenerates more slowly than

one taken nearer the mouth.

Mitotic activity occurs throughout the stalk, but the rate of production of new cells is especially high in the region just below the hypostome. Because the tips of the tentacles, basal disc, and buds are sites of cell death or cell loss (bud), there is a gradual migration of cells down the stalk and out along the tentacles from the mitotic zone (Fig. 4–11). Within a period of several weeks, all the cells in the body of an individual are replaced. Thus, hydras never grow old!

Sexual reproduction in hydras occurs chiefly in the fall, to produce eggs, which are a means by which the species survives the winter. Most hydras are dioecious. As in all cnidarians, the germ cells originate for the most part from interstitial cells, which aggregate in the stalk to form ovaries or testes (Fig. 4–24C). A single egg is produced in each ovary, with the other interstitial cells of the ovarian aggregate merely serving as food in the egg's formation. As the egg enlarges, a rupture occurs in the overlying epidermis, exposing the egg. The testis is a conical swelling with

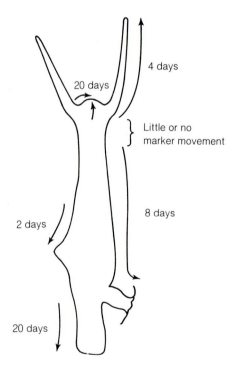

FIGURE 4–11 Tissue movement in *Hydra* as a result of mitotic activity. Arrows indicate direction of movement, and numbers indicate days required for cells to move along path of arrow. *(From Campbell, R. D. 1967. J. Morphol. 121(1):19–28)*

a nipple through which the sperm escape. Sperm liberated from the testes into the surrounding water penetrate the exposed surface of the egg, which is thus fertilized *in situ.*

The egg then undergoes cleavage and simultaneously becomes covered by a chitinous shell. When shell formation is complete, the encapsulated embryo drops off the parent and remains in its protective casing through the winter. With the advent of spring the shell softens, and a young hydra emerges. Because each individual may bear several ovaries, a number of eggs may be produced each season.

The reproductive pattern described for hydras is not typical of most marine cnidarians. Eggs, as well as sperm, may be liberated into the seawater, where fertilization occurs. Cleavage is complete and usually radial. The primitive blastula is hollow (coeloblastula), but a yolky solid blastula (stereoblastula) has evolved in many species. A **planula larva** forms following gastrulation. The planula larva is elongated and radially symmetrical but with anterior (apical) and posterior ends. The surface is covered by monociliated cells representing the future epidermis; the interior is composed of endodermal cells that will form the gastrodermis. A mouth is sometimes present at the posterior end of the larva, and the gastrovascular cavity may develop before settling. Where settling occurs, the larva attaches by the anterior end.

SUMMARY

1. Cnidarians are aquatic, radially symmetrical animals with tentacles encircling the mouth at one end of the body. The mouth is the only opening into the gut cavity.

2. Cnidarians exhibit two body forms: the medusa, which is adapted for a pelagic existence, and the polyp, which is adapted for an attached, benthic existence. Colonial organization has evolved in many polypoid groups.

3. The body wall consists of an outer epidermis, an inner gastrodermis, and an intervening mesoglea. The latter may be thin basal lamina or a thick, acellular or cellular connective tissue.

4. Cnidarians are primitive in their lack of organs, the predominance of myoepithelial cells, and the diploblastic origin of the adult body.

5. Most feed on zooplankton, although some utilize larger animals and some are suspension feeders of fine particulate matter. Prey is caught with the tentacles and immobilized by explosive cells, called cnidocytes, which are unique to the phylum. Digestion is initially extracellular, then intracellular.

6. The neurons are usually arranged as nerve nets at the base of the epidermal and gastrodermal layers, and impulse transmission tends to be radiating. Synaptic junctions are commonly nonpolarized.

7. A ciliated, free-swimming stereogastrula, called the planula larva, occurs in the life cycle of most cnidarians.

CLASS HYDROZOA

The class Hydrozoa contains about 2700 species, but because of their small size and plantlike appearance, most people are largely unaware of their existence. Much of the growth attached to rocks, shells, and wharf pilings, usually dismissed as "seaweed," is frequently composed of hydrozoan cnidarians. The few known freshwater cnidarians belong to the class Hydrozoa and include the hydras and some small, freshwater jellyfish.

Hydrozoans display either the polypoid or the medusoid structure, and some species pass through both forms in their life cycle. Three characteristics unite the members of this class. The mesoglea lacks cells; the gastrodermis lacks cnidocytes; and the gonads are epidermal, or if gastrodermal, the eggs and sperm are shed directly to the outside and not into the gastrovascular cavity.

Hydroid Structure

Although some hydrozoans display only the medusoid form, most species possess a polypoid stage in their life cycle. Some species, such as the hydras, exist as solitary polyps, but the vast majority are colonial. In hydras, buds form on the stalk as simple evaginations of the body wall. The distal end of the bud forms a mouth and a circle of tentacles (Fig. 4–2A); then the whole bud drops off to form a new individual. In the development of colonies, the buds remain attached, in turn producing buds so that each polyp is connected to the others. Such a collection of

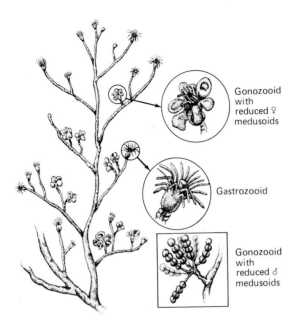

Gonozooid with reduced ♀ medusoids

Gastrozooid

Gonozooid with reduced ♂ medusoids

FIGURE 4–12 *Eudendrium,* a genus of hydroids having an arborescent growth form. The medusoids, called gonophores, are reduced and never liberated from the parent colony.

palmata, which is found on intertidal mud flats along the California coast, may reach a height of 14 cm when submerged and erect.

Probably because of the increased size resulting from a colonial organization, most hydroids are, at least in part, surrounded by a supporting, nonliving,

A

B

FIGURE 4–13 A, Scanning electron micrograph of a gastrozooid of the thecate hydroid *Gonothyrae loveren.* The vaselike structure is the skeleton around the hydranth. Diatoms are growing on it. **B,** *Pennaria,* a hydroid with a pinnate growth form. *(A, Micrograph courtesy of C. Ostman. B, Photograph courtesy of Betty M. Barnes.)*

polyps is known as a **hydroid colony** (Figs. 4–12 and 4–15). The three body layers in a hydroid colony—epidermis, mesoglea, and gastrodermis—and the gastrovascular cavities are all continuous.

In describing hydrozoan individuals or colonies, it is convenient to use the term **hydranth** to refer to the oral end of the polyp bearing the mouth and tentacles and the term **hydrocaulus** to refer to the stalk of the polyp.

In most species of hydrozoans the colony is anchored by a horizontal, rootlike stolon, called the **hydrorhiza,** which grows over the substratum (Fig. 4–15). From the stolon arise single upright polyps or branches of polyps. The branches develop through different growth and budding patterns and vary in form. They may be arborescent (bushlike) (Fig. 4–12) or pinnate (feather-like) (Fig. 4–13B), or single polyps may arise from basal stolons (Fig. 4–14B).

Most hydroid colonies are 5 to 15 cm high. The individual polyps are usually very small and inconspicuous, but the genus *Branchiocerianthus,* a deep-sea giant with solitary polyps, may reach a length of over 2 m, and the shallow-water *Corymorpha*

chitinous envelope (a cuticle) secreted by the epidermis. Such a cylinder is known as the **perisarc** (Fig. 4–15), and the living tissue that it surrounds is called the **coenosarc.** The perisarc may be confined to the hydrocaulus (Fig. 4–14A), but often it encloses the hydranth itself in a casing known as the **hydrotheca,** as in *Obelia* and *Campanularia.* The hydrotheca may be bell-like and open (Figs. 4–13A and 4–15), or its opening may be covered by a lid of one to several pieces. The lid opens when the polyp is extended and feeding and closes when the polyp contracts. Hydroids with a hydrotheca surrounding the polyp proper are said to be **thecate;** those without the hydrotheca, **athecate.**

The hydras and the solitary polyps of other hydrozoans have no such external skeleton, nor is a skeleton very extensive in some colonial species in which all polyps arise from hydrorhizae. For example, in *Hydractinia echinata,* which lives on snail shells inhabited by hermit crabs, the densely packed,

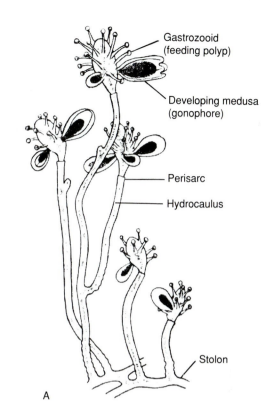

A

B

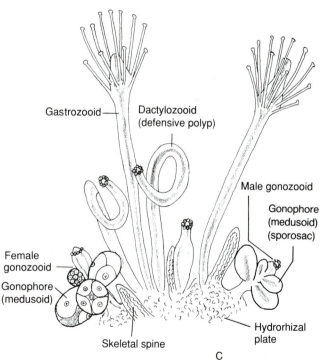

C

FIGURE 4–14 A, *Coryne,* a dimorphic hydroid in which medusoids are formed directly on gastrozooids. **B** and **C,** *Hydractinia echinata.* This hydroid lives only on snail shells occupied by hermit crabs. Polyps arise separately from a mat of stolons, or hydrorhizae. The skeleton is limited to the covering of stolons, and fusion of adjacent skeletons forms a plate. The colony is tetramorphic; the four kinds of zooids are shown in C. *(A, Modified from Neumov. B, from Ruppert, E. E. and Fox, R. S. 1988. Seashore Animals of the Southeast. Univ of So. Carolina Press, Columbia, SC. C, After Hyman, L. H. 1940. The Invertebrates. Vol. I. McGraw-Hill Book Co., New York)*

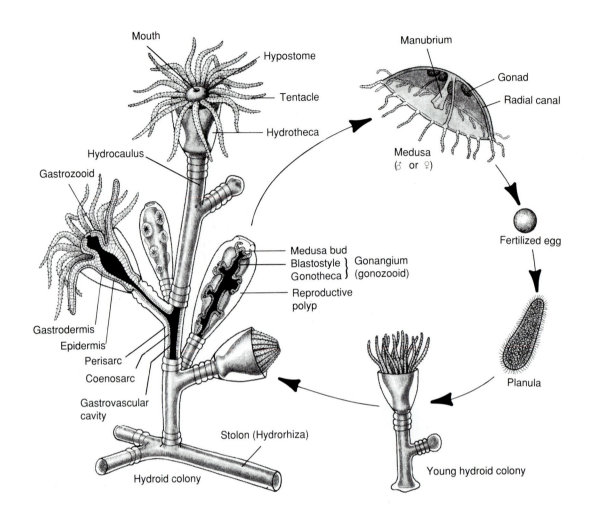

FIGURE 4–15 Life cycle of *Obelia*, showing structure of hydroid colony.

naked polyps arise directly from hydrorhizae. The hydrorhizae are provided with skeletons, which are fused to form a more or less continuous spiny plate anchoring the colony to the shell (Fig. 4–14B).

Polymorphism is another characteristic of hydroids associated with their colonial organization. Most hydroid colonies are at least dimorphic; that is, the colony consists of at least two structurally and functionally different types of members. The most numerous and conspicuous type of zooid is the nutritive, or feeding, polyp, called a **gastrozooid** (or **trophozooid**) which looks like a short hydra (Figs. 4–14 and 4–15). The gastrozooids capture and ingest prey, and thus provide nutrition for the colony. Most

hydrozoans feed on any zooplankton that is small enough to be handled by the gastrozooids. Studies on *Hydractinia echinata* showed that each gastrozooid in the colony takes a mean of 4.3 prey items each 24 h, and each item remains in the gastrovascular cavity for no longer than about 5 h.

Extracellular digestion takes place in the gastrozooid itself, after which the partially digested broth passes into the common gastrovascular cavity of the colony, where intracellular digestion occurs. Distribution is probably facilitated by rhythmic pulsations and contraction waves, which have been observed in many hydroids. Some hydroids contain zooxanthellae within the gastrodermal cells (see p. 148).

FIGURE 4–16 The hydroid *Aglaophenia*. **A,** Branch of colony. **B,** Three successive gastrozooids, each of which is surrounded by small defensive polyps.

A   B

In most species the gastrozooids also fulfill the defensive functions of the colony, but in some hydroids there are special defensive polyps **(dactylozooids).** The defensive polyps assume a variety of forms but are frequently club-shaped structures, well supplied with cnidocytes and adhesive cells (Fig. 4–14B). Defensive polyps commonly are located around the gastrozooid and also contribute to prey capture (Fig. 4–16).

Part of the colony of all hydroids are the reproductive members produced as asexual medusoid buds from some part of the colony (Fig. 4–15). The medusoids may either develop into free medusae capable of active swimming (e.g., *Obelia*), or be retained on the colony (e.g., *Hydractinia*); in either case, they produce the gametes to complete the sexual phase of the life cycle. The name **gonophore** ("gonad bearer") is commonly used to refer to hydroid medusoid members.

Medusoids of different species of hydrozoans assume a variety of shapes and locations. They may arise from the hydrocaulus, from the hydrorhiza, from the gastrozooid stalk, or frequently from the body of the gastrozooid itself (Figs. 4–14A and 4–17). In some species medusoids are produced only on certain polyps, called **gonozooids,** which are often reduced, lacking mouth and tentacles. Such a reduced gonozooid is called a **blastostyle.** In thecate hydroids,

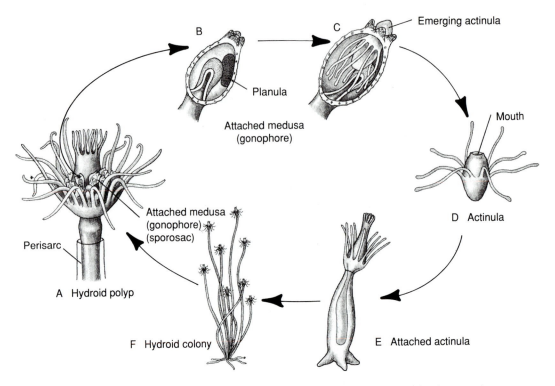

FIGURE 4–17 Life cycle of *Tubularia*. **A,** Skeleton is restricted to stalk of polyp; medusae are formed on gastrozooid and remain attached. **B,** Egg develops into planula within attached parent medusa. **C,** Actinula larva is released from medusa (**D**) and eventually settles to the bottom and develops into a new hydroid colony (**E** and **F**). *(After Allman from Bayer and Owre.)*

such as *Obelia* and *Campanularia,* the medusoids are restricted to a type of gonozooid called a **gonangium,** consisting of a central blastostyle surrounded by an extension of the perisarc, called the **gonotheca** (Fig. 4–15). Although free medusae are produced in some species, such as the well-known *Obelia* (Fig. 4–15), the majority of hydroids do not release their medusae.

An unusual group of hydroids consists of members of the genera *Porpita* (blue buttons) and *Velella* (by-the-wind sailors). Floating on or near the surface, these planktonic species, which range from 2 to 10 cm in diameter, are considered a colony. Each is suspended from a rather flattened, chambered, chitinous float (Fig. 4–18). At the center of the colony is a large gastrozooid. Gonozooids bearing gonophores hang down between the mouth of the gastrozooid and the marginal, tentacle-like defensive polyps. Found throughout the world's oceans, these floating colonies are often washed onto beaches. They are sometimes placed in a separate hydrozoan order, the Chondrophora.

Medusa Structure

Unlike the medusae of the Scyphozoa, hydrozoan medusae, commonly referred to as **hydromedusae,** are usually small, ranging from 0.5 cm to 6 cm in diameter (Fig. 4–19). The upper surface of the bell, typically covered with flattened cells, is called the **exumbrella,** and the lower surface, the **subumbrella.** The margin of the bell projects inward to form a shelf called a **velum** (characteristic of most hydromedusae). The tentacles that hang down from the margin of the bell are richly supplied with cnidocytes.

The mouth opens at the end of a tubelike extension called the **manubrium,** which hangs down from the center of the subumbrella and corresponds to the

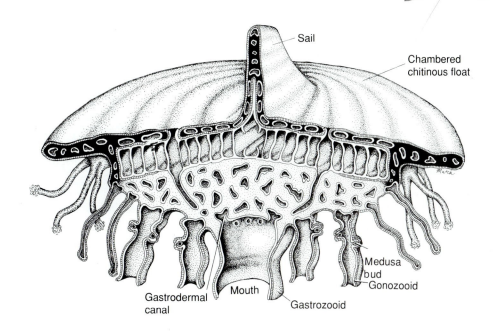

FIGURE 4–18 *Velella* (by-the-wind sailor), a pelagic hydroid. These pelagic hydroids float near the surface, and the aboral surface bears a sail.

hypostome of polyps. The manubrium also possesses cnidocytes and is often lobed or frilled with the inner surface lined with gastrodermis. The mouth leads into a central stomach, from which, typically, extend four **radial canals** lined with gastrodermis. The radial canals join with a **ring canal** running around the margin of the umbrella.

As in all medusoid forms, the mesoglea is thick and gelatinous and constitutes the bulk of the animal. The mesoglea of hydromedusae is devoid of cells but does contain elastic fibers. The muscular system of hydromedusae is best developed around the bell margin and subumbrella surface, where epidermal contractile fibers are organized as circular and radial striated sheets. The contractions of these fibers produce rhythmic pulsations of the bell, driving water out from beneath the subumbrella (Fig. 4–20). The compressed mesoglea with its elastic fibers provides the antagonistic force to restore the bell shape between contractions. "Joints" of soft mesoglea are commonly present (Figs. 4–20A and B). The velum, which is

most effective in deeper bell-shaped medusae, reduces the subumbrella aperture and thus increases the force of the water jet. Although the medusa often turns when swimming, the general direction is more or less vertical, and following a series of pulsations that drive the animal upward, it slowly sinks. Horizontal movement largely depends on water currents. There are a few species of hydromedusae, such as *Gonionemus,* that crawl about over the bottom, attaching to vegetation with their tentacles.

The nervous system of the medusa (Fig. 4–21) is more highly specialized than that of the polyp. In the margin of the bell, the epidermal nerve cells are usually organized and concentrated into an inner and an outer nerve ring. These nerve rings, which represent one of the highest levels of nervous organization in cnidarians, connect with fibers innervating the tentacles, the musculature, and the sense organs. The inner ring contains large motor neurons to the swimming muscles and is the center of rhythmic pulsation; that is, it contains the pacemakers.

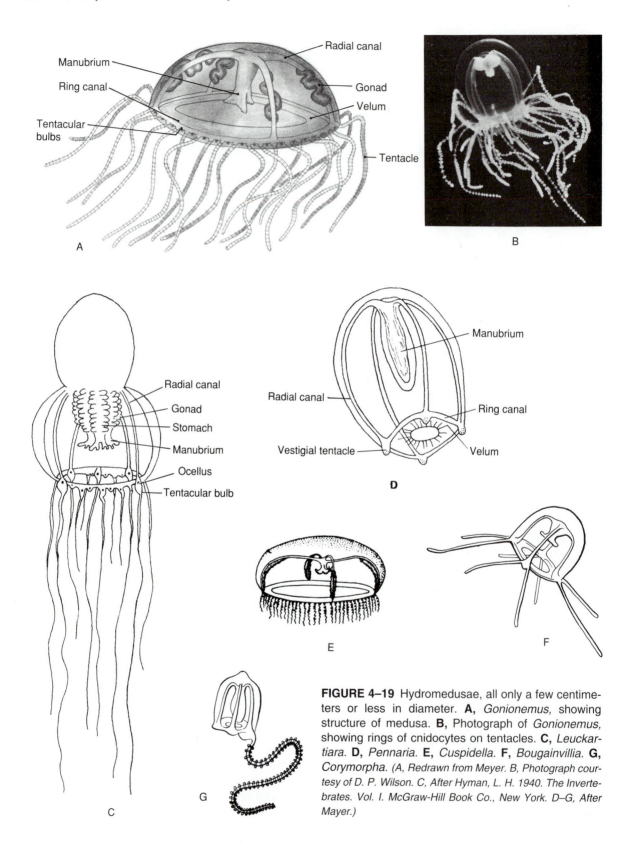

FIGURE 4–19 Hydromedusae, all only a few centimeters or less in diameter. **A,** *Gonionemus,* showing structure of medusa. **B,** Photograph of *Gonionemus,* showing rings of cnidocytes on tentacles. **C,** *Leuckartiara.* **D,** *Pennaria.* **E,** *Cuspidella.* **F,** *Bougainvillia.* **G,** *Corymorpha.* (A, Redrawn from Meyer. B, Photograph courtesy of D. P. Wilson. C, After Hyman, L. H. 1940. The Invertebrates. Vol. I. McGraw-Hill Book Co., New York. D–G, After Mayer.)

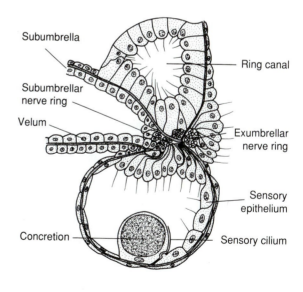

FIGURE 4–20 Swimming pulsations of hydromedusae. **A** and **B,** Aboral views of the relaxed and contracted phases of *Euphysa flammea,* showing action of the joints. **C,** Lateral views of *Bougainvillia multitentaculata* during swimming. *(A–C, From Gladfelter, W. G. 1973. A comparative analysis of the locomotory systems of medusoid Cnidaria. Helgol. wiss. Meeresunters. 25:228–272.)*

In addition to scattered receptor cells, the bell margin contains two types of true sense organs, light-sensitive **ocelli** and **statocysts.** The ocelli consist of patches of pigment and photoreceptor cells organized within either a flat disc or a pit on the outer side of the tentacular bulbs. Statocysts in the form of pits, vesi-

cles, or pendent clubs, are located between the tentacles or associated with the tentacular bulb at the tentacle base (Fig. 4–22). Stimulation of statocysts appears to inhibit muscular contractions on that side of the bell, and the opposite side throws water beneath the tilted margin to bring about righting.

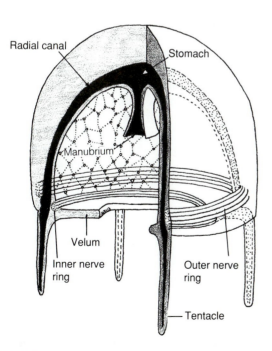

FIGURE 4–21 Nervous system of a hydromedusa. *(After Bütschli from Kaestner.)*

FIGURE 4–22 Closed vesicular statocyst of a hydromedusa in horizontal position. *(From Singla, C. L. 1975. Statocysts of hydromedusae. Cell Tissue Res. 158:391–407.)*

Like polyps, medusae are carnivorous, feeding on other planktonic animals, including small fish that come in contact with the tentacles.

Reproduction and Life Cycle

Hydrozoans may be hermaphroditic or dioecious, and all medusae reproduce sexually, although a few (e.g., *Aequorea*) can also reproduce asexually. The eggs and sperm arise from interstitial cells that have aggregated in specific locations in the epidermis, where a "gonad" is formed. The gonads are usually located beneath the radial canals in the epidermis of the subumbrella, as in *Gonionemus* (Fig. 4–19A).

Fertilization may be external in the sea water, on the surface of the manubrium, or internal, with the eggs beginning development in the gonad. Cleavage is complete and leads to a hollow blastula and then usually to a stereogastrula. The endoderm forms the future gastrodermis, and the ectoderm forms the epidermis. The stereogastrula rapidly elongates to become a ciliated, free-swimming planula larva (p. 113). After a free-swimming existence lasting from several hours to several days, the planula larva attaches to an object by the anterior end and develops into a hydroid colony.

Such a life cycle, with a free-swimming medusoid generation as well as a colonial polypoid stage, is displayed by *Obelia* (Fig. 4–15), *Pennaria, Syncoryne,* and other hydrozoan genera. However, most hydroids, such as *Tubularia, Sertularia,* and *Plumularia,* do not produce a free-swimming medusa. Instead, the medusa remains attached to the parent (Fig. 4–18), and varying degrees of arrested development are seen in different species. Despite this incomplete development, the medusa remains a sexually reproducing individual. In some hydroids the attached medusa is represented by only the gonadal tissue. Such an incomplete medusa is often called a **sporosac** (Figs. 4–14B and 4–17). Finally, in the hydras no vestige remains of a medusoid generation. Gonads form directly in the epidermis of the polyp stalk (See Fig. 4–24C). As is the case in free-swimming medusae, the eggs of attached medusoids may be retained and pass through their early embryonic stages while in the gonad. In some species, such as *Orthopyxis* (Fig. 4–23), the egg may develop through gastrulation within a sporosac, and the embryo escapes as a planula larva. In *Tubularia* even the plan-

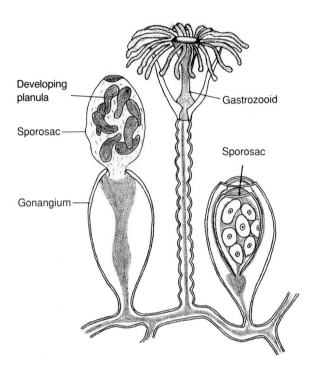

FIGURE 4–23 *Orthopyxis* with uneverted and everted sporosacs. *(After Nutting from Hyman.)*

ula stage is passed in the gonophore, and an **actinula larva** is eventually released. An actinula looks like a stubby hydra and creeps about on its tentacles (Fig. 4–17). The lack of a larval stage in hydras is probably correlated with their freshwater existence where larval stages are commonly absent.

From the discussion thus far it might appear that all hydrozoans are polypoid colonies, but this is by no means true. The medusoid form is dominant in some hydrozoans. In *Liriope* and *Aglaura,* there is no polypoid stage. The planula forms a planktonic actinula larva that transforms directly into a medusa (Fig. 4–24A). In *Gonionemus* and the freshwater *Craspedacusta* a polypoid stage is present, but the polyp is small and solitary, and medusae bud off from the sides (Fig. 4–24B).

The appearance of hydroid colonies and hydromedusae is usually highly seasonal and, in temperate regions at least, is closely correlated with water temperatures. Hydrozoans that produce free medusae may liberate enormous numbers of them in a short time. A 7-cm colony of *Bougainvillia* released 4450 medusae over a three-day period; as a result of such production, hydromedusae may often make a dramatic appearance in the plankton. The life span of a medusa ranges from a few days to many months.

Hydrozoan Evolution

The evolutionary significance of the life cycle of hydrozoans is a fascinating problem. Which came first, the polyp or the medusa? In 1886, W. K. Brooks worked out a theory of cnidarian evolution that is still supported by many zoologists today. According to Brooks's theory, the ancestral cnidarian form was medusoid. The tendency among hydrozoans has been to suppress the medusoid stage, so that in such forms as *Hydra* the medusa has completely disappeared. The polyp, on the other hand, represents an evolutionary retention and development of the polyp-like actinula larva.

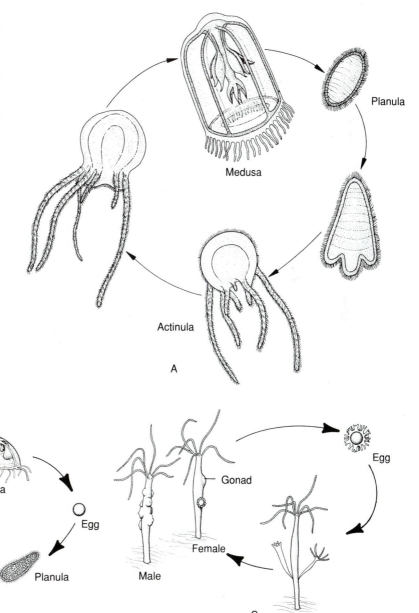

FIGURE 4–24 Some hydrozoan life cycles. **A,** *Aglaura,* a hydrozoan that has no polypoid state. Planula larva develops into an actinula, which develops directly into a medusa. **B,** *Craspedacusta.* (Life cycle of *Gonionemus* is similar.) Polyp is solitary. **C,** *Hydra,* a freshwater hydrozoan in which the medusoid state has disappeared and planula larva is suppressed. *(A, From Bayer and Owre. C, From Bayer, F., and Owre, H. B. 1968. The Free-Living Lower Invertebrates. Macmillan Co., New York.)*

Medusa

Planula

Actinula

A

Medusa

Egg

Planula

Polyp

B

Gonad

Egg

Female

Male

C

The evolutionary sequence may have occurred in the following manner:

Medusa→Egg→Planula→Actinula→Medusa

The ancestral cnidarians were medusoid, and development led through a planktonic planula larva and a later planktonic actinula larva. The actinula, in which the polypoid characters first make their appearance, developed into the adult medusa.

Medusa→Planula→Attached Actinula→Medusa

In some groups of such ancestral medusoid cnidarians, the actinula took up an attached benthic existence to give rise to a polyp. Such an attached condition could have been an adaptation to exploit a new food supply, to extend larval life, or to facilitate asexual reproduction of additional polyps by budding. Hydromedusae may have first developed by direct transformation of the attached actinula polyp.

Medusa→Planula→Polyp→Medusa

The evolution of gonophores (medusoid budding) would have made possible the formation of numerous medusae from one polyp.

Attached Medusa→Planula→Polyp→Attached Medusa

The medusa, which primitively was free, became retained.

Polyp→Planula→Polyp

Then the attached medusa was gradually suppressed until it disappeared completely. Living hydrozoans display life cycles that illustrate different stages in this evolutionary sequence.

The hydroid colony evolved through the retention of budded polyps. Correlated with the evolution of colonial organization was the development of skeletons and polymorphism, but these events probably occurred independently in different evolutionary lines. Medusoid suppression was probably also repeated numerous times, and carried to various degrees, throughout the Hydrozoa. For example, it is

A

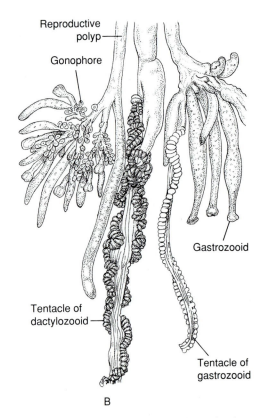

Reproductive polyp

Gonophore

Gastrozooid

Tentacle of dactylozooid

Tentacle of gastrozooid

B

FIGURE 4–25 A, *Physalia,* the Portuguese man-of-war, a siphonophoran that has a large, horizontal float and no swimming bell. **B,** Part of a Portuguese man-of-war colony. *(A, Photograph courtesy of the New York Zoological Society. B, After Lane.)*

unlikely that the hydras evolved from a colonial hydroid; they are probably derived from a line of hydrozoans, similar to *Gonionemus* and *Craspedacusta,* in which polyps were always solitary but originally budded off free-swimming medusae.

We will return to this theory of polypoid origins when we examine the other classes of cnidarians.

Specialized Hydrozoan Orders

Siphonophora

Members of the order Siphonophora, which includes the familiar *Physalia* (the Portuguese man-of-war, Fig. 4–25), are large polymorphic pelagic colonies composed of modified polypoid and medusoid members. The members of the colony are commonly attached to a long stem (Fig. 4–26). A conspicuous feature of many species is a gas-filled float, containing mostly carbon monoxide. The large float of the Por-

tuguese man-of-war may attain a length of 30 cm, keeps the colony at the surface, and functions as a sail. Other siphonophores possess mouthless, pulsating swimming bells (**nectophores**) in addition to or instead of a gas-filled float.

Most siphonophores live below the surface and may migrate vertically in the water by secreting or releasing gas from the floats. The Pacific *Nanomia bijuga,* which possesses both a float and swimming bells, has been reported to move 300 m in less than an hour in vertical migration.

Feeding is carried on by gastrozooids with terminal mouths and a single, long, fishing tentacle, which is contractile and armed with batteries of cnidocytes (Figs. 4–25 and 4–26). Siphonophores also have defensive polyps with long tentacles. The 4 to 150 polyps and tentacles form an effective fishing net for catching various planktonic invertebrates and small fish. Many siphonophores interrupt their fishing with a short interval of swimming, after which their "nets" are let out in a new location. The Gulf of California siphonophore *Rhizophysa eysenhardti,* which reaches a length of somewhat less than a meter and contains up to 28 gastrozooids, was found to consume about nine fish larvae (5–15 mm long) per day. This siphonophore fed only during the day and only on fish larvae, each of which took about 8 min to ingest and 3 to 7 h to digest.

Siphonophores are largely tropical and semitropical, but numbers of Portuguese men-of-war (*Physalia*) are often seen on North Atlantic shores, both in Europe and North America, especially following storms that have blown them in from the Gulf Stream. An accidental encounter with the tentacles of this siphonophore, which may hang several meters below the float (Fig. 4–25), can be a painful and even a dangerous experience for a swimmer.*

Hydrocorals

The two small groups of **hydrocorals,** sometimes placed in the separate orders Milleporina (*Millepora*) and Stylasterina (*Stylaster*), secrete a calcareous skeleton (Fig. 4–27). Both are colonial polypoid hydrozoans with either an encrusting or an upright growth form. Both may attain considerable size. Gas-

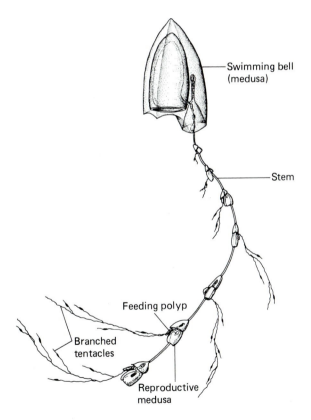

FIGURE 4–26 *Muggiaea,* a submergent, more typical siphonophore than *Physalia.* Clusters of feeding polyps and reproductive medusae are connected by a long stem hanging beneath a swimming bell, which is about 1 cm long.

*Meat tenderizer, commonly carried in the first-aid box of snorklers and scuba divers, can be an effective pain remedy if quickly rubbed on the sting of this or other cnidarians. The protease in the meat tenderizer breaks down the protein toxin.

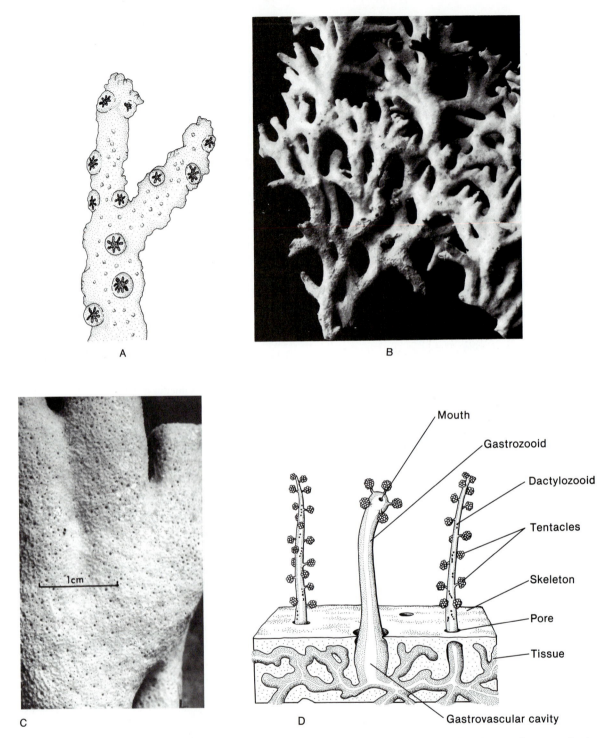

Mouth

Gastrozooid

Dactylozooid

Tentacles

Skeleton

Pore

Tissue

A

B

C

D — Gastrovascular cavity

FIGURE 4–27 Hydrocorals. **A,** *Allopora* of the order Stylasterina. These hydrocorals are usually more finely branched than the Milleporina. The gastrozooids and surrounding defensive polyps are located in the large, notched cups. *Millepora* of the order Milleporina: **B,** Part of a colony. **C,** Pores in skeleton through which polyps emerge. Gastrozooids emerge through large pores, defensive polyps from surrounding small pores. **D,** Defensive polyp (dactylozooid) and gastrozooids. *(A, After Moseley from Bayer, F. M., and Owre, H. B. 1968. The Free-Living Lower Invertebrates. Macmillan Co., New York. B, Courtesy of the Encyclopedia Britannica. D, Modified from de Kruijf, H. A. M. 1975. General morphology and behavior of gastrozooids and dactylozooids in two species of Millepora. Mar. Behav. Physio. 3:181–192. © Gordon and Breach Science Publishers, Inc.)*

trozooids and defensive polyps emerge from pores in the skeleton, which are sometimes arranged in rings (Fig. 4–27). The defensive polyps are numerous and prominent; in the Milleporina, sometimes called fire corals, their sting can be severe. The polyps are connected by canals beneath the surface of the skeleton. Although the skeleton is covered by the tissues of the colony, it is actually external to the epidermis. The Milleporina produce minute medusae. *Millepora* is a common component of coral reefs and is yellow-brown, like most stony corals, because of symbiotic zooxanthellae. The order Stylasterina contains many species, which are found in both temperate and tropical seas.

SUMMARY OF HYDROZOA

1. Members of the class Hydrozoa are medusoid or polypoid or exhibit both forms in their life cycle. The mesoglea lacks cells, cnidocytes are restricted to the epidermis, and gametes develop in the epidermis. Hydrozoans may be the most primitive of the three classes of cnidarians.

2. The planktonic hydromedusae are small and have a velum and manubrium.

3. The most primitive hydrozoans are probably medusoid species in which the pelagic actinula develops directly into an adult medusa. Such a life cycle may also be primitive for the phylum.

4. The polypoid form may have arisen in some medusoid species in which the actinula passed through a period of attachment prior to development into a pelagic adult; that is, the attached actinula was the first polyp.

5. Early polypoid stages, including the attached actinula, probably reproduced asexually by budding. Persistent attachment of the buds led to colonial polypoid species, called hydroids, which now compose the majority of hydrozoans.

6. The evolution of a skeleton (support) and polymorphism (division of labor) have been associated with colonial organization.

7. Naked solitary species, such as hydras and the *Gonionemus* polyp, probably stem from early polypoid forms that were not colonial.

8. Suppression of the medusa through attachment to the polyp and subsequent reduction have evolved independently in different hydrozoan lines, and living species exhibit all degrees of reduction in the medusoid form.

CLASSES SCYPHOZOA AND CUBOZOA

Scyphozoans and cubozoans are the cnidarians most frequently referred to as **jellyfish.** In these classes the medusa (Figs. 4–28A and 4–29) is the dominant and conspicuous individual in the life cycle; the polypoid form is restricted to a small, sessile stage. Scyphomedusae and cubomedusae are generally larger than hydromedusae, most having a bell diameter ranging from 2 to 40 cm; some species are even larger. The bell of *Cyanea capillata,* for example, may reach 2 m in diameter. Cubomedusae are somewhat smaller than scyphomedusae. Coloration is often striking—the gonads and other internal structures (which may be deep orange, pink, or other colors) are visible through the transparent or more delicately tinted bell.

There are some 200 described species of scyphozoans, living in cold to tropical oceans, some at great depths. In contrast, there are only 15 described species of cubozoans, all of which are semitropical or tropical. Unless otherwise indicated, the following discussion focuses on the more numerous and frequently encountered scyphozoans.

Species of both groups inhabit coastal waters and may be a nuisance on bathing beaches. Their large size and their nematocysts make them unpleasant and sometimes dangerous swimming companions. The so-called sea nettles, which inhabit the North Atlantic coast in large numbers during late summer, are members of this class. Some Indo-Pacific cubozoans, called sea wasps, of which *Chironex fleckeri* is the most notorious (Fig. 4–30), produce such virulent toxins that they are extremely dangerous. The bell may reach 17 cm in height with trailing tentacles 2 m in length, but their transparency makes them difficult to see from the water's surface. Many lethal and nonlethal stings have been recorded from northeastern, northern, and northwestern Australian coasts. Death, if it occurs, takes place 3 to 20 min after stinging. Lesions from nonlethal stings may be severe and slow to heal. Swimming enclosures and warning signs mark many Queensland beaches (Fig. 4–30C).

In general, scyphozoan medusae are similar to hydromedusae. The bell varies in shape from a shallow saucer to a deep helmet, but the margin is typically scalloped to form lobes called **lappets** (Fig. 4–29). A velum is absent. The manubrium of many common coastal species (Semaeostomeae) is drawn out into four or eight often frilly, **oral arms,** which

(Text continues on page 130)

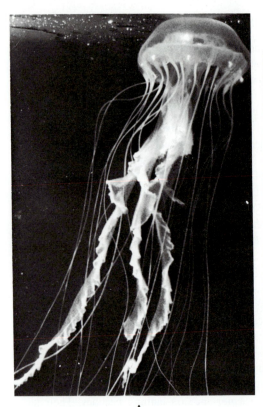

A

B

FIGURE 4–28 A, The sea nettle, *Chrysaora quinquecirrha,* a common scyphozoan along the Atlantic coast. **B,** Two species of sessile scyphomedusae (stauromedusae) attached to an alga. The specimen on the left, *Haliclystus auricula,* has a diameter of less than 1 in. The specimen on the right is *Craterolophus convolvulus. (A, Photograph courtesy of William H. Amos. B, Photograph courtesy of D. P. Wilson. In Buchsbaum, R. M. and Milne, L. J. 1961. The Lower Animals. Chanticleer Press, New York.)*

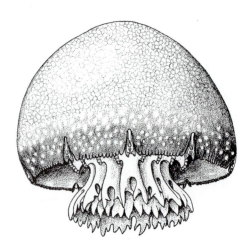

FIGURE 4–29 *Stomolophus meleagris,* sometimes called the cannonball jellyfish, is found along the southeastern coast of the United States and in the Caribbean. This species is a rapid swimmer, and its hemispherical, relatively rigid bell may reach the size of a football. *Stomolophus* is a rhizostome medusa. The members of this order lack tentacles around the bell margin and have multiple secondary mouths in the oral arm area. *(After Mayer.)*

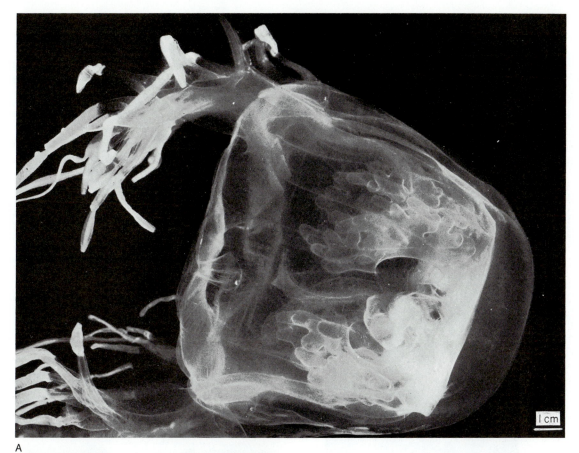

A

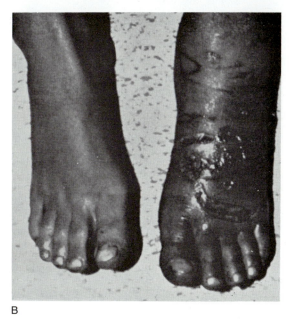

B

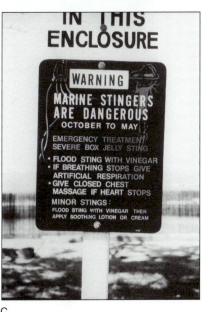

C

FIGURE 4–30 A, The Indo-Pacific sea wasp, *Chironex fleckeri,* a dangerous scyphozoan of the order Cubome-dusae. The tentacles have been cut to permit handling of the specimen. **B,** Five-day-old lesions on the victim's left foot produced by a small sting of this sea wasp. **C,** Sea wasp warning sign on beach in Queensland, Australia. *(B, Photograph courtesy of J. H. Barnes. In Rees, W. J. (Ed.). 1966. The Cnidaria and Their Evolution. Zoological Society of London. C, From Dorit, R. L. Walker, W. F., Jr., and Barnes, R. D. 1991. Zoology. Saunders College Publishing, Philadelphia.)*

bear cnidocytes and aid in the capture and ingestion of prey. The tentacles around the bell margin vary from four to many but are absent in rhizostomes (Fig. 4–29). In *Aurelia,* the type most often studied in introductory courses, the tentacles are small and form a short fringe around the margin (Fig. 4–31); but in other species the tentacles are much longer (Fig. 4–28A). One order, the Stauromedusae, is sessile (Fig. 4–28B). In this group, the exumbrellar surface is drawn out into a stalk by which the animal attaches to algae and other objects. They are thus polypoid in general structure.

The cubomedusae, commonly called box jellies, are somewhat cuboidal in shape, as the name sug-

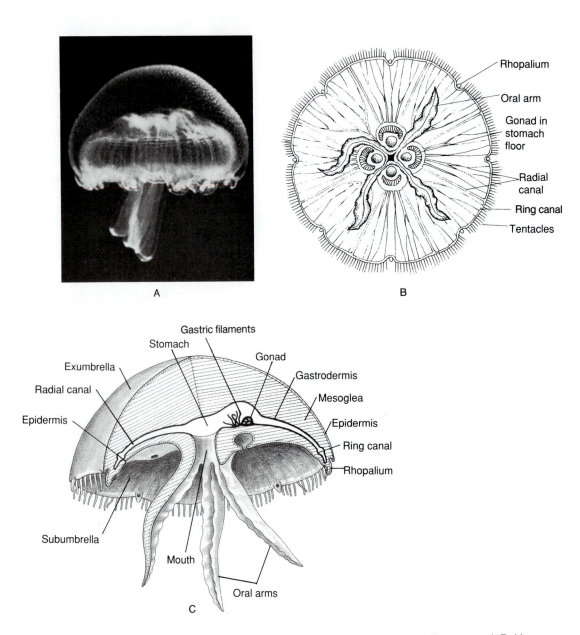

FIGURE 4–31 *Aurelia,* a scyphozoan medusa. **A,** Young specimen with bell contracted. **B,** Ventral view. **C,** side view in section. *(A, Photograph courtesy of D. P. Wilson.)*

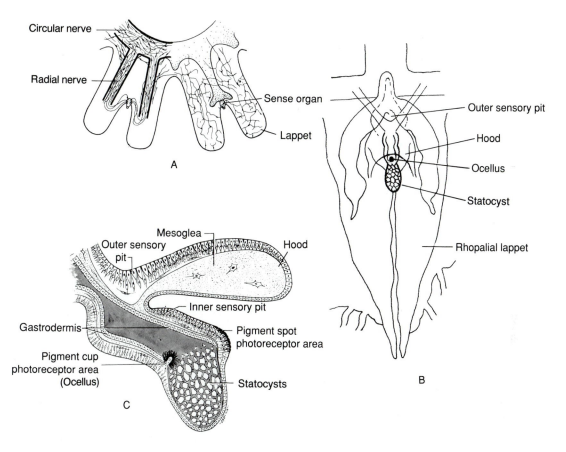

FIGURE 4–32 Marginal sense organs of scyphozoans. **A,** Diagram of a section of bell margin showing nerve net, lappets, and position of sense organ, or phopalium. **B,** Rhopalium of *Aurelia* (aboral view). **C,** Section through a rhopalium showing the hood and the various sensory areas. *(B, After Hyman, L. H. 1940. The Invertebrates, Vol. I. McGraw-Hill Book Co., New York. C, Modified from Hyman In Schewiakoff.)*

gests. The bell margin is not scalloped, possesses a velum-like structure **(velarium),** and bears four tentacles or tentacle clusters (Fig. 4–32A).

The mesoglea of scyphozoans and cubozoans is thick, gelatinous, and fibrous, but unlike that of the hydrozoans, it contains ameboid cells, which originate from the epidermis.

A band of powerful circular muscle fibers (the coronal muscle) and radial fibers in the subumbrella produce swimming pulsations similar to those of hydromedusae. The mesoglea provides elastic recoil, and some species have areas around the bell that "fold" during contraction. Although scyphomedusae drift with currents or waves, most can swim vertically and horizontally, and some tropical cubozoans and rhizostome medusae (Figs. 4–29 and 4–30A) are very

rapid swimmers. The velum-like flap of a cubomedusa greatly increases the force of the water jet leaving the subumbrellar cavity.

The plan of the scyphozoan gastrovascular system is somewhat different from that seen in the hydromedusae. The mouth opens through the manubrium into a central **stomach,** from which extend four **gastric pouches** (Fig. 4–33). Between the pouches are **septa,** each of which contains an opening to help circulate water. Thus, all four pouches are in lateral communication with each other. The margin of the septum, which faces the central portion of the stomach, bears a large number of **gastric filaments** containing cnidocytes and gland cells. Radiating canals typically run from the stomach pouches to the bell margin. A ring canal may be present or absent.

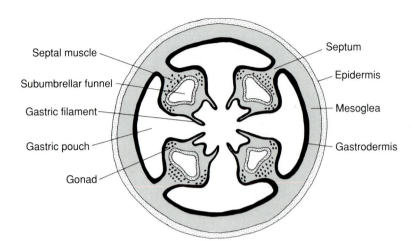

Septal muscle

Subumbrellar funnel

Gastric filament

Gastric pouch

Gonad

Septum

Epidermis

Mesoglea

Gastrodermis

FIGURE 4–33 Section through a scyphozoan with gastric pouches.

There is considerable modification of this plan in different groups. Some, such as *Aurelia,* lack the gastric pouches. Nevertheless, filaments are present and are attached interradially to the periphery of the stomach floor. *Aurelia* has an extensive system of branched and unbranched canals extending from the stomach to a ring canal in the bell margin (Fig. 4–30). Adult scyphozoans feed on all types of small animals, especially crustaceans. Some scyphozoans feed on fish; however, larval fish of a number of species swim with certain species of scyphozoans for protection. As the medusa gently swims or slowly sinks, prey is cap-

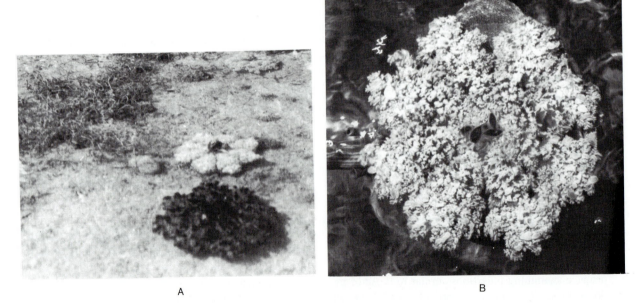

A

B

FIGURE 4–34 *Cassiopea,* a common tropical genus of rhizostome scyphozoans, which live upside down on the bottom in quiet shallow water. **A,** Specimens resting on the bottom, viewed through water. **B,** A single specimen about 21 cm in diameter, which has been lifted to the surface. *(A and B, Photographs courtesy of Betty M. Barnes.)*

tured on contact with the tentacles or oral arms of the manubrium. The tentacles may contract to bring the prey to the vicinity of the manubrium. Some species, including *Aurelia,* are actually suspension feeders, trapping plankton in mucus on the ciliated subumbrellar surface. Cilia then sweep the food to the bell margin, where it is scraped off by the oral arms. Ciliated grooves on the oral arms carry the food to the mouth and stomach. *Cassiopea,* a jellyfish common to Florida and the Caribbean, lies upside down on the bottom in the quiet shallow water of mangrove embayments (Fig. 4–34). This genus, like other members of the order Rhizostomeae, possesses many small secondary mouths that open into the stomach by way of canals in the oral arms. Small animals trapped on the surface of the frilly oral arms are carried into the mouths within mucous cords. However, *Cassiopea* possesses symbiotic algae (zooxanthellae) in its mesoglea, and in adequate light it can survive and grow largely on the products of the algal photosynthesis.

Digestion is essentially as described in hydras. The gastric filaments are the source of extracellular enzymes, and the gastrodermal cnidocytes are probably used to quell prey that is still active.

A nerve ring occurs in the cubozoans. In the scyphozoans the control of bell pulsations is centered in marginal **rhopalia,** which are concentrations of neurons and sensory organs housed in a little club-shaped structure. The rhopalia are located around the bell margin between lappets and number four or multiples of four. Each rhopalium is flanked by a pair of small specialized lappets called **rhopalial lappets** and is covered by a hood (Fig. 4–32).

Each rhopalium contains two sensory pits, a statocyst, and sometimes an ocellus (Fig. 4–32B). Cubozoans have complex eyes containing a lens and a retina-like arrangement of sensory cells and can orient to small sources of light. Many scyphozoans display distinct phototaxis. They come to the surface of the water during cloudy weather and at twilight but move downward in bright sunlight and at night.

With few exceptions, scyphozoans and cubozoans are dioecious, and the gonads are gastrodermal, in contrast to the usual epidermal gonad in hydrozoans. In septate groups with gastric pouches, the eight gonads are located on both sides of the four septa (Fig. 4–33). In semaeostome medusae, which lack septa, four horseshoe-shaped gonads lie on the floor of the stomach (Fig. 4–31). When mature, the gametes usually break into the gastrovascular cavity and exit through the mouth. In some species, such as *Cyanea,* the eggs are fertilized within the gonad and are released at the gastrula stage, whereas in others, including *Aurelia,* the eggs become lodged in pits on the oral arms and this temporary brood chamber is the site of fertilization and early development to the planula larva (Fig. 4–35). After a brief free-swimming existence, the planula settles to the bottom, becomes attached by its anterior end, and develops into a little polypoid **scyphistoma.** The scyphistoma looks very much like a hydra (Fig. 4–35) but has the four internal gastrodermal septa characteristic of primitive scyphomedusae. It feeds and produces additional scyphistomae by asexual budding, either directly from the midcolumn wall or from stolons (in *Aurelia*). At certain periods of the year, under the influence of hormonal and environmental factors, the scyphistoma produces young medusae. In the one species of cubozoans for which the life cycle is known, the scyphistoma transforms directly into a small medusa (Fig. 4–36). Note that this life cycle illustrates one of the evolutionary stages described for hydrozoans (p. 123). In other scyphozoans, medusa formation is accomplished by transverse fission of the oral end of the scyphistoma, a process called **strobilation.** Medusae may form one at a time (monostrobilation) or by multiple budding (polystrobilation) so that the immature medusae, called **ephyrae,** are stacked up like saucers at the oral end of the body stalk (Fig. 4–35). As each ephyra is formed, it breaks away from the oral end of the scyphistoma.

After strobilation, the scyphistoma may resume its polypoid existence until the following year, when formation of ephyrae is repeated. A scyphistoma may live for one or several years. The ephyra is almost microscopic and has a deeply incised bell margin (Fig. 4–35). Some ephyrae take two years to grow into sexually-reproducing adult medusae; others are relatively short lived. The ephyrae of *Aurelia aurita* on the west coast of the United States are produced in March and reach sexual maturity by June.

Species of *Stephanoscyphus* and *Nausithoe* (order Coronatae) have branching colonial polyps with a supportive skeletal tube (Fig. 4–37). Some even have reduced medusoid stages, with precocious gamete production occurring during or immediately after strobilation.

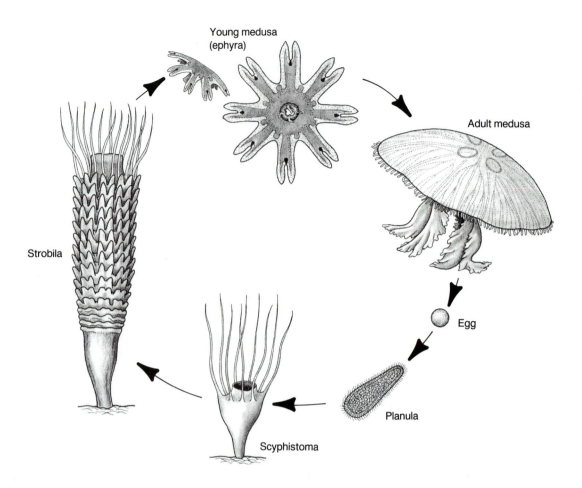

FIGURE 4–35 Life cycle of the scyphozoan *Aurelia.* The polypoid stage is a larva and produces medusae by transverse budding.

SUMMARY OF SCYPHOZOA AND CUBOZOA

1. Members of the classes Cubozoa and Scyphozoa are pelagic cnidarians in which the medusa is the dominant and conspicuous form. A polypoid form, equivalent to an actinula, follows the planula.

2. Within the Scyphozoa, specialization has led to complexities in medusoid structure, as evidenced by the following features: larger size than most hydromedusae, more highly developed manubrium, mesoglea containing cells, septate stomach or at least a stomach with gastric filaments, gastrodermal cnidocytes, and some development of sense organs.

3. The gonads are gastrodermal, and the eggs, which are shed through the mouth, develop into planula larvae. Following settling, the planulae develop into polyps, which feed and may reproduce asexually.

4. In at least some species of cubozoans the polypoid stage transforms directly into a young medusa. In scyphozoans, young medusae are budded off transversely from the oral end of the polypoid stage.

CLASS ANTHOZOA

Anthozoans are either solitary or colonial polypoid cnidarians in which the medusoid stage is completely absent. Many familiar animals, such as sea anemones, corals, sea fans, and sea pansies, are members of this

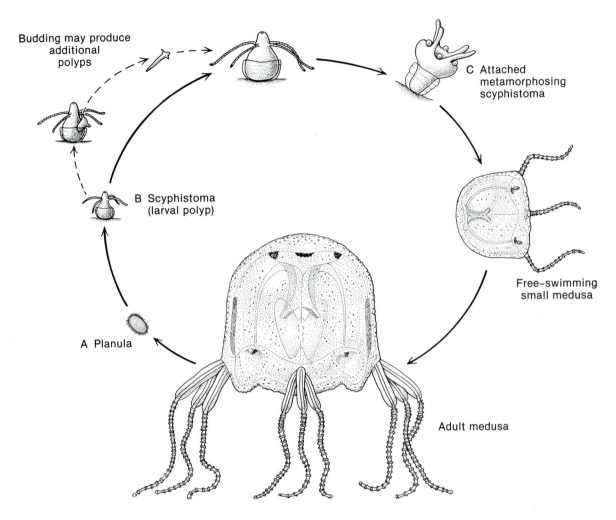

Budding may produce
additional
polyps

B Scyphistoma
(larval polyp)

C Attached
metamorphosing
scyphistoma

Free-swimming
small medusa

A Planula

Adult medusa

FIGURE 4–36 Life cycle of the cubomedusa *Tripedalia cystophora.* The planula (**A**) develops into a solitary attached larval polyp (**B**), which eventually metamorphoses directly (**C**) into the medusa. Additional polyps may be derived by budding. *(From Werner, B. 1973. New investigations on systematics and evolution of the class Scyphozoa and the phylum Cnidaria. Proc. 2nd Internat. Symp. Cnidaria, Publ. Seto Mar. Biol. Lab. 20:35–61.)*

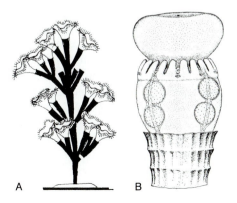

FIGURE 4–37 *Stephanoscyphus racemosus* (order Coronatae). **A,** Colonial scyphistomae. **B,** Medusa, just released from strobila, already possesses ripe eggs. *(From Werner, B. 1973. New investigations on systematics and evolution of the class Scyphozoa and the phylum Cnidaria. Proc. 2nd Internat. Symp. Cnidaria, Publ. Seto Mar. Biol. Lab. 20:35–61.)*

A   B

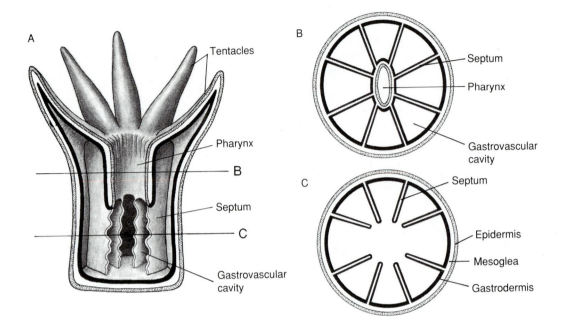

FIGURE 4–38 Structure of an anthozoan polyp. **A,** Longitudinal section. **B,** Cross section at the level of the pharynx. **C,** Cross section below the pharynx.

class, which is the largest of the cnidarian classes, containing over 6000 species.

Although the anthozoans are polypoid, they differ considerably from hydrozoan polyps. The mouth leads into a tubular **pharynx** that extends more than halfway into the gastrovascular cavity (Fig. 4–38). The gastrovascular cavity is divided by longitudinal **septa,** or **mesenteries,** into radiating compartments, and the edges of the septa bear nematocysts. The gonads, as in the scyphozoans, are gastrodermal, and the fibrous mesoglea contains mesenchymal cells. The nematocysts, unlike those of hydrozoans and scyphozoans, do not possess an operculum. Some anthozoan nematocysts have a three-part tip that folds back on expulsion; in others, the thread appears to rupture directly through the end of the capsule.

To simplify the survey of this class, we concentrate on the sea anemones, the stony corals, and the octocorallian corals separately.

Sea Anemones

Sea anemones are solitary polyps and are considerably larger and heavier than the polyps of hydrozoans (Fig. 4–39). Most sea anemones range from 1.5 to 10 cm in length and from 1 to 5 cm in diameter, but specimens of *Tealia columbiana* on the North Pacific coast of the United States and *Stichodactyla* on the Great Barrier Reef of Australia may grow to a diameter of more than a meter at the oral end. Sea anemones are often brightly colored and may be white, green, blue, orange, or red, or a combination of colors. Sea anemones inhabit deep or coastal waters throughout the world but are particularly diverse in tropical oceans. They commonly live attached to rocks, shells, and submerged timbers, although some species burrow in mud or sand. A few species are commensal on other animals, such as the shells of hermit crabs (see p. 115). The shape of the body is

A

B

C

D

FIGURE 4–39 Sea anemones. **A** and **B,** *Anthopleura,* a genus of common sea anemones found in shallow water along the Pacific coast of the United States. The column of the partially closed specimen in B is covered with shell fragments, which adhere to epidermal tubercles. **C,** A West Indian species of *Actinia.* **D,** The venomous *Alicia mirabilis* from the Caribbean. The column is covered with berry-like evaginations containing nematocyst batteries. *(A, Photograph courtesy of Turtox News. B and C, Photographs courtesy of Betty M. Barnes. D, Photograph courtesy of Charles Arneson.)*

FIGURE 4–40 Relationship of body form to habitat in Caribbean reef-dwelling sea anemones. **A,** *Rhodactis sanctithomae,* a surface-dwelling form with short column and tentacles. **B,** *Phymanthus crucifer,* and **C,** *Bartholomea annulata,* sand pocket forms. **D,** *Heteractis lucida,* a hole dweller with long column and tentacles. *(From Sebens, K. 1976. The ecology of Caribbean sea anemones in Panama: Utilization of space on a coral reef. In Mackie, G. O. (Ed.): Coelenterate Ecology and Behavior. Plenum Press, New York, p. 69.)*

often related to the habitat in which the sea anemone lives (Fig. 4–40).

The major part of the sea anemone is a thick **column** (Fig. 4–41), which may be smooth, or bear tubercles or even tentacle-like outgrowths. The common intertidal *Anthopleura elegantissima* from the west coast of the United States attaches shells and pebbles to the tubercles on its column, providing itself with a protective cover when the tide is out. At the aboral end of the column there is a flattened **pedal disc** for attachment. At the oral end, the column flares slightly to form the **oral disc,** which bears eight to several hundred hollow tentacles, and in some species is drawn out into lobes. In the center of the oral disc is the slit-shaped mouth, bearing at one or both ends a ciliated groove called a **siphonoglyph,** which provides for the entry of water into the gastrovascular cavity. This current of water functions to maintain an internal fluid, or hydrostatic, skeleton against which the muscular system can act, and also provides for the exchange of gases through the gastrodermal surface.

When a sea anemone contracts, the upper surface of the column of most species is pulled over and covers the oral disc. In many sea anemones, including the familiar *Metridium* of the northern Atlantic and Pacific coasts, the column bears a circular fold, or **collar,** at its junction with the oral disc. The collar covers the oral surface when the animal is contracted (Fig. 4–40A).

The mouth leads into a flattened pharynx, which extends approximately two thirds of the way into the column (Fig. 4–40). The pharynx is formed from an infolding of the body wall and therefore, both consist of the same tissue layers: an inner ciliated epidermis, an outer gastrodermis, and between the two, a layer of mesoglea. The pharynx is kept closed and flat by the water pressure in the gastrovascular cavity. The siphonoglyph is kept open by an especially thick mesoglea and very heavy epidermal cells.

As in all anthozoans, the gastrovascular cavity of the sea anemone is partitioned by longitudinal, radiating septa, which increase the internal surface area. In the sea anemones, there are usually two types of septa—complete and incomplete. Complete septa are connected to the body wall on one side and to the wall of the pharynx on the other (Fig. 4–39B). Incomplete septa are connected only to the body wall and extend only partway into the gastrovascular cavity. The septa, both complete and incomplete, are arranged in adjacent pairs. The pairs at each end of the tapered pharynx are called **directives.** The septa usually occur in multiples of 12. When only 12 septa are present, as in the primitive *Halcampoides,* they are complete and are called the primary cycle. The addition of a cycle of 12 secondary septa between the primary ones brings the total to 24. A tertiary cycle brings the number to 48. Only the first cycle may be complete, with subsequent cycles incomplete and successively smaller. Many exceptions exist to the numerical symmetry just described. Moreover, asexual reproduction produces considerable irregularity, particularly in *Metridium.*

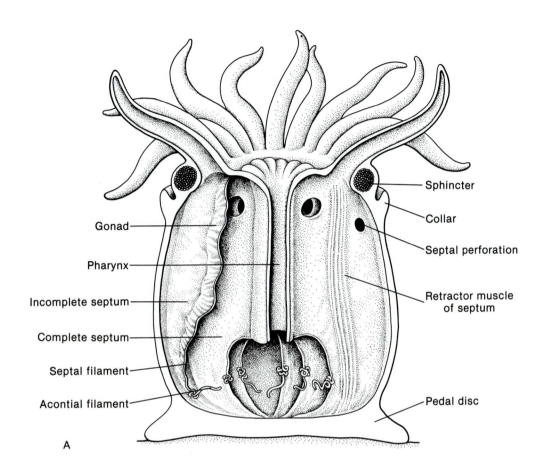

Sphincter

Collar

Septal perforation

Gonad

Pharynx

Retractor muscle
of septum

Incomplete septum

Complete septum

Septal filament

Acontial filament

Pedal disc

A

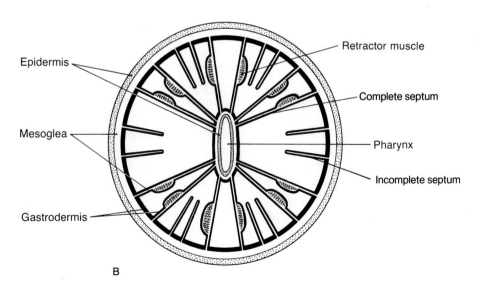

Retractor muscle

Epidermis

Complete septum

Pharynx

Mesoglea

Incomplete septum

Gastrodermis

B

FIGURE 4–41 Structure of a sea anemone. **A,** Longitudinal section. **B,** Cross section at
the level of the pharynx.

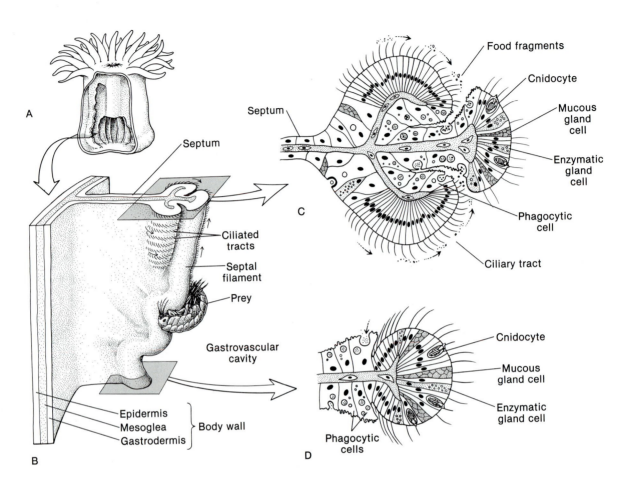

FIGURE 4–42 Structure of a sea anemone septal filament. **A,** An opened sea anemone, showing the location of the section of body wall, septum, and septal filament depicted in **B.** The upper part of the septal filament is trilobed; the lower part bears a single lobe. **C,** Section of trilobed filament. **D,** Section of single-lobed filament. *(Adapted from Van-Praet, M. 1985. Nutrition of sea anemones. Adv. Mar. Biol. 22:65–99.)*

In the upper part of the pharyngeal region, the septa are perforated by openings that facilitate water circulation (Fig. 4–41A). Below the pharynx, the complete septa have free margins and recurve toward the body wall. The hollow tentacles, which contain an extension of the coelenteron within their interior, are always located between septa (Fig. 4–41).

Histologically, each septum consists of two layers of gastrodermis separated by a layer of mesoglea. Both the gastrodermis and the mesoglea are continuous with their corresponding layers in the body wall and also in the pharynx when the septa are complete.

The free edge of the septum is trilobed and is called a **septal filament** (Fig. 4–42B). The septal filament is longer than the septum and thus tends to be somewhat convoluted. The lateral lobes of the filament, which are restricted to the upper part of the septum, are composed of ciliated cells and aid in water circulation. The middle lobe contains cnidocytes and enzymatic-gland cells along the entire length of the septum. In some sea anemones, including *Metridium,* the middle lobe continues beyond the base of the septum as a thread called an **acontium,** which projects into the gastrovascular cavity.

The epidermis of sea anemones may be ciliated and in some species is covered with a cuticle. The mesoglea is much thicker than that of hydrozoan polyps and contains a large number of fibers, as well as amebocytes. Associated with the tip of at least some anthozoan nematocysts is a cnidocil-like cilium and microvilli complex that is involved in the reception of stimuli for discharge (p. 108). In addition to nematocysts, sea anemones also possess **spirocysts,** which have a capsule with a single wall and a long adhesive thread. The spirocysts function in the capture of prey that have hard surfaces, such as small crustaceans, and in attachment to the substratum (Fig. 4–10E). In the tentacles spirocysts are usually more prevalent than nematocysts.

A few sea anemones occur in European and American waters whose nematocysts can produce a severe toxic reaction in humans. They include the berried sea anemones *(Alicia mirabilis)* (Fig. 4–39D), whose column is covered with berry-like clusters of nematocyst batteries, and the Caribbean sea anemone *Lebrunia danae.* The West Australian *Dofleina armata* is believed to be the most toxic sea anemone.

Sea anemones generally feed on various invertebrates, but large species can capture fish. Species with delicate tentacles lodge their bodies in protective crevices. Many large intertidal species feed on crabs and bivalves washed down from higher intertidal levels by waves or predators. The prey is caught by the tentacles, paralyzed by nematocysts, and carried to the mouth. The mouth is opened by radial muscles in the septa, and the prey is swallowed.

When the prey passes from the pharynx, which is richly supplied with gland cells, into the lower part of the gastrovascular cavity, it is brought into contact with the free edges of many septa and also, when present, the acontia. The septal filaments form, in effect, a closely fitting bag around the food. They produce the enzymes for extracellular digestion of proteins and fats. These septal filaments also appear to be the principal site of intracellular digestion and absorption.

Some large sea anemones with short tentacles, such as *Stichodactyla* and *Radianthus,* feed on fine particles. Planktonic organisms are trapped on the surface of the column and tentacles. Cilia on the surface of the column beat toward the oral disc, and cilia on the tentacles beat toward the tentacle tips. The tentacles then bend over and deposit the food in the mouth.

In the Indo-Pacific, little fish of the genus *Amphiprion* (clown fish or anemone fish) live symbiotically among the tentacles of large sea anemones. Juvenile fish are recruited to their sea anemone home

and remain there by species-specific attractant compounds released by the sea anemone. The mucous coat on the surface of the fish does not contain the compounds that activate the cnidocyte receptors of the sea anemone, making it possible for the fish to live in an otherwise lethal habitat. The sea anemone provides protection and some food scraps; the fish in turn may bring some food to the sea anemone, protects the sea anemone from some predators, removes necrotic tissue, and by its swimming and ventilating movements, reduces fouling of the anemone by sediment of various sorts. Other commensals of sea anemones include, amphipods, cleaning shrimps (p. 696), snapping shrimps, arrow crabs, brittle stars, and various fish.

Symbiotic zooxanthellae or zoochlorellae or even both are found in the gastrodermal cells, especially in the tentacles and the oral disc of many sea anemones. The color variation of *Anthopleura elegantissima* is determined in large part by the predominance of zoochlorellae or zooanthellae in the tissues. The relationship is similar to that described for corals on page 148.

The muscular system in sea anemones is much more specialized than that in the other cnidarian classes. The longitudinal, epidermal fibers of the column and pharynx have largely disappeared, except in primitive species. They are present, however, in the tentacles and oral disc. Thus, the muscular system is primarily gastrodermal. Bundles of longitudinal fibers in the septa form retractor muscles for shortening the column (Fig. 4–41). Circular muscle fibers in the columnar gastrodermis are well developed. The presence of complete septa may faciliate the function of the internal hydraulic skeleton by limiting the maximum extent of the diameter of the column when the retractor muscles contract. Also important are radial muscles in the complete septa that, on contraction, open the pharynx. During contraction of the column, water can be released from pores at the tip of the tentacles, through the mouth, or through a pore at the base of the column in some burrowing species.

Although sea anemones are essentially sessile animals, many species are able to change locations by slow gliding on the pedal disc, by crawling on the side of the column, or by walking on the tentacles. A few species can detach the pedal disc and swim briefly with lashing motions of the column or tentacles. Such swimming is used to escape predators, such as sea stars. Burrowing sea anemones slowly insert the body column into sand or mud by peristaltic contractions, which change the column diameter. Members of the genus *Minyas* are planktonic and hang upside down from a chitinous float secreted by the pedal disc.

The nervous system exhibits the typical cnidarian pattern, and no specialized sense organs are present.

Some species of sea anemones exhibit aggressive behavior toward members of other clones or toward other species. Cnidocytes on specialized searching tentacles or on column projections (**acrorhagi**) are fired on contact with the "foreign" sea anemone. There is some withdrawal between combatants, and one or both parties may suffer tissue damage. Such aggressive behavior apparently provides for some spatial separation between species or clones.

Asexual reproduction is common in sea anemones. One method is by **pedal laceration**, in which parts of the pedal disc are left behind as the animal moves. In some instances the disc puts out lobes that pinch off. These detached portions then regenerate into small sea anemones. Many sea anemones reproduce asexually by longitudinal fission, and a few species do so by transverse fission.

Most sea anemones are hermaphroditic but produce only one type of gamete during any one reproductive period. The gonads are located in the gastrodermis on all or certain of the septa in the form of longitudinal, bandlike cushions behind the septal filament (Fig. 4–41A).

The eggs may be fertilized in the gastrovascular cavity, with development taking place in the septal chambers, or fertilization may occur outside the body in the sea water. The planula larva may be plank-totrophic (feeding) or lecithotrophic (getting nutrition from yolk) and has a variable larval life span. Septa develop from the column wall and grow toward the pharynx. There are still no tentacles, and the young sea anemone lives as a ciliated ball, unattached and free-swimming. With further development, the young polyp settles, attaches, and forms tentacles. Studies indicate that a New Zealand intertidal sea anemone (*Actinia tenebrosa*) requires 8 to 66 years to reach a column diameter of 40 mm and has an average longevity of 50 years.

Ceriantharians and Zoanthideans

Two small orders of anemone-like anthozoans contain a number of commonly encountered species. The order Ceriantharia includes the "tube-dwelling" sea anemones. These large, solitary anthozoans are adapted for life in soft bottoms. The body is lodged in a heavy, secreted tube, which is buried within the substratum. This tube is formed of mucus and fired threads and capsules of **ptychocysts**, nematocyst-like organelles. When feeding, the animal projects the tentacles and oral disc from the surface or elevated opening of the tube (Fig. 4–43).

Members of the order Zoanthidea are largely tropical, and some are common reef inhabitants. Most are 1 to 2 cm in diameter, and the majority are colonial, with a connecting stolon or a common mat. The body may be columnar, but many species are rather short and button-like (Fig. 4–44). A short fringe of tentacles surrounds the broad oral disc. The column and mat are covered by a thick cuticle, and many species have sand or other debris embedded in the

FIGURE 4–43 *Cerianthus,* a large, burrowing anthozoan. The animal secretes a tube into which the body can be retracted. The members of this order (Ceriantharia) are similar to sea anemones in size and general appearance. *(Photograph courtesy of Buchsbaum, R. M., and Milne, L. J. 1961. The Lower Animals. Chanticleer Press, New York.)*

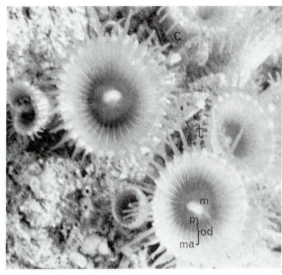

A   B

FIGURE 4–44 Order Zoanthidea, colonial and semicolonial anemone-like anthozoans. **A,** Contracted individuals of *Palythoa*. In the tropics species of *Palythoa* commonly carpet rocks in shallow water. **B,** Expanded specimens of *Palythoa psammophilia*. The letters indicate mouth (m), oral disc (od), peristome (p), marginal area (ma), and tentacles (t). *(A, Photograph courtesy of Betty M. Barnes. B, From Reimer, A. A. 1971. Specificity of feeding chemoreceptors in Palythoa psammophilia. Comp. Gen. Pharm., Pergamon Press. 2(8):383–396.)*

surface layer. Some reef species pave rocks or form large encrusting masses and harbor zooxanthellae (Fig. 4–44). Some are commensal.

Stony, or Scleractinian, Corals

Closely related to sea anemones are the stony, or scleractinian, corals (also called madreporarian corals), which constitute the largest order of anthozoans. In contrast to sea anemones, stony corals produce a calcium carbonate skeleton. Some corals, such as the Indo-Pacific reef-inhabiting *Fungia* and some deep-sea species, are solitary and have polyps as large as 25 cm in diameter (See Fig. 4–47E), but the majority are colonial with small polyps averaging 1 to 3 mm in diameter (Figs. 4–45 and 4–49D). However, the entire colony may become large. Coral polyps are very similar in structure to sea anemones but do not possess siphonoglyphs.

The skeleton is composed of calcium carbonate and is secreted by the epidermis of the lower half of the column as well as by the basal disc. This secreting process produces a skeletal cup, within which the polyp is fixed. The floor of the cup contains thin, radi-

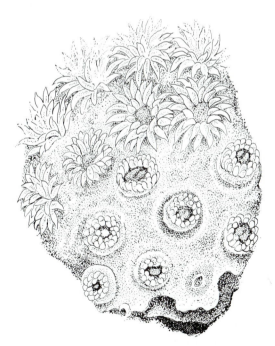

FIGURE 4–45 Part of a coral colony. The polyps, some contracted and some expanded, are connected by a lateral sheet of tissue.

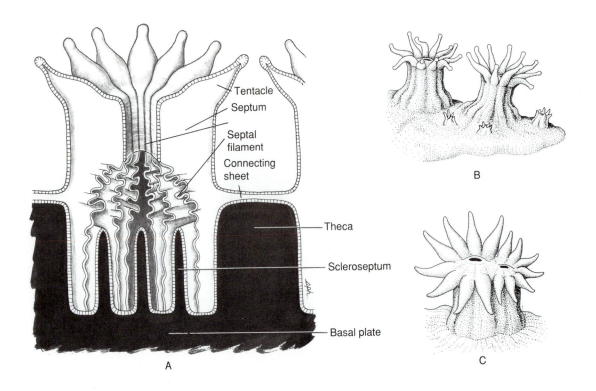

FIGURE 4–46 A, A coral polyp in its theca (longitudinal section). **B,** Extratentacular budding. **C,** Intratentacular budding. *(A, Modified after Hyman, L. H. 1940. The Invertebrates. Vol. I. McGraw-Hill Book Co., New York.)*

ating, calcareous **sclerosepta** (Figs. 4–46A and 4–47C and F). Each scleroseptum projects upward into the base of the polyp, folding the basal layers and inserting them between a pair of internal tissue septa. As long as a colony is alive, calcium carbonate is deposited beneath the living tissues.

In addition to providing a uniform substratum on which the living colony can attach, the skeleton (and especially the scleorsepta) also serves as protection. When contracted, the polyps project little above the skeletal platform and are difficult for most fish and other predators to remove.

The polyps of colonial corals are all interconnected by lateral folds of the column wall folds, which connect with similar folds of adjacent polyps.

Thus, all the members of the colony are connected by a horizontal sheet of tissue (Figs. 4–45 and 4–46). Because this sheet represents a fold of the body wall, it contains an extension of the gastrovascular cavity as well as an upper and lower layer of gastrodermis and epidermis. The lower epidermal layer secretes the part of the skeleton that is located between the cups in which the polyps lie. The living coral colony, therefore, lies entirely above the skeleton and completely covers it.

The skeletal configurations of various species of corals are due in part to the growth pattern of the colony and in part to the arrangement of polyps in the colony (Figs. 4–47 and 4–49). Some species form flat or rounded skeletal masses; others have an upright

FIGURE 4–47 A, Oral surface of a living brain coral. Note the row of mouths. **B,** Skeleton of the brain coral *Diploria.* Note the arrangement of scleorsepta. **C,** Skeleton of *Montastrea.* In life the polyps are in the large, distinct cups. **D,** Skeleton of lettuce coral, *Agaricia,* in which the polyps are arranged in rows. **E,** Skeleton of *Fungia,* a solitary coral of the Indo-Pacific. The skeleton of this very large polyp is limited to scleorsepta projecting from the basal plate. There is no wall, hence no cup. **F,** Solitary deep-water coral. *(A, Photograph courtesy of Catala, R. L. 1964. Carnival under the Sea. R. Sicard, Paris. © R. L. Catala. B–E, Photographs courtesy of Betty M. Barnes. F, Photograph courtesy of Katherine E. Barnes.)*

FIGURE 4–48 Part of a colony of a species of *Seriatopora,* an Indo-Pacific genus of delicate branching corals. Specimen is 6 cm across. *(Photograph courtesy of Katherine E. Barnes.)*

A

B

C

D

FIGURE 4–49 Scleractinian, or stony, corals. **A,** Brain coral on the Great Barrier Reef, Australia. **B,** Staghorn coral from the Great Barrier Reef. **C,** Knobbed and lettuce coral on a Bahamian reef. **D,** Cup corals. *(A, Photograph courtesy of Fritz Goro. B, Photograph courtesy of Allen Keast. From Buchsbaum, R. M., and Milne, L. J. 1961. The Lower Animals. Chanticleer Press, New York. C, Photograph courtesy of John Storr. D, Photograph courtesy of D. P. Wilson.)*

and branching growth form. Some are large and heavy; others are small and delicate (Fig. 4–48). When the polyps are well separated, the coral skeleton has a pitted appearance, as in *Oculina,* the eyed coral, in *Astrangia,* one of the corals living along the North Atlantic coast of the United States, and in the reef coral *Montastrea* (Fig. 4–47C). The polyps of brain coral are arranged in rows (Figs. 4–47A and B, 4–49A). The rows are well separated, but the polyps in each row are fused together so that their cups are confluent. As a result, the skeleton of the colony has the appearance of a human brain, containing troughs or valleys separated by skeletal ridges.

The coral colony expands by the budding of new polyps from the bases of old polyps (Fig. 4–46B) or from the oral discs of old polyps. In the latter case, the oral disc of the parent lengthens in one direction (Fig. 4–46C). Gradually, the oral disc constricts, and the separation extends down the length of the column to form two new polyps. The budding process is accompanied by simultaneous changes in the deposition of the underlying sclerosepta. Brain corals arise by intratentacular budding, in which the oral discs and columns never constrict after new mouths are formed. Thus, the polyps in a row of brain coral share a common oral disc bearing many mouths (Fig. 4–47A).

Calcium carbonate is continually deposited by the basal epidermis of the living colony that rests on it. In many corals the polyps periodically lift their bases and secrete a new floor to their cup. This closes off a minute chamber in the skeleton (Fig. 4–50A). The growth rate varies greatly, depending on the species and water temperature. Many dome and plate corals grow only 0.3 to 2 cm a year through vertical or linear deposition of calcium carbonate. Some branched corals, on the other hand, grow rapidly, increasing in the linear direction as much as 10 cm per year.

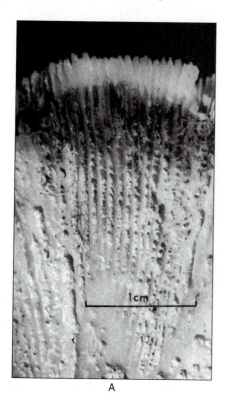

A

B

FIGURE 4–50 A, Minute chambers in coral skeleton that result from the lifting of the polyp and base and subsequent secretion of a new underlying calcareous plate. **B,** X-ray of a radial section through a 15-year-old specimen of *Porites lobata* from the Great Barrier Reef, Australia. One year of growth is represented by a dark band (low-density skeleton, formed in winter) and a light band (high-density skeleton, formed in summer.) *(A, Photograph courtesy of Katherine E. Barnes. B, From Isdale, P. 1977. Variation in growth rate of hermatypic corals in a uniform environment. Proceedings of the 3rd International Coral Reef Symposium, University of Miami, Miami, FL. 2(Geology):406.)*

The density of the secreted calcium carbonate is not the same throughout the year, the change being governed by seasonal shifts of temperature and light. Thus, many coral skeletons exhibit seasonal growth bands like tree rings, which can show up on x-radiographs and be used to determine the age and growth rate of the coral (Fig. 4–50B).

Although coral species are restricted to various parts of a reef depending on environmental conditions, at any particular zone there is often competition for space. If a coral touches another species, it may extrude its septal filaments and digest the intruder's tissues. Or, like certain sea anemones, it may use specialized tentacles. Aggressive responses depend on the species, some being very aggressive, others only slightly so.

Corals feed like sea anemones, and depending on the size of the polyps, the prey ranges from small fish down to small zooplankton. When expanded, the outstretched tentacles of adjacent polyps present a broad, continuous mesh that prey might touch. In addition to capturing zooplankton, many corals also collect fine particles in mucous films or strands, which are then driven by cilia to the mouth. Some, such as the foliaceous agaricids (Fig. 4–47D), which have reduced or no tentacles, are entirely mucus suspension feeders. Corals produce large quantities of mucus. *Acropora acuminata* has been estimated to release in mucus as much as 40% of the daily fixed carbon received from its symbionts.

Although there are many exceptions, most Caribbean corals feed at night and are contracted during the day (Fig. 4–51), and some species display a persistent rhythm of expansion, even when kept in constant darkness or light. Many Indo-Pacific corals feed during both day and night.

Over 60 genera of corals contain symbiotic zooxanthellae within the gastrodermal cells. Deepwater and some cold-water corals lack zooxanthellae, but virtually all reef-dwelling (hermatypic) corals possess them. The algae may reach such concentrations as to account for 50% of the protein nitrogen of the coral. The algal symbionts give most of their coral host a yellow-brown to dark brown color. Coral bleaching,

A   B

FIGURE 4–51 *Favia favus* from the Great Barrier Reef, Australia during the day (**A**) and at night (**B**). *(From Vernon, J. E. N., Pichon, M., and Wijsman-Best, M. 1977. Scleractinia of Eastern Australia. Part II. Australian Institute of Marine Science Monograph. Series 3, p. 29.)*

when corals expel their zooxanthellae, has been observed in some areas in recent years. In localities where bleaching is taking place, every coral or even an entire colony may not expel its symbionts, but the condition may be visibly conspicuous. Coral bleaching has been shown to be correlated with seasonal elevated water temperatures and may be evidence of a global rise in temperature. Although the coral is stressed by the loss of its symbionts, it can recover if the condition is not prolonged.

Our knowledge of the physiological relationship between hermatypic corals and their symbiotic algae has grown considerably in recent years. The nutritive needs of the coral are supplied in part by the prey on which it feeds and in part by its algal symbionts. A large portion of the carbon fixed by the algae in photosynthesis is passed to the coral, largely in the form of glycerol but including glucose and alanine. The food caught by the coral probably supplies both coral and algae with nitrogen, which is then cycled back and forth between the two. The zooxanthellae can also obtain inorganic nutrient ions directly from the sea water, but in nutrient-poor water, which characterizes most coral reefs, the contribution to the zooxanthellae from coral feeding is probably important. The symbiosis also facilitates the deposition of the coral skeleton, because corals that are deprived of their algae or kept in the dark deposit calcium carbonate at a much slower rate than under normal conditions. The degree to which the coral depends on the algae varies by species and even within species populations.

Sexual reproduction is similar to that in the sea anemones, and there are both dioecious and hermaphroditic species. The planula, which is produced by sexual reproduction, attaches and the subsequent first polyp (which develops by asexual budding) becomes the parent of all other members of the colony.

Corals are subject to injury or death from storms, extremely low tides, predation, and disease. The living colony can regenerate about 1 cm of destroyed tissue but not much more. In some branching corals, however, such as the Caribbean elkhorn coral *(Acropora palmata),* whole colonies may regenerate from broken fragments. White band and black band disease is caused by microorganisms and can produce tissue death that is more permanent. Disease is exacerbated where pollution raises nutrient levels in the surrounding water.

Octocorals

Sea anemones and corals possess 12 or more tentacles and septa among other similarities and are grouped in the subclass Hexacorallia, or Zoantharia. Most of the remaining anthozoans, including such common marine forms as sea pens, sea pansies, sea fans, whip corals, and pipe corals, form the subclass Octocorallia. The Octocorallia possess a number of distinctive features. Octocorallians always have eight tentacles, which are **pinnate,** that is, they possess side branches, as does a feather (Fig. 4–52). There are always eight complete septa, one on each side of a tentacle base, and this may be the primitive anthozoan condition. Only one siphonoglyph is present.

The octocorallians are colonial cnidarians, and the polyps are usually small and similar to those of stony corals. The polyps of an octocorallian colony are connected by a mass of tissue called a **coenenchyme** (Fig. 4–52). This tissue consists of a thick mass of mesoglea, perforated by gastrodermal tubes that are continuous with the gastrovascular cavities of the polyps. The surface of the entire fleshy mass is covered by a layer of epidermis, which joins the epidermis of the polyp column. Only the upper portion of the polyp projects above the coenenchyme (Fig. 4–52).

The amebocytes of the mesoglea secrete calcareous skeletal material that supports the colony. Thus, the skeleton of the Octocorallia is internal and is an integral part of the tissue. This arrangement is in sharp contrast to that of the stony corals, whose skeletons are entirely external. The octocorallian skeleton may be composed of separate or fused calcareous **spicules** or of a horny material. The skeleton functions in support, but in some species, such as soft corals, it can also deter predators.

Among the most familiar of the octocorallians are the gorgonian, or horny, corals of the order Gorgonacea (Fig. 4–53), which include the whip corals, sea feathers, sea fans, and precious red coral *(Corallium).* Gorgonians are common and conspicuous members of reef faunas, especially in the West Indies (Fig. 4–52). The body of most gorgonian corals contains a central **axial rod** composed of an organic substance called **gorgonin** (a tanned collagen). The axial rod is commonly impregnated with calcium carbonate, and in some species a calcified section alternates with a noncalcified section, so the rod is jointed.

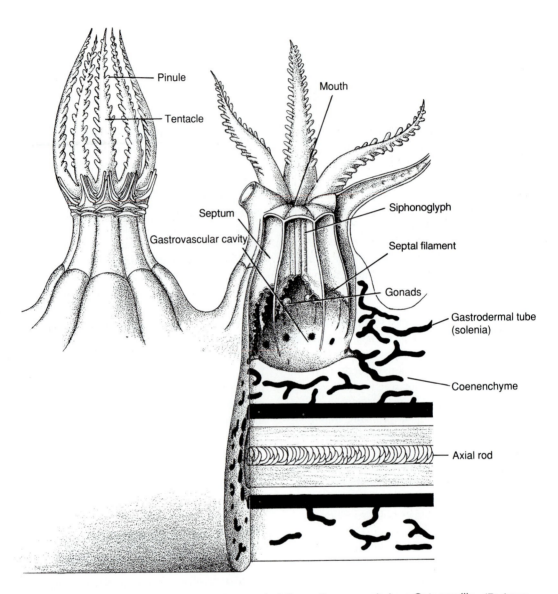

FIGURE 4–52 Structure of a gorgonian coral of the anthozoan subclass Octocorallia. *(Redrawn from Bayer, F. M. 1956. Octocorallia. In Moore, R. C., Treatise on Invertebrate Paleontology. Courtesy of Geological Society of America and University of Kansas, Lawrence. p. F169.)*

Around the axis is a cylinder of coenenchyme and polyps (Fig. 4–52). The coenenchyme contains embedded calcareous ossicles or spicules of different shapes and colors (Fig. 4–54). It is the color of the calcareous skeletal components that accounts for the yellow, orange, or lavender color of some species. The yellow-brown color of many reef species results from the presence of symbiotic zooxanthellae. The

colonies of most gorgonian corals are erect branching rods and are thus rather plantlike (Fig. 4–53). Whip corals consist of slightly branched, long, cylindrical filaments about 5 mm in diameter. Many gorgonians are branched only in one plane, and in sea fans the branches may be connected by cross bars to form a lattice (Figs. 4–53B and C). Sea fans are usually oriented at right angles to the water current. In general,

A

B

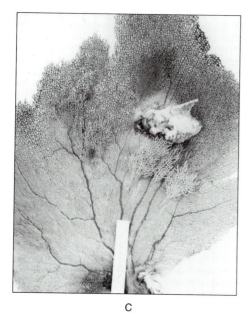

C

FIGURE 4–53 Gorgonian corals (order Gorgonacea): **A,** Sea rods, the most common gorgonian growth form. **B,** A species of sea fan in which there is extensive branching in one plane but branches do not interconnect. **C,** *Gorgonia ventilina,* the common West Indian sea fan, in which the branches anastomose to form a lattice. Rule is 15 cm. A winged oyster *(Pteria)* is attached to the fan. *(A, Photograph courtesy of T. Parkinson. B and C, Photographs courtesy of Katherine E. Barnes.)*

the design of most gorgonian corals requires a small amount of surface area for attachment but provides a large surface area for feeding, as a consequence of the branching vertical development. The flexible skeleton permits bending in currents. Depending on the species, gorgonians reach the maximum size their architecture will permit in 10 to 30 years.

Gorgonian polyps are more commonly expanded during the day than during the night. They have been assumed to feed on zooplankton like most scleractinian corals, but studies on a number of gorgonians and other octocorallians indicate that they have few cnidocytes and feed on smaller particles than zooplankton.

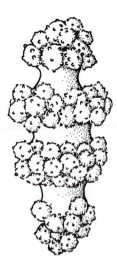

FIGURE 4–54 Spicule of a gorgonian coral.

Gorgonians harbor many symbiotic animals that are either attached to or crawl over the gorgonian surface—colonial tunicates, barnacles, bivalves, snails, and gobies. Some take on the colors of their gorgonian host.

The tropical Indo-Pacific organ-pipe coral, *Tubipora,* which belongs to another group of octocorallians, is differently organized (Fig. 4–55). Its long, parallel polyps are encased in calcareous tubes of fused spicules. The tubes are connected at intervals by transverse, calcareous plates, or platforms.

FIGURE 4–55 Skeleton of organ-pipe coral, a member of the small tropical octocoral order Stolonifera. *(Photograph courtesy of Betty M. Barnes.)*

The sea pens and sea pansies (order Pennatulacea) are inhabitants of soft bottoms and are quite different from coral-like members of the Octocorallia in that they have a large, primary polyp with a stemlike base, which is anchored in the sand. The body is fleshy, although the coenenchyme contains calcareous spicules, or ossicles. The upper part of this primary polyp gives rise to secondary polyps, the more typical and conspicuous of which are termed **autozooids.** Highly modified polyps, called **siphonozooids,** pump water into the interconnected gastrovascular cavities, thereby keeping the colony turgid and erect. In the sea pens, the primary polyp is elongated and cylindrical. In the sea pansy, *Renilla* (Fig. 4–56A), which is common along the Atlantic, Gulf, and southern California coasts of the United States, the primary polyp is leaflike, with the secondary polyps limited to the upper surface. The flattened horizontal surface reduces resistance to the turbulent near-shore waters in which these colonies live. Moreover, they are able to uncover themselves when buried by shifting sediment and reanchor themselves when dislodged. There are also deep-sea species, such as *Umbellula,* which lives on the Atlantic abyssal plain, and looks like a pinwheel (secondary polyps) mounted at the end of a long stalk (primary polyp) embedded in the substratum (Fig. 4–56B).

Members of the order Alcyonacea are known as soft corals, or leather corals. These octocorallians possess soft fleshy or leathery colonies that may reach a large size, 1 m in diameter in the case of *Sarcophyton.* The colonies are irregular in shape, some encrusting and some massive, often with lobate or finger-like projections (Fig. 4–57). Separate calcareous spicules are embedded in the coenenchyme. Alcyonacean corals are often conspicuous on Indo-Pacific reefs. In some respects they resemble scleractinian corals, with the fleshy coenenchymal mass acting as the substratum for the colony instead of calcium carbonate. There are also some cold-water species.

SUMMARY OF ANTHOZOA

1. Members of the class Anthozoa are polypoid cnidarians; the medusoid stage is entirely lacking.

2. The anthozoan polyp is more specialized than that of hydrozoans, and its cellular mesoglea, septate gastrovascular cavity, cnidocytes in gastric

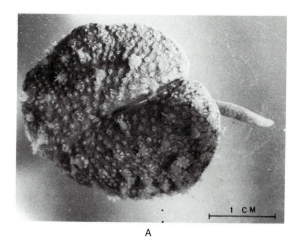

FIGURE 4–56 Order Pennatulacea. **A,** Sea pansy, *Renilla.* **B,** *Umbellula,* a strange deep-sea octocoral related to sea pansies. Photograph was taken at a depth of over 5000 m some 350 miles off the west coast of Africa. Stalk is estimated to be about a meter in length. *(A, Photograph courtesy of Betty M. Barnes. B, Photograph courtesy of Walter Jahn, U. S. Naval Oceanographic Service.)*

FIGURE 4–57 Soft corals of the order Alcyonacea. **A,** Soft corals (pale) and scleractinian corals (dark) from Fiji. Mound of corals is about 2 m across and about 2 m below the water surface. **B,** Part of a colony of *Capnella gaboensis* from Sidney Harbour, Australia, showing polyps. *(Both photographs courtesy of Penny Farrant.)*

filaments, and gastrodermal gonads indicate a closer phylogenetic relationship with the Scyphozoa than with the Hydrozoa.

3. The difference in the body form of the Scyphozoa and the Anthozoa (medusa versus polyp) may be reconciled if the anthozoans are derived through the polypoid stage of scyphozoans.

4. The two subclasses, the Zoantharia and the Octocorallia, reflect different levels of structural evolution within the Anthozoa. The Octocorallia have retained an arrangement of eight complete septa and eight tentacles, which may be the primitive anthozoan condition. Colonial organization is characteristic of almost all octocorallians, and the polyps are interconnected through a complex mass of mesoglea and gastrodermal tubes. The Zoantharia display a more complex system of septa, arranged in multiples of six (usually at least 12). There are many solitary forms, and colonial species are connected by more or less simple outfoldings of the body wall.

5. Sea anemones are the principal group of solitary anthozoans, and perhaps because of their solitary condition, many species have evolved a larger size than most other anthozoan polyps. The number and complexity of their septa, providing a large surface area of gastric filaments, may be related to the utilization of larger prey.

6. The majority of anthozoans are colonial, and this type of organization has evolved independently a number of times within the class. Although colonies may reach a large size, the individual polyps are generally small. There are some groups with polymorphic colonies, but this condition is not as widespread as in the hydrozoans.

7. Scleractinian corals, although similar to sea anemones, are largely colonial and are unique in their secretion of an external calcareous skeleton. The skeleton provides the colony with a uniform substratum on which the living colony rests. The sclerosepta may contribute to the adherence of the polyps within the thecal cups and provide some protection against grazing predators when the polyps are withdrawn.

8. The majority of scleractinian corals are tropical reef inhabitants (hermatypic) and harbor zooxanthellae. Zooxanthellae are found in many other anthozoans as well as some scyphomedusae and some hydrozoans.

9. The colonial alcyonaceans, or soft corals, which are most abundant on Indo-Pacific reefs, in many ways parallel the scleractinian corals, for the massive coenenchymal mass forms the substratum from which the individual polyps arise.

10. The branching, rodlike colonies of gorgonian corals are adapted for exploiting the vertical water column while using only a small area of the substratum for attachment. Flexible support is provided by a central, organic axial rod and separate calcareous spicules embedded in the coenenchyme.

11. The pennatulaceans, which include sea pens, sea feathers, and sea pansies, are adapted for life on soft bottoms. A large, primary polyp, which determines the form of the colony, not only provides anchorage in the sand but also acts as the substratum from which the small, secondary polyps arise.

12. A planula larva is characteristic of most anthozoans and develops into the polyp. Colonial forms are derived by budding from the first polyp.

THE EVOLUTION OF CNIDARIANS

Scyphozoans and anthozoans display so many similar features that they are believed to be more closely related to each other than either is to the hydrozoans. Scyphozoans and anthozoans both have a mesoglea containing cells, gastrodermal cnidocytes, gastrodermal gonads, and a septate gastrovascular cavity (at least in the scyphozoan polyp). How then can we reconcile the medusoid form of the scyphozoan adults with the polypoid form of the anthozoans? It is probable that the anthozoans are derived from an ancestral stock common to both scyphozoans and anthozoans. The scyphozoans retained the medusa as the adult form. In the line leading to the anthozoans, the medusa became suppressed. Although the coronate schypozoans are probably not the ancestors of anthozoans, certain species that have minute sexually precocious medusae (Fig. 4–37), illustrate a step in the kind of suppression that may have occurred. Note the importance that can be made of comparative life cycles as evidence for cnidarian evolution.

The most primitive cnidarian life cycle, as we postulated at the end of the section on the hydrozoans, could have been one in which medusae were the sexually reproducing adults, and the late larval stage (actinula) was planktonic and developed directly into the medusa. Some hydrozoans possess such a life cycle (Fig. 4–24A). The hydrozoans as a whole exhibit a number of features that could be argued are more primitive than those found in scyphozoans and anthozoans: the mesoglea is acellular and in polypoid forms is a basal lamina on which a thicker, cellular mesoglea is believed to have been elaborated. Cnidocytes are limited to the epidermis, and the gastrovascular cavity is simple, not septate.

Let us assume, as do some zoologists, that the anthozoan polyp is the most primitive cnidarian body form. What does this premise demand? Hydrozoans become the most advanced cnidarian class and the simpler features we have just enumerated would all have to be secondarily derived (i.e., by reduction). The primitive symmetry of cnidarians would be biradial, not radial. The medusoid form would have to have evolved independently in hydrozoans, scyphozoans, and perhaps cubozoans. Just how this would have occurred to bring about the transfer of gamete-producing tissue to a medusa is not clear. However, even if the medusa evolved only once, it is difficult to explain the presence of gametes within a medusa where a polyp is also present in the life cycle. This evolution would have had to have been followed by subsequent medusoid suppression, because such suppression has clearly occurred in various cnidarians. Finally, a polypoid starting point begs the question of cnidarian origins. Motility is a basic animal feature. What would have been the origin of a sessile polypoid cnidarian ancestor?

SYSTEMATIC RÉSUMÉ OF THE PHYLUM CNIDARIA

Class Hydrozoa. Cnidarians having a polypoid, medusoid, or both forms in the life cycle. Mesoglea acellular. Cnidocytes confined to the epidermal layer. Gonads epidermal, or if gastrodermal, gametes do not escape through the coelenteron and mouth.

Order Trachylina. Medusoid hydrozoans lacking a polypoid stage. Medusa develops directly from an actinula. This order contains perhaps the most primitive members of the class. *Liriope, Aglaura.*

Order Hydroida. Hydrozoans with a well-developed polypoid generation. Medusoid stage present or absent. The majority of hydrozoans belong to this order.

Suborder Limnomedusae. Mostly freshwater hydrozoans possessing small solitary polyps and free medusae. The marine *Gonionemus;* the freshwater *Craspedacusta.*

Suborder Anthomedusae. Skeletal covering, when present, does not surround hydranth (athecate). Free medusae, which are tall and

bell-shaped, are often present. *Tubularia, Pennaria, Syncoryne, Eudendrium, Hydractinia, Polyorchis, Branchiocerianthus,* the freshwater hydras; the hydrocoral *Millepora.*

Suborder Leptomedusae. Hydranth surrounded by a skeleton (thecate). Free medusae are commonly absent, but when present, they are more or less flattened. *Obelia, Campanularia, Abietinaria, Sertularia, Plumularia, Aglaophenia, Aequorea.*

Suborder Chondrophora. Pelagic, polymorphic, polypoid colonies. (These cnidarians can also be interpreted as large, single, inverted polyps.) *Velella* (by-the-wind sailor), *Porpita.*

Order Actinulida. Tiny, solitary hydrozoans resembling actinula larvae. Unlike other mature polypoid and medusoid hydrozoans, the epidermal cells are ciliated. No medusoid stage present. Interstitial inhabitants. *Halammohydra, Otohydra.*

Order Siphonophora. Pelagic hydrozoan colonies of polypoid and medusoid individuals. Colonies with floats or large swimming bells. Largely in warm seas. *Physalia* (Portuguese man-of-war), *Stephalia, Nectalia.*

Order Stylasterina. Hydrocorals having a thick layer of tissue overlying the skeleton. Defensive and feeding polyps located within star-shaped openings on the skeleton. *Stylaster, Allopora.*

Class Scyphozoa. Cnidarians in which the medusoid form is dominant; the polypoid form is small. Velum absent and stomach is primitively tetraseptate. Mesoglea is cellular; some cnidocytes are gastrodermal; gametes are gastrodermal.

Order Stauromedusae, or Lucernariida. Sessile polypoid scyphozoans attached by a stalk on the aboral side of the trumpet-shaped body. Chiefly in cold littoral waters. *Haliclystus, Craterolophus, Lucernaria.*

Order Coronatae. Bell of medusa with a deep encircling groove or constriction, the coronal groove, extending around the exumbrella. Many deep-sea species. *Periphylla, Stephanoscyphus, Nausithoe, Linuche, Atolla.*

Order Semaeostomeae. Scyphomedusae with bowl-shaped or saucer-shaped bells having scalloped margins. Manubrium divided into four oral arms. Gastrovascular cavity with radial canals or channels extending from central stomach to bell margin. Occur throughout the oceans of the world, especially along coasts. *Cyanea, Pelagia, Aurelia, Chrysaora, Stygiomedusa.*

Order Rhizostomeae. Bell of medusa lacking tentacles. Oral arms of manubrium, branched and bearing deep folds into which food is passed. Folds, or "secondary mouths," lead into arm canals of manubrium, which pass into stomach. Original mouth lost through fusion of oral arms, except in *Stomolophus.* Mostly tropical and subtropical shallow-water scyphozoans. *Cassiopea, Rhizostoma, Mastigias, Stomolophus.*

Class Cubozoa. Medusoid cnidarians with bells having four flattened sides. Bell margin simple and bearing a velum and four tentacles or tentacle clusters. An attached polypoid stage arises from the planula. The Cubozoa were formerly considered a class of Scyphozoa, however, the presence of a velum, the possession of a type of nematocyst (stenotele) found only in hydrozoans, the lack of a notched bell margin and rhopalia are considered evidence that they are not closely related to the other scyphozoans and should be placed within a separate class. Tropical and subtropical oceans. *Carybdea, Chiropsalmus, Chironex.*

Class Anthozoa. Cnidarians having only the polypoid form in their life history. Body wall around mouth infolded to form a sleevelike pharynx. Coelenteron partitioned by gastrodermal septa. Mesoglea cellular; some cnidocytes are gastrodermal; gonads gastrodermal.

Subclass Octocorallia, or Alcyonaria. Polyp with eight pinnate tentacles and eight septa. Almost entirely colonial, and polyps usually connected by coenenchyme.

Order Stolonifera. No coenenchymal mass; polyps arising from a creeping mat or stolon. Skeleton of calcareous tubes of separate or fused calcareous spicules, or horny external cuticle. Tropical and temperate oceans in shallow water. *Tubipora* (organ-pipe coral), *Clavularia.*

Order Telestacea. Lateral polyps on simple or branched stems. Skeleton of calcareous spicules. *Telesto.*

Order Alcyonacea. Soft corals. Coenenchyme forming a rubbery mass. Colony may have a massive mushroom shape or an encrusting growth form. Skeleton of separate calcareous spicules. Largely tropical. *Alcyonium, Gersemia, Sarcophyton.*

Order Helioporacea. Contains the Indo-Pacific blue coral, *Heliopora,* having a massive, blue calcareous skeleton.

Order Gorgonacea. Horny corals or gorgonian corals. Common tropical and subtropical octocorallian cnidarians having a largely, upright, plantlike growth form and an axial skeleton of a horny organic material. Separate or fused calcareous spicules may also be present. *Gorgonia* (sea fan), *Leptogorgia* (sea whip), *Corallium* (precious red coral), *Muricea* (sea rod).

Order Pennatulacea. Sea pens. Colony having a fleshy, flattened or elongated body, or rachis. Skeleton of calcareous spicules. *Stylatula* (sea pen), *Veretillum, Renilla* (sea pansy), *Umbellula.*

Subclass Hexacorallia, or Zoantharia. Polyps with more than eight tentacles and septa, typically in cycles of 12. Tentacles rarely pinnate. Solitary or colonial.

Order Zoanthidea. Small, anemone-like anthozoans having one siphonoglyph and no skeleton. Solitary or colonial. *Palythoa* and *Zoanthus.* Some, such as *Epizoanthus* and *Parazoanthus,* epizoic on other invertebrates.

Order Actiniaria. Sea anemones. Solitary anthozoans with no skeleton, with septa in hexamerous cycles, and usually with two siphonoglyphs. *Halcampoides, Edwardsia, Metridium, Epiactis, Stichodactyla.*

Order Scleractinia, or Madreporaria. Stony corals. Mostly colonial anthozoans secreting a heavy, external, calcareous skeleton. Sclerosepta arranged in hexamerous cycles. *Fungia, Acropora, Porites, Astrangia, Oculina.* Many fossil species.

Order Corallimorpharia. Solitary and some colonial species with tentacles often in radiating rows. Resemble true corals but lack skeletons. *Corynactis, Discosoma, Ricordea.*

Order Ceriantharia. Anemone-like anthozoans with greatly elongated bodies adapted for living within secreted tubes buried in sand or mud. One siphonoglyph; numerous septa all complete. *Cerianthus, Ceriantheopsis.*

Order Antipatharia. Black or thorny corals. Gorgonian-like species with upright, plantlike colonies. Polyps arranged around an axial skeleton composed of a black, horny material and bearing thorns. Largely in deep water in tropics. *Antipathes.*

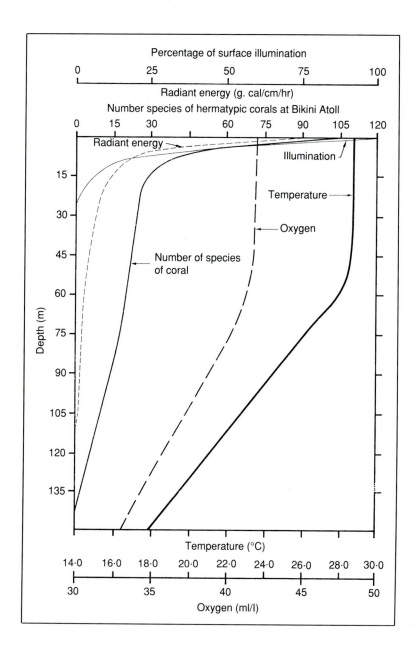

Percentage of surface illumination

Radiant energy (g. cal/cm/hr)

Number species of hermatypic corals at Bikini Atoll

FIGURE 4–58 Relationship between the vertical distribution of hermatypic corals at Bikini Atoll and environmental factors. *(After Wells and Motoda. In Stoddard, D. R. 1969. Biol. Rev. 44:433–498. Copyrighted and reprinted by permission of Cambridge University Press.)*

CORAL REEFS

Coral reefs are tropical, shallow-water, calcareous structures supporting a diverse association of marine plants and animals. A unique characteristic of coral reefs is that they are formed by certain of the plants and animals that inhabit them. Of all of the organisms secreting calcium carbonate that contribute to modern reef formation, the scleractinian corals are the most important. Not only do they deposit calcium carbonate, but the environmental demands of these animals also describe the limits of reef distribution.

Coral reefs occur in shallow water, ranging to depths of 60 m. Reef-building, or hermatypic, corals contain gastrodermal symbiotic algae (zooxanthellae), which require light for photosynthesis. Thus, the vertical distribution of living reef corals is restricted to the depth of light penetration. That relationship is dramatically reflected in the graph in Figure 4–58, which shows the 150 species of corals on Bikini Atoll plotted against depth. The number of species declines rapidly in deeper water, the curve closely following that for light elimination.

Because of their dependence on light, reef corals require clear water. Thus, coral reefs are found only where the surrounding water contains relatively small amounts of suspended material, that is, in water of low turbidity and low productivity. Thus, the great

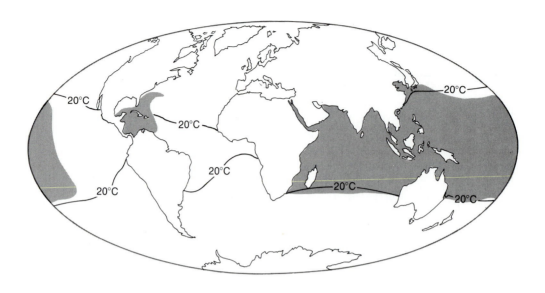

FIGURE 4–59 Distribution of coral reefs today (heavy shading).

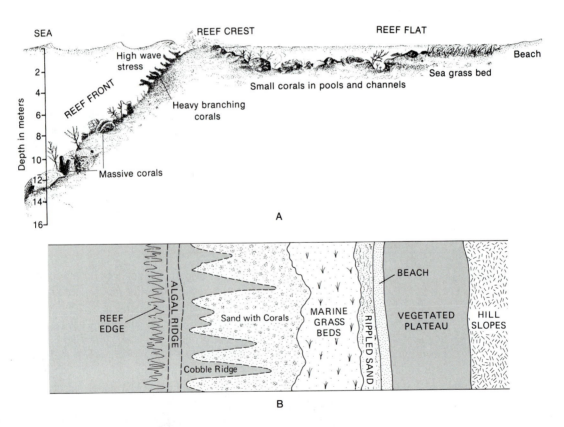

FIGURE 4–60 A, Generalized profile of a fringing reef. **B,** Overview of a reef on the Seychelles in the Indian Ocean. *(B, After Lewis and Taylor. In Stoddard, D. R. 1973. Coral reefs of the Indian Ocean. In Jones, O. A. and Endean, R. (Eds.): Biology and Geology of Coral Reefs, 1(Geology 1):51–92. Academic Press, New York.)*

river systems disgorging silt off the west coast of Africa and the east coast of South America prevent reef development along these coasts.

Coral reefs are further restricted by water temperature and occur only in tropical and semitropical seas, where the average minimum water temperature is not less than 20°C. Existing reefs are restricted to the Caribbean, the Indian Ocean, and the tropical Pacific (Fig. 4–59).

Thus, the two important physical parameters of coral reef ecosystems are warm water (20–28°C) and light (clear water).

REEF STRUCTURE

Three general types of reefs can be recognized on the basis of their structure and the underlying substratum. **Fringing reefs** project seaward directly from the shore. They surround islands as well as border continental land masses and are the most common type of reef. **Barrier reef** platforms are separated from the adjacent land mass by a lagoon. The Great Barrier Reef of Australia is the longest in the world, stretching over 1000 miles along Australia's northeast coast. Another long barrier reef is located in the Caribbean off the coast of Belize, between Mexico and Guatemala.

Atolls, the third type of reef, rest on the summits of submerged volcanos (See Fig. 4–65). They are usually circular or oval with a central lagoon, and parts of the reef platform may emerge as one or more islands. Breaks in the reef provide passes into the central lagoon. Over 300 atolls are present in the Indo-Pacific.

The term **patch reef** refers to small circular or irregular reefs that rise from the floor of lagoons behind barrier reefs or within atolls. They are rather typical lagoon features and are often numerous.

All three reef types, fringing, barrier, and atoll, show certain similarities in profile (Fig. 4–60). The seaward side of the reef, the reef front or fore-reef slope, rises from lower depths to a level just below or just at the water surface. The inclination of the reef front varies from gentle to steep and is often interrupted by a terrace. In some reefs, the lower part of the front is a vertical wall, referred to as the drop-off.

The reef front does not present a flat, wall-like barrier to incoming wave energy. Instead, there are finger-like, seaward projections of the reef that alternate with deeper, sand-filled pockets from the sea (Fig. 4–60B). This so-called spur-and-groove formation disperses wave energy.

Behind the reef crest is a reef flat, or back-reef (Fig. 4–60), which is highly variable in character. The flat may be short or it may extend back several hundred meters (Fig. 4–61), and it may be dissected by channels several meters deep. Further back, the flat usually becomes shallower and may even include

FIGURE 4–61 Reef flat fringing the island of Atutaki (Cook Islands) in the South Pacific.

intertidal areas. The flat may be paved with rock or sand or strewn with coralline boulders. The sand often supports beds of sea grass. Extensive reef flats are rarely uniform in character and exhibit zones (Fig. 4–60B). The reef flat ends at the shore on fringing reefs and descends into the lagoon of atolls and barrier reefs.

The profile just described clearly indicates differing environments that support different species of corals and other animals. On the reef front, coral populations extend to depths between 10 and 60 m. On the intermediate slopes there is a rich zone of massive dome and columnar corals. Below this coral head zone, platelike species predominate. Higher up, where wave stress is greater, still other species are found. In the Caribbean this is commonly the zone of *Acropora palmata,* or elkhorn coral (Fig. 4–62), whose heavy, spreading branches may form a seaward-projecting thicket. Wave stress is an important factor determining the species of corals or other organisms that occupy the reef crest.

Behind the protected reef front, pools, channels, and other areas of deeper, less turbulent water support more delicate corals, for instance, in the Caribbean, *Acropora cervicornis* (staghorn coral), and in the Indo-Pacific there are many species of small, fragile, branching corals (Fig. 4–48). Other species are characteristic of quiet, shallow water farther from the reef front.

The Indo-Pacific contains a very rich and diverse scleractinian fauna. In contrast to 20 known Caribbean genera, there are 80 in the Indo-Pacific. There are only three species of *Acropora* in the Caribbean but 150 in the Indo-Pacific and over 200 species in the Great Barrier Reef of Australia.

The reef platform supports a great array of other animals and plants, some of which can be important contributors to calcium carbonate deposition and cementation. Gorgonian corals, the branching sea rods and sea fans, are a conspicuous component of Caribbean reefs, particularly on the shallow fore-reef. In the Indo-Pacific, gorgonians are usually less conspicuous; but large, massive, or encrusting alcyonacean (soft) corals are often abundant.

Besides cnidarians many other attached animals live on reefs, including clams, tunicates, and bryozoans. There are also certain sponges, clams, and various worms that bore into exposed coral surfaces. Reef topography is highly irregular (much more so than the diagram in Fig. 4–60A suggests) and con-

tains innumerable holes and passages of all sizes, in which dwell a great cryptic fauna of shrimp, crabs, snails, worms, fish, and other animals.

Reef Ecosystem

Much of the blue wavelength of light passing through the photic zone of the sea is scattered by water molecules. Suspended material in productive seas, those with large amounts of plankton and detritus, tends to reflect yellow wavelengths, which accounts for the typically green color of productive seas. Light transmission in productive seas is reduced, and light disappears rapidly below the surface.

Clear water, water of low productivity, is blue. Indeed, blue has been called the color of ocean deserts. Such impoverished water bathes coral reefs, yet paradoxically, coral reefs are among the most productive marine environments. Reef organisms are largely benthic and include not only many attached algae but also an enormous population of symbiotic zooxanthellae. These photosynthetic organisms are at the base of the food chain and constitute such an important energy source that reefs are generally considered to be autotrophic—meaning that resident primary production is the source of most energy flow through reef food webs.

Not all reef energy has a resident origin, however. The plankton in the surrounding water, although scarce compared with most nontropical seas, is still an important food source for the many suspension-feeding animals, including corals. Some, or most, of this plankton develops off the reef and is washed to the platform by currents.

To varying degrees the reef platform constitutes an energy and nutrient trap. Rather than being lost to deep-water sediments, some organic compounds and nutrients (e.g., nitrogen and phosphorus compounds) are retained on the platform and recycled.

Reef Formation

The building of the reef platform is not simply a matter of deposition of new calcium carbonate on top of old. The building involves constructive and destructive phases. Scleractinian corals and the larger skeletons of some other organisms form the framework

FIGURE 4–62 A, *Acropora palmata,* elkhorn coral, on a reef off the coast of Belize in the Caribbean. One "horn" may cover an arc of over a meter. **B,** Coral diversity and cover on Great Barrier Reef off the coast of northeast Australia. Coral is growing in shallow protected water of reef flat. *(A, Photograph courtesy of K. Ruetzler and I. G. Macintyre. B, Photograph courtesy of the Australian News and Information Bureau.)*

material, or the "bricks," of the reef platform; finer skeletal material forms the "mortar" (Fig. 4–63).

The destructive phase may begin long before the death of a living coral colony. Any exposed surface of the coral is quickly attacked by boring organisms, particularly boring sponges and bivalves (Fig. 4–63). Living corals are attacked from the underside, which sometimes leads to the toppling of a large dome. When a living colony dies, regardless of the cause of death, the underlying skeleton also becomes riddled with boring organisms. Eventually, even a large coral is fragmented or reduced to rubble.

The duration of the destructive phase of reef formation is determined by the deposition of fine, excavated material produced by the borers. This material

accumulates; eventually the borers are smothered in their own debris and boring stops. The debris produced by boring, as well as fine skeletons and shells of gorgonian corals, calcareous algae, sponges, molluscs, and other organisms, settles into crevices and holes, slowly filling the spaces between the larger framework pieces with a calcareous mortar. The fusion, or lithification, of this material occurs within 10 to 15 cm of the surface. A core taken from a living reef reveals all phases of platform formation (Fig. 4–63). The reef may grow seaward, but vertical growth is limited by light and water depth. Yet core sampling on most modern reefs reveals a platform thickness of 6 to more than 1000 m. Moreover, the lower parts of many platforms are located below the photic zone (See Fig. 4–65). How can this be explained?

Extensive vertical growth of reef platforms is a result of changes in sea level or subsidence of the substratum. Virtually all modern reefs reflect some growth associated with the most recent sea level rise. During the last Pleistocene glacial period, when enormous quantities of sea water were locked in ice over much of the present north temperate regions, the sea level dropped 120 m below the present level (Fig. 4–64). Coastal areas now covered by water were exposed; islands were larger or connected to mainlands; and submerged platforms, such as the Great Bahamas Bank, were out of water. Fringing and other reef types developed in favorable areas, as is true today.

Glacial melting and sea level rise toward modern levels began about 18,000 years before the present (B.P.). Some modern reefs followed this rise, but most made their appearance later. Upward growth rates of modern reefs vary from 3 to 15 m per 1000 years.

Reef platforms of great thickness are the result of substratum subsidence. Such movement has been especially important in the formation of atolls and certain barrier reefs, such as the Great Barrier Reef of Australia. Atolls are typically located on top of volcanic seamounts. Slow subsidence of the seamount has been matched by compensating upward growth of the reef platform (Fig. 4–65).

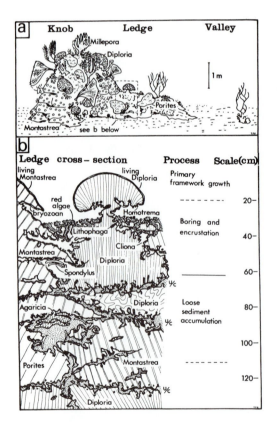

FIGURE 4–63 Section (**B**) through a Bermuda patch reef (**A**) showing phases of reef formation. *(A and B, From Scoffin, T. P. 1972. Fossilization of Bermuda patch reefs. Science, 178:1280–1282. © 1972 by the American Association for the Advancement of Science.)*

SUMMARY OF CORAL REEFS

1. Coral reefs are calcareous rock formations supporting a great array of marine plants and animals, and certain of these reef organisms deposit the cal-

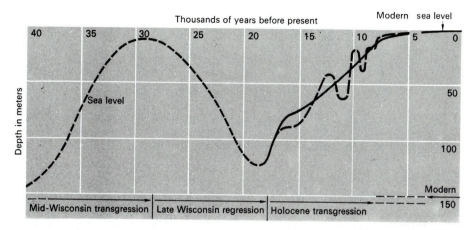

FIGURE 4–64 Sea level changes during the last Pleistocene interglacial and glacial periods. Transgression refers to encroachment by the sea as a result of glacial melting and sea level rise; regression refers to the reverse conditions. *(From Curray, J. H. 1965. Late Quaternary history, continental shelves of the United States. In The Quaternary of the United States. Princeton University Press, Princeton. p. 725.)*

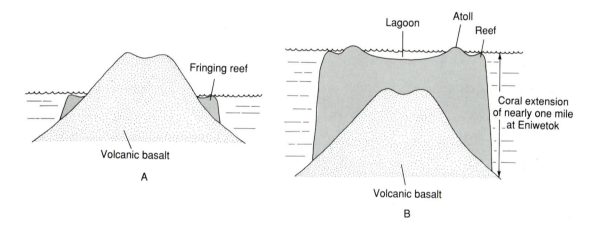

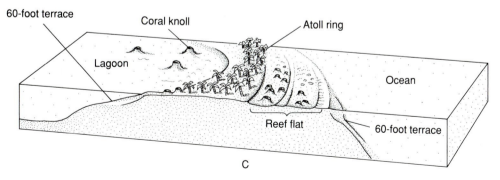

FIGURE 4–65 Formation of an atoll. **A,** Fringing reef around an emergent volcano. **B,** Continuous deposition of coral as the volcanic cone subsides, leading to the formation of a great coralline cap; the emergent part of the cap is atoll. **C,** Section through part of the atoll. *(After Ladd.)*

cium carbonate that forms the underlying reef formation. The distribution of coral reefs is restricted to clear tropical seas, permitting the penetration of light required by symbiotic zooxanthellae

   2.   The principal types of reefs are fringing reefs, barrier reefs, and atolls. Most exhibit a somewhat similar cross-sectional profile, consisting of a seaward-facing and sloping reef front, which receives most of the wave energy, a shallow or just-emergent reef crest, or ridge, and a shallow reef flat behind the ridge. Because environmental conditions differ across the reef, there is a corresponding zonation of reef organisms.

   3.   The reef structure grows upward and seaward by the accumulation of calcium carbonate. The principal animal framework builders are scleractinian corals. Intervening spaces between chunks of framework builders are filled with the excavated debris and the fine skeletal material of other organisms. This "brick and mortar" is cemented, or lithified.

   4.   The great thickness of many reefs resulted from sea level rise or subsidence of the platform on which the reef rests, or from both.

THE CTENOPHORES

The Ctenophora constitutes a small phylum of planktonic, nearly transparent, marine animals that are commonly known as sea walnuts or comb jellies. The phylum contains approximately 50 known species, some of which are abundant in coastal waters, and many others of which are oceanic. The group is still poorly known, even though a great deal has been learned about them over the last ten years. Ctenophores are so delicate that they are difficult to collect with plankton nets. Much of our recent knowledge of oceanic ctenophores is based on specimens collected by scuba divers with hand-held jars.

   Ctenophores were thought to be an offshoot from some medusoid cnidarian because the general body plan is somewhat similar to that of a medusa. The gastrovascular cavity has a canal system, and the thick, middle body layer is comparable to the cnidarian mesoglea. However, there are fundamental differences and the similarities may represent more convergence than common ancestry. We will examine the comparisons at the end of the chapter.

   The more primitive ctenophores, such as the common coastal *Pleurobrachia* (Fig. 4–66), are

spherical or ovoid in shape and range in size from that of a pea to that of a golf ball. Some deep-water species the size of watermelons, however, have been observed from submersibles. Most ctenophores are transparent, but some are brightly colored or possess spots of bright pigment.

   The body wall is composed of an outer epidermis containing sensory cells and considerable mucous gland cells. A layer of true smooth muscle cells lies beneath the epidermis. The thick mesoglea, a connective tissue (p. 70), is composed of a jelly-like material strewn with fibers and amebocytes derived from mesoderm (p. 103). The mesoglea of ctenophores is also crossed by large muscle cells, which are arranged as an anastomosing network.

   The biradial body can be divided into two hemispheres (Figs. 4–66 and 67). The mouth, on one side, forms the oral pole; the diametrically opposite point on the body bears an apical organ and marks the aboral pole. The body is further divided into equal sections by eight ciliated bands. These bands, called **comb rows,** are characteristic of ctenophores and are the structures from which the name of the phylum is derived (*ktenes* in Greek means combs and *ophora,* carrying). Each band extends about four fifths of the distance from the aboral pole to the oral end of the body and is made up of short transverse plates of long, fused cilia called **combs.** The combs are arranged in succession, one behind the other, to form a comb row.

   The combs provide the locomotive power in ctenophores, although lobate forms can also swim by contractions of the lobes. The ciliary beat functions in waves beginning at the aboral end of the row. The effective sweep of each comb is toward the aboral pole, so the animal is driven with the mouth, or oral end, forward, but the ciliary beat can be reversed.

   From each side of the aboral hemisphere is suspended a long, solid, branched, contractile tentacle, which emerges from the bottom of a deep, ciliated, epidermal canal called the tentacular sheath, or pouch. There are two pouch openings, located between comb rows on opposite sides of the body, each at an approximately 45-degree angle from the aboral pole.

   The tentacular epidermis possesses peculiar adhesive cells called **colloblasts.** A colloblast cell is somewhat pear-shaped, with the narrowed end anchored in the tentacular mesoglea and having a synaptic junction with a nerve cell. An intracellular

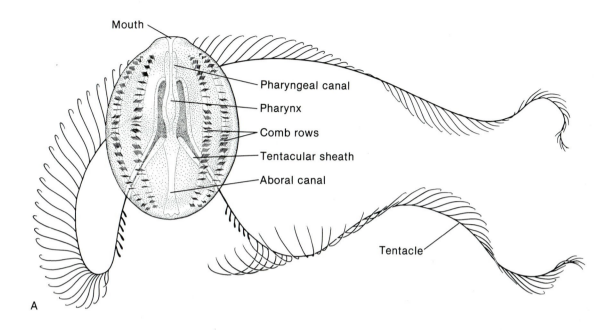

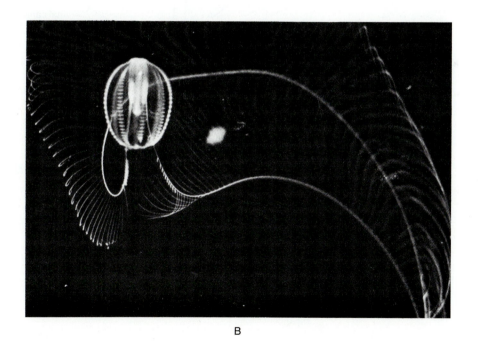

B

FIGURE 4–66 A, *Pleurobrachia.* **B,** *Pleurobrachia pileus* with expanded tentacles. *(A, Adapted from Hyman. B, From Greve, W. 1976. Publikation wiss. Film Gött. 9(1):53–62.)*

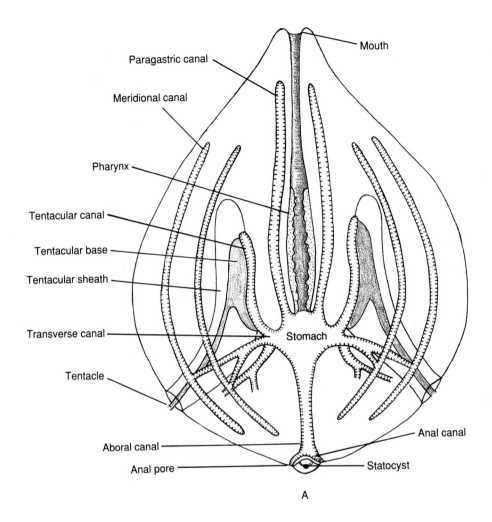

Mouth

Paragastric canal

Meridional canal

Pharynx

Tentacular canal

Tentacular base

Tentacular sheath

Transverse canal

Tentacle

Stomach

Anal canal

Aboral canal

Anal pore

Statocyst

A

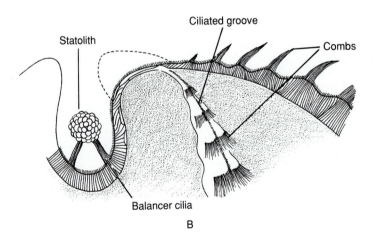

Statolith

Ciliated groove

Combs

Balancer cilia

B

FIGURE 4–67 A, Digestive system of a cydippid ctenophore. **B,** Part of anterior end of a ctenophore showing statocyst, ciliated groove, and combs. *(A, After Hyman, L. H. 1940. The Invertebrates, Vol. I. McGraw-Hill Book Co., New York. B, After Horridge.)*

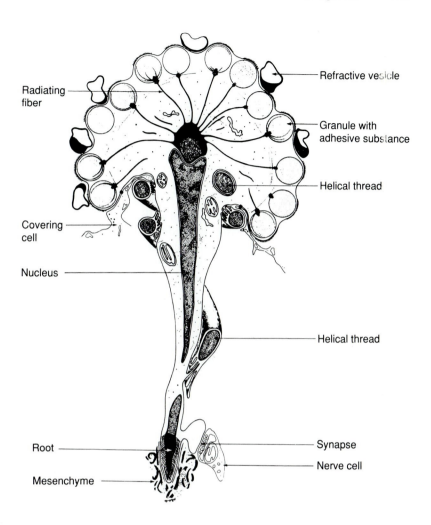

Radiating
fiber

Refractive vesicle

Granule with
adhesive substance

Helical thread

Covering
cell

Nucleus

Helical thread

Root

Synapse

Nerve cell

Mesenchyme

FIGURE 4–68 Colloblast.
(From Franc, J. 1978. Organization and function of ctenophore colloblasts: An ultrastructural study. Biol. Bull. 155:527–541.)

helical thread winds around the long axis of the cell and at its distal end gives rise to a large number of smaller radiating fibers (Fig. 4–68). Each radiating fiber terminates in a granule filled with an adhesive mucoid material, which is liberated on contact with prey. The digestive system is composed of an elaborate, biradial system of canals that arise from a central stomach (or infundibulum) (Fig. 4–67A). The mouth leads into a long, flattened ciliated pharynx, which extends along the polar axis to the stomach.

Ctenophores are carnivorous, feeding on other planktonic animals. Tentaculate ctenophores, such as *Pleurobrachia,* fish with their branched tentacles, which form a large net when fully expanded, up to 30 cm in length. Prey, especially copepods, are caught on the adhesive colloblasts, hauled in by the tentacle retraction, and wiped into the mouth and pharynx. The body is rotated to bring the mouth near the tenta-

cles. Colloblasts are only used once and usually the entire tentacular branch is lost and replaced. The lobate ctenophores, such as *Mnemiopsis* and *Leucothea,* use both the tentacles and the mucus-covered oral surfaces of the lobes to capture prey, especially small crustaceans. The cylindrical *Beroe,* which lacks tentacles, feeds on other ctenophores. Contact of the large mouth with the prey causes an inward gulp, and the prey is swallowed. One ctenophore, *Euchlora rubra,* feeds on jellyfish and conserves the nematocysts, which are then transported by gastrodermal cells to the tentacles and utilized by the ctenophore instead of colloblasts. Digestion is both extracellular and intracellular, and indigestible wastes are passed out through anal pores and mouth.

The digestive canal contains numerous pores that open into the mesoglea. Each pore is surrounded by a rosette of ciliated cells that may function to regulate the fluid content and density of the mesoglea as part of buoyancy regulation.

Ctenophores are noted for their luminescence, which is characteristic of all species. The light-producing photocytes are located in the walls of the meridional digestive canals, so externally the light appears to emanate from the comb rows.

The nervous system of ctenophores is a subepidermal nerve network that is particularly well developed beneath the comb rows. The only sense organ is an apical organ containing a statolith, which rests on four tufts of balancer cilia in a deep pit (Fig. 4–67B). When the animal is tilted, the pressure exerted by the statolith on the respective balancer cilia can change the rate and direction of the beat of the comb row cilia. The change is transmitted by way of the ciliated grooves to the corresponding comb rows, and the animal turns. The transmission of metachronal ciliary waves along a comb row results from the hydrodynamic interaction between adjacent combs.

Almost all members of the phylum are hermaphroditic. The gonads are in the form of two bands located in the thickened wall of each meridional canal. One band is an ovary and the other a testis. The eggs and sperm are shed to the exterior through gonoducts, and fertilization takes place in the sea water, except in the few species that brood their eggs.

Cleavage is total, biradial, and highly determinate (p. 195). The mesoglea and muscles are derived from certain micromeres given off at the oral side of the embryo during gastrulation. Some authors have held that these are ectodermal cells, but there are good reasons for believing that these cells represent mesoderm, in which case ctenophores are triploblastic animals. The gastrula develops into a free-swimming, **cydippid larva** that closely resembles the adult of ctenophores with the more ovoid or spherical body structure described previously. The flattened species of ctenophores also possess a spherical cydippid larva that undergoes a more extensive transformation to attain the adult structure. This general existence of a spherical larva in ctenophores seems to substantiate the belief that the primitive shape was spherical or ovoid.

The spherical shape is restricted to some species *(Pleurobrachia)* of one order. Both *Mertensia* and *Mnemiopsis* are moderately flattened laterally, and in *Mnemiopsis* and *Leucothea* the middle of the body has become constricted along the tentacular plane, leaving the expanded outer portions in the form of large lobes (Fig. 4–69A). The resulting shape is somewhat similar to that of a clam. In the lobate ctenophores, tentacles are short, lack sheaths, and have moved to a position near the mouth.

Velamen (Fig. 4–69B) and *Cestum*, a genus known as Venus's girdle, have become so expanded and flattened that they look like transparent belts. One species of *Cestum* reaches a length of over 1 m. These

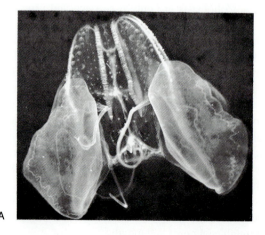
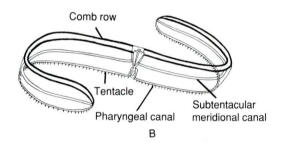

FIGURE 4–69 A, *Leucothea multicornis*, an oceanic lobate ctenophore. **B,** *Velamen. (A, From Harbison, G. R., Madin, L. P., and Swanberg, N. R. 1978. On the natural history and distribution of oceanic ctenophores. Deep-Sea Research, Pergamon Press, 25:233–256. B, After Mayer from Hyman.)*

animals swim not only by means of the comb rows but also by muscular undulations.

The Beroida *(Beroe)* are conical but somewhat flattened along the oral/aboral axis (Fig. 4–70).

The evolutionary position of ctenophores is very obscure. Their spherical, gelatinous character make them superficially similar to medusae, and they have long been considered related to cnidarians. However, there are many fundamental differences. The triploblastic origin of their tissues and determinate cleavage is like that of some bilateral phyla, not cnidarians. The ciliated cells of ctenophores always bear numerous cilia in contrast to the monociliated condition of cnidarians, and ctenophores have true muscle cells instead of the cnidarian myoepithelium. If ctenophores are bilateral animals, their nearest relatives are uncertain.

SYSTEMATIC RÉSUMÉ OF PHYLUM CTENOPHORA

Class Tentaculata.   Ctenophores with tentacles.

Order Cydippida.   Body rounded or oval; tentacles branched and retractable into pouches. *Mertensia, Pleurobrachia.*

Order Lobata.   Body moderately compressed, with two large oral lobes to either side of tentacular plane. Small tentacles not in pouches. *Mnemiopsis, Bolinopsis, Leucothea.*

Order Cestida.   Body ribbon-shaped and greatly compressed along tentacular plane. The two principal tentacles and two of the four comb rows reduced. *Cestum, Velamen.*

Order Platyctenida.   Body greatly flattened as a result of a reduction in the oral/aboral axis. Adapted for creeping. Comb rows reduced or absent in adult. *Ctenoplana, Coeloplana.*

Class Nuda. Ctenophores without tentacles.

Order Beroida.   Body conical or cylindrical, somewhat flattened along the tentacular plane. Mouth muscular. *Beroe.*

SUMMARY OF CTENOPHORES

1. Members of the phylum Ctenophora, or comb jellies, are pelagic marine animals that are superficially similar to medusae in their globose form, their jelly-like mesoglea, and their transparency. However, the similarities to medusae may represent convergence rather than a close evolutionary relationship to cnidarians.

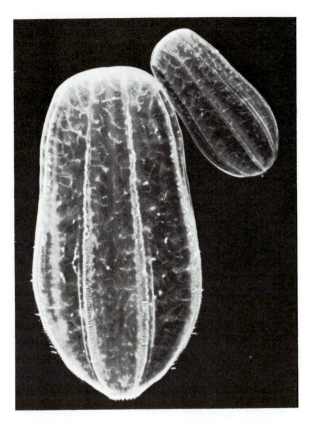

FIGURE 4–70 Lateral view of *Beroe cucumis*. (From Greve, W. 1976. Publikation wiss. Film Gött. 9(1):53–62.)

2. Ctenophores are distinguished by eight meridional ciliary comb rows that propel the body in swimming, oral end forward.

3. Ctenophores are predatory on other pelagic invertebrates. Many species use two branched retractile tentacles that bear special adhesive structures (colloblasts) to capture prey. Some ctenophores have reduced tentacles and capture prey directly with the mouth.

4. Primitive ctenophores have spherical biradial bodies, but many species have become flattened to varying degrees, compressed or depressed.

5. Ctenophores are hermaphroditic. In the development of flattened species, the body is initially spherical (cydippid larva).

REFERENCES

The literature included here is restricted in large part to the cnidarians and ctenophores. The introductory references on page 6 include many general works and field guides that contain sections on these animals.

Epithelia

Darnell, J., Lodish, H. and Baltimore, D. 1990. Molecular Cell Biology. Scientific American Books, Inc., New York. 1105 pp.

Lane, N. J. and Chandler, H. J. 1980. Definitive evidence for the existence of tight junctions in invertebrates. J. Cell Biol. *86*:765–774.

Larsen, W. J. 1983. Biological implications of gap junction structure, distribution, and composition. Tissue & Cell. *15*:645–671.

Lord, B. A., and diBona, D. R. 1976. Role of the septate junction in the regulation of paracellular transepithelial flow. J. Cell Biol. *71*:967–972.

Mackie, G. O. 1984. Introduction to the diploblastic level. *In* Bereiter-Hahn, J., Matoltsky, A. G. and Richards, K. S. (Eds.): Biology of the integument. Vol. 1. Springer-Verlag, Berlin. pp. 43–46.

Welsch, U. and Storch, V. 1976. Comparative Animal Cytology and Histology. University of Washington Press, Seattle. 243 pp.

Skeletons

Clark, R. B. 1964. Dynamics in Metazoan Evolution. Oxford University Press, London. 313 pp.

Koehl, M. A. R. 1982. Mechanical design of spicule-reinforced connective tissue: stiffness. J. Exp. Biol. *98*:239–267.

Wainwright, S. A. 1988. Axis and Circumference. Harvard University Press, Cambridge, MA. 132 pp.

Vogel, S. 1988. Life's Devices. The physical world of animals and plants. Princeton University Press, Princeton. 367 pp.

General

Boardman, R. S., Cheetham, A. H., and Oliver, W. A. (Eds.): 1973. Animal Colonies. Dowden, Hutchinson and Ross, Stroudsburg, PA. 603 pp.

Campbell, R. D. 1974a. Cnidaria. *In* Giese, A. C., and Pearse, J. S. (Eds.): Reproduction of Marine Invertebrates I. Academic Press, New York. pp. 133–200.

Cook, C. B. 1983. Metabolic interchange in algae-invertebrate symbiosis. Internat. Rev. Cytology, Suppl. *14*:177–209.

Gladfelter, W. G. 1973. A comparative analysis of the locomotory system of medusoid Cnidaria. Helgol. wiss. Meeresunters. *25*:228–272.

Harrison, F. W. (Ed.): 1991. Microscopic Anatomy of Invertebrates, Vol. 2. Placozoa, Porifera, Cnidaria, and Ctenophora. Alan Liss, New York. 436 pp. (This volume contains chapters on each of the classes of cnidarians.)

Hessinger, D. A., and Lenhoff, H. M. (Eds.):1988. The Biology of Nematocysts. Academic Press, San Diego. 600 pp.

Holstein, T., and Tardent, P. 1984. An ultrahigh-speed analysis of exocytosis: Nematocyst discharge. Science. *223*:830–833.

Hyman, L. H. 1940. The Invertebrates: Protozoa Through Ctenophora. Vol. 1. McGraw-Hill, New York. pp. 365–696. (An old but still useful account.)

Lenhoff, H. M., and Muscatine, L. (Eds.): 1971. Experimental Coelenterate Biology. University Press of Hawaii, Honolulu. 288 pp.

Mackie, G. O. (Ed.): 1976. Coelenterate Ecology and Behavior. Plenum Press, New York. (Papers presented at the 3rd International Symposium on Coelenterate Biology at the University of Victoria, B. C.)

Muscatine, L., and Lenhoff, H. M. (Eds.): 1974. Coelenterate Biology. Academic Press, New York. 501 pp.

Muscatine, L., Pool, R. R., and Trench, R. K. 1975. Symbiosis of algae and invertebrates: Aspects of the symbiont surface and the host-symbiont interface. Trans. Am. Micros. Soc. *94(4)*:450–469.

Rees, W. J. (Ed.): 1966. The Cnidaria and Their Evolution. Academic Press, New York.

Taylor, D. L. 1973. The cellular interactions of algal-invertebrate symbiosis. Adv. Mar. Biol. *11*:1–56.

Tokioka, T., and Nishimura, S. 1973. Proceedings of the Second International Symposium on Cnidaria. Seto Mar. Biol. Lab. Japan. (53 papers presented at a 1972 symposium held in Japan.)

Werner, B. 1973. New investigations on systematics and evolution of the class Scyphozoa and the phylum Cnidaria. Proceedings of the Second International Symposium on Cnidaria, Publ. Seto Mar. Biol. Lab. *20*:35–61.

Wyttenbach, C. R. (Ed.): 1974. The developmental biology of the Cnidaria. Am. Zool. *14(2)*:540–866. (Papers presented at a 1972 symposium held in Washington, DC.)

Young, R. E. 1983. Oceanic bioluminescence: An overview of general functions. Bull. Mar. Sci. *33(4)*:829–847.

Hydrozoans

Benos, D. J., and Prusch, R. D. 1972. Osmoregulation in fresh-water hydra. Comp. Biochem. Physiol. [A], *43A*:165–171.

Biggs, D. C. 1977. Field studies of fishing, feeding and digestion in siphonophores. Mar. Behav. Physiol. *4*:261–274.

Calder, D. R. 1975. Biotic census of Cape Cod Bay: Hydroids. Biol. Bull. *149*:287–315.

Christensen, H. E. 1967. Ecology of *Hydractinia echinata*. I. Feeding biology. Ophelia. *4*:245–275.

Fraser, C. 1954. Hydroids of the Atlantic Coast of North America. University of Toronto Press, Toronto.

Gierer, A. 1974. Hydra as a model for the development of biological form. Sci. Am. *231(6)*:44–54.

Mackie, G. O., Pugh, P. P., and Purcell, J. E. 1987. Siphonophore biology. Adv. Mar. Biol. *24*:97–262.

McNeil, P. L. 1981. Mechanisms of nutritive endocytosis. I. Phagocytic versatility and cellular recognition in *Chlorohydra* digestive cells, a scanning electron microscope study. J. Cell Sci. *49*:311–339.

Pardy, R. L., and White, B. N. 1977. Metabolic relationships between green hydra and its symbiotic algae. Biol. Bull. *153*:228–236.

Petersen, K. W. 1990. Evolution and taxonomy in capitate hydroids and medusae. Zool. J. Linn. Soc. *100*:101–231.

Purcell, J. E. 1981. Feeding ecology of *Rhizophysa eysenhardti*, a siphonophore predator of fish larvae. Limnol. Oceanogr. *26(3)*:421–432.

Purcell, J. E. 1984. The functions of nematocysts in prey capture by epipelagic siphonophores. Biol. Bull. *166*:310–327.

Satterlie, R. A., and Spencer, A. D. 1983. Neuronal control of locomotion in hydrozoan medusae: A comparative study. J. Comp. Physiol. *150(2)*:195–206.

Singla, C. L. 1975. Statocysts of Hydromedusae. Cell Tiss. Res. *158*:391–407.

Scyphozoans and Cubozoans

Calder, D. R., and Peters, E. C. 1975. Nematocysts of *Chiropsalmus quadrumanus* with comments on the systematic status of the Cubomedusae. Helgol. wiss. Meeresunters. *27(3)*:364–369.

Fancett, M. S. 1988. Diet and prey selectivity of scyphomedusae from Port Phillip Bay, Australia. Mar. Biol. *98*:503–509.

Kramp, P. L. 1961. Synopsis of the medusae of the world. Mar. Biol. Assoc. U.K. *40*:1–469.

Larson, R. J. 1976a. Cnidaria: Scyphozoa. Marine Flora and Fauna of the Northeastern United States. NOAA Tech. Rep. NM FS circular 397. U.S. Government Printing Office, Washington, DC. 18 pp.

Larson, R. J. 1976b. Cubomedusae: Feeding-functional morphology, behavior and phylogenetic position. *In* Mackie, G. O. (Ed.): Coelenterate Ecology and Behavior. Plenum Press, New York. pp. 237–245.

Sandrini, L. R., and Avian, M. 1989. Feeding mechanism of *Pelagia noctiluca*, laboratory and open sea observations. Mar. Biol. *102*:49–55.

Shih, C. T. 1977. A guide to the Jellyfish of Canadian Atlantic Waters. Natural History Museum of Canada. University of Chicago Press, Chicago. 90 pp.

Werner, B. 1975. Structure and life history of the polyp of *Tripedalia cystophora* (Cubozoa, class. nov., Carybdeidae) and its importance for the evolution of the Cnidaria. Helgol. wiss. Meeresunters. *27(4)*:461–504.

Anthozoans

Bak, R. P. M., and Steward-Van Es, Y. 1980. Regeneration of superficial damage in the scleractinian corals *Aqaricia aqaricites*, f. purpurea and *Porites asteroides*. Bull. Mar. Sci. *30(4)*:883–887.

Bayer, F. M. 1961. The shallow-water Octocorallia of the West Indian Region. Martinus Nijhoff, Hague.

Benson, A. A., and Lee, R. F. 1975. The role of wax in oceanic food chains. Sci. Am. *232(3)*:76–86.

Bigger, C. H. 1982. The cellular basis of the aggressive acrorhagial response of sea anemones. J. Morph. *173(3)*:259–278.

Cairns, S. 1976. Guide to the commoner shallow-water gorgonians of Florida, the Gulf of Mexico, and the Caribbean region. Sea Grant Field Guide Series #6, University of Miami, Miami, FL.

Clayton, W. S., Jr., and Lasker, H. R. 1982. Effects of light and dark treatment of feeding by the reef coral *Pocillopora damicornis*. J. Exp. Mar. Biol. Ecol. *63(3)*:269–280.

Cook, C. B., D'Elia, C. F., and Muller-Parker, G. 1988. Host feeding and nutrient sufficiency for zooxanthellae in the sea anemone *Aptasia pallida*. Mar. Biol. *98*:253–262.

Cook, C. B., Logan, A., Ward, J. et al. 1990. Elevated temperatures and bleaching on a high latitude coral reef: The 1988 Bermuda event. Coral Reefs. *9*:45–49.

Fadlallah, Y. H. 1983. Sexual reproduction, development and larval biology in scleractinian corals. A review. Coral Reefs. *2(3)*:129–150.

Falkowski, P. G., Dubinsky, Z., Muscatine, L. et al. 1984. Light and the bioenergetics of a symbiotic coral. BioSci. *34(11)*:705–709.

Fautin, D. G. 1992. A shell with a new twist. Natu. Hist. April, 1992. (Sea anemone-hermit crabs symbiosis.)

Francis, L. 1973. Intraspecific aggression and its effect on the distribution of *Anthopleura elegantissima* and some related sea anemones. Biol. Bull. *144*:73–92.

Gladfelter, E. H., Monohan, R. K., and Gladfelter, W. G. 1978. Growth rates of five species of reef-building corals in the northeastern Caribbean. Bull. Mar. Sci. *28(4)*:728–734.

Godknecht, A., and Tardent, P. 1988. Discharge and mode of action of the tentacular nematocysts of *Anemonia sulcata*. Mar. Biol. *100*:83–92.

Isdale, P. 1977. Variation in growth rate of hermatypic corals in a uniform environment. Proceedings Third International Coral Reef Symposium, University of Miami, Miami, FL. *2*:403–408.

Jokiel, P. L., and York, R. H., Jr. 1982. Solar ultraviolet photobiology of the reef coral *Pocillopora damicornis* and symbiotic zooxanthellae. Bull. Mar. Sci. *32(1)*:301–315.

Kastendiek, J. 1976. Behavior of the sea pansy *Renilla kollikeri* Pfeffer and its influence on the distribution and biological interactions of the species. Biol. Bull. *151*:518–537.

Knutson, D. W., Buddemeier, R. W., and Smith, S. V. 1972. Coral chronometers: Seasonal growth bands in reef corals. Science. *177*:270–272.

Lang, J. 1973. Interspecific aggression by scleractinian corals: 2. Why the race is not only to the swift. Bull. Mar. Sci. *23(2)*:260–279.

Lasker, H. R. 1981. A comparison of particulate feeding abilities of three species of gorgonian soft corals. Mar. Ecol. Prog. Ser. *5*:61–67.

Lewis, D. H., and Smith, D. C. 1971. The autotrophic nutrition of symbiotic marine coelenterates with special reference to hermatypic corals. Proc. Roy. Soc. London [Biol.]. *178*:111–129.

Lewis, J. B., and Price, W. S. 1975. Feeding mechanisms and feeding strategies of Atlantic reef corals. J. Zool. [London]. *176*:527–544.

Lewis, J. B., and Price, W. S. 1976. Patterns of ciliary currents in Atlantic reef corals and their functional significance. J. Zool. [London]. *178*:77–89.

Manuel, R. L. 1981. British Anthozoa. Synopses of the British Fauna No. 18. Academic Press, London. 250 pp.

Mariscal, R. N. 1970. Nature of symbiosis between Indo-Pacific anemone fishes and sea anemones. Mar. Biol. *6*:58.

Mariscal, R. N., Conklin, E. J., and Bigger, C. H. 1977. The ptychocyst, a major new category of cnida used in tube construction by a cerianthid anemone. Biol. Bull. *152*:392–405.

Mariscal, R. N., McLean, R. B., and Hand, C. 1977. The form and function of cnidarian spirocysts. 3. Ultrastructure of the thread and the function of spirocysts. Cell Tiss. Res. *178*:427–433.

Murata, M., Miyagawa-Kohshima, K., Nakanishi, K. et al. 1986. Characterization of compounds that induce symbiosis between sea anemone and anemone fish. Science. *234*:585–586.

Ottaway, J. R. 1980. Population ecology of the intertidal anemone, *Actinia tenebrosa*: 4. Growth rates and longevities. Aust. J. Mar. Freshw. Res. *31(3)*:385–396.

Porter, J. W. 1976. Autotrophy, heterotrophy, and resource partitioning in Caribbean reef-building corals. Am. Nat. *110*:731–742.

Purcell, J. E. 1977. Aggressive function and induced development of catch tentacles in the sea anemone *Metridium sinile*. Biol. Bull. *153*:355–368.

Richardson, C. A., Dustan, P., and Lang, J. C. 1979. Maintenance of living space by sweeper tentacles of *Montastrea cavernosa*, a Caribbean reef coral. Mar. Biol. *55(3)*:181–186.

Rinkevich, B., and Loya, Y. 1984. Does light enhance calcification in hermatypic corals? Mar. Biol. *80*:1–6.

Shick, J. M. 1991. A Functional Biology of Sea Anemones. Chapman and Hall, London.

Smith, F. G. W. 1971. Atlantic Reef Corals. University of Miami Press, Coral Gables, FL. 164 pp.

Steele, R. D., and Goreau, N. I. 1977. The breakdown of symbiotic zooxanthellae in the sea anemone *Phyllactis flosculifera*. J. Zool. [London]. *181*:421–437.

Thorington, G. U., and Hessinger, D. A. 1990. Control of cnida discharge: III. Spirocysts are regulated by three classes of chemoreceptors. Biol. Bull. *178*:74–83.

Van-Praet, M. 1985. Nutrition of sea anemones. Adv. Mar. Biol. *22*:65–99.

Velimirov, B. 1975. Wachstum und Altersbestimmung der Gorgoni *Eunicella carolinii*. Oecologia. *19*:259–272.

Veron, J. E. N., Pichon, M., and Wijsman-Best, M. 1976–1984: Scleractinia of Eastern Australia. Pts. I–IV. Australian Institute of Marine Sciences Monograph Series. (This superb treatise on the scleractinian corals of the Australian Great Barrier Reef is of great value in the study of corals of other parts of the tropical Pacific.)

Watson, G. M., and Hessinger, D. A. 1989. Cnidocyte mechanoreceptors are tuned to the movements of swimming prey by chemoreceptors. Science. *243*:1589–1591.

Wellington, G. M., and Glynn, P. W. 1983. Environmental influences on skeleton banding in eastern Pacific corals. Coral Reefs. *1*:215–222.

Williams, R. B. 1978. Some recent observations on the acrorhagi of sea anemones. J. Mar. Biol. Assoc. U.K. *58(3)*:787–788.

Wineberg, S., and Wineberg, F. 1979. The life cycle of a gorgonian: *Eunicella singularis*. Bijdr. Dierjunde. *48(2)*:127–140.

Wood, E. M. 1983. Corals of the World. T. F. H. Publications, Neptune City, NJ. 256 pp.

Ctenophores

Franc, J.-M. 1978. Organization and function of ctenophore colloblasts: An ultrastructural study. Biol. Bull. *155*:527–541.

Greve, W. 1975. Ctenophora. Fich. Ident. Zooplancton 146. 6 pp.

Greve, W. 1976. Die Rippenquallen der sudlichen Nordsee und ihre interspezifischen Relationen. (The comb jellies of the southern North Sea and their interspecific relations). Publ. wiss. Film Gött. (1976) Bd. *9(1)*:53–62.

Harbison, G. R. 1985. On the classification and evaluation of the Ctenophora. *In* Conway Morris, S., George, J.

D., Gibson, R. et al. (Eds.): The Origins and Relationships of Lower Invertebrates. Systematics Association Special Volume No. 28. Clarendon Press, Oxford. 394 pp.

Harbison, G. R., Madin, L. P., and Swanberg, N. R. 1978. On the natural history and distribution of oceanic ctenophores. Deep-Sea Res. *25*:233–256.

Reeve, M. R., and Walter, M. A. 1978. Nutritional ecology of ctenophores—a review of recent research. Adv. Mar. Biol. *15*:249–287.

Siewing, R. 1977. Mesoderm in ctenophores. Z. Zool. Syst. Evolutions forsch. *15(1)*:1–8.

Coral Reefs

Adey, W. H. 1977. Shallow water Holocene biotherms of the Caribbean Sea and West Indies. Proceedings of the Third International Coral Reef Symposium, University of Miami. Vol. 2 (Geology):xxi–xxiv.

Adey, W. H. 1978. Coral reef morphogenesis: A multidimensional model. Science. *202*:831–837.

Goreau, T. F., Goreau, N. I., and Goreau, T. J. 1979. Corals and coral reefs. Sci. Am. *241(2)*:124–136.

Kaplan, E. H. 1982. A Field Guide to Coral Reefs of the Caribbean and Florida. Peterson Field Guide Series. Houghton Mifflin Co., Boston. 289 pp.

Kuhlmann, D. 1985. Living Coral Reefs of the World. Arco. Publications, New York.

Newell, N. D. 1972. The evolution of reefs. Sci. Am. *226(6)*:54–65.

Scoffin, T. P. 1972. Fossilization of Bermuda patch reefs. Science. *178*:1280–1282.

Smith, S. V., and Kinsey, D. W. 1976. Calcium carbonate production, coral reef growth, and sea level change. Science. *194*:937–939.

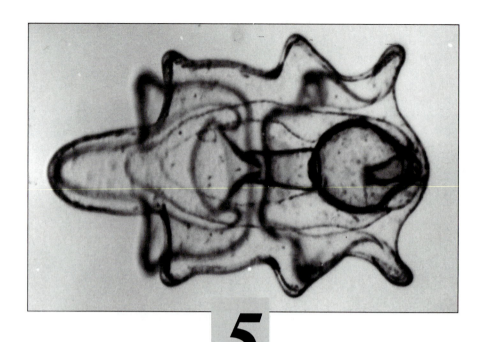

5

BILATERAL ANIMALS

PRINCIPLES AND EMERGING PATTERNS

Bilateral Symmetry and Movement

Compartmentation

Reproductive Biology of Bilateral
   Animals

Relationships of Bilateral Animals

PRINCIPLES AND EMERGING PATTERNS

All of the animals to be discussed in the remainder of the book show **bilateral symmetry,** that is, the left and right halves of the body are mirror images of each other. Correlated with the evolution of bilaterality is the organization of neurons in longitudinal cords and the concentration of receptors and integrating systems at the anterior end of the body. Most bilateral animals also have well-developed, mesodermally derived tissues and organs, which create regulated extracellular compartments, allowing for physiological specialization. Bilateral animals are sometimes referred to as **triploblastic** because of the importance of mesodermal derivatives in their body organization. The bodies of cnidarians, in contrast, are said to be **diploblastic** in structure because their predominant tissues are derived only from ectoderm and endoderm. Finally, the gut of bilateral animals often has a mouth and anus, and is regionally specialized for particular functions. The distinctions between bilateral animals and the radially symmetrical cnidarians and ctenophores—bilaterality, highly organized nervous systems, mesoderm, and a complete gut—however, are not absolute. They appear in diploblastic animals also but not to the same degree as in triploblastic invertebrates. This higher level of structural organization and its functional consequences are the subjects of this chapter.

BILATERAL SYMMETRY AND MOVEMENT

Although cnidarian larvae have elongated bodies and swim with the anterior pole forward, all of the parts of the body around the longitudinal axis (anterior-to-posterior pole) are similar. In bilateral animals, the upper side of the body (dorsal side) is different from the lower (ventral side) so that only one plane of symmetry divides the body into two mirror-image halves (Fig. 5–1). Bilateral symmetry is related to

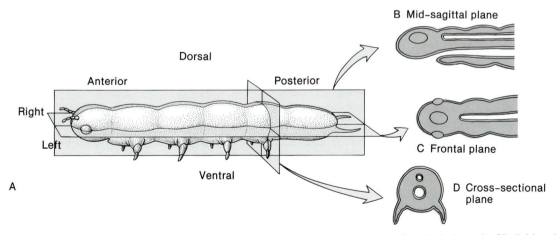

FIGURE 5–1 Bilateral symmetry is typical of animals in which only one plane (vertical plane, in **A**) divides the body into two mirror-image halves. When describing the orientation of a bilateral animal, anterior refers to the head end, posterior to the tail, lateral to the left and right sides, dorsal to the upper surface, and ventral to the belly side. A plane that divides the body into left and right sides is called a sagittal plane. When a sagittal plane passes through the midline of the body, it is called a midsagittal plane **(B).** The midsagittal plane is the only plane that produces mirror-image body halves. A plane that divides across the long axis of the body, for example, as you would slice a loaf of bread, is the cross-sectional plane **(D).** If the body is divided into dorsal and ventral halves, as you would slice a submarine sandwich, the surfaces are cut in the frontal plane **(C).**

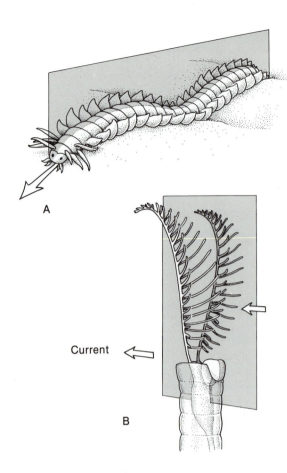

A

Current

B

FIGURE 5–2 Bilateral symmetry is correlated with locomotion, as shown in **A** for a segmented worm, or with an unidirectional input of food to a sessile suspension-feeding animal, such as the pterobranch hemichordate shown in **B**. In motile bilaterians **(A)**, cephalization is apparent, whereas sessile bilaterally symmetrical animals **(B)** are usually not cephalized.

the tendency of these animals when swimming or creeping to keep the same end of the body forward and the same surface downward toward the substratum (Fig. 5–2A). When an animal consistently moves with the same part of the body forward, there is a tendency for its sensory organs and nervous system to become concentrated at the forward end where environmental conditions are first being met. Such a concentration is called **cephalization** (literally, "head development") and has evolved to various degrees in most, although not all, bilateral animals. The mouth usually is located at the leading end where it can ingest food that has been detected by the sensory organs on the head. Neurons become organized as longitudinal cords for the rapid transmission of information up and down the body. Because there is a concentration of sensory organs at the anterior end, there also tends to be a concentration of nervous tissue for the integration of these sensory signals and their translation into motor commands. The brains typical of most bilateral animals could be defined as centers of sensory integration and motor command.

If all animals but sponges, cnidarians, and ctenophores are bilateral, you might well wonder about starfish, or sea stars. Indeed, adult sea stars are radially symmetrical. But the larvae of sea stars are bilateral, good evidence that sea stars evolved from bilateral ancestors. Moreover, when the sea star larva changes to the adult symmetry, it attaches to the bottom during the transformation. In general, bilateral animals that adopt a sessile existence commonly exhibit some shift toward radial symmetry, which better enables them to meet their environment from all directions (Fig. 5–3). The shift may be slight, as in the case of acorn barnacles, where only the protective circular wall plates are arranged radially. Or the shift may be profound as

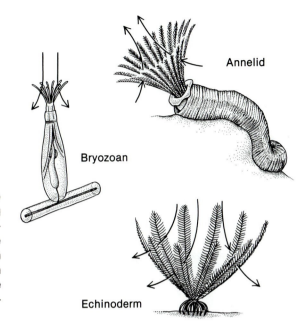

FIGURE 5–3 Many sessile bilateral animals have adopted radial symmetry as an adaptation to meeting the environment equally in all directions. In these examples, suspended food particles are distributed more or less uniformly in all directions and a radial collection system, which shows no directional preference, is an effective adaptation for gathering food. Although these animals mask their bilateral symmetry, it is nevertheless evident in many larval and some adult structures.

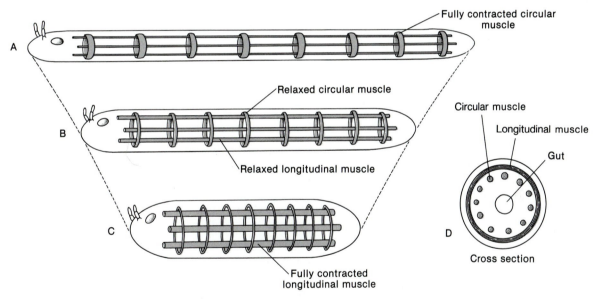

FIGURE 5–4 The basic arrangement of body wall muscles in soft-bodied bilateral animals, as shown in **B** and the cross section **(D)**, is an outer circular and an inner longitudinal musculature. These two layers have antagonistic actions: contraction of the circular musculature causes elongation of the body **(A)**, whereas contraction of the longitudinal musculature causes shortening **(C)**. The circular body wall muscles are typically positioned outside of the longitudinal muscles because the effectiveness of their action (elongation or peristalsis) depends on compression of the bodily tissues, including the longitudinal musculature.

in the case of sea stars. Some early fossil relatives of sea stars were sessile. Living sea stars may have become detached secondarily but retain the radial symmetry of their more immediate sessile ancestors.

The typical body wall musculature of bilateral animals is arranged in circular and longitudinal layers. When the body is viewed in cross section, the outer circular musculature almost always encloses the inner longitudinal musculature (Fig. 5–4, B, D).

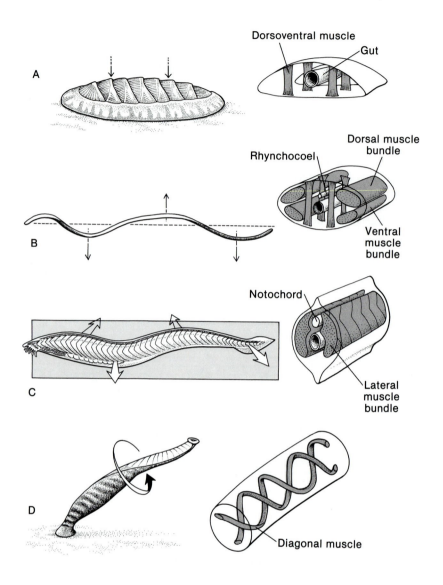

FIGURE 5–5 In addition to a circular and a longitudinal musculature, bilateral animals have a variety of other specialized muscles. Some of the more common ones are illustrated here. Dorsoventral muscles **(A, B)** often contribute to dorsoventral flattening of the body in animals such as chitons **(A)** and nemerteans **(B).** Flattening expands the surface in contact with the substratum in creeping animals and thus promotes adhesion. In addition, some nemerteans and leeches flatten their bodies while swimming to expand their thrusting surfaces **(B).** A longitudinal musculature can produce swimming undulations if bundles on opposite body sides contract alternately and if an axial structure is present to prevent shortening of the body. Under these conditions, if contractions alternate between dorsal and ventral muscle bundles, then vertical undulations occur **(B)**; if they alternate between lateral bundles, horizontal undulations occur, as in cephalochordates **(C).** The incompressible axial rod in cephalochordates is the notochord whereas, in nemerteans, the rhynchocoel may serve the same function. In leeches **(D),** twisting movements of the body are accomplished using helical muscles.

The alternate contraction of these muscles results in two commonly occurring and opposing body movements. Contraction of the circular musculature causes elongation, and contraction of the longitudinal musculature results in shortening (Fig. 5–4A–C). Both the circular and longitudinal muscles are equally represented in animals, such as earthworms, that use peristalsis for burrowing. In many soft-bodied bilaterians that do not use peristalsis, the longitudinal muscles dominate or exist exclusively in the absence of circular muscles. A longitudinal musculature alone is used by some tube worms and crevice dwellers (polychaete annelids) to retract into their refuges for safety, and by lancelets (cephalochordates) to produce swimming undulations (Fig. 5–5C).

Many bilateral animals have specialized muscles that change the body shape (or move appendages) for particular functions. A vertical dorsoventral musculature, for example, flattens the body and lowers its profile in many animals that creep over surfaces, such as flatworms and chitons (Fig. 5–5A). In other animals, such as ribbon worms (nemerteans) and leeches, the dorsoventral musculature flattens the body to improve thrust while swimming (Fig. 5–5B). Dorsoventral muscles are typically arranged in a series along the length of the body (Fig. 5–5A). A few bilateral animals, such as echiuran worms, flatworms, and leeches, have an oblique diagonal musculature, which winds around the body in two helices, one left-handed and one right-handed. These muscles are used to produce twisting movements of the body (Fig. 5–5D).

COMPARTMENTATION

Evolution of Compartmentation

A milestone in the evolution of metazoans was the organization of cells into epithelia. As a general rule, the function of an epithelium is to separate and physiologically regulate different extracellular body compartments. Before epithelia evolved in metazoans, steady-state regulation (homeostasis) was more or less limited to regulation within separate cells, such as probably occurs in sponges. Homeostatic regulation of extracellular compartments, such as fluid-filled cavities and the connective tissue, was probably a precondition for the evolution of specialized muscular, nervous, digestive and other systems.

The cnidarians evolved two epithelia, the epidermis and gastrodermis, and use them to regulate two extracellular compartments. The gastrodermis encloses and regulates the composition of the gastrovascular cavity, whereas the epidermis and gastrodermis together enclose and regulate the mesoglea (Fig. 5–6A, B). The epidermal and gastrovascular cell layers control the entry and exit of nutrients, gases, ions, and other substances to and from the mesoglea. These epithelia apparently establish an extracellular milieu in the mesoglea that is more appropriate to such special functions as muscle operation, nerve conduction, and maturation of gametes than is present in the epithelium itself. Probably because of this, there is a trend in cnidarians and ctenophores for the germ, nerve, and muscle cells to lose their association with the surface epithelia and to invade the regulated mesoglea. The ctenophores, as we have seen, have already evolved such a mesodermal musculature, the cells of which lie entirely within the mesoglea. By the same token, the regulatory gastrodermis controls the concentration of substances, such as digestive enzymes, in the coelenteron thus permitting the extracellular digestion of large foods.

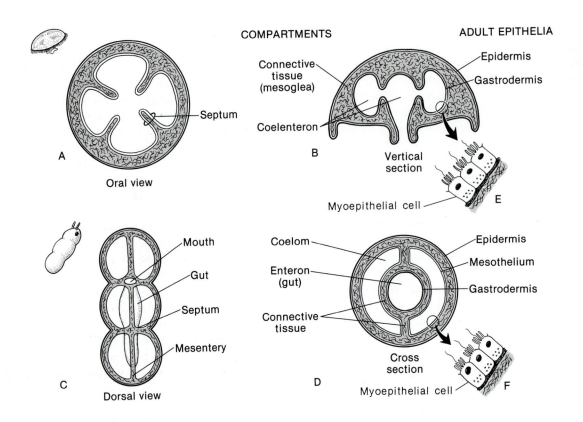

FIGURE 5–6 The body of radial animals **(A, B)** consists of two epithelia, the outer epidermis and the inner coelenteron lining (gastrodermis), and an intermediate layer of jelly-like connective tissue (mesoglea). This arrangement of epithelia allows for physiological regulation and specialization of two extracellular body compartments: the mesoglea and the coelenteron. In contrast to cnidarians and ctenophores, most bilateral animals specialize and regulate three extracellular compartments, the connective tissue, the gut lumen, and the coelom. The additional compartment is made possible by a body design that consists of three, rather than two, epithelia, the epidermis, gastrodermis, and mesothelium **(C, D).** The fact that the myoepithelial musculature of cnidarians is identical to the coelomic myoepithelial musculature of bilaterians **(E, F)** suggests that bilaterian muscles in general evolved from a coelomic myoepithelium.

Extracellular digestion, a specialized function in an extracellular compartment, first appears in radial animals, primarily because epithelia are present.

Although cnidarians were first to evolve a cavity for extracellular digestion, the coelenteron is limited in its freedom to become further specialized for digestion by the need to satisfy other important functions as well. Besides digestion and absorption, these other functions are internal transport (circulation), hydrostatic support, excretion, and reproduction. The coelenteron could not evolve into a regionally specialized gut, as we generally think of it, until it divested itself of circulatory, hydrostatic, excretory, and reproductive functions. This divestiture occurred in large bilateral animals as the multifunctional coelenteron lining was replaced by two new epithelia that delineate a total of three new compartments: the **gut** cavity and its specialized lining, which function primarily in digestion and absorption (Fig. 5–6C, D); the **coelom** and its lining **mesothelium** for hydrostatic support, circulation, reproduction, and excretion (Fig. 5–6C, D); and a specialization of the connective tissue called the **blood-vascular system,** which is important in circulation (Figs. 5–8, 5–10, 5–11).

Gut Compartment

Although there are many exceptions, the gut of bilateral animals is typically a complete tube that opens to the exterior at both ends. It consists of a **mouth,** a **foregut (pharynx),** a **midgut,** a **hindgut,** and an **anus** (Fig. 5–7). Because the gut is open at both ends, food can move unidirectionally through it, permitting regions to specialize, like a (dis)assembly line, for different steps in the digestive process. The mouth and pharynx are specialized for ingestion of food but the pharynx may also grind the food and secrete digestive enzymes. The midgut is the site of digestion and absorption, and the hindgut is usually specialized for the formation and egestion of feces through the anus. A few bilaterians, such as flatworms, do not have an anus or hindgut, but even these have specialized foreguts and midguts.

Foreguts and hindguts develop from embryonic ectoderm (the stomodeum and proctodeum, respectively) that unites with the midgut to form a complete digestive tube. In animals, such as crustaceans and roundworms, that have a body cuticle secreted by ectoderm, the foregut and hindgut are also lined by cuticle. These gut cuticles may be specialized as grinding, piercing, or filtering teeth in the foregut, and as valves and other structures in the hindgut. The midgut develops from embryonic endoderm, and the adult epithelium lacks a cuticle, adapting the midgut for absorption of nutrients. Many invertebrate zoologists restrict their use of the term **gastrodermis,** which means literally, "digestive skin," to the lining of the coelenteron of cnidarians and ctenophores, but it applies equally well, perhaps better, to the midgut epithelium of bilateral animals, in which its function is more or less limited to digestion and absorption.

Coelomic and Vascular Compartments

Most bilateral animals of large body size have transport systems for the mass flow of internal fluids. These transport systems are the coelom and the blood-vascular system (**hemal system**), and they often occur jointly in one organism (Fig. 5–8C).

FIGURE 5–7 The gut of many bilateral animals is divided into three general regions. Foregut and hindgut often develop from invaginated ectoderm, whereas the midgut originates from endoderm.

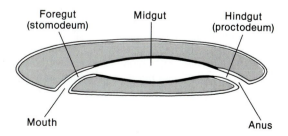

Small or flat bilaterians that rely on simple diffusion for internal transport do not have a coelom (or blood system) and are said to be **acoelomate** in organization (Fig. 5–8E, F).

Animals having a coelom are known as **coelomates.** In all cases, the coelom is lined by an epithelium of cells derived from embryonic mesoderm (Fig. 5–6D, F), and this lining epithelium separates and presumably regulates the composition of

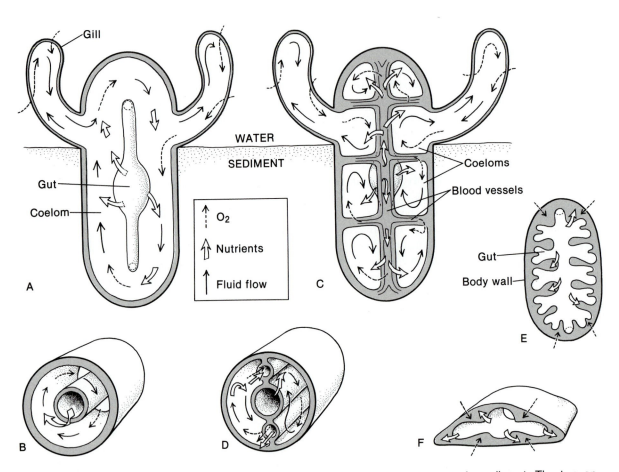

FIGURE 5–8 A and **C** are of large bilateral animals that are partly buried in anoxic sediment. The two appendages are gills that are extended into the oxygen-rich water overhead. Because these are large animals, simple diffusion is not rapid enough to transport oxygen (broken arrows) from the gills and nutrients from the gut (open arrows) to the consuming tissues elsewhere in the body. The animals instead must rely on a circulatory system for internal transport. Some bilaterians, as in **A** (cross section in **B)** have an unpartitioned coelom that is continuous throughout the body. In these, the coelomic fluid reaches all tissues and is the sole circulatory system. In most bilateral animals, however, the coelom is divided by septa and mesenteries, as shown in **C** (cross section in **D),** and because of them, the coelomic fluid can only circulate locally. For whole-body transport, these animals have evolved a blood-vascular system, which consists of fluid-filled channels in the connective tissue (shaded). On the other hand, in very small or flat animals, as shown in **E** and **F,** coelomic and blood-vascular circulatory systems are absent. Because diffusion distances are very short, gases diffuse across the body wall and nutrients diffuse from the gut to the consuming tissues. In the special case of very flat and wide animals, the distance from a central tubular gut to the body margins often exceeds the range of simple diffusion. To compensate for this, the gut develops branches that transport nutrients to the periphery. Such transport in bilateral animals is fundamentally similar to gastrovascular transport in cnidarians and ctenophores.

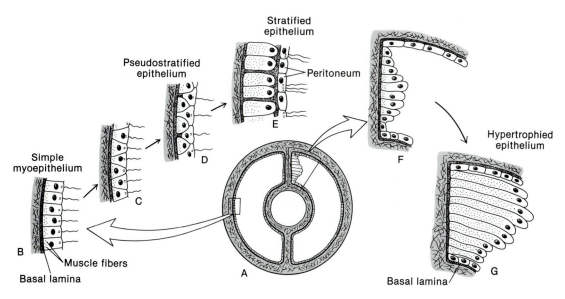

FIGURE 5–9 The embryonic mesothelium of large bilateral animals becomes the body wall musculature. In the simplest case **(A, B)**, the epithelium is simple, forming a single layer of myocytes (lophophorates, some echinoderms, pterobranch hemichordates, some annelids). In others **(C–D)**, noncontractile epithelial cells arch up over the myocytes and separate them from direct contact with the coelomic fluid. This arrangement is termed a pseudostratified epithelium **(D)** because, although the cells are in two layers, all cells remain attached to a common basal lamina (many annelids, echinoderms, enteropneust hemichordates). In a stratified epithelium **(E)**, the noncontractile cells again separate the myocytes from the coelomic fluid but have now lost contact with the original basal lamina and have secreted a new one beneath themselves. The new epithelium formed by the noncontractile cells is called peritoneum (sipunculans, echiurans, chaetognaths, vertebrates). A hypertrophied epithelium **(A, F, G)** is another variation on a simple myoepithelium. In such an epithelium, the myocytes enlarge and eventually fill the coelomic cavity. This results in the formation of a powerful musculature (hemichordates, some annelids; cephalochordates and vertebrate myomeres). © *(A–E, Modified and redrawn from Rieger, R. M. and Lombardi, J. 1987. Ultrastructure of coelomic lining in echinoderm podia: Significance for concepts in the evolution of muscle and peritoneal cells. Zoomorphology. 107:191–208)*

the coelomic cavity independent of adjacent connective tissue and gut compartments. In vertebrates, including humans, the large cavities enclosing the viscera, lungs, and heart are all coeloms, and in earthworms the paired cavities in each segment along the length of the animal are also coeloms. The epithelial lining of the vertebrate coelom is a thin, noncontractile layer, called **peritoneum** (Fig. 5–9E), which separates the coelomic fluid from the underlying (retroperitoneal) muscle and connective tissue. The presence of peritoneum is usually part of most textbook definitions of a coelom, but few invertebrate coelomates actually have a vertebrate-like peritoneum.

In coelomate invertebrates, the simplest and probably most primitive coelom, like that of some segmented worms (annelids) and acorn worms (hemichordates), is lined by a simple epithelium composed largely of myoepithelial cells. This single epithelium is simultaneously the body musculature and the coelomic lining; there is no musculature beneath the epithelium (Fig. 5–6D, F; 5–9A, B). Such a simple coelomic lining is given the general name, mesothelium, which means, "middle epithelium." Because a vertebrate-like peritoneum and a retroperitoneal musculature are not universal in invertebrate coelomates, the definition of coelom should be—*any cavity lined by a mesodermally derived epithelium.*

Just how the vertebrate peritoneum and retroperitoneal musculature could have evolved from a coelom lined only with a myoepithelial mesothelium is unknown, but recent evidence suggests one possible pathway. According to data and ideas developed by R. M. Rieger and his colleagues, the primitive, simple mesothelium became stratified into two or more layers of cells. The innermost layer, nearest the coelomic fluid, became the noncontractile peritoneum and the deeper layers lost their epithelial characteristics and became the body wall muscles (Fig. 5–9A–E). Thus, the mesothelium of primitive coelomates may have evolved into both the peritoneum and some of the body wall muscles of higher coelomates.

In other coelomates, however, the specialized body musculature arises directly from the mesothelium by simple enlargement of some of its myoepithelial cells (Figs. 5–9A, F, G). In animals, such as acorn worms (hemichordates) and lancelets (cephalochordates), these muscle cells enlarge at the expense of the coelomic cavity itself and may nearly obliterate it. This process results in a powerful musculature, such as the swimming muscles (myomeres) of lancelets (Fig. 5–5C).

Coeloms are usually paired and arranged in a longitudinal series as, for example, one pair in each of the many segments of an earthworm. The left member of a pair is separated from the right where their walls abut above and below the gut. Together, these two walls form a vertical partition in the midsagittal plane, called a **mesentery** (Fig. 5–6C). Where a pair of coeloms in a longitudinal series abuts the next pair in the series, the region of contact is a double-walled partition in the cross-sectional plane. Such a partition, which resembles a bulkhead in a ship or an airplane, is called a **septum** (Fig. 5–6C). Septa and mesenteries divide the coelom into separate fluid-filled compartments each of which may be regulated, by virtue of its lining epithelium, for a particular function. Thus, the evolution of a segmented coelom not only contributed to internal transport and hydrostatic support but also created a potential for further functional specialization—as one coelomic cavity became specialized for a particular function, a second cavity became specialized for another. Specialized coelomic cavities, for example, often enclose gonads as genital coeloms (echinoderms, hemichordates) or enclose hearts as pericardia (molluscs, urochordates). The pleural, pericardial, and abdominal cavities of mammals are also specialized coelomic cavities.

Circulation of coelomic fluid is often an important internal transport mechanism. Coelomic fluid is typically circulated by cilia on the coelomic lining cells, although ciliary circulation may be augmented by muscular contractions of the body. In animals, such as peanut worms (sipunculans), in which the perivisceral coelom is undivided, the coelom is the sole circulatory system and its fluid flows unimpeded throughout the entire body (Fig. 5–8A, B). In most other animals, the coelom is divided into few or many segmental compartments by septa and mesenteries. As a result of these subdivisions, fluid is usually confined to one compartment only. This means that any physiologically important substances (oxygen, nutrients) that enter one coelomic compartment are not circulated to another (Fig. 5–8C, D). Instead, such substances have to diffuse or be actively transported across the septa and mesenteries to other coelomic compartments for distribution throughout the body. Because diffusion is slow and because septa and mesenteries are barriers to the continuous flow of fluid, large animals that have a segmented coelom usually distribute substances throughout the body using another noncoelomic circulation. This second circulatory system is called a blood-vascular, or hemal, system (Fig. 5–8C, D).

The blood-vascular system of invertebrates is a network of channels and spaces in the connective tissue, which is continuous around and between all epithelial layers of the body (Figs. 5–8C; 5–10A; 5–11B). The blood itself, although part of the connective tissue, lacks the fibers and molecules typical of the adjacent structural connective tissue. As a result, the blood is fluid and it flows (Fig. 5–10A). Each blood channel or space passes *between* epithelia—in the body wall between epidermis and mesothelium, in the gut wall between the gastrodermis and mesothelium, or in mesenteries and septa between adjacent coelomic mesothelia (Figs. 5–8C; 5–10A; 5–11B). Because the connective tissue is continuous throughout the body, the blood, unlike the cavity-bound coelomic fluid, can circulate freely and unimpeded to within a few millimeters of all living cells in an animal's body. For this

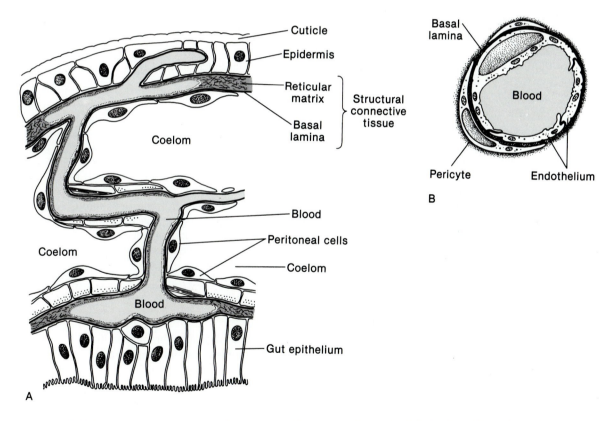

FIGURE 5–10 Invertebrate and vertebrate blood vessels are both fluid-filled channels in the connective tissue compartment. **A,** The blood of most invertebrates is in direct contact with structural connective tissue (basal lamina and reticular matrix). **B,** The blood of vertebrates, cephalopod molluscs, and a few other groups is separated from the structural connective tissue by a secondary endothelium, which lines the vessels, controls transport across the vessel wall, and facilitates flow within. © *(A, Modified from Nakao, T. 1974. An electron microscopic study of the circulatory system in Nereis japonica. J. Morphol. 144:217–236. B, Modified from Welsch, U. and Storch, V. 1976. Comparative Animal Cytology and Histology. University of Washington Press, Seattle. 343 pp.)*

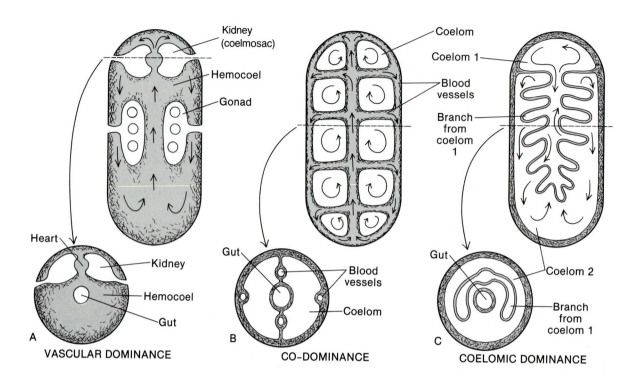

FIGURE 5–11 Most large bilateral animals have codominant coelomic and blood-vascular systems. Codominant coelomic and blood-vascular systems occur, for example, in most annelids, echiurans, phoronids, and hemichordates. In animals with vascular dominance, such as molluscs, arthropods, rotifers, and urochordates, the connective tissue compartment enlarges at the expense of the coelom to form a hemocoel (or pseudocoel). The reduced coelom typically persists only as the pericardium/kidney and the gonadal cavities. Coelomic dominance is common in sipunculans, bryozoans, and echinoderms. In these, the coelom may be continuous throughout the body, thus providing whole-body transport in the absence of blood vessels. Alternatively, the coelom may be segmented, in which case rudimentary but functional blood vessels may bridge the partitions between adjacent cavities, or outgrowths from one coelom may push into another to form an exchange surface (some sipunculans, echinoderms).

reason, the blood-vascular system typically circulates substances throughout the body. When they occur jointly with a blood system, coelomic cavities usually function in more regional transport (Figs. 5–8, 5–11).

Invertebrate blood channels range widely in size. Small tubular channels are called **vessels;** large saclike spaces are **sinuses;** and in cases where the connective tissue compartment enlarges to form a voluminous, blood-filled cavity, as in most arthropods and many molluscs, the cavity is called a **hemocoel.**

Contrary to the situation in vertebrates, most invertebrate blood vessels are not lined by their own epithelium. Instead, they are lined by the basal lamina of the adjacent epithelial sheets (Fig. 5–10A). Blood vessels that pass through mesenteries and septa are overlain by mesothelium, which is often composed of myocytes bearing circularly arranged fibers (Fig. 5–13). When these cells contract, they pressurize the blood, causing it to circulate. When the myocytes are distributed widely over the surface of a blood vessel, their rhythmic contraction produces peristalsis, which pushes the blood along in waves. Often, however, the myocytes are localized and thickened into one or more specialized **hearts.** As a general rule, muscular contraction circulates blood, whereas cilia circulate coelomic fluid.

The Pseudocoel: Another Internal Compartment?

A third sort of fluid-filled body cavity, called a **pseudocoel,** occupies the space between the body wall and gut in a few phyla of small animals called aschelminths. Like a blood space, a pseudocoel lacks an epithelial lining. It is usually defined as "a persistent blastocoel," and most textbooks still provide this definition. There are a number of problems associated with the concept of a pseudocoel, however, but only one will be discussed here.

As noted in Chapter 4, the blastocoel is either a fluid- or gel-filled cavity between the embryonic germ layers. It has long been suspected (and now is confirmed for a few species in several phyla of invertebrates: annelids, hemichordates) that the blood-vascular system develops from either a persistent blastocoel or in a position that was formerly occupied by it (Fig. 5–18). This being so, then either the blood-vascular system is a kind of pseudocoel, or the pseudocoel is but one form of a blood-vascular system, namely a *hemocoel.* Typically, a heart and vessels are associated with a hemocoel but are absent in animals having a pseudocoel.

Relative Development of Coelomic and Vascular Compartments

In animals that have elaborate and specialized hydrostatic systems, a coelom is usually the dominant fluid-filled compartment (Fig. 5–11C). Among echinoderms, for example, the coelomic compartments are not only highly specialized for locomotion and skeletal support, but have largely superseded the blood-vascular system for internal transport.

On the other hand, in many bilateral animals that have evolved rigid-framework skeletons, such as shells and exoskeletons, the role of the coelom in hydrostatic support is diminished, and the blood-vascular system (and other connective tissues) are expanded to dominate the middle body layer (Fig. 5–11A). The expansion of the blood-filled cavity (hemocoel), a characteristic of arthropods and molluscs, largely displaces the coelom and appropriates from it the functions of circulation and, in some cases, hydrostatic support. The coelom generally persists, however, though much reduced in size, as gonadal cavities, pericardial cavities, or as specialized cavities for the receipt of primary urine from blood, often called **coelomosacs.** These cavities are specialized for functions, such as excretion and reproduction, that have been retained by the coelom and for which the coelom remains responsible.

Composition of Blood and Coelomic Fluids

Invertebrate blood, like that of vertebrates, consists of noncellular **plasma,** which contains water, ions, and proteins, and, sometimes, cellular **corpuscles (hemocytes).** The blood may be colorless or pigmented because of the presence of respiratory proteins. Respiratory proteins, such as hemoglobin, are sometimes small, intracellular molecules in **erythrocytes** but often are large extracellular molecules carried in the plasma. Extracellular respiratory proteins (like the plasma proteins) not only increase the capacity of the blood to transport oxygen but are also important in osmosis. As small, osmotically active particles, the proteins (in combination with other solutes) increase the osmotic concentration of the blood above that of the surrounding coelomic fluid. This establishes an osmotic gradient that causes the blood to take up water from the coelom (and probably also across the gut and body walls; Fig. 5–12). The uptake of water by blood offsets the loss of water from the blood to the coelom during ultrafiltration (Fig. 5–13). In some bilateral animals, such as arthropods, ascidians, and vertebrates, the blood may contain many functionally distinct corpuscles.

Invertebrate coelomic fluid has not been as well analyzed as blood but, in general, contains water, ions, low-molecular-weight solutes, and cells. Overall, it differs from blood in having a low, rather than a high, protein concentration. Circulating cells, which may or may not contain a respiratory protein, are termed **coelomocytes.** When respiratory proteins occur in coelomic fluid (and in tissue cells), they are almost always intracellular and are not free in solution. Invertebrate coelomocytes, like vertebrate blood cells, can have many different functions. In fact, in some invertebrates, even the developing germ cells are released from the gonad into the perivisceral coelomic fluid, where they circulate as coelomocytes. The solutes in coelomic fluid, like those of blood, are osmotically active. Even in marine animals, the overall osmotic concentration of the coelomic fluid is slightly

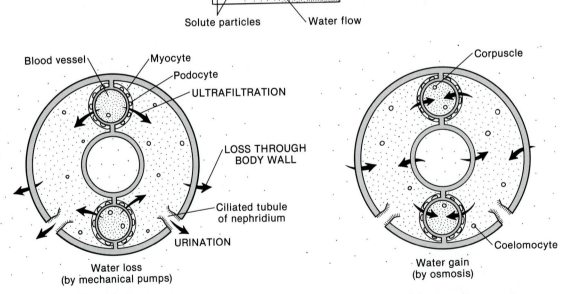

FIGURE 5–12 Water balance in a generalized bilateral animal is the result of losses by ultrafiltration and urination that are offset by osmotic water gains and by drinking. The water losses result from the action of muscular (blood vessels, body wall musculature) and ciliary action (nephridia). Water gains result from osmotic concentrations of solutes in the body fluids that are higher than those in the surrounding environment. In general, the concentration of the coelomic fluid exceeds that of the environmental water, and the concentration of the blood is greater than that of the coelomic fluid. As a result, osmosis across the body wall and gut replaces loss of coelomic fluid during urination, and the uptake of water by blood from coelomic fluid replaces losses resulting from ultrafiltration.

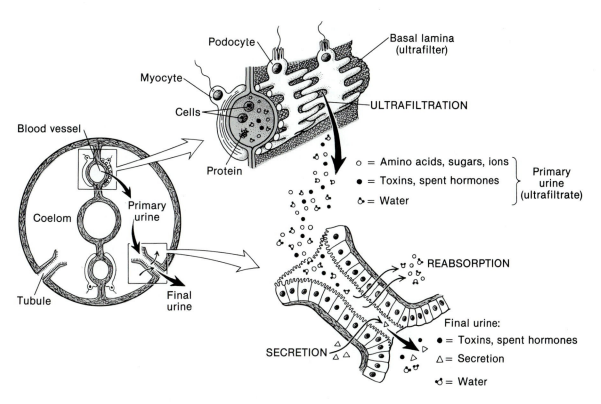

Podocyte

Basal lamina
(ultrafilter)

Myocyte

ULTRAFILTRATION

Cells

Blood vessel

Protein

o = Amino acids, sugars, ions
• = Toxins, spent hormones
♂ = Water

Primary
urine
(ultrafiltrate)

Coelom

Primary
urine

REABSORPTION

Tubule

Final
urine

Final urine:
• = Toxins, spent hormones
△ = Secretion
♂ = Water

SECRETION

FIGURE 5–13 A metanephridial excretory system consists of an ultrafiltration site on a blood vessel (as indicated by podocytes), a coelom, and a tubule that leads to the exterior. As in all filtration kidneys, excretion is a two-step process consisting of: (1) ultrafiltration to form primary urine and (2) modification to form final urine, which is released to the exterior. Modification involves two processes that alter the composition of the primary urine: selective recovery, or reabsorption, of useful metabolites and secretion of toxins from blood or coelomic fluid into the forming urine. The pressure for ultrafiltration results from muscular contractions of the vessel wall.

greater than that of the surrounding seawater. As a result, water tends to diffuse inward from the sea, across the permeable body and gut walls, and into the coelom (Fig. 5–12). This inflow of water may help to maintain the volume of the hydrostatic skeleton.

Blood-Coelom Interplay: Excretion

Blood must be pressurized for it to flow through a blood vessel. As a result of pressurization, there is a tendency for the blood to leak across the vessel wall into the surrounding coelom. Most large animals actually encourage such leakiness at specialized sites called **filtration sites** (Fig. 5–13). The mesothelium at such sites does not form a tight pavement over the vessel but rather is broken by numerous minute gaps. The gaps are created by special mesothelial cells, called **podocytes** (Fig. 5–13), each of which resembles a foot with long splayed-apart toes that interdigitate with other toes. If it were not for the continuous unbroken basal lamina on the vessel wall beneath the podocyte, pressurized blood would probably flow freely from the vessel into the coelom. The basal lamina, however, prevents such an uncontrolled flow because of its fibrous meshlike structure, but it is not a tight seal. In those invertebrates in which it has been studied, the basal lamina does seal in large particles, such as proteins and cells, but is permeable to water and small solutes, for example ions, glucose, amino acids, fatty acids, and probably many toxins. The basal lamina is thus a selective filter. Filters that retain protein-sized particles are called ultrafilters, and the filtration process is referred to as ultrafiltration.

Ultrafiltration is a more or less continuous process by which a protein- and cell-free filtrate of the blood enters the coelom (Fig. 5–12). Among coelomates having a blood-vascular system, the coelomic fluid, therefore, is partly an ultrafiltrate of the blood. The ultrafiltrate itself contains water, ions, glucose, and other useful metabolites as well as unwanted toxic byproducts of cellular metabolism. It may also contain hormones that have completed their tasks in internal communication and must then be eliminated. These useful and useless or toxic metabolites are separated from each other by a process of selective recovery called reabsorption.

Reabsorption is the active, cellular uptake of metabolites from the ultrafiltrate for return or storage elsewhere in the body. Apparently, the reabsorbing cells do not recognize toxic metabolites, which may have been ingested with food, and do not reabsorb them. If the ultrafiltrate containing these substances is allowed to trickle *slowly* from the coelom through tubules leading to the exterior, the useful metabolites will have time to be reabsorbed and recycled and the ignored toxic substances will pass to the outside in the **urine** (Fig. 5–13). The tubules through which the ultrafiltrate passes from the coelom to the exterior are termed **nephridia.**

As indicated previously, the organization of cells into epithelia is a means of regulating the constancy of an animal's internal environment. It follows, then, that any substance that enters the body, whether through the skin or across the gut wall *changes* the composition of the stable internal milieu. **Excretion** maintains the body's internal constancy (homeostasis) by the elimination of excess substances (or their metabolic byproducts) that enter the body. Such substances may be water (which was imbibed by the animal and passed across the gut or entered the body by osmosis), salts, and byproducts of cellular activity, especially nitrogenous wastes. In aquatic invertebrates, most protein nitrogen is excreted as ammonia, which is soluble in water, readily diffuses across the body surface (often across gills), and is carried away by the surrounding water. Most invertebrate nephridia are, therefore, unimportant in the excretion of nitrogenous wastes. Their importance is more often associated with osmoregulation and the excretion of other metabolites. Excretion does *not* include the **defecation (egestion)** of undigested fecal material from the gut because these wastes have not crossed the gut epithelium and become incorporated into the body.

Filtration kidneys (nephridia) characteristically use the processes of ultrafiltration and reabsorption to produce urine. In many animals, however, the composition of the ultrafiltrate may also be modified by the active **secretion** of a specific waste into it (Fig. 5–13), for example, uric acid (a breakdown product of nucleic acids). The sites of waste secretion are not well known in most invertebrates, but in some groups, such as some molluscs and crustaceans, secretion of urea and organic acids occurs in the kidney tubule itself. Because secretion, unlike ultrafiltration, moves a specific metabolite and very little water from blood to urine, it is often employed as the predominant or *sole* mode of urine formation, replacing ultrafiltration, in animals that need to conserve water. Such kidneys are called **secretion kidneys.** They are typical in insects and tardigrades, for example.

By contrast, filtration kidneys are potentially wasteful of water because water comprises most of the ultrafiltrate and is lost in the final urine unless it is reabsorbed. (Such reabsorption, when it occurs, is accomplished by osmosis.) Although nearly all marine and freshwater animals (including the vertebrates) have filtration kidneys, their ability to eliminate water is especially significant to animals that live in fresh water. In these, the high salt concentration of the tissues in relation to that of the environment causes water to enter the body by osmosis. This water must then be excreted, like any other excess substance, and a filtration kidney is well designed to accomplish the task.

Metanephridial systems and protonephridia are common filtration kidneys of bilateral animals. A **metanephridial system** consists of a vascular filtration site or sites, a coelom containing the ultrafiltrate, and a ciliated tubule (**metanephridium**) leading from the coelom to the exterior (Fig. 5–13). Because filtration occurs across the wall of a blood vessel into the coelom in metanephridial systems, such excretory systems occur primarily in animals that have both a blood-vascular system and a coelom. These animals are usually relatively large, for example, most annelids, hemichordates, phoronids, molluscs, crustaceans, and vertebrates (Fig. 5–14).

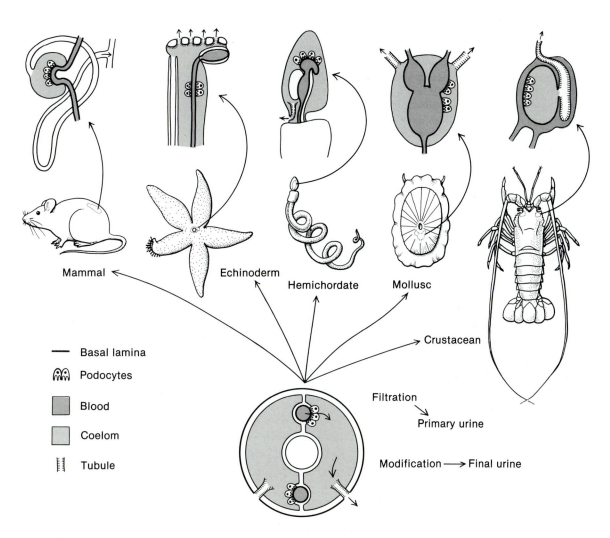

FIGURE 5–14 Metanephridial systems occur in large animals that have blood-vascular systems. © *(From Ruppert, E. E. and Smith, P. R. 1988. The functional organization of filtration nephridia. Biol. Rev. 63:231–258)*

Protonephridia occur in animals that lack blood vessels, a coelom, or both. These are typically small animals (e.g., small annelids, flatworms, larvae, rotifers) or, occasionally, large animals in which only one body cavity (the coelom) is continuous and unpartitioned (e.g., the bloodworm, *Glycera*) (Fig. 5–16). Because only a single body cavity is present, the ultrafilter is positioned on the inner end of the nephridial tubule where it separates the fluid in the body cavity (or tissue spaces) from fluid in the interior of the protonephridial tubule (Fig. 5–15). Protonephridial filtration occurs as extracellular body fluid crosses the ultrafilter and enters the nephridial tubule. The filtration sites, like those of metanephridia, are gaps in an epithelium, but a protonephridial epithelium often consists of few cells, sometimes only one. Each gap-forming cell resembles a single metanephridial podocyte, and may be evolutionarily related to it, but is designated as a **terminal cell** because it is attached to the inner terminal end of the nephridial tubule (Fig. 5–15). Contractile blood vessels are usually absent in animals having protonephridia and, consequently, blood pressure cannot provide the filtration force. Terminal cells, however, are flagellated, with one or more flagella each, and the flagellar motion probably creates the pressure difference that drives ultrafiltration. This is, then, another example of the use of cilia to motivate a process (ultrafiltration) in small animals, whereas muscles are used for the same process in large animals (compare Figs. 5–13 and 5–15).

A protonephridial terminal cell that bears one flagellum is called a **solenocyte.** These are common among the segmented worms and their larvae. A terminal cell that has many flagella, which beat synchronously and resemble a minute flickering flame, is called a flame cell or a flame bulb. A **flame bulb** has its nucleus offset to

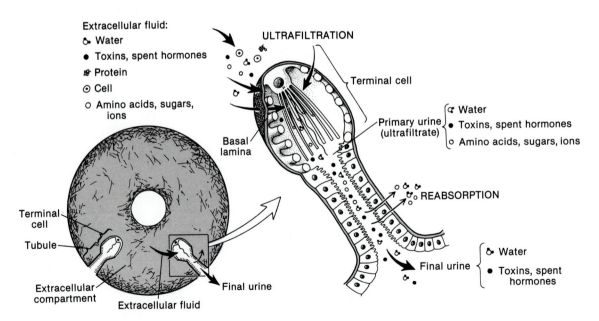

FIGURE 5–15 As in metanephridial systems, protonephridial excretion is a two-step process involving ultrafiltration and modification to form a final urine. In protonephridia, ultrafiltration occurs across the wall of a terminal cell, which is like a metanephridial podocyte attached to the inner end of the nephridial tubule. The force driving fluid across the terminal cell wall is not muscular in origin but results instead from the action of flagella.

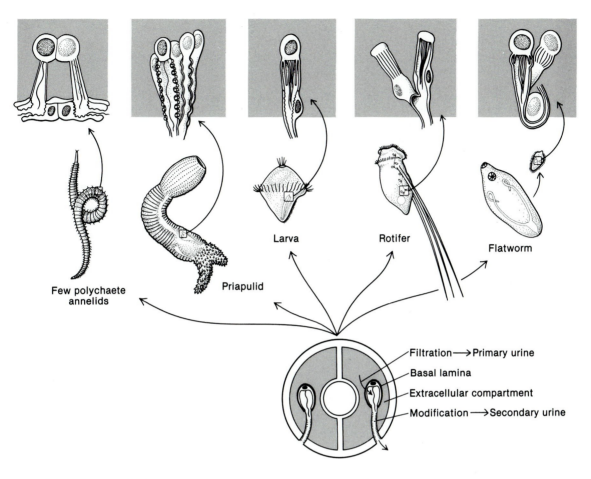

FIGURE 5–16 Bilateral animals that lack blood vessels, coeloms, or both have protonephridia instead of a metanephridial system. Typically, these are either very small or flat animals that rely on simple diffusion for internal transport. Some, on the other hand, are large animals that have only *one* fluid-filled body cavity, typically an unpartitioned coelom or hemocoel. © *(From Ruppert, E. E. and Smith, P. R. 1988. The functional organization of filtration nephridia. Biol. Rev. 63:231–258)*

one side of the flame and is typical of rotifers. A **flame cell** bears its nucleus at the base of the flame and can be found in many flatworms.

REPRODUCTIVE BIOLOGY OF BILATERAL ANIMALS

Sexual Reproduction

In Chapter 3, it was noted that adaptations evolved in many different groups of animals to increase the likelihood that sperm and eggs are released synchronously and in close proximity. Internal fertilization is a common and widespread adaptation among many, but by no means all, bilateral animals and has evolved independently numerous times. The deposition of sperm within the female reproductive tract not only greatly increases the proximity of eggs and sperm for fertilization, but in many species also makes possible storage of sperm for the fertilization of eggs over an extended period. In bilateral animals that use external fertilization, such as many sea stars and sea urchins, the duct from the gonad to the exterior (**gonoduct**) is, typically, a simple tube for transport. In animals that use internal fertilization, the gonoduct has become modified for a variety of functions. In the male system, there are usually glands that produce secretions that nourish and transport sperm, a **seminal vesicle** that stores sperm prior to transmission, and a **penis** or some other type of copulatory structure that aids in the transmission of sperm to the female. In the female system, copulatory **bursa**, **seminal receptacle**, and **spermatheca** are names given to pouches or chambers that receive and store sperm for short, intermediate, and long periods, respectively. When fertilization is internal, the fertilized eggs are usually deposited within some sort of protective **capsule** or **case.**

Species with intracapsular development may release either a larva or a juvenile from the capsule. In either case, the hazards of *early* development in the plankton are avoided and a smaller number of eggs need be produced, although their yolk content must be greater if development is direct. The production of egg capsules or cases requires glands associated with the female reproductive tract for their construction. Although reproductive systems may be simple in species that free spawn their gametes into the water, they can be extraordinarily complex when internal fertilization is the rule, and especially among hermaphrodites.

The amount of yolk in each egg not only affects the cleavage pattern in developing embryos (p. 71 and Fig. 3–3) but also the pattern of larval development in indirect developing species. Embryos developing from small, relatively nonyolky eggs often become long-lived larvae that feed on plankton (**planktotrophic** larvae). Their metamorphosis is often complex and involves loss of many larval structures and extensive remodeling of others. Embryos developing from large, yolky eggs may become larvae that do not feed on plankton but instead rely on yolk reserves (**lecithotrophic** larvae), or they may develop directly into juveniles. In some groups, such as clams, development is predominantly indirect; in others, such as most free-living flatworms, it is largely direct; in some large groups, such as gastropod molluscs, all types of developmental patterns are encountered.

Cleavage Patterns in Early Development

In all bilateral animals, the late embryo (postgastrula) is bilaterally symmetrical, but this is usually preceded by radial symmetry, which may reflect the original metazoan pattern of symmetry. This initial radial symmetry results from one of two predominant cleavage patterns: **radial** or **spiral** (Fig. 5–17). During radial cleavage, which occurs in some cnidarians, phoronids (lophophorates), echinoderms, and hemichordates (deuterostomes), cleavage planes are oriented parallel or at right angles to the polar axis of the egg. The tiers of blastomeres arrange themselves radially around the polar axis, and corresponding blastomeres of different tiers are located directly above and below each other. Spiral cleavage, on the other hand, occurs in many flatworms, annelids, and molluscs (protostomes). In spiral cleavage, the cleavage planes are oriented obliquely to the polar axis of the egg. Successive tiers of blastomeres arrange themselves radially about the polar axis, but corresponding blastomeres of different tiers are offset with respect to those above and below. A blastomere in an upper tier rests in the furrow between two blastomeres in the next lower tier. With gastrulation, the radial symmetry of cleavage is replaced by bilateral symmetry. Some invertebrate embryos, such as those of roundworms

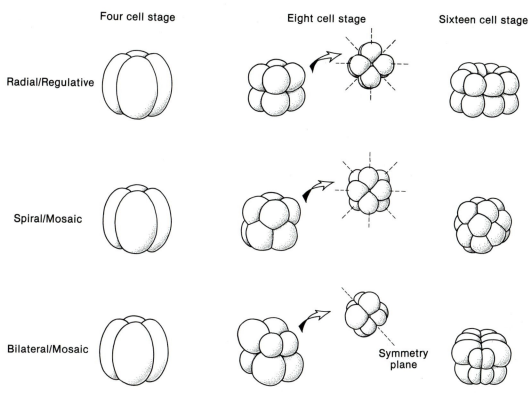

Four cell stage   Eight cell stage   Sixteen cell stage

Radial/Regulative

Spiral/Mosaic

Bilateral/Mosaic

Symmetry
plane

FIGURE 5–17 The three basic cleavage patterns of bilateral animals shown in 4-, 8- and 16-cell stages (lateral views). Arrows point to eight-cell stages in animal pole view. Dashed lines indicate mirror-image symmetry planes that establish radial symmetry in early radial and spiral embryos, and bilateral symmetry in bilaterally cleaving embryos. Later in development, the embryos of all bilaterians adopt bilateral symmetry. © *(Modified and redrawn from Siewing, R. 1969. Lehrbuch der vergleichenden Entwicklungsgeschichte der Tiere. Verlag Paul Parey, Hamburg, Berlin. pp. 46, 60; and Conklin, E. G. 1905. The organization and cell-lineage of the ascidian egg. J. Am. Nat. Soc. 13:1–119)*

(nematodes), and sea squirts (urochordates), are bilaterally symmetrical from the beginning of cleavage and never show radially symmetrical or spiral patterns. Such embryos have a **bilateral cleavage** pattern (Fig. 5–17). In cases in which the egg accumulates large amounts of yolk (which has occurred in various species among almost all groups of animals), the original cleavage pattern becomes modified and obscured. For example, arthropods have yolky eggs, like those of birds, that cleave superficially (meroblastic cleavage).

Because all animals develop from embryos composed of undifferentiated cells into adults composed of differentiated cells in specific locations, at some point during development the fate of embryonic blastomeres must be established. The specification of cell fate during development results from two general processes, which are roughly correlated with the cleavage patterns previously described. One of these processes is called **mosaic development** and is associated with spiral and bilateral cleavage patterns. In such embryos, cell fate is determined early in development and results primarily from the action of specific factors that are unevenly distributed, like pieces of a mosaic, in the cytoplasm of the uncleaved egg. As the egg divides, some blastomeres incorporate the determining factors, and others do not. A clear example of mosaic development occurs in annelids, molluscs, and their relatives. In the early embryos of these animals, an unidentified cytoplasmic factor is segregated into one blastomere, the **mesentoblast,** whose presence there causes the cell and its progeny to form mesoderm (Fig. 5–19). (The mesentoblast is sometimes called the "4d" cell in a formal system of labeling each blastomere.) Thus, the fate of the mesentoblast is fixed early in development by inheritance of a maternal mesodermalizing factor that is already present in the cytoplasm of the uncleaved egg.

Another form of embryonic fate determination, called **regulative development,** predominates in animals that have radial cleavage. Cell fates in these animals (e.g., sea urchins) are determined by a network of cellular communication in the embryo. The fate of a particular cell results from its position in the network and not from a cytoplasmic determinant that specifies its fate regardless of embryonic location. Because cell fate determination in regulative embryos requires the presence of several interacting cells, it tends to occur later in development (after a few more cleavages) than in mosaic embryos. It must be noted, finally, that although mosaic development predominates in spirally and bilaterally cleaving embryos and regulative development predominates in radially cleaving embryos, both forms of determination occur simultaneously, but in different proportions, in the development of *all* animals.

Gastrulation

Origins of Tissues and Extracellular Compartments

All bilateral zygotes gastrulate to form embryos that have three germ cell layers, which are from outside to inside, **ectoderm, mesoderm,** and **endoderm.** A three-layered (triploblastic) embryo is not unique to bilateral animals, however—ctenophores also have three layers—but bilateral animals, in general, diversify and specialize the layers to a greater extent than do their radial ancestors. In many bilateral animals, each of the embryonic germ layers differentiates into an adult epithelium: ectoderm becomes **epidermis,** mesoderm becomes **mesothelium,** and endoderm becomes the **gastrodermis.** Epidermis encloses and covers the body; secretes the cuticle (exoskeleton); and may contain sensory, glandular, ciliated, and other cells, including absorptive cells. Much of the nervous system arises from and is often restricted to the epidermal layer. Mesothelium lines the coelom and forms the musculature, septa, mesenteries and, together with the gametes, constitutes the gonads. The mesothelium often contains muscle and ciliated, secretory, absorptive, and storage cells. It also gives rise to nonepithelial blood corpuscles, coelomocytes, and connective tissue cells. The gastrodermis lines the midgut and contains ciliated, secretory, absorptive, and storage cells. When present, the embryonic **blastocoel** may develop partly into blood spaces, as in acorn worms (hemichordates; Fig. 5–18E–I), or become occupied by skeletal connective tissue (as in sea cucumbers).

Modes of Mesoderm Segregation

Although ectoderm and endoderm are usually epithelial tissues (as are their adult derivatives), mesoderm may be epithelial or nonepithelial in organization. Epithelial mesoderm usually originates as outfolds of the wall of the archenteron, as in

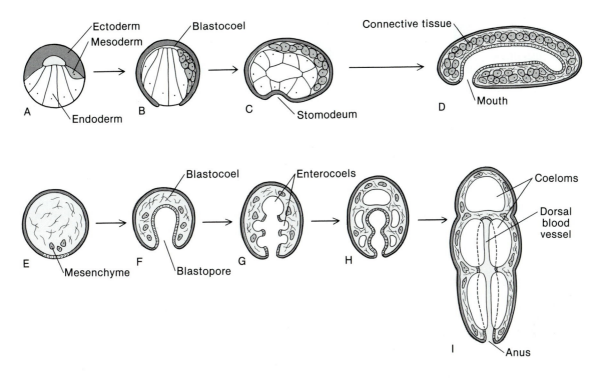

FIGURE 5–18. Two basic patterns of morphogenesis of the connective tissue and other internal compartments occur in bilateral animals. In **A–D** (an acoelomate flatworm), gastrulation occurs by epiboly **(A, B)**, and a stereogastrula **(B)** results. The endoderm hollows to form the gut and joins the stomodeum to form the mouth **(C, D)**. Mesoderm is segregated as mesenchyme into the space previously occupied by the blastocoel **(B, C)** and differentiates into the gonads, body musculature, and connective tissue **(D)**. In a coelomate acorn worm **(E–I)**, gastrulation occurs by invagination. Enterocoelous outpockets from the archenteron pinch off to form the coelomic cavities, and mesenchyme invades the blastocoel **(F–I)**. The blood and other connective tissues are derived from the blastocoel.

many echinoderms. Each of these outfolds (the number varies) is thus an epithelium from the outset. Eventually each outpocket pinches free of the archenteron and becomes an independent coelomic compartment. This process is referred to as **enterocoely** (Figs. 5–18, 5–19).

Nonepithelial mesoderm, which may arise from either ectoderm or endoderm, is said to be **mesenchyme.** Mesenchymal cells are independent nonepithelial cells that are located in the blastocoel, or, when a blastocoel is absent, push in between the ectoderm and endoderm. They are often motile, using a form of ameboid locomotion, and may or may not reorganize into an epithelium later in development. When they do not later become an epithelium but remain as mesenchyme, they form the cellular part of the adult connective tissue. Such is the case, for example, in flatworms, gastrotrichs, gnathostomulidans, and other acoelomate animals in which a perivisceral coelom does not form and muscle and many other mesodermal cells are nonepithelial in organization (Fig. 5–18).

In animals, such as annelids and vertebrates, mesoderm destined to form a coelomic lining usually originates as embryonic mesenchyme but later reorganizes

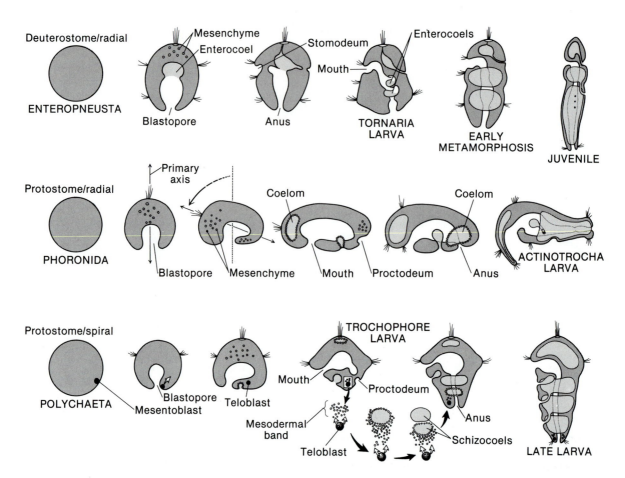

FIGURE 5–19 Three of the many patterns of larval development illustrate the various modes of mesoderm segregation and coelom formation in bilateral animals. In deuterostomes, such as the acorn worm (Enteropneusta), mesoderm originates as mesenchyme (column 2) that forms some of the larval muscles and one small coelom, whereas the major coeloms arise by enterocoely (columns 3 and 4). In phoronids, most or all of the mesoderm originates as mesenchyme, which later becomes epithelial and forms the coeloms. Finally, in polychaetes, some larval mesoderm and the head coelom arise from mesenchyme (columns 3 and 4), but most of the mesoderm and the segmented coeloms originate from a single cell, the mesentoblast, whose fate is determined early in development (column 1). Before or as the mesentoblast enters the blastocoel (column 2), it divides to form two teloblast cells, which locate themselves near the posterior end of the body. The two teloblasts divide repeatedly and proliferate cells in two apical bands (columns 4 and 5). These solid mesodermal bands hollow sequentially in a process called schizocoely, to form the segmental coeloms.

into the coelomic epithelium. This kind of mesenchymal-to-epithelial transition is called **schizocoely** because the mesenchymal cells aggregate initially as solid bands that later, during epithelialization, split apart and hollow out to form a coelomic cavity (Figs. 5–18, 5–19).

Fates of the Blastopore

The blastopore of radially cleaving embryos usually becomes the adult anus, and the mouth forms as a new structure (e.g., hemichordates). In spirally cleaving embryos, the blastopore either becomes the mouth (anus forms anew; e.g., molluscs in general) or gives rise to both mouth *and* anus (e.g., some molluscs, polychaetes, onychophorans). When the latter occurs, the blastopore first forms a long furrow.

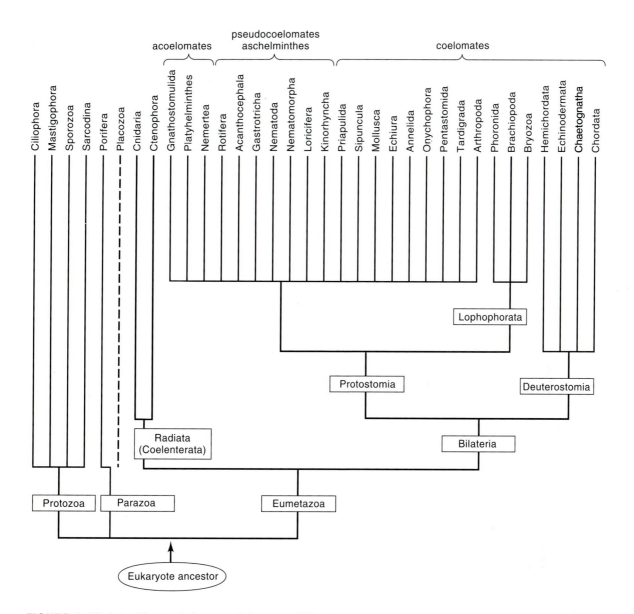

FIGURE 5–20 A traditional phylogeny of the animal kingdom separates the Eumetazoa from the sponges, or Parazoa. The Eumetazoa are then divided into two sister groups, the radial animals (cnidarians and ctenophores) and the bilaterally symmetrical animals. Within the Bilateria, a deep division occurs between the protostomes and deuterostomes. The evolutionary relationships of phyla within the protostomes and deuterostomes is far from certain, however, and much current evolutionary research centers on establishing a clearer understanding of relationships within these two broad divisions.

Later, the furrow margins come together and fuse between the two extremes, one of which remains open as the mouth and the other as the anus.

RELATIONSHIPS OF BILATERAL ANIMALS

Traditional and modern techniques in morphology, which now extend to the molecular level, and limited fossil evidence present a consistent outline of the *general* relationships among bilateral animals (Fig. 5–20). In this outline, bilateral animals belong to one of two main lines of evolution, each containing several closely related phyla. A phylum that belongs to one line is more closely related to other phyla in that same line than it is to a phylum in another line. Nevertheless, the evolutionary order within a line—which phylum is the most primitive and which is least primitive—is not yet established with anything close to certainty. Consequently, the sequence of phyla presented in this book could be changed to accommodate various interpretations.

The two lines of evolution, or groupings, are called **Protostomia** and **Deuterostomia.** Deuterostomes include hemichordates, echinoderms, chordates, and chaetognaths; whereas protostomes include the remaining bilateral animals, such as molluscs, arthropods, annelids, and flatworms. In protostomes, cleavage is generally spiral, the embryonic blastopore usually persists to become the adult mouth (or mouth and anus), and the mesothelium is typically schizocoelous in origin. In deuterostomes, cleavage is usually radial, the blastopore typically becomes the anus, the mouth forms anew from the surface ectoderm, and the mesothelium arises by enterocoely.

In this book, the protostome/deuterostome classification (Fig. 5–20) is adopted because it is familiar and allows for an orderly development of conceptual themes. In adopting it, we are not suggesting that protostomes, as a group, are more primitive than deuterostomes. The opposite may, in fact, be true. Neither is it suggested that the presentation order of the phyla within these groups is the order in which they evolved, although it may have been. Those who believe that the sequence of phyla adopted is inappropriate can easily change the chapter order. Animal evolution will be discussed in the last chapter (Chapter 20).

SUMMARY

1. The hallmark of bilateral animals, grouped together as the **Bilateria,** is bilateral symmetry, symmetry in which only one plane divides the body into mirror-image halves. Its occurrence is correlated with locomotion. Radial symmetry often is superimposed on bilateral symmetry in sessile suspension-feeding animals.

2. Common trends in bilaterian body design include cephalization of the nervous system and structural compartmentation of extracellular spaces.

3. In soft-bodied bilaterians, body wall muscles are almost always arranged as an outer circular musculature, which causes elongation on contraction, and an inner longitudinal musculature for shortening the body. Commonly occurring specialized muscles include dorsoventral muscles (for flattening) and diagonal muscles (for twisting).

4. Extracellular compartmentation in bilaterians, as in other metazoans, is made possible by the regulatory capacity of tissue sheets called epithelia. Cnidarians and ctenophores have two adult epithelia that regulate the mesogleal and gastrovascular compartments. Most large bilaterians have three primary epithelia, epidermis, gastrodermis, and mesothelium. These allow for regulation of digestive, coelomic, and connective tissue compartments.

5. Bilateral animals, with some exceptions, have a complete gut and many also have a mesothelial coelom and a fluid transport system in the connective tissue known as a blood-vascular system.

6. A coelom is a cavity lined by a mesodermal epithelium and located between epidermis and gut. It provides hydrostatic support and also may function in excretion, reproduction, and circulation. Because it is often segmented into regional compartments, its role in circulation is usually limited to local rather than whole-body transport. Coelomic fluid may contain coelomocytes, some of which may contain a respiratory pigment. Its lining epithelium, or mesothelium, has several functions, one of which is to provide the body musculature.

7. A blood-vascular system is a system of fluid-filled channels in the connective tissue compartment. Because the connective tissue extends around and between epithelia and other tissues, it forms an uninterrupted track for whole-body fluid transport. Respiratory pigments occur in erythrocytes or, frequently, are extracellular in the plasma.

8. In very small or flat bilaterians, both coelomic and blood-vascular systems are usually reduced or absent, and their bodies are acoelomate in organization. Such animals use diffusion rather than convection for internal transport of substances.

9. The excretion of toxic and excess substances from blood and coelomic fluid occurs by way of filtration nephridia. In animals that have blood vessels and coeloms, these are metanephridial systems; in those that lack one or the other, or both, they are protonephridia.

10. Both external and internal fertilization are widespread among bilaterians; internal fertilization is often accompanied by specialized and complex male and female accessory reproductive organs.

11. Three general cleavage patterns characterize the early development of bilaterians. Spiral-mosaic cleavage is typical of most protostomes; radial-regulative cleavage of many deuterostomes; and bilateral-mosaic cleavage of some protostomes and deuterostomes.

12. During gastrulation, mesoderm destined to form the coelom is segregated as epithelial outpockets (enterocoely) in deuterostomes or as independent cells (mesenchyme) in protostomes. Mesenchymal cells differentiate into solid mesodermal bands, which organize into epithelia and become hollow during a process called schizocoely (protostomes). In acoelomate protostomes, schizocoely does not occur, and a coelom does not form.

13. In most large bilaterians, three embryonic germ layers give rise to three adult epithelial layers: most ectoderm becomes epidermis, some mesoderm becomes mesothelium, and endoderm becomes gastrodermis.

14. Planktotrophic larvae feed on plankton and are long-lived, whereas lecithotrophic larvae feed on yolk and are short-lived. At the completion of their larval lives, indirect developers metamorphose into juveniles.

15. The classification adopted here divides bilaterians into two broad groups of phyla called Protostomia and Deuterostomia.

REFERENCES

Bartolomaeus, T. and Ax, P. 1992. Protonephridia and metanephridia—Their relation within the Bilateria. Z. zool. Syst. Evolut.-forsch. *30*:21–45.

Clark, R. B. 1964. Dynamics in Metazoan Evolution. The Origin of the Coelom and Segments. Clarendon Press, Oxford.

Costello, D. P. and Henley, C. 1976. Spiralian development: a perspective. Am. Zool. *16*:277-291.

de Beauchamp, P. 1961. Generalites sur les metazoaires triploblastiques. *In* Grassé, P.-P. (Ed.): Traité de Zoologie. Vol. 4. Masson et Cie, Paris. pp. 1–19.

Fransen, M. E. 1988. Coelomic and vascular system. *In* Westheide, W. and Hermans, C. O. (Eds.): Ultrastructure of the Polychaeta. Mikrofauna Marina *4*:199–213.

Goodrich, E. S. 1945. The study of nephridia and genital ducts since 1895. Q. J. Micros. Sci. *86*:113–392.

Greenberg, M. J. 1985. Ex bouillabaisse lux: The charm of comparative physiology and biochemistry. Am. Zool. *25*:737–749.

Hanson, J. 1949. The histology of the blood system in Oligochaeta and Polychaeta. Biol. Rev. *24*:9–173.

Hyman, L. H. 1951. The Invertebrates: Platyhelminthes and Rhynchocoela. The Acoelomate Bilateria. Vol. 2. Mc-Graw-Hill Book Co., Inc., New York. pp. 1–51.

Jägersten, G. 1972. Evolution of the Metazoan Life Cycle. A Comprehensive Theory. Academic Press, New York. 282 pp.

Kirschner, L. B. 1967. Comparative physiology: Invertebrate excretory organs. Ann. Rev. Physiol. *29*:169–196.

Korschelt, E. and Heider, K. 1895. Text-book of the Embryology of Invertebrates. Swan Sonnenschein & Co., London. 4 volumes.

Kume, M. and Dan, K. (Eds.):1968. Invertebrate Embryology. NOLIT, Belgrade. 605 pp.

Kümmel, G. 1975. The physiology of protonephridia. Fortschr. Zool. *23*:18–32.

Kümmel, G. 1977. Der gegenwartige Stand der Forschung zur Funktionsmorphologie excretorischer Systeme. Versuch einer vergleichenden Darstellung. (The current status of the functional morphology of excretory systems. Attempt at a comparative presentation.) Verhandlung Deutsche zoologische Gesellschaft 1977. pp. 154–174.

LaBarbera, M. 1990. Principles of design of fluid transport systems in zoology. Science *249*:992–1000.

LaBarbera, M. and Vogel, S. 1982. The design of fluid transport systems in organisms. Am. Sci. *70*:54–60.

Lankester, E. R. 1900. The Enterocoela and the Coelomocoela. *In* Lankester, E. R. (Ed.): A Treatise on Zoology. Part 2. The Porifera and Coelentera. Adam and Charles Black, London. pp. 1–37.

MacBride, E. W. 1914. Text-book of Embryology. Vol. 1. Invertebrata. MacMillan and Co., Ltd., London. 692 pp.

Mackie, G. O. 1984. Introduction to the diploblastic level. *In* Bereiter-Hahn, J., Matoltsky, A. G. and Richards, K. S. (Eds.): Biology of the Integument. Vol. 1. Invertebrates. Springer-Verlag, Berlin, Heidelberg. pp. 43–46.

Oglesby, L. C. 1981. Volume regulation in aquatic invertebrates. J. Exp. Zool. *215*:289–301.

Pantin, C. F. A. 1959. Diploblastic animals. Proc. Linn. Soc. London *171*:1–14.

Rähr, H. 1981. The ultrastructure of blood vessels of *Branchiostoma lanceolatum* (Pallas) Cephalochordata: I. Relations between blood vessels, epithelia, basal laminae, and connective tissue. Zoomorphology *97*:53–74.

Riedl, R. 1970. Water movement. Animals. *In* Kinne, O. (Ed.): Marine Ecology: A Comprehensive, Integrated Treatise on Life in Oceans and Coastal Waters. Wiley-Interscience, London, New York. pp. 1085–1150.

Riegel, J. A. 1972. Comparative Physiology of Renal Excretion. Hafner Publ. Co., New York.

Rieger, R. M. and Lombardi, J. 1987. Ultrastructure of coelomic lining in echinoderm podia: Significance for concepts in the evolution of muscle and peritoneal cells. Zoomorphology. *107*:191–208.

Ruppert, E. E. 1991. Introduction to the aschelminth phyla: A consideration of mesoderm, body cavities, and cuticle. *In* Harrison, F. W. and Ruppert, E. E. (Eds.): Microscopic Anatomy of Invertebrates. Vol. 4. Wiley-Liss, New York. pp. 1–17.

Ruppert, E. E. and Carle, K. J. 1983. Morphology of metazoan circulatory systems. Zoomorphology. *103*:193–208.

Ruppert, E. E. and Smith, P. R. 1988. The functional organization of filtration nephridia. Biol. Rev. *63*:231–258.

Schmidt-Nielsen, K. 1983. Animal Physiology: Adaptation and Environment. Cambridge University Press, Cambridge.

Siewing, R. 1981. Problems and results of research on the phylogenetic origin of coelomata. Atti dei Convegni Lincei. *49*:123–160.

Smith, P. R. 1986. Development of the blood vascular system in *Sabellaria cementarium* (Annelida, Polychaeta): An ultrastructural investigation. Zoomorphology. *105*:67–74.

Stearns, S. C. 1976. Life history tactics: A review of the ideas. Q. Rev. Biol. *51*:3–47.

Strathmann, R. R. 1989. Existence and functions of a gel filled primary body cavity in development of echinoderms and hemichordates. Biol. Bull. *176*:25–31.

Strathmann, R. R. and Strathmann, M. F. 1982. The relationship between adult size and brooding in marine invertebrates. Am. Nat. *119*:91–101

Vogel, S. 1988. Life's Devices. The Physical World of Animals and Plants. Princeton University Press, Princeton. 367 pp.

Willmer, P. 1990. Invertebrate Relationships. Patterns in Animal Evolution. Cambridge University Press, Cambridge. 400 pp.

Wilson, R. A. and Webster, L. A. 1974. Protonephridia. Biol. Rev. *49*:127–160.

6

PLATYHELMINTHS, GNATHOSTOMULIDS, AND MESOZOANS

PRINCIPLES AND EMERGING PATTERNS

Utilization of Dissolved Organic Nutrients in Seawater
Regeneration of Missing or Damaged Parts
Marine Interstitial Animals
Reproductive Adaptations to Life in Fresh Water
Hermaphroditic versus Dioecious Reproductive Patterns
Parasitism

PLATYHELMINTHS

Class Turbellaria
Class Trematoda
Class Monogenea
Class Cestoidea
Origins of the Parasitic Flatworms

GNATHOSTOMULIDA

MESOZOANS: PHYLUM ORTHONECTIDA AND PHYLUM RHOMBOZOA (CLASSES DICYEMIDA AND HETEROCYEMIDA)

PRINCIPLES AND EMERGING PATTERNS

UTILIZATION OF DISSOLVED ORGANIC NUTRIENTS IN SEAWATER

Seawater contains considerable amounts of dissolved organic material (DOM), including amino acids. To what extent might animals supplement nutrients taken up through the gut by direct absorption of dissolved organic material through the skin? Using radioactive tracers, uptake of DOM across body surfaces has been demonstrated in a number of invertebrate groups. Aquatic animals that lack a cuticle and have microvilli on the surface of the epidermis, such as many cnidarians, free-living flatworms, nemerteans, hemichordates, and echinoderms, are likely candidates, but filter feeders such as bivalves, which move large volumes of water over their gills or other surfaces, have been most commonly investigated. The uptake involves active transport, because the dissolved substances must be moved across a steep concentration gradient; however, there is also some leakage of amines and other substances in the outward direction. So is there any net influx? A net influx has been demonstrated in mussels (*Modiolus* and *Mytilus*), the polychaete *Nereis diversicolor*, and the sand dollar *Dendraster excentricus*. Most attention has been given to the movement of free amines. Less is known about sugars and very little about lipids. Assuming that a net influx occurs in most marine invertebrates, how important is the influx in the animal's nutrition? Absorbed DOM coupled with organic materials supplied by algal symbionts appears significant in the total energy budget of one soft coral studied (*Heteroxenia fuscesens*), but we are still a long way from understanding the physiological or ecological significance of DOM uptake for most invertebrates.

REGENERATION OF MISSING OR DAMAGED BODY PARTS

Wound healing and regeneration of missing organs are widespread among invertebrates. Invertebrates that extend their head and other body parts from protective tubes and burrows and do not retract them quickly enough, frequently have them bitten off by passing predators. Such losses of body parts typically do not result in death, and nearly always the lost parts are regenerated. The phenomenon of regeneration has been intensively studied in the freshwater planarians. As long ago as 1825, biologists realized that a planarian whose head had been bisected longitudinally with a razor would soon heal the cut surfaces and regenerate two complete heads. Similarly, if an animal was severed into two pieces, either transversely or longitudinally, each of the two halves would regenerate the missing parts and become a whole worm. Similar experiments continue to provide an introduction to planarian regeneration, but contemporary research centers on answering two questions: What is the source of the regenerating cells? and What controls regeneration?

When a planarian is cut or wounded, the adjacent epidermis spreads over and seals the wound. A dome-shaped mass of unspecialized cells, called a **blastema**, then forms beneath the epidermis. Eventually, the missing parts of the body differentiate from cells within the blastema.

What is the source of the blastema? The undifferentiated cells of the blastema may arise from differentiated cells, such as muscle cells, by a process of dedifferentiation, or reversion of the cell to its totipotent, embryonic, undifferentiated state or

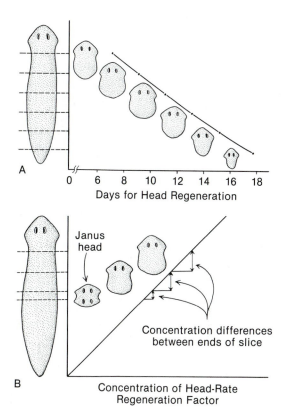

FIGURE 6–1 A, Regeneration rate of a planarian head from different body levels. **B,** Thickness of the slice affects maintenance of head/tail polarity. Very thin slices lose polarity and form "Janus" heads. *(A, Adapted from Dubois, F. 1949. Contribution a l'étude de la régénération chez planaires dulcicoles. Bull. Biol. 83:213–283. B, Adapted from Goss, R. 1969. Principles of regeneration. Academic Press, New York.)*

from a permanent pool of totipotent, undifferentiated cells, such as the archeocytes of sponges or the interstitial cells of cnidarians. At present, there is uncertainty over which of these sources predominates in a regenerating turbellarian flatworm.

The control of regeneration is a complex and lengthy topic and only one example is given here. One of the earliest controls to be studied was regulation of head/tail polarity in regenerating transverse slices. If an intact worm is cut transversely into a series of slices each of equal length, a new head always regenerates on the front end of each slice but the new heads do not regenerate at the same rate on all the slices. Heads regenerate faster on slices from the anterior end of the worm than from the posterior end. Such experiments suggest that the factor (or factors) that controls the rate of head regeneration is distributed along the worm in an anterior-to-posterior concentration gradient (Fig. 6–1A), with more of the factor present in the head and less in the tail. The concentration difference between any two points along the length of the worm may be the control of head/tail polarity. As long as a significant difference in concentration exists between the two ends of the slice, a head will regenerate on the end having the higher concentration. Therefore, there should be some minimum concentration difference, below which head/tail polarity is lost. This minimum difference should be reflected in a minimum distance that can be established experimentally by cutting progressively thinner transverse slices. Eventually, a slice results that is too thin to contain a minimum concentration difference, polarity is lost, and a head is regenerated at both ends (Fig. 6–1B). Such two-headed monsters are called Janus heads, after the Roman god of doors, who is depicted as a head with two opposing faces.

MARINE INTERSTITIAL ANIMALS

Marine sands harbor a complex community of animals, called interstitial fauna (or meiofauna), that lives in the water-filled spaces between grains of sand. The interstitial fauna includes representatives from nearly every invertebrate phylum, and many occur in great numbers. The spaces, or interstices, form a complex three-dimensional labyrinth through which the animals move, feed, and reproduce, but the interstices are small and the sands shift as water flows over and through the bottom.

The animals of this dynamic Lilliputian world are protozoans and metazoans that are similar in being small, typically 100 to 200 μm in diameter (in order to fit in the voids) but elongated (because there is no limitation on length) (Fig. 6–2). Because there is a niche for a small worm-shaped body, animals such as nematodes (roundworms), turbellarians, small segmented worms, and wormlike crustaceans called harpacticoid copepods, are dominant groups. Many of these small interstitial species belong to phyla composed entirely of small animals, but others, such as some polychaete annelids and crustaceans, often resemble the larvae of their larger macrofaunal cousins. This fact suggests that some interstitial species may have evolved from large-bodied species by **progenesis,** that is, the early onset of sexual maturity in a larval stage, and there is good evidence for this form of evolution within annelids and crustaceans. The high frequency of progenesis among intersti-

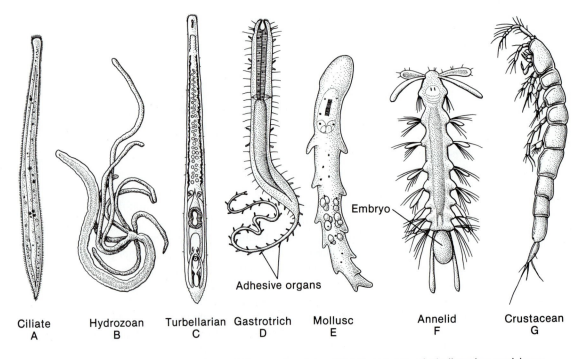

Embryo

Adhesive organs

Ciliate
A

Hydrozoan
B

Turbellarian
C

Gastrotrich
D

Mollusc
E

Annelid
F

Crustacean
G

FIGURE 6–2 All representatives of the marine interstitial fauna are of similar size and have worm-shaped bodies. **A,** The ciliate *Tracheloraphis remanei.* **B,** The hydrozoan *Halammohydra vermiformis.* **C,** The turbellarian *Coelogynopora biarmata.* **D,** The gastrotrich *Urodasys viviparus.* **E,** The opisthobranch mollusc *Pseudovermis papillifer.* **F,** The polychaete annelid *Nerillidium troglochaetoides.* **G,** The harpacticoid copepod *Cylindropsyllis laevis. (Combined from Swedmark, B. 1964. The interstitial fauna of marine sands. Biol. Rev. 39:1–42.)*

tial animals has prompted the idea that whole phyla of small-bodied meiofaunal animals, such as the Platyhelminthes, may have evolved in this manner from large-bodied ancestors (see "Turbellarian Origins," later in this chapter).

Adult interstitial animals are in the size range of plankton and must be under constant risk of being washed out of the sand and into the water, especially in turbulent intertidal areas or on substrata scoured by currents. To prevent this from happening, most interstitial animals have **adhesive organs** with which they make temporary, but tenacious, attachments to sand grains (Figs. 6–2D; 6–3A, B). The sand grains, of course, may be cast into suspension by waves or currents, but because of their density they settle out rapidly, carrying their riders with them. The most common adhesive organs, called duo-gland adhesive organs, are composed of two different kinds of glands. One, designated as the viscid gland, secretes the adhesive and cements the animal to the substratum, and the other, called the releasing gland, secretes the deadhesive, the substance that breaks the attachment (Fig. 6–3B). Using groups of these duo-gland adhesive organs, interstitial animals, such as turbellarians, gastrotrichs, and annelids, can securely and rapidly stick to and release from sand grains.

Another parallel adaptation, related to small body size, is internal fertilization and direct benthic development in interstitial animals. In many large animals, the high mortality associated with free-spawned eggs is tolerated because myriads of eggs are released, and enough survive to replace their parents. But in tiny interstitial species, which produce few eggs, free spawning is out of the question. Interstitial

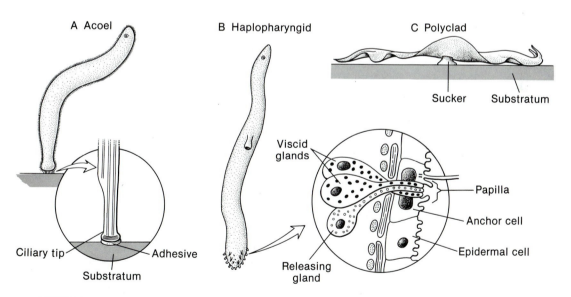

FIGURE 6–3 A, Temporary adhesion in turbellarians is achieved by adhesive cilia in the acoel *Paratomella rubra.* **B,** A duo-gland adhesive papilla (inset) in *Haplopharynx* and many other turbellarians. **C,** An adhesive sucker in the polyclad *Oligoclado floridanus* and its relatives. *(A, Modified from Tyler, S. 1973. An adhesive function for modified cilia in an interstitial turbellarian. Acta Zool. 54:139–151. B, Modified and redrawn from Tyler, S. 1976. Comparative ultrastructure of adhesive systems in the Turbellaria. Zoomorphologie 84:1–76.)*

animals, by and large, copulate and store sperm to ensure fertilization of their few eggs. Once fertilized, the zygotes are usually cemented to sand grains to prevent them from being washed out of the environment. A few remarkable species (Fig. 6–2F) brood their young, for example, the polychaete *Nerillidium*; and at least one, the gastrotrich *Urodasys*, is viviparous (Fig. 6–2D).

REPRODUCTIVE ADAPTATIONS TO LIFE IN FRESH WATER

Compared with the ocean, fresh water is an extreme environment. The concentration of salts in fresh water is generally much lower than in the tissues of the animals. As a result, freshwater animals are like leaky ships and must either protect their tissues from dilution or expend metabolic energy to bail out water. They must also accumulate salts. Freshwater habitats, especially puddles and small ponds and streams, may become dry at certain seasons of the year, and the seasonal range of temperature may be wide in temperate regions. Under these conditions, either the adult itself or a developmental stage, usually the egg, must be physiologically adapted to survive the extremes. Food, in the form of algae, is often only available for a few months of the year and, as a result, mechanisms must exist for rapid population growth and exploitation of the temporary resource. Finally, fresh water in streams, creeks and rivers flows downstream only, and organisms must avoid being washed out of their favored habitat.

One of the most noticeable adaptations to life in fresh water is the near absence of pelagic larvae, which, if released, might suffer lethal osmotic stress (because of an unfavorably large surface-area-to-volume ratio) or be washed downstream out of the adult habitat; floating eggs are also uncommon. Freshwater invertebrates typically have direct development and produce heavy, yolky eggs, which sink to the bottom or are cemented to stones or debris. In temperate waters, these eggs may be of two forms: a rapidly developing summer (subitaneous) egg and an overwintering resting egg. The production of subitaneous eggs is one adaptation for rapid population growth during favorable periods. Resting eggs, on the other hand, are often encapsulated in thick shells and are resistant to desiccation and extremes of temperature that may occur during the winter or at other times of the year. Typically, they germinate with the return of moisture or the warming of water in the spring.

Asexual reproduction is common and well developed in many freshwater invertebrates and provides another means of rapidly increasing the number of individuals available to exploit a temporarily abundant resource. Asexual reproduction may occur by budding or fission as, for example, in turbellarian flatworms or *Hydra*. In some cases, buds may be dormant and resistant to environmental extremes as, for example, with the gemmules of sponges. Often, however, **parthenogenesis** is a common pattern. Parthenogenesis is the embryonic development of unfertilized, usually diploid, eggs. As long as conditions remain more or less constant, the females continue to reproduce by parthenogenesis, often producing huge numbers of similar individuals. Under adverse conditions, such as a decline in food or water temperature, typically in the fall of the year in temperate areas, males appear in the population and mate with the females. The males typically develop from unfertilized haploid eggs, and thus the female shifts from the production of diploid eggs to haploid eggs under the influence of photoperiod, population density, and dietary cues, among other factors. Fertilization of the haploid eggs by the males' sperm results in dormant and desiccation-resistant eggs.

HERMAPHRODITIC VERSUS DIOECIOUS REPRODUCTIVE PATTERNS

The general topic of reproductive strategies, of which these two conditions are a part, is complex and controversial, and only one line of reasoning is offered here. Speculation on this topic can be investigated further by examining references in the bibliography at the end of the chapter.

Hermaphroditism is not, as it might seem, an adaptation for self-fertilization. In fact, most hermaphrodites have structural and physiological barriers that prevent self-fertilization. Instead, hermaphroditism generally prevails under conditions in which potential mates meet infrequently. Such conditions may occur when population densities are low and contacts are infrequent (for example in many flatworms) or where the adults are sessile and are within mating range of relatively few adjacent individuals, such as in barnacles. The security of hermaphroditism is that contact between *any* two individuals can result in successful mating. This seems like a great advantage, and it must therefore be asked–Why aren't all animals hermaphrodites? There must be some subtle cost inherent in hermaphroditism that is absent when the sexes are separate. This additional cost is believed to be in the need for both sets of accessory reproductive organs.

If a hermaphrodite and a dioecious individual have an equal energy allotment for reproduction, both invest it in gametes (especially eggs) and accessory reproductive organs (e.g., ducts, penes, vaginas, seminal receptacles, seminal vesicles). The hermaphrodite must invest energy in the construction of *both* male and female accessory organs (a total of two sets: 1 male, 1 female). A dioecious individual, on the other hand, has either male *or* female organs (a total of one set: either male or female). Thus, a dioecious pair has surplus reproductive energy that can be invested in gametes, theoretically increasing its fecundity over that of a hermaphroditic pair.

An example of this principle can be found in this chapter. Virtually all flatworms are hermaphrodites, and many, but not all, occur in low population densities. Nevertheless, a group of endoparasitic flatworms, the schistosomes, is dioecious. An adult schistosome male and female are physically joined throughout life, thus eliminating the condition of infrequent contact and giving them the freedom to be dioecious. Being dioecious may increase their fecundity, and with it, the chance that their life cycle will be successfully completed.

PARASITISM

The most common form of symbiosis between animals is **parasitism**, a relationship in which the host is harmed by the presence of the symbiont, or parasite. Parasites that live on the outside of their host are said to be **ectoparasites**; those that live on the inside, **endoparasites**. A few parasites can utilize several closely related or ecologically similar species as hosts but most show a high degree of host specificity. Some parasites require two or more hosts to complete their life cycles; in these cases the host for the larval or developmental stage is termed the **intermediate host**; the host for the adult stage is termed the **definitive**, or **primary host**. Many parasites, however, have only one host.

Nutrition is the usual benefit derived by a parasite from its relationship with the host. The parasite feeds on the host's tissues or body fluids or utilizes the food ingested by the host. When the parasite is enclosed by some part of the host's body, there may be other advantages, such as protection.

The establishment and maintenance of a parasitic relationship, however, pose problems and are not without cost to the parasite. The host species represents a small and discontinuous habitat, and a primary problem for parasites is that of locating and infecting new hosts. Most hosts are not without defenses and respond to invading eukaryotic parasites as to such other invaders as viruses and bacteria. Any tissue damage produces initial inflammation and phagocytic reactions of leucocytes or amoebocytes. The host's immune system is also stimulated by the antigenic proteins on the parasite's surface. Parasites that overcome these host defenses may yet be harmlessly isolated or walled off within the host's body. In the vertebrates, hydrochloric acid and proteases present in the stomach are a formidable barrier that few potential parasites entering the gut can pass. The high body temperatures of birds and mammals and the low level of oxygen in the gut contents of many animals are also barriers to parasitism.

From these generalizations about barriers to parasitism, we can anticipate many of the adaptations encountered in parasites. A variety of structural adaptations may be present depending on whether the parasite lives on or within the host. Ectoparasites that feed infrequently, such as leeches and ticks, may have a part of the gut modified for storage. On the other hand, some endoparasites, such as tapeworms, utilize digested food of the host and have lost the gut entirely. The problems of penetration and attachment to the host have resulted in the evolution of a variety of structures, such as suckers, hooks, and teeth. Enzymes may facilitate penetration in some species. The most common point of entry for endoparasites is through the mouth of the host. The parasite may remain in the gut of the host or, in some species, break through the gut wall to reach other organs.

The problem of reaching new hosts has been met in most parasites through the production of enormous numbers of eggs or other developmental stages. The species is perpetuated if only a few individuals reach the proper host and survive to adulthood. The reproductive system of parasites is highly developed for the production of great numbers of gametes. There are some parasites in which most of the body is concerned with reproduction.

The general structure of endoparasites usually reflects the less rigorous demands of the limited and uniform environment in which they live. Locomotor processes may be reduced, and the sensory organs found in free-living relatives are reduced or absent. The nervous system in turn is greatly reduced.

Nevertheless, parasites have evolved various specializations for dealing with the defenses of the host. A common one is analogous to storming a castle: there is great loss of invading troops, but a few get through. Having successfully entered, some parasites, particularly among the protists, can multiply in the body of the host. To cope with the host's immune system, some parasites simply avoid it by residing in sites, such as the gut, that are largely out of reach of the immune system. The developmental stages of some parasites are protected by cyst walls. Blood flukes have evolved the neat trick of concealing themselves immunologically by absorbing the host's antigens onto their surface so that the host's immune system does not recognize them as "foreign." Flagellate trypanosomes, which produce sleeping sickness and other diseases, keep changing their surface antigens so that the host's immune system cannot get a "fix" on them.

A "successful" parasite usually does not kill its host, for a dead host results in a dead parasite. If the parasite does kill the host, the timing must be such that it or its eggs escape before the host's death. In many cases, a host may carry a small population of parasites without any obvious serious consequences. Where the stress of a parasitic infection is manifested in the host, the condition is recognized as a disease. Parasitic disease may cause the destruction of host cells and tissues. Blood vessels, ducts, or the gut may be clogged. The parasite may produce wastes or substances that have a toxic or allergic effect on the host.

What have been the avenues along which parasitism evolved? Ectoparasites may have evolved from species that were initially commensals or occasional occupants of the host's surface. Some parasites may have evolved from ancestors that were preadapted for a parasitic existence. For example, certain parasitic flies that lay their eggs in the skin wounds of cattle are closely related to species that lay their eggs in dead animal tissues. The host's mouth has clearly been an important avenue for the evolution of many endoparasites. The ancestors were accidentally ingested and survived. Survival would have led to rapid selection of characteristics that enhanced the parasitic relationship. In parasites that have intermediate hosts, the intermediate host, at least in some cases, was probably the original host, for example, digeneans in snails. When the first host was eaten by the second host, the parasite survived in the new host, but the adaptations that had already evolved required the first host stage to be preserved in the life cycle of the parasite.

PLATYHELMINTHS

Of all the bilateral phyla, the free-living flatworms of the phylum Platyhelminthes have long been considered the most primitive. Although this assumption is supported by preliminary analyses of gene sequences, some zoologists question this view, in part because several flatworm attributes that were previously thought to be indicative of the phylum's primitiveness are now believed to be the result of small body size, a general characteristic of flatworms. Such attributes include the absence of circulatory systems, the use of cilia rather than muscles for locomotion, and the presence of protonephridia. The phylum Platyhelminthes embraces four classes of worms. Three are entirely parasitic and include the Trematoda, Monogenea, and the Cestoidea (tapeworms). The fourth class, the Turbellaria, is free-living and is certainly the group from which arose the ancestors of the three parasitic classes. Members of the phylum may be tiny (less than a millimeter to a few millimeters), longer than they are wide, and more or less cylindrical; or larger (centimeters, rarely meters) and dorsoventrally flattened—accounting for the name flatworm. In either case, diffusion distances are short, and as a result the typical fluid transport systems (coelom and blood-vascular systems) are absent and the body is compact (acoelomate). The connective tissue compartment, called the parenchyma in flatworms, has mesenchymal cells, as well as bodies of cells whose apices form part of the body and gut walls, such as gland cells. Mesenchymal cells and ex-

tracellular matrix are especially well developed in the larger species. The mouth is typically the only opening to the digestive tract, when a digestive tract is present. Protonephridia are usually present, and the reproductive system is generally hermaphroditic.

CLASS TURBELLARIA

Turbellarians vary in shape from oval to elongated and, like the other flatworm representatives, are usually dorsoventrally flattened. In general, the larger the worm, the more pronounced the flattening. Marine polyclads, some of which have a body width of a few centimeters, are usually less than 1 mm in thickness and are quite fragile [e.g., *Oligoclado* (Fig. 6–5E)]. Head projections are present in some species. These may be in the form of short marginal or dorsal tentacles (Figs. 6–4H, 6–5A, E) or lateral projections of the head called auricles (Figs. 6–5B, 6–20E), which are common in freshwater planarians. Coloration is mostly in shades of black, brown, and gray, although some marine turbellarians have brightly colored patterns. Turbellarians are mostly small but range in size from microscopic to more than 60 cm long (the Lake Baikal triclad, *Rimacephalus arecepta).*

Turbellarians are primarily aquatic, and the great majority are marine. Although there are a few pelagic species, most are bottom dwellers that live in sand or mud, under stones and shells, or on seaweed. Many species are common members of the interstitial fauna (p. 206). Freshwater turbellarians live in lakes, ponds, streams, and springs, where they occupy benthichabiitats. Some species have become terrestrial, but these are confined to humid areas and usually hide beneath logs and leaf mold during the day, emerging only at night to feed. They are mostly large tropical species, but a few, such as the North American *Bipalium adventitium* (Fig. 6–5A) and related species, live in temperate regions. Some 3000 species of turbellarians have been described.

Despite their external simplicity, turbellarians exhibit considerable internal complexity, and the class is composed of a relatively large number of diverse groups. Those orders that are mentioned fre-

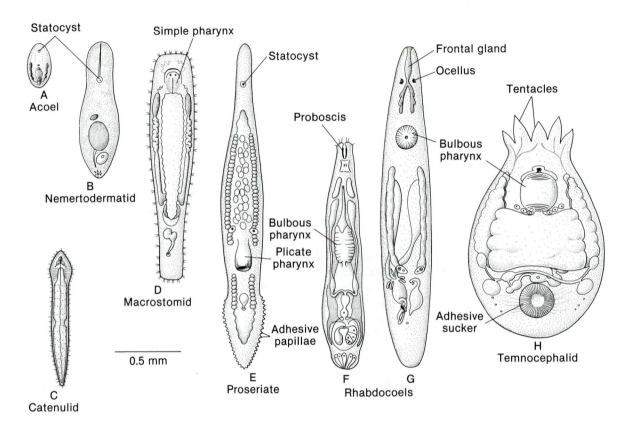

FIGURE 6–4 Representative small turbellarians. **A,** The acoel *Pseudactinoposthia parva.* **B,** The nemertodermatid *Flagellophora.* **C,** The catenulid *Stenostomum virginianum.* **D,** The macrostomid *Macrostomum appendiculatum.* **E,** The proseriate *Monocelis galapagoensis.* **F,** The kalyptorhynch rhabdocoel *Karkinorhynchus tetragnathus.* **G,** The typhloplanoid rhabdocoel *Ceratopera bifida.* **H,** The temnocephalid *Temnocephala geonoma.* *(All figures redrawn from original sources. A, From Ehlers, U., and Dörjes, J. 1979. Interstitielle Fauna von Galapagos. XXIII. Acoela (Turbellaria). Mikrofauna Meeresboden. 72:1–74. B, From Sterrer, W. 1966. New polylithophorous marine Turbellaria. Nature. 210:436. C, From Nuttycombe, J. W. 1931. Two new species of Stenostomum from the southeastern United States. Zool. Anz. 97:80–85. D, From Ferguson, F. F. 1937. The morphology and taxonomy of Macrostomum virginianum n. sp. Zool. Anz. 119:25–32. E, From Ax, P., and Ax, R. 1977. Interstitielle Fauna von Galapagos. XIX. Monocelididae (Turbellaria, Proseriata). Mikrofauna Meeresboden. 64:1–40. F, From Ax, P., and Schilke, K. 1971. Karkinorhynchus tetragnathus nov. spec. ein Schizorhynchier mit zweigeteilten Russelhaken (Turbellaria, Kalyptorhynchia). Mikrofauna Meeresboden. 5:1–10. G, From Ehlers, U., and Ax, P. 1974. Interstitielle Fauna von Galapagos. VIII. Trigonostominae (Turbellaria, Typhloplanoida). Mikrofauna Meeresboden. 30:1–33. H, From Williams, J. B. 1980. Morphology of a species of Temnocephala (Platyhelminthes) ectocommensal on the isopod Phreatoicopsis terricola. J. Nat. Hist. 14:183–199.)*

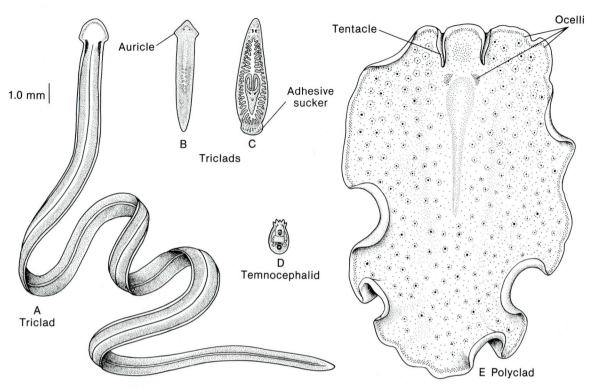

FIGURE 6–5 Representative large turbellarians. The temnocephalid **(D)**, from Figure 6–4, is shown for comparison of scale. **A,** The terrestrial triclad *Bipalium adventitium.* **B,** The freshwater triclad *Dugesia tigrina.* **C,** The horseshoe crab commensal triclad *Bdelloura candida.* **D,** the temnocephalid from Figure 6–4H, for scale. **E,** The polyclad *Oligoclado floridanus. (A and C From Hyman; B and E from Life.)*

quently in the following discussion can be more easily kept in mind if they are grouped and designated at the outset. The members of the orders Nemertodermatida, Acoela, Catenulida, Macrostomida, and Polycladida have a more primitive type of reproductive system (referred to as archoophoran). All five are marine, but the Macrostomida and Catenulida are also represented in freshwater habitats. The Nemertodermatida, Acoela, Catenulida, and Macrostomida are small to microscopic in size, but members of the Polycladida are large, some species reaching 5 cm or more in length (Fig. 6–5E). The Proseriata, Tricladida, and Rhabdocoela are among those turbellarians that have a more advanced type of reproductive system (referred to as neoophoran). The proseriates are marine exclusively, but triclads and rhabdocoels are found both in the sea and in fresh water. The proseriates and rhabdocoels are generally small, but triclads, which include the freshwater and terrestrial species

called planarians, may attain a moderately large size. The turbellarian orders may be summarized as follows:

Archoophoran Level of Organization*
    Order Nemertodermatida (Fig. 6–4B). Marine. Small.
    Order Acoela (Fig. 6–4A). Marine. Small.
    Order Catenulida (Fig. 6–4C). Mostly freshwater. Small.
    Order Macrostomida (6–4D). Marine and freshwater. Small.
    Order Polycladida (Fig. 6–5E). Marine. Large.

Neoophoran Level of Organization
    Order Proseriata (Fig. 6–4E). Marine. Small.

*This synopsis is abbreviated. See the systematic résumé on page 237 for a complete classification.

Order Tricladida (Figs. 6–5A–C). Marine, freshwater and terrestrial. Large.

Order Rhabdocoela (Figs. 6–4F, G). Marine and freshwater. Small.

Order Temnocephalida (Figs. 6–4H, 6–11I). Freshwater (ectosymbionts). Small.

Body Wall

The body of a turbellarian is covered by a simple **ciliated epidermis** in which each cell bears many cilia (Fig. 6–6). The swirling motion of microscopic particles close to the ciliated surface of free-living flatworms is responsible for the name Turbellaria, which means "whirlpool." In some triclads and polyclads, cilia are confined to the ventral surface. Short microvilli cover the epidermal surface between the cilia, but a cuticle is absent. In a few turbellarians the epidermis is syncytial, and cellular boundaries are only

partially present or absent (Fig. 6–6D). Beneath the epidermis is a basal lamina [except in the orders Acoela and Nemertodermatida (Fig. 6–6A)].

In the absence of a cuticle, turbellarians typically use the basal lamina and intracellular fibers to support the body wall. The most common intracellular skeleton is a weblike sheet of actin filaments (terminal web) within the epidermis itself (Fig. 6–6). This sheet apparently helps to bear the stresses on the epidermis. Clear evidence for use of the basal lamina for body wall support is found in some unusual marine turbellarians in which calcareous spicules occur in this layer and form an elaborate skeleton (Fig. 6–7). Ciliary rootlets, the fibrous structures anchoring each cilium within the cell, may also bear stress; and the interjoining of such rootlets into a mesh in acoels and nemertodermatids perhaps compensates for their lack of a basal lamina. The role of the parenchyma in skeletal support is discussed later.

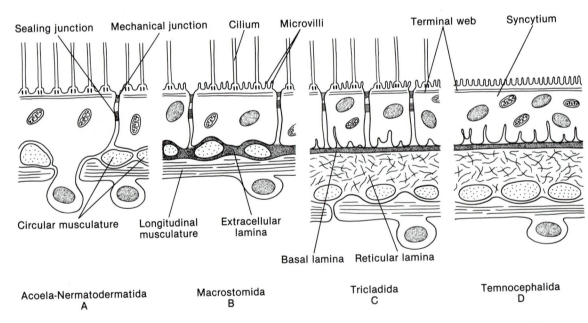

FIGURE 6–6. Sections show the body wall organization in acoels and nemertodermatids **(A)**, macrostomids **(B)**, triclads **(C)**, and temnocephalids **(D)**. The amount of extracellular matrix tends to increase with body size from **A** to **D**. A syncytial epidermis occurs in some rhabdocoels but is typical of the ectocommensal temnocephalids **(D)**. A similar syncytial tegument is characteristic of trematodes, monogeneans, and tapeworms. *(Adapted from Tyler, S. 1984. Turbellarian platyhelminths. In Bereiter-Hahn, J., Matoltsky, A. G., and Richards, K. S. (Eds.): Biology of the Integument. Vol. 1. Invertebrates. Springer-Verlag, Berlin. pp. 112–131. And Rieger, R. M. 1981. Morphology of the Turbellaria at the ultrastructural level. Hydrobiologia. 84:213–229.)*

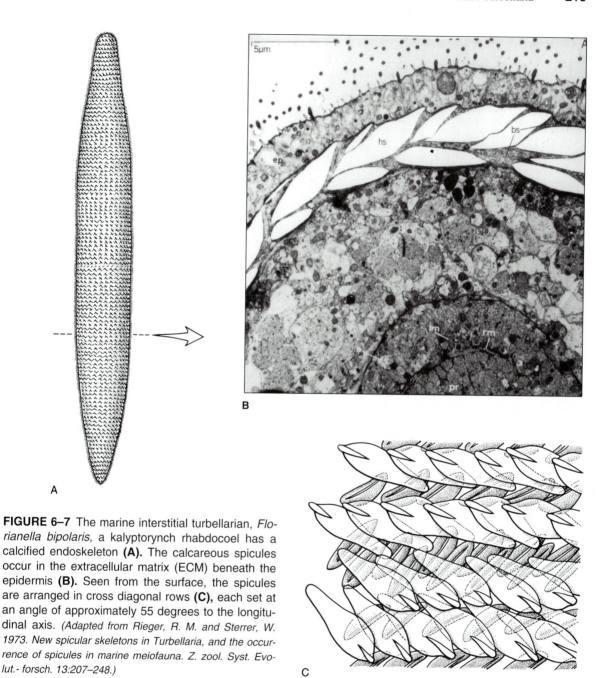

FIGURE 6–7 The marine interstitial turbellarian, *Florianella bipolaris,* a kalyptorynch rhabdocoel has a calcified endoskeleton **(A).** The calcareous spicules occur in the extracellular matrix (ECM) beneath the epidermis **(B).** Seen from the surface, the spicules are arranged in cross diagonal rows **(C),** each set at an angle of approximately 55 degrees to the longitudinal axis. *(Adapted from Rieger, R. M. and Sterrer, W. 1973. New spicular skeletons in Turbellaria, and the occurrence of spicules in marine meiofauna. Z. zool. Syst. Evolut.- forsch. 13:207–248.)*

A characteristic feature of turbellarians is the presence of numerous gland cells (Figs. 6–3, 6–8, 6–9, 6–13). The gland cells may be entirely within the epidermis but are commonly submerged into or below the muscle layers, with only the neck of the gland penetrating the epidermis [or gastrodermis (Fig.

6–9)]. The glands may provide for adhesion, mucus secretion, and other secretory functions. Temporary adhesion to the substratum is made possible by adhesive glands, adhesive cilia, or muscular suckers (Fig. 6–3). Many interstitial marine species adhere to the sand grains of their environment with glandular adhe-

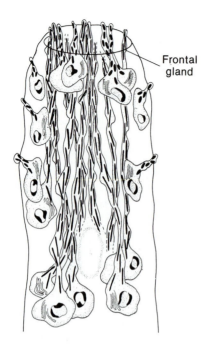

FIGURE 6–8 Frontal gland of the macrostomid *Paramalostomum coronum*. The individual gland cells opening on the sides of the head are not part of the frontal gland. *(From Klauser, M. D., and Tyler, S. 1987. Frontal glands and frontal sensory structures in the Macrostomida (Turbellaria). Zool. Scripta. 16:95–110.)*

sive organs known as **duo-gland organs** (p. 207), each of which may project from the surface as a papilla (Figs. 6–3B, 6–4E, 6–11E). Large species, such as many polyclads and the temnocephalids, ectocommensal on crustaceans, attach to surfaces using a welldeveloped ventral sucker (Figs. 6–3C, 6–4H, 6–5C). At least one tiny acoel *(Paratomella rubra)* is able to stick to surfaces using the flattened tips of specialized cilia (6–3A).

An anterior aggregation of secretory cells, called a **frontal gland,** is characteristic of many turbellarians and is believed to be a primitive turbellarian feature (Figs. 6–4G, 6–8). Its function, however, is unknown, although roles in defense, slime production for locomotion, and adhesion (in larvae) have been suggested. Typical of almost all turbellarians are numerous membrane-bounded, rod-shaped secretions known as rhabdoids that are released to the surface, where they swell and dissolve to form mucus. The most common kind of rhabdoid is the **rhabdite,** characterized by a specific, layered ultrastructure (Fig. 6–9). Rhabdites are secreted by epidermal gland cells, usually submerged below the epidermis, and are found largely in archoophoran macrostomids and polyclads, as well as in many neoophorans. Other rhabdoids of different ultrastructure are found in gland cells in acoels and in the epidermal (epithelial) cells of planarians. Some turbellarians have glands at the posterior end (Fig. 6–4F). In *Bdelloura*, which lives as a commensal on the book gill of the Atlantic

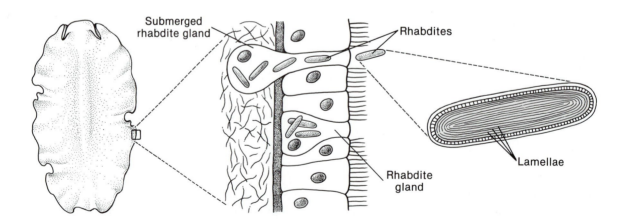

FIGURE 6–9. One of the characteristic turbellarian features is a rhabdite, a rod-shaped secretion, microscopically composed of successive lamellae. *(Enlarged granule drawn from Smith, J. P. S., III, Tyler, S., Thomas, M. B. et al. 1982. The morphology of turbellarian rhabdites: Phylogenetic implications. Trans. Am. Micros. Soc. 101:209–228.)*

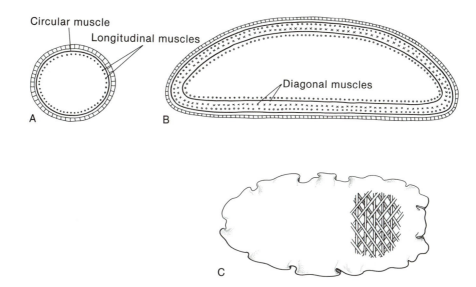

FIGURE 6–10 In most small turbellarians, the body wall musculature consists of the usual outer circular and inner longitudinal layers **(A)**. The musculature of many large turbellarians, however, has additional diagonal (55 degrees in respect to longitudinal axis), circular, and longitudinal layers **(B, C)**. Often, dorsoventral muscles are also present (not shown). *(B, C, Redrawn from Lang, A. 1894. Die Polycladen des Golfes von Neapel und der angrenzenden Meeresabschnitte. Verlag Wilhelm Engelmann. Leipzig. 688 pp.)*

horseshoe crab, the glands form an adhesive plate (Fig. 6–5C).

In most small turbellarians, the musculature consists of the typical outer circular and inner longitudinal layers (Fig. 6–10A), but diagonal and dorsoventral muscles also may be present, although these are best developed in large species (Fig. 6–13B). Larger turbellarians often have additional layers of circular and longitudinal muscles (Figs. 6–10B, C). Turbellarian muscle fibers usually are smooth.

Locomotion

Turbellarians have a wide variety of locomotory adaptations that range from ciliary gliding over surfaces and through water to muscular creeping, swimming, peristalsis, and other movements, such as twisting, turning, retraction and extension, and somersaulting (Fig. 6–11). In general, small aquatic turbellarians use cilia for locomotion (Fig. 6–11A), whereas large turbellarians use muscular movements (perhaps aided by cilia) of the entire body or along a specialized ventral sole (Figs. 6–11C, D). In some

small species, body wall peristalsis is common but is not used in locomotion. Instead, these movements may be important in mixing the contents of the gut, which often has a weakly developed musculature (Fig. 6–11G). Some polyclad species swim using dorsoventral undulations of the lateral body margins (Fig. 6–12).

Parenchyma

The connective tissue compartment between the body wall musculature and gut is called the **parenchyma.** Like the typical connective tissue described in Chapter 3, the parenchyma of *large* flatworms is composed of mesenchymal cells in a fibrous extracellular matrix (Fig. 6–13B). Departures from this organization, however, occur in two directions. First, the parenchyma of *small* flatworms contains little extracellular matrix, and in acoels an extracellular matrix is absent and the parenchyma is entirely cellular (Fig. 6–13A). Second, in many freshwater catenulids, the extracellular matrix part of the parenchyma is well developed and fluid rather than fibrous and forms a pseudocoel (Fig.

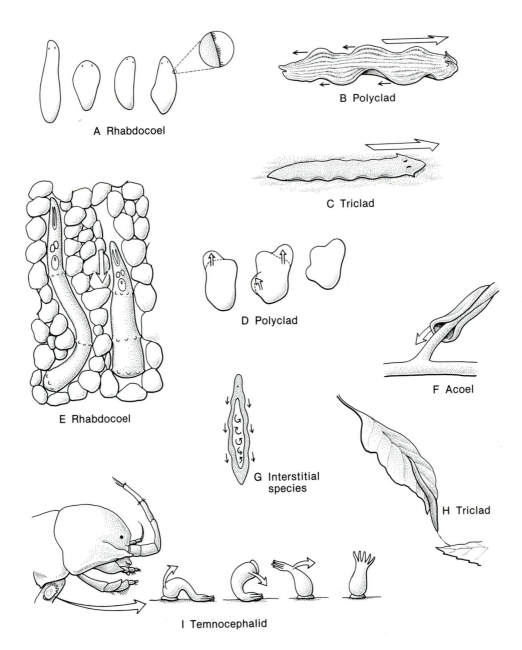

A Rhabdocoel

B Polyclad

C Triclad

D Polyclad

E Rhabdocoel

F Acoel

G Interstitial species

H Triclad

I Temnocephalid

6–13C). The pseudocoel may play a role in internal transport and as a hydrostatic skeleton.

Flatworm parenchymal cells are diverse; only a few of whose functions are established and are mentioned here. **Epidermal replacement cells** migrate from the parenchyma to the body surface and replace any damaged or destroyed epidermal cells. This unusual means of replacement is necessitated by an absence of mitosis in the adult epidermis. The epidermal replacement cells are situated immediately below the body wall and each contains a cluster of centrioles (which later become the ciliary basal bodies). Many turbellarians have a population of totipotent cells called **neoblasts** that are important in wound healing and regeneration (Fig. 6–13B). Another common cell of the parenchyma is the **fixed parenchymal cell,** a large branched cell that makes contacts with and interjoins other cells and tissues (Fig. 6–13B).

(Text continues on page 221)

◀**FIGURE 6–11** Ciliary locomotion is common in small turbellarians, such as the rhabdocoel *Kytorhynchella meixneri* **(A)** and the acoel *Convoluta convoluta* **(F)**. The large polyclad *Pseudoceros crozieri* **(B)** swims using undulations of the body margins. Planarians produce muscular waves along their ventral surface to move over substrata **(C)**, whereas polyclads, such as *Corondena mutabilis,* shuffle forward ditaxically **(D)** like some snails. Retraction movements are common among turbellarians, shown here for the interstitial rhabdocoel *Proschizorhynchus anophthalmus* **(E)**, which not only withdraws from a disturbance but simultaneously anchors itself using girdles of duo-gland adhesive papillae. While stationary, some interstitial species produce peristaltic waves along the body surface, presumably to help mix the gut contents **(G)**. The terrestrial triclad *Rhynchodemus terrestris* casts a mucous thread and then uses it as a suspension bridge to cross between two leaves **(H)**. Some temnocephalids, which are ectocommensals on crus- taceans, can somersault by alternately attaching the posterior sucker and anterior tentacles **(I)**. *(A, Modified from Rieger, R. M. 1974. A new group of Turbellaria-Typhlo- planoida with a proboscis and its relationship to Kalyp- torhynchia. In Riser, N. W. and Morse, M. P. (Eds.): Biology of the Turbellaria. McGraw-Hill Book Co., New York. pp. 23–62. E, Redrawn and modified from L'Hardy, J. -P. 1965. Turbellaries Schizorhynchidae des sables de Roscoff. II. Le genre Proschizorhynchus. Cah. Biol. Mar. 6:135–161. F, Re- drawn from Apelt, G. 1969. Fortpflanzungsbiologie, Entwick- lungszyklen und vergleichende Frühentwicklung acoeler Turbellarian. Mar. Biol. 4:267–325. H, Redrawn and modified from Reisinger, E. 1923. Turbellaria. Strüdelwurmer. Biol. Tiere Deutsch. Lief. 6:1–64. I, Redrawn and modified from Williams, J. B. 1980. Morphology of a species of Temno- cephala (Platyhelminthes) ectocommensal on the isopod Phreatoicopsis terricola. J. Nat. Hist. 14:183–199, and after Haswell from de Beauchamp, P. 1961: Classe des Turbellar- ies. In Grassé, P. -P. (Ed.): Traité de Zoologie. Vol. 4. Mas- son et Cie, Paris. pp. 216.)*

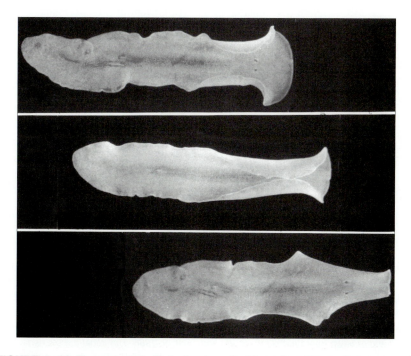

FIGURE 6–12 The polyclad, *Stylochoplana floridana,* is a 20-mm long worm that swims rapidly (15 mm/s) by flapping its anterior body margins in the manner of a sea hare. *(Ponce de Leon Inlet, Florida. Photograph courtesy of Paul Clem.)*

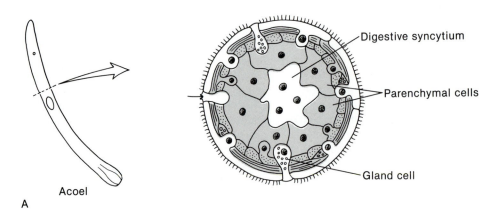

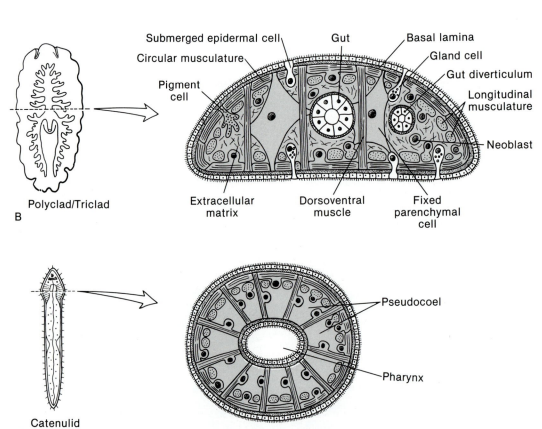

FIGURE 6–13 Turbellarian parenchyma (connective tissue compartment) can have three grades of organization. **A,** In acoels, such as *Diopisthoporus longitubus,* the parenchyma is entirely cellular and lacks an extracellular matrix. **B,** In most other turbellarians, including the polyclads and triclads, the parenchyma is like typical connective tissue, containing both cells and extracellular matrix. **C,** The extracellular matrix of some catenulids is fluid and forms a pseudocoel. *(A, B, Modified and redrawn from Smith, J. P. S. III., and Tyler, S. 1985. The acoel turbellarians: Kingpins of metazoan evolution or a specialized offshoot? and Rieger, R. M. 1985. The phylogenetic status of the acoelomate organization within the Bilateria: A histological perspective. Both in Morris, S. C., George, J. D., Gibson, R. et al. (Eds.) The Origins and Relationships of Lower Invertebrates. Clarendon Press, Oxford. pp. 123–142, 101–122.)*

Gap junctions, forming low-resistance pathways for intercellular transport of metabolites, are often associated with these contacts, suggesting that fixed parenchymal cells have an integrative role in turbellarian physiology. Some planarians have been found to possess parenchymal pigment cells (Fig. 6–13B) and **chromatophores,** which can cause the animal to lighten when the pigment in the chromatophore is concentrated toward the nucleus. Control is apparently under the direction of the brain, for the posterior half of a bisected dark worm does not lighten until it has regenerated a brain. Lastly, in at least one interstitial acoel *(Paratomella rubra),* parenchymal cells containing hemoglobin color the body red and probably function as an oxygen store for use when the animal wanders into oxygen-poor layers of the sand.

Digestive System and Nutrition

The digestive cavity, or gut, of turbellarians is typically a blind sac, and the mouth is used for both ingestion and egestion (Fig. 6–14). An anus or multiple anuses occur only in some very long worms and in some turbellarians with highly branched guts. In these, the normal return of undigested wastes to the mouth is apparently complicated by the extreme length or complex branching of the gut. The wall of the gut is single-layered and composed of phagocytic and gland cells. The primitive gut in the Macrostomida, Catenulida, and some polyclads is ciliated, but in most others it lacks cilia.

The form of the gut is in part related to the size of the worm (Fig. 6–14). Small turbellarians, such as the macrostomids, catenulids, and rhabdocoels, have a gut that is a simple unbranched sac. The members of the order Acoela lack a permanent gut cavity, and the gut is usually a syncytial mass enclosed by a common cell membrane (Fig. 6–14F). It is to the lack of a gut cavity that the name *Acoela* refers.

The larger turbellarians have guts with lateral diverticula, which increase the surface area for digestion and absorption and compensate for the absence of an internal nutrient transport system. In polyclads, the intestine consists of a central tube, from which a great many lateral branches arise (Fig. 6–14E). These, in turn, are subdivided and may anastomose with other branches. The members of the Tricladida, to which the planarians belong, have an intestine composed of three principal branches—one anterior and two posterior (Figs. 6–14D, 6–20A). Each of these branches, in turn, has many lateral diverticula. The three branches join in the middle of the body, anterior to the mouth and pharynx. The names *Polycladida* and *Tricladida* refer to the branching intestine of these groups of turbellarians.

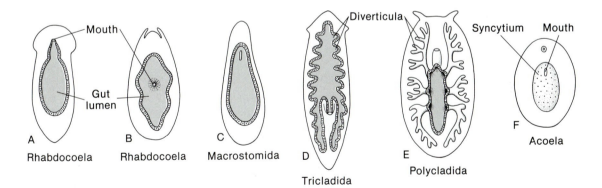

FIGURE 6–14 Turbellarian midguts are more or less simple sacs in small turbellarians **(A–C)** such as most rhabdocoels **(A, B)** and macrostomids **(C).** In the acoels **(F),** the gut is usually a lumenless syncytium, as shown here. The larger turbellarians, such as triclads **(D)** and polyclads **(E),** have vessel-like diverticula that increase the area of the gut and distribute nutrients to all parts of the body. *(A–F, Modified and redrawn after Hyman from Rieger, R. M., Tyler, S., Smith, J. P. III. et al. 1991. Platyhelminthes: Turbellaria. In Harrison, F. W. and Bogitsh, B. (Eds.): Microscopic Anatomy of Invertebrates. Vol. 3. Wiley-Liss., New York. pp. 7–140.)*

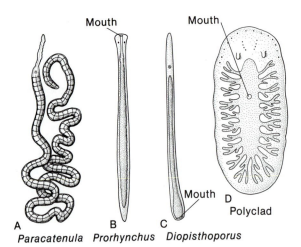

Mouth Mouth Mouth D Polyclad

A B C
Paracatenula Prorhynchus Diopisthoporus

FIGURE 6–15 The presence and position of the mouth is variable in turbellarians. In at least one rare marine interstitial species **(A)**, the catenulid *Paracatenula urania,* the mouth is entirely absent, and the animal derives its nutrition directly from prokaryotic symbionts found in its rudimentary gut. When present, the mouth ranges from anterior in the prolecithophoran *Prorhynchus stagnalis* **(B)**, posterior in the acoel *Diopisthoporus gymnopharyngeus* **(C)**, to midventral in many others, such as this polyclad **(D)**. *(A, Redrawn from Sterrer, W. and Rieger, R. M. 1974. Retronectidae—A new cosmopolitan marine family of Catenulida (Turbellaria). In Riser, N. W. and Morse, M. P. (Eds.): Biology of the Turbellaria. McGraw-Hill Book Co., New York. pp. 63–82. B, Redrawn from Hyman, L. H. 1951. The Invertebrates: Platyhelminthes and Rhynchocoela. The Acoelomate Bilateria. Vol. II. McGraw-Hill Book Co., New York. p. 152. C, Drawn from a photograph In Smith, J. P. S. III., and Tyler, S. 1985. Fine-Structure and Evolutionary Implications of the Frontal Organ in Turbellaria Acoela. 1. Diopisthoporus gymnopharyngeus sp. n. Zool. Scripta 14:91–102.)*

The mouth is commonly located on the midventral surface but may be situated anteriorly, posteriorly, or anywhere along the midventral line (Fig. 6–15). The pharynx shows increasing complexity within the class. Most acoels lack a pharynx, and the mouth opens directly into the digestive syncytium. The pharynx is a simple, ciliated tube (**simple pharynx**) in the small, primitive Macrostomida and Catenulida *(Stenostomum)* (Figs. 6–16A, 6–17). In

higher turbellarians (polyclads and all neoophorans) the pharynx is a more complex ingestive organ. A folded, or **plicate, pharynx** is characteristic of the polyclads and triclads, which are large turbellarians with branched intestines. The folded condition is believed to have evolved from a simple pharynx and has resulted in a muscular pharyngeal tube lying within a pharyngeal cavity (Fig. 6–16B). The free end of the tube can be protruded from the mouth during feeding (Fig. 6–18). The pharynx may project backward (Fig. 9–20D), as in the common freshwater planarians, or the pharynx may be attached to the cavity posteriorly and extend forward. A **bulbous pharynx,** encountered in the Rhabdocoela, is believed to be derived from the plicate condition through reduction of the outer pharyngeal cavity and separation by a special muscular septum from the parenchyma of the body (Figs. 6–4F–H, 6–16C). This separation isolates the action of the pharynx from the influence of the inertial parenchyma. The sucking muscular bulb, which characterizes this type of pharynx, can be protruded from the mouth in many species. Many predatory and most commensal and parasitic species have a bulbous pharynx.

Turbellarians are largely carnivorous and prey on various invertebrates that are small enough to be captured, as well as on the dead bodies of animals that sink to the bottom (Fig. 6–17). Feeding behavior is elicited, at least in some species (e.g., planarians), by substances emitted from the potential food source. Protozoa, rotifers, insect larvae, small crustaceans (e.g., water fleas, copepods, amphipods, isopods), snails, and small annelid worms are common prey, but there are marine species that feed on sessile animals, such as bryozoans and small tunicates. The polyclad *Stylochus frontalis* feeds on living oysters and is nicknamed the "oyster leech;" *Stylochus triparitus* preys on barnacles; and the West Indian *Pseudoceros crozieri* feeds on the colonial tunicate, *Ecteinascidia turbinata* (Fig. 6–18). The ectocommensal triclad *Bdelloura,* which lives on the book gill of horseshoe crabs, shares in the food of its host.

Not all turbellarians are predators. Some acoel, macrostomid, and polyclad species feed on algae, especially diatoms, and other predaceous species feed on diatoms as juveniles before switching over to animal food. Several acoel species harbor green zoochlorellae or golden zooxanthellae or diatoms in their parenchyma, and at least one of these species

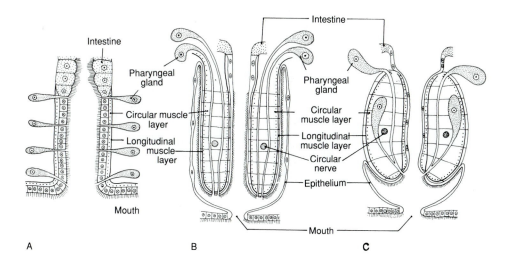

FIGURE 6–16 The three basic designs of the turbellarian pharynx include a simple pharynx **(A)** in macrostomids, catenulids, and a few acoels; a plicate pharynx **(B)** in polyclads and planarians; and a bulbous pharynx **(C)** in rhabdocoels and temnocephalids. *(Modified from Ax, P. 1963. Relationships and phylogeny of the Turbellaria. In Dougherty, E. C. (Ed.): The Lower Metazoa. University of California Press, Berkeley. pp. 191–224.)*

(Convoluta roscoffensis) relies on them for nutrition because the acoel does not ingest food as long as the algal symbionts are present. Species of the rhabdocoel family, Fecampiidae, are endoparasitic in the hemocoel of crustaceans. These turbellarians lack a digestive tract and absorb host nutrients across the body wall (p. 204).

Many turbellarians capture living prey by wrapping themselves around it, entangling it in slime, and pinning it to the substratum by means of the adhesive organs. Species of *Mesostoma* paralyze their prey with mucus. A few species are known to stab prey with a penis, which terminates in a hardened stylet and projects from the mouth; the interstitial kalyp-

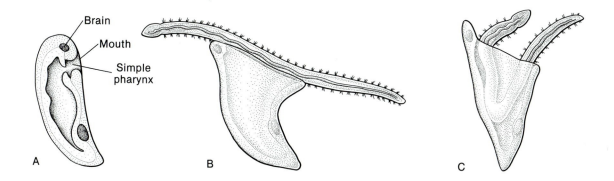

FIGURE 6–17 The freshwater, *Microstomum caudatum* (Macrostomida) uses a simple pharynx and mouth, which undergo a surprising dilation, to engulf an oligochaete worm. *(All lateral views. Modified and redrawn from Kepner, W. A. and Helvestine, F., Jr. 1920. Pharynx of Microstoma caudatum. J. Morphol. 33:309–316.)*

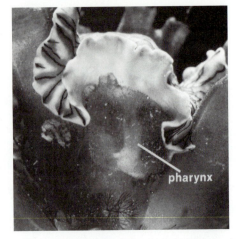

FIGURE 6–18 The polyclad, *Pseudoceros crozieri* crawls over (left photograph) and preys on (right photograph) one zooid of the colonial tunicate, *Ecteinascidia turbinata*. While feeding, the worm thrusts its everted plicate pharynx into one of the sea squirt's siphons (lower photograph). *(Big Pine Key, Florida.)*

torhynch rhabdocoels possess an anterior, raptorial proboscis that may have either a sticky tip or grasping hooks (Fig. 6–19). The proboscis of these species is not connected with the mouth and bulbous pharynx.

Prey is swallowed whole by those turbellarians with a simple pharynx (Fig. 6–17), by those with a protrusible bulbous pharynx, and even by the polyclads, which have a plicate pharynx. In the triclads, the pharyngeal tube is extended from the mouth and inserted into the body of the prey or the carrion. The exoskeleton of crustaceans is penetrated at thin points, such as the articulations between body segments. Penetration by the pharynx and ingestion of

body tissues of the prey are aided by proteolytic enzymes (endopeptidases) produced by pharyngeal glands that open onto the tip of the pharynx. The partially digested and liquified contents are then pumped into the gut by peristaltic action.

In studies on the acoel *Convoluta paradoxa*, small prey is captured and engulfed by the digestive syncytium, which is partially everted through the mouth. Larger prey is pressed into the mouth and swallowed. In both of these cases, prey probably enter the syncytium by phagocytosis.

Digestion is first extracellular. Disintegration of the ingested food is initiated by pharyngeal enzymes,

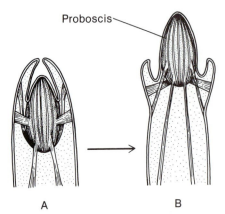
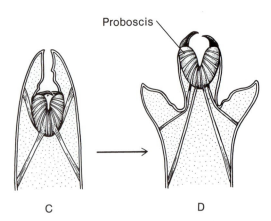

FIGURE 6–19 An eversible proboscis occurs in several groups of turbellarians but is particularly well developed in the kalyptorhynch rhabdocoels, shown here. One type **(A, B)** is a muscular mass covered by a sticky surface to which prey adhere. Another form **(C, D)** is a grasping organ that often bears hooks. *(Redrawn and modified from de Beauchamp, P. 1961. Classe des Turbellaries. In Grassé, P. -P. (Ed.): Traité de Zoologie. Vol. 4. Masson et Cie, Paris. p. 172.)*

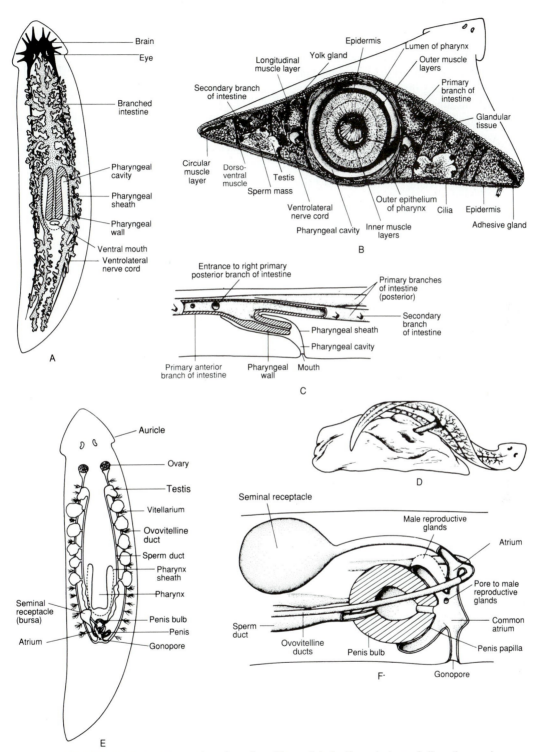

FIGURE 6–20 Anatomy of a planarian *(Dugesia).* **A,** Dorsal view of digestive and nervous systems. Commissures between cords are not shown. **B,** Cross section of the body at the level of the pharynx and gut. **D,** Lateral view of pharynx protruded in food mass. **E,** Dorsal view of reproductive system. **F,** Sagittal view of reproductive system.

and additional endopeptidase is supplied by gland cells of the intestine. The resulting food fragments are then engulfed by the phagocytic cells, where digestion by endopeptidases continues in an acid medium. About 8 to 12 h following ingestion the phagocytic vacuole becomes alkaline, which marks the appearance of exopeptidases, lipases, and carbohydrases necessary to complete digestion. During the course of intracellular digestion, the vacuole sinks more deeply into the phagocytic cell and eventually disappears.

Freshwater planarians are able to withstand prolonged periods of starvation experimentally. In extreme cases they resorb and utilize part of the gut and all of the parenchyma and reproductive system tissues. In fact, the body volume may be reduced to as little as 1/300 of the original.

Although parasitism in flatworms is usually associated with trematodes and cestodes, there are a number of commensal and parasitic turbellarians. They are largely freshwater and marine rhabdocoels (suborder Dalyellioida) as well as the freshwater order Temnocephalida (Fig. 6–11I). Commensals include species that live within the mantle cavities of molluscs and on the gills of crustaceans. Parasitic forms inhabit the guts and body cavities of molluscs, crustaceans, and echinoderms, as well as the skin of fishes.

Internal Transport

Turbellarians are either tiny and more or less cylindrical or larger and dorsoventrally flattened. In either case, the diffusion distance for gas exchange is short, and oxygen is absorbed across the general body wall. In small turbellarians, nutrients diffuse from the central gut to the nearby tissues, but in large *flat* worms the distance from the central gut to the lateral body margin exceeds the range of simple diffusion. As a result, large flat turbellarians have branches from the gut that transport nutrients to the peripheral parts of the body. In this functional sense, the branched digestive tract of large turbellarians is similar to the gastrovascular canal system of some cnidarians and ctenophores. Nutrients may also be transported intracellularly from gastrodermal phagocytes to fixed parenchymal cells in a manner similar to nutrient transport from a choanocyte to an archaeocyte in sponges. Actual circulation of fluid may occur in the fluid-filled pseudocoel of freshwater catenulids.

Excretion

Turbellarians eliminate protein nitrogen in the form of ammonia, which diffuses across the general body surface, but they release excess water and probably other waste metabolites using protonephridia that

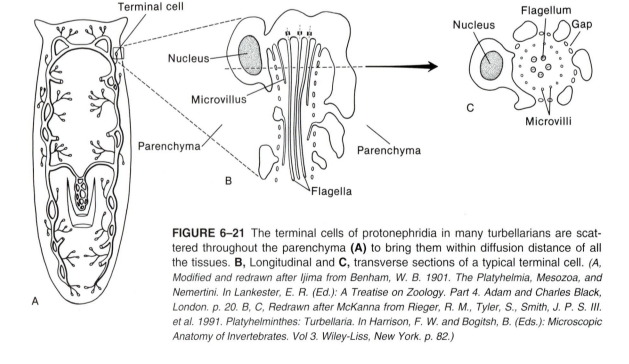

FIGURE 6–21 The terminal cells of protonephridia in many turbellarians are scattered throughout the parenchyma **(A)** to bring them within diffusion distance of all the tissues. **B,** Longitudinal and **C,** transverse sections of a typical terminal cell. *(A, Modified and redrawn after Ijima from Benham, W. B. 1901. The Platyhelmia, Mesozoa, and Nemertini. In Lankester, E. R. (Ed.): A Treatise on Zoology. Part 4. Adam and Charles Black, London. p. 20. B, C, Redrawn after McKanna from Rieger, R. M., Tyler, S., Smith, J. P. S. III. et al. 1991. Platyhelminthes: Turbellaria. In Harrison, F. W. and Bogitsh, B. (Eds.): Microscopic Anatomy of Invertebrates. Vol 3. Wiley-Liss, New York. p. 82.)*

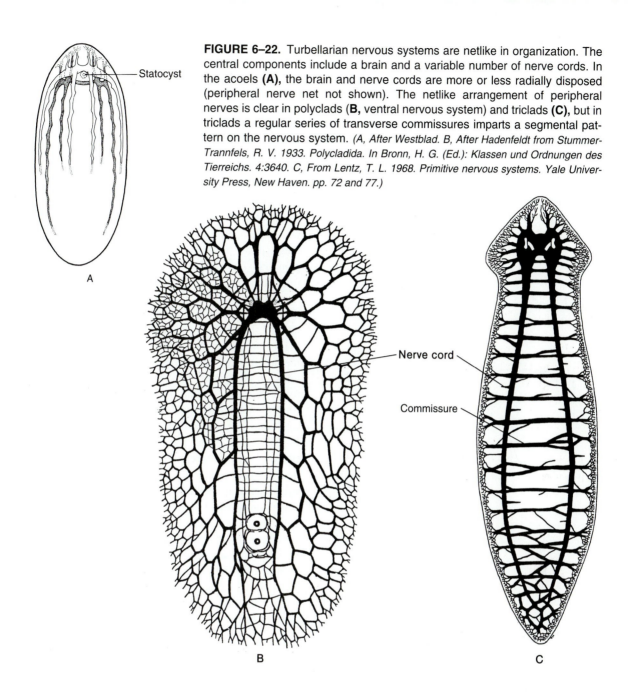

Statocyst

FIGURE 6–22. Turbellarian nervous systems are netlike in organization. The central components include a brain and a variable number of nerve cords. In the acoels **(A)**, the brain and nerve cords are more or less radially disposed (peripheral nerve net not shown). The netlike arrangement of peripheral nerves is clear in polyclads (**B,** ventral nervous system) and triclads **(C),** but in triclads a regular series of transverse commissures imparts a segmental pattern on the nervous system. *(A, After Westblad. B, After Hadenfeldt from Stummer-Trannfels, R. V. 1933. Polycladida. In Bronn, H. G. (Ed.): Klassen und Ordnungen des Tierreichs. 4:3640. C, From Lentz, T. L. 1968. Primitive nervous systems. Yale University Press, New Haven. pp. 72 and 77.)*

Nerve cord

Commissure

A

B

C

bear multiciliated terminal cells (Figs. 6–21B, C). Because there is no circulatory system to deliver excess water and other wastes from all parts of the body to a centralized kidney, turbellarian terminal cells typically are scattered widely to position them within diffusion distance of all tissues (Fig. 6–21A). The terminal cells then feed into anastomosing and sometimes ciliated ducts, which eventually open to the exterior at one or more pores, depending on the species. Unlike all other turbellarians, acoels lack any recognized excretory system.

Nervous System and Sensory Organs

The general design of the turbellarian nervous system is variable, especially with respect to the number and arrangement of nerve cords. In general, and perhaps primitively, it consists of a subepidermal ringlike brain from which one to several nerve cords, depending on the species, extend posteriorly through the body. When several pairs of longitudinal nerves extend from the brain, they are usually equidistant from each other and impart radial symmetry on the nervous system (Fig. 6–22A). The longitudinal nerve cords join a nerve net located internal to the body wall musculature. This submuscular net, in turn, joins two other more peripheral nets—one between the epidermis and musculature and another within the epidermis. Although some nervous tissue is concentrated into brain and longitudinal cords, the nervous system as a whole is diffuse, netlike, and reminiscent of that of cnidarians (Fig. 6–22B). A specialized net may be associated with the pharynx and midgut of some turbellarians.

A common evolutionary trend within the turbellarian nervous system is a departure away from the nerve net toward a concentration of nerves into a well-developed bilateral brain and two ventrolateral, longitudinal nerve cords, as occurs in the common planarian *Dugesia*. Although *Dugesia* retains a peripheral nerve net, the longitudinal cords are joined at regular intervals by transverse commissures, which together with the cords give the nervous system a segmented, ladder-like appearance (Fig. 6–22C). Such a highly organized nervous system suggests a level of hierarchic order and control not present in netlike systems. In all turbellarians, the nervous system is relatively primitive in that it lacks ganglia, except in the brain, but typical types of neurons are present. Pigment-cup eyes are common in most turbellarians (Fig. 6–23B). Two is the usual number (Fig. 6–4G), but two or three pairs are not uncommon. In the polyclads and the land planarians, there may be a great many eyes located in clusters over the brain, in tentacles, or distributed around the body margin (Fig. 6–23B). The eyes function largely in orienting to light, and most turbellarians are negatively phototactic. Strong light directed onto a swimming polyclad, for example, will stop it immediately, causing it to sink to the bottom, where it will crawl away in search of cover.

Other than eyes, statocysts are the most conspicuous of turbellarian sensory organs (Fig. 6–23A), but

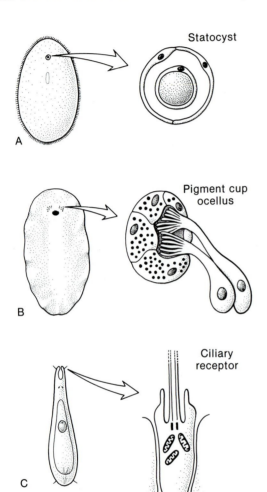

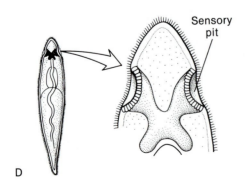

◀**FIGURE 6–23** Examples of turbellarian sensory organs are a statocyst of an acoel **(A),** pigment cup ocellus of a polyclad **(B),** simple ciliary receptor of a rhabdocoel, probably a mechanoreceptor **(C),** and a sensory pit of a catenulid, probably a chemoreceptor **(D).** *(A–C, Modified and redrawn from Rieger, R. M., Tyler, S. III et al. 1991. Platyhelminthes: Turbellaria. In Harrison, F. W. and Bogitsh, B. (Eds.): Microscopic Anatomy of Invertebrates. Vol. 3. Wiley-Liss, Inc., New York. pp. 7–140, D, Modified and redrawn from Nuttycombe, J. W. and Waters, A. J. 1938. The American species of the genus Stenostomum. Proc. Am. Philos. Soc. 79:213–300.)*

they only occur in a few orders, such as the acoels, nemertodermatids, proseriates, and some catenulids. Turbellarian statocysts are unpaired and located medially near the brain. Each statocyst consists of a capsule that encloses a fluid-filled cavity and a central concretion called a statolith. Sometimes, more than one statolith may be present; nemertodermatids, for example, typically have two. Because of their general similarity to the statocysts of some cnidarian medusae and ctenophores, turbellarian statocysts are presumed to be gravity receptors. Unlike the statocysts of diploblastic animals, however, turbellarian statocysts lack sensory cilia, and the statolith makes contact with the unspecialized wall of the capsule. The mechanism of sensory reception is unknown.

Dispersed ciliary receptors (Fig. 6–23C), many of which are probably mechanoreceptors, are distributed over the entire body but are particularly concentrated on tentacles, auricles, and body margins. Cilia or microvilli are located in pits or grooves in the head region and probably contain chemoreceptors (Fig. 6–23D), which may be used in locating food.

Asexual Reproduction

Many turbellarians, especially freshwater species, reproduce asexually by means of budding or transverse fission. Budding occurs in the genera *Catenula, Stenostomum,* and *Microstomum,* whose species are all small. The buds, called zooids, differentiate along the length of the parent's body and form chains before fission separates them into new individuals (Fig. 6–24A). This form of reproduction, which is called **paratomy,** resembles strobilation in scyphomedusae. The large freshwater planarians divide asexually by transverse fission and regeneration (p. 204) occurs after separation (Fig. 6–24B). The fission plane usually forms behind the pharynx, and separation appears to depend on locomotion: the posterior end of the worm clings to the substratum, and the anterior half continues to move forward until the two regions pull apart. Each half then regenerates missing structures to form a complete worm.

A few species of freshwater planarians, such as members of the genus *Phagocata* and some land pla-

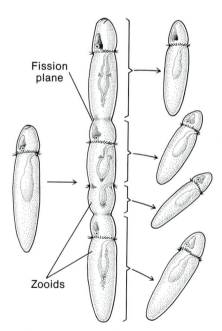

Fission plane

Zooids

A PARATOMY

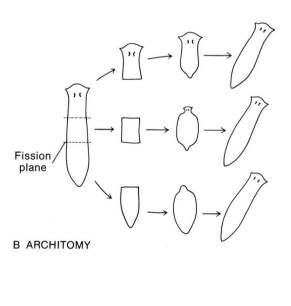

Fission plane

B ARCHITOMY

FIGURE 6–24 A, Paratomy in the catenulid, *Catenula lemnae.* **B,** Architomy in a planarian. *(B, After Marcus from de Beauchamp, P. 1961. Classe des Turbellaries. In Grassé, P. -P. (Ed.): Traité de Zoologie. Vol 4. Masson et Cie, Paris. p. 193.)*

narians, simultaneously fragment the body into several pieces. This form of fission is called **architomy** (Fig. 6–24B). In *Phagocata,* each piece forms a cyst in which regeneration takes place and from which a small worm emerges.

Reproduction may be controlled by day length and temperature. Freshwater planarians, for example, which are almost all inhabitants of temperate regions, reproduce asexually by fission during the summer and sexually during the fall under the stimulus of shorter day lengths and lower temperatures. Asexually reproducing laboratory cultures of *Catenula* have been maintained for as long as six years without the occurrence of sexual reproduction. Parthenogenesis is also an important asexual reproductive strategy in most species of Catenulida.

The common planarian *Dugesia dorotocephala,* which is easily maintained in laboratory culture and has been studied extensively, undergoes fission only at night. During the day the brain produces some sub-

stance that inhibits fission, and the production of the inhibitor appears to be under photoperiodic control.

Sexual Reproduction

All turbellarians, except for a few specialized parasitic species, are **hermaphrodites** (p. 209), reproducing by way of copulation and **internal fertilization.** Because turbellarians are small animals and because eggs have a lower size limit of approximately 50 μm, turbellarians produce relatively few eggs, which are never recklessly spawned. Development is lecithotrophic and direct, but in many large marine polyclads successive batches of numerous eggs develop indirectly into planktotrophic larvae.

Except for acoels and a few others (Fig. 6–25A), the gonads are separated by an epithelium from the surrounding parenchyma (Fig. 6–25B). The male and female systems are complex and variable, but can be described as follows: the male part of the system,

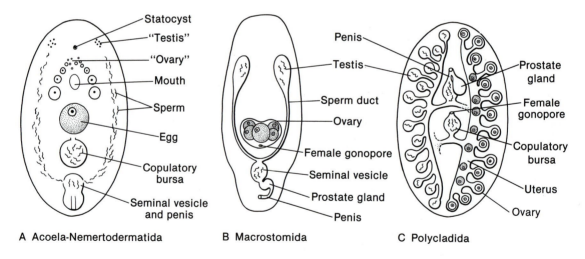

A Acoela-Nemertodermatida   B Macrostomida   C Polycladida

FIGURE 6–25 The archoophoran reproductive system. **A,** Acoel-nemertodermatid. Ovaries and testes are absent and germ cells mature in parenchyma. **B,** Macrostomid. Small animals with ovaries, testes, and well-developed accessory reproductive organs. **C,** Polyclad. Multiple ovaries and testes scattered among branches of gut. Accessory organs are well developed. *(A, Modified and redrawn from Rieger, R. M., Tyler, S., Smith, J. P. S. III. et al. 1991. Platyhelminthes: Turbellaria. In Harrison, F. W. and Bogitsh, B. (Eds.): Microscopic Anatomy of Invertebrates. Vol. 3. Wiley-Liss, New York. p. 98. B, Modified and redrawn from Schmidt, P. and Sopott-Ehlers, B. 1976. Interstitielle Fauna von Galapagos. XV. Macrostomum O. Schmidt, 1848 und Siccomacrostomum triviale nov. gen. nov. spec. (Turbellaria, Macrostomida). Mikrofauna Meeresboden. 57:1–44. C, Redrawn and modified after von Graff from Benham, W. B. 1901. The Platyhelmia, Mesozoa, and Nemertini. In Lankester, E. R. (Ed.): A Treatise on Zoology. Part 4. Adam and Charles Black, London. p. 37.)*

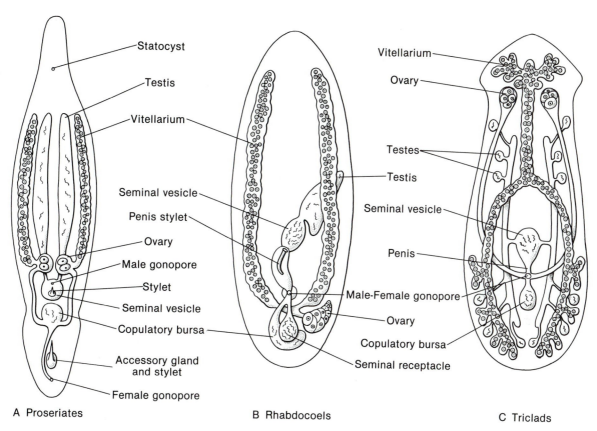

FIGURE 6–26 Neoophoran reproductive systems of proseriate **(A)**, rhabdocoel **(B)**, and triclad **(C)** turbellarians. In all three examples, the vitellaria, which contain the yolk-forming cells, lie near the wall of the gut and its branches. Vitellaria, with their large volume and small yolk cells, may be a specialization for rapid yolk synthesis. *(A, Modified and redrawn from Ax, P. and Ax, R. 1977. Interstitielle Fauna von Galapagos. XIX. Monocelidae (Turbellaria, Proseriata). Mikrofauna Meeresboden. 64:1–44. B, Modified and redrawn from Noldt, U. and Reise, K. 1987. Morphology and ecology of the kalyptorhynch Typhlopolycystis rubra (Plathelminthes) an inmate of lugworm burrows in the Wadden Sea. Helgolander Meeresunters. 41:185–199. C, Modified and redrawn from de Beauchamp, P. 1961. Classe des Turbellaries. In Grassé, P. -P. (Ed.): Traité de Zoologie. Vol. 4. Masson et Cie, Paris. p. 83.)*

transporting sperm *out,* consists of paired **testes,** each leading into a **sperm duct,** a storage sac or **seminal vesicle,** and a **penis** (copulatory organ). The penis may be armed with a stylet and often receives secretions from a prostate gland. The female part of the system, through which sperm move *in* from the partner, is specialized as a **gonopore** (vagina) **copulatory bursa** and **seminal receptacle,** for short and long-term sperm storage, respectively. The female system also produces eggs and transports them *out* from paired **ovaries** via an **oviduct** to the gonopore. Such a generalized hermaphroditic system is essentially similar to that in *Macrostomum. Macrostomum,* however, lacks a seminal receptacle and has cement glands for

the attachment of eggs to the substratum (Figs. 6–25B, 6–34). Some turbellarians have more than two testes or ovaries.

The ovaries of archoophoran turbellarians, such as *Macrostomum,* are like those of other animals and produce eggs in which yolk material is an integral part of the cytoplasm of the egg cell (entolecithal eggs) (Fig. 6–27E). The ovary of many other turbellarians, however, has evolved into two specialized regions, the ovary proper for the production of female gametes, and a **vitellarium,** sometimes called a **yolk gland,** for the production of yolk-filled nurse cells (Figs. 6–26B, 6–27F). The ovary and vitellarium may be united into an ovovitellarium, or they may be sepa-

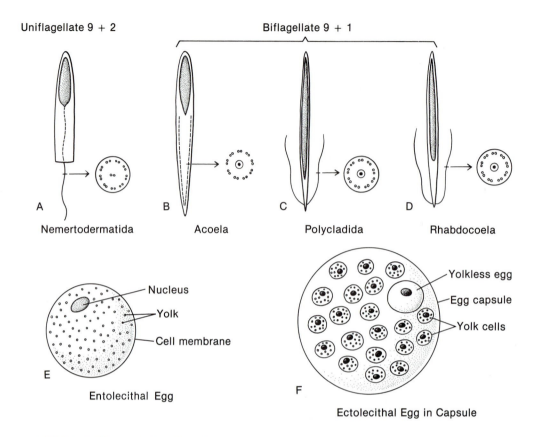

FIGURE 6–27 A, Uniflagellate 9×2+2 sperm of the Nemertodermatida. **B,** Sperm of some Acoela without free flagella (two 9×2+1 axonemes incorporated into cell). **C, D,** Biflagellated sperm of Polycladida and Rhabdocoela, both with 9×2+1 axonemes. **E,** Entolecithal (yolk-containing) egg of archoophoran turbellarians. **F,** Ectolecithal egg of neoophoran turbellarians. In some examples, the capsule may contain several eggs and hundreds of yolk cells. *(A–D, Redrawn from Hendelberg, J. 1986. The phylogenetic significance of sperm morphology in the Platyhelminthes. Hydrobiologia. 132:53–58.)*

rate organs linked together only by a common duct (Fig. 6–26B). In either case, the egg, after being released from the ovary, is surrounded by several yolk cells (ectolecithal egg), and together they are enclosed in an **egg capsule** (Fig. 6–27F; 6–30C, D).

The division of labor in the neoophoran ovary may be an adaptation to increase the rate of yolk synthesis and ultimately, fecundity. The small molecules used for yolk synthesis enter the cell across its membrane. The rate at which the precursors enter must therefore be determined, in part, by the surface area of the cell. Because many small cells have a larger total surface area than one large cell of equal volume, the small cells (vitellocytes) should be able to synthe-size yolk at a greater rate than one large cell. Furthermore, by scattering the many yolk cells throughout the body in vitellaria (Fig. 6–26C), the yolk-forming cells have access to nutrients along the length of the midgut.

The two types of ovaries are the basis for distinguishing the two levels of organization in the Turbellaria, namely the archoophora and the neoophora. The generally more primitive archoophoran turbellarians, in which there is a simple ovary, include the acoels, the macrostomids, the catenulids, and the polyclads. The neoophoran flatworms, in which there is an ovary and vitellaria, include the proseriates, rhabdocoels, and the triclads. The parasitic classes of flatworms, containing the trematodes, monogeneans, and tapeworms, also fall within the neoophoran group and have specialized ovaries with separate vitellaria.

In addition to the copulatory bursa and seminal receptacle, another female accessory reproductive organ, a **uterus,** serves as a temporary storage sac for ripe eggs. The uterus may be a blind sac, as is the case in some rhabdocoels, or it may be merely a dilated part of the oviduct, as is true for the polyclads (Fig. 6–25C). However, most turbellarians lack uteri because only a few eggs are laid at a time.

In acoels and catenulids, the female system is less well developed than in other turbellarians. Some possess no female ducts at all, not even a gonopore (Fig. 6–25A). In others, there are no oviducts, but a short blind vagina for receiving the penis leads from a female gonopore. Some zoologists believe that this condition in acoels is a reduction from the more developed condition described for *Macrostomum.* Others consider it to be primitive. Separate male and female gonopores are characteristic of most macrostomids, the acoels, and the polyclads, and this is probably the primitive turbellarian condition. Many turbellarians, however, including the common planarian, possess a single gonopore and genital atrium into which both male and female systems open.

Most turbellarians have biflagellate sperm with a 9-1 axoneme (or 9-0 in some acoels), a condition that does not seem primitive (Figs. 6–27B–D). Perhaps the original condition is reflected in *Nemertoderma,* which has conventional uniflagellate sperm (Fig. 6–27A).

Sperm transfer in turbellarians involves copulation and is usually reciprocal. In most turbellarians, the penis is inserted into the female gonopore or common gonopore of the partner. During copulation the worms orient themselves in a variety of ways with the ventral surfaces around the genital region pressed together and elevated (Fig. 6–28A, B). Hypodermic impregnation occurs in some acoels, macrostomids, rhabdocoels, and polyclads. The penis, which bears stylets, is pushed through the body wall of the copulating partner, depositing sperm into the parenchyma (Fig. 6–28B, C). The sperm then migrate to the ovaries. Self-fertilization probably occurs in some exceptional species.

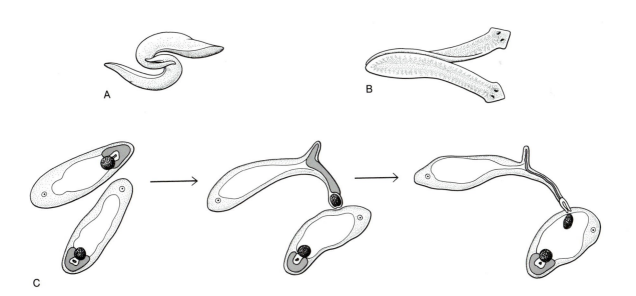

FIGURE 6–28 Copulation in the acoel, *Polychoerus carmelensis* **(A),** the freshwater triclad, *Dugesia* **(B),** and the acoel, *Pseudaphanostoma psammophilum* **(C).** *(A, Redrawn from Costello, H. M. and Costello, D. P. 1938. Copulation in the acoelous turbellarian Polychoerus carmelensis. Biol. Bull. 75:85–98. C, Redrawn from Apelt, G. 1969. Fortpflanzungsbiologie, Entwicklungszyklen und vergleichende Frühentwicklung acoeler Turbellarien. Mar. Biol. 4:267–325.)*

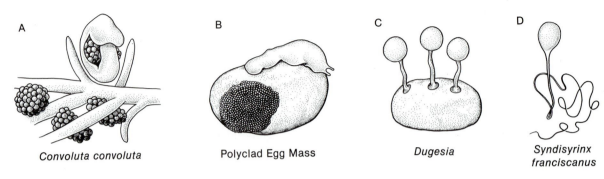

A   Convoluta convoluta

B   Polyclad Egg Mass

C   Dugesia

D   Syndisyrinx franciscanus

FIGURE 6–29 A, Egg masses attached to algae by the acoel, *Convoluta convoluta.* **B,** Egg mass attached to a stone by a polyclad. **C,** Egg capsules attached to a stone by the triclad, *Dugesia.* **D,** Egg capsule released into the intestine of a sea urchin by the parasitic rhabdocoel, *Syndisyrinx franciscanus. (A, Redrawn from Apelt, G. 1969. Fortpflanzungsbiologie, Entwicklungszyklen und vergleichende Frühentwicklung acoeler Turbellarien. Mar. Biol. 4:267–325. C, Redrawn from Pennak. D, Drawn from a photograph from Shinn, G. L. and Cloney, R. A. 1986. Egg capsules of a parasitic flatworm: Ultrastructure of hatching sutures. J. Morphol. 188:15–28.)*

Turbellarians that have oviducts but lack vitellaria release only a small number of eggs at any one time (Fig. 6–29A), except for the polyclads. The polyclads lay their numerous eggs in gelatinous strings or masses (Fig. 6–29B), and one individual may lay several egg masses in succession. Although vitellaria are absent, the fecundity of polyclads probably is related to large body size and division of the ovary into numerous small follicles, each of which is within diffusion distance of a branch from the midgut (Fig. 6–25C). Acoels that have no gonoducts release their eggs through the mouth or by temporary rupture of the body wall (Fig. 6–29A).

Proseriates, rhabdocoels, and triclads possess vitellaria, and as a result egg production is somewhat modified. As the fertilized eggs pass through the oviduct, they are accompanied by yolk cells from the vitellaria. On reaching the atrium, one or several eggs, along with many yolk cells, are enclosed by a hard capsule that is cemented to the substratum. In many forms, including some freshwater triclads, such as the common *Dugesia,* each oval brownish capsule has a slender stalk for attachment to rocks and other debris. Such attached capsules resemble little balloons (Figs. 6–29C, D). One worm can produce a number of capsules, in each of which several embryos may develop.

Some freshwater turbellarians often produce two types of eggs (p. 208): summer eggs (subitaneous eggs), which are enclosed in a thin capsule and hatch in a relatively short period, and autumn eggs (resting eggs), which have a thicker and more resistant capsule (Fig. 6–30). Resting eggs remain dormant during

FIGURE 6–30 Thick-walled resting eggs in freshwater rhabdocoel, *Mesostoma lingua. (From Mac-Fira, V. 1974. The turbellarian fauna of the Romanian littoral waters of the Black Sea and its annexes. In Riser, N. W., and Morse, M. P. (Eds.): Biology of the Turbellaria. McGraw-Hill Book Co., New York. p. 263.)*

the winter, can resist freezing and drying, and hatch in the spring with the rise in water temperature. In *Mesostoma ehrenbergii,* the generation time is between 16 and 75 days, depending on water temperature, and one individual produces about 15 summer eggs or 45 resting eggs. The life span is between 65 and 140 days. Parthenogenesis is characteristic of some turbellarians, and there are some parthenogenic species in which maleness is unknown (p. 208).

Early development of the entolecithal eggs of archoophoran turbellarians involves spiral cleavage like that occurring in other protostome phyla (Fig. 6–31A, B; see Chapter 5). Gastrulation is by epiboly and produces a stereogastrula. The mouth and pharynx form from an ectodermal invagination near the original site of the blastopore. This invagination connects with the archenteron, which has formed from an entodermal mass and become hollow. Mesenchyme originates from both ectoderm and endoderm, migrates into the region between the two primary germ layers, and later gives rise to the musculature, parenchyma, and germ cells (Figs. 6–31A, B).

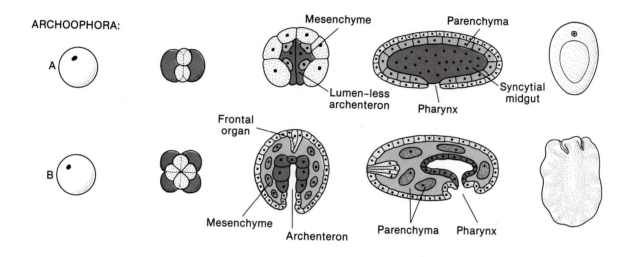

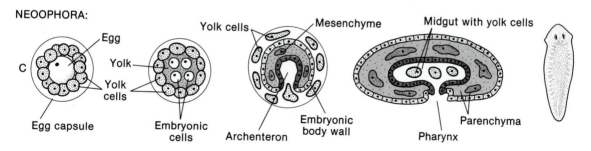

FIGURE 6–31 More or less typical spiralian development is found in the archoophoran acoels **(A)** and polyclads **(B).** In the neoophorans, on the other hand, cleavage patterns and other aspects of early development are highly modified because of the presence of numerous yolk cells. These yolk cells are incorporated into the developing embryo in a variety of ways, depending on the species. One example is given here for a triclad, *Dendrocoelum* **(C),** in which some of the yolk cells liberate their yolk into the embryonic mass (later the parenchyma), and others are ingested by the archenteron. *(A, Modified and redrawn after Bresslau from de Beauchamp, P. 1961. Classe des Turbellaries. In Grassé, P. -P. (Ed.): Traité de Zoologie. Vol. 4. Masson et Cie, Paris. p. 183. B, Modified and redrawn after Kato from Thomas, M. B. 1986. Embryology of the Turbellaria and its phylogenetic significance. Hydrobiologia. 132:105–115. C, Modified and redrawn after Hallez from Korschelt, E. and Heider, K. 1895. Text-book of the Embryology of Invertebrates. Part 1. Sawn Sonnenschein & Co., New York. pp. 170–172.)*

Turbellarians in general have **direct development,** but some polyclads produce free-swimming larvae that feed on plankton. There are two kinds of these larvae, one with four arms, or lobes, called Goette's larva and another with eight arms, called Mueller's larva (Fig. 6–32). The arms bear a band of long cilia used in locomotion and feeding, like the prototroch of trochophore larvae. Anteriorly, there is a ciliary tuft on an apical organ. The larva swims about for a few days, gradually resorbs its arms, and then settles to the bottom as a young worm.

The pattern of embryonic development in neoophoran flatworms probably evolved from spiral cleavage, but in most species the presence of external yolk cells has altered the cleavage pattern, disguising its origin (Fig. 6–31). There is no free-swimming larval stage in the neoophoran orders of Turbellaria; development is direct and the young worms emerge from the capsule in a few weeks.

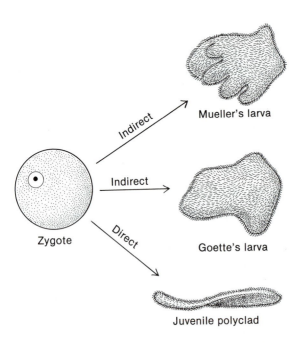

FIGURE 6–32 Although development is often direct, many polyclad species have indirect development resulting in planktotrophic larvae. During metamorphosis, these free-swimming larvae resorb their arms and soon resemble the juvenile at the bottom of the figure. *(Adapted from Ruppert, E. E. 1978. A review of metamorphosis of turbellarian larvae. In Chia, F. -S. and Rice, M. E. (Eds.): Settlement and Metamorphosis of Marine Invertebrate Larvae. Elsevier/North-Holland, Inc., New York. pp. 65–82.)*

Turbellarian Origins

Few evolutionary questions have prompted more speculation and controversy than the evolutionary origin of the Turbellaria. Although several theories propose to trace their ancestry, two are actively discussed by current researchers. The first is the planula theory, which was devised by von Graff in the nineteenth century and championed in the United States by the late Libbie Hyman (Fig. 6–33A). It states that the turbellarians and cnidarians arose from a common ancestor, the planuloid, which resembled a cnidarian planula larva. The planuloid ancestor would have had an outer epidermal epithelium, an inner gastrodermal epithelium, and a connective tissue (mesoglea) in between. According to this theory, the flatworm gut and parenchyma correspond to the coelenterate coelenteron and mesoglea, respectively.

An alternative idea is that turbellarians are not primitive bilaterians but instead have evolved from coelomate animals by anatomical simplification. This hypothesis receives indirect support from the complexity of the turbellarian reproductive system and epidermis, neither of which has a counterpart in diploblastic animals. The ciliated epidermal cells of turbellarians, for example, always bear many cilia each, whereas those of diploblastic animals primitively and predominantly have only one cilium each.

Although it is relatively easy to imagine how an organism like a planula larva could evolve into a simple turbellarian by the addition of complexity (Fig. 6–33A), it is more difficult to see how a coelomate animal, such as a segmented worm (annelid) or an acorn worm (hemichordate) could, by simplification, transform into an acoelomate animal. Modern proponents of this latter theory, such as A. Remane and R. M. Rieger, believe that reduction in body size could have been the crucial factor in such a transformation. It has been noted already that small body size is correlated with ciliary rather than muscular modes of locomotion, an absence of internal fluid transport systems (including a coelom), and the occurrence of protonephridia (Chapter 5; p. 192). The actual disappearance of a coelom could occur, as seen in hemichordates and cephalochordates (Fig. 6–9F, G) and especially some annelids, as described by M. Fransen, by an infilling of the coelomic cavity with transformed lining cells to form a cellular parenchyma. More recently, R. M. Rieger has suggested that turbellarians could have evolved not from an adult

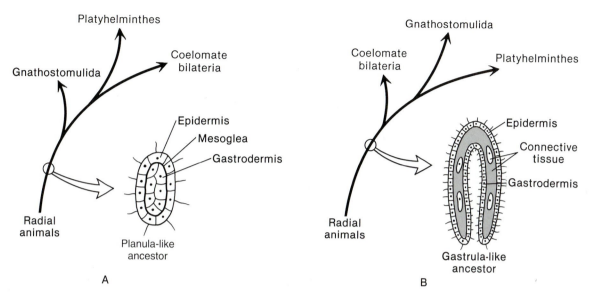

FIGURE 6–33 Two alternative theories for the evolution of flatworms. In **A,** the flatworms are believed to have evolved from an ancestor that resembled a cnidarian planula larva. Mesoderm would have evolved as cells migrated into the space between ectoderm and endoderm and differentiated into germ cells, muscles, and various parenchymal cells. According to this theory, the flatworms and gnathostomulids are the most primitive bilateral animals. In **B,** the flatworms and gnathostomulids are viewed as having evolved by progenesis from a developmental stage, such as a gastrula, of a large coelomate animal. In this case, the embryonic mesenchyme and blastocoel, which has a parenchyma-like organization, becomes the body wall musculature, gonads, and adult parenchyma. According to this theory, flatworms and gnathostomulids are not primitive bilaterians. At present, it is not possible to choose between these two alternatives.

coelomate ancestor but rather by progenesis (p. 206) from a developmental stage before morphogenesis of the coelom had occurred; in most coelomates, such a stage would have to be a gastrula or early larva (Fig. 6–33B). It is well known that progenesis occurs within phyla of many invertebrates and especially in salamanders among the vertebrates.

On the other hand, Peter Ax, an advocate of the planula theory, has attempted to rationalize some of the structural discrepancies between turbellarians and diploblastic animals by suggesting that the acoelomate and monociliated Gnathostomulida form an evolutionary bridge between the two groups. A recent comparison of ribosomal RNA gene sequences by J. M. Turbeville and colleagues supports the planula theory.

Whether the turbellarians arose from a diploblastic or a triploblastic ancestor, turbellarians show typi-cal spiral cleavage suggesting that their closest bilateral relatives are likely to be found among the protostome line of evolution. Their position within the protostomes, whether primitive or evolutionarily derived, is at present uncertain.

SYSTEMATIC RÉSUMÉ OF CLASS TURBELLARIA

The modern classification of the turbellarians divides them into three major groups, which may or may not be closely related to each other (Fig. 6–34). These are the Acoelomorpha (Nemertodermatida and Acoela), Catenulida (Catenulida), and Rhaditophora (remaining orders) (see Rieger et al. 1991). The first two groups (Acoelomorpha, Catenulida) have archoophoran organization, as do the primitive members of the Rhabditophora (i.e., the Macrostomida, Haplopharyngida, and Polycladida); the rest of the Rhabditophora have the neoophoran organization.

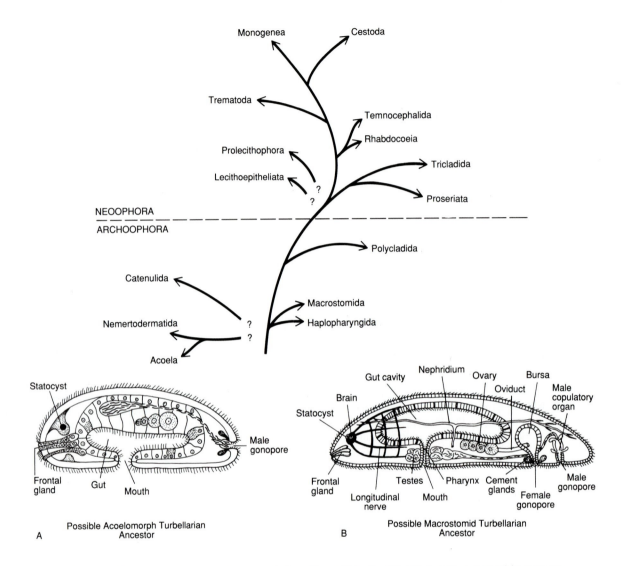

FIGURE 6–34 Possible evolutionary relationships within the Platyhelminthes based in part on the ideas of Ehlers (1985b) Smith, J. P. III et al. (1985) and Llewellen (1965). A turbellarian ancestor **(B)** similar to living macrostomids (Ax, 1963) is more widely accepted at the present time than is one similar to the acoels **(A)**, although the latter view was long held in the past. *(A, After Smith, J. P. III et al., 1985.)*

Archoophoran Turbellarians. Orders that reflect a more primitive level of organization. Vitellaria absent; eggs entolecithal; cleavage spiral.

Order Nemertodermatida. Small marine species similar to acoels but possessing uniflagellate sperm and an epithelial digestive tract. Statocyst usually with two statoliths. *Nemertoderma.*

Order Acoela. Small marine flatworms, usually less than 2 mm in length. Statocyst with one statolith. Mouth and sometimes a simple pharynx present, but no digestive cavity. Protonephridia absent. Gonads often not bounded by a cellular wall. Oviducts absent. *Amphiscolops, Anaperus, Afronta, Polychoerus, Convoluta, Archaphanostoma, Pseudaphanostoma, Diopisthoporus,*

Paratomella. A few species are commensal within the intestine of various echinoderms.

Order Catenulida. Mostly small, freshwater species having a simple pharynx; ciliated, saclike intestine; and unpaired gonads, with the male gonopore dorsal above the pharynx. Statocyst sometimes present, usually with one statolith. No female gonoducts. *Stenostomum, Catenula.*

Order Haplopharyngida. Small marine species similar to macrostomids but possessing a proboscis and a temporary anus. *Haplopharynx.*

Order Macrostomida. Small marine and freshwater species having a simple saclike ciliated intestine, a simple pharynx, and one pair of ventrolateral nerve cords. *Macrostomum, Microstomum.*

Order Polycladida. Marine flatworms of moderate size, averaging 3 to 20 mm in length, with a greatly flattened and more or less oval shape. A pair of anterior marginal or dorsal tentacles may be present. Many are brightly colored. Intestine elongated and centrally located, with many highly branched diverticula. Plicate pharynx either an anteriorly directed tube or pendant from the roof of the pharyngeal cavity. Eyes numerous. *Gnesioceros, Leptoplana, Notoplana, Stylochus, Prostheceraeus, Pseudoceros.*

Neoophoran Turbellarians. Orders that reflect an advanced level of organization. Vitellaria present; eggs ectolecithal; and development greatly modified from the spiral pattern.

Order Lecithoepitheliata. Marine and freshwater species in which ovary produces eggs surrounded by follicle-like yolk cells. Mouth and complex pharynx at anterior end of body, intestine simple. *Prorhynchus.*

Order Prolecithophora. Usually small. Marine and freshwater species having a plicate or bulbous pharynx and a simple intestine. Ovary produces eggs and follicle-like yolk cells. *Plagiostomum, Hydrolimax.*

Order Proseriata. Small, mostly marine turbellarians, including many interstitial forms. Statocyst with one statolith. Pharynx is plicate and tubular, but gut is not branched. *Otoplana, Monocelis, Nemertoplana.*

Order Temnocephalida. Commensal and parasitic on crustaceans, molluscs, and turtles. Posterior ventral surface provided with an adhesive disc, and anterior margin bears finger-like projections, by which worm moves leechlike on its host. *Temnocephala.*

Order Rhabdocoela. A large group of small marine and freshwater turbellarians, having a bulbous pharynx, a simple intestine, and one pair of nerve cords.

Suborder Typhloplanoida. Mouth usually located in the middle and pharynx oriented at right angles to long axis of body. Contains marine and freshwater, free-living species. *Mesostoma.*

Suborder Dalyellioida. Mouth typically at anterior end of body and pharynx oriented parallel to long axis of body. Contains marine and freshwater species, some of which are commensal and parasitic on and within snails, clams, sea urchins, and sea cucumbers. *Anoplodiera, Syndesmis, Kronborgia.*

Suborder Kalyptorhynchia. Mouth anterior to middle of body. Anterior protrusible, muscular proboscis often bears cuticularized hooks or teeth. Contains mostly marine interstitial species. *Gyratrix, Gnathorhynchus.*

Order Tricladida. Relatively large, marine, freshwater, and terrestrial turbellarians. Pharynx is plicate, tubular, and posteriorly directed; gut has three branches. Among marine species, *Bdelloura* is commensal on the book gills of horseshoe crabs. The freshwater species are known as planarians and include *Planaria, Dendrocoelum, Procerodes, Dugesia, Phagocata, Polycelis, Procotyla.* Land planarians include *Bipalium, Orthodemus, Geoplana.*

(**Neodermata.** Group including the parasitic platy-helminths, if they were classified according to strict cladistic principles.)

SUMMARY

1. Free-living members of the Platyhelminthes are grouped in the class Turbellaria. Turbellarians generally are small, bilaterally symmetrical animals with a low level of cephalization and an acoelomate grade of body construction.

2. The majority of turbellarians are marine, but there are freshwater species and a few terrestrial forms in humid environments. Turbellarians are benthic animals, living on or beneath stones, algae, and other objects. They are common members of the interstitial fauna.

3. Most turbellarians move using cilia; large species (polyclads) are markedly flattened and augment ciliary motion with muscular undulations. Adhesive organs make possible temporary attachment in many species.

4. Most turbellarians are predators and scavengers. A few are herbivores, commensals, or parasites. Digestion is initially extracellular and then intracellular. Small species have a simple saclike intestine with a simple or bulbous pharynx. Large species have a branched intestine and a plicate pharynx, usually tubular.

5. Mucus produced by epidermal glands plays an important role in the life of turbellarians, coating the substratum over which the animal crawls. Mucus also aids in prey trapping and swallowing.

6. The small size, flattened shape and branched gut (in larger forms) make unnecessary special systems for internal transport and gas exchange. Protonephridia are present in most flatworms.

7. A radial arrangement of several pairs of longitudinal nerve cords is probably primitive, and arrangements with lesser numbers have likely evolved through reduction. The nerve cords are associated with peripheral nerve nets. Pigment cup ocelli, which may be numerous, and statocysts are the principal sensory organs.

8. Turbellarians are simultaneous hermaphrodites, and the reproductive system is adapted for internal fertilization and egg deposition.

9. At the primitive, archoophoran level the eggs are entolecithal, cleavage is spiral, and there is sometimes a free-swimming larva. In most archoophoran species, however, development is direct.

10. Many turbellarians have evolved an ovarian division of labor between egg production and yolk production, leading to ectolecithal eggs (neoophoran level). Spiral cleavage has been lost, and development is always direct.

Class Trematoda[†]

The class Trematoda includes two subclasses of closely related parasitic flatworms, the Digenea, a large, economically and medically important taxon;

[†]Coverage of the remaining platyhelminths, the parasitic flatworms, is designed to accommodate those courses that cover the parasites in their survey of the invertebrates. It is of necessity much briefer than the account of the turbellarians.

and the Aspidogastrea, a small taxon of no medical or economic importance.

Subclass Digenea

Digeneans, which are sometimes called flukes, are common endoparasites of all classes of vertebrates: fishes, amphibians, reptiles, birds, and mammals. Some cause debilitating disease in livestock and humans. There are 11,000 species of digeneans, more than all other flatworms combined, and they are second in diversity among metazoan parasites only to the roundworms. Development is indirect, and the life cycle always includes at least two infective stages, which accounts for the name Digenea, meaning "two generations." Two or more hosts are infected before the life cycle is complete. The first intermediate host is typically a gastropod snail; if a **second intermediate host** is involved, it is usually an arthropod, and the definitive host is a vertebrate (Fig. 6–36). Having successfully invaded the first intermediate host, digenean developmental stages undergo two rounds of asexual divisions, which greatly increase their numbers, and thus their chances, to complete the life cycle. Thus, one egg can give rise to dozens of sexual adults in the definitive host.

The life cycle of digeneans varies, depending upon the species, but it is possible to describe a general pattern (Fig. 6–35, 6–36). An adult fluke releases eggs in capsules that must pass out of the host. Eggs are generally voided in the feces, although in some species, they exit in the urine or sputum. The first cleavage divides each zygote into two cells: a **somatic cell** and a **germinal cell.** Somatic cells become the body of the parasite, and germinal cells are reproductive. After the egg leaves the body of the host, it may be ingested accidently by the first intermediate host (usually a terrestrial event), or in aquatic situations, a ciliated **miracidium** eventually hatches from the capsule and actively seeks and penetrates the skin of the gastropod host. Once in the hemocoel of the snail, the miracidium metamorphoses into a saclike **sporocyst,** containing a number of germinal cells. Each germinal cell divides repeatedly and gives rise either to another generation of sporocysts or to a **redia,** which also has numerous germinal cells in its interior. Each germinal cell develops into a **cercaria.** The cercaria, a fourth developmental stage, possesses a digestive tract, suckers, and a tail. The cercaria

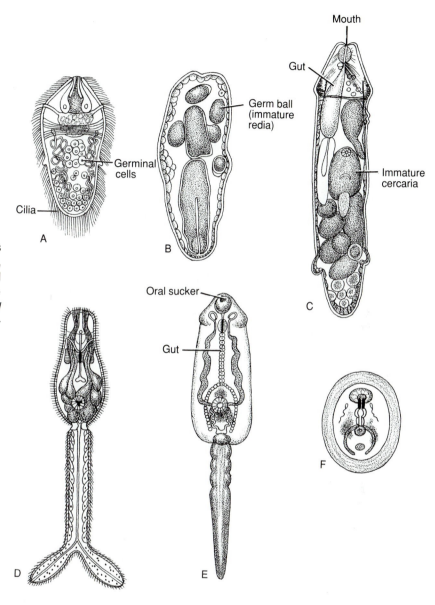

FIGURE 6–35 Larval stages of Digenea. **A,** Miracidium. **B,** Sporocyst. **C,** Redia. **D** and **E,** Cercariae. **F,** Metacercaria. *(From the U.S. Naval Medical School Laboratory Manual.)*

leaves the snail host and is generally freeswimming. Where there is a second intermediate host, it may be an invertebrate (commonly an arthropod) or a vertebrate (commonly a fish), in which the cercaria encysts, casting off its tail. The encysted resting stage is called a **metacercaria** (Fig. 6–35F). If the host arthropod and its metacercaria are eaten by the final vertebrate host, the metacercaria escapes from its cyst, migrates, and grows into the adult within a characteristic location in the final host. In many digeneans, the cercariae encyst on aquatic vegetation and are ingested by the definitive host.

A great many digeneans infect the gut or gut derivatives of their definitive host. Lungs, bile ducts, pancreatic ducts, and intestines are common sites.

Anatomy and Physiology

Adult digeneans range in size from approximately 0.2 mm to 6.0 cm in length. They are typically dorsoven-

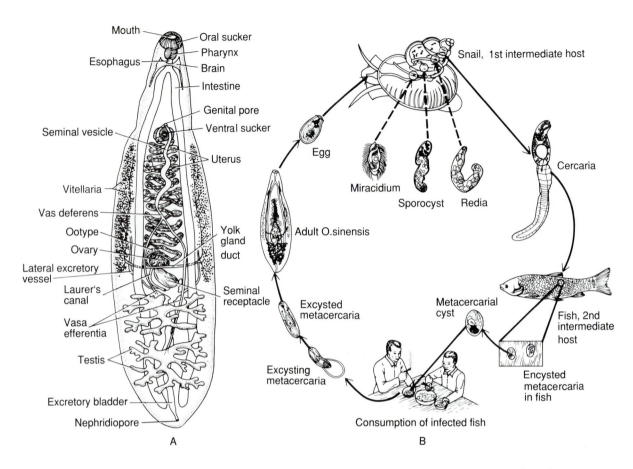

FIGURE 6–36 The Chinese liver fluke, *Opisthorchis sinensis.* Millions of Asians are believed to be infected with liver flukes transmitted from uncooked fish. One worm can live as long as 8 years. **A,** Dorsal view of adult worm. **B,** Life cycle. *(A, After Brown from Noble and Noble. B, After Yoshimura from Noble and Noble, 1982: Parasitology. 5th Edition Lea and Febiger, Philadelphia.)*

trally flattened, but some are thick and fleshy, whereas others are long and threadlike. An **oral sucker** surrounds the mouth, and many, but not all, have a midventral or posterior **ventral sucker** (acetabulum)(Fig. 6–37). The suckers are important organs that prevent dislodgment and aid in feeding (oral sucker).

In contrast to the ciliated epidermis of the turbellarians, the digenean body is covered by a nonciliated cytoplasmic syncytium, the **tegument,** overlying consecutive layers of circular, longitudinal, and diagonal muscle. The syncytium is continuous with the cytoplasm surrounding the nuclei, which are submerged into the parenchyma (Fig. 6–38).

The tegument plays a vital role in the physiology of digeneans. It provides protection, especially against the host's enzymes in gut-inhabiting species. Nitrogenous wastes diffuse to the exterior through the tegument (nephridia are used also), and it is the site of gas exchange. In endoparasites, the tegument, as well as the gut, absorbs glucose and some amino acids. The protein synthesis involved in digenean egg production and in asexual reproduction in developmental stages places especially heavy demands on the amino acid supply.

The mouth is surrounded by a powerful, muscular oral sucker, which is used for attachment and aids the transport of food into the mouth. The bulbous pharynx ingests cells and cell fragments, mucus, tissue fluids, or blood of the host on which the parasite feeds. The pharynx passes into a short esophagus leading into two blind intestinal ceca that extend pos-

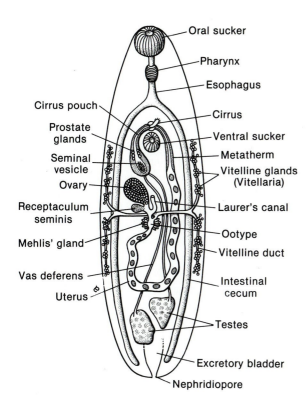

FIGURE 6–37 A diagram of a generalized trematode. *(After Cable, 1949, from Smyth, J. D. 1976. Introduction to Animal Parasitology. 2nd Edition. Hodder & Stoughton, London. 466 pp.)*

teriorly along the length of the body (Fig. 6–37). Digestion is primarily extracellular in the ceca.

Digeneans are facultative anaerobes. The amount of oxygen utilized in respiration depends on the location within the host and also on the developmental stage of the parasite.

Digeneans, like other flatworms, have protonephridia. Typically, there is a pair of longitudinal collecting ducts, which empty into a single posterior bladder and nephridiopore (Fig. 6–36). The nephridia excrete water and waste metabolites, such as unwanted iron from hemoglobin in blood-feeding worms.

The nervous system is essentially like that of turbellarians. There is a pair of anterior cerebral ganglia from which longitudinal nerve cords extend posteriorly, of which the ventral pair is, typically, the most highly developed. The fluke body surface has a variety of sensory papillae, and ocelli occur in many miracidia and some cercariae.

Reproduction

The reproductive system is similar to that of neoophoran turbellarians (Fig. 6–37), but egg production is estimated to be 10,000 to 100,000 times greater than in the free-living flatworms. There are

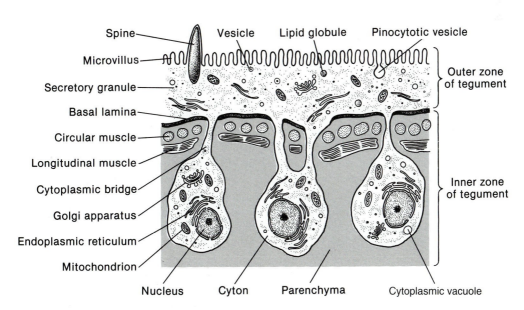

FIGURE 6–38 Section through the tegument and body wall of a generalized digenean. *(From Cheng, T. C. 1973. General Parasitology. Academic Press, New York. 965 pp.)*

usually two testes, and their position is of taxonomic importance. Sperm ducts, one from each testis, unite anteriorly and may expand into an external seminal vesicle before entering into the **cirrus sac.** This sac contains the internal seminal vesicle, prostate glands, and an eversible copulatory **cirrus,** which protrudes into a common **genital atrium.** Sperm, on leaving the testes, are stored in the seminal vesicle. The single gonopore opens from the common genital atrium and is usually located midventrally anterior to the ventral sucker. There are many variations of the general plan just described.

The female part of the system consists of, usually, a single ovary and an oviduct, which leads into a small sac, called the **ootype** (Figs. 6–36A, 6–37). En route to the ootype, the oviduct receives a duct from the seminal receptacle and a common duct from the right and left vitellaria. The ootype is surrounded by conspicuous unicellular gland cells important in egg capsule formation, called collectively **Mehlis's gland.** Leaving the ootype is the uterus, which runs anteriorly to the genital atrium and gonopore. In most digeneans, there is a short, inconspicuous canal (**Laurer's canal),** perhaps a vestigial vagina, that extends from the duct of the seminal receptacle to the dorsal surface of the worm, where it may open at a minute pore.

During copulation, sperm exchange is mutual, and cross fertilization is the general rule, although self-fertilization is known to occur in rare cases. As copulation proceeds, the cirrus of one worm is inserted, via the gonopore, into the uterine opening of the other worm, and sperm are ejaculated. The prostate gland provides semen for sperm survival. Sperm travel up the uterus, through the ootype, to be stored in the seminal receptacle.

Once released from the ovary, the eggs are fertilized either in the oviduct or within the ootype. Because eggs of digeneans are ectolecithal, as are those of neoophoran turbellarians, the yolk is supplied by yolk cells, which also carry the material needed in egg capsule formation. The egg capsules pass along the entire length of the uterus, where embryonic development usually begins, and eventually are released from the gonopore.

Life Cycle Examples

The Chinese liver fluke, *Opisthorchis* (or *Clonorchis*) *sinensis,* infects the bile ducts of over 20 million east Asians (Fig. 6–36). Adult flukes reach 2.5 cm in length, and may live as long as 8 years, each fluke producing as many as 4000 eggs per day for up to 6 months of the year.

Chinese liver fluke eggs pass in feces from humans and domesticated animals, especially dogs, into water where they are consumed by certain species of aquatic gastropod snails, the first intermediate host. Once ingested by the snail, the miracidium penetrates the wall of the intestine and metamorphoses into a sporocyst within the tissues of the host. The sporocyst produces the redia generation, which gives rise to cercariae that escape into the water. Each cercaria swims to the surface using the muscular tail and then sinks slowly through the water until it contacts an inert object or is disturbed by turbulence in the water. When this happens, the cercaria is stimulated to return to the surface where it again sinks down. This repetitive swim-and-sink behavior keeps the cercaria off the bottom, where it is most likely to encounter the second intermediate host, a fish. On contact with a fish, the cercaria attaches with its suckers, sheds its tail, and bores through the skin of the fish. Within the subcutaneous tissue or the musculature, the cercaria encysts to form a metacercaria. When humans, or other mammals, ingest the uncooked fish, young worms emerge from their cysts in the small intestine and migrate into the bile ducts.

The number of flukes infecting a person is generally low, but even modest infections (20–200 flukes per liver) can cause jaundice, gallstones, general debilitation, and perhaps liver cancer. In one unfortunate case, over 21,000 flukes were found at autopsy. Although drugs are partially effective against the disease, prevention is the best approach, which unfortunately is hampered by cultural and economic restrictions. Transmission of the disease in rural areas is the result of building toilet facilities over ponds. In urban areas, especially in Asia, consumption of uncooked fish, which is sometimes glistening with metacercarial cysts, is considered to be a delicacy, but in poor areas it can be a necessity because fuel for cooking is either unavailable or prohibitively expensive.

There are three families of digeneans that inhabit the blood of their hosts, but certainly the best known blood flukes belong to the family Schistosomatidae, which contains species producing the human disease schistosomiasis. *Schistosoma mansoni,* one of several species parasitic in humans, occurs in Africa and tropical areas of the New World (Fig. 6–39). The adult, like that of other species of schistosomes, in-

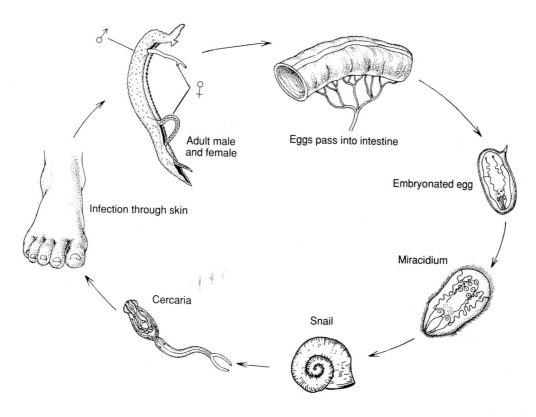

FIGURE 6–39 Life cycle of *Schistosoma mansoni.*

habits the intestinal veins. The members of this family, in contrast to most other flatworms, are dioecious with a male and female permanently paired throughout life. The male is 6 to 10 mm in length and 0.5 mm in width. A ventral groove extends along most of the length of the male, into which the longer but more slender female fits. When laying eggs, the female protrudes from the groove of the male worm and deposits the eggs in small intestinal venules. Eventually, they break through using spines and enzymes into the lumen of the intestine and are released in the feces of the host. If defecation occurs in the water, the thin-walled egg capsules hatch, and the miracidia escape. The miracidia are well supplied with sensory receptors and seek out a particular species of freshwater snails. After penetration, they transform into sporocysts, which eventually give rise to a second generation of sporocysts. The sporocysts produce cercariae without an intermediate redia stage. The cercariae leave the molluscan hosts and, on contact with human skin, penetrate, using enzymes and muscular boring

movements. The now tailless "schistosomules" are carried by the bloodstream first to the lungs, then to the liver, and finally to the veins of the intestine or bladder. During this period the schistosomules gradually transform into adults.

The two other species that attack humans are the Asian *Schistosoma japonicum,* which has a life cycle similar to that of *Schistosoma mansoni,* and *Schistosoma haematobium* of North Africa, Turkey, and Portugal. The eggs of *Schistosoma haematobium,* which occurs in the veins of the urinary bladder, leave the primary host via the bladder and urine.

Schistosomiasis is a seriously debilitating disease and can be lethal. Egg penetration through the intestinal wall and bladder, aberrant lodging of eggs in various organs, and the developmental stages in the lung and liver can result in inflammation, necrosis, or fibrosis, depending on the degree of infection. Pathogenic response to the eggs is generally more serious than that to the schistosomules or adults. The percentage of the population infected in endemic areas is enormous, and globally some 300 million people are estimated to be infected by one of the three *Schistosoma* species. Schistosomiasis, malaria, and hookworm infections are the three great parasitic scourges of humankind.

Other members of the Schistosomatidae infect various birds and mammals, including domestic species. "Swimmer's itch," which occurs commonly in North America, is an irritation produced by the incomplete penetration into human skin by cercariae of blood flukes of birds.

Subclass Aspidogastrea

The subclass Aspidogastrea includes a small group of trematodes that show similarities to other parasitic flatworms but are more closely related to the Digenea and thus are included in the Trematoda. The distinguishing feature is the adhesive organ, which is either a single, septate sucker covering the entire ventral surface or a longitudinal row of suckers (Fig. 6–40). The digestive tract contains a single intestinal cecum. The reproductive system is essentially like that of the Digenea, but there is typically only one testis.

The aspidogastreans are mostly endoparasites in the gut of fish and reptiles and in the pericardial and renal cavities of bivalve molluscs. Their life cycles involve one or two hosts, depending on species.

CLASS MONOGENEA

The some 1100 species of monogeneans are mostly ectoparasites of aquatic vertebrates, especially fishes, but amphibians and reptiles are also hosts. Because they are attached to the skin of a fast-moving host, monogeneans are dorsoventrally flattened and have a large, posterior attachment organ, the **haptor,** which bears hooks and suckers, allowing the parasite to cling tenaciously to the host (Fig. 6–41). The monogenean life cycle is different from that of the Digenea,

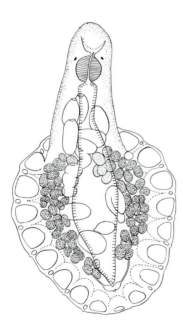

FIGURE 6–40 Dorsal view of *Cotyaspis insignis,* an aspidogastrean fluke parasitic in freshwater mussels. Note the large ventral sucker with subdivisions. *(After Hendrix and Short.)*

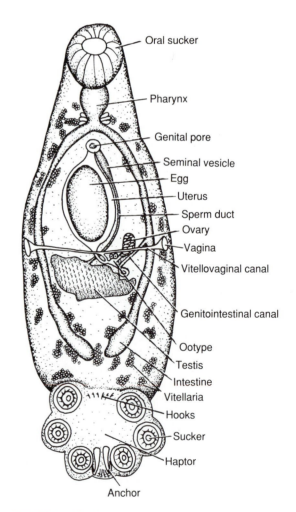

Oral sucker

Pharynx

Genital pore

Seminal vesicle

Egg

Uterus

Sperm duct

Ovary

Vagina

Vitellovaginal canal

Genitointestinal canal

Ootype

Testis

Intestine

Vitellaria

Hooks

Sucker

Haptor

Anchor

FIGURE 6–41 *Polystomoidella oblongum,* a monogenean parasitic in the urinary bladder of turtles. *(After Cable.)*

as there is no intermediate host, and one egg [by way of a ciliated larva, the **oncomiracidium** (Fig. 6–42)] gives rise to only one adult worm, hence the name *monogenea,* meaning "one generation."

The body is composed of a head, trunk, and haptor (Fig. 6–41). Although the head lacks an oral sucker, adhesive glands (head organs) are present. Alternate attachment of the head organs and haptor enable some monogeneans to creep like inchworms; others may be attached permanently to the host. A syncytial tegument covers the body.

The digestive system is similar to that of digeneans, but the pharynx of some monogeneans secretes a protease that digests the host's skin, allowing the parasite to ingest blood and cellular debris. Many digeneans, however, graze exclusively on mucus and superficial cells of the host skin.

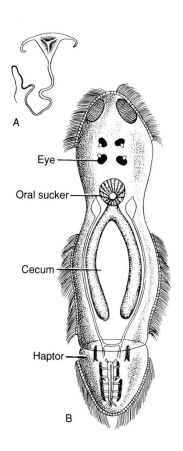

FIGURE 6–42 **A,** Egg capsule and **B,** oncomiracidium of *Benedenia melleni,* a monogenean parasitic on fish. *(Both after Jahn and Kuhn.)*

Monogeneans have inconspicuous protonephridia that consist of scattered terminal cells and their collecting tubules. The latter open to the exterior at two dorsolateral pores. In some species, a urinary bladder is associated with each pore.

In contrast to the endoparasitic digeneans, monogenean ectoparasites have aerobic metabolism.

Reproduction

Monogeneans are neoophoran flatworms with complex hermaphroditic reproductive systems (Figs. 6–41, 6–44). The male reproductive system usually has only one circular or oval testis, but a few species have two or more. The sperm duct usually leads into the base of the copulatory organ, which is a protrusible penis-like structure that may be armed with hooks. During copulation, there is a mutual exchange of sperm with seminal fluid being passed into the single or paired vagina of the partner. Sperm are generally stored in a seminal receptacle situated near the ovary and oviduct. There is a single ovary and extensive vitellaria.

Life Cycle Examples

Polystoma integerrimum is found in the bladders of frogs and toads and is an example of remarkable synchronization of parasite life cycle with that of the amphibian host (Fig. 6–43). The egg capsules are produced and stored until the frog or toad returns to the water to breed, at which time the capsules are released. The oncomiracidium attaches to the gills of the tadpoles. When the tadpoles metamorphose, the parasite leaves the gill chamber, crawls over the host's belly, and enters the bladder. When the tadpole is very young, some of the larvae may attain a precocious sexual maturity and produce egg capsules. This ectoparasitic generation dies when metamorphosis occurs.

The various species of *Dactylogyrus* are common ectoparasites on the gills of various freshwater fishes (Fig. 6–44). *Dactylogyrus* can be a serious problem in fish hatcheries, causing high mortality in the young fish from secondary infection, smothering by excess mucus production, or loss of blood. The egg capsules are released and drop to the bottom. Eventually, the capsules hatch to liberate the oncomiracidia. On contacting a suitable host fish, each larva transforms into the adult worm. Egg production increases with rise of

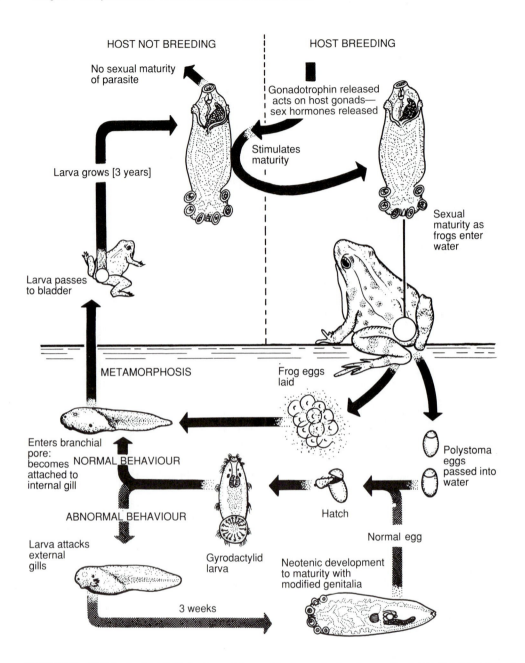

FIGURE 6–43 Life cycle of *Polystoma integerrimum,* a monogenean found in the bladder of frogs. Diagram also shows a progenetic population parasitic on the gills of the tadpole. *(From Smyth, J. D. 1976. Introduction to Animal Parasitology. 2nd Edition. Hodden and Stoughton Educational, Kent. p. 139.)*

water temperature so that the population of monogeneans builds up over the summer. Some egg capsules can overwinter to begin new cycles of infestation in the spring, and adult *Dactylogyrus* can persist on their hosts over the winter.

CLASS CESTOIDEA

The Cestoidea are the most evolutionarily–derived of the flatworm classes. All 3400 species of the Cestoidea are endoparasites with a body covered by a

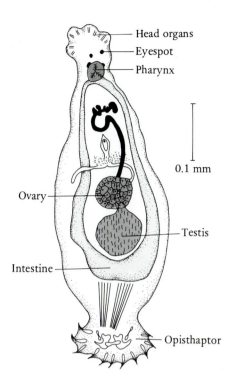

FIGURE 6–44 *Dactylogyrus vastator,* a monogenean ectoparasite on the gills of freshwater fishes. *(After Bychowsky, from Schmidt, G. D. and Roberts, L. S. 1989. Foundations of Parasitology, 41 Edition. Times Mirror/Mosby College, St. Louis.)*

syncytial tegument as in the Digenea and Monogenea, but cestoideans have special tegumental modifications associated with nutrient uptake because they lack a digestive tract. Most members of the Cestoidea can be included in the subclass Eucestoda, which is considered in the present book.

Subclass Eucestoda

Structure and Physiology

Cestoideans are generally known as tapeworms. The body of an adult tapeworm is unlike that of the other flatworms in that there is an anterior head region, the **scolex,** which is adapted for adhering to the host. Behind the scolex is a narrow **neck,** which is proliferative, and gives rise to the body, sometimes called the **strobila** (Fig. 6–45). The strobila consists of linearly arranged segment-like sections, called **proglottids,** that constitute the greater part of the worm (Fig. 6–48). Tapeworms are generally long; some species can reach lengths of 15 m. The scolex, neck, and strobila are regarded as a single individual and not a colony.

Compared with the mature proglottids, the scolex is often minute. It generally is a four-sided knob provided with suckers or hooks for attachment to the host gut wall. Although there are generally four large suckers arranged around the sides of the scolex in

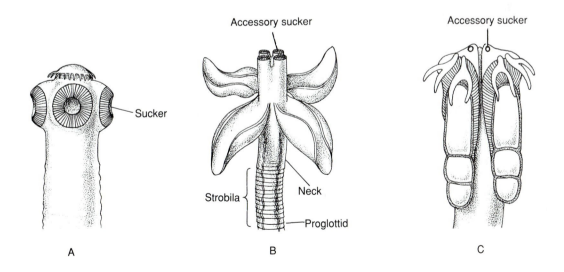

FIGURE 6–45 Scoleces of three tapeworms, showing the four principal suckers or sucker-like adhesive structures, small accessory suckers, and the hooks. **A,** *Taenia.* **B,** *Myzophyllobothrium.* **C,** *Acanthobothrium. (A and C, After Southwell. B, After Shipley and Hornell.)*

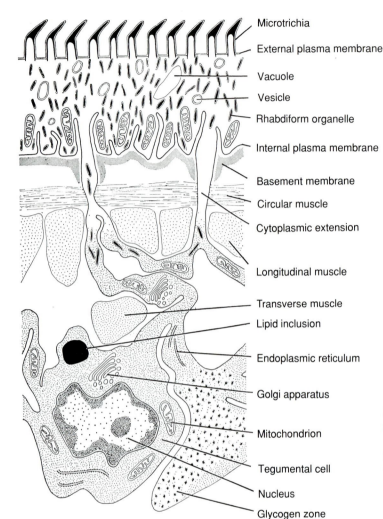

Microtrichia

External plasma membrane

Vacuole

Vesicle

Rhabdiform organelle

Internal plasma membrane

Basement membrane

Circular muscle

Cytoplasmic extension

Longitudinal muscle

Transverse muscle

Lipid inclusion

Endoplasmic reticulum

Golgi apparatus

Mitochondrion

Tegumental cell

Nucleus

Glycogen zone

FIGURE 6–46 Section through the tegument and body wall of the tapeworm *Caryophyllaeus.* Other than the folding of the surface membrane, note the similarity to the digenean body wall shown in Figure 6–38.*(After Beguin from Smyth, J. D. 1969. The Physiology of Cestodes. W. H. Freeman and Co., San Francisco.)*

tapeworms such as *Taenia* (Fig. 6–45A), the scolex is often a more complex structure in other representatives (Fig. 6–45B, C). The main attachment organs may be leaflike or ruffled, and there may be terminal accessory suckers in place of or in addition to hooks.

The neck is a short region behind the scolex, which produces each proglottid by mitotic growth followed by a transverse constriction (Fig. 6–45B). The youngest proglottids are thus at the anterior end immediately behind the head; they increase in size and maturity as they are displaced toward the posterior end of the strobila.

The tapeworm tegument plays a vital role in active transport of food molecules of various carbohy-drates and amino acids. It is also important in the evasion of the host immune response. The outer plasma membrane of the syncytial tegument is thrown into projections, the **microtrichia,** a type of microvilli (Fig. 6–46). Anaerobic metabolism apparently predominates in tapeworms but is not their exclusive mode of metabolism.

The muscle layers of tapeworms consist of the usual circular and longitudinal layers, but in addition there is musculature of longitudinal, transverse, and dorsoventral fibers, which encloses the interior parenchyma.

Extending through the chain of proglottids are the nervous and protonephridial systems and the longitudinal musculature. An anterior nerve mass lies in the scolex, and two lateral, longitudinal cords extend posteriorly through the strobila (Fig. 6–48). There may also be a dorsal and ventral pair of cords and, quite commonly, accessory lateral cords. Ring commissures connect the longitudinal cords in each proglottid.

Protonephridial terminal cells and tubules in the parenchyma drain into four lateral longitudinal collecting canals, two of which are dorsolateral and two ventrolateral (Fig. 6–48). The ventral canals are usually connected by a transverse canal in the posterior end of each proglottid. After the proglottids have started to be shed, the collecting ducts open to the exterior through the last terminal proglottid.

Reproduction

A complete reproductive system occurs within each proglottid and makes up a major part of each of these body sections. As shown in Figure 6–48, the tapeworm reproductive system is basically like that of the Digenea. There is usually a common male and female atrium and gonopore. However, the tapeworm differs from many digeneans in that a separate vaginal canal extends between the genital atrium and seminal receptacle close beside the ootype. The uterus is usually a blind sac extending out of the ootype into the parenchyma of the proglottid and functions solely in storage of the maturing eggs.

Cross-fertilization is probably the rule where there is more than one worm in the host's gut, but self-fertilization between two proglottids in the same strobila or even within the same proglottid is known to occur.

At copulation the cirrus of the male is everted into the vaginal opening of the proglottid of an adjacent worm. Sperm are stored in the seminal receptacle of the female system and then liberated for the fertilization of the eggs in the oviduct before entering the ootype together with a number of yolk cells. The zygotes remain in the blind-ended uterus, where development begins and the egg capsule is hardened. In many cases, terminal proglottids, packed with egg capsules, break away from the strobila. The capsules are freed from the proglottid, either within the host's intestine or after they leave with the feces.

Life Cycles

Tapeworms are endoparasites in the guts of all classes of vertebrates. Their life cycles require one, two, or

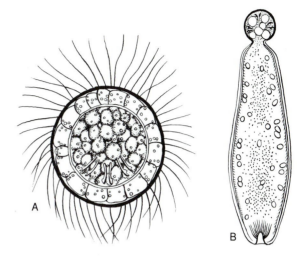

FIGURE 6–47 Some stages in the life cycle of the fish tapeworm, *Diphyllobothrium latum.* **A,** Ciliated oncosphere. **B,** Mature procercoid. *(A, After Vergeer. B, After Brumpt.)*

sometimes more intermediate hosts, which are usually arthropods and vertebrates. Two life cycles are included here: in one transmission is through an aquatic food chain; in the other transmission is terrestrial.

Diphyllobothrium spp. are widely distributed in northern areas and are parasitic in the gut of many fish-eating carnivores, including humans (*Diphyllobothrium latum*). If the egg capsules are deposited with feces in water, a ciliated, free-swimming **oncosphere (coracidium),** which bears six small hooks, hatches after approximately ten days of development (Fig. 6–47A). The coracidium is ingested by certain copepod crustaceans. It penetrates the intestinal wall and develops within the hemocoel into a **procercoid,** which retains the coracidial hooks (Fig. 6–47B). When the copepod is eaten by a freshwater fish, the procercoid penetrates the gut of the fish and migrates to the striated muscles where it transforms into a metacestode, known in this case as the **plerocercoid** stage. The plerocercoid, which looks like a small unsegmented tapeworm complete with a miniature scolex, develops into an adult tapeworm when ingested by a suitable warm-blooded definitive host.

Species of the family Taeniidae are among the best known tapeworms. *Taenia saginatus,* the beef

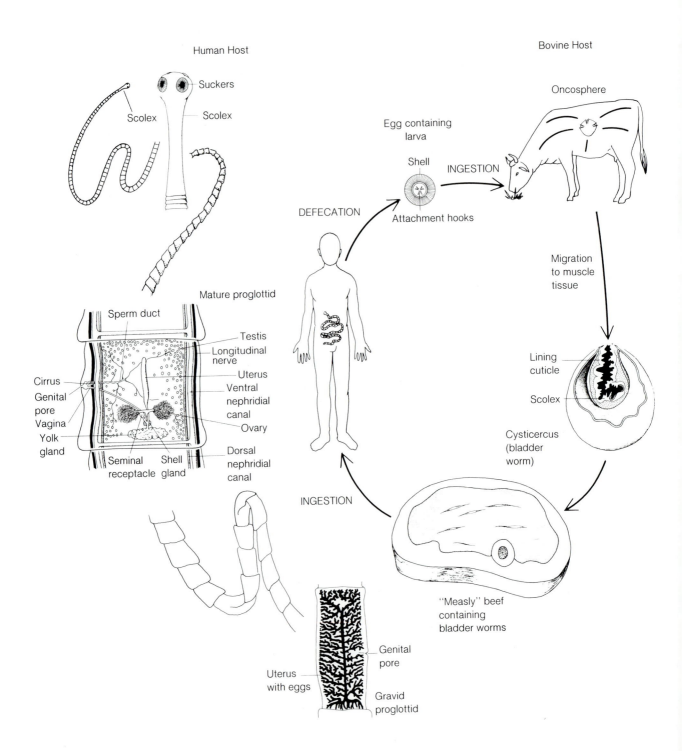

FIGURE 6–48 Structure and life cycle of the beef tapeworm *Taenia saginatus.*

tapeworm, is cosmopolitan and one of the most commonly occurring species in humans, where it lives in the intestine and may reach a length of over 3 m (Fig. 6–48). Proglottids containing embryonated eggs are eliminated through the anus, usually with feces. If an infected person defecates on a pasture, the egg capsules may be eaten by grazing cattle. On hatching in the gut of the intermediate host, the oncosphere, bearing three pairs of hooks, bores into the intestinal wall, where it is picked up by the circulatory system and transported to striated muscle. Here the oncosphere develops into the metacestode, in this case called a **cysticercus.** The cysticercus, sometimes called a bladder worm, is an oval worm about 10 mm in length, with an invaginated scolex. If raw or insufficiently cooked beef is ingested by humans, the cysticercus is freed, and the scolex evaginates, proliferating numerous proglottids to develop into an adult worm in the gut.

Taenia solium, the pork tapeworm, is also a parasite of humans, but the intermediate host is the pig and the cysticercus is obtained from pork. *Taenia pisiformis* occurs in cats and dogs, with rabbits as the intermediate hosts. This order (Cyclophyllidea) contains tapeworms that are largely parasitic in birds and mammals. Vertebrates, insects, mites, annelids, and molluscs serve as their intermediate hosts.

A severe infection of adult tapeworms may cause diarrhea, weight loss, and reactions to the toxic wastes of the worm. The worms can be eliminated easily with drugs. Much more serious are the "accidental" cysticercus infections of humans in the role of intermediate host. In the case of the pork tapeworm, *Taenia solium,* and for the dog tapeworm, *Echinococcus granulosus,* serious disease may result in humans. The adult *Echinococcus* is minute, with only a few proglottids present at any one time. Many different mammals, including humans, can act as intermediate hosts, although herbivores are the most important in completing the life cycle. The bladder worm, or hydatid, of *Echinococcus* develops mostly in the lung or liver but can develop in many other sites as well. The cysticerci of the pork tapeworm develop in subcutaneous connective tissue and in the eye, brain, heart, and other organs. The bladder worms of both species are life threatening when growing in such places as the brain and can do much damage elsewhere. Hydatid cysts can reach a large size and contain a great volume of fluid (up to many liters), which if released

into the host can cause severe reactions. Bladder worm cysts can be removed only by surgery.

ORIGIN OF THE PARASITIC FLATWORMS

By virtue of their neoophoran reproductive systems, which are essentially variations on a turbellarian theme, it is likely that all parasitic platyhelminths evolved from an ancestor, or ancestors, shared with neoophoran turbellarians (Fig. 6–34). Cestoideans and monogeneans probably evolved independently of the digeneans from a rhabdocoel-like ancestor. The close relationship of monogeneans and cestodes is based, in part, on the great similarity of their attachment hooks (haptor in monogenean oncomiracidia, oncosphere in cestoideans) in the larvae of the taxa.

SUMMARY

1. Three classes of flatworms, Trematoda, Monogenea, and Cestoidea, are entirely parasitic. In contrast to the turbellarians, all have a syncytial, nonciliated body covering, or tegument.

2. Most parasitic flatworms are hermaphrodites, and the neoophoran reproductive systems are adapted for copulation, internal fertilization, ectolecithal development, and the formation of an egg capsule.

3. Digenean trematodes are endoparasites of vertebrates and constitute the largest taxon of parasitic flatworms. The oval to elongated body is flattened and provided with a ventral sucker. A gut is present, and the anterior mouth is provided with an oral sucker.

4. The digenean life cycle is indirect and involves two to four different hosts and a number of developmental stages, including three or four types of larvae (miracidium, sporocyst, redia, cercaria). The definitive host is always a vertebrate, and gastropod snails are common first intermediate hosts. Species of blood flukes (*Schistosoma*) are among the most widespread and serious groups of human parasites.

5. Most monogeneans are ectoparasites of fishes, but they also parasitize amphibians and reptiles. A gut is present, but the mouth lacks a sucker. They have a large posterior haptor with which they hold onto the integument of the host. The life cycle is

direct, with adults laying eggs from which the oncomiracidium emerges to locate and attach to a new host of the same species.

6. Cestoidea, or tapeworms, are gut parasites of vertebrates. They have a scolex, usually with hooks and suckers for attachment, a proliferative neck, and a strobila, which consists of a chain of proglottids budded off from the neck region.

7. A gut is absent in tapeworms, and the reproductive system, which is somewhat similar to that of digeneans and monogeneans, is repeated in each proglottid.

8. The tapeworm life cycle is indirect and involves a larva, the oncosphere, and one or more intermediate hosts.

GNATHOSTOMULIDS

The Gnathostomulida is a small phylum of minute, acoelomate worms that live in the interstitial spaces of fine marine sands, especially those that have little or no oxygen and reek of hydrogen sulfide. Like many other interstitial groups, the Gnathostomulida was discovered in the latter half of this century (in 1956 by Peter Ax). Since that time, more than 80 species and 18 genera have been found worldwide, concentrated especially along the east coast of North America, where they have been extensively sought after.

Gnathostomulids may reach 3 mm but most are between 0.5 and 1.0 mm in length and only 50 μm in diameter. All are elongated, and some species are threadlike (Fig. 6–49A). Commonly, the cylindrical body consists of an anterior head separated from the trunk by a slightly constricted neck (Fig. 6–49B). The trunk tapers posteriorly into a tail.

The gnathostomulid epidermis lacks a cuticle and is ciliated, but each epithelial cell has only one cilium instead of many. Each of these cilia is surrounded by a low collar of microvilli. The epidermal cilia are used for locomotion. The body wall musculature, which is cross-striated, underlies the epidermal basal lamina and is not used for locomotion but rather to turn or shorten the body. It consists of weakly developed outer circular fibers and usually three pairs of longitudinal muscles. Gnathostomulids have no connective tissue, and as a result, muscles, protonephridia,

and reproductive organs are simply sandwiched between the epidermis and gut (6–49C). The nervous system is intraepidermal and consists of an anterior cerebral ganglion, a buccal ganglion innervating the pharyngeal bulb, and one to three pairs of longitudinal cords. The sensory organs are ciliary pits and sensory cilia, which are especially well developed on the head. There are two to five pairs of monociliated protonephridia, each with its own duct to the exterior.

Gnathostomulids have a blind tubular or saclike gut although, as is the case in some turbellarians, a poorly developed anus is sometimes present. The mouth and muscular pharynx, which are located ventrally behind the head, bear a comblike basal plate on the ventral lip and a pair of toothed lateral **jaws** embedded in the walls of the pharyngeal bulb (Figs. 6–49D, E). Some of the pharyngeal muscles are myoepithelial cells. The remainder of the gut is a simple epithelial tube lined with microvilli. The food appears to be bacteria and fungi, which are scraped up by the comb on the ventral lip and passed into the gut by snapping movements of the jaws (Fig. 6–49E).

Gnathostomulids are generally hermaphrodites. The female reproductive system usually consists of a single ovary and associated bursa, or sperm storage sac. A vagina is present in some species. The male system consists of one or a pair of posterior testes and usually a copulatory organ, which in some species bears a stylet. The male gonopore is at the posterior end. Copulation has not yet been observed, but transfer of the uniflagellate or aflagellate sperm probably occurs by hypodermic impregnation of the body wall or attachment of the male gonopore to the integument of the partner or by injection into the vagina. A single large egg is laid at each oviposition. The egg ruptures through the body wall (which quickly heals) and adheres to the bottom. Cleavage is spiral, and development is direct.

Gnathostomulids are often classified with the aschelminth phyla (Chapter 8), but their lack of cuticle, ciliated surface, and acoelomate body organization indicate a closer alliance with the platyhelminths. Furthermore, the presence of monociliated cells, medusa-like cross-striated muscles, and ectodermal myoepithelial cells in the pharynx are all reminiscent of cnidarian tissues. For these reasons, the gnathostomulids are believed by some to form an evolutionary bridge between a cnidarian-like ancestor and the turbellarians (see "Turbellarian Origins," p. 236).

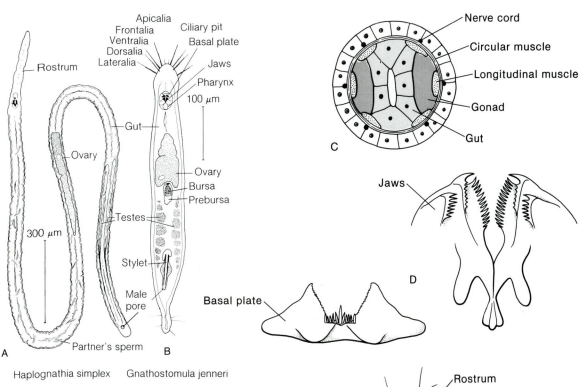

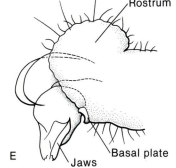

FIGURE 6–49 Gnathostomulids *Haplognathia simplex* **(A)**, *Gnathostomula jenneri* **(B)**. **C**, Simplified cross section of a gnathostomulid trunk. **D**, Jaws and basal plate of *Gnathostomula mediterranea*. **E**, Jaws and basal plate of *Problognathia minima* in feeding position *(A, B, D, from Sterrer, W. 1972. Systematics and evolution within the Gnathostomulida. Syst. Zool. 21:151–173. E, From Sterrer, W. and Farris, R. A. Problognathia minima n. g., n. sp., A representative of a new family of Gnathostomulida, Problognathiidae n. fam. from Bermuda. Trans. Am. Micros. Soc. 94:357–367.)*

MESOZOANS: PHYLUM ORTHONECTIDA AND PHYLUM RHOMBOZOA (CLASSES DICYEMIDA AND HETEROCYEMIDA)

Mesozoans comprise some 50 species of small parasitic worms that have simple structure but complex life cycles (Figs. 6–50, 6–51). Mesozoans parasitize and absorb dissolved nutrients from the tissues and internal cavities of marine invertebrates. Members of the phylum Orthonectida have been found in flat-worms, nemerteans, polychaetes, bivalve molluscs, brittle stars, and other groups. The dioecious adults are unattached within the host, and the microscopic, wormlike body, which lacks organs, consists of an outer layer of naked multiciliated cells enclosing an internal mass of either sperm or egg cells. In some orthonectids, circular and longitudinal muscles occur beneath the epidermis (Fig. 6–50D). Parenchymal cells are absent, but some extracellular matrix is present. The feeble movements are primarily the result of cilia and not muscular contraction, although the musculature produces flexing movements of the body.

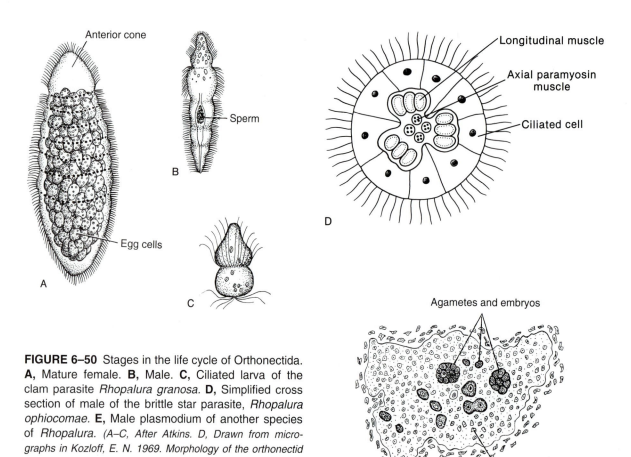

FIGURE 6–50 Stages in the life cycle of Orthonectida. **A,** Mature female. **B,** Male. **C,** Ciliated larva of the clam parasite *Rhopalura granosa.* **D,** Simplified cross section of male of the brittle star parasite, *Rhopalura ophiocomae.* **E,** Male plasmodium of another species of *Rhopalura.* (*A–C, After Atkins. D, Drawn from micrographs in Kozloff, E. N. 1969. Morphology of the orthonectid Rhopalura ophiocomae. J. Parasitol. 55:171–195. E, After Caullery and Mesnil.*)

Adult females and males (Fig. 6–50A, B) are released from the host simultaneously. Fertilization occurs outside of the host as sperm from the males penetrate the bodies of the females and fertilize the eggs. Cleavage leads to a ciliated larva that is released from the female parent and infects a new host (Fig. 6–50C). Within host tissues the larva loses its ciliated cells and reportedly gives rise to an amoeboid syncytial **plasmodium** (Fig. 6–50E). The infection spreads, often sterilizing the host, as the plasmodium divides asexually and the fragments are transported elsewhere in the host's body. Some of the nuclei within the plasmodium, called **agametes,** divide mitotically and give rise to sexual adults.

Rhombozoans in the class Dicyemida live attached within the nephridial cavities of squids, cuttlefish, and octopods and presumably absorb metabolites from the primary urine of the host, but they do not cause any obvious pathology. Although this phylum was known before the orthonectids, knowledge of its life cycle is still incomplete. The adult dicyemid is called a **nematogen** and is 0.5 to 7 mm long. The body, composed of only 20 to 30 cells, consists of a long, central **axial cell** surrounded by a single layer of ciliated cells. The anterior cells are used for attachment. The population of individuals within their hosts, especially young cephalopods, increases through the production of young having the same form as the parent. Such young are called **vermiform embryos** and are formed within the axial cell of the parent (Fig. 6–51).

In mature cephalopods, the nematogens become **rhombogens,** which have a similar structure to nematogens except that the axial cell of the parasite gives rise to another type of larva, called an **infusoriform larva.** The axial cell first forms a structure that

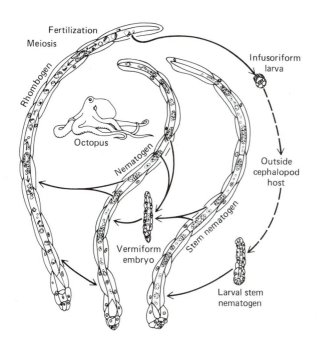

FIGURE 6–51 Life cycle of the dicyemid, *Dicyemmenea*, which lives in the kidney of *Octopus*. *(From Hochberg, F. G. 1982. The "kidneys" of cephalopods: A unique habitat for parasites. Malacologia. 23:121–134.)*

is interpreted as being a hermaphroditic gonad (infusorigen). An egg produced by this structure and fertilized by sperm produced in the same "gonad" gives rise to the infusoriform larva. All of this takes place within the axial cell of the parent. The infusoriform larvae are small with posteriorly directed cilia (Fig. 6–51). They are passed out with the urine of the host, and although their fate is unknown, it is now believed that they are picked up by their cephalopod host without an intermediate host. Heterocyemid rhombozoans are rare, poorly known parasites of *Octopus* and cuttlefish kidneys.

Because of their simple structure, the phylogenetic position of mesozoans has been difficult to determine. Originally believed to have only two germ layers—ectoderm and endoderm (the latter not organized into a gut)—they were interpreted as being intermediate between Protozoa and Metazoa. It was from this old idea that the name Mesozoa, meaning "middle animals," was coined. Now, however, it is clear that the inner layer is actually mesoderm (it forms muscle and germ cells), and that endoderm is probably absent secondarily (as in tapeworms). Consequently, the absence of a gut and the occurrence of few cells and tissue layers may be regarded as specializations for endoparasitism, rather than as hallmarks of a primitive body organization. Assuming that mesozoans are bilateral animals, their evolutionary relationship to other bilaterians is far from clear. The multiciliated epidermal cells and the occurrence of unusual smooth muscles (paramyosin muscles) of mesozoans, both suggest that they are not primitive bilaterians, but their nearest relations are unclear. They are grouped with the Platyhelminthes because of the general similarity of their body walls.

SUMMARY

1. The phylum Gnathostomulida is a small group of acoelomate, monociliated, interstitial worms. They have a blind gut and a muscular pharynx bearing jaws.

2. Gnathostomulids are usually viewed as primitive bilaterians forming a bridge between platyhelminths and cnidarians.

3. Mesozoans constitute two small phyla of minute endoparasites of invertebrates. The phylum inhabiting the excretory organs of squids, cuttlefish, and octopods is the best known.

4. The simple, elongated, ciliated mesozoan body lacks the usual cell types and typical organs. Mesozoans may share a common ancestor with flatworms or may be an offshoot from some early, free-living metazoan. The two mesozoan phyla may have had separate origins.

REFERENCES

Uptake of Dissolved Organics

General discussions of the role of dissolved organics in invertebrate nutrition are provided by

Stevens, G. C. et al. 1982. The role of uptake of organic solutes in the nutrition of marine organisms. Am. Zool. *22*:611–733 (ten papers from a symposium).

Stewart, M. G. 1979. Absorption of dissolved organic nutrients by marine invertebrates. Oceanogr. Mar. Biol Ann. Rev. *17*:163–192.

Regeneration

General accounts of regeneration in flatworms and other organisms may be found in

Bronsted, D. 1969. Planarian Regeneration. Pergamon Press, Oxford, New York. 216 pp.

Chandebois, R. 1976. Histogenesis and morphogenesis in planarian regeneration. Monogr. Devel. Biol. *11*:1–182.

Mattson, P. 1976. Regeneration. Bobb-Merrill Co., Inc., Indianapolis. 178 pp.

Rose, S. M. 1970. Regeneration. Appleton-Century-Crofts, New York.

Interstitial Animals

The following references are general accounts and provide an introduction to the interstitial fauna. An indispensible reference is the volume edited by:

Higgins, R. P., and Thiel, H. (Eds.): 1988. Introduction to the Study of Meiofauna. Smithsonian Institution Press, Washington, DC. 488 pp.

Older but still valuable papers include those of:

McIntyre, A. D. 1969. Ecology of marine meiobenthos. Biol. Rev. *44*:245–290.

Swedmark, B. 1964. The interstitial fauna of marine sand. Biol. Rev. *39*:1–42.

The subject of temporary adhesion in turbellarians and other animals is described in papers by:

Hermans, C. O. 1983. The duo-gland adhesive system. Oceanogr. Mar. Biol. Ann. Rev. *21*:283–339.

Tyler, S. 1976. Comparative ultrastructure of adhesive systems in the Turbellaria. Zoomorphologie *84*:1–76.

A discussion of the principles of adhesion in general can be found in:

Denny, M. W. 1988. Biology and Mechanics of the Wave-Swept Environment. Princeton University Press, Princeton. pp. 228–237.

The importance of progenesis in meiofaunal evolution is discussed by:

Westheide, W. 1984. The concept of reproduction in polychaetes with small body size: Adaptations in interstitial species. Fortschr. Zool. *29*:265–287.

and in general by:

Gould, S. J. 1977. Ontogeny and Phylogeny. Harvard University Press, Cambridge, MA. pp. 303–351.

Reproductive Adaptations to Life in Fresh Water

Pennak, R. W. 1989: Freshwater Invertebrates of the United States: Protozoa to Mollusca. 3rd Edition. John Wiley, New York. pp. 628.

Thorp, J. H., and Covich, A. P. 1991. Ecology and Classification of North American Freshwater Invertebrates. Academic Press, New York. pp. 911).

Hermaphroditic and Dioecious Reproductive Patterns

Ghiselin, M. T. 1969. The evolution of hermaphroditism among animals. Q. Rev. Biol. *41*:189–208.

Heath, D. J. 1977. Simultaneous hermaphroditism; Cost and benefit. J. Theor. Biol. *64*:363–373.

Williams, G. C. 1975. Sex and Evolution. Princeton University Press, Princeton. pp. 200.

General Parasitology Textbooks

The following general parasitology texts provide information on trematodes and cestodes.

Baer, J. G. 1971. Animal Parasites. McGraw-Hill Book Co., New York. 256 pp. (An excellent introduction to general parasitology)

Cox, F. E. G. 1982. Modern Parasitology. Blackwell Scientific Publications, Oxford, London. 346 pp.

Chappell, L. H. 1980. Physiology of Parasites. John Wiley and Sons, New York. 230 pp.

Cheng, T. C. 1973. General Parasitology. Academic Press, New York. 965 pp.

Noble, E. R., and Noble, G. A. 1982. Parasitology. 5th Edition. Lea & Febiger, Philadelphia. 522 pp.

Olsen, O. W. 1967. Animal Parasites: Their Biology and Life Cycles. Burgess Publishing Co., Minneapolis. 431 pp.

Read, C. P. 1970. Parasitism and Symbiology. Ronald Press, New York. 316 pp. (A text that examines host/parasite relationships rather than surveying parasitic groups.)

Schmidt, G. D., and Roberts, L. S. 1989. Foundations of Parasitology. 4th Edition. The C. V. Mosby Co., St. Louis. 750 pp.

Smyth, J. D. 1976. Introduction to Animal Parasitology. 2nd Edition. John Wiley and Sons, New York. 466 pp.

Whitfield, P. J. 1979. The Biology of Parasitism: An Introduction to the Study of Associating Organisms. University Park Press, Baltimore. 277 pp.

Flatworms

The introductory references on page 6 include many general works and field guides that contain sections on flatworms.

Apelt, G. 1969. Fortpflanzungsbiologie, Entwicklungszyklen und vergleichende Frühentwicklung acoeler Turbellarien. Mar. Biol. *4(4)*:267–325. (Reproductive biology, developmental cycles, and comparative early development of acoel turbellarians.)

Arne, C. 1983. Aspects of tapeworm physiology. J. Biol. Ed. *17(4)*:352–357. (An excellent brief review of tapeworm physiology.)

Arne, C., and Pappas, P. W. (Eds.): 1983. Biology of the Eucestoda. Vols. I and II. Academic Press, London.

Ax, P. 1963. Relationships and phylogeny of the Turbellaria. *In* Dougherty, E. C. (Ed.) The Lower Metazoa. University of California Press, Berkeley. pp. 191–224.

Ax, P. 1977. Life cycles of interstitial Turbellaria from the eulittoral of the North Sea. Acta Zool. Fennica *154*:11–20.

Ax, P. 1985. The position of the Gnathostomulida and Platyhelminthes in the phylogenetic system of the Bilateria. *In* Morris, S. C., George, J. D., Gibson, R. et al. (Eds.): The Origins and Relationships of Lower Invertebrates. Clarendon Press, Oxford. pp. 168–180.

Ax, P., and Apelt, G. 1965. Die "Zooxanthellen" von *Convoluta convoluta* (Turbellaria, Acoela) entstehen aus Diatomen. Naturwissenschaften, *52(15)*:444–446.

Ax, P., and Borkett, H. 1968. Organisation und Fortpflanzung von *Macrostomum romanicum* (Turbellaria, Macrostomida). Zool. Anz. Suppl. *32*:344–347.

Benham, W. D. 1901. The Platyhelmia, Mesozoa, and Nemertini. *In* Lankester, E. R. (Ed.): A Treatise on Zoology. Adam & Charles Black, London. pp. 204. (A classic but still valuable reference.)

Bush, L. 1975. Biology of *Neochildia fusca* n. gen., n. sp. from the northeastern coast of the United States. Biol. Bull. *148(1)*:35–48.

Crezée, M. R. 1984. Turbellaria. *In* Parker, S. P. (Ed.): Synopsis and classification of living organisms. Vol 1. McGraw-Hill, New York. pp. 718–740.

de Beauchamp, P., Caullery, M., Euzet, L. et al. 1961. Plathelminthes et Mesozaires. *In* Grassé, P. (Ed.): Traité de Zoologie. Vol. 4 pt. 1. Masson et Cie, Paris. pp. 1–729.

Doe, D. A. 1981. Comparative ultrastructure of the pharynx simplex in Turbellaria. Zoomorphologie. *97*:133–193.

Eakin, R. M., and Brandenberger, J. L. 1981. Unique eye of probable evolutionary significance. Science. *211*:1189–1190.

Ehlers, U. 1985a. Das phylogenetisches System der Platyhelminthes. Verlag Paul Parey, Stuttgart, New York. pp. 317.

Ehlers, U. 1985b. Phylogenetic relationships within the Platyhelminthes. *In* Morris, S. C., George, J. D., Gibson, R. et al. (Eds.): The Origins and Relationships of Lower Invertebrates. Clarendon Press, Oxford. pp. 143–158.

Erasmus, D. A. 1972. The Biology of Trematodes. Crane, Russak & Co., New York. 312 pp.

Faisst, J., Keenan, C. L., and Koopowitz, H. 1980. Neuronal repair and avoidance behavior in the flatworm, *Notoplana acticola*. J. Neurobiol. *11*:483–496.

Fournier, A. 1984. Photoreceptors and photosensitivity in Platyhelminthes. *In* Ali, M. A. (Ed.): Photoreception and Vision in Invertebrates. Plenum Press, New York. pp. 217–240.

Freeman, R. 1973. Ontogeny of cestodes and its bearing on their phylogeny and systematics. Adv. Parasitol. *11*:481–557.

Harrison, F. W., and Bogitsh B. J. (Eds.): 1991. Microscopic anatomy of invertebrates. Vol. 1. Platyhelminthes and Nemertinea. pp. 1–347. (The most recent summary of platyhelminth structure and ultrastructure.)

Heitkamp, U. 1977. The reproductive biology of *Mesostoma ehrenbergii*. Hydrobiologia. *55*:21–32.

Henley, C. 1974. Platyhelminthes (Turbellaria). *In* Giese, A. C., and Pearse, J. S. (Eds.): Reproduction of Marine Invertebrates. Vol. I. Acoelomate and Pseudocoelomate Metazoans. Academic Press, New York. pp. 267–343.

Highnam, K. C., and Hill, L. 1977. The Comparative Endocrinology of the Invertebrates. 2nd Edition. University Park Press, Baltimore.

Hyman, L. H. 1951. The Invertebrates: Platyhelminthes and Rhynchocoela. Vol. 2. McGraw-Hill, New York. pp. 52–219.

Jennings, J. B. 1957. Studies on feeding, digestion, and food storage in free-living flatworms. Biol. Bull. *112*:63–80.

Jennings, J. B. 1968. Nutrition and digestion in Platyhelminthes. *In* Florkin, M., and Scheer, B. T. (Eds.): Chemical Zoology, Vol. 2. Academic Press, New York. pp. 305–327.

Jennings, J. B. 1974. Digestive physiology of the Turbellaria. *In* Riser, N. W., and Morse, M. P. (Eds.): Biology of the Turbellaria. McGraw-Hill Book Co., New York. pp. 173–197.

Karling, T. G. 1974. On the anatomy and affinities of the turbellarian orders. *In* Riser, N. W., and Morse, M. P. (Eds.): Biology of the Turbellaria. McGraw-Hill Book Co., New York. pp. 1–16.

Keenan, C. L., Coss, R., and Koopowitz, H. 1981. Cytoarchitecture of primitive brains: Golgi studies in flatworms. Jour. Comp. Neurobiol. *195*:697–716.

Kenk, R. 1972. Freshwater planarians (Turbellaria) of North America. Biota of Freshwater Ecosystems. Identification Manual No. 1. Environmental Protection Agency, Washington, DC. 81 pp.

Koopowitz, H. 1974. Some aspects of the physiology and organization of the nerve plexus in polyclad flatworms. *In* Riser, N. W., and Morse, M. P. (Eds.): Biology of the Turbellaria. McGraw-Hill Book Co., New York. pp. 198–212.

Kozloff, E. N. 1972. Selection of food, feeding and physical aspects of digestion in the acoel turbellarian *Otocelis luteola*. Trans. Am. Microsc. Soc. *91*:556–565.

Lauer, D. M., and Fried, B. 1977. Observations on nutrition of *Bdelloura candida*, an ectocommensal of *Limulus polyphemus*. Am. Midl. Nat. *97*:240–247.

Llewellyn, J. 1965. The evolution of parasitic platy-helminths. *In* Taylor, A. E. R. (Ed.): Evolution of Parasites. 3rd Symposium for the British Society for Parasitology. Blackwell, London.

Martin, G. G. 1978. A new function of rhabdites: Mucus production for ciliary gliding. Zoomorphologie. *91*:235–248.

Moraczewski, J. 1977. Asexual reproduction and regeneration of *Catenula.* Zoomorphologie. *88*:65–80.

Morita, M., and Best, J. B. 1984. Effects of photoperiods and melatonin on planarian asexual reproduction. J. Exp. Zool. *231*:273–282.

Muscatine, L., Boyle, J. E., and Smith, D. C. 1974. Symbiosis of the acoel flatworm *Convoluta reoscoffensis* with the alga *Platymonas convolutae.* Proc. R. Soc. Lond. [Biol.] *187*:221–234.

Nentwig, M. R. 1978. Comparative morphological studies after decapitation and after fission in the planarian *Dugesia dorotocephala.* Trans. Am. Micros. Soc. *97*:297–310.

Palladini, G., Medolago-Albani, L., Margotta, V. et al. 1979. The pigmentary system of planaria: 2. Physiology and functional morphology. Cell Tissue Res. *199*:203–211.

Prusch, R. D. 1976. Osmotic and ionic relationships in the freshwater flatworm *Dugesia dorotocephala.* Comp. Biochem. Physiol. *54A*:287–290.

Reisinger, E. 1968. *Xenoprorhynchus* ein Modellfall für progressiven Funktionswechsel. Z. Zool. Syst. Evolut.-forsch. *6*:1–55.

Reisinger, E. 1972. Die Evolution des Orthogons der Spiralier und der Archicoelomatenproblem. Z. Zool. Syst. Evolut.-forsch. *10*:1–43.

Reynoldson, T. B., and Sefton, A. D. 1976. The food of *Planaria torva,* a laboratory and field study. Freshwater Biol. *6(4)*:383–393.

Rieger, R. M. 1981. Morphology of the Turbellaria at the ultrastructural level. Hydrobiologia. *84*:213–229.

Rieger, R. M. 1985. The phylogenetic status of the acoelomate organization with the Bilateria: A histological perspective. *In* Morris, S. C., George, J. D., Gibson, R. et al. (Eds.): The Origins and Relationships of Lower Invertebrates. Clarendon Press, Oxford. pp. 101–122.

Rieger, R. M., Tyler, S., Smith, J. P. III et al. 1991. Turbellaria. *In* Harrison, F. W. and Bogitsh, B. J. (Eds.) Microscopic Anatomy of Invertebrates. Vol. 3. Platyhelminthes and Nemertinea. Wiley-Liss, Inc., New York. pp. 7–140.

Riser, N. W., and Morse, M. P. (Eds.): 1974. Biology of the Turbellaria. McGraw-Hill Book Co., New York. 530 pp. (Papers from a 1970 symposium in Chicago sponsored by the American Society of Zoologists.)

Ruppert, E. E. 1978. A review of metamorphosis of turbellarian larvae. *In* Chia, F. S. and Rice, M. (Eds.): Settle-ment and Metamorphosis of Marine Invertebrate Larvae. Elsevier and North Holland Biomedical Press. pp. 65–81.

Schell, S. C. 1970. How to Know the Trematodes. W. C. Brown Co., Dubuque, IA. 355 pp.

Schockaert, E. R., and Ball, I. R. (Eds.): 1981. The Biology of the Turbellaria. Dr. W. Junk Publishers, The Hague. (Papers from a symposium on the Turbellaria.)

Shinn, G. L. 1987. Phylum Platyhelminthes with emphasis on marine Turbellaria. *In* Strathmann, M. F. Reproduction and Development of Marine Invertebrates of the Northern Pacific Coast. University of Washington Press, Seattle. pp. 114–128.

Shinn, G. L. and A. M. Christensen. 1985. *Kronborgia pugettensis* sp. nov. (Neorhabdocoela: Fecampiidae), an endoparasitic turbellarian infesting the shrimp *Heptacarpus kincaidi* (Rathbun), with notes on its life history. Parasitology *91*:431–447.

Slais, J. 1973. Functional morphology of cestode larvae. Adv. Parasitol. *11*:395–480.

Smith, J. P. S. III, and Tyler, S. 1985. The acoel turbellarians: Kingpins of metazoan evolution or a specialized offshoot? *In* Morris, S. C., George, J. D., Gibson, R. et al. (Eds.): The Origins and Relationships of Lower Invertebrates. Clarendon Press, Oxford. pp. 123–142.

Smith, J. P. S. III, Tyler, S., Rieger, R. M. 1985. Is the Platyhelminthes polyphyletic? *In* Research in Turbellarian Biology. Hydrobiologia. *132*:13–21.

Smith, J. P. S. III, Tyler, S., Thomas, M. B. et al. 1982. The morphology of turbellarian rhabdites: Phylogenetic implications. Trans. Amer. Micros. Soc. *101*:209–228.

Smyth, J. D., and Halton, D. W. 1983. The Physiology of Trematodes. 2nd Edition. Cambridge University Press, Cambridge. 446 pp.

Tempel, D., and Westheide, W. 1980. Uptake and incorporation of dissolved amino acids by interstitial Turbellaria and Polychaeta and their dependence on temperature and salinity. Mar. Ecol. Prog. Ser. *3*:41–50.

Thomas, M. B. 1986. Embryology of the Turbellaria and its phylogenetic significance. Hydrobiologia. *132*:105–115.

Turbeville, J. M., Field, K. G. and Raff, R. A. 1992. Phylogenetic position of phylum Nemertini inferred from 18S rRNA sequences: Molecular data as a test of morphological character homology. Mol. Biol. Evol. *9*:235–249.

Tyler, S. (Ed.) 1986. Advances in the biology of turbellarians and related platyhelminths. Hydrobiologia, *132*:1–357.

Tyler, S. (Ed.) 1991. Turbellarian biology. Hydrobiologia, *227*:1–398.

Tyler, S., and Rieger, R. M. 1975. Uniflagellate spermatozoa in *Nemertoderma* and their phylogenetic significance. Science. *188*:730–732.

Wright, C. A. 1971. Flukes and snails. *In* Carthy, J. D., and

Sutcliffe, J. F. (Eds.): The Science of Biology. Series 4. Allen and Unwin, Publishers, London.

Gnathostomulida

Lammert, V. 1991. Gnathostomulida. *In* Harrison, F. W., and Ruppert, E. E. (Eds.): Microscopic Anatomy of Invertebrates. Vol. 4. Aschelminthes. Wiley-Liss, New York. pp. 19–39

Mainitz, M. 1983. Gnathostomulida. *In* Adiyodi, K. G., and Adiyodi, R. G. (Eds.): Reproductive Biology of Invertebrates. Vol. 4. John Wiley, New York. pp. 169–180.

Rieger, R. M., and Mainitz, M. 1977. Comparative fine structure of the body wall in Gnathostomulida and their phylogenetic position between Platyhelminthes and Aschelminthes. Z. zool. Syst. Evolut.-forsch. *15*:9–35.

Sterrer, W. 1972. Systematics and evolution within the Gnathostomulida. Syst. Zool. *21*:151–173.

Sterrer, W., Mainitz, M., and Rieger, R. M. 1986. Gnathostomulida: Enigmatic as ever. *In* Morris, S. C., George, J. M., Gibson, R. et al. (Eds.): The Origins and Relationships of Lower Invertebrates. Clarendon Press, Oxford. pp. 181–199.

Mesozoans

Kozloff, E. N. 1969. Morphology of the orthonectid *Rhopalura ophiocomae*. J. Parasitol. *55*:171–195.

Kozloff, E. N. 1971. Morphology of the orthonectid *Ciliocincta sabellariae*. J. Parasitol. *57*:585–597.

Matsubara, J. A., and Dudley, P. L. 1976a. Fine structural studies of the dicyemid mesozoan *Dicyemmenea californica* McConnaughey. I. Adult stages. J. Parasitol. *62*:377–389.

Matsubara, J. A., and Dudley, P. L. 1976b. Fine structural studies of the dicyemid mesozoan, *Dicyemmenea californica* McConnaughey. II. The young vermiform stage and the infusoriform larva. J. Parasitol. *62*:390–409.

McConnaughey, B. H. 1963. The Mesozoa. *In* Dougherty, E. C. (Ed.): The Lower Metazoa. University of California Press, Berkeley. pp. 151–168.

Lapan, E. A., and Morowitz, H. 1972. The Mesozoa. Sci. Am. *227*:94–101.

Stunkard, H. W. 1954. The life history and systematic relations of the Mesozoa. Q. Rev. Biol. *29*:230–244.

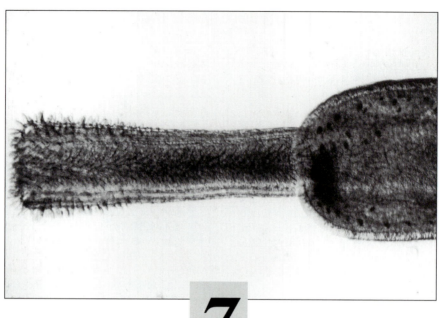

7

NEMERTEANS

PRINCIPLES AND EMERGING PATTERNS

SKELETONS 2

Recall from the initial discussion of hydrostatic skeletons (Chapter 4, p. 96) that animal hydrostats are cylinders whose walls are reinforced by a helical array of fibers that toughens the wall and helps prevent kinks and aneurisms. The angle between any fiber and a line parallel to the longitudinal axis of the body at rest is often just less than 55 degrees. What is the significance of this unusual angle? To provide an answer to this question and explore additional properties of hydrostats, it is necessary to consider the properties of an ideal, fiber-wound cylinder (Fig. 7–1).

Any increase or decrease in the length of an ideal cylinder wound with helical fibers causes a change in its volume. Such changes can be visualized by mentally extending a cylinder until it becomes a straight line, or shortening it until it becomes a circle—both two-dimensional shapes of zero volume. Approximately midway between these two extremes, the cylinder reaches a maximum volume. Moreover, as the shape and volume of the cylinder change between the extremes, so does the angle of the helical fibers with respect to the longitudinal axis. For example, when the cylinder is extended fully as a straight line of zero volume, the fiber angle is 0

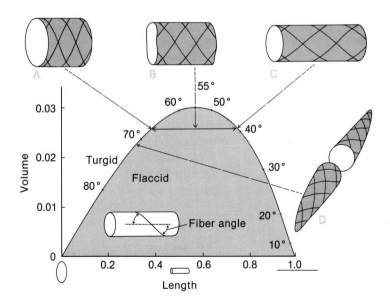

FIGURE 7–1 How volume and fiber angle vary with changes in length of a helically wound cylinder. The fully extended cylinder is arbitrarily given a length of 1 unit. Beneath the curve, the cylinder is underinflated and flattened. Any horizontal line beneath the curve represents the range of body length change permissible for a constant-volume hydrostat. The lowest lines permit the greatest range of extensibility and require the hydrostat to be flat at 55 degrees. A nemertean is shown at two extremes **(A, B)** and midpoint **(C)** of its range of extension. The nematode, *Ascaris,* is illustrated at a point corresponding to its body shape at rest **(D)**. *(Modified and redrawn from Vogel, S. 1988. Life's Devices. Princeton University Press, Princeton. p. 221.)*

degrees; at the opposite extreme, it is 90 degrees; but when the cylinder is at maximum volume, the fiber angle is approximately 55 degrees, exactly the value measured in the body wall of many animals.

It should now be noted that, unlike such an ideal cylinder, the fiber-wound body of real animals does not change volume as its length varies; most soft-bodied animals are constant-volume hydrostats. But, if a helically wound cylinder must change volume when it changes length, how do cylindrical animals change length without changing volume? This apparent contradiction is explained by the fact that, at rest, most animals are flat, underinflated cylinders of a lower volume than that of a circular, fully inflated cylinder. Because an animal body is underinflated, its length can change until it becomes circular in cross section. At that point, any further change in length must be accompanied by a change in volume, which is impossible for a constant-volume hydrostat. Thus, it follows that the flattest, most underinflated hydrostats should be able to undergo the greatest degree of length change, whereas the nearly cylindrical, inflated hydrostats, should change the least. Some species of nemertean ribbon worms [e.g., *Cerebratulus lacteus* (Fig.7–2D)] fulfill the requirements of a highly extensible, constant-volume hydrostat. At rest *Cerebratulus* is a flat, underinflated worm, which can range from a contracted length of approximately 20 cm to an extended length of 2 m.

The 55-degree fiber angle is significant because it corresponds to the midpoint (maximum volume) of the length/volume curve (Fig. 7–1). A body at rest that has a 55-degree fiber angle can lengthen or shorten to an equal extent before becoming cylindrical and ceasing to be able to change in length. Furthermore, for animals that are nearly cylindrical at rest, the *rate* of change in volume as length varies is least about the 55-degree point on the curve (Fig. 7–1).

Although the body of many animals is underinflated and extensible, that of a few invertebrates is a fully inflated, fiber-wound cylinder. Invertebrates with such hydrostats either change both length and volume or change neither length nor volume. For example, anemones, especially burrowing species, are animals that significantly elongate and shorten their cylindrical bodies. Changes in body length, however, are accompanied by changes in body volume, as water flows in or out of the coelenteron via the mouth. Similarly, many sea cucumbers (Holothuroidea) among the echinoderms have an inflated, cylindrical body capable of changes in length. Sea cucumbers, like anemones, may change the volume of their hydrostat by allowing water to enter and leave the coelom by way of ducts in the wall of the cloaca. Nematode roundworms are also cylindrical animals, but unlike sea anemones and sea cucumbers they maintain a constant body volume. As a result, nematodes do not move by shortening or elongating their bodies, rather they bend one way and then another, producing whiplike movements.

NEMERTEANS

The phylum Nemertea, or Rhynchocoela, comprises approximately 900 species of elongated and often flattened worms, called ribbon worms. They are sometimes also known as proboscis worms because of the presence of a remarkable proboscis apparatus used in capturing food. Most nemerteans (sometimes called nemertines) are benthic marine animals, but there are some deepwater pelagic species. The many common shallow-water nemerteans live beneath shells and stones or in algae, or burrow in mud and sand. Some species form semipermanent burrows lined with mucus or even distinct cellophane-like tubes. There are three freshwater genera, and six genera containing terrestrial species are confined primar-

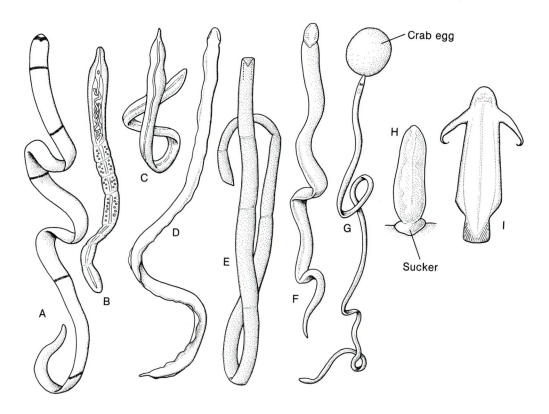

FIGURE 7–2 A,_Tubulanus rhabdotus,_ a tube-dwelling paleonemertean. **B,** An annulated sand-dwelling paleonemertean. **C,** _Carinoma tremaphoros,_ a peristaltic burrowing paleonemertean. **D,** _Cerebratulus lacteus,_ a large burrowing heteronemertean. **E,** _Lineus socialis,_ a crevice-dwelling heteronemertean. **F,** _Paranemertes peregrina,_ a burrowing hoplonemertean. **G,** _Carcinone-mertes carcinophilia,_ a hoplonemertean parasitic on crabs (shown feeding on egg). **H,** _Malacob-della grossa,_ a bdellonemertean commensal of bivalves. **I,** _Nectonemertes mirabilis,_ a pelagic hoplonemertean. _(F, Redrawn from Coe, W. R. 1901. The nemerteans. Wash. Acad. Sci. Proc. 3:1–110. H, Redrawn from Coe, W. R. 1951. The nemertean faunas of the Gulf of Mexico and southern Florida. Bull. Mar. Sci. 1:149–186. I, Redrawn from Coe, W. R. 1926. The pelagic nemerteans. Mem. Mus. Comp. Zool. 49:1–244.)_

ily to the tropics and subtropics. A few species live as ectosymbionts on crabs, in the mantle cavity of bivalve molluscs, or in the atrium of tunicates.

External Structure

In appearance, nemerteans resemble flatworms but tend to be larger, thicker-bodied, and more elongated (Fig. 7–2). A few species are annulated and give the appearance of being segmented, but this is only superficial (Fig. 7–2A, B, E). The anterior end is com-monly pointed or shaped like a spatula. Although most species are less than 20 cm in length and some are only a few millimeters long, there are a few genera, such as the burrowing, shallow-water _Cerebratulus_ and _Lineus,_ which are ribbon-shaped and may measure several meters or more. The greatest reported length is approximately 30 m for some specimens of the European _Lineus longissimus,_ which lie irregularly coiled beneath stones. Most nemerteans are pale, but some are brightly colored with patterns of yellow, orange, red, and green. Many bathypelagic species that live below the photic zone are bright red, orange, or yellow, but the colors appear black at such depths.

BODY WALL AND LOCOMOTION

The body wall (Fig. 7–3C) is composed of an epidermis, a layer of connective tissue, and a thick musculature. The epidermis lacks a cuticle, and each ciliated cell bears many cilia and microvilli. Beneath the epidermis is a basal lamina and a layer of connective tissue, which is continuous with the general connective tissue of the body. Beneath the connective tissue

layer, the body wall musculature is composed of circular and longitudinal smooth muscle, of which the longitudinal is the most developed (Fig. 7–3C). Dorsoventral and radial muscles may also be present. In some nemerteans, especially pelagic nemerteans, jelly-like parenchymal tissue is said to occupy the space between the digestive tract and the muscle layers, but in most it is absent.

Most nemerteans use their epidermal cilia to glide over the substratum on a trail of slime, much of which is secreted by **cephalic glands** on the head (Fig. 7–3A, B). Nemerteans that burrow in sediments, such as *Cerebratulus*, *Carinoma* and *Zygeupolia*, do

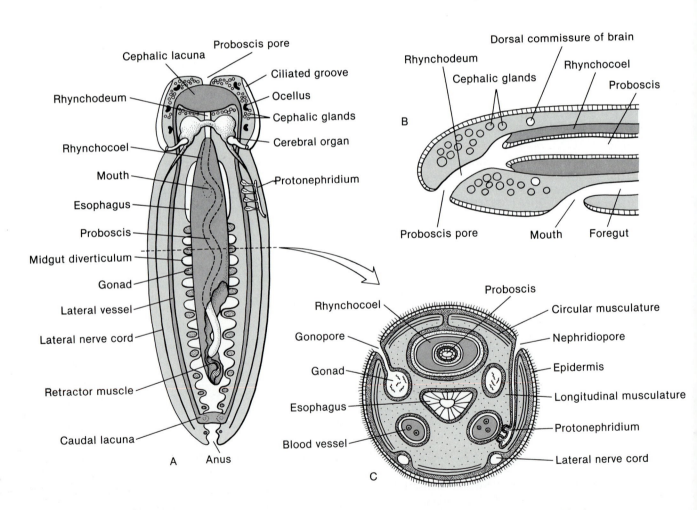

FIGURE 7–3 Body organization of unarmed nemerteans. **A,** Body in dorsal view with rhynchocoel wall partly broken away to show proboscis within. **B,** Sagittal section of head as it might appear in a paleo- or heteronemertean. **C,** Cross section of trunk of a paleonemertean.

not use cilia for locomotion but instead move using peristalsis (Fig. 7–4A). The peristaltic waves are most powerful anteriorly, where the fluid-filled rhynchocoel of the proboscis apparatus functions as a hydrostat. Burrowing nemerteans often have extremely muscular body walls. Indeed, some nemerteans are unique among metazoans in having extensions of the musculature into the epidermis, in which they form extra muscle layers. Other ribbon worms, particularly certain pelagic species but also the benthic *Cerebratulus,* are able to swim and use dorsoventral undulations exclusively. In these, the dorsoventral musculature is particularly well developed.

NUTRITION AND DIGESTIVE SYSTEM

The characteristic feature of the phylum is the **proboscis apparatus** (Fig. 7–3A, B). Although, in other phyla, a proboscis may be present and associated with the digestive tract, the proboscis of nemerteans is separate from the digestive tract, except secondarily in some species. The opening of the proboscis apparatus is through a pore at or near the anterior tip of the worm. This **proboscis pore** leads into a short canal known as the **rhynchodeum,** which extends approximately to the level of the brain (Fig. 7–3A, B). The tissues that make up the wall of the rhynchodeum, as well as those of the proboscis itself, are similar to the tissues of the body wall because both rhynchodeum and proboscis develop as ectodermal invaginations.

The lumen of the rhynchodeum is continuous with that of the proboscis proper. The **proboscis** consists of a long tube, often coiled, lying free in a fluid-filled cavity called the **rhynchocoel,** which is a coelom both in structure and developmental origin. The posterior end of the proboscis is blind and is attached to the back of the rhynchocoel by a **retractor muscle** (Fig. 7–3A).

In some nemerteans (class Anopla), the proboscis is a simple tube (Fig. 7–5A) (or branched tubes) (Fig. 7–5B), but in others (class Enopla) the proboscis has become more specialized and is armed with a heavy, calcareous barb called a **stylet** (sometimes many stylets), which is anchored in the proboscis wall via a bulbous secreted structure, called the **basis** (Figs. 7–5C, D; 7–6A, C, D). Where the stylet is attached, approximately two thirds of the way from the anterior of the animal, the proboscis wall is swollen, particularly on one side, nearly occluding the lumen of the tube. Accessory, or **reserve stylets,** are present on

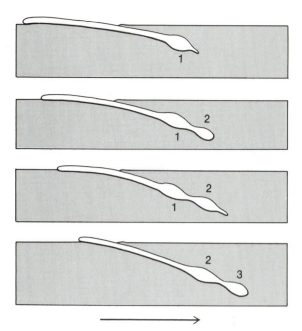

FIGURE 7–4 Peristaltic burrowing in *Carinoma tremaphoros.* Peristaltic waves (numbered) originate anteriorly and progress rearward as the animal burrows. *(Modified and redrawn from Turbeville, J. M. and Ruppert, E. E. 1983. Epidermal muscles and peristaltic burrowing in Carinoma tremaphoros: Correlates of effective burrowing without segmentation. Zoomorphology. 103:103–120.)*

each side of the active central stylet, providing replacement as the animal increases in size or when the main stylet is lost during feeding (Fig. 7–6A, C, D). The mouth is ventral and located at the anterior end of the body near the level of the brain. It opens into a foregut, which is often subdivided into a buccal cavity, an esophagus, and a glandular stomach. The foregut opens into a long intestine, which has lateral diverticula and, in some species, extends anteriorly beyond the junction with the foregut as a cecum. The intestine opens at the **anus** located at the tip of the tail (Figs. 7–3A, 7–6A).

In many armed nemerteans the mouth has disappeared, and the esophagus opens into the rhynchodeum (Fig. 7–6B); in commensal bdellonemerteans the rhynchodeum has disappeared, and the proboscis opens into the anterior part of the gut (Fig. 7–2H). In all other nemerteans, the digestive system is completely separate from the proboscis apparatus.

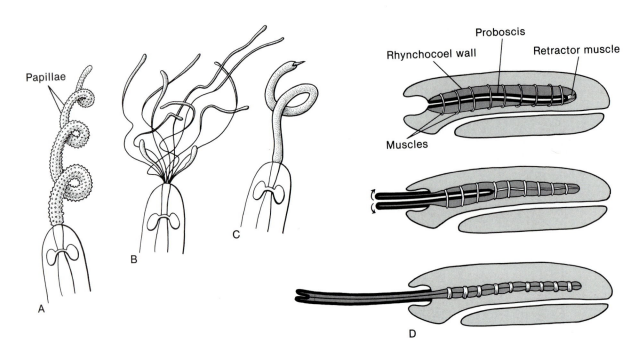

FIGURE 7–5 A, Paleo- and heteronemerteans have a proboscis that lacks a calcified stylet and is typically covered with sticky papillae. **B,** A branched proboscis occurs in species of the heteronemertean, *Gorgonorhynchus.* In these species, the proboscis erupts from its pore like a mass of sticky spaghetti and entangles the prey. **C,** An armed proboscis of a hoplonemertean. **D,** Contraction of muscles in the rhynchocoel wall raises the rhynchocoelic fluid pressure, causing the proboscis to evert. Proboscis eversion stretches the proboscis retractor muscle, which on contraction retracts the proboscis into the rhynchocoel. *(B, Redrawn from Sterrer, W. 1986. Marine Fauna and Flora of Bermuda. John Wiley and Sons, Inc., New York p. 210.)*

Nemerteans are entirely carnivorous and feed primarily on annelids and crustaceans. The proboscis, which is used to capture prey, shoots out of the body under pressure as it everts from the fluid-filled rhynchocoel. Because the proboscis is attached to the posterior wall of the rhynchocoel by a retractor muscle, which is not infinitely extensible, the proboscis cannot be completely everted (Fig. 7–5D). In the armed nemerteans (Hoplonemertea), the proboscis is everted only as far as the stylet, which then is exposed on the everted tip of the proboscis (Fig. 7–6D). The force for eversion results from muscular pressure on the rhynchocoelic fluid, and retraction is caused by the retractor muscle. The proboscis coils around the prey, and sticky toxic secretions from the anterior region of the proboscis aid in holding and immobilizing it (Fig. 7–7). In hoplonemerteans, the stylet is used to stab the prey repeatedly, which in turn, allows the neuro-toxic secretions to enter the body of the prey. Some additional toxin may be pumped (from the posterior proboscis chamber) into the wound by contraction of the muscular stylet bulb (Fig. 7–6C, D). The immobilized prey is either swallowed whole, or its tissues simply are sucked into the mouth.

The large unarmed *Cerebratulus lacteus* of the East coast of the United States enters the burrow of the razor clam *Ensis directus* from below and swallows the anterior end of the clam. *Paranemertes peregrina,* an armed, intertidal nemertean found on the Pacific coast of the United States, feeds on polychaete annelids. The nemertean leaves its burrow to feed and can follow the mucous trails of prey, but it must touch the polychaete for the feeding response to be initiated. Once contact occurs, the proboscis everts and wraps around and stabs the prey repeatedly with the stylet. The prey is quickly paralyzed by injected toxin and then swallowed as the retracting proboscis pulls the prey toward the mouth. After feeding, the nemertean relocates the burrow by following its own mucous trail. Other armed nemerteans feed on small crustaceans, such as amphipods. They kill the prey with a piercing strike of the stylet on the ventral exoskeleton and then force their head through the opening. The esophagus is everted and the contents of the prey are sucked out and digested (Fig. 7–7). Species of *Carcinonemertes* are economically important predators on eggs brooded by female crabs. When the short pro-

boscis is everted, the stylet extends to just beyond the proboscis pore, where it punctures the eggshell and allows the contents to be sucked out (Fig. 7–2G).

Digestion, which takes place in the intestine, is initially extracellular but is concluded intracellularly in phagocytic cells.

NERVOUS SYSTEM AND SENSORY ORGANS

The nervous system consists of an anterior brain of four ganglia that surround the rhynchodeum or ante-

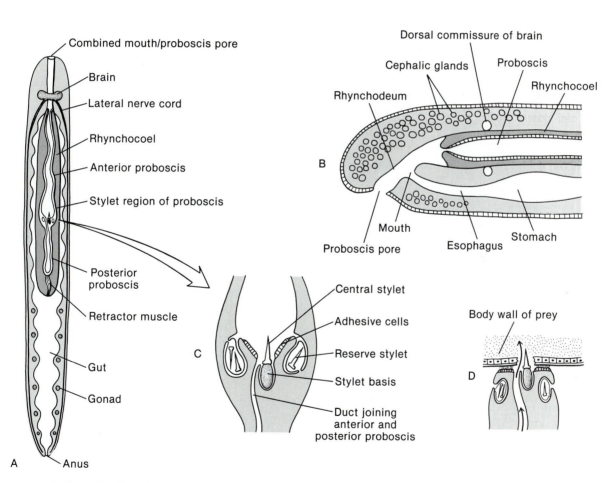

FIGURE 7–6 Body organization of armed nemerteans. **A,** Body organization in dorsal view. **B,** Sagittal section of head showing union of the foregut and rhynchodeum. **C–D,** Detail of the stylet apparatus before **(C)** and after **(D)** eversion. Much of the toxin that enters the prey originates in the anterior proboscis, but some also may enter the wound from the posterior proboscis (arrows). *(B, Modified and redrawn from Stricker, S. A. 1982. The morphology of Paranemertes sanjuanensis sp. n. (Nemertea, Monostilifera) from Washington, U.S.A. Zool. Scripta 11:107–115. C, Modified and redrawn from Stricker, S. A., and Cloney, R. A., 1981. The stylet apparatus of the nemertean Paranemertes peregrina: Its ultrastructure and role in prey capture. Zoomorphology. 97:205–223.)*

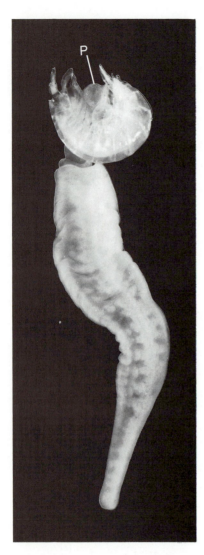

FIGURE 7–7 The armed nemertean *Nipponemertes pulcher* attacking an amphipod crustacean. The proboscis (P) is wrapped around the amphipod from behind and is striking the ventral surface. *(From McDermott, J. J. 1984. The feeding biology of Nipponemertes pulcher, with some ecological implications. Ophelia 23:1–21.)*

rior rhynchocoel and a pair of large, lateral, nonganglionated nerve cords (Figs. 7–3A, C; 7–6A). A number of other longitudinal nerves are frequently present, including a dorsal cord, and may represent vestiges of a primitive, radial arrangement. The nerve cords and especially the brain of nemerteans commonly contain hemoglobin and are accordingly pinkish or red. The pigment stores oxygen for use by the nervous system when the animals must be active under anoxic environmental conditions.

Sensory organs consist of sensory epidermal pits, pigment-cup eyes, ciliated cephalic slits and grooves, cerebral organs, and eversible frontal organs (Fig. 7–8A–D). The last three of these are probably chemoreceptors. The **cerebral organs** are a pair of ciliated canals associated with the brain (Fig. 7–8A, C). The external openings of the canals are in the cephalic slits or grooves, or in a pair of pits over the brain area. Bidirectional water currents, created by cilia in the canals, appear to be activated in the presence of food. The cerebral organs also appear to have a neuroendocrine role in osmoregulation.

INTERNAL TRANSPORT AND EXCRETION

Because of their large size and thick, muscular body wall, nemerteans, like other large animals, use fluid circulation rather than simple diffusion to transport substances throughout their bodies. Nemerteans have two circulatory systems, and in both, muscular contractions circulate the fluid. One circulatory system is the rhynchocoel, whose fluid bathes and probably supplies nutrients to the proboscis. Thus, the rhynchocoel has at least two biological roles: it is a hydrostat used in proboscis eversion and burrowing, and it is a circulatory system. The second nemertean circulatory system is called the **blood system** (Figs. 7–3A, C; 7–9). In its simplest form there are only two vessels, one on each side of the gut but joined together anteriorly and posteriorly. In many nemerteans, this basic plan has been considerably elaborated on with the development of additional longitudinal and transverse vessels.

Nemertean blood vessels are lined by a continuous endothelium, which occurs nowhere else among invertebrates except in cephalopod molluscs and a few specialized taxa in other phyla. In a few nemerteans, the endothelial cells are myoepithelial and may even bear a rudimentary cilium (Fig. 7–9D, E). The occurrence of an endothelium and its resemblance to the coelomic linings of coelomate invertebrates has prompted some biologists to suggest that nemertean blood vessels are actually specialized coelomic channels (p. 274).

The vessels are contractile, although blood flow depends on both the contraction of the blood vessels

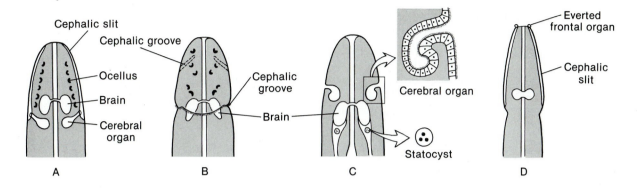

FIGURE 7–8 Many nemerteans have an impressive battery of cephalic sensory organs, but their functions, except for the ocelli, are not well understood. Many heteronemerteans have lateral cephalic slits **(A, D)**, whereas hoplonemerteans have transverse or oblique cephalic grooves **(B)**. Photoreceptive ocelli **(A, B)** are common in nemerteans and range in number from one to many pairs. Most nemerteans also have complex cerebral organs—most likely chemoreceptors—closely associated with the brain **(A–C** and inset). A few tiny nemerteans that occupy the interstices of marine sands have statocysts **(C** and inset), which are gravity receptors. Eversible frontal organs **(D)** occur sporadically in nemerteans. **A,** *Lineus socialis.* **B,** *Amphiporus ochraceous.* **C,** *Ototyphlonemertes pallida.* **D,** *Cerebratulus lineolatus.* *[C, Adapted from Mock, H. 1978. Ototyphlonemertes pallida (Keferstein, 1862). Mikrofauna Meeresbodens. 67:559–570.]*

FIGURE 7–9 The circulatory systems of **A,** *Cephalothrix* spp. **B,** *Amphiporus cruentatus,* and *Cerebratulus* spp. Blood flows anteriorly in the median dorsal vessel of *Amphiporus* and posteriorly in the two lateral vessels. The cross sections of the vessels in *Cephalothrix* show myoepithelial **(D)** and ciliated **(E)** lining cells. Such cells are found nowhere else in blood vessels but are of common occurrence in coelomic linings. These facts are among several that suggest that the nemertean circulatory system is, like the rhynchocoel, a modified coelom. The "blood" of *Amphiporus cruentatus* is red because of the presence of hemoglobin-containing erythrocytes. *(Cross sections modified and redrawn from Turbeville, J. M. 1991. Nemertinea. In Harrison, F. W. and Bogitsh, B. J. (Eds.): Microscopic Anatomy of Invertebrates. Vol. 3. Wiley-Liss, Inc., New York, pp. 285–328)*

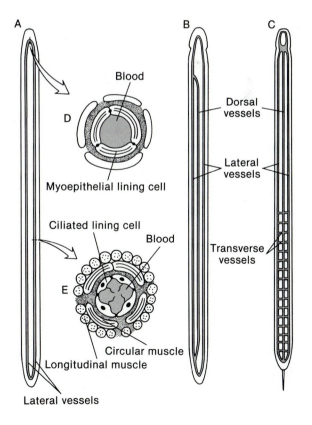

and the body wall musculature. Circulation is intermittent and usually does not follow a defined circuit because the blood may ebb and flow in each of the two lateral vessels. Nevertheless, in at least one species, *Amphiporus cruentatus,* which has lateral vessels and a dorsal longitudinal vessel, blood flows anteriorly in the dorsal vessel and posteriorly in the lateral vessels (Fig. 7–9B).

Nemertean blood is usually colorless, but in many species it contains corpuscles bearing yellow, red (because of hemoglobin in erythrocytes), orange, or green pigments of uncertain function. In addition to the pigmented corpuscles, the blood and rhynchocoel also contain amebocytes.

Communication between the blood and rhynchocoelic circulations occurs across specialized exchange sites called **vascular plugs.** A vascular plug is a blind vessel from the blood system that distends the rhynchocoel wall and bulges into the rhynchocoel cavity. In at least one species, the plug is overlain with podocytes across which an ultrafiltrate of the blood presumably enters the rhynchocoel, perhaps supplying nutrients to the proboscis.

The excretory system consists of two or more protonephridia, each bearing many terminal cells (Fig. 7–3A). A nephridiopore, or pores, is located on each side of the foregut, and a tubule extends anteriorly from the opening of each. The terminal cells of most nemerteans project into the wall of the lateral blood vessel (Fig. 7–10). In a few cases, the vessel lining is interrupted at the sites of contact so that the basal lamina around the terminal cells is bathed directly in blood. The protonephridia are probably important in osmoregulation because semiterrestrial and freshwater species have many more terminal cells per nephridium than do their marine counterparts. The physiology of excretion, however, is not known.

REGENERATION AND REPRODUCTION

Nemerteans, especially the larger species, display a marked tendency to fragment when irritated. Collecting large, intact specimens is therefore usually difficult. Frequently the proboscis becomes detached when everted. The proboscis soon regenerates, but the ability of the body fragments to regenerate varies greatly, depending on the species. Some species, including certain members of the genus *Lineus,* reproduce asexually by fragmentation, and even posterior

sections of the body are capable of total regeneration (Fig. 7–11), which takes place within a mucous cyst.

The majority of nemerteans are dioecious, and the reproductive system is simple. Gametes develop from stem cells that aggregate and become enclosed in an epithelium to form a gonad. Such gonads alternate with the intestinal diverticula to form a regular row on each side of the body (Figs. 7–3A; 7–6A).

After maturation of the gametes, a short duct grows from the gonad to the outside, allowing the gametes to escape. Each ovary produces 1 to 50 eggs, depending on the species. The shedding of eggs or sperm does not necessarily require contact between two worms, although some species aggregate at the time of spawning or a pair of worms may occupy a common burrow or secreted cocoon (Fig. 7–12). Fertilization is external in most nemerteans, and the eggs are either shed and dispersed into the seawater or deposited within the burrow, or tube, or in gelatinous strings (Fig. 7–12).

DEVELOPMENT

Cleavage in nemerteans is spiral. All orders except heteronemerteans have direct development or a short-lived larval stage. Many heteronemerteans pass through a free-swimming and planktotrophic larval stage called a **pilidium,** which possesses an apical

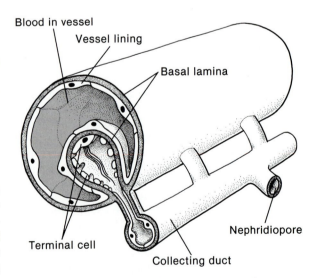

FIGURE 7–10 The nemertean excretory system consists of protonephridia in close association with the blood vessels.

Blood in vessel

Vessel lining

Basal lamina

Nephridiopore

Terminal cell

Collecting duct

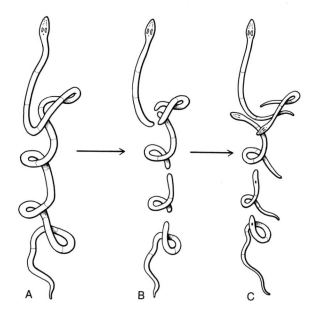

FIGURE 7–11 During asexual reproduction in *Lineus*, as well as some other nemerteans, the adult body can break along preformed lines, which appear as annulations on the body surface **(A)**. After fragmentation **(B)**, each fragment can regenerate missing parts and become a complete worm **(C)**. *(Modified and redrawn from Coe, W. R. 1943. Biology of the nemerteans of the Atlantic coast of North America. Trans. Conn. Acad. Arts Sci. 35:129–328.)*

tuft of cilia and is somewhat helmet-shaped (Fig. 7–13). After a free-swimming existence, the pilidium undergoes a complex metamorphosis into a young worm (Fig. 7–13E, F). *Paranemertes peregrina* has a life span of about one and a half years. Spawning occurs in spring and summer, and adults die in the winter. Juveniles resulting from the spring and summer spawn attain sexual maturity the following spring and summer.

PHYLOGENY

Traditionally, nemerteans are believed to have evolved from a common acoelomate ancestor shared with flatworms, and there are many zoologists who hold to such a relationship. The evidence for this interpretation is the common occurrence in flatworms and nemerteans of a ciliated epidermis lacking a cuticle, parenchyma, protonephridia, rhabdites, and an eversible proboscis (present in a few flatworms).

Thus, zoologists generally accept that nemerteans are the first animals to have a coelom (albeit an unusual one in the form of a rhynchocoel), a blood-vascular system, and a close association between the blood and the excretory systems.

Results from recent work in the areas of development, as well as ultrastructural and molecular systematics, consistently point to another evolutionary view—that nemerteans are not advanced acoelomate animals but rather are derived coelomates. This interpretation is supported by an array of new facts. For example, nemertean rhabdites are now known to be unlike and unrelated to those of flatworms; nemerteans probably lack parenchyma; the flatworm proboscis, where it occurs, is unrelated to that of nemerteans, and both a ciliated epidermis and pro-

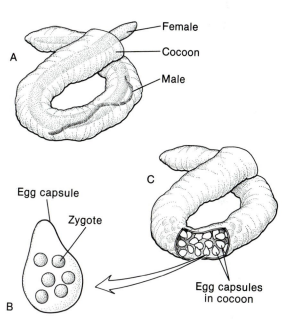

FIGURE 7–12 Preceding fertilization in *Lineus viridis*, the female constructs a tubular cocoon (which resembles the permanent tube of the paleonemertean, *Tubulanus*) around herself and the smaller male **(A)**. After an hour or two, the male releases sperm and these are believed to enter the female and fertilize her eggs internally. She then releases fertilized eggs in capsules **(B)** that are attached to the tube lining and surrounded by mucus **(C)**. *(Modified and redrawn from Bartolomaeus, T. 1984. Zur Fortpflanzungsbiologie von Lineus viridis (Nemertini). Helgol. wiss. Meeresunters. 38:185–188.)*

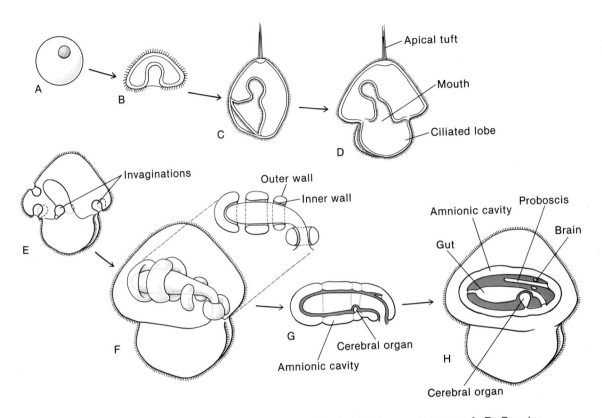

FIGURE 7–13 Larval development and metamorphosis in heteronemerteans. **A–D,** Development of the pilidium larva **(D).** Metamorphosis is initiated when seven invaginations of larval ectoderm **(E)** pinch off into the blastocoel to form seven hollow vesicles (imaginal discs). These vesicles surround the larval endoderm **(F)** and eventually fuse together **(G)** to form a continuous sheath over the larval gut. The inner wall of the sheath becomes the epidermis of the developing worm, and the outer wall and the space between it and the epidermis form an amniotic cavity **(G, H).** Eventually, the juvenile worm ruptures from the amnion and larval body to adopt an independent existence. *(A–D, Modified and redrawn from Iwata, F. 1985. Foregut formation of the nemerteans and its role in nemertean systematics. Am. Zool. 25:23–36.)*

tonephridia are found in coelomate and acoelomate animals, not in flatworms and nemerteans exclusively. Perhaps the most provocative new interpretation is that the nemertean blood system is not a blood-vascular system at all but rather is a modified coelom like the rhynchocoel. This idea, as studied by J. M. Turbeville, is based on several features that nemertean blood vessels share with coeloms, including the lateral position of the vessels, a continuous epithelial lining (absent in invertebrate blood vessels), the occurrence of cilia on, and muscle filaments in some of the vessel linings, and the observation that the blood vessels develop by schizocoely from solid mesodermal bands. His comparison of ribosomal RNA gene sequences of flatworms, nemerteans, and various coelomate invertebrates points to the same conclusion—that nemerteans are closely related to the coelomates and only remotely to flatworms.

SYSTEMATIC RÉSUMÉ OF THE PHYLUM NEMERTEA (RHYNCHOCOELA)

Class Anopla. Unarmed nemerteans. Mouth located below or posterior to the brain.

Order Paleonemertea. Primitive neritic nemerteans. Body wall musculature two, three, or more layers. Nerve cords subepidermal outside of muscle layers or within longitudinal muscle layer. *Carinina, Carinoma, Cephalothrix, Tubulanus.*

Order Heteronemertea. Body wall musculature is three-layered: circular layer between two longitudinal layers (but a reduced, fourth, outermost circular layer is probably present). Nerve cords within outer longitudinal muscle layer. *Cerebratulus, Micrura, Lineus, Zygeupolia.*

Class Enopla. Armed nemerteans. Mouth located anterior to brain. Nerve cords located inside of body wall musculature, composed of outer circular and inner longitudinal layers.

Order Hoplonemertea. All of the armed, stylet-bearing nemerteans belong to this order, which contains several interesting ecological groups, such as the floating, or pelagic, nemerteans (e.g., *Nectonemertes*), which live in the open ocean, many at depths below 1000 m; the freshwater nemerteans (e.g., *Prostoma*); the tropical and subtropical terrestrial nemerteans (e.g., *Geonemertes*); the crab parasites (*Carcinonemertes*), and many marine shallow-water forms, such as *Amphiporus, Nemertopsis, Paranemertes, Zygonemertes.*

Order Bdellonemertea. This small order contains but one genus, *Malacobdella*, with four species: three are commensal in the mantle cavity of marine clams and one is commensal in the mantle cavity of a freshwater snail. Although the proboscis is unarmed, it has probably been derived from the armed type. The proboscis and esophagus open into a common chamber. Bdellonemerteans feed on bacteria, algae, and protozoa removed from the ventilating current of the host.

SUMMARY

1. Nemerteans are small to large, often dorsoventrally flattened worms. Typically, the body is slender, elongated, and muscular. The epidermis is ciliated and glandular and lacks a cuticle.

2. Nemerteans live in rocky crevices, beneath stones, within algal mats, which may include sessile animals, or in soft bottoms. They move by cilia or muscular peristalsis, or by a combination of the two.

3. Nemerteans prey on other invertebrates, which they capture with a unique proboscis apparatus. The proboscis is explosively everted by a build up of hydrostatic pressure in the rhynchocoel, and the prey is trapped by the coiling proboscis or stabbed by a proboscis stylet. The proboscis apparatus is not generally connected to the gut although in some groups it unites with it secondarily. Digestion is initially extracellular but is concluded intracellularly.

4. Gas exchange occurs across the surface of the body wall, and both the rhynchocoel and blood system are important in internal transport. Hemoglobin occurs in the blood of a few nemerteans and in the nervous system of many species. Protonephridia are intimately associated with the lateral blood vessels.

5. The rhynchcoel is a specialized coelom, unique to nemerteans. The blood system may also be a specialized coelom unrelated to the blood-vascular system of other invertebrates.

6. The nervous system consists of a ganglionated brain and two primary, lateral longitudinal cords. Additional longitudinal cords are often present. Complex sensory organs include ocelli and cerebral organs.

7. The sexes are usually separate with simple, serially arranged gonads, each with its own gonoduct and pore. Fertilization is external, although individuals may be in contact or close together at the time of spawning. Eggs are shed into the seawater or encased in gelatinous masses. Development is typically direct but may be indirect with a pilidium larva.

8. Traditionally, nemerteans have been compared to platythelminths. Recent data, however, suggest that nemerteans may be more closely related to coelomate animals, such as annelids, than to the acoelomate flatworms.

REFERENCES

Hydrostatic Skeletons

Clark, R. B. and Cowey, J. B. 1958. Factors controlling the change of shape of certain nemertean and turbellarian worms. J. Exp. Biol. *35*:731–748.

Vogel, S. 1988. Life's Devices. Princeton University Press, Princeton. 367 pp.

Wainwright, S. A. 1988. Axis and Circumference. Harvard University Press, Cambridge, MA. 132 pp.

Nemerteans

The literature listed here deals almost exclusively with nemerteans. The introductory references on page 6 include many general works and field guides that contain sections on nemerteans.

Amerongen, H. M. and Chia, F.-S. 1982. Behavioral evidence for a chemoreceptive function of the cerebral organs in *Paranemertes peregrina* Coe (Hoplonemertea: Monostilifera). J. Exp. Mar. Biol. Ecol. *64*:11–16.

Amerongen, H. M. and Chia, F.-S. 1987. Fine structure of the cerebral organs in hoplonemertines (Nemertini) with a discussion of their function. Zoomorphology. *107*:145–159.

Cantell, C. E. 1969. Morphology, development, and biology of the pilidium larvae from the Swedish west coast. Zool. Bidrag. Fran Uppsala. *38*:61–111.

Gibson, R. 1972. Nemerteans. Hutchinson University Library, London. 224 pp. (An excellent general account of the nemerteans.)

Gibson, R. 1974. Histochemical observations on the localization of some enzymes associated with digestion in four species of Brazilian nemerteans. Biol. Bull. *147*:352–368.

Gibson, R., and Moore, J. 1976. Freshwater nemerteans. Zool. Linn. Soc. *58*:117–218.

Hyman, L. H. 1951. The Invertebrates: Platyhelminthes and Rhynchocoela. Vol. 2. McGraw-Hill, New York pp. 459–531. (An old but still valuable general treatment of the nemerteans.)

Jespersen, A. and Lutzen, J. 1987. Ultrastructure of the nephridio-circulatory connections in *Tubulanus annulatus* (Nemertini, Anopla). Zoomorphology. *107*;181-189.

McDermott, J. J. 1976a. Predation of the razor clam *Ensis directus* by the nemertean worm *Cerebratulus lacteus*. Chesapeake Sci. *17*:299–301.

McDermott, J. J. 1976b. Observations on the food and feeding behavior of estuarine nemertean worms belonging to the order Hoplonemertea. Biol. Bull. *150*:57–68.

McDermott, J. J. and Roe, P. 1985. Food, feeding behavior and feeding ecology of nemerteans. Am. Zool. *25*:113–126.

Moore, J. and Gibson, R. 1985. The evolution and comparative physiology of terrestrial and freshwater nemerteans. Biol. Rev. *60*:267–312.

Riser, N. W. 1974. Nemertinea. *In* Giese, A. C., and Pearse, J. S. (Eds.): Reproduction of Marine Invertebrates. Vol. 1. Academic Press, New York pp. 359–389.

Roe, P. 1970. The nutrition of *Paranemertes peregrina*. I. Studies on food and feeding behavior. Biol. Bull. *139*:80–91.

Roe, P. 1976. Life history and predator–prey interactions of the nemertean *Paranemertes peregrina*. Biol. Bull. *150*:80–106.

Stricker, S. A. 1985. The stylet apparatus of monostiliferous hoplonemerteans. Am. Zool. *25*:87–97.

Stricker, S. A. 1986. An ultrastructural study of oogenesis, fertilization, and egg laying in a nemertean ectosymbiont of crabs, *Carcinonemertes epialti* (Nemertea, Hoplonemertea). Can. J. Zool. *64*:1256–1269.

Stricker, S. A. and Cavey, M. J. 1986. An ultrastructural study of spermatogenesis and the morphology of the testis in the nemertean worm *Tetrastemma phyllospadicola* (Nemertea, Hoplonemertea). Can. J. Zool. *64*:2187–2202.

Stricker, S. A. and Cloney, R. A. 1982. Stylet formation in nemerteans. Biol. Bull. *162*:387–403.

Stricker, S. A. and Reed, C. G. 1981. Larval morphology of the nemertean *Carcinonemertes epialti* (Nemertea, Hoplonemertea). J. Morphol. *169*:61–70.

Turbeville, J. M. 1986. An ultrastructural analysis of coelomogenesis in the hoplonemertine *Prosorhochmus americanus* and the polychaete *Magelona* sp. J. Morphol. *187*:51–60.

Turbeville, J. M. 1991. Nemertinea. *In* Harrison, F. W. and Bogitsh, B. J. (Eds.): Microscopic Anatomy of Invertebrates. Wiley-Liss, Inc., New York. pp. 285–328. (The most recent review of nemertean anatomy and ultrastructure.)

Turbeville, J. M., Field, K. G. and Raff, R. A. 1992. Phylogenetic position of phylum Nemertini inferred from 18s rRNA sequences: Molecular data as a test of morphological character homology. Mol. Biol. Evol. *9*:235–249.

Turbeville, J. M. and Ruppert, E. E. 1983. Epidermal muscles and peristaltic burrowing in *Carinoma tremaphoros*: Correlates of effective burrowing without segmentation. Zoomorphology *103*:103–120.

Turbeville, J. M. and Ruppert, E. E. 1985. Comparative ultrastructure and the evolution of nemertines. Amer. Zool. *25*:53–72.

ASCHELMINTHS

PRINCIPLES AND EMERGING PATTERNS
Monociliated Cells in Metazoans
Syncytial Epithelia
Cryptobiosis

THE ASCHELMINTHS

PRINCIPLES AND EMERGING PATTERNS

MONOCILIATED CELLS IN METAZOANS

There is increasing evidence that the primitive ciliated epidermal cell of animals bore only one cilium, which was surrounded by a ring of microvilli, as in a sponge choanocyte (Fig. 8–1). Monociliated cells are characteristic of most lower animals (sponges, placozoans, cnidarians), a few protostomes (gnathostomulids, some gastrotrichs, some annelid larvae, and phoronids), and many deuterostomes (brachiopods, pterobranch hemichordates, echinoderms, cephalochordates) and undoubtedly gave rise to multiciliated cells by replication of the cilium, or flagellum. This is important evidence for the evolution of metazoans from colonial flagellates, which typically have only one (or two) flagellum, rather than from ciliates, which have many.

Additional clues to the evolutionary origin of metazoans are found in the anchoring structures of each cilium. In metazoan monociliated cells such structures are organized into diplosomes. A **diplosome** consists of two centrioles, one of which is attached to the cilium as the basal body and the other of which, the accessory centriole, is free. One or more proteinaceous rootlet fibers anchor the diplosome within the cell. Curiously, the accessory centriole is always at a 90-degree angle to the basal body and never produces a cilium. Among the flagellated protists, however, both centrioles often become basal bodies and produce cilia. In such cases, the result is a biflagellated cell, such as *Euglena* or *Chlamydomonas*, but in these cells, the angle between the two centrioles is not 90 degrees. Only one group of nonphotosynthetic protists has its two centrioles set at a right angle—the choanoflagellates—providing another clue that an organism resembling a choanoflagellate (p. 69) was the most likely ancestor to the metazoans.

The derivation of the multiciliated cells of some gastrotrichs from the monociliated condition of primitive species is depicted in the right column in Figure 8–1 (from bottom to top). The possible evolution of the ciliary rootlet system of the multiciliated flatworms, nemerteans, and annelids is depicted at the top.

SYNCYTIAL EPITHELIA

The epithelia of many aschelminth animals are either partly or wholly syncytial in organization (Fig. 8–2A). In some groups the gut and muscular and gonadal tissues are also organized as syncytia. During development, a syncytium arises from a typical cellular epithelium (p. 70; Fig. 8–2B) as the lateral membranes between adjacent cells break down and the cytoplasm becomes continuous. As a result of the lateral fusion of cells, the syncytial epithelium may be viewed as single multinucleated cell.

Because invertebrates other than protozoans are multicellular animals, we are conditioned to believe in the central importance of multicellular tissues because cellular compartmentation allows for specialization and division of labor among cells. We may ask therefore, what is the functional significance of a reversion to a syncytium? Although there is much that remains to be learned about the functions of syncytia, two ideas are suggested here.

Syncytia may be less freely permeable to substances than cellular epithelia because materials cannot cross the epithelium by slipping between cells. There is some evidence, for example, that epidermal syncytia have low water permeability and could therefore protect animals living in osmotically stressful environments, such as in fresh water or, among parasites, in the host's gut. Osmotic stress may be acute in small animals, like many aschelminths, because small body size means an unfavorably large surface in relation to volume.

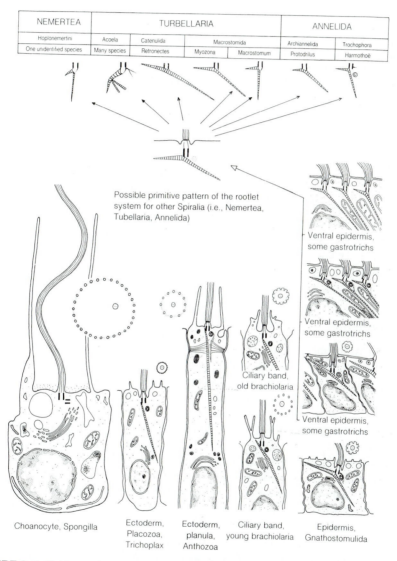

FIGURE 8–1 Evidence for the evolution of metazoans from protozoan flagellates receives support from electron microscopy of ciliated cells. Primitive animals, such as sponges and cnidarians, have exclusively monociliated cells, and the 90-degree arrangement of the basal body and accessory centriole (seen clearly in the choanocyte figure) is identical to that of choanoflagellates. The evolution of multiciliated cells from a cell with only one cilium is illustrated in the right column (read from bottom to top) and is based on a comparative study of species of gastrotrichs. *(From Rieger, R. M. 1976. Monociliated epidermal cells in Gastrotricha: Significance for concepts of early metazoan evolution. Z. Zool. Syst. Evolut.-forsch. 14:198–226.)*

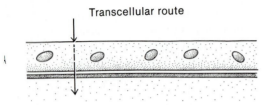

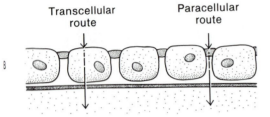

FIGURE 8–2 **A,** In a syncytial epithelium all substances that cross the epithelium must pass through the cell where direct regulation can occur. **B,** In a cellular epithelium on the other hand, there is a possibility of unregulated diffusion between cells.

Another explanation for the occurrence of syncytia may be uniformity of function. Cellular epithelia are commonly composed of cells such as secretory, ciliated, and transportive cells, that have broadly different functions. Syncytia, however, may be specialized for one predominant function, such as absorption of nutrients or yolk synthesis. Under such circumstances, the avoidance of cellular organization, which may be energetically costly and unnecessary, could be an advantage. Thus, specialization for absorption may explain the syncytial epidermis of endoparasitic tapeworms and spiny-headed worms.

CRYPTOBIOSIS

Metazoans that inhabit mosses, lichens, temporary ponds, and damp soils periodically face the threat of dehydration and other physical extremes. Often, such animals have evolved reproductive adaptations, such as resistant eggs (p. 208), which allow their progeny to survive the extreme conditions. This same ability to resist and survive adverse conditions occurs in the adults of some remarkable species of roundworms, rotifers, and tardigrades. As their surrounding environment deteriorates, the animals enter a deathlike state of suspended animation, called **cryptobiosis** (meaning "hidden life") and are resurrected when conditions again become favorable. Cryptobiotic animals can remain dormant for a period of months to years—some for even as long as a century—before reviving. While dormant, they survive desiccation and temperature extremes that seem incompatible with life. There are records of cryptobiotic animals surviving temperatures ranging from 150°C (for a few minutes) to −200°C (for days). In one experiment, a rotifer and a tardigrade recovered after a brief exposure to near absolute zero (0.008° K). Clearly the ability to become cryptobiotic is advantageous to animals that live in transiently uninhabitable environments.

To successfully enter cryptobiosis, the animal must dry slowly, usually over several days, as would be typical in natural habitats. As the animals dry, they change their body shape to minimize surface area and, perhaps, to retard the rate of desiccation. Many nematodes tightly coil their bodies, rotifers retract head and foot and adopt a spherical shape, and tardigrades contract into the shape of a barrel (Fig. 8–3).

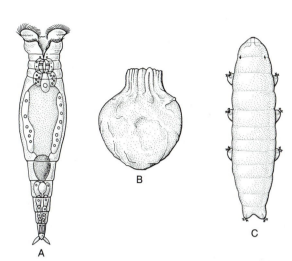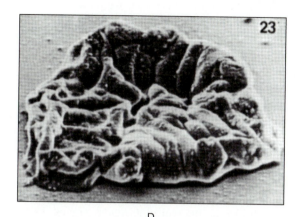

FIGURE 8–3 Cryptobiosis in adult aschelminths: **A,** Active hydrated and **B** cryptobiotic form of the rotifers, *Philodina roseola* **(A)** and *Habrotrocha rosa* **(B). C,** Normal hydrated adult and **D** desiccated cryptobiotic "tun" of the tardigrade, *Macrobiotus hufelandi. (A, Redrawn after Hickernell from Pennak, R. W. 1953. Fresh-Water Invertebrates of the United States. Ronald Press, New York. B, Redrawn from Wallace, R. L., and Snell, T. W. Rotifera. In Thorp, J. H., and Covich, A. P. 1991. Ecology and Classification of North American Freshwater Invertebrates. Academic Press, New York. pp. 187–248. C, Redrawn after Higgins, R. P. from Nelson, D. R. 1982. Developmental biology of the Tardigrada. In Harrison, F. W., and Cowden, R. R. (Eds.): Developmental Biology of Freshwater Invertebrates. Alan R. Liss, New York, p. 368. D, Drawn from a micrograph from Greven, H. 1971. On the morphology of Tardigrades: a stereoscan study of Macrobiotus hufelandi and Echiniscus testudo. Forma et Functio. 4:283–302.)*

Cryptobiosis is truly a deathlike state. Metabolism is almost immeasurably low, 0.01% of normal, or is absent entirely, and the water content of the body decreases to less than 1%. In the absence of metabolism and near total dehydration, cryptobiotic animals are essentially dead—but revivable. If metabolic processes actually cease, then somehow the structural order essential to life must be preserved in a sort of stable crystallized condition until water returns and life resumes. Two molecules, glycerol and the disaccharide trehalose, appear to be important in the formation and maintenance of the crystalline state. Both are synthesized as the animals slowly enter cryptobiosis. Glycerol is believed to protect the tissues against oxidation and to replace water bound to biological macromolecules. Trehalose replaces water in membranes. In substituting for water, glycerol and trehalose act as spacers to preserve macromolecular organization during dehydration.

ASCHELMINTHS

The aschelminths are a heterogeneous assemblage of marine and freshwater animals, such as gastrotrichs, rotifers, and roundworms (nematodes). Although previously recognized as classes within the phylum Aschelminthes, each of the seven or eight former classes is now designated as a distinct phylum. The informal name aschelminths, however, is still a convenient term of reference for the entire assemblage. Historically, the Priapulida (Chapter 9) have been included within the Aschelminthes. The tardigrades, or water bears, traditionally have not been grouped with the aschelminth phyla but rather with the arthropods. Their evolutionary relationship to arthropods is unclear, however, and aspects of their biology are similar to those of some aschelminths. For these reasons, the tardigrades are described in this chapter.

The majority of free-living aschelminths are tiny worm-shaped animals, ranging from microscopic size to a centimeter in length, and their small body size is correlated with several familiar design features. For example, a well developed coelom and blood-vascular system are absent. Cilia are used for locomotion in several groups, but in those that have a thick cuticle, muscular forms of locomotion predominate. Protonephridia are the typical excretory organs. Minute body size also means that the body itself is composed of relatively few cells—some aschelminths have as few as 1000 somatic cells. The rotifer *Epiphanes senta,* for example, has 958 nuclei, and the male of the roundworm *Caenorhabditis elegans* has 1031. A correlation between small body size and few cells is to be expected, but it is remarkable that many species of aschelminths, including the previous two examples, have an invariant and genetically fixed number of cells—a phenomenon called **eutely.** In eutelic animals, mitosis commonly ceases following embryonic development, and growth continues only through increase in the size of cells.

Although the anterior of the body bears the mouth and sense organs, there is no well-formed head. An external cuticle is present in some phyla but absent in others. In general, those that have a thin flexible cuticle or that lack a cuticle altogether have antagonistic circular and longitudinal muscles acting on a hydrostatic skeleton. Circular musculature is absent in aschelminths that have a thick cuticle, and the action of longitudinal musculature is antagonized by the elastic cuticle. Adhesive glands in various numbers are characteristic of many species and often open to the outside of the body via projecting cuticular tubes. The digestive tract is usually a complete tube bearing a mouth and anus, and a specialized pharyngeal region is almost always present. Most aschelminths are dioecious, and cleavage is strongly determinate.

Historically, the aschelminthes were described as pseudocoelomate animals, meaning that they had a fluid-filled cavity, unlined by a coelomic epithelium, that arose as a persistent embryonic blastocoel. Ultrastructural and embryological research over the past 30 years, however, has shown that many aschelminths have no body cavity at all, that is, they are acoelomate animals, as would be expected from their small body size. Thus, gastrotrichs, most nematodes, and kinorhynchs, for example, are acoelomate protostomes. Nevertheless, a few aschelminths, such as rotifers and large roundworms, do indeed have an expanded, fluid-filled, connective-tissue cavity. Traditionally, this cavity is called a pseudocoel, but it should be recalled (pp. 185–187) that fluid connective tissue is blood, or hemolymph, and the cavity is a blood sinus similar to a hemocoel as found, for example, in arthropods and molluscs. In those few aschelminths, such as acanthocephalans and tardigrades, in which the body cavity is voluminous, it does not develop as a persistent blastocoel (a blastocoel is absent in the development of these taxa) but rather forms anew as an enlarged space in the connective tissue, exactly as the hemocoel devel-

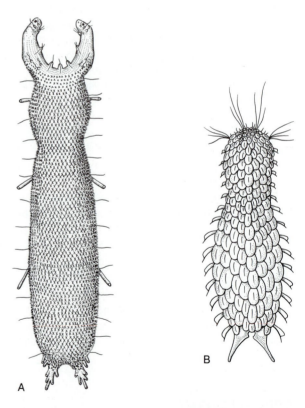

FIGURE 8–4 A, A marine macrodasyid gastrotrich, *Pseudostomella* and **B** a marine chaetonotid, *Halichaetonotus. (A, From Ruppert, E. E. 1970. On Pseudostomella Swedmark 1956 with descriptions of P. plumosa nov. spec., P. cataphracta nov. spec. and a form of P. roscovita Swedmark 1956 from the West Atlantic coast. Cah. Biol. Mar. 11:121–143.)*

ops in arthropods and molluscs. The pseudocoelic fluid functions as a hydrostatic skeleton and as a medium for internal transport, but a heart is absent and body movements circulate the fluid irregularly.

Phylum Gastrotricha

The gastrotrichs are a small phylum of some 430 marine and freshwater aschelminths that inhabit the interstitial spaces of bottom sediments and superficial detritus, the surfaces of submerged plants and animals, and the water films of soil particles. Many gastrotrichs are common animals of ponds, streams, and lakes. The intertidal zone of marine beaches may harbor 40 or more species, and as many as 1000 individ-

uals have been reported in 20 ml of sand. The phylum is divided into two groups designated as orders: the marine Macrodasyida and the marine and freshwater Chaetonotida (Fig. 8–4).

External Structure

Most gastrotrichs are microscopic in size. They range in length from 50 to 1000 μm, although members of some species may approach 4 mm. The tenpin- or strap-shaped body is flattened ventrally and arched dorsally. There is an anterior head bearing the sensory organs, brain, and pharynx, as well as an elongated trunk housing the midgut and reproductive organs. Posteriorly, the trunk usually bears two or more **adhesive organs.** (Figs. 8–4, 8–5).

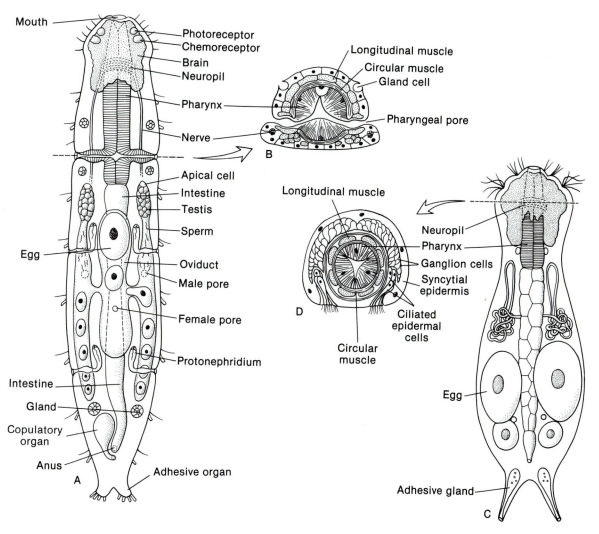

FIGURE 8–5 A,B, Generalized macrodasyid gastrotrich. **C,D,** Generalized chaetonotid gastrotrich.

The locomotory cilia are always restricted to the flattened ventral surface of the trunk and head. The entire ventral surface may be ciliated, or the cilia may be arranged in longitudinal bands, transverse rows, or patches or grouped into cirri, like those of hypotrichous ciliates (Fig. 8–6). The primitive groups of gastrotrichs have monociliated epidermal cells, a feature shared only with the gnathostomulids among the acoelomate protostomes (p. 279).

Adhesive tubes provide temporary adhesion to the substratum (Figs. 8–4, 8–5, 8–7) and contain duogland systems, that is, a viscid gland coupled with a releasing gland (p. 207) as in the turbellarians. The adhesive tubes may be numerous and located beneath the head, along the sides of the body, or on a terminal appendage (macrodasyids), or there may be only one pair restricted to the tail fork at the posterior end of the body (chaetonotids) (Fig. 8–5).

Internal Structure

The body wall is composed of an external **cuticle,** an epidermis, and underlying bands of circular and longitudinal muscle fibers. There is no body cavity, and the worms are acoelomate. The cuticle consists microscopically of a fibrous basal layer and an outer epicuticle composed of bilayers that resemble cell membranes. The flexible bilayers surround the entire body, including each of the cilia, and probably function as a physiological barrier (Fig. 8–8B). In some gastrotrichs the basal layer is locally thickened and specialized to form scales, spines, and hooks (Figs. 8–7, 8–8).

The **ventrally ciliated epidermis** can be cellular, as in macrodasyids, or partly syncytial, as in most chaetonotids. In some chaetonotids, the ciliated cells

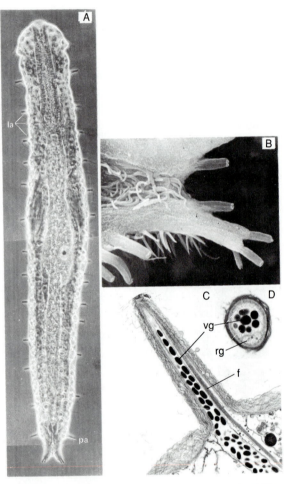

FIGURE 8–7 Adhesive tubes of *Turbanella.* **A,** Photograph of living animal, showing lateral and posterior adhesive tubes. Each lateral adhesive tube is accompanied by a sensory cilium. **B,** Enlarged view of posterior adhesive tubes. **C** and **D,** Longitudinal and cross sections of an adhesive tube. (pa = posterior adhesive tube; la = lateral adhesive tube; f = fiber; vg = viscid gland; rg = releasing gland) *(From Tyler, S., and Rieger, G. E. 1980. Adhesive organs of the Gastrotricha. I. Duo-gland organs. Zoomorphologie. 95:1–15.)*

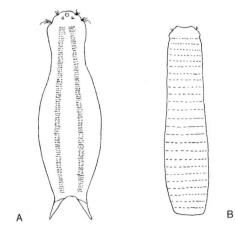

FIGURE 8–6 Ventral ciliary patterns in gastrotrichs. **A,** *Chaetonotus.* **B,** *Thaumastoderma. (After Remane, A. 1936. Gastrotricha. In Klassen Ordn. Tierreichs. 4:1–242.)*

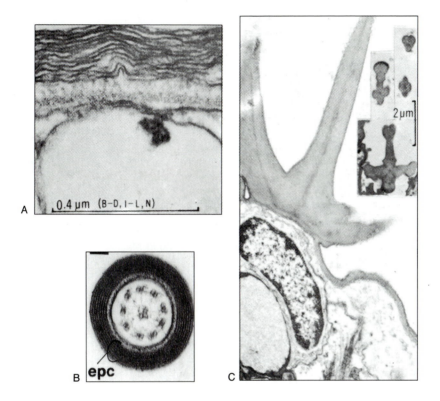

FIGURE 8–8 A, Cuticle and part of epidermis of *Turbanella ocellata,* showing a fuzzy endocuticle overlying a large epidermal vesicle, and the outer epicuticle composed of numerous bilayers. **B,** Cross section of a locomotory cilium of *Urodasys nodostylis,* showing exocuticle (epc) around cilium. **C,** Cuticular spines of *Tetranchyroderma* spp. Inset shows cross sections of spines at various levels. *(A,C, from Rieger, G. E., and Rieger, R. M. 1977. Comparative fine structure of the gastrotrich cuticle and aspects of cuticle evolution within the Aschelminthes. Z. zool. Syst. Evolut.-forsch. 15:81–124. B, From Ruppert, E. E. 1991. Gastrotricha. In Harrison, F. W., and Ruppert, E. E. (Eds.): Microscopic Anatomy of Invertebrates. Vol. 4. Wiley-Liss, New York. pp. 41–109.)*

are arranged in two longitudinal rows that bulge inward to form thickened ridges, or cords (Fig. 8–5).

The musculature, which may be cross-striated or obliquely-striated depending on the species, is usually arranged as hoops of outer circular muscle and bands of inner longitudinal muscle. The two ventrolateral longitudinal bands are strongly developed and extend the length of the body, enabling it to shorten and curl ventrally. The circular musculature allows for reextension of the body in most species of Macrodasyida, which have a soft flexible cuticle, but in many Chaetonotida a circular body wall musculature is absent, and the thick cuticle antagonizes the longitudi-

nal muscles. The innervation of gastrotrich muscle fibers, like that of roundworms and some other invertebrates (p. 290), is accomplished by the extension of noncontractile processes of muscle cells to the nerve cords rather than extension of nerves to the muscles.

Forward movement in gastrotrichs is the result of smooth ciliary gliding. Muscular action is important, however, in specialized movements, such as escape responses, searching activity, and copulation (Fig. 8–9). An escape response can be a rapid rearward withdrawal of the head and trunk to an attachment made by the posterior adhesive organ or a series of inchworm-like retreating movements. During copula-

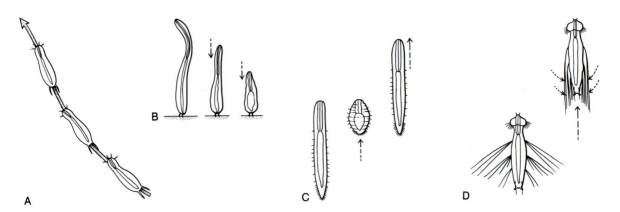

FIGURE 8–9 **A,** Ciliary gliding over a surface in *Chaetonotus.* **B,** Retraction toward anchored posterior adhesive organs in *Neodasys.* **C,** Forward creeping in *Macrodasys.* **D,** Skipping loco-motion using muscle-actuated spines in the pelagic *Stylochaeta. (A, C, D Redrawn from Remane, A. 1936. Gastrotricha. In Bronn, H. G. (Ed.): Klassen Ordn. Tierreichs 4:1–242.)*

tion, two animals use muscles to twist their posterior ends together like interlocking the tips of your index fingers.

The terminal mouth opens directly into the pharynx or, in some chaetonotids, into a small, cuticle-lined buccal cavity. The **myoepithelial pharynx** is an elongated, glandular, muscular tube (Fig. 8–5A, C). The pharyngeal wall is composed largely of myoep-ithelial cells, that is, cells that not only have a lining function but also possess radial myofibrils. The lumen is triangular and lined with cuticle (Fig. 8–5B, D). In macrodasyids, the pharynx first opens to the exterior through a pair of pores before joining the midgut where a specialized nonreturn valve is present (Figs. 8–5A, B). The midgut is a cellular cylindrical tube that tapers to join a posterior and ventral anus. The walls of the midgut generally lack cilia, but a brush border of microvilli is borne on the simple lining ep-ithelium (Fig. 8–5).

Gastrotrichs feed on small, dead or living organic particles, such as bacteria, diatoms, and small proto-zoa, all of which are sucked into the mouth by the pumping action of the muscular pharynx. In the ma-rine macrodasyids the paired **pharyngeal pores** per-mit the release of excess water ingested with food. Digestion takes place in the midgut and is probably a combination of extracellular and intracellular processes.

A single pair of protonephridia occurs in the an-terior trunk of marine and freshwater chaetonotids, but several pairs are arranged serially along the length of the body in the marine macrodasyids (Fig. 8–5). They are probably primarily osmoregulatory in func-tion (see p. 192). The nephridiopore or pores open on the ventrolateral surface of the body.

The brain is composed of two ganglionic masses (Fig. 8–5), one on each side of the anterior of the pharynx; the two masses are connected dorsally and ventrally by a commissure. Each lateral ganglion of the brain gives rise to an intraepidermal nerve cord that extends laterally along the length of the body. The nerve cords are adjacent to the thickened epider-mal cords, when present, and the principal pair of lon-gitudinal muscles. The sense organs include cerebral structures as well as sensory bristles located over the general body surface. The cerebral sensory organs consist of ciliary bristles and tufts (mechanorecep-tors), ciliated pits, fleshy appendages (chemorecep-tors), and simple ciliary ocelli (photoreceptors) (Fig. 8–5A).

Reproduction

In contrast to other aschelminths, gastrotrichs are her-maphrodites. In the marine *Macrodasys,* which prob-ably approximates the primitive plan, there is a pair of hermaphroditic gonads, each with an anterior testis and a posterior ovary (Fig. 8–5A). Sperm ducts carry sperm, which may be formed into spermatophores (sperm packets), to a pair of ventral male gonopores located about two-thirds of the body length from the anterior end; a copulatory organ is located at the posterior end of the body. Immediately prior to copulation, the copulatory organ is brought into contact with the male gonopores and loaded with sperm. The copulatory organ then functions as a penis to transfer the sperm to the copulating partner (Fig. 8–10). The sperm are then stored within a seminal receptacle. The internally fertilized eggs are released from the body by rupture. All species have complex accessory reproductive organs.

The male system of the freshwater chaetonotids has become so degenerate that all individuals are functionally female and reproduce parthenogenetically (Fig. 8–5C). Recently, however, sperm have been discovered in the common and widespread *Lepi-*

dodermella squamata and in many other species, but it is uncertain whether the sperm are functional.

In the freshwater parthenogenetic chaetonotids, two types of eggs are produced and attached to the substratum. One type is the dormant, or resting, egg, which, like those of freshwater flatworms, can withstand desiccation and low temperatures; the other type hatches in one to four days. The eggs are enormous, occupying most of the trunk and are produced and laid one at a time. Cleavage is bilateral but determinate. Young gastrotrichs have most of the adult structures on hatching and reach sexual maturity in about three days. Growth results primarily in elongation of the trunk and morphogenesis of the reproductive system.

Laboratory studies on *Lepidodermella* indicate a maximum life of about 40 days, during the first 10 of which a female lays four or five eggs.

SYSTEMATIC RÉSUMÉ OF PHYLUM GASTROTRICHA

As is the case for the Kinorhyncha (discussed later in the chapter), the two orders of Gastrotricha actually represent classes but are not formally recognized as such. Originally, the Macrodasyida and Chaetonotida were designated as orders in the class Gastrotricha (phylum Aschelminthes), and the traditional rank of order has been retained although the class Gastrotricha has been elevated to phylum status.

Order (Class) Macrodasyida. Body usually wormlike, elongated, and dorsoventrally flattened. Adhesive tubes located on anterior and posterior ends and on lateral sides of body. Pharyngeal pores present. Hermaphrodites. Marine and estuarine species. *Macrodasys, Urodasys, Turbanella, Dactylopodola, Tetranchyroderma, Pseudostomella.*

Order (Class) Chaetonotida. Freshwater and marine gastrotrichs. Body usually tenpin-shaped. Adhesive tubes restricted to the posterior end. Pharyngeal pores absent. Most freshwater species are parthenogenetic females. *Chaetonotus, Halichaetonotus, Lepidodermella, Xenotrichula.*

PHYLUM NEMATODA

The nematodes, called roundworms, form the largest aschelminth phylum (12,000 described species, but there are probably many more undescribed than described species) and include some of the most widespread and numerous of all multicellular animals.

FIGURE 8–10 Copulating pair of *Macrodasys* sp. during the process of sperm transfer. *(From Ruppert, E. E. 1978. The reproductive system of gastrotrichs. II. Insemination in Macrodasys: A unique mode of sperm transfer in Metazoa. Zoomorphologie. 89:207–228.)*

Free-living nematodes are found in the sea, in fresh water, and in the soil, and there are many parasitic species. They occur from the polar regions to the tropics in all types of environments, including deserts, high mountain elevations, and great ocean depths. Nonparasitic nematodes are benthic animals and live in interstitial spaces of algal mats and especially aquatic sediments and soil. They are often present in enormous numbers. One square meter of bottom mud off the Dutch coast has been reported to contain as many as 4,420,000 nematodes. An acre of good farm soil has been estimated to contain several hundred million to billions of terrestrial nematodes. A single decomposing apple on the ground of an orchard has yielded 90,000 roundworms belonging to a number of species.

Some unusual aquatic habitats of nematodes include hot springs, in which the water temperature may reach 53°C, and the water in tropical, epiphytic bromeliads. In large lakes, there is often a distinct zonation of the nematode species from the shoreline to deeper water.

Terrestrial species live in the film of water that surrounds each soil particle, and they are therefore actually aquatic. In fact, some species are found in both soil and fresh water. Although nematodes exist in enormous numbers in the upper soil, the population decreases rapidly at greater depths. Moreover, the numbers are greater in the vicinity of plant roots. In addition to the more typical terrestrial habitats, nematodes have also been reported from accumulations of detritus in leaf axils and in the angles of tree branches. Mosses and lichens maintain a characteristic nematode fauna, and these and many soil species are able to withstand periodic desiccation. During such times the worm passes into a state of cryptobiosis (p. 280).

That nematodes are ubiquitous and numerous animals was noted long ago in a now famous aphorism of an early twentieth-century nematologist, N. A. Cobb, "If all the matter in the universe except the nematodes were swept away, our world would still be dimly recognizable, and if, as disembodied spirits, we could investigate it, we should find its mountains, hills, vales, rivers, lakes and oceans represented by a thin film of nematodes. . . . " In addition to free-living species, there are many parasitic nematodes. The parasites display all degrees of parasitism and attack virtually all groups of plants and animals. The numer-ous species that infest food crops, domesticated animals, and humans make this phylum one of the most important of the parasitic animal groups. The phylum also contains one of the most intensely studied laboratory animals, *Caenorhabditis elegans,* whose every cell has been traced throughout the course of development, and whose genome is one of the best known of any organism.

The initial discussion of this phylum emphasizes the free-living members, and the parasitic forms will be dealt with at the end of the section.

External Structure

The size and form of nematodes are important adaptations for living in interstitial spaces. They have slender, elongated bodies with both ends gradually tapered in most species (Fig. 8–11A). The majority of free-living nematodes are less than 2.5 mm in length, most are approximately 1 mm, and many are microscopic. However, some soil nematodes are as long as 7 mm, and some marine species attain a length of 5 cm.

Nematodes have a perfectly cylindrical body; hence the name *roundworm.* A cuticle encloses the body and lines the pharynx, hindgut, and other body openings. The mouth is located at the somewhat rounded anterior end and is surrounded by lips and sensilla of various sorts. In many marine nematodes (Fig. 8–12A), which include the most primitive members of the phylum, six liplike lobes border the mouth, three on each side, but as a result of fusion, there are often only three lips in terrestrial and parasitic species. Primitively, the lips and the anterior surface outside the lips bear 18 sensilla (Fig. 8–12A) and may also carry a variety of cuticular projections. As in other aschelminths, the cuticle of the general body surface is often sculptured or ornamented in different ways (Fig. 8–11C–F). Members of the marine interstitial family Stilbonematidae have body surfaces clothed with a symbiotic blue-green alga and appear to be hairy.

A **caudal gland,** also called a spinneret, is typical of many free-living nematodes, including most marine species. The gland opens at the posterior tip of the body, which is sometimes drawn out to resemble a tubelike tail. The caudal gland of nematodes opens on a projecting cuticular tube, as in several other aschelminths, and in at least some nematodes, it is a duo-gland system.

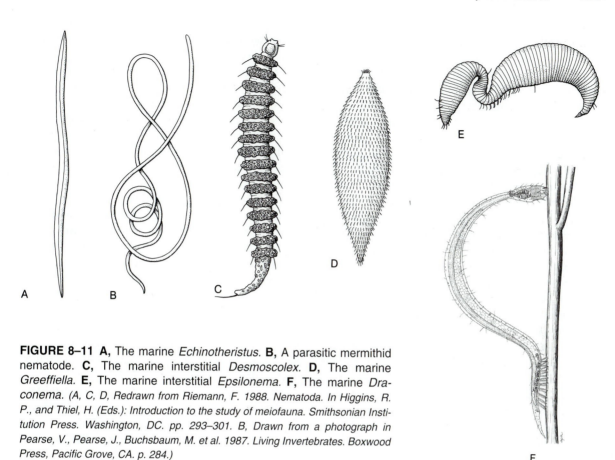

FIGURE 8–11 A, The marine *Echinotheristus.* **B,** A parasitic mermithid nematode. **C,** The marine interstitial *Desmoscolex.* **D,** The marine *Greeffiella.* **E,** The marine interstitial *Epsilonema.* **F,** The marine *Draconema. (A, C, D, Redrawn from Riemann, F. 1988. Nematoda. In Higgins, R. P., and Thiel, H. (Eds.): Introduction to the study of meiofauna. Smithsonian Institution Press. Washington, DC. pp. 293–301. B, Drawn from a photograph in Pearse, V., Pearse, J., Buchsbaum, M. et al. 1987. Living Invertebrates. Boxwood Press, Pacific Grove, CA. p. 284.)*

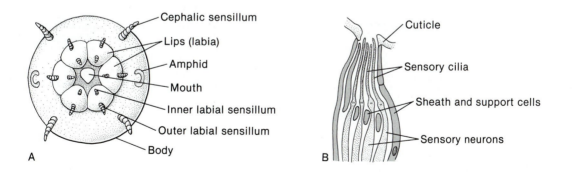

FIGURE 8–12 A, Oral view of a generalized nematode, showing typical sensory structures. **B,** Longitudinal section of an amphid of *Caenorhabditis elegans. (A, Modified after Jones from Lee, D. L., and Atkinson, H. J. 1977. Physiology of nematodes. Columbia University Press, New York. p. 161. B, Modified from Ward, S., Thomson, J. G., and Brenner, S. 1975. Electron microscopical reconstruction of the anterior sensory anatomy of the nematode Caenorhabditis elegans. J. Comp. Neurol. 160:313–338.)*

Body Wall

The nematode cuticle is more complex than that of most other aschelminths. It has three main layers, which contain collagen, as well as other compounds, and an epicuticle (Fig. 8–13). The thin epicuticle, which may exhibit quinone tanning, covers the outer cuticular layer (cortex) externally. The outer layer is typically annulated (ringed) and sometimes divided into outer and inner parts (Fig. 8–13). The median layer variously has a uniform granular structure or struts, skeletal rods, fibrils, or canals, depending on the species. The basal layer may be striated or laminated or contain fibers in a crossed helical arrangement (p. 263).

Growth in nematodes is accompanied by four **molts** of the cuticle. Beginning at the anterior end, the old cuticle separates from the underlying epidermis, and a new cuticle is secreted, at least in part. The old cuticle is shed, sometimes in fragments (Fig. 8–13). Molting does not occur after the worm becomes adult, but the cuticle continues to expand as the animal grows.

The epidermis, also called the hypodermis because it lies beneath a substantial cuticle, is usually cellular but may be syncytial in some species. A constant feature of the nematode epidermis is its inward expansion to form ridges along the middorsal, midventral, and midlateral lines of the body (Figs. 8–14C and D). The four ridges, called **longitudinal cords,** extend the length of the body. The epidermal nuclei are commonly restricted to these cords and are typically arranged in rows. The epidermis secretes the cuticle, stores nutrients, bears attachment fibers that link the musculature to the cuticle (Fig. 8–15B) and, in some endoparasitic species, is an important surface for the absorption of host nutrients.

The muscle layer of the body wall is composed entirely of longitudinal fibers, which are located in four quadrants between the longitudinal cords (Figs. 8–14C, D; 8–15). The fibers may be relatively broad and flat or tall and narrow, depending on the species. Each nematode muscle fiber, like those of gastrotrichs, possesses a slender arm that extends from the fiber to either the dorsal or the ventral longitudinal nerve cord, where innervation occurs (Figs. 8–14C, D; 8–15). In addition to the body wall musculature, nematodes have muscles associated with the reproductive organs and sometimes the intestine.

The nematode pseudocoel is small or nonexistent in most small free-living species but may be voluminous in large forms, such as the parasitic *Ascaris.* When present, the cavity extends from the muscula-

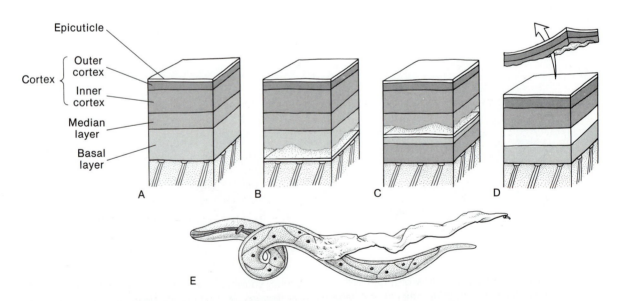

FIGURE 8–13 Although details vary in different species of nematodes, molting often occurs in the sequence shown. **A,** Intact cuticle. **B,** Secretion of new epicuticle and digestion of old basal layer. **C,** Secretion of new outer and inner cortex and partial digestion of old median layer. **D,** Secretion of new basal layer to complete the new cuticle. The remaining old cuticle is shed **(D, E).**

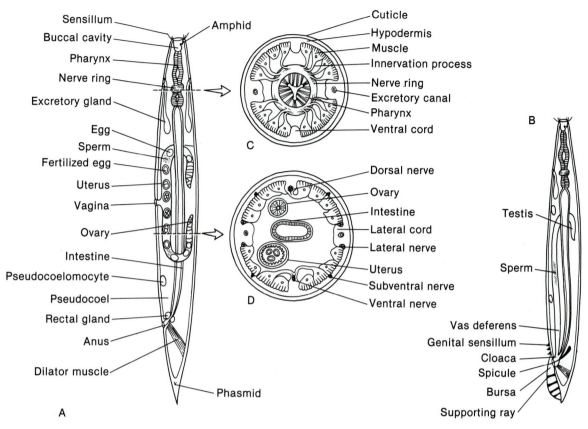

FIGURE 8–14 Generalized female nematodes in lateral view **(A)** and cross sections **(C, D). B,** Generalized male nematode in lateral view. The volume of the pseudocoel has been exaggerated for clarity. *(Redrawn from Lee, D. L., and Atkinson, H. J. 1977. Physiology of Nematodes. Columbia University Press, New York. pp. 2, 3.)*

ture to the gut wall and surrounds the reproductive organs. Fluid in the cavity is pressurized and functions as a hydrostat. The fluid contains a variety of organic metabolites, including hemoglobin in some species, but no circulating cells. A few phagocytic cells, however, are permanently attached to the walls of the cavity and are important in internal defense.

Locomotion

Most nematodes move forward and rearward using sinuous eellike undulations of the body. The undulations are in the dorsoventral plane, rather than side-to-side as in eels, and are produced by the alternate contraction of dorsal and ventral longitudinal muscles (Fig. 8–16A). Annulations in the cuticle of many nematodes may improve flexibility, whereas crossed

helical fibers prevent kinking and herniation as the body flexes and the hydrostatic pressure rises (pp. 96 and 263).

When nematodes are removed from sediment, as is the case during laboratory observations, their movement becomes undirected and of a whipping and thrashing nature. Keep in mind that free-living nematodes are largely interstitial inhabitants, and their undulatory movements are effective for progression only when applied against substratum particles or the surface tension of water films. Nematodes move through soil pores of 15 to 45 μm in diameter, and the pore size that allows for optimum movement is about 1.5 times the worm's diameter.

Many nematodes may swim intermittently for short distances. This is true, for example, of moss-inhabiting species when the plant is flooded. A few

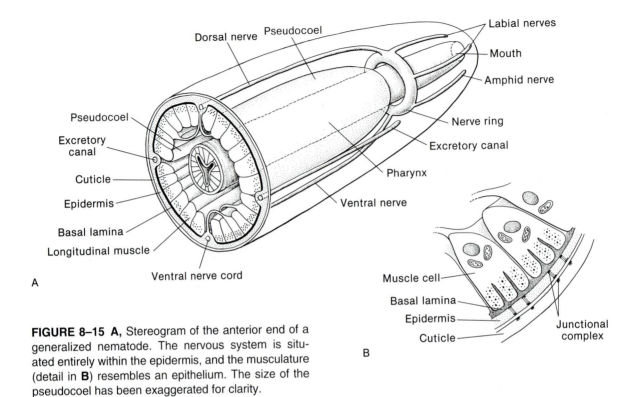

FIGURE 8–15 A, Stereogram of the anterior end of a generalized nematode. The nervous system is situated entirely within the epidermis, and the musculature (detail in **B**) resembles an epithelium. The size of the pseudocoel has been exaggerated for clarity.

species can crawl. The cuticle is sculptured to aid in gripping a surface. In one species, which possesses a ringed cuticle, crawling is similar to that in earthworms (the cuticle functions to oppose the longitudinal musculature because there is no circular musculature in nematodes); others crawl like caterpillars; and still others move like inchworms (Figs. 8–16D, F). The caudal gland, or spinneret, which is present in most marine nematodes, provides for temporary attachment (Fig. 8–16B) but may also be used for tail-anchored retreat movements and in springing (Fig. 8–16B and C).

A great many freshwater and terrestrial nematodes have a cosmopolitan distribution. Birds and other animals, as well as floating debris to which small amounts of mud adhere, are undoubtedly important agents in the spread of nematodes. Many of the saprophagous nematodes that inhabit dung utilize dung insects to move from one habitat to another.

Nutrition

Many free-living nematodes are carnivorous and feed on small metazoan animals, including other nematodes. Other species are phytophagous. Many marine and freshwater species feed on diatoms, algae, fungi, and bacteria. Algae and fungi are also important food sources for many terrestrial species. A large number of terrestrial nematodes pierce the cells of plant roots and suck out the contents. Such nematodes are responsible for approximately $5 billion worth of damage to commercial plants annually in the United States. There are also many deposit-feeding marine, freshwater, and terrestrial species that ingest particles of substratum. Deposit feeders and the many nematodes that live on dead organic matter, such as dung or on the decomposing bodies of plants and animals, digest only the associated bacteria and fungi, however. This is true of the common vinegar eel, *Turbatrix aceti,* which lives in the sediment of nonpasteurized vinegar. Nematodes are the largest and most ubiquitous group of organisms feeding on fungi and bacteria and are of great importance in the food chains leading from decomposers. Although many nematodes feed on fungi, there are fungi that prey on nematodes. The worms are caught when they pass through special hyphal (threadlike) loops, which close on stimulation, snaring and digesting the prey.

The mouth of the nematode opens into a **buccal cavity,** or stoma (Figs. 8–14A; 8–17A), which is

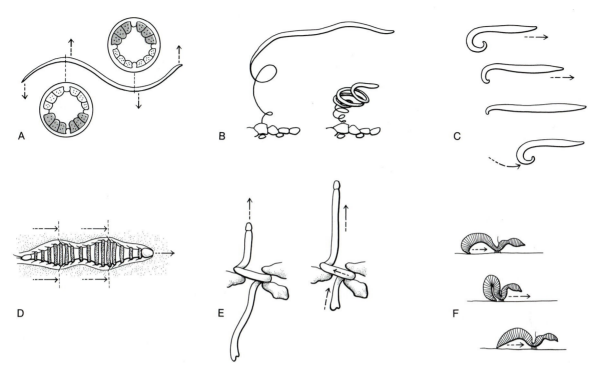

FIGURE 8–16 A, Typical dorsoventral undulatory locomotion in a generalized nematode. Cross sections show contracted muscles (shaded) used to bend the body in the indicated regions. **B,** Retraction movement similar to that found in *Vorticella* shown here in the marine semisessile nematode, *Trefusia.* **C,** Springing or thrusting, locomotion in the marine *Theristus caudasaliens.* This nematode forcibly flicks its posterior end and springs forward. **D,** Earthworm-like progression in the marine genus, *Desmoscolex.* **E,** Single spiral wave used by the gut parasite *Nippostrongylus* to move among the intestinal villi. **F,** Inchworm-like locomotion in the marine interstitial *Epsilonema. Epsilonema* alternates attachment of its adhesive toe and stilt bristles on the trunk. *(B, Redrawn from Riemann, F. 1974. On hemisessile nematodes with flagelliform tails living in marine soft bottoms and on micro-tubes found in deep sea sediments. Mikrofauna Meeresbodens. 40:1–15. C, Redrawn from Adams, P.J.M., and Tyler, S. 1980. Hopping locomotion in a nematode: Functional anatomy of the caudal gland apparatus of Theristus caudasaliens sp. n. J. Morphol. 164:265–285. E, Redrawn from Lee, D. L., and Atkinson, H. J. 1977. Physiology of Nematodes. Columbia University Press, New York. p. 156.)*

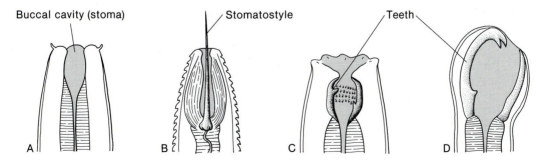

FIGURE 8–17 Nematode buccal cavities: **A,** The bacteriovore, *Rhabditis.* **B,** The plant root parasite, *Criconemoides.* **C,** The protozoan and micrometazoan carnivore, *Mononchus.* **D,** The intestinal parasite, *Ancylostoma. (A, C, D, redrawn from Lee, D. L., and Atkinson, H. J. 1977. Physiology of Nematodes. Columbia University Press, New York. p. 30. B, Redrawn from Nicholas, W. L. 1984. The Biology of Free-Living Nematodes. Clarendon Press, Oxford, p. 28.)*

somewhat tubular and lined with cuticle. The cuticular surface is often strengthened with ridges, rods, or plates, or it may bear teeth (Fig. 8–17C, D). The structural details of the buccal cavity are correlated with feeding habits and are of primary importance in the identification of nematode species. Teeth are especially typical of carnivorous nematodes; they may be small and numerous or limited to a few, large, jaw-like processes. The toothed terrestrial nematode, *Mononchus papillatus,* which has a large dorsal tooth opposed by a buccal ridge, consumes as many as 1000 other nematodes during its life-span of approximately 18 weeks. In feeding, this nematode attaches its lips to the prey and makes an incision in it with the large tooth. The contents of the prey are then pumped out by the pharynx.

In some carnivores, as well as in many species that feed on the contents of plant cells, the buccal cavity carries a long hollow or solid cuticular

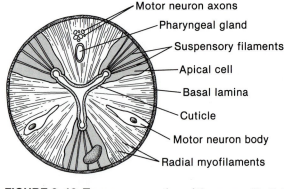

FIGURE 8–19 Transverse section of the myoepithelial pharynx of *Caenorhabditis elegans (Drawn from an electron micrograph from Albertson, D. G., and Thomson, J. N. 1976. The pharynx of Caenorhabditis elegans. Philos. Trans. R. Soc. Lond., B. Biol. Sci. 275:299–325.)*

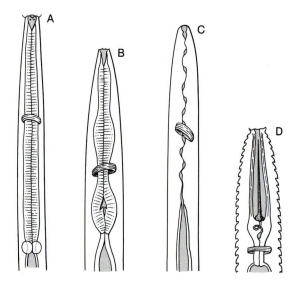

FIGURE 8–18 Pharyngeal diversity in nematodes: **A,** *Monhystera,* a marine and freshwater genus whose species feed on bacterial aggregates. **B,** *Rhabditis,* marine and terrestrial species that feed on bacteria. **C,** Juvenile *Gastromermis* species parasitize the hemocoel of insects, absorbing nutrients directly through the body wall rather than with the rudimentary pharynx. **D,** The stylet-bearing *Criconemoides,* which punctures plant roots and sucks out the contents. *(A, B, D, Modified and redrawn from Nicholas, W. L. 1984. The Biology of Free-Living Nematodes. Clarendon Press, Oxford. p. 35.)*

stylet (Fig. 8–17B), which can protrude from the mouth. Both kinds of stylet are used to puncture prey, and the hollow stylet may act as a tube through which the contents of the victim are pumped out by the pharynx. In a stylet-bearing herbivore, it is used to penetrate the root cell walls, being thrust rapidly forward and backward. Species with either kind of stylet secrete pharyngeal enzymes that initiate digestion of the prey or plant cell contents and that may even aid in the penetration of the plant cell wall.

The buccal cavity leads into a tubular pharynx, referred to as the esophagus by nematologists (Figs. 8–14A, C; 8–18). The pharyngeal lumen is triradiate in cross section and lined with cuticle (Fig. 8–19). The wall is composed of myoepithelial and gland cells, as in gastrotrichs. Frequently, the pharynx contains more than one muscular swelling or bulb. The pharynx or pharyngeal bulbs act as pumps and bring food from the mouth into the intestine. Valves are frequently present.

From the pharynx a long tubular intestine composed of a single layer of epithelial cells extends the length of the body (Fig. 8–14A, B). It is primitively ciliated, as in species of *Eudorylaimus* (Fig. 8–20A), but cilia are absent in other nematodes and the epithelium has instead a brush border of microvilli (Fig. 8–20B, C). A valve located at each end of the intestine prevents food from being forced out of the intestine by the fluid pressure of the pseudocoel. A short,

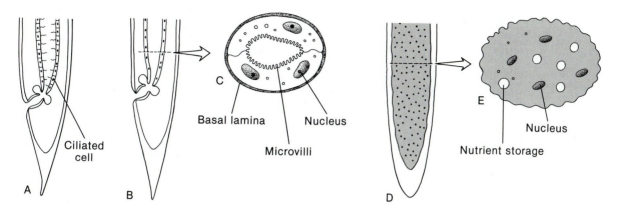

FIGURE 8–20 A, Although rare in nematodes, the midgut is ciliated in *Eudorylaimus*. The midgut of most nematodes **(B, C)** lacks cilia and is lined only by a brush border of microvilli. **D,E,** In the mermithid nematodes, the midgut is syncytial and lacks a lumen. Mermithids absorb host nutrients directly across the body wall, the gut playing no role in ingestion or digestion. Instead, it functions as a nutrient storage organ, called a trophosome. *(A, After Zmoray, I., and Guttekova, A. 1969. Ecological conditions for the occurrence of cilia in intestines of nematodes. Biol. Bratisl. 24:97–112. B, Redrawn after Lee, D. L., and Atkinson, H. J. 1977. Physiology of Nematodes. Columbia University Press, New York. p. 37. D, E, Modified and redrawn after Batson, B. S. 1979. Ultrastructure of the trophosome, a food-storage organ in Gastromeris boophthoorae (Nematoda: Mermithidae). Int. J. Parasitol. 9:505–514.)*

cuticle-lined rectum connects the intestine with the anus, which is on the midventral line just in front of the posterior tip of the body (Fig. 8–14A). In mermithid nematodes, which are parasites of invertebrates, the midgut is syncytial, lacks a lumen, and does not function in digestion (Fig. 8–20D, E). Instead, nutrients that are absorbed through the body wall are stored in the gut syncytium, which thus functions as a liver.

Digestive enzymes are produced by the pharyngeal glands (Fig. 8–19) and the intestinal epithelium. Digestion begins extracellularly within the intestinal lumen but is completed intracellularly. The intestine is also an important organ of nutrient storage and yolk synthesis for developing oocytes. In *Caenorhabditis elegans,* the intestine synthesizes yolk proteins and exports them, via the pseudocoel, to the ovary.

Excretion and Osmoregulation

Nematodes excrete nitrogenous wastes in the form of ammonium ions that diffuse across the body wall. Osmoregulation, ionic regulation and perhaps the excre-

tion of other waste metabolites seem to be associated with specialized, excretory structures unique to nematodes, either an excretory gland cell or cells, an excretory canal system, or both (Fig. 8–21). When gland and canal systems occur together, they share a common outlet pore. Neither of these resembles a filtration nephridium, such as a protonephridium or metanephridial system, but rather may be unique secretion kidneys. A few nematodes lack any excretory organs.

The **excretory gland cell,** also called a ventral or renette cell, occurs alone in the class Adenophorea, which includes most marine and freshwater nematodes, and in conjunction with a canal system in some members of the class Secernentea. This large cell protrudes into the pseudocoel and is provided with a necklike duct that opens midventrally at a pore (Fig. 8–21). Its role in excretion is uncertain, and several alternative functions have been suggested or demonstrated for it. These suggested functions include secretion of the gelatinous matrix around the eggs in the case of the citrus nematode, *Tylenchulus semipenetrans;* secretion of the outer glycoprotein coat on the cuticle in the case of the root-knot nematode,

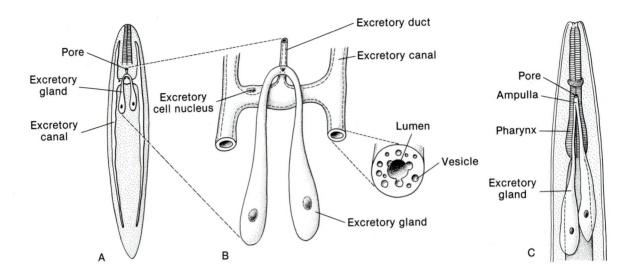

FIGURE 8–21 A, Generalized anatomy of the binucleate excretory gland cell and excretory canal system of secernentean nematodes; ventral view. **B,** Enlargement of the system in **A. C,** The excretory gland of the adenophorean nematode, *Rhabdias. (A, B, Based chiefly on Caenorhabditis elegans. Modified from Nelson, F. K., Albert, P. S., and Riddle, D. L. 1983. Fine structure of the Caenorhabditis elegans secretory–excretory system. J. Ultrastruct. Res. 82:156–171. C, Redrawn after Hyman, L. H. 1951. The Invertebrates. Vol. 3. McGraw-Hill Book Co., New York. p. 241.)*

Meloidogyne javanica; and production of exoenzymes to initiate digestion of host tissues in the case of some animal parasitic nematodes.

All members of the class Secernentea, which includes many terrestrial species, have a hollow **excretory canal system,** often in addition to the excretory gland cell. The entire canal system lies within a single elaborate cell, typically the largest cell in the animal's body. Generally the cell is laid out in the form of an H, the two long canals are situated in the lateral epidermal cords and are joined together by a transverse canal. A short stem leads anteriorly from the transverse canal to a midventral outlet pore in the pharyngeal region of the body (Fig. 8–21A, B). In some nematodes, the stem is enlarged to form a bladder-like ampulla, which fills and empties rhythmically. The canal system has been shown to have an osmoregulatory function in one nematode, *Caenorhabditis elegans,* but the mechanism by which water enters the canals from the body is unknown, although osmosis may be part of the process. Cilia and muscles are absent from the canals.

Nervous System

The entire nematode nervous system is intraepithelial in position, being located within the epidermis, pharynx, and hindgut. The brain is a circumpharyngeal nerve ring (Figs. 8–14C; 8–15). From the brain, nerves extend anteriorly, innervating the cephalic sensilla and the amphids (sense organs, which are described later). Dorsal, lateral, and ventral nerves extend posteriorly from the brain and run within the longitudinal cords. Muscle cell tails make contact with the dorsal and ventral nerves; lateral nerves are associated with the excretory canals (Fig. 8–15). The largest of these nerves is the ventral nerve, which contains motor neurons that innervate both the ventral and dorsal quadrants of longitudinal muscles. Motor neurons that innervate the dorsal musculature leave the ventral nerve, pass around the circumference of the body to the dorsal midline, and then extend anteriorly or posteriorly within the dorsal nerve.

The principal sensilla of nematodes are papillae, setae, amphids, and phasmids, all of which bear a cili-

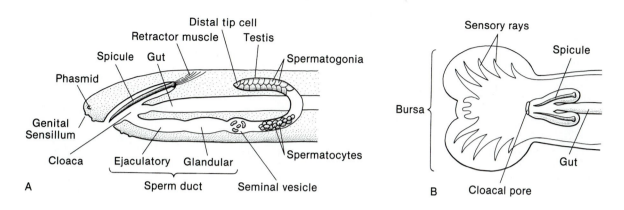

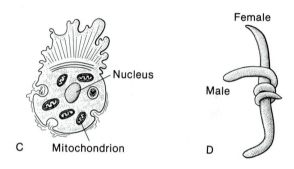

FIGURE 8–22 **A,** Lateral view of the male reproductive system of a generalized nematode with one testis. **B,** Copulatory bursa of a parasitic trichostrongyle nematode. During copulation, the bursa wraps around the body of the female like a glove and secures the male firmly to the surface of her body. **C,** Amoeboid sperm of *Caenorhabditis elegans.* **D,** Nematode copulation. *(A, B, Redrawn after Lee, D. L., and Atkinson, H. J. 1977. Physiology of Nematodes. Columbia University Press, New York. pp. 116, 118. C, Redrawn after Ward, S., Argon, Y., and Nelson, G. A. 1981. Sperm morphogenesis in wild-type and fertilization-defective mutants of Caenorhabditis elegans. J. Cell Biol. 91:26–44.)*

ated dendrite enclosed in a specialized part in the body wall (Fig. 8–12). The labial and cephalic papillae are low projections of the cuticle on the lips and head, and the setae are elongated bristles on both the head and body. The setae are touch receptors that, when stimulated, cause the animal to withdraw from the stimulus. Outer labial papillae and cephalic papillae are also believed to be mechanoreceptors. The remaining sensilla, the inner labial papillae, the paired amphids, and the paired phasmids (Figs. 8–12A, 8–14A; 8–22A) all open to the exterior via a small cuticular pore, thus exposing the sensory cilia directly to external environment. These sensilla are believed to be chemosensory in function but each, perhaps, responds to a different set of chemical cues.

The **amphid** sensilla, which reach their highest development in the aquatic nematodes, especially in the marine species, are blind, pouchlike or tubelike invaginations of the cuticle. One amphid is situated on each side of the head, often immediately posterior to the cephalic setae (Fig. 8–12A and B). In the tail region of some groups of nematodes (e.g., Secernentea) is a pair of unicellular glands, called phasmids (Figs. 8–14A; 8–22A), which open separately on either side of the tail. They reach their best development in parasitic nematodes.

A pair of simple ocelli is located one on each side of the pharynx of some marine and freshwater nematodes, but the function is uncertain. Stretch receptors have been found in the epidermal cords of nematodes and probably regulate locomotor movements.

Reproduction

Most nematodes are dioecious, but hermaphrodites, such as the well-studied *Caenorhabditis elegans,* are not uncommon. Males are often smaller than females, and the posterior of the male may be curled like a hook or broadened into a fan-shaped copulatory aid,

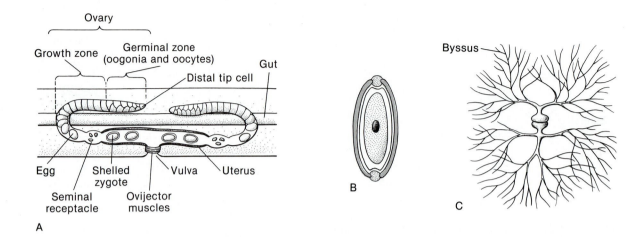

FIGURE 8–23 A, Lateral view of the reproductive anatomy of a generalized female nematode. **B,** Shelled egg of *Trichuris.* **C,** Egg of the mermithid nematode, *Mermis,* whose juveniles parasitize grasshoppers. Adults live in soil and, when ripe, the female attaches her eggs to grass blades and other vegetation using the proteinaceous byssus. *(A, B, Redrawn and modified from Lee, D. L., and Atkinson, H. J. 1977. Physiology of Nematodes. Columbia University Press, New York. pp. 121, 123. C, Redrawn after Christie from Hyman, L. H. 1951. The Invertebrates. Vol. 3. McGraw-Hill Book Co., Inc., New York. p. 253.)*

called a **bursa** (Fig. 8–22B). The paired reproductive system opens at a single pore, the **cloaca** (which also receives intestinal wastes) in males (Fig. 8–22A) and the **vulva** in females (Fig. 8–23A). The tubular gonads are typically paired, but they are not bilaterally positioned within the body. Instead, one gonad is usually oriented anteriorly, and the other, posteriorly (Fig. 8–23A). In many species, each gonad bends back on itself in the shape of a (Fig. 8–22A), but in many parasitic species, each gonad is long and coiled.

The innermost end of each gonad contains a syncytium consisting of an axial core of common cytoplasm (rachis) and a rind of germ cell nuclei, which are partly separated from each other by lateral membranes but are in cytoplasmic continuity with the central rachis. As the germ cells mature, they move proximally along the gonad and separate from each other and the rachis. The innermost (distal) cell of each gonad, the **distal tip cell,** secretes a mitosis-promoting substance that causes the proliferation of germ cell nuclei (Figs. 8–22A, 8–23A). As these nuclei migrate proximally toward the gonopore, they escape the influence of the mitogenic substance and then undergo meiosis in preparation for fertilization.

In males there may be one or two tubular testes, each of which passes more or less imperceptibly into a long sperm duct. Each sperm duct eventually widens to form a long seminal vesicle. A muscular ejaculatory duct, containing prostatic glands, connects the seminal vesicles with the cloaca (Fig. 8–22A). The prostatic secretions are adhesive and supposedly aid in copulation. The wall of the cloaca is evaginated to form two pouches, which join before they open into the cloacal chamber. Each pouch contains a **spicule,** which is usually short and shaped somewhat like a pointed, curved blade (Fig. 8–22A). Special muscles cause the spicules to protrude through the cloaca and out of the anus or vent. In many nematodes, the dorsal walls (and sometimes even the ventral and lateral walls) of the pouch bear special, cuticular pieces (the gubernaculum) that guide the spicules through the cloacal chamber.

There may be one or, more typically, two ovaries, which are usually oriented in opposite directions (Fig. 8–23A). Each ovary grades into a tubular oviduct and then into a much widened, elongated uterus. Each of the two uteri opens into a short, common, muscular tube called the vagina, which leads to the outside through the vulva. The vulva is located on the midventral line, usually in the midregion of the body. The upper end of the uterus usually functions as a seminal receptacle.

The females of some nematodes are known to produce a pheromone that attracts males—a mechanism that is probably widespread. In copulation the curved posterior of the male nematode is usually coiled around the body of the female in the region of the genital pores (Fig. 8–22D). The copulatory spicules of the male are extended through the cloaca and anus and are used to hold open the female gonopore during the transmission of the sperm into the vagina. Nematode sperm are peculiar in that they lack flagella, and some move in a more or less ameboid manner (Fig. 8–22C).

After copulation the sperm migrate to the upper end of the uterus, where fertilization takes place. The fertilized egg secretes a thick fertilization membrane, which hardens to form the inner part of the shell. To this inner shell is added an outer layer, which is secreted by the uterine wall. The outer surface of nematode eggs is sculptured in species-specific ways. Eggs passed in the feces of nematode hosts can therefore be identified (Fig. 8–23B, C). Nematode eggs may be stored in the uterus prior to deposition, and not infrequently embryonic development begins while the eggs are still in the female. Many parasitic nematodes and some free-living species, such as the vinegar eel, are viviparous.

Some terrestrial nematodes, particularly the rhabditoids (including *Caenorhabditis elegans*), are hermaphrodites. Sperm develop before the eggs within the same gonad (ovotestis) and are stored. Self-fertilization takes place after the formation of the eggs. Although hermaphrodites, like *Caenorhabditis elegans*, do not cross-fertilize, low numbers of males arise periodically in hermaphroditic populations and cross-fertilize the hermaphrodites. Parthenogenesis also occurs in some nematodes, especially terrestrial species, and there are some parthenogenetic nematodes for which males are unknown.

The deposition of the eggs of free-living nematodes is still not well known. Marine species rarely produce more than 50 eggs, which are often deposited in clusters. Terrestrial species may produce up to several hundred eggs, which are deposited in the soil.

Development

Early cleavage is not spiral or radial but does follow a fixed asymmetrical pattern that soon becomes bilateral in form (Fig. 8–24). Development is determinate, and there is an early separation of future germ cells

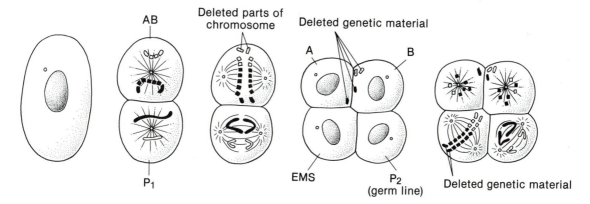

FIGURE 8–24 Chromosome diminution in *Parascaris aequorum* during early development. During the first cleavages of some nematodes, the chromosomes in all cells, except the stem cell of the germinal line (P$_2$), fragment and cast off some of the pieces, which degenerate. The deleted fragments presumably contain genetic information that is essential only to the germ line and not the somatic cells. Thus, in nematodes, such as *Parascaris,* there is genetic differentiation as early as the two-cell stage. Blastomeres A and B proliferate primarily ectodermal cells whereas blastomere EMS produces endodermal, mesodermal and stomodeal cells. *(Adapted from Boveri, T. 1904. Ergebnisse über die Konstitution der chromatischen Substanz des Zellkerns. Verlag Gustav Fischer, Jena, Germany. p. 27.)*

from somatic cells. As in some other aschelminths, the various organs of the body contain a relatively fixed number of cells, and these cells have been largely produced by the time hatching takes place. For example, in certain *Rhabditis* adults there are 200 nerve cells and 120 epidermal cells, and the digestive tract is composed of 172 cells. In adult *Caenorhabditis elegans,* there are exactly 959 somatic nuclei. Although there is a limited increase in the number of cells during juvenile stages, most growth in the size of nematodes results from increases in cell size.

Juveniles, sometimes inappropriately called larvae, have almost all of the adult structures when they hatch, except for certain parts of the reproductive system. Growth is accompanied by four molts of the cuticle, the first two of which may occur within the shell before hatching (as in Secernentea). Adults do not molt, but many continue to increase in size.

Parasitism

The great numbers of parasitic species of nematodes attack virtually all groups of animals and plants and display all degrees and types of parasitism. Moreover, some of the major groups of nematodes contain both free-living and parasitic species. All of these facts suggest that parasitism has evolved many times within the phylum. Some nematode radiation was coupled with the evolution of flowering plants, insects, and higher vertebrates, all of which contain important nematode host species. The types of nematode host/parasite relationships have been outlined by Hyman (1951). A modification of that outline is followed here:

1. *Completely free-living.* Life cycle is direct, and all stages are free-living.

2. *Ectoparasites of plants.* Worms feed on the external cells of plants by puncturing the cell wall with stylets and sucking out the cell contents. (These nematodes could just as readily be called herbivores.)

3. *Endoparasites of plants.* Worms in juvenile stage enter the plant body and feed on the living cells, producing gall-like structures or causing tissue death. Reproduction takes place within the host, and the new generation of juveniles migrates to other plants. *Heterodera* contains many such species.

Most of the plant parasites, both ectoparasites and endoparasites, are members of the orders Tylenchida and Aphelenchida. All parts of a plant are attacked, depending on the species. Because of the

damage they can cause, many of these nematodes are of great economic importance.

4. *Saprophagous type of zooparasitism.* Adults and juveniles are free-living in soil, but worms in late juvenile stage enter an invertebrate. The host is not injured, and the worms feed on the dead tissues when the host dies.

5. *Zooparasitic juvenile stages only.* The juveniles are parasitic in an animal host, usually an invertebrate, for at least part of the early life cycle. Adults are free-living.

6. *Phytoparasitic juveniles and zooparasitic female adults.* The female worm produces juveniles within a plant-feeding insect host. When the insect punctures the plant tissue, the juvenile worms enter the plant and remain as endoparasites. When mature and after copulation, the female enters the larva of the insect host, which lives on the same plant. The larval host metamorphoses into an adult and deposits a new generation of juvenile worms. *Heterotylenchus aberrans,* a parasite of onion flies, illustrates this type of life cycle.

7. *Zooparasitic juveniles and phytoparasitic adults.* Early stages of development take place within an invertebrate host. Later, worms in juvenile stages leave the host and enter the plant on which the host feeds. Worms complete development and reproduce as phytoparasites. The new generation of young then enters the animal host.

8. *Zooparasitism in adult females only.* The young become adults in the soil. After copulation, the male dies, and the female enters an invertebrate host to produce the next generation.

9. *Adult zooparasites with one host.* Adult worms of both sexes are parasitic within a vertebrate or invertebrate host. Transmission from one host to another is by eggs or newly hatched young, which may be free-living for a part of their development. Many economically or medically important nematode parasites possess this type of life history; a few of the more interesting or important examples follow.

The ascaroid nematodes, which feed on the intestinal contents of humans, dogs, cats, pigs, cattle, horses, chickens, and other vertebrates, include the largest species of nematodes. They are entirely parasitic within a single host, and the life cycle typically involves transmission by the ingestion of eggs or juveniles passed in the feces of another host. The juvenile stages usually penetrate the intestinal wall to enter the circulatory system, where they are carried to the lungs. Here they break into the alveoli and migrate back to the intestine via the trachea and esophagus.

The human ascaroid, *Ascaris lumbricoides,* reaches a length of 49 cm and is one of the best known parasitic nematodes (Fig. 8–25). The species is widely distributed throughout the world, including the southeastern United States, particularly in children. The developing eggs are notoriously resistant to adverse environmental conditions and may remain viable in soil for 10 years. The very closely related species in pigs probably had a common evolutionary origin with the human species, the common ancestor being confined to one host or the other prior to human domestication of the pig.

Physiological studies suggest that *Ascaris* produces enzyme inhibitors that protect the worm from the host's digestive enzymes. If this is true, such a mechanism is probably utilized by most other gut-inhabiting nematodes. The ascarids feed on the host's intestinal contents.

Toxocara canis and *Toxocara cati* are two small ascaroid species common in dogs and cats. It is these species for which puppies and kittens are usually wormed.

The hookworms are another group of parasites of the digestive tract of vertebrates. Most members of this group feed on the host's blood. The mouth region is usually provided with cutting plates, hooks, teeth,

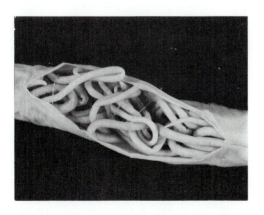

FIGURE 8–25 Adult specimen of *Ascaris suum* within the small intestine of a pig. *Ascaris lumbricoides,* which affects humans, is similar. *(From Schmidt, G. D., and Roberts, L. S. 1985. Foundations of Parasitology. 3rd Edition. Times Mirror/Mosby College Publishing, St Louis. p. 491.)*

or combinations of these structures for attaching to and lacerating the gut wall (Fig. 8–17D).

An infection of more than about 25 worms produces symptoms of hookworm disease, and a heavy infection can produce serious danger to the host through loss of blood and tissue damage. An adult worm may live as long as two years in the intestine. Hookworm infection is widespread in humans. It is estimated that over 380 million people are infected with *Necator americanus,* the most important species throughout the tropical regions of the world (despite the species' name).

The life cycle of hookworms involves an indirect migratory pathway by the juveniles, as in ascaroids. The fertilized eggs leave the host in its feces and hatch outside the host's body on the ground. The juvenile gains reentry by penetrating the host's skin (feet in humans) and is carried in the blood to the lungs. From the lungs the juvenile migrates to the pharynx, where it is swallowed and passes to the intestine.

Oxyurid nematodes, known as pinworms, have a simpler life cycle. These small nematodes are parasitic in the gut of vertebrates and invertebrates. Infection usually occurs through the ingestion of eggs passed in feces. The eggs hatch, and juveniles develop within the gut of the new host. The human pinworm, *Enterobius vermicularis,* affects children throughout the world. The female worm deposits eggs at night in the perianal region. Itching is caused by the migration of the female depositing her eggs, and scratching by the child contaminates the fingernails and hands with eggs. The eggs thus easily spread to other children or reinfect the same child.

Trichinelloids are also parasites of the alimentary tracts of vertebrates, especially birds and mammals. The whipworms, *Trichuris,* which infect human beings, dogs, cats, cattle, and other mammals, are relatively small (the human whipworm, *Trichuris trichura,* is about 4 cm) and have a life cycle similar to that of pinworms. Certainly the most familiar of the trichinelloids is *Trichinella spiralis,* which infects mammals and is the cause of the disease trichinosis. The minute worm, which lives in the intestinal wall, is viviparous, and its juveniles are carried in the blood to the striated muscles. There the juveniles form calcified cysts and, if infection is high, can produce pain and stiffness (Fig. 8–26). Transmission to another host can occur only if flesh containing encysted juveniles is ingested. Thus, in some animals, such as

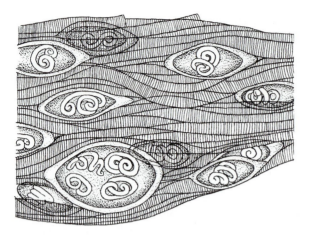

FIGURE 8–26 "Larvae" of *Trichinella spiralis* within calcareous cysts in striated muscle tissue of the host. *(After Chandler, A. C., and Read, C. P. 1961. Introduction to Parasitology. John Wiley and Sons, New York.)*

the microfilariae enter with the host's blood. Development within the intermediate host involves a migration through the gut to the thoracic muscles and after a certain period into the proboscis. From the proboscis the microfilariae are introduced into the definitive host when the mosquito feeds. In severe filariasis the blocking of the lymph vessels by large numbers of worms results in serious short-term lymphatic inflammation marked by pain and fever. Over a long period, increase in the volume of connective tissue affected may result in terrible enlargement of the legs, breast, and scrotum. Such enlargement is called elephantiasis (Fig. 8–27), but extreme cases are no longer common.

the rat, this can be a one-host parasite; in others, such as man and the pig, it would normally require two hosts.

10. *Zooparasites with one intermediate host.* Varying degrees of juvenile development take place within an intermediate host, after which there is reinfection of the definitive host, where reproduction occurs.

This type of life cycle is illustrated by a number of nematode parasites, including the familiar filarioids and dracunculoids. The filarioids are threadlike worms that inhabit the lymphatic glands and some other tissues of the vertebrate host, especially birds and mammals. The female is viviparous, and the larvae are called microfilariae. Blood-sucking insects, such as fleas, certain flies, and especially mosquitoes, are the intermediate hosts. A number of species parasitize man, producing filariasis.

The chiefly African and Asian *Wuchereria bancrofti* illustrates the life cycle. (The male is 40 mm × 0.1 mm, and the female is about 90 mm × 0.24 mm.) The adults live in the ducts adjacent to the lymph glands. The microfilariae are found in the blood and are present in the peripheral bloodstream. A distinct periodicity of larval migration to the peripheral circulation, coinciding with the activity of certain species of mosquitoes, has been demonstrated by a number of investigators. When the mosquito bites the next host,

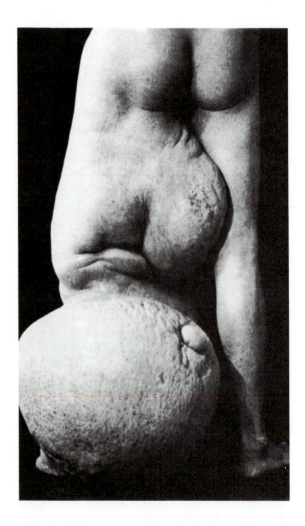

FIGURE 8–27 A victim of elephantiasis, which results from severe filariasis. *(Photograph courtesy of the Mayo Clinic, Rochester, MN.)*

Dirofilaria immitis, the heartworm, which live as adults in the hearts or pulmonary arteries of dogs, wolves, and foxes, is also transmitted by mosquitoes.

Loa loa, the African eye worm, lives in the subcutaneous tissues of humans and baboons. The worm migrates about in the tissue and sometimes passes across the eyeball, accounting for the name eye worm (Fig. 8–28).

The dracunculoids are also rather threadlike worms. They live in the connective tissues and body cavities of vertebrate hosts. The most notable example is the guinea worm, *Dracunculus medinensis,* which parasitizes humans and many other mammals, especially in Asia and Africa. The female is about 1 mm in diameter and up to 120 cm in length. After a period of development in the body cavity and connective tissue of the host, the gravid female migrates to the subcutaneous tissue and produces an ulcerated opening to the exterior. If the ulcerated area of the host comes in contact with water, larvae are released. After a short free-living stage, the larvae are ingested by species of copepod crustaceans *(Cyclops)* and continue their development in the host's hemocoel. When the definitive host swallows copepods in drinking water, the nematode larvae are released and penetrate the intestinal wall to reach the coelom or subcutaneous tissue. The worms can be removed surgically, but the ancient method, still practiced, is to slowly wind them out on a small stick (Fig. 8–29). Breaking of the worm causes severe inflammation. The dracunculid and its removal from the human is the basis of the caduceus, or symbol of the medical profession. The symbol, however, has been misinterpreted as a staff with snakes instead of a nematode wrapped around a stick.

FIGURE 8–29 A guinea worm, *Dracunculus medinensis,* being removed from an ulcerated opening of the arm by slow winding on a match stick. *(Photograph courtesy of the Institute of Public Health Research, Teheran University School of Public Health.)*

Classification

The phylum Nematoda is divided into two classes and 14 orders. The class Adenophorea, or Aphasmida, contains nematodes that have variously shaped amphids behind the lips. Most species are free-living; some are parasitic. The free-living species include terrestrial and freshwater forms and almost all of the marine forms. To this class belong the orders Enoplida, Isolaimida, Mononchida, Dorylaimida, Trichocephalida, Mermithida, Muspiceida, Chromadorida, Desmodorida, Desmoscolecida, Monhysterida, and Araeolaimida.

The class Secernentea, or Phasmida, contains nematodes that usually possess porelike amphids in the lateral lips. Many parasitic forms are members of this class, and the free-living species largely inhabit soil. To this class belong the orders Rhabditida, Strongylida, Ascaridida, Spirurida, Camallanida, Diplogasterida, Tylenchida, and Aphelenchida.

FIGURE 8–28 The eye worm *Loa loa* in the cornea. *(From Chandler, A. C., and Read, C. P. 1961. Introduction to Parasitology. John Wiley and Sons, New York.)*

PHYLUM NEMATOMORPHA

Superficially resembling the nematodes and perhaps closely related to them is a small group (about 320 species) of long worms known as horsehair worms or hairworms, which constitutes the phylum Nematomorpha. The adults are free-living and short-lived, and the larvae, which parasitize arthropods (and

rarely, humans), are the dominant stage in the life cycle. Most hairworms, namely of the class Gordioida, live as adults in fresh water and damp soil. The single genus *Nectonema*, which makes up the class Nectonematoida, is pelagic in marine waters (Fig. 8–30E). The body of nematomorphs (Fig. 8–30A, E) is threadlike—extremely long (up to 100 cm) and slender (1–2 mm), without a distinct head

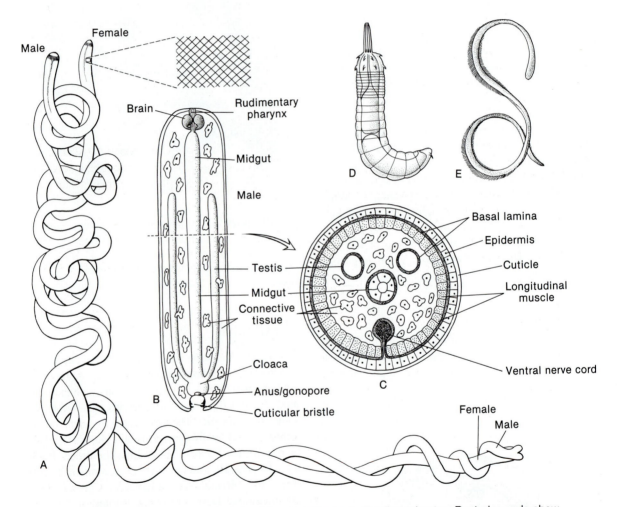

FIGURE 8–30 A, Male and female of the nematomorph *Gordius robustus.* Posterior ends show typical copulatory position. Inset shows crossed helical fibers in the cuticle. The fiber angle with respect to the longitudinal body axis is approximately 55 degrees. **B,** Generalized anatomy of a gordioid nematomorph. **C,** Cross section of B. **D,** Larva of the gordioid, *Paragordius.* **E,** The marine nectonematoid worm, *Nectonema agile,* showing lateral swimming setae. *(A, Modified from May, H. G. 1919. Contributions to the life histories of Gordius robustus Leidy and Paragordius varius (Leidy). Illinois Biol. Monogr. 5:1–118. C, Redrawn after Bresciani, J. 1991. Nematomorpha. In, Harrison, F. W., and Ruppert, E. E. (Eds.): Microscopic Anatomy of Invertebrates. Vol. 3. pp. 197–218. D, After Montgomery from Hyman, L. H. 1951. The Invertebrates. Vol. 3. McGraw-Hill Book Co., New York. p. 468. E, After Fewkes from Hyman, L. H. 1951. The Invertebrates. Vol. 3. McGraw-Hill Book Co., New York. p. 467.)*

(Fig. 8–31). Lengths of 5 to 10 cm or more are typical. The hairlike nature of these worms is so striking that they were formerly thought to arise spontaneously from the hairs of a horse's tail. This interpretation was reinforced by the frequent occurrence of worms in horses' drinking troughs where grasshoppers, which are common intermediate hosts of some hairworm species, happened to fall.

The nematomorph body wall consists of a thick, outer, white-to-black iridescent cuticle composed of collagen fibers in crossed helical layers (Fig. 8–30A, inset). Beneath the cuticle is a cellular epidermis and a sheath of obliquely striated longitudinal muscle; a circular musculature is absent. The myofilaments include paramyosin filaments, which are especially thick and long, like those in the closing muscles of clams, and like them may be important in sustaining muscular contraction. Adults coil tightly around streamside vegetation, and each other, for extended periods of time. Adult nematomorphs, especially males, swim using whiplike undulations of the body. During copulation, a male forms and sustains tight coils around the female posterior end (Fig. 8–30A). If crowded together, nematomorphs coil together in masses that resemble the proverbial gordian knots. The nematomorph connective tissue compartment ("mesenchyme") is voluminous (Fig. 8–30C) and contains loose cells in a fibrous collagenous matrix similar to the parenchyma of large flatworms. In postreproductive adults, however, the cavity is an open fluid-filled pseudocoel. The digestive tract is vestigial, and the adults do not feed. The nervous system is composed of an anterior nerve ring and a non-ganglionated ventral cord. Excretory organs are absent, but some researchers have speculated that the vestigial midgut may function as a secretion kidney (p. 190), like an insect malpighian tubule (p. 840). The sexes are separate, and two long, cylindrical gonads extend the length of the body. As in nematodes, the sperm ducts empty into a cloaca, but there are no copulatory spicules (Fig. 8–30B). The oviducts also empty into the cloaca.

Hairworms live in all types of freshwater habitats in temperate and tropical regions of the world. Although some species copulate, others may deposit sperm on the body surface of the female rather than internally. Females deposit an egg string in the water or among plant roots on the shore. On hatching, the nematomorph larva has a protrusible proboscis armed with spines (Fig. 8–30D), which it uses to penetrate the host.

After hatching, the larvae actively penetrate or are ingested by an arthropod host living in water or along the water's edge. Common hosts are beetles, cockroaches, crickets, grasshoppers, centipedes, and millipedes. Recently, leeches have been reported as hosts. The rare human infections are probably accidental, but worms have, nevertheless, been recovered from human digestive and urogenital tracts, and larval hairworms have infected facial tissue and produced orbital tumors. The larvae of *Nectonema,* the only marine hairworms, parasitize hermit crabs, true crabs, and some shrimps. The larvae either penetrate the body wall of the host or are ingested as cysts and penetrate the gut wall. The young then enter the host's hemocoel, where development is completed. Their nutrition as parasites is apparently accomplished by direct absorption of food materials through the body wall. After several weeks to several months of development, during which a number of molts occur, the worms attain the form of adults. They emerge only when the host is near or in water. Sexual maturity is shortly attained during the free-living adult phase of the life cycle.

FIGURE 8–31 Three female gordioid nematomorphs, or horsehair worms. *(From Pennak, R. W. 1978. Freshwater Invertebrates of the United States. 2nd Edition. John Wiley and Sons, New York.)*

SYSTEMATIC RÉSUMÉ OF THE PHYLUM NEMATOMORPHA

Class Nectonematoida. Marine nematomorphs of the genus *Nectonema*. The 20-cm long body has lateral swimming setae and, internally, dorsal and ventral nerve cords and a fluid-filled pseudocoel. Larvae and juveniles parasitize decapod crustaceans.

Class Gordioida. Freshwater and semiterrestrial adults; juveniles parasitize insects. Body cavity as parenchyma in recently emerged adults (later a pseudocoel). Ventral nerve cord only. *Gordius, Paragordius, Chordodes.*

PHYLUM ROTIFERA

The phylum Rotifera contains the common animals known as rotifers, which, with protozoans and small crustaceans, dominate the freshwater zooplankton and are important in nutrient recycling in aquatic systems. Although some marine species exist and some species live in mosses, the majority inhabit fresh water. In many freshwater systems, rotifers commonly reach densities of 1000 individuals per liter, or more. Over 1500 species have been described, and most have a widespread distribution.

Most rotifers are .1 to 1 mm in length, only a little longer than ciliated protozoa. The body is composed of about 1000 cells, and the organ systems are eutelic. Most are solitary, free-swimming or crawling animals, but there are sessile (Fig. 8–32G, H) as well as some colonial species (Fig. 8–33E). Colonies, however, are essentially aggregations of solitary individuals, which have arisen as a result of parthenogenic reproduction. The body is usually transparent, although some rotifers appear green, orange, red, or brown, owing to coloration of the digestive tract.

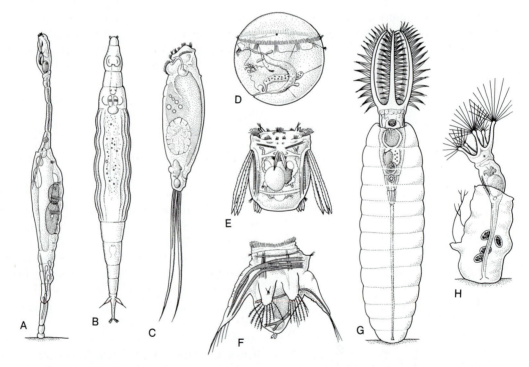

FIGURE 8–32 A, The seisonidean rotifer, *Seison.* **B,** The bdelloid rotifer, *Dissotrocha.* **C–H,** Monogonont rotifers: The rotifer *Monommata* **(C),** planktonic rotifers *Trochosphaera* **(D),** *Polyarthra* **(E),** *Hexarthra* **(F);** sessile rotifers *Stephanoceros* **(G)** and *Collotheca* **(H).** *(A, Redrawn from Hyman, L. H. 1951. The Invertebrates. Vol. 3. McGraw-Hill Book Co., New York. p. 106. B, Redrawn from Donner, J. 1966. Rotifers. Frederick Warne & Co., Ltd., London. p. 61. C–H, Redrawn from Ruttner-Kolisko, A. 1974. Plankton Rotifers. In Elster, H. -J., and Ohle, W. (Eds.): Die Binnengewässer 26 (Suppl.):1–146.)*

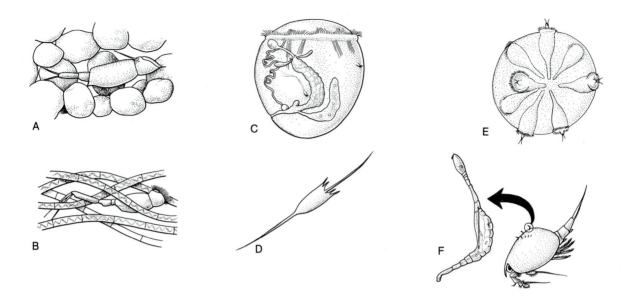

FIGURE 8–33 A, B, Interstitial rotifers, *Bryceella tenella* (**A,** in sand) and *Scaridium longicaudum* (**B,** in algae). Both have long slender bodies and well-developed adhesive toes. **C–E,** Planktonic rotifers. The voluminous pseudocoel of *Trochosphaera solstitialis* (**C**) and the gelatinous cuticle of *Conochilus unicornis* (**E**) may be adaptations to lower the body density and improve buoyancy. In *Kellicottia longispina* (**D**), on the other hand, long spines retard the rate of sinking. The marine ectoparasite, *Seison annulatus* (**F**), attaches securely to the carapace of certain shrimps, using a well developed toe. *(A–E, Redrawn from Ruttner-Kollisko, A. 1974. Plankton rotifers. In Elster, H.-J., and Ohle, W. (Eds.): Die Binnengewässer 26 (Suppl.):1–146.)*

External Structure

The elongated or saccate body, which is relatively cylindrical, can be divided in many species into a short anterior region, a large trunk composing the major part of the body, and a terminal foot (Fig. 8–34A, B). The body is often distinctly ringed, sculptured, or ornamented in various ways.

The broad or narrow anterior end bears a ciliated organ called the **corona,** which is characteristic of all members of the phylum. The corona is used in feeding and in swimming. The primitive corona is believed to have consisted of a large, ventral, ciliated area called the **buccal field** (Fig. 8–35), which surrounded the mouth. If rotifers evolved from a small, ciliated, creeping ancestor, as some believe, then the buccal field may represent a vestige of the ancestral ventral ciliation. From the buccal field, cilia extended around the anterior margins of the head to form a crownlike ring called the **circumapical band.** The

area devoid of cilia inside the ring is called the **apical field.**

The different types of coronas characteristic of different groups of rotifers are believed to have evolved from this basic plan. Various parts of the buccal field and the circumapical band have either been lost or become more highly developed. Not infrequently, certain cilia have become modified to form cirri, membranelles, or bristles. A few of the more common types of coronal ciliation are described here.

The buccal field in *Collotheca* and related species is modified into a funnel with the mouth at the bottom, and ciliation is reduced (Fig. 8–32H). The edges of the buccal field have become expanded, forming a varying number of lobes bearing long bristles, or setae, which may be arranged in bundles or tufts and are probably ciliary in structure.

In *Polyarthra* and related forms (Fig. 8–32E, F) the corona is derived entirely from the circumapical

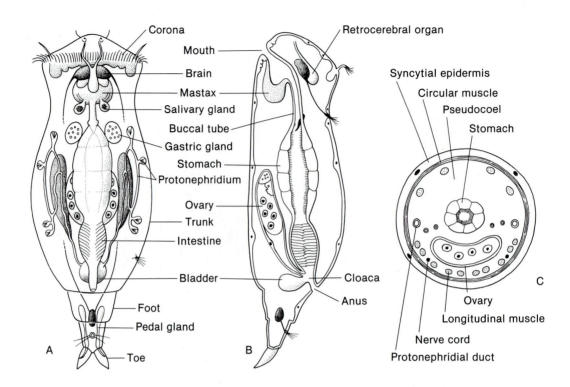

FIGURE 8–34 Rotifer anatomy: **A,** Dorsal view. **B,** Lateral view. **C,** Cross section. *(A, B, Redrawn from Remane, A. 1929–33. Rotatoria. In Bronn, H. G. (Ed.): Klassen Ordn, Tierreichs 2:1–576. C, Redrawn from de Beauchamp, P. 1965. Classè des Rotifères. In Grassé, P.-P. (Ed.): Traité de Zoologie. Vol. 4. Masson et Cie., Paris. p. 1235.)*

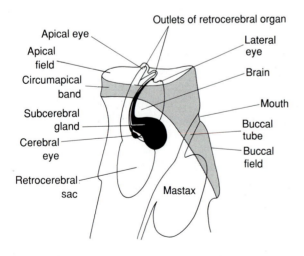

FIGURE 8–35 Primitive rotifer corona (lateral view). *(After de Beauchamp, from Hyman.)*

band, which has been transformed into two close circlets of modified cilia—an anterior band called the **trochus** and a posterior band called the **cingulum,** which passes below the mouth. The corona of bdelloid rotifers, which include many common species, also possess these two ciliated bands, but the anterior circlet of cilia (the trochus) is raised on a pedestal and divided into two discs, called **trochal discs** (Fig. 8–3A). The posterior circlet passes around the base of the pedestals and runs beneath the mouth. In living species, the beating membranelles of the trochal discs resemble two rotating wheels at the anterior of the body, and it is this type of corona from which the names *rotifer* and *wheel animalcule* are derived. The trochal discs function in swimming and in feeding, and the pedestals can be retracted when the discs are not in use.

The head in bdelloid and notommatid rotifers carries a middorsal projection called the **rostrum**

(Fig. 8–39). This little projection bears cilia and sensory bristles at its tip, and it is also adhesive. Other anterior structures in rotifers include the eyes (which vary in number and location), one or two short antennae, and the mucus-secreting retrocerebral organ (Figs. 8–34 and 8–35).

The trunk forms the major part of the body. The epidermis frequently is stiffened into a distinctive armor, called a **lorica** (Fig. 8–36). The lorica may be divided into distinct plates or ringlike sections and is usually ornamented with ridges, spines, or articulated appendages. The spines may be long, and in some rotifers they are movable. In *Polyarthra* the appendages are long, flat, skipping blades that are grouped in four clusters of three each (Fig. 8–32E).

The terminal portion of the body, or foot, is considerably narrower than the trunk region (Figs. 8–32A and 8–34A, B). The epidermis is frequently ringed, and in many bdelloids (Fig. 8–32A, B) the resulting segments or joints of the foot are able to telescope into similar larger joints of the trunk. Even the head may be retracted in this manner. The end of the foot usually bears one to four projections, called **toes.** In both the crawling and the sessile rotifers, the foot is used as an attachment organ; in these groups, the foot contains **pedal glands** that open by ducts at the tips of the toes or other parts of the foot. These pedal glands produce an adhesive substance for temporary attachment. Duo-gland organs (p. 207) have not yet been found in rotifers.

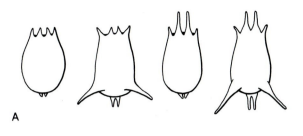

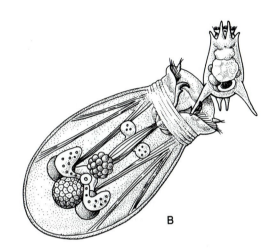

FIGURE 8–37 Cyclomorphosis in rotifers. **A,** Seasonal variability in spines of the planktonic rotifer *Brachionus,* a favorite prey of *Asplanchna* **(B),** An unidentified substance released by *Asplanchna* induces the offspring of *Brachionus* to grow long defensive spines. The spines increase the body size of *Brachionus* and make it difficult for *Asplanchna* to swallow them. The *Asplanchna* in **B** is engaged in a futile attempt to ingest a long-spined *Brachionus. (Modified and redrawn from Koste, W. 1978. Rotatoria. Gebrüder Bornträger, Berlin.)*

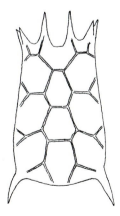

FIGURE 8–36 Lorica of *Keratella quadrata. (From Koste, W. 1978. Rotatoria. Gebrüder Bornträger, Berlin.)*

In a number of common bdelloids, such as *Philodina* (Fig. 8–3A) and *Rotaria,* the pedal glands open onto the ends of two long and diverging conical spurs located near the end of the foot. Functionally, the spurs replace the toes, which are very small in these genera. In planktonic rotifers, the foot is usually reduced or has disappeared altogether.

Many pelagic species undergo seasonal changes in body shape or proportions, a phenomenon known as **cyclomorphosis** and one that also takes place in small crustaceans (Fig. 8–37A). For example, individuals of certain species during one season of the year have spines that are longer or shorter than those developed by their descendents during another season. In *Brachionus calyciflorus,* spines in subsequent generations can be induced by starvation, low temperature, and by some substance produced by the predatory rotifer *Asplanchna.* Elongated spines protect *Brachionus* from being eaten by this predator (Fig. 8–37B).

Body Wall and Pseudocoel

Rotifers generally lack a cuticle. Instead, the epidermis is supported by a dense intracellular web of actin-like protein fibers. All of the so-called cuticular structures—spines, stiff body sections, and the lorica—are stiffened solely by the intracellular web (Fig. 8–38). Only the extracellular jelly coats and tubes of certain rotifers, such as *Stephanoceros* and *Collotheca,* are true cuticles (Fig. 8–32G, H). The epidermis is thin and syncytial and always possesses a species-specific number of nuclei (Fig. 8–34C). Secretory pores are scattered over the epidermal surface. Beneath the epidermis are the cross-striated body

muscles. Although some of the muscle fibers are circular (ring muscles) and some are longitudinal (retractor muscles) (Fig. 8–39), the body wall musculature is not organized into complete circular and longitudinal sheaths. A more or less spacious fluid-filled pseudocoel lies beneath the body wall and surrounds the gut and other internal organs (Fig. 8–34C).

Locomotion

Rotifers move by creeping over substrata, ciliary swimming (Fig. 8–40A), or skipping through water using specialized appendages (Fig. 8–40B). Most species of freshwater rotifers inhabit the substratum or live on submerged vegetation and other objects (Fig. 8–33). Some of these benthic species never swim, but many both creep and swim. In the common bdelloids, for example, the corona is retracted when the animal creeps, and the foot adheres to the substratum using the adhesive secretion produced by the pedal glands. The animal then extends the body, attaches the rostrum, and detaches the foot to move forward and again grip the substratum (Fig. 8–40C).

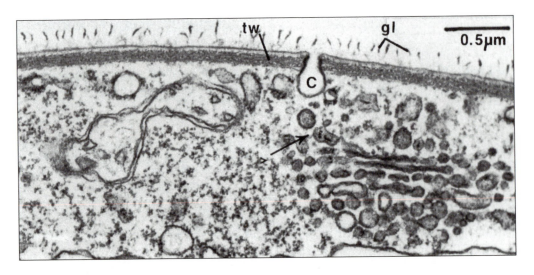

FIGURE 8–38 Electron micrograph of the syncytial epidermis of *Asplanchna sieboldi* showing the intracellular terminal web (tw) and one surface pore (C). The terminal web forms the skeleton of the body wall and is much thickened in loricate species, such as *Keratella.* (gl, surface coat of body wall) *(After Koehler from Ruppert, E. E. 1991. Introduction to the aschelminth phyla: A consideration of mesoderm, body cavities and cuticle. In Harrison, F. W., and Ruppert, E. E. (Eds.): Microscopic Anatomy of Invertebrates. Vol. 4. Wiley-Liss, New York. p. 14.)*

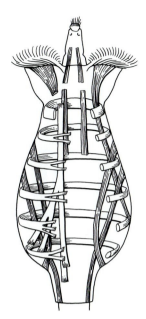

FIGURE 8–39 Musculature of *Rotaria* in ventral view. *(Redrawn after Brakenhoff from Hyman, L. H. 1951. The Invertebrates. Vol. 3. McGraw-Hill Book Co., New York. p. 83.)*

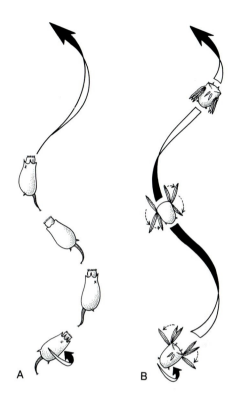

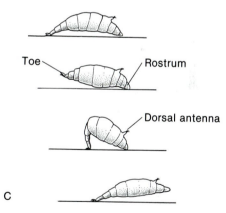

During swimming, which is only for short distances, the corona is extended, and the foot is retracted. Swimming speeds are typically 1 mm/s, but a few species, such as *Polyarthra* and *Hexarthra,* skip forward at velocities up to 35 mm/s using rapid appendage movements.

Many species of sessile rotifers attach to vegetation and display a remarkable restrictiveness, not only to the species of alga or plant to which they attach but also to the site of attachment on the plant. The primitive *Seison* occurs only on the gill cover of the marine crustacean *Nebalia* (Figs. 8–32A; 8–33F). Many species of the order Flosculariaceae live in vaselike tubes, commonly composed of foreign particles embedded within a secreted material (Fig. 8–32G, H).

Pelagic rotifers swim continually. Usually, the body is globose, the body wall thin and flexible, and the pseudocoelomic volume large (Fig. 8–32D); oil droplets may occur to further lower the density of the animal. Long spines may be present to retard the rate of sinking, and the foot and toes may be absent (Fig. 8–33D). Among the many strictly pelagic species are a few colonial rotifers, such as *Conochilus,* whose members resemble trumpets radiating from a common center (Fig. 8–33E). The combined ciliary action of the coronae propels the colony through the water.

FIGURE 8–40 A, Ciliary swimming in *Brachionus pala* at approximately 1 mm/s. **B,** Springtail-like skipping using muscular movements of appendages in *Polyarthra.* Each skip moves the animal 12 body lengths at a velocity of 35 mm/s. **C,** Inchworm-like locomotion in a bdelloid rotifer. Alternate attachments are made with the toe and rostrum. *(A, After Viaud from Hyman, L. H. 1951. The Invertebrates. Vol. 3. McGraw-Hill Book Co., New York. p. 143. B, Greatly modified from Gilbert, J. J. 1985. Escape response of the rotifer Polyarthra: A high-speed cinematographic analysis. Oecologia. 66:322–331.)*

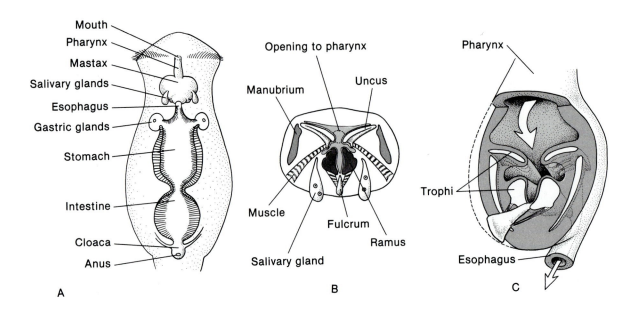

A

B

C

FIGURE 8–41 A, Anatomy of the digestive system in dorsal view. **B,** Enlargement of mastax in **A. C,** Three-dimensional view of mastax and trophi. *(A, Redrawn after Remane from Ruttner-Kolisko, A. 1974. Plankton Rotifers. In Elster, H.-J., and Ohle, W. (Eds.): Die Binnengewässer 26 (Suppl):8.)*

Nutrition

The mouth of the rotifer is typically ventral and is usually surrounded by some part of the corona (Fig. 8–34B). The mouth may open directly into the pharynx, or a ciliated buccal tube may be situated between the mouth and the pharynx, as in suspension feeders (Fig. 8–41A). The pharynx, or **mastax** (Figs. 8–34A, B and 8–41), is characteristic of all rotifers, and its structure is a distinguishing feature of the phylum. The mastax is usually oval or elongated and highly muscular. The inner epithelium bears seven large, interconnected, projecting pieces, or **trophi,** composed of an acid mucopolysaccharide material (Fig. 8–41B, C). It is not certain whether the trophi are intracellular or extracellular structures. The mastax is used both in capturing and in triturating food, and its structure therefore varies considerably, depending on the type of feeding behavior involved (Fig. 8–42).

Most rotifers are either suspension or raptorial feeders, although the latter group is rather omnivorous. The suspension feeders, of which the bdelloids are the most notable examples, feed on minute organic particles that are brought to the mouth in the water current produced by the coronal cilia. In the bdelloids, which have trochal discs, the larger, preoral cilia (trochus) produce the principal water current, which is directed backward, and function in both swimming and feeding. Food particles brought in by the water current are swept by both preoral and postoral cilia into a food groove that lies between them. Food groove cilia then carry the particles around to the mouth. The mastax of suspension feeders (Figs. 8–42B) is adapted to grinding; two of the pieces are extremely large, platelike, and ridged.

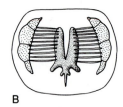

A B

FIGURE 8–42 A, Incudate trophi of *Asplanchna* used for seizing prey. **B,** Malleoramate trophi of *Filinia* used for grinding. *(Redrawn from Grassé.)*

The two plates oppose each other, and the ridges form a surface for grinding. The mastax of suspension feeders probably also acts as a pump, sucking in particles that have collected at the mouth. Food intake can be regulated in various ways. In *Brachionus,* for example, the ciliated buccal field can be screened or uncovered by certain large coronal cirri, the buccal field's ciliary beat can be reversed, or the mastax can reject particles.

The carnivorous species, which feed on protozoa, rotifers, and other small metazoan animals, capture their prey by trapping or suction. The forceps-like trophi of suction feeders are used to hold or manipulate prey once it is in the mastax cavity (Fig. 8–42A). After the prey is broken up, the indigestible parts of its body are commonly discarded.

Rotifers, such as *Collotheca* and other forms that possess funnel-like buccal fields (Fig. 8–32H), capture prey in a manner similar to the insectivorous flower, Venus's flytrap. When small protozoa accidentally swim into the funnel, the setae-bearing lobes of the funnel fold inward, preventing escape. The captured organisms are then sucked into the foregut by the mastax, which functions as a pump and is called the proventriculus. The mastax of a trapping rotifer is often very much reduced.

Some rotifers, especially notommatids, feed on the cell contents of filamentous algae. In the bdelloid, *Henoceros,* the mastax and trophi can be extended out from the mouth to puncture algal cells.

A number of epizoic and parasitic rotifers live primarily on small crustaceans, particularly on the gills. Examples are the previously mentioned *Seison,* (Fig. 8–33F); *Zelinkiella,* which attaches to the gills of the polychaete, *Amphitrite,* and the tube feet of some brittle stars. Endoparasitic species inhabit snail eggs, heliozoans, the interior of *Volvox,* and the intestine and coelom of earthworms, freshwater oligochaetes, and slugs. One genus, *Proales,* is parasitic within the filaments of the freshwater alga, *Vaucheria,* and produces gall-like swellings. In parasitic rotifers either the foot or the mastax becomes modified as an attachment organ, and the corona is reduced.

Located in the mastax walls of most rotifers are enzymatic glandular masses called **salivary glands** (Fig. 8–41A), which open through ducts just in front of the mastax proper. A tubular esophagus connects the pharynx with the stomach. At the junction of the esophagus and stomach is a pair of enzyme-secreting **gastric glands** (Fig. 8–41A), each of which opens by a pore into each side of the digestive tract. The digestive and absorptive stomach is a large sac or tube that passes into a short intestine. It is cellular and ciliated in the monogonont rotifers but syncytial, with or without a permanent lumen, in the bdelloids. The excretory organs and the oviduct also open into the terminal end of the intestine, which thus functions as a cloaca. The anus opens to the dorsal surface near the posterior of the trunk (Figs. 8–34B, 8–41A).

An intestine and anus are absent in the large predatory species of *Asplanchna.* In some of the sessile tube-dwelling species, such as *Collotheca* and *Stephanoceros* (Fig. 8–32G, H), the anus has shifted anteriorly to allow egestion of wastes over the lip of the tube.

Excretion

Rotifers are ammonotelic animals. There are typically two protonephridia in the pseudocoel, one on each side of the body. Each protonephridium has few to many terminal cells, which discharge into a collecting tubule. The two collecting tubules empty into a bladder (Fig. 8–34A), which opens into the ventral side of the cloaca; in the bdelloids, there is a constriction between the stomach and the intestine, and the somewhat bulbous cloaca acts as a bladder (Fig. 8–3A). The contents of the bladder or cloaca are emptied through the anus by contraction. The pulsation rate is often between one and four times per minute.

The protonephridia of rotifers function in osmoregulation and ionic regulation. The excreted fluid is hypoosmotic to the fluid of the pseudocoel, and the rate of bladder discharge is determined by the ionic content of the environmental medium. The high rate of discharge from the bladder suggests that some fluid may be entering the body through the mouth in feeding.

Most terrestrial rotifers are associated with soil, leaf mold, mosses, and lichens and are active only during the short periods when these surfaces are filled with water. During this time these terrestrial rotifers swim about in the water films. They are capable of undergoing desiccation, usually without the formation of cysts, and can remain in a dormant, highly resistant state (cryptobiosis) for as long as three to four years (p. 280; Fig. 8–3A, B).

Nervous System

The brain consists of a dorsal ganglionic mass lying over the mastax (Fig. 8–43), which gives rise to a varying number of nerves that extend to the anterior sense organs and to other parts of the body. The sense organs consist of sensory bristles in various parts of the ciliated crown (Fig. 8–43), an often conspicuous dorsal antenna (Fig. 8–43), and one or two cerebral eyes or a pair of anterior eyes or both (Fig. 8–35). The eyes are simple pigment-cup ocelli composed of one to two photoreceptor cells plus an accessory, red-pigmented cell. The eyes provide for locomotor orientation to light and for photic regulation of reproduction.

Reproduction

The great ecological success of rotifers is correlated in large part with their reproductive adaptations. Rotifers are either dioecious or parthenogenetic females. Among the dioecious species, the males are always smaller than the females (except in Seisonidea), and their nonreproductive organs are often degenerate.

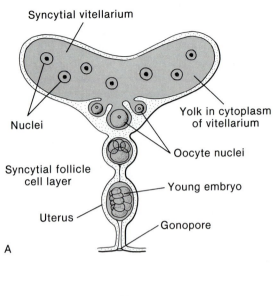

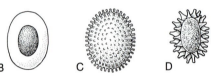

FIGURE 8–44 A, Germovitellarium (ovary) of *Asplanchna brightwelli.* **B,** Rapidly developing summer egg of *Ploeosoma hudsoni.* **C,** Resting egg of *Rhinoglena frontalis.* **D,** Resting egg of *Ploeosoma truncatum. (A, Modified and redrawn from Bentfield, M. E. 1971. Studies of oogenesis in the rotifer Asplanchna. I. Fine structure of the female reproductive system. Z. Zellforsch. 115:165–183. B–D, redrawn from Ruttner-Kolisko, A. 1974. Plankton Rotifers. In Elster, H.-J., and Ohle, W. (Eds.): Die Binnengewässer. 26 (Suppl.):59, 107.)*

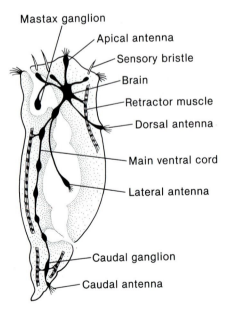

FIGURE 8–43 Nervous system and sensory structures (lateral view). *(Redrawn after Remane from Ruttner-Kolisko, A. 1974. Plankton Rotifers. In Elster, H.-J., and Ohle, W. (Eds.): Die Binnengewässer 26 (Suppl):10.)*

Parthenogenesis is characteristic of most groups of rotifers. Among most parthenogenic species, males are present in the population only at certain times; in the bdelloids no males have ever been reported. Thus, females make up all or most of the population of rotifers usually encountered.

The female reproductive system in the majority of species consists of one or two ovaries located anteriorly in the pseudocoel. Each syncytial ovary, called a **germovitellarium,** consists of a germinal region, where the oocyte nuclei are located, and a yolk-producing vitellarium (Fig. 8–44A). After yolk accumulates around an oocyte nucleus, the nucleus and its yolk reserve pinch off from the syncytium as a mature egg and pass through an oviduct into the cloaca or to a genital pore if there is no intestine.

The male is short-lived, and the gut is vestigial or absent. A single, saclike testis and a ciliated sperm duct (Fig. 8–45A) are present. Because a gut is absent in the male, the sperm duct runs directly to a gonopore that is homologous to the anus in the female and has the same position. Two or more glandular masses called accessory (prostate) glands are associated with the sperm duct, and the end of the sperm duct is usually modified to form a copulatory organ. Each sperm has an anterior flagellum, which pulls the cell forward as it moves (Fig. 8–45C).

Copulation is by hypodermic impregnation or by insertion through the cloaca (Fig. 8–45B). In the planktonic *Asplanchna*, the penis adheres to the female epidermis, and the male may be pulled about by the female. The mechanism by which penetration takes place is still uncertain. In *Brachionus*, penetration occurs only in the softer coronal region. In some species copulation occurs within a few hours of hatching, when the female's epidermis is still relatively thin and unstiffened.

Each ovarian nucleus becomes incorporated into one egg, and because only some 8 to 20 germinal nuclei exist in most species, there is a corresponding limit to the number of eggs produced in the lifetime of a particular female. Each egg is surrounded by a shell and a number of egg membranes, all of which are secreted by the egg itself. The eggs may be free-floating, attached to objects on the substratum, or attached to the body of the female. A few rotifers, such as *Asplanchna* and *Rotaria*, brood their eggs internally.

In the monogonont rotifers, several types of eggs are produced (Fig. 8–44B). One type, called an **amictic egg,** is thin-shelled, cannot be fertilized, and develops by parthenogenesis into amictic females. Typical meiosis does not take place in maturation, and the eggs are diploid. A second type of egg, called a **mictic egg,** is also thin-shelled but is haploid. If these eggs are not fertilized, they produce haploid males parthenogenetically. If they are fertilized, they secrete heavy, resistant shells. Such fertilized eggs are called resting, or **dormant eggs** (Fig. 8–44C, D); in contrast to the thin-shelled, unfertilized amictic and mictic eggs, which hatch in several days, these dormant

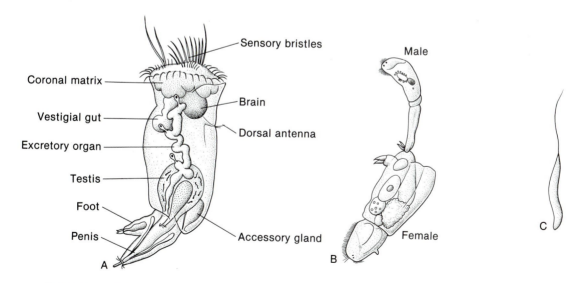

FIGURE 8–45 A, Anatomy of dwarf male of *Brachionus calyciflorus* in lateral view. **B,** Cloacal copulation in *Cephalodella catellina*. **C,** Typical sperm with leading flagellum (top). *(A, Redrawn after de Beauchamp from Ruttner-Kollisko, A. 1974. Plankton Rotifers. In Elster, H.-J., and Ohle, W. (Eds.): Die Binnengewässer 26 (Suppl.):14. B, Redrawn after Wulfert from de Beauchamp, P. 1965. Classe des Rotifères. In, Grassé, P.-P. (Ed.): Traité de Zoologie. Vol. 4. Masson et Cie., Paris, p. 1266.)*

eggs are capable of withstanding desiccation and other adverse conditions and may not hatch for several months or even years. Dormant eggs hatch into females. At any one time, a female may produce amictic or mictic eggs, but not both, thus the type of egg produced appears to be determined at the time of oocyte development.

Bdelloid rotifers, in which males are unknown and development is parthenogenetic, produce egg nuclei mitotically, and the eggs hatch into females only.

Parthenogenesis and the production of two kinds of eggs are probably adaptations for life in fresh water, especially temporary ponds and streams (p. 208).

The reproductive pattern of such rotifers tends to be cyclic. After spring rains, with the advent of warmer temperatures, dormant eggs that have passed through the winter hatch into amictic females. These females produce a number of generations of parthenogenetic females, each having a life-span of one to two weeks. Some species can double their population every two days. In the late spring or early summer, when this population reaches a peak, mictic eggs are produced and males appear (Fig. 8–46). Dormant eggs carry the species through until the next season and, if the pond or stream dries up, can be dispersed by birds or blown with dust. Rotifers inhabiting large, permanent bodies of fresh water may display a number of cycles or population peaks during the warmer months or may be present during the whole year. The production of mictic eggs is induced by specific environmental factors, such as high population density or changes in the amount or kind of food, photoperiod, and temperature, but the importance of different factors varies from species to species.

Development

Cleavage in rotifers is a kind of spiral cleavage and is determinate, but it is modified from the typical spiralian pattern. Nuclear division is completed early in development and never occurs again. In free-moving species, no larval development takes place. When the females hatch, they have all the adult features and attain sexual maturity after a growth period of a few days. For example, the common bdelloid rotifer, *Philodina roseola,* has an average life span of 48 days, and the adult is 28 times heavier than a newly hatched individual. A female lays 45 eggs, and the

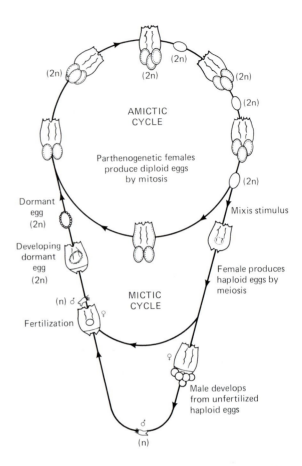

FIGURE 8–46 Life cycle of a monogonont rotifer. This particular species, *Brachionus leydigi,* lives in temporary ponds. Dormant eggs hatch with melting snows and spring rains to begin a first amictic cycle. Stagnating water stimulates the production of mictic eggs, and dormant eggs carry the species through the summer, when the pond dries. With autumn rains, there is a second amictic cycle. Frost stimulates the production of mictic eggs, and dormant eggs carry the species over the winter. *(Modified from Koste, W. 1978. Rotatoria. Gebrüder Bornträger, Berlin, p. 34.)*

generation time is four days. The smaller male rotifers do not undergo a growth period but are sexually mature when they leave the egg. The sessile rotifers hatch as free-swimming "larvae" that are structurally very similar to free-swimming species. After a short period they settle down, attach, and assume the characteristics of the sessile adults.

SYSTEMATIC RÉSUMÉ OF PHYLUM ROTIFERA

Class Seisonidea. A single genus of marine rotifers commensal on certain crustaceans. Elongated body with reduced corona. Little sexual dimorphism; gonads paired in both sexes; ovaries lack vitellaria. *Seison.*

Class Bdelloidea (Digononta). Anterior end retractile and usually bearing two trochal discs. Mastax adapted for grinding, with one pair of flattened trophi. Germovitellaria paired. Telescopic cylindrical body. Swimming and creeping species. Parthenogenesis only; males absent. *Philodina, Embata, Rotaria, Adineta.*

Class Monogononta. Rotifers with one ovary. Parthenogenesis common, but periodic sexual reproduction does occur involving dwarf, nonfeeding males. Mastax, if adapted for grinding, not designed as in the bdelloids.

> **Order Ploima.** Swimming rotifers. Body with or without a lorica, often short, sometimes saclike. This order contains the majority of rotifers. *Notommata, Proales, Polyarthra, Synchaeta, Chromogaster, Gastropus, Asplanchna, Brachionus, Euchlanis, Keratella.*
>
> **Order Flosculariacea.** Sessile, many tubicolous; or free-swimming, toeless rotifers. Corona with a double wreath of cilia. *Conochilus, Floscularia, Hexarthra, Testudinella.*
>
> **Order Collothecacea.** Mostly sessile rotifers. Mouth at the bottom of a shallow concavity. Anterior end often surrounded by arms or bundles of setae. *Stephanoceros, Collotheca.*

PHYLUM ACANTHOCEPHALA

The acanthocephalans are a phylum of some 1150 species of parasitic, wormlike aschelminths. All are endoparasites requiring two hosts to complete the life cycle. The juveniles are parasitic within crustaceans and insects; the adults live in the digestive tracts of vertebrates, especially fish. The body of the adult is elongated and composed of a **trunk** and a short, anterior spiny **proboscis** joined to the trunk by a neck (Fig. 8–47A). Most acanthocephalans are white, but some range from red to brown in color. They are commonly a few millimeters in length, although one species, *Macracanthorhynchus hirudinaceus,* the spiny-headed worm of pigs, may attain a length of 80 cm. The proboscis is covered with recurved spines (Fig. 8–47B), hence the name *Acanthocephala* meaning "spiny head."

The acanthocephalan proboscis and neck can be retracted into a muscular **proboscis sac** in the anterior of the trunk. The trunk is frequently covered with spines and in some is divided into superficial segments (8–47A). The retractable proboscis and anchoring spines provide the means of attachment in the host's gut and also enable the acanthors (larvae) to move within the host. To either side of the neck an invagination of the body wall projects posteriorly into the trunk pseudocoel. These two invaginations, called **lemnisci** (Fig. 8–47C), are filled with fluid and function as part of the hydraulic system in proboscis eversion, or in fluid transport to and from the proboscis.

The acanthocephalan epidermis is a thick syncytial epithelium. A cuticle is absent, but, as in rotifers, there is a well developed web of protein filaments within the epidermis that helps to support and toughen the integument. The web is penetrated by numerous branches of pore ducts. The cytoplasm beneath the web and pore ducts surrounds the large basal nuclei and contains bundles of vertical fibers that anchor the epidermis to the underlying basement membrane (Fig. 8–48). The proboscis hooks (Fig. 8–47D) originate from and remain attached to the epidermal basement membrane. The epidermal nuclei of some acanthocephalans are numerically constant, few in number, and very large (up to 5 mm in diameter), and they often produce conspicuous bulges in the body wall (Fig. 8–47C, D).

A unique circulatory system, the **lacunar canal system,** is located within the epidermis (Fig. 8–47A, D). The lacunar system occurs throughout the epidermis and the lemnisci, but the major canals are dorsal and ventral longitudinal canals (in most acanthocephalans) or lateral longitudinal canals (in eoacanthocephalans). The canals are believed to be filled with fluid, but the composition of the fluid and its manner of circulation are unclear. In electron micrographs of the lacunae, each canal is simply an organelle-free part of the epidermal cytoplasm. The contents of the lacunar canal is thus fluid cytoplasm. Respiratory pigments are absent from the lacunar fluid.

Beneath the epidermis is a basal lamina and connective tissue dermis (a circular and a longitudinal musculature), both of which are composed of smooth

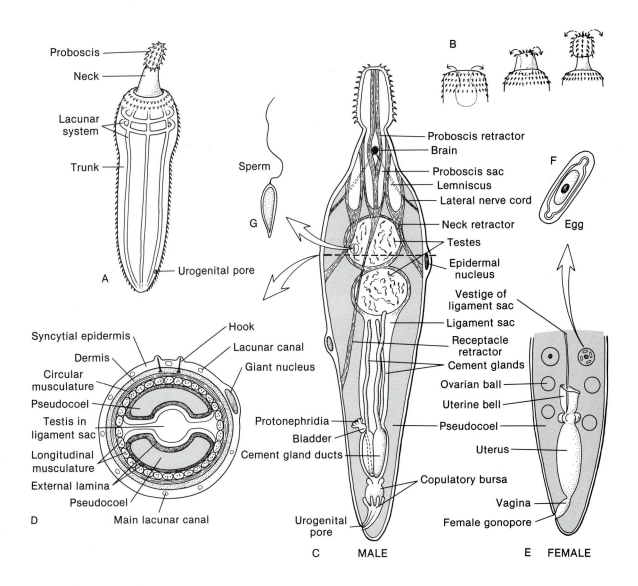

FIGURE 8–47 A, External view of a generalized acanthocephalan showing lacunar canals in the epidermis. **B,** Eversion of the spiny proboscis. **C,** Anatomy of a male acanthocephalan. **D,** Cross section through the trunk of a male. **E,** Posterior anatomy of a female. **F, G,** Egg and sperm (flagellum is anterior).

muscles (Fig. 8–47D). Both muscle layers are embedded in a thick fibrous extracellular matrix, which lines the fluid-filled body cavity—a pseudocoel. A hollow, elongated sac, the **ligament sac,** is suspended in the pseudocoel from an anterior attachment on or near the proboscis sac to the posterior accessory reproductive organs. There is one median sac in paleacanthocephalans (Fig. 8–47D) and two (a **dorsal ligament sac** and a **ventral ligament sac**) in archi- and eoacantho-

cephalans. The wall of the ligament sac is composed of connective tissue. The ligament sacs house the gonads and open posteriorly into the cavities of the accessory reproductive organs (Fig. 8–47C, D).

Acanthocephalans lack a digestive system; food is absorbed directly through the body wall from the host. In the absence of a gut, dorsal and ventral body surfaces are difficult to define with certainty. The brain and gonopore are viewed as ventral by

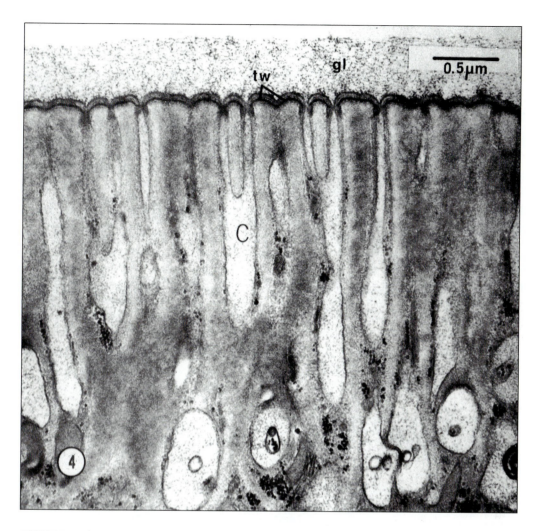

FIGURE 8–48 Electron micrograph of the outer part of the syncytial epidermis of the acantho-cephalan, *Moniliformis dubius.* As in rotifers, the epidermis is supported apically by a terminal web (tw) of proteinaceous filaments and bears a number of invaginations (C), which open, via pores, onto the surface of the body. Lacunar canals are not visible in this photograph (gl, extra-cellular surface coat on epidermis). *(After Byram and Fisher from Ruppert, E. E. 1991. Introduction to the aschelminth phyla: A consideration of mesoderm, body cavities, and cuticle. In Harrison, F. W., and Rup-pert, E. E. (Eds.): Microscopic Anatomy of Invertebrates. Wiley-Liss, New York. p. 14.)*

most specialists—an interpretation that seems un-usual considering that the brain is dorsal in virtually all other bilaterally symmetrical animals. Two pro-tonephridia, which open into the reproductive ducts, are present in one family (Fig. 8–47C). The nervous system is composed of an anterior brain, two princi-pal lateral nerves and many small specialized nerves.

The sexes are separate, and the gonopore is lo-cated at the end of the trunk. The male system in-

cludes a protrusible penis, and fertilization is internal following copulation. Acanthocephalan sperm, like those of rotifers, have an anterior leading flagellum (Fig. 8–47G). In females, the ovary breaks up into fragments, called **ovarian balls,** which are free in the pseudocoel (Fig. 8–47E). Development of the egg takes place within the female pseudocoel and results in the formation of encapsulated larvae. Each larva has an anterior crown or rostellum with hooks. The larvae are removed from the pseudo-coel by a funnel-shaped structure, called a **uterine bell,** are released into the vertebrate host's intestine, and are passed to the exterior in the host's feces. If the capsules are eaten by certain insects, such as roaches or grubs, or by aquatic crustaceans, such as amphipods (Fig. 8–49), isopods, or ostracods, the larva emerges from the capsule, bores through the gut wall of the host, and becomes lodged in this interme-diate host's hemocoel. The larva, called an **acanthor,** possesses a rostellum with hooks that are used in pen-etrating the host's tissues. Once in the intermediate host, the acanthor passes through an **acanthella** stage and becomes an encysted **cystacanth.** When the in-termediate host is eaten by a fish, bird, or mammal, the cystacanth excysts and the juvenile worm attaches to the intestinal wall of the vertebrate host by using the spiny proboscis. Compared with tapeworms and roundworms, acanthocephalans are rare parasites, but some vertebrate hosts nevertheless have heavy para-

site loads that can do considerable damage to the in-testinal wall. As many as 1000 acanthocephalans have been reported in the intestine of a duck, and 1154 in the intestine of a seal.

Although all modern acanthocephalans are para-sites, one extinct species, *Ottoia prolifica,* from the mid-Cambrian Burgess shale may have been a free-living predator or scavenger. The assignment of *Ot-toia* to this phylum, however, is controversial and many paleontologists regard it as a fossil priapulid.

SYSTEMATIC RÉSUMÉ OF THE PHYLUM ACANTHOCEPHALA

Class Archiacanthocephala. Parasites of birds and mammals; centipedes, millipedes, and insects are intermediate hosts. Main lacunar canals dorsal and ventral. Ligament sacs break down in repro-ductive females. Protonephridia present in one family. *Macracanthorhynchus, Moniliformis, Oli-gacanthorhynchus.*

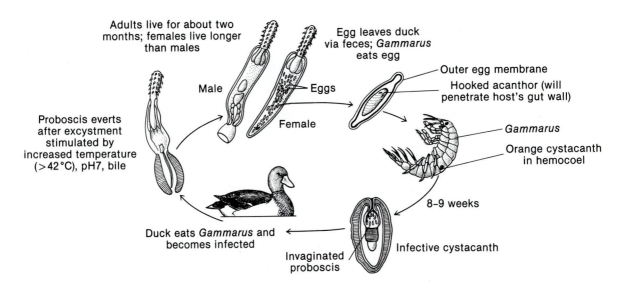

FIGURE 8–49 Life cycle of *Polymorphus minutus,* an acanthocephalan that parasitizes the gut of wildfowl and domestic ducks. The eggs of *Polymorphus* pass out of the duck with feces and enter the water. The eggs are then eaten by the amphipod, *Gammarus.* A larval acanthor hatches from the egg in the shrimp's gut and bores through the gut wall into the hemocoel. After passing through an acanthella stage, the parasite becomes an encysted cystacanth, which is bright orange and can be seen through the transparent body wall of the amphipod. Ducks are re-infected when they eat the amphipods. *(From Lyons, K. M. 1978. The Biology of Helminth Parasites. Edward Arnold Publishers, Ltd., London, p. 7.)*

Class Eoacanthocephala. Parasites of fishes and some amphibians and reptiles; crustaceans are intermediate hosts. Main lacunar canals are dorsal and ventral. Ligament sacs persist in reproductive females. *Neoechinorhynchus.*

Class Palaeacanthocephala. Largest and most diverse group of acanthocephalans. Parasites of all vertebrate classes; intermediate hosts usually crustaceans. Main lacunar canals lateral. Ligament sac breaks down in reproductive females. *Acanthocephalus, Echinorhynchus, Leptorhynchoides, Polymorphus.*

PHYLUM KINORHYNCHA

The phylum Kinorhyncha, formerly known as Echinoderida, consists of a small group of some 150 described marine aschelminths that burrow in the surface layer of marine mud or live in the interstitial spaces of marine sand. They have been found from the intertidal zone to depths of several thousand meters. The members of this phylum are usually less than 1 mm in length. The short grublike body is flattened ventrally, like that of gastrotrichs, but kinorhynchs lack locomotory cilia and except for the lack of paired appendages resemble interstitial copepod crustaceans, with which they are often confused by beginners.

A distinguishing feature of the phylum is the division of the chitinous cuticle into clearly defined **segments** (zonites). The **head** composes the first segment, the **neck** makes up the second segment, and the trunk in all adults contains 11 segments (Fig. 8–50A). In the males of *Pyncnophyes* and *Kinorhynchus,* a pair of adhesive tubes is commonly found on the ventral surface of segment four, but in *Echinoderes* they occur in males and females. In other species additional adhesive tubes may be present ventrally and laterally on various trunk segments.

The anteriorly located mouth is situated at the end of a protrusible cone, which is surrounded at the tip by styles and at the base by circlets of spines (scalids) (Fig. 8–50A). The entire head can be withdrawn either into the neck or into the first trunk segments, hence the name *Kinorhyncha,* meaning "movable snout." In the former case, the cuticular plates of the neck (placids) are adapted for closing over the retracted head. A similar closing apparatus is present on the first trunk segments in those species in which both head and neck retract into the trunk.

A thin cellular epidermis underlies and secretes the cuticle and bears segmental mucous glands. The musculature is located beneath the epidermal basal lamina. It consists of segmental bundles of longitudinal, diagonal, and dorsoventral muscles, all of which are cross-striated, like those of rotifers and arthropods. A network of circular and longitudinal muscles surround the esophagus and midgut. A fluid-filled pseudocoel is reduced and may be absent entirely in many species, but amoeboid cells of unknown function occur between some of the internal organs (Fig. 8–50B).

A kinorhynch burrows by an alternate eversion and retraction of its spiny head. The body moves forward during *eversion* of the head and remains stationary as it is retracted. As the head everts, the spines unfurl rearward and pull the animal forward (Fig. 8–50C). When the head is fully everted, a **mouth cone,** bearing a terminal mouth surrounded by styles, is protruded into the sediment.

Kinorhynchs feed either on diatoms or fine organic material, or both. The foregut is lined with a cuticle and consists of a filtering buccal cavity, a sucking pharynx composed of separate muscular and nonmuscular epithelial layers, and a short **esophagus,** which joins the midgut. The tubular midgut leads into a short, cuticle-lined hindgut before opening to the exterior via a terminal anus. The physiology of digestion is unknown.

Two protonephridia open to the lateral surface of the 11th segment (Fig. 8–50A). The nervous system, which is within the epidermis, consists of a ganglionated nerve ring around the anterior of the pharynx, and from this ring arises a midventral nerve cord with paired ganglia in each segment (Fig. 8–50A, B). The sense organs consist of the scalids, which contain ciliary receptors, and sensory bristles located over the general body surface, especially the trunk. A few species bear anterior ocelli of unusual structure.

Kinorhynchs are dioecious, with a pair of ovaries or testes and two gonopores on the last segment (Fig. 8–50A, B). The end of each sperm duct usually carries two to three penial spines or spicules. Copulation has never been observed, but species in two genera have been found to extrude spermatophores, which are directed toward a female with the penial spines. Sperm are stored in the female's seminal receptacles,

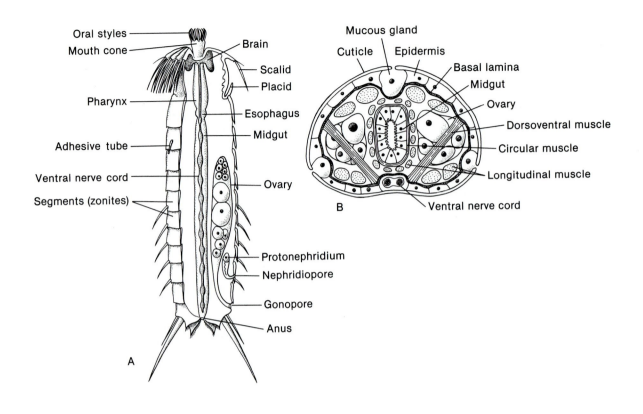

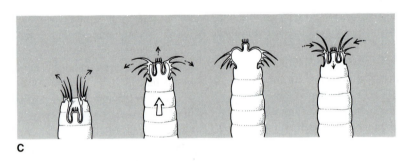

FIGURE 8–50 A, Anatomy of a generalized kinorhynch. **B,** Cross section of the trunk of a female kinorhynch. **C,** Introvert eversion and locomotion. A kinorhynch moves forward through sediment as the introvert is everted and the spines (scalids) curve rearward. The body does not move when the introvert retracts. *(A, B, Modified and redrawn from Kristensen, R. M., and Higgins, R. P. 1991. Kinorhyncha. In Harrison, F. W., and Ruppert, E. E. (Eds.): Microscopic Anatomy of Invertebrates. Wiley-Liss, New York. pp. 378, 379.)*

and fertilization is probably internal. The young look very much like adults but have fewer segments and may possess spine arrangements somewhat different from those of the adult. Periodic molts of the cuticle occur in the attainment of adulthood, after which molting typically does not occur.

SYSTEMATIC RÉSUMÉ OF THE PHYLUM KINORHYNCHA

Kinorhynchs are divided into two orders, rather than classes, for historical reasons. As in the gastrotrichs, the orders were defined at a time when kinorhynchs were believed to be a class of the phylum Aschelminthes. Although still designated as orders, each actually holds class rank within the Kinorhyncha.

Order (Class) Cyclorhagida. Most widely distributed and diverse group of kinorhynchs. Fourteen to 16 closing plates in the neck region of most species. Trunk spiny and circular to triangular in cross section. Common in marine muds and subtidal sands. *Echinoderes.*

Order (Class) Homalorhagida. Relatively large (up to 1 mm) kinorhynchs. Six to eight closing plates in the neck region. Trunk with few spines and distinctly triangular in cross section. Common in marine muds. *Pycnophyes.*

Phylum Loricifera

The Loricifera, described in 1983 by the Danish zoologist R. M. Kristensen, is the most recently discovered phylum of animals and bring to three the number of phyla discovered in this century—Pogonophora (1914), Gnathostomulida (1956), and Loricifera (1983). The loriciferan *Nanaloricus mysticus* is a minute animal, about a quarter of a millimeter in length, that lives in the interstitial spaces of shelly, marine gravel. Although specimens have been found off the coasts of North Carolina and Greenland and in the Coral Sea, the phylum Loricifera originally was based on a species of *Nanaloricus* collected from bottom samples taken off the coast of Roscoff, France, at a depth of 25 m. None of the nine loriciferan species have ever been observed alive, and current knowledge of them is based solely on preserved specimens.

The animals are believed to adhere so tightly to the gravel substrate that they cannot be extracted by conventional methods, which probably accounts for their late discovery. Most loriciferans have been collected by submerging the bottom sample briefly in fresh water. This shocks the animals osmotically, and they release.

Nanaloricus mysticus looks like a cross between a rotifer and a kinorhynch (Fig. 8–51). The major part of the body, called the abdomen, is encased within a cuticular **lorica** composed of a dorsal, a ventral, and two lateral plates. It is to the lorica that the name of the phylum refers (lorica bearer). The cone-shaped anterior end, or introvert, of the animal bears many recurved spines (scalids) on its lateral surface. The field of spines is continuous with that of the necklike thorax, which joins the introvert with the abdomen.

Both introvert and thorax can be retracted into the anterior end of the lorica (Fig. 8–51A).

Eight stylets surround the telescopic mouth cone at the end of the introvert, but their tips open to the side of the mouth and not within it. A buccal tube leads into a bulbous pharynx composed of myoepithelial cells, and a long midgut region forms most of the gut tube. A terminal rectum opens to the outside through a posterior anal cone (Fig. 8–51C). A large brain lies within the introvert, but little is known about the remainder of the nervous system. The sexes are separate, each with a pair of gonads. A protonephridium is associated with each gonad, reminiscent of the priapulids (p. 357), and it empties into the urogenital duct (Fig. 8–51C). At present little is known about the habits and physiology of loriciferans.

A larval stage, called a **Higgins larva,** is somewhat similar to the adult except that the buccal cone lacks stylets and the thorax lacks spines (Fig. 8–51B). The thorax can enclose the introvert but cannot itself be retracted into the lorica. The posterior end of the larval trunk carries a pair of toes, which are believed to function in swimming.

Phylum Tardigrada

The tardigrades are a group of very tiny but highly specialized animals known commonly as water bears. Some reach 1.2 mm, but the majority are no longer than 0.3 to 0.5 mm. Although not uncommon, tardigrades are seldom encountered unless looked for because of their minute size and rather specialized habitats.

A few marine interstitial tardigrades have been collected from both shallow and deep water. Also, there are some freshwater species that live in bottom detritus or on aquatic algae and mosses, and there are even species that live in the water films of soil and forest litter. Many tardigrades, however, live in the water films surrounding the leaves of terrestrial mosses and lichens, especially those that grow on stones and trees. This latter habitat is shared primarily with some bdelloid rotifers, and the two groups present many parallel adaptations.

Approximately 600 species of tardigrades have been described, many of which are cosmopolitan. A single fossil tardigrade has been described from Cretaceous amber.

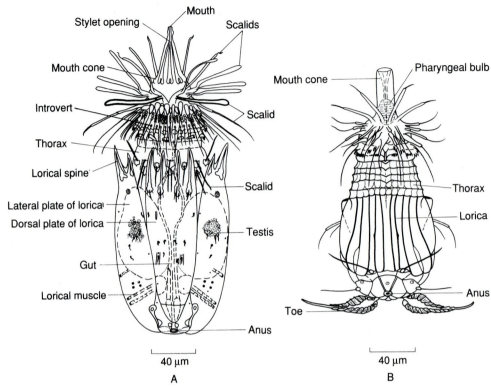

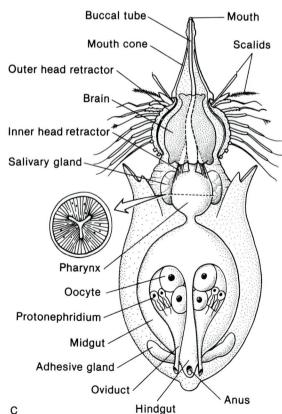

FIGURE 8–51 Organization of the Loricifera *(Nanaloricus mysticus)*. **A,** Dorsal view of adult male. **B,** Dorsal view of the Higgins larva. **C,** Internal anatomy of the adult (inset shows cross section of myoepithelial pharynx). *(A, B, From Kristensen, R. M. 1983. Loricifera, a new phylum with Aschelminthes characters from the meiobenthos. Z. zool. Syst. Evolut.-forsch. 21:163–180. C, Modified and redrawn from Kristensen, R. M. 1991. Lorcifera. In Harrison, F. W., and Ruppert, E. E. (Eds.): Microscopic Anatomy of Invertebrates. Wiley-Liss, New York. pp. 334, 335.)*

External Structure

The bodies of tardigrades are short, plump, and cylindrical, and there are four pairs of ventral, stubby legs (Figs. 8–52 and 8–53). Each leg terminates in four to eight claws or discs (many interstitial species) (Fig. 8–53). The body is covered by either a smooth or an ornamented cuticle, which in some tardigrades, such as *Echiniscus* and *Megastygarctides,* is divided into symmetrical segmental plates (Fig. 8–53B).

The cuticle contains chitin, mucopolysaccharides, proteins, and lipids, composing three primary layers (Fig. 8–54B). Periodically, the old cuticle is

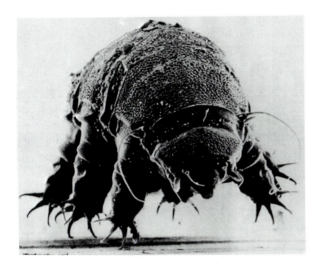

FIGURE 8–52 Anterior view of the eutardigrade, *Echiniscus arctomys*. Mouth is located at the end of the snoutlike, anterior projection. *(From Crowe, J. H., and Cooper, A. F. 1971. Cryptobiosis. Sci. Amer. 225:30–36.)*

shed, including the pharyngeal stylets, claws, and linings of the fore- and hindguts. The epidermis secretes a new body cuticle, but the stylets are replaced by the salivary gland and the claws by the claw glands (Fig. 8–54A). During molting, the body contracts, pulling away from the old cuticle, which is then slipped off and left behind as a relatively intact casing.

Internal Structure and Physiology

The musculature consists of separate muscle bands, each composed of a single, smooth or striated muscle cell, extending from one subcuticular point of attachment to another (Fig. 8–54A). The actual muscular attachment with the cuticle is via intraepidermal tonofilaments. Tardigrades move about slowly, crawling on their telescopic legs and using the hooks at the ends of the legs to grasp the substratum. The first three pairs of legs are used in forward locomotion, and the last pair function in retreat or prehension (Fig. 8–55).

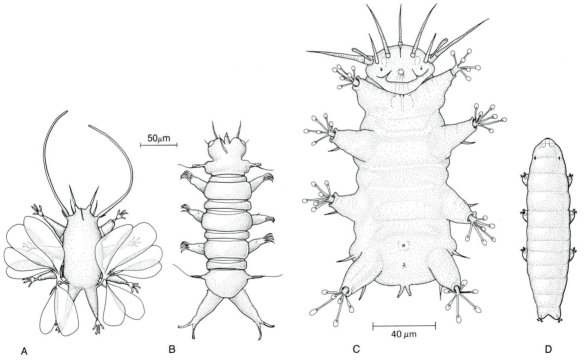

50μm

40 μm

A   B   C   D

FIGURE 8–53 A–C, The marine heterotardigrades, *Tanarctus velatus* **(A)**, *Megastygarctides orbiculatus* **(B)**, *Batillipes noerrevangi* **(C). D,** The eutardigrade, *Macrobiotus hufelandi*. *(A, B, From McKirdy, D., Schmidt, P., and McGinty-Bayly, M. 1976. Interstitielle Fauna von Galapagos. XVI. Tardigrada. Mikrofauna Meeresbodens. 58:1–43. C, From Kristensen, R. M. 1981. Sense organs of two marine arthrotardigrades. Acta Zool. 62:27–41; D, Redrawn after Higgins, R. P. from Nelson, D. R. 1982. Developmental Biology of the Tardigrada. In Harrison, F. W., and Cowden, R. R. (Eds.): Developmental Biology of Freshwater Invertebrates. Alan R. Liss, New York.)*

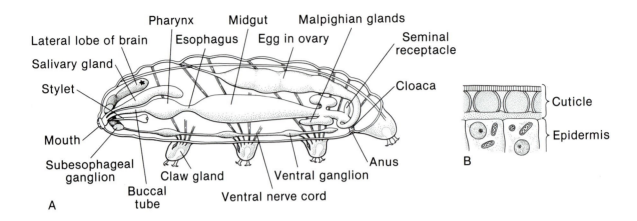

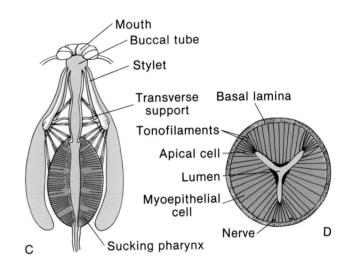

FIGURE 8–54 A, Lateral view of the eutardigrade, *Macrobiotus hufelandi*. **B,** Section of the cuticle and epidermis of *Batillipes noerrevangi*. **C,** Dorsal view of foregut apparatus of a tardigrade. **D,** Cross section of a tardigrade myoepithelial pharynx. *(A, Modified and redrawn from Cuénot, L. 1949. Les Tardigrades. In Grassé, P.-P. (Ed.): Traité de Zoologie. Vol. 6. Masson et Cie., Paris. B, Modified from Kristensen, R. M. 1976. On the fine structure of Batillipes noerrevangi Kristensen 1976. Zool. Anz., Jena 197: 129–150. C, Combined from Pennak, R. W. 1978. Fresh-Water Invertebrates of North America, 2nd Edition. John Wiley and Sons, New York, and Kristensen, R. M. 1982. The first record of cyclomorphosis in Tardigrada based on a new genus and species from Arctic meiobenthos. Z. zool. Syst. Evolut.-forsch. 20:249–270.)*

A fluid-filled pseudocoel extends between the muscle bands and the other internal organs (Fig. 8–54A). The pseudocoelomic fluid is colorless, but hemocytes circulate as the body moves. Correlated with their size is the lack of specialized gas exchange organs in tardigrades.

The majority of tardigrades feed on the contents of plant cells, which are pierced with a **stylet apparatus** resembling that of herbivorous nematodes and rotifers (Fig. 8–54C). Soil tardigrades feed on algae and probably detritus, and some are predators on nematodes and other minute soil animals, including other

tardigrades. The anterior mouth opens into a chitin-lined **buccal tube** and a bulbous pharynx composed of myoepithelial cells (Fig. 8–54C, D). The pointed ends of two stylets project into the anterior end of the buccal tube.

During feeding, the mouth is placed against the plant cells, and the stylets are projected to puncture the cell wall. The contents of the cell are then sucked out by the pharynx.

The pharyngeal bulb passes into the tubular esophagus, which opens in turn into a large midgut, or intestine (Fig. 8–54A, C), and in which digestion and absorption take place. The end of the intestine leads into a short cuticle-lined hindgut (rectum), which opens to the outside through the terminal anus.

At the junction of the intestine and rectum of some tardigrades are three large glands, sometimes called **malpighian glands,** that are thought to be excretory in function (Fig. 8–54A).

The nervous system consists of a dorsal brain and circumbuccal commissures, which join a pair of ventrolateral ganglionated nerve cords. Each pair of ganglia innervates a pair of appendages (Fig. 8–54A). The clear segmentation of the nervous system suggests that tardigrades are composed of a head and five

segments. The subesophageal ganglion probably innervates the stylets, which may be the modified claws of the original first segment and are now fused with the head. Tardigrade sensilla contain cilia and consist of bristles and spines, especially on the head. Most tardigrades also possess a pair of simple eyespots, each containing a single red- or black-pigmented cell.

Reproduction

Tardigrades are largely dioecious (three species are known to be hermaphrodites) and have a single saccular gonad, testis or ovary, located above the intestine. In the male, two sperm ducts open through a single median gonopore just in front of the anus. In the female, the single oviduct opens either above the anus or into the ventral side of the rectum, which then serves as a cloaca (Fig. 8–54A). In the latter case a small, adjacent seminal receptacle also opens into the rectum.

Females are often more numerous than males, and in some genera, such as *Echiniscus,* males are unknown. Mating and egg laying occur at the time of a molt (Fig. 8–56A). In some aquatic species sperm are deposited into an old female cuticle containing eggs; in most terrestrial forms, sperm are deposited into the female tract before the cuticle is completely shed, and fertilization occurs in the ovary.

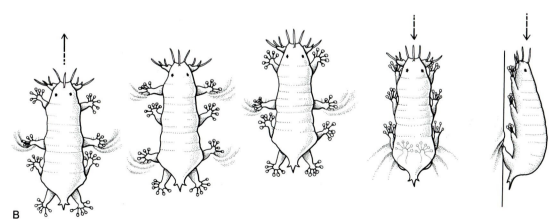

FIGURE 8–55 A, Tardigrades climbing on an algal filament. **B,** Locomotion over a surface (microscope slide) by a marine interstitial tardigrade *(Batillipes). (A, From Marcus, E. 1929. In Bronn, H. G. (Ed.): Klassen Ordn. Tierreichs. 5:156.)*

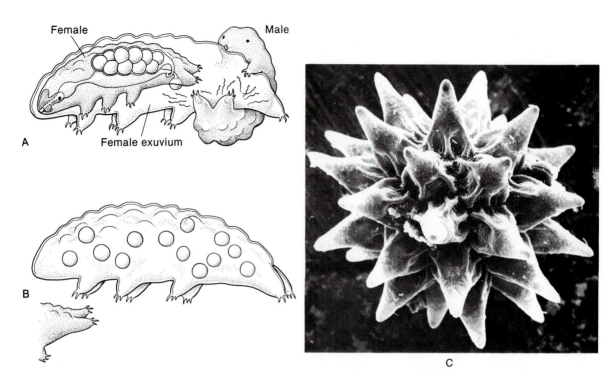

FIGURE 8–56 A, Copulation in the eutardigrade, *Hypsibius nodosus.* **B,** Female *Hypsibius* ovipositing in her own exuvium. **C,** Scanning electron micrograph of an attached dormant egg of *Macrobiotus tonollii. (A, B, Redrawn from Marcus, E. 1929. Tardigrada. In Bronn, H. G. (Ed.): Klassen Ordn. Tierreichs. 5:105, 108. C, From Nelson, D. R. 1982. Developmental biology of the Tardigrada. In Harrison, F. W., and Cowden, R. R. (Eds.): Developmental Biology of Freshwater Invertebrates. Alan R. Liss, New York. p. 382.)*

One to 30 eggs are laid at a time, depending on the species. Aquatic tardigrades either deposit them in the old cuticle or attach them singly or in groups to various objects (Fig. 8–56B). Like rotifers and gastrotrichs, some aquatic tardigrades produce thin-shelled eggs when environmental conditions are favorable and thick-shelled eggs when environmental conditions are adverse. The eggs of terrestrial species typically possess a thick, sculptured shell that can resist the frequent periods of desiccation to which mosses are subjected (Fig. 8–56C). Parthenogenesis is common.

Development is direct and rapid. Cleavage is holoblastic but does not follow a typical spiral or radial pattern. Development is completed within 14 days or less, and the little tardigrades hatch by breaking the shells with their stylets. Further growth is attained by increase in the size of cells rather than by the addition of cells, but mitoses have been reported in the adults. As a result, tardigrades may not be strictly eutelic. As many as 12 molts may take place in the life of a tardigrade, which has been estimated to last 3 to 30 months.

Moss-dwelling adult tardigrades, like their nematode and rotifer counterparts, enter cryptobiosis as the habitat dries (p. 280; Fig. 8–3C, D). Tardigrades can remain dormant for seven years or more, and after repeated periods of cryptobiosis, may lengthen their life span up to 70 years.

Phylogeny

The phylogenetic position of tardigrades is uncertain. The segmented cuticle, appendages, musculature, and nervous system, as well as the malpighian tubules and pseudocoel (equated with a hemocoel), are all reminiscent of those of arthropods. On the other hand, a pseudocoel also occurs in many aschelminths, and the tardigrade pharynx (myoepithelium) is organized like that of gastrotrichs, nematodes, and loriciferans and not like that of arthropods and many other phyla in which there are separate muscular and epithelial layers. Many of the tardigrade structural peculiarities are undoubtedly specializations for living in their restricted habitats.

SYSTEMATIC RÉSUMÉ OF THE PHYLUM TARDIGRADA

Class Heterotardigrada.   "Armored" marine and terrestrial tardigrades. Thickened dorsal cuticle divided into segmental plates. Conspicuous sensilla on head. Stubby legs terminate in claws or adhesive discs. Anus and gonopore separate. *Batillipes, Echiniscus, Stygarctus, Tanarctus.*

Class Eutardigrada.   "Naked" freshwater and terrestrial tardigrades. Thin, undivided dorsal cuticle. Head lacks conspicuous sensilla. Legs terminate in claws. Intestine and gonad open into a common cloaca. *Macrobiotus, Milnesium.*

PHYLOGENETIC RELATIONSHIPS OF ASCHELMINTHS

The aschelminths are clearly a diverse assemblage, and no distinctive features bind them together as a whole. For this reason, most zoologists are opposed to placing them within a single phylum. It was once believed that all aschelminths shared similar adhesive organs, an extracellular cuticle, and, especially, a body cavity called a pseudocoel. The pseudocoel was interpreted as an evolutionary grade between the complete absence of a fluid-filled cavity (acoelomate grade) and a cavity lined by a mesothelium (coelomate grade). Because of this interpretation, the aschelminths were seen as evolutionary intermediates between the primitive acoelomates and the evolutionarily derived coelomate animals. Nowadays, the close relationship of all aschelminths is doubted by many researchers, and their evolutionary standing in relation to other animals is less clear-cut primarily because of differences between the body cavities and cuticles of the aschelminths. In no case is the adult body cavity a persistent embryonic blastocoel, and in some cases, for example the gastrotrichs, a body cavity of any sort is absent. An extracellular cuticle may be well represented in phyla, such as nematodes, gas-

trotrichs, and kinorhynchs, but it is absent altogether in rotifers and acanthocephalans.

Within the aschelminth assemblage, the closest evolutionary ties appear to be between gastrotrichs and nematodes (and perhaps nematomorphs) on the one hand, and the rotifers and acanthocephalans on the other. Gastrotrichs and nematodes share several features, including a similar developmental pattern, pharynx, nervous system, muscle innervation, cuticle, and epidermal cords. Rotifers and acanthocephalans show similarities in sperm structure, early development, and organization of the epidermis. Both phyla lack a cuticle, and the syncytial epidermis is supported instead by a dense apical web of cytoplasmic fibers. The outer epidermal membrane of both phyla has permanent invaginations that lead blindly into a pocket or crypt. In acanthocephalans, the pores and crypts are sites of absorption of certain host nutrients. They are believed to be secretory in rotifers.

The relationships among the remaining aschelminths are unclear, but links are sometimes made between kinorhynchs, loriciferans, priapulids (p. 357), acanthocephalans, and sometimes nematomorphs based on the common occurrence of a retractile spiny head (larval nematomorphs). Differences in structural details of the heads, however, speak against some of the comparisons. For example, although all these phyla have spiny heads, the spines of acanthocephalans are not cuticle but develop instead from the basement membrane of the epidermis. The loriciferan body is reminiscent of that of larval priapulids (Figs. 9–20C and 8–51A), raising the possibility that loriciferans evolved via progenesis (p. 206) from the Priapulida. Because of their segmentation, the kinorhynchs and tardigrades could perhaps be considered as minor lines within the Arthropoda. With the possible exception of the kinorhynchs and tardigrades, the central evolutionary problem posed by the aschelminths is that no clues have yet been uncovered that firmly establish their relationship to any of the major phyla of protostomes or deuterostomes.

SUMMARY

1. With many exceptions, the aschelminth assemblage of phyla share such features as a cuticle (except rotifers and acanthocephalans), adhesive tubes, protonephridia (except nematodes and tardigrades), and small body size. When an extracellular body cavity is present, as in the nematodes, it is never completely lined by cells, and is occupied by connective tissue fluid (pseudocoel) or a stiff collagenous matrix (some nematomorphs) or is absent altogether (gastrotrichs, kinorhynchs).

2. Gastrotrichs are acoelomate inhabitants of fresh water and the sea, the marine species being largely interstitial. They glide over the bottom on ventral cilia and temporarily anchor by means of adhesive tubes that contain duo-gland organs. Most feed on bacteria, which they ingest using a sucking pharynx that is nearly identical to that of nematodes, tardigrades, and loriciferans. They are hermaphrodites with internal fertilization and direct development.

3. Free-living nematodes (roundworms) occur in large numbers in the sea, in fresh water, and in soil. Their long, tapered, cylindrical bodies are adapted for living in minute spaces, especially interstitial spaces of aquatic sediments and algal mats. Terrestrial species live in the water films around soil particles and are actually aquatic.

4. The cylindrical body shape in combination with longitudinal body wall muscles, the complex elastic cuticle, and the hydrostatic pressure of the pseudocoelomic fluid make possible the undulatory movements that drive nematodes through interstitial spaces.

5. Nematodes exhibit a wide range of feeding habits, most of which involve a muscular sucking pharynx for ingestion. Teeth or stylets are commonly present in carnivorous and plant-feeding nematodes. The large number of parasitic nematodes includes many species of economic and medical importance, such as hookworms, ascarid worms, pinworms, *Trichinella,* and filarial worms. Most nematodes are dioecious with internal fertilization and direct development. Parasitic species vary greatly in their life cycles.

6. Members of the phylum Nematomorpha, called horsehair worms, have threadlike bodies as long as 100 cm. The free-living adults of most species live in fresh water or in damp soil. The larval stages are parasitic in arthropods. They are dioecious, and fertilization is internal. A gordian larva bearing anterior stylets enters the host.

7. Rotifers are distinguished by a ciliated corona, and mastax-bearing trophi. A cuticle is absent.

8. Rotifers are largely inhabitants of fresh water. Benthic species swim intermittently by using a ciliated corona and temporarily attach by using adhesive glands that open at the ends of the toes on the feet. Planktonic rotifers swim continually and have reduced feet and enlarged pseudocoels. The trophi are adapted for raptorial or suspension feeding. Reproduction is largely by parthenogenesis; development is direct.

9. Members of the phylum Acanthocephala are parasites of the gut of aquatic and terrestrial vertebrates. Each elongated parasite hooks to the gut wall with an anterior, retractile proboscis bearing recurved spines. A cuticle is absent. There is no digestive tract, and food is absorbed through the complex body wall. They are dioecious animals, and the life cycle requires a crustacean or insect as an intermediate host.

10. The phylum Kinorhyncha is composed of a small number of marine interstitial species. The cuticle and body wall are segmented, and the mouth is at the end of a spiny, anterior, retractile head. They are dioecious, development is direct, and molting occurs with growth.

11. The Loricifera is the most recently discovered phylum of animals. They are minute, marine, interstitial species with a spiny, cone-shaped anterior region (introvert) that can be telescopically retracted into a lorica-encased posterior region (abdomen).

12. The tardigrades, or water bears, are tiny interstitial inhabitants of beaches, soils, mosses, and lichens. The body bears four pairs of telescoping stubby legs, each of which is tipped with claws or adhesive discs. A pair of piercing stylets is used in combination with a sucking pharynx for feeding. They are dioecious, development is direct, and molting occurs with growth.

13. Nematodes, rotifers, and tardigrades that inhabit the high-stress environments of water films around soil particles and mosses are capable of remarkable degrees of cryptobiosis.

14. The aschelminth assemblage appears to represent multiple evolutionary lines. One line includes the gastrotrichs, nematodes, and, perhaps, the nematomorphs; another may include the rotifers and acanthocephalans. Loriciferans may be linked to the priapulids, and kinorhynchs and tardigrades appear to be arthropod relatives.

REFERENCES

Monociliated Cells and Evolution

Barnes, R. D. 1985. Current perspectives on the origins and relationships of lower invertebrates. *In* Conway Morris, S., George, J. D., Gibson, R. et al. (Eds.): The Origins and Relationships of Lower Invertebrates. Clarendon Press, Oxford. pp. 360–367.

Lyons, K. M. 1973. Collar cells in planula and adult tentacle ectoderm of the solitary coral *Balanophyllia regia* (Anthozoa, Eupsammidae). Z. Zellforsch. *145:*57–74.

Nørrevang, A., and Wingstrand, K. G. 1970. On the occurrence and structures of choanocyte-like cells in some echinoderms. Acta Zool. *51:*249–270.

Rieger, R. M. 1976. Monociliated epidermal cells in Gastrotricha: Significance for concepts of early metazoan evolution. Z. zool. Syst. Evolut.-forsch. *14:*198–226.

Syncytia

Podesta, R. B. 1983. Syncytial epithelia: Transport in the absence of paracellular pathways. J. Exp. Biol. *106:*195–204.

Cryptobiosis

Crowe, N. H., and Maden, K. A., 1974. Anhydrobiosis in tardigrades and nematodes. Trans. Am. Micros. Soc. *93:*513–524.

Wright, J. C., Westh, P., and Ramlov, H. 1992. Cryptobiosis in Tardigrada. Biol. Rev. *67:*1–29.

The literature included here is restricted to aschelminths. The introductory references on page 6 include many general works and field guides that contain sections on the aschelminths. General parasitology texts, which contain much additional information on parasitic nematodes and on nematomorphs and acanthocephalans, are listed at the beginning of the references for flatworms, Chapter 6, page 258.

General Aschelminths

Grassé, P. (Ed.): 1965. Nemathelminthes, Rotifères, Gastrotriches, et Kinorhynques. In Traité de Zoologie, Vol. 4. Pts. 2 and 3. Masson et Cie, Paris. (A detailed general account of the pseudocoelomate invertebrates.)

Higgins, R. P., and Thiel, H. (Eds.): 1988. Introduction to the Study of Meiofauna. Smithsonian Institution Press, Washington, DC. 488 pp.

Hyman, L. H. 1951. The Invertebrates: Acanthocephala, Aschelminthes, and Entoprocta. Vol. 3. McGraw-Hill, New York. (Out of date in many areas, but still useful.)

Lorenzen, S. 1985. Phylogenetic aspects of pseudocoelomate evolution. *In* Conway Morris, S., George, J. D., Gibson, R. et al. (Eds.): The Origins and Relationships of Lower Invertebrates. Clarendon Press, Oxford. pp. 210–223.

Malakhov, V. V. 1980. Cephalorhyncha, a new type of animal kingdom uniting Priapulida, Kinorhyncha, Gordiacea, and a system of Aschelminthes worms. Zool. Zh. *59:*485–499. (In English.)

Remane, A. 1963. The systematic position and phylogeny of the Pseudocoelomata. *In* Dougherty, E. C. (Ed.): The Lower Invertebrates. University of California Press, Berkeley. pp. 247–255.

Storch, V. 1984. Minor Pseudocoelomates. *In* Bereiter-Hahn, J., Matoltsy, A. G., and Richards, K. S. Biology of the Integument, Vol. 1. Springer-Verlag, New York. pp. 242–268.

Gastrotricha

Colacino, J. M., and Kraus, D. W. 1984. Hemoglobin-containing cells of *Neodasys* (Gastrotricha, Chaetonotida)—II. Respiratory significance. Comp. Biochem. Physiol. *79A:*363–369.

d'Hondt, J. L. 1971. Gastrotricha. Oceanogr. Mar. Biol. Ann. Rev. *9:*141–192.

Hummon, M. R. 1984. Reproduction and sexual development in a freshwater gastrotrich. 3. Postparthenogenic development of primary oocytes and the x-body. Cell Tiss. Res. *236:*629–636.

Hummon, M. R. 1986. Reproduction and sexual development in a freshwater gastrotrich. 4. Life history traits and the possibility of sexual reproduction. Trans. Am. Microsc. Soc. *105:*97–109.

Hummon, W. D. 1971. Biogeography of sand beach Gastrotricha from the northeastern United States. Biol. Bull. *141*(2):390.

Hummon, W. D., and Hummon, M. R. 1983. Gastrotricha. *In* Adiyodi, K. G., and Adiyodi, R. G. (Eds.): Reproductive Biology of Invertebrates. Vol. I:211–221; Vol. II: 195–205.

Rieger, G. E., and Rieger, R. M. 1977. Comparative fine structure study of the gastrotrich cuticle and aspects of cuticle evolution within the Aschelminthes. Z. Zool. Syst. Evolut.-forschung. *15*(2):81–124.

Rieger, R. M. 1976. Monociliated epidermal cells in Gastrotricha: Significance for concepts of early metazoan evolution. Z. Zool. Syst. Evolut.-forschung. *14*(3):198–226.

Ruppert, E. E. 1978. The reproductive system of gastrotrichs. II. Insemination in *Macrodasys:* A unique mode of sperm transfer in Metazoa. Zoomorphologie. *89:*207–228.

Ruppert, E. E. 1982. Comparative ultrastructure of the gastrotrich pharynx and the evolution of myoepithelial foreguts in Aschelminthes. Zoomorphologie. *99:*181–220.

Ruppert, E. E. 1988. Gastrotricha. *In* Higgins, R. P., and Thiel, H. (Eds.): Introduction to the Study of Meiofauna. Smithsonian Institution Press, Washington,

DC. pp. 302–311. (Includes a pictorial key to the genera of marine and freshwater Gastrotricha.)

Ruppert, E. E. 1991. Gastrotricha. *In* Harrison, F. W., and Ruppert, E. E. (Eds.): Microscopic Anatomy of Invertebrates. Vol. 4. Aschelminthes. Wiley-Liss, New York. pp. 41–109.

Strayer, D. L., and Hummon, W. D. 1991. Gastrotricha. *In* Thorp, J. H., and Covich, A. P. (Eds.): Ecology and Classification of North American Freshwater Invertebrates. Academic Press, Inc., New York. pp. 173–185.

Teuchert, G. 1968. Zür Fortpflanzung und Entwicklung der Macrodasyoidea (Gastrotricha). Z. Morphol. Tiere. *63*:343–418. [On the reproduction and development of the Macrodasyoidea (Gastrotricha)]

Tyler, S., and Rieger, G. E. 1980. Adhesive organs of the Gastrotricha. I. Duo-gland organs. Zoomorphologie. *95*:1–15.

Weiss, M. J., and Levy, D. P. 1979. Sperm in "parthenogenetic" freshwater gastrotrichs. Science. *205*:302–303.

Nematoda

Adams, P. J. M., and Tyler, S. 1980. Hopping locomotion in a nematode: Functional anatomy of the caudal gland apparatus of *Theristus caudasaliens* sp. n. J. Morphol. *164*:265–285.

Bennet-Clark, H. C. 1976. Mechanics of nematode feeding. *In* Croll, N. A. (Ed.): The Organization of Nematodes. Academic Press, New York. pp. 313–342.

Bird, A. F. 1971. The structure of Nematodes. Academic Press, New York. 318 pp.

Chitwood, B. G., and Chitwood, M. B. 1974. Introduction to Nematology. University Park Press, Baltimore. 334 pp.

Croll, N. A. 1970. The Behavior of Nematodes. St. Martin's Press, New York. 117 pp.

Croll, N. A., and Matthews, B. E. 1977. Biology of Nematodes. John Wiley and Sons, New York. 201 pp. (A general account of the phylum.)

Croll, N. A., and Smith, J. M. 1974. Nematode setae as mechanoreceptors. Nematologica. *20*(3):291–296.

Croll, N. A., and Viglierchio, D. R. 1969. Osmoregulation and uptake of ions in a marine nematode. Proc. Helminthol. Soc. Wash. *36*(1):1–9.

Deutsch, A. 1978. Gut ultrastructure and digestive physiology of two marine nematodes, *Chromadorina germanica* (Butschli, 1874) and *Diplolaimella* sp. Biol. Bull. *155*:317–355.

Doncaster, C. C., and Seymour, M. K. 1973. Exploration and selection of penetration site by Tylenchida. Nematologica. *19*(2):137–145.

Dropkin, V. H. 1980. Introduction to Plant Nematology. Wiley-Interscience, New York. 293 pp.

Ferris, V. R. 1971. Taxonomy of the Dorylaimida. *In* Zuckerman, B. M., Mai, W. F., and Rohde, R. A. (Eds.): Plant Parasitic Nematodes. Vol. 1. Academic Press, New York. pp. 163–189.

Ferris, V. R., Ferris, J. M., and Tjepkema, J. P. 1973. Genera of freshwater nematodes (Nematoda of eastern North America). Biota of Freshwater Ecosystems. Ident. Manual 10. EPA. U.S. Government Printing Office, Washington, DC.

Golden, A. M. 1971. Classification of the genera and higher categories of the order Tylenchida (Nematoda). *In* Zuckerman, B. M., Mai, W. F., and Rohde, R. A. (Eds.): Plant Parasitic Nematodes. Vol. 1. Academic Press, New York. pp. 191–232.

Goodey, J. B. 1951. Soil and Freshwater Nematodes. John Wiley and Sons, New York.

Hope, W. D., and Murphy, D. G. 1972. A Taxonomic Hierarchy and Checklist of the Genera and Higher Taxa of Marine Nematodes. Smithson. Contrib. Zool. *137*:1–101.

Jennings, J. B., and Colam, J. B. 1970. Gut structure, digestive physiology and food storage in *Pontonema vulgaris* (Nematoda: Enoplida). J. Zool. London. *161*:211–221.

Jensen, P. 1982. Reproductive behaviour of the free-living marine nematode *Chromodorita tenuis*. Mar. Ecol. Prog. Ser. *10*:89–95.

Lee, D. L., and Atkinson, H. J. 1977. Physiology of Nematodes. 2nd Edition. Columbia University Press, New York. 215 pp.

Levine, N. D. 1968. Nematode Parasites of Domestic Animals and of Man. Burgess Publishing Co., Minneapolis. 600 pp.

Maggenti, A. 1981. General Nematology. Springer Verlag, New York. 372. pp. (Emphasis on parasitic nematodes.)

Mai, W. F., and Lyon, H. H. 1975. Pictorial Key to the Plant-Parasitic Nematodes. 4th Edition. Comstock Publishing Associates, Cornell University Press, Ithaca.

Nicholas, W. L. 1984. The Biology of Free-Living Nematodes. 2nd Edition. Oxford University Press, London. (An excellent introductory account of free-living nematodes. including sections on methods of collection and preparation of material.)

Platt, H. M., and Warwick, R. M. 1983. Freeliving Marine Nematodes. Part I: British Enoplids—Pictorial key to world genera and notes for identification of British species. Synopses of the British Fauna, No. 28. Cambridge University Press, Cambridge. 307 pp.

Poinar, G. O. 1983. The Natural History of Nematodes. Prentice-Hall, Inc., Englewood Cliffs, NJ. 323 pp.

Poinar, G. O. 1991. Nematoda and Nematomorpha. *In* Thorp, J. H. and Covich, A. P. (Eds.): Ecology and

Classification of North American Freshwater Invertebrates. Academic Press, Inc., New York. pp. 249–283.

Riemann, F. 1972. *Kinochulus sattleri* n.g.n. sp., an aberrant free-living nematode from the lower Amazonas. Veroff. Inst. Meersforsch. Bremerh. *13:*317–326.

Roggen, D. R., Raski, D. J., and Jones, N. O. 1966. Cilia in nematode sensory organs. Science. *152:*515–516.

Schaefer, C. 1971. Nematode radiation. Syst. Zool. *20:*(1)77–78.

Siddiqui, I. A., and Viglierchio, D. R. 1977. Ultrastructure of the anterior body region of marine nematode *Deontostoma californicum.* J. Nematol. *9*(1):56–82.

Somers, J. A., Shorey, H. H., and Gastor, L. K. 1977. Sex pheromone communication in the nematode, *Rhabditis pellio.* J. Chem. Ecol. *3*(4):467–474.

Steiner, G., and Heinly, H. 1922. Possibility of control of Heterodera radicola by means of predatory nemas. J. Washington Acad. Sci. *12:*367–396.

Tarjan, A. C., Esser, R. P., and Chang, S. L. 1977. An illustrated key to nematodes found in freshwater. J. Water Pollut. Control Fed. *49*(11):2318–2337.

Tietjen, J. H. 1969. The ecology of shallow water meiofauna in two New England estuaries. Oecologia. *2:*251–291.

Wallace, H. R. 1970. The movement of nematodes. *In* Fallis, A. M. (Ed.): Ecology and Physiology of Parasites. University of Toronto Press, Toronto. pp. 201–212.

Wood, W. B. (Ed.): 1988. The Nematode, *Caenorhabditis elegans.* Cold Spring Harbor, New York. 667 pp.

Wright, K. A. 1991. Nematoda. *In* Harrison, F. W. and Ruppert, E. E. (Eds.): Microscopic Anatomy of Invertebrates. Vol. 4. Aschelminthes. Wiley-Liss, New York. pp. 11–195.

Yeats, G. W. 1971. Feeding types and feeding groups in plant and soil nematodes. Pedobiologia. *11*(2):173–179.

Zuckerman, B. M., Mai, W. F., and Rohde, R. A. 1971. Plant Parasitic Nematodes: Cytogenetics, Host Parasite Interactions and Physiology. Vol. 2. Academic Press, New York. 347 pp.

Nematomorpha

Bresciani, J. 1991. Nematomorpha. *In* Harrison, F. W. and Ruppert, E. E. (Eds.): Microscopic Anatomy of Invertebrates. Vol. 4. Aschelminthes. Wiley-Liss, New York. pp. 197–218.

Eakin, R. M., and Brandenburger, J. L. 1974. Ultrastructural features of a Gordian worm (Nematomorpha). J. Ultrastruct. Res. *46:*351–374.

May, H. G. 1919. Contributions to the life histories of *Gordius robustus* Leidy and *Paragordius varius* (Leidyi). Illinois Biol. Monogr. *5:*1–119.

Rotifera

Aloia, R. C., and Moretti, R. L. 1973. Mating behavior and ultrastructural aspects of copulation in the rotifer *Asplanchna brightwelli.* Trans. Am. Microsc. Soc. *92:*371–380.

Clément, P., and Wurdak, E. 1991. Rotifera. *In* Harrison, F. W. and Ruppert, E. E. (Eds.): Microscopic Anatomy of Invertebrates. Vol. 4. Aschelminthes. Wiley-Liss, New York. pp. 219–297.

Dumont, H. J., and Green, J. 1980. Rotatoria. Proceedings of 2nd International Rotifer Symposium. Hydrobiologia. 73. W. Junk, The Hague. 263 pp.

Gilbert, J. J. 1974. Dormancy in rotifers. Trans. Amer. Micros. Soc. *93:*490–513.

Gilbert, J. J. 1980. Female polymorphism and sexual reproduction in the rotifer, *Asplanchna:* Evolution of their relationship and control by dietary tocopherol. Amer. Nat. *116*(3):409–431.

Gilbert, J. J., and Starkweather, P. L. 1977. Feeding in the rotifer *Brachionus calyciflorus:* I. Regulatory mechanisms. Oecologica. *28*(2):125–132.

Gilbert, J. J., and Stemberger, R. S. 1985. Prey capture in the rotifer *Asplanchna girodi.* Verh. Internat. Verein. Limnol. *22:*2997–3000.

Koehler, J. K. 1965. A fine-structural study of the rotifer integument. J. Ultrastruct. Res. *12:*113–134.

Koste, W. 1978. Rotatoria. Die Rädertiere Mitteleuropas. I. Text, II. Illustrations. Gebrüder Bornträger, Berlin. (The wide distribution of most species of rotifers makes this work a very important reference for identification, but the section on bdelloids is not yet completed.)

Lebedeva, L. I., and Gerasimova, T. N. 1981. Growth, breeding and production of *Philodina roseola* upon individual cultivation. Zool. Zh. *60*(11):1614–1620.

Pontin, R. M. 1964. A comparative account of the protonephridia of *Asplanchna* (Rotifera) with special reference to the flame bulbs. Proc. Roy. Soc. Lond. *142:*511–525.

Pontin, R. M. 1966. The osmoregulatory function of the vibratile flames and the contractile vesicle of *Asplanchna* (Rotifera). Comp. Biochem. Physiol. *17:*1111–1126.

Pourriot, R. 1979. Soil rotifers. Rev. Ecol. Biol. Sol. *16*(2):279–312.

Salt, G. W., Sabbadini, G. F., and Commins, M. L. 1978. Trophi morphology relative to food habits in six species of rotifers (Asplanchnidae). Trans. Amer. Microsc. Soc. *97*(4):469–485.

Thane, A. 1974. Rotifera. *In* Giese, A. C., and Pearse, J. S. (Eds.): Reproduction of Marine Invertebrates. Vol. I. Acoelomate and Pseudocoelomate Metazoans. Academic Press, New York. pp. 471–484. (This review covers freshwater as well as marine forms.)

Wallace, R. L. 1980. Ecology of sessile rotifers. Hydrobiologia. *73:*181–193.

Wallace, R. L., and Snell, T. W. 1991. Rotifera. *In* Thorp, J. H., and Covich, A. P. (Eds.): Ecology and Classification of North American Freshwater Invertebrates. Academic Press, Inc., New York. pp. 187–248. [An excellent modern account of the biology, ecology, and classification of rotifers.]

Warner, F. D. 1969. The fine structure of the protonephridia in the rotifer *Asplanchna.* J. Ultrastruct. Res. *29:*499–524.

Acanthocephala

Abele, L. G., and Gilchrist, S. 1977. Homosexual rape and sexual selection in acanthocephalan worms. Science. *197:*81–83.

Asaolu, S. O. 1989. Morphological studies on the testis of adults of *Moniliformis* (Acanthocephala). Acta Zool. (Stockholm) *70:*65–69.

Baer, J. C. 1961. Acanthocephales. *In* Grassé P. (Ed.): Traité de Zoologie. Vol. 4. pt. 1. Masson et Cie, Paris. pp. 733–782.

Byram, J. E., and Fisher, F. M. 1973. The absorptive surface of *Moniliformis dubius* (Acanthocephala). I. Fine structure. Tissue and Cell. *5:*553–579.

Byram, J. E., and Fisher, F. M. 1974. The absorptive surface of *Moniliformis dubius* (Acanthocephala). II. Functional aspects. Tissue and Cell. *6:*21–42.

Conway Morris, S., and Crompton, D. W. T. 1982. The origins and evolution of the Acanthocephala. Biol. Rev. *57:*85–115.

Crompton, D. W. T. 1970. An Ecological Approach to Acanthocephalan Physiology. Cambridge University Press, New York. 136 pp.

Crompton, D. W. T., and Nichol, B. B. (Eds.): 1985. Biology of Acanthocephala. Cambridge University Press, New York. 500 pp.

Dunagan, T. T., and Miller, D. M. 1991. Acanthocephala. *In* Harrison, F. W., and Ruppert, E. E. (Eds.): Microscopic Anatomy of Invertebrates. Vol. 4. Aschelminthes. Wiley-Liss, New York. pp. 299–332.

Nicholas, W. L. 1967. The Biology of the Acanthocephala. Adv. Parasitol. *5:*205–246.

Nicholas, W. L. 1973. The Biology of the Acanthocephala. Adv. Parasitol. *11:*671–706.

Pratt, I. 1969. The biology of the Acanthocephala. *In* Florkin, M., and Scheer, B. T. (Eds.): Chemical Zoology. Vol. 3. Academic Press, New York. pp. 245–252.

Van Cleave, H. J. 1941. Relationships of the Acanthocephala. Am. Nat. *75:*31–47.

Whitfield, P. J. 1971. Phylogenetic affinities of Acanthocephala: An assessment of ultrastructural evidence. Parasitology. *63*(1):49–58.

Whitfield, P. J. 1984. Acanthocephala. *In* Bereiter-Hahn, J., Matoltsy, A. G., and Richards, K. S.: Biology of the Integument, Vol. I. Invertebrates. Springer-Verlag, New York. pp. 234–241.

Kinorhyncha

Brown, R. 1983. Spermatophore transfer and subsequent sperm development in homalorhagid kinorhynchs. Zool. Scripta. *12*(3):257–266.

Higgins, R. P. 1971. A historical overview of kinorhynch research. *In* Hulings, N. C. (Ed.): Proceedings of the First International Conference on Meiofauna. Smithson. Contrib. Zool. *76:*25–31.

Kozloff, E. N. 1972. Some aspects of development in *Echinoderes* (Kinorhyncha). Trans. Amer. Microsc. Soc. *91:*119–130.

Kristensen, R. M., and Hay-Schmidt, A. 1989. The protonephridia of the arctic *Echinoderes aquilonius* (Cyclorhagida, Echinoderidae). Acta Zool. (Stockholm) *70:*13–27.

Kristensen, R. M., and Higgins, R. P. 1991. Kinorhyncha. *In* Harrison, F. W., and Ruppert, E. E. (Eds.): Microscopic Anatomy of Invertebrates. Vol. 4. Aschelminthes. Wiley-Liss, New York. pp. 377–404.

Loricifera

Higgins, R. P., and Kristensen, R. M. 1988. Loricifera. *In* Higgins, R. P., and Thiel, H. (Eds.): Introduction to the Study of Meiofauna. Smithsonian Institution Press, Washington, DC. pp. 319–321.

Kristensen, R. M. 1983. Loricifera, a new phylum with Aschelminthes characters from the meiobenthos. Z. f. zool. Syst. Evoluts.-forsch. *21:*163–180. (See also the short report in Science. Oct. 14, 1983, p. 149.)

Kristensen, R. M. 1991. Loricifera. *In* Harrison, F. W., and Ruppert, E. E. (Eds.): Microscopic Anatomy of Invertebrates. Vol. 4. Aschelminthes. Wiley-Liss, New York. pp. 351–375.

Kristensen, R. M. 1991. Loricifera—A general biological and phylogenetic overview. Verh. dtsch. zool. Ges. *84:*231–246.

Tardigrada

Bussers, J. C., and Jeuniaux, C. 1973. Structure et composition de la cuticle de *Macrobiotus* et de *Milnesium tardigradum.* Ann. Soc. R. Zool. Belg. *103:*271–279.

Cuénot, L. 1949. Les Onychophores, Les Tardigrades, et Les Pentastomides. *In* Grassé, P. (Ed.): Traité de Zoologie. Vol. 6. Masson et Cie, Paris. pp. 3–75.

Greven, H. 1984. Tardigrada. *In* Bereiter-Hahn, J., Matoltsy, A. G., and Richards, K. S.: Biology of the Integument, Vol. I. Invertebrates. Springer-Verlag, New York. pp. 714–727

Kristensen, R. M. 1981. Sense organs of two marine arthro-tardigrates (Heterotardigrada, Tardigrada). Acta Zool. *62:27–41.*

Marcus, E. 1929. Tardigrada. *In* Bronn, H. G. (Ed.): Klassen und Ordnungen des Tierreichs. Bd. 5, Abt. IV. Akademische Verlagsgesellschaft, Frankfurt.

Morgan, C. I., and King, P. E. 1976. British Tardigrades. Synopses of the British Fauna No. 9. Academic Press, London. 133 pp.

Nelson, D. R. 1975. The hundred-year hibernation of the water bear. Nat. Hist. *84:62–65.*

Nelson, D. R. 1982a. Developmental biology of the Tardigrada. *In* Harrison, F. W., and Cowden, R. R. Developmental Biology of Freshwater Invertebrates. Alan R. Liss, New York. pp. 363–398.

Nelson, D. R. (Ed.): 1982b. Proceedings of the Third International Symposium on the Tardigrada. East Tennessee State University Press, Johnson City.

Nelson, D. R. 1991. Tardigrada. *In* Thorp, J. H., and Covich, A. P. (Eds.): Ecology and Classification of North American Freshwater Invertebrates. Academic Press, Inc., New York. pp. 501–521.

Pollock, L. W. 1976. Tardigrada. Marine flora and fauna of the northeastern U.S. NOAA Technical Reports NMFS Circular 394 U.S. Government Printing Office, Washington, DC. 25 pp.

Renaud-Mornant, J. 1988. Tardigrada. *In* Higgins, R. P., and Thiel, H. (Eds.): Introduction to the Study of Meiofauna. Smithsonian Institution Press, Washington, DC. pp. 357–364. (Includes a pictorial key to the marine, freshwater, and moss-dwelling tardigrades.)

Riggin, G. T., Jr. 1962. Tardigrada of southwest Virginia, with the addition of a description of a new marine species from Florida. Virginia Agr. Exp. Sta. Tech. Bull. *152:1–145.* (This paper is a good source for taxonomic work on North American tardigrades.)

Walz, B. 1978. Electron microscopic investigation of cephalic sense organs of the tardigrade *Macrobiotus hufelandi* C.A.S. Schultze. Zoomorphologie. *89:1–19.*

9

NONSEGMENTED COELOMATE WORMS

PRINCIPLES AND EMERGING PATTERNS

BURROWING MECHANICS

Burrowing in moist or submerged sediments is a common behavior in invertebrates, from sea anemones and nemerteans to earthworms and the animals discussed in this chapter. The principles of burrowing are common and simple: any animal that burrows must anchor itself posteriorly in order to push the sediment aside without backslipping. This kind of anchor is called a **penetration anchor** because it holds one part of the body in place as another part penetrates and advances into the sediment (Fig. 9–1A). Once the leading end of the body has pushed forward some distance into the sediment, it too must anchor itself so that the trailing end can be drawn forward into the burrow. Such an anchor at the leading end of an animal is called a **terminal anchor** (Fig. 9–1B). Continuous burrowing thus requires a sustained and coordinated alternation of penetration and terminal anchors.

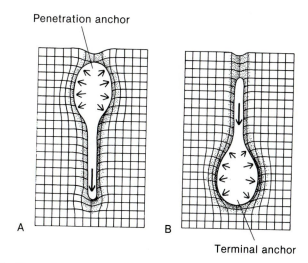

FIGURE 9–1. Burrowers must establish a penetration anchor **(A)** to prevent back-slipping while pushing the body into sediment. Once the body enters the sediment, a terminal anchor **(B)** allows the trailing part of the body to be pulled forward into the burrow.

Peristalsis is one of the simplest ways of coordinating penetration and terminal anchors (Fig. 9–2). As a peristaltic wave passes along the length of a burrowing worm, longitudinal and circular muscles contract alternately. At points where the longitudinal musculature is fully contracted, the body shortens and widens to form both the penetration and terminal anchors. At points where the longitudinal musculature is only partly contracted, the body shortens without anchoring, thus pulling a section of the body toward a terminal anchor. In body regions where the circular musculature is contracted, the body elongates pushing forward into the sediment. Peristaltic burrowing is common in annelids, echiurans, nemerteans, and a few anemones.

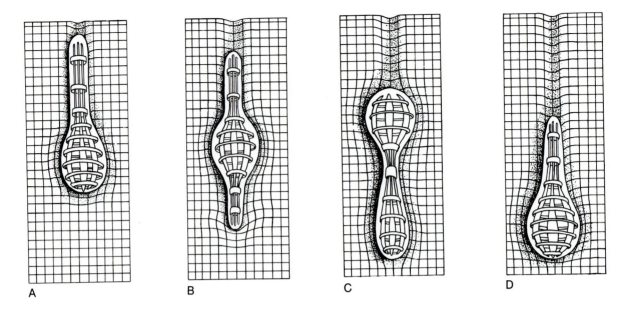

FIGURE 9–2. Peristalsis is a common means of coordinating penetration and terminal anchors while burrowing. During peristaltic burrowing, alternating waves of longitudinal and circular muscle contraction pass rearward along the body. In this figure, one complete wave of longitudinal and circular muscle contraction is shown.

Not all worms use whole-body peristalsis for burrowing. Instead, some use **eversion/retraction** cycles of an eversible body part, such as a proboscis, pharynx, or introvert (Figs. 9–7; 9–19). As burrowing occurs in these animals, the trunk usually forms the penetration anchor while the eversible proboscis thrusts forward and enters the sediment. The flow of body fluid into the proboscis during eversion causes the advancing tip to swell into a terminal anchor. As the proboscis is retracted, usually by contraction of specialized retractor muscles, the trunk is pulled forward toward the terminal anchor. Often, the tip of the eversible body part is adorned with spines or setae, which help to anchor it in the substratum. This sort of burrowing occurs in sipunculans, priapulids (Fig. 9–19), a few annelids, and perhaps some nemerteans.

DEPOSIT FEEDING

Small organic particles are continually settling to the bottom in marine and freshwater environments. The particles are derived from the decomposing bodies of algae, aquatic plants, animals, and planktonic organisms. The fecal pellets of many different animals also are a major source of organic particulate matter. All of this deposited material, which becomes mixed with mineral particles on the bottom, is an important source of food for deposit-feeding animals. **Deposit feeders** may be **nonselective** or **selective.** Nonselective deposit feeders ingest both organic and mineral particles and then digest out some of the organic material, especially bacteria that have colonized the surface of sediment particles. The remaining organic and mineral matter is defecated and is commonly visible as continuous piles, called **castings.** By way of a variety of different mechanisms selective deposit feeders are capable of separating the organic particles from mineral particles and ingesting only the organic material.

RESPIRATORY PIGMENTS

Three different respiratory pigments appear sporadically in bilateral animals, and two are found in nonsegmented protostomes. The three pigments are **hemoglobin, hemerythrin,** and **hemocyanin.** Hemoglobin is widespread among bilateral animals, hemocyanin is limited to molluscs and arthropods, and hemerythrin has a quirky distribution, being found only in sipunculans, priapulids, inarticulate brachiopods, and one family (Magelonidae) of polychaete annelids.

A hemoglobin molecule consists of two parts, a porphyrin ring with an iron atom at its center, called *heme,* and a protein part, called *globin* (Fig. 9–3A). An important function of globin is to prevent oxygen from binding too tightly to the heme; when globin is present, oxygen binds reversibly to heme and can be released to the tissues. Hemoglobin occurs within cells, which is probably the primitive condition, and also extracellularly in blood plasma. The extracellular hemoglobins are typically large molecules of high molecular weight compared with the relatively small sizes of intracellular hemoglobins. The extracellular occurrence of few large molecules rather than many small ones may be necessary to keep the osmotic concentration of the blood within reasonable limits, perhaps also to prevent loss of the mole-

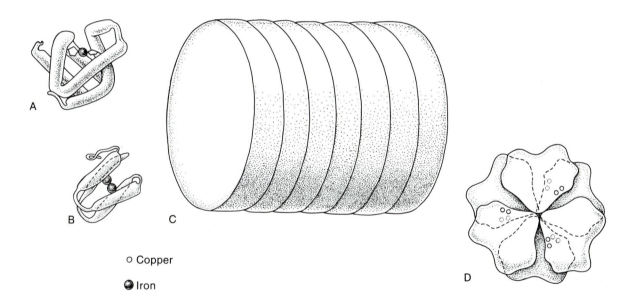

○ Copper

● Iron

FIGURE 9–3. Respiratory pigments are all complexes of metal (iron or copper) and protein. Although oxygen actually binds to the metal and not the protein, the protein regulates the tightness of the bond and, therefore, the delivery of oxygen to the tissues. Hemoglobin **(A)** and hemerythrin **(B)** both contain iron and are similar in size and shape (myohemerythrin and myoglobin shown). In hemoglobin, one iron atom is bound to heme (pentagons), whereas in hemerythrin two iron atoms are bound directly to the protein. The amino acid sequences of the two molecules are also different. Both molluscs **(C)** and arthropods **(D)** have large copper-based hemocyanins, but the pigments are not homologous in the two phyla. Molluscan hemocyanin is a large cylindrical molecule, whereas arthropodan hemocyanin is a smaller (but still large) hexamer (shown) or dodecamer. *(A, Redrawn from Fig. 2.7 of Dickerson, R. E., and Geis, I., 1983. Hemoglobin. The Benjamin Cummings Publ. Co., Inc., Menlo Park, NJ. 176 pp. B, Modified and redrawn from Fig. 3. of Klippenstein, G. L., 1980. Structural aspects of hemerythrin and myohemerythrin. Am. Zool. 20:39–51. C, Drawn from micrograph. In Mangum, C. P., 1985. Oxygen transport in invertebrates. Am. J. Physiol. 248:R505–R514. D, Redrawn from Fig. 3A of Linzen, B. et al., 1985. The structure of arthropod hemocyanins. Science. 229:519–524.)*

cules during ultrafiltration. Most hemoglobin is red when oxygenated, but one variant (chlorocruorin) is green.

A molecule of hemerythrin, like hemoglobin, uses iron atoms to bind oxygen, but unlike hemoglobin the two iron atoms are bound directly to the protein and not to a heme (Fig. 9–3B). Hemerythrin is always found within cells and never free in plasma or coelomic fluid. It is pink or violet when oxygenated and colorless when deoxygenated.

Hemocyanin binds a molecule of oxygen between a pair of copper atoms. The many pairs of copper atoms are bound directly to the protein part of the molecule and a heme is absent. Hemocyanin only occurs as large *extra*cellular molecules and is never found in cells (Fig. 9–3C, D). It is light blue when oxygenated and colorless when deoxygenated.

Because they chemically bind oxygen, respiratory pigments increase the capacity of blood, coelomic fluid, or tissue to transport or store oxygen. An enhancement of oxygen transport is important in situations in which metabolic oxygen demand is not met by a supply of oxygen via simple diffusion or transport in physical solution. Animals that have a high oxygen demand or that occupy oxygen-poor habitats usually have a respiratory pigment.

Functionally different forms of the same respiratory pigment may occur in different body compartments of the same animal. For example, there may be one hemoglobin in the tissues, one in the coelom, and one in the blood-vascular system, each with a different protein component (globin) that determines the affinity of the hemoglobin for oxygen. The oxygen affinity of the pigment increases from the blood compartment near the environmental source to an inner compartment, such as muscle tissue, containing the metabolic sink. In between, for example in the coelom, the oxygen affinity of the hemoglobin is intermediate between that of the blood and muscle hemoglobins. This stepwise arrangement of affinities, from low to high as one moves from outside to inside the animal, creates a cascade down which oxygen is transferred from blood to coelom to tissues.

DEVELOPMENTAL PATTERNS OF EVOLUTIONARY SIGNIFICANCE AND TROCHOPHORE LARVAE

A peculiar but characteristic arrangement of blastomeres, which may be of evolutionary significance, appears at the animal pole of many embryos undergoing spiral cleavage at the 64-cell stage (after six cleavages; Fig. 9–4). The pattern consists of four pairs of cells, each pair radiating outward from the animal pole like a spoke on a bicycle wheel, and a central hub composed of an additional four cells. The four radial spokes are equidistant from each other and together form an X called the "cross." The four "hub" cells are also arranged tetraradially directly above the animal pole and are called the "rosette." In annelids, the arms of the cross extend outward along interradial lines passing *between* the rosette cells, and the resultant pattern is called the **annelidan cross** (Fig. 9–4A, B). In molluscs, on the other hand, the four arms of the cross extend outward along radial lines drawn *through* the rosette cells. This figure is known as the **molluscan cross** (Fig. 9–4C, D). Rosette and cross figures have been noted in embryos of only two phyla of animals other than annelids and molluscs. These are the peanut worms (Sipuncula) and spoon worms (Echiura), which have an uncertain evolutionary relationship to each other and to other phyla of animals. The sipunculans, however, show a molluscan cross

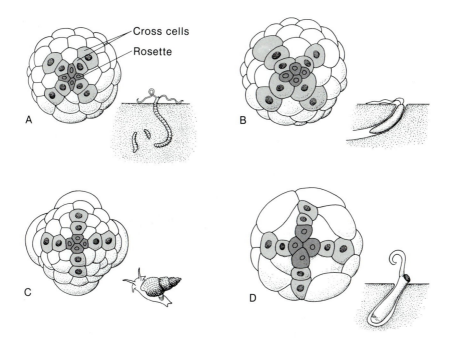

FIGURE 9–4. The embryos of several phyla having spiral cleavage show "rosette and cross" patterns at the animal pole. In annelids **(A)** and echiurans **(B),** the four arms of the cross are interradial with respect to the rosette, whereas in molluscs **(C)** and sipunculans **(D),** the cross is radial in position. *(A, Modified and redrawn from Wilson, E. B. 1892. The cell-lineage of Nereis. J. Morphol. 6:361–466. B, After Newby, W. W. 1932 from Kume, M., and Dan, K. 1968. NOLIT Publishing House, Belgrade. 605 pp. C, Modified and redrawn from McBride, E. W. 1914. Text-book of embryology. Vol. I. Invertebrata. Macmillan and Co., Ltd., London. 692 pp. D, After Gerould, J. H. (1906) from Rice, M. E. 1985. Sipuncula: Developmental evidence for phylogenetic inference. In Conway Morris, S., George, J. D., Gibson, R. et al. The Origins and Relationships of Lower Invertebrates. Oxford University Press, Oxford. pp. 274–296.)*

during development, whereas echiurans have an annelidan cross. These occurrences suggest that although peanut worms and spoon worms are similar to each other in general form and habit, sipunculans may be more closely related to molluscs, but echiurans may share a common ancestor with annelids. The developmental fates of the rosette and cross cells do not shed further light on the evolutionary relationships of these two phyla: in both annelids and molluscs, the rosette cells become the apical plate, and the cross cells join other cells to form the pretrochal epidermis of the trochophore larva.

The characteristic larva of sipunculans, echiurans, polychaetes, molluscs, and many other protostomes is a **trochophore** (Fig. 9–5). A typical trochophore larvae is top-shaped and bears a tuft of cilia at the apical end. The distinguishing feature is a conspicuous girdle of cilia, called the **prototroch,** which rings the body approximately one third to one half the distance from the apical tuft. The gut is a complete tube, and the mouth opens posterior to the prototroch. In the trochophore of many polychaetes and especially in sipunculan larvae, a second girdle of cilia, called the **metatroch,** develops below the mouth, and a third, the **telotroch** (well developed in echiuran trochophores), forms just before the anus at the posterior end (Fig. 9–5A).

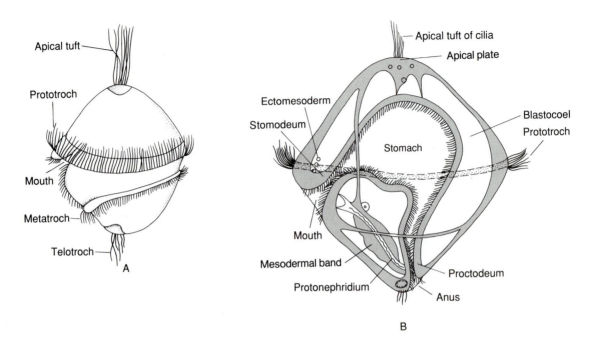

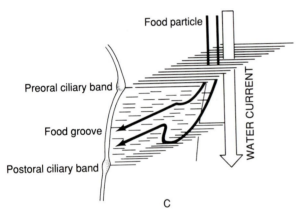

FIGURE 9–5. A, External view of a trochophore larva showing characteristic ciliary bands. **B,** Internal anatomy of a trochophore. **C,** Food collection mechanism of some trochophores. The preoral and postoral ciliary bands correspond to the prototroch and metatroch, respectively, in **A. A,** Trochophore of the annelid *Polygordius.* **B,** An annelid trochophore. **C,** Two-band suspension feeding system. *(A, After Dawydoff. B, After Shearer from Hyman. C, From Strathmann, R. R., Jahn, T. L., and Fonesca, J. R. 1972: Suspension feeding by marine invertebrate larvae: Clearance of particles by ciliated bands of rotifer, pluteus and trochophore. Biol. Bull. 142:505:519.*

Internally, the old blastocoel persists as a large, gelatinous, connective tissue layer lying between the gut and outer ectoderm (Fig. 9–5B). Larval muscle bands cross the blastocoel, and within is a pair of protonephridia, one on each side of the gut. There is a pair of ventrolateral mesodermal bands from which most adult mesodermal structures are derived. The cells of the apical plate contain the primordium of the cerebral ganglia.

The fully developed trochophore larva can be divided into three regions: the **pretrochal region** consisting of the apical plate, prototroch, and area about the mouth; the **pygidium,** consisting of the telotroch and anal area behind it; and the **growth zone,** which includes all of the larva between the mouth and telotroch. In polychaetes, the growth zone eventually forms all of the trunk segments.

Trochophores may be planktotrophs and have long planktonic lives, or they may be lecithotrophs with a short planktonic existence. The prototroch is the swimming organ and, in trochophores that feed, it also collects suspended food particles. In feeding trochophores a food groove lies between the preoral prototroch and the postoral metatroch. The longer cilia of the preoral band drive water and suspended food particles backward. The shorter cilia of the postoral band beat in the opposite direction, and suspended particles are driven into the food groove between the two bands. The short cilia of the food groove transport the particles to the mouth (Fig. 9–5C)

NONSEGMENTED COELOMATE WORMS

This chapter covers three small phyla of coelomate protostomes, all of which lack segmentation and are similar in general appearance. Although probably not closely related to each other, each may preserve features associated with ancestral members of different evolutionary lines.

PHYLUM SIPUNCULA

The sipunculans are a group of approximately 320 species of marine animals, sometimes called peanut worms because of the resemblance of their contracted body to a shelled peanut, or starworms because of how their extended tentacles radiate from the head (Figs. 9–6; 9–7; 9–8C). Sipunculans are rather drab-colored worms that range in length from 2 mm to more than 72 cm, although most are less than 10 cm long. All are benthic animals, the majority living in shallow water. Some (e.g., *Sipunculus*) live in sand and mud and burrow actively, whereas others occupy mucus-lined excavations. Still others live in coral crevices, in empty mollusc shells *(Phascolion)* or annelid tubes, and in other sorts of protective retreats. Several species bore into coralline rock, and at least one bores into wood. Borers direct their anterior end toward the opening of the burrow from which they extend to feed. Densities as high as 700 individuals per square meter of coralline rock have been reported from Hawaii. Apparently, rock boring requires both mechanical abrasion and chemical softening. In parts of the tropical Indo-Pacific, sipunculans are consumed as human food.

External Structure

The cylindrical body of sipunculans is divided into a slender anterior section, called the **introvert,** and a swollen posterior **trunk** (Fig. 9–6). The introvert is the anterior end of the body, but it can be retracted

into the trunk (Fig. 9–7). The anterior end of the introvert bears the mouth, which is located below or is surrounded, at least in part, by a scalloped fringe, lobes, or tentacles. Most of these projections are ciliated, and each bears a deep ciliated groove on its

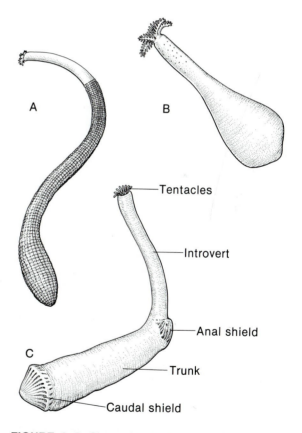

FIGURE 9–6. Sipunculan body forms. **A,** The sand burrower, *Sipunculus nudus* (10 cm). **B,** The crevice dweller, *Themiste lageniformis* (1 cm). **C,** The rock borer, *Paraspidosiphon klunzingeri* (2 cm). *(C, Redrawn from Sterrer, W. (Ed.). 1986. Marine Fauna and Flora of Bermuda. John Wiley & Sons, New York. plate 70, p. 227)*

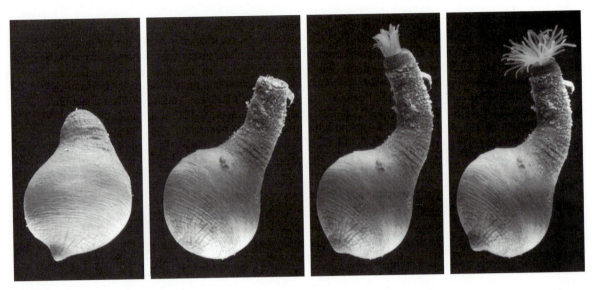

FIGURE 9–7. Introvert eversion in the sipunculan, *Themiste lageniformis* (1–2 cm).

inner side (Fig. 9–8C). Behind the anterior end, the surface of the introvert is typically covered with spines, tubercles, or other cuticular elements.

The trunk is cylindrical, and in some rock-boring sipunculans, the cuticle of the trunk is thickened and sometimes calcified at the anterior end to form a dorsal or collar-like **anal shield.** The shield is used to block the entrance of the burrow, like the closing plate on a snail's shell, when the introvert is invaginated (Fig. 9–9D).

Internal Organization and Physiology

The sipunculan body wall consists of a thick collagenous cuticle, a glandular epidermis, circular and longitudinal muscle layers, and a noncontractile ciliated peritoneum that encloses the coelom. One, two, or four **introvert retractor muscles,** depending on the species, originate on the ventral lining of the trunk coelom, span the cavity, and insert on the wall of the esophagus (Fig. 9–8A).

The locomotion of adult sipunculans is best developed in the burrowing species of the family Sipunculidae, at least one of which can also swim (Fig. 9–9A). In these, burrowing is accomplished by thrusting movements of the introvert and peristalsis of the trunk (Fig. 9–9B). The movements of rock, wood, and crevice dwellers, on the other hand, are largely restricted to extension and retraction of the introvert (Fig. 9–9C, D).

The nervous system is subepidermal and consists of a dorsal brain above the esophagus and circumesophageal connectives that join the brain to the ventral nerve cord. The unpaired nonganglionated nerve cord gives off a series of lateral branches that innervate the body wall muscles. In *Sipunculus nudus* and other species, the cord is pink and may contain hemerythrin. Sensory cells are particularly abundant on the end of the introvert that is used to probe the surrounding environment. They are also present elsewhere on the body. Many species of *Golfingia* have a pair of ciliated pits (nuchal organs) at the end of the introvert. A pair of pigmented eyes is typically embedded dorsally in the brain (Fig. 9–8A).

Sipunculans are apparently nonselective, suspension or deposit feeders, and particulate material is collected on their ciliated tentacles. In burrowing species, such as *Sipunculus,* the worm ingests the sand and silt through which it burrows. Some rock borers, such as *Phascolosoma antillarum,* spread the tentacular crown at the opening of the gallery while suspension feeding. Other hard-bottom species extend the introvert over the rock surface and ingest deposited material. The mechanism of ingestion is not certain, but apparently, food collected on the tentacles is ingested as the introvert and tentacles are retracted into the trunk.

The digestive tract is J-shaped, and the tubular intestine is long and complexly coiled. The mouth is at the tip of the introvert, and the anus is middorsal on the anterior end of the trunk except in *Onchnesoma,* in which it opens on the introvert. Sipunculans occupy blind galleries, and the anterior placement of the anus at the gallery opening promotes household sanitation. Behind the mouth, the esophagus descends into the trunk where it joins a long intestine wound into a double helix consisting of proximal descending and distal ascending coils. A longitudinal **spindle muscle** is in the axis of the coils, and when contracted, it compresses them. The intestinal coils an-

chor to the body wall by numerous radial **fixing muscles,** the functions of which are to suspend the gut and to help stir the intestinal contents. The ascending coil of the intestine joins a rectum before opening to the exterior at the anus. The inner surface of the intestine is folded to form a longitudinal **ciliary groove.** Digestion in sipunculans is extracellular in the intestinal lumen, and the function of the ciliary groove is probably to shunt water, which could dilute the intestinal enzymes, past the digestive region of the gut (Fig. 9–8A).

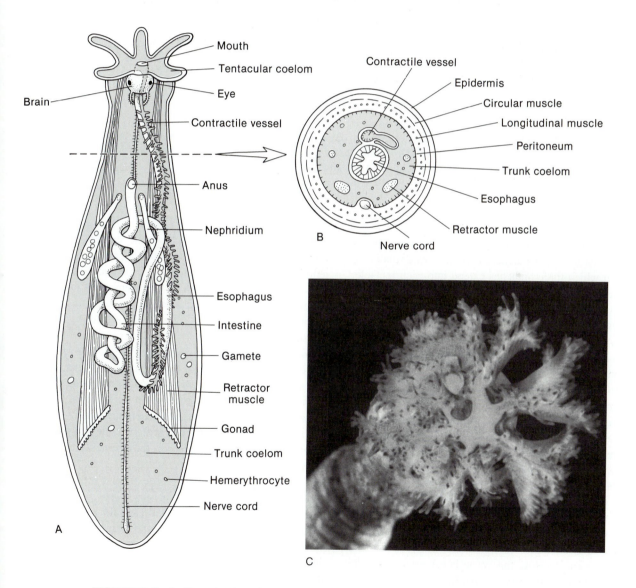

FIGURE 9–8. A, Organization of a generalized sipunculan that has a well-developed contractile vessel. **B,** Cross section of the body. **C,** Expanded tentacles of *Themiste alutacea.*

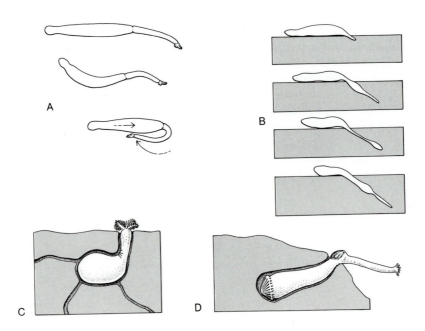

FIGURE 9–9. Swimming **(A)** and burrowing **(B)** by *Sipunculus nudus*. **(C)** *Themiste lageniformis,* a crevice and rock dweller. **D,** *Lithacrosiphon,* a rock-boring sipunculan having a calcareous anal shield that caps the opening when the introvert is retracted.

Sipunculans have two coelomic cavities. One is at the base of the tentacles where it forms a small ring from which three branches extend into each tentacle. The other, which is separated from the first by an elaborate septum, is large and voluminous and occupies the trunk (Fig. 9–8A). The coelomic fluid in both cavities is kept in circulation by cilia on some of the peritoneal cells and by contractions of the muscular body. Hemerythrocytes are the most common and conspicuous of the numerous coelomocytes. In addition to these, some sipunculans, such as *Sipunculus,* have microorgans called **ciliated urns.** These may be attached ("fixed") to the coelomic lining or may circulate ("free") in the coelomic fluid. Free urns swim using a band of cilia and resemble microscopic medusae (Fig. 9–10A). Because of their appearance, they were first thought to be symbionts rather than specialized structures functioning in internal defense. Swimming urns secrete mucus, which trails behind them and entangles foreign material and exhausted coelomocytes. How the waste-laden urns leave the body, if indeed they do, is unknown.

Sipunculans lack a blood-vascular system, and as a result, transport between the two coelomic cavities must occur across the tentacular/trunk septum. To facilitate exchange between the two compartments, the tentacular coelom has developed one or two diverticula, the **contractile vessels** (compensation sacs), which project posteriorly along the sides of the esophagus into the trunk coelom (Fig. 9–8A, B). These increase the surface area of the septum and thereby the rate of exchange between the two coelomic compartments. In some sipunculans, such as *Themiste,* the contractile vessels branch and may occupy much of the trunk coelom. In such cases, the contractile vessels superficially resemble blood vessels of other animals. The contractile vessels may also function as fluid reservoirs for the tentacles when they are retracted.

In the absence of blood vessels, coelomic fluid is the sole transport medium for gases and nutrients alike. The tentacles are important respiratory surfaces in rock and crevice dwellers, but the entire body wall is important for gas exchange in sand-burrowing species. In burrowers like *Sipunculus,* coelomic fluid flows from the trunk coelom into a series of longitudinal canals located immediately below the thin epidermis where gas exchange occurs. In *Xenosiphon,* outgrowths of these canals form slender gills that project outward from the body surface. Hemerythrin in corpuscles is found in both coelomic cavities. In *Themiste,* species of which are rock borers or crevice dwellers, the oxygen affinity of the trunk hemerythrin is higher than that of the tentacles. As a result, environmental oxygen bound at the tentacles and transported to the trunk in the contractile vessels is removed by the trunk hemerythrin and transported to the tissues of the body.

Sipunculans have one *(Phascolion)* or two elongated saclike nephridia in the anterior part of the trunk (Fig. 9–8A). Each has a ciliated funnel at its anterior end (Fig. 9–10B) and a ventrolateral pore, also at the anterior end. It is not certain whether sipunculan nephridia are filtration or secretion kidneys. Podocytes have been noted on the walls of the contractile vessels of *Themiste,* suggesting that filtration occurs there and that sipunculans have a metanephridial system. On the other hand, podocytes have also been observed on the outer surface of the

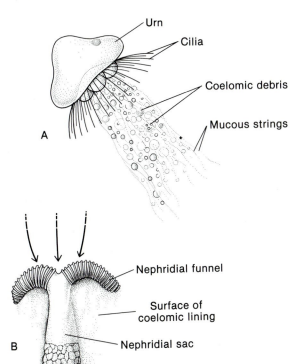

A

Urn

Cilia

Coelomic debris

Mucous strings

Nephridial funnel

Surface of coelomic lining

B

Nephridial sac

FIGURE 9–10. A, Ciliated urn of *Sipunculus nudus.* **B,** Cilia on the coelomic lining of *Siphonosoma cumanense* direct fluid toward the ciliated funnel of each nephridium. The lower lip of the funnel is attached to the coelomic lining, and the free upper lip (shown) arches upward like a scoop.

nephridium itself, an arrangement reminiscent of protonephridia. Whatever the arrangement, however, sipunculan nephridia function in osmoregulation *and* in gamete storage and maintenance prior to spawning.

Reproduction

The sexes are separate in most sipunculan species, and inconspicuous gonads are situated at the base of the introvert retractor muscles. Gametes are released from the gonads at an early stage and undergo maturation in the trunk coelom (Fig. 9–8A). When mature, they are removed from the coelomic fluid by the nephridial funnel and stored in the nephridial sacs (Fig. 9–10B).

Sipunculans spawn their gametes into seawater, and fertilization is external. The developing zygotes undergo spiral cleavage, and at the 64-cell stage show a typical molluscan cross (Fig. 9–4D). Embryos gastrulate by invagination or epiboly and may develop directly into small worms as in *Phascolion cryptus,* indirectly via a lecithotrophic trochophore as in *Phascolion strombus,* or indirectly via a secondary larva, called a **pelagosphera** (Fig. 9–11D).

The pelagosphera larva develops from the trochophore (with one exception), is usually planktotrophic and long-lived, and is the agent of long-range dispersal in sipunculans. The pelagosphera swims and feeds using a well developed metatrochal band of cilia that replaces the earlier prototroch. The larval head and metatroch can be retracted into the trunk. Metamorphosis of the pelagosphera involves lengthening of the trunk, loss of the terminal organ, lengthening and remodeling of the head to form the introvert and developing feeding tentacles.

Phylogeny

The evolutionary relationships of sipunculans are uncertain, but most investigators link them to either annelids or molluscs, a few to echinoderms. Although similarities to annelids include a worm-shaped body, ventral nerve cord, spiral cleavage, and a trochophore larva, these characteristics are widespread among protostomes and are not unique to sipunculans and annelids. The hallmark of annelids is segmentation (see Chapter 11), a trait that is absent in sipunculans.

A relationship to echinoderms is suggested by similarities between sipunculan tentacles and those of sea cucumbers (holothuroids, see Chapter 18). Other

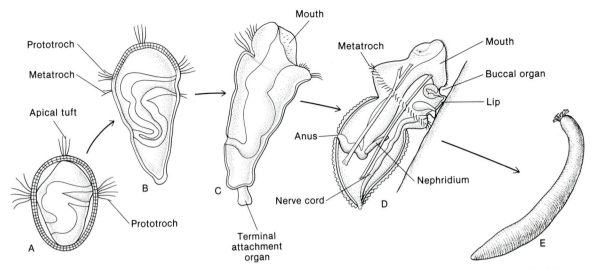

FIGURE 9–11. In sipunculans, such as *Golfingia misakiana,* a feeding larva called a pelagosphera **(C, D)** succeeds the earlier nonfeeding trochophore **(A, B).** The pelagosphera eventually settles and crawls over the bottom on its lip. As it does so, the eversible buccal organ probes the substratum and may dislodge food particles to be ingested by the mouth. The head, lip, and metatrochal region **(D)** of the pelagosphera can be retracted into the trunk for protection. The pelagosphera metamorphoses into a young sipunculan **(E).** Metamorphosis in this species requires approximately two weeks to complete. *(All redrawn from Rice, M. E. 1978. Morphological and behavioral changes at metamorphosis in the Sipuncula. In Chia, F. S., and Rice, M. E. (Eds.): Settlement and Metamorphosis of Marine Invertebrate Larvae. Elsevier North Holland, Inc., New York pp. 83–102.)*

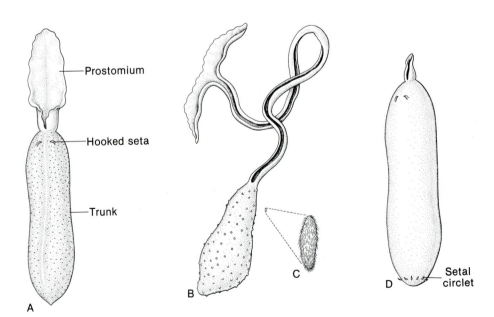

FIGURE 9–12. **A,** The echiuran, *Thalassema hartmani,* from the southeastern coast of the United States. **B–C,** The European, *Bonellia viridis.* **B,** Female. **C,** Dwarf male. **D,** The filter feeder, *Urechis caupo,* from the west coast of the United States. *(B, C, Modified and redrawn from MacGinitie, G. E., and MacGinitie, N. 1968. Natural History of Marine Animals, McGraw-Hill Book Co., New York. 523 pp.)*

than the tentacles, however, there are few correspondences between echinoderms and sipunculans.

The relationship to molluscs seems to be more firmly established by the appearance of molluscan features during the development of sipunculans. These include a molluscan cross shared uniquely with molluscs (Fig. 9–4C, D), and an eversible, scraping buccal organ and a muscular creeping lip in the pelagosphera, which are reminiscent of the molluscan buccal mass and foot. If the sipunculans are the nearest living relatives of the molluscs, however, the divergence between them must have been early indeed, certainly before the adoption of many typical molluscan features.

PHYLUM ECHIURA

Echiurans, or spoonworms, are sausage-shaped marine animals that resemble sipunculans in size and general habit (Fig. 9–12). Many species such as *Thalassema, Urechis,* and *Ikeda,* occupy burrows in sand and mud, whereas others live in rock and coral crevices (Fig. 9–13C). *Lissomyema mellita,* which is found along the southeastern coast of the United States, inhabits the test of dead sand dollars and fissures in discarded mollusc shells. When the worm is small, it enters the test and later becomes too large to leave (Fig. 9–13B). Echiurans range in size (trunk length) from approximately 1 cm *(Lissomyema)* to over 50 cm *(Urechis).* The majority of spoonworms

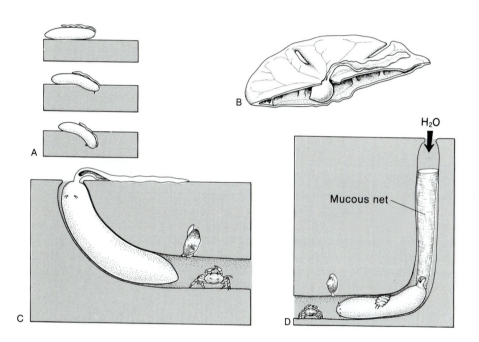

FIGURE 9–13 A, *Thalassema hartmani* uses its anterior setae to scoop away sand as the trunk undergoes a slow peristalsis. Sometimes, but not always, the ciliated prostomium is used to gain initial entry into the sand. **B,** The small *Lissomyema mellita* is a crevice dweller and may often be found among mollusc shells and in dead sand dollar tests. **C,** *Thalassema hartmani,* from the east coast of the United States, constructs a U-shaped burrow from which it extends its deposit-feeding prostomium. The security and resources within the burrow attract a minute clam, *Paramya subovata,* and a small filter-feeding crab, *Pinnixa lunzi.* On the west coast of the United States, the filter-feeding *Urechis caupo,* constructs a U-shaped burrow through which it pumps water and strains out food particles. Its burrow houses at least three commensals, the clam, *Cryptomya californica,* the crab, *Pinnixa franciscanus,* and the scaleworm, *Hesperonoe adventor. (D, Modified and redrawn from MacGinitie, G. E., and MacGinitie, N. 1968. Natural History of Marine Animals. McGraw-Hill Book Co., New York. 523 pp.)*

live in shallow water, but there are also deep-water species. In all, approximately 140 species have been described. Echiurans may be an important food in the diet of some fishes. In one dietary study of leopard sharks caught along the California coast, *Urechis* was the most important single food. Apparently, the sharks use suction to slurp the large spoonworms from their burrows.

External Structure

The echiuran body consists of two distinct regions, a nonsegmented cylindrical **trunk** and a flat, ribbon-like, anterior **prostomium,** or proboscis (Fig. 9–12). The whitish prostomium, which is situated anterior to the mouth, is ciliated ventrally and glandular else-where (Figs. 9–14; 9–15B). It is highly mobile and is usually capable of extensions that are at least ten times its retracted length. In the Japanese echiuran, *Ikeda,* a specimen with a trunk length of 40 cm, a prostomium of 1.5 m has been recorded. Similarly, a female *Bonellia* (Fig. 9–12B) that has an 8-cm long trunk can have a prostomium that extends to over 2 m. On the other hand, the Pacific echiuran, *Urechis,* has a prostomium that is always much shorter than its trunk (Fig. 9–12D).

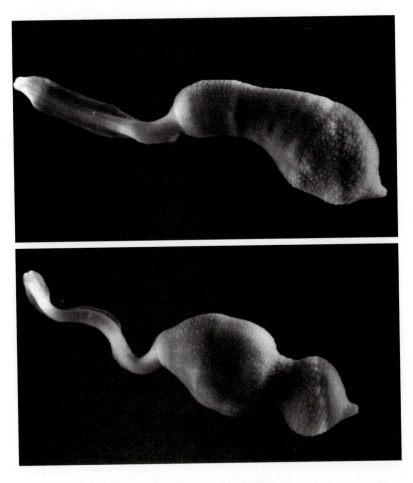

FIGURE 9–14. During extension, the prostomium is flattened against the substratum as it creeps slowly outward on its ventral cilia (upper photograph). Once extended, regions of the prostomium curl ventrally (lower photograph) to form an inverted gutter through which food is transported to the mouth at the anterior end of the trunk. While in the burrows, echiurans gener-ate slow peristaltic waves, which pass from front to rear along the trunk (lower photograph). Such movements create a ventilating water flow over the body. *(Lissomyema mellita, Beaufort, NC.)*

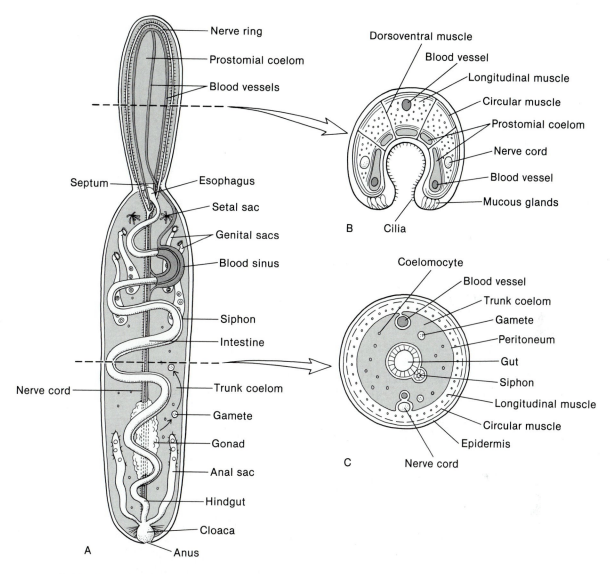

FIGURE 9–15. **A,** Organization of a generalized echiuran. **B,** Cross section of prostomium that is curled ventrally in feeding position. **C,** Cross section of the trunk.

The echiuran trunk may be gray, reddish-brown, rose, or red. Adult and larval *Bonellia* are greenish because of **bonellin,** a dermal pigment that may have antibiotic properties. A pair of short, hooked, chitinous setae occurs ventrally on the anterior part of the trunk (Fig. 9–12). When they are extended from their setal sacs, they curve posteriorly and are used for digging as the animal burrows (Fig. 9–13A). In addition to the anterior setae, some echiurans, such as *Echiurus* and *Urechis,* have one or two circles of hooked setae around the posterior extremity of the

trunk (Fig. 9–12D). These setae are used for burrow maintenance and, perhaps, for anchorage.

Internal Organization and Physiology

The body wall of the trunk consists of a thin, poorly developed cuticle, glandular epidermis, a musculature consisting of at least three layers, and a ciliated peritoneum that lines the coelom (Fig. 9–15B, C). The wall of the prostomium differs from that of the trunk in being ciliated ventrally, lacking a spacious coelom, and having especially well developed longitudinal and dorsoventral muscles (Fig. 9–15B). These muscles are responsible for retracting and flattening the prostomium, respectively. Echiurans are slow, sluggish animals whose movements are predominantly a slow peristalsis of the trunk (Fig. 9–14), used for burrowing and burrow ventilation, and extension/retraction movements of the prostomium, used in feeding and probably to escape predators.

The subepidermal nervous system consists of a nerve ring around the perimeter of the prostomium and a nonganglionated ventral nerve cord that extends through the trunk (Fig. 9–15A). The nerve cord gives off numerous lateral branches into the musculature and is often pink from the presence of hemoglobin. There are no specialized sensory organs in adult echiurans.

Most echiurans are deposit feeders but at least one, *Urechis caupo* (Fig. 9–13D), which lives along the California coast, filter feeds like the polychaete, *Chaetopterus* (Chapter 11). *Urechis* casts a mucous net from a glandular girdle on its anterior trunk to its burrow wall and then pumps water through it at a rate of approximately 1 l per hour. Virtually all particles, including typical plankton, are trapped on the net as the water passes through. When loaded with food, the net is detached from the body, seized by the prostomium, and swallowed. *Urechis* is often called the "Innkeeper" because it provides a protected ventilated burrow and is a harmless landlord to several commensals, including a polychaete scaleworm, a tiny clam, a filter-feeding crab, and a goby (Fig. 9–13D). Similar commensals are common in the burrows of other species of spoonworms (Fig. 9–13C).

Typical deposit feeders, such as *Thalassema* and *Bonellia*, extend the prostomium onto the surface of sediment above their burrows. The prostomium glides outward from the trunk on its ventral ciliated surface (Fig. 9–14). When extended, the prostomium curls ventrally to form a gutter along which deposited particles are transported to the mouth at its base. Any unusual disturbance to the prostomium causes its rapid retraction.

The digestive tract is extremely long and coiled and loosely suspended in the coelom (Fig. 9–15A). In one specimen of *Thalassema hartmani,* for example, the uncoiled and outstretched gut was ten times longer than the trunk. The mouth is situated ventrally at the base of the prostomium, and the anus opens on the posterior end of the trunk. The intestine, which constitutes most of the tract, is the site of digestion (probably extracellular). An accessory gut, or **siphon,** originates from the anterior intestine, runs parallel to the intestine, and rejoins it posteriorly where the intestine joins the hindgut (9–15A, C). The siphon is an intestinal bypass that probably transports water ingested with food around the digestive region of the gut. It is thus functionally similar to the intestinal ciliary groove of sipunculans and the intestinal siphon of sea urchins (Chapter 18). The hindgut widens into a cloaca, which receives the excretory tubules, before opening to the exterior at the anus.

Echiurans, like sipunculans, have two coelomic cavities that are separated from each other by a septum (Fig. 9–15A). The trunk coelom is voluminous and unpartitioned except by partial mesenteries and radial muscles between the body wall and gut. Fluid is circulated throughout the coelom by muscular contractions of the body and by the cilia on the coelomic lining and contains erythrocytes, amebocytes, and germ cells. The prostomial coelom, a small compartment restricted to the ventral part of the prostomium (Fig. 9–15B), contains fluid only; coelomocytes are absent.

A blood-vascular system is present in all echiurans, except *Urechis*. The system consists of a blood sinus around the foregut from which a dorsal median vessel transports blood into the prostomium. Blood returns to the trunk in the lateral prostomial vessels, which unite in the trunk to form a ventral longitudinal vessel. Branches from the ventral vessel unite with the gut sinus to complete the circuit (Fig. 9–15A). A respiratory pigment is absent and the blood is colorless, thus a role in gas exchange seems unlikely. It is more probable that the blood transports nutrients from the gut primarily to the prostomium, nervous system, and musculature.

Gas exchange in echiurans presumably occurs across the general body wall of both trunk and prostomium. While in the burrow, peristaltic ventilation movements of the body (Figs. 9–14; 9–15) facilitate exchange across the trunk. Because the proboscis is flat and ribbonlike, oxygen transport to it is probably by simple diffusion. The trunk, however, is large, and oxygen that diffuses across the body wall is transported internally within coelomic erythrocytes. Body wall gas exchange in the large-bodied *Urechis* is augmented by spasmodic intake of water through the anus, with the cloaca and anal sacs functioning as a water "lung."

Excretion in echiurans is accomplished by two specialized organs known as **anal sacs** (Fig. 9–15A). Each anal sac is a thin-walled, hollow diverticulum from the wall of the cloaca that extends into the trunk coelom. The anal sac surface is covered with numerous (sometimes hundreds) valved ciliated funnels through which coelomic fluid enters the sac. It is not known whether the coelomic fluid contains an ultrafiltrate of the blood, and it is therefore uncertain whether or not the anal sacs are part of a filtration excretory system. In any case, the anal sacs produce urine and then discharge it through the cloaca and anus by contraction of their muscular walls.

Reproduction

The sexes are separate in echiurans and, in the Bonelliidae, sexual dimorphism is pronounced (Fig. 9–12B, C). The echiuran gonad is unpaired and attached to the wall of the ventral blood vessel (Fig. 9–15A). Gametocytes are released from the gonad into the trunk coelom, where they mature into gametes. The gametes are then removed from the coelomic fluid by ciliated funnels on paired storage organs, the **genital**

sacs (Fig. 9–15A). The sacs and their funnels resemble metanephridial ducts in annelids and sipunculans and may be homologous to them. However, in echiurans, they do not function in excretion but are involved solely in the storage of gametes. When the animals spawn, the genital sacs contract and squeeze the gametes out through ventral pores. Depending on the species, echiurans can have one *(Bonellia)* to hundreds *(Ikeda)* of pairs of genital sacs but most have two pairs.

Gametes are shed from the genital sacs into seawater, where fertilization occurs. The zygote undergoes spiral cleavage, and a typical annelid cross (Fig. 9–4A) appears early in development. Development is usually indirect and leads to a planktotrophic trochophore larva (often with two eyespots) in which both prototroch and telotroch are well developed and sometimes subdivided into additional trochs around the larva. During metamorphosis, the pretrochal region becomes the prostomium, and the growth zone and pygidium become the trunk (Fig. 9–16).

Reproduction in *Bonellia* differs from that of other echiurans. A dwarf male (Fig. 9–12C) permanently inhabits the unpaired genital sac of the female. The male not only fertilizes the eggs, but at oviposi-

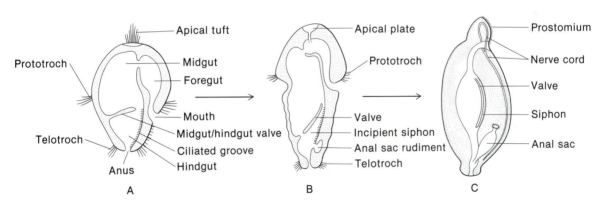

FIGURE 9–16. A, The echiuran trochophore has two well-developed ciliary bands, the prototroch and telotroch. **B–C,** During metamorphosis, the preoral region of the trochophore becomes the prostomium, and the postoral region becomes the trunk. The larval midgut/hindgut valve, a sheet of tissue, shifts ventrally during metamorphosis and roofs over the ciliated intestinal groove to form the intestinal siphon. Anal sacs develop from hindgut diverticula, and the nervous system, as in most protostomes, develops from the larval apical plate. *(After Newby, W. W. 1940. From Pilger, J. F. 1978. Settlement and metamorphosis in the Echiura: A review. In Chia, F-S, and Rice, M. E. (Eds.): Settlement and Metamorphosis of Marine Invertebrate Larvae. Elsevier North Holland Inc. Biomedical Press, New York. pp. 103–112.)*

tion, secretes gelatinous material used to bind the eggs together. The yolky eggs develop into lecithotrophic trochophores. If the short-lived larvae settle on an adult female prostomium, most become males; if they settle apart from a female, they metamorphose into juvenile females. Larvae induced to become males develop an adhesive organ with which they attach to the prostomium of the female. Later, they find their way into the genital sacs, where they reside until required to fertilize the eggs. Male bonelliids are necessarily much smaller than females; do not have a prostomium, mouth, anus, or blood-vascular system; and meet their metabolic needs by exchange with female fluids in which they are bathed. The small ciliated body consists of a gonad, seminal vesicle, and a pair of protonephridia.

Phylogeny

The particular combination of features found in echiurans, for example, an elongate prostomium and a nonsegmented trunk, clearly sets them apart from other worms as a distinct phylum. Nevertheless, characteristics of their early development, such as a trochophore larva and an annelidan cross, suggest a close alliance with annelids. In fact, it is only the absence of segmentation in echiurans that distinguishes them from the Annelida. It therefore seems probable that echiurans and annelids evolved along separate lines from a nonsegmented (or oligomeric) ancestor.

PHYLUM PRIAPULIDA

The phylum Priapulida consists of only 16 living and 11 fossil species of cucumber-shaped or wormlike marine animals. Priapulids are benthic invertebrates that live buried in sand and mud in shallow and deep water. Most of the large-bodied species live in cold water, ranging northward from Massachusetts and California off the North American coast but occurring also in the Baltic Sea off Siberia, and a few species have been taken from Antarctic waters. Tiny species of *Tubiluchus* and other genera are distributed widely, including in the tropics.

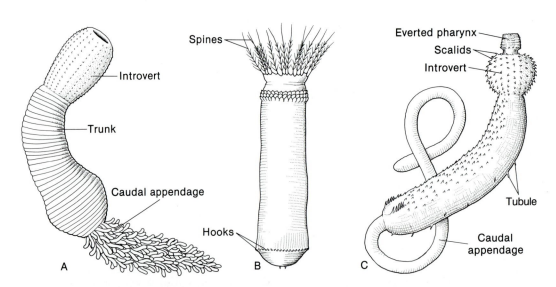

FIGURE 9–17. A, The large (8 cm) cold-water *Priapulus caudatus*. **B,** The small (2–3 mm) tube-dwelling *Maccabeus tentaculatus* from the Mediterranean. This animal apparently uses its spiny tentacles as a spring trap to snare small prey. **C,** The tiny (less than 2 mm) *Tubiluchus corallicola* burrows in tropical coralline sands. *(All modified and redrawn. A, After Theel from Hyman. B, Combined from Por, F. D. 1972. Priapulida from deep bottoms near Cyprus. Israel J. Zool. 21:525–528; and Calloway, C. B. 1982. Priapulida. In Parker, S. P. (Ed.): Synopsis and Classification of Living Organisms. McGraw-Hill Book Co., New York. pp. 941–944. C, From drawing by Brian Marcotte.)*

External Structure

The cylindrical body of priapulids ranges in length from 0.5 mm to 30 cm and is divided into an anterior **introvert** (proboscis) and a posterior **trunk** (abdomen) (Fig. 9–17A). The introvert, which constitutes the anterior part of the animal, is somewhat barrel-shaped and bears longitudinal rows of riblike, conical projections called **scalids** (Fig. 9–17C). In one species, the mouth is surrounded by a crown of long-branched, tentacle-like spines in addition to the normal scalids (Fig. 9–17B). The introvert invaginates into the anterior region of the trunk, which is covered with small variable projections. In the little *Tubiluchus corallicola,* the trunk bears a long, terminal tail, that is probably used to anchor the body in the surrounding sediment (Fig. 9–17C). In the much

larger *Priapulus,* however, the posterior end of the trunk bears one or two appendages, each consisting of a hollow stalk from which extend many short, finger-like diverticula (Fig. 9–17A). A gas exchange or chemoreceptive function has been suggested for these structures, but neither has yet been demonstrated. Setae-like hooks on the posterior end of the body of *Maccabeus tentaculatus* and *Meiopriapulus fijiensis* help grip the sediment as the animals burrow tail first (Fig. 9–17B).

Internal Organization and Physiology

The body wall of the priapulids is composed of a well-developed chitinous cuticle, a nonciliated epidermis that contains the nervous system, and a layer of circular and longitudinal muscles (Fig. 9–18). The

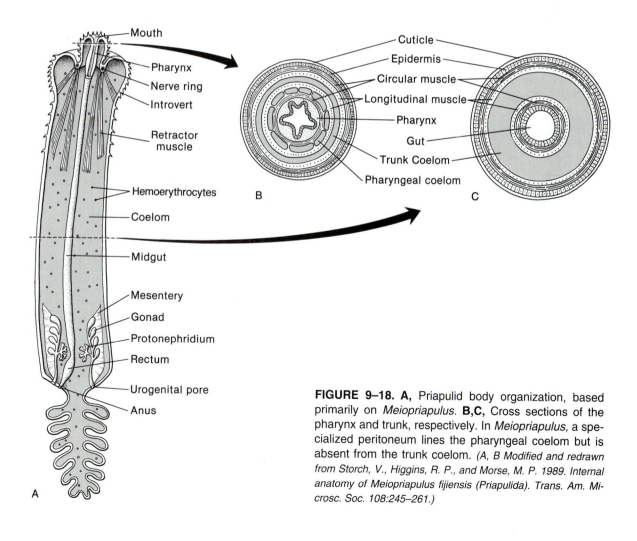

FIGURE 9–18. A, Priapulid body organization, based primarily on *Meiopriapulus.* **B,C,** Cross sections of the pharynx and trunk, respectively. In *Meiopriapulus,* a specialized peritoneum lines the pharyngeal coelom but is absent from the trunk coelom. *(A, B Modified and redrawn from Storch, V., Higgins, R. P., and Morse, M. P. 1989. Internal anatomy of Meiopriapulus fijiensis (Priapulida). Trans. Am. Microsc. Soc. 108:245–261.)*

cuticle is molted periodically as the animal grows. The body cavity of the trunk is undivided and bounded on the body wall and gut sides by musculature. A peritoneum is absent, except in a specialized pharyngeal cavity found in *Meiopriapulus fijiensis* (Fig. 9–18B). The body cavity fluid contains phagocytic amebocytes and respiratory hemerythrocytes, both of which are circulated by action of the body wall muscles. The body cavity is the sole circulatory system; it also functions as a hydrostatic skeleton.

Historically, the priapulid body cavity has been regarded as either a coelom or a pseudocoel (see Chapter 5). Although the issue remains controversial, it should be recalled that primitive coeloms are lined with a muscular epithelium but lack a peritoneum. Only evolutionarily derived coelomic cavities, like the pharyngeal cavity of *Meiopriapulus fijiensis*, have a peritoneum. Thus, it is possible that the priapulid trunk cavity is a coelom with a primitive grade of organization.

The larger priapulids burrow through sediments using a combination of body wall peristalsis and eversion/retraction movements of the introvert (Fig. 9–19).

Large priapulids, such as *Priapulus* (Fig. 9–17A) and *Halicryptus,* are thought to be carnivores, whereas some small species, such as *Tubiluchus* (Fig. 9–17C) and *Meiopriapulus,* are probably deposit-feeding bacteriovores. A few, for example *Maccabeus,* may be suspension feeding carnivores (Fig. 9–17B). The carnivores feed on soft-bodied, slow-moving invertebrates, particularly polychaete worms, and sometimes each other. The mouth and pharyngeal regions can be everted during feeding (Fig. 9–17C), and oral and pharyngeal teeth may be used to seize prey.

The terminal mouth leads into a muscular pumping pharynx that is lined with cuticle, part of which is modified as teeth. The straight tubular midgut (intestine) lacks cuticle and cilia but is lined with a brush border of absorptive microvilli. The midgut is surrounded by a sheath of circular and longitudinal mus-

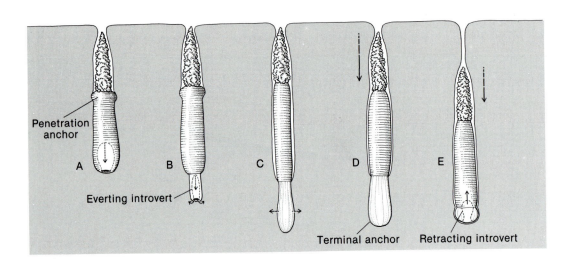

FIGURE 9–19. Priapulids, such as *Priapulus caudatus* (shown), burrow primarily by using eversion/retraction movements of the introvert. In **A** and **B,** the introvert pushes downward into the sediment as the expanded body forms a penetration anchor. The introvert then swells to form a terminal anchor **(C)** as the trunk is drawn forward **(D).** The trunk continues to advance downward as the introvert is retracted **(E)** in preparation for the next burrowing cycle. *(Modified and redrawn from Elder, H.Y., and Hunter, R.D. 1980. Burrowing of Priapulus caudatus (Vermes) and the significance of the direct peristatic wave. J. Zool Lond. 191:333–351.)*

cles. A combination of extracellular and intracellular digestion occurs in the midgut. The midgut joins a short terminal rectum that, like the pharynx, is lined with cuticle. The anus is located at the terminal end of the trunk (Fig. 9–18A).

The priapulid nervous system is intraepidermal in position and consists of a nerve ring around the anterior end of the pharynx and a single, midventral, nonganglionated cord that terminates in a rectal ganglion. Sensory papillae on the trunk of some species are called **tubuli** (Fig. 9–17C).

The urogenital system consists of two elongated ducts, each suspended in a mesentery beside the posterior region of the intestine and joined anteriorly by a protonephridial tubule and laterally by the gonads. The protonephridial tubule bears many monociliated terminal cells that are suspended in the coelom but separated from it by a basal lamina. Each urogenital duct opens to the exterior at a pore situated on the posterior end of the trunk (Fig. 9–18A).

Reproduction

The sexes are separate, and fertilization is external in large-bodied species, but internal fertilization probably occurs in most small meiofaunal species. The egg is reported to cleave radially, and a nonciliated sterogastrula is the hatching stage (Fig. 9–20A, B). The priapulid larva, which inhabits bottom mud as does the adult, has an introvert that can be retracted into the trunk (Fig. 9–20C, D). The trunk cuticle is thick and is called a lorica. The lorica disappears at metamorphosis.

Phylogeny

The evolutionary relationships of the Priapulida are poorly understood primarily because the phylum is only now beginning to be studied using modern techniques and because priapulid development requires further study. Nevertheless, current opinions regarding their nearest relatives seem to fall into two camps, and it is at present impossible to choose between them. One suggests that priapulids are closely related to phyla, such as kinorhynchs and loriciferans, because of the shared occurrences of a molted cuticle, an eversible anterior end that bears hooks, and, at least in loriciferans and priapulids, a urogenital system with protonephridia. The second opinion is that priapulids and sipunculans are close relatives and that

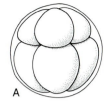
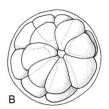

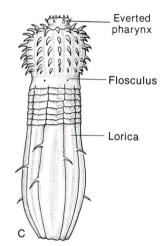
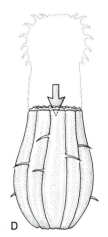

Everted pharynx

Flosculus

Lorica

FIGURE 9–20. Priapulid early development and larval morphogenesis are imperfectly known. The few observations of early development **(A,B)** indicate that cleavage is radial, holoblastic, and nearly equal. The larva **(C,D;** stage unknown) is dorsoventrally flattened and has a thick trunk cuticle, the lorica, into which the introvert can be retracted. *(A,B, Redrawn after Ginkine from Grassé.)*

both share a primitive position within the protostomes. This idea is supported by the occurrence in both phyla of an introvert with hooks, molting (limited in sipunculans), a similar nervous system and musculature, and hemerythrin as a respiratory pigment.

SUMMARY

1. The phylum Sipuncula comprises approximately 320 species of nonsegmented marine worms that burrow in sand and mud or live in coral or wood excavations or in old mollusc shells.

are the mouth and a partial or complete circlet of tentacles or lobes that are used in deposit and suspension feeding.

3. Sipunculans have two coelomic cavities, one in the tentacles and one in the trunk. A blood-vascular system is absent. The respiratory pigment is hemerythrin, which occurs in coelomocytes of both coeloms, in some muscles, and in some nerves.

4. Sipunculans have paired, saclike metanephridial ducts that function in excretion and as genital ducts.

5. The sipunculan subepidermal nervous system consists of a ganglionated circumesophageal brain and a nonganglionated ventral nerve cord.

6. Cleavage is spiral, and some species have a trochophore larva, but the pelagosphera, which succeeds the trochophore, is the characteristic sipunculan larva. Developmental features suggest an evolutionary relationship with molluscs.

7. The phylum Echiura contains some 140 species of marine worms that burrow in sand and mud or live in rock or coral crevices.

8. The cylindrical trunk has an anterior flattened prostomium (or proboscis) that bears the mouth at its base. A single pair of large, ventrolateral setae is located near the anterior end of the trunk.

9. Echiurans are deposit or suspension feeders on detritus that is collected in mucus and transported to the mouth by the prostomium.

10. Echiurans have a small coelomic cavity in the prostomium and a large one in the trunk. The latter contains erythrocytes with hemoglobin. A blood-vascular system, which contains colorless blood, extends throughout the body.

11. Echiuran nephridia, or genital sacs, function only in gamete storage. The excretory organs are anal sacs.

12. The echiuran subepidermal nervous system consists of a nonganglionated anterior nerve ring and a nonganglionated ventral nerve cord.

13. Cleavage is spiral, and there is a trochophore larva. Developmental features, especially the presence of an annelidan cross, suggest a close relationship between annelids and echiurans.

14. The phylum Priapulida contains 16 species of marine worms that live in mud and sand.

15. The cylindrical trunk bears a swollen, spiny, retractile introvert (or proboscis) that is used in prey capture and burrowing. The mouth is at the tip of the introvert.

16. Priapulids have one (sometimes two) coelomic cavity, which occupies the trunk and introvert and lacks a peritoneum. Hemerythrin-containing coelomocytes occur in the body cavity. A blood-vascular system is absent.

17. Priapulids have two protonephridia and two gonads that share a common duct leading to the exterior.

18. The priapulid nervous system is located within the epidermis. It is composed of a nonganglionated circumpharyngeal brain and a single nonganglionated ventral nerve cord.

19. Early development is poorly known, but cleavage is said to be radial. A loricate larva occurs in most species. Priapulids molt their cuticles as they grow.

Anat. *103:*389–423. (In German but with an English abstract and good illustrations.)

REFERENCES

Burrowing

Trueman, E. R., and Ansell, A. D. 1969. The mechanisms of burrowing into soft substrata by marine animals. Oceanogr. Mar. Biol. Ann. Rev. *7:*315–366.

Respiratory Pigments

Klippenstein, G. L. 1980. Structural aspects of hemerythrin and myohemerythrin. Am. Zool. *20:*39–51.

Mangum, C. P. 1985. Oxygen transport in invertebrates. Am. J. Physiol. *248:*R505–R514.

Mangum, C. P. 1992a. Respiratory function of the red blood cells hemoglobins of six animal phyla. Adv. Comp. Environ. Physiol. *13:*117–149.

Mangum, C. P. 1992b. Physiological function of the hemerythrins. Adv. Comp. Environ. Physiol. *13:*173–192.

Developmental Patterns and Trochophore Larvae

MacBride, E. W. 1914. Text-Book of Embryology. MacMillan and Co., Ltd., London. 692 pp.

Salvini-Plawen, L. v. 1980. What is a trochophora? An analysis of protostomian marine larval types. Zool. Jb.

Anat. *103:*389–423. (In German but with an English abstract and good illustrations.)

Strathmann, R. R., John, T. L., and Fonseca, J. R. C. 1972. Suspension feeding by marine invertebrate larvae: Clearance of particles by ciliated bands of a rotifer, pluteus, and trochophore. Biol. Bull. *142:*505–519.

General

Anderson, D. T. 1973. Embryology and Phylogeny in Annelids and Arthropods. Pergamon Press, New York. 495 pp. (Detailed accounts of the embryology of onychophorans and uniramian arthropods and the phylogenetic implications of the embryonic patterns.)

Clark, R. B. 1969. Systematics and phylogeny; Annelida, Echiura, Sipuncula. *In* Florkin, M., and Scheer, B. T. (Eds.): Chemical Zoology. Vol. 4. Academic Press, New York. pp. 1–68.

Dawydoff, C. 1959. Classes des Echiuriens et Priapuliens. *In* Grassé, P. (Ed.): Traité de Zoologie. Vol. 5. Pt. 1. Masson et Cie, Paris. pp. 855–926.

Giese, A. C., and Pearse, J. S. (Eds.): 1975. Reproduction of Marine Invertebrates. Vol. 2. Entoprocts and Lesser Coelomates. Vol. 3. Annelids and Echiurans. Academic Press, New York. (Vol 2 includes tardigrades, priapulids, sipunculids, and pogonophorans.)

Hulings, N. C. (Ed.): Proceedings of the First International Conference on Meiofauna. Smithson. Contrib. Zool. *76:*25–31.

Hyman, L. H., 1951. The Invertebrates: Acanthocephala, Aschelminthes, and Entoprocta. Vol. 3. McGraw-Hill, New York. (Out of date in many areas, but still useful.)

Kohn, A. J., and Rice, M. E. 1971. Biology of Sipuncula and Echiura. BioSci. *21:*583–584. (A brief review of a symposium on the biology of these two phyla.)

Malakhov, V. V. 1980. Cephalorhyncha, a new type of animal kingdom uniting Priapulida, Kinorhyncha, Gordiacea, and a system of Aschelminthes worms. Zool. Zh. *59:*485–499. (In English.)

Rice, M. E., and Todororic, M. (Eds.): 1970. Proceedings of the International Symposium on the Biology of the Sipuncula and Echiura. Vols. 1. (355 pp.) and 2 (254 pp.). Published by the Institute for Biological Research, Yugoslavia and Smithsonian Institution, Washington, DC.

Storch, V. 1984. Minor Pseudocoelomates. *In* Bereiter-Hahn, J., Matoltsy, A. G., and Richards, K. S. Biology of the Integument. Vol. 1. Springer-Verlag, New York. pp. 242–268.

Sipuncula

Cutler, E. B. 1973. Sipuncula of the western North Atlantic. Bull. Am. Mus. Nat. Hist. *152:*103–204. (Key and detailed descriptions of sipunculans on the western side of the North Atlantic.)

Cutler, E. B. 1977. Sipuncula. Marine flora and fauna of the northeastern U.S. NOAA Technical Report NMFS Circular 403. U.S. Government Printing Office, Washington, DC.

Gibbs, P. E. 1977. British Sipunculans. Synopses of the British Fauna No. 12. Academic Press, London.

Hyman, L. H., 1959. The Invertebrates. Vol. 5. The Smaller Coelomate Groups. (Covers sipunculans, pp. 610–696.) McGraw-Hill, New York.

Mangum, C. P., and Burnett, L. E. 1987. Response of sipunculid hemerythrins to inorganic ions and CO_2. J. Exper. Zool. *244:*59–65.

Moya, J., and Serrano, T. 1984. Podocyte-like cells in the nephridial tube of sipunculans. Cuad. Invest. Biol. (Bilbao) *5:*33–37.

Pilger, J. F. 1982. Ultrastructure of the tentacles of *Themiste lageniformis* (Sipuncula). Zoomorphology. *100:*143–156.

Rice, M. E. 1969. Possible boring structures of sipunculids. Am. Zool. *9:*803–812.

Rice, M. E. 1970. Asexual reproduction in a sipunculan worm. Science. *167:*1618–1620.

Rice, M. E. 1985. Sipuncula: Developmental evidence for phylogenetic inference. *In* Conway Morris, S., George, J. D., Gibson, R. et al. (Eds.): The Origins and Relationships of Lower Invertebrates. Oxford University Press, Oxford. pp. 274–296.

Rice, M. E. 1985. Description of a wood dwelling sipunculan, *Phascolosoma turnerae,* new species. Proc. Biol. Soc. Wash. *98:*54–60.

Rice, M. E. 1986. Larvae adrift: Patterns and problems in life histories of sipunculans. Amer. Zool. *21:*605–619.

Rice, M. E. 1989. Comparative observations of gametes, fertilization, and maturation in sipunculans. *In* Ryland, J. S., and Tyler, P. A. (Eds.): Reproduction, Genetics and Distributions of Marine Organisms. Olsen & Olsen, Fredensborg, Denmark. pp. 167–182.

Ruppert, E. E., and Rice, M. E. 1983. Structure, ultrastructure and function of the terminal organ of a pelagosphera larva (Sipuncula). Zoomorphology. *102:*143–163.

Walter, M. D. 1973. Feeding and studies on the gut content in sipunculids. Helgol. Wiss. Meeresunters. *25:*486–494.

Williams, J. A., and Margolis, S. U. 1974. Sipunculid burrows in coral reefs: Evidence for chemical and mechanical excavation. Pac. Sci. *28:*357–359.

Echiura

Harris, R. R., and Jaccarini, V. 1981. Structure and function of the anal sacs of *Bonellia viridis.* J. Mar. Biol. Assoc. U.K., *61:*413–430.

Jaccarini, V., Agius, L., Schembri, P. J. et al. 1983. Sex determination and larvae sexual interaction in *Bonellia viridis*. J. Exp. Mar. Biol. Ecol. *66:*25–40.

Jaccarini, V., and Schembri, P. J. 1977. Feeding and particle selection in the echiuran worm *Bonellia viridis*. J. Exp. Mar. Biol. Ecol. *28:*163–181.

Pilger, J. F. 1978. Settlement and metamorphosis in the Echiura: A review. *In* Chia, F-S., and Rice, M. E. Settlement and Metamorphosis of Marine Invertebrate Larvae. Elsevier North Holland Inc., New York. pp. 103–112.

Schembri, P. J., and Jaccarini, V. 1977. Locomotory and other movements of the trunk of *Bonellia viridis*. J. Zool. *182:*477–494.

Schuchert, P., and Rieger, R. M. 1990. Ultrastructural observations on the dwarf male of *Bonellia viridis*. Acta Zool. *71:*5–16.

Pripaulida

Higgins, R. P., and Storch, V. 1991. Evidence for direct development in *Meiopriapulus fijiensis* (Priapulida). Trans. Am. Microsc. Soc. *110*:37–46.

McLean, N. 1984. Amoebocytes in the lining of body cavity and mesenteries of *Priapulus caudatus*. Acta Zool. *65:*75–78.

Por, F. D. 1983. Class Seticoronaria and phylogeny of the phylum Priapulida. Zool. Scripta. *12:*267–272.

Storch, V. 1991. Priapulida. *In* Harrison, F. W., and Ruppert, E. E. (Eds.): Microscopic Anatomy of Invertebrates. Wiley-Liss Publishers, New York. pp. 333–350.

Storch, V., Higgins, R. P., and Morse, M. P. 1989. Internal anatomy of *Meiopriapulus fijiensis* (Priapulida). Trans. Am. Micrsoc. Soc. *108:*245–261.

10

THE MOLLUSCS

PRINCIPLES AND EMERGING PATTERNS

EYES AND IMAGE FORMATION

The widely distributed pigment-cup eyes described on p. 100 are adapted for animals that monitor light only for orientation, for detecting the shadow of a possible predator, or for determining photoperiod and setting circadian rhythms. Many animals that move rapidly and search for prey or other objects have evolved eyes capable of some degree of image formation and the neural circuitry to permit object discrimination. In the simplest terms, this requires that each photoreceptor or group of photoreceptors be able to determine the light intensity of some small part of the visual field. When various intensities across the field are differentiated and projected, as are the dots on a television screen, the differences register as contrast in an image. The quality of image formation among invertebrates varies greatly, but in general it is poor. The variation depends on external environmental factors as well as limitations in eye structure. There have been few complete studies of vision in invertebrates, and as a result, functional hypotheses are based largely on anatomical information.

The ability of an eye to produce a good image depends on the number of photoreceptors it has—a factor that varies enormously. There are about 200 photoreceptors in the pigment-cup eye of *Planaria maculata,* which measures 300 μm across the aperture (approximately 3000 receptors/mm²), compared with 70,000 photoreceptors/mm² in the eye of an octopus. But the large octopus eye is an exception. Most invertebrate eyes have far fewer photoreceptors, in large part because the animals are small, and receptor cells have a minimal size. As a result, the eyes of small animals would have to be disporportionately large to accomodate enough receptors to form an image (see Fig. 11–43). A lens improves the quality of the image by bending light (focusing) onto the retinal receptors so that all light emanating from one point in the visual field converges at one point on the retina. Light is bent, or refracted, when it passes from one medium into another. In aquatic animals, the cornea plays a minimal role in focusing because its index of refraction is not much different from that of water. The lens plays the principal role in focusing only if it is composed of a material that has a higher index of refraction than the medium from which the light is entering. Many invertebrates, such as many polychaete annelid worms and most prosobranch molluscs, have a "lens" of soft material that fills up much of the interior of the eye, but it contributes little to the eye's focusing ability. True lenses are usually spherical. A spherical lens produces aberrations, however, because light striking different parts of the curved surface do not all meet at the same point, blurring the image. This problem has been solved in some animals by a gradient of density from the outside of the lens to the center. A good quality focus also requires space between the back of the lens and the retina. The eyes of very few aquatic invertebrates meet all these conditions: a large number of photoreceptors; a highly developed, dense lens; and a space between the lens and the retina. The notable examples are the pelagic alciopid polychaetes, the heteropod gastropods, and the cephalopods, all of which are raptorial, swimming carnivores. On land, where light passes through air, the cornea plays a primary role in refraction, and the lens is secondary. In humans and other terrestrial mammals, for example, the lens functions largely for focal adjustment, the primary task having been accomplished by the cornea. Among terrestrial invertebrates, some spiders with simple eyes and certain crustaceans and insects with compound eyes achieve the highest degree of object discrimination. Their eyes will be discussed later. However, some periwinkles *(Littorina)* and land snails *(Helix)* appear to be able to recognize vertical bars (marsh grass and other vegetation).

Active animals that are nocturnal or live in dimly lighted habitats must capture the maximum amount of light over a given unit of time. Light capture is facilitated by an eye with a large pupil and large photoreceptors. The larger the eye the better, but because the small size of the animal limits the size of the eye, image resolution ability is usually inversely related to light-capturing ability. Some animals, such as scallops, wolf spiders, and some insects, possess a reflecting layer of pigment (tapetum) behind the retina, which bounces light back to the photoreceptors, improving the performance of the eye under conditions of dim light.

Light waves coming from the sun across the sky oscillate at right angles to the path of the light rays. Sky light is therefore said to be polarized. A few animals, such as certain spiders, amphipods, bees, and sea hares, have the photosensitive pigment in the membrane of certain specific receptors oriented so as to be sensitive to light only when aligned with the direction of polarization. Such animals can use polarized sky light to determine the sun position even when the sun is hidden. The sun position provides navigational orientation to habitat, food sources, and so forth.

CHEMOSYNTHETIC SYMBIONTS

An assortment of free-living invertebrates—pogonophoran worms, a few segmented worms (oligochaetes), and a small number of bivalve molluscs—are gutless or have a greatly reduced gut. For some of these animals, the gutless condition has been long known and very puzzling. How do they feed? Are they living on dissolved organic substances absorbed from sea water? The puzzle has been solved in recent years. They depend on symbiotic chemosynthetic bacteria. All of these gutless animals live in habitats where the hydrogen sulfide content of the surrounding water is very high—a characteristic of intertidal or subtidal black sediments that contain a high level of organic matter and are anoxic. The sediments smell a little like rotten eggs and account for the sulfur taste of coastal drinking water in many places. Deep-sea hydrothermal vents are also high in hydrogen sulfide. Here seawater circulates through subsurface channels and crevices in areas of sea floor spreading and is heated. It emerges through vents, like hot springs, laden with hydrogen sulfide along with various metallic salts. Various deep-sea animals have become adapted for living in the vicinity of these thermal vents. Both the gutless animals of hydrothermal vents and those living in anoxic sands harbor symbiotic bacteria that are capable of oxidizing the hydrogen sulfide to sulfur. The energy liberated in the oxidation is used by the bacteria to fix carbon, that is, they synthesize organic compounds from CO_2. The bacteria are located in different places within the host animal, such as in the gills of bivalves; share the chemosynthate with the host; and at least in some cases, have hydrogen sulfide delivered to them by the host's blood. Chemosynthesis forms the base of the food chain within the assemblage of animals living around the thermal vents and, in part, within the food chain of animals in anaerobic sands. Thus, the mutualistic relationship of chemosynthetic bacteria and their host invertebrates is a striking parallel to the mutualism between photosynthetic zooxanthellae and zoochlorellae and their animal hosts (Chapter 4).

SPERMATOPHORES

By keeping sperm and eggs concentrated, internal fertilization enhances the likelihood of fertilization and conserves gametes. This concentration is usually accomplished by retaining eggs in the body of the females and by transferring sperm directly from the male gonopore into the female gonopore, following which sperm are commonly stored within a seminal receptacle. In the absence of direct transfer of sperm from male system to female system, some animals package sperm and transfer the package, called a spermatophore, thereby avoiding the waste of broadcasting sperm into the surrounding water. Spermatophores commonly have shapes characteristic of the species. For example, in squids and octopods, the spermatophore looks like a baseball bat (Fig. 10–111A). Primitively, in aquatic animals, the spermatophores are transferred to some part of the female body outside of the reproductive tract but within easy access of the eggs at discharge. In some groups, however, spermatophores have come to be directly transferred into the female tract. After transfer, the spermatophore disintegrates and releases sperm. Spermatophores evolved numerous times within various groups of animals, and the conditions of origin and transfer are somewhat different in each. As we shall see, spermatophores appear to have been the initial solution of terrestrial arthropods to the problem of sperm transfer on land (p. 619).

LARVAL SETTLEMENT

As we have already seen, many benthic marine animals possess a larval stage in their life history. Indeed, there are at least some representatives of most phyla with marine larvae. A major advantage of a larval stage is dispersal, especially when the adults are slow moving (many polychaetes) or sessile (barnacles). Some species with long-lived feeding larvae may be carried great distances on ocean currents. The larvae of certain marine snails, for example, are believed to be transported for thousands of kilometers. The larvae of many other marine species are planktonic for only hours or a few days. Nevertheless, within this short time they may be moved considerable distances by currents produced by tides and waves.

Settlement is a crucial stage in the life of the animal because most species can live only in relatively restricted types of habitats. Because intrinsic developmental factors ultimately cause larvae to settle, many larvae perish before they reach a suitable habitat for the adult mode of life. Two adaptive responses, however, have evolved among many marine invertebrates that increase the chances of favorable settlement. Commonly, the larva responds only to very specific features of the surface on which it might settle and, in the absence of such cues, can delay settlement. Together these responses enable the larva to select for optimal settlement conditions.

The larva responds to favorable surface signals by attachment and metamorphosis to the adult body form. Light, surface texture, and chemistry can be important signals. For example, many larvae require a bacterial film over the surface in order for attachment and metamorphosis to be triggered. Those invertebrates that live on kelps (large marine algae) have larvae that respond to certain chemicals that are produced by the kelp. The juveniles of the red abalone of California, *Haliotis rufescens,* as well as many other abalone species, graze on coralline red algae. The larvae of these molluscs stop swimming on contact with a small protein on the surface of the coralline algae. In the absence of this chemical signal, settling can be delayed for as long as a month. On the other hand, the larvae must undergo at least seven days of larval life, because the larval receptors for the algal protein have not developed before this time. The planula larvae of corals of the genus *Agaricia* also settle on coralline algae, but the different species of *Agaricia* are restricted to certain species of coralline algae, responding to a surface oligosaccharide. Larvae of invertebrates that live in aggregations, such as barnacles and reef-building tubeworms *(Phragmatopoma),* are stimulated to settle by surface proteins produced by adults of the species. There can be a combination of factors influencing settlement. The sea slug *Rostanga pulchra* feeds on an encrusting sponge that lives in rocky crevices and overhangs. The sponge provides the settling signal for the sea slug larva, but the larva is also negatively phototaxic, a characteristic that increases the chances that it will make contact with the sponge.

MOLLUSCS

Members of the phylum Mollusca are among the most conspicuous and familiar invertebrate animals and include such forms as clams, oysters, squids, octopods, and snails. In abundance of species, molluscs constitute the largest invertebrate phylum aside from the arthropods. Over 50,000 living species have been described. In addition, some 35,000 fossil species are known because the phylum has had a long geological history, and the animals' mineral shells, which increase the chances of preservation, have resulted in a rich fossil record that dates back to the Cambrian.

Despite the striking differences among snails, clams, and squids, molluscs are built on the same fundamental plan. To understand the basic design, we begin by examining a generalized mollusc. Many features of this hypothetical animal, such as the structure and function of the digestive tract and gills, are encountered among primitive living species of different classes. Yet, as we shall see, not all features of this generalized mollusc are primitive molluscan characteristics. Nevertheless, they provide a model for understanding the body organization of the major classes of molluscs.

A generalized mollusc is an aquatic animal that moves over and grazes on the surface of a hard substratum. Its body is bilaterally symmetrical and several centimeters in length, and has a somewhat ovoid shape (Fig. 10–1A). The ventral surface is flattened and muscular to form a creeping sole, or **foot**. The dorsal surface is covered by an oval, convex, shield-like **shell** that protects the underlying internal organs, or **visceral mass**. The underlying epidermis, called the **mantle** (or **pallium**), secretes the animal's shell, and the most active secretion occurs around the edge of the mantle, although some new material is added to the older portions of the shell. Thus, the shell increases in diameter and thickness at the same time.

Pairs of **pedal retractor muscles** enable the animal to pull its shield-shaped shell down against the substratum on which it lives. Each retractor muscle is attached to the inner surface of the shell and is inserted into each side of the foot.

The periphery of the shell, as well as its underlying mantle, overhangs the body only slightly, except toward the posterior, where the overhang is so great that it creates a chamber called the **mantle cavity.** Within this protective chamber are pairs of **gills**, as well as openings from pairs of nephridia.

Each gill (**ctenidium**) consists of a long, flattened **axis** projecting from the anterior wall of the mantle cavity and contains blood vessels, muscles, and nerves (Fig. 10–1A). To each side of the broad surface of the axis are attached flattened, triangular **filaments** that alternate in position with those filaments on the opposite side of the axis (Fig. 10–1C, D). Such a gill is said to be **bipectinate.** Many living molluscs, however, have **monopectinate** gills in which the filaments occur on only one side of the axis, like teeth on a comb. The gills are located on opposite sides of the mantle cavity and are held in position by a ventral and a dorsal membrane. Water enters the lower part of the mantle cavity from the posterior, passes upward between gill filaments, and then moves posteriorly back out of the cavity.

Propulsion of water through the mantle cavity is largely effected by the beating of a powerful band of **lateral cilia** located on the gills just behind the frontal margin (which has first contact with the inhalant water stream). Sediment brought in by water currents and trapped by mucus on the gills is carried upward

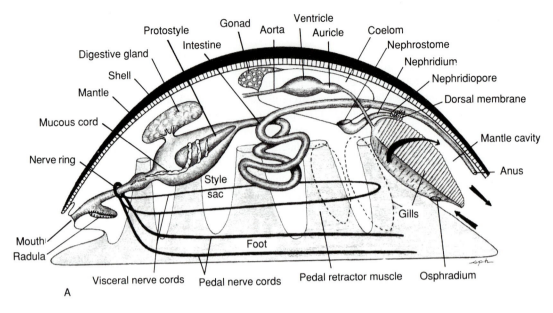

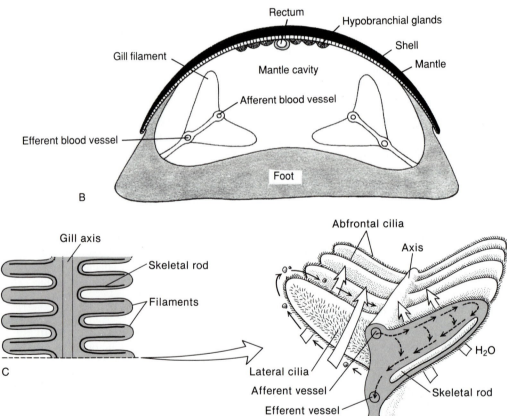

FIGURE 10–1 A, Generalized mollusc (lateral view). Arrows indicate path of water current through mantle cavity. **B,** Transverse section through the body of an ancestral mollusc at the level of the mantle cavity. **C,** Frontal section through the primitive gill, showing alternating filaments and supporting chitinous rods. **D,** Transverse section through the gill of the primitive gastropod *Haliotis.* Large outlined arrows indicate the direction of water current over gill filaments; small solid arrows indicate the direction of cleansing ciliary currents; small broken arrows indicate the direction of blood flow within gill filaments. *(B–D, Modified and redrawn from Yonge, C.M. 1947. The pallial organs of the aspidobranchiate Mollusca, a functional interpretation of their structure and evolution. Phil. Trans. R. Soc. Lond. B. 443–518.)*

first by **frontal cilia** and then by **abfrontal cilia** toward the axis, where it is swept out by the exhalant current. On the mantle roof are two patches of mucus-secreting epithelium, called **hypobranchial glands** (Fig. 10–1B). They lie downstream to each gill and trap sediment in the exiting water current.

Two blood vessels run through the gill axis. The **afferent vessel,** which carries blood into the gill, runs just within the abfrontal margin. The **efferent vessel,** which drains the gill, runs along the frontal margin. Blood flows through the filaments from the afferent to the efferent vessel (Fig. 10–1D) and thus constitutes a countercurrent to the external water stream flowing from the frontal to the abfrontal margin. Such countercurrent flow maximizes the uptake of oxygen by the gill.

In most living molluscs, not only the mantle epidermis but also the epidermis of the remainder of the exposed body parts, including the foot, are covered by cilia and contain mucous gland cells. Mucous glands are especially prevalent on the foot, where they lubricate the substratum to facilitate locomotion.

We assume that our generalized mollusc, like many living molluscs, is a grazer of fine algae and other organisms growing on rocks. The anterior mouth opens into a cuticle-lined buccal cavity. The floor of the buccal cavity is thickened by an elongated, muscular, cartilagenous mass, the **odontophore** (Fig. 10–2A). A membranous belt, the **radula,** bears transverse rows of teeth and extends medially over the odontophore and around its anterior end. The radula arises from a deep outpocket, called

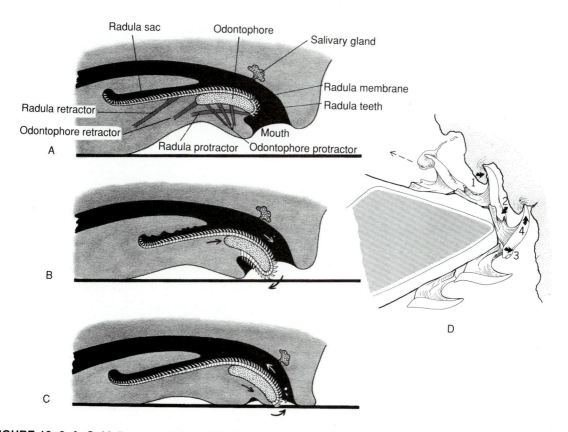

FIGURE 10–2 A–C, Molluscan radula. **A,** Mouth cavity, showing radula apparatus (lateral view). **B,** Protraction of the radula against the substratum. **C,** Retracting movement, during which substratum is scraped by radula teeth. **D,** The cutting action of radula teeth when they are erected over the end of the odontophore during radula retraction. *(D, From Solem, A. 1974. The Shell Makers: Introducing Mollusks. Reprinted by permission of John Wiley and Sons, New York. pp. 135 and 150.)*

the **radula sac,** from the posterior wall of the buccal cavity (Fig. 10–2D).

Not only can the odontophore be projected out of the mouth, but also the radula can move to some extent over the odontophore. Within the sac the lateral margins of the radula tend to roll up, but as the odontophore is projected out of the mouth over the substratum, the changing tension causes the radula belt to flatten as it bends around the odontophore tip. The flattening in turn brings about the erection of the teeth. In living species that are grazers, the radula is a scraper and collector (Fig. 10–2B–D). Because the radula teeth recurve posteriorly, the effective scraping stroke is forward and upward when the odontophore is retracted. Imagine pressing your lips against a surface with your mouth slightly open and then licking the surface with your tongue. Your tongue would be moving somewhat like a radula.

As a result of the hard wear caused by scraping, there is a gradual loss of membrane and teeth at the anterior end of the radula. To compensate for this loss, new teeth are continuously secreted at the posterior end. The radula slowly grows forward over the odontophore at a rate of one to five rows of teeth per day.

At least one pair of **salivary glands** opens onto the anterior dorsal wall of the buccal cavity. These glands secrete mucus, which lubricates the radula and entangles the ingested food particles. Food in mucous strings passes from the buccal cavity into a tubular **esophagus,** from which it is moved posteriorly toward the **stomach** (Fig. 10–3A). In primitive living molluscs, the stomach is shaped like an ice cream cone with a broad, hemispherical anterior end, into which the esophagus opens, and a tapered posterior end, which leads into the intestine. The anterior region of the stomach is lined with chitin except for a ciliated, ridged **sorting region** and the entrance point for two ducts from a pair of lateral **digestive glands** (liver), or diverticula. The posterior conical region of the stomach, called the **style sac,** is ciliated.

The contents of the stomach are rotated by cilia of the style sac. The rotation winds up the mucous food strings, drawing them along the esophagus and into the stomach (Fig. 10–1A). The relatively stiff rotating mucous mass is called a **protostyle.** The size and consistency of particles within the string vary greatly, and the chitinous lining of the anterior part of the stomach protects the wall from damage by sharp surfaces (Fig. 10–3A). The acidity of the stomach

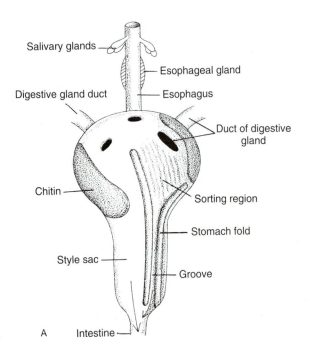

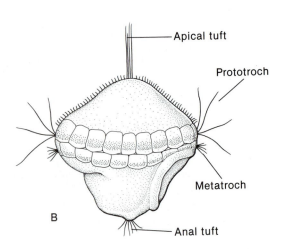

FIGURE 10–3 A, Diagram of a primitive molluscan stomach. **B,** Trochophore larva of the gastropod *Notoacmaea scutum. (A, Modified from Owen. B, Redrawn from a sketch by G.L. Shinn.)*

fluid (pH 5–6 in living molluscs) decreases the viscosity of the mucus and aids in freeing the contained particles. Such particles are eventually swept against the sorting region, in which they are graded by size. Lighter and finer particles are driven by the cilia of the ridges to the duct openings of the digestive diverticula. Heavier and larger particles are carried in the grooves between the ridges to a large groove running along the floor of the stomach to the intestine. Particles utilized as food pass into the ducts of the digestive glands, and digestion occurs within the cells of the distal tubules. Although digestion appears to be mostly intracellular in primitive molluscs, at least some extracellular digestion occurs within the stomach cavity of most living species.

The long, coiled **intestine** functions largely in the formation of fecal pellets. The anus opens middorsally at the posterior margin of the mantle cavity, and wastes are swept away by the exhalant current (Fig. 10–1A).

The relatively small **coelomic cavity,** or **pericardium,** is located in the middorsal region of the body (Fig. 10–1A); it surrounds the **heart** dorsally and a portion of the intestine ventrally. The heart consists of a pair of posterior **auricles** and a single anterior **ventricle.** The auricles drain blood from each gill via the efferent vessels and then pass it into the muscular ventricle, which pumps it anteriorly through a single **aorta.** The aorta branches into smaller blood vessels that deliver the blood into tissue spaces and sinuses of the hemocoel. From the tissue sinuses, return flow to the heart is by way of the nephridia and the gills. This is an oversimplified picture of molluscan circulation, and in living molluscs there is considerable variation in the flow pattern. Like vertebrates, and unlike all other invertebrates, the blood of squids and octopods is completely enclosed within vessels lined by endothelium. This is presumably because the vascular system of these animals, like that of vertebrates, is a high-pressure system, important in supporting their high activity. The blood contains amebocytes as well as the respiratory pigment called **hemocyanin** (see p. 339).

The excretory organs, usually called **kidneys,** are organized into a metanephridial system. In living molluscs, there are commonly one or two kidneys (Fig. 10–1A). Although a typical kidney tubule has one end connected to the pericardial cavity and the other opening to the outside through a nephridiopore, in most molluscs the connection with the pericardial cavity (**renopericardial canal**) and the nephridiopore are at the same end of the nephridium (Fig. 10–50A). The nephridium is thus a blind sac. The nephridiopore opens at the back of the mantle cavity. The pericardial coelom receives blood ultrafiltrate through the auricular wall of the heart, which contains podocytes. The pericardial fluid then passes through the nephrostome into the kidney tubule. Here secretion of wastes from the blood, as well as some selective reabsorption, may occur through the tubule wall, and the final urine is discharged through a pair of renal pores into the mantle cavity.

The ground plan of the molluscan nervous system consists of a circumesophageal **nerve ring** from the underside of which two pairs of nerve cords extend posteriorly. Because there are four nerve cords, molluscs are said to show **tetraneury.** The ventral pair of cords, called the **pedal cords,** innervates the muscles of the foot; the dorsal pair, called the **visceral cords,** innervates the mantle and visceral organs. Transverse connections give each pair of cords the appearance of a ladder.

The sense organs of many living molluscs include **tentacles,** a pair of **eyes,** a pair of **statocysts** in the foot, and **osphradia.** The osphradia are patches of sensory epithelium located on the posterior margin of the ventral mesentery that supports each gill (Fig. 10–1A); they function as chemoreceptors and also determine the amount of sediment in the inhalant current.

The generalized mollusc is dioecious and has a pair of anterior, dorsolateral **gonads.** When ripe, the eggs or sperm break into the coelomic cavity and are transported to the outside through the kidneys. Fertilization occurs externally in the seawater. Following a period of spiral cleavage, a gastrula develops into a free-swimming **trochophore larva** (Fig. 10–3B).

In most of the molluscan classes, the trochophore passes into a more highly developed **veliger larva,** in which the foot, shell, and other structures make their appearance (Fig. 10–53). Characteristically, a veliger larva has two ciliated flaps, called a **velum,** with which it swims and feeds. At the end of larval life, the larva sinks to the bottom and metamorphoses to assume the benthic habit of the adult.

SUMMARY

1. Members of the phylum Mollusca, one of the largest phyla of animals, are found in the sea, in fresh water, and on land. They are distinguished by a muscular foot, a calcareous shell (the mantle) secreted by the underlying body wall, and a feeding organ (the radula).

2. A generalized mollusc possesses a flat, creeping foot; a dorsal, shield-shaped shell; and a poorly developed head.

3. Several pairs of gills are housed within a mantle cavity created by the overhanging mantle and shell. The gills (ctenidia) are composed of numerous flattened filaments that extend alternately from each side of a supporting axis. Each filament bears lateral cilia, which create the ventilating current, and frontal cilia and abfrontal cilia, which remove particulate matter.

4. The molluscan radula, a belt of recurved chitinous teeth stretched over a muscular cartilage base, functions as a scraper in feeding.

5. The primitive stomach is adapted for processing fine particles of food (especially algae) scraped from hard surfaces by the radula. A rotating, mucous mass in the style sac acts like a windlass, pulling in a food-laden mucous string from the esophagus. Particles are separated over a sorting region, and fine particles are sent up the ducts of the surrounding digestive glands, where intracellular digestion occurs.

6. The blood-vascular system is a hemocoel. Blood flows from the gills into one (or more) pair of auricles. From each auricle blood passes into the central ventricle, which pumps it out through the aorta for distribution to the tissue sinuses.

7. The heart is surrounded by a coelomic cavity (pericardial cavity) that receives an ultrafiltrate of the blood. The metanephridial excretory tubules, called kidneys, drain the pericardial cavity, modify the ultrafiltrate, and empty into the mantle cavity.

8. The ground plan of the nervous system consists of a circumesophageal nerve ring, from which extend a pair of pedal nerve cords innervating the foot and a pair of visceral cords innervating the mantle and visceral mass. Typical sense organs are tentacles, eyes, statocysts, and one or two osphradia in the mantle cavity.

9. The generalized mollusc is dioecious, with a pair of gonads in the visceral mass adjacent to the pericardium. Ripe gametes are released into the pericardial coelom, and the kidneys function as gonoducts. In such primitive molluscs fertilization is external, and development is planktonic.

10. Cleavage is spiral, a trochophore is the first larval stage, and a veliger is the second.

CLASS MONOPLACOPHORA

Of the seven classes of living molluscs, the monoplacophorans are a good starting point for our survey because they are most like the generalized mollusc we have just described, and a monoplacophoran-like mollusc may have been the ancestor of the snails, bivalves, squids, and octopods.

In 1952, ten living specimens of *Neopilina,* a group of molluscs previously known only from Cambrian and Devonian fossils, were dredged from a deep ocean trench off the Pacific coast of Costa Rica. Since this discovery, specimens belonging to eleven species and three genera have been collected from the shelf edge and from deep water, 2000 to 7000 m, in various parts of the world—the North and South Atlantic, the Indian Ocean, and in a number of places in the eastern Pacific.

As the name implies, a monoplacophoran possesses a single, symmetrical shell, as in some snails. The shell varies in shape from a flattened, shieldlike plate to a short cone. Because of the limpet-like shell, monoplacophorans had been classified with the gastropods, but a gastropod shell usually has two pedal retractor muscle scars, whereas the monoplacophoran shell has three to eight pairs, depending on the species. In this respect, monoplacophorans resemble chitons, with which they have also been classified.

The living specimens are 3 mm to little more than 3 cm long and externally resemble limpets (Fig. 10–4B). The apex of the shell is directed anteriorly. A pallial groove (the mantle cavity) separates the edge of the broad flat foot from the mantle on each side. The mouth is located in front of the foot, and the anus is located in the pallial groove at the posterior of the body. In front of the mouth is a preoral fold, or velum (homology with the larval organ is uncertain), which extends laterally on each side as a rather large, ciliated, palplike structure. Another fold lies behind the mouth and projects to either side as a pair of postoral tentacles.

Compared with most living molluscs, the striking feature of monoplacophorans is the repetition of parts. The pallial groove contains five or six pairs of monopectinate gills (Fig. 10–4D). Internally, usually eight pairs of pedal retractor muscles are present.

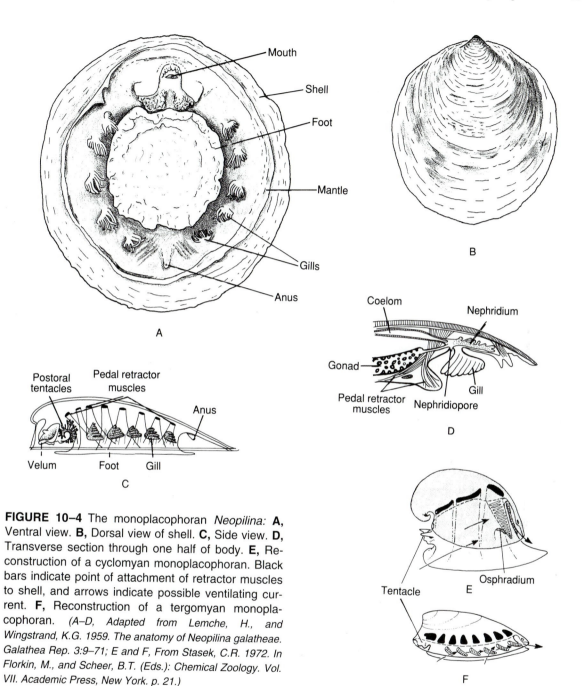

FIGURE 10–4 The monoplacophoran *Neopilina:* **A,** Ventral view. **B,** Dorsal view of shell. **C,** Side view. **D,** Transverse section through one half of body. **E,** Reconstruction of a cyclomyan monoplacophoran. Black bars indicate point of attachment of retractor muscles to shell, and arrows indicate possible ventilating current. **F,** Reconstruction of a tergomyan monoplacophoran. *(A–D, Adapted from Lemche, H., and Wingstrand, K.G. 1959. The anatomy of Neopilina galatheae. Galathea Rep. 3:9–71; E and F, From Stasek, C.R. 1972. In Florkin, M., and Scheer, B.T. (Eds.): Chemical Zoology. Vol. VII. Academic Press, New York. p. 21.)*

There are six pairs of kidneys located on each side of the body, and each nephridium probably opens into a dorsal coelom that may join the pericardial coelom. The nephridiopores open into the pallial groove. Two pairs of auricles open into two ventricles, one on each side of the rectum (Fig. 10–5), and the heart is surrounded by a paired pericardial coelom.

The digestive system includes a radula and a subradular organ within the buccal cavity. The stomach contains a style sac and a crystalline style (p. 436);

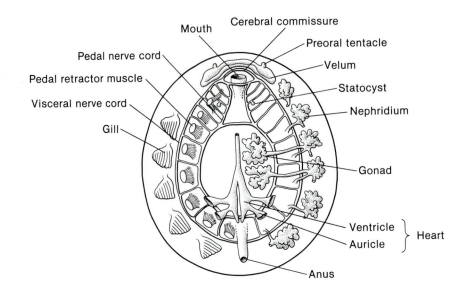

FIGURE 10–5 Internal anatomy of *Neopilina*, ventral view. *(Redrawn from Lemche, H., and Wingstrand, K.G. 1959. The anatomy of Neopilina galatheae. Galathea Rep. 3:9–71.)*

the intestine is greatly coiled. Stomach contents consist of diatoms, forams, and sponge spicules.

The nervous system consists a pair of cerebral ganglia, and a circumoral nerve ring from which emerge a pair of visceral nerve cords into the mantle fold and a pair of pedal nerve cords to the foot.

The sexes are separate, and two pairs of gonads are located in the middle of the body. Each gonad is provided with a separate gonoduct that connects to one of the pairs of nephridia in the middle of the body. Fertilization must occur externally. Development is still unknown.

The survival of a few species of monoplacophorans is probably correlated with their living at great depths, where perhaps they escaped competition, predation, or other factors that led to the extinction of most members of the class. Fossil species appear to have evolved along two lines. In one group (subclass Cyclomya), there was an increase in the dorsoventral axis of the body, leading to a planospiral shell and a reduction of gills and retractor muscles (Fig. 10–4E). Although they disappeared from the fossil record in the Devonian, this group may have been ancestral to the gastropods (snails) and cephalopods (squids and octopods). The other line (subclass Tergomya) retained a flattened shell with five to eight pairs of retractor muscles (Fig. 10–4F). Although this group was thought to have become ex-

tinct in the Devonian when it disappeared from the fossil record, the living species may represent survivors.

SUMMARY

1. The small number of living species of the class Monoplacophora are deep-water relicts of a much larger and more widespread group of molluscs that dates back to the Cambrian.

2. The repetition of both external and internal structures—gills, retractor muscles, auricles, and kidneys—is a distinctive feature of living monoplacophorans. Fossil species show only multiple muscle scars.

3. Many monoplacophoran features, such as the shield-shaped shell, flat creeping foot, slight cephalization, multiple gills and retractor muscles, radula, and cone-shaped stomach, are thought to be primitive, and the class is believed by many malacologists to be ancestral to the gastropods, bivalves, and cephalopods.

CLASS POLYPLACOPHORA

The class Polyplacophora contains the **chitons.** Although some features of their structure and development are primitive, chitons have become highly

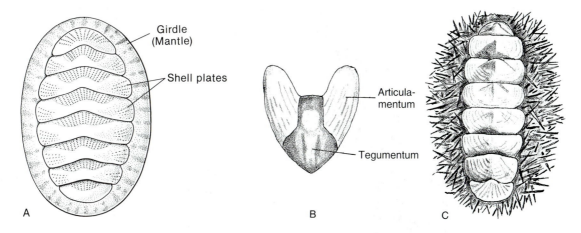

FIGURE 10–6 A, Common American Atlantic coast chiton *Chaetopleura apiculata,* which is only a few centimeters long. **B,** Single shell plate of *Katharina.* **C,** *Chiton.* *(A, After Pierce.)*

adapted for adhering to rocks and shells. The oval body is greatly flattened dorsoventrally (Fig. 10–6) and is covered not by a single shell plate but by **eight overlapping plates.** Later we will attempt to reconcile this significant difference from our generalized mollusc (p. 379). Chitons have no cephalic eyes or tentacles, and the head is indistinct. The mantle is thick, and the foot is broad and flat to facilitate adhesion to hard substrata. There are approximately 800

existing species of chitons, of which many are found along the west coast of North America. The fossil record, which dates to the late Cambrian period, is rather sparse. Some 350 fossil chitons are known.

Chitons range in size from 3 mm to 40 cm, the largest being the giant Pacific gumshoe chiton, *Cryptochiton stelleri* (Fig. 10–7A). However, most species are 3 to 12 cm in length. Chitons are commonly drab shades of red, brown, yellow, or green.

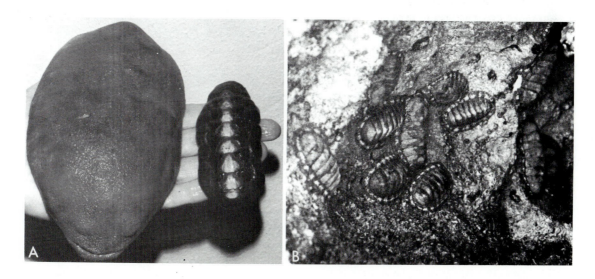

FIGURE 10–7 A, Two species of chitons from the northwest Pacific coast. The shell plates of the larger species *(Cryptochiton)* are completely covered by the mantle; those of the smaller species *(Katharina)* are partially covered. **B,** The West Indian chiton *Chiton tuberculatus* exposed at low tide. *(Both photographs courtesy of Betty M. Barnes.)*

Shell and Mantle

The most distinctive characteristic of a chiton is the shell, which is divided into eight overlapping transverse plates. From the nature of the shell is derived the name of the class, Polyplacophora, meaning "bearer of many plates" (Fig. 10–6). Except for the overlapping posterior edge, the margins of each plate are covered by mantle tissue, but the degree of coverage varies among different species. In many species, most of the plate width is exposed (Fig. 10–6A). However, in *Katharina* only the midsection of each plate is uncovered, and in *Cryptochiton* the shell is completely covered by the mantle (Fig. 10–7A). The peripheral area of the mantle, called the **girdle,** is thick and stiff and extends a considerable distance beyond the lateral margins of the plates. The girdle surface is covered by a thin cuticle and may be smooth or bear scales, bristles, or calcareous spicules (Fig. 10–6C).

Foot and Locomotion

The broad, flat foot occupies most of the ventral surface and functions in adhesion as well as in locomotion (Fig. 10–8A). Chitons creep very slowly by pedal muscular waves in the same manner as snails (p. 395). The division of the shell into transverse plates and their articulation with one another enable chitons to move over and adhere to a sharply curved surface. Chitons roll up into a ball if dislodged, and although this may be a defense mechanism, it also enables the animal to right itself.

Adhesion is brought about by both the foot and the girdle. The foot is responsible for ordinary adhesion, but when a chiton is disturbed, the girdle is also employed. The girdle is clamped down tightly against the substratum, and the inner margin is then raised. This creates a vacuum that enables the animal to grip the substratum with great tenacity.

Chitons are common rocky intertidal inhabitants, and like limpets, most species are motionless at low tide. When the rock surface is submerged or splashed, they move about to feed. They are usually negatively phototactic and thus tend to locate themselves under rocks and ledges. They are most active at night if they are submerged by the tide. Like limpets, some species exhibit homing behavior.

Water Circulation and Gas Exchange

The mantle cavity of chitons consists of a trough, or groove, on each side of the body between the foot and the mantle edge (Fig. 10–8A). This condition is correlated with the dorsoventral flattening of the body, one of the adaptations of chitons for life on rocky surfaces. Six to 88 pairs of bipectinate gills are arranged in a linear series within the two mantle troughs. The number of pairs varies among species and even within a single species, depending on the size of the animal.

On each side of the body toward the anterior end of the mantle groove, the mantle margin is raised to form an inhalant opening through which water enters (Fig. 10–8A). As the water flows along the course of the groove, it passes through the gills before it flows out of the groove posteriorly in an exhalant stream. The exhalant opening is created by a local elevation of the mantle margin.

Nutrition

Most chitons are microphagous feeders on fine algae and other organisms that they scrape from the surface of rocks and shells with the radula. The gut contents of three species from the Maine coast included remains from 14 different algal and animal organisms, and about three quarters of the contents consisted of sediment. Some species, however, feed on coarser algae, and one genus, *Placiphorella,* on the west coast of the United States, uses its raised and flaring anterior end to trap small crustaceans and other invertebrates. The radula, which is very long, bears 17 teeth in each transverse row, and certain teeth are capped with magnetite, an iron-containing mineral. The great length of the radula and its hardened teeth are both adaptations to the wear imposed by almost continuous grazing on the surface of rocks. The mouth opens into the chitin-lined buccal cavity (Fig. 10–8C). A long radula sac projects posteriorly from the back of the buccal cavity, as does a smaller, more ventral evagination called the subradula sac. The latter contains a cushion-shaped chemosensory structure, the **subradula organ,** hanging from the roof.

When a chiton feeds, the subradula organ is first protruded and applied against the rock. If food is present, the odontophore and its radula project from the mouth and scrape. Periodically the subradula organ is protruded and tests the substratum again. From the

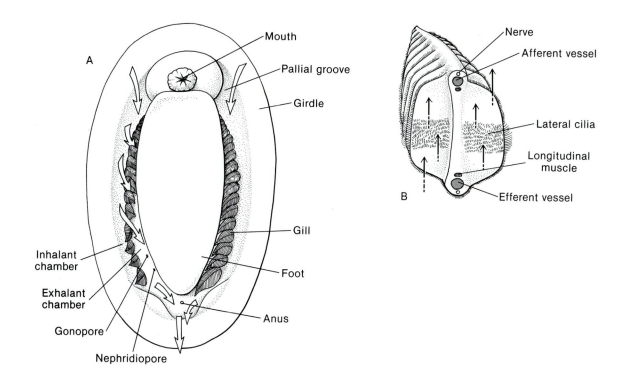

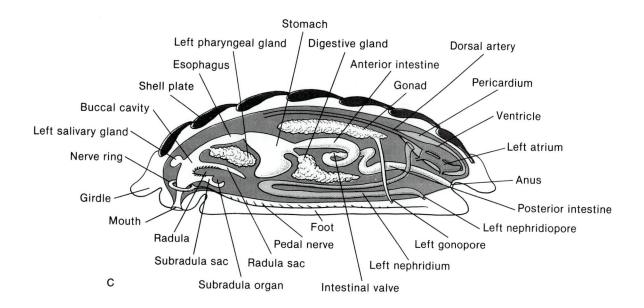

FIGURE 10–8 The chiton *Lepidochitona cinerea*. **A,** Mantle groove, showing direction of water currents (ventral view). **B,** Transverse section through gill axis. Arrows indicate direction of water current over filament surface. **C,** Buccal region and digestive tract (lateral view). *(A and B, After Yonge, C.M. 1939. On the mantle cavity and its contained organs in the Loricata. Q. J. Micro. Sci. 81:367–390.)*

buccal cavity, mucus from a pair of salivary glands entangles food particles as they enter the esophagus and are carried along a ciliated food channel toward the stomach (Fig. 10–8C). During this passage, the food particles are mixed with amylase secreted by a pair of large pharyngeal glands (sugar glands), the ducts from which open at the beginning of the esophagus.

The esophagus opens into a stomach, where the food is further mixed with proteolytic secretions from the digestive gland. Digestion is almost entirely extracellular and takes place in the digestive gland, in the stomach, and in the anterior intestine.

The anterior intestine loops and then joins a large, coiled, posterior intestine, where fecal pellets are formed. The anus opens at the midline just behind the posterior margin of the foot, and the egested fecal pellets are swept out with the exhalant current.

Internal Transport, Excretion, and Nervous System

The pericardial cavity is large and located beneath the last two shell plates. A single pair of auricles collects blood from all of the gills. Each of the two large, U-shaped kidneys connects with the pericardial cavity, and the nephridiopore opens into the pallial groove (Fig. 10–8A).

The nervous system is like that described for the generalized mollusc (Fig. 10–9B). Ganglia are lacking or at least poorly developed. The chief sense organs are the subradula organ, girdle hairs, and the **esthetes**. Esthetes, which are unique to chitons, are mantle cells lodged within minute vertical canals in the upper layer of the shell plate (Fig. 10–10). The canals and sensory endings terminate beneath a cap on the shell surface. The density is very great; 1750 terminate on 1 mm of shell surface in *Lepidochitona cinerea*. Although the structure of esthetes has been studied in detail and although it is clear that the shell plates are involved in the light response, the function of esthetes is still uncertain. Secretory structures are usually more conspicuous than sensory elements in esthetes, but photoreceptors occur in some.

In one family of chitons (Chitonidae) distinct eyes are lodged in the shell canals. These ocelli may number in the thousands per individual and are especially concentrated on the anterior shell plates.

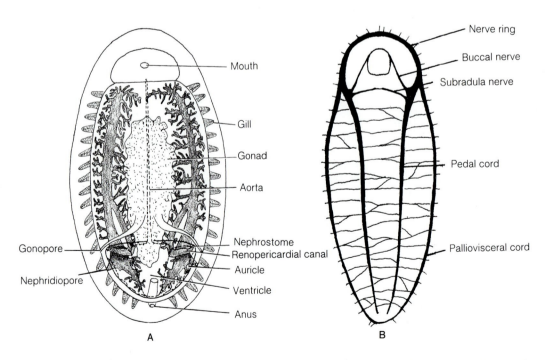

FIGURE 10–9 A, Internal structure of a chiton. **B,** Nervous system of a chiton. (*A, After Lang and Haller. B, After Thiele from Parker and Haswell.*)

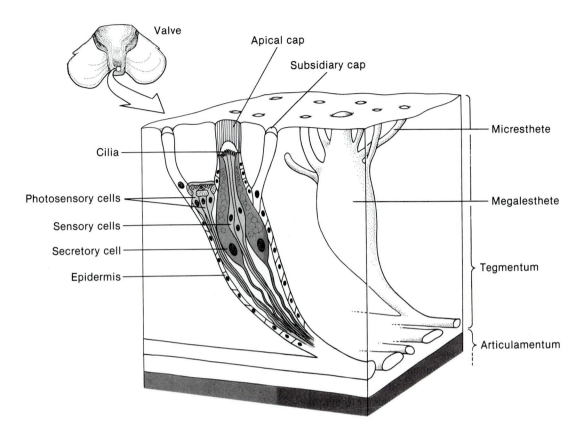

FIGURE 10–10 Chiton esthetes. Section of shell showing esthete canals and surface terminations. The larger canal houses a megalesthete; the smaller, a micresthete. *(Modified and redrawn from Boyle, P.R., 1974: The aesthetes of chitons. II. Fine Structure in* Lepidochitona cinereus. *Cell Tiss. Res., 153:383–398 and other sources.)*

Reproduction and Development

Most chitons are dioecious. A single median gonad (paired early in development) is located in front of the pericardial cavity beneath the middle shell plates (Fig. 10–9A). The gametes are transported to the outside by two gonoducts, instead of by the nephridia. A gonopore is located in each pallial groove in front of the nephridiopore (Fig. 10–8A).

Chitons do not copulate. Instead, males release sperm in the exhalant respiratory currents, and fertilization occurs in the sea or within the mantle cavity of the female. The usual gregariousness of chitons facilitates fertilization. The eggs, which are enclosed within a spiny envelope, are usually shed into the sea either singly or in strings, but in some species the eggs are brooded within the mantle cavity.

There is a free-swimming trochophore, except in some of those forms that brood their eggs, but a veliger is absent (Fig. 10–11A). In the metamorphosis of the chiton trochophore, the posttrochal region elongates to form the major part of the body (Fig. 10–11B). The prototroch degenerates, and the animal sinks to the bottom as a young chiton. The larval eyes are retained for some time after metamorphosis.

SUMMARY

1. Chitons, members of the class Polyplacophora, are adapted for living on hard surfaces, especially in the intertidal zone. The foot and girdle provide for gripping; the low profile reduces water resistance; and the eight articulating shell plates provide protection while permitting folding of the large body across angles on the substratum.

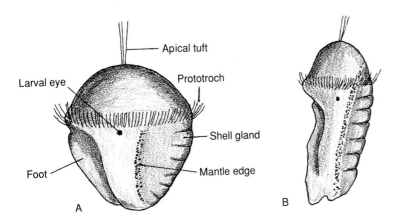

FIGURE 10–11 Development of chitons. **A,** Trochophore of *Ischnochiton.* **B,** Metamorphic stage of *Ischnochiton. (A and B, After Heath from Dawydoff.)*

2. Chitons feed on algae and other organisms that encrust the rocks and shells on which they live. Food is removed by a scraping radula.

3. The poorly developed head, nature of the shell, multiple gills, and lack of a veliger larva suggest that chitons may have diverged early from the main line of molluscan evolution.

CLASS APLACOPHORA

The class Aplacophora comprises some 288 species of strange, small, wormlike molluscs. They are found throughout the oceans of the world to depths of 7000 m. Although there are shallow-water species, they are most numerous between 200 and 3000 m. Some (Chaetodermomorpha) (Fig. 10–12A) burrow into the bottom, and others (Neomeniomorpha) (Fig. 10–12B) creep on the bottom or on hydroids and alcyonarian corals. Most specimens have been collected by dredging, and the biology of the group is still poorly known.

Aplacophorans are usually less than 5 mm in length. The head is poorly developed, and the typical molluscan shell is absent (Fig. 10–12). However, the integument (mantle) is covered by a cuticle and contains embedded calcareous scales or spicules. The body is vermiform because the mantle margins have

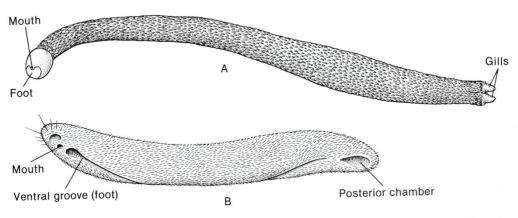

FIGURE 10–12 External organization of chaetoderm **(A)** and neomeniomorph **(B)** aplacophorans. *(Modified and redrawn from Salvini-Plawen, L. v. 1972. Zur Morphologie und Phylogenie der Mollusken: Die Beziehungen der Caudofoveata und der Solenogastres als Aculifera, als Mollusca und als Spiralia. Z. wiss. Zool. 184:205–394.)*

rolled inward ventrally. The burrowing species have a reduced foot (Fig. 10–12A), but the others have a midventral groove on which they creep and which is homologous to the foot of other molluscs (Fig. 10–12B). The posterior end of the body contains a cavity into which the anus empties. The posterior cavity is believed to represent a mantle cavity, and in the chaetoderms, it houses a pair of bipectinate gills.

Most creeping species feed on cnidarians; the burrowers feed on small organisms and deposited material. A radula may or may not be present. The excretory system is not yet understood fully, but pericardial glands and pericardial ducts, both of which have roles in excretion in other molluscs, have been described. Most aplacophorans (neomeniomorphs) are hermaphrodites, and the gonoducts extend to the mantle cavity either directly from the gonad or more usually from the pericardial cavity. In the very small numbers of species studied, the eggs are brooded and develop directly or are spawned and develop into a trochophore larva.

Molluscan Origins

Aplacophorans lack a shell; chitons have eight shell plates; and monoplacophorans have a large single shell. We have described our generalized mollusc as being similar to a monoplacophoran. How then can we account for the condition of chitons and aplacophorans? What was the nature of the first molluscs? The earliest molluscan fossils are shells from the Cambrian. But that does not mean that the earliest molluscs had shells because it is the shell that is usually preserved as a fossil. What evidence do we have from the comparative anatomy of living molluscs? The aplacophorans are certainly a mix of primitive and specialized features. Among the latter is the reduction of the foot, the loss of the radula in many, and loss of the gills in most. The nervous system is distinctly primitive and there is no evidence that aplacophorans *ever had a shell*. Thus, the first molluscs were probably shell-less, although calcareous spicules or scales may have been present in the mantle.

Could chitons have evolved from an extinct monoplacophoran ancestor, or monoplacophorans from a chiton-like ancestor? This seems improbable given the nature of the shells—there is neither fossil nor developmental evidence for the evolution of a single shell from an ancestral eight shells or eight from an ancestral one. Thus, it appears that the poly-

placophoran condition and the monoplacophoran condition evolved independently. The first molluscs were probably without shells, the dorsal surface being covered by a cuticle and perhaps, as in aplacophorans, integumental calcareous spicules. The aplacophorans were probably an early offshoot of the ancestral molluscs and have retained the shell-less condition. Chitons and monoplacophorans may have diverged from some common shell-less molluscan ancestor, because both have eight pairs of pedal retractors. Shells evolved independently within each line. Because all of the remaining molluscs—gastropods (snails), bivalves (clams), scaphopods (tooth shells), and cephalopods (squids and octopods)—theoretically can be derived from a monoplacophoran-like ancestor, all of these molluscs, including the monoplacophorans, are sometimes referred to as **conchiferans,** meaning "shell-bearers." The generalized mollusc described in the introduction to the chapter is a conchiferan archetype rather than a hypothetical molluscan ancestor. It is with the conchiferans, which make up the great bulk of molluscan species, that the rest of this chapter will be concerned.

CLASS GASTROPODA

The class Gastropoda is the largest class of molluscs. About 30,000 existing species have been described, and to this total should be added some 15,000 fossil forms. The class has had an unbroken fossil record beginning with the early Cambrian period and has undergone the most extensive adaptive radiation of all the major molluscan groups. Considering the wide variety of habitats the gastropods have invaded, they are certainly the most successful of the molluscan classes. Marine species have become adapted to life on all types of bottoms as well as to a pelagic existence. They have invaded fresh water, and the pulmonate snails and several other groups have conquered land by eliminating the gills and converting the mantle cavity into a lung.

Origin and Evolution

The evolution of gastropods involved four major changes from the organization of a generalized mollusc: (1) development of a head, (2) dorsoventral elongation of the body, (3) conversion of the shell from a shield to a deep protective retreat, and (4) torsion. Although gastropods retain the flat, creeping ancestral foot, most are relatively active, mobile animals and are more highly cephalized than such molluscs as chitons and bivalves. Primitively, the **head** bears a pair of tentacles with an eye at each tentacle base.

The shell of most gastropods is an asymmetrical spiral that functions as a portable retreat. The animal therefore does not depend on clamping against a hard substratum for protection. The change in shell design involved an increase in height and a decrease in aperture, thus changing the shape from a shield to a cone (Fig. 10–13). However, a cone not only would be unwieldy to carry but would also make it difficult for the animal to exploit crevices and holes for food and shelter. The problem was avoided by the spiraling of the shell over the head as it became higher and more conical (Fig. 10–13C). The early shell was probably a planospiral; that is, it was bilaterally symmetrical with each spiral, or **whorl,** located completely outside of the one preceding it and in the same plane, like a hose coiled flat on the ground. Reduction of the shell aperture and resulting limitation of space within the mantle cavity perhaps account for the reduction of gills, retractor muscles, and nephridia to a single pair, which is the maximum number in any gastropod. The evolution of an asymmetrical shell is a later event to which we will return shortly.

The most distinctive modification of gastropods is the twisting, or **torsion,** that the body has undergone. Torsion is not the coiling of the shell; all evidence indicates that a planospiral shell evolved before torsion. Torsion and the spiraling of the shell were therefore separate evolutionary events. Torsion was a much more drastic change than the spiraling of the shell. When viewed dorsally most of the body behind the head, including the visceral mass, mantle, and mantle cavity, was twisted 180 degrees counterclockwise (Fig. 10–13A, B). The mantle cavity, gills, anus, and two nephridiopores were now located in the anterior part of the body behind the head. Internally, the digestive tract was looped, and the nervous system was twisted into a figure 8. The head and foot retained the original untorted bilateral symmetry, and the shell remained a symmetrical spiral.

Torsion is not merely an evolutionary event because it appears in the ontogeny of living gastropods. The larva is at first bilaterally symmetrical and then quite suddenly undergoes twisting as a result of muscular contraction and differential growth. No widely accepted explanation of the evolutionary significance of torsion has yet been advanced, despite many contributions. A number of researchers have postulated that torsion represents a larval adaptation for protec-

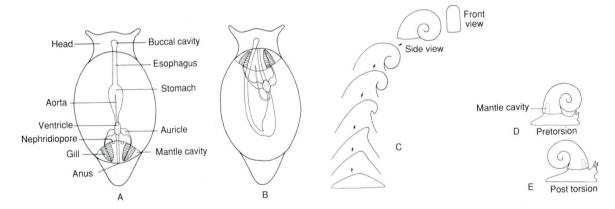

FIGURE 10–13 Dorsal views of ancestral gastropod. **A,** Prior to torsion. **B,** After torsion. **C,** Evolution of a planospiral shell. Height of the shieldlike shell of a hypothetical ancestral mollusc increases and a peak forms. The peak is pulled forward and coiled under. The aperture is reduced, and the animal can withdraw into a spiral shell, which is more compact and less awkward to carry than a straight conical shell. Note that the shell is bilaterally symmetrical. **D,** Hypothetical pretorsion gastropod with a planospiral shell. **E,** Posttorsion gastropod. Torsion does not affect the planospiral shell except to place coils of shell posteriorly. The mantle cavity is now anterior. *(A and B, Modified from Graham.)*

tion of the head. Others believe that torsion was an adult adaptation, having the principal advantages of head protection or utilization of the oncoming water stream by the gills or sensory organs in the mantle cavity.

Up to this point we have considered a gastropod with a symmetrical, planospiral shell. Such a form is not entirely hypothetical, because there are early fossil species (the Bellerophontacea; Fig. 10–14A, B) with a symmetrical, planospiral shell bearing a cleft along the anterior, middorsal edge. The cleft indicates the midline of the mantle cavity and a corresponding cleft in the mantle (Fig. 10–14C). Such a cleft shell and mantle, which is also found in some living gastropods, reflects a modified ventilating current, which helps to prevent fouling now that the anus is above the head (p. 384).

All existing gastropods possess asymmetrical shells, or if the shells are symmetrical, this symmetry has been secondarily derived. The planospiral shell had the disadvantage of not being very compact; because each coil lies completely outside of the preceding one, the diameter of the shell could become relatively great (Fig. 10–14C). The problem was solved with the evolution of asymmetrical coiling, in which the coils are laid down around a central axis called the **columella** and each coil lies to some extent beneath the preceding coil (Fig. 10–14D). Such a shell is more compact and stronger than a planospiral shell.

The new asymmetrical conical shell obviously could not be carried like the old planospiral shell, because all the weight would hang on one side of the body (Fig. 10–14D). To bring the center of gravity of the shell/visceral mass over the head/foot, the shell position shifted so that the axis of the spiral slanted upward and somewhat posteriorly. The shell was eventually carried obliquely to the long axis of the body, as in most living gastropods. The changes in shell symmetry and carriage would have occurred simultaneously.

The new position of the shell along with the coiling restricts the mantle cavity to the left side of the body because the mantle cavity on the right side is occluded by the bulging whorl of the visceral mass (Fig. 10–14F). This occlusion has had profound effects, resulting in the decrease in size, or the complete loss, of the gill, auricle, and nephridium on the right side of the body.

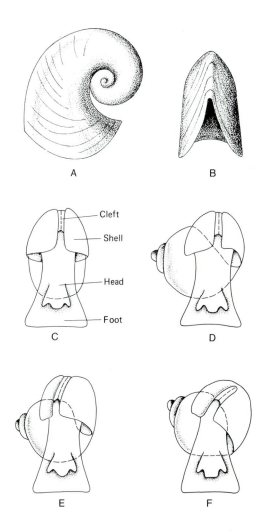

FIGURE 10–14 Side view (**A**) and front view (**B**) of the planospiral shell of *Strepsodiscus,* a genus of the fossil Bellerophontacea, possibly the earliest known archaeogastropod. **C–F,** Evolution of the asymmetrical gastropod shell. Slot in shell for exhalant water current marks the location of mantle cavity. **C–E,** Hypothetical stages. **C,** Ancestral posttorsion gastropod with planospiral shell. **D,** Apex of the spiral is drawn out, producing a more compact shell. **E,** Position of shell over the body is shifted, providing more equal distribution of weight. **F,** Final position of the shell over the body, typical of most living gastropods. Axis of the shell is oblique to the long axis of body, and the mantle cavity is on the left side. The right side is compressed by the shell. *(C–F, After Yonge.)*

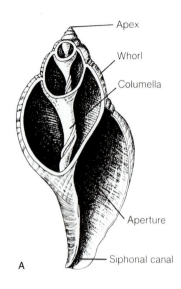

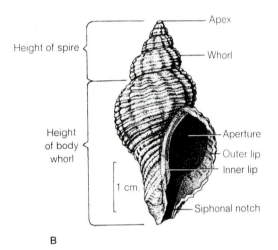

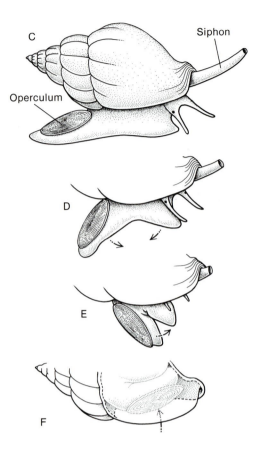

FIGURE 10–15 Gastropod shells. **A,** Longitudinal section through a shell. **B,** Shell of the oyster drill *Urosalpinx cinerea,* showing commonly designated features. **C,** Gastropod with an operculum. **D–F,** Withdrawal into the shell and closure by the operculum. *(B, After Turner.)*

The gastropod head and foot are withdrawn into the shell by a retractor muscle, the **columella muscle,** that arises in the foot and is inserted onto the columella of the shell. Primitively, two retractor muscles are present (See Fig. 10–18B); this paired arrangement is found in a few living species, although the left muscle is usually very small. In most gastropods, the left muscle has disappeared, and only the right one remains. The foot of most prosobranch gastropods (and the larvae of some opisthobranchs and pulmonates) bears a horny disc, called the **operculum,** on its posterior dorsal surface (Fig. 10–15C).

Unlike monoplacophorans and chitons, which clamp themselves tightly onto the substratum for pro-

tection, most gastropods withdraw the well-developed head and foot through the narrow aperture of the shell. This is accomplished partly by compressing and partly by folding the soft tissues as they are withdrawn. As the foot withdraws, the posterior half folds forward, like a hinged lid, to close the aperture with the operculum (Fig. 10–15D, E).

For other ideas about the evolution of gastropods, see Solem (1974) and Linsley (1978).

With this background of possible gastropod evolutionary origins, we must now consider the manner in which existing gastropods are classified. Gastropods are divided into three subclasses. The first, known as the **Prosobranchia,** includes all gastropods that respire by gills and in which the mantle cavity, gills, and anus are located at the anterior of the body—in other words, those gastropods in which torsion is clearly evident. There are about 18,000 species. From the Prosobranchia evolved the two other subclasses: the **Pulmonata** and **Opisthobranchia.** The Opisthobranchia displays detorsion. The shell and mantle cavity are usually either reduced in size or absent, and many species have become secondarily bilaterally symmetrical. The sea hares and the sea slugs (nudibranchs) are perhaps the most familiar members of this subclass. Members of the subclass Pulmonata, which includes the land snails, are also detorted, and the gills have disappeared; the mantle cavity has been modified into a lung.

Shell and Mantle

The typical gastropod shell is a conical spire composed of tubular whorls and containing the visceral mass of the animal (Fig. 10–15A). Starting at the **apex,** which contains the smallest and oldest whorls, successively larger whorls are coiled about a central axis (the columella); the last and largest whorl (the **body whorl**), eventually terminates at the opening, or **aperture,** from which the head and foot of the living animal protrude. The whorls above the body whorl constitute the **spire.** A shell may be spiraled clockwise or counterclockwise or, as it is more frequently stated, displays a right-handed (dextral) or left-handed (sinistral) spiral. A spiral is right-handed when the aperture opens to the right of the columella (if the shell is held with the spire up and the aperture facing the observer) and left-handed when it opens to the left. Most gastropods are right-handed, a few are left-handed, and some species have both right-handed and left-handed individuals.

A gastropod shell typically consists of four layers. The outer **periostracum** is composed of a quinone-tanned, horny protein material called **conchiolin,** or **conchin** (Fig. 10–16). Although usually thin, the periostracum may be absent, as in the cowries, or thick and hairy, as in some whelks. The inner shell layers consist of calcium carbonate. The outermost calcareous layer is generally **prismatic;** that is, the mineral is deposited as vertical crystals, each surrounded by a thin protein matrix. The inner calcareous layers, usually two but sometimes more, are laid down as sheets (**lamellae**), over a thin organic matrix. The sheets are usually oriented perpendicular to the surface, but in

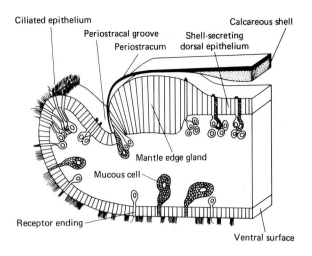

FIGURE 10–16 Diagrammatic section (not to scale) through mantle edge of an aquatic pulmonate snail (*Heliosoma*). Periostracum is secreted initially at the bottom of the periostracal groove and is thickened as it passes over the mantle edge gland. Secretion of the calcareous portion of the shell begins to the inner side of the mantle edge gland. *(Modified slightly from Jones, G.M., and Saleuddin, A.S.M. 1978. Cellular mechanisms of periostracum formation in Physa spp. Can. J. Zool. 56:2299–2311.)*

some gastropods the sheets of the innermost layer are parallel to the surface, an arrangement called **nacreous,** or **nacre,** that makes the inner surface smooth and lustrous.

The color of the shell results from pigments in the periostracum or in the calcareous layers. The shell is enlarged by the addition of mineral from the outer edge of the mantle to the lips of the aperture. A constant difference in the rate of mineral deposition along the inner and outer lips results in the characteristic (logarithmic) spiral of the shell (Fig. 10–20). Growth is usually not continuous, and the intervals can often be determined by interval growth lines, as in bivalves (p. 426), and by the sculpturing of the shell surface. In most gastropods shell growth declines with age.

Gastropod shells display an infinite variety of colors, patterns, shapes, and sculpturing, but at this point only two of the more radical modifications in shell form will be mentioned. In a considerable number of gastropods, the shell is conspicuously spiraled only in the juvenile stages. The coiled nature disappears with growth, and the adult shell represents a single, large, expanded body whorl. In the abalone, *Haliotis* (Fig. 10–18A), and in the slipper snails, *Crepidula,* the shell remains asymmetrical, but in the **limpets,** of which there are a number of unrelated groups, the shell has become secondarily symmetrical and looks like a Chinese straw hat (Fig. 10–18D). Bilaterality has been derived in a very different way in the beautiful cowrie shells. Here the last whorl of the shell has completely overgrown the previous whorls, and the aperture is greatly narrowed (Fig. 10–17).

The second modification is shell reduction and shell loss, a condition that has occurred many times in the history of gastropods. When the shell is greatly reduced, it often becomes buried within the mantle tissues. Other shell modifications will be described later in connection with ventilation, movement, and habitation.

Water Circulation and Gas Exchange: The Evolution of the Gastropod Groups

The great diversity of gastropods reflects adaptive radiation at various points in their evolutionary history. The main lines of this evolution are reflected in the modifications for enhancement of water circulation and gas exchange. We will focus attention on these two processes as a way of gaining an initial overview of the class. Such an overview has been diagrammatically depicted in Figure 10–21, which can be used in following the discussion below.

Prosobranchs

The most primitive type of gill structure and water circulation occurs in those prosobranchs with cleft or perforated shells, which may have been the primitive gastropod solution to possible sanitation problems caused by torsion, that is, having the exhalant water stream located over the head. These more or less primitive prosobranchs belong to the order **Archaeogastropoda** and include slit snails, abalones, and the keyhole limpets. In all three groups, two bipectinate gills are present[*]. The rectum and anus are removed from the edge of the mantle cavity and open beneath the shell perforation, or cleft. The ventilating current produced by the action of the lateral cilia of

---

[*]The Archaeogastropoda is also sometimes called the Aspidobranchia because of the bipectinate gills or called the Diotocardia because there are two auricles (although one may be reduced).

FIGURE 10–17 Cowrie shells. The shells of these tropical gastropods are superficially bilateral because each whorl completely encloses the previous whorls. The specimen on the right is viewed from the aperture side. The specimens on the left and in the middle are viewed from the side opposite the aperture, and the one on the left has been cut away to show the younger whorls. *(Photographs courtesy of Betty M. Barnes.)*

the gills enters the mantle cavity at the anterior of the body; it passes between the gill filaments and then continues upward and out through the shell cleft.

The slit snails (*Scissurella* and the deep-water *Pleurotomaria*) have typical spiral shells, but the anterior margin of the shell and the underlying mantle are deeply cleft (Fig. 10–18C). The abalones (*Haliotis*) and the keyhole limpets are intertidal and shallow-water inhabitants of wave-swept rocks. The broad shells of both groups are designed for minimum water resistance and as protective shields when the animal is clamped against rock. They are perforated instead of slotted, avoiding the structural weakness of a deep notch.

The low, shieldlike shell of the abalone is asymmetrical and constitutes in large part a single expanded whorl (Fig. 10–18A). As in most other prosobranch gastropods, the mantle cavity is displaced to the left side of the body. In *Haliotis tuberculata* the shell above the cavity contains a line of five holes (Fig. 10–18A). The mantle is cleft along the line of shell perforations, and the edges of the mantle fit together and project into the shell openings to form a lining for each hole. The ventilating current enters the mantle cavity through the anterior two holes and beneath the shell to the left of the head. It leaves through the last three holes (Fig. 10–18A). The anus and nephridial openings lie beneath one of the poste-

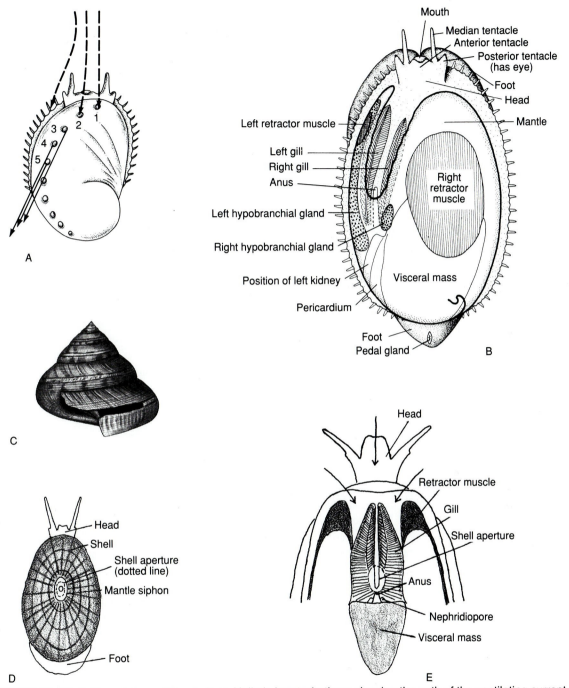

FIGURE 10–18 A, Dorsal view of the abalone *Haliotis kamtschatkana* showing the path of the ventilating current. Part of the inhalant water stream enters the mantle cavity beneath the shell to the left of the head, and part enters the anterior through shell perforations. The exhalant water stream leaves the mantle cavity through perforations 3, 4, and 5. Old, posterior, sealed perforations appear as a line of short tubercles on the shell. **B,** Dorsal view of *Haliotis* with the shell removed and the mantle cavity exposed. **C,** *Pleurotomaria,* a gastropod with a slotted shell for exhalant current. **D,** The keyhole limpet, *Diodora,* a dorsal view. **E,** Dorsal view of the exposed mantle cavity of *Diodora,* showing paired, bipectinate gills. *(A, Based on the work of Voltzow, J. 1983. Flow through and around the abalone Haliotis kamtschatkana. Veliger. 26(1):18–21. B, After Bullough. C, Drawn from a photograph by Abbott. D and E, After Yonge.)*

rior perforations. A succession of holes develop as the shell grows. Each hole arises as a notch at the front margin of the shell and eventually becomes sealed at the rear.

The keyhole limpet has a conical, secondarily symmetrical shell, which has either a cleft at the anterior margin or a hole at the apex (Fig. 10–18D, E). The opening arises as a notch along the shell margin during early stages of development. The notch then becomes enclosed, and through differential growth it gradually assumes a position at the apex of the shell. Water enters the mantle cavity anteriorly, flows over the gills, and issues as a powerful stream from the opening at the shell apex. The anus and urogenital openings are located just beneath the posterior margin of the shell opening.

The Patellacea, now commonly separated from the archeogastropods as a separate order, Patellogastropoda, contain the largest assemblage of limpets. They are found from shallow water to the deep sea, and many are common inhabitants of the rocky intertidal zone. Their shells evolved independently from those of the keyhole limpets, and shell openings or clefts are lacking. In *Acmaea, Notoacmaea, Collisella,* and *Lottia,* which include the common limpets on the west coast of North America, there is only a left gill, which projects to the right side of the body. As in all limpets, the mantle and shell overhang produces a distinct groove on each side between the foot and the mantle edge (Fig. 10–19A). The inhalant ventilating current enters the mantle cavity anteriorly on the left side and exits on the right. In some species the exiting current or part of it may flow posteriorly in the mantle grooves. *Patella,* another genus of widespread intertidal limpets, lacks a gill in the mantle cavity. Instead, mantle folds form secondary gills, which project into the pallial groove along each side of the body (Fig. 10–19B).

The remaining archaeogastropods—the Trochacea (top snails and turban snails) and the Neritacea—possess only left gills, and their ventilating current enters the mantle cavity on the left side of the head and exits on the right (Fig. 10–21). The anus opens at the right edge of the mantle cavity, and wastes are carried away in the exhalant water stream. Such an oblique water current must be an efficient solution to ventilation because it is found in most prosobranchs.

The neritaceans include many common, rocky, intertidal species, such as the semitropical and tropi-

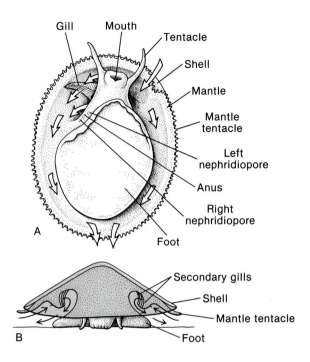

FIGURE 10–19 Patellacean limpets. **A,** *Acmaea* (ventral view) **B,** Cross section through a patellid limpet, showing secondary gills and ventilating currents (arrows).

cal species of *Nerita* (Fig. 10–20C). The gill, if present, is secondary. Members of this group have also invaded fresh water (*Theodoxus*), and perhaps from some freshwater stock evolved a family of tropical land snails, the Helicinidae (Fig. 10–21).

Of the some 18,000 species of prosobranchs, the majority are not archaeogastropods but **mesogastropods** and **neogastropods.** The archaeogastropods are largely restricted to the surfaces of rock and kelps. The great adaptive diversity of the mesogastropods and neogastropods is in part correlated with their ability to exploit other types of habitats, especially soft bottoms. This ability may be related to a major change in the structure of the gills. The dorsal and ventral membranes that suspend the bipectinate gills of the archaeogastropods present considerable surface areas that could be fouled by sediment carried within the ventilating current, and this perhaps accounts for the restriction of these primitive prosobranchs to the cleaner water over rocky bottoms. In the mesogastropods and neogastropods the membranous suspension of the gills has disappeared, and the gill axis is

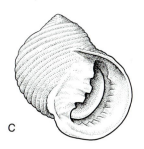

FIGURE 10–20 Archaeogastropods with a single gill (left) and an oblique ventilating current. **A,** Top snail shells (Trochacea). The large specimen is *Trochus niloticus,* a common species around South Pacific islands. **B,** Turban snail (Trochacea), *Turbo,* a genus of common tropical Pacific and Indian Ocean species. The heavy, calcareous operculum of the South Pacific species is often washed up on beaches in large numbers and is called a cat's eye. **C,** *Nerita tesselata* from south Florida.

attached directly to the mantle wall (Fig. 10–21). The filaments on the side of the attachment have disappeared; those on the opposite side project into the mantle cavity. The gill of mesogastropods and neogastropods is thus monopectinate.

A further modification associated with the ventilating current of many species of these two higher orders of prosobranchs, especially the neogastropods, is the development of an inhalant **siphon** by the extension and inward rolling of the mantle margin (Fig. 10–22B). In many species the anterior margin of the shell aperture is notched (Fig. 10–41A) or drawn out as a **siphonal canal** to house the siphon (Fig. 10–40B). The siphon may provide access to surface water in some species that burrow, or by making possible the selection of restricted areas of water, the mobile siphon may function as a sense organ, especially in carnivores. There has also been a tendency in neogastropods to direct the exhalant current toward the rear.

The ventilating current and monopectinate gill are relatively uniform among mesogastropods and neogastropods[†], and the diversity of these higher prosobranchs results from variations in adaptations for locomotion, habitation, feeding, and other functions. Only their immigrations from the sea need be mentioned here. Mesogastropods are well represented in fresh water as a result of several independent invasions (Fig. 10–21). The majority are tropical, but there are many genera, such as *Goniobasis, Pleurocera, Viviparus, Campeloma,* and *Valvata,* that contain temperate species. All are operculate in contrast to the freshwater pulmonates, which lack an operculum.

There are two large families of mesogastropod land snails, the Cyclophoridae and the Pomatiasidae (Fig. 10–21). Like the archaeogastropod Helicinidae, they are largely tropical and operculate, they have no gill, and gas exchange occurs across a vascularized mantle wall within the mantle cavity (lung). A notch

---

[†]The two groups are often treated as suborders of a single order, Caenogastropoda or Pectinibranchia or Monotocardia (referring to the single monopectinate gill or the single auricle of the heart).

(Text continues on page 390)

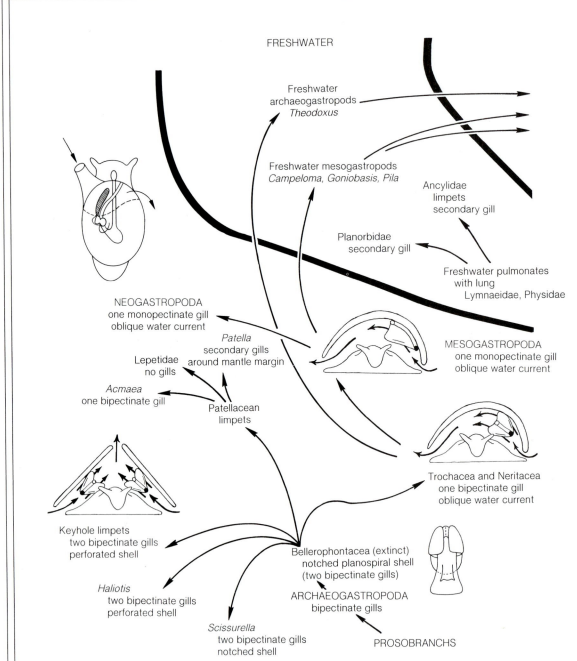

FIGURE 10–21 Evolution of water circulation and gas exchange in gastropods. Diagram reflects phylogenetic relationships of the subclasses and orders. Families and genera listed represent only examples. *(Figures adapted from Graham, Hyman, Morton, and Yonge.)*

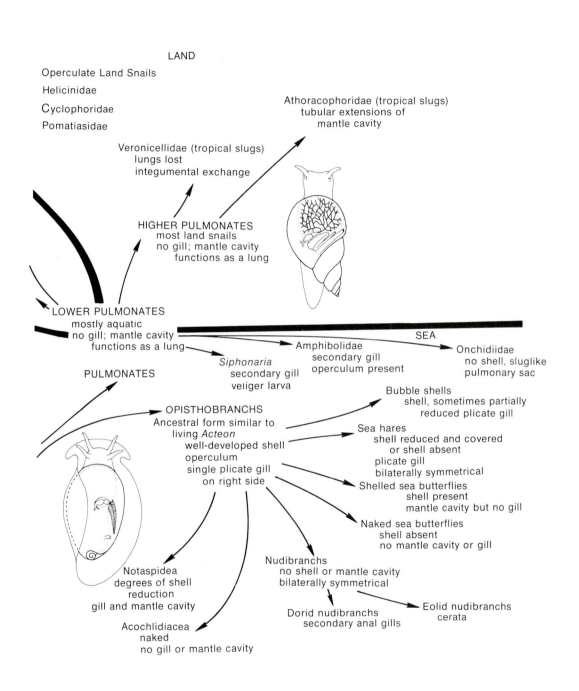

LAND

Operculate Land Snails

Helicinidae

Cyclophoridae

Pomatiasidae

Athoracophoridae (tropical slugs)
tubular extensions of
mantle cavity

Veronicellidae (tropical slugs)
lungs lost
integumental exchange

HIGHER PULMONATES
most land snails
no gill; mantle cavity
functions as a lung

LOWER PULMONATES
mostly aquatic
no gill; mantle cavity
functions as a lung

SEA

Siphonaria
secondary gill
veliger larva

Amphibolidae
secondary gill
operculum present

Onchidiidae
no shell, sluglike
pulmonary sac

PULMONATES

OPISTHOBRANCHS
Ancestral form similar to
living *Acteon*
well-developed shell
operculum
single plicate gill
on right side

Bubble shells
shell, sometimes partially
reduced plicate gill

Sea hares
shell reduced and covered
or shell absent
plicate gill
bilaterally symmetrical

Shelled sea butterflies
shell present
mantle cavity but no gill

Naked sea butterflies
shell absent
no mantle cavity or gill

Notaspidea
degrees of shell
reduction
gill and mantle cavity

Nudibranchs
no shell or mantle cavity
bilaterally symmetrical

Eolid nudibranchs
cerata

Acochlidiacea
naked
no gill or mantle cavity

Dorid nudibranchs
secondary anal gills

A

B

FIGURE 10–22 Prosobranchs. **A,** Shell of a cyclophorid land snail, showing the breathing tube, which permits gas exchange when the animal is withdrawn and the aperture is sealed by the operculum. **B,** Anterior end of the neogastropod *Mitra*, showing the folded origin of the siphon. *(A, After Rees from Purchon, R.D. 1968. Biology of the Mollusca. Pergamon Press, New York. B, Based on a photograph by Paul Zahl.)*

or a breathing tube in the shell aperture of some species permits the entrance of air when the operculum is closed (Fig. 10–22A).

Opisthobranchs

The remaining two subclasses of gastropods, the Opisthobranchia and the Pulmonata, are probably both derived from a prosobranch-like ancestor that possessed only the left gill. The some 2000 highly diverse species of opisthobranchs are largely marine and are characterized by the placement of the mantle cavity and its organs on the right side of the body (Fig. 10–21). On the assumption that ancestral opisthobranchs had an anterior mantle cavity, as in prosobranchs, the opisthobranchs are usually said to be detorted. There is, however, neither fossil nor developmental evidence for detorsion in opisthobranchs. During their morphogenesis, they simply tort 90 degrees, instead of 180 the degrees in prosobranchs. Another consideration in the study of opisthobranch evolution is the fact that the gill is plicate, or folded, rather than filamentous and may not be homologous with the prosobranch gill (Fig. 10–23C).

A primitive opisthobranch is asymmetrical and possesses the more or less typical, coiled, gastropod shell (Fig. 10–21), although an operculum is usually lacking in adults. Throughout the subclass, however, perhaps correlated with detorsion but also with the adoption of chemical defenses, there has been a tendency toward shell reduction and loss, reduction of the mantle cavity and associated loss of the original gill, and attainment of a secondary bilateral symmetry. Characteristic of most opisthobranchs is a second pair of tentacles, called **rhinophores,** that are located behind the first pair, and are commonly surrounded at the base by a collar-like fold (Figs. 10–21 and 10–23L).

The Cephalaspidea, containing the bubble snails *(Acteon, Scaphander, Hydatina, Bulla),* is the largest order of opisthobranchs, and the one to which the more primitive members of the subclass belong (Fig. 10–23A). In *Acteon,* the most primitive known opisthobranch, the nervous system is still twisted, and the shell is closed by an operculum. Many bubble snails burrow or crawl on the surface of soft bottoms, and the lateral, skirtlike folds of the foot reduce fouling of the mantle cavity and other parts of the body.

The order Anaspidea contains the sea hares, which reach the largest size (40 cm) of any opisthobranch. The reduced shell is buried in the mantle or completely lost, and the body is bilaterally symmetrical. The mantle cavity and gill are still present, and the posterior edge of the mantle can be rolled to form an exhalant siphon (Fig. 10–23B). When disturbed, many sea hares release a defensive purple ink derived from the pigments of the red algae on which they feed.

The pteropods comprise two orders of small, swimming, pelagic opisthobranchs. The shelled pteropods (order Thecosomata) possess a shell, often with an operculum (Fig. 10–23D). The naked pteropods (order Gymnosomata) lack a shell. A gill is absent from most pteropods, and the naked forms have no mantle cavity. Gas exchange occurs across the general body surface.

The sea slugs, members of the order Nudibranchia, rank among the most spectacular and beautiful molluscs. Shell, mantle cavity, and original gill have disappeared, and the body is secondarily bilaterally symmetrical, with the anus at the rear. The dorsal body surface is greatly increased in many nudibranchs by numerous projections called **cerata** (Fig. 10–23I–L). The cerata can be filamentous, club-shaped, branched, or look like a cluster of grapes, or resemble numerous variations on these themes. Cerata are lacking in some nudibranchs, but these sea slugs have secondary gills arranged in a circle around the posterior anus (Fig. 10–23J). The cerata, as well as other parts of the nudibranch body, are usually brilliantly colored and commonly are red, yellow, orange, blue, green, or a combination of colors. The sea slugs, as well as some other opisthobranchs, have evolved other defenses in the absence of well-developed shells. Escape swimming is a common ability. Many have skin glands that produce sulfuric acid or a nonacidic noxious substance that repels potential predators, especially fish. Some utilize nematocysts from the prey on which they feed (p. 408). Some have spicules embedded in the mantle. The flamboyant coloration of some species of sea slugs probably represents warning coloration; for others it may be camouflage (see review by Todd, 1981). Several other small orders—Acochlidioidea, Notaspidea, and Sacoglossa—contain sluglike forms, some with cerata (Fig. 10–23G, H).

Pulmonates

The subclass Pulmonata contains the highly successful land snails as well as many freshwater forms, and the more than 16,000 described species are widely distributed in both tropical and temperate regions throughout the world. Pulmonates show the same sort of detorsion as the opisthobranchs, but the distinctive feature from which the group takes its name is the conversion of the mantle cavity into a **lung**. The edges of the mantle cavity have become sealed to the back of the animal except for a small opening on the right side called the **pneumostome** (Figs. 10–21 and 10–12A). The gill has disappeared, and the roof of the mantle cavity has become highly vascularized. Ventilation is facilitated by the arching and flattening

of the mantle cavity floor (actually the back of the animal). The pneumostome generally remains open at all times, or opens and closes with the ventilating cycle. Gas exchange by diffusion through the pneumostome is probably important in most pulmonates and predominant in small species.

The first pulmonate land snails appeared in the Carboniferous, but their origin is obscure. They probably evolved from some group of operculate prosobranchs that had a single gill. These ancestral forms perhaps inhabited estuarine marshes and mud flats, and the pulmonate condition could have evolved as a means of gas exchange when the animals were confined to small, stagnant puddles or to wet but exposed surfaces. These conditions are similar to those postulated for the origin of amphibians.

There are a few primitive, marine species, all of which live at the edge of the sea on intertidal rocks or in estuarine habitats. Tropical limpets of the genus *Siphonaria* and the temperate *Melampus* of salt marshes and drift are among the few pulmonates that possess a veliger larva (p. 419), indicating that the marine habit does not represent a secondary return to the sea. *Amphibola*, another marine pulmonate, has a typical shell but is unusual in possessing an operculum. In all other pulmonates the operculum is lost during the course of development. This is a distinguishing characteristic. All freshwater and terrestrial prosobranchs are operculate; pulmonates, which are found in the same habitats, are not.

The lower pulmonates (order Basommatophora), with one pair of tentacles and with eyes at the tentacle bases, include the few marine and all the freshwater forms. Many freshwater species, such as the cosmopolitan *Lymnaea* and *Physa*, come to the surface to obtain air for gas exchange. In *Lymnaea*, the edges of the mantle cavity can be extended as a long tube for this purpose. The pneumostome is closed when the animal is submerged, and submergence may last from 15 min to more than an hour, depending on the time of year and other conditions. However, some deep-lake lymnaeids have abandoned air breathing and fill the mantle cavity with water. A secondary gill (pseudobranch) has evolved in other aquatic pulmonates as folds of the mantle near the pneumostome. Such a secondary gill is found in many planorbids and in the ancylids. The latter are limpets adapted for life in fast-running streams.

Various species of freshwater pulmonate snails are important hosts for certain human parasites. The

(Text continues on page 394)

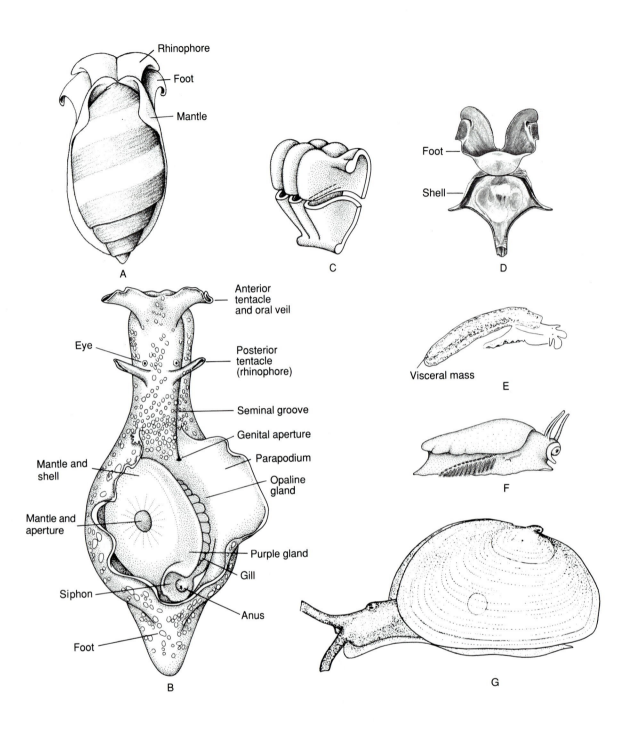

FIGURE 10–23 Opisthobranchs. **A,** The bubble snail *Hydatina,* a cephalaspidean. **B,** Diagrammatic dorsal view of a sea hare, an anaspidean. **C,** Part of gill of *Aplysia* cut horizontally to show folded condition. **D,** *Cavolina,* a shelled sea butterfly (pteropod; Thecosomata). **E,** *Microhedyle* (Acochlidioidea). **F,** Sluglike *Pleurobranchus* (Notaspidea). **G,** *Berthelinia,* an opisthobranch with a bivalved shell (Sacoglossa).

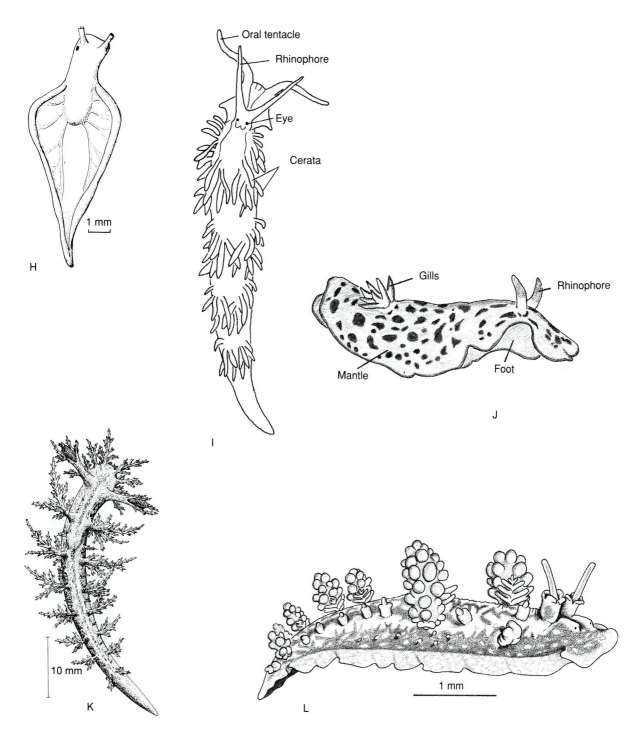

FIGURE 10–23 *(continued)* **H,** *Elysia viridis,* a sluglike sacoglossan. **I,** *Eolidia papillosa,* a nudibranch with cerata (dorsal view). **J,** *Glossodoris,* a nudibranch with secondary anal gills. **K,** *Dendronotus frondosus,* a nudibranch with branched cerata. **L,** *Doto chica,* a nudibranch with cerata that look like clusters of grapes. *(B, After Guiart from Kandel, E.R. 1979. Behavioral Biology of Aplysia. W.H. Freeman and Co., San Francisco. C, Modified after Carew et al. D After a photograph by Abbott. E, After Odhner from Hyman. F, After Vayassiere and Hyman. G, After Kawaguti from Hyman. H, After Gascoigne, T. 1975. Methods of mounting sacoglossan radulae. Microscopy. 32:513. I, After Pierce. J, After a photograph by P. Zahl. K, From Thompson, T.E. and Brown, G.H. 1976. British Opisthobranch Molluscs. Academic Press, London. p. 67.)*

African genus *Bulinus,* for example, is the principal host for trematodes causing schistosomiasis.

The higher pulmonates (order Stylommatophora) include the terrestrial species and are a considerably larger group than that containing the aquatic forms. These pulmonates have two pairs of tentacles (the anterior lower pair may be inconspicuous), and their eyes are mounted at the top of the upper pair (Fig. 10–24B). The usually calcareous shell of these terrestrial pulmonates is not as heavy as those of many marine gastropods, although some variation may result from the availability of calcium in the soil. The periostracum protects the calcareous layers from humic acid and may also function as a water repellent. In many small species the aperture of the shell is partially occluded by teeth or ridges, which keep out such predators as insects but allow the soft body of the snail to protrude. The largest shells, 23 cm in height, belong to members of the African species *Achatina fulica,* but the South American strophocheilids are also large, reaching 15 cm in height. Many species of land snails have shells that measure

A

B

C

D

FIGURE 10–24 Pulmonates. **A,** A terrestrial slug. Opening into lung seen at lower edge of saddle-like mantle. **B,** A land pulmonate (Stylommatophora), showing the eyes mounted at the ends of the tentacles. **C,** *Testacella,* a land slug with reduced shell. **D,** Shell of specimen of the large South American pulmonate *Strophocheilus,* compared with two common species of temperate American land snails, *Polygyra* and *Retinella* (the smallest is on ruler). *(Photographs of A, B, and D courtesy of Betty M. Barnes. C, After Baker.)*

less than 1 cm. Although such species are found throughout the world in leaf mold and beneath bark and stones, they are especially abundant on oceanic islands, as in the Pacific.

Shell reduction or loss has occurred independently a number of times within the higher pulmonates, and such naked species are called **slugs** *(Arion, Philomycus, Limax)*. The shell is generally absent or reduced and buried within the mantle, but in *Testacella* a little shell is perched on the back (Fig. 10–24C). The pneumostome of slugs is usually a conspicuous opening on the right side of the body (Fig. 10–24A). The evolution of the slug form is perhaps an adaptive response to low availability of calcium because their original centers of distribution are restricted to areas of high humidity and low soil calcium. Slugs have now been introduced into many parts of the world from which they were originally absent.

It should be remembered that not all land snails are pulmonates. The operculate land snails, although a smaller group (4000 species of terrestrial prosobranchs, compared with some 20,000 species of terrestrial pulmonates), are very common in the tropics, and their adaptation for life on land parallels that of the pulmonates in many ways.

Locomotion and Habitation

The typical gastropod foot is a flat, creeping sole, but it has become adapted for locomotion over a variety of substrata. Typically, the sole is ciliated and provided with numerous gland cells or, in the pulmonates, with a large pedal gland. The glands of the foot elaborate a mucous trail over which the animal moves. Some very small snails, as well as species that live on sand and mud bottoms, move by ciliary propulsion. Most hard-bottom gastropods, terrestrial pulmonates, and even large soft-bottom species, when moving rapidly, are propelled by waves of fine muscular contraction that sweep along the foot. The sole of the foot is firmly anchored to the substratum by gelatinous mucus except in the region of a wave, where the foot slides forward over liquefied mucus. In species in which the waves move anteriorly (see the discussion below), the mucus changes very rapidly from gel to sol as a consequence of the shearing force produced by muscle contraction (Fig. 10–25). Thus, each wave performs a small step. In some species a

wave extends across the entire width of the foot **(monotaxic)** (Fig. 10–26A, B), but in many forms, a wave involves only half of the width of the foot, and the waves on the right side move alternately to those on the left **(ditaxic)** (Fig. 10–26C, D).

The waves may be **direct,** progressing in the same direction as the movement of the animal, that is, from back to front (Fig. 10–25A). Or the waves may be **retrograde,** passing from front to back in the opposite direction of the animal's movement (Fig. 10–25B). Direct waves involve contraction of longitudinal and dorsoventral musculature beginning at the posterior end of the foot. Successive sections of the foot are in effect pushed forward (Fig. 10–26A). Retrograde waves involve contraction of transverse muscles, which along with blood pressure extends the front of the foot forward. This backward-moving wave of elongation is followed by the contraction of longitudinal muscles, and successive areas of the foot are pulled forward (Fig. 10–25B). Direct and indirect waves may be associated with either the monotaxic or the ditaxic condition; that is, the pattern may be direct monotaxic, direct ditaxic, and so on. The most common pattern among prosobranchs is retrograde ditaxic. Most pulmonates exhibit direct monotaxic waves. In general, pedal waves, whether direct or indirect, occupy about one-third of the foot length and only one or two waves are present simultaneously. Commonly, retrograde ditaxic waves are oblique (Fig. 10–26D).

Some prosobranchs and cephalaspid opisthobranchs that live on soft sand bottoms have become adapted for burrowing. In the moon shells *Natica* and *Polinices,* the front of the foot, called the **propodium,** acts like a plough and anchor, and a dorsal flaplike fold of the foot covers the head as a protective shield (Fig. 10–27A).

The conch *Strombus* crawls over sand in a very different fashion from other gastropods (Fig. 10–27B). The large, clawlike operculum digs into the sand, and the animal then "poles" foreward by rapid extension of the foot.

Shell shape, movement, and habitation are correlated. In general, shells with low spires are more stable and better adapted for carriage upside down or on the vertical surfaces of rocks and vegetation. Shells with long spires are carried horizontally or even dragged over soft bottoms. Spines and other shell projections and sculpturing may contribute to shell strengthening, to protection, to stabilization in soft

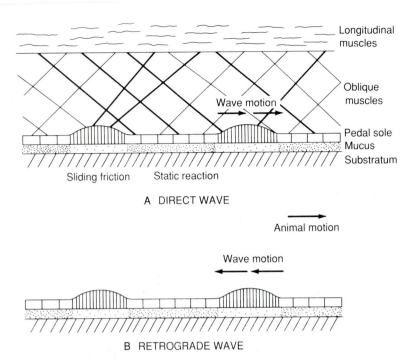

A DIRECT WAVE

Animal motion

Wave motion

B RETROGRADE WAVE

FIGURE 10–25 Pedal muscular waves in gastropods. **A,** Direct waves. Waves of pedal contraction sweep over the foot in the same direction as the animal is moving. In the region of the wave (close vertical lines), the sole slides forward over liquefied mucus. **B,** Retrograde waves. Waves sweep from front to back of the foot opposite to the direction of animal movement. *(Modified after Denny, M.W. 1980. A quantitative model for the adhesive locomotion of the terrestrial slug, Ariolimax columbianus, J. Exp. Biol. 91:195–218; and After Trueman, E.R., and Jones, H.D. 1977. Crawling and burrowing. In Alexander, R.M., and Goldspink, G. (Eds.): Mechanics and Energetics of Animal Locomotion. Chapman and Hall, London.)*

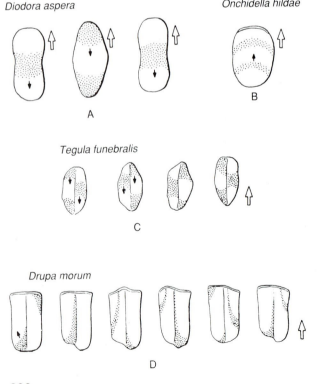

Diodora aspera

Onchidella hildae

A

B

Tegula funebralis

C

Drupa morum

D

FIGURE 10–26 Patterns of pedal waves in gastropods. **A,** Retrograde monotaxic waves in the keyhole limpet *Diodora aspera*. **B,** Direct monotaxic waves in the intertidal pulmonate slug *Onchidella hildae*. **C,** Retrograde ditaxic waves in the archaeogastropod *Tegula funebralis*. **D,** Direct ditaxic waves in the neogastropod *Drupa morum*. Wave sequence is from left to right in each case. Large outlined arrows indicate the direction of the animal's movement, small black arrows the direction of the waves. *(From Miller, S.L. 1974. The classification, taxonomic distribution, and evolution of locomotor types among prosobranch gastropods. Proc. Malacol. Soc. London. 41:233–272.)*

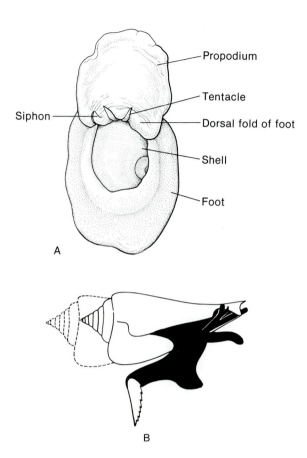

Propodium

Tentacle

Siphon

Dorsal fold of foot

Shell

Foot

A

B

FIGURE 10–27 A, *Polinices,* a burrowing gastropod (dorsal view). **B,** Lateral view of *Strombus.* The black section represents the foot, to which the toothed, bladelike operculum is attached. Dotted outline indicates the length of one "leap." *(B, From Morton, J.E. 1964. In Wilbur, K.M., and Yonge, C.M. (Eds.): Physiology of Mollusca. Vol. I. Academic Press, New York.)*

in the rock surface, over which the edges of the shell have come to fit very tightly, constitutes the limpet's "home" (Fig. 10–28A). At high tide the animal may wander 10 to 150 cm away in feeding, depending on the species, but then returns to its home. The experi-

A

B

FIGURE 10–28 A, Photograph of a small area of intertidal rock on the west coast of Scotland. Three specimens of the neogastropod *Thais* are feeding on barnacles. Four patellacean limpets belonging to the genus *Patella* can be seen to the right and above the snails. Numerous barnacles have settled on the limpets' shells. **B,** Ventral view of the keyhole limpet *Fissurella,* showing the sensory tentacles on the mantle margin. *(Photographs courtesy of Betty M. Barnes.)*

bottoms, to burrowing, or even to landing right side up if the animal is knocked off a rock.

Limpets, abalones, and slipper snails are especially adapted for clinging to rocks and shells. All have low, broad shells that can be pulled down tightly and offer less resistance to waves and currents. The large, mucus-covered foot functions as an adhesive organ as well as for movement, and the surrounding skirtlike mantle margin, which may bear tentacles, serves as an important sense organ (Fig. 10–28B).

The homing ability of many intertidal limpets—*Collisella, Patella, Fissurella, Siphonaria*—has been the subject of numerous studies. A slight depression

mental studies to date indicate that homing ability depends primarily on chemical cues in the mucus laid down by the limpet as it makes its feeding excursions. Homing may reduce intraspecific competition by establishing a grazing territory and may reduce desiccation and predation because of the tight fit into the home site. Especially remarkable is the discovery by Connor and Quinn that the mucus of at least some homing species, but not of nonterritorial migratory species, stimulates algal growth in their territory.

There are a number of other marine snails, such as *Ilyanassa obsoleta* and certain species of *Nerita*, that use mucous trails to follow other individuals of the same species. On striking a trail they are able to determine from the mucous components which is the forward direction of the trail. Many terrestrial snails and slugs are able to return to shelters beneath logs and stones, especially when the shelter is occupied by other individuals of the same species. The homing cue appears to be a pheromone in mucous trails or airborne pheromones from fecal pellets.

A small number of gastropods are adapted for a sessile existence. The worm snails, members of three unrelated mesogastropod families (Vermetidae, Turritellidae, and Siliquariidae), have typical larval and juvenile shells, but as the animal grows older, the whorls become completely separated, and the adult shell looks like a corkscrew or is completely irregular (Fig. 10–29). Worm snails attach their shells to sponges, other shells, or rocks, and the separated whorls provide greater surface area for attachment. The foot is reduced, but an operculum is present (Fig. 10–29B).

A pelagic existence has been adopted by some mesogastropods (heteropods) and by some opisthobranchs, notably the pteropods, or sea butterflies. In most of these groups, the foot has become modified as an effective finlike swimming organ. The Heteropoda are laterally compressed, and the foot is transformed into a ventral fin, even though these animals swim upside down (Fig. 10–30D).

The swimming foot of opisthobranchs is modified differently. Two fins, called parapodia, arise as lateral projections from the side of the foot. In the sea hares, which swim intermittently, the fins arise from the middle of the body and are very broad (Fig. 10–31). The pteropods, or sea butterflies, have anteriorly located parapodia, which function as oars (Fig. 10–30A–C), and the animals swim upside down.

Space permits only brief mention of the still other modes of existence that might be described. The

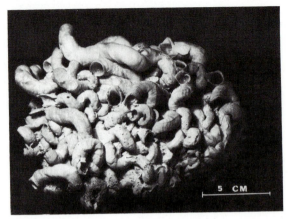

A

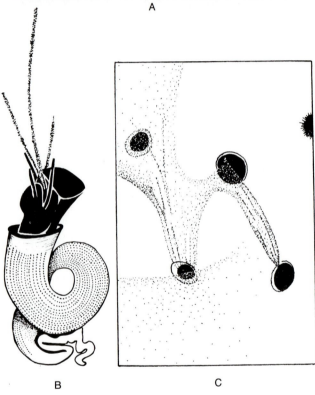

B   C

FIGURE 10–29 Worm snails. **A,** A mass of vermetid worm snails. **B,** The worm snail *Serpulorbis* laying down mucous threads. The shell has been removed. **C,** Interconnecting mucous nets extending from apertures of adjacent worm snails. *(A, Photograph courtesy of Betty M. Barnes. B, After Morton from Hyman, L.H. 1967. The Invertebrates. Vol. VI Mollusca. Pt. I. McGraw-Hill Book Co., New York. C, From Hughes, R.N., and Lewis, A.H. 1974. On the spatial distribution, feeding and reproduction of the vermetid gastropod Dendropoma maxima. J. Zool., London. 172(4):539.)*

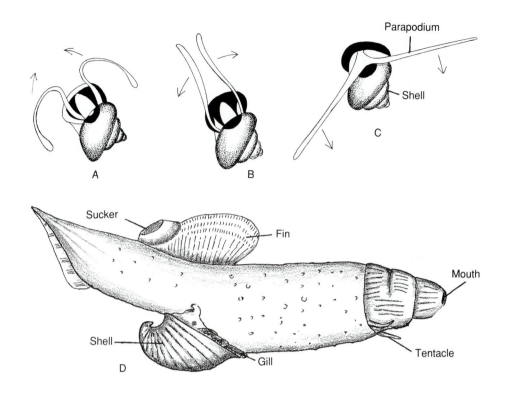

FIGURE 10–30 A–C, Swimming in the shelled sea butterfly, *Lomacina.* Arrows indicate the direction of movement of the parapodia. **A** shows the recovery stroke; **B,** the beginning of the effective stroke; and **C,** the middle of the effective stroke. **D,** *Carinaria,* a pelagic prosobranch. *(A–C, After Morton, J.E. 1967. Molluscs. 4th Edition. Hutchinson University Library, London. D, After Abbot.)*

common but minute prosobranch species of *Caecum,* which live in sand, have short, tusklike shells; the prosobranch violet snails *(Janthina)* float beneath a raft of bubbles secreted by the foot (Fig. 10–32A); planktonic sea slugs *(Glaucus* and *Glaucilla)* stay afloat by means of a bubble of air held in the stomach; the canal boring Coralliophilidae, with often strangely shaped shells, bore into coral (Fig. 10–32B); the carrier snail, *Xenophora,* attaches foreign objects, including other gastropod and bivalve shells, to its own shell with a foot secretion (Fig. 10–32C); there are also a few minute, naked, interstitial opisthobranchs (Fig. 10–23E); and a group of algae-inhabiting, sacoglossan opisthobranchs have secondarily derived bivalve shells (Fig. 10–23G).

Nutrition

Virtually every type of feeding habit is exhibited by gastropods. There are herbivores, carnivores, scav-engers, deposit feeders, suspension feeders, and para-sites. Despite great differences in feeding habits, it is possible to make a few generalizations:

1. A radula is usually employed in feeding.

2. Digestion is always at least partly extra-cellular.

3. With few exceptions, the enzymes for extra-cellular digestion are produced by the salivary glands, esophageal pouches, the digestive diverticula, or a combination of these structures.

4. The stomach is the site of extracellular digestion, and the digestive diverticula are the sites of absorption and of intracellular digestion, if such digestion takes place.

5. As a result of torsion, the stomach has been rotated 180 degrees, so the esophagus enters the stomach posteriorly, and the intestine leaves anteriorly (Fig. 10–13B). In the higher gastropods there has been a tendency for the esophageal opening to mi-

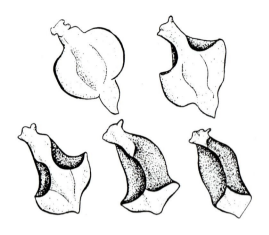

FIGURE 10–31 Swimming in the sea hare *Aplysia*. The lateral swimming fins are foot folds, or parapodia. *(After Pruvot-Fol from Farmer, W.M. 1970. Swimming gastropods. Veliger. 13(1):73.)*

grate forward again toward the more usual anterior position.

In most gastropods the radula has become a highly developed feeding organ, acting as a grater, rasp, brush, cutter, grasper, or conveyor. The total number of teeth varies from 16 to thousands and are always arranged in a longitudinal series of transverse rows. Usually, each row bears a median tooth, and to each side of the median tooth, lateral and marginal teeth appear in succession (Fig. 10–33B). The median, lateral, and marginal teeth usually differ from one another in shape and structure and have different functions. The character and form of the radula teeth are often important in classification.

Primitive Gastropods

The most primitive feeding habit and digestive tract are found in the archaeogastropods. The marine archaeogastropods are largely microphagous animals,

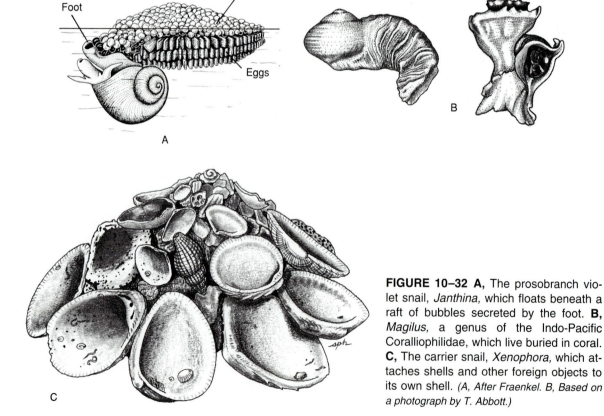

FIGURE 10–32 A, The prosobranch violet snail, *Janthina,* which floats beneath a raft of bubbles secreted by the foot. **B,** *Magilus,* a genus of the Indo-Pacific Coralliophilidae, which live buried in coral. **C,** The carrier snail, *Xenophora,* which attaches shells and other foreign objects to its own shell. *(A, After Fraenkel. B, Based on a photograph by T. Abbott.)*

grazing on fine algae, sponges, or other organisms growing on rocks or kelps. Except in some patellacean limpets the radula usually bears many teeth, at least 12 in each transverse row (Fig. 10–33B, C), acting as a broad scraper. When the radula is retracted, the numerous, fanlike, marginal teeth direct the ingested particles into the center of the gutter produced by the lateral folding of the radula ribbon. The food-laden mucous string formed within the gutter is pulled into the esophagus and then the stomach by the rotating mucous mass within the style sac.

The grazing activity of large populations of intertidal limpets and chitons may greatly limit the growth of algae. Patellacean limpets have a rasping radula with 6 to 20 teeth in a transverse row; the teeth are stout and impregnated with iron and silicon.

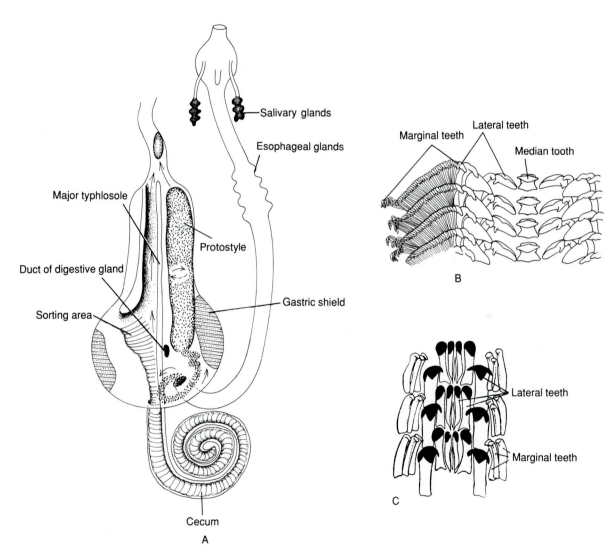

FIGURE 10–33 A, Diagram of digestive tract of a primitive prosobranch (Trochidae). Arrows show ciliary currents and rotation of mucous mass (protostyle) within the style sac. **B,** Part of the radula (rhipidoglossan type) of the abalone *Haliotis* (archaeogastropod), with transverse rows of many teeth. Marginal teeth on one side have been omitted. **C,** Part of the radula (docoglossate type) of the archaeogastropod *Patella*, showing three transverse rows of 12 teeth each. *(A, After Owen. B and C, From Fretter, V., and Graham, A. 1962. British Prosobranch Molluscs. Ray Society, London. p. 171.)*

With the exception of the patellacean limpets, the stomach (Fig. 10–33A) is like that described for the generalized mollusc (p. 368). The large surface area of the sorting region is accommodated within a cecum (Fig. 10–33A). Digestion is partly extracellular, by enzymes that are elaborated by glands in the esophageal region, and partly intracellular within the digestive diverticula.

Higher Gastropods

Perhaps because of the colonization of soft bottoms and other habitats, the diets and feeding habits of higher gastropods became extremely diverse, especially among the mesogastropods and opisthobranchs (Fig. 10–34). Many higher gastropods are macrophagous animals. Digestion has become entirely extracellular and takes place in the stomach, which has lost most of its primitive features—chitinous lining, sorting area, style sac—and become more or less a simple sac (Fig. 10–35C). Enzymes are supplied by the digestive diverticula or by glands associated with the esophagus or buccal region.

The highly adaptable mesogastropod radula bears seven teeth in a transverse row (Fig. 10–35A); the marginal teeth are hook-shaped. The neogastropods, which are mostly carnivores, have radulae with only

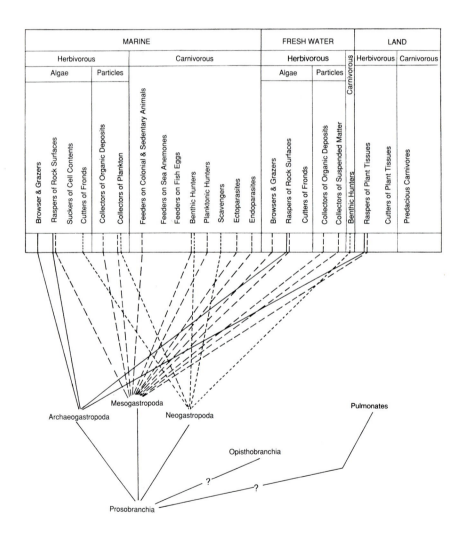

FIGURE 10–34 Diagram illustrating the adaptive radiation in prosobranch feeding habits. *(From Purchon, R.D. 1968. Biology of the Mollusca. Pergamon Press, London. p. 73.)*

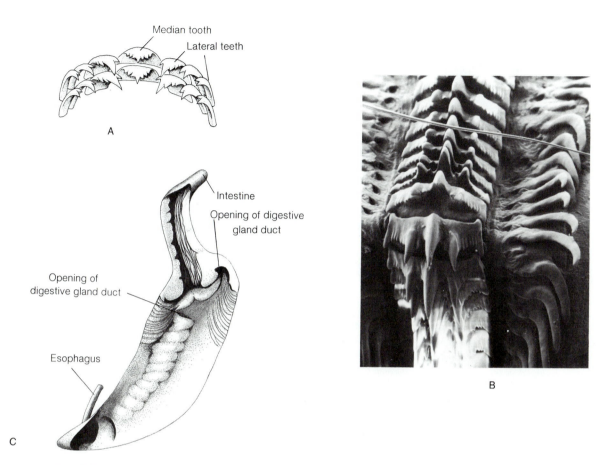

FIGURE 10–35 A, Two transverse rows of radula teeth of the mesogastopod *Lanistes* (tae-nioglossate radula). **B,** Scanning electron micrograph of the radula of *Urosalpinx,* a carnivorous drilling neogastropod. There are only three teeth to a transverse row, but the middle tooth bears several cusps (stenoglossate radula). **C,** Saclike stomach of a carnivorous prosobranch *(Natica).* The stomach has been opened dorsally. *(A, Modified from Turner. B, From Carriker, M.R. 1969. Excavation of boreholes by the gastropod Urosalpinx. Am. Zool. 9:917–933. C, From Fretter, V., and Graham, A. 1962. British Prosobranch Molluscs. Ray Society, London. p. 226.)*

three teeth (sometimes only one tooth) per transverse row, but the teeth are heavy and usually bear several cusps (Fig. 10–35B). The outer, hooked-shaped teeth collect torn or detached particles and bring them into the center when the radula is retracted. The largely herbivorous pulmonates have radulae with the largest number of teeth of any gastropod: up to 750 small teeth per transverse row (Fig. 10–36). The opistho-branch radula is highly variable. The efficiency of the radula results not only from the adaptive design of particular teeth but also from the complex ways in which the teeth interact with each other.

Feeding is also facilitated in many gastropods, including most opisthobranchs and pulmonates, by jaws, which are thickened cuticular pieces in the front of the buccal cavity.

Herbivores

The many herbivorous gastropods include some marine prosobranchs, the freshwater prosobranchs, the operculate land snails, a variety of opisthobranchs, and the majority of the pulmonates. Most marine species feed on fine algae that can be rasped from a rock or other surfaces, or on large algae, such as

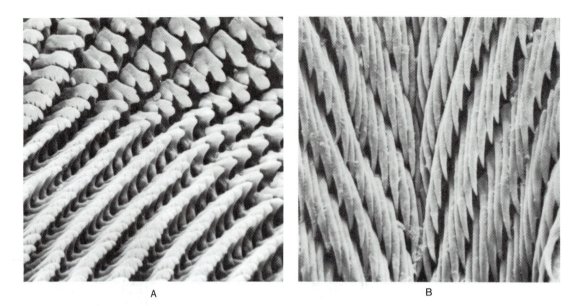

FIGURE 10–36 Scanning electron micrographs of pulmonate radulae. **A,** The herbivorous snail *Diastole conula.* **B,** The carnivorous *Euglandina rosea.* Each tooth is a long cone, which is erected when the radula is in a functional position. *(From Solem, A. 1974. The Shell Makers: Introducing Mollusks. Reprinted by permission of John Wiley and Sons, New York. pp. 163 and 168.)*

kelps, that can carry the weight of the snail. Freshwater and land forms also consume the tender parts of aquatic and terrestrial vascular plants, decaying vegetation, or fungi. A few terrestrial snails and slugs are serious agricultural pests. The giant African snail *Achatina fulica,* introduced into Hawaii and the United States, can be very destructive, and considerable effort has been expended to prevent its spread.

Members of the mesogastropod family Littorinidae, called periwinkles, are found on rocky shores, mangroves, and even marsh grasses throughout the world. These common and often abundant snails live in the intertidal zone, each species occupying a characteristic level between the low-tide mark and the high-splash region. At low tide the animal withdraws into its shell behind the operculum, attaching the lip of the shell to the substratum with mucus. At high tide it emerges to graze on fine algae, including endolithic forms that penetrate just below the surface. The gut contents include large quantities of mineral material.

The sacoglossan opisthobranchs are specialized herbivores. In these tiny, sluglike gastropods the radula is reduced to a single, longitudinal row of teeth, used to slit open algal cells (Fig. 10–37). The contents of the cells are then sucked out. Sacoglossans tend to be rather restrictive in the species of algae they use for food. Members of the family Elysidae are remarkable in incorporating the chloroplasts of their food into their own digestive gland cells, where photosynthesis then occurs. Some nudibranch sea slugs are now known to harbor zooxanthellae, which they obtain from the octocorallians on which they feed. Limpets living near hydrothermal vents harbor a mat of filamentous sulfide bacteria on their gills, which is utilized in their nutrition.

In many herbivorous species, the esophagus or anterior part of the stomach is modified as a crop and gizzard. The gizzard may be lined with cuticle (as in the sea hares, which feed on large pieces of algae) or contain sand grains (as in many freshwater snails). Amylases and cellulases are produced by the esophageal glands or the digestive glands. Terrestrial pulmonates possess a powerful array of digestive enzymes, and digestion is initiated in the large crop, which to a great extent replaces the stomach (Fig. 10–38). There is no gizzard. Cellulase activity has been reported from among the gut digestive enzymes

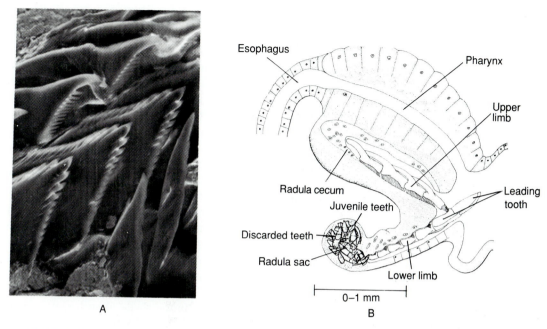

A

B

FIGURE 10–37 A, Scanning electron micrograph of the radula of the aeolidacean nudibranch *Coryphella*. **B,** Lateral section of the radula and buccal mass of the bivalve sacoglossan *Limapontia*. The single row of radula teeth is adapted for puncturing algal cells. *(A, From Cowen, R.K., and Laur, D.R. 1977. A new species of Coryphella from Santa Barbara, California. Veliger. 20(3):292–294. B, From Gascoigne, T., and Sartory, P.K. 1974. The teeth of three bivalved gastropods and three other species of the order Sacoglossa. Proc. Malacol. Soc. London. 41:11.)*

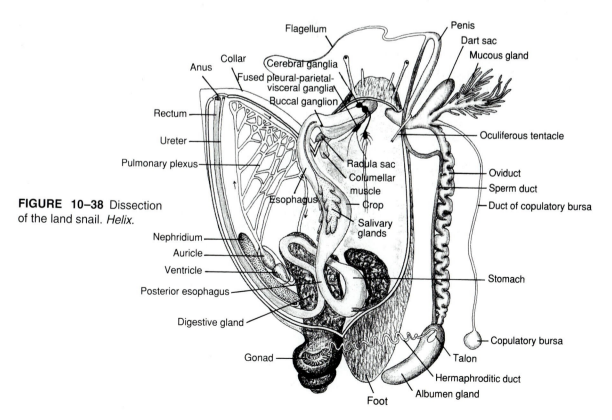

FIGURE 10–38 Dissection of the land snail. *Helix.*

and also appears to result from bacterial action in the crop and intestine.

Carnivores

Although a few of the many carnivorous gastropods are pulmonates, feeding on earthworms or other snails and slugs, the majority of carnivorous families are prosobranchs and opisthobranchs. The radula of these marine families is variously modified for cutting, grasping, tearing, scraping, or conveying. Jaws are sometimes present. The most common adaptation of carnivorous prosobranchs is a highly extensible **proboscis,** which enables the animal to reach and penetrate vulnerable areas of the prey. The proboscis is part of the alimentary tract (contains the esophagus, buccal cavity, and radula) and lies within a proboscis sac, or sheath (Fig. 10–39). The proboscis should not be confused with the tubular siphon, mentioned above, which is part of the mantle and is used as a snorkel to conduct water to the osphradium and gills. In feeding, the proboscis is projected out of the opening of the proboscis sac by blood pressure (Figs. 10–39A and 10–40). Specific proteins liberated by prey or carrion, once detected in the ventilating current, aid in locating the food source and elicit protrusion and search with the proboscis. The whelk *Buccinum undatum* can locate a food source 30 m upstream, but it takes several days to reach it.

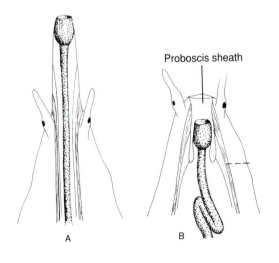

FIGURE 10–39 Diagram showing the proboscis of a prosobranch protracted **(A)** and retracted **(B)** *(After Fretter, V., and Graham, A. 1962. British Prosobranch Molluscs. Ray Society, London.)*

Bottom-dwelling carnivores, especially large species of neogastropods and some mesogastropods, commonly feed on bivalve molluscs, other gastropods, sea urchins, sea stars, polychaetes, crustaceans, and even fish. Many burrow into the sand to reach their prey. Bonnets *(Phalium),* tuns *(Tonna),* and tritons *(Cymatium)* narcotize their prey with salivary secretions containing sulfuric acid or another toxin. Using sulfuric acid secretions, helmets *(Cassis)* can cut a hole into a sea urchin within 10 min. Some olives and volutes smother victims with their feet. A whelk *(Buccinum, Busycon, Fasciolaria, Murex)* may grip the bivalve with its foot, pulling or wedging the two valves apart with the edge of its shell or siphonal canal (Fig. 10–41A). To accomplish the wedging, the gastropod may first erode the valve margin with the lip of its own shell.

A number of prosobranchs are adapted for drilling holes in the shells of such prey as limpets, barnacles, and especially bivalves. The two best known such families are the neogastropod Muricidae *(Urosalpinx, Murex, Thais, Nucella, Eupleura)* and the mesogastropod Naticidae (moon snails, *Natica, Polinices).* The mechanism has been most extensively studied in the Muricidae, and particularly in *Urosalpinx,* which causes great damage in oyster beds. Both the American drill, *Urosalpinx,* and the Japanese drill, *Rapana,* have been introduced into other parts of the world with shipments of oysters. The anterior sole of the foot contains an eversible gland (accessory boring organ), which is applied to the area to be drilled. The acidic secretion produced by this gland reduces the organic framework and demineralizes the shell. Penetration is primarily a result of glandular activity rather than the radula. The animal drills with the radula for about a minute and then applies the eversible gland for about 30 to 40 min, repeating the cycle until the bivalve shell is penetrated. Approximately 8 h are required to penetrate a shell 2 mm thick, and penetration to a depth of 5 mm has been recorded. The beveled sides readily identify the hole as having been drilled by a gastropod. When drilling is completed, the proboscis is extended through the hole, and the soft tissues of the prey are torn by the radula and ingested. In the naticids the shell-softening gland is located at the proboscis tip rather than on the foot.

One of the most remarkable groups of carnivores is the toxoglossan neogastropods containing the turrids, auger, and cone snails. Cone snails (Conidae) are tropical and subtropical and are found mainly in the

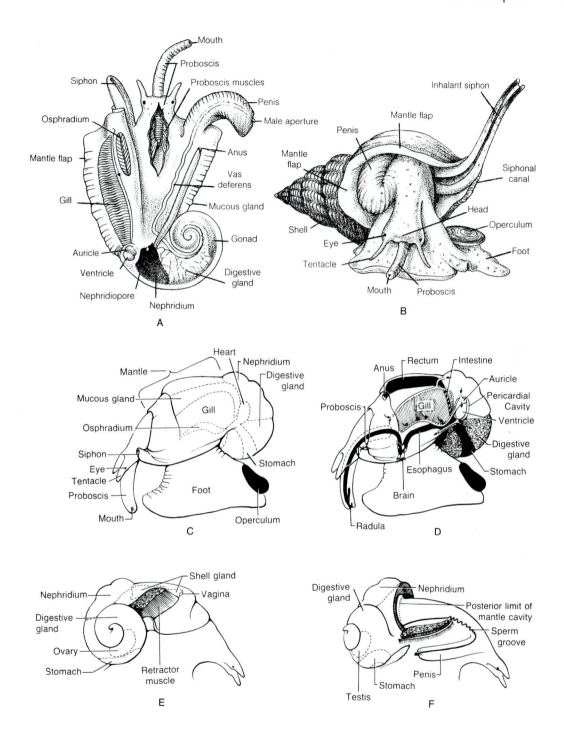

FIGURE 10–40 Anatomy of two marine prosobranchs (neogastropods). **A** and **B**, *Buccinum undatum*. **A,** Dorsal view, with shell removed and wall of mantle cavity cut and reflected. **B,** Animal crawling with proboscis protruded. **C–F,** *Busycon canaliculatum* with the shell removed. The shell is similar to that of *Buccinum* and is carried in the same position. **C,** Left side, showing external organs and internal organs visible through the integument. **D,** Same view with digestive, respiratory, circulatory, and nervous systems indicated. **E,** Female, showing portion of the right side. **F,** Male, showing portion of the right side with mantle and retractor muscle cut short. In **E** and **F,** the proboscis is withdrawn. *(A and B, After Cox, L.R. 1960. Gastropoda. In Moore, R.C. (Ed.): Treatise on Invertebrate Paleontology. Vol. I (pt. 1). Geological Society of America and University of Kansas Press, Lawrence. pp. 189 and 191.)*

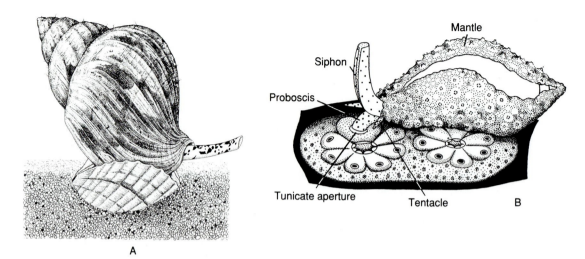

Siphon

Mantle

Proboscis

Tunicate aperture

Tentacle

B

A

FIGURE 10–41 A, *Buccinum* using the edge of its shell to pry open a cockle. **B,** The cowrie *Erato voluta* feeding on a colonial tunicate. The proboscis of the snail is thrust into the buccal opening of the tunicate. The shell of the cowrie is partially covered by the reflexed mantle; the erect structure is the siphon. *(A, From Nielsen, C. 1975. Observations on Buccinum undatum attacking bivalves and on prey responses, with a short review on attack methods of other prosobranchs. Ophelia. 13:87–108. B, From Fretter, V. 1951. Proc. Malacol. Soc. London. 29:15.)*

western Atlantic and Indo-Pacific oceans. They feed primarily on polychaete worms, other gastropods, or fish, which they stab and poison with their radular teeth. The odontophore has disappeared, and the radula is greatly modified (Fig. 10–42A). The teeth, which function singly, are long, grooved, and barbed at the end (Fig. 10–42B), and they are attached to the radula membrane by a slender cord of tissue. A large muscular bulb functions as an injector and is connected to the buccal cavity by a duct, which secretes the poison.

The cone snail has a long, highly maneuverable proboscis. When the proboscis is projected, the barbed end of a single radula tooth slips out of the radula sac into the buccal cavity and is then thrust into the prey. In those species that feed on snails, the tooth is freed from the proboscis, but in forms that feed on polychaetes and fish (Fig. 10–42C), the cone lies buried in the sand and does not strike until the fish pauses over the bottom. The harpoon is then thrust into the soft underbelly of the victim, and the cone retains a grip on the end of the tooth. The victim is very quickly immobilized by the neurotoxic peptide poison, which enters the wound through the hollow cavity of the tooth. The bite, or sting, of a number

of South Pacific species is highly toxic to humans; a few deaths have been reported, in one case within 4 h.

The possible adaptive evolution of the carnivorous prosobranchs is illustrated in Figure 10–43.

There are many carnivorous opisthobranchs, the principal raptorial groups of which are the naked sea butterflies and many bubble snails. The former prey on shelled sea butterflies, and the latter on bivalves and gastropods, which are seized with the hooked teeth of the radula and swallowed whole.

The nudibranchs are grazing carnivores that feed on sessile animals, such as hydroids, sea anemones, soft corals, bryozoans, sponges, ascidians, barnacles, and fish eggs. Each family of nudibranchs is generally restricted to one type of prey. There usually is no proboscis, but jaws are commonly present. In the Aeolidiidae, most of which feed on hydroids and sea anemones, the pair of bladelike jaws are used to cut small pieces of tissue from the prey.

The most remarkable feature of some nudibranchs with cerata is their utilization of the prey's nematocysts. Ciliary tracts in the stomach and in the ducts from the digestive diverticula carry the undischarged, even immature nematocysts to the cerata, where they are engulfed but not digested. The undis-

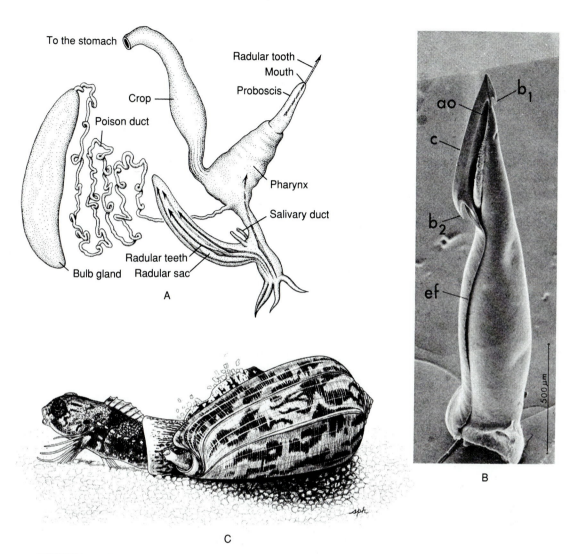

FIGURE 10–42 Feeding mechanism of the cone snails *(Conus)*. **A,** Buccal structures of *Conus striatus*. **B,** Scanning electron micrograph of the harpoon-like radula tooth of *Conus imperialis*. Note the folded structure and barbed end. **C,** A cone snail swallowing a fish. *(A, Modified after Clench. B, From Kohn, A.J. et al. 1972. Science. 176:49–51. Copyright 1972 by the American Association for the Advancement of Science. C, Based on a photograph by Robert F. Sisson and Paul Zahl.)*

charged nematocysts are moved to the distal tips of the cerata, called **cnidosacs,** which open to the exterior. There the nematocysts are used by the nudibranchs for defense, although the discharge mechanism is not yet understood. Nematocysts are replaced in 3 to 12 days, and most nudibranchs utilize only certain types of the various kinds of nematocysts present in their cnidarian prey.

The prosobranch counterparts of the predacious sea butterflies are the pelagic heteropods, which feed on other small, pelagic invertebrates; the cowries (mesogastropods), like the nudibranchs, are grazers that feed on sessile invertebrates (Fig. 10–41B).

Scavengers and Deposit Feeders

A scavenging habit has been adopted by numerous gastropods, of which *Nassarius* and allies are notable examples. The feeding habits of these little neogastropods range all the way from a carnivorous habit to deposit feeding. On quiet, protected beaches along the

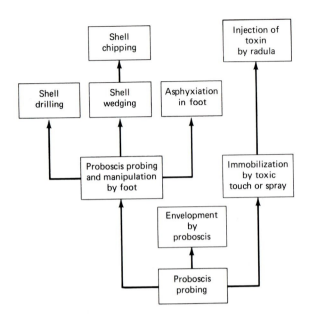

FIGURE 10–43 Possible evolutionary relationships of carnivorous feeding modes in prosobranchs. See the text for descriptions. *(From Taylor, J.D., Morris, N.J., and Taylor, C.N. 1980. Food specialization and the evolution of predatory prosobranch gastropods. Palaeontology. 23(2):375–409.)*

Atlantic coast of the eastern United States, *Ilyanassa obsoleta* may occur in enormous numbers at low tide, feeding on organic material deposited in the intertidal zone (Fig. 10–44A). This same species, however, is also a facultative carrion feeder and consumes the flesh of fresh fish. Along the coasts of Britain and Europe the tiny deposit-feeding mesogastropod *Hydrobia* also occurs in enormous numbers (as many as $30,000/m^2$) on muddy, intertidal flats.

Other deposit feeders include many common species of horn snails *(Cerithium, Cerithidea, Batillaria)*. The conch *Strombus* and the related burrower *Aporrhais* are deposit feeders or grazers on fine algae, with a large, mobile proboscis (under the protective canopy of the flaring shell aperture) that sweeps across the bottom like a vacuum cleaner.

Suspension Feeders

Suspension feeding has evolved a number of times within the Gastropoda. In the slipper snail *(Crepidula)*, a filtering suspension feeder, the gill filaments have been tremendously lengthened, increasing the surface area that traps plankton on a mucous sheet (Fig. 10–45A, B). Many of the sessile worm snails (Turritellidae and Vermetidae) use the gills as a food-trapping surface. Others secrete a net or veil of mucous threads in the vicinity of the shell opening. The net is produced by the pedal gland, laid

A

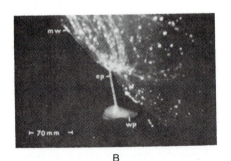

B

FIGURE 10–44 A, *Ilyanassa obsoleta* feeding at low tide. This intertidal prosobranch scavenger and deposit feeder occurs in enormous numbers on protected beaches on the east coast of the United States. **B,** The sea butterfly *Gleba cordata* feeding from its large, delicate, mucous web. The long process labeled ep is the proboscis. The label wp indicates one of the winglike extensions of the foot. *(A, Photograph courtesy of Betty M. Barnes. B, From Gilmer, R.W. 1972. Science. 176:1240. Copyright 1972 by the American Association for the Advancement of Science.)*

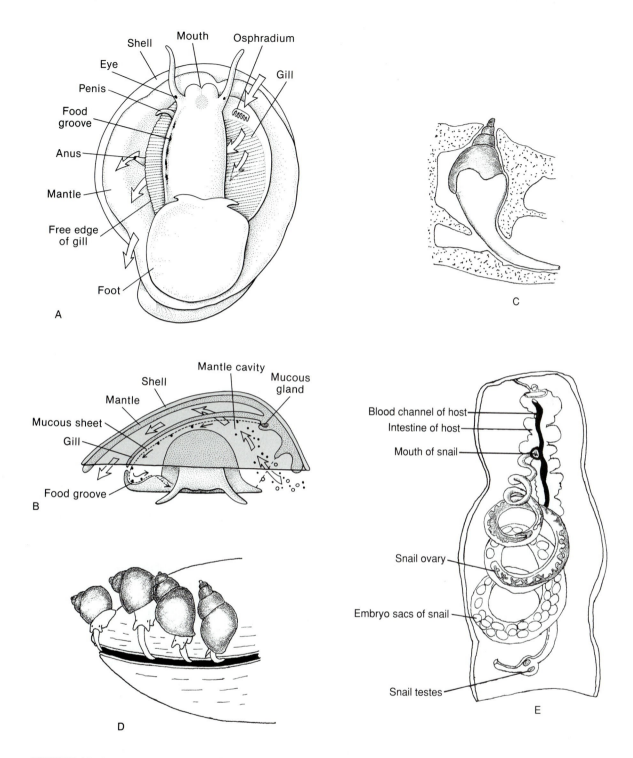

FIGURE 10–45 A, The slipper snail *Crepidula fornicata,* a ciliary feeder (ventral view). **B,** *Crepidula,* showing direction of water current (large outlined arrows) through mantle cavity. Ciliary currents (small arrows) carry food particles. Dashed lines indicate filtering sheet of mucus. Coarse particles are removed by the mucous web at the entrance to the mantle cavity; fine particles are trapped by the mucous sheet on the gill. The latter is driven by the frontal gill cilia to a food-collecting groove along the side of the body, where it is rolled up into a string and periodically pulled into the mouth by the radula. **C,** The parasite *Stilifer* embedded in the body wall of a sea star. **D,** *Brachystomia,* an ectoparasite, shown feeding on the body fluids of a clam. **E,** The endoparasite *Entoconcha* within the body cavity of a sea cucumber. *(B, Modified from Morton. C and D, After Abbott. E, After Baur from Hyman.)*

down by pedal tentacles, and spread by wave action (Fig. 10–29B, C), covering as much as 50 cm in *Serpulorbis squamigerous* of the American Pacific coast. The Red Sea *Dendropoma maxima* takes 2 min to haul in the net with the radula, which it does every 13 min.

The shell-bearing sea butterflies are suspension feeders, trapping food particles in the mucus covering the parapodia on their mantle cavities. Some sea butterflies, such as *Gleba* and *Corolla,* secrete enormous, floating mucous nets as big as 2 m in diameter (Fig. 10–44B). The animal hangs beneath the net by its extended proboscis. There is no radula, and food particles trapped in the net are pulled in by proboscis cilia.

Crystalline styles are found in many suspension and deposit feeders and some grazers—*Crepidula, Struthiolaria,* many worm snails, *Strombus,* species of *Nassarius, Cerithium,* and some sea butterflies among the opisthobranchs. This structure is typical of bivalve molluscs and is described on page 436. The presence of a crystalline style in gastropods is associated with the more or less continuous feeding of the animal on a diet of phytoplankton or organic detritus.

Parasites

Parasitism has evolved in a number of gastropods, which along with certain free-living forms present an interesting adaptive series leading from an epizoic to an ectoparasitic existence and thence to endoparasitism. The ectoparasites are modified chiefly with respect to the nature of the buccal region and digestive system. Members of the opisthobranch family Pyramidellidae possess chitinous jaws, stylets, and a pumping pharynx for sucking blood from bivalve molluscs and polychaetes (Fig. 10–45D). The remaining commensals and parasites belong to a related group of small families (Eulimacea) that live on or within echinoderms (Fig. 10–45C). The most modified are members of the wormlike Entoconchidae, which live within sea cucumbers (Fig. 10–45E).

Excretion and Water Balance

Many archaeogastropods possess two kidneys[‡] but in all other gastropods the right kidney has disappeared, except for a small section that contributes to the reproductive duct. Also, as a result of torsion, the kidney (nephridium) is located anteriorly in the visceral mass (Fig. 10–40). The structure of the kidney is that of a blind sac, with the walls greatly folded to increase the surface area for secretion. At the end of the sac, near the nephridiopore, the kidney connects with the pericardial cavity via a renopericardial canal (Fig. 10–50A). The nephridiopore of both the prosobranchs and the lower opisthobranchs opens at the rear of the mantle cavity, on the downstream side of the gills, and wastes are removed by the circulating water current (Fig. 10–21). Such an arrangement is not possible in terrestrial pulmonates because the mantle cavity functions as a lung. As a result, the pulmonate tubule has lengthened along the right wall of the mantle and opens to the front of the mantle cavity or to the outside near the anus and the pneumostome.

Freshwater gastropods maintain a rather low level of blood salts, and the nephridia are capable of excreting a hyposmotic urine by reabsorption of salts. Freshwater species expel large amounts of water through the nephridia. Terrestrial pulmonates and operculate land snails conserve water by converting ammonia to relatively insoluble uric acid. Perhaps as an adaptation for water conservation, the renopericardial canal of pulmonates has only a small orifice opening into the pericardial cavity, and pericardial glands (sites of ultrafiltration) are absent. The urine is thus largely a result of nephridial secretion. This is also true of some intertidal prosobranchs, such as the littorinids.

Terrestrial pulmonates lose considerable water through evaporation in the production of the slime trail for crawling. Many pulmonates, however, can survive extensive desiccation: *Helix* can survive a water loss equal to 50% of body weight, and the slug *Limax* can survive an 80% loss. As would be expected, the majority of pulmonates nevertheless require a humid environment. They either are nocturnal or live in damp places, such as beneath logs or in leaf mold on forest floors. Some pulmonates inhabit dry, rocky areas, dunes, and even deserts, but such species are active only at night or following rains. During dry periods in the tropics and in deserts and during the winter in temperate regions, snails become inactive, either estivating or hibernating. They may first burrow into humus or soil, or climb into vegetation and attach the aperture edge of the shell with dried mucus. Estivation in elevated positions is probably an adaptation to avoid ground-dwelling predators, such as mice, reptiles, and beetles. Terrestrial prosobranchs close the aperture with the operculum, but in pul-

---

[‡]The left nephridium is reduced in patellacean limpets.

monates the edges of the mantle are drawn together in front of the shell aperture, and a protective mucous membrane that hardens when dry (or a thin calcareous membrane) is secreted over the opening. Several such **epiphragms** may be secreted as the snail withdraws further into the shell. Freshwater snails also estivate when ponds dry up and hibernate when the water is frozen. Estivation may last a number of months, and there are records of snails estivating for several years. Reactivation commonly results from temperature change, rise in humidity, or jarring, as from rain drops beating on the shell. Handling can also cause reactivation.

Internal Transport

As a result of torsion, the heart of a gastropod is located anteriorly in the visceral mass (Fig. 10–13B). The primitive archaeogastropods have retained two auricles. In all other gastropods, however, the right auricle either has become vestigial or, in most instances, has disappeared as a result of the loss of the right gill, which supplied it with blood (Figs. 10–21 and 10–40). The ventricle gives rise to a posterior aorta supplying the visceral mass and an anterior aorta supplying the head and foot, or there may be a single short aorta, which then divides into an anterior and a posterior artery. In the abalone, which has been studied in some detail, there is an anterior aorta supplying the dorsal part of the mantle and a posterior aorta that supplies the rest of the body. Moreover, in the head/foot region, where blood has a hydraulic function in addition to transport, the arterial system opens into sinuses, but in the region of the visceral mass, where the system is largely involved in transport, blood is confined to small vessels.

From the arterial sinuses, blood eventually collects in a large cephalopedal vein (Fig. 10–46). Much of this blood passes through the kidney before entering the branchial circulation, but some may return directly to the heart. In the return flow to the heart in pulmonates, all blood from the venous sinuses passes through the pulmonary capillary network in the roof of the lung (Fig. 10–38).

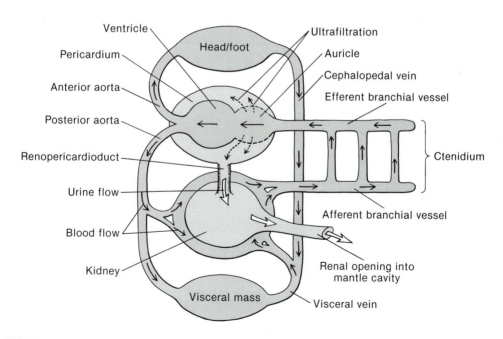

FIGURE 10–46 Pattern of blood circulation in the periwinkle, *Littorina littorea*. The outlined arrows indicate urine flow, the solid arrows blood flow, and the dashed arrows ultrafiltration. *(Simplified and redrawn from Andrews, E.B., and Taylor, P.M. 1988. Fine structure, mechanism of heart function and haemodynamics in the prosobranch gastropod mollusc Littorina littorea (L.) J. Comp. Physiol. B. 158:247–262.)*

Prosobranchs and pulmonates possess the respiratory pigment hemocyanin, which is dissolved in the plasma. The freshwater Planorbidae (pulmonates) possess hemoglobin instead of hemocyanin in the plasma. The opisthobranch gas transport provisions are poorly known, but some species of sea hares *(Aplysia),* which have been extensively studied, possess hemocyanin and some lack it.

Nervous System

The nervous system of gastropods may be more easily understood if the ground plan is first described as if torsion had not taken place (Fig. 10–47A). A pair of adjacent cerebral ganglia lie over the posterior of the esophagus and give rise to nerves that connect anteriorly to the eyes, tentacles, statocysts, and a pair of buccal ganglia, which are located in the back wall of the buccal cavity. The buccal ganglia innervate the muscles of the radula and other structures in this vicinity. A nerve cord issues ventrally from each cerebral ganglion on each side of the esophagus. These are the two pedal connectives, which extend ventrally to a pair of ganglia located in the midline of the foot. The pedal ganglia innervate the foot muscles.

A pair of connectives arise from the cerebral ganglia and extend back to a pair of pleural ganglia, which supply the mantle and the columella muscle. One pair of connectives joins the pleural and pedal

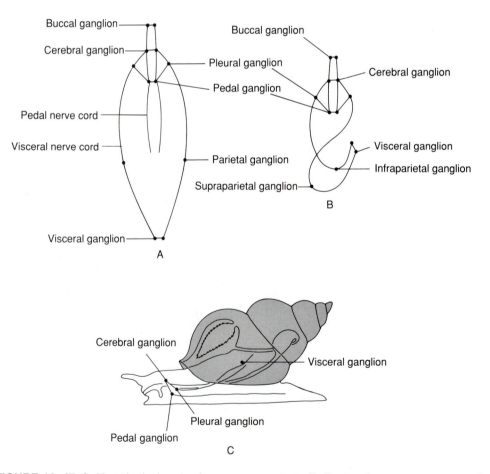

FIGURE 10–47 A, Hypothetical pretorsion nervous system. **B,** Posttorsion nervous system. **C,** Lateral view of gastropod showing position of nervous system. *(After Naef from Kandel, E.R. 1979. Behavioral Biology of Aplysia. W.H. Freeman and Co., San Francisco.)*

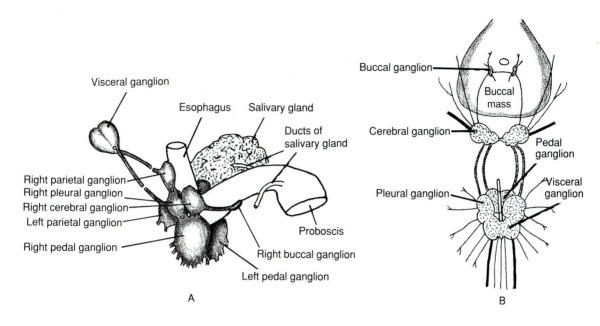

Visceral ganglion

Esophagus   Salivary gland

Ducts of
salivary gland

Right parietal ganglion
Right pleural ganglion
Right cerebral ganglion
Left parietal ganglion

Right pedal ganglion

Proboscis

Right buccal ganglion

Left pedal ganglion

A

Buccal ganglion

Buccal
mass

Cerebral ganglion

Pedal
ganglion

Pleural ganglion

Visceral
ganglion

B

FIGURE 10–48 A, Concentrated nervous system in *Busycon.* **B,** Secondarily symmetrical nervous system of the pulmonate *Helix. (A, After Pierce, M. In F.A. Brown, Jr. 1950. Selected Invertebrate Types. John Wiley and Sons, New York. B, After Bullough.)*

ganglia. A second pair of cords leaves the pleural ganglion and extends posteriorly until it terminates in a pair of visceral ganglia that is located in the visceral mass and that supplies organs in this region. A pair of parietal, or intestinal, ganglia innervates the gills and osphradium. These ganglia are located along the length of the visceral nerves.

As a result of torsion, the visceral cords are twisted to resemble a figure 8 (Fig. 10–47B), and one intestinal ganglion (supraparietal ganglion) is now located higher in the visceral mass than the other.

The twisted condition is one distinctive feature of the primitive gastropod nervous system. Another is the separation of the ganglia by nerve cords, as previously described. Such a primitive nervous system is found, with some modifications, in many prosobranchs. However, in most gastropods the original arrangement is obscured because of two evolutionary tendencies. First, there has been a tendency toward both concentration and fusion of ganglia with a consequent shortening of the connectives between ganglia. Second, there has been a tendency for the ganglia and cords to adopt a secondary bilateral symmetry (Fig. 10–48A, B).

The ganglia of molluscs are organized in the same way as those of annelids. Because their neurons are large and identifiable, sea hares and some other opisthobranchs have been the principal molluscan subjects of neurophysiological investigations of specific neuronal circuitry and its control of behavioral reflexes. Knowledge of endocrine control, especially for reproduction, has been increasing.

Sense Organs

The sense organs of gastropods include eyes, tentacles, osphradia, and statocysts. The primitive eye, as in *Patella,* is a simple pit lined by photoreceptor and pigment cells (Fig. 10–49A), but in most higher gastropods the pit has become closed over and differentiated into a single, chambered, vesicular eye with a lens (Fig. 10–49B). The eyes of most gastropods appear to detect only changes in general light intensity (p. 362).

A pair of closed statocysts is located in the foot near the pedal ganglia of gastropods. Statocysts are absent from sessile forms, such as slipper snails and worm snails. The evolution of the osphradium of gastropods closely parallels that of the gills. In the primi-

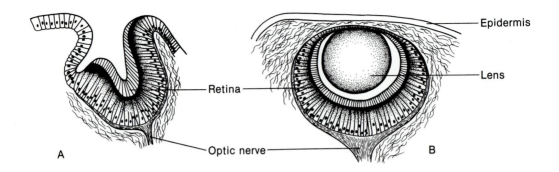

FIGURE 10–49 Eyes of two marine prosobranchs. **A,** *Patella,* a limpet. **B,** *Murex. (Both After Helger from Parker and Haswell.)*

tive Archaeogastropoda an osphradium is present for each gill. In the other prosobranchs, which possess but one gill, there is also only one osphradium, which is located on the mantle cavity wall anterior and ventral to the attachment of the gill (Figs. 10–40A, C; 10–45A). In most cases the osphradium has become either filamentous or folded, thereby increasing its chemoreceptive surface. This organ is highly developed in prosobranch carnivores and scavengers, which can locate carrion, animal juices, or prey from a considerable distance, as much as 2 m in the case of scavenger *Cominella.* Many gastropods continually wave their siphons about as they move over the bottom, selecting and monitoring water from various parts of the environment.

Reproductive System

Most prosobranchs are dioecious. The single gonad, either ovary or testis, is located in the spirals of the visceral mass near the digestive gland (Fig. 10–40A, E, F). The gonoduct ranges from a very simple to a highly complex structure, but in all cases it has developed in close association with the right kidney. In most Archaeogastropoda both kidneys are functional, and the right kidney provides an outlet for either the sperm or the eggs (Fig. 10–50A). The gametes pass through a short duct that extends from the gonad and opens into the distal part of the kidney; the gametes are then conducted by the kidney into the mantle cavity through the nephridiopore. The genital duct is thus formed from two elements—the gonoduct proper and the right kidney. In this type of reproductive system,

the eggs are provided, at most, with gelatinous envelopes produced by the ovary or the terminal part of the kidney. There is no copulation, and fertilization takes place in the sea water after the eggs are swept out of the mantle cavity.

In all the other gastropods there is copulation and internal fertilization and a correspondingly complex reproductive system. The right kidney has degenerated except for a very small portion that functions solely as part of the genital duct. Furthermore, the genital duct has become considerably lengthened by a third addition, derived from the mantle; as a result, the genital pore is located at the opening of the mantle cavity. This third section, which might be called the **pallial duct,** probably first arose as a ciliated groove extending from the nephridiopore, but in at least the females of existing gastropods the groove has become closed over to form a distinct tube (Fig. 10–50B). It is the pallial portion of the genital duct that has undergone elaboration or differentiation to provide for sperm storage and egg membrane formation. Certainly the freeing of the right kidney from excretory functions and the subsequent development of a complex reproductive system has contributed to the success of the mesogastropods and neogastropods.

In the male of these gastropods, the entire gonoduct consists of a coiled tube (vas deferens) leading from the testis and functions in sperm storage and transport. The pallial vas deferens, containing a prostate, runs in the floor of the mantle cavity and out to a tentacle-like penis located behind the right cephalic tentacle. In some prosobranchs (neritids, littorinids, cerithiids) this pallial portion is represented

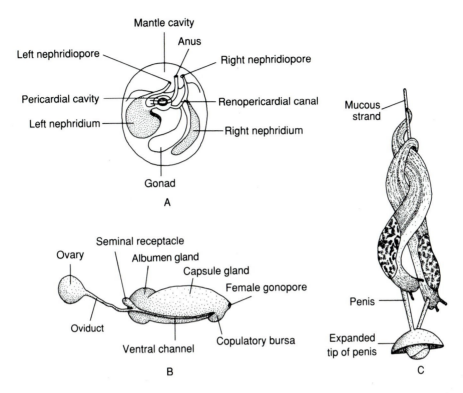

FIGURE 10–50 A, Diagrammatic dorsal view of an archaeogastropod (Trochacea) showing the relationship of the gonad to the nephridia. Note that the gonad opens into the renopericardial canal of the right nephridium. **B,** Diagram of the reproductive system of *Littorina,* a mesogastropod. **C,** Copulating pair of pulmonate slugs, *Limax. (A and B, From Fretter, V., and Graham, A. 1962. British Prosobranch Molluscs. Ray Society, London. pp. 283 and 360.)*

by a ciliated groove instead of a duct. In various genera of all classes of gastropods the sperm are delivered in spermatophores. For example, the sessile worm snails are dioecious and transfer sperm by rafting spermatophores, which are caught by the mucous nets of other individuals.

In the female the pallial section of the oviduct is modified to form both an albumen gland and a large jelly gland or capsule gland (Fig. 10–50B). Species such as *Cerithium* and *Littorina* embed the eggs in jelly masses produced by the jelly gland, but in many of the higher prosobranchs, the eggs are enclosed in a capsule. There is a seminal receptacle or bursa or both for the storage of sperm. Where both are present, the bursa handles the initial reception and short-term storage of sperm (Fig. 10–50B). The outgoing eggs pass through the dorsal portion of the oviduct, where

membranes are applied. In neogastropods, which tend to have tougher egg cases, the soft egg capsules from the female gonopore are conducted down to a pedal gland in the foot, where the egg case is molded, hardened and then attached. The oyster drill *(Urosalpinx),* for example, deposits 7 to 96 such cases containing 4 to 12 eggs each during a breeding season.

A small number of prosobranchs are hermaphrodites (Fig. 10–51). Protandric hermaphrodites include many patellacean limpets and the slipper snails *(Crepidula).* In the more or less sessile slippers, individuals of some species tend to live stacked up on one another (Fig. 10–52A). The right shell margins are adjacent, thereby permitting the penis of the upper individual to reach the female gonopore of the individual below. Young specimens of such aggregating species are always males. This initial male phase is

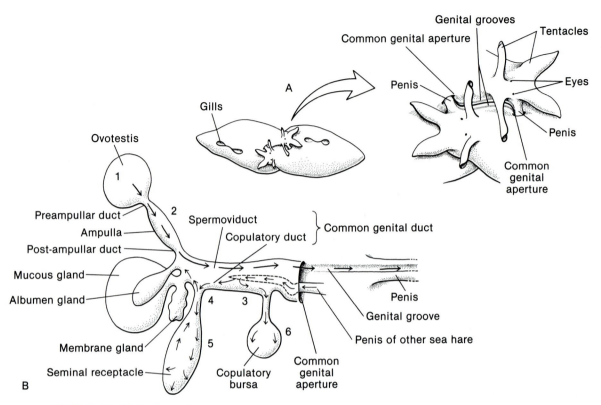

FIGURE 10–51 Reciprocal sperm transmission in the hermaphroditic sea hare, *Phyllaplysia tay-lori.* **A,** Dorsal view of copulating pair. **B,** Sperm movements within one individual. Heavy arrows indicate path of the emitted sperm; light arrows, the path of the sperm received; numbers, the sequential location of sperm before and after emission. *(From Beeman, R.D. 1970. An autoradiographic study of sperm exchange and storage in a sea hare, Phyllaplysia taylori, a hermaphroditic gastropod. J. Exp. Zool. 175(1):130.)*

followed by a period of transition in which the male reproductive tract degenerates; the animal then develops into a female or another male. The sex of each individual is influenced by the sex ratio of the association, probably by pheromones. An older male will remain male longer if it is attached to a female. If such a male is removed or isolated, it will develop into a female. The scarcity of females influences certain of the males to become females. The actual change is directed by the nervous system through control of hormone secretion. Once the individual becomes female, it remains in that state.

Pulmonates and opisthobranchs are mostly simultaneous hermaphrodites, although the **ovotestis** may not produce sperm and eggs at the same time. Copulation with reciprocal sperm transfer is typical. The reproductive systems are very complicated and

display endless variations (Figs. 10–38 and 10–51B). In land pulmonates the sperm are usually exchanged in spermatophores, and copulation is commonly preceded by a courtship involving circling, oral and tentacular contact, and intertwining of the bodies. Bizarre sexual behavior occurs in some species, particularly slugs. In the shelled Helicidae *(Helix),* the vagina contains an oval dart sac, which secretes a calcareous spicule. When two snails are intertwined, one drives its spicule dart into the body wall of the other. Copulation follows this rather drastic form of courtship. The dart shooter is stimulated to copulate and to bite less, whereas the behavior of the dart receiver appears relatively unmodified. The process may have evolved as a means of inducing one individual to act as a male and the other as a female. Copulating limacid slugs hang intertwined from a cord of mucus attached to a tree trunk or branch, and in the process of exchanging sperm, the penes are unrolled to a length of 10 to 25 cm (sometimes as much as 85 cm) and are twisted together at the tips (Fig. 10–50C). Spermatophores are exchanged by the penial tips and carried back to the body by the retracting penes for storage prior to fertilization.

Development

Egg deposition in gelatinous strings, ribbons, or masses is characteristic of most mesogastropods and opisthobranchs, but neogastropods and some mesogastropods embed their eggs in an albumen mass surrounded by a capsule or case, which is usually attached to the substratum. The size and shape of the case, the nature of the wall (which may be leathery or gelatinous), and the number of cases attached together are extremely variable and are characteristic of the species (Fig. 10–52B).

Aquatic pulmonates deposit their eggs in gelatinous capsules. Terrestrial species produce a relatively small number of eggs, each enclosed with albumen within a separate capsule. The eggs are usually laid in a heap in soil. Among the largest eggs—up to 16 mm in diameter—are laid by a 15-cm South American species of *Strophocheilus* (Fig. 10–52H).

In over half of the terrestrial pulmonate families and even in some prosobranch land snails, the capsule wall contains calcite crystals embedded in mucus or consists of a calcareous shell. The shell not only supports and protects the egg contents but serves as a source of calcium for the shell of the developing snail within. A contractile, vesicle-like extension of the embryonic foot, called a **podocyst,** functions in absorption of albumen as well as in gas exchange and excretion. The podocyst is thus somewhat analogous to the allantois of reptiles and birds.

A free-swimming trochophore larva is found only in archaeogastropods that shed their eggs directly into the sea water; in all the other gastropods, the trochophore stage is suppressed and is passed before hatching.

More characteristic of marine gastropods is a free-swimming veliger larva. The veliger larva is derived from a trochophore but represents a later, more developed stage (Fig. 10–53). The characteristic feature of the veliger is the swimming organ, called a velum, which consists of two large, semicircular lobes bearing long cilia. The velum forms as an outward extension of the prototroch of the trochophore. The foot, eyes, and tentacles differentiate from the body of the embryo. The shell, called the **protoconch,** develops spirally in the veliger and may remain at the apex of the adult shell for some time until it eventually disappears through differential growth and erosion. In the sea slugs a shell appears in the veliger and is later cast off during metamorphosis.

Some gastropods have feeding (planktotrophic) veligers with a larval life that may last as long as three months; others have short-lived, yolk-laden, nonfeeding (lecithotrophic) veligers. The long cilia of the velum function not only in locomotion but also in suspension feeding. The beating of the long velar cilia brings fine plankton in contact with the shorter cilia of a subvelar food groove. Within the food groove, particles become entangled in mucus and are conducted to the mouth (Fig. 10–53C).

During the course of the veliger stage, torsion occurs and the shell and visceral mass twist 180 degrees in relation to the head and foot. Torsion may be very rapid (only about 3 min in the marine limpet *Acmaea*), or it may be a gradual process. As development proceeds, the veliger reaches a point at which not only can it swim by means of the velum, but also the foot is sufficiently formed to allow creeping. Settling and metamorphosis occur. The velum is lost, and the final features of the adult form are attained. Settling sites are of critical importance in the survival of the larvae, and many species can delay metamorphosis until specific types of substrata can be reached (p. 364).

Some marine prosobranchs, especially the Neogastropoda, nearly all freshwater prosobranchs, and almost all pulmonates have no free-swimming larvae. At hatching, a tiny snail emerges from the protective shell or case.

The course of development may vary greatly even within a closely related group. For example, among the common intertidal periwinkles *(Littorina)*, some release pelagic egg cases from which veliger larvae hatch; some attach their egg cases, with or without larvae; and some brood their eggs, giving birth to larvae or to little snails. The condition is related in part to the particular intertidal level occupied by the species, and the timing of reproductive events may be tied to tidal cycles.

Many gastropods reach adult size and sexual maturity at 6 months to 2 years, but slow growth continues and larger species may not reach maximum size

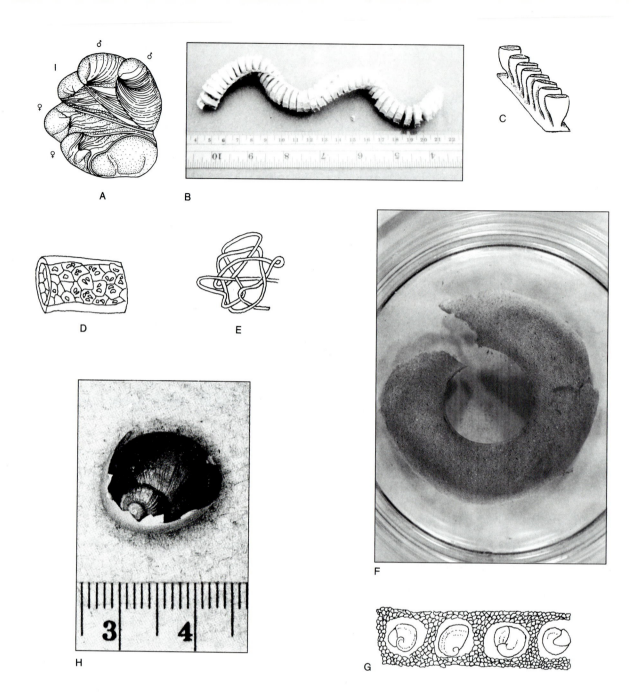

FIGURE 10–52 A, A stack of slipper snails, *Crepidula fornicata,* in life position. **B,** Parchment egg cases of the whelk, *Busycon carica.* The many cases, each shaped like a pill box, are connected by a cord. Development is direct, and the little whelks emerge from a hole at the margin of each case. Although often washed up on beaches, the cases are produced while the whelk is buried in the sand, so for a time at least the string is anchored in sand. **C,** Parchment egg cases of *Conus.* **D–E,** Gelatinous egg string of a sea hare, *Aplysia;* **E,** Enlarged. **F,** Egg case, or "sand collar," of the moon snail, *Polinices duplicatus.* **G,** Section through the egg case of a moon snail. **H,** The very large opened egg (about 16 mm) of the pulmonate, *Strophocheilus,* showing an enclosed young snail. *(A, From Hoagland, K.E. 1979. The behavior of three sympatric species of Crepidula from the Atlantic, with implications for evolutionary ecology. Nautilus. 94(4):143–149. C–E, After Abbott, R.T. 1954. American Sea Shells. Van Nostrand Co., Princeton, NJ. B and F, Photographs courtesy of Betty M. Barnes. H, From Tompa, A. 1980. A method for the demonstration of pores in calcified eggs of vertebrates and invertebrates. J. Micros. 118(4):477–482.)*

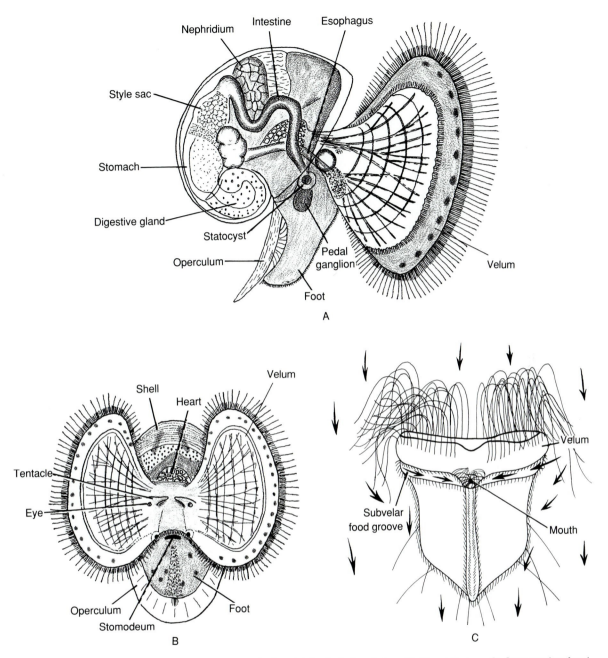

FIGURE 10–53 Veliger larva of the slipper snail, *Crepidula*. **A,** Lateral view. **B,** Frontal view. **C,** Suspension feeding in the early veliger of *Archidoris*. Arrows indicate direction of movement of food particles. *(A and B, After Werner from Raven. C, After Thompson.)*

for many years. In general, growth is more rapid in tropical species than in temperate ones. The life span is highly variable: 5 to 16 years for the limpet, *Patella vulgata;* 4 to 10 years for the periwinkle, *Littorina littorea;* 1 to 2 years for many freshwater pulmonates; 5 to 6 years for the land snail, *Helix aspersa;* and only 1 year for many nudibranchs. Few juvenile gastropods survive long enough to grow to sexual maturity. Of the 1000 eggs produced each year by the muricid prosobranch, *Thais,* not more than ten juveniles ever reach an age of one year.

SUMMARY

1. The evolution of the class Gastropoda involved four important changes: greater cephalization, dorsoventral body elongation, development of an asymmetrical spiral shell, and torsion.

2. All evidence indicates that shell spiraling preceded torsion and was probably related to the conversion of the shell from a shield to a retreat into which the animal could withdraw.

3. The evolutionary significance of torsion is uncertain but may have made possible the withdrawal of the head into the shell before the foot, as is true of living gastropods.

4. Living gastropods possess asymmetrical spiral shells, which have the advantage of compactness over planospiral shells and greater resistance to crushing. The shell of modern forms is carried obliquely across the body with the spire apex directed to the right, posteriorly, and upward.

5. The asymmetrical shell and its carriage cause some occlusion of the mantle cavity on the right side, which in turn has brought about the reduction or loss of the right gill and associated auricle. The left pedal retractor muscle is reduced or lost.

6. Most gastropods move by waves of muscle contraction that sweep along the length of the broad ventral surface of the foot. Cilia are important in the locomotion of juveniles and forms that live on soft bottoms.

7. The prosobranch archaeogastropods are believed to be the most primitive of living forms. They have retained the bipectinate condition of the gill. The most primitive archaeogastropods are probably the slit snails, keyhole limpets, and abalones, in which shell notches or perforations permit the exit of the ventilating current and waste products. They are also the only living gastropods to retain the right gill and auricle and the left pedal retractor.

8. The patellacean limpets (archaeogastropods) have a single anterior gill with a laterally or posteriorly directed ventilating current, or the original gill has been lost, with or without replacement by laterally paired gills in the mantle groove.

9. The trochaceans and neritaceans exhibit an oblique water current, which enters the mantle cavity on the left side of the head and exits on the right side. This ventilating current has been retained by all higher prosobranchs.

10. The archaeogastropods are largely restricted to firm substrata, where they graze on fine algae and other organisms found on rock surfaces. Their microphagous habit is reflected in the nature of the radula and the presence of a protostyle and sorting region within the stomach.

11. The gametes of archaeogastropods exit by the right kidney, which continues to serve in excretion. The reproductive circumstances of most archaeogastropods—no copulation, external fertilization, planktonic eggs or eggs that are weakly enveloped when deposited—are correlated with the limitations in gonoduct specialization. A trochophore larva precedes the veliger.

12. The single left gill of higher prosobranchs (mesogastropods and neogastropods) is monopectinate rather than bipectinate. The shift to the monopectinate condition, with the gill axis directly attached to the mantle wall, perhaps had the advantage of reducing surfaces that might be fouled with sediment in the ventilating current. In any event, these prosobranchs are not confined to hard surfaces, and many species live on soft bottoms.

13. Some mesogastropods invaded freshwater and land and evolved many species, although the land snails are largely confined to the tropics. Freshwater and terrestrial prosobranchs are operculate.

14. The great diversity of mesogastropods and neogastropods is reflected in their modes of feeding, locomotion, and habitation. Macrophagous snails typically have a simple saclike stomach. The various feeding styles are correlated with radula design. A proboscis occurs in many carnivorous species.

15. The right kidney of mesogastropods and neogastropods has been freed of an excretory function and contributes only to the gonoduct. Another contribution from the wall of the mantle cavity has extended the gonoduct length and potential for complexity of function. Most species copulate, fertilize eggs internally, and deposit the eggs within well-developed envelopes or cases. The eggs hatch into veligers, or development is direct.

16. Pulmonates show 90 degrees of torsion and may have evolved from mesogastropod-like ancestors, which lived in some shallow-water marine habitat subjected to frequent, periodic reduction in standing water. The gill was lost, and the mantle cavity was converted to a lung for gas exchange in air. Most lower pulmonates inhabit fresh water, where they continue to use the lung for gas exchange or have evolved secondary gills. Higher pulmonates are terrestrial and may be readily distinguished from mesogastropod land snails by the absence of an operculum (true of all pulmonates). Higher pulmonates differ from lower pulmonates in having two pairs of tentacles, with the eyes mounted at the top of the second pair.

17. Most pulmonates are herbivores, although there are some carnivorous species. The radula bears a large number of teeth in each transverse row. Higher pulmonates display an array of structural, physiological, and behavioral adaptations for life on land.

18. Pulmonates are hermaphrodites, with copulation and mutual sperm transfer. The eggs are deposited within envelopes, and development is direct, except in the few marine species.

19. Opisthobranchs are marine gastropods that exhibit 90 degrees of torsion and are said to be detorted. The gill and mantle cavity are located on the right side. The operculum has been lost except in a single primitive family. Primitively, there is a shell and gill, but in many opisthobranchs reduction or loss of shell, mantle cavity, and gill has occurred. The numerous, bilaterally symmetrical, sluglike forms, of which the colorful nudibranchs are the most notable examples, represent the extreme products of this change. Opisthobranchs are very diverse in feeding modes, habitation, and locomotion.

20. Opisthobranchs are simultaneous or protandric hermaphrodites, with copulation, internal fertilization, and egg deposition. The veliger is the hatching stage in indirect development.

CLASS BIVALVIA

The class Bivalvia, also called Pelecypoda or Lamellibranchia, includes such common animals as clams, oysters, and mussels. Bivalves are laterally compressed and possess a shell composed of two valves, hinged dorsally, that completely enclose the body. The foot, like the remainder of the body, is laterally compressed, hence the origin of the name Pelecypoda, meaning "hatchet foot." The head is very poorly developed. The mantle cavity is the most capacious of any class of molluscs, and the gills are usually very large, having assumed in most species a food-collecting function in addition to that of gas exchange. Most of these characteristics represent modifications that enabled bivalves to become soft-bottom burrowers, for which the lateral compression of the body is well suited. Although modern bivalves have invaded other habitats, the original adaptations

for burrowing in mud and sand have taken bivalves so far down the road of specialization that they have become largely chained to a sedentary existence.

Bivalves are believed by many malacologists to have evolved from an extinct class of molluscs called **rostroconchs,** which in turn were probably derived from laterally compressed monoplacophorans. Like bivalves, the body of rostroconchs was enclosed within two shell valves, but there was no dorsal hinge; that is, the two halves of the shell were continuous with each other across the dorsal surface. In many species the ventral shell margins touched except for an anterior gape for the foot, and commonly there was a tubular rostrum-like posterior extension of the shell through which circulating water may have passed.

The class Bivalvia contains three major groups, distinguished by the nature of their gills: protobranchs, lamellibranchs, and septibranchs. These groups were formerly considered to be subclasses. Although only the protobranchs have any formal taxonomic status, these three are still very useful groupings and we will make continual reference to them. The protobranchs are generally believed to be the most primitive of existing bivalves. The septibranchs are highly specialized. The lamellibranchs include the majority of the bivalve species.

Shell, Mantle, and Foot

A typical bivalve shell consists of two similar, more or less oval, usually convex **valves,** which are attached and articulate dorsally with each other (Fig. 10–54A, B). Each valve bears a dorsal protuberance called the **umbo,** which rises above the line of articulation and is the oldest part of the shell. The two valves are attached by a noncalcified, elastic, protein band called the **hinge ligament,** which is covered above with the periostracum. The hinge ligament bridges the two valves and, together with them, forms the shell. The ligament is so constructed that when the valves are closed, the dorsal or outer part (lamellar layer) is stretched and the ventral or inner part (fibrous layer) is compressed (Fig. 10–54A). Thus, when the adductor muscles relax, the natural elasticity of the ligament causes the valves to open. Numerous hinge specializations have evolved. In some bivalves, the hinge has grown upward, acquiring the form of a horseshoe in cross section (Fig. 10–55A).

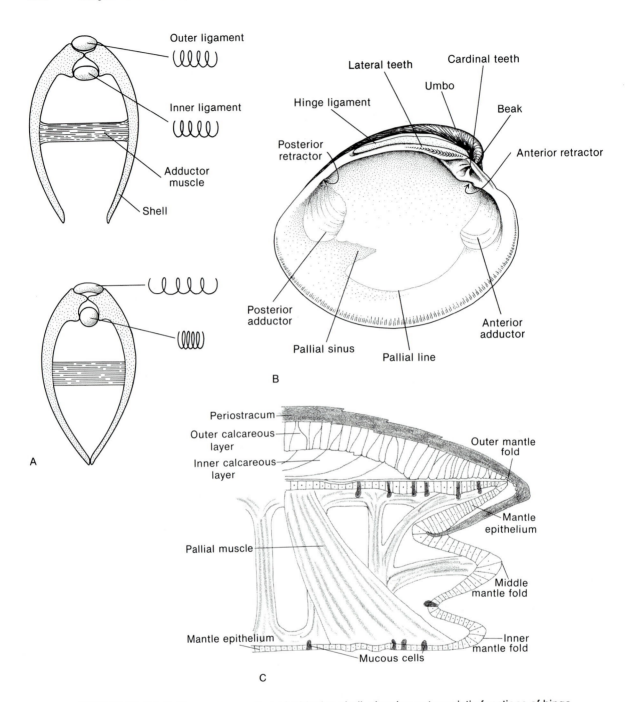

FIGURE 10–54 A, Transverse section of a bivalve shell, showing antagonistic functions of hinge ligaments and adductor muscles. When the valves are closed by the adductor muscles, the outer hinge ligament is stretched and the inner ligament is compressed. **B,** Inner surface of the left valve of the marine clam *Mercenaria.* **C,** Transverse section through the margin of shell and mantle of a bivalve, showing mantle lobes and points of shell secretion. Although the periostracum is shown pressed against the middle mantle lobe, this layer is actually secreted by the inner side of the outer mantle lobe. *(After Kennedy W.J. et al. 1969. Environmental and biological controls on bivalve shell minerology. Biol. Rev. 44:499–530. Copyrighted and reprinted by permission of Cambridge University Press.)*

When the valves are pulled together by the adductor muscles, the arms of the horseshoe are pressed toward each other; they spring back, opening the valves when the muscles relax. There are also bivalves in which each valve bears a ventrally directed flange in the hinge area (**chondrophore**) that provides a large surface area for the attachment of the inner hinge ligament (Fig. 10–55B).

To prevent lateral slipping, the two valves in most species are locked together by **teeth** or ridges and apposing sockets or grooves, located on the hinge line of the shell beneath the ligament (Fig. 10–54B). However, a dorsal flange of tissue still lies between the tooth and socket of the two apposing valves.

The valves of the shell are pulled together by two large dorsal muscles, called **adductors,** which act antagonistically to the hinge ligament (Fig. 10–54A, B). An anterior and a posterior adductor extend transversely between the valves, and scars on the inner surfaces of the valves indicate where these muscles are attached. The adductors of most bivalves contain both striated and smooth fibers, facilitating rapid and sustained closing of the valves, respectively.

The bivalve shell is composed of an outer periostracum covering two to four calcareous layers. The periostracum may be very thick, as in many large, freshwater clams, or very thin, as in the edible marine quahog *(Mercenaria)*. An important function of the periostracum is related to shell secretion (described later), but it also protects the underlying calcium carbonate from dissolution and may contribute to a tight seal when the edges of the valves are brought together on closure. The calcareous layers may be entirely aragonite (primitive) or a mixture of aragonite and calcite, and they may be deposited as prisms or as minute laths or tablets arranged in sheets (nacre), lenses, or more complex forms. It is nacre that gives the lustrous inner surface to many shells. The basic elements—prisms, tablets, and so forth—are always deposited within an organic framework, which together with the periostracum may account for 12 to 72% of the dry weight of the shell. Although shell structure is not uniform for the class, it is constant and characteristic for different groups of bivalves.

The shells of bivalves exhibit a great variety of sizes, shapes, surface sculpturing, and colors. Where the rate of shell addition is the same all around the margin, the shell is equilateral; where the rates are different, shell shape changes. As in gastropods, surface sculpturing may contribute to traction, protec-

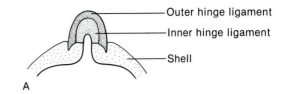

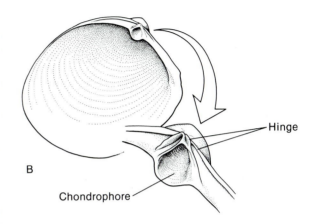

FIGURE 10–55 A, Cross section of a horseshoe-shaped type ligament (C spring) of such bivalves as *Cardium* and *Tellina*. **B,** Shell flange (chondrophore) bearing the attachment surface for the inner hinge ligament of *Raeta*. *(A, From Moore, R.C. (Ed.): 1969. Treatise on Invertebrate Paleontology. Part N. Vol. I. PN61. Geological Society of America and the University of Kansas.)*

tion, or shell strengthening. The familiar ribbing, or corrugations, of cockle and some scallop shells, for example, increases the shell strength. Bivalves range in size from the tiny seed shells of the freshwater family Sphaeriidae, some of which do not exceed 2 mm in length, to the giant clam *Tridacna* (Figs. 10–74 and Fig. 10–75B) of the South Pacific, which attains a length of over a meter and may weigh over 1100 kg.

Like the shell, the mantle greatly overhangs the body, and it forms a large sheet of tissue lying beneath the valves. The edge of the mantle bears three **folds:** an inner, a middle, and an outer fold (Fig. 10–54C). The innermost fold contains radial and circular muscles, the middle fold is sensory in function, and the outer fold secretes the shell. Molluscan shell secretion has been most studied in bivalves. The inner surface of the outer fold (periostracal groove) lays down the periostracum, and the outer surface secretes the first calcareous layer. The entire mantle surface secretes the remaining calcareous portion. There is a minute, extrapallial space between mantle and shell, except at points of muscle attachment. It is into this space that shell materials are first secreted; from the extrapallial fluid of this space both the calcareous elements and the surrounding organic framework are then deposited. Because the mantle epithelium is in contact with the shell surface at the periostracal groove, where the periostracum is secreted, the periostracum plays an important role in sealing off the extrapallial space from the external aqueous medium (Fig. 10–54C).

In recent years much has been learned about the various lines and bands produced in shell growth. They provide information about the age of the animal and the environmental conditions under which shell growth took place. Some of the lines are visible on the outside of the shell, but most are only seen when the shell is sectioned radially, that is, from hinge to ventral margin. The shells of many species, including the hard-shell clam, *Mercenaria,* and the mussel, *Mytilus edulis,* that live in intertidal habitats show fine lines between microgrowth increments, reflecting daily tidal cycles as well as the semilunar cycle of spring tides. The line is believed to be produced when the valves are closed. At that time the production of organic acid from anaerobic respiration causes a slight dissolution of calcium carbonate, leaving a preponderance of organic framework material, which forms the line.

Annual growth increments, which are present in most bivalve shells, can be detected by the seasonal thickness of microgrowth increments or seasonal differences in the density of the shell (Fig. 10–56B). In some bivalves annual growth increments are recorded in concentric lines or checks on the shell surface, resulting from a winter break in the periostracum and outer calcified layer. The previous year's layers may overlap the new like a shingle or be interrupted by a groove (Fig. 10–56A). However, similar lines can be produced by environmental disturbances, such as storms.

The mantle is attached to the shell at the points of insertion of muscle fibers of the inner mantle lobe. These describe a semicircular line a short distance from the shell edge. The line of mantle attachment is impressed on the inner surface of the shell as a scar, called the **pallial line** (Fig. 10–54B).

Despite the attachment of the mantle, occasionally some foreign object, such as a sand grain or a parasite, lodges between the mantle and the shell. The object then becomes a nucleus around which are laid concentric layers of nacreous shell. This is how a pearl is formed. If the object is enfolded within the mantle and moved about during secretion, the pearl becomes spherical or ovoid. Commonly, however, the developing pearl adheres to or even becomes completely embedded in the shell.

Pearls can be produced by most shell-bearing molluscs, but only those with shells having an inner nacreous layer produce pearls of commercial value. The finest natural pearls are produced by the pearl oysters, *Pinctada margaritifera* and *Pinctada mertensi,* which inhabit most of the warmer Pacific areas. There are two methods of producing cultured pearls. Most cultured pearls are started with a microscopic globule of liquid or ground "seeds" of unionid (freshwater clam) shells, placed into a pearl oyster. The resulting year-old seed pearl is then transplanted into another oyster. A pearl of marketable size is obtained three years after transplantation. Bead pearls are produced more quickly by starting with a spherical shell fragment only slightly smaller than the finished pearl. This fragment is enclosed in a bag of mantle tissue and transplanted into the gonadal tissue of another pearl oyster. A calcareous veneer approximately 1 mm thick is then laid down around the bead.

The foot of most bivalves has become compressed, bladelike, and directed anteriorly as an adaptation for burrowing. Foot movement, which will be described in detail later in connection with burrowing, is effected by a combination of blood pressure and muscle actions of **pedal protractors** and **retractors**. These latter muscles, which are homologous to the pedal retractors of other molluscs, extend from each side of the foot to the opposite valve, where they are usually attached to the shell near the adductor muscles (Fig. 10–54B).

Other specializations of the bivalve shell, mantle, and foot will be described later in connection with different adaptive groups.

Evolution of Bivalve Feeding

It is generally agreed that the early bivalves were shallow burrowers in soft substrata. They belonged to the protobranch group, which is represented by some of the oldest fossil forms (Ordovician, perhaps Cam-

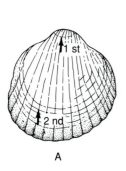

A

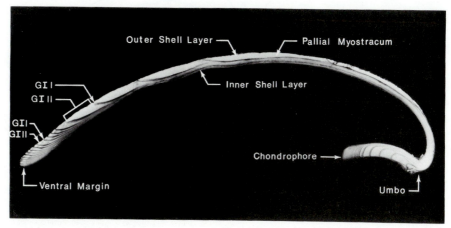

B

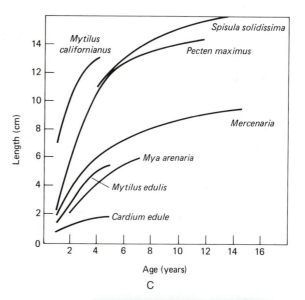

C

FIGURE 10–56 A, First and second annual growth rings in a European cockle. The ring is produced by winter interruption of shell deposition. Thus, shell produced during the second year lies between the faint first annual ring and the more conspicuous second ring. **B,** Radial cross section of a surf clam shell *(Spisula solidissima)* showing repeating annual growth layers. Each pair of light and dark bands (GL I and GL II) represent one year's growth. Note that most of the growth occurs in the early years of the life of the clam. **C,** Growth curves of shell length for seven bivalves. The approximate age of a specimen can be determined by measuring the shell length and locating the corresponding point on the curve. **D,** Anterior view of the large hard clam, *Mercenaria campechiensis.* The growth of each valve can be visualized by its growth lines. The oldest part of the shell is above at the umbo of each valve, and the youngest part is below at the gape. Notice that each valve curves in an open spiral (arrows) in much the same manner as the shell of gastropods. The form of the juvenile shell can be seen (dotted lines), again thanks to the growth lines, on each side of the hinge. The configuration of the juvenile valves indicates that, as the clam grows, the valves are pushed apart to accommodate the growing soft parts within. *(A, From Richardson, C.A. et al. 1980. The use of tidal growth bands in the shell of Cerastoderma edule to measure seasonal growth rates under cool temperate and subarctic conditions. J. Mar. Biol. Assoc. U.K. 60:977–989. Copyrighted and reprinted by permission of Cambridge University Press. B, From Jones, D.S. 1981. Repeating layers in the molluscan shell are not always periodic. Paleontology. 55(5):1076–1082.)*

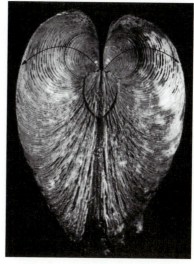

D

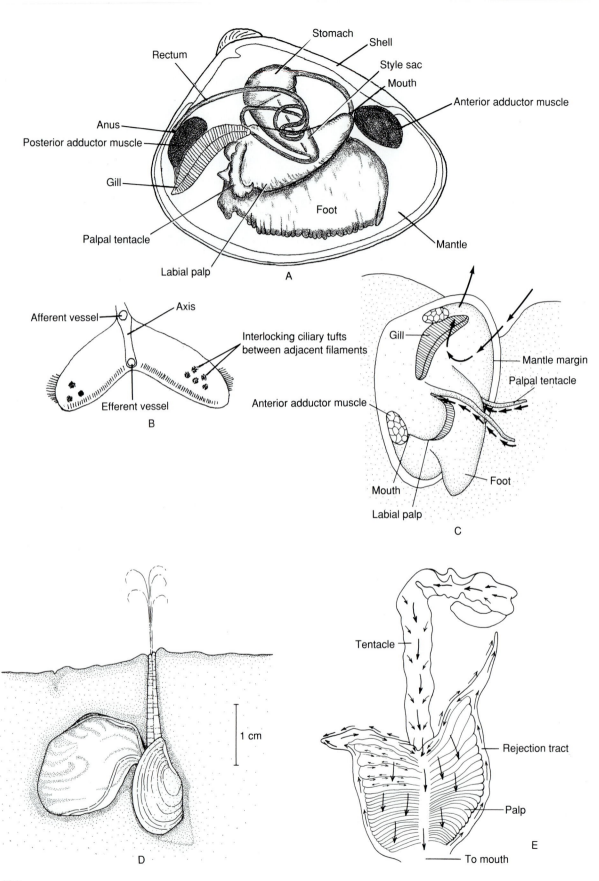

A

Stomach

Shell

Style sac

Mouth

Anterior adductor muscle

Rectum

Anus

Posterior adductor muscle

Gill

Palpal tentacle

Labial palp

Foot

Mantle

B

Afferent vessel

Axis

Interlocking ciliary tufts
between adjacent filaments

Efferent vessel

C

Gill

Mantle margin

Palpal tentacle

Anterior adductor muscle

Mouth

Labial palp

Foot

D

1 cm

E

Tentacle

Rejection tract

Palp

To mouth

brian) as well as by some living species—*Nucula, Nuculana, Yoldia, Solemya,* and *Malletia.* Most extant species of protobranchs live in the substratum with the anterior end directed downward and the posterior end directed toward the surface. They possess a single pair of posterolateral bipectinate gills, from which the name *protobranch,* meaning "first gills" is derived. In most protobranchs, as well as in almost all other bivalves, the ventilation current enters the mantle cavity through the shell gape posteriorly and ventrally, passes up through the gills, and exits posteriorly and dorsally (Fig. 10–57C). Lateral gill cilia create the water current, and frontal cilia remove sediment trapped on the gill surface.

Most living protobranchs are selective deposit feeders, and this is believed by many malacologists to have been the mode of feeding of the early and extinct members of the group. In the ancestral molluscs, the mouth rested against the hard bottom over which the animal crawled. However, when bivalves became adapted for burrowing in sand or mud, the mouth was lifted above the substratum as a result of both the lateral compression of the body and the greatly increased height of the dorsoventral axis. The radula disappeared. Protobranchs maintain contact with the substratum by a pair of tentacles, elongations of the margins of the mouth. Each tentacle is associated with a large fold composed of two flaps, called a **labial palp,** one located to either side of the mouth (Fig. 10–57C, E). During feeding the tentacles are extended into the bottom sediments (Fig. 10–57D). Deposit material adheres to the mucus-covered surface of the tentacle and then is transported by cilia back to the palps, which function as sorting devices. The inner apposing surfaces of each palp are ridged and ciliated (similar to Fig. 10–57E). Light particles are carried by crest cilia to the mouth; heavy particles are carried by groove cilia to the palp margins, where they are ejected into the mantle cavity.

Solemya velum, which is common along the east coast of the United States and in other parts of the world, does not feed in the typical protobranch manner. It lacks palpal tentacles and has greatly reduced labial palps. Members of this genus, including some gutless, deep-sea forms, rely on carbon fixed by chemosynthetic bacteria living in their gills (p. 363). Most remarkable is the discovery that *Solemya reidi,* itself, can oxidize sulfide to obtain energy for ATP synthesis.

Although some protobranchs, such as *Nucula, Yoldia,* and *Solemya,* live in shallow water, the group is more abundantly represented on deep ocean floors.

In some group of early protobranch bivalves, filter feeding evolved. An explosive evolution followed this development, and the filter feeders, called **lamellibranchs,** came to dominate the bivalve fauna. The gills and ventilating current of protobranchs preadapted them for filter feeding. As the lamellibranchs evolved, detrital particles and microorganisms in the ventilating current came to be utilized as a source of food, the gills became filters, and the gill cilia that originally served to keep the gills clean became adapted for the transport of particles trapped in mucus from the filter to the labial palps and mouth.

The principal modification of the gills for filtering was the lengthening and folding of the gill filaments, which greatly increased their surface area (Fig. 10–58). Many filaments were added to the gills so that they extended anteriorly, reaching the palps. Each gill filament on each side of the axis became folded, or U-shaped. The arm of the U that is attached to the axis of the gill is called the **descending limb,**

◀ **FIGURE 10–57** Protobranchs. **A,** Body of *Nucula* with right valve and mantle removed (lateral view). **B,** Gill of *Nucula,* showing lateral filaments (transverse section). Note the similarity to the gill of the primitive gastropod *Haliotis,* shown in Figure 10–18B. **C,** A generalized protobranch, illustrating shallow burrowing and deposit feeding. Small arrows indicate the path of food particles along palpal tentacles and labial palps; large arrows show the direction of water current. **D,** The protobranch *Yoldia limatula* deposit feeding within a chamber excavated below the surface. Siphons come to the surface. This species may also extend palpal tentacles to surface and feed on surface particles. **E,** One labial palp and palpal tentacle of *Nuculana minuta.* Apposing flaps of the palp are pulled back, and arrows indicate the direction of ciliary currents. *(A and B, After Yonge, C.M. 1939. The protobranchiate Mollusca, a functional interpretation of their structure and evolution. Phil. Trans. Roy. Soc. London B. 230:79–147. D, From Bender, K., and Davis, W.R. 1984. The effect of feeding by Yoldia limatula on bioturbation. Ophelia. 23(1):91–100. E, After Atkins.)*

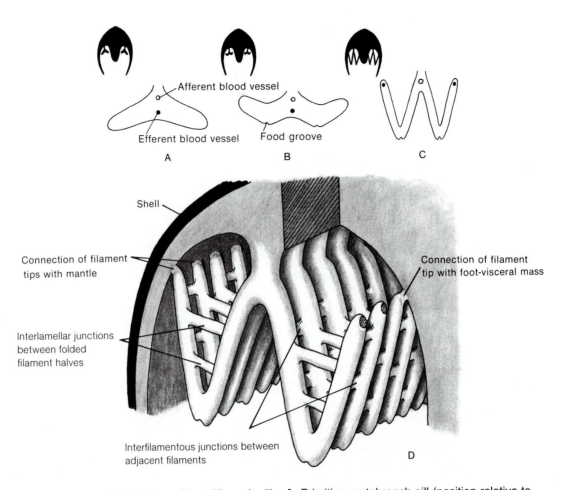

FIGURE 10–58 Evolution of lamellibranch gills. **A,** Primitive protobranch gill (position relative to the foot/visceral mass and mantle indicated in cross section). **B,** Development of the food groove in a hypothetical intermediate condition. **C,** Folding of filaments at the food groove to produce the lamellibranch condition. **D,** Tissue connections that provide support for the folded lamellibranch filaments.

and the arm next to the mantle or visceral mass, the **ascending limb.** The net result of the filaments on both sides of the axis becoming folded has been to transform each original single gill into what appears to be a pair of gills, or **demibranchs;** the original outer filaments form one member of the pair, and the original inner filaments form the other (Fig. 10–58). The lengthened, folded filaments and their attachment to one another give the gill a sheetlike form, hence the name of these bivalves, lamellibranchs, meaning "sheet gill." Four large, broad, filtering surfaces (lamellae) are present, two on each demibranch. At the angle of flexure, the frontal surface of each

filament has developed an indentation, or notch, which, when lined up with the notches of adjacent filaments, forms a food groove that extends the length of the underside of the gill. These modifications in gill structure have necessitated a change in ciliation. The frontal cilia carry food particles trapped on the gill surface vertically to the **food grooves** (Figs. 10–60B and 10–62). The abfrontal cilia, now inside, are usually lost from most filaments. Lateral cilia still produce the water current through the gills. On each side of the filaments, between the lateral and the frontal cilia, is a new ciliary tract composed of laterofrontal cirri. Each cirrus is a bundle of many ad-

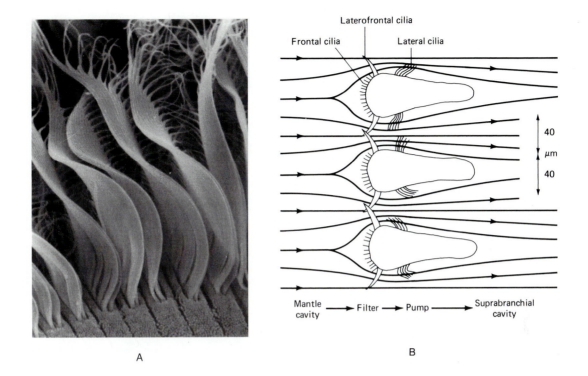

FIGURE 10–59 A, Laterofrontal cirri of *Venus casina.* Each cirrus is composed of two rows of fused cilia on a single cell. The distal ends of the cilia are unfused. **B,** Diagrammatic cross section through three gill filaments of *Mytilus.* The frontal cilia beat toward and away from the viewer. The effective stroke of the cirri is to the left, and the effective stroke of the lateral cilia is to the right. Arrows indicate the direction of the feeding/ventilating current produced by the lateral cilia. *(A, From Owen, G. 1978. Classification and the bivalve gill. Phil. Trans. Roy. Soc. London B. 284:377–386. B, From Silvester, N.R., and Sleigh, M.A. 1984. Hydrodynamic aspects of particle capture by Mytilus. J. Mar. Biol. Assoc. U.K. 64:859–879. Copyrighted and reprinted by permission of Cambridge University Press.)*

hering cilia (Fig. 10–59A). Opposing cirri form a fine mesh that filters particles from the water entering the gill; the cirri then move the particles onto the frontal cilia. The pressure of the water stream generated by the lateral cilia is more than sufficient to overcome the resistance offered by the cirri (Fig. 10–59B).

The inhalant, feeding/ventilating current enters the lower part of the mantle cavity (**infrabranchial chamber**) at the posterior end of the animal, flows between the filaments, and then moves up between the two lamellae. From the interlamellar spaces, water passes into the exhalant, or **suprabranchial,** chamber and finally flows out through the posterior exhalant opening.

Support for the long, folded filaments is provided by three kinds of new tissue connections at various points within the gill: (1) cross connections, called **interlamellar junctions,** between the folded filament halves, or lamellae; (2) connections, called **interfilamentous junctions,** between adjacent filaments; (3) connections between the tips of the filaments and the mantle or foot (Fig. 10–58D). The extent of these connections varies in different groups of lamellibranchs and accounts for several types of lamellibranch gills.

If the individual filaments are still more or less separate, the gill is known as a **filibranch gill** (Fig. 10–60A, B). Bars of tissue, the interlamellar junctions, attach the two limbs of each U at intervals, and there may be tissue union of adjacent filaments at the bottoms or tops of the lamellae. But throughout most of their length, filaments are attached to adjacent

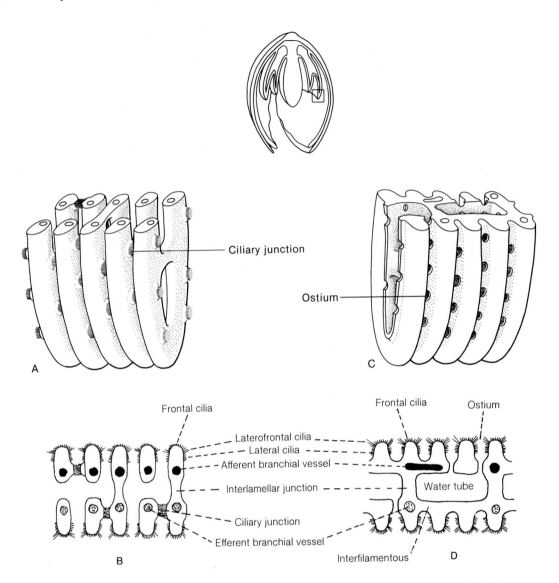

FIGURE 10–60 A–B, Filibranch gill. A, Five adjacent filaments (surface 3-D view). B, Frontal section. C–D, Eulamellibranch gill. C, Five fused, adjacent filaments (surface 3-D view). D, Frontal section.

filaments only by specialized ciliary junctions. Filibranch gills are found in such bivalves as arks *(Arca),* scallops *(Pecten),* and mussels *(Mytilus).* In the oysters and pen clams, which are said to have a **pseudolamellibranch** type of gill, the filaments are bound together with some (though not extensive) interfilamentous tissue junctions.

The most specialized lamellibranch gill is known as a **eulamellibranch gill** in which the union of filaments has developed further, so that the lamellae

actually consist of solid sheets of tissue (Fig. 10–60C, D). Furthermore, the interlamellar junctions have increased in number and extend the length of the lamellae (dorsoventrally). Thus, the interlamellar space is partitioned into vertical **water tubes.** The tips of the ascending limbs have become fused with the upper surface of the mantle on the outside and the foot on the inside, morphologically separating the inhalant chamber from the suprabranchial chamber. Ciliation remains the same because the frontal edges

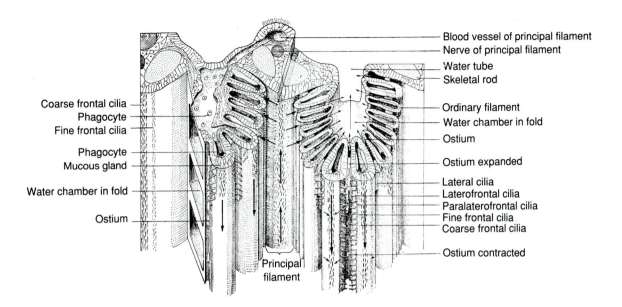

Coarse frontal cilia
Phagocyte
Fine frontal cilia

Phagocyte
Mucous gland

Water chamber in fold

Ostium

Principal filament

Blood vessel of principal filament
Nerve of principal filament
Water tube
Skeletal rod

Ordinary filament
Water chamber in fold
Ostium

Ostium expanded

Lateral cilia
Laterofrontal cilia
Paralaterofrontal cilia
Fine frontal cilia
Coarse frontal cilia

Ostium contracted

FIGURE 10–61 Stereodiagram of part of a plicate gill *(Crassostrea)*. *(After Nelson from Jorgensen, C.B. 1966. Biology of Suspension Feeding. Pergamon Press, London. p. 71.)*

of the filaments are not involved in the interfilamentous fusion. Thus, a frontal section of the eulamellibranch gill exhibits a lamella with a ridged outer surface, each ridge representing one of the original filaments (Fig. 10–60C, D).

Water in the inhalant chamber circulates between the ridges and enters the water tubes through numerous pores (**ostia**) in the lamellae. Oxygenation of the blood takes place as the water moves over the surface of the gill and dorsally in the water tubes. From the water tubes, water flows into the suprabranchial cavity and out the exhalant opening.

In the primitive gills of protobranchs, the efferent, or drainage, vessel ran within the axis of the filament beneath the afferent vessel, as in the ancestral molluscan gill. With the elongation and folding of the filament, the old drainage vessel dropped out, and a new drainage vessel formed at the junction of the ends of the fused ascending limbs of each filament with the mantle wall or visceral mass (Fig. 10–58C). Blood therefore flows in but one direction through each filament.

In many pseudolamellibranchs and eulamellibranchs, the surface area of the lamellae has been increased further by folding along the length of the gill, and the **plicate gill** surface of these bivalves presents an undulated appearance (Fig. 10–61).

Most lamellibranchs feed on fine plankton and suspended detritus. Food particles, in some cases as small as 1 μm, are removed from the water currents passing between filaments or entering the ostia. The particles are then passed onto the frontal cilia, where they are entangled in mucus and moved up or down the margin of the filament to a food groove.

The primitive lamellibranch has five food grooves transporting particles anteriorly to the palps. Three of the grooves are located at the top of the gills between and outside the demibranchs; the other two are located ventrally, one along the margin of each demibranch. The frontal cilia are divided into separate tracts of coarse and fine cilia, one carrying particles upward and one downward. Such a two-way vertical tract system with five food grooves is found in oysters and scallops. From such a primitive condition, the great variation in number and location of food grooves and direction of vertical tracts encountered in other lamellibranchs is believed to be derived by deletion. A few of these variations are illustrated in Figure 10–62, in which the orally directed food grooves are indicated by black circles. Note that the dual-tract system has disappeared in most lamellibranchs.

Various mechanisms provide for some prepalpal sorting by the gills and for coping with sediment-

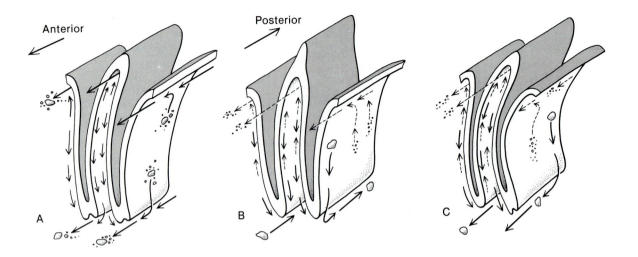

Anterior

Posterior

A

B

C

FIGURE 10–62 Transverse sections of different lamellibranch gills, showing direction of frontal cilia beat and position of anteriorly moving food tracts. In **B** and **C,** broken arrows indicate the fine frontal cilia carrying food particles upward; solid arrows indicate the coarse frontal cilia carrying particles ventrally. The inner demibranch, or gill, is on the right in all cases. **A,** Mytilidae and Pinnidae. **B,** Arcidae and Anomiidae. **C,** Ostreidae and Pectinidae. *(All modified and redrawn after Atkins, D. 1937. Q. J. Micros. Sci. 79.)*

laden water short of halting ventilation. In the families Arcidae and Anomiidae the heavy particles that are carried to the ventral margin of the gills are transported posteriorly and ejected onto the mantle. In a number of families that have plicate gills, such as the Ostreidae (oysters), Pectinidae (scallops), and Solenidae (razor clams), the filament that lies between the folds, called the **principal filament** (Fig. 10–61), carries only light particles upward. When the water is relatively clean, the gills are expanded, and the upward-moving tracts are largely in operation. When there is a lot of sediment in the water, the gills are stimulated to contract, placing the principal filament deep within the folds (Figs. 10–63A, B). The downward moving tracts of the other filaments drive much of the trapped material to the ventral grooves, which if too heavily laden will drop their loads into the bottom of the mantle cavity, where they can be released to the outside.

The lamellibranch palpal lamellae supposedly have the same sorting and conveying function as in the Protobranchia (Fig. 10–64). Particles are said to be sorted by size and weight. Small, light particles are retained for ingestion and are carried up the palpal surface across the crests of the ciliated ridges (Fig. 10–64). Large particles, destined to be rejected, are carried to the edge of the lamellae in the grooves between ridges and fall to the mantle or foot. Studies on the scallop, *Placopecten magellanicus,* and the oyster, *Crassostrea gigas,* raise questions about various aspects of this classical account. The palps do not receive free particles from the gills but rather a cord of particles bound in mucus that travels in the oral groove at the junction between the two palpal lamellae. In the normal position in which the two lamellae are appressed, there does not appear to be any particle selection by the crests of the ridges.

Rejected materials, called **pseudofeces,** from both the palps and the gills leave the mantle cavity most commonly by the inhalant aperture. The particles are carried posteriorly along a ventral ciliated tract of the mantle to accumulate behind the inhalant aperture. When the valves are closed periodically, water is forced out of the inhalant opening, taking these accumulated wastes with it.

The animal can regulate water flow by changing the size of the apertures into the mantle cavity and by gill contraction or expansion, which permits less or more water to pass between the filaments.

From some protobranch or lamellibranch ancestor evolved still another group of bivalves, the septibranchs (*Poromya* and *Cuspidaria*). The members of

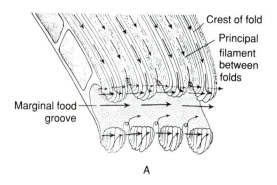 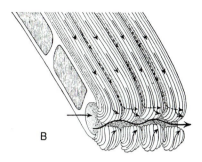

FIGURE 10–63 A, Ventral margin of a section of the plicate gill of *Pinna fragilis* when relaxed and expanded. **B,** Same gill as in **A** when contracted. Note that contraction buries the principal filaments that are located between the folds (plicae) and closes the marginal food groove. In members of this family, most frontal ciliary tracts carry particles downward. (See Fig. 10–62A.) *(From Atkins, D. 1937. Q. J. Micros. Sci. 79:348.)*

this small group have become carnivores or scavengers and are more common in the deep sea than in shallow water, where filter-feeding bivalves predominate. The gills of septibranchs have been modified to form a pair of perforated muscular septa, which separate the exhalant, suprabranchial chamber from the inhalant, infrabranchial chamber. By muscular contractions the septum moves up and down, forcing water into the inhalant chamber and out of the exhalant chamber. The force of the pumping septa is

sufficient to bring small animals, such as crustaceans and worms, into the mantle cavity. These prey are then seized by the reduced but muscular labial palps and carried to the mouth. Vibrations of small moving crustaceans are detected by siphonal tentacles. The siphon is then shot out in the direction of the prey, which is quickly sucked up by the simultaneous inhalant water current *(Cuspidaria),* or the siphon is everted as a hoodlike trap over the prey *(Poromya)* (Fig. 10–65).

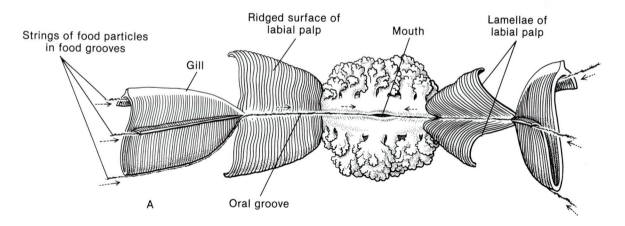

FIGURE 10–64 Labial palps of the deep sea scallop *Placopecten magellanicus.* **A,** Diagram of labial palps each with its two lamellae turned back, showing relationship to the gills (one demibranch) and mouth. The mouth is surrounded by bushy folds. *(From Benninger, Auffret, M., and Le Pennec, M. 1990. Peribuccal organs of Placopecten magellanicus and Chlamys varia: structure, ultrastructure and implications. Mar. Biol. 107:215–223.)*

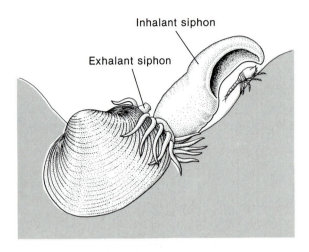

FIGURE 10–65 The septibranch *Poromya granulata* capturing a crustacean with its hoodlike, inhalant siphon. *(Modified and redrawn from Morton, B. 1981. Prey capture in the carnivorous septibranch Poromya granulata. Sarsia. 66:241–256.)*

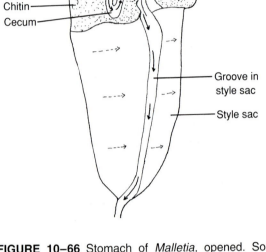

FIGURE 10–66 Stomach of *Malletia,* opened. Solid arrows indicate food tracts; dashed arrows indicate direction of beat of style-rotating cilia. *(After Yonge.)*

Digestion

The structure and physiology of the digestive tract of some of the deposit-feeding protobranchs have retained a number of primitive features (p. 367) and are similar to those of archaeogastropods (Fig. 10–66). Digestion in most protobranchs is extracellular in the stomach, and absorption occurs in the digestive gland.

The use of finer particles as food in the filter-feeding Lamellibranchia is reflected in a number of stomach modifications. The girdle of chitin, present in the protobranchs, is reduced to a small plate, the **gastric shield** (Fig. 10–67A). A style sac is present, but the mucus has become consolidated into a very compact and often long rod, the **crystalline style.** In addition to the protein matrix of the style itself, the style sac secretes enzymes that are absorbed onto the style in its formation. The projecting anterior end of the style is rotated against the platelike gastric shield by cilia in the style sac, and in the process, the style end is abraded and dissolved, releasing various carbohydrate-splitting enzymes and lipase, depending on the species. Similar enzymes are released from the stomach wall. Thus, starches and lipids are digested at least in part extracellularly. Most protein digestion

occurs intracellularly within the digestive gland. The rotation of the style also aids in mixing the enzymes with the stomach contents and acts as a windlass to pull food-laden mucous strings from the esophagus into the stomach (Fig. 10–67). The lower pH of the stomach facilitates the dislodgment of particles from the mucous strings.

The length of the style varies, but it is remarkably long considering the size of the animal. In many bivalves, the style is approximately 3 cm long, but a 12-cm *Tagelus* may have a 5-cm style, and a 1-m *Tridacna* may have a 36-cm style.

The mixing of stomach contents by the crystalline style continually throws partially digested particulate material against a ciliated sorting area. Coarse, heavy particles are segregated and sent to the intestine along a deep intestinal groove. Fine particles and fluid containing digestion products are retained by the cilia of the sorting ridges and directed toward the numerous apertures of the digestive glands. In many lamellibranchs, these apertures, along with a typhlosole bordering the intestinal groove, are located in one cecum or, in higher forms, in two ceca of the

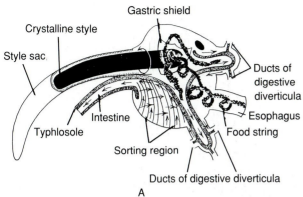

A

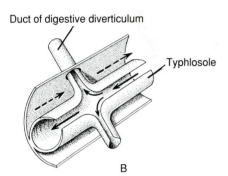

B

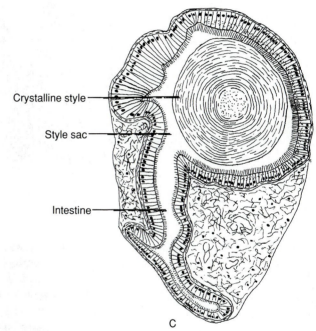

C

FIGURE 10–67 A, Stomach of a lamellibranch, showing rotation of the crystalline style and winding in of the mucous food string. Arrows indicate the ciliary pathways. **B,** Style sac and intestine of the freshwater clam, *Lampsilis anodontoides* (transverse section). **C,** Typhlosole within the cecum of the stomach, showing extensions into the ducts of the digestive diverticula. Solid arrows indicate inhalant ciliary currents; dashed arrows indicate exhalant currents. **D,** Diagram of a section of digestive diverticulum, showing the absorption and intracellar digestion of material passed inward from the stomach (solid arrows) and the outward passage of wastes (dashed arrows). *(A, After Morton, J.E. 1967. Molluscs. Hutchinson University Library, London. B, After Nelson from Yonge. C and D, After Owen, G. 1974. Feeding and digestion in the Bivalvia. Adv. Comp. Physiol. Biochem. 5:1–35.)*

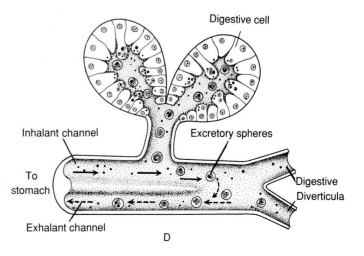

D

stomach (Fig. 10–67A). Within the main ducts of the digestive diverticula, there is a continuous two-way flow of materials entering for intracellular digestion and absorption and of cell fragments and wastes leaving (Fig. 10–67C). The deep intestinal groove by which wastes leave the stomach and pass into the intestine or the actual separation of the style sac from the intestine prevents wastes from becoming incorporated into the style matrix (Fig. 10–67B).

In intertidal bivalves, such as *Crassostrea virginica, Lasaea rubra,* and *Cardium edule,* that feed only at high tide, the different digestive processes display a tidal rhythm. The crystalline style, which in these forms lies in an intestinal groove, is dissolved at low tide when the animal is not feeding and is reformed as the tide comes in. A circadian feeding rhythm appears to be present in some species of freshwater clams. It appears that in most shallow-water bivalves, feeding and extracellular digestion in the stomach and extracellular digestion in the digestive diverticula are to varying degrees rhythmic or phasic.

The muscular stomach of the carnivorous septibranchs is lined with chitin and acts as a crushing gizzard. Proteases from the digestive gland initiate extracellular digestion in the stomach. Material digested this way is conveyed into the ducts of the digestive diverticula, where further digestion occurs intracellularly. The style is very reduced and barely projects into the lumen of the stomach.

Studies on energy expenditure in *Mytilus edulis* reveal that the cost of water pumping and food transport by the gills is slight. At high food intake levels about half of the energy cost is in general body maintenance and the remainder is partitioned between the cost of digestion and absorption and the cost of growth. A considerable part of a large food intake into the gut may be expelled undigested.

Adaptive Radiation of Bivalves

The evolution of filter feeding freed lamellibranchs from dependence on deposit material and made possible the colonization of many habitats that were uninhabitable for their protobranch ancestors. The success of this adaptive radiation is reflected in the fact that of the some 7700 described species and 75 families of bivalves, most are lamellibranchs. It must be emphasized that the adaptive groups described below do not necessarily constitute closely related species. Colonization and adaptation for a particular type of habitat have been achieved independently by a number of bivalve families or superfamilies.

Soft-Bottom Burrowers (Infauna)

It should not be thought that all lamellibranchs departed from the ancestral habitat. Most species inhabit soft bottoms, exploiting the protection offered by a subterranean life in marine sand and mud while utilizing food suspended in water brought in from above the surface. Some live just beneath the surface, many are adapted for deep burrowing, which offers greater protection from predators, some move between the surface and lower levels, and some are especially adapted for shallow, rapid burrowing in a shifting environment. In addition to the many marine forms, soft-bottom-dwelling lamellibranchs also include most of the freshwater clams.

The mechanism of burrowing involves an alternation of penetration and terminal anchors. A penetration anchor is formed by the slightly gaped valves, which wedge the shell in the sediment. Ridges or other sculpturing of the shell surface may increase traction. Against this anchor, the foot is extended by contraction of pedal protractor muscles (Fig. 10–68). The projecting foot probes and pushes into the surrounding sand. As the foot is protruded, the valves begin closing by contraction of the adductor muscles, reducing the profile of the shell and expelling some water from the mantle cavity, thereby loosening the sand or mud and facilitating the movement of the foot. The water remaining in the mantle cavity and the blood act as hydrostatic skeletons. The pressure of these two fluids is elevated by the adducted valves. Blood from the visceral mass is forced down into the pedal hemocoel, causing the foot to dilate, forming a terminal anchor in the substratum (Fig. 10–68D). With the foot anchored, an anterior pair and a posterior pair of pedal retractor muscles contract, pulling the shell downward. In many species, retraction by the anterior pedal muscle occurs before that of the posterior muscle. The effect is to rock the shell, which facilitates its movement through the substratum. Following pedal retraction, relaxation occurs, and the valves gape. To return to the surface or to a higher level within the substratum, most bivalves back out, pushing against the anchored end of the foot.

In the primitive protobranchs, such as *Nucula, Solemya, and Yoldia,* the foot bears a flattened sole (Fig.

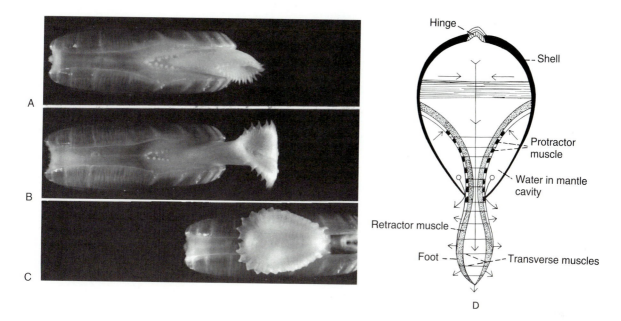

FIGURE 10–68 Operation of foot of *Solemya velum,* a protobranch. **A,** End of the foot folded and partly extended. **B,** Foot fully extended, opened, and anchored. **C,** Body advanced by movement of the foot. **D,** Diagrammatic cross section of a bivalve, showing hydrostatic forces that produce dilation of the foot. Central vertical arrow indicates the flow of blood into the foot. *(D, Modified after Trueman, 1966.)*

10–68B). However, the two sides of the sole can be folded together, producing a bladelike edge; in this condition, the foot is thrust into the mud or sand. The sole then opens and serves as an anchor.

The common cockles (*Cardium* and related genera), which have rounded, convex, ribbed shells, are shallow burrowers (Fig. 10–56A). They use the foot not only for conventional burrowing but also for escape. When threatened by a predator, such as a starfish, the cockle leaps by rapidly extending the foot from a folded position against the substratum.

A major problem arising from burrowing in soft bottoms is the sediment brought in by water currents. Blood engorgement within the mantle margin enables the mantle edges to be appressed, even when the valves gape slightly (the muscles of the inner fold provide for retraction), but because circulation of water through the mantle cavity is necessary for gas exchange and, in most species, for feeding, at least the posterior part of the mantle must be opened for water to enter, and some sediment invariably enters.

There has been a tendency to seal the mantle edges morphologically where openings are not neces-

sary (Fig. 10–69). The most frequent point of fusion of the mantle edges is at the posterior end of the cavity between the inhalant and exhalant openings. This fusion forms a distinct dorsal exhalant aperture. Often, a second point of fusion occurs immediately below the ventral inhalant opening, forming a permanent inhalant aperture. Bivalves that lack a permanent fusion usually have a temporary opening that forms as the left and right margins of the mantle are pressed together below the inhalant aperture.

Still further mantle fusion has occurred in some species, especially deep burrowers, and most of the ventral margin anterior to the inhalant aperture has become sealed. Thus, three apertures remain: the inhalant and exhalant apertures and an anterior pedal aperture through which the foot protrudes. Extensive mantle fusion not only reduces fouling of the mantle cavity but is of primary importance in maintaining the hydraulic pressure within the mantle cavity necessary for burrowing.

Commonly, when structural inhalant and exhalant apertures are present as a result of fusion, the mantle edges surrounding the apertures have become

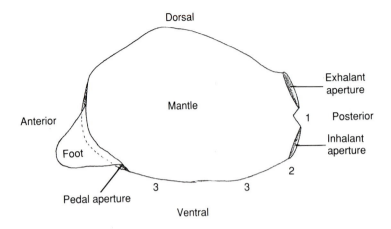

FIGURE 10–69 Areas of mantle fusion seen in various bivalves: (1) Between inhalant and exhalant apertures or siphons, the most common point of fusion. (2) Fusion below inhalant aperture or siphon. (3) Fusion between inhalant aperture and foot, leaving only a pedal aperture for the extension of the foot.

elongated to form tubular siphons of varying lengths (Fig. 10–70). With siphons the animals can be completely buried in the mud, and only the siphon tips need project above the bottom. The siphons are extended by blood pressure or by water pressure within the mantle cavity when the valves are closed and are withdrawn by siphon retractor muscles derived from the muscle tissue of the innermost mantle fold. Considerable variation exists in siphon length and degree of fusion between inhalant and exhalant siphons. Where there are separate inhalant and exhalant siphons, siphon formation involves only the muscular inner fold of the mantle.

Well-developed siphons are indicated by scars on the inner face of the valve. The pallial line impression is recurved sharply inward just below the posterior adductor and represents the point of attachment of the siphon retractor muscles. This bay in the pallial line is called the **pallial sinus** (Fig. 10–54B). Most modern soft-bottom-dwelling filter feeders possess siphons. Those that do not are rather sluggish shallow burrowers. The geoducks, *Panopea generosa,* of the Pacific coast of the United States are among the deepest burrowers, going down more than a meter. They have siphons so large that they can no longer be retracted between the shell valves (Fig. 10–71B).

Many bivalves that burrow deeply (i.e., to depths greater than the lengths of their bodies) tend to have semipermanent or even permanent burrows. The walls of the burrows become coated with mucus, which also reduces sediment fouling.

The primitive form of lamellibranch soft-bottom burrowers is considered to be one in which the adductor muscles are more or less equal **(isomyarian),**

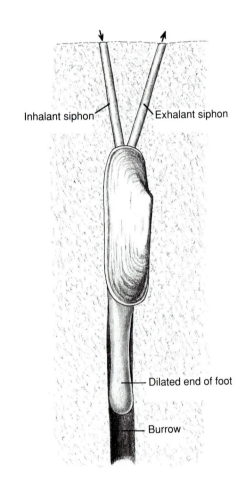

FIGURE 10–70 Razor clam *Tagelus plebius* from the southeastern coast of the United States, where it may occur in enormous numbers.

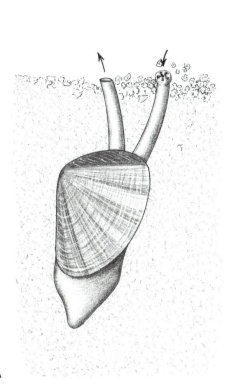

FIGURE 10–71 A, *Donax variabilis,* a common inhabitant of surf beaches. Rapid burrowing is facilitated by the thin, pointed foot. Opening of the inhalant siphon is frilled, preventing entrance of sand grains. **B,** The geoduck, *Panopea generosa,* a giant Californian bivalve, the body and siphon of which cannot be enclosed within valves. *(From Milne and Milne, 1959. Animal Life. Prentice-Hall, Englewood Cliffs, NJ.)*

the mantle is unfused, and the valves are equal and circular in outline. We have already described modifications involving mantle fusion. In many groups, the valves are also modified, becoming more streamlined for burrowing by being flattened or elongated. Species of *Donax,* which are inhabitants of surf beaches, have shells that are pointed anteriorly and blunt posteriorly. They back out on incoming waves and reburrow with great rapidity as the wave recedes. The margins of the inhalant siphon are fringed with infolded tentacles, which keep out swirling sand grains (Fig. 10–71A).

The razor clams *Ensis* and *Tagelus,* two unrelated but similar groups, have greatly elongated valves and an elongated foot, which enable them to move rapidly within their more or less permanent burrows (Fig. 10–70). *Tagelus* has long siphons, each of which has a separate opening to the surface. *Ensis,* which has short siphons, comes to the surface to feed from the deeper, more protected part of its burrow.

Two further specializations of soft-bottom burrowers should be mentioned. Some tellinaceans have reverted to deposit feeding. *Scrobicularia,* for example, extends its inhalant siphon above the surface at low tide and, like a vacuum cleaner, sucks in deposit material (Fig. 10–72), which is then sorted on the gills. Deposit feeding is generally an addition to, rather than a substitute for, filter feeding.

The lucines, rather small bivalves, with rounded shells, and members of the superfamily Lucinacea

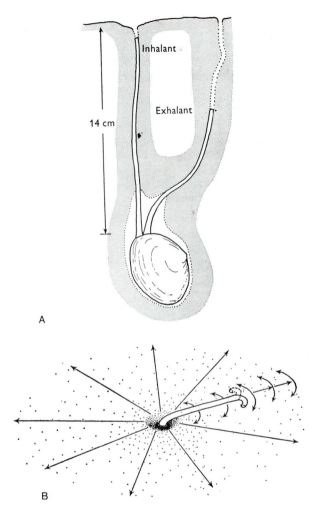

A

B

FIGURE 10–72 Deposit feeding in *Scrobicularia*. **A,** Animal in burrow with inhalant siphon withdrawn. **B,** Feeding movements of inhalant siphon at low tide. *(From Hughes, R.N. 1969. J. Mar. Biol. Assoc. U.K. 49:807. Copyrighted and reprinted by permission of Cambridge University Press.)*

(*Linga, Codakia, Lucina, Divaricella, Thyasira*), are remarkable in using the foot not only for burrowing but also for constructing a mucus-lined canal back up to the surface. The feeding/ventilating current enters anteriorly by this passageway and leaves by the usual exhalant opening. A number of lucines (*Lucina, Lucinoma, Loripes*) possess chemosynthetic bacteria in their gills, upon which they rely at least in part for their nutrition (p. 363).

Attached Surface Dwellers (Epifauna)

A number of evolutionary lines of lamellibranchs have invaded firm substrata—peat; wood; shell; coral; rock; or manufactured sea walls, jetties, and pilings. Attachment is provided either by a byssus or by fusion of one valve to the substratum. A **byssus** is a tough protein secretion produced by a gland in the foot and commonly has the form of threads. In mytilids, which have been most studied, a groove runs from a deep byssus gland to the anterior tip of the foot (Fig. 10–73B, C); a byssal thread byssus is formed in the groove by a mixture of secretions produced by the byssus gland and various glands along the groove. After a few minutes, when the thread has hardened by tanning, the foot is withdrawn, leaving the thread anchored at one end to the substratum and at the other to the byssus gland (Fig. 10–73A). A byssus retractor muscle may enable the animal to pull against its anchorage.

Among living surface dwellers attached by byssal threads, the widely distributed mussels (Mytilidae) are perhaps the most familiar. They live attached to wharf pilings, sea walls, and rocks or among oysters, often in great numbers. The threads, laid down by the little finger-like foot, often radiate outward like guy wires (Fig. 10–73A). Young individuals even use byssal threads to climb walls. Mytilids are widely used as food in many parts of the world and are sometimes "farmed" on ropes suspended from rafts.

A mytilid and the large white *Calyptogena magnifica* live attached to rocks near hydrothermal vents at the Galapagos Rift. Both depend on carbon fixed by sulfide-oxidizing bacteria located in the gills (p. 363). *Calyptogena* obtains the sulfides by means of its highly vascularized foot, which it wedges into cracks through which sulfide-laden water is emerging. The sulfides are transported to the gill bacteria bound to special extracellular blood proteins.

Other surface inhabitants attached by byssal secretions include many of the heavy-bodied arks (Arcidae) (Fig. 10–73C), which are very common on tropical coralline substrates; mangrove oysters (*Isognomon*) (Fig. 10–73F), which hang in clusters from mangrove roots; and winged oysters (pteriids), which live attached to sea fans and other gorgonian corals (Fig. 10–73E). The sole group of freshwater bivalves that have byssal attachment threads are the dreissenids, which include the zebra mussel, *Dreissena polymorpha.*

Some members of the Tridacnidae, which includes the giant clams *(Tridacna)* of the Indo-Pacific, are also surface dwellers. The smallest of the six species is only 10 cm long; the largest, *Tridacna gigas,* reaches 1.37 m. All live vertically oriented with the hinge side down and are initially attached by byssal threads. Some retain the byssus; some lose it and rest on the bottom by the weight of the shell; *Tridacna maxima* bores into coral or coralline rock so that the valve margins are flush with the substratum surface. The gape of all tridacnids is directed upward with the large mantle surface (actually the siphons) protruded across the fluted shell for maximum exposure to light (Figs. 10–74; 10–75B). Blood sinuses within the mantle contain extracellular, symbiotic zooxanthellae that provide the clam with an auxilliary source of nutrition. The mantle tissue also contains pigments of brilliant green, blue, red, violet, or brown that probably reduce light intensity. Large populations of tridacnids, such as the little boring *Tridacna maxima,* can be a spectacular addition to the beauty of an Indo-Pacific coral reef.

Utilization of byssal threads by adults of some 18 bivalve superfamilies is believed to represent a persistent larval adaptation, for the larvae of many unattached burrowing forms produce a byssus for initial temporary anchorage on settling (Fig. 10–76A). Adults of a few living species use byssal threads to anchor rootlike in soft bottoms. The mytilid genera *Modiolus* and *Geukensia* live partially buried in peat or coarse sediments (Fig. 10–73H), and the pen shells, *Pinna* and *Atrina* (Fig. 10–73G) occupy a similar position in sand, attaching the byssal threads to small stones.

Surface-dwelling bivalves that are attached by cementation lie on one side, fixed to the substratum by either the right or the left valve, depending on the species. Such sessile bivalves include at least eight evolutionary lines, of which the oysters are the most familiar. However, the name *oyster* is applied to a wide variety of species, some of which attach by byssal threads. In the family Ostreidae, which contains the edible American east coast oyster, *Crassostrea virginica,* and the European *Ostrea edulis,* the metamorphosing veliger is initially anchored with an organic adhesive produced first by the foot and then by the mantle. The mantle margin then attaches the left valve to the substratum in the process of shell secretion (Fig. 10–73A).

Shell attachment has led to varying degrees of inequality in the size of the two valves, the lower being larger or smaller (as in the Ostreidae) than the upper one. In *Chama,* the tropical jewel boxes, the upper valve forms a lid over the boxlike lower valve (Fig. 10–77A). The extreme condition was reached in the extinct Mesozoic rudists, in which the lower valve was shaped like a tube or horn (Fig. 10–77B). The rudists often occurred in reeflike aggregations.

The common jingle (or toenail) clams, members of the family Anomiidae, possess features of both the byssally attached and the cemented bivalves. They lie on one side but are actually anchored by a large, calcified byssal thread that passes through the attached valve (Fig. 10–77D).

Attached bivalves share a number of features. As would be expected, the foot is reduced to varying degrees, and it is completely absent in those bivalves, such as the oysters, that are attached by one valve. There has also been a tendency for the anterior end to become smaller, leading to a reduction of the anterior adductor muscle (**anisomyarian**) (Fig. 10–73B) or its loss (**monomyarian**) (Figs. 10–78; 10–79B). It has been suggested that anterior reduction in mussels (Mytilidae) is perhaps an adaptation to elevate the posterior end, thereby reducing the likelihood of obstruction that the dense aggregations of these bivalves might create. Many sessile bivalves, such as various scallops and thorny oysters, have a well-developed middle mantle lobe with tentacles and eyes.

Mantle fusion and siphon formation have not occurred in epifaunal bivalves; they live above the surface and generally on hard substrata, where sedimentation is less of a problem. However, oysters and mussels, which occur in dense beds, depend on the cleansing action of tidal currents to prevent them from becoming completely buried in their own feces and pseudofeces. The young *Mytilus edulis* actually cleans the exterior of its shell with its precociously elongated foot.

Unattached Surface Dwellers

Among the small number of bivalves that live free on the surface, some scallops (Pectinidae) (Fig. 10–79A) and some file clams (Limidae) (Fig. 10–79C) are the

(Text continues on page 449)

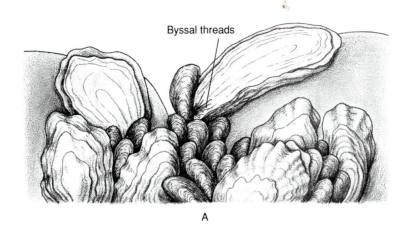

A

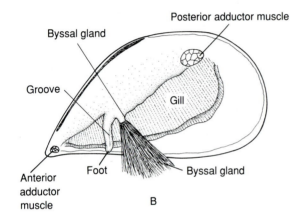

B

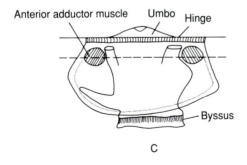

C

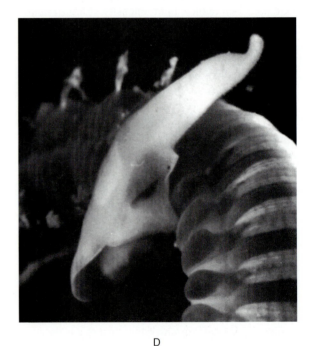

D

FIGURE 10–73 Sessile bivalves attached by byssal threads. **A,** Mussels *(Mytilus)* anchored among oysters (*Crassostrea*). **B,** Lateral view of a mytilid with the valve removed. **C,** Diagrammatic lateral view of *Arca.* **D,** The commensal bivalve *Entovalva,* attached by its foot to the body of the polychaete *Notomastus.* The clam's shell is reduced to a pair of small white valves seen by transparency through the extensive mantle. The extended mantle, which gives the clam a sluglike appearance, houses the ctenidia and is a brood

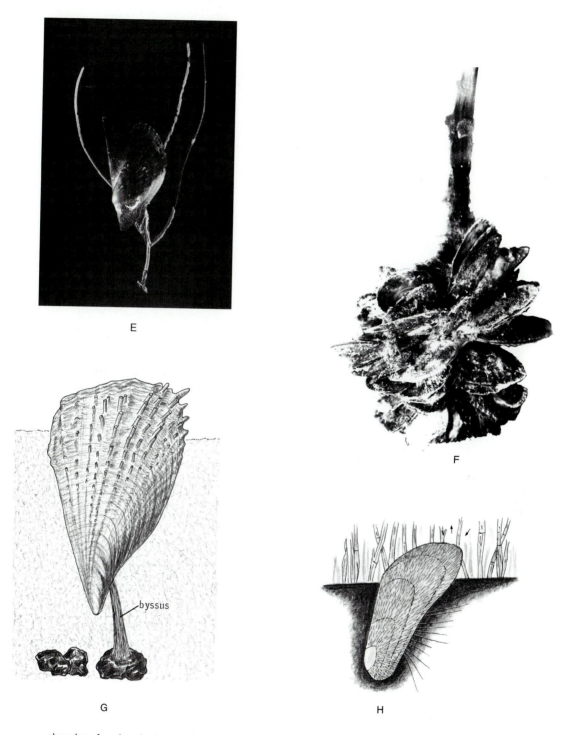

E

F

G

byssus

H

chamber for developing embryos. **E,** *Pteria,* a winged oyster, attached to a sea fan. **F,** *Isognomon,* a mangrove oyster. **G,** A pen clam, *Atrina rigida.* **H,** *Modiolus (Geukensia)* partially buried among intertidal marsh grass. *(A–C, Modified from Yonge, C.M. 1953. Phil. Trans. R. Soc. London B. 237:365. F, Photograph courtesy of Betty M. Barnes. H, After Yonge from Stanley, S.M. 1972. J. Paleontol. 46(2):165–212.)*

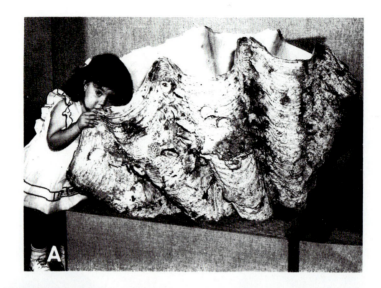

FIGURE 10–74 The giant clam, *Tridacna*. **A,** Shell. **B,** Looking down at expanded specimen of *Tridacna derasa*. The mantle extends over the shell fluting. The conical aperture is the exhalant siphon. *(A, Photograph courtesy of Cranbrook Institute of Science. B, Photograph courtesy of the British Museum.)*

FIGURE 10–75 Comparison of the structures of a cockle, which is a shallow burrower in soft bottoms, with the related but aberrant giant clam *(Tridacna)*, which is byssally attached and oriented with the hinge side down. *(From Yonge, C.M. 1975. Giant clams. Sci. Am. 232(4):96–105.)* ▶

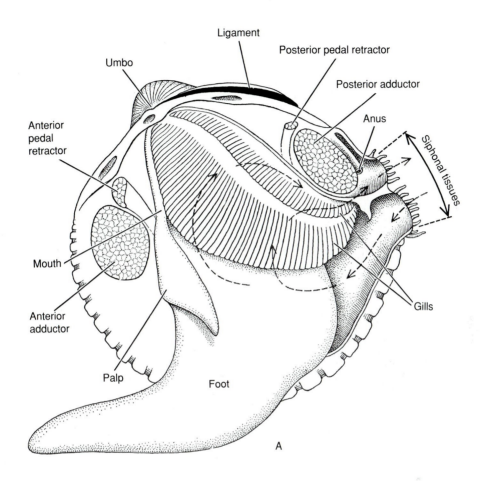

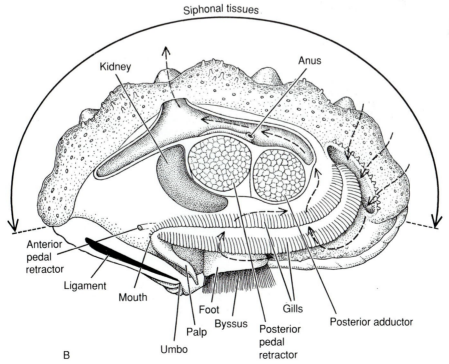

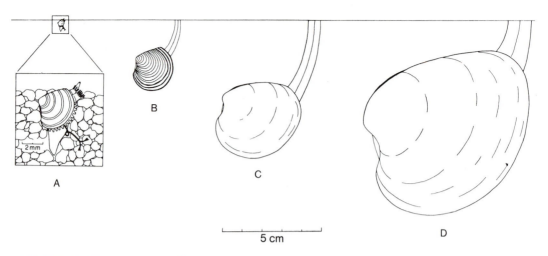

FIGURE 10–76 Habitation of *Mercenaria mercenaria* at different ages. **A,** A newly settled clam anchored in sand by byssal threads. **B,** First-year individual with heavy, stabilizing, concentric ridges on shell. **C** and **D,** Older individuals, showing changes in siphon length. *(From Stanley, S.M. 1972. J. Paleontol. 46(2):165–212.)*

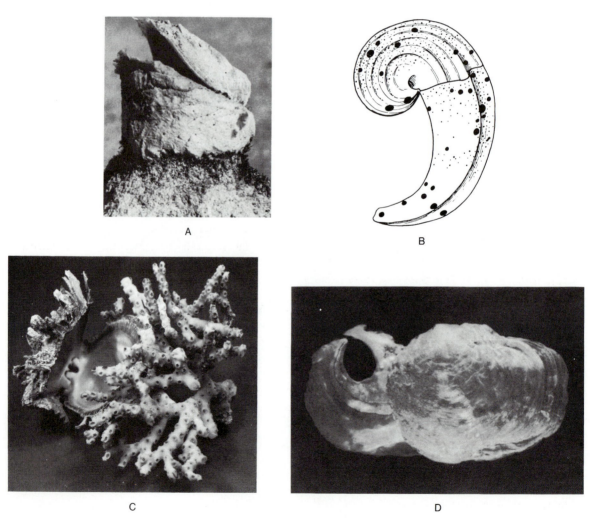

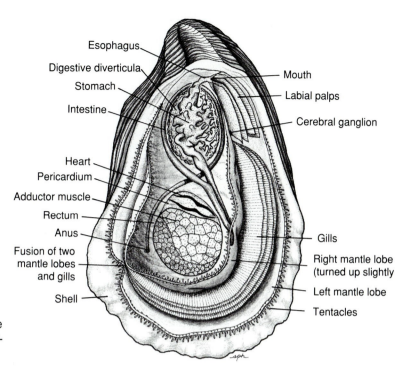

FIGURE 10-78 Anatomy of the American oyster, *Crassostrea virginica. (Modified after Galtsoff.)*

Labels: Esophagus, Digestive diverticula, Stomach, Intestine, Mouth, Labial palps, Cerebral ganglion, Heart, Pericardium, Adductor muscle, Rectum, Anus, Fusion of two mantle lobes and gills, Shell, Gills, Right mantle lobe (turned up slightly), Left mantle lobe, Tentacles

most familiar. There are members of both families that live anchored by byssal threads, but others are unattached or only weakly attached. They lie on their sides, and in many scallops the bottom valve (right) is flattened, the foot is reduced and used in cleaning the mantle cavity, and the anterior adductor muscle has disappeared. Free-living file clams and scallops have evolved the ability to swim by clapping the valves, which forces water from the mantle cavity in a jet that propels the animal away from its predator. The solitary, posterior adductor muscle has shifted to a more central position and is divided into smooth and striated sections (Fig. 10-79B). The rapid contraction of the striated fibers provides for swimming, and the sustained contraction of the smooth fibers provides for prolonged closure of the valves. The muscular inner lobe of the mantle margin, when appressed against the lobe on the opposite mantle surface, controls the direction of the water jet, permitting it (in scallops) to exit on either side of the hinge line or opposite the hinge line.

The swimming ability of scallops and file clams is used primarily to escape predators or other sudden disturbing conditions. For example, if a predatory starfish, or even one tube foot of such a starfish, touches the mantle margins of a scallop, a swimming response is evoked, and the scallop will swim a meter or so away. Some scallops use the water jets to blow out a depression in the sand surface into which they settle. File shells typically nest in crevices beneath stones and swim only when disturbed.

Boring Bivalves

The ability to penetrate and live beneath the surface of firm substrata—peat, clay, sandstone, shell, coral, coralline rock, and wood—has evolved in seven su-

◀FIGURE 10-77 **A,** The jewel box, *Chama,* a common, tropical, sessile bivalve, whose upper valve forms a lid over the lower valve. **B,** A rudist, a Mesozoic bivalve in which a caplike upper valve covered a hornlike or tubelike lower valve. **C,** A species of thorny oyster, *Spondylus,* attached to a piece of coral *(Oculina).* **D,** A jingle shell, or toenail shell (Anomiidae). The attached valve contains a large hole for the calcareous peduncle, homologous to the byssus of other bivalves. *(A and C, Photographs courtesy of Betty M. Barnes. B, From Kauffman, E.G., and Sohl, N.F. 1973. Verh. Nat. Ges. D, Photograph courtesy of Gates, J.B., from Andrews, J. 1971. Sea Shells of the Texas Coast. University of Texas Press, Austin. p. 168.)*

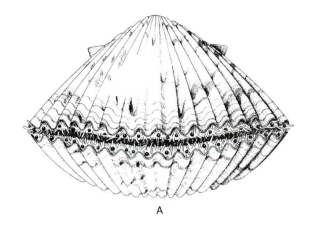

A

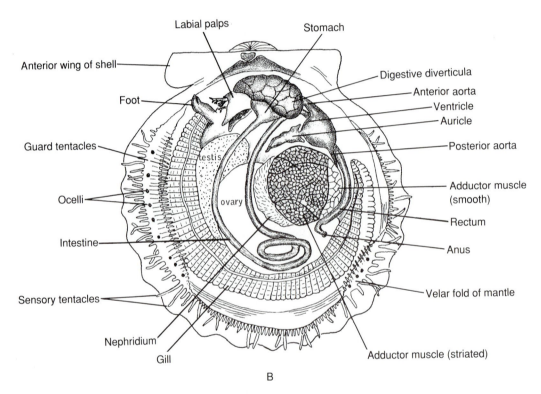

Anterior wing of shell

Labial palps

Stomach

Digestive diverticula

Anterior aorta

Ventricle

Auricle

Posterior aorta

Foot

Guard tentacles

testis

ovary

Adductor muscle (smooth)

Ocelli

Rectum

Anus

Intestine

Sensory tentacles

Velar fold of mantle

Nephridium

Gill

Adductor muscle (striated)

B

C

FIGURE 10–79 The scallop, *Pecten.* **A,** Gape view, showing eyes and tentacles on the sensory fold of the mantle margin. **B,** Internal structure, viewed from the left side. The anterior is on the viewer's left. **C,** The file clam, *Lima pellucida. (A and B, After Pierce, M. 1950. In Brown, F.A. (Ed.): Selected Invertebrate Types. John Wiley and Sons, New York.)*

perfamilies of lamellibranchs, of which the Pholadacea, containing the piddocks, are among the most conspicuous examples. The ancestors of some of these boring bivalves were soft-bottom inhabitants that evolved the ability to burrow in successively firmer substrata. The ancestors of others were probably surface inhabitants attached to hard substrata by byssal threads.

Each boring bivalve begins excavation after the larva settles and slowly enlarges and deepens the burrow with growth. The animal is forever locked within its burrow, and only the siphons project to the small surface opening. If a boring bivalve is removed and placed on the surface, it cannot excavate a new chamber.

Boring bivalves with epifaunal ancestors hold to the side of the burrow by byssal threads. Boring species with burrowing ancestors usually attach by the foot, which has developed a sucker-like ventral surface. In the great majority of species, drilling is a mechanical process, and the anterior ends of the valves, which are frequently serrated, are the abrading surfaces (Fig. 10–80A). The drilling movements are commonly adaptations of the locomotory movements found in their nonboring ancestors.

Drilling rates vary. Over an 18-month period *Penitella penita* and *Chaceia ovoidea* excavated soft shale at rates of 2.6 mm and 11.4 mm per month, respectively.

Some boring bivalves rotate within the burrow; that is, change position, and as a result the burrow cross section is round. Others remain attached in one place, and the burrow tube takes the shape of the shell. Such forms may even have two openings to the surface if the ends of the siphons secrete some material between them (*Gastrochaena*) (Fig. 10–80D). Much of the sediment produced in drilling is taken into the mantle cavity and then ejected with the pseudofeces through the inhalant siphon.

Lithophaga bisulcata, a very common, cigar-shaped, byssally attached borer in shell and coral, excavate chemically; others, such as *Lithophaga nigra,* excavate mechanically. A mucoprotein-chelating agent secreted by glands in the middle fold of the mantle margin softens the calcareous substratum, which is then scraped away with the valves. Those species that bore into live coral have, in addition, glands that prevent the coral from depositing calcium carbonate over the bivalve's bore hole and glands that inhibit the firing of the coral's nematocysts.

Many boring bivalves inhabit wood. The wood-boring pholadids, such as *Martesia* and *Xylophaga,* are adapted in much the same way as the many rock- and shell-boring members of the same family. Wood panels planted 1830 m deep on the sea bottom were completely riddled by *Xylophaga* and a related genus when recovered 104 days later. The most specialized wood borers are the shipworms, members of the family Teredinidae. The natural habitats of the some 60 species of this widely distributed family are mangrove roots and timber swept into the sea by rivers, and they play an important ecological role in the reduction of sea-borne wood. They are destructive animals in piers, pilings, and other wooden structures placed by humans in the sea, and much expense and research have been devoted to their control. Timbers can become completely riddled with tunnels (Fig. 10–81C).

The body of the shipworm is greatly elongated and cylindrical (Fig. 10–81A). The shell is reduced to two small, anterior valves. Cutting of the wood is effected by opening and rocking motions of the valves while the anterior end of the body is attached to the burrow by the small foot. The mantle, enclosing the greater part of the body behind the valves, produces a calcareous lining within the tunnel. The long, delicate siphons open at the surface of the wood, and the burrow entrance is plugged by calcareous pallets when the siphons are retracted. Burrow size increases with the growth of the shipworm that fills it, and may reach a length of 18 cm to 2 m, depending on the species. The life span is one to several years, again depending on the species.

Shipworms use the excavated sawdust for food, although many also filter feed. The stomach is provided with a cecum for sawdust storage, and a section of the digestive gland is specialized for handling wood particles. Symbiotic bacteria housed within a special organ that opens into the esophagus not only provide for cellulose digestion but also, by fixing nitrogen, compensate for the low-protein diet.

Commensals and Parasites

A small number of bivalves have evolved commensal and parasitic relationships. Most commensals are related to free-living epibenthic forms, such as the little *Kellia* and *Lasaea*, which nestle in crevices. Most attach by byssus threads (Fig. 10–73D), but some crawl on the foot like a snail. The hosts are usually burrowing echinoderms, such as heart urchins; brittle stars;

(Text continues on page 454)

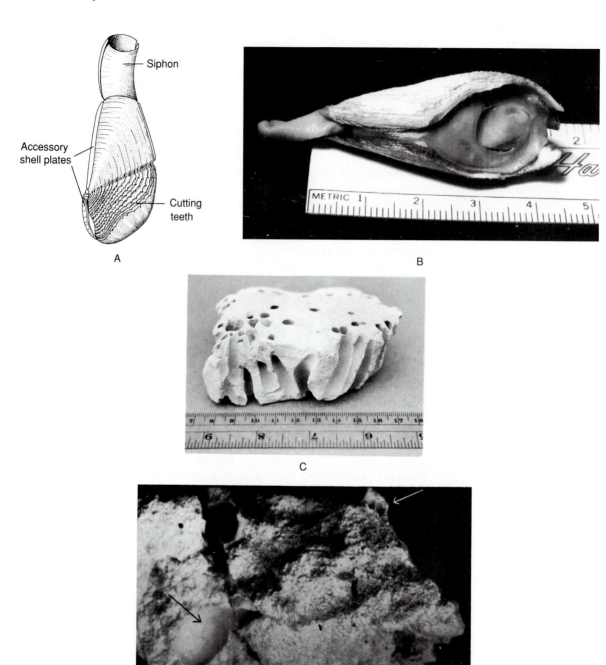

FIGURE 10–80 A, Structure of a pholadid. **B,** Ventral view of a pholadid, showing the sucker-like foot in the pedal aperture. Siphon is to the left. **C,** Surface openings and part of the burrows of pholadid bivalves in a piece of hard clay. **D,** The separate siphon openings of *Gastrochaena hians,* a common West Indian borer in coralline rock. This bivalve does not turn in its burrow. *(B, Photograph courtesy of Katherine E. Barnes.)*

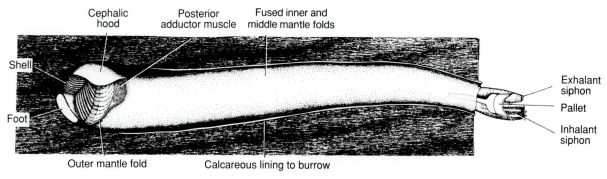

Cephalic hood

Posterior adductor muscle

Fused inner and middle mantle folds

Shell

Foot

Outer mantle fold

Calcareous lining to burrow

Exhalant siphon

Pallet

Inhalant siphon

A

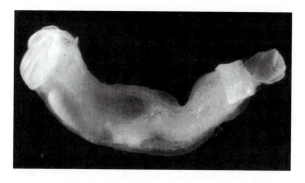

B

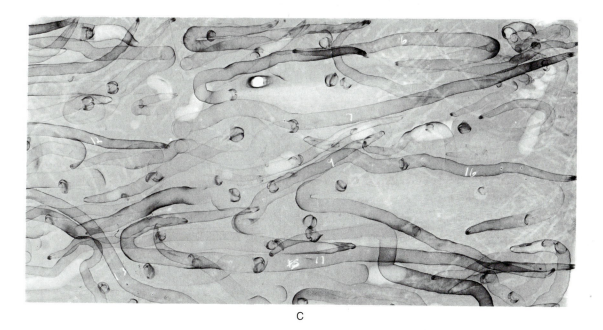

C

FIGURE 10–81 A, A shipworm, a wood-boring bivalve. **B,** Photograph of a living *Nototeredo knoxi* extracted from driftwood. **C,** X-ray photograph of a marine timber section showing shipworms. *(A, Drawing by Brian Morton. C, X-ray photograph courtesy of Lane, C.E. Sci. Am. February, 1961.)*

sea cucumbers; and burrowing, shrimplike crustaceans. A species of *Entovalva,* which lives in the gut of sea cucumbers, is the only known parasitic bivalve.

Internal Transport and Gas Exchange

In the majority of bivalves, the ventricle of the heart is folded around the gut (rectum), so the pericardial cavity encloses not only the heart but also a short section of the digestive tract (Fig. 10–82B). The contractions of the ventricle can be easily observed in a large clam from which one of the valves has been carefully removed. Pulsations are slow, about 20 per minute in *Anodonta.*

An anterior aorta issues from the ventricle, and in the eulamellibranchs there is a posterior aorta as well (Figs. 10–79B; 10–82B).

The typical molluscan circulatory route—heart, tissue sinuses, nephridia, gills, heart—is exhibited in the bivalves, although some modifications of this circuit have taken place in different species (Fig. 10–83). In all bivalves, there is a more or less well-developed circulatory pathway through the mantle,

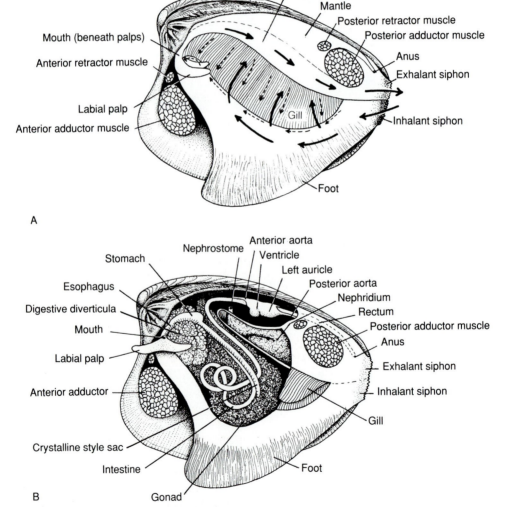

FIGURE 10–82 Anatomy of *Mercenaria mercenaria.* **A,** Interior of the left side. **B,** Partial dissection, showing some of the internal organs.

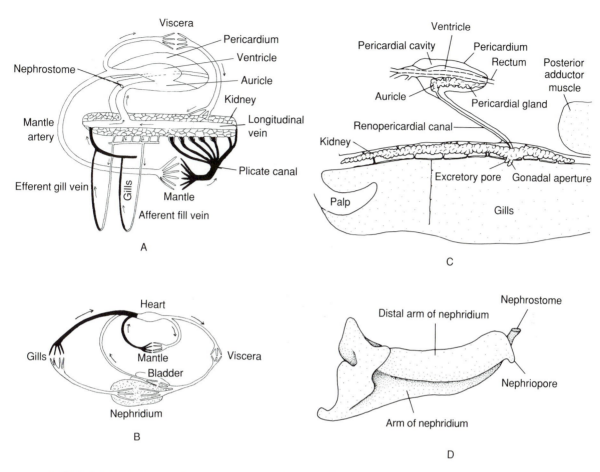

FIGURE 10–83 A, Circulatory system of *Mytilus*. **B,** Circulatory system of the freshwater clam, *Anodonta*. **C,** Lateral view of heart and one nephridium of *Mytilus*. **D,** Lateral view of nephridium of *Anodonta*. The nephridium is bent like a hairpin and lies between the pericardial cavity above and the mantle cavity below. *(C, After Pirie, B.J.S., and George, S.G. 1979. Ultrastructure of the heart and excretory system of Mytilus edulis. J. Mar. Biol. Assoc. U.K. 59:819–829. Copyrighted and reprinted by permission of Cambridge University Press. D, Modified after Fernau.)*

which is an additional site of gas exchange. Depending on the species, some blood may be returned from the mantle or from the kidney directly to the heart.

Gas exchange takes place as water moves over and within the gills. The amount of oxygen removed from the water current is low compared with that in other molluscs (2.5–6.8% for the scallop compared with 48–70% for the abalone, a gastropod). This low oxygen uptake is correlated with the large gill size, which is greater than the respiratory needs of the animal but is required for filter feeding. Moreover, at least in the intertidal clam, *Geukensia demissa,* there is considerable direct passage of oxygen across the body wall, because in highly oxygenated water, blockage of the circulatory system reduces oxygen consumption by less than 15%. When out of water at low tide, the clam utilizes oxygen in air, but deep tissues respire anaerobically.

The blood of most bivalves lacks any respiratory pigment, but in some 21 species, including arks *(Noetia, Arca, Anadara)* and *Calyptogena* of deep hydrothermal vents, the blood contains intracellular or extracellular hemoglobin. In some species, such as *Tellina alternata,* the hemoglobin (neuroglobin) is associated with the ganglia. Hemoglobin and muscle myoglobin may give the mantle and other tissues a bright red color. In *Noetia,* at least, the blood hemoglobin functions both in oxygen transport and in oxygen storage.

Recently, hemocyanin has been found in two groups of protobranchs. This is probably additional evidence that hemocyanin is the primitive molluscan respiratory pigment but has been lost in most bivalves.

Excretion

The two nephridia of bivalves are located beneath the pericardial cavity and above the gills. The nephridium of freshwater clams of the genus *Anodonta,* among the best studied species, is shaped like a hairpin with the two ends close together (Fig. 10–83D). One end opens into the pericardial cavity by way of the nephrostome and the other into the mantle cavity by way of the nephridiopore. The arm of the nephridium associated with the nephrostome has highly folded interior walls. In *Mytilus* the nephridium is a long, branched tube connected to the pericardium by a pericardial canal (Fig. 10–83C). The canal joins the nephridium very near the nephridiopore. The walls of the auricles and pericardial glands are composed of cells like vertebrate podocytes and are believed to be sites of ultrafiltration. Selective reabsorption and secretion probably occur in the sections of the nephridium with folded walls.

Except for random representatives of different marine families, such as the oysters, which can tolerate low salinities and have invaded brackish estuaries and marshes, the freshwater bivalves are members of nine families variously represented in different parts of the world. North American freshwater bivalves are primarily members of the Unionidae and the Sphaeriidae (fingernail clams). Many species of fingernail clams are adapted for living in temporary bodies of fresh water. One Asian *Corbicula* has been introduced into many North American drainage systems, but the most serious invader is the Asian zebra mussel, *Dreissena polymorpha,* which has been introduced into many parts of the world in the ballast water of cargo ships (Fig. 10–84). The zebra mussel is believed to have entered the Great Lakes region of North America in 1986 via ballast discharge and has since undergone an explosive population increase. The mussel attaches by byssal threads and forms mats containing as many as 700,000 individuals per square meter. In addition to altering the ecosystem by re-

FIGURE 10–84 The zebra mussel, *Dreissena polymorpha. (From a Sea Grant Marine Biotechnology Brochure.)*

moving a large amount of plankton and overgrowing and smothering native clams, the zebra mussel clogs the intake pipes of municipal water plants and various industrial systems.

Like gastropods, freshwater bivalves excrete large amounts of water through the nephridia. The urine is very hyposmotic, and the blood salts are maintained at a very low level. The mantle and gills pick up salts from the respiratory water stream.

Nervous System and Sense Organs

The nervous system is bilateral with three pairs of ganglia and two pairs of long nerve cords (Fig. 10–85). On each side of the esophagus is located a cerebropleural ganglion, which is connected to its opposite member by a short commissure dorsal to the esophagus. From each cerebropleural ganglion arise two major posteriorly directed nerve cords. The upper pair of nerve cords (one from each ganglion) extends directly back through the viscera and terminates in a pair of closely adjacent visceral ganglia on the anteroventral surface of the posterior adductor muscle. The second pair of cords arising from the cerebropleural ganglia extend posteriorly and ventrally into the foot and connect with a pair of pedal ganglia. Foot movement and the anterior adductor muscle are under the control of the pedal and cerebral ganglia, but the visceral ganglia control the posterior adductor muscles and the siphons. Coordination of pedal and valve movements is a function of the cerebral ganglia.

The margin of the mantle, particularly the middle fold, is the principal location of most of the bivalve

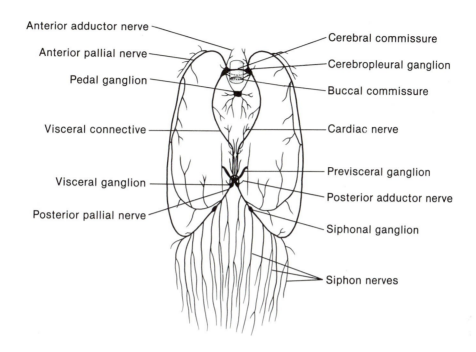

FIGURE 10–85 Dorsal view of the nervous system of *Pholas*, a siphonate eulamellibranch that bores into hard substrata. *(After Forster from Bullock, T.H., and Horridge, G.A. 1965. Structure and Function in the Nervous Systems of Invertebrates. Vol. II. W.H. Freeman, San Francisco. p. 1391.)*

sense organs. In many species the mantle edge bears pallial tentacles, which contain mechano- and chemoreceptor cells. The entire margin may bear tentacles, as in the swimming *Pecten* and *Lima* (Fig. 10–79); more commonly, tentacles are restricted to the inhalant or exhalant apertures or siphons, or they may even fringe the pedal aperture.

A pair of statocysts is usually found in the foot near or within the pedal ganglia, but innervated from the cerebral ganglion. The statocysts of some attached forms, such as oysters, are reduced.

Ocelli may be present along the mantle edge or even on the siphons in some bivalves. They enable the surface-dwelling clam to detect sudden changes in light intensity, like the shadow of a predator. The exposed mantle tissue of giant clams contains several thousand ocelli. In most cases, bivalve ocelli are of simple pigment-spot or pigment-cup type. In arks, however, the ocelli are compound, and in scallops and thorny oysters, the ocelli are well developed (Fig. 10–79), with a cornea and lens.

Scallop *(Pecten)* eyes are peculiar in having two layers of photoreceptors, one rhabdomeric and one of the ciliary type (p. 100). Behind the retina is a layer of reflecting pigment (tapetum), which reflects the incoming light back to the ciliary receptor layer, where the image is formed. The function of the rhabdomeric receptor layer is uncertain.

Cephalic eyes present in some filibranch bivalves, including some Mytilidae, persist from the larval stage and are probably homologous to the eyes of gastropods. A cephalic eye is located at the anterior end of each gill axis.

Immediately beneath the posterior adductor muscle in the exhalant chamber is a patch of sensory epithelium, usually called an osphradium. The osphradium has been considered an organ of chemoreception for monitoring the water passing through the mantle cavity. The position of this sensory tissue in the exhalant chamber makes it doubtful that it is actually homologous with the osphradium of a gastropod.

Reproduction

The majority of bivalves are dioecious. The two gonads encompass the intestinal loops and are usually so close to one another that the paired condition is difficult to detect (Fig. 10–82B). The gonoducts are always simple because there is no copulation. In the protobranchs and the more primitive lamellibranchs, the short gonoduct opens into the nephridium, and sperm and eggs exit by way of the nephridiopores. In most lamellibranchs the gonoducts open into the mantle cavity, close to the nephridiopore.

The hermaphroditic bivalve species include shipworms, the freshwater Sphaeriidae, a few Unionidae, and some species of cockles, oysters, and scallops. In hermaphroditic scallops the gonad is divided into a ventral ovary and a dorsal testis, both of which lie on the anterior side of the adductor muscle (Fig. 10–79B). The European oyster, *Ostrea edulis,* like other species of *Ostrea,* is a protandric hermaphrodite (species of *Crassostrea,* including the American east coast oyster, *Crassostrea virginica,* are mostly dioecious). *Ostrea edulis* not only shifts from male to female but also changes back from female to male. An

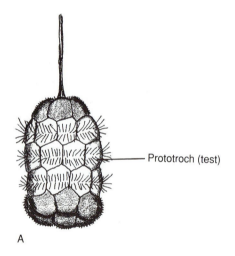

A

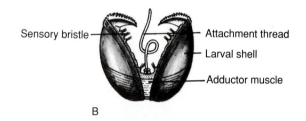

B

C

D

FIGURE 10–86 A, Trochophore of *Yoldia limatula.* **B,** Glochidium of the freshwater clam, *Anodonta.* **C,** Scanning electron micrograph of the interior of the marsupial gill of *Anodonta cataracta* showing glochidia larvae (G). Interlamellar septa (ILS), secondary water tubes (SWT), and a gill filament (F) are also visible. **D,** A fish lure produced by an extension of the inner lobe of the mantle of brooding females of *Lampsilis ventricosa.* Presumably, the lure attracts predaceous fish to the clam, which then releases its glochidia larvae for attachment to the fish's gills. *(A, After Drew from Dawydoff. B, Redrawn from Harms. C, From Tankersley, R.A., and Dimock, R.V. Jr. 1992. Quantitative analysis of the structure and function of the marsupial gills of the freshwater mussel Anodonta cataracta. Biol. Bull. 182:145–154. D, From Gilbert, S.F. 1988. Developmental Biology. 2nd edition. Sinauer Associates, Inc., Sunderland, MA.)*

individual may exhibit active male and female phases each year.

Development

In most bivalves fertilization occurs in the surrounding water; the gametes are shed into the suprabranchial cavity and then swept out with the exhalant current. Some bivalves brood their eggs within the suprabranchial cavity, as in some shipworms, or within the gills, as in *Ostrea edulis* and the freshwater Unionidae and Sphaeriidae. Brooded eggs are fertilized by sperm brought in with the ventilating current.

The development of a free-swimming trochophore, succeeded by a veliger larva, is typical of marine bivalves (Fig. 10–86). The veliger is bilaterally symmetrical and eventually becomes enclosed within the two valves characteristic of bivalves (Fig. 10–87).

Like gastropods, some marine bivalves have long-lived, planktotrophic (feeding) veligers; others have short-lived, lecithotrophic (nonfeeding) veligers.

It is speculated that the larvae of oysters (Ostreidae), for example, are capable of dispersion over a distance as great as 1300 km. Some bivalves, such as *Ostrea* and species of shipworms, are larviparous, releasing the veligers following an initial period of brooding (eight days in *Ostrea edulis*).

Metamorphosis is characterized by a sudden shedding of the velum. Settling may involve considerable testing of the substratum and delayed metamorphosis. *Ostrea edulis* swims upward and attaches to the shaded underside of objects. Shipworms settle only on wood.

With the exception of *Dreissena* and *Nausitoria*, which have free-swimming veliger larvae, freshwater bivalves exhibit modified development. Direct development is characteristic of the freshwater Sphaeriidae, which brood the eggs in marsupial sacs that develop between the gill lamellae. At the completion of development, the young clams are shed from the gills.

The freshwater mussels (Unionacea and Mutelacea) display an indirect but very specialized development. As in the sphaeriids, the eggs are

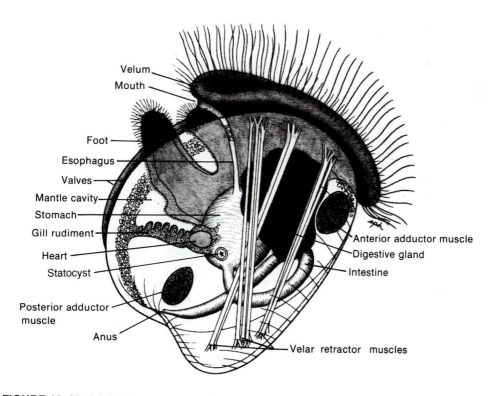

FIGURE 10–87 A fully developed veliger larva of an oyster. *(After Galtsoff.)*

brooded between the gill lamellae, where they develop through the veliger stage. However, in the worldwide Unionacea, the African Mutelidae, and the South American Mycetopodidae, the veliger, called a **glochidium, haustorium,** and **lasidium,** respectively, has become highly modified for a parasitic existence on fish. The fish during this period disperse the rather sedentary freshwater bivalves.

The glochidium larva is enclosed by two valves, each edge of which may bear a hook, as in *Anodonta* (Fig. 10–86B). The shell valves cover a larval mantle, which bears clusters of sensory bristles. A rudimentary foot is present, to which is attached a long adhesive thread. There is neither mouth nor anus, and the digestive tract is rather poorly developed.

When mature, the glochidia range in size from 0.5 mm to 5 mm, depending on the species. In *Unio* and *Anodonta,* the glochidia leave the gills (Fig. 10–86C) through the suprabranchial cavity and exhalant aperture. In *Lampsilis,* the glochidia emerge directly from gills through temporary openings, and dispersal into the surrounding water is aided by the movement of special mantle flaps near the exhalant aperture (Fig. 10–86D). The way the glochidia are released is related to the habits of the host. In some species the glochidia are dispersed over the bottom and are picked up by species of fish that have benthic nesting habits. Other clams release their glochidia in the form of colored masses that look like worms. The mass is eaten by the fish host, and the glochidia attach to the gills. Most clam species parasitize more than one species of fish, and a species of fish is usually the host for a number of species of clams.

The hooked glochidia of *Anodonta* immediately clamp onto the fins and other parts of the body surface of the fish. A long, sticky thread aids in initial contact and adhesion, and clasping is a response to certain molecules in the fish mucus. Hookless glochidia are picked up by the respiratory currents of the fish and attach to the gills. In either case the tissue of the fish in the vicinity of the attached glochidium is stimulated to grow around the parasite and form a cyst. The larval mantle contains phagocytic cells that feed on the tissues of the host and obtain nutrition for the developing clam. During this parasitic period, which lasts 10 to 30 days, many of the larval structures (sensory bristles on the mantle, the adhesive thread, the larval adductor muscle, the larval mantle) disappear, and the adult organs begin developing. Eventually, the immature clam breaks out of the cyst,

falls to the bottom, and burrows in the mud. Here the remainder of development is completed, and the adult habit is gradually assumed.

Some of the larger freshwater mussels may produce as many as 3,000,000 glochidia each, and a single fish has been reported to contain 3000 glochidia. Adult fish are apparently not harmed by the parasitic glochidia, but young fry may die from secondary infection.

Growth and Life Span

As in gastropods, the rates of growth and life span of bivalves vary greatly. The common mussel of the California coast *(Mytilus californianus)* may reach a length of 86 mm within one year. In general, most bivalves grow most rapidly during their early years. Ages of 20 to 30 years are now known to be common for bivalves, and for certain species there are records of 150-year-old individuals.

The growth stages of commercial species are well known. Oysters (Ostreidae), for example, reach marketable size in one to three years depending on the species, latitude, and various environmental conditions. Newly settled oysters, called *spat,* are collected on tiles, twigs, or other objects and allowed to grow to a few centimeters in length. These seed oysters are then distributed over a managed bed, where they grow until harvested. In a natural oyster reef, the average life span is uncertain. Certainly some live longer than ten years.

Small scallops of the genus *Argopecten* have a life span of only one to two years, but the deep-sea scallops *(Placopecten)* are about ten years old when they reach a maximum size of 15 cm.

SUMMARY

1. The general characteristics of the class Bivalvia, a bivalved shell and a reduced head, largely reflect adaptations for burrowing in soft substrata, although many species have secondarily colonized other epibenthic habitats. A primary feature, to which many others are related, is lateral compression of the body. A radula is absent.

2. Primitive bivalves (protobranchs) possess bipectinate gills and usually a ventilating current that enters and leaves from the posterior. Most are selective deposit feeders, utilizing a pair of palpal tentacles in obtaining and transporting deposit material. The stomach contains a protostyle and sorting region.

3. The great majority of bivalves are filter-feeding lamellibranchs. They probably evolved from some group of protobranchs that were preadapted for filter feeding. The gills were used as the filter, the ventilating current became the filtering current, and frontal cilia were employed for vertical transport of trapped food particles. Changes necessary for the lamellibranch condition involved lengthening and folding of the gill filament for greater filtering surface and formation of food grooves for horizontal transport.

4. In the more primitive lamellibranch condition (filibranchs) connections between the folded gill filaments are not substantial. In the specialized condition (eulamellibranchs), however, the filaments are fused by a variety of tissue junctions, and the channel for blood leaving the gill runs adjacent to the tips of the ascending limbs of the gill filaments.

5. The stomach of lamellibranchs has retained a number of primitive features associated with a diet of fine particles, but the protostyle has become consolidated into a gelatinous crystalline style, which liberates enzymes as it is eroded.

6. The evolution of filter feeding led to an explosive evolution of lamellibranchs because they were no longer chained to deposit material as a food source. The majority of lamellibranchs have remained in soft bottoms, for which varying degrees of mantle fusion and siphons are important adaptations.

7. Many lamellibranchs live on hard substrata, the result of a number of separate invasions of this habitat. Anchorage is by byssal threads or by cementation of one valve to the substratum. Reduction or loss of the foot and anterior adductor muscle is common in attached bivalves. Those species that habitually lie on one valve tend to exhibit valve inequality.

8. Ability to drill into hard substrata—rock, shell, coral, and wood—evolved in several groups of lamellibranchs. Most drill mechanically, using the anterior shell margins as cutting tools.

9. A few bivalves live unattached on the surface. Of these, scallops and file clams are capable of escape swimming.

10. The lack of respiratory pigment in most bivalves is correlated with their relatively sluggish habits and their large gill surface.

11. Most bivalves are dioecious. Gametes exit by nephridia in protobranchs and by special gonod-

ucts in other bivalves. Fertilization is usually external and development planktonic, with a trochophore and veliger in marine species.

CLASS SCAPHOPODA

The class Scaphopoda contains about 350 species of burrowing marine molluscs that are popularly known as tusk, or tooth, shells. These names are derived from the shape of the shell, which is an elongated cylindrical tube usually shaped like an elephant's tusk (Fig. 10–88). Both ends of the tube are open. The shells of most scaphopods average 3 to 6 cm in length, but *Cadulus mayori,* found off the Florida coast, does not exceed 4 mm in length. *Dentalium vernedei,* found off Japan, is the largest living species, reaching a length of 15 cm. A fossil species of *Dentalium,* however, has a shell 30 cm long with a maximum diameter of well over 3 cm. The shells of most scaphopods are white or yellowish, but an East Indian species of *Dentalium* is a brilliant jade green.

The majority of scaphopods burrow in sand in water depths greater than 6 m. The living animals are therefore not frequently encountered, but judging from the number of shells washed up on beaches in some areas, scaphopods are not rare.

The body of the scaphopod is greatly elongated along the anterior/posterior axis (Fig. 10–88C). The head and foot project from the larger and anterior aperture of the shell. When the shell is slightly curved, the concavity lies over the dorsal surface. Scaphopods live buried in sand, head downward with the body steeply inclined; only the small, posterior aperture projects above the surface.

Adapted to a burrowing habit, the head is reduced to a short, conical projection, or proboscis, bearing the mouth. The cone-shaped foot is projected into the sand, and lateral flanges expand and anchor the tip as in protobranch bivalves.

The scaphopod mantle cavity is large and extends the entire length of the ventral surface. The posterior aperture serves for both inhalant and exhalant water currents. Water slowly enters the mantle cavity as a result of mantle ciliary action and perhaps foot protraction. After 10 to 12 min of inhalation, a violent muscular contraction (probably foot retraction) expels the water from the same opening it entered. There are no gills; exchange of gases takes place through the mantle surface.

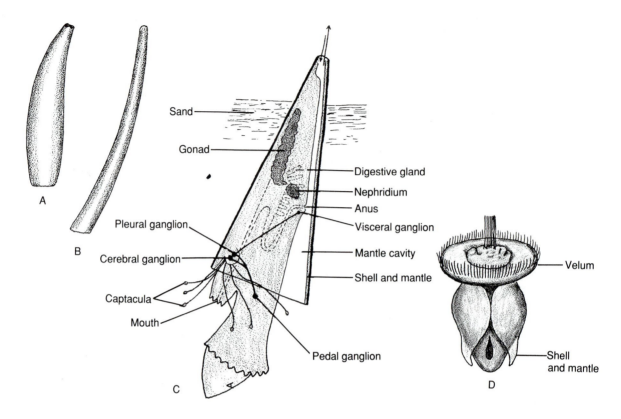

FIGURE 10–88 Scaphopoda. **A,** Shell of *Cadulus.* **B,** Shell of *Dentalium.* **C,** Structure of *Dentalium.* Arrows indicate water current direction through mantle cavity. **D,** Veliger larva. *(A and B, After Abbott, R.T. 1954. American Sea Shells. Van Nostrand Co., Princeton, NJ. D, After Lacaze Duthiers from Dawydoff.)*

Scaphopods feed on microscopic organisms, such as forams, juvenile bivalves, and kinorhynchs, in the surrounding sand and water. Two lobes, located above the head, each bear a large number of unique threadlike tentacles called **captacula** (Fig. 10–88C). Each captaculum has a ciliated knob at the tip and extends by ciliary creeping into the sand to capture food. A selected particle adheres to the knob by adhesive secretions that empty into a pit on one side of the knob. The pit surface is always in contact with the particle, and the secretions appear to be products of a duogland system. Contractions of muscles within the captacular stalk bring the particle back to the mouth, where it enters by means of oral cilia. The buccal cavity contains a median jaw and a well-developed radula with large, flattened teeth. The radula breaks up particles and moves food back to the esophagus. Some species of *Dentalium* can collect fine sediment by means of a ciliated tract on the captacular stalk.

Digestion is extracellular in the stomach. The intestine empties through the anus into the mantle cavity.

The circulatory system is reduced to a system of blood sinuses, and there may be no heart. The nervous system exhibits the typical molluscan plan. There are no eyes, sensory tentacles, or osphradia. A pair of nephridia is present, and the nephridiopores are located near the anus.

Scaphopods are dioecious. The unpaired gonad fills most of the posterior part of the body, and the sperm or eggs reach the outside by way of the right nephridium. The eggs are shed singly and are planktonic. Fertilization is external.

Scaphopod development is very similar to that of the marine bivalves. There is a free-swimming trochopore larva, succeeded by a bilaterally symmetrical veliger (Fig. 10–88D). As in bivalves, the larval mantle and shell in scaphopods are at first bilobed, but then the mantle lobes fuse along their ventral margins (Fig. 10–88D). This fusion thus results in a cylindrical mantle and shell that remain open at each end.

Scaphopods are clearly related to bivalves and perhaps like bivalves may have evolved from rostroconchs. Many rostroconchs had shells in which the ventral margins were in contact. If they fused, the shell would have become tubular with a posterior aperture (rostral) and an anterior aperture (pedal). Thus, scaphopods and bivalves may be sister groups. The reduction of the head, the burrowing habit, the symmetrical veliger, and the embryonic bilobed mantle and shell are strikingly similar to the respective bivalve characteristics.

SUMMARY

1. Members of the class Scaphopoda are burrowing molluscs with cylindrical, tusk-shaped shells open at each end. The animal lives buried in soft bottoms, with the larger, anterior end downward and the small, posterior end, through which the ventilating current enters and leaves, near the surface of the substratum. They burrow like protobranch bivalves.

2. Scaphopods feed on interstitial organisms collected by means of small tentacles (captacula) and ingested with a radula.

3. The mantle surface, rather than gills, provides for gas exchange.

4. Scaphopods are dioecious, fertilization is external, and development is planktonic. There is both a trochophore and a veliger larva.

CLASS CEPHALOPODA

The class Cephalopoda contains the nautili, cuttlefish, squids, and octopods and fossil ammonoids. Although some cephalopods, such as the octopus, have secondarily assumed a bottom-dwelling habit, the class as a whole is adapted for a swimming existence and contains the most highly organized and active of all molluscs. The head projects into a circle of large, prehensile **tentacles,** or **arms,** which are homologous to the anterior of the foot of other molluscs. In the evolution of the cephalopods, the body became greatly lengthened along the dorsoventral axis, and as a result of a change in the manner of locomotion, this axis became the functional anterior/posterior axis (Fig. 10–89). The circle of tentacles is thus located at the anterior of the body, and the visceral hump is posterior. The original posterior mantle cavity is now ventral.

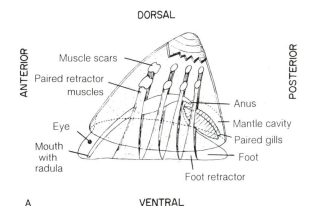

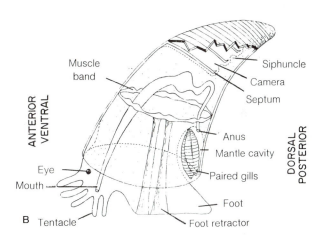

FIGURE 10–89 Evolution of the cephalopod form. **A,** Reconstruction of a monoplacophoran with a cap-shaped shell. Note that the apex tilts slightly posteriorly. Some high cone monoplacophorans, from which the cephalopods may have evolved, had septate shells but lacked a siphuncle. **B,** Reconstruction of an early cephalopod, such as the late Cambrian *Plectronoceras. (From Yochelson, E.L., Flower, R.H., and Webers, G.F. 1973. The bearing of the new Late Cambrian monoplacophoran genus Knightoconus upon the origin of the Cephalopoda. Lethaia. 6:275–310.)*

The cephalopods have attained the largest size of any invertebrate. Although the majority range between about 6 and 70 cm in length, including the arms and tentacles, some species reach giant proportions. The largest cephalopods are the giant squids, *Architeuthis* (Fig. 10–95); one specimen was reported to have measured 16 m, including the tentacles. The tentacles alone were 6 m long, and the circumference of the body was 4 m. Giant octopods with arms 10 to 15 m long have been observed by divers in the Sea of Japan, but no specimens have been taken. The dorsal mantle length of *Octopus dofleini* of the Pacific coast, one of the largest known octopods, does not usually exceed 36 cm, though its rather slender arms may be five times the body length. The record is a specimen with a 9.6-m arm span.

There are only about 600 living species of cephalopods but more than 7500 fossil forms. The class first appeared in the Cambrian period and then, once during the Paleozoic era and once in the Mesozoic era, underwent great periods of evolutionary development with the formation of many species.

Shell

A completely developed external shell is found only in the fossil representatives of the class and the four living species of *Nautilus* found in the tropical Indo-western Pacific. In squids and cuttlefish the shell is reduced and internal, and in the octopods it is completely lacking. The shell of *Nautilus* is coiled over the head in a bilaterally symmetrical planospiral (Fig. 10–90). Only the last two whorls are visible, since they cover the inner whorls. The shell of *Nautilus*, like all cephalopod shells, is divided by transverse **septa** into internal chambers, and only the last chamber is occupied by the living animal. As the animal grows, it periodically moves forward, and the posterior part of the mantle secretes a new septum.

Each septum is perforated in the middle, and through the opening extends a thin calcareous tube that houses a slender cord of tissue, called the **siphuncle,** which is an outgrowth of the visceral mass. The siphuncle secretes gas through the porous wall of the tube into the empty chambers, making the shell buoyant and allowing the animal to swim. The shell is composed of an outer porcelaneous layer, containing prisms of calcium carbonate in an organic matrix, and an inner nacreous layer.

Cephalopods may have evolved from high-cone monoplacophorans (Fig. 10–89), some of which were septate. Such septation, however, was probably a spatial adaptation, and only later did the siphuncle evolve as a true cephalopod innovation. The apical chambers were perhaps first filled with fluid, but with the evolution of the siphuncle, gas production became possible. Initially, such gas-filled chambers may only have aided in keeping the shell upright as the animal

moved about over the bottom. Swimming and invasion of the pelagic environment were probably later cephalopod developments.

The first cephalopod shells are believed to have been curved cones and, from them, there has been an evolutionary toward straight and coiled shells. Later, however, some groups of coiled cephalopods reevolved straight and curved shells. The straight shells of some species from the Ordovician period exceeded 5 m in length, and the aperture was 36 cm in diameter. Most coiled shells were planospirals, but one fossil group (heteromorph ammonoids) had asymmetrically or highly irregular coiled shells adapted for planktonic life (Fig. 10–91E, F). The largest fossil species with a coiled shell was *Pachydiscus seppenradensis* from the Cretaceous period, which had a shell diameter of 3 m. However, many fossil cephalopods were small species with shells only 3 cm in diameter.

One of the most important characteristics for the classification of fossil cephalopods is the nature of the internal **suture**—the junction between the septum and the wall of the shell. As the chambers filled with sediment, the details of the suture pattern, once internal, were beautifully preserved on the outer surface of the fossil. The simplest suture lines were straight or slightly waved, as in *Nautilus* (Fig. 10–90A), but one large group, the ammonoids, developed elaborate sutures that were zigzagged, or, more frequently, minutely crinkled (Fig. 10–91B). Such sutures indicate a corresponding complexity in the nature of the septal junction and probably represented an adaptation providing for greater strength to compensate for the somewhat thinner ammonoid shell.

The current system of cephalopod classification separates those forms with complete shells into two subclasses: the Nautiloidea and the Ammonoidea. The Nautiloidea is characterized by straight or coiled shells and simple sutures. The Nautiloidea first appeared in the Cambrian period and is represented today by *Nautilus*. All members of the Ammonoidea were coiled and displayed complex septa and sutures. They appeared in the Silurian period after the nautiloids and disappeared at the end of the Cretaceous period.

We know little concerning the soft parts of fossil ammonoids and nautiloids and can only assume that they were somewhat similar to those of *Nautilus*.

Except for *Nautilus*, all living cephalopods belong to the subclass Coleoidea, in which the shell is

A

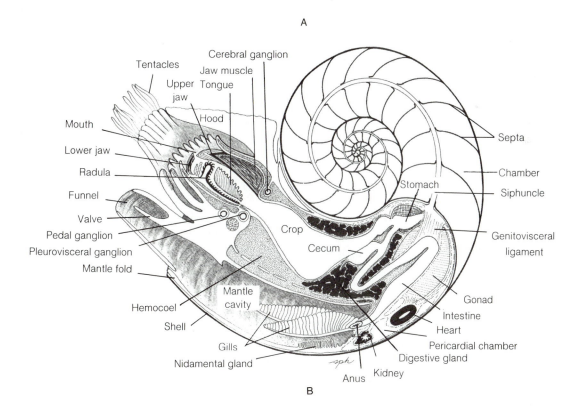

B

FIGURE 10–90 A, Side view of *Nautilus* in swimming position. **B**, Sagittal section. *(B After Stenzel.)*

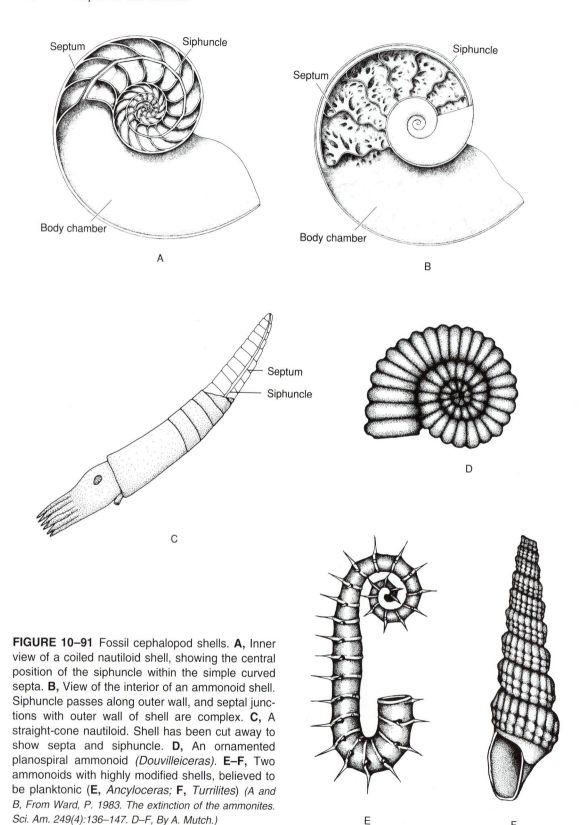

FIGURE 10–91 Fossil cephalopod shells. **A,** Inner view of a coiled nautiloid shell, showing the central position of the siphuncle within the simple curved septa. **B,** View of the interior of an ammonoid shell. Siphuncle passes along outer wall, and septal junctions with outer wall of shell are complex. **C,** A straight-cone nautiloid. Shell has been cut away to show septa and siphuncle. **D,** An ornamented planospiral ammonoid *(Douvilleiceras).* **E–F,** Two ammonoids with highly modified shells, believed to be planktonic (**E,** *Ancyloceras;* **F,** *Turrilites*) *(A and B, From Ward, P. 1983. The extinction of the ammonites. Sci. Am. 249(4):136–147. D–F, By A. Mutch.)*

reduced and internal or lacking altogether. These cephalopods are believed to have evolved from some early straight-shelled nautiloid, whose shell became completely enclosed by the mantle.

From some primitive form the coleoid shell evolved in four directions, all leading to modern species and all involving a reduction in the weight of the shell. This evolutionary development is illustrated in Figure 10–92. In *Spirula,* a common, worldwide, deep-water cuttlefish, the shell has become coiled and is similar to that of *Nautilus* (Fig. 10–93B). In the evolutionary line leading to the squids, the largest group of cephalopods (including *Loligo pealei,* the common squid of the western North Atlantic), the shell is reduced to a long, flattened, chitinous plate, called a **pen,** or **gladius.**

A third evolutionary line, represented by the European cuttlefish, *Sepia,* has a chalky, platelike shell

with only a trace of the original septa (Fig. 10–93A). Finally, in *Octopus* the shell has disappeared completely.

Locomotion and Adaptive Diversity

Most cephalopods swim by jet propulsion, rapidly expelling water from the mantle cavity. The mantle contains both radial and circular muscle fibers. During the inhalant phase of water circulation, the circular fibers relax and the radial muscles contract. This action increases the volume of the mantle cavity, and water rushes in laterally between the anterior margin of the mantle and the posterior end of the head.

During the exhalant phase the contraction of the circular muscles not only increases the water pressure within the cavity but also locks the edges of the mantle tightly around the head. Flap valves seal the man-

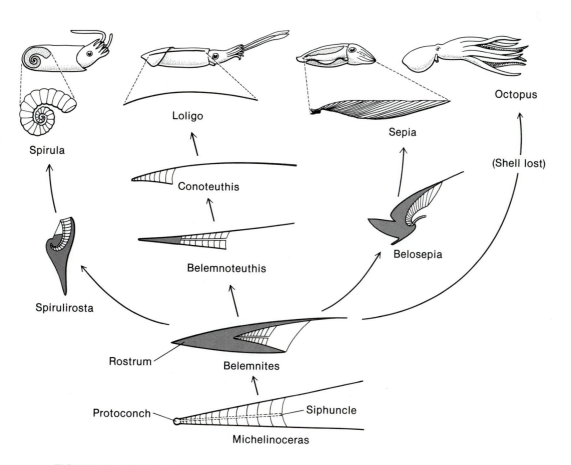

FIGURE 10–92 Evolution of shells of the Coleoidea. *(After Shrock and Twenhofel.)*

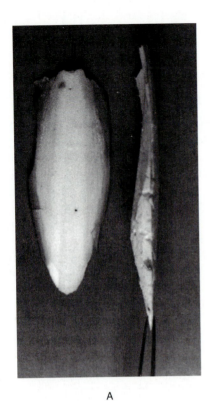

A

B

FIGURE 10–93 Internal shells of coleoids. **A,** Cuttlefish *Sepia*. Right shell is viewed from the side; the left shell is viewed from the upper surface. **B,** *Spirula,* side (left) and end (right) views. Note the siphuncle at the bottom of the septum in the end view. The shell of *Sepia* is located on the dorsal side of the body; in *Spirula* it is the posterior end of the body. *(Photographs courtesy of Katherine E. Barnes.)*

tle cavity ventrally, and thus water is forced to leave through the ventral tubular **funnel.** The force of water leaving the funnel propels the animal in the opposite direction. The funnel is highly mobile and can be directed anteriorly or posteriorly, resulting in either forward or backward movement. The fastest movement is achieved in backward escape swimming, when powerful contractions of the mantle eject water from the anteriorly directed funnel.

Squids possess a number of features that increase the efficiency of this jet propulsion system. The powerful contractions of escape swimming are produced by one type of circular muscle fiber; the rhythmic, less powerful contractions of ventilation and slow swimming are produced by another. The radial muscles function only during escape swimming, when they actually hyperinflate the mantle cavity, that is, they increase its volume beyond the usual ventilating capacity. In slow swimming, the mantle cavity is expanded by elastic recoil of stretched collagen fibers extending through the mantle wall, and some additional propulsive power is provided by undulations of the fins.

The squids and cuttlefish are best adapted for swimming by water jet. They can hover, perform subtle swimming movements, slowly cruise, or dart rapidly. Squids attain the greatest swimming speeds of any aquatic invertebrate, up to 40 km/h. The body of a typical squid is long and tapered posteriorly and has a pair of posterior lateral fins, which function as stabilizers, rudders, and even for propulsion at low speeds (Fig. 10–94). The "flying squids" (Onycoteuthidae), which have long, tapered bodies and highly developed fin vanes and funnels, shoot out of the water during escape swimming and glide for some distance. There have been reports of squids accidentally leaping onto the decks of ships 12 ft above the water's surface.

The giant architeuthid squids may inhabit depths of 300 to 600 m over the continental slopes (Fig. 10–95) and are not rapid swimmers. Two other interesting groups of deep-water squids are the bathypelagic chiroteuthids and the cranchiids. The chiroteuthids have very long, slender bodies and long, whiplike tentacles (Fig. 10–96A, B). The cranchiids are planktonic and are often strangely shaped (Fig.

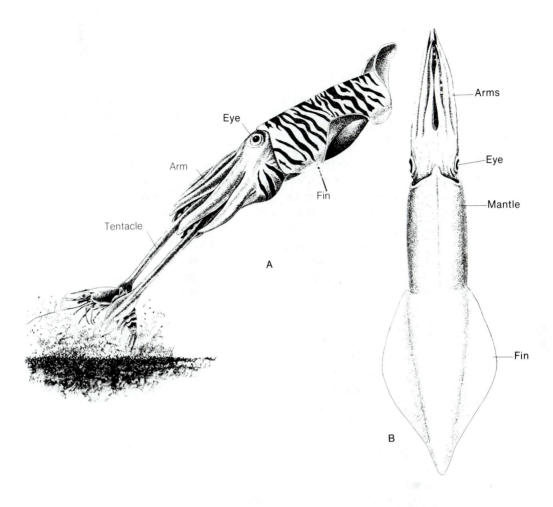

Eye

Arm

Tentacle

Fin

A

Arms

Eye

Mantle

Fin

B

FIGURE 10–94 A, The cuttlefish, *Sepia,* seizing a shrimp with its tentacles. **B,** Dorsal view of the squid, *Loligo,* in swimming position. Tentacles and arms are held together, acting as a rudder.

10–96C). Two thirds of the entire body volume of a cranchiid is provided by the enormous, fluid-filled coelom, which functions as a buoyancy chamber. Most of the seawater cations are replaced in the coelomic fluid with lighter ammonium ions derived from the metabolic wastes. Ammonium is used to achieve neutral buoyancy in about half of the 26 families of oceanic squids, including the chiroteuthids and the giant squids, but is located in special body tissues instead of the coelom.

The body of a cuttlefish, such as *Sepia* and *Spirula,* tends to be short, broad, and flattened (Fig. 10–94A). Cuttlefish are versatile swimmers but not as fast as the more streamlined squids.

The shell of cuttlefish *(Sepia),* despite its reduction from the ancestral form, still functions in providing buoyancy. Spaces between the thin septa contain fluid and gas (air). By regulating the relative amounts of fluid and gas, the degree of buoyancy can be varied. Light is an important factor controlling the regulating mechanism. During the day the cuttlefish lies buried in the bottom; at night the animal becomes active, swimming and hunting for food. Buoyancy decreases when the animal is exposed to light and increases in the dark.

The deep-water *Spirula,* which lives down to about 1000 m, swims with the tentacles directed downward. When the animal dies, the little gas-filled

(Text continues on page 472)

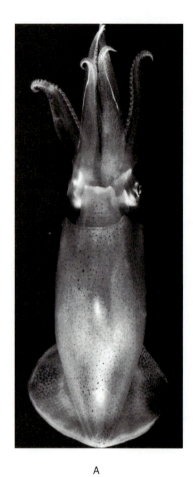

A

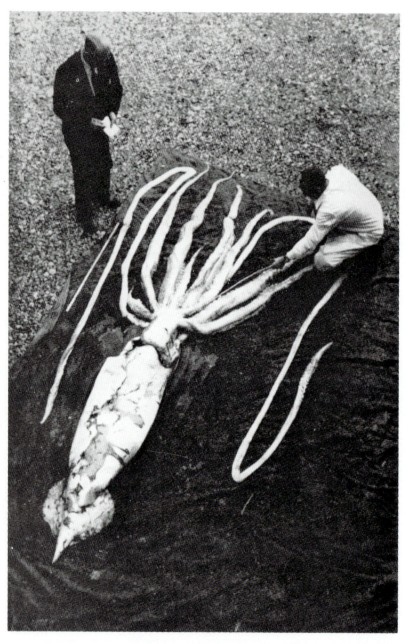

B

FIGURE 10–95 A, The small squid, *Lolliguncula brevis,* of tidal creeks in the southeastern United States (ventral view). **B,** A giant squid of the genus *Architeuthis* stranded at Rahneim, Norway, in 1954. *(B, Photograph courtesy of Clark, M.R. 1966. Adv. Mar. Biol. 4:103. Copyright by Academic Press, Inc., London.)*

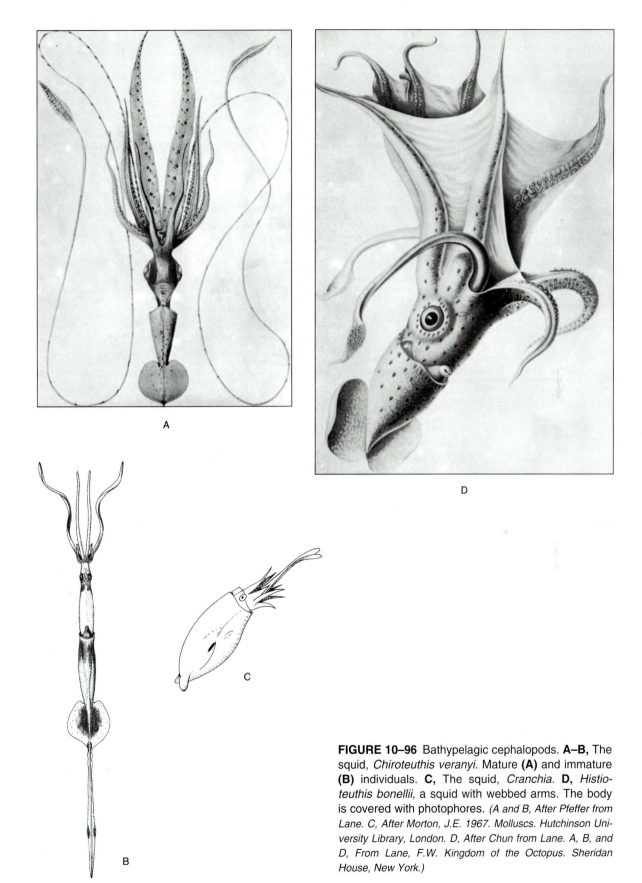

FIGURE 10–96 Bathypelagic cephalopods. **A–B,** The squid, *Chiroteuthis veranyi.* Mature **(A)** and immature **(B)** individuals. **C,** The squid, *Cranchia.* **D,** *Histioteuthis bonellii,* a squid with webbed arms. The body is covered with photophores. *(A and B, After Pfeffer from Lane. C, After Morton, J.E. 1967. Molluscs. Hutchinson University Library, London. D, After Chun from Lane. A, B, and D, From Lane, F.W. Kingdom of the Octopus. Sheridan House, New York.)*

471

shell floats to the surface and is commonly washed ashore (Fig. 10–93B). The large number that can often be found on tropical and semitropical beaches attests to the abundance of this cephalopod.

The smallest cuttlefish are species of the genus *Idiosepius*, which are about 15 mm long. They live in tide pools and possess a dorsal disc on the mantle for attaching to algae.

Species of *Nautilus* are mobile epibenthic animals, found from the surface to depths as great as 600 m. Because they prefer cool water temperatures, they are not found above 100 m where surface water temperatures are warm. Moreover, they ascend at night and descend during the day. Using radio telemetry, one specimen tagged in Palau was found to ascend to 150 to 100 m at night and then move down to 250 to 350 m during the day. When resting, they attach to rubble or the walls of crevices with their appendages. Whether the animal is swimming or resting, the gas-filled chambers keep the shell upright and provide buoyancy that counteracts the shell and body weight.

The regulatory mechanism is essentially the same in *Nautilus, Spirula,* and *Sepia.* An osmotic mechanism accounts for gas and fluid exchange between the siphuncle and chambers (Fig. 10–97). The siphuncular tissue actively pumps ions from chamber water into the blood causing the blood to be saltier than the chamber water. Following the osmotic gradient, water diffuses from the chamber into the blood. As water is removed from the chambers, gas dissolved in the blood diffuses into the chamber. Fluid exchange is restricted to the more anterior chambers. In fact, fluid fills the space between the last septum and the posterior end of the animal when a new septum is being secreted. Fluid is retained in the chamber until the septum is sufficiently strong to withstand pressure changes. In *Nautilus,* chamber fluid is gradually removed to compensate for the gradual growth of the shell. There is no regulation with short-term changes in depth.

Except when feeding, *Nautilus* swims backward, at about the same speed as a person doing a slow

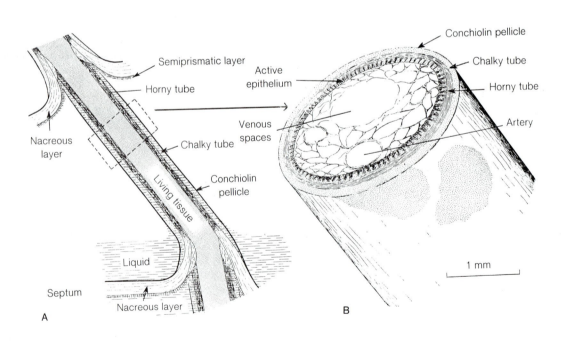

FIGURE 10–97 Siphuncle of *Nautilus.* **A,** Longitudinal section between two septa. **B,** Cross section. *(From Denton, E.J. 1974. On buoyancy and the lives of modern and fossil cephalopods. Proc. Roy. Soc. London B. 185:273–299.)*

breast stroke. The locomotor mechanism is the same as that of squids, although the ejection of water through the funnel results from the retraction of the body and contraction of the muscles of the funnel rather than by the mantle. The water jet must have been generated in a similar manner in fossil forms, for as in *Nautilus,* enclosure by the shell would have inhibited mantle contraction. Unlike that of coleoids, the funnel of nautiloids is a rolled sheet and not a complete tube, although it functions as a tube. It can be extended and directed up or down and from side to side. Lift is obtained by directing the funnel downward.

Living cephalopods that still use gas-filled shells to maintain neutral buoyancy—*Nautilus, Spirula,* and *Sepia*—are restricted to depths at which the shell strength can withstand the pressure. Below that depth the shell implodes (caves in). Fossil cephalopods would have been limited in the same manner, and it is therefore reasonable to assume they were all inhabitants of relatively shallow water.

During the late Paleozoic and Mesozoic cephalopods were the dominant and most highly developed pelagic animals, until they were forced to give way to competition from fish. The speedier coleoid design perhaps evolved as a result of competition with fish and may account for the many convergent features, such as complex eyes, the two groups share. With further modification, the coleoid design permitted invasion of the deep and open ocean, from which the ammonoids and nautiloids were excluded.

Pelagic cephalopods of the open ocean, which includes 46% of the genera, inhabit various levels of the vertical water column. The greatest number live in the upper 100 m and near the boundary between the mesopelagic and bathypelagic zones, but there are species that live at deeper levels. Many epipelagic and mesopelagic cephalopods exhibit diurnal vertical migration, moving upward during the night and inhabiting lower levels during the day. Four sperm whales caught off Chile and Peru contained in their stomachs the beaks of 1000 cephalopods, which indirectly reflects the population of pelagic cephalopods and their importance in the pelagic food web.

The octopods have reverted to more benthic habits. The body is globular and baglike, and there are no fins (Fig. 10–98). The mantle edges are fused dorsally and laterally to the head, resulting in a much more restricted aperture into the mantle cavity. Although octopods are capable of swimming by water

FIGURE 10–98 Lateral view of an octopus.

jets, with arms trailing, they more frequently crawl about over the rocks around which they live. The arms, which are provided with adhesive suction discs, are used to pull the animal along or anchor it to the substratum. Species of *Octopus* usually occupy a den or retreat, from which they make feeding excursions.

A very different mode of existence is exhibited by a number of families of deep-water octopods, some bathypelagic and some abyssalbenthic. These animals have webbed, umbrella-like arms and swim somewhat like jellyfish, by pulsations of the arms as well as by water jets from the funnel (Fig. 10–99). Many have rather gelatinous bodies.

Gas Exchange

The circulation of water through the mantle not only produces the power for locomotion but, as in other molluscs, also provides oxygen for the gills. The nautiloids have four gills, but the coleoids have only two. Muscular contractions of the mantle create a concurrent, rather than a countercurrent, flow over the gills. The loss of countercurrent exchange efficiency is compensated for by the rapid muscle-generated movement of water, rather than by ciliary movement (cilia are absent), and by increased folding of the gill surface. In many cephalopods that swim with webbed arms, the gills are vestigial, and gas exchange takes place through the general body surface.

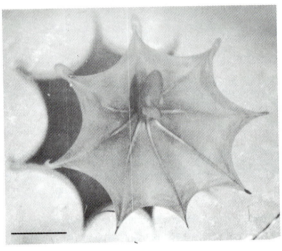

FIGURE 10–99 Deep-sea photographs of a cirrate octopod, *Cirroteuthis*, which lives near the bottom. **A,** Animal perhaps at the beginning of a downward stroke, with arm web closed and fins folded dorsally. **B,** View onto interbrachial web. Cirri on top of the arms may have a sensory function. The scale mark equals 30 cm. Photographs taken with a deep-sea camera at 3000 m in the Virgin Islands Basin. *(From Roper, C.E.F., and Brundage, W.L. 1972. Cirrate octopods with associated deep sea organisms: New biological data based on deep benthic photographs. Smithson. Contrib. Zool. 21:1–46.)*

Nutrition

Cephalopods are highly adapted for raptorial feeding and a carnivorous diet. Prey is located with the highly developed eyes, and capture is effected by the tentacles or arms. *Nautilus* possesses some 90 tentacles arranged around the head (Fig. 10–90). Many are chemosensory and tactile in function, but some of those around the buccal mass are prehensile and used in bringing food to the mouth. These tentacles lack adhesive suckers or discs but have transverse ridges. The bases of the tentacles are united and form a cephalic sheath that encircles the buccal area. Dorsally the cephalic sheath is continuous with a large, leathery, protective hood that acts like an operculum and covers the aperture of the shell when the animal withdraws.

Squids and cuttlefish possess only ten appendages, arranged in five pairs around the head (Figs. 10–94A; 10–102A). Eight are short and heavy and are called arms; the fourth pair down from the dorsal side are larger and are called tentacles. The inner surface of each arm is flattened and covered with stalked, cup-shaped, adhesive discs that function like suction cups. The rim of the sucker is commonly horny and toothed, and the inner wall sometimes has hooks (Fig. 10–100). Suckers are present only on the flattened spatulate ends of the highly mobile tentacles, which are shot out with great rapidity to seize prey. The arms aid in holding the prey after capture (Fig. 10–94A). Some squids have curved hooks instead of suckers.

Octopods have only eight arms, which are similar to the arms of squids except that the suckers are stalkless and lack horny rings and hooks.

A radula is present in cephalopods (Fig. 10–101), but more important is the pair of powerful, beaklike jaws in the buccal cavity. The beak can bite and tear off large pieces of tissue, which are then pulled into the buccal cavity by a tonguelike action of the radula and finally swallowed (Fig. 10–102A). The location of the buccal mass within a blood sinus permits the animal to turn the entire buccal apparatus and use the jaws with great dexterity. In coleoids two pairs of salivary glands empty into the buccal cavity. The posterior salivary glands of *Sepia* and octopods secrete poison and, at least in *Octopus,* proteolytic enzymes. The poison (a glycoprotein) can enter the tissues of the prey through the wound inflicted by the jaws. The little blue-ringed octopus, *Hapalochlaena maculosa,* which feeds on crustaceans in shallow water in the Indo-Pacific, is extremely venomous, and its bite has resulted in a few human fatalities.

The diet of cephalopods depends on the habitats in which they live. Pelagic squids, such as *Loligo* and *Alloteuthis,* feed on fish, crustaceans, and other squids. *Loligo* darts into a school of young mackerel,

A

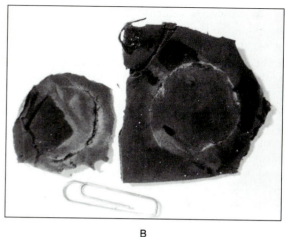

B

FIGURE 10–100 A, Toothed suckers of the squid, *Lolliguncula brevis.* **B,** Scars from the suckers of the giant squid, *Architeuthis,* on the skin of a sperm whale. Sperm whales are the principal predators of the giant squid. *(B, Photograph courtesy of C. Roper.)*

FIGURE 10–101 Radula teeth of *Octopus briareus.* Only half of a series of seven transverse rows is shown. *(From Solem, A., and Roper, C.F.E. 1975. Structures of recent cephalopod radulae. Veliger. 18(2):127–133.)*

seizes a fish with the tentacles, and quickly bites out a chunk behind the head or bites off the head. The fish is devoured with small bites of the beak until only the gut and tail remain. These parts are then dropped to the bottom.

Cuttlefish swim over the bottom and feed on surface-inhabiting invertebrates, especially shrimps and crabs. *Sepia* may rest on the bottom, lying in wait for passing prey.

Octopods live in dens located in crevices and holes. They make excursions in search of food or lie in wait near the entrances of their lairs. Clams, snails, and crustaceans are common prey, and the diet of a particular species may include as many as 55 different prey items, although a few items predominate. Some of the prey are eaten during the foraging excursion, and some are taken back to the den and eaten. Thus, octopods commonly have midden heaps of shells around their dens.

Octopus dofleini, found along the west coast of the United States, is most active at night, with foraging areas around the den of no more than 250 m. The reef-inhabiting, Indo-Pacific *Octopus cyanea* is active during the day. The latter may make hour-long hunting excursions and cover a distance of 100 m. Movement is by combined swimming and crawling, and the animal may perch on a rock for a while during the course of a trip. The octopus leaps on motile prey, such as a crab, enveloping it in the outstretched arm web. It also leaps on algal clumps and other objects and then feels beneath the web for a possible catch. The octopus may take several animals, all paralyzed by salivary toxin, home for consumption.

In contrast to cuttlefish and squids, which tear prey with the jaws, the feeding habit of octopods is rather like that of spiders. The prey is injected with poison, with or without a bite of the jaws, and then, while held, is flooded with enzymes. The partially digested tissues pass into the gut, and the indigestible remains are eventually discarded. To remove gastropods from their shells, octopods drill a hole through the shell with the radula or the toothed salivary papilla and inject poison directly into the occupant. A chemical softening agent is secreted in the drilling process.

A few deep-sea octopods are suspension feeders and convergent with jellyfish.

Nautilus is a scavenger and predator over the bottom, and decapod crustaceans, especially hermit crabs, appear to be common prey. When feeding, it swims forward, searching with extended tentacles. Baited traps capture nautiloids only when situated on the bottom, never when suspended in the water column.

The esophagus of cephalopods is muscular and conducts food by peristaltic action into the stomach or, as in *Nautilus* and *Octopus,* into the crop, which is an expansion of one end of the esophagus. The stomach is very muscular, and attached to its anterior is a large cecum (Figs. 10–102A, B; 10–103). The digestive gland in cephalopods is divided into a small, spongy, diffuse portion, sometimes called the pancreas, and a large solid "liver" (paired in *Sepia*), which is probably homologous to the digestive gland of the other molluscs. In squids the two divisions of the digestive gland are morphologically separated from each other, and the pancreas empties into the liver duct. Digestion is entirely extracellular. Enzymes from both digestive gland divisions empty through a common duct into the junction between the stomach and cecum. Secretions from the two glands, which may be released in separate phases, can be sent to the stomach or cecum through regulation of a groove.

Absorption takes place in the cecal walls (*Loligo*) or in the digestive gland (*Sepia, Octopus*). Large, indigestible residues in the stomach can be passed directly into the intestine. Some intestinal absorption takes place. Wastes leave the anus near the funnel and are carried away with the exhalant water jet. The digestive modifications of cephalopods, particularly squids, probably represent an adaptation for rapid digestion correlated with an active pelagic life and a carnivorous diet.

Excretion

There are two nephridia in coleoids and four in *Nautilus*. The conspicuous part of the cephalopod nephridium is a large renal sac (Fig. 10–104). Each sac opens to the mantle cavity through a nephridiopore and communicates with the pericardium by way of a renopericardial canal. The renal sac receives pericardial filtrate via the renopericardial canal and also secretions from large, contractile renal appendages, which are evaginations of the wall of the branchial vein crossing the sac. The wall separating the pericardial coelom and an appendage of the branchial heart contains podocytes, through which a filtrate passes into the pericardial fluid. Selective reabsorption begins within the coelomic cavity even before the filtrate enters the renal sacs.

Internal Transport

The circulatory system of cephalopods is an extensive system of vessels and capillaries, lined by endothelium, a condition shared with few animals other than vertebrates. Blood within the vessels follows more or less the same route through the body as in other molluscs (Fig. 10–104), but in addition to a systemic heart, the circuit includes two **branchial hearts,** which pump blood through the gills.

The structure and physiology of the cephalopod circulatory system are closely correlated with the higher metabolic rate of these animals compared with other molluscs and, perhaps, the concurrent water flow over the gills. The existence of capillaries, some contractile arteries, and the branchial hearts increases the blood pressure and the speed of blood flow. The contraction of the branchial hearts, which receive unoxygenated blood from all parts of the body, boosts the pressure of the blood, sending it through the capillaries of the gills. The two auricles of the heart drain blood from the gills and then pass it into the median ventricle. The ventricle pumps blood out to the body through both an anterior and a posterior aorta and eventually through smaller vessels. The blood of cephalopods contains hemocyanin that loads oxygen at the gills and unloads at the tissues at relatively high partial pressures of oxygen.

Nervous System and Sense Organs

The high degree of development of the cephalopod nervous system is unequaled among other inverte-

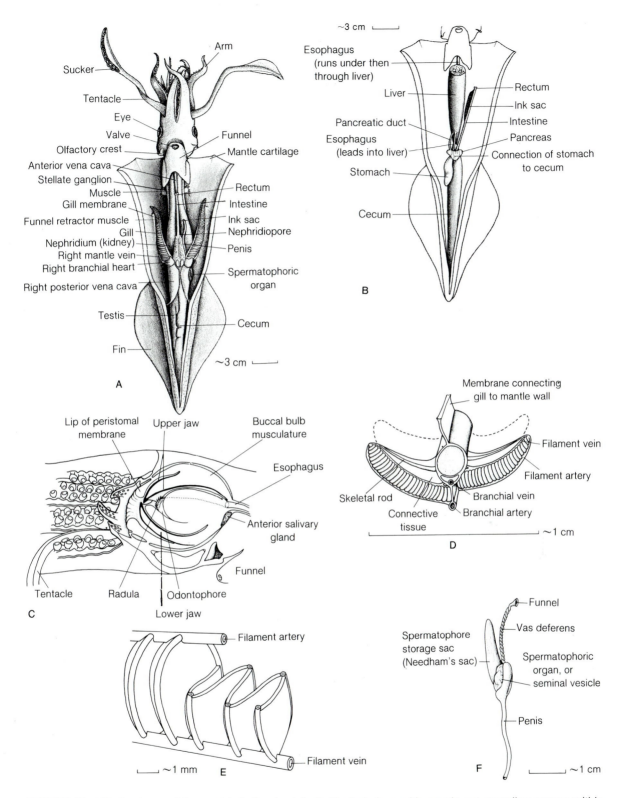

FIGURE 10–102 Anatomy of the squid, *Loligo* (male). **A,** Ventral view, with mantle cut revealing organs within mantle cavity. **B,** Digestive system. The obscuring organs have been removed, the pancreas is drawn as semi-transparent to show the digestive structures on opposite side, and the liver is cut to show the esophagus. **C,** Longitudinal section through the buccal mass, showing the base of the arms on the right side. **D,** Section of the gill axis, showing two filaments on either side; only the outlines of the posterior two are shown. **E,** Enlarged view of one filament showing folding. **F,** Male reproductive system. Testis is shown in **A.** *(By Mary Ann Nelson.)*

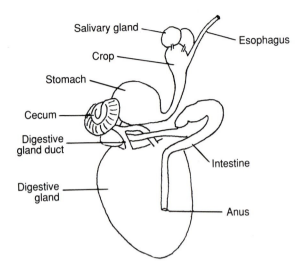

FIGURE 10–103 Digestive tract of *Octopus vulgaris.* (*After Masao.*)

brates and is correlated with the locomotor dexterity and carnivorous habit of these animals. There is great cephalization. All of the typical molluscan ganglia are concentrated and more or less fused to form a brain that encircles the esophagus and is encased in a cartilaginous cranium (Fig. 10–105A).

In the subesophageal regions of the brain, the pedal ganglia supply nerves to the funnel, and anterior divisions of the pedal ganglia, called **brachial ganglia,** send nerves to each of the arms. The innervations by the pedal and brachial ganglia are evidence that the arms and the funnel of cephalopods are homologous to the foot of other molluscs.

A large pair of nerves from the visceral ganglion supplies the mantle. Ordinary swimming and ventilating contractions of the mantle musculature result from impulses conveyed through a system of many small motor neurons radiating from two **stellate ganglia,** one in each side of the mantle wall. The rapid escape movements of swimming cephalopods, such as squids, result from a highly organized system of giant motor fibers that bring about powerful and synchronous contractions of the circular muscles of the mantle.

The command center of the system is a pair of very large, first-order **giant neurons** that lie in the median ventral lobe of the fused visceral ganglia (Fig. 10–105B). These neurons are probably fired by a barrage of impulses from the sense organs. They run to another center within the visceral ganglia, where each makes a connection to a second-order giant neuron, which traverses the mantle nerve to the stellate ganglion. Here connections are made with third-order

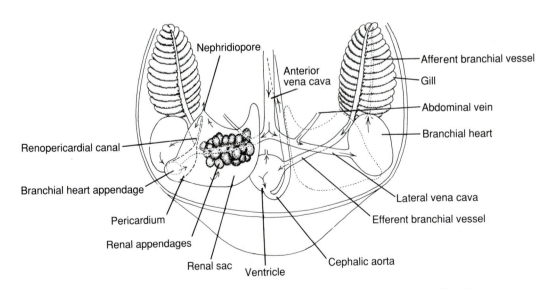

FIGURE 10–104 Excretory and circulatory systems of *Octopus dofleini.* (*After Potts.*)

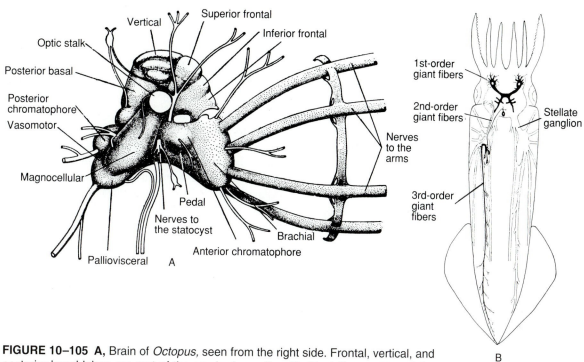

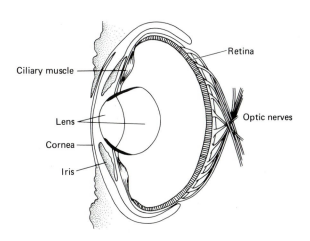

FIGURE 10–105 A, Brain of *Octopus,* seen from the right side. Frontal, vertical, and posterior basal lobes are part of the cerebral ganglion. **B,** Giant fiber system of the squid, *Loligo. (A, After Boycott and Young from Wells, M.J. 1962. Brain and Behaviour in Cephalopods. Stanford University Press, Stanford, CA. p. 93. B, From Young, J.Z. 1936. Cold Spring Harbor Symposia on Quantitative Biology. 4:4.)*

giant neurons supplying the circular muscle fibers. The diameters of the third-order fibers are not uniform. The greater the distance between the stellate ganglion and the muscle terminal, the greater the diameter of the fiber, thus ensuring that all impulses arrive at the muscle fibers simultaneously and produce a powerful, synchronous contraction of the mantle musculature.

The eyes of coleoid cephalopods are highly developed and are strikingly similar in structure to those of fishes (Fig. 10–106). A spherical housing containing cartilaginous plates fits into a sort of orbit, or socket, of cartilages associated with those surrounding the brain. The lens, which is suspended by a ciliary muscle, is a rigid sphere with a fixed focal length, and in front of the lens is an iris diaphragm, which can control the amount of light entering the eye through the slit-shaped pupil. The retina contains closely packed, long, rodlike photoreceptors that are directed toward the source of light. The eye is thus of

FIGURE 10–106 Eye of *Octopus. (After Wells, M.J. 1961. What the octopus makes of it; our world from another point of view. Adv. Sci. Lond. 20:461–471.)*

the direct type, instead of the indirect type common in vertebrates. The photoreceptors are connected to retinal cells that send fibers back to an optic ganglion.

The cephalopod eye undoubtedly forms an image, and the optical connections appear to be especially adapted for analyzing vertical and horizontal projections of objects in the visual field. Experimental studies indicate that *Octopus* can discriminate objects as small as 0.5 cm from a distance of 1 m, which is a considerable advantage in catching prey.

Functioning in an aquatic environment, the cornea of the cephalopod eye contributes little to focusing, because there is almost no light refraction at the corneal surface, as there is in an air/corneal surface (see p. 362). Accommodation takes place by forward and backward movement of the lens, as in fish.

The cephalopod eye can accommodate itself to light changes both by modifications in the pupil's size and by the migration of pigment in the retina. Squids of the family Histioteuthidae, which live in the mesopelagic zone, have dimorphic eyes. The large left eye is directed upward and responds to faint light from the surface, and the little right eye is directed forward and responds to bioluminescent light.

The eyes of *Nautilus* are large and are carried at the end of short stalks. Although the eye contains a large number of photoreceptors, strangely it lacks a lens and is open to the external sea water through a small aperture, the pupil. Supposedly, the eye functions like a pinhole camera, but without a lens the resolving power and light-gathering capacity must be very limited. The long evolutionary persistence of such an inefficient eye is puzzling.

Statocysts are found in nautiloids but are particularly well developed in coleoids, in which they are large and are embedded in the cartilages located on each side of the brain. They not only provide information about static spatial orientation (i.e., body position in relation to gravitational pull) but are so constructed that, like the semicircular canals of vertebrates, they inform the animal of changing positions in motion, such as turning. Without the statocysts a cephalopod can neither keep the pupil slits of the eyes horizontal nor discriminate between horizontal and vertical surfaces.

Osphradia are present only in *Nautilus*. The arms, and especially the sucker epithelium, are liberally supplied with tactile cells and chemoreceptor cells, particularly in the benthic hunting octopods. Textural and chemical discrimination of a surface by the arms has been demonstrated in *Octopus*, but these animals cannot determine shape with the tentacles, perhaps because the arms contain no divisions or structures that could function as a unit of measurement, as for example the distance between thumb and forefinger in primates or leg joints in arthropods.

Cephalopod behavior is an area of great interest. Learning, memory, tactile and visual discrimination, and localization of motor and sensory functions within the brain have been studied with rewarding results. Among the many investigators of cephalopod behavior, the British zoologists, J. Z. Young, M. J. Wells, J. Wells, and B. B. Boycott, are especially prominent. For a more detailed account of the cephalopod nervous system and cephalopod behavior, refer to the excellent résumés by M. J. Wells (1978, 1979) and J. Z. Young (1972).

Chromatophores, Ink Gland, and Luminescence

The unusual coloration of cephalopods (most other than *Nautilus*) is caused by the presence of **chromatophores** in the integument. The expansion of these cells results from the action of small muscle cells attached to the periphery of the chromatophore. When the muscles contract, the chromatophore is drawn out to form a large, flat plate; when the muscles relax, the pigment is concentrated and less apparent. Particular species possess chromatophores of several colors—yellow, orange, red, blue, and black—and the chromatophores of a particular color may occur in groups or layers. The chromatophore effect is enhanced by deeper layers of **iridocytes,** or reflector cells, which differentially reflect light (Fig. 10–107). The coloration of the skin at any particular time is thus a result of the light passing through the chromatophore filters, the particular chromatophores that are expanded, and the iridocyte filters. The chromatophores are controlled by the nervous system and probably by hormones, with vision being the principal initial stimulus. The degree to which these animals can change color and the stimulus for color change vary considerably. The cuttlefish, *Sepia officinalis,* displays complex color changes and may simulate the background hues of sand, rock, and so forth, but most changes appear to be correlated with behavior. Many species exhibit color change when alarmed. For example, the littoral squid, *Loligo vulgaris,* is generally very pale and only darkens when

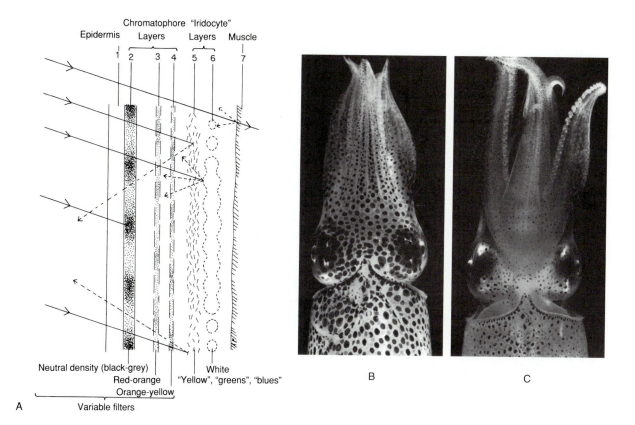

Chromatophore "Iridocyte"
Epidermis Layers Layers Muscle

1 2 3 4 5 6 7

Neutral density (black-grey)
Red-orange
Orange-yellow

White
"Yellow", "greens", "blues"

Variable filters

A

B

C

FIGURE 10–107 A, Diagram of the layers of chromatophores and iridocytes in the skin of *Octopus* and their interaction. Arrows indicate possible absorption and reflection of light coming from the left. See the text for further explanation. **B,** Expanded and contracted chromatophores of the squid, *Lolliguncula brevis. (A, From Packard, A., and Hochberg, F.G. 1977. Skin patterning in Octopus and other genera. In Nixon, M., and Messenger, J.B. (Eds.): The Biology of Cephalopods. Symposia of the Zoological Society of London. 38, Academic Press, London. p. 197.)*

disturbed; others, such as the Caribbean reef squid, *Sepioteuthis sepioidea,* lighten (Fig. 10–108). When alarmed, an octopus may flatten its body and present an elaborate "defensive" color display, including color changes flowing over the body and large dark spots around the eyes. Color displays also are associated with courtship in many cephalopods and are described in the next section.

In cephalopods other than *Nautilus* and some deep-water species, a large **ink sac** opens into the rectum just behind the anus (Fig. 10–102B). The gland secretes a brown or black fluid containing a high concentration of melanin pigment, which is stored in the reservoir. When an animal is alarmed, the ink is released through the anus, and the cloud of inky water

forms a "dummy," confusing the predator. It is also believed that the alkaloid nature of the ink may be objectionable to predators, particularly fish, in which it may anesthetize the chemoreceptive senses.

Many of the mid- and deep-water squids are bioluminescent, with luminescent **photophores** arranged in various patterns over the body, even on the eyeball (Fig. 14–2A). In some species, such as *Sepiola,* the luminescence is due to symbiotic bacteria, but in most it is intrinsic (p. 681).

Reproduction

Cephalopods are dioecious, and the single gonad is located at the posterior of the body (Fig. 10–102A).

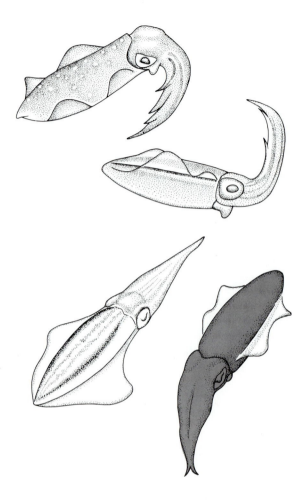

FIGURE 10–108 Four chromatophoric color patterns of the Caribbean reef squid, *Sepioteuthis sepioidea.* Most of these color patterns, as well as the body and tentacle positions, are probably alarm responses. *(From Moynihan, M., and Rodaniche, A.F. 1982. The Behaviour and Natural History of the Caribbean Reef Squid [Sepioteuthis sepioidea]. Paul Parey, Publishers, Hamburg.)*

Fertilization may take place within the mantle cavity or outside, but in either case it involves copulation. One of the arms of the male has become modified as an intromittent organ, called a **hectocotylus.** The degree of modification varies. In *Sepia* and *Loligo* several rows of suckers are smaller and form an adhesion area for the transport of spermatophores (Fig. 10–109). In *Octopus* the tip of the arm carries a spoonlike depression; in *Argonauta* and others there is actually a cavity or chamber where the spermatophores are stored.

Before copulation a male cephalopod performs various displays that serve to identify it to the female. The cuttlefish, *Sepia,* presents a striped color pattern and establishes a temporary bond with a female, swimming above her. The display is also directed toward intruding males, and the weaker male departs. In pelagic cephalopods copulation occurs while the animals are swimming, the male seizing the female head-on (Fig. 10–110). During copulation the hectocotylus receives spermatophores from the male's funnel or plucks a mass of spermatophores from the storage sac. The male hectocotylus is then inserted into the mantle cavity of the female and deposits the spermatophores on the mantle wall near the openings from the oviducts or, in *Octopus,* into the genital duct itself. In some octopods the male inserts his copulatory arm into the mantle cavity of the female without grasping her body.

In *Loligo* and other genera of the same family, the spermatophores may be received by the mantle cavity as just described, or the hectocotylus may be inserted into a horseshoe-shaped seminal receptacle, located in a fold beneath the mouth. The buccal membrane alone receives the spermatophores in *Sepia.*

In the male the highly coiled vas deferens conducts the sperm from the testis anteriorly to a seminal vesicle, which has ciliated grooved walls. Here in the seminal vesicle the sperm are rolled together and encased in a very elaborate **spermatophore.** From the seminal vesicle spermatophores pass into a large storage sac (**Needham's sac**) (Fig. 10–102F), which opens into the left side of the mantle cavity.

In females the oviduct terminates in an oviductal gland. In octopods and some oceanic squids two oviducts are present.

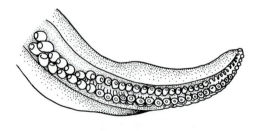

FIGURE 10–109 Hectocotylus of *Loligo roperi. (From Nesis, K.N. 1987. Cephalopods of the World. T.F.H. Publications, Inc., Neptune City, NJ, 351 pp.)*

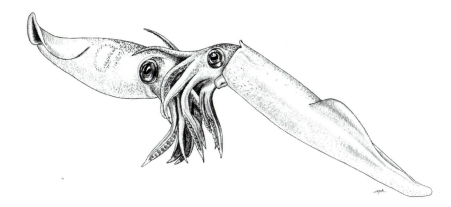

FIGURE 10–110 Copulation in the squid, *Loligo. (Based on a photograph by Robert F. Sisson.)*

The spermatophore is shaped like a baseball bat and consists of an elongated sperm mass, a cement body, a coiled, springlike, ejaculatory organ, and a cap (Fig. 10–111A). With the cap removed as a result of traction in the transfer process or from water uptake, the ejaculatory organ is everted, pulling out the sperm mass. The cement body adheres to the seminal receptacle or mantle wall, and the sperm mass disintegrates.

As the eggs are being discharged from the oviduct, each is enveloped by a paired membrane or capsule in the oviduct gland. Additional protective covering is produced by a pair of nidimental glands in most groups, which are located on the ventral side of the visceral mass and open independently into the mantle cavity. In *Loligo* secretions from the nidimental glands surround the eggs in a gelatinous mass. After leaving the mantle cavity, the egg string is held by the arms, and the eggs may be fertilized by sperm from the seminal receptacle under the mouth. The female then attaches the fertilized eggs to the substratum in a cluster of 10 to 50 elongated strings, each containing as many as 100 eggs (Fig. 10–112). The gelatinous covering of each mass hardens on exposure to sea water, and the individual egg capsules swell to several times the original diameter. Large numbers of *Loligo* come together to copulate and spawn at the same time, and a "community pile" of egg strings may be formed on the bottom. Death of the adult usually follows soon after spawning.

Sepia deposits its eggs singly but attaches them by a stalk to seaweed or other objects. *Octopus* forms egg clusters that resemble a bunch of grapes and are attached within rocky recesses. Female benthic octopods remain to ventilate the eggs after they are deposited. The females die after brooding their eggs. The egg masses of deep-water and some pelagic squids, such as the oegopsids, are commonly free-floating rather than attached.

A remarkable adaptation for egg deposition occurs in the pelagic genus *Argonauta,* commonly known as the paper nautilus. The two dorsal arms of the female secrete a beautiful, calcareous, bivalved case into which the eggs are deposited. The case is carried about and serves as a brood chamber. The posterior of the female usually remains in the case; when disturbed, she withdraws completely into the retreat.

The reproductive system of *Nautilus* is somewhat similar to that of coleoids. Copulation is head-on and spermatophores are transferred by means of a copulatory organ derived from four fused modified tentacles. The spermatophores are attached to the ventral surface of the sheath bearing the tentacles surrounding the buccal region. Some 12 large eggs (25–35 mm) are deposited and attached to the substratum singly within rather elaborate capsules.

Cephalopod eggs are much more heavily yolk-laden than other mollusc eggs, particularly the eggs of *Sepia* and *Ozaena* are very large and may reach 15 mm in diameter.

Reproduction is under hormonal control, although endocrinological studies have been largely restricted to octopods. Hormones are produced by a pair

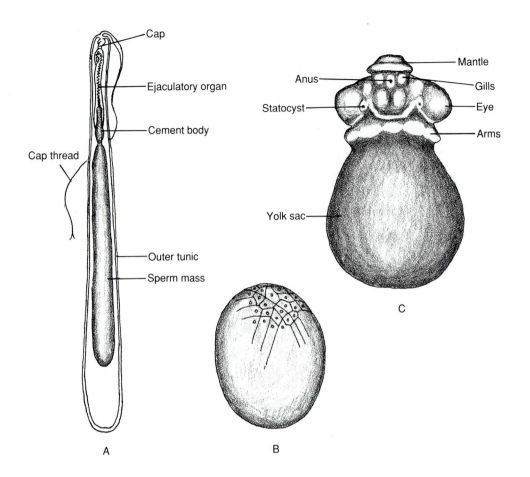

FIGURE 10–111 A, Spermatophore of *Loligo*. **B,** Discoidal meroblastic cleavage in *Loligo*. **C,** Embryo of *Loligo*. *(B, After Watase from Dawydoff. C, After Naef from Dawydoff.)*

of spherical optic glands associated with the optic tracts. The secretions not only regulate the production of eggs and sperm but also, following spawning, cause the female to cease feeding and to brood her eggs. Death follows the reproductive period in both sexes. If the optic glands of brooding females are removed, they stop brooding, resume feeding, and the life span is extended. In contrast to coleoids, *Nautilus* can breed annually for a number of years.

Development

Cleavage, which is meroblastic, results in the formation of a germinal disc, or cap, of cells at the animal pole (Fig. 10–111B). Here the embryo forms (Fig. 10–111C). The margin of the disc grows down and around the yolk mass and forms a yolk sac; the yolk is gradually absorbed during development.

Although development is direct in that there is no trochophore or veliger larva, the little cephalopods may be planktonic for a while following hatching. This is true of octopods that do not take up a benthic existence until they have reached a larger size (Fig. 10–113). Even in some pelagic squids, the juveniles live at higher levels than do the adults. The shell of newly hatched *Nautilus* is about 25 mm in diameter and contains seven septa.

Many cephalopods have short lives. Squids *(Loligo)* live only one to three years, depending on the species, and usually die after a single spawning. *Octopus vulgaris* also dies after one brood, when the animal is two years old. *Nautilus,* however, may have

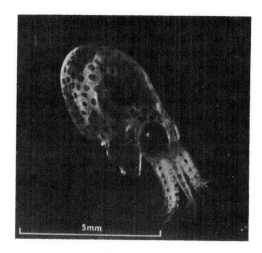

FIGURE 10–113 Newly hatched planktonic young of the octopus, *Eledone cirrosa*. *(From Boletzky, S.V. 1977. Post-hatching behaviour and mode of life in cephalopods. In Nixon, M., and Messenger, J.B. (Eds.): The Biology of Cephalopods. Symp. Zool. Soc. London. 38, Academic Press, London, p. 562.)*

FIGURE 10–112 Egg cluster of the Caribbean reef squid, *Sepioteuthis sepioidea*. One of the young has just hatched, and others can be seen within the eggs. *(Photograph courtesy of Olga F. Linares. In Moynihan, M., and Rodaniche, A.F. 1982. The Behaviour and Natural History of the Caribbean Reef Squid [Sepioteuthis sepioidea]. Paul Parey, Publishers, Hamburg.)*

a life span of 15 or more years, taking as long as 10 years to reach sexual maturity, at which time there is little subsequent growth of the shell.

SUMMARY

1. The primary design of the class Cephalopoda is one adapted for a pelagic, raptorial existence.

2. The gas-filled, chambered shells of *Nautilus, Sepia,* and *Spirula* provide buoyancy. Regulation occurs by changing the volume of fluid within the chambers via the siphuncle.

3. Cephalopods swim by a water jet produced by expulsion of water from the mantle cavity through the funnel. In *Nautilus,* which is a slow swimmer, force is generated by contraction of the funnel and re-

traction of the body. This was probably also true of the many fossil cephalopods.

4. Most living cephalopods (coleoids) generate the water jet by contracting the mantle wall, making possible greater force and speed. The shell is reduced and internal, or even lost altogether, freeing the mantle wall for the water pumping function.

5. Octopods have taken up a secondary crawling, benthic existence and use jet propulsion only for escape and intermittent swimming.

6. Most cephalopods seize prey with a pair of prehensile tentacles and hold it with the eight arms, all provided with suckers. Octopods lack the tentacles, perhaps because of their crawling habits. *Nautilus* possesses about 90 nonsuckered tentacles, only some of which are prehensile.

7. Prey is dispatched with a horny, parrot-like beak and a pair of poison glands (modified salivary glands). The radula functions as a tongue, pulling in pieces of tissue torn off by the beak. The digestive system is adapted for rapid digestion.

8. Many cephalopod features are directly or indirectly related to their active life and attendant higher metabolic rate. Such features parallel those of vertebrates and include

    a. Secondary folded gills

    b. Absence of gill cilia

    c. Blood-vascular system with arteries, veins, and capillaries

    d. Accessory branchial hearts

    e. Presence of hemocyanin

    f. Highly developed eyes

    g. Complex nervous system and behavior

    h. Chromatophores

    i. Ink gland

  9. In cephalopod copulation, one of the male arms transfers spermatophores to the female. Before being released, the eggs are fertilized in the oviduct, in the mantle cavity, or from spermatophores deposited by the male in a receptacle around the mouth of the female. The encased eggs are either deposited on the bottom or shed into the sea water. Development is direct in that there is neither trochophore nor veliger larva, although juveniles ("larvae") of some species are planktonic.

THE ORIGIN OF THE MOLLUSCS

The striking similarity between the development of molluscs and that of the polychaete annelids has long been recognized. Both kinds of animals exhibit spiral cleavage and virtually identical trochophore larvae. In both, the 4d cell (a cleavage blastomere—p. 196) divides to form two mesoteloblast cells, each of which produces a mesodermal band. Within each band the coelom arises by schizocoely (p. 198). All of these similarities have been the principal evidence to support the view that the annelids and molluscs arose from some common ancestral stock.

From the standpoint of adult anatomy, annelids and molluscs are unique in possessing metanephridial systems that also may function as gonoducts.

The discovery of living monoplacophorans contributed fresh views not only to a molluscan/annelidan relationship but also to the nature of the ancestral molluscs. The distinctive feature of monoplacophorans is the linear repetition of structure—eight pairs of pedal retractor muscles, five pairs of gills, and six pairs of nephridia. Chitons also display a repetition of certain features of the nervous system and possess eight pedal retractors and eight shell plates. There are, therefore, good reasons for believing that the ancestral molluscs had a body plan of eight "segments." If so, this provides additional support for a

common ancestry with annelids. In annelids segmentation came to involve the coelom and was exploited as a hydrostatic skeleton in burrowing. The early molluscs, on the other hand, were surface dwellers. Segmentation was more limited and never involved the coelom. As other life styles and body designs evolved, the original molluscan segmentation was lost.

SYSTEMATIC RÉSUMÉ OF THE PHYLUM MOLLUSCA

Class Aplacophora. Wormlike molluscs with no shell and having the foot absent or reduced to a ventral ridge.

  Subclass Chaetodermomorpha (=Caudofoveata). Burrowing footless aplacophorans. A pair of bipectinate gills located in the posterior mantle cavity. *Chaetoderma, Scutopus.*

  Subclass Neomeniomorpha (=Solenogastres). Aplacophorans that live on the surface of cnidarians. Foot reduced to a ventral ridge. Gills absent.

Class Polyplacophora. Molluscs with more or less oval dorsoventrally flattened bodies, having the dorsal surface covered by eight shell plates. Ventral surface occupied by a broad creeping foot. Multiple gills located in the pallial groove lying between the foot and mantle edge (girdle).

  Order Paleoloricata. Paleozoic and mesozoic chitons in which the shell plates lack the middle calcareous layer (articulamentum).

  Order Neoloricata. Fossil and living chitons in which the shell plates possess a middle calcareous layer.

    Suborder Lepidopleurina. Small chitons in which the shell plates lack insertion plates. Only a small number of gills present and located posteriorly. *Lepidopleurus.*

    Suborder Ischnochitonina. Chitons with shell plates having insertion plates with teeth. A relatively large part of shell plate exposed. Most chitons belong to this suborder. *Chiton, Chaetopleura, Mopalia.*

    Suborder Acanthochitonina. Chitons with toothed insertion plates. A relatively small part of the shell plates are exposed or shell plates are completely covered. *Acanthochitona, Cryptochiton* (gumshoe).

Class Monoplacophora. Bilateral molluscs having the body covered by a shieldlike shell. Paired pedal retractors, gills, and nephridia repeated along the length of the body. Eleven species in three genera, all from relatively deep water. *Neopilina, Vema, Monoplacophorus.*

Class Gastropoda.   Cephalized molluscs in which the visceral mass is twisted counterclockwise (torsion) on a flat creeping headfoot. Shell is in one piece and asymmetrically spiraled.

Subclass Prosobranchia.   Marine, freshwater, and terrestrial forms in which the mantle cavity and contained organs are located anteriorly. Aquatic species possess one or two gills within the mantle cavity. Shell and usually an operculum present. Mostly dioecious.

Order Archaeogastropoda (Diotocardia, Aspidobranchia).   Primitive forms in which there are usually two bipectinate gills, two auricles, and two nephridia. The right gill may be reduced or absent, but even if only one gill is present, it is bipectinate. Osphradium is simple (ridgelike).

Superfamily Pleurotomariacea.   Shell with notch, slit, or row of holes; two gills. The deep-water slit snails: *Pleurotomaria, Perotrochus; Scissurella.* The abalones: *Haliotis.*

Superfamily Fissurellacea.   Keyhole limpets. Shell with single apical hole. Two gills. *Emarginula, Diodora, Fissurella.*

Superfamily Patellacea (Docoglossa).   Limpets without a hole or notch in the shell. Single auricle; single bipectinate gill, secondary gills, or gills absent. *Acmaea, Notoacmaea, Collisella, Patelloida, Lottia, Patella, Cellana, Lepeta.*

Superfamily Trochacea.   Conical shells with an operculum; single bipectinate gill. Top snails: *Margarites, Calliostoma, Trochus, Tegula, Monodonta, Gibbula.* Turbans: *Turbo.* Star snails: *Astraea.*

Superfamily Neritacea.   Globose operculate snails. The presence of a single nephridium and a complex reproductive system excludes these snails from the archaeogastropods in some systems; however, they possess a single bipectinate gill. The common tropical intertidal *Nerita* and *Neritina;* the freshwater *Theodoxus;* the terrestrial Helicinidae.

Order Mesogastropoda (Taenioglossa, or suborder Taenioglossa within the order Monotocardia or Pectinibranchia).   Single monopectinate gill, one auricle, and one nephridium; osphradium is simple (ridgelike); complex reproductive system, usually with a penis; radula taenioglossate, that is, seven teeth in a transverse row. Chiefly marine but with many freshwater and terrestrial genera.

Superfamilies Cyclophoracea and Vivaparacea.   Terrestrial, gill-less, operculate Cyclophoridae (absent from the Western Hemisphere); freshwater *Viviparus, Campeloma, Pomacea, Pila.*

Superfamily Littorinacea.   The intertidal Littorinidae: *Littorina, Tectarius; Lacuna.* The terrestrial Pomatiasidae: *Pomatias.*

Superfamily Rissoacea.   A large assemblage of small marine, freshwater, and terrestrial snails with conical shells. *Hydrobia, Bulimus, Rissoina, Alvania, Cingula.*

Superfamily Cerithiacea.   Marine and freshwater snails with high shells. Turret snails (Turritellidae): *Turritella.* Worm snails (Vermetidae): *Vermetus, Serpulorbis, Petaloconchus, Dendropoma,* and (Siliquariidae) *Siliquaria;* the scaphopod-like *Caecum;* the freshwater *Goniobasis, Pleurocera;* marine *Cerithium, Bittium; Batillaria.*

Superfamily Epitoniacea.   The pelagic violet snails. *Janthina.*

Superfamily Eulimacea.   Predators, commensals, and parasites of echinoderms. *Eulima, Stilifer,* and the Entoconchidae, endoparasites of sea cucumbers—*Entoconcha, Entocolax,* and *Enteroxenos.*

Superfamily Calyptraeacea.   Protandric snails with mostly caplike or limpet-like shells. *Capulus; Calyptraea.* Slipper snails: *Crepidula.*

Superfamily Strombacea.   Mostly large gastropods having shells with siphonal canals and a flaring lip. Carrier snails *(Xenophora)* Xenophoridea; *Struthiolaria; Aporrhais;* the conchs, *Strombus, Lambis.*

Superfamily Cypraeacea.   Cowries. Spire enclosed within the last whorl of the shell. *Cypraea, Trivia.*

Superfamily Heteropoda (Atlantacea).   Pelagic snails with finlike foot and reduced shell. *Atlanta, Carinaria.*

Superfamily Naticacea.   Moon snails. Burrowing snails with globose shells and a drilling apparatus. *Natica, Polinices.*

Superfamily Tonnacea. Marine snails with heavy and often large shells. Helmet snails: *Cassis, Cassidarius.* Bonnets: *Phalium.* Tritons: *Cymatium.* Tuns: *Tonna.*

Order Neogastropoda (Stenoglossa, or suborder Stenoglossa within the order Monotocardia or Pectinibranchia). Members of this order are similar to the Mesogastropoda in having a single monopectinate gill, one auricle, and one nephridium, and in the complexity of their reproductive system. They differ in having a radula with three teeth to a transverse row (rachiglossate) and a complex osphradium (with bipectinate folds). All marine.

Superfamily Muricacea. Heavy conical sculptured shells with a long siphonal canal. The canal-boring Coralliophilidae. The drills (Muricidae): *Murex, Urosalpinx, Rapana, Eupleura, Purpura, Thais, Nucella.*

Superfamily Buccinacea. A large assemblage of forms having shells with a short siphonal canal. Whelks (Buccinidae): *Buccinum, Neptunea,* and (Melongenidae) *Busycon* (an exception in having a long siphonal canal). Tulip snails (Fasciolariidae): *Fasciolaria.* Mud snails (Nassariidae): *Nassarius, Ilyanassa.*

Superfamily Volutacea. Shell surface is usually smooth, and spire is conical to low. Olives (Olividae): *Oliva* and *Olivella.* Miter snails (Mitridae): *Vexillum, Mitra.* Harp snails (Harpidae): *Harpa.* Volutes (Volutidae): *Voluta, Cymbium.*

Superfamily Conacea (Toxoglossa). Predatory species with poison gland and highly modified radula or no radula. Cone snails (Conidae): *Conus.* Turret snails (Turridae): *Polystira.* The high-spired Terebridae: *Terebra.*

Subclass Opisthobranchia. Have one gill, one auricle, and one nephridium but display detorsion. Reduction and loss of shell and mantle cavity common. Many are secondarily bilaterally symmetrical. Head commonly bears two pairs of cephalic tentacles. Buccal cavity with a pair of jaws. Hermaphroditic. Mostly marine.

Order Cephalaspidea. Bubble snails. Dorsal surface of the head is shieldlike. Shell generally present but reduced, absent in some. *Acteon, Hydatina, Philine, Scaphander, Bulla.*

Order Pyramidellacea. Ectoparasites of bivalve molluscs and polychaetes. Shell and operculum present. Proboscis contains a stylet instead of a radula. *Boonea, Pyramidella, Odostomia, Brachystomia.*

Order Acochlidioidea. Small naked species with no gills and having the visceral mass sharply set off from the rest of the body. Some members of this order found in fresh water. *Acochlidium.*

Order Anaspidea, or Aplysiacea. Sea hares. Large opisthobranchs with more or less bilaterally symmetrical external form. Reduced shell buried in the mantle; gill and mantle cavity present; foot with lateral parapodia. *Aplysia, Bursatella, Akera.*

Order Notaspidea. Shelled or naked opisthobranchs. Gill present. *Pleurobranchus.*

Order Sacoglossa. Shelled and sluglike opisthobranchs with a radula bearing a single row of teeth adapted for suctorial feeding on algae. *Elysia, Alderia;* the bivalved *Berthelinia.*

Order Thecosomata. Shelled pteropods, or sea butterflies. Shelled pelagic species with large parapodia. *Limacina, Cavolina, Spiratella, Clio, Cymbulia, Gleba.*

Order Gymnosomata. Naked pteropods. Pelagic species with no shell or mantle cavity. Parapodial fins present. *Pneumoderma, Cliopsis.*

Order Nudibranchia. Nudibranchs, or sea slugs. Shell and mantle cavity absent, and body secondarily bilaterally symmetrical. Doridaceans, with secondary gills around the anus: *Doris, Chromodoris, Glossodoris, Jorunna, Onchidoris.* Dendronotaceans, with simple to branched cerata: *Doto, Tritonia, Dendronotus.* Arminaceans, with platelike gills beneath mantle edge or cerata: *Armina.* Aeolidaceans, with simple cerata: *Aeolidia, Glaucus, Glaucilla.*

Subclass Pulmonata. One auricle and nephridium. Gills absent. Mantle cavity, on the right side, is converted into a vascularized chamber for gas exchange in air or secondarily in water. Nervous system concentrated and symmetrical.

Shell usually present, but operculum lacking. Hermaphroditic.

Order Systellommatophora. Slugs with anus located at posterior end of the body, instead of laterally as in other pulmonates. The intertidal Onchidiidae, having a posterior pulmonary sac and the tropical Veronicellidae, which have lost the lung.

Order Basommatophora. Pulmonates with one pair of tentacles; eyes located near the tentacle base. Primarily freshwater forms, a few marine. Marine limpets: *Siphonaria, Otina.* Marine *Melampus, Amphibola,* the only operculate pulmonate. Freshwater snails: *Lymnaea, Planorbis, Helisoma, Bulinus, Physa.* Freshwater limpets: *Ancylus, Ferrissia.*

Order Stylommatophora. Pulmonates with two pairs of tentacles, the upper pair bearing eyes at the tip. Terrestrial. *Partula, Achatina, Retinella, Polygyra, Helix.* Land slugs: *Arion, Limax, Deroceras, Phylomycus, Testacella.*

Class Bivalvia. Laterally compressed molluscs in which the entire body is enclosed within two lateral shells (valves) that are hinged dorsally. Foot is bladelike in burrowing species and reduced in attached forms. There is no cephalization, and a radula is absent.

The relationship of the older subclasses Lamellibranchia and Septibranchia to the system below is readily determined. Taxa containing the protobranchs and septibranchs are indicated; all others contain lamellibranchs.

Subclass Protobranchia. Bivalves with bipectinate gills. Most are deposit feeders in soft bottoms and possess a pair of palpal tentacles. Usually isomyarian.

Order Nuculoida. (=Subclass Palaeotaxodonta) Shell valves equal and taxodont (row of short teeth along hinge). *Nucula, Yoldia, Malletia.*

Order Solemyoida (=Subclass Cryptodonta). Awning clams. Valves thin, equal, somewhat elongated, and without hinge teeth. Palpal tentacles lacking; possess chemosynthetic bacteria in gills. *Solemya.*

Subclass Pteriomorphia. Epibenthic bivalves, most attached by byssus threads or cemented to the substratum, but some secondarily free. Unfused mantle margins.

Order Arcoida. Arks. Hinge straight and usually with vertical plications. Iso- or anisomyarian; filibranch gills. *Arca, Barbatia, Anadara, Noetia, Glycymeris.*

Order Mytiloida. Mussels. Bivalves attached by byssal threads. Hinge usually without teeth. Anterior adductor muscle reduced. Includes only the family Mytilidae. *Mytilus, Brachidontes, Modiolus, Geukensia, Lithophaga.*

Order Pterioida. Mostly byssally attached bivalves. Valves more or less equal. Aniso- or monomyarian. Pinnidae (pen shells): *Pinna, Atrina.* Pteriidae (winged oysters): *Pteria, Pinctada; Isognomon; Malleus.*

Order Ostreoida. Bivalves that lie on one side, free or attached. Valves commonly unequal. Aniso- or monomyarian. Pectinidae (scallops): *Pecten, Chlamys, Hinnites, Aequipecten, Argopecten* (Atlantic bay scallop), *Placopecten* (Atlantic deep-sea scallop); *Spondylus* (thorny oysters). Anomiidae (jingles, or Venus' toenails): *Anomia;* Ostreidae (oysters): *Ostrea, Crassostrea, Lopha; Placuna* (window pane clams).

Order Limoida. Free or byssally attached bivalves. Valves equal. Monomyarian. Mantle margin with long tentacles. *Lima* (File clams).

Subclass Palaeoheterodonta. Equivalve with a few hinge teeth, in which the elongated lateral teeth, when present, are not separated from the large cardinal teeth. An inner nacreous layer is present. Usually without siphons.

Order Unionoida. Valves round to elongated. Freshwater bivalves. Unionidae: *Margaritifera, Elliptio, Anodonta, Lampsilis.* The African Mutelidae: *Mutela.* Etheriidae (freshwater oysters).

Order Trigonioida. Valves triangular. A single Australian genus *Trigonia.*

Subclass Heterodonta. Equivalve with a few large cardinal teeth separated by a toothless space from the elongated lateral teeth. Shell without nacreous layer. Siphons usually present; eulamellibranch gills.

Order Veneroida. Usually equivalve and isomyarian. Lucinidae: *Lucina, Codakia, Linga, Myrtea, Divaricella, Thyasira.* Chamidae (jewel boxes). Lasaeidae (commensal bivalves): *Erycina, Lasaea* (free-living); *Kellia* (free-living); *Lepton; Montacuta* (commensal); *Entovalva* (commensal and parasitic). Cardiidae (cockles): *Cardium, Trachy-cardium, Dinocardium, Laevicardium,* Tridacnidae (giant clams); *Mactra, Spisula.* Solenidae (razor clams): *Solen, Ensis, Siliqua.* The deposit feeders: *Tellina, Macoma, Scrobicularia; Donax* (coquina clams); *Abra.* Solecurtidae (razor clams): *Solecurtus, Tagelus;* the freshwater Sphaeriidae; the freshwater and estuarine Dreissenidae: *Dreissena* (zebra mussel). Veneridae (Venus clams): *Venus, Mercenaria, Chione, Callista, Macrocallista, Dosinia, Gemma Petricoliduc: Petricola* (rock borer); Corbiculidae: *Corbicula* (Asiatic clam).

Order Myoida. Thin-shelled burrowers with well-developed siphons. One or no cardinal teeth. Shell without nacreous layer. *Mya* (soft-shell clam); *Corbula;* the borers *Hiatella, Gastrochaena; Panopea* (the geoducks). The boring Pholadidae: *Barnea, Pholas, Zirphaea, Martesia* (in wood), *Xylophaga* (in wood). Teredinidae (wood-boring shipworms): *Teredo, Bankia.*

Order Hippuritoida. Valves unequal and one valve cemented to substratum. Chamidae (jewel boxes).

Subclass Anomalodesmata. Equivalve with no hinged teeth. Isomyarian. Mantle margins fused and siphonate. Mostly filter-feeding burrowers but also includes members of former subclass Septibranchia. *Lyonsia; Pandora; Clavagella* (watering-pot clams). The septibranch superfamily Poromyacea: *Poromya, Cuspidaria.*

Class Scaphopoda. Burrowing molluscs having a tubular, tusklike shell, open at each end. The large anterior opening contains the conelike foot and buccal region. Adhesive knobbed tentacles (captacula) surround the mouth and are used in feeding. Circulating water enters and leaves through the smaller posterior opening of the shell. Gills absent. *Dentalium, Cadulus.*

Class Cephalopoda. Raptorial pelagic molluscs having the foot divided into muscular arms located around the buccal area. Water pumped through the mantle cavity and out a ventral funnel provides the force for swimming. Shell absent or reduced and covered by the mantle in most species. Where the shell is complete, it is partitioned into gas-filled chambers; where external, the living animal occupies only the chamber opening through the shell aperture.

Subclass Nautiloidea. Possess external shells, which may be coiled or straight; sutures are not complex. Living species possess many slender, suckerless tentacles. Two pairs of gills and two pairs of nephridia are present. The class has been in existence since the Cambrian period, but all members are extinct except *Nautilus.*

Subclass Ammonoidea. Fossil forms with coiled external shells having complex septa and sutures. Silurian to Cretaceous. Includes *Ceratites, Scaphites,* and *Pachydiscus.*

Subclass Coleoidea. Shells are internal, reduced, or absent. The eight or ten appendages bear suckers. One pair of gills and one pair of nephridia. Mississippian to the present.

Order Belemnoidea. Extinct species. Shell internal, chambered but with a posterior solid rostrum and a dorsal, shieldlike extension. *Belemnites, Belemnoteuthis.*

Order Sepioidea. Cuttlefish and sepiolas. Eight arms and two tentacles. Shell with septa, or shell greatly reduced or lost. Body mostly short and broad or saclike. *Spirula, Sepia, Idiosepius, Sepiola, Rossia.*

Order Teuthoidea. Squids. Shell or pen a flattened blade or vane. Body mostly elongated with eight arms and two long tentacles.

Suborder Myopsida. Squids with a transparent corneal membrane over the eye. *Loligo, Lolliguncula, Sepioteuthis.*

Suborder Oegopsida. Squids without a transparent corneal membrane but with eyelids. Pupil circular. Contains most squids, many of which live in deep water. *Architeuthis, Abralia, Abraliopsis, Gonatus, Onychoteuthis, Ctenopteryx, Histioteuthis, Bathyteuthis, Illex, Ommastrephes, Chiroteuthis, Cranchia.*

Order Vampyromorpha. Vampire squids. Small, deep-water, octopod-like forms with eight arms united by a web, but two small filaments also present. *Vampyroteuthis.*

Order Octopoda. The octopods. Possess eight arms; body globular.

Suborder Cirrata. Finned octopods. Mantle with a pair of fins. Arms, with finger-like cirri, connected by a broad web. Mostly deep-sea species. *Cirrothauma, Opisthoteuthis.*

Suborder Incirrata. Octopods without fins. *Octopus, Ozaena, Eledonella, Vitreledonella, Amphitretus, Argonauta.*

REFERENCES

The literature included here is restricted to molluscs. The introductory references on page 6 include many general works and field guides that contain sections on molluscs.

More than 15 journals are devoted to studies on molluscs: *American Malacological Bulletin; Archiv für Molluskenkunde* (German); *Basteria* (Dutch); *Bolletino Malacologico* (Italian); *Bulletin of Malacology of the Republic of China; Haliotis* (French); *Journal of Conchology* (British); *Journal of the Malacological Society of Australia; Journal of Molluscan Studies* (British); *Lavori della Societa Malacologica Italiana* (Italian); *Malacologica* (U.S.); *Malacological Review* (U.S.); *Malakologische Abhändlungen* (German); *Nautilus* (U.S.); *Veliger* (U.S.); *Venus* (Japanese).

References for Identification

Abbott, R. T. 1974. American Seashells. 2nd Edition. Van Nostrand Reinhold Co., New York 663 pp. (A complete and excellent guide to the marine molluscs of the Atlantic and Pacific coasts of North America.)

Andrews, J. 1971. Sea Shells of the Texas Coast. University of Texas Press, Austin. 298 pp. (An excellent illustrated systematic account of the molluscs of the Texas coast; useful for much of the Gulf Coast and West Indies.)

Behrens, D. W. 1980. Pacific Coast Nudibranchs, a Guide to the Opisthobranchs of the Northeastern Pacific. Western Marine Enterprises, Ventura, CA. 112 pp.

Bertsch, H., and Johnson, S. 1981. Hawaiian Nudibranchs. Oriental Publishing Co., Honolulu. 112 pp.

Bouchet, P. 1979. Seashells of Western Europe. American Malacologists, Inc., Melbourne, FL. 156 pp.

Burch, J. B. 1962. How to Know the Eastern Land Snails. W. C. Brown Co., Dubuque, IA. 214 pp.

Burch, J. B. 1973. Freshwater Unionacean Clams of North America. Biota of Freshwater Ecosystems. Identification Manual No. 11. 176 pp. U.S. Government Printing Office.

Burch, J. B. 1975. Freshwater Unionacean Clams of North America. Revised Edition. Malacological Publications, Hamburg, MI. 204 pp.

Burch, J. B. 1982. North American freshwater snails: Identification keys, generic synonymy, supplemental notes, glossary, references, index. Walkerana. *4*:217–365.

Burch, J. B., and van Devender, A. S. 1980. Identification of eastern North American land snails: The Prosobranchia, Opisthobranchia and Pulmonata (Actophila). Walkerana. *2*:33–80.

Burch, J. B., and Tottenham, J. L. 1980. North American freshwater snails: Species list, ranges, and illustrations. Walkerana. *3*:81–215.

Cameron, R. A. D., and Redfern, M. 1976. British Land Snails: Synopses of the British Fauna No. 6. Academic Press, London. 64 pp.

Clarke, A. H. 1981. The Freshwater Molluscs of Canada. National Museum of Natural Science, National Museums of Canada. University of Chicago Press, Chicago. 446 pp.

Ellis, A. E. 1978. British Freshwater Bivalve Mollusks: Synopses of the British Fauna No. 11. Academic Press, London. 109 pp.

Graham, A. 1971. British Prosobranch and Other Operculate Gastropod Molluscs. Synopses of the British Fauna No. 2. Academic Press, New York. 112 pp.

Keen, A. M., and McLean, J. H. 1971. Sea Shells of Tropical West America. 2nd Edition. Stanford University Press, Stanford, CA. 1080 pp. (Excellent manual for marine molluscs from San Diego to Peru.)

Lalli, C. M., and Gilmer, R. W. 1989. Pelagic Snails. The Biology of Holoplanktonic Gastropod Molluscs. Stanford University Press, Stanford, CA 259 pp. (A contemporary and beautifully-illustrated account of the lives of planktonic snails.)

Lindner, G. 1978. Field Guide to Seashells of the World. Van Nostrand Reinhold Co., New York. 271 pp. (Guide to the common shelled molluscs of the world. Translation of a 1975 German work.)

Marcus, E., and Marcus, E. 1967. American Opisthobranch Mollusks. Studies in Tropical Oceanography Series: No. 6. University of Miami Press, Coral Gables, FL. 256 pp.

Morris, P. A. 1973. A Field Guide to Shells of the Atlantic and Gulf Coasts and the West Indies. 3rd Ed. Houghton Mifflin Co., Boston. 330 pp.

Pilsbry, H. A. 1939–1946: Land Mollusca of North America. Vol. 1, Pts. 1 and 2; Vol. 2, Pts. 1 and 2. Academy of Natural Sciences, Monograph No. 3, Philadelphia. (Although old, this is a valuable taxonomic work on the North American land snails.)

Roper, C. F. E., Sweeney, M. J., and Nauen, C. E. 1984. FAO species catalogue. Vol. 3. Cephalopods of the world. An annotated and illustrated catalogue of species of interest to fisheries. FAO Fisheries Synopsis. (*125*) 3. 277 pp.

Roper, C. F. E., Young, R. E., and Voss, G. L. 1969. An illustrated key to the families of the order Teuthoidea (Cephalopoda). Smithson. Contrib. Zool. *13:*1–32.

Thompson, T. E., and Brown, G. H. 1976. British Opisthobranch Molluscs. Synopses of the British Fauna No. 8. Academic Press, London. 204 pp.

Willan, R. C., and Colemann, N. 1984. Nudibranchs of Australasia. Sea Australia Productions, Sydney, Australia.

General

Bourne, G. B., Redmond, J. R., and Jorgensen, D. D. 1990. Dynamics of the molluscan circulatory system: Open versus closed. Physiol. Zool. *63*(1):140–166.

Burke, R. D. 1983. The induction of metamorphosis of marine invertebrate larvae: Stimulus and response. Can. J. Zool. *61:*1701–1719.

Chia, f.-S., and Rice, M. E. (Eds.): 1978. Settlement and Metamorphosis of Marine Invertebrate Larvae. Elsevier, New York. 290 pp.

Childress, J. J., Felbeck, H., and Somero, G. N. 1987. Symbiosis in the deep sea. Sci. Am. *256*(5):115–120.

Cronin, T. W. 1986. Photoreception in marine invertebrates. Am. Zool. *26*(2):403–415.

Edwards, J. S. 1988. Life in the allobiosphere. Trends in Ecology and Evolution. *3*(5):111–114.

Feder, H. M. 1972. Escape responses in marine invertebrates. Sci. Am. *227:*93–100.

Ghiselin, M. T. 1988. The origin of molluscs in the light of molecular evidence. *In* Harvey, P. H., and Partridge, L. (Eds.): Oxford Surveys in Evolutionary Biology. Oxford University Press. Vol. 5. pp. 66–95.

Giese, A. C., and Pierce, V. S. (Eds.): 1977 and 1979. Reproduction of Marine Invertebrates. Vol. 4 (1977) (Gastropods and Cephalopods). Vol. 5 (1979) (Chitons and Bivalves). Academic Press, New York.

Graham, A. 1949. The molluscan stomach. Trans. R. Soc. Edinburgh. Pt. 3., *61*(27):737–778. (A comparative treatment of the molluscan stomach; it deals primarily with gastropods and bivalves.)

Grassé, P. P. (Ed.): 1968. Traité de Zoologie. Vol. V (Pt. II) Introduction to the Mollusks, the Chitons, the Monoplacophorans, and the Pelecypods. (Pt. III) Gastropods and Scaphopods. Masson et Cie, Paris.

Grassle, J. F. 1986. The ecology of deep-sea hydrothermal vent communities. Adv. Mar. Biol. *23:*301–362.

Harrison, F. W. (Ed.): 1992. Microscopic Anatomy of Invertebrates. Vols. 5 and 6. Molluscs. Wiley-Liss, New York.

Heller, J. 1990. Longevity in molluscs. Malacologia. *31*(2):259–295.

Hyman, L. H. 1967. The Invertebrates. Vol. 6. Mollusca. I. McGraw-Hill, New York. (This volume, the last of Dr. Hyman's contributions to the Invertebrate Series, covers a part of the phylum Mollusca. Four classes are treated: Aplacophora, Polyplacophora, Monoplacophora, and Gastropoda.)

Land, M. F. 1981. Optics and vision in invertebrates. *In* Autrum, H. (Ed.): Handbook of Sensory Physiology. Vol. III/6B. Springer-Verlag, Berlin. pp. 471–592.

Land, M. F. 1984. Molluscs (Eyes). *In* Alii, M. A. Photoreception and Vision in Invertebrates. Plenum Press, New York. pp. 699–725.

Lutz, R. A., and Rhoads, D. C. 1980. Growth patterns within the molluscan shell, an overview. *In* Rhoads, D. C., and Lutz, R. A. (Eds.): Skeletal Growth of Aquatic Organisms. Plenum Press, New York.

Mangum, C. P. 1985. Oxygen transport in invertebrates. Am. J. Physiol. *248:*505–514.

Moore, R. C. (Ed.): 1957–1971. Treatise on Invertebrate Paleontology. Mollusca. Vol. I–N. Geological Society of America and University of Kansas Press, Lawrence. (In addition to descriptions of fossil species, the introductory sections cover various aspects of the biology of molluscs.)

Morse, A. N. C. 1991. How do planktonic larvae know where to settle? Am. Sci. *79:*154–167.

Morton, J. E. 1967. Molluscs. 4th Edition. Hutchinson University Library, London.

Nakazima, M. 1956. On the structure and function of the midgut gland of Mollusca, with a general consideration of the feeding habits and systematic relations. Jap. J. Zool. 2(4):469–566. (A good comparative account of the digestive diverticula of molluscs.)

Nilsson, D. E. 1989. Vision optics and evolution. BioSci. *39*(5):298–307.

Potts, W. T. W. 1967. Excretion in molluscs. Biol. Rev. *42*(1):1–41.

Price, T. J., Thayer, G. W., Lacroix, M. W. et al. 1974. The organic content of shells and soft tissues of selected estuarine gastropods and pelecypods. Proc. Natl. Shellfish Assoc. *65:*26–31.

Purchon, R. D. 1977. The Biology of the Mollusca. 2nd Edition. Pergamon Press, New York. 596 pp. (Good general account of many aspects of molluscan biology, especially gastropod and bivalve habitation and feeding.)

Runnegar, B., and Pojeta, J. 1974. Molluscan phylogeny: The paleontological viewpoint. Science. *186:*311–317.

Solem, A. 1974. The Shell Makers: Introducing Mollusks. John Wiley and Sons, New York. 289 pp. (This little book is largely concerned with gastropods, especially land snails. A good complement to the volume by Yonge and Thompson.)

Stasek, C. R. 1972. The molluscan framework. *In* Florkin, M., and Scheer, B. T. (Eds.): Chemical Zoology. Vol. 3. Academic Press, New York. pp. 1–44.

Trueman, E. R. 1983. Locomotion in molluscs. *In* Wilbur, K. M. (Ed.): The Mollusca. Vol. 4, Pt. 1. Academic Press, New York. pp. 155–198.

Vagvolgyi, J. 1967. On the origin of mollusks, the coelom, and coelomic segmentation. Syst. Zool. *16:*153–168.

Wilbur, K. M. (Ed.): 1983–1985. The Mollusca. Academic Press, New York. (Ten volumes covering the biochemistry, physiology, development, ecology, and evolution of molluscs.)

Wright, W. G. 1988. Sex change in the Mollusca. Trends in Ecology and Evolution. *3*(6):137–140.

Yochelson, E. L. 1978. An alternative approach to the interpretation of the phylogeny of ancient mollusks. Malacologia. *17*(2):165–191.

Yochelson, E. L. 1979. Early radiation of Mollusca and mollusc-like groups. *In* House, M. R. (Ed.): The Origins of Major Invertebrate Groups. Systematics Association Special Volume No. 12. Academic Press, London. pp. 323–358.

Yonge, C. M., and Thompson, T. E. 1976. Living Marine Molluscs. Collins, London. 288 pages. (A brief but excellent general treatment of marine molluscs, with special emphasis on the British fauna. Solem's book provides good coverage of the land snails.)

Monoplacophorans, Aplacophorans, Polyplacophorans, Scaphopods

Bilyard, G. R. 1974. The feeding habits and ecology of *Dentalium entale stimpsoni.* Veliger. *17*(2):126–138.

Boyle, P. R. 1972. The aesthetes of chitons. I. Role in the light response of whole animals. Mar. Behav. Physiol. *1:*171–184.

Boyle, P. R. 1974. The aesthetes of chitons. II. Fine structure in *Lepidochitona cinereus.* Cell Tiss. Res. *153:*383–398.

Boyle, P. R. 1977. The physiology and behavior of chitons. Oceanog. Mar. Biol., Ann. Rev. *15:*461–509.

Fischer, v.-F. P. 1978. Photoreceptor cells in chiton esthetes. Spixiana. *1*(3):209–213.

Gainey, L. F. 1972. The use of the foot and captacula in the feeding of *Dentalium.* Veliger. *15*(1):29–34.

Langer, P. 1983. Diet analysis for three subtidal coexisting chitons from the northwestern Atlantic. Veliger. *25*(4):370–377.

Mook, D. 1983. Homing in the West Indian chiton *Acanthopleura granulata.* Veliger. *26*(2):101–105.

Poon, P. 1987. The diet and feeding behavior of *Cadulus tolmiei.* Nautilus. *101:*88–91.

Scheltema, A. H. 1978. Position of the class Aplacophora in the phylum Mollusca. Malacologia. *17*(1):99–109.

Shimek, R. L. 1988. The functional morphology of scaphopod captacula. Veliger. *30*(3):213–231.

Smith, S. Y. 1975. Temporal and spatial activity patterns of the intertidal chiton *Nopalia muscosa.* Veliger. *18*(Supp.):57–62.

Trueman, E. R. 1968. The burrowing process of *Dentalium.* J. Zool. London. *154:*19–27.

Warren, A., and Hain, S. 1992. *Laevipilina antarctica* and *Micropilina arntzi,* two new monoplacophorans from the Antarctic. Veliger. *35*(3):165–176. (This paper reviews the distribution of all living species of monoplacophorans.)

Wingstrand, K. G. 1985. On the anatomy and relationships of recent Monoplacophora. Galathea Rep. *16:*7–94.

Gastropods

Branch, G. M. 1981. The biology of limpets: Physical factors, energy flow and ecological interactions. Oceanogr. Mar. Biol., Ann. Rev. *19:*235–380.

Brandley, B. K. 1984. Aspects of the ecology and physiology of *Elysia* cf. *furvacauda.* Bull. Mar. Sci. *34*(2):207–219.

Carefoot, T. H. 1987. *Aplysia:* Its biology and ecology. Oceanogr. Mar. Biol., Ann. Rev. *25:*167–284.

Carriker, M. R. 1981. Shell penetration and feeding by natacean and muricacean predatory gastropods: A synthesis. Malacologia. *20*(2):403–422.

Cather, J. N., and Tompa, A. S. 1972. The podocyst in pulmonate evolution. Malacol. Rev. *5:*1–3.

Chung, D. J. D. 1987. Courtship and dart shooting behavior of the land snail *Helix aspersa.* Veliger. *30*(1):24–39.

Conklin, E. J., and Mariscal, R. N. 1977. Feeding behavior, ceras structure, and nematocyst storage in the aeolid nudibranch, *Spurilla neapolitana.* Bull. Mar. Sci. *27*(4):658–667.

Connor, V. M., and Quinn, J. F. 1984. Stimulation of food species growth by limpet mucus. Science. *225:*843–844.

Cook, A. 1979. Homing in the Gastropoda. Malacologia. *18:*315–318.

Croll, R. P. 1983. Gastropod chemoreception. Biol. Rev. *58:*293–319.

Day, R. M., and Harris, L. G. 1978. Selection and turnover of coelenterate nematocysts in some aeolid nudibranchs. Veliger. *21*(1):104–109.

Denny, M. W. 1981. A quantitative model for the adhesive locomotion of the terrestrial slug *Ariolimax columbianus*. J. Exp. Biol. *91:*195–218.

Fretter, V., and Graham, A. 1962. British Prosobranch Molluscs. Ray Society, London.

Fretter, V., and Peake, J. (Eds.): 1975. Pulmonates. Vol. 1. Functional Anatomy and Physiology. Academic Press, London. 417 pp.

Garstang, W. 1928. Origin and evolution of larval forms. Rep. Brit. Assoc. Sec. D. p. 77.

Ghiselin, M. T. 1966. The adaptive significance of gastropod torsion. Evolution. *20:*337–348.

Greenwood, P. G., and Mariscal, R. N. 1984. Immature nematocyst incorporation by the aeolid nudibranch *Spurilla neapolitana*. Mar. Biol. *80:*35–38.

Gurin, S., and Carr, W. E. 1971. Chemoreception in *Nassarius obsoletus:* The role of specific stimulatory proteins. Science. *174:*293–295.

Hamilton, P. V. 1977. Daily movements and visual location of plant stems by *Littorina irrorata*. Mar. Behav. Physiol. *4:*293–304.

Haszprunar, G. 1985. The fine morphology of the osphradial sense organs of the Mollusca. I. Gastropoda. Prosobranchia. Phil. Trans. R. Soc. London B. *307:*457–496.

Hickman, C. S. 1983. Radular patterns, systematics, diversity, and ecology of deep-sea limpets. Veliger. *26*(2):73–92.

Hickmann, C. S. 1988. Archaeogastropod evolution, phylogeny and systematics: A re-evaluation. Malacol. Rev. Suppl. *4:*17–34.

Himmelman, J. H. 1988. Movement of welks (*Buccinum undatum*) towards a baited trap. Mar. Biol. *97:*521–531.

Hinde, R. 1983. Retention of algal chloroplasts by molluscs. *In* Goff, L. J. (Ed.): Algal Symbiosis. Cambridge University Press, Cambridge, p. 97.

Hoagland, K. E. 1978. Protandry and the evolution of environmentally mediated sex change: A study of the Mollusca. Malacologia. *17*(2):365–391.

Hughes, R. N. 1986. A Functional Biology of Marine Gastropods. The John Hopkins Press, Baltimore. 245 pp.

Hughes, R. N., and Lewis, A. H. 1974. On the spatial distribution, feeding and reproduction of the vermetid gastropod *Dendropoma maximum*. J. Zool. (London). *172*(4):531–548.

James, M. J. 1984. Comparative morphology of radular teeth in *Conus:* Observations with scanning electron microscopy. J. Moll. Stud. *46:*116–128.

Jensen, K. R. 1980. A review of sacoglossan diets, with comparative notes on radular and buccal anatomy. Malacol. Rev. *13:*55–77.

Kaelker, H., and Schmekel, L. 1976. Structure and function of the cnidosac of the Aeolidoidea. Zoomorphol. *86*(1):41–60.

Kandel, E. R. 1979. Behavioral Biology of *Aplysia*. W. H. Freeman and Co., San Francisco, 463 pp.

Kennedy, D., Selverton, A. I., and Remler, M. P. 1969. Analysis of restricted neural networks. Science. *164:*1488–1496.

Lindberg, D. R. 1988. The Patellogastropoda. Malacol. Rev. Suppl. *4:*35–63.

Lindberg, D. R., and Dwyer, K. R. 1982. The topography, formation and role of the home depression of *Collisella scabra*. Veliger. *25*(3):229–233.

Linsley, R. M. 1978. Shell form and the evolution of gastropods. Am. Sci. *66*(4):432–441.

Mackenstedt, U., and K. Markel. 1987. Experimental and comparative morphology of radula renewal in pulmonates. Zoomorphology. *107:*209–239.

Miller, S. L. 1974a. Adaptive design of locomotion and foot form in prosobranch gastropods. J. Exp. Mar. Biol. Ecol. *14:*99–156.

Miller, S. L. 1974b. The classification, taxonomic distribution, and evolution of locomotor types among prosobranch gastropods. Proc. Malac. Soc. London. *41:*233–272.

Pender, W. F. 1973. The origin and evolution of the Neogastropoda. Malacologia. *12*(2):295–338.

Pennington, B. J., and Currey, J. D. 1984. A mathematical model for the mechanical properties of scallop shells. J. Zool. *202*(2):239–264.

Pennington, J. T., and Chia, F.-S. 1985. Gastropod torsion: A test of Garstang's hypothesis. Biol. Bull. *169:*391–396.

Ponder, W. F. (Ed.): 1986. Prosobranch Phylogeny. Malacol. Rev. 4 (Suppl.).

Pratt, D. M. 1974. Attraction to prey and stimulus to attack in the predatory gastropod *Urosalpinx cinerea*. Mar. Biol. *27*(1):37–45.

Rollo, C. D., and Wellington, W. G. 1981. Environmental orientation by terrestrial Mollusca with special reference to homing behavior. Can. J. Zool. *59:*225–239.

Rudman, W. B. 1981. The anatomy and biology of alcyonarian-feeding aeolid opisthobranch molluscs and their development of symbiosis with zooxanthellae. Zool. J. Linn. Soc. *72:*219–262.

Runham, N. W., and Hunter, P. J. 1970. Terrestrial Slugs. Hutchinson University Library, London. 184 pp. (A general biology of pulmonate slugs.)

Smith, C. R. 1977. Chemical recognition of prey by the gastropod *Epitonium tinetum*. Veliger. *19*(3):331–340.

Solem, A., and Yochelson, E. L. 1979. North American Paleozoic land snails, with a summary of other Paleozoic non-marine snails. U.S. Geological Survey. Professional Paper 1072.

Spight, T. M. 1975. On a snail's chances of becoming a year old. Oikos. 26(1):9–14.

Steneck, R. S., and Watling, L. 1982. Feeding capabilities and limitations of herbivorous molluscs: A functional approach. Mar. Biol. 68:299–319.

Taylor, J. D., and Norris, N. J. 1988. Relationships of neogastropods. Malacol. Rev. Suppl. 4:167–179.

Taylor, J. D., Morris, N. J., and Taylor, C. N. 1980. Food specialization and the evolution of predatory prosobranch gastropods. Paleontology. 23(2):375–409.

Thiriot-Quievreux, C. 1973. Heteropoda. Oceanogr. Mar. Biol. Ann. Rev. 11:237–261.

Thompson, T. E. 1976. Biology of Opisthobranch Mollusca. Vol. I. The Ray Society, London, 206 pp.

Thompson, T. E., and Brown, G. H. 1984. Biology of Opisthobranch Mollusca. Vol. II. The Ray Society Monographs. No. 156. The Ray Society, London. 229 pp.

Tillier, S. 1989. Comparative morphology, phylogeny and classification of land snails and slugs. Malacologia. 30(1–2):1–303.

Todd, C. D. 1981. The ecology of nudibranch molluscs. Oceanogr. Mar. Biol. Ann. Rev. 19:141–234.

Tompa, A. S. 1980. Studies on the reproductive biology of gastropods: Part III. Calcium provision and the evolution of terrestrial eggs among gastropods. J. Conch. 30:145–154.

Underwood, A. J. 1979. The ecology of intertidal gastropods. Oceangr. Mar. Biol., Ann. Rev. 16:111–210.

Willows, A. O. D. 1971. Giant brain cells in mollusks. Sci. Am. 224:69–76.

Bivalves

Ansell, A. D., and Nair, N. B. 1969. A comparison of bivalve boring mechanisms by mechanical means. Am. Zool. 9:857–868.

Bayne, B. L. 1976. Marine Mussels: Their Ecology and Physiology. Cambridge University Press, New York. 506 pp.

Beninger, P. G., Auffret, M., and Le Pennec, M. 1990. Peribuccal organs of Placopecten magellanicus and Chlamys varia: Structure, ultrastructure and implications for feeding. Mar. Biol. 107(2):215–223.

Beninger, P. G., Le Pennec, M., and Salaun. M. 1988. New observations of the gills of Placopecten magellanicus and implications for nutrition. I. General anatomy and surface microanatomy. Mar. Biol. 98(1):61–70.

Booth, C. E., and Mangum, C. P. 1978. Oxygen uptake and transport in the lamellibranch mollusc Modiolus demissus. Physiol. Zool. 51(1):17–32.

Carpenter, E. J., and Culliney, J. L. 1975. Nitrogen fixation in marine shipworms. Science. 187:551–552.

Cary, S. C., Fisher, C. R., and Felbeck, H. 1988. Mussel growth supported by methane as sole carbon and energy source. Science. 240:78–80.

Childress, J. J., Fisher, C. R., Brooks, J. M. et al. 1986. A methanotrophic marine molluscan bivalve symbiosis: Mussels fueled by gas. Science. 233:1306–1308.

Childress, J. J., Fisher, C. R., Favuzzi, J. A. et al. 1991. Sulfide and carbon dioxide uptake by the hydrothermal vent clam, Calyptogena magnifica, and its chemoautotrophic symbionts. Physiol. Zool. 64(6):1444–1470.

Deaton, L. E., and Mangum, C. P. 1976. The function of hemoglobin in the arcid clam Noetia ponderosa. II. Oxygen uptake and storage. Comp. Biochem. Physiol. 53A:181–186.

Freadman, M. A., and Mangum, C. P. 1976. The function of hemoglobin in the arcid clam Noetia ponderosa. I. Oxygenation in vitro and in vivo. Comp. Biochem. Physiol. 53A:173–179.

Galtsoff, P. S. 1964. The American Oyster. U.S. Fisheries Bull. 64:1–480.

Haderlie, E. C. 1981. Growth rates of Penitella penita, Chaceia ovoidea and other rock boring marine bivalves in Monterey Bay, California, U.S.A. Veliger. 24(2):109–114.

Heard, W. H. 1977. Reproduction of fingernail clams (Sphaeriidae: Sphaerium and Musculium. Malacologica. 16(2):421–456.

Herry, A., Diouris, M., and Le Pennec, M. 1989. Chemoautotrophic symbionts and translocation of fixed carbon from bacteria to host tissues in the littoral bivalve Loripes lucinalis. Mar. Biol. 101:305–312.

Jones, D. S. 1983. Sclerochronology: Reading the record of the molluscan shell. Am. Sci. 71:384–391.

Jorgensen, C. B. 1966. Biology of Suspension Feeding. Pergamon Press, New York.

Jorgensen, C. B. 1974. On gill function in the mussel Mytilus edulis. Ophelia. 13(1/2):187–232.

Kat, P. W. 1984. Parasitism and the Unionacea. Biol. Rev. 59:189–207.

Knudsen, J. 1979. Deep-sea bivalves. In van der Spoel, S. et al. Pathways in Malacology. Bohn, Scheltema and Holkema, Utrecht. pp. 195–224.

Kristensen, J. H. 1972. Structure and function of crystalline styles in bivalves. Ophelia. 10(1):91–108.

Mangum, C. P., Miller, K. I., Scott, J. L. et al. 1987. Bivalve homocyanin: Structural, functional, and phylogenetic relationships. Biol. Bull. 173:205–221.

Mason, J. 1972. Cultivation of the European mussel,

Mytilus edulis. Oceanogr. Mar. Biol. Ann. Rev. *10:*437–460.

Mathers, N. F. 1973. Carbohydrate digestion in *Ostrea edulis.* Proc. Malacol. Soc. London. *40*(5):359–367.

McKee, P. M., and Mackie, G. L. 1981. Life history adaptations of the fingernail clams *Sphaerium occidentale* and *Musculium securis* to ephemeral habits. Can. J. Zool. *59:*2219–2229.

McKee, P. M., and Mackie, G. L. 1983. Respiratory adaptions of the fingernail clams *Sphaerium occidentale* and *Musculium securis* to ephemeral habitats. Can. J. Fish. Aquat. Sci. *40*(6):783–791.

Morton, B. 1970. The tidal rhythm and rhythm of feeding and digestion in *Cardium edule.* J. Mar. Biol. Assoc. U.K. *50:*499–512.

Morton, B. 1978a. The diurnal rhythm and the processes of feeding and digestion in *Tridacna crocea.* J. Zool. (London). *185:*371–387.

Morton, B. 1978b. Feeding and digestion in shipworms. Oceanogr. Mar. Biol. Ann. Rev. *16:*107–144.

Morton, B. 1981. Prey capture in the carnivorous septibranch *Poromya granulata.* Sarsia. *66:*241–256.

Morton, B. 1983. Feeding and digestion in Bivalvia. *In* Wilbur, K. M. (Ed.): The Mollusca. Vol. 5. Pt. 2. Academic Press, New York. pp. 65–147.

Morton, B., and Scott, P. J. B. 1980. Morphological and functional specializations of the shell, musculature and pallial glands in the Lithophaginae. J. Zool. *192:*179–203.

Neves, R. J., and Moyer, S. N. 1988. Evaluation of techniques for age determination of freshwater mussels. Am. Malacol. Bull. *6*(2):179–188.

Newell, N. D. 1969. Classification of Bivalvia. *In* Moore, R. C. (Ed.): Treatise on Invertebrate Paleontology, Vol. N. Pt. 1. Mollusca 6. Bivalvia. pp. N205–N224.

Owen, G. 1974. Feeding and digestion in the Bivalvia. Adv. Comp. Physiol. Biochem. *5:*1–35.

Palmer, R. E. 1979. A histological and histochemical study of digestion in the bivalve *Arctica islandica.* Biol. Bull. *156:*115–129.

Pirie, B. J. S., and George, S. G. 1979. Ultrastructure of the heart and excretory system of *Mytilus edulis.* J. Mar. Biol. Assoc., U.K. *59:*819–829.

Pojeta, J., and Runnegar, B. 1974. *Fordilla troyensis* and the early history of pelecypod molluscs. Am. Sci. *62:*706–711.

Powell, M. A., and Somero, G. N. 1986. Hydrogen sulfide oxidation is coupled to oxidative phosphorylation in mitochondria of *Solemya reidi.* Science. *233:*563–566.

Reed-Miller, C., and Greenberg, M. J. 1982. The ciliary junctions of scallop gills: The effects of cytochalasins and concanavalin. A. Biol. Bull. *163:*225–239.

Reid, R. G. B., and Bernard, F. R. 1980. Gutless bivalves. Science. *208:*609–610.

Reid, R. G. B., and Reid, A. M. 1974. The carnivorous habit of members of the septibranch genus *Cuspidaria.* Sarsia. *56:*47–56.

Richardson, C. A. 1989. An analysis of the microgrowth bands in the shell of the common mussel *Mytilus edulis.* J. Mar. Biol. Assoc. U. K. *69:*477–491.

Roberts, L. 1990. Zebra mussel invasion threatens U.S. waters. Science. *249:*1370–1372.

Rosen, M. D., Stasek, C. R., and Hermans, C. D. 1978. The ultrastructure and evolutionary significance of the cerebral ocelli of *Mytilus edulis,* the bay mussel. Veliger. *21*(1):10–18.

Silvester, N. R., and Sleigh, M. A. 1984. Hydrodynamic aspects of particle capture by *Mytilus.* J. Mar. Biol. Assoc. U.K. *64:*859–879.

Stanley, S. M. 1968. Post-Paleozoic adaptive radiation of infaunal bivalve molluscs—a consequence of mantle fusion and siphon formation. J. Paleontol. *42*(1):214–229.

Stanley, S. M. 1970. Relation of Shell Form to Life Habits of the Bivalvia (Mollusca). Geological Society of America, Mem. 125. 296 pp.

Stanley, S. M. 1972. Functional morphology and evolution of byssally attached bivalve mollusks. J. Paleontol. *46*(2):165–212.

Stanley, S. M. 1975. Why clams have the shape they have: An experimental analysis of burrowing. Paleobiology. *1*(1):48.

Taylor, J. D. 1973. The structural evolution of the bivalve shell. Palaeontology. *16*(3):519–534.

Trench, R. K., Wethey, D. S., and Porter, J. W. 1981. Observations on the symbiosis with zooxanthellae among the Tridacnidae. Biol. Bull. *161:*180–198.

Trueman, E. R. 1966. Bivalve mollusks: Fluid dynamics of burrowing. Science. *152:*523–525.

Turner, R. D. 1973. Wood-boring bivalves, opportunistic species in the deep sea. Science. *180:*1377–1379.

Waite, J. H. 1983. Adhesion in byssally attached bivalves. Biol. Rev. *58:*209–231.

Waterbury, J. B., Calloway, C. B., and Turner, R. D. 1983. A cellulolytic nitrogen-fixing bacterium cultured from the gland of Deshayes in shipworms. Science. *221:*1401–1403.

Widdows, J., and Hawkins, A. J. S. 1989. Partitioning of rate of heat dissipation by *Mytilus edulis* into maintenance, feeding, and growth components. Physiol. Zool. *62*(3):764–784.

Wood, E. M. 1974. Some mechanisms involved in host recognition and attachment of the glochidium larva of *Anodonta cygnea.* J. Zool. *173:*15–30.

Yonge, C. M. 1941. The protobranchiate Mollusca: A functional interpretation of their structure and evolution. Phil. Trans. R. Soc. Lond. B. *230:*79–147.

Yonge, C. M. 1977. Form and evolution in the Anomiacea—*Pododesmus, Anomia, Patro, Enigmonia* (Anomiidae): *Placunanomia, Placuna* (Placunidae Fam. Nov.). Phil. Trans. R. Soc. Lond. B. *276:*453–527.

Yonge, C. M. 1979. Cementation in the bivalves. *In* van de Spoel, S. et al (Eds.): Pathways in Malacology. Bohn, Scheletema and Holkema, Utrecht. pp. 83–106.

Zwarts, L., and Wanink, J. 1989. Siphon size and burying depth in deposit- and suspension-feeding benthic bivalves. Mar. Biol. *100:*227–240.

Cephalopods

Bone, Q., Pulsford, A., and Chubb, A. D. 1981. Squid mantle muscle. J. Mar. Biol. Assoc. U.K. *61:*327–342.

Boyle, P. R. (Ed.): 1983. Cephalopod Life Cycles. Vols. I and II. Species Accounts. Academic Press, London. 475 pp.

Clarke, M. R., Denton, E. J., and Gilpin-Brown, J. B. 1979. On the use of ammonium for buoyancy in squids. J. Mar. Biol. Assoc. U.K. *59:*259–276.

Clarke, M. R., Macleod, N., and Paliza, O. 1976. Cephalopod remains from the stomach of sperm whales caught off Peru and Chile. J. Zool. (London). *180:*477–493.

Cloney, R. A., and Brocco, S. L. 1983. Chromatophore organs, reflector cells, iridocytes and leucophores in cephalopods. Am. Zool. *23:*581–592.

Denton, E. J. 1974. On buoyancy and the lives of modern and fossil cephalopods. Proc. R. Soc. Lond. B. Bio. Sci. *185(1080):*273–299.

Denton, E. J., and Gilpin-Brown, J. B. 1973. Flotation mechanisms in modern and fossil cephalopods. Adv. Mar. Biol. *11:*197–264.

Ghiretti-Magaldi, A., Ghiretti, F., and Salvato, B. 1977. The evolution of haemocyanin. *In* Nixon, M., and Messenger, J. B. (Eds.): The Biology of Cephalopods. Symp. Zool. Soc. London. *38:*513–523. Academic Press, London.

Gilbert, D. L., Adelman, W. J., and Arnold, J. M. (Eds.): 1990. Squid as Experimental Animals. Plenum Press, New York. 516 pp.

Gilpin-Brown, J. B. 1972. Buoyancy mechanisms of cephalopods in relation to pressure. Symp. Soc. Exp. Biol. *26:*251–259. (Good review, with bibliography.)

Gosline, J. M., and DeMont, M. E. 1985. Jet-propelled swimming in squids. Sci. Am. *252(1):*96–103.

Hartwick, B., Tulloch, L., and MacDonald, S. 1981. Feeding and growth of *Octopus dofleini.* Veliger. *24(2):*129–138.

Herring, P. J. 1977. Luminescence in cephalopods and fish. *In* Nixon, M., and Messinger, J. B. (Eds.): The Biology of Cephalopods. Symp. Zool. Soc. Lond., 38. 127–159, Academic Press, London.

Holme, N. A. 1974. The biology of *Loligo forbesi* Steen-

strup in the Plymouth area. J. Mar. Biol. Assoc. U.K. *54:*481–503.

Lehmann, U. 1981. The Ammonites. Cambridge University Press, Cambridge, U. K. 246 pp.

Macy, W. K. 1982. Feeding patterns of the long-finned squid, *Loligo pealei,* in New England waters. Biol. Bull. *162:*28–38.

Mather, J. 1991. Foraging, feeding and prey remains in middens of juvenile *Octopus vulgaris.* J. Zool. *224:*27–39.

Mather, J. A., Resler, S., and Cosgrove, J. 1985. Activity and movement patterns of *Octopus dofleini.* Mar. Behav. Physiol. *11:*301–314.

Moynihan, M. 1985. Communication and Noncommunication by Cephalopods. Indiana University Press, Bloomington. 160 pp.

Moynihan, M., and Rodaniche, A. F. 1982. The Behavior and Natural History of the Caribbean Reef Squid (*Sepioteuthis sepioidea*). Paul Parey, Publishers, Hamburg.

Nixon, M., and Messenger, J. B. (Eds.): 1977. The Biology of Cephalopods. Symp. Zool. Soc. Lond. 38. Academic Press, London. 616 pp. (Papers presented at a symposium on cephalopods.)

Packard, A. 1972. Cephalopods and fish: The limits of convergence. Bio. Rev. *47:*241–307.

Roper, C. F. E., and Boss, K. J. 1982. The giant squid. Sci. Am. *246(4):*96–105.

Roper, C. F. E., Lu, C. C., and Hochberg, F. G. (Eds.): 1983. Proceedings of the workshop on the biology and resource potential of cephalopods, Melbourne, Australia, 1981. Mem. Nat. Mus. Victoria. *44:*311.

Roper, C. F. E., and Young, R. E. 1975. Vertical distribution of pelagic cephalopods. Smithson. Contrib. Zool. *209:*1–51.

Saunders, W. B. 1983. Natural rates of growth and longevity of *Nautilus belauensis.* Paleobiology. *9(3):*280–288.

Saunders, W. B. 1984. The role and status of *Nautilus* in its natural habitat: Evidence from deep-water remote camera photosequences. Paleobiology. *10(4):*469–486.

Ward, P. 1983. The extinction of the Ammonites. Sci. Am. *249(4):*136–147.

Ward, P. 1987. The Natural History of *Nautilus.* Allen and Unwin, Boston, 267 pp.

Ward, P., Greenwald, L., and Greenwald, O. E. 1980. The buoyancy of the chambered Nautilus. Sci. Am. *243:*190–203.

Wells, M. J. 1978. *Octopus:* Physiology and Behaviour of an Advanced Invertebrate. Chapman and Hall, London. 417 pp.

Wells, M. J. 1979. The world of a mollusc; brain and behaviour in *Octopus vulgaris. In* van der Spoel, S. et al:

Pathways in Malacology. Bohn, Scheltema and Holkema, Utrecht. pp. 139–156.

Wodinsky, J. 1977. Hormonal inhibition of feeding and death in *Octopus:* Control by optic gland secretion. Science. *198*:948–951.

Yarnell, J. L. 1969. Aspects of the behaviour of *Octopus cyanea* Gray. Animal Behav. *17*(4):747–754.

Yochelson, E. L., Flower, R. H., and Webers, G. F. 1973. The bearing of the new late Cambrian monoplacophoran genus *Knightoconus* upon the origin of the Cephalopoda. Lethaia. *6*:275–310.

Young, J. Z. 1972. The Anatomy of the Nervous System of *Octopus vulgaris.* Oxford University Press, New York. 690 pp.

Young, R. E. 1975. Function of the dimorphic eyes in the midwater squid *Histioteuthis dofleini.* Pac. Sci. *29*(2):211–218.

Young, R. E., and Roper, C. F. E. 1976. Bioluminescent counter-shading in midwater animals: Evidence from living squid. Science. *191*:1046–1048.

Young, R. E., and Roper, C. F. E. 1977. Intensity regulation of bioluminescence during countershading in living midwater animals. Fishery Bull. *75*(2):239–252.

![11]

ANNELIDS AND POGONOPHORANS

PRINCIPLES AND EMERGING PATTERNS

SIGNIFICANCE OF SEGMENTATION

Segmentation, or **metamerism,** is characteristic of animals whose bodies are divided into a longitudinal series of similar repeated units, or **segments.** Segmentation is widespread among animals, occurring in annelids, kinorhynchs, arthropods, and most chordates, but many zoologists believe that it evolved independently at least three times. Segmented animals typically have a specialized anterior **acron** and posterior **pygidium** (or **telson**) and a varying number of intermediate segments. In examples of near-perfect segmentation, appendages, muscles, nerves, blood vessels, coeloms, and excretory and reproductive systems are faithfully replicated in each segment. The organization of many annelids approaches such ideal segmentation, but in the vertebrates segmentation is usually apparent only in the axial skeleton, muscles, and nerves, although primitively other organs and tissues are also segmented. The initial stages of growth in segmented animals result from the addition of segments, elongating the body. Among invertebrates, the growth region may be localized, like a shoot tip in many plants, to the region immediately in front of the pygidium. Thus, the newest and youngest segment in the body is located posteriorly, and the oldest is situated immediately behind the head.

The serial repetition of segments in segmented animals is reminiscent of the replication of zooids in colonial animals, and segmented animals share some of the adaptive advantages associated with colonial organization (pp. 103, 997). Perhaps the most important of these advantages is that segmentation divides a body into a series of compartments, each of which can be regulated more or less independently of others. As such, the evolution of segmentation, like the evolution of multicellularity or coloniality, provided a **framework for specialization.** In colonies, such specialization is apparent in the polymorphism of zooids, whereas in segmented animals, it results from **regional specialization** of segments. The simplest examples of regional specialization are found in annelids in which the body may be weakly divided into a head, thorax, and abdomen (or trunk) (Fig. 11–1). These same three divisions are more prominent in insects and many crustaceans. In extreme cases, for example in the higher vertebrates, even the segmental arrangement of muscles is obscured by regional specialization.

Regional specialization is the result of three processes. The first is the *restriction* of certain segmental structures to only a few segments. For example, gonads are often restricted to a few specialized genital segments. Other segmental structures may be retained in all segments but may structurally diverge from each other and adopt different functions. Such *divergence* is common among segmental appendages, some of which may be specialized for locomotion, whereas others function, for example, in grasping, chewing, and gas exchange. Specialization also results from the *fusion* of segments. Although fusion may occur anywhere along the length of a segmented animal, it is commonly expressed as anterior fusion of one or more segments with the acron to form a complex head. The head of the common polychaete, *Nereis,* for example, consists of an acron and at least two segments, whereas that of the fruitfly, *Drosophila,* is composed of five segments.

Segmentation is also an important event in the orderly development of form in many animals and has been studied best in insects. As a fruitfly develops from an egg to an adult, a sequence of genes is activated that first defines the anatomical axes—anterior/posterior, dorsal/ventral—of the bilateral embryo, then forms a se-

FIGURE 11–1 Regional specialization in the polychaete annelid, *Americonuphis magna.* The body of this tube-dwelling worm is specialized into a *head* bearing long slender sensory appendages, a *thorax* with anteriorly directed limbs, and an *abdomen* bearing branched gills.

ries of equivalent segments, and finally specifies specializations within each of the segments. The genes that control morphogenesis at each of these levels of development are called maternal-effect genes, segmentation genes, and homeotic genes, respectively.

While considering annelid locomotion, the English zoologist, R. B. Clark, developed another explanation for the significance of segmentation. His idea, which may be called the **burrowing theory,** states that segmentation of the hydraulic coelom enabled worms to burrow more efficiently than their nonsegmented relatives.

The burrowing theory may be elucidated by comparing two coelomate worms that are identical except that one has segmental septa that partition its coelom into a series of isolated compartments, and the coelom of the other is undivided and continuous throughout the body. Both worms burrow using alternating contractions of longitudinal and circular muscles to generate a wave of peristalsis along the body (Fig. 11–2). At any instant, the peristaltic waves of both worms are identical in appearance, but the musculature of the nonsegmented worm is more active than that of the segmented worm and requires more energy to maintain the proper body shape. Along the body of the nonsegmented worm, where circular or longitudinal muscles are contracted maximally (regions of minimum and maximum body diameter), the pressure of the coelomic fluid is at a maximum. Because the coelom is unpartitioned, the elevated fluid pressure is transmitted throughout the coelom and must be antagonized by the action of body wall muscles to prevent aneurisms and other deviations from the proper, peristaltic, wave shape (Fig. 11-2B). Segmented animals, on the other hand, isolate changes in coelomic fluid pressure to individual segments or groups of segments. As a result, body regions between contracted segments do not experience high fluid pressures and need not contract fully, or at all, to maintain the preferred shape of the body (Fig. 11-2C).

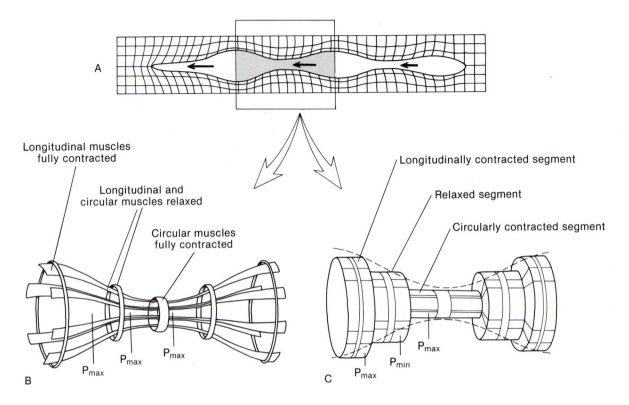

FIGURE 11–2 Peristaltic burrowing (**A**) with (**C**) and without (**B**) segmentation. The peristaltic burrowing wave (**A**) can be sustained in segmented worms (**C**) with less muscular involvement than in nonsegmented worms (**B**) because fluid pressure maxima are localized to a few segments. P_{max}, maximal pressure in hydrostat; P_{min}, minimal pressure in hydrostat. *(Modified and redrawn from Turbeville, J. M. and Ruppert, E. E. 1983. Epidermal muscles and peristaltic burrowing in Carinoma tremaphoros (Nemertini): Correlates of effective burrowing without segmentation. Zoomorphology. 103:103–120.)*

ORIGIN OF COELOMATES AND ACOELOMATES

Annelids, like nearly all large invertebrates, possess a type of body cavity called a coelom. Although the coelom may be reduced or absent in the adult, as in arthropods, the coelom is still present in the embryo. The coelom arises during embryonic development as a cavity within mesoderm, either by the splitting of a mass of mesodermal cells (schizocoely) or by outpocketing of the archenteron in the initial separation of mesoderm from endoderm (enterocoely) (p. 196 and Fig. 5–19). The inner mesodermal cells differentiate to form an epithelium, called mesothelium, that lines the coelom and encloses the coelomic fluid. Typically, the coelom is composed of two or more paired chambers (right and left), depending on the animal design. Annelids, for example, being segmented (metameric), possess numerous pairs separated by transverse septa. The fluid that fills the coelom serves a variety of functions in different coelomate groups. It transports gases and nutritive materials, provides fluid for processing excretory wastes, functions as a hydrostatic skeleton, and as a site for gamete maturation and brooding of embryos. Most or all of these functions are performed by the coelomic fluid of annelids.

The internal transport function is important for most coelomates and may have been a primary factor in the evolution of the coelom. The evolution of the coelom is uncertain. One theory postulates that coelomates evolved from an ancestral acoelomate, such as some flatworm, by a hollowing out of mesenchymal parenchyma, some cells of which would form mesothelium. This schizocoel theory of coelom origin claims as supporting evidence the embryonic development of annelids and molluscs, whose coeloms form in this way. The enterocoel theory, on the other hand, argues that the coelom evolved from the gastric pouches of some cnidarian ancestor, such as schyphozoans or anthozoans, and claims as supporting evidence the enterocoelous mode of mesoderm and coelom formation in the development of echinoderms, hemichordates, and chordates. Depending on the theory, then, the acoelomate condition is either ancestral to or derived from the coelomate condition.

According to the schizocoel theory, the acoelomate body plan is primary and ancestral to the coelomate plan. The acoelomate flatworms are the stem group in the evolution of bilateral animals. The enterocoel theory proposes that all bilateral animals are basically coelomate and that acoelomate forms, such as the flatworms, are secondarily derived from coelomate ancestors by loss of the cavity. Although both theories date from the end of the nineteenth century, the enterocoel theory has never gained much acceptance, at least in part because it is difficult to postulate functional steps that would have led to a change in both design (from coelomate to acoelomate) and symmetry (from bilateral to radial). In contrast, the ancestral position of flatworms among the Bilateria and the primitive nature of the acoelomate body plan has been widely accepted in the literature, including some previous editions of this text. Recent ultrastructural research on flatworms and annelids suggests, however, that the acoelomate condition may indeed be secondary. Studies on the parenchymal cells of turbellarian flatworms clearly indicate that this "filling" is not a uniform, mesenchyme-like tissue and its composition differs among turbellarian groups (p. 217 and Fig. 6–13). This lack of uniformity suggests an independent origin of the acoelomate condition either by invasion of cells or by retention of larval features in the adult. Studies on the mesothelial lining of the coelom also raise questions about the likelihood of a coelom evolving within an acoelomate. Myoepithelium, cells that both line surfaces and contract, was once thought to be limited to the cnidarians, but it is now known to occur in many animals, including polychaetes, where it may contribute to or form much of the coelomic lining. This is not the sort of cell that would be expected to line the coelom if the mesothelium differentiated from the mesenchymal parenchyma of an acoelomate ancestor, but it would be expected if the coelom were derived from gastric pouches of some cnidarian ancestor, where the myoepithelial gut lining (monociliated nutritive muscle cells) became the lining of the coelom.

Still other clues about the origin of acoelomate organization come from some strange little worms that were recently discovered in the marine interstitial fauna by the Austrian zoologist, R. M. Rieger. If flatworms were derived from some annelidan, coelomate ancestor, these worms appear to be somewhat intermediate. Called *Lobatocerebrum* and *Jennaria,* members of these genera are completely ciliated, similar to turbellarians (Fig. 11–3). But like an annelid, *Lobatocerebrum* and *Jennaria* have a cuticle, a pair of ganglionated ventral nerve cords, and segmentally arranged protonephridia that open separately to the exterior. The digestive tract

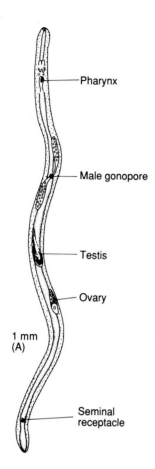

1 mm
(A)

FIGURE 11–3 The evolutionarily enigmatic worm, *Lobatocerebrum*, which shares characteristics with platyhelminths and annelids. *(From Rieger, R. M. 1980. A new group of interstitial worms, Lobatocerebridae, nov. fam. (Annelida) and its significance for metazoan phylogeny. Zoomorphologie. 95:41–84.)*

exits by way of a posterior anus. On the other hand, members of both genera lack setae, septa, and a coelom; they are acoelomate. The parenchyma, however, is composed of baglike, fluid-filled cells with very little intercellular matrix, and in *Jennaria,* the cells form an epithelium. On the whole, both genera display more annelidan than turbellarian features and are therefore tentatively assigned to the Annelida. *Lobatocerebrum* is placed in the class Oligochaeta because it is hermaphroditic and has a glandular region a little like a clitellum. *Jennaria* has not yet been assigned to a class. Besides *Lobatocerebrum* and *Jennaria,* there are some small interstitial polychaetes (*Dinophilus,* protodrilids) that are ciliated and display coelom reduction, the cavity being partially filled by muscle cells or large, baglike cells. Many of the peculiarities of these interstitial worms may be related to the assumption of a life among sand grains, but Rieger believes that just such changes may have been the route by which turbellarians evolved from the coelomate annelids. As our knowledge of comparative ultrastructure and the interstitial fauna grows, evidence may enable us to trace the major pathways that led to the evolution of lower invertebrates—or at least to know which were likely and which were improbable.

CUTICLE EVOLUTION

A **cuticle** may be defined as a nonliving, extracellular secretion of the epidermis that tightly covers the body. Although a cuticle is generally absent in animals, such as cnidarians, gnathostomulids and most flatworms, their bodies are nevertheless covered with a permanent film, called a **glycocalyx,** which is seen between and around the microvilli on the surface of the epidermal cells. A glycocalyx is believed to be present on the exposed surface of epithelia in general and is not regarded as cuticle but may be the framework for its morphogenesis. A relatively simple flexible cuticle, consisting of a mesh of proteinaceous fibers (usually collagen and sometimes polysaccharides), occurs in annelids, nematomorphs, sipunculans, and several other groups of animals. The fibers are often arranged in left- and right-handed spirals around the body and probably evolved to toughen the body wall and to prevent kinks and aneurisms during movements (p. 96). A rigid **exoskeleton** probably evolved from an organic cuticle in one of two ways: by chemically cross-linking the fibers together (sclerotization), as occurs in insects, or by the addition of inorganic minerals (calcification).

Figure 11–4 supports the idea mentioned above of a progressive evolution of a cuticle from a glycocalyx. The possibility exists, however, that the absence of a cu-

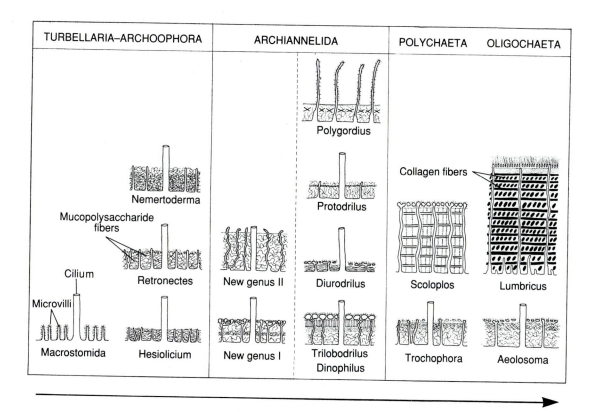

FIGURE 11–4 Evolution of the cuticle in flatworms, nemerteans, and annelids. The cuticle of turbellarians is composed mucopolysaccharide fibers deposited between microvilli. The tips of the microvilli are commonly enlarged as vesicles. Primitive flatworms, like living macrostomids, may have had microvilli but not cuticle. Between the microstomid-like primitive condition and the complex cuticle of some annelids, such as *Scoloplos* and *Lumbricus,* all degrees of cuticle exist as a result of deposition of fibers between microvilli. Rieger believes that the cuticle was originally an adaptation to regulate the absorption of dissolved organic compounds and secondarily assumed a protective function. The cuticle of aschelminths may have had a different origin. *(From Rieger, R. M. and Rieger, G. E. 1976. Fine structure of the archiannelid cuticle and remarks on the evolution of the cuticle within the Spiralia. Acta Zool. (Stockh.) 57:53–68.)*

ticle in certain groups of animals may be secondary rather than primary, particularly if animals rely on epidermal uptake of nutrients to any significant degree (p. 204). Ample evidence for the regressive evolution of a cuticle can be found among some endoparasitic species of predominantly free-living groups of animals. Among the barnacles for example, certain species parasitize crabs, echinoderms, and other invertebrates. Although the parasites have evolved from ancestral barnacles with an exoskeleton, they now nearly lack one and absorb host nutrients through their skin. Some parasitic barnacles clearly have evolved via progenesis, the same phenomenon that may have given rise to some of the small, acoelomate groups of animals (p. 206).

THE ANNELIDS

The phylum Annelida comprises the segmented worms and includes the familiar earthworms and leeches, plus a number of marine and freshwater species of which most people are completely unaware. A shovelful of muddy sand taken from the shore along a coastal sound at low tide usually brings to light a much richer and far more spectacular collection of "worms" than could be found in a backyard garden.

In general, the annelids attain the largest size of any of the wormlike invertebrates. The smallest annelids are interstitial polychaetes in the size range of a few tenths of a mm and the largest are the giant 3-m-long earthworms of Australia. A distinguishing characteristic of the phylum is segmentation (metamerism), the division of the body into similar parts, or segments, which are arranged in a linear series along the anteroposterior axis. The segmented part of the body is always limited to the trunk. The head, or acron, is represented by the **pros-** **tomium** and contains the brain, whereas the pygidium is the terminal part of the body that bears the anus. The prostomium and pygidium are not regarded as segments because they do not arise from the segmental growth zone, but rather from the pretrochal and pygidial regions of the larval body, respectively (Fig. 9–5; p. 342). In all segmented animals there has been a tendency for anterior trunk segments to fuse, in varying degrees, with the unsegmented prostomium or acron. This fusion gives rise to a secondary, compound "head." The formation of new segments in a segmented animal always takes place just in front of the pygidium. The oldest body segments are therefore anterior, and the youngest are posterior.

In annelids the primary segmental structures are the coelomic compartments created by partitioning of the coelom with **transverse septa.** Each septum is composed of two layers of peritoneum, one derived from the segment in front and one from the segment behind and a layer of connective tissue sandwiched in between (Figs. 11–5; 11–6A,B). As an accommoda-

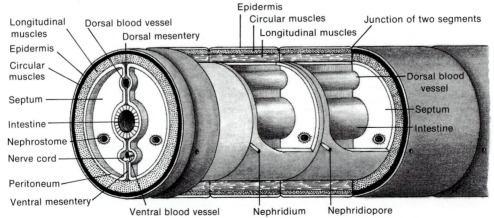

FIGURE 11–5 Annelid segments and anatomy. *(After Kaestner, A. 1967. Invertebrate Zoology, Vol. 1. Interscience Publishers, New York.)*

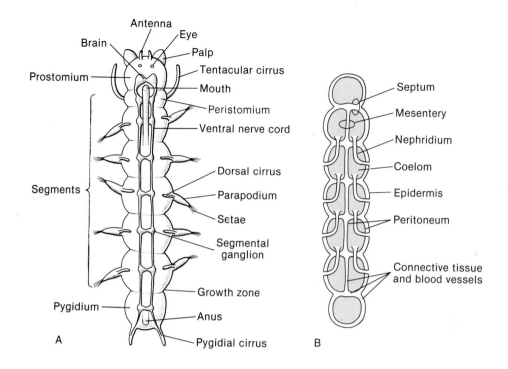

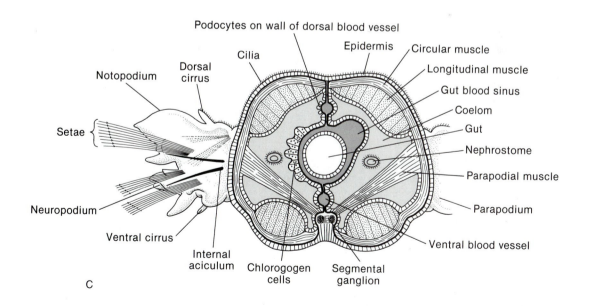

FIGURE 11–6 Polychaete organization. **A, B,** Dorsal views. **C,** Cross section of trunk *(Based primarily on Nereis).*

tion to serve the primary segmentation of the coelom, the lateral nerves to the body wall musculature, blood vessels, and excretory organs are also segmentally arranged.

Coelomic fluid functions as a hydraulic skeleton against which the muscles act to change the body shape. Contraction of the longitudinal muscles causes the coelomic fluid to exert a laterally directed force, and the body widens. Contraction of the circular muscles causes the coelomic fluid to exert an anteriorly and posteriorly directed force, and the body elongates (Chapter 5, Fig. 5–4). Powerful force can be generated, enabling the animal to thrust the anterior end through the substratum to move rapidly through a previously excavated burrow. Chitinous, paired, lateral bristles, or **setae** (chaetae), on each segment increase traction with the substratum. The functional significance of the compartmentation of the coelom by transverse septa is discussed on page 501 (Fig. 11–2).

The typical annelid body wall consists of a fibrous collagenous cuticle, a glandular **epidermis** in which the nerve fibers are situated, and a connective tissue **dermis** (cutis) of varying thickness. The cuticular fibers, which are usually arranged in a crossed helical pattern (Fig. 11–7E), toughen the body wall, resist bulges, and often impart an iridescent sheen to the body. A few of the tube-dwelling polychaetes lack a cuticle, but their secreted tubes resemble the cuticle in structure and composition (Fig. 11–7F).

Annelids possess a more or less **straight digestive tract** running from the anterior mouth to the posterior anus. The gut is suspended within the coelom by longitudinal mesenteries and by the septa, through which the gut penetrates (Fig. 11–5). Digestion is extracellular. Excretion takes place by means of **filtration nephridia** (metanephridial systems and protonephridia), which characteristically occur as a single pair per segment. There is usually a well-developed **blood-vascular system,** in which the blood is usually confined to small vessels, but larger sinuses may also occur. The nervous system consists of an anterior, dorsal ganglionic mass, or brain; a pair of anterior connectives surrounding the gut; and a long double or single ventral nerve cord with ganglionic swellings and lateral nerves in each segment.

The phylum contains over 12,000 described species, which are placed into three classes: Polychaeta, Oligochaeta, and Hirudinea. The ancestral annelids were probably marine animals burrowing in the bottom sand and mud of shallow coastal waters. The class Polychaeta contains most of the living marine species. The class Oligochaeta, which includes the freshwater annelids and the terrestrial earthworms as well as many marine species, may have stemmed from some early polychaetes, but more likely the oligochaetes evolved independently from the ancestral annelids. The class Hirudinea, the leeches, clearly arose from some stock of freshwater oligochaetes.

SUMMARY

1. Members of the phylum Annelida are vermiform (wormlike) segmented animals. Segmentation, or metamerism, which is a distinguishing feature, may have evolved as an adaptation for peristaltic burrowing in soft substrata.

2. The primary segmental structures are the coelomic compartments, which locally isolate pressure changes and, as a result, minimize muscle involvement during peristaltic burrowing.

3. Segmental setae increase traction with the substratum.

4. A fibrous cuticle covers and protects the body.

5. The nervous, circulatory, and excretory systems are segmented, for as maintenance systems, they accommodate the primary segmentation of the coelomic compartments.

6. The gut is typically a straight tube extending through the body between the anterior mouth and posterior anus.

CLASS POLYCHAETA

Polychaete worms are very common marine animals, but their secretive habits result in their being overlooked by casual observers. Within the 8000 described species, a great diversity of body forms and lifestyles is reflected in the establishment of 86 families of polychaetes. The majority are less than 10 cm long with a diameter ranging from 2 to 10 mm, but some interstitial forms are less than 1 mm, and body lengths of 70 cm or more are not uncommon; species of *Eunice* and *Nereis* may attain a length of greater than 1 m. Many polychaetes are strikingly beautiful, colored red, pink, or green or a combination of colors.

The generalized polychaete is perfectly segmented, with identical cylindrical body segments, each bearing a pair of lateral, fleshy, paddle-like ap-

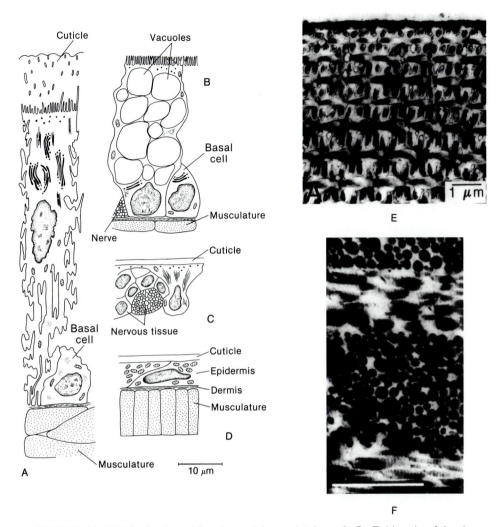

FIGURE 11-7 Polychaete epidermis, cuticle, and tubes. **A–D,** Epidermis of *Lanice conchilega* (**A**), *Spiochaetopterus typicus* (**B**), *Plakosyllis quadrioculata* (**C**), and *Streblospio benedicti* (**D**). **E,** Crossed helical collagen fibers in the cuticle of *Arabella iricolor*. **F,** Orthogonal collagen fibers in the tube of *Chaetopterus variopedatus*. (*A–E, From Storch, V. 1988. Integument. In Westheide, W., and Hermans, C. O. (Eds.): The Ultrastructure of Polychaeta. Mikrofauna Marin. 4:13–36. F, From Gaill, F., and Hunt, F. 1988. Ibid. pp. 61–70.*)

pendages called **parapodia** (Fig. 11–6A). At the anterior end of the worm is a well-developed prostomium, which bears sense organs. The mouth is situated on the ventral side of the body between the prostomium and a postoral region called the **peristomium,** the first true segment. The terminal, unsegmented region, the pygidium, carries the anus. Few polychaetes possess such a typical structure, however. The different lifestyles of the worms of this class

have led to various degrees of modification in the basic plan.

Polychaetes can be errant (free-moving) or sedentary, but the distinction is not always sharp. The errant polychaetes include some species that are strictly pelagic, some that crawl about beneath rocks and shells, some that are active burrowers in sand and mud, but also some that occupy stationary tubes. Many sedentary species construct and live in stabilized burrows, galleries, or tubes of various degrees of complexity. The obligate tubicolous species usually cannot leave the tubes and can only project their heads from the tube openings.

Polychaete Structure

The generalized polychaete body plan is most nearly attained in the errant families. The dorsal, preoral prostomium is well developed and bears numerous sensory structures. The prostomial sense organs usually consist of **eyes, antennae,** and ventrolateral **palps** (Figs. 11–6A, 11–8A). The prostomium projects like a shelf over the mouth. Behind it is the buccal segment, or peristomium, which forms the lateral and ventral margins of the mouth. Often the peristomium bears sensory **tentacular cirri** or two long feeding appendages, called **tentacular palps** (for example, the spionids).

The most distinguishing feature of polychaetes is the presence of parapodia, the paired, lateral ap-

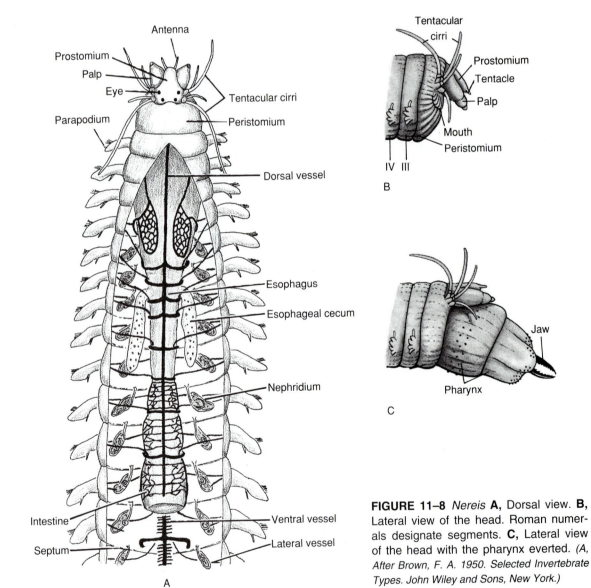

FIGURE 11–8 *Nereis* **A,** Dorsal view. **B,** Lateral view of the head. Roman numerals designate segments. **C,** Lateral view of the head with the pharynx everted. *(A, After Brown, F. A. 1950. Selected Invertebrate Types. John Wiley and Sons, New York.)*

FIGURE 11–9 Dorsal views of the anterior ends of four polychaetes. **A,** The polynoid scaleworm, *Lepidasthenia varia,* which occupies the burrow of its host, the giant capitellid polychaete, *Notomastus lobatus.* **B,** The green phyllodocid, *Phyllodoce fragilis,* a common inhabitant of oyster reefs and other fouling communities. **C,** The arabellid "opalworm," *Arabella iricolor,* a burrower in fine sands. **D,** The eunicid, *Marphysa sanguinea,* an inhabitant of muddy oyster reef communities in the southeastern United States and also found in holes in calcareous rock.

pendages extending from the body segments. A typical parapodium is a fleshy projection extending from the body wall and is more or less laterally compressed (Fig. 11–6C). The parapodium is basically biramous, consisting of an upper division, the **notopodium,** and a ventral division, the **neuropodium.** Each division is supported internally by one or more chitinous rods, each called an **aciculum.** A tentacle-like process (**cirrus**) projects from the dorsal base of the notopodium and from the ventral base of the neuropodium. The notopodia and neuropodia assume various shapes in different families and may be subdivided into several lobes or even greatly reduced (Figs. 11–9; 11–10A, B; 11–20B).

Several families of errant polychaetes, collectively called scaleworms, have modified some of their dorsal cirri into peculiar platelike scales, or **elytra,** which are carried by short stalks on the dorsal side of the body (Figs. 11–9A; 11–15B). A few scaleworms burrow (sigalionids), others are secretive and occupy tight crevices beneath stones (polynoids), a few

(polyodontids) live in secreted tubes, but most are commensals (polynoids) in the burrows and tubes of other invertebrates or the shells of hermit crabs. Because most live in tight quarters, the scales may provide a protective channel for the ventilating current. The elytra also bear a variety of sensory structures. In the scale worm, *Aphrodita* (called a sea mouse), the entire dorsal surface, including the elytra, is covered by hairlike "felt," composed of setae that arise from the notopodia and trail back over the dorsal surface of the animal (Fig. 11–16A).

The parapodial lobes (rami) contain pockets, or **setal sacs,** from which many chitinous bristles, or setae, project. Each simple seta is secreted by a single cell at the base of the setal sac and usually projects a considerable distance beyond the end of the parapodial lobe. New setae are continually produced by the setal sac as older setae are lost (Fig. 11–11). They assume a great variety of shapes, and the setal bundles of a particular species may be composed of more than one type of seta (Fig. 11–10B). Most setae are used

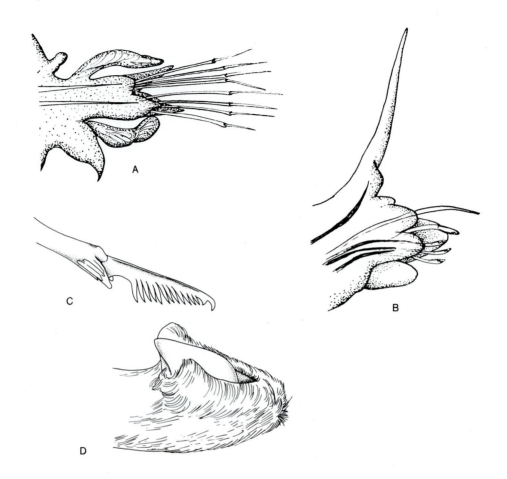

FIGURE 11–10 **A,** Parapodium and setae of *Glycera dibranchiata,* a burrowing poly-chaete. **B,** Parapodium and setae of *Diopatra cuprea,* a tube dweller. **C,** Jointed seta from *Typosyllis pulchra,* a crawler in fouling communities. **D,** A hooded, hooked seta from the spionid, *Tripolydora spinosa,* a sedentary worm. *(A,B, From Renaud, J.C. 1956. A report on some polychaetous annelids from the Miami-Bimini area. Am. Mus. Novit. 1812:1–40. C, Based on an SEM by A. E. Heacox. D, Based on an SEM by Blake, J. A., and Woodwick, K. H. 1981. The morphology of Tripolydora spinosa: An application of scanning electron microscopy to polychaete systematics. Proc. Biol. Soc. Wash. 94:352–364.)*

for locomotion, but some spatulate setae (paleae) are used for digging (Fig. 11–25G), and others, which re-semble Velcro hooks, are used to grip the inner walls of tubes and burrows (Figs. 11–14B; 11–77B)

The members of the largely tropical family Am-phinomidae, which live in coral and beneath stones, have brittle, tubular, calcareous setae containing poi-son (Fig. 11–13B). The setae are used for defense. The amphinomids are commonly known as fireworms because of the pain produced by the poison liberated when the setae break after easily penetrating the skin. These worms are avoided by fish.

The body segments of polychaetes are generally similar, but in some burrowers and tube dwellers there has been a tendency for the trunk to become re-gionally specialized into distinct regions (thorax and abdomen) as a result of variations in the parapodia or the presence or absence of gills (Fig. 11–1). The num-ber of segments ranges from fewer than 10 to over 200, depending on the species.

The polychaete epidermis is composed of a sin-gle layer of cuboidal or columnar epithelium, which is covered by a thin, fibrous, collagenous cuticle (see Fig. 11–4, and p. 504 for the evolution of cuticle).

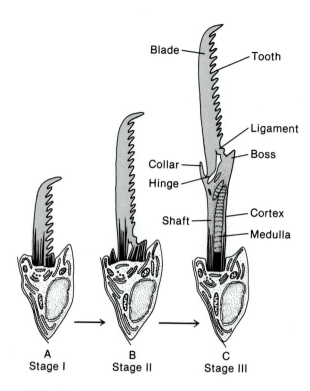

FIGURE 11–11 Morphogenesis of setae in the larva of *Nereis vexillosa*. The setae are gradually secreted on the surface of chaetoblast cells. The shape of the developing seta is directed by specialized microvilli (black in drawings), some of which fabricate the blade and teeth (**A**), and others, arising later in morphogenesis, secrete the shaft and its medulla. *(From O'Clair, R. M., and Cloney, R. A. 1974. Patterns of morphogenesis mediated by dynamic microvilli: Chaetogenesis in Nereis vexillosa. Cell Tissue Res. 151:141–157).*

Mucus-secreting gland cells are a common component of the epithelium. Beneath the epithelium lie, in order, a layer of circular muscle fibers, a much thicker layer of longitudinal muscle fibers, and a thin layer of peritoneum (Fig. 11–7). Although the muscles of the body wall essentially comprise two sheaths, the longitudinal fibers typically are broken up into four bundles—two dorsolateral and two ventrolateral (Fig. 11–6C).

Within the spacious coelom, the gut is suspended by septa and mesenteries (Figs. 11–5; 11–6C). Thus, each coelomic compartment is divided into right and left halves, at least primitively. The septa, however, have partially or completely disappeared in many polychaetes.

Locomotion

Polychaetes have a wide variety of locomotory adaptations depending on their lifestyle and body design. Peristaltic burrowing, as described on p. 337, is common among families of polychaetes having elongated bodies, reduced parapodia and head appendages, and many similar segments. The circular musculature is usually well developed, and the septa are often complete, restricting the coelomic fluid to individual segments. Actively burrowing polychaetes typically resemble earthworms (which also burrow using peristalsis) and are members of families, such as the Arabellidae, Lumbrineridae, Capitellidae, and Orbiniidae (Figs. 11–9C; 11–20B).

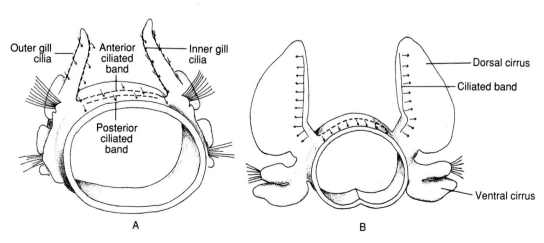

FIGURE 11–12 Surface ciliation in two polychaetes. Arrows indicate direction of water currents. Flows on trunk surface are directed posteriorly. **A,** *Scolelepis squamata.* **B,** *Phyllodoce laminosa. (After Segrove.)*

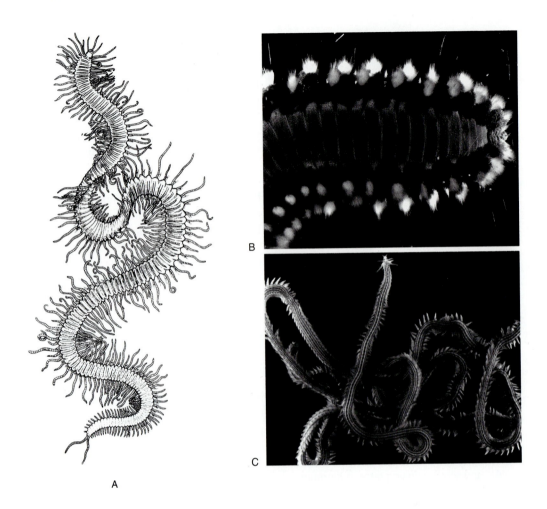

FIGURE 11–13 Three surface-dwelling polychaetes, all with well-developed heads and parapodia. **A,** A syllid, *Trypanosyllis zebra,* with long dorsal cirri. **B,** An amphinomid fireworm, *Hermodice carunculata,* from a coral reef in the Florida Keys. When disturbed, these polychaetes shed some of their calcareous setae, which contain an irritant. **C,** The phyllodocid, *Eulalia myriacyclum,* from beneath stones in southern Florida. *(A, After McIntosh.)*

Many other burrowers do not use peristalsis but have adopted other means of locomotion. The bloodworm, *Glycera,* and lugworm, *Arenicola,* use an eversible pharynx to punch and anchor in the sediment and then, upon retraction, to pull themselves forward (Figs. 11–14A; 11–21). Some members of the family Nephtyidae, called shimmy worms, and the amphioxus-like, *Armandia* and *Ophelia* (Opheliidae), produce rapid lateral undulations of the body and literally swim through loosely consolidated sand (Fig. 11–20A). Very small sediment dwellers, such as

members of the interstitial Dinophilidae and Diurodrilidae, use cilia for locomotion.

A large number of polychaetes, such as nereids, phyllodocids, and hesionids, crawl over surfaces using well-developed parapodia and setae. In these worms, the prostomium is equipped with eyes and other sensory organs, the parapodia are well developed, and the body segments are generally similar (Figs. 11–8A; 11–9B; 11–13C; 11–17). Movement is the result of the combined action of the parapodia, the body wall musculature, and to some extent the

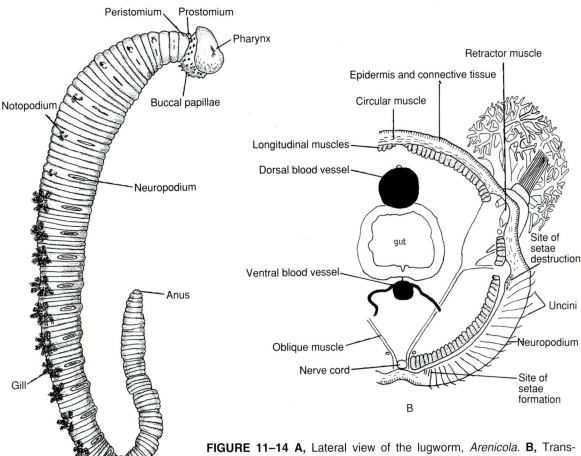

FIGURE 11–14 A, Lateral view of the lugworm, *Arenicola*. **B,** Transverse section of a setigerous segment of *Arenicola marina*. *(A, After Brown, F. A. 1950. Selected Invertebrates Types. John Wiley and Sons, New York. B, Modified from Wells.)*

coelomic fluid. The longitudinal muscle layer is better developed than the circular layer, and the septa tend to be incomplete.

During slow crawling (Fig. 11–17A), the parapodia and setae push against the substratum like legs. Parapodial movement involves an effective backward stroke in which the acicula and setae are extended and the parapodium is in contact with the substratum. After the backward stroke of the parapodium, the acicula and setae are retracted, and the parapodium lifts off the substratum and moves forward. The combined sweeps of the numerous parapodia propel the worm forward.

As soon as one parapodium begins its effective stroke, the preceding parapodium sweeps forward and

starts its effective stroke. The waves of activation thus move forward rather than backward. Activation takes place on both sides of the body, but the waves alternate with each other; when a particular parapodium on one side is executing an effective stroke, the parapodium on the opposite side is executing the recovery stroke (Figs. 11–17A; 11–18A). The overall pattern is similar to that of a crawling millipede.

Rapid crawling in worms, such as nereids or hesionids, involves not only the parapodia but also lateral body undulations, which are produced by waves of contraction in the longitudinal muscles of the body wall (Fig. 11–17B). These waves of contraction move

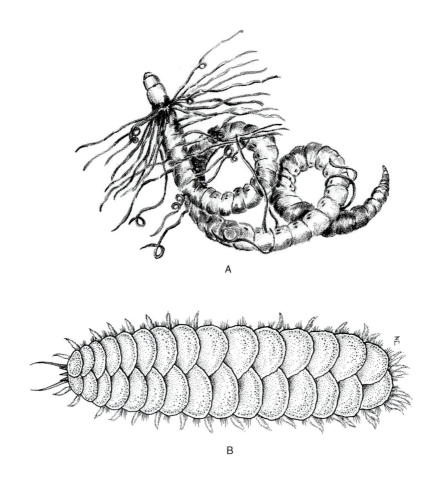

FIGURE 11–15 A, *Cirratulus cirratus,* a polychaete with long, threadlike, dorsal cirri (gills). **B,** A free-living polynoid scaleworm, *Harmothoe aculeata,* that is common among rocks in the eastern United States. The elytra bear many low sensory papillae on their upper surface. *(A, After McIntosh from Fauvel. B, From Ruppert, E. E., and Fox, R. S. 1988. Seashore Animals of the Southeast. University of South Carolina Press, Columbia, SC. 429 pp.)*

forward along the body and coincide with the alternating waves of parapodial activity just described. The backward power thrust of the parapodia occurs when they are at the crest of an undulatory wave and in contact with the substratum. Additional force is transmitted to the parapodia in contact with the substratum by contraction of the longitudinal muscles on the opposite side of the body in the trough of the advancing wave. As these muscles compress the body on the trough side of the wave, they spread it on the opposite side, creating a backthrust against the para-

podia planted into the substratum, thus pushing the animal forward (Figs. 11–18B,C).

Many benthic polychaetes, such as *Nereis,* can swim using the movements employed during rapid crawling. The same movements are used for swimming in six families of polychaetes that are exclusively planktonic or pelagic. They resemble crawling polychaetes but tend to be transparent, as are many other planktonic animals. The Alciopidae, which have enormous eyes (Fig. 11–19B), and the Tomopteridae, which have lost the setae and possess membranous

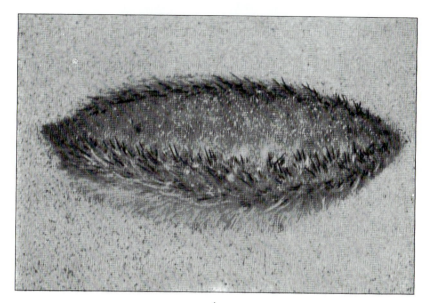

A

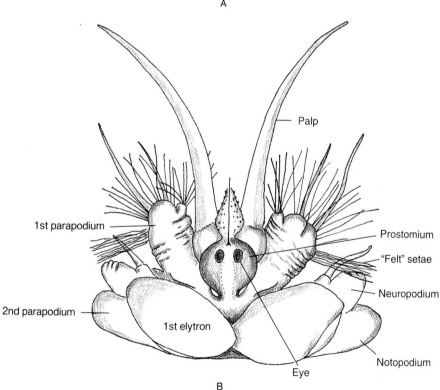

B

FIGURE 11–16 The sea mouse, *Aphrodita aculeata*. **A,** Dorsal view. The dorsal and lateral surfaces are covered by felt setae. **B,** Anterior end, including the first pair of dorsal scales, or elytra. Sea mice occur in Europe and on the northwest and northeast coasts of the United States. They live offshore where they burrow in soft bottom muds, leaving only their posterior ends exposed at the surface. The ventral surface is free of felt and forms a flat, muscular, creeping sole. They ventilate their burrow by elevating the ventral surface, which pumps surface water into the burrow and moves it anteriorly along the sole. The sole is then depressed, forcing the water posteriorly in a channel formed by the dorsum and its covering scales. Movement of the scales also helps create the exhaust flow. *(A, Photograph courtesy of D. P. Wilson. B, After Fordham.)*

A   B

FIGURE 11–17 Slow and fast crawling in polychaetes. **A,** The nereid, *Ceratonereis irritabilis* (Beaufort, NC) showing parapodial movements while crawling slowly. **B,** The phyllodocid, *Eulalia macroceros* (Ft. Pierce, FL) combining parapodial movements with a progressive undulatory wave while crawling rapidly.

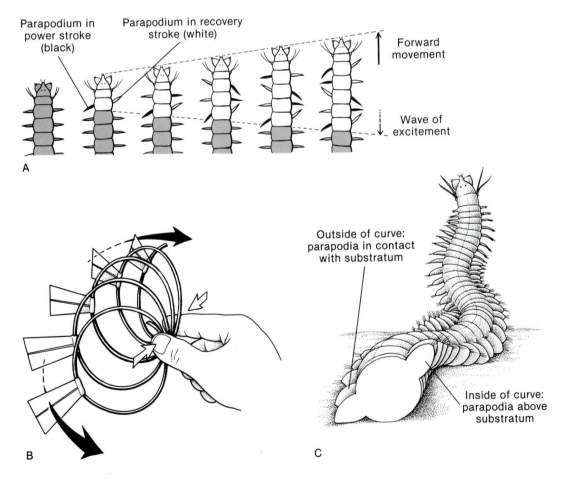

Parapodium in power stroke (black)

Parapodium in recovery stroke (white)

Forward movement

Wave of excitement

A

Outside of curve: parapodia in contact with substratum

Inside of curve: parapodia above substratum

B

C

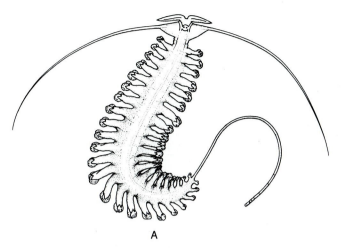

A

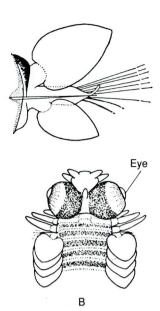

Eye

B

FIGURE 11–19 Pelagic polychaetes. **A,** *Tomopteris renata,* a tomopterid. **B,** Dorsal view of *Rhynchonerella angelina,* an alciopid. **C,** A parapodium of *Rhynchonerella. (From Day, J. 1967. Polychaeta of South Africa. British Museum, London.)*

parapodial pinnules, are the most highly specialized of the pelagic families (Fig. 11–19A).

Burrow- and Tube-Dwelling Polychaetes

Sedentary burrowers occupy more or less fixed, simple vertical or U-shaped burrows excavated in the substratum. These sedentary burrowers include members of such families as the Arenicolidae (the lugworms, Figs. 11–14; 11–22) and the Terebellidae (spaghetti worms, Fig. 11–23). The prostomial sensory appendages are generally absent, but the head region may carry specialized feeding structures (Fig. 11–23). Movement through the burrow is usually by peristaltic contractions. The parapodia are greatly reduced and are in part represented by transverse ridges provided with setae modified into hooks, called **uncini,** which grip the sides of the burrow. In terebellids like *Amphitrite,* which occupies a U-shaped burrow, the hooks are oriented oppositely on the anterior and posterior ends of the animal. As a result, attempts to pull the worm from either head or tail ends of its burrow are opposed by the uncini.

A tube-dwelling habit has evolved in many families of tubicolous polychaetes. The tube may serve the worm as a protective retreat or as a lair for catching passing prey. It may provide access to clean, oxygenated water above a muddy or sandy bottom. It may permit the worm to inhabit hard, bare surfaces such as rock, shell, or coral. Polychaete tubes may be composed of secreted material or sand grains cemented together or a combination of secreted and foreign material. Tube secretions are commonly produced by glands on the ventral surfaces of the segments.

Some errant tubicolous worms, including members of the families Eunicidae and Onuphidae, are commonly carnivorous and extend from the opening

◄ **FIGURE 11–18** Slow and fast crawling in nereid polychaetes. **A,** Slow crawling showing alternation of parapodial power and recovery strokes on opposite sides of the same segment, and the retrograde wave of activation along the body. **B,** Fast crawling superimposes a progressive undulatory wave of the body on the stepping movements of parapodia. Parapodia on the wave crests are undergoing power strokes which are amplified by the contraction of longitudinal muscles in the troughs. As the trough side of the body shortens, it creates a backthrust on the crest side through the active parapodia in contact with the substratum (**C**). *(A, Modified from Clark, R. B. 1964. Dynamics in Metazoan Evolution. Clarendon Press, Oxford. 313 pp.)*

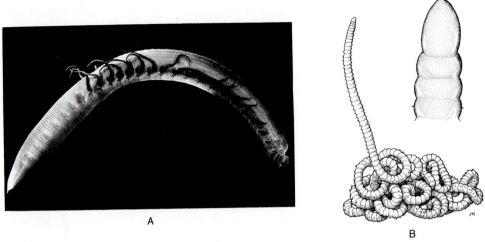

A

B

FIGURE 11–20 Burrowing polychaetes. **A,** Lateral view of *Ophelia denticulata*. Note the small pointed prostomium. The long projections are gills. **B,** The threadworm, *Drilonereis magna*, a burrower, and its conical prostomium. *(A, Photograph courtesy of C. R. Gilmore. B, From Ruppert, E. E., and Fox, R. S. 1988. Seashore Animals of the Southeast. University of South Carolina Press, Columbia, SC. 429 pp.)*

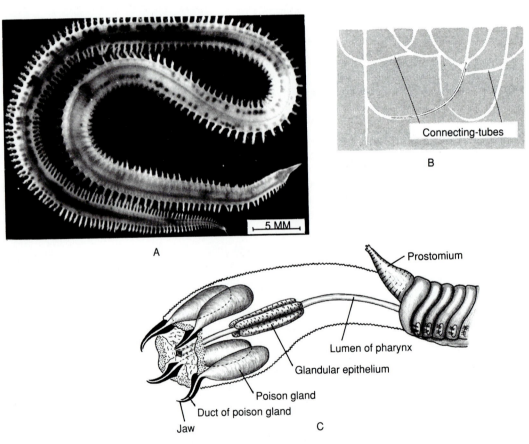

FIGURE 11–21 A, The bloodworm, *Glycera americana,* a common burrowing polychaete found along the east coast of the United States. Note the pointed prostomium. **B,** Burrow system of *Glycera alba,* showing worm lying in wait for prey. *(A, Photograph courtesy of G. M. Moore. B, After Ockelmann and Vahl. C, After Michel.)*

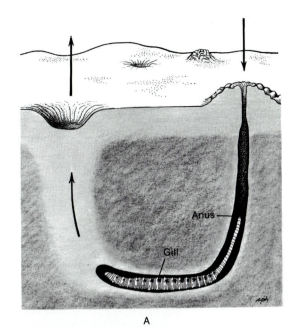

FIGURE 11–22 **A,** The lugworm, *Arenicola,* in its burrow. Arrows indicate the direction of water flow produced by the worm. The worm ingests the column of sand on the left, through which water is filtered. The pile of sand at the burrow opening is defecated castings. **B,** Tracing of activity cycles of *Arenicola* over a period of six hours. The downstroke reflects the worm backing up to the burrow opening to defecate; the sharp upstroke reflects the worm moving back down to the head of the burrow and vigorously resuming ventilation contractions and deposit feeding. Intervals between defecations are about 40 min. *(B, After Wells. 1959. From Newell, R. C. 1970. Biology of Intertidal Animals. American Elsevier Co., New York.)*

A

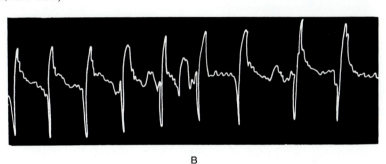

B

of the tube to seize passing prey. They are not greatly different from the crawling polychaetes. Prostomial sensory appendages are well developed, and the parapodia, which are used in crawling through the tube, are not markedly reduced (Figs. 11–1; 11–9D; 11–24A). The onuphids, *Diopatra* and *Onuphis,* build heavy, conspicuous, membranous tubes that may occur in great numbers in intertidal areas. The projecting chimney of the tube is bent over and flares at the end like a ship's ventilator funnel (Fig. 11–24B). The chimneys are covered with bits of shell, seaweed, and other debris that the worm collects and places in position with its jaws. The ornamentation probably provides a cryptic refuge and aids in the detection of possible predators or prey by more readily transmitting disturbances in the surrounding water or from contact with the tube.

The majority of tube-dwelling species are sedentary tubicolous polychaetes that are highly specialized

for a tube-dwelling existence. Like the sedentary burrowers, the prostomial sensory appendages are reduced or absent, and special anterior feeding structures are common. The worms usually move within the tube by peristaltic contractions, and the parapodia are reduced and provided with uncinate setae for gripping the tube wall. Commonly the body shows regional specialization (p. 500).

The sand-grain tubes of some members of the Maldanidae, called bamboo worms, are common in the intertidal zone. These worms, which live anterior end downward in their tubes, have truncate heads and parapodia that are reduced to ridges having the appearance of cane joints (Fig. 11–25C, D).

Both the Sabellariidae and the Pectinariidae construct sand-grain tubes and have highly modified heads bearing heavy, conspicuous setae. In the Sabellariidae, two fused segments have grown forward and dorsally to form an operculum for blocking the tube

(Text continues on page 524)

521

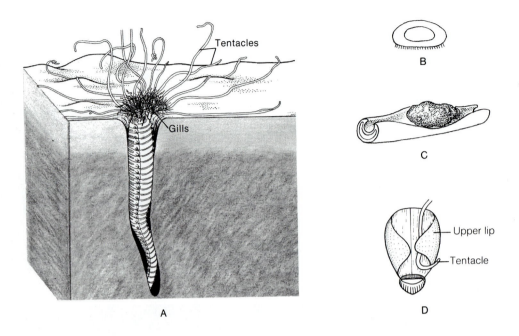

FIGURE 11–23 *Amphitrite* at the aperture of its U-shaped burrow with tentacles outstretched over the substratum. **B,** Cross section through a tentacle of *Terebella lapidaria,* creeping over the substratum. **C,** Section of a tentacle of *Terebella lapidaria* rolled up to form a ciliary gutter, transporting deposit material. **D,** Tentacle being wiped by one of the lips. *(B–D, After Dales.)*

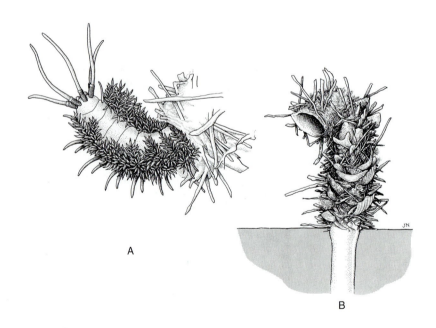

FIGURE 11–24 **A,** *Diopatra cuprea,* a common inhabitant of intertidal mudflats, and its characteristic downturned, shaggy tube (**B**). *(From Ruppert, E. E., and Fox, R. S. 1988. Seashore Animals of the Southeast. University of South Carolina Press, Columbia, SC. 429 pp.).*

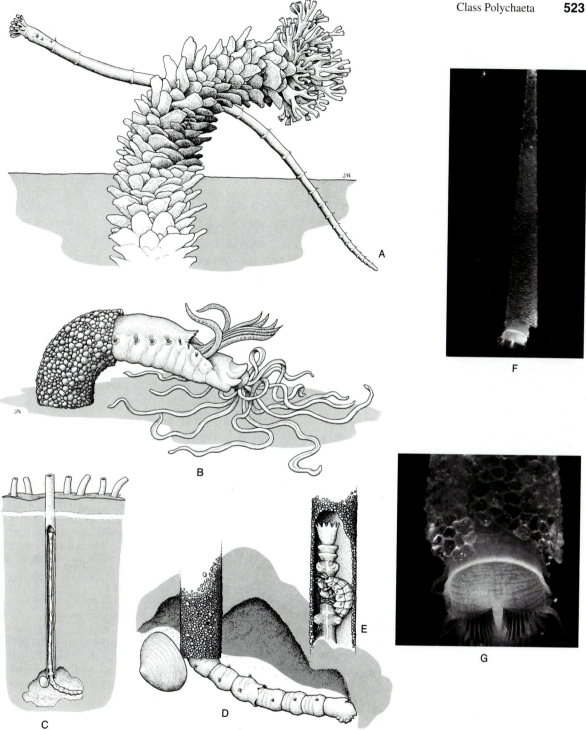

FIGURE 11–25 Tubicolous deposit-feeding polychaetes. **A,** *Owenia fusiformis,* which also suspension feeds, shown in its tube of flat shell fragments (foreground) and removed from its tube. **B,** The ampharetid, *Melinna maculata,* surface deposit feeding using its oral tentacles. **C–G,** Subsurface deposit feeders. **C,** The bamboo worm (Maldanidae), *Clymenella torquata,* and **F,** the pectinariid ice-cream-cone worm, *Pectinaria (Cistenides) gouldii.* Bamboo and ice-cream-cone worms live head-down in their tubes and excavate subsurface caverns. **G,** *Pectinaria* digs with its stout setae (paleae). **D, E,** Two commensals, a tiny clam, *Aligena elevata* (**D**), and an amphipod, *Listriella clymenellae* (**E**), share the protective retreat created by *Clymenella.* (*A–E, From Ruppert, E. E., and Fox, R. S. 1988. Seashore Animals of the Southeast. University of South Carolina Press, Columbia, SC. 429 pp.*)

entrance (Fig. 11–26A, B). Where the water is turbulent, species of *Sabellaria* and related genera build their tubes on top of each other, creating honeycomb-like aggregations (Fig. 11–26C, D). Such colonies may be composed of millions of individuals and assume reeflike proportions.

The sand-grain tube of the pectinariids, or ice-cream-cone worms, is conical, with the smaller end opening at the surface (Fig. 11–25F, G). The head of this worm bears rows of large, conspicuous setae that are used in digging in soft sand or mud.

The tube of *Owenia* is composed of an inner, membranous, secreted lining and an outer layer of sand grains. The animal builds the tube of flat shell fragments, attached at one edge and overlapping adjacent fragments (Fig. 11–25A; 11–33). The free edges of each fragment are directed upward and together they anchor the tube in the substratum. Because of its flexibility, the tube is like a loose-fitting cuticle that changes its shape as the worm moves. *Owenia* is said to be able to burrow through sand while occupying its tube. Sediment particles are collected during feeding, and those fragments suitable for tube construction are stored in a ventral pouch situated beneath the mouth. During construction of the tube, the pouch projects outward and downward and fastens a fragment to the margin of the tube. The membranous lining is secreted by paired glands in each of the first seven trunk segments. The secretion is applied by parapodial setae as the worm revolves.

Among the most beautiful of the sedentary polychaetes are the fan worms, or feather dusters or Christmas tree worms, of the families Sabellidae, Ser-

pulidae, and Spirorbidae. In these groups the prostomial palps have developed to form a funnel-shaped or spiral crown consisting of a few to many pinnate processes called **radioles** (Figs. 11–27). The radioles are rolled up or closed together when the worm withdraws its anterior end into the free end of the tube. Sabellids build membranous or sand-grain tubes. Serpulids and spirorbids secrete calcareous tubes that are attached to rocks, shells, or algae and thus can live on an otherwise inhospitable hard substratum. The most dorsal radiole on one or both sides of a serpulid or spirorbid is modified into a long, stalked knob called an **operculum** (Fig. 11–27B,C), which acts as a protective plug at the end of the tube when the crown is withdrawn. Sabellids lack an operculum but when the animal withdraws into its flexible tube, the opposite walls of the opening come together and seal. Sometimes, the compressed tube end folds over.

In all fanworms, the peristomium is folded back to form a distinct collar, which fits over the tube opening and is the principal structure used to mold additions on the end of the tube. In a serpulid two large glands secreting calcium carbonate open beneath the collar folds. Crystals of calcium carbonate are added to an organic matrix material secreted by the ventral surface (shields) of the anterior segments. When additions are made to the tube, secretions flow out between the collar and the body wall. This space then acts as a mold in which the secretion hardens and is simultaneously fused as a new ring on the end of the tube.

The fan worm, *Sabella*, constructs a tube of sand grains embedded in mucus. The worm sorts detritus

FIGURE 11–26 Sabellariid polychaetes. **A,** Ventral view of *Sabellaria floridana* removed from its sand-grain tube. ▶ Its suspension-feeding tentacles are expanded, and behind them are two lobes bearing golden setae that form a closing operculum when the animal withdraws into its tube. The soft tubular structure in the animal's midline is the reduced posterior end of the body, which bends anteriorly. It encloses only the intestine and is a provision for releasing feces at the mouth of the tube. **B,** Opercular bundles of setae projecting from the tube mouth of *Sabellaria vulgaris*. Like *Sabellaria floridana*, this species builds sand-grain tubes attached to shells, stones, pilings, and other objects along much of the Atlantic coast of North America. Farther south and on parts of the California coast, other species build reefs as shown in **C. C,** An intertidal rock on the Cornish coast of England encrusted with the tubes of *Sabellaria alveolata*, a species that forms colonies. Tubes are oriented at right angles to the substratum and are attached to each other like cells in a honeycomb. The scale is 15 cm long. **D,** Enlarged view of a small area of crust surface showing the tube openings. *(C and D, From Wilson, D. P. 1974. Sabellaria colonies at Duckpool, North Cornwall, 1971–1972, with a note for May 1973. J. Mar. Biol. Ass. U.K. 54:393–436.)*

A

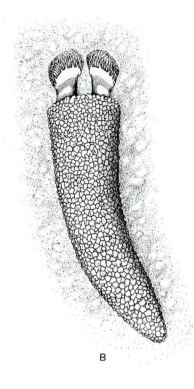

B

C

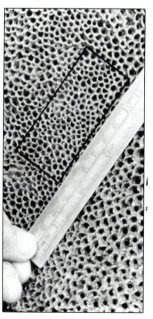

D

A

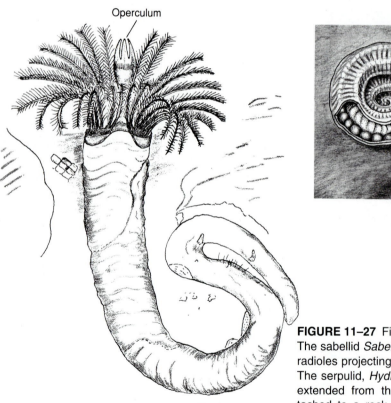

Operculum

B

C

FIGURE 11–27 Filter-feeding fan, or peacock, worms. The sabellid *Sabella pavonina,* showing the expanded radioles projecting from the apertures of the tubes. **B,** The serpulid, *Hydroides,* with radioles and operculum extended from the end of the calcareous tube attached to a rock. **C,** *Spirorbis,* a common spirorbid with a snaillike tube found attached to a variety of substrata, including algae and seagrass. Cutaway of shell shows eggs being brooded in the tube. *(A, Photograph courtesy of D. P. Wilson)*

collected by the ciliated radioles, and sand grains of suitable size for the tube construction are stored in a pair of opposing ventral sacs situated below the mouth. The walls of the sacs produce mucus, which is mixed with the sand grains. To make additions at the end of the tube (Fig. 11–35A), the ventral sacs deliver a ropelike string of mucus and sand grains to the collar folds below, which are divided midventrally, like the front of a shirt collar. The string of building materials is received at the collar folds. The worm rotates slowly in the tube, and the collar folds act like a pair of hands, molding and attaching the rope to the end of the tube. The operation is quite similar to an Indian method of making pottery.

In fanworms, the ventral surface of each segment bears a pair of large, mucus-secreting pads, or glandular shields. When the worm rotates, these glands lay down a mucous coating on the inner surface of the tube.

Boring Polychaetes

Representatives of a number of different polychaete families bore into the calcareous shells of dead or living molluscs and the skeleton of corals to form protective retreats. On coral reefs the brightly colored, spiral radioles of Christmas tree worms (the serpulid, *Spirobranchus*) can often be seen on the surface of living coral heads. Boring is begun by a newly settled young worm, but the mechanisms are still largely unknown. Species of the spionid, *Polydora,* excavate chemically in the shells of live oysters. The worm partially fills the excavation with debris, leaving room for a U-shaped burrow. When boring takes place between the mantle edge and shell, the oyster attempts to wall out the worm with new shell, creating unsightly "mud blisters" and reducing its market value.

Symbiotic Polychaetes

In their commensal relationship with other animals, polychaetes may be hosts or guests. As might be expected, the role of host is played primarily by the noncarnivorous tube dwellers and the burrowing polychaetes who provide a protected, ventilated retreat. Their guests include scale worms, bivalves, and crustaceans, particularly species of little crabs. The giant (80 cm) capitellid, *Notomastus lobatus,* the

Maitre d' worm, lives in clay sediment along the southeastern coast of the United States. It occupies a permanent helical burrow that it shares with no fewer than eight commensals, including one scaleworm and two other polychaetes, three clams, a crab, and an amphipod. Commensal polychaetes, some of which are facultative, are distributed in numerous families throughout the class, but the scale worms (Polynoidae) include the largest number. They live in tubes and the burrows of other polychaetes and crustaceans, with hermit crabs, on echiuran worms, on corals, in the ambulacral grooves of sea stars, and on sea urchins and sea cucumbers as well as other animals. Many display colors similar to those of the host, are small or dorsoventrally flattened, and have setae modified for clinging.

Parasitism is not common among polychaetes. *Labrorostratus* and other arabellids live in the coelom of other polychaetes and may be almost as big as their host. Polychaete ectoparasites include the blood-sucking Ichthyotomidae, which attach to the fins of marine eels.

The myzostomes, sometimes placed in a separate class, the Myzostomida, are a strange group of commensal and parasitic polychaetes. These little worms are rarely more than 5 mm long and resemble flatworms, but when disturbed they scoot rapidly away on stubby legs (Fig. 11–28A). The body is oval and greatly flattened, and the five pairs of parapodia are carried on the undersurface. They are found only on echinoderms, especially crinoids.

Nutrition

The feeding methods of polychaetes are closely correlated with the various life habits of the class.

Raptorial Feeders

Raptorial feeders include members of many families of surface-dwelling species, many pelagic groups, tubicolous eunicids and onuphids, and active gallery dwellers like the glycerids and nephtyids. The prey consists of various small invertebrates, including other polychaetes, which are usually captured by means of an eversible pharynx (proboscis). The pharynx commonly bears two or more horny jaws composed of cross-linked (tanned) protein (Fig. 11–29). The pharynx is rapidly everted, placing the jaws at the anterior of the body and causing them to open.

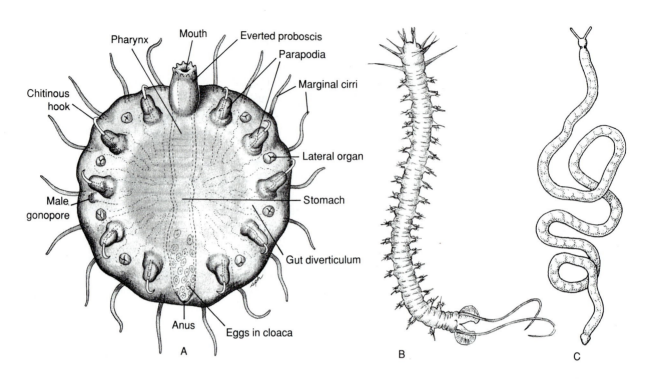

FIGURE 11–28 Commensal and interstitial polychaetes. **A,** *Myzostoma,* a commensal on crinoids. **B,** *Hesionides arenaria,* a minute polychaete, no more than 2 mm in length, that inhabits interstitial spaces of beach sands in northern Europe. The ventrally directed parapodia are adapted for crawling between sand grains. **C,** *Polygordius neopolitanus,* an interstitial polychaete. Segmentation is poorly indicated externally. *(B, From Ax, P. 1966. Veröffentlungen des Institut für Meeresforschung Bremerhaven. Suppl. II. pp. 15–66. C, After Fraipoint.)*

The food is seized by the jaws, and the pharynx is retracted. Although protractor muscles may be present, an increase in coelomic pressure resulting from the contraction of body wall muscles is an important factor in the eversion of the pharynx. When pressure on the coelomic fluid is reduced, the pharynx is withdrawn by the retractor muscles, which extend from the body wall to the pharynx (Fig. 11–8B,C).

Raptorial tube dwellers may leave the tube partially or completely when feeding, depending on the species. *Diopatra* uses its hood-shaped tube as a lair (Fig. 11–24A). Chemoreceptors monitor the ventilating current of water passing into the tube, and when approaching prey is detected, the worm partially emerges from the tube opening and seizes the victim with a complex pharyngeal armature of teeth. During feeding the prey may be clasped with the enlarged anterior parapodia. Species of *Diopatra* may also feed on dead animals, algae, organic debris, and small or-

ganisms, such as forams, that are in the vicinity of the tube or become attached to it.

Some raptorial feeders, such as syllids and glycerids, have a long, tubular pharynx. Species of *Glycera* live within a gallery system constructed in muddy bottoms. The system contains numerous loops that open to the surface (Fig. 11–21B). Lying in wait at the bottom of a loop, the worm uses its four tiny antennae to detect the surface movements of prey, such as small crustaceans and other invertebrates. It slowly moves to the burrow opening and then seizes the prey with its pharynx.

When the pharynx is retracted, it occupies approximately the first 20 body segments. At the back of the pharynx are four jaws arranged equidistantly around the wall. The proboscis is attached to an S-shaped esophagus. No septa are present in these anterior segments, and the proboscis apparatus lies free in the coelom. Just prior to eversion of the proboscis the

FIGURE 11–29 Ventral view of the head of the eunicid, *Marphysa sanguinea,* showing its complex jaws. Within the expanded beaklike parts (white) is a pair of opposing mandibles (black).

longitudinal muscles contract violently, sliding the proboscis forward and straightening out the esophagus. The proboscis is then everted with explosive force, and the four jaws emerge open at the tip (Fig. 11–21C). Each jaw contains a canal that delivers poison from a gland at the jaw base.

Herbivores, Omnivores, Scavengers, and Browsers

Not all errant polychaetes that possess jaws are carnivores. Scavenging and omnivorous habits have evolved in many polychaetes. The jaws may, for example, be used to tear off pieces of algae. These polychaetes generally belong to the same families as do the carnivores, and they are similarly adapted.

Studies on *Nereis* by H. Goerke demonstrate the diversity of feeding habits that may exist even within a genus. Species of *Nereis* possess a muscular, eversible pharynx with a pair of heavy jaws (Fig. 11–8C). Some, such as *Nereis pelagica, Nereis virens,* and *Nereis diversicolor,* are omnivorous and feed on algae, other invertebrates, and even detritus. *Nereis succinea* and *Nereis longissima* feed primarily on detritus material in the substratum. However, *Nereis fucata,* a commensal in hermit crab shells, is carnivorous. *Nereis brandti,* which occurs on the northwest coast of the United States and reaches 1.8 m in length, feeds primarily on green algae.

Nonselective Deposit Feeders

Some nonselective, deposit-feeding polychaetes consume sand or mud directly by applying the mouth against the substratum. Ingestion is generally facilitated by means of a simple, nonmuscular pharynx, which is everted by elevated coelomic fluid pressure (Figs. 11–30; 11–31; and 11–32). Among polychaetes, such deposit feeders include burrowers and tube dwellers. The less stationary burrowers include some capitellids and opheliids, both of which ingest the substratum through which they burrow.

Population counts were taken by zoologists of the little burrowing opheliid, *Euzonus (Thoracophelia) mucronatus,* which is about 25 mm long and not more than 2 mm in diameter. These worms inhabit the intertidal zone on the Pacific coast of the United States and form colonies that occupy extensive stretches of protected (low-energy) beaches. In such colonies the number of worms averages 2500 to 3000 per square foot. Worms occupying a typical strip of beach 1 mi long, 10 ft wide, and 1 ft thick ingest approximately 14,600 tons of sand each year.

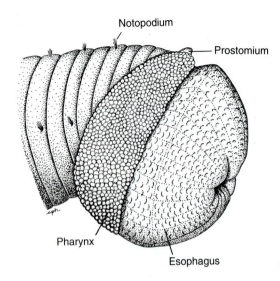

FIGURE 11–30 Lateral view of the anterior end of the capitellid *Notomastus,* a burrowing, nonselective deposit feeder. The large everted pharynx obscures the tiny prostomium. *(After Michel.)*

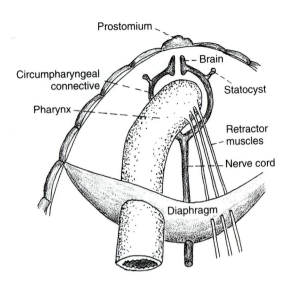

FIGURE 11–31 Anterior of *Arenicola* (dorsal dissection). *(After Ashworth from Brown.)*

The lugworms, members of the family Arenicolidae, are common deposit feeders. *Arenicola* (and *Abarenicola*) lives in an L-shaped burrow whose vertical part opens to the surface (Fig. 11–22). The head of the worm is directed toward the blind, horizontal part of the burrow, where sand is continually ingested by means of a simple pharynx. The ingested sand is rich in surface organic material which tumbles into the surface depression above the animal's head and slumps downward to the worm. The worm irrigates the burrow by peristaltic contractions that drive water into the burrow opening. Water leaves the burrow by percolating up through the sand. At cyclic intervals the worm backs up to the surface to defecate mineral material (castings).

Bamboo worms, the Maldanidae, are examples of deposit-feeding tube dwellers. The worm lives upside down and ingests the substratum at the bottom of the sand-grain tube. Cilia within the everted pharynx drive loose particles into the gut, and there is actually some particle selection (Fig. 11–25D). Following a distinct rhythm, the feeding halts and the worm backs up to the top of the tube to defecate the mineral particles passed through the gut.

Selective Deposit Feeders

Selective deposit feeders lack a specialized pharynx. Special head structures extend over or into the substratum. Deposit material adheres to mucous secretions on the surface of these feeding structures and is then conveyed to the mouth along ciliated tracts or grooves. These polychaetes thus select organic deposit material from between sand particles.

The prostomial ingestive organs of the terebellids, for example, *Amphitrite* and *Terebella* and the ampharetids, are formed of large clusters of contractile tentacles, which stretch over the surface of the substratum by ciliary creeping (Figs. 11–23; 11–25B). Surface detritus adheres to the mucus secreted by the tentacular epithelium. Particles are moved down a ciliated gutter formed by the rolled tentacle, and food accumulates at the base of the tentacles, each of which is wiped over the upper lip bordering the mouth. Cilia on the lip then drive the food into the mouth. The tentacles of some large tropical species, such as *Eupolymnia crassicornis,* reach a meter or more in length and stream over the surface of the sand like active strands of spaghetti.

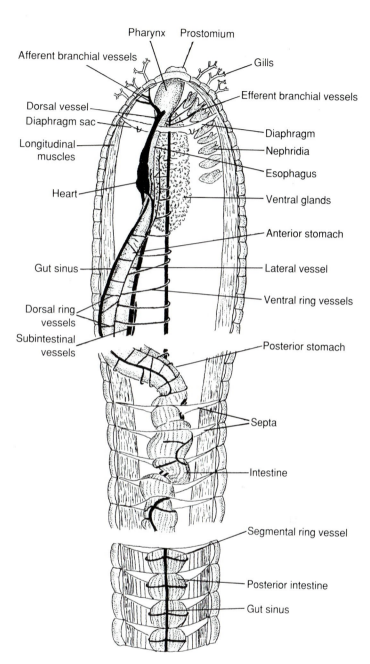

FIGURE 11–32 *Amphitrite* (dorsal dissection). *(After Brown F. A. 1950. Selected Invertebrate Types. John Wiley and Sons, New York.)*

The tubicolous *Owenia* feeds in a somewhat similar manner but utilizes a prostomial crown of flattened, branched, ribbon-like filaments, each of which ends in a bifid lobe (Figs. 11–25A; 11–33). The worm also suspension feeds.

The tentaculate, deposit-feeding cone worm, *Pectinaria (Cistenides),* was studied by zoologist R. B. Whitlatch. The animal lives buried in the sand head down (Fig. 11–25F, G), selects organic aggregates and mineral particles encrusted with organic

material, and by selection concentrates the organic matter from about 32% (in sediment) to 42% (in gut). Of the ingested organic fraction, about 30% is utilized, and the remainder passes through the gut. By differentially staining the organic matter, it is possible to estimate the relative amounts of proteins and carbohydrates present and utilized.

The tubicolous spionids are suspension feeders that probably depend on particles stirred up from the bottom. They possess two long, tentacular palps that

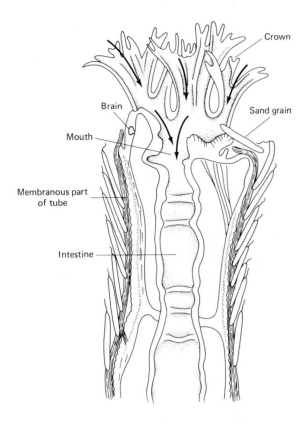

FIGURE 11–33 *Owenia* adding a sand grain to its tube (longitudinal section). *(Modified from Watson.)*

lash the water or project from the tube opening and extend over the bottom (Fig. 11–34E, F). Particles that adhere to the surface are propelled toward the mouth down the length of the palps in a ciliated channel.

Filter Feeders

Many of the sedentary burrowers and tubicolous polychaetes are filter feeders. The head is usually equipped with special feeding processes that collect detritus and plankton from the surrounding water. The particles adhere to the surface of the feeding structures and are then conveyed to the mouth along ciliated tracts.

The crownlike, bipinnate radioles of serpulid, sabellid, and spirorbid fan worms form a funnel of one or two spirals when expanded outside the end of the tube (Fig. 11–35). Beating of the cilia situated on the pinnules produces a current of water that flows

through the radioles into the funnel and then flows upward and out. Particles are trapped on the pinnules and are driven by the cilia into a groove running the length of each radiole. The particles are carried along the groove down to the base of the radiole, where a rather complex sorting process takes place (Fig. 11–35B). The largest particles are rejected, and fine material is carried by ciliated tracts into the mouth. Many sabellids sort particles into three grades and store the medium grade for use in tube construction.

The feeding mechanism of the chaetopterids differs from that of the other filter feeders (Fig. 11–34A, C). *Chaetopterus,* which lives in U-shaped parchment tubes, has a highly modified body structure. The notopodia on the twelfth segment are extremely long and aliform (winglike), and the epithelium is ciliated and richly supplied with mucous glands. The notopodia on segments 14 to 16 are modified and fused, forming semicircular fans that project like piston rings against the cylindrical wall of the tube (Fig. 11–34C). The beating of the fans produces a current of water that enters the chimney of the U-shaped tube near the anterior end of the worm, flows through the tube, and then flows out of the opposite chimney.

The paired, aliform notopodia are stretched out around the walls of the tube, and a sheet of mucus is secreted between them. The mucous film is continuously secreted from each notopodium, and so the sheet assumes the shape of a bag. The posterior of the bag is grasped by a ciliated cup on the middorsal side of the worm a short distance behind the aliform notopodia. Water brought into the tube by the rhythmic beating of the fan parapodia passes through the mucous bag, which strains suspended detritus and plankton.

Large objects brought into the tube by the water current are detected by peristomial cilia; the aliform notopodia then are pulled back to let the large objects pass by. The food-laden mucous bag is continuously being rolled up into a ball by the dorsal cupule. When the ball reaches a certain size, the bag is cut loose from the notopodia and rolled up with the ball. The cupule then projects forward and deposits the mucous food ball onto a ciliated middorsal groove, which extends to the anterior of the worm, and the ball is carried to the mouth. An 18- to 24-cm long specimen of *Chaetopterus* may produce mucous film for the bag at the rate of approximately 1 mm/s, with food balls averaging 3 mm in diameter.

The other members of the Chaetopteridae build straight, vertical tubes but utilize mucous bags for filter feeding. The number of mucous bags and the site of their formation vary—as many as 13 are formed at one time in *Spiochaetopterus.* In several genera the water current is activated by cilia rather than by pumping (Fig. 11–34D).

The Alimentary Canal

Typically, the alimentary canal of polychaetes is a straight tube extending from the mouth at the anterior end of the worm to the anus situated in the pygidium. The canal is commonly differentiated into a pharynx (or buccal cavity if the pharynx is absent), short esophagus, stomach (in sedentary species), intestine, and rectum (Fig. 11–8A). However, in many species, these regions can be detected only histologically, and the gross appearance of the digestive tract behind the pharynx is that of a simple, uniform tube. The stomach or anterior intestine elaborates enzymes for extracellular digestion. The intestine is the site of absorption, and not infrequently the walls are folded, increasing the intestinal surface area. In *Nereis* two large, glandular ceca open into the esophagus (Fig. 11–8A). They, along with the anterior end of the intestine, secrete digestive enzymes.

The egested wastes from a worm living in a tube with double openings, such as *Chaetopterus,* are readily removed by water currents. Such flushing, however, is less efficient when the tube is deeply buried in mud and sand, or is secreted with only one opening, as in serpulid fanworms and sabellariids. Many polychaetes turn around in the tube or burrow and thrust the pygidium out of the opening during defecation. In sabellariids, the posterior half of the worm folds back on the anterior half, bending the body into a U and venting the anus at or near the tube opening (Fig. 11–26A). Some species produce fecal pellets or strings, which reduce the risks of fouling. A fan worm has a ciliated groove, which carries fecal pellets from the anus anteriorly out of the tube.

Gas Exchange

Gills are common among the polychaetes, but they vary greatly in both structure and location, indicating that they have arisen independently within the class a number of times. They are never enclosed within protective chambers; many species that possess gills are already protected, since they live in tubes and burrows. Gills are lacking in polychaetes that are very small or that possess long, threadlike bodies, such as many burrowing Lumbrineridae, Arabellidae, and Capitellidae.

In the scaleworms gas exchange is largely restricted to the dorsal body surface, which is roofed over by the elytra. Cilia on the dorsal surface create a current of water flowing posteriorly beneath the elytra. The felt-covered sea mouse *(Aphrodita)* lacks cilia (Fig. 11–16), but a similar dorsal water current is produced by the animal, which tilts the elytra upward and then rapidly brings them down in sequence.

Most commonly the gills are associated with the parapodia and in many cases are modified parts of the parapodium. The notopodium may possess a flattened branchial lobe, which acts as a gill, as in nereids (Fig. 11–6C). Commonly, the dorsal cirrus of the parapodium is modified to serve as a gill (Figs. 11–12A, B; 11–36C, D), or the gills arise from the base of the dorsal cirrus. Cirratulids have long, contractile, threadlike gills, each attached to the base of the notopodium (Fig. 11–36A).

The gills are not always associated with the parapodia. Many sedentary species have gills at the anterior ends near the opening of the tubes or burrows. For example, the gills of some terebellids, such as *Amphitrite,* are arborescent and are situated on the dorsal surface of the anterior segments (Fig. 11–36B). The bipinnate radioles composing the fans serve as sites of gas exchange in the fan worms.

Ventilation may be provided by gill cilia or by gill contractions (see Fig. 11–12, page 513), but many burrowing and tube-dwelling polychaetes drive water through their burrows or tubes by undulating or peristaltic contractions of the body. Worms that ventilate by muscular activity typically exhibit a spontaneous ventilating rhythm in which a period of ventilation alternates with a period of rest. According to research conducted by C. P. Mangum, ventilation activity increases the worm's oxygen requirement as much as 15-fold, but there is approximately a 20-fold increase in oxygen uptake.

Internal Transport

Circulation in most polychaetes results from fluid movement in both the blood-vascular system and the coelom. A common variation on this pattern occurs in many polychaetes that have reduced septa, for example the glycerid bloodworms. In these, the coelomic

(Text continues on page 536)

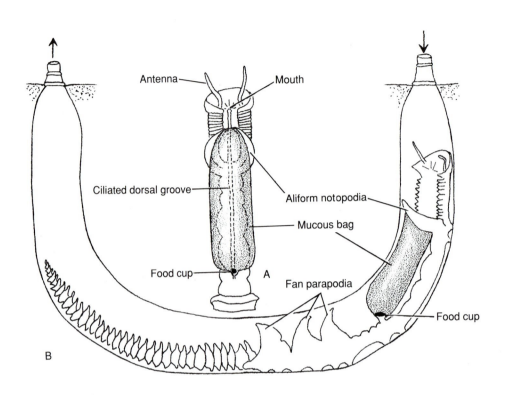

FIGURE 11–34 *Chaetopterus* during feeding. **A,** Anterior part of the body (dorsal view). **B,** Worm in tube (lateral view). Arrows indicate direction of water current through the tube. **C,** Side views of the parapodial fans of *Chaetopterus* during one pumping cycle. Worm is in glass tube. **D,** Dorsal view of three segments of middle body region of *Spiochaetopterus,* showing the position of mucous bags and the formation of food balls. Arrows indicate the direction of water currents. **E,** Palps of a spionid polychaete projecting from its sand-grain tube. Ciliated groove of the palp conveys detritus material picked up from the substratum or suspended in the water to the mouth. **F,** Ventral view of anterior end of *Spio pettiboneae* showing the base of palps and mouth. *(A, B, After MacGinitie. C., From Brown, S. C. 1975. Biomechanics of water-pumping by Chaetopterus variopedatus Renier. Skeletomusculature and kinematics. Biol. Bull. 148:136–150. D, From Barnes.)*

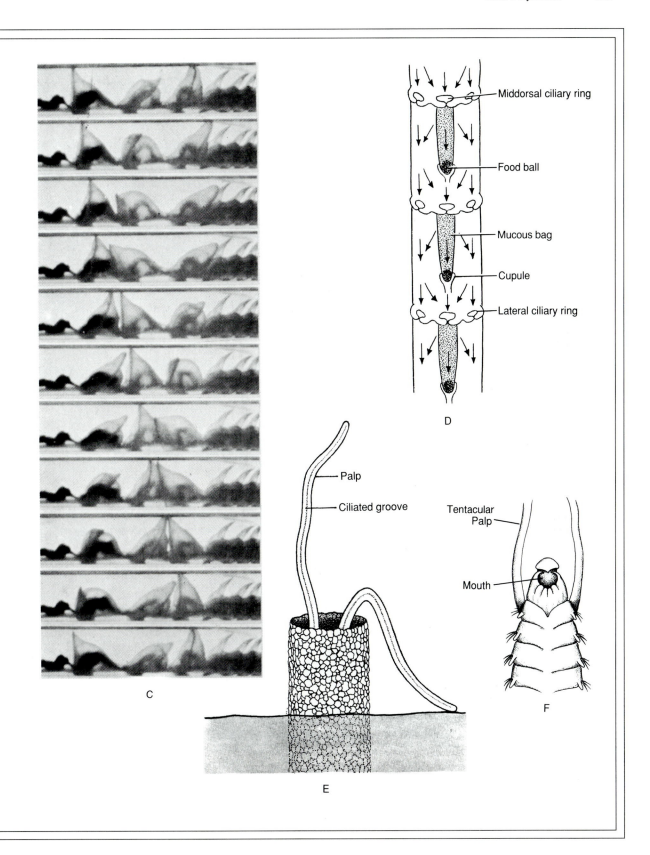

C

Middorsal ciliary ring

Food ball

Mucous bag

Cupule

Lateral ciliary ring

D

Palp

Ciliated groove

E

Tentacular
Palp

Mouth

F

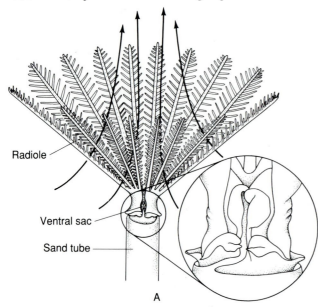

FIGURE 11–35 A, Anterior end of the fan worm, *Sabella,* showing the filter-feeding currents and tube building. **B,** Filter feeding in *Sabella,* showing water current (large arrows) and ciliary tracts (small arrows) over a section of one radiole. The letters A, B, and C indicate the different sizes of the particles sorted. *(A and B, Modified after Nicol from Newell, R. C. 1970. Biology of Intertidal Animals. American Elsevier Co. New York.)*

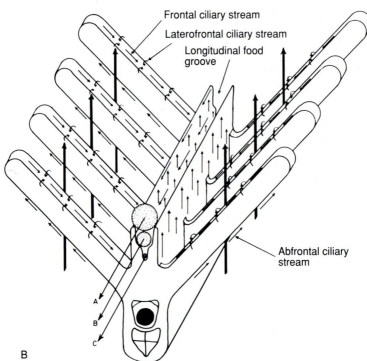

system replaces the blood system and transports substances throughout the body. Very small species also typically lack a blood system and sometimes also the coelomic cavities (Fig. 11–28B, C).

In most polychaetes there exists a well-developed blood-vascular system, in which the blood is enclosed within vessels. In a typical blood system, blood flows anteriorly in a **dorsal vessel** situated over the digestive tract; at the anterior end of the body, the

dorsal vessel is connected to a **ventral vessel** by one to several vessels or by a network of vessels passing around the gut. The ventral vessel carries blood posteriorly beneath the alimentary tract (Fig. 11–37A).

In each segment the ventral vessel gives rise to one pair of ventral, **parapodial vessels,** which supply the parapodia, the body wall, and the nephridia, and to several ventral, **intestinal vessels,** which supply the gut (Fig. 11–37A). The dorsal vessel, in turn, re-

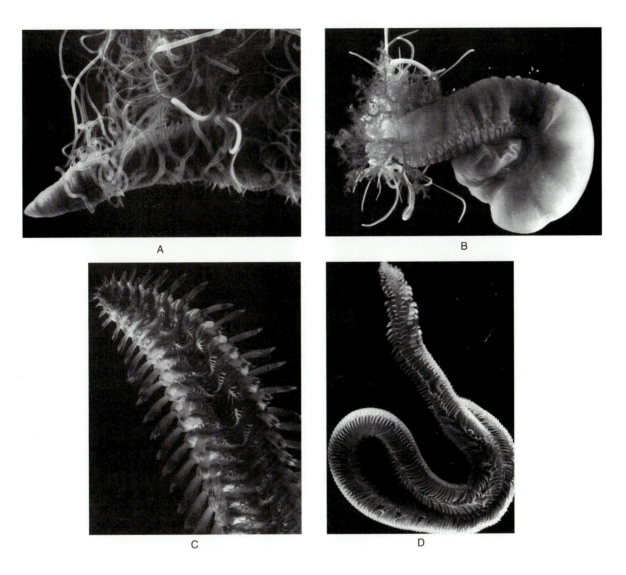

FIGURE 11–36 A, Threadlike gills of *Cirriformia cirriformia.* **B,** Arborescent gills of the terebellid polychaete, *Amphitrite ornata.* **C,** Featherlike gills of the amphinomid, *Chloeia viridis.* **D,** The fingerlike gills (dorsal cirri) of the orbiniid burrower, *Scoloplos rubra,* arch over the animal's dorsum and form a semienclosed ventilatory canal.

ceives a corresponding pair of dorsal parapodial vessels and a dorsal intestinal vessel. The dorsal and ventral parapodial vessels and the dorsal and ventral intestinal vessels are interconnected by a network of smaller vessels.

Polychaete blood is confined to small- and large-diameter vessels, and in some species, large-volume sinuses, which typically occur on the wall of the gut (Fig. 11–6C). Both vessels and sinuses lack a contin-

uous cellular endothelium and are lined instead only by the basal lamina of overlying cells (Fig. 11–38). The gills are usually provided with afferent and efferent vascular loops permitting a two-way flow. This is true, for example, of the gills of lugworms and the branchial, notopodial lobes of nereids (Fig. 11–37A). On the other hand, the radioles of fan worms, which function in both food gathering and gas exchange, contain only a single vessel, within which blood flows

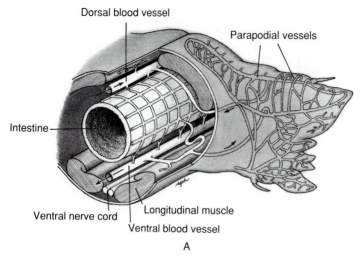

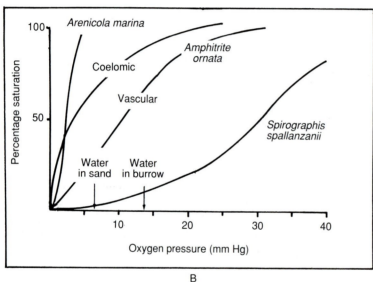

FIGURE 11–37 A, Vascular system within a segment of *Nereis virens.* Arrows indicate the direction of blood flow. B, Oxygen dissociation curves of the respiratory pigments of three species of polychaetes. *Arenicola ornata* has two hemoglobins, one vascular and one coelomic. The respiratory pigment of *Spirographis spallanzanii* is chlorocruorin. (See p. 539 of text for the significance of the positions of the curves.) *(In part after Jones from Dales.)*

tidally, in and out. In many polychaetes, such as glycerids, the gills are irrigated with coelomic fluid and not blood.

In general, blood is driven by peristaltic waves of contraction that sweep over the blood vessels, particularly the dorsal vessels. The vessel wall in *Magelona,* for example, consists of a single layer of myoepithelial cells that contain striated myofibrils arranged in a circular direction or in both circular and longitudinal directions. Many polychaetes have accessory, heartlike pumps situated in various places within the blood-vascular system.

The blood contains few cells compared with coelomic fluid. In small polychaetes it is usually colorless, but in larger species and those that burrow in soft bottoms, the blood contains respiratory pigments dissolved in the plasma. In fact, within the Polychaeta are found three of the four respiratory pigments of animals. Hemoglobin is the most common of these pigments, but chlorocruorin is characteristic of the blood of the serpulid, spirorbid, and sabellid fan worms (the blood of *Serpula* contains both hemoglobin and chlorocruorin) and also of the Flabelligeridae and Ampharetidae. **Chlorocruorin** is a kind of hemoglo-

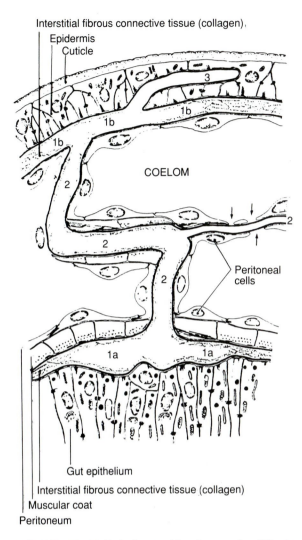

Interstitial fibrous connective tissue (collagen)
Epidermis
Cuticle

3

1b

1b

1b

COELOM

2

2

2

2

Peritoneal cells

1a

1a

Gut epithelium

Interstitial fibrous connective tissue (collagen)

Muscular coat

Peritoneum

FIGURE **11–38** Polychaete blood vessels *(Nereis japonica),* like those of most other invertebrates, are specialized fluid-filled spaces (numbered) in the connective tissue of the body. **1a** a gut sinus; **2** a coelomic vessel; **1b** a body wall vessel; **3** an epidermal vessel. *(From Nakao, T. 1974. An electron microscopic study of the circulatory system in Nereis japonica. J. Morphol. 144:217–235.)*

bin, but a slight difference in side chains gives it a green rather than a red color. *Magelona* has a blood-vascular system with enucleated corpuscles containing a third iron-bearing, but nonheme (not a porphyrin), protein pigment called hemerythrin (p. 339), found elsewhere only in Sipuncula, Priapulida, and inarticulate brachiopods.

When present in polychaetes respiratory pigments are typically small molecules in corpuscles in the coelom and large extracellular molecules in the blood plasma. The extracellular molecules in the blood plasma not only carry oxygen but also contribute to the osmotic concentration of the blood and thus help to maintain the plasma volume of the blood. Packaging of respiratory pigments within coelomocytes may be a means of controlling the osmotic concentration of coelomic fluid intermediate between that of seawater and that of the blood (p. 108).

The coelomic fluid is circulated by muscular contractions of the body wall or by both muscular contractions and cilia on the coelomic lining. The coelomic fluid may be colorless or contain a corpuscular hemoglobin. Sometimes, as in capitellid polychaetes, the pigment occurs only in the coelom and not in the blood system, but in others, such as many terebellids and opheliids, two different hemoglobins are present, one in the blood and the other in the coelom. Some polychaetes, including *Glycera* and most scaleworms, have hemoglobin (neuroglobin) in glial cells surrounding their nerve cords, which are bright red. Myoglobin occurs in the muscles of many scaleworms.

The coelomic hemoglobin of *Amphitrite* has a greater affinity for O_2 at low-oxygen tensions (dissociation curve to the left) than does the blood-vascular hemoglobin. This difference facilitates the passage of oxygen from the blood-vascular system to the coelomic fluid, which is the principal source of oxygen for internal tissues (Fig. 11–37B).

In the majority of polychaetes, the respiratory pigments function in oxygen transport, although supplying only a part of the oxygen consumed. When the blood from the gills does not become mixed with unoxygenated blood before delivering its oxygen load to the target tissues, the oxygen affinity of the hemoglobin is relatively low (oxygen dissociation curve to the right). This is the situation for polychaetes like *Amphitrite* and the fan worms (e.g., *Sabella*), in which the gills are at the anterior end. In worms with segmental gills, in which the blood from the gills is mixed with unoxygenated blood en route to the target tissues, the oxygen affinity of the hemoglobin is high; that is, the hemoglobin holds on to its oxygen at relatively low oxygen tensions (Fig. 11–37B). This ensures that oxygen is not lost from the pigment during its delivery to the target tissues.

In some polychaetes, such as the bloodworm *Glycera,* the hemoglobin may also store oxygen during the resting periods between ventilation or at low tide, when the oxygen tension of the water in the burrow and surrounding sand is considerably decreased. The amount of hemoglobin present would seem to permit an oxygen reserve lasting only a few minutes. Like many invertebrates, polychaetes are however, oxyconformers; that is, oxygen consumption is regulated in part by the amount of oxygen available in the surrounding environment. The physiologist C. P. Mangum has suggested that the decreasing metabolic demands for oxygen during stagnation enable the worm to extend over a significant period of time what would otherwise be an inadequate store of oxygen provided by the hemoglobin. According to research by E. Ruby and D. Fox, the intertidal burrower, *Euzonus mucronatus* (p. 529), uses its oxyhemoglobin storage to carry it through regular 2- to 4-h periods of anoxia during low tides. If subjected to longer periods of anoxia, the worm switches over to anaerobic respiration, on which it can survive as long as 20 days.

Excretion

Polychaete excretory organs are filtration nephridia (p. 189), which in general are distributed as one pair per segment, but reduction to few or even one pair for the entire worm has occurred in some families. The anterior end of the nephridial tubule is situated in the coelom of the segment immediately anterior to that from which the nephridiopore opens (Figs. 11–6B; 11–8A). The tubule penetrates the posterior septum of the segment, extends into the next segment, where it may be coiled, and then opens to the exterior in the region of the neuropodium. Both the preseptal portion of the nephridium and the postseptal tubule are covered by a reflected layer of peritoneum from the septum.

Depending on their body design, polychaetes have either protonephridia or metanephridial systems. Polychaetes that lack a blood-vascular system (nine families and all larvae), or in which the blood system is reduced (five families), have protonephridia. (The exception is the Nephtyidae, which have both a blood system and protonephridia.) All remaining 81 families have blood-vascular and metanephridial systems. The correlation of blood vessels with metanephridial tubules and their absence with protonephridia is an indication of how ultrafiltration, the first step in urine formation, occurs in the two broad groups. It is believed that ultrafiltration of blood occurs across the vessel wall in vascularized polychaetes (Fig.

11–39A). Then the ultrafiltrate is modified and swept to the exterior through a ciliated duct, the metanephridial tubule (metanephridium) (Fig. 11–39B–D). In the absence of blood vessels, ultrafiltration is believed to occur as the coelomic fluid is drawn across the walls of protonephridial terminal cells (Fig. 11–40C, D). Reabsorption occurs in the protonephridial tubule, which is identical histologically to a metanephridium. The protonephridia terminal cells, called **solenocytes,** have a single flagellum and a long tubular filtration collar.

The preseptal end of polychaete metanephridia, for example in *Nereis,* possesses an open ciliated funnel, the **nephrostome.** The nephrostome has an outer investment of peritoneum, and the interior is densely ciliated. The postseptal canal, which extends into the next successive segment, becomes greatly coiled to form a mass of tubules, which are enclosed in a thin, saclike covering of peritoneal cells. Coiling is probably an adaptation that increases the surface area for tubular secretion or reabsorption. The nephridiopore opens at the base of the neuropodium on the ventral side. The entire lining of the tubules is ciliated.

The metanephridia of most other polychaetes differ only in minor details (Fig. 11–39B–D) but may display various degrees of regional restriction in the more specialized families. In the fan worms, where only one pair of functional nephridia remains, the two nephridia join at the midline to form a single median canal, which extends forward to open through a single nephridiopore on the head (Fig. 11–39C). Urine is released directly outside, and fouling of the tube is avoided.

In polychaetes there may or may not be a direct association of the blood vessels with the metanephridial tubules. The fan worms and the arenicolids lack a well-developed nephridial blood supply, and the coelomic fluid must be the principal route for waste removal from the blood to the metanephridia. In other polychaetes the nephridia are surrounded by a network of vessels. In the nereids the nephridial blood supply is greater in those species that live in brackish water.

Many polychaetes, particularly nereids, can tolerate low salinities and have become adapted to life in brackish sounds and estuaries. The gill (notopodial lobe) of *Nereis succinea* contains cells specialized for absorbing ions. A small number of species live in fresh water. The sabellid, *Manyunkia speciosa,* for example, occurs in enormous numbers in certain regions of the Great Lakes, such as around the mouth of

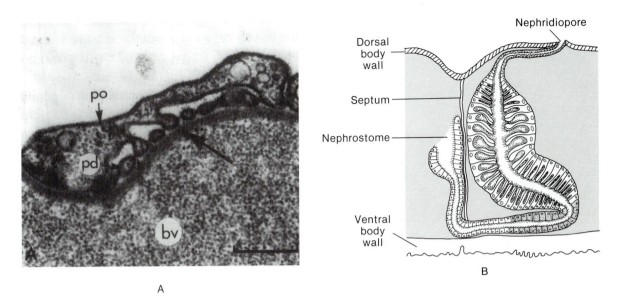

A

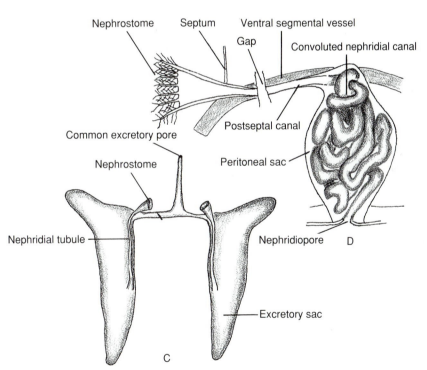

FIGURE 11–39 Polychaete metanephridial systems. **A,** Podocytes *(po)* on the wall of a segmental blood vessel *(bv)* of the spionid, *Spio setosa.* **B–D,** Metanephridia of three polychaetes: **B,** The spionid, *Polydora* (this nephridium also fabricates spermatophores); **C,** The serpulid fan worm, *Pomatoceros;* **D,** *Nereis vexillosa. (A, From Smith, P. R., and Ruppert, E. E. 1988. Nephridia. In Westheide, W., and C. O. Hermans (Eds.): Ultrastructure of Polychaeta. Microfauna Marin. 4:231–262, Gustav Fischer Verlag, Stuttgart and New York. B, From Rice, S. A. 1980. Ultrastructure of the male nephridium and its role in spermatophore formation in spionid polychaetes (Annelida). Zoomorphologie. 95:181–194. C, After Thomas. D, After Jones.)*

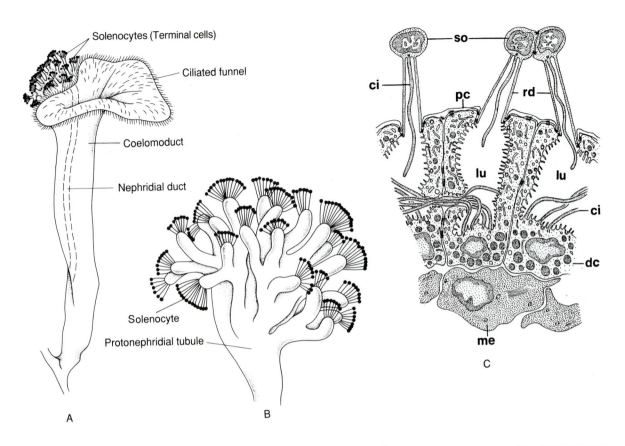

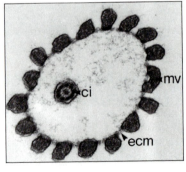

FIGURE 11–40 A, Protonephridium and coelomoduct (gonoduct) of *Phyllodoce paretti*. **B,** Branched end of protonephridium of *Phyllodoce paretti*. **C,** Ultrastructure of three terminal cells and the collecting protonephridial tubules (lu) of the bloodworm, *Glycera dibranchiata*. The terminal cells (so) project from the surface of the nephridium into the coelomic fluid and provide a filtration cylinder (cross section in D) composed of microvilli (rd = mv). Coelomic fluid is ultrafiltered across the cylinder wall, and reabsorption occurs within the tubule (lu). Reabsorbed metabolites are digested intracellularly and, eventually, stored as glycogen in nearby cells (me). *(A, B, From Goodrich. C, From Smith, P. R. 1992. Polychaeta: Excretory System. In Harrison, F. W., and Gardiner, S. L. (Eds.): Microscopic Anatomy of Invertebrates. Vol. 7: Annelida, pp. 71–108. D, from an unpublished poster by P. R. Smith.)*

the Detroit River. There are a few terrestrial polychaetes, all tropical Indo-Pacific nereids, that burrow in soil or live in moist litter.

Chloragogen tissue, coelomocytes, and the intestinal wall may play accessory roles in excretion. **Chlorogogen tissue** is composed of brown or greenish peritoneal cells situated on the wall of the intestine or on various blood vessels. Chloragogen tissue, which has been studied much more extensively in earthworms (see p. 560), is an important site of nutrient synthesis and storage and hemoglobin synthesis (perivasal cells).

Nervous System

The polychaete brain, usually bilobed, lies in the prostomium beneath the dorsal epithelium (Figs. 11–41; 11–42A). Depending on the degree of devel-

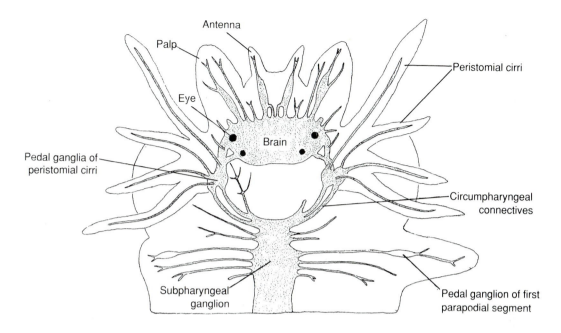

FIGURE 11–41 Anterior part of *Nereis* nervous system. *(After Henry from Kaestner.)*

opment of sense organs, the brain supplies nerves to the palps, antennae, eyes, and nuchal organs. Typically, a pair of circumpharyngeal or circumesophageal connectives surround the anterior gut and interconnect the brain and the ventral nerve cord.

The primitive ventral nerve cord is completely double and ladder-like throughout with transverse commissures between the separate ganglia (Fig. 11–6A), as in fan worms, but in most polychaetes, the two cords are fused in varying degrees, and in the tube-dwelling *Owenia,* an unpaired, nonganglionated nerve cord is found within the epidermis. There is typically one ganglionic swelling per segment, and from each ganglion usually three or four pairs of lateral nerves emerge that innervate the body wall of that segment.

The pattern of innervation of body muscles in the polychaetes is like that of the arthropods (Chapter 12). Each muscle fiber (cell) receives more than one neuronal ending, and the rate of fiber contraction depends on the summed effects of all the neurons. The degree and rate of muscular contraction in polychaetes are thus a result of differential activity of the multiple neurons contacting each individual fiber. In the vertebrates, on the other hand, each muscle fiber receives only one neuron to which it responds by contracting all or none. Control of muscular contraction is largely by recruitment of more or fewer fibers.

An important defense of most polychaetes against their many predators is the ability to contract very rapidly. The rapid end-to-end contraction reflex is particularly well developed in the tube dwellers, which project from their protective housing to feed. Correlated with this ability to contract rapidly is the presence of giant axons in the ventral nerve cords. The enlarged diameter of the axon increases the rapidity of conduction and therefore makes possible simultaneous contractions of the segmental muscles (Fig. 11–42B, C). For example, the single giant fiber of the fanworm, *Myxicola* (1.7 mm in diameter; largest in the animal kingdom), can be fired at any level along the length of the body and conducts an impulse in either direction. Conduction along the giant fiber occurs at a speed of 20 m/s, compared with about 0.5 m/s along ordinary longitudinal tracts. Efferent branches from the giant axon run to the longitudinal muscles. The giant fiber is not involved in the conduction of ordinary locomotor impulses, for if the fiber is severed, locomotion is not inhibited; only rapid contraction is blocked.

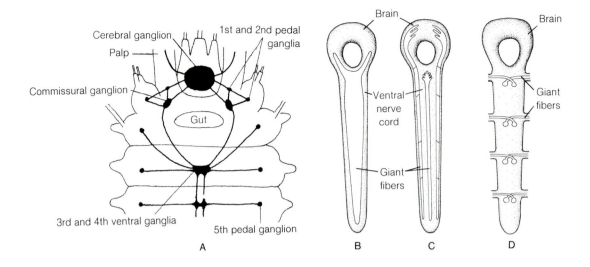

FIGURE 11–42 A, Nervous system of the scaleworm, *Lepidasthenia.* **B–D,** Nervous systems of three polychaetes, showing types of giant axons: **B,** *Eunice,* with a single giant axon; **C,** *Nereis,* with medial and lateral giant fibers; **D,** *Thalanessa,* with intrasegmental giant axons. *(A, After Storch from Fauvel. B and C, After Nicol. D, After Rhodes from Nicol.)*

Sense Organs

The principal specialized sense organs of polychaetes are eyes, nuchal organs, and statocysts. The eyes, which are best developed in errant polychaetes, are found on the surface of the prostomium in two, three, or four pairs (Fig. 11–8; 11–9B–D; 11–17A). In general, the polychaete eye is of the retinal-cup variety, the wall of which is composed of rodlike photoreceptors, pigment, and supporting cells (Fig. 11–43).

The eyes of most polychaetes can probably determine only light intensity and light source, but the huge bulging eyes of the pelagic, raptorial Alciopidae are capable of image formation (Fig. 11–43C, D) (see p. 362).

The radioles of some sabellids bear eye spots, and those of serpulids have dispersed photoreceptors. Fan worms are very sensitive to sudden light reduction and immediately withdraw into their tubes when such a reduction in light occurs. This behavior probably represents a protective adaptation against passing predators.

Nuchal organs consist of a pair of ciliated sensory pits, or slits, often eversible, situated in the head region of most polychaetes (Fig. 11–44A, B). These sense organs are important for detecting food and attain their greatest development in the predatory species, such as some of the amphinomid fireworms, in which they expand to form a convoluted, brainlike crown (caruncle) on the upper surface of the head (Fig. 11–13B).

Statocysts are found in many sedentary burrowers or tube dwellers. The statocysts of *Arenicola* are situated within the body wall of the head with a canal that opens to the outer, lateral body surface (Fig. 11–44C). The statocysts of *Arenicola* contain spicules, diatom shells, and quartz grains, all covered with a chitinoid material. *Arenicola* always burrows head downward, and if an aquarium containing a worm is tilted 90 degrees, the worm makes a compensating 90-degree turn in burrowing. If the statocysts are destroyed, this compensating ability is lost.

Regeneration

Polychaetes have relatively great powers of regeneration. Tentacles, palps, and even heads ripped off by predators are soon replaced. Such replacement is a common occurrence in burrowers and tube dwellers. In general, the potential for regeneration is somewhat greater in worms with undifferentiated trunks than in those with thoracic and abdominal regions but some polychaetes, such as *Chaetopterus* and *Dodecaceria,* can regenerate the entire body from a single segment. Cells for regeneration are supplied by the remains of whatever tissues have been lost. Experimental studies indicate that the nervous system plays an important inductive role in regeneration and that the neuroendocrine system is involved in some way. If the nerve cord alone is severed, a new head forms where the cut is made. A lateral secondary head forms if the severed end of a cord, cut just behind the subesophageal ganglion, is pulled through a hole in the lateral body wall.

Reproduction

Asexual reproduction is known in some polychaetes, including cirratulids, syllids, sabellid fan worms, and

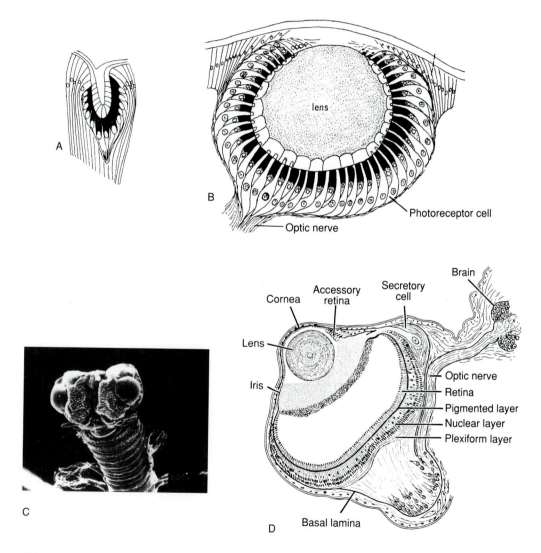

FIGURE 11–43 A, Simple eye of *Mesochaetopterus.* **B,** Eye of *Nereis.* **C,** The head and bulging eyes of the pelagic alciopid, *Vanadis formosa.* **D,** Section through the eye of *Vanadis formosa. (A and B, After Hesse from Fauvel. C, From Rice, S. A. 1987. Reproductive biology, systematics, and evolution in the polychaete family Alciopidae. Biol. Soc. Wash. Bull. 7:114–127. D, After Hesse from Hermans, C. O., and Eakin, R. M. 1974. Fine structure of the eyes of an alciopid polychaete Vanadis tagensis. Z. Morphol. Tiere. 79:245–267.)*

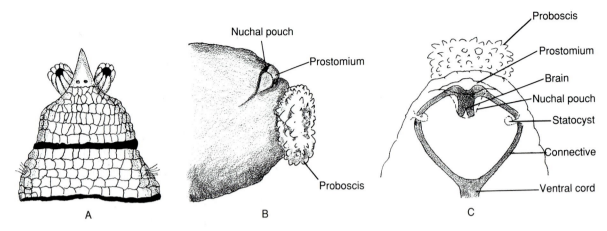

FIGURE 11–44 Anterior of *Notomastus latericeus* with everted nuchal organs. **B,** Head of *Arenicola* (side view). **C,** Statocysts and anterior part of nervous system of *Arenicola*. *(A, After Rullier from Fauvel. B and C, After Wells.)*

spionids; it takes place by budding or division of the body into two parts or into a number of fragments (Fig. 11–45C).

As far as we know, most polychaetes reproduce only sexually, and the majority of species are dioecious. Polychaete gonads are usually distinct organs but, depending on species, vary in position and number. In general, they occur in the connective tissue associated with such structures as septa, blood vessels, and the lining of the coelom (Fig. 11–46).

In the primitive state most of the segments produce gametes, and this is true of many polychaetes, but the gonads may be restricted to **genital segments** in some species. When there are distinct thoracic and abdominal regions, the gonads are usually limited to the abdomen. Among the few hermaphroditic polychaetes, some fan worms have anterior abdominal segments that produce eggs and posterior ones that produce sperm.

The gametes often are shed into the coelom as gametogonia or primary gametocytes, and maturation takes place in the coelomic fluid (Fig. 11–46A). When the worm is mature, the coelom is packed with eggs or sperm; in species in which the body wall is thin or not densely pigmented, the gravid condition is easily apparent. For example, the abdomen of a ripe male *Pomatoceros* appears white, and that of a female, bright pink or orange, because of the color of the sperm and eggs respectively. The blue eggs of the

red, white, and blue worm, *Proceraea fasciata* (Syllidae), show through the red-banded, whitish body.

There is considerable diversity in the ways gametes reach the exterior. A few polychaetes, such as the capitellids, have separate **coelomoducts,** or gonoducts. They develop at the time of sexual maturity, one pair per segment, somewhat like nephridia, and possess a ciliated funnel that receives the eggs or sperm. In many species a coelomic funnel, which is specialized for gathering gametes from the coelomic fluid, joins the nephridium so that the gametes leave the body through the nephridiopores. The nephridia alone serve as gonoducts in such groups as syllids and spionids, the organs enlarging with gamete development (Fig. 11–39B). In many male nereids the sperm may exit through special anal apertures.

The escape of gametes through a rupture in the body wall is perhaps a specialized condition. Rupturing is found among nereids, syllids, and eunicids, all of which become pelagic at sexual maturity, and in nephtyids, which do not. After rupture of the body wall, the adults die.

Epitoky

Epitoky is a reproductive phenomenon characteristic of many polychaetes and is especially well known in nereids, syllids, and eunicids. Epitoky is the formation of a pelagic reproductive individual, or **epitoke,** that is adapted for leaving bottom burrows, tubes, and

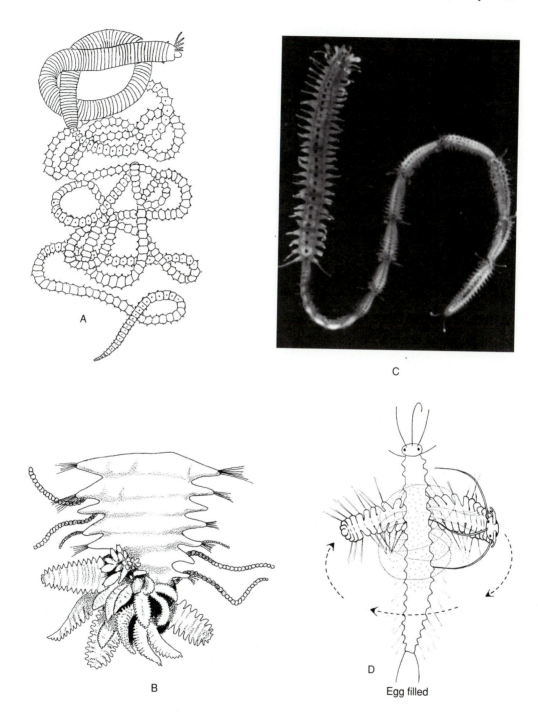

FIGURE 11–45 A, *Palola viridis,* the Samoan palolo worm, with posterior, epitokal region. **B,** Posterior of *Trypanosyllis,* showing cluster of budding epitokes. **C,** Budding of epitokes (stolonization) in the syllid, *Myrianida pachycera.* **D,** Syllid polychaetes during swarming. Male is swimming around female and releasing sperm. *(A, After Woodworth from Fauvel. B, After Potts from Fauvel. D, After Girdholm.)*

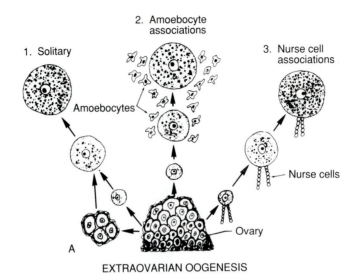

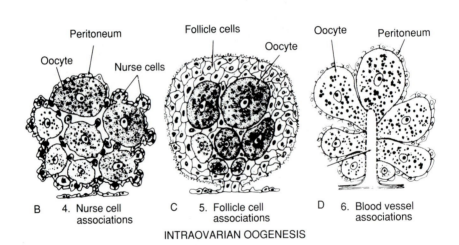

FIGURE 11–46 Polychaete ovaries and testes are compact organs situated beneath the peritoneum. In ovaries, shown here, oogenesis may occur after eggs are released from the ovary (**A**), or it may be completed within the ovary (**B–D**). *(From Eckelbarger, K. J. 1983. Evolutionary radiation in polychaete ovaries and vitellogenic mechanisms: Their possible role in life history patterns. Can. J. Zool. 61:487–504.)*

other habitations. Epitokal modifications include changes in the formation of the head, the structure of the parapodia and setae, the size of the segments, and the segmental musculature.

Epitokous individuals arise from a nonreproductive **atoke** either by direct transformation of the entire individual, as in nereids, or by transformation and separation of the posterior end from the atoke, as in eunicids and syllids. Syllid epitokes usually arise as buds at the caudal end of the atoke (Fig. 11–45A–C).

Often the gamete-bearing segments of the epitoke are the most strikingly modified, and the body of the worm appears to be divided into two markedly different regions. For example, the epitoke of *Nereis irrorata* (and *Nereis succinea*) has large eyes and reduced prostomial palps and tentacles (Fig. 11–47). The anterior 15 to 20 trunk segments are not greatly modified, but the remaining segments, forming the epitokal region and packed with gametes, are much enlarged; their parapodia contain fans of long, spatulate swimming setae. In *Palola (Eunice) viridis,* the Samoan palolo worm, the anterior end of the worm is unmodified, and the epitokal region consists of a chain of egg-filled segments (Fig. 11–45A).

Swarming

Usually, epitokous polychaetes swim to the surface during the shedding of the eggs and sperm. This synchronous behavior, known as **swarming,** congregates sexually mature individuals in a relatively short time and increases the likelihood of fertilization. (See p. 72 for discussion of the general importance of synchrony and proximity.) Experimental evidence indicates that the female produces a pheromone that attracts the male and stimulates shedding of the sperm. The sperm in turn stimulate the shedding of the eggs. The male syllid, *Autolytus,* for example, swims in circles around the female, touching her with his antennae and releasing sperm (Fig. 10–45D).

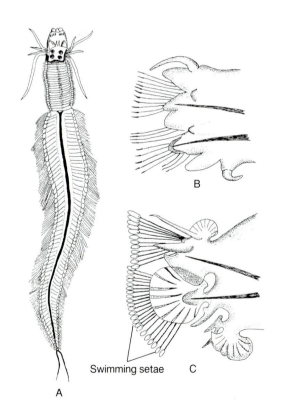

Swimming setae

FIGURE 11–47 A, Epitokous male of *Nereis irrorata.* **B** and **C,** Parapodia of atoke (**B**) and epitoke (**C**) of *Nereis irrorata* male. *(A, After Rullier from Fauvel. B and C, After Fauvel.)*

Swarming is induced in *Autolytus edwardsi* by changes in light intensity, and the worms leave the bottom and swim to the surface at dawn and dusk. Swarming often coincides with lunar periods. *Odontosyllis enopla* in the West Indies and Bermuda swarms in the summer about 50 to 60 min after sunset up to 12 days following a full moon. The worms luminesce when they reach the surface, and when males and females swim around each other, releasing gametes, they create small circles of light. Striking examples of swarming lunar periodicity are displayed by the so-called palolo worms. The name *palolo* originally referred to the Samoan species of the eunicid genus *Palola,* but it is now applied to other species as well. The Samoan palolo worm occupies rock and coral crevices below the low-tide mark and releases epitokes in October or November at the beginning of the last lunar quarter. The natives, who consider the epitokes a great delicacy, eagerly await the predicted night of swarming and scoop up great numbers of the worms from the ocean surface.

The West Indian palolo, *Eunice schemacephala,* lives in habitats similar to those of the Samoan palolo. The worm is negatively phototactic and emerges from the burrow to feed only at night. Swarming takes place in July near the last quarter of a lunar cycle. At three or four o'clock in the morning during such a period, the worm backs out of its burrow, and the caudal, sexual epitokal region breaks free. The epitoke swims to the surface, where it makes spiral motions. By dawn the ocean surface is covered with sexual bodies, and at the rising of the sun the epitokes burst. Fertilization immediately follows rupture, and acres of eggs may cover the sea. A ciliated larval stage is attained by the next day, and in three days the larvae sink to the bottom.

Control of Reproduction

Polychaete reproductive events are regulated by hormones. The hormones are neurosecretions produced by the brain or, in the case of syllids, by nervous elements of the sucking foregut. In worms, such as nereids and syllids, which reproduce only once and then die, the hormone regulates the entire reproductive state, that is, both the production of gametes and the development of epitokal features. In worms that breed more than once, a hormone is required for gamete development, especially for eggs, and the hormone effect is largely limited to that process.

The precise mechanisms that control swarming and the relationship between swarming and normal control of reproduction are still poorly understood. The relation of lunar phases to swarming periods differs among species, and swarming even occurs on cloudy nights. This makes any hypothesis based merely on light intensity difficult to support.

Egg Deposition

Many polychaetes shed their eggs freely into the sea water, where they become planktonic. Some polychaetes, however, retain the eggs within the tubes or burrows or lay them in mucous masses that are attached to tubes or to other objects. For example, *Axiothella* (Fig. 11–48A), a bamboo worm, produces a small ovoid egg mass that is attached to the chimney of the tube.

Many polychaetes brood their eggs. There are tubicolous species, such as some spionids and serpulids, that brood their eggs within the tubes. Some species of *Spirorbis* brood their eggs in the cavity of the operculum, and *Autolytus* broods its eggs within a secreted sac attached to the ventral surface of the body (Fig. 11–48B). A few species, such as *Nereis limnicola,* brood their eggs within the coelom.

Development

The polychaete egg contains a variable amount of yolk, depending on the species, and cleavage is spiral and holoblastic. A displaced blastocoel is usually pre-

sent, but a stereoblastula develops in *Nereis, Capitella,* and others. Gastrulation takes place by invagination, epiboly, or both.

The Trochophore

After gastrulation, the embryo rapidly develops into a top-shaped trochophore larva (Figs. 11–49A; 11–50A, B; p. 340). The greatest development of larval structures is attained in planktotrophic trochophores, those that feed on plankton (e.g., *Owenia, Polygordius,* phyllodocids, serpulid fan worms; Fig. 11–49E). The trochophores of many species, however, are lecithotrophic, that is, yolky and nonfeeding (e.g., nereids and eunicids), and their short larval existence is spent near the bottom.

Metamorphosis

Polychaete metamorphosis transforms the trochophore into the juvenile body form (Fig. 11–50). The most conspicuous feature of metamorphosis is the gradual lengthening of the growth zone—the region between the mouth and the telotroch—as trunk segments form and develop (Fig. 11–50). The segments develop from anterior to posterior, and the germinal region remains just in front of the terminal pygidium. Thus, in adult polychaetes the oldest segments are those closest to the head of the worm. In the pretrochal region, which originally formed the major part of the body of the trochophore, the cells of the apical plate form the prostomium and the brain.

Metamorphosis may result in the immediate termination of a planktonic existence, but more often the elongated, metamorphosing larvae remain planktonic for varying lengths of time. The metamorphosing stages of spionids, sabellariids, and oweniids even possess greatly enlarged, erectile, anterior setae that serve as flotation or protective devices.

In many polychaetes the trochophore stage is passed in the egg prior to hatching, which occurs at various times during advanced development. In such species metamorphosis is more direct because larval structures are never greatly developed to begin with. There may still be a free-swimming, post-trochophoral, larval stage. For example, in *Autolytus* an elongated larva breaks free from the brood sac of the mother. On the other hand, *Axiothella mucosa* and *Scoloplos armiger* have no planktonic stages and assume the adult mode of existence on emerging from the jelly egg case.

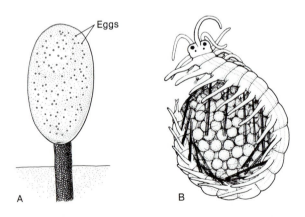

FIGURE 11–48 A, *Axiothella mucosa* egg jelly mass attached to the end of the tube. **B,** The syllid, *Autolytus,* carrying the egg mass beneath its body. *(A, B, After Throson.)*

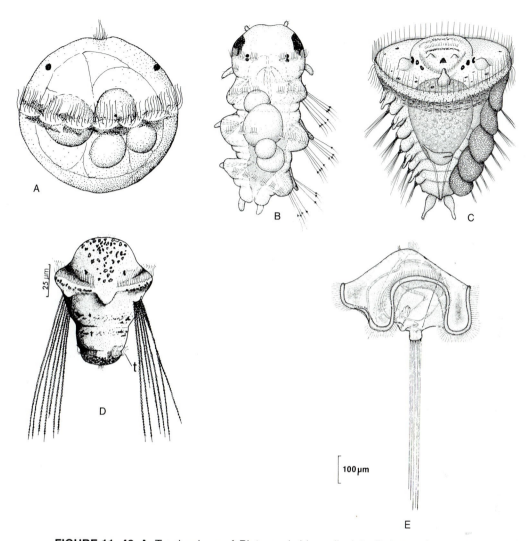

FIGURE 11–49 A, Trochophore of *Platynereis bicanaliculata*. **B,** Later, three-setiger stage of *Platynereis*. **C,** Late metatrochophore of the scaleworm, *Halosydna brevisetosa*. **D,** Larva of the sabellariid, *Phragmatopoma* (t = telotroch). **E,** Larva, called a mitraria, of *Owenia*. *(A–C, From Blake, J. D. 1975. The larval development of Polychaeta from the northern California coast. III. Eighteen species of Errantia. Ophelia. 14:23–84. D, From Eckelbarger, K. J. 1976. Larval development and population aspects of the reef-building polychaete Phragmatopoma lapidosa from the east coast of Florida. Bull. Mar. Sci. 26:117–132. E, After Wilson from Smith, P. R., Ruppert, E. E., and Gardiner, S. L. 1987. A deuterostome-like nephridium in the mitraria larva of Owenia fusiformis (Polychaeta, Annelida). Biol. Bull. 172:315–323.)*

According to the polychaete specialist K. Fauchald, the developmental patterns of the relatively small number of polychaetes that have been studied appear to fall into three categories. Annual species, those that live only one or two years and spawn only once, producing a large number of relatively small eggs. They have well-developed feeding larvae that are planktonic for a week or more. Perennial species, which live and breed for more than one year, produce a small number of large, yolky eggs and nonfeeding, benthic larvae. Multiannual species, those with such short life spans that several generations can be pro-

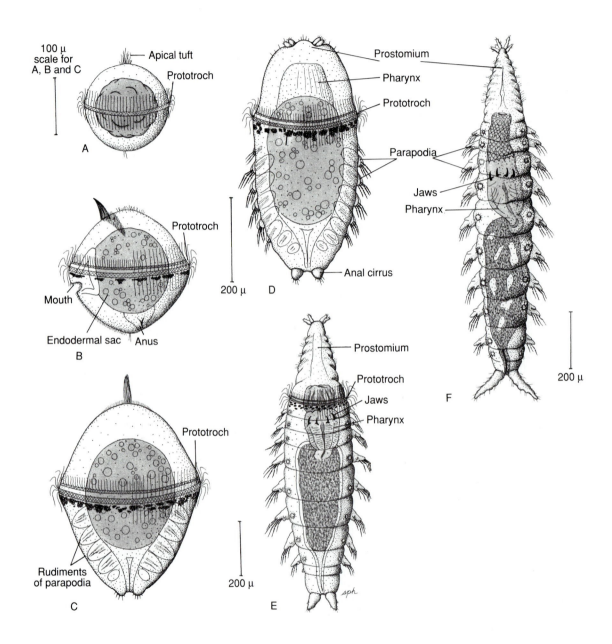

FIGURE 11–50 Larval stages of *Glycera convoluta*. **A,** Early trochophore (15 h). **B,** Later trochophore (10 days). **C,** Young metatrochophore (4 weeks). **D,** Metatrochophore at 7 weeks. Although still a swimming stage, it frequently comes to rest on the bottom. **E,** Postlarva at 8 weeks. Metamorphosis follows metatrochophore illustrated in **D.** Larva becomes benthnic and raptorial. **F,** Young worm at 2 months. Compare with Figure 11–21**A**. *(All after Cazaux, C. 1967. Vie et Milieu. 18:559–571.)*

duced in one year, produce numerous small batches of large, yolky eggs that hatch nonfeeding, benthic larvae.

Population Biology

Burrowing and tubicolous polychaetes commonly occur in enormous numbers on the ocean floor and compose a major part of the soft-bottom infauna. A study by S. Santos and J. Simon in Tampa Bay, Florida, reported the average density of polychaetes to be 13,425 individuals per square meter, that is, living in the sediment beneath that area. They belonged to some 37 species. On the upper continental slope and the deep ocean floor polychaetes compose 40 to 80% of the infauna.

In general, such populations do not appear to be limited by the resources available, at least not in shallow water. Predation and other pressures usually prevent annelidan, molluscan, and other infaunal populations from ever reaching the carrying capacity of the habitat. When areas in the York River estuary of the Chesapeake Bay were protected from fish and crabs by means of wire cages, over half of the species in the polychaete population increased from two to many times their numbers in unprotected conditions.

SUMMARY

1.   The evolution of the class Polychaeta from the ancestral burrowing annelids is perhaps correlated with a shift to a crawling, surface existence. Various lines then diverged to invade other habitats and to assume other modes of existence, including burrowing in soft bottoms.

2.   Surface-dwelling, errant polychaetes possess well-developed parapodia and heads (prostomium) with sense organs. They crawl with the parapodia, using them as leglike appendages. Most are predaceous, but some are herbivorous or scavengers. They typically possess an eversible pharynx equipped with jaws.

3.   Gallery-dwelling and burrowing polychaetes show some convergence with earthworms. The prostomium is usually small and more or less conical with poorly developed sense organs. The parapodia are smaller than those of surface dwellers

and provide anchorage in peristaltic movement through the galleries. Gallery dwellers are predators or deposit feeders.

4.   The more sedentary burrowers, which live in simple vertical or U-shaped excavations, move by peristaltic contractions and possess parapodial ridges with hooked setae for gripping the mucus-lined burrow walls. Those species that are nonselective deposit feeders have a small prostomium without conspicuous sense organs; those that are selective deposit feeders usually possess head appendages, such as tentacles or long palps, which are used in feeding.

5.   Tubicolous polychaetes live in tubes composed of secreted materials or of sand grains or shell fragments cemented together. The majority are selective or nonselective deposit feeders or filter feeders and are adapted much like sedentary burrowers. The few that are predatory are similar to errant surface dwellers.

6.   Most large polychaetes possess gills (thin-walled evaginations with an interior vascular supply), or some part of the parapodium is especially modified as a gas exchange surface.

7.   Internal transport is provided by a blood-vascular system, by coelomic fluid, or by a combination. Gas transport frequently involves respiratory pigments, of which hemoglobin is the most common, but some polychaetes possess chlorocruorin or hemerythrin. Blood-vascular hemoglobin and chlorocruorin are always extracellular and large molecules; blood-vascular hemerythrin and coelomic hemoglobin are always intracellular and small molecules.

8.   Most polychaetes possess paired, segmental metanephridial systems, in which each nephrostome opens into the coelomic compartment that is anterior to the one housing the tubule. Polychaetes that lack a blood-vascular system possess protonephridia, which may be primitive for the phylum.

9.   Primitively, polychaetes have a ladder-like ventral nervous system.

10.   The sexes are separate in most polychaetes. The gonads are in the connective tissue but the gametes are often released into the coelom for maturation and storage. After maturing in the coelom, they exit by coelomoducts, coelomoducts joined to nephridia, nephridia alone, or rupture of the body wall. In primitive polychaetes, gametes are associated with most segments.

11. Copulation is rare, and synchronous emission of sperm and eggs is important. Epitoky and swarming bring a dispersed benthic population together for a brief pelagic existence, when gametes are shed and the likelihood of fertilization is increased.

12. A trochophore larva is the basic larval stage of polychaetes.

CLASS OLIGOCHAETA

The class Oligochaeta contains some 3100 species of annelids, including the familiar earthworms and many species that live in fresh water. Some freshwater oligochaetes burrow in bottom mud and silt; others live among submerged vegetation. About 200 marine species have been described, even some that inhabit sediments of the deep sea. Oligochaetes approximate the polychaetes in size; however, the giant earthworms of Australia and other parts of the world may exceed 3 m in length (Fig. 11–51). In general, aquatic species are smaller than earthworms.

According to some authorities, oligochaetes are believed to have evolved directly from the burrowing ancestral marine annelids, independently of the polychaetes. The first oligochaetes were probably burrowers in freshwater sediment. They may have given rise in one direction to the strictly freshwater species that invaded loose bottom debris and in another direction to the earthworms that invaded successively drier sediments.

External Anatomy

In the oligochaetes segmentation is well developed, parapodia are absent, and the prostomium is usually a small, rounded lobe or a small cone without sensory appendages (Figs. 11–52; 11–56). In a few genera, such as *Stylaria* (Fig. 11–52C), the prostomium is drawn out into a tentacle.

Oligochaetes have no parapodia, but with few exceptions, they have setae, although not as diverse as those of polychaetes. Oligochaete setae commonly have tips that are simple, bifid, pectinate, or in other ways different from the shaft (Fig. 11–52B). Genital setae may be more complex.

In general, the longer setae are characteristic of aquatic species (Fig. 11–52C); the setae of earthworms project only a short distance beyond the integument and are commonly sigmoid. On each side of a segment there are setal sacs, in which the setae are secreted and from which they emerge as groups or bundles. Two of the groups are ventral, and two are ventrolateral or dorsolateral. The number of setae per bundle varies from 1 to 25 (Fig. 11–52C). In any

FIGURE 11–51 An Australian giant earthworm. *(Photograph courtesy of Globe photos.)*

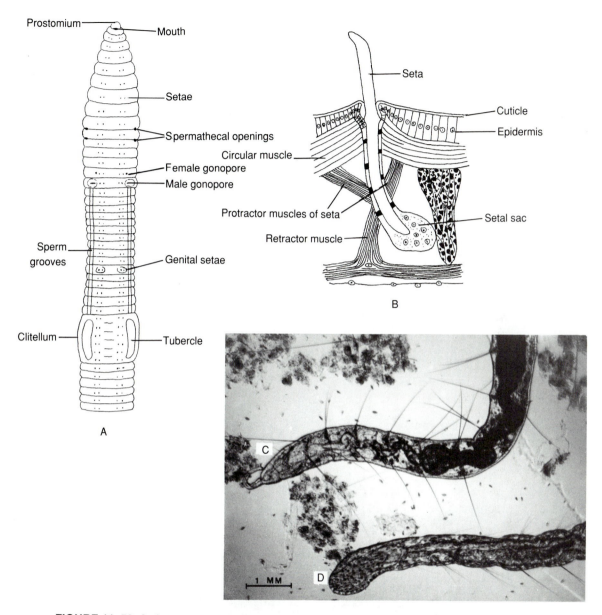

FIGURE 11–52 A, Anteroventral surface of the earthworm, *Lumbricus terrestris.* **B,** Body wall of the earthworm, *Pheretima* (transverse section). **C** and **D,** Anterior of two freshwater olio-gochaetes, *Stylaria* (**C**) and *Aeolosoma* (**D**). *Stylaria* has a long tentacular prostomium. *(A, After Stephenson from Avel. B, After Bahl from Avel. C, D, Photograph courtesy of Betty M. Barnes.)*

case, they are generally less numerous in these worms than in polychaetes, hence the origin of the name *Oligochaeta,* meaning "few setae."

In most earthworms, such as *Lumbricus,* and in some aquatic families, the setae are limited to eight with two setae forming each group (Fig. 11–52A). Attached to the base of each seta are protractor and

retractor muscles that allow the seta to be extended or withdrawn (Fig. 11–52B).

In mature oligochaetes certain adjacent segments in the anterior half of the body are thickened and swollen by glands that secrete mucus for copulation and also secrete the cocoon. The glandular area of these segments, collectively called the **clitellum,** partially or completely covers the segments and often forms a conspicuous girdle around the body (Fig. 11–52A). The presence of a clitellum and hermaphroditism, egg-laying in a **cocoon,** and the restriction of gonads to a few genital segments, distinguish the oligochaetes from the polychaetes.

Body Wall and Coelom

The structure and histology of the oligochaete body wall, especially in terrestrial species, is essentially like that of burrowing polychaetes. A thin cuticle overlies an epidermal layer, which contains mucus-secreting gland cells. The circular muscles are well developed, and the septa partitioning the coelom are relatively complete. Earthworms, which have the best developed septa, may possess sphincters around septal perforations to control the flow of coelomic fluid from one segment to another.

In most earthworms each coelomic compartment, except at the extremities, is connected to the outside by a middorsal pore situated in the intersegmental furrows and provided with a sphincter. These pores exude coelomic fluid, which aids in keeping the integument moist. When disturbed, some giant earthworms squirt fluid several centimeters.

Locomotion

Oligochaetes move by peristaltic contractions, as described for burrowing annelids (p. 513, Fig. 11–2). Earthworm locomotion has been studied extensively by several authorities. (Fig. 11–53). Circular muscle contraction and the consequent elongation of segments are most important in crawling and always generate a coelomic fluid pressure pulse. Longitudinal muscle contraction is more important in burrowing, dilating the burrow, or anchoring the segments against the burrow wall.

Setae are extended during the longitudinal contraction and retracted during circular contraction. Each segment moves forward in steps of 2 to 3 cm at

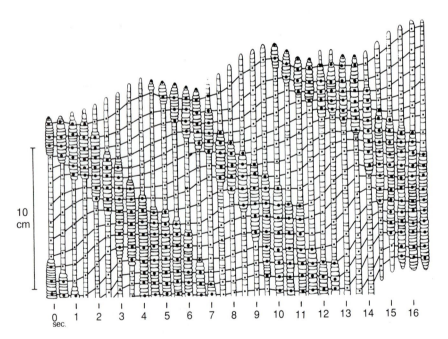

FIGURE 11–53 Diagram showing the mode of locomotion of an earthworm. Segments undergoing longitudinal muscle contraction are marked with the larger dot and drawn twice as wide as those undergoing circular muscle contraction. The forward progression of a segment during the course of several waves of circular muscle contraction is indicated by the horizontal lines connecting the same segments. *(After Gray and Lissman).*

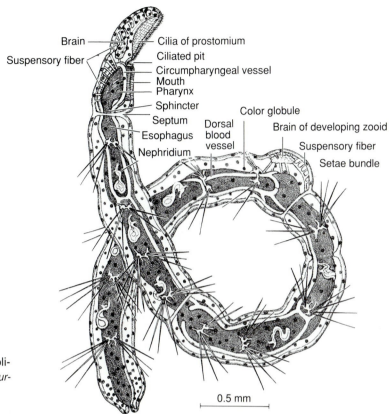

Brain
Suspensory fiber
Cilia of prostomium
Ciliated pit
Circumpharyngeal vessel
Mouth
Pharynx
Sphincter
Septum
Esophagus
Nephridium
Dorsal blood vessel
Color globule
Brain of developing zooid
Suspensory fiber
Setae bundle

0.5 mm

FIGURE 11–54 The freshwater oligochaete, *Aeolosoma. (Drawing courtesy of R. Singer.)*

the rate of seven to ten steps per minute (Fig. 11–53). The direction of contraction waves can be reversed, thus enabling the worm to crawl backward.

Freshwater species move through bottom debris and algae in the same manner as earthworms, but the microscopic aeolosomatids swim by means of a ciliated prostomium (Fig. 11–54).

Habitation and Ecological Distribution

Oligochaetes live in all types of freshwater habitats, where they usually burrow in bottom debris. Only a small number construct tubes. Oligochaetes are most abundant where the water is shallow, although several families have benthic representatives in deep lakes. Abundance of different species of aquatic oligochaetes can be a good indication of water pollution.

Oligochaetes have reinvaded the sea, and some 200 marine species have been described. Most belong to the families Enchytraeidae and Tubificidae and are chiefly inhabitants of the supratidal and intertidal zones, but subtidal species are known even from abyssal and hadal depths. Marine oligochaetes are members of the interstitial fauna, are shallow burrowers, or live beneath intertidal rocks or in algal drift.

Many species are amphibious or transitional between a strictly aquatic and a strictly terrestrial environment. These worms live in marshy or boggy land and on the margins of ponds and streams.

There are ten families of earthworms, of which four contain large numbers of species: Glossoscolecidae, Lumbricidae, Megascolecidae, Moniligastridae. All are burrowers and are found everywhere except in deserts. They often occur in tremendous numbers. As many as 8000 enchytraeids and 700 lumbricids have been reported from a square meter of meadow soil.

Soils containing considerable organic matter, or at least a layer of humus on the surface, maintain the largest fauna of worms, but other soil factors are important to the distribution of terrestrial species. Acid soils are favorable habitats for most earthworms.

A cross section through soil reveals a distinct vertical stratification of worms. The tunnels of larger species, such as *Lumbricus terrestris,* range from the surface to several meters deep, depending on the nature of the soil. Young worms and small species are restricted to the few centimeters of upper humus; others have a wider vertical distribution but are still limited to the upper level of the soil that contains some organic matter.

The terrestrial oligochaete constructs its burrow by forcing its anterior end through crevices and by swallowing soil. The egested material and mucus are plastered against the burrow wall, forming a distinct lining. Some egested material is removed from the burrow as castings. According to P. J. Bolton and J. Phillipson, the gut turnover time in the lumbricid, *Allolobophora rosea,* is 1 to 2.5 h. The burrows may be complex, with two openings and horizontal and vertical ramifications. *Lumbricus terrestris* always plugs its burrow with debris pulled into the hole. It is not yet clear why large numbers of *Lumbricus terrestris* leave their burrows after heavy rains, for high mortality certainly results. Moreover, many species of earthworms, including *Lumbricus terrestris,* can survive in submerged ground for months.

As first demonstrated by Charles Darwin, the activities of earthworms have a beneficial effect on the soil. The extensive burrows increase soil drainage and aeration, but more important are the mixing and churning of the soil. Deeper soil is brought to the surface as castings, and organic material is moved to lower levels. Some tropical species produce enormous castings. For example, the tower-like castings of *Hyperoidrilus africanus* may reach 8 cm in height and 2 cm in diameter.

Earthworms' ability to churn soil can be demonstrated in a container filled halfway with sand and halfway with potting soil. Five worms will thoroughly mix 500 ml of sand with 500 ml of soil in several months.

In the tropics a number of arboreal species have been reported. Such terrestrial forms live in accumulated humus and detritus in leaf axils and branches of trees. Aquatic species inhabit the water reservoirs of bromeliad epiphytes, living on the trunks and branches of tropical trees.

Members of the major aquatic families occur throughout the world wherever suitable habitats exist. Many terrestrial species, however, have very limited geographical distribution. Humans have certainly been an important agent in the spread of earthworms. For example, the earthworm fauna of the larger Chilean cities consists solely of European species, which have displaced the endemic forms.

Nutrition

The majority of oligochaete species, both aquatic and terrestrial, are scavengers and feed on dead organic matter, particularly vegetation. Earthworms feed on decomposing matter at the surface and may pull leaves into the burrow. They also deposit feed, ingesting organic material obtained from mud or soil while burrowing. The food source and feeding habits of earthworms are related to the species zonation described in the previous section.

Fine detritus, algae, and other microorganisms are important food sources for many tiny, freshwater species. The common, minute *Aeolosoma* collects detritus with its prostomium (Fig. 11–54). The ciliated ventral surface of the prostomium is placed against the substratum, and the center is elevated by muscular contraction. The partial vacuum dislodges particles, which are then swept into the mouth by cilia. Members of the genus *Chaetogaster,* small oligochaetes that are commensals on freshwater snails, are raptorial and catch amebas, ciliates, rotifers, and trematode larvae by a sucking action of the pharynx.

The digestive tract is straight and relatively simple (Fig. 11–55). The mouth, situated beneath the prostomium, opens into a small buccal cavity, which in turn opens to a more spacious pharynx. The dorsal wall of the pharyngeal chamber is muscular and glandular and forms a bulb or pad, which is the principal ingestive organ. In aquatic forms the pharynx is everted and the mucus-covered muscular disc collects particles on an adhesive pad (Fig. 11–56). In earthworms the pharynx acts as a pump. Pharyngeal glands produce a salivary secretion containing mucus and enzymes.

The pharynx opens into a narrow, tubular esophagus, which may be modified at different levels to form a gizzard or, in lumbricid earthworms, a crop. In some forms there are two to ten gizzards, each occupying a separate segment. The **gizzard,** which is used for grinding food particles, is lined with cuticle and is very muscular. The **crop** is thin-walled and is a storage chamber.

A characteristic feature of the oligochaete gut is the presence of **calciferous glands** in certain parts of the esophageal wall. When highly developed, the glandular region becomes completely separated from

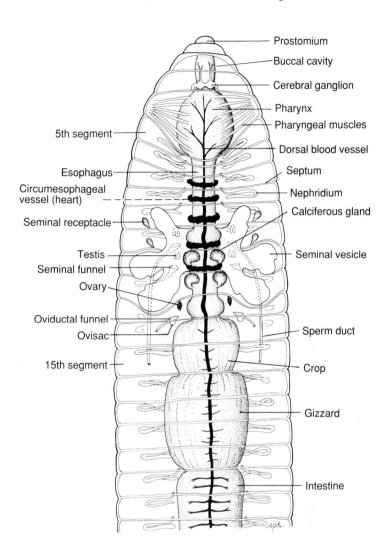

Prostomium
Buccal cavity
Cerebral ganglion
Pharynx
Pharyngeal muscles
Dorsal blood vessel
Septum
Nephridium
Calciferous gland
Seminal vesicle
Sperm duct
Crop
Gizzard
Intestine

5th segment
Esophagus
Circumesophageal vessel (heart)
Seminal receptacle
Testis
Seminal funnel
Ovary
Oviductal funnel
Ovisac
15th segment

FIGURE 11–55 Dorsal view of the anterior internal structures of the earthworm, *Lumbricus.*

the esophageal lumen and may appear externally as lateral or dorsal swellings (Fig. 11–55). The calciferous glands secrete calcium carbonate, in the form of calcite crystals, into the esophagus. The crystals are then transported along the gut, but are not reabsorbed, and eventually pass out of the body with the feces. The calciferous glands do not play a role in digestion and their function is uncertain, but two hypotheses have been suggested. According to aquatic chemists, soil CO_2 levels can be several hundred times higher than atmospheric levels because of bacterial respiration. When earthworms encounter such surroundings, elimination of their own respiratory CO_2 by diffusion may be hampered by an unfavorable concentration gradient. To overcome this difficulty, CO_2 (as bicarbonate ions) in the blood and other tissues may combine with calcium ions in the calciferous glands to form calcite, thus eliminating CO_2 indirectly via the gut. It has also been suggested that the glands function to eliminate excess calcium taken in with food.

The intestine forms the remainder of the digestive tract and extends as a straight tube through all but the anterior quarter of the body. The anterior half of the intestine is the principal site of secretion and digestion, and the posterior half is primarily absorptive. In addition to the usual classes of digestive enzymes, the intestinal epithelium of earthworms also secretes cellulase and chitinase. The absorbed food materials

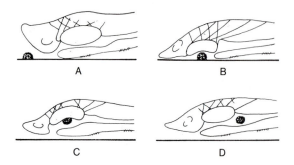

A

B

C

D

FIGURE 11–56 Mechanism of ingestion in *Aulophorus carteri,* which possesses a dorsal, padlike pharynx. *(After Marcus and Avel.)*

are passed to blood sinuses that lie between the mucosal epithelium and the intestinal muscles. The surface area of the intestine is increased in many earthworms by a ridge or fold, called a **typhlosole,** which projects internally from the middorsal wall.

Surrounding the intestine and investing the dorsal vessel of oligochaetes is a layer of yellowish peritoneal cells, called chlorogogen cells, which play a vital role in intermediate metabolism, similar to the role of the liver in vertebrates. Chloragogen tissue is the chief center of glycogen and fat synthesis and storage. Storage and detoxification of toxins, hemoglobin synthesis, protein catabolism and formation of ammonia, and synthesis of urea also take place in these cells. In terrestrial species silicates obtained from food material and the soil are removed from the body and deposited in the chloragogen cells as waste concretions.

Gas Exchange

Gas exchange in almost all oligochaetes, both aquatic and terrestrial, takes place by the diffusion of gases through the general body integument, which in the larger species contains a capillary network within the outer epidermal layer.

True gills occur in only a few oligochaetes. Species of the aquatic genera *Dero* (Fig. 11–57B) and *Aulophorus* have a circle of finger-like gills at the posterior end of the body. A tubificid, *Branchiura* (Fig. 11–57A), has filamentous gills situated dorsally and ventrally in the posterior quarter of the body.

The larger oligochaetes usually have hemoglobin dissolved in the plasma. The hemoglobin of *Lumbri-*

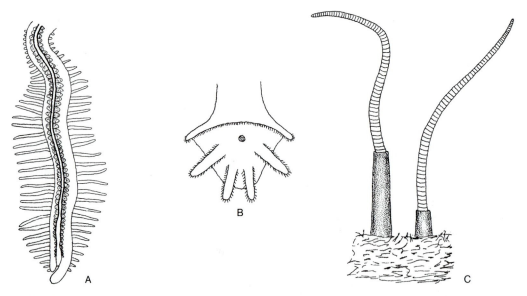

A

B

C

FIGURE 11–57 A, Posterior of *Branchiura sowerbyi,* showing dorsal and ventral filamentous gills. **B,** Posterior of *Dero,* showing circlet of gills around the anus. **C,** Posterior of the body of *Tubifex* projected from the tube and waved about in the water, facilitating gas exchange. *(A, After Beddard from Avel. B, C, After Pennak, R. W. 1978. Freshwater Invertebrates of the United States, 2nd Edition. John Wiley and Sons, New York.)*

cus transports 15 to 20% of the oxygen utilized under ordinary burrow conditions, in which the partial pressure of oxygen is about the same as that in the atmosphere above ground. When the partial pressure drops, the hemoglobin compensates by increasing its carrying capacity.

Many aquatic oligochaetes tolerate relatively low oxygen levels and, for a short period, even a complete lack of oxygen. Members of the family Tubificidae, which live in stagnant mud and lake bottoms, are notable examples. There are members of this family, such as *Tubifex tubifex,* that die from long exposure to ordinary oxygen tensions. *Tubifex* ventilates in stagnant water by exposing its posterior end out of the mud and waving it about (Fig. 11–57C).

Internal Transport

The circulatory system of oligochaetes (Fig. 11–55) is basically similar to that of polychaetes. Branches from the segmental vessels send blood into capillaries in the integument (best developed in earthworms) and supply the various segmental organs. The vessels lack an endothelium and are lined only by the basal lamina of the overlying peritoneum, as in polychaetes. The dorsal vessel is contractile and is the principal means by which the blood is propelled. The vessels in oligochaetes (commonly referred to as hearts) are certain anterior, commissural vessels that are conspicuously contractile and function as accessory organs for blood propulsion. The number of such hearts varies. Five are present in *Lumbricus* and surround the esophagus (Fig. 11–55). Only one pair of hearts is present in *Tubifex,* and this pair is circumintestinal. Valves in the form of folds in the vessel walls are present in the hearts and may also be found in the dorsal vessel at junctions containing segmental vessels.

Excretion, Water Balance, and Quiescence

Adult oligochaetes have a metanephridial system (Fig. 11–58), and typically, there is one pair of metanephridial tubules per segment except at the extreme anterior and posterior ends. In the segment following the nephrostome, the tubule is greatly coiled, and in some species, such as *Lumbricus,* there are several separate groups of loops or coils (Fig. 11–58B). Before the nephridial tubule opens to the outside, it is sometimes dilated to form a bladder. The nephridiopores are usually situated on the ventrolateral surfaces of each segment.

In contrast to the majority of oligochaetes, which possess in each segment a single, typical pair of nephridia called **holonephridia,** many earthworms of the families Megascolecidae and Glossoscolecidae are peculiar in possessing additional nephridia, which are multiple or branched. Either typical or modified nephridia may open to the outside through nephridiopores, or they may open into various parts of the digestive tract, in which case they are termed **enteronephric.** A single worm may possess a number of different types of these nephridia, each being restricted to certain parts of the body.

Earthworms excrete urea, but they are less perfectly ureotelic than are other terrestrial animals. Although urea is present in the urine of *Lumbricus* and other earthworms, ammonia remains an important excretory product. The level of urea depends on the condition of the worm and the environmental situation.

Salt and water balance, of particular importance in freshwater and terrestrial environments, is regulated in part by the nephridia (Fig. 11–58C). The urine of both terrestrial and freshwater species is hypoosmotic, and considerable reabsorption of salts must take place as fluid passes through the nephridial tubule. Some salts are also actively absorbed by the integument.

In the terrestrial earthworms water absorption and loss occur largely through the integument. Under normal conditions of adequate water supply, the nephridia excrete a copious hypoosmotic urine. It is not certain whether reabsorption by the ordinary nephridia is of importance in water conservation, but the enteronephric nephridia do appear to represent an adaptation for the retention of water. By passing the urine into the digestive tract, much of the remaining water can be reabsorbed as it goes through the intestine. Worms with enteronephric systems can tolerate much drier soils or do not have to burrow so deeply during dry periods.

A few aquatic oligochaete species are capable of encystment during unfavorable environmental conditions. The worm secretes a tough, mucous covering that forms the cyst wall. Some species form summer cysts for protection against desiccation; others form winter cysts when the water temperature becomes low.

During dry seasons or during the winter, earthworms migrate to deeper levels of the soil, down to 3 m in the case of certain Indian species. After moving to deeper levels, an earthworm often undergoes a pe-

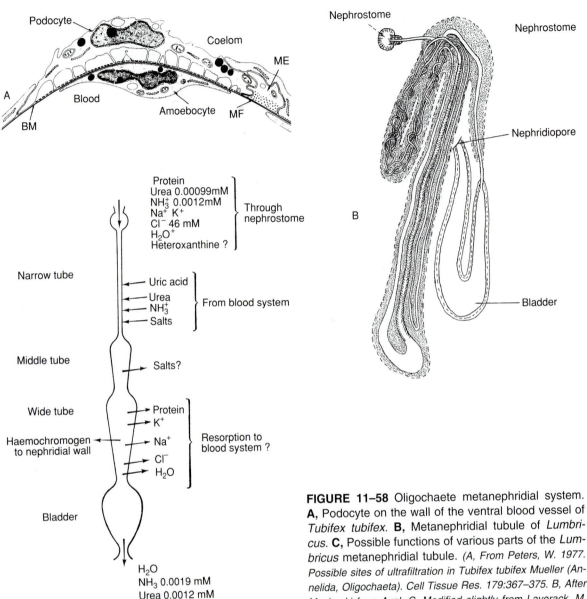

FIGURE 11–58 Oligochaete metanephridial system. **A,** Podocyte on the wall of the ventral blood vessel of *Tubifex tubifex*. **B,** Metanephridial tubule of *Lumbricus*. **C,** Possible functions of various parts of the *Lumbricus* metanephridial tubule. *(A, From Peters, W. 1977. Possible sites of ultrafiltration in Tubifex tubifex Mueller (Annelida, Oligochaeta). Cell Tissue Res. 179:367–375. B, After Maziarski from Avel. C, Modified slightly from Laverack, M. S. 1963. The Physiology of Earthworms. Pergamon Press, Oxford, p. 67.)*

riod of quiescence and in dry periods may lose as much as 70% of its water. Balance is restored and activity resumed as soon as water is again available.

Nervous System

In most oligochaetes, there are two fused, ventral nerve cords situated inside the muscle layers of the body wall. The oligochaete brain has shifted posteri- orly and in lumbricids lies in the third segment, above the anterior margin of the pharynx (Fig. 11–59).

Like polychaetes, oligochaetes have giant axons. The earthworms possess five giant nerve fibers. Three are quite large and are grouped at the middorsal side of the ventral nerve cord. The other two are less conspicu- ous; they are situated midventrally and are rather widely separated. The middorsal fiber is fired by ante- rior stimulation, and the two dorsolateral fibers are fired by posterior stimulation. They make connection in each ganglion with giant motor neurons that innervate the longitudinal muscles.

The basic peristaltic locomotor rhythm of body wall contraction appears to arise from the ventral nerve cord, but contraction can be initiated indirectly by way of reflex arcs involving sensory neurons in the body wall. A wave of longitudinal muscular contraction exerts a pull on the following segments. This pull apparently stimulates the sensory neurons of those following segments, and a reflex action is initiated that causes the contraction of the circular muscle layer and the elongation of the segments. This traction-stimulated reflex was illustrated by B. Friedlander, who in 1888, severed a worm and loosely connected the two parts by a thread. Although the nerve cord was cut, a peristaltic wave continued from one part of the worm to the other, resulting from the pull of the thread on the severed posterior half and from the initiation of the traction reflex.

The subpharyngeal ganglion is the principal center of motor control and vital reflexes and dominates the succeeding ganglia in the chain. All movement ceases when the subpharyngeal ganglion is destroyed. Motor control continues normally following removal of the brain, but the worm loses its ability to correlate movement with external environmental conditions. The relation of the subpharyngeal ganglion to the brain is thus somewhat analogous to the relation of the medulla to the higher brain centers in vertebrates.

Sense Organs

Oligochaetes lack eyes, except for a few aquatic forms that have simple pigment-cup ocelli. The integument however, is well supplied with dispersed photoreceptors situated in the inner part of the epidermis, especially at the anterior end. The other receptors have a more or less general distribution in the integument. Clusters of sensory cells that form a projecting tubercle, with sensory processes extending above the cuticle, appear to be chemoreceptors. The **tubercles** form three rings around each segment and are particularly numerous on the more anterior segments and especially on the prostomium, where (in *Lumbricus*) there may be as many as 700/mm².

Reproduction and Development

Asexual Reproduction

Asexual reproduction is very common among many species of aquatic oligochaetes, particularly the aeolosomatids and the naidids. In fact, there are many asexually reproducing naidids in which sexual individuals are rare or have never been observed; a clone

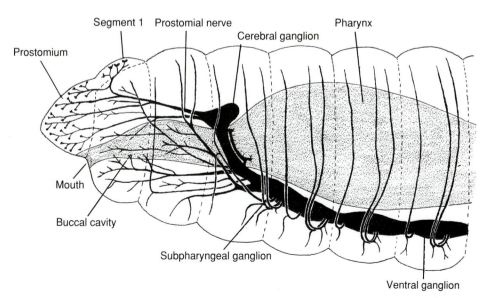

FIGURE 11–59 *Lumbricus's* nervous system (lateral view). *(After Hess from Avel.)*

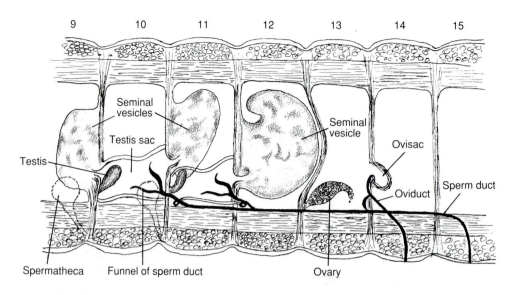

FIGURE 11–60 Reproductive segments of the earthworm, *Lumbricus* (lateral view). *(After Hess from Avel.)*

of *Aulophorus furcatus* was traced through 150 generations for a three-year period with no appearance of sexual individuals and with no diminishing of the fission rate. Other oligochaetes reproduce asexually in the summer and sexually in the fall. Asexual reproduction always involves a transverse division of the parent worm into two or more new individuals. Regeneration commonly precedes the separation of the daughter individuals, and not infrequently, as in species of *Nais,* a new fission zone forms before division has occurred in an old one. Such delayed divisions (paratomy) produce chains of individuals, or zooids, similar to those in certain turbellarian flatworms that also undergo paratomy.

The Reproductive System

The reproductive system of oligochaetes differs from that of polychaetes in a number of striking respects. Oligochaetes are all hermaphrodites, they possess distinct gonads, and the number of reproductive segments is limited to a few genital segments. The oligochaete arrangement is undoubtedly a specialized one, and the polychaete condition is probably primitive. Two oligochaete specialists, R. Brinkhurst and B. Jamieson, have suggested that the ancestral annelids might have resembled oligochaetes but had polychaete reproductive systems.

In most aquatic groups there is usually one ovarian segment followed by one testicular segment; in terrestrial families, two male segments may be present. The gonad-containing segments are situated in the anterior half of the worm, and the female segment or segments are situated behind the male segments (Fig. 11–60). The exact location of genital segments along the trunk varies in different families.

The ovaries and testes, both of which are typically paired, are situated in the genital segments on the lower part of the anterior septum and project into the coelom. Although maturation of the gametes is completed in the coelom, it is typically restricted to special coelomic pouches called **seminal vesicles** and **ovisacs.** Both arise as outpocketings of the septa of the reproductive segments, but the number, size, and position vary.

The reproductive segments, whether male or female, are each provided with a pair of **sperm ducts** or **oviducts** for the exit of sperm or eggs. The ducts extend backward and pass through one or more segments before opening on the ventral surface of the body. In earthworms the two pairs of sperm ducts on each side of the body usually become confluent before opening to the outside through a single, common, male genital pore, which has a raised border or lips. In many aquatic species the chamber (atrium) within the common gonopore contains a penis or an eversible area of the body wall or atrium tip (pseudopenis) (Fig. 11–61C). Glandular tissues, called **prostate glands,** are commonly associated with the male gon-

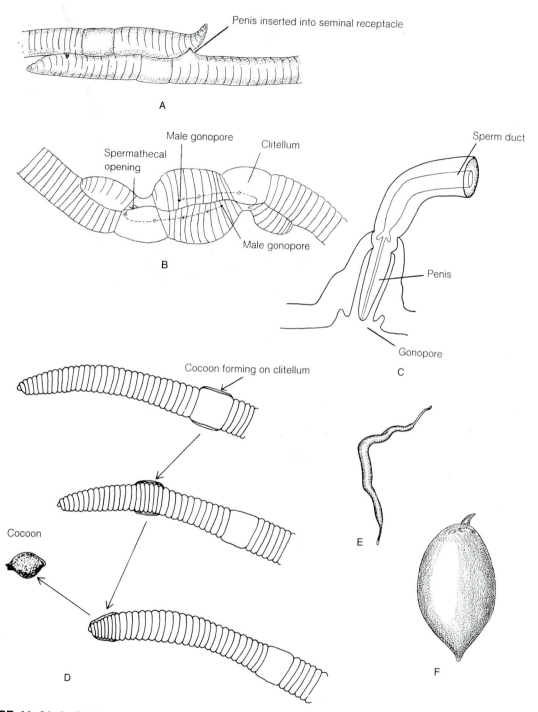

FIGURE 11–61 A, Copulation by direct sperm transmission in the earthworm, *Pheretima communissima.* **B,** Copulation by indirect sperm transmission in the lumbricid, *Eisenia foetida.* Arrows indicate path of sperm from male gonopores to openings of seminal receptacles. **C,** Penis of the lumbriculid, *Rhynchelmis.* **D,** Cocoon formation in a lumbricid earthworm. **E,** Cocoon of the glossocolecid, *Alma nilotica.* **F,** Cocoon of the lumbricid, *Allobophora terrestris. (A, After Oishi from Avel. B, After Grove and Cowley from Avel. C and E, From Brinkhurst, R., and Jamieson, B. G. M. 1972. Aquatic Oligochaeta of the World. Toronto University Press, Toronto. D, Modified from Tembe and Dubash. In Edwards, C. A., and Lofty, J. R. 1972. Biology of Earthworms. Chapman and Hall, London, p. 64. F, After Avel.)*

oducts. In some megascolecid earthworms, the prostates are not connected to the vas deferens, and they open separately onto the ventral surface of segments adjacent to those bearing the male gonopore. Prostates are absent from most lumbricids.

Forming a part of the female reproductive system, but completely separated from the female gonoducts, are the **spermathecae (seminal receptacles).** These storage chambers are simple pairs of sacs, usually opening into the ventral intersegmental groove adjacent to the segment containing them. The number of spermathecae ranges from one to many pairs, each pair commonly in a separate segment. Although they are usually situated in certain segments anterior to the ovarian segment, the exact position is variable.

The position of reproductive structures in *Lumbricus* is illustrated in Figure 11–60. The male segment in this genus is partitioned so that the testes, the sperm duct funnel, and the opening to the seminal vesicles are enclosed in a special ventral compartment called a **testis sac.** The testis sac is completely separated from the larger, remaining portion of the coelomic cavity. For this reason, the testes are not visible in the usual dorsal dissection.

The general plan of the oligochaete reproductive system is relatively uniform, but the numbers of various structures, the segments in which they are situated, and the segments onto which the genital pores open are extremely variable. This variation is of considerable importance in the taxonomy of oligochaetes.

The only exception to the oligochaete pattern appears in the Aeolosomatidae, which, although hermaphroditic, are similar to polychaetes in that they lack distinct gonads and have a large number of segments capable of producing gametes. Also, there is no gonoduct, and the sperm, at least, use the nephridia to exit from the body. However, some zoologists question whether these little worms are really oligochaetes.

The Clitellum

The clitellum is a reproductive structure characteristic of oligochaetes. It consists of certain adjacent segments in which the epidermis is greatly swollen with unicellular glands that form a girdle, partially or almost completely encircling the body from the dorsal side downward (Figs. 11–52A; 11–61A). The number of segments composing the clitellum varies considerably; there are 2 clitellar segments in many aquatic

forms, 6 or 7 in *Lumbricus,* and as many as 60 in certain Glossoscolecidae. In aquatic species and megascolecid earthworms, the clitellum is often situated in the same region as the genital pores; in the lumbricids, the clitellum is considerably posterior to the gonopores. The degree of development of the clitellum varies from group to group. In aquatic species the clitellum may be only one cell thick, whereas in many earthworms it forms a thick girdle. The development of the clitellum also varies from season to season. It generally coincides with sexual maturity, but there are some worms in which the clitellum becomes conspicuous only during the breeding season.

The glands of the clitellum produce mucus for copulation, secrete the wall of the cocoon, and secrete the albumin in which the eggs are deposited within the cocoon. In the earthworms the glands performing each of these three functions form three distinct layers with the large, albumen-secreting glands forming the deepest and thickest (Fig. 11–62).

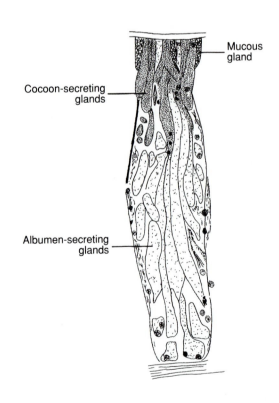

FIGURE 11–62 Section through the clitellum of *Lumbricus terrestris. (After Grove from Avel.)*

Copulation

Most oligochaetes breed semicontinually in contrast to the noncontinuous breeding of most polychaetes. Copulation is the rule, and mutual sperm transfer occurs. The ventral anterior surfaces of a pair of copulating worms are in contact, with the anterior of one worm directed toward the posterior of the other. In most oligochaetes except the lumbricids, the male genital pores of one worm directly appose the spermathecae of the other. The two worms are held in position by an enveloping mucous coat secreted by the clitella; they may also be hooked together by genital setae. The genital setae are modified ventral setae generally situated in the region of the male gonopore or the spermathecae. Transmission of sperm into one pair of spermathecae in the earthworm, *Pheretima communissima,* takes more than 1.5 h and then is repeated for each of the other two pairs of receptacles (Fig. 11–61A).

In the lumbricids the male genital pores do not appose the spermathecae during copulation, so sperm must travel a considerable distance from one opening to the other (Fig. 11–61B). During copulation the posteriorly placed clitellum of one worm attaches to the segments containing the spermathecae of the opposite worm. Attachment is accomplished by means of an adhesive slime tube and by the genital setae. The intervening region between the two clitella is less rigidly attached, although each worm is covered by a slime tube. At the emission of the sperm, certain muscles in the body wall of the segments posterior to the male gonopores contract and form a pair of ventral sperm grooves, which extend posteriorly to the clitellum. Because the grooves are roofed over by the enveloping slime tube, the sperm are actually passed through an enclosed channel.

The movement of sperm down the sperm groove is effected by a greater contraction of the muscles producing the groove. A pit is thus formed that travels the length of the groove, carrying small amounts of semen. When the semen reaches the region of the clitellum, it passes over to the other worm and enters the spermathecae. The mucous tube obviously must be incomplete in this region. The emission of semen may or may not be accomplished simultaneously in both members of the copulating pair. Copulation in *Lumbricus* continues for 2 to 3 h.

The Cocoon

A few days after copulation, a cocoon is secreted for the deposition of the eggs. First a mucous tube is secreted around the anterior segments, including the clitellum. Then the clitellum secretes a tough, encircling, chitin-like material; this material forms the cocoon. The deeper layer of clitellar glands secretes albumen into the space between the wall of the cocoon and the clitellum.

When the cocoon is completely formed, it slips forward over the anterior end as the worm pulls backward (Fig. 11–61D). Eggs are discharged into the cocoon from the female gonopores as it passes forward from the clitellum over these openings. Sperm are deposited in the cocoon as it passes over the spermathecae. There is external fertilization in the cocoon. As the cocoon slips over the head of the worm and is freed from the body, the mucous tube quickly disintegrates, and the ends of the cocoon constrict and seal themselves. The cocoons of terrestrial species are left in the soil, and the cocoons of aquatic species are left in the bottom debris or mud or are attached to vegetation.

The cocoons are yellowish in color and ovoid in shape (Fig. 11–61D–F). Cocoons of *Tubifex* are 1.60 mm $\times$ 0.85 mm; those of *Lumbricus terrestris* are approximately 7 mm $\times$ 5 mm. The largest cocoons, 75 mm $\times$ 22 mm, are produced by the giant Australian earthworms, *Megascolides australis.* A cocoon contains anywhere from 1 to 20 eggs, depending on the species. A succession of cocoons may be produced. Under favorable conditions, lumbricids may mate continually during the spring, and cocoons are formed every three or four days. Tubificids and lumbriculids generally reproduce only once a year. Their reproductive systems are then resorbed and are reformed the following year.

There is increasing evidence that growth and reproduction in oligochaetes are regulated by neurosecretions of the brain, as is true of polychaetes. In contrast to polychaetes, however, the hormone produced by the brain appears to stimulate rather than inhibit reproduction.

Development

In general, the eggs of aquatic groups, particularly the primitive families, contain relatively large amounts of yolk. On the other hand, terrestrial species have much smaller eggs with much less yolk; the abundant albumen in the cocoon supplies most of the nutritive needs of the embryo. In both aquatic and terrestrial groups, there is direct development, there are no larval stages, and all development takes place within the cocoon.

The cleavage pattern, although retaining some traces of the spiral character of the cleavage pattern of the polychaetes, is considerably modified in oligochaetes, especially in earthworms.

The oligochaete young emerge from the end of the cocoon after eight days to several months of development. *Lumbricus* hatches in 12 to 13 weeks. Usually, only some of the eggs deposited in the cocoon hatch; in *Lumbricus terrestris,* only one egg develops.

Many oligochaetes live several years. Earthworms in aquaria have been known to live for six years, but their life span is probably much shorter in natural, unprotected conditions. Lumbricid earthworms reach sexual maturity in six months to a year, depending on environmental conditions and the species. *Lumbricus terrestris* requires at least 200 days. Some aquatic oligochaetes have shorter generation times. At elevated temperatures enchytraeids inhabiting sewage percolating filters reach sexual maturity in 13 to 28 days, depending on the species, and the generation time between cocoons ranges from one to two months. On the other hand, some tubificids have a two-year life span, breeding once and then dying.

SUMMARY

1. Members of the class Oligochaeta are believed to have evolved independently of the Polychaeta from the ancestral annelids. Most move using peristalsis and have well-developed segmentation and a simple prostomium.

2. The first oligochaetes were probably inhabitants of freshwater sediments. Some lines then successively invaded drier substrata to give rise to the earthworms. Others remained aquatic but became adapted for living in loose debris and algae.

3. The digestive tract of oligochaetes is adapted for a diet of decomposing organic matter, largely plant material.

4. Gills are generally absent, but cutaneous vascular networks are well developed in larger species, especially earthworms.

5. Oligochaetes have a metanephridial system, like that in polychaetes, consisting of segmentally arranged tubules.

6. Many of the adaptations of earthworms for life on land are behavioral, but others include the presence of calciferous glands and enteronephric

nephridia, the production of urea, encystment, and egg deposition in cocoons.

7. Oligochaetes are hermaphrodites, with well-developed reproductive systems limited to a few segments. There is copulation and reciprocal sperm transfer. Fertilization and direct development occur within a cocoon secreted by the clitellum.

CLASS HIRUDINEA

The class Hirudinea contains over 500 species of marine, freshwater, and terrestrial worms, commonly known as leeches. Although they are all popularly considered to be bloodsuckers, a large number of leeches are not ectoparasites.

As a group, the leeches are certainly the most specialized annelids, and most of the distinguishing characteristics of the class have no counterpart in the other two annelid groups. But they do display many oligochaete features. Both leeches and oligochaetes lack parapodia and head appendages; both are hermaphrodites, with gonads and gonoducts restricted to a few segments; and both possess direct development within cocoons secreted by a clitellum. These similarities clearly suggest a common ancestry, and the two groups are commonly considered subclasses or classes within the class or subphylum Clitellata. Similar to oligochaetes, leeches are basically a freshwater group, with some invasions of land and a secondary invasion of the sea.

Leeches are never as small as many polychaetes and oligochaetes. The smallest leeches are 1 cm in length, but most species are 2 to 5 cm long. Some species, including the medicinal leech *(Hirudo medicinalis),* may attain a length of 12 cm, but the giant of the class is the Amazonian *Haementeria ghiliani,* which reaches 30 cm. Black, brown, olive green, and red are common colors, and striped and spotted patterns are not unusual.

According to some authorities, the class Hirudinea may be divided into three subclasses, the Branchiobdellida (p. 581) which occurs on crayfishes, the oligochaete-like Acanthobdellida, and the Euhirudinea, which constitutes the majority of leeches. The Euhirudinea contains two orders, the Rhynchobdellida and the Arhynchobdellida. Rhynchobdellid leeches include the marine fish and turtle leeches and freshwater leeches that feed on vertebrate and invertebrate hosts using an eversible proboscis. Rhynchobdellid leeches have blood vessels and often lay large yolky eggs. Arhynchobdellid leeches have a non-eversible pharynx often bearing jaws with which they suck blood or ingest prey. They lack blood vessels and lay small eggs that develop into larvae within the cocoon. Both the European and American medicinal leeches are members of the Arhynchobdellida.

External Anatomy

The anatomy of leeches is remarkably uniform. The body is typically **dorsoventrally flattened** and frequently tapered at the anterior end (Fig. 11–63). The segments at both extremities have been modified to form **suckers.** The anterior sucker is usually smaller than the posterior one and frequently surrounds the mouth. The posterior sucker is disc-shaped and turned ventrally. Segmentation is very much reduced. Unlike other annelids, leeches have a **fixed number of segments,** 32 (or 34 according to some authorities who claim that two segments have fused with the prostomium), but secondary external **annulations** obscure the original segmentation. Setae are absent.

The head, or cephalic region, contains the reduced prostomium (and perhaps two fused segments) plus four body segments (Fig. 11–64). Dorsally, the head bears a number of eyes; ventrally, it bears the **anterior sucker** surrounding the mouth (Figs. 11–64; 11–65B). The trunk consists of 21 segments, including the preclitellar region, clitellum, and postclitellar region. The clitellum, covering three segments, is never conspicuous, except during reproductive periods. The large, ventral **posterior sucker** is derived from seven segments. The anus opens *dorsally* on or near the last trunk segment.

The number of annulations per segment varies not only in different regions of the body but also in different species. The best means of determining the primary segmentation of leeches is by study of the nervous system and the innervation of the annulations by segmental nerves. The occurrence of a ring of sensory papillae around the first annulation of each segment, the serial repetition of color patterns, and the placement of the ventral nephridiopores, however, give good indications of the segmentation.

Body Wall, Coelom, and Locomotion

The body wall consists of a typical annelidan cuticle and epidermis, but unlike polychaetes and most oligochaetes, the fibrous connective tissue beneath the epidermis is very thick and occupies much of the interior of the body. The musculature and other tissues occupy this **expanded connective tissue** region. Some of the cell bodies of the enlarged epidermal glands have submerged into the peripheral connective tissue layer (dermis). Below the dermis is a layer of circular musculature followed by diagonal muscles and a powerful longitudinal musculature. Dorsoventral muscles are also present (Fig. 11–66).

Expansion of the connective tissue compartment of leeches is correlated with a **reduction of the coelom and septa,** a striking difference between leeches and other annelids. Only in the five anterior segments of the primitive *Acanthobdella* are there coelomic compartments and separating septa. Interestingly, this same leech lacks an anterior sucker, and the anterior segments with coelomic compartments bear setae, the only known exception to the absence of setae in leeches. *Acanthobdella* provides additional evidence linking leeches and oligochaetes. In all other leeches septa have disappeared, and the coelom has become reduced and specialized into a circulatory system composed of interconnected sinuses and channels (Figs. 11–66; 11–67).

The loss of septa, setae, and coelomic compartments in leeches is correlated with a change from peristaltic burrowing to new modes of locomotion; leeches crawl or swim but do not burrow using peristalsis. While crawling, only the anterior and posterior suckers anchor to the substratum. When the posterior sucker is attached, a wave of circular contraction sweeps over the animal, and the body is lengthened and extended forward. The diagonal muscles may be postural, keeping the animal rigid while it is raised and attached posteriorly. The anterior sucker then attaches, and the posterior sucker releases. A wave of longitudinal contraction then occurs, shortening the animal and moving the posterior sucker forward (Fig. 11–68). The diagonal muscles may also enable leeches to twist their raised bodies while attached posteriorly. When a leech swims, the body is flattened by contraction of the dorsoventral musculature, and waves of contraction along the longitudinal muscles produce vertical undulations.

Ecological Distribution

Although some leeches are marine, most aquatic species live in freshwater. Relatively few species tolerate rapid currents; most prefer the shallow, vegetated water-bordering ponds, lakes, and sluggish

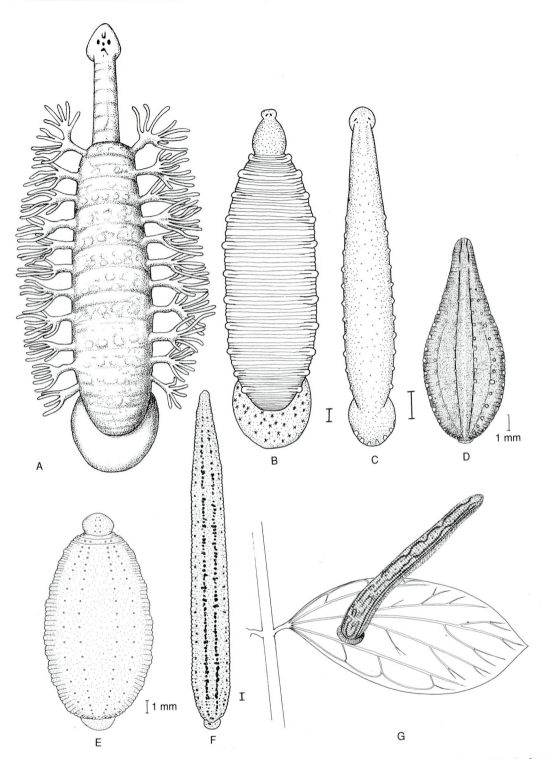

FIGURE 11–63 External dorsal views of different species of leeches. **A–C.** Fish leeches (Piscicolidae). **A,** *Ozobranchus,* showing lateral gills. **B,** *Cystobranchus.* **C,** *Piscicola.* **D** and **E,** Glossiphoniid leeches. **D,** *Glossiphonia complanata,* a common European and North American leech that feeds on snails. **E,** *Theromyzon,* a cosmopolitan genus of leeches that attacks birds. **F,** *Erpobdella punctata* (Erpobdellidae), a common North American scavenger and predatory leech. **G,** *Haemadipsa* (Haemadipsidae), a blood-sucking terrestrial leech of south Asia poised on a leaf. *(A, After Oka from Mann. B–F, from Sawyer, R. J. 1972. North American Freshwater Leeches. University of Illinois Press, Champaign, IL. G, Adapted from Keegan, et al.)*

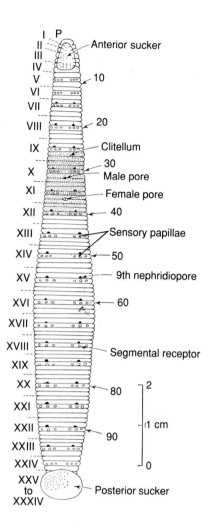

FIGURE 11–64 External, ventral surface of the medicinal leech, *Hirudo medicinalis. (From Mann, K. H. 1962. Leeches. Pergamon Press, Elmsford, New York.)*

United States. Complete terrestriality has been attained in the Haemadipsidae, which inhabit humid jungles of southern Asian and Australian regions. Although leeches are found through the world, they are most abundant in north temperate lakes and ponds. Much of the North American leech fauna is shared with Europe.

Internal Transport and Gas Exchange

Among the euhirudineans, the rhynchobdellids have retained the blood-vascular system of oligochaetes (Fig. 11–67), but the coelomic sinuses act as a supplemental circulatory system. In the arhynchobdellids the ancestral circulatory system has disappeared, and the coelomic sinuses and fluid have become the sole internal transport system, now termed the **hemocoelomic system** (Fig. 11–66A). Because the hemocoelomic system evolved from the original coelom and not the blood-vascular system, all of its vessels and channels are lined by an endothelium, the peritoneum. Much of this peritoneum, especially in the capillaries, is specialized into large nutrient storage cells, the chlorogogen tissue in rhynchobdellids and the **botryoidal tissue** in arhynchobdellids (Fig. 11–66B). The hemocoelomic fluid is propelled by the muscular contractions of the lateral longitudinal channels.

Gills are found only in the Piscicolidae (fish leeches), the general body surface providing for gas exchange in other leeches. The piscicolid gills are lateral leaflike or branching outgrowths of the body wall (Fig. 11–63A).

Respiratory pigment (extracellular hemoglobin) is found only in the arhynchobdellid leeches and is responsible for about one half of the oxygen transport.

Nutrition

Leeches possess either an eversible **proboscis** or a noneversible sucking pharynx and jaws. The proboscis (order Rhynchobdellida) is an unattached tube lying within a proboscis cavity, which is connected to the ventral mouth by a short, narrow canal (Figs. 11–65A; 11–69B). The proboscis is highly muscular, has a triangular lumen, and is lined internally and externally with cuticle. Ducts from large, unicellular salivary glands open into the proboscis. When feeding, the animal extends the proboscis out of the mouth, forcing it into the tissue of the host.

streams. In favorable environments, often high in organic pollutants, overturned rocks may reveal an amazing number of individuals; more than 10,000 individuals per square meter have been reported from Illinois. Some species estivate during periods of drought by burrowing into the mud at the bottom of a pond or stream and can survive a loss of as much as 90% of their body weight.

There is a tendency toward amphibious habits in the arhynchobdellid leeches. For example, there is a terrestrial representative, *Haemopsis terrestris,* which is occasionally plowed up in fields in midwestern

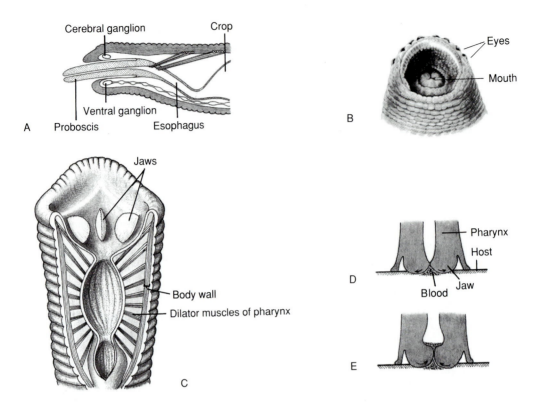

FIGURE 11–65 A, Sagittal section through anterior end of *Glossiphonia,* a rhynchobdellid leech, showing protruded tubular proboscis. (Fig. 11-66B shows nonprotruded state.) **B–E,** Arhynchobdellid leeches. **B,** Ventral view of oral region of a terrestrial, blood-sucking, haemadipsid leech. Jaws are not exposed. **C,** Ventral dissection of anterior end of *Hirudo,* showing three jaws. **D** and **E,** Ingestion by *Hirudo.* Outward movement of teeth (**D**), followed by medial movement of teeth and dilation of pharynx (**E**). *(A, Modified after Scribin. B, From Keegan, H. L., et al. 1968. 406th medical laboratory special report. U.S. Army Med. Command, Japan. C, Modified after Pfurtscheller. D, E, After Herter.)*

In the arhynchobdellid leeches, which lack a proboscis, the mouth is situated in the anterior sucker (Fig. 11–65B, C). Just within the mouth cavity of most species are three large, oval, bladelike **jaws,** each bearing a large number of small teeth along the edge. The three jaws are arranged in a triangle, one dorsally and two laterally. When the animal feeds, the anterior sucker is attached to the surface of the prey or host, and the edges of the jaws slice through the integument (Fig. 11–65D, E). The jaws swing toward and away from each other, activated by muscles attached to their bases. Salivary glands secrete an anticoagulant called **hirudin.**

Immediately behind the teeth, the buccal cavity opens into a muscular, pumping pharynx. The erpobdellids, also have a pumping pharynx, but the jaws are replaced by muscular folds.

The remainder of the digestive tract is relatively uniform throughout the class. A short esophagus either opens directly into a relatively long stomach, or first expands into a crop. The stomach may be a straight tube, as in the erpobdellids, but more commonly it is provided with 1 to 11 pairs of lateral **ceca** (Figs. 11–66A; 11–69). Following the stomach is an intestine, which may be a simple tube or, as in the rhynchobdellids, may have four pairs of slender lateral ceca. The intestine opens into a short rectum, which empties to the outside through the dorsal anus, situated in front of the posterior sucker.

Many leeches are predaceous, but about three

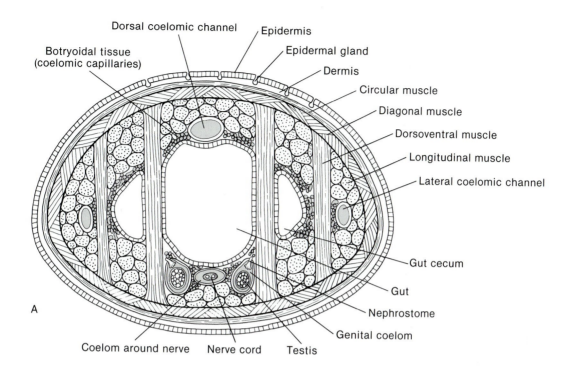

A

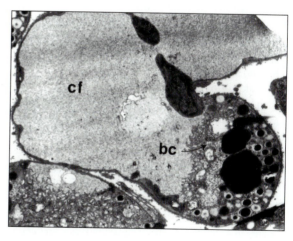

B

FIGURE 11–66 A, Transverse section through the arhynchobdellid leech, *Hirudo*. In arhynchobdellid leeches, the blood-vascular system has been completely replaced by the modified coelomic circulatory system. **B,** Electron micrograph of a section through a botryoidal capillary of the leech, *Macrobdella decora,* showing a botryoidal lining cell (bc). (cf = hemocoelomic fluid.) *(A, Modified and redrawn from Bourne, A. G. 1884. Contributions to the anatomy of the Hirudinea. Quart. J. Microsc. Sci. 24:419–506 plus plates.)*

fourths of the known species are blood-sucking ectoparasites. However, in many cases the difference lies only in the size of the host. The arhynchobdellid Hirudinidae, which includes the American medicinal leech *(Macrobdella decora),* especially demonstrate a gradation from predation to parasitism. The Erpobdellidae contain the greatest number of predaceous leeches, but this type of feeding habit is found in other families as well. Predatory leeches always feed on invertebrates. Prey includes worms, snails, and in-

sect larvae. Feeding is relatively frequent, and the prey is usually swallowed whole. Many glossiphoniids (Rhynchobdellida) suck all the soft parts from their hosts and are best regarded as specialized predators. In laboratory studies by W. H. Cross, *Erpobdella punctata* consumed 1.78 tubificids (oligochaete worms) per day and *Helobdella stagnalis* consumed 0.57 per day.

The blood-sucking leeches attack a variety of hosts. Some, primarily species of *Glossiphonia* and

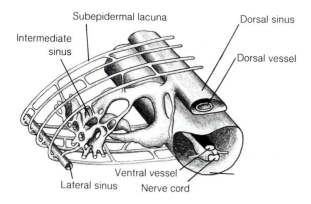

FIGURE 11–67 Section of blood-vascular (black) and coelomic (grey) circulatory systems of the rhynchobdellid, *Placobdella costata* (Glossiphonidae). *(After Oka from Harant and Grassé.)*

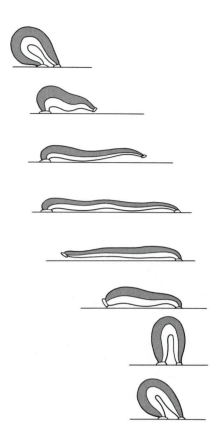

FIGURE 11–68 Leech locomotion. *(Modified and redrawn from Fretter, V., and Graham, A. 1976. A Functional Anatomy of Invertebrates. Academic Press, London and New York.)*

Helobdella, feed only on invertebrates, such as snails, oligochaetes, crustaceans, and insects, but vertebrates are hosts for most species. Piscicolidae are parasites of both freshwater and marine fish, sharks, and rays (Fig. 11–63A, B). The glossiphoniids feed on amphibians, turtles, snakes, alligators, and crocodiles. Species of the cosmopolitan glossiphoniid genus, *Theromyzon,* attach to the nasal membranes of shore and water birds. The aquatic Hirudinidae and the terrestrial Haemadipsidae feed primarily on mammals, including humans (Fig. 11–63G).

Parasitic leeches are rarely restricted to one host, but they are usually confined to one class of vertebrates. For example, *Placobdella* feeds on almost any species of turtles and even alligators, but they rarely attack amphibians or mammals. On the other hand, mammals are the preferred hosts of *Hirudo.* Furthermore, some species of leeches that are exclusively bloodsuckers as adults are predaceous during juvenile stages.

The mammalian bloodsuckers, such as *Hirudo,* on contacting a thin area of the host's integument, attach the anterior sucker very tightly to this area, and then slit the skin. The jaws of *Hirudo* make about two slices per second. The incision is anesthetized by a substance of unknown origin. The pharynx provides continual suction, and the secretion of hirudin prevents coagulation of the blood. Penetration of the host's tissues is not well understood in the many jawless, proboscis-bearing species that are bloodsuckers. The proboscis becomes rigid when extended, and it is possible that penetration is aided by enzymatic action.

Leech digestion is peculiar in a number of respects. The gut secretes no amylases, lipases, or endopeptidases. The presence of only exopeptidases perhaps explains the fact that digestion in blood-sucking leeches is so slow. Also characteristic of the leech gut is a symbiotic bacterial flora that is important in nutrition. In both the blood-sucking medicinal leech of Europe, *Hirudo medicinalis,* and the predaceous *Erpobdella octoculata,* the gut bacteria are responsible for a considerable part of digestion; they may be significant in the digestion of all leeches. The bacterium, *Pseudomonas hirudinicola,* of *Hirudo medicinalis* breaks down high-molecular-weight proteins, fats, and carbohydrates, and the bacterial population increases significantly following the ingestion of blood by the leech. The bacteria may also produce vitamins and other compounds that are used by the leech host.

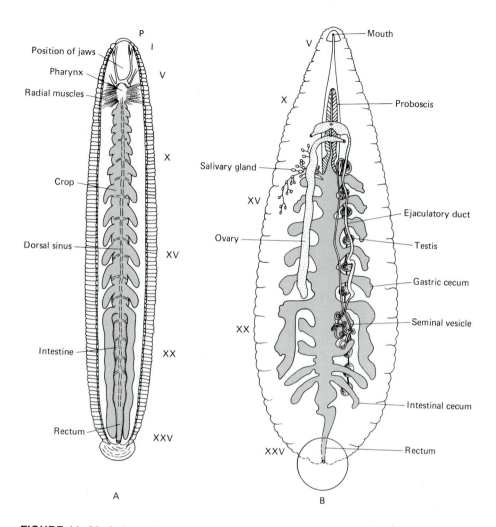

FIGURE 11–69 A, Internal structure of the arhynchobdellid, *Hirudo medicinalis* (dorsal view). **B,** Internal structure of the rhynchobdellid, *Glossiphonia complanata* (ventral view). *(A, From Mann, K. H. 1962. Leeches. Pergamon Press, Elmsford, New York. B, After Harding and Moore from Pennak.)*

Bloodsuckers feed infrequently, but when they do, they can consume an enormous quantity of blood. *Haemadipsa* may ingest ten times its own weight, and *Hirudo* two to five times its own weight. Following ingestion, water is removed from the blood and excreted through the nephridia. The digestion of the remaining blood cells then takes place very slowly. These leeches can then tolerate long periods of fasting. Medicinal leeches have been reported to have gone without food for one and one half years, and because they may require 200 days to digest a meal, they need not feed more than twice a year in order to grow.

Excretion

Leeches contain 10 to 17 pairs of metanephridial tubules, situated in the middle third of the body, one pair per segment (Fig. 11–66A). As a result of the coelom reduction and the loss of septa in the leech body, the nephridial tubules are embedded in connective tissue, and the nephrostomes project into the coelomic channels. Each ciliated nephrostome opens into a nonciliated capsule (Fig. 11–70).

In most leeches the cavity of the capsule does not open into the nephridial canal, and the two parts of the nephridium may lose all structural connection in

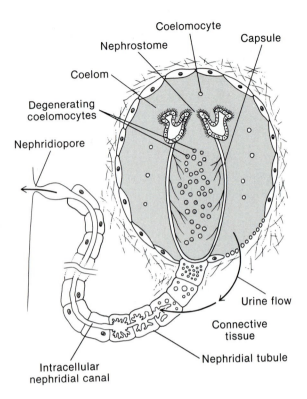

Coelomocyte

Nephrostome

Capsule

Coelom

Degenerating
coelomocytes

Nephridiopore

Urine flow

Connective
tissue

Nephridial tubule

Intracellular
nephridial canal

FIGURE 11–70 Nephridium of the arhynchobdellid leech, *Trocheta*. The nephrostome and capsule do not communicate with the tubule but rather collect and break down spent coelomocytes. Urine formation (arrows) apparently results from ultrafiltration of coelomic fluid across the wall of the botryoidal capillary into the connective tissue, followed by secretion into the nephridial tubule.

some species. The nephridial tubule consists of a main duct that receives numerous branched ductules (canaliculi). The main duct typically expands into a **urinary bladder** before opening to the exterior at a ventrolateral nephridiopore. The epithelial lining of the coelomic channels in the region of the nephrostomes is porous (the lining cells resemble podocytes), and thus the initial phase of urine formation is believed to occur as the hemocoelomic fluid is pressure filtered across the wall of the channel into the connective tissue around the canaliculi. The canalicular cells secrete salts into the canalicular lumen, and water enters from the connective tissue by osmosis. As water and salts flow outward, salt is reabsorbed by the cells lining the main duct (which must therefore be imper-

meable to water), and a watery hypoosmotic urine accumulates in the bladder before discharge to the exterior. The nephridia are thus important organs of osmoregulation.

The function of the structurally isolated nephrostomes and nephridial capsules has changed from excretion to internal defense. The capsules, similar to lymph nodes, contain a population of amoeboid phagocytes that engulf foreign material and cellular debris; some of these cells may be released into the hemocoelomic circulation as coelomocytes. Particulate waste is also picked up by botryoidal and vasofibrous tissue of the hirudinid leeches and by pigmented and coelomic epithelial cells of glossiphoniids and piscicolids. In the rhynchobdellids, cilia on the nephridial funnels beat inwards, carrying particulates into the capsule. The funnels of the arhynchobdellids, on the other hand, have cilia that beat outward away from the capsule and are believed to promote the circulation of coelomic fluid. The conversion of a nephridial or coelomic funnel into a site for phagocytosis and coelomocyte destruction is also of common occurrence in the polychaetes.

Nervous System and Sense Organs

The nervous system of leeches is similar to that of other annelids, but the anterior and posterior ganglia are concentrated because of the segmental modifications forming the suckers. The brain consists of a mass of neural tissue in the prostomium, the supraesophageal ganglion, and the fused first four pairs of ventral segmental ganglia, forming the subesophageal ganglion. A series of 21 ganglia form the ventral nerve cord of the trunk, and the last 7 ganglia are fused together to form the caudal ganglion associated with the posterior sucker (Fig. 11–71A). The entire central nervous system is enclosed in the unpaired ventral coelomic channel (Fig. 11–66).

The relatively small number and large size of the neurons have made leeches, along with sea hares and crayfish, favorite invertebrate subjects of neuroanatomists and neurophysiologists. Each of the 21 segmental ganglia in *Hirudo* contains 175 pairs of neuron cell bodies arranged bilaterally around a central mass of nerve fibers (neuropil) where synaptic junctions are made. The cell bodies are large enough to be probed with electrodes and mapped (Fig. 11–71B).

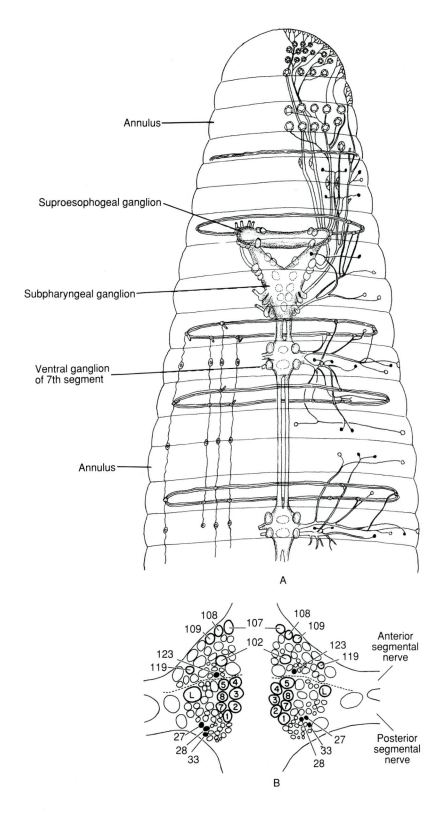

FIGURE 11–71 A, Nervous system of *Erpobdella punctata.* **B,** Dorsal view of a segmental ganglion of a leech. Circles represent cell bodies; solid black circles are interneurons, and those enclosed by heavy lines are motor neurons. *(A, After Bristol from Harant and Grassé. B, From Stent, G. S., et al. 1978. Neuronal generation of the leech swimming movement. Science. 200:1348–1356.)*

Some rhynchobdellid leeches can dramatically change color as a result of pigment movements in large **chromatophores,** which are under neurohumoral control. The significance of the color change is not certain, for these leeches do not adapt to background coloration.

The specialized sense organs in leeches consist of two to ten pigment-cup eyes and **sensory papillae.** The sensory papillae are small, projecting discs arranged in a dorsal row or in a complete ring around one annulation of each segment (Fig. 11–64). Each papilla consists of a cluster of many sensory cells and supporting epithelium.

Despite the lack of highly organized, concentrated sense organs, leeches can detect low levels of many types of stimuli, and the sensitivity is often an adaptation for finding prey or a host. Fish leeches respond to moving shadows and water pressure vibrations. Both predatory and blood-sucking leeches will attempt to attach to an object smeared with various host or prey substances, such as fish scales, tissue juices, oil gland secretions, or sweat. *Hirudo,* which is a bloodsucker of warm-blooded animals, will swim into waves, which may be generated by a possible host. The same leech is also attracted by body secretions and elevated temperatures and has been reported to swim toward a man standing in water. Supposedly, the terrestrial blood-sucking Haemadipsidae of the tropics, which are attracted by passing warm-blooded mammals, will move over vegetation and converge on a person standing in one place.

Reproduction

Unlike many other annelids, leeches do not reproduce asexually, nor can they regenerate lost parts. Like oligochaetes, all leeches are hermaphrodites, but they are protandric, not simultaneous hermaphrodites. The reproductive system is similar to that of oligochaetes. There are, however, no separate seminal receptacles (spermathecae), and there is internal fertilization (Fig. 11–72).

Sperm transfer in the hirudinids, most of which possess a penis, is similar to direct sperm transmission in earthworms (Fig. 11–73A). The ventral surfaces of the clitellar regions of a copulating pair come together, with the anterior end of one worm directed toward the posterior end of the other. Thus, the male gonopore of one worm apposes the female gonopore of the other. The penis is everted into the female gonopore, and sperm are introduced into the vagina, which probably also acts as a storage center.

Sperm transfer in most glossiphoniids, piscicolids, and erpobdellids, all of which lack a penis, is by hypodermic impregnation. The two copulating worms commonly intertwine and grasp each other with their anterior suckers. The ventral clitellar regions are in apposition, and by muscular contraction of the atrium, a spermatophore is expelled from one worm and penetrates the integument of the other. The site of penetration is usually in the clitellar region, but spermatophores may be inserted some distance away. As soon as the head of the spermatophore has penetrated the integument, perhaps by a combination of expulsion pressure and a cytolytic action by the spermatophore itself, the sperm are discharged into the tissues (Fig. 11–73F). Following liberation from the spermatophore, the sperm are carried to the ovisacs in the coelomic channels or by a special tissue pathway where there is a restricted region of integument for the reception of the spermatophore.

Eggs are laid from two days to many months after copulation. At this time, the clitellum becomes conspicuous and in most families secretes a cocoon, as in the oligochaetes (Fig. 11–73B, C). The cocoon is filled with a nutritive albumen produced by certain of the clitellar glands. The cocoon then receives the one to many fertilized eggs as it passes over the female gonopore. The eggs are secondarily small and relatively yolkless. Beginning in May, *Erpobdella punctata* in Michigan, for example, produces some ten cocoons, each with five eggs, which hatch three to four weeks later. The cocoons are affixed to submerged objects or vegetation. Some piscicolids attach their cocoons to their hosts. Terrestrial species place them in damp soil beneath stones and other objects, and the hirudinids, such as *Hirudo* and *Haemopsis,* leave the water to deposit their cocoons in damp soil.

The glossiphoniids brood their eggs. In some species the cocoons are attached to the bottom and covered and ventilated by the ventral surface of the worm. In others, the cocoons are membranous and transparent and are attached to the ventral surface of the parent (Fig. 11–73E). During the course of development the embryonic leeches break free of the cocoon and attach themselves directly to the ventral surface of the parent.

A larva (cryptolarva) develops within the cocoon of arhynchobdellid leeches. It has a pair of provi-

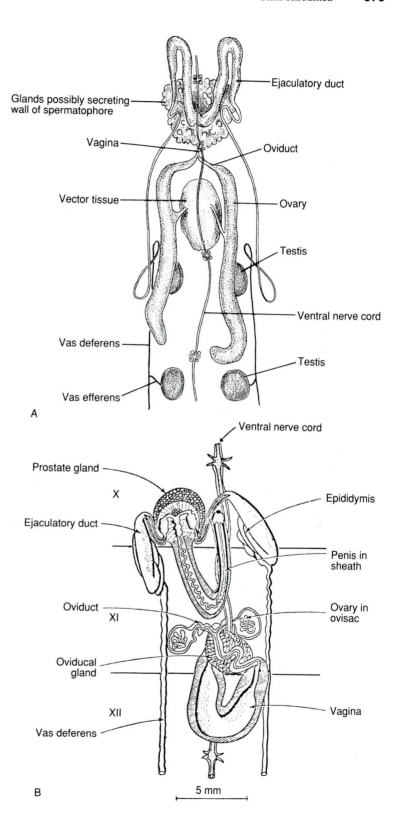

FIGURE 11–72 A, Reproductive system of *Piscicola geometra*. **B,** Reproductive organs of *Hirudo*. Testes are associated with vas deferens in more posterior segments. *(A, After Brumpt from Harant and Grassé. B, After Leuckart and Brandes from Mann.)*

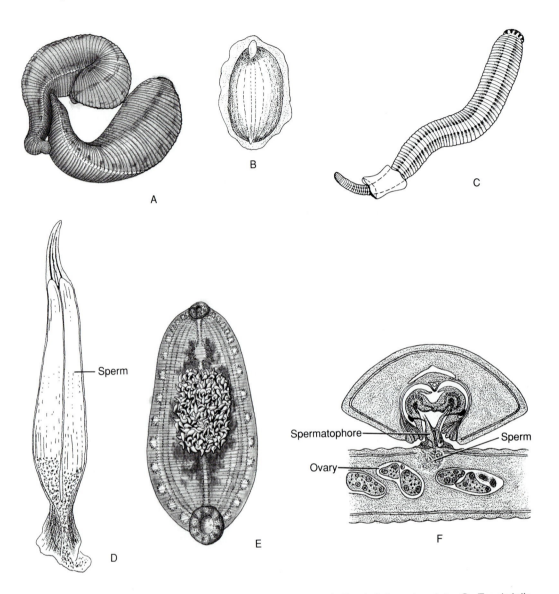

FIGURE 11–73 A, *Hirudinaria* copulating. **B,** Cocoon of *Erpobdella octoculata*. **C,** Erpobdella withdrawing its body from a cocoon. **D,** Spermatophore of the rhynchobdellid, *Haementeria*. **E,** Glossiphoniid brooding young. **F,** Section through two copulating *Erpobdella* individuals. Upper leech is injecting spermatophores into lower individual. *(A, After a photograph by Keegan, et al. B, After Pavlovsky from Harant and Grassé. C, From Nagao, Z. 1957. J. Fac. Sci. Hokkaido Univ. Ser. VI, Zool. 13:192–196. D, After Pavlovsky from Harant and Grassé. E, F, After Brumpt.)*

sional protonephridia and a mouth with which it imbibes the nutritive fluid within the cocoon.

Most leeches have an annual or two-year cycle, breeding in the spring or summer and maturing by the following year. Life cycles are correlated in part with feeding habits. Some, such as *Hirudo,* are associated with the host only during actual feeding; others, such as the marine fish leech, *Hemibdella,* never leave the host. But most leeches leave the host at least to breed. Less than three months is required to complete the entire life cycle in *Hemibdella.*

SUMMARY

1. The presence of a clitellum, the similarity of reproduction, and the existence of a single genus with anterior setae and coelomic compartments are evi-

dence that members of the class Hirudinea evolved from an ancestor shared in common with the oligochaetes.

2. Leeches possess an anterior and a posterior sucker; they lack setae, external segmentation, and septa; they have expanded connective tissue and a reduced coelom that is specialized into a circulatory system. All of these features are probably correlated with a change from peristaltic locomotion to movement involving extension and contraction of the entire body and attachment with the suckers.

3. The shift in the mode of movement is probably related to the assumption of predaceous and ectoparasitic feeding habits. The latter is characteristic of about three quarters of the species of leeches.

4. A proboscis or jaw with a pumping pharynx is utilized in both predaceous and ectoparasitic feeding.

5. Vertebrates are the principal hosts of ectoparasitic, or blood-sucking, leeches. Digestion of blood, which is slow, depends on exopeptidases produced by the leech and a symbiotic bacterial flora.

6. Segmentation is reflected internally in the nephridia and the ganglia of the ventral nerve cord. Gills are generally absent, and the blood-vascular system is similar to that of other annelids or has been replaced by coelomic channels.

7. Reproduction is similar to that of oligochaetes, but some species transfer sperm by hypodermic impregnation with spermatophores.

BRANCHIOBDELLIDA

The branchiobdellidans are annelids that are parasitic or commensal on freshwater crayfish. They show similarities to both oligochaetes and leeches and have been included with both groups, but authorities now believe that they diverged early from the base of the common oligochaete/hirudinean stock and justify being placed within a separate class or subclass (Branchiobdellida).

All branchiobdellidans are very small (1–10 mm) and are composed of only 17 segments (Fig. 11–74). The head is modified into a sucker with a circle of finger-like projections. The mouth is terminal, and the buccal cavity contains two teeth. The posterior segments (15–17) are also modified to form a sucker, and all of the segments lack setae. The anus is dorsal on segment 14. Branchiobdellidans have a segmented coelom, a more or less typical annelidan blood-vascu-

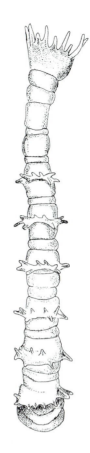

FIGURE 11–74 *Stephanodrilus,* a member of the leechlike Branchiobdellida. These worms are parasitic and commensal on freshwater crayfish. *(After Yamaguchi from Avel.)*

lar system, and two pairs of metanephridia. Some species are ectoparasitic on the gills of crayfish; others live on the outer surface of the exoskeleton and graze on accumulated organic debris and microorganisms. Branchiobdellidans are hermaphrodites with internal fertilization. Zygotes are deposited in cocoons, which are attached to the host. The embryos develop into cryptolarvae, as in arhychobdellid leeches.

THE POGONOPHORANS

The Pogonophora was unknown prior to the twentieth century. Their discovery awaited the development of oceanographic vessels capable of sampling deep-water habitats. The first specimen was dredged from Indonesian waters in 1900. Since that time more than

80 species have been described and more are being discovered. The Pacific northwest has yielded the richest pogonophoran fauna. It is now becoming apparent that the phylum is widespread in the world's seas, especially along the continental slopes and deep-sea spreading centers. In the North Atlantic, pogonophorans have been collected from European waters, from Nova Scotia to Florida, and from the Gulf of Mexico.

Pogonophorans are exclusively marine animals and usually occur at depths exceeding 100 m. They are sessile, living in secreted tubes that are composed of a mixture of protein and chitin. Tubes are usually stiff and fixed upright in bottom ooze, but the giant tube worm from the Pacific, *Riftia pachyptila*, produces a flexible tube that is attached to a hard substratum or to the tube of an adjacent worm. Pogonophorans often occur in dense aggregations, as many as 200/m^2. There are a few species that construct their tubes in decaying wood or other debris.

Although there is disagreement among zoologists concerning the taxonomic status of the phylum, the present account considers the phylum to comprise two groups, the perviate pogonophores and the obturate (vestimentiferan) pogonophores; this distinction is based primarily on differences between their anterior ends.

The long, transparent, hairlike body of most species of perviate pogonophores is slender and ranges in length from 5 to 85 cm and contrasts with the thicker opaque bodies of vestimentiferans. In 1977, the research submersible, *Alvin,* while investigating the fauna around the warm-water vents on the 2600-m deep floor of the Galapagos Rift, discovered clusters of large vestimentiferans 1.5 m in length and almost 4 cm in diameter (Fig. 11–75).

The body bears one to many (up to 200,000) slender **tentacles** anteriorly that are the distinguishing feature of the phylum (Pogonophora means "beard bearer"). In the vestimentiferans, the tentacles arise from and enclose two anteriorly directed, supportive structures **(obturacula),** which are fused medially, and the individual tentacles are fused along their lengths to a lesser or greater extent to form sheets. The tentacles of perviate pogonophores, on the other hand, are unfused and borne dorsally on the **cephalic lobe,** which is the anterior extremity of a glandular region, called the **forepart** (Fig. 11–76B). The forepart of perviate pogonophores may correspond to the collar-like **vestimentum,** which is situated immediately

FIGURE 11–75 The giant vestimentiferan tube worm, *Riftia pachyptila,* from the Galapagos undersea rift. *(From Jones, M. L. 1984. The giant tube worms. Oceanus. 27:47–52.)*

behind the tentacular crown of vestimentiferans (Fig. 11–76A). The vestimentum and the posterior part of the forepart secrete the animals' tube. All pogonophores have a trunk that constitutes most of the body length and bears girdles of setae in the perviate pogonophores. Each girdle seta terminates in numerous hooklike teeth, some of which are directed anteriorly and some posteriorly (Fig. 11–77A). The posterior end of the body is an **opisthosoma,** composed of numerous (up to 95) segments (Fig. 11–76A, B). One half or more of the opisthosomal segments bear setae. Those of vestimentiferans resemble the girdle setae of perviate pogonophores, whereas the latter group have toothed, peglike setae on some of their opisthosomal segments. The opisthosoma anchors the worm in its tube, allowing it to withdraw rapidly. It may also aid in burrowing

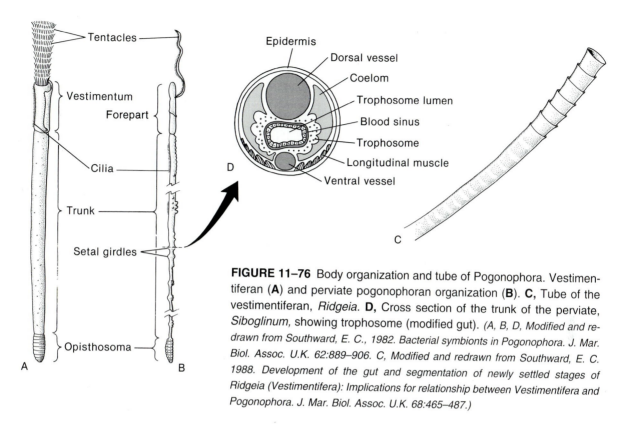

FIGURE 11–76 Body organization and tube of Pogonophora. Vestimentiferan (**A**) and perviate pogonophoran organization (**B**). **C,** Tube of the vestimentiferan, *Ridgeia*. **D,** Cross section of the trunk of the perviate, *Siboglinum,* showing trophosome (modified gut). *(A, B, D, Modified and redrawn from Southward, E. C., 1982. Bacterial symbionts in Pogonophora. J. Mar. Biol. Assoc. U.K. 62:889–906. C, Modified and redrawn from Southward, E. C. 1988. Development of the gut and segmentation of newly settled stages of Ridgeia (Vestimentifera): Implications for relationship between Vestimentifera and Pogonophora. J. Mar. Biol. Assoc. U.K. 68:465–487.)*

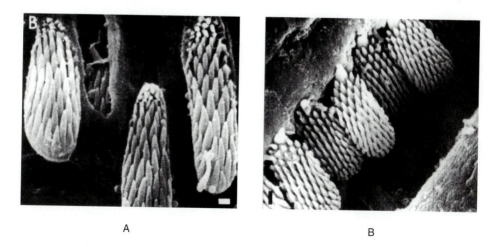

FIGURE 11–77 Comparison of the girdle setae of the perviate pogonophoran, *Siboglinum* (**A**), and the toothed setae of the sabellid polychaete, *Oriopsis* (**B**). The heads (shown) of the setae bear rasplike teeth, some of which oppose each other in the pogonophoran but not in the polychaete. *(From George, J. D., and Southward, E. C. 1973. A comparative study of the setae of Pogonophora and polychaetous Annelida. J. Mar. Biol. Assoc. U.K. 53:403–424.)*

in the bottom ooze below the lower end of the tube. The opisthosoma is fragile and easily breaks away from the rest of the body when the animal is pulled. As a result, most specimens dredged from the bottom lack this region, and it has been only since 1964 that the opisthosoma has been known to exist.

Internally, there are coelomic compartments in each of the body divisions, and there are extensions of the coelom into the tentacles. In the opisthosoma there are septa between the segmental coelomic compartments, and in vestimentiferans a median mesentery divides each coelomic space into two compartments (Fig. 11–78A). A well-developed blood-vascular system is present, and each tentacle is supplied with two vessels. The respiratory pigment, hemoglobin, usually occurs in the vascular system

and sometimes also in the fluid of some coelomic spaces. There is a nerve plexus at the base of the epidermis and an intraepidermal, ventral nerve cord.

A remarkable feature of adult pogonophores is the absence of a mouth and normal digestive tract. In fact, zoologists who examined the first specimen thought that part of the body was missing. In the absence of a digestive tract, the mode of nutrition in these animals was puzzling. The large size of the vestimentiferans from the Galapagos Rift and elsewhere made possible the discovery that in the trunk of the worm is a central mass of tissue, called the **trophosome,** that is packed with symbiotic bacteria (Fig. 11–76D). The bacteria oxidize sulfur-containing compounds and use the resultant energy to fix carbon. The pogonophoran host obtains its nutrition from the pro-

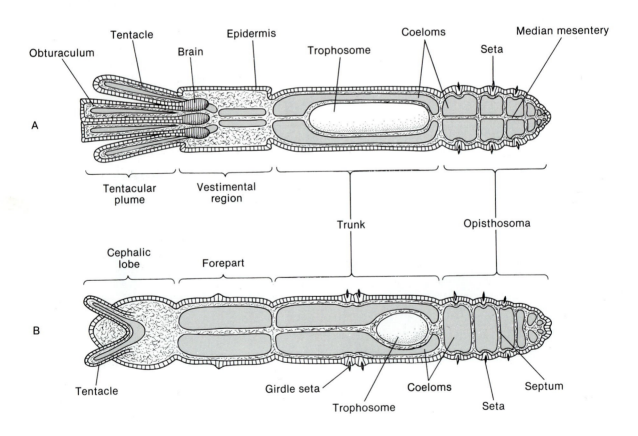

FIGURE 11–78 Pogonophoran coelomic organization. **A,** Vestimentiferan; **B,** Perviate. *(Modified and redrawn from Southward, E. C.1988. Development of the gut and segmentation of newly settled stages of Ridgeia (Vestimentifera): Implications for relationship between Vestimentifera and Pogonophora. J. Mar. Biol. Assoc. U.K. 68:465–487.)*

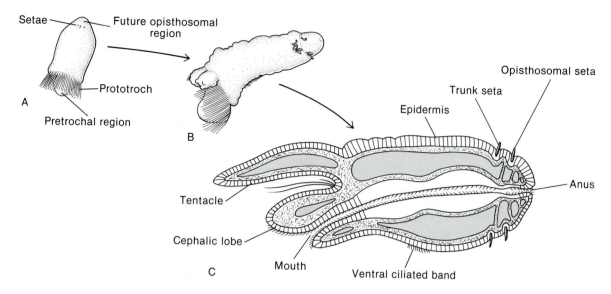

FIGURE 11–79 A, Trochophore of the vestimentiferan, *Ridgeia*, in lateral view. **B,** An early juvenile stage of the Ridgeia. **C,** Sagittal section of juvenile of *Ridgeia* showing the presence of a typical annelidan gut. Later in development, the gut and associated tissues become modified to form the trophosome. *(A, After a sketch provided by S. L. Gardiner; B, C, Redrawn from Southward, E. C. 1988. Development of the gut and segmentation of newly settled stages of Ridgeia (Vestimentifera): Implications for relationship between Vestimentifera and Pogonophora. J. Mar. Biol. Assoc. U.K. 68:465–487.)*

duction of excess organic compounds by the bacteria and by lysis and subsequent absorption of bacterial components. The vascular hemoglobin, which accounts for the red color of the tentacular crown of vestimentiferans, is important in delivering the large amounts of oxygen and sulfur-containing compounds required by the bacteria. Similar trophosomal tissue was subsequently found in perviate pogonophores living in water around Scandinavia, in the Bay of Biscay and elsewhere. Varying degrees of supplementary nutrition may, however, be possible through absorption of organic compounds dissolved in the surrounding sea water (p. 204). Dependence on symbiotic bacteria for nutrition is not unique to pogonophores. This mode of nutrition is found in certain gutless clams (p. 442) and some other groups, including some oligochaetes and polychaetes.

In 1988, E. Southward working in England and M. Jones and S. Gardiner working in the United States simultaneously discovered a complete annelid-like gut in the juveniles of vestimentiferans (Fig. 11–79B). This discovery enabled the scientists to

confirm the ventral position of the nerve cord and to demonstrate that the trophosome, at least in part, is an adult modification of the embryonic midgut.

Pogonophores have a pair of nephridia situated in the forepart or the vestimental region, but the nature of these organs is not yet clear.

Most perviate pogonophores are dioecious, and two cylindrical gonads are situated one on each side in the trunk coelom. The two male gonopores are situated at the anterior part of the trunk, and the terminal part of the sperm duct packages the sperm into spermatophores. In the female the two oviducts open through separate gonopores farther back on the trunk. Sperm transfer, fertilization, and egg deposition have not been observed, but one specimen of *Siboglinum* was seen to move spermatophores to the mouth of the tube with its tentacles. The spermatophores perhaps reach the tubes of neighboring females by floating.

Vestimentiferans are dioecious, and the gonads are situated in the trunk coelom. In both sexes, paired gonoducts lead to two gonopores situated on the dorsal surface of the vestimentum near its posterior margin. Externally, males are distinguished from females by paired ciliated ridges that extend anteriorly from the gonopores along the dorsal surface of the vestimentum. The anterior part of each sperm duct organizes sperm into masses, but true spermatophores are absent.

Because the eggs of many perviate pogonophores are brooded within the tube, we have some knowledge of pogonophoran development. Cleavage follows a bilateral pattern, which could perhaps be derived from either a spiral or a radial type. Descriptions of mesoderm and coelom formation are contradictory. The larvae of the vestimentiferan, *Ridgeia*, are considered to be trochophores and each has a single ciliary girdle (prototroch). Embryos of perviate pogonophores are collected from tubes, and late embryos of other species are wormlike and yolk-laden and have ciliary girdles (Fig. 11–79A). Whether the larvae are dispersed by currents or quickly sink to the bottom is uncertain.

The occurrence of a complete gut in early developmental stages, at least in vestimentiferans, indicates that pogonophorans probably evolved from animals having a more conventional body organization. The segmented setiferous terminal part of the body, the similarity of the setae to those of annelids, the segmentation of the mesoderm, and a ventral nerve cord indicate a protostome position of the Pogonophora and suggest that they are closely related to the Annelida, especially the Polychaeta. Although this view is held by most zoologists, there are others who place pogonophorans between deuterostomes and protostomes or on a separate line of evolution.

SUMMARY

1. The Pogonophora is a phylum of 80 or more species of wormlike, marine animals that live in chitinous tubes fixed upright in the bottom sediments at depths of 100 m or more.

2. The elongated body of pogonophorans is composed of three sections. The anterior section bears from one to thousands of tentacles, depending on the species, and the posterior section is segmented and bears setae. A typical gut is lacking in the adult, but a trophosome bearing symbiotic bacteria is usually present.

SYSTEMATIC RÉSUMÉ OF CLASS POLYCHAETA

Polychaete families are in many ways more distinctive and useful taxa than are the orders. The following list includes the more common families to which references have been made in the preceding discussion. Lack of space permits mention of only the distinguishing characteristics. See Fauchald (1977) and Parker (1982) for complete definitions of orders and families and see Barnes, Harrison and Gardiner (1992) for a shortened account of polychaete classification.

Errant Polychaetes. Polychaetes with most of the following characteristics: segments numerous and similar; parapodia well developed, with acicula and setae; head with sensory structures; pharynx with jaws or teeth. Swimming, crawling, burrowing, and tube-dwelling worms.

Family Aphroditidae. Sea mice. Long setae forming a feltlike covering of the dorsal surface. *Aphrodita, Laetmonice.*

Families Polynoidae, Sigalionidae, and Polyodontidae. Scale worms. Dorsal surface bears elytra. Chiefly crawling polychaetes. *Lepidonotus, Harmothoe, Polynoe, Lepidasthenia, Lagisca, Iphinoe, Arctonoe, Euthalenessa, Sigalion, Sthenelais, Thalanessa, Polyodontes.*

Family Phyllodocidae. Uniramous parapodia with flattened, leaflike cirri. Crawling polychaetes. *Notophyllum, Phyllodoce, Eulalia.*

Family Amphinomidae. Crawling polychaetes with brittle, poisonous setae. Fireworms. *Hermodice, Amphinome, Chloeia.*

Family Alciopidae. Planktonic worms with transparent bodies and two large eyes. *Alciopa, Vanadis.*

Family Tomopteridae. Planktonic worms with membranous pinnules in place of setae. *Tomopteris.*

Family Hesionidae. Crawling polychaetes with well-developed prostomial sense organs and long, dorsal parapodial cirri. *Podarke, Ophiodromus.*

Family Syllidae. Small crawling worms with long, delicate bodies and uniramous parapodia. *Syllis, Odontosyllis, Pionosyllis, Autolytus, Proceraea, Trypanosyllis, Brania (Grubia), Myrianida.*

Family Nereididae. Four eyes and four pairs of peristomial cirri. Pharynx contains a pair of jaws. Large crawling species. *Nereis, Platynereis.*

Family Nephtyidae. Rapid-crawling worms with well-developed prostomial sense organs situated between parapodial lobes. *Nephtys.*

Family Glyceridae. Errant burrowing worms. Conical prostomium and long proboscis with four jaws. *Glycera.*

Superfamily Eunicea. Elongated worms with proboscis armature of at least two pieces. This large superfamily contains a number of ecologically diverse but closely related families. The Eunicidae and Onuphidae, with eyes and antennae, contain many tubicolous species. *Eunice, Marphysa, Palola, Lysidice; Onuphis, Diopatra, Hyalinoecia, Nothria.* The Lysaretidae are crawling polychaetes. *Oenone.* The Arabellidae and Lumbrineridae are threadlike, errant burrowers with reduced prostomial sensory structures. *Arabella, Drilonereis, Labrorostratus; Lumbrineris.*

Family Histriobdellidae. Ectoparasitic. *Histriobdella.*

Family Ichthyotomidae. Ectoparasitic. *Ichthyotomus.*

Family Myzostomidae. Greatly flattened commensals and parasites of echinoderms, particularly crinoids. *Myzostoma.* Sedentary Polychaetes. Polychaetes with most of the following characteristics: body regions commonly differentiated, parapodia reduced, without acicula or compound setae; prostomium without sensory appendages, but head commonly provided with palps, tentacles, and other structures for feeding; no teeth or jaws.

Family Orbiniidae. Sedentary burrowers with conical or globular prostomium, without appendages. *Orbinia, Scoloplos.*

Family Spionidae. Tubicolous polychaetes with two long prostomial palps. *Spio, Scolelepis (Nerine), Polydora, Mala.*

Family Magelonidae. Sedentary burrowers with one pair of long papillate palps. *Magelona.*

Family Chaetopteridae. Tubicolous polychaetes with one pair of long palps. *Chaetopterus, Phyllochaetopterus, Spiochaetopterus, Mesochaetopterus (Ranzania).*

Family Cirratulidae. Segments bear long, threadlike gills. *Cirratulus, Cirriformia (Audouinia).*

Family Flabelligeridae. Sedentary burrowers with cephalic gills that can be retracted with the prostomium into anterior segments. *Flabelligera, Piromis.*

Family Opheliidae. Errant burrowers with conical prostomium. Number of segments, usually relatively low, are fixed within each species. *Ophelia, Armandia, Polyophthalmus, Euzonus (Thoracophelia).*

Family Capitellidae. Errant burrowers with conical prostomium and long body. *Dasybranchus, Capitella, Notomastus.*

Family Arenicolidae. Sedentary burrowers without head appendages. Lugworms. *Arenicola, Abarenicola.*

Family Maldanidae. Tubicolous polychaetes with small prostomium fused to peristomium, without head appendages. The bamboo worms. *Clymenella, Maldane, Axiothella.*

Family Oweniidae. Tubicolous species without prostomial appendages, or with foliaceous prostomial crown. *Owenia.*

Family Sabellariidae. Tubicolous polychaetes. Setae of anteriorly projecting segments are modified to form an operculum for closing the tube. *Sabellaria, Phragmatopoma.*

Family Pectinariidae, or Amphictenidae. Tubicolous polychaetes constructing conical tubes. Head with heavy, forward-directed, golden setae used as an operculum and for digging. *Pectinaria (Cistenides).*

Family Ampharetidae. Tubicolous polychaetes with retractile buccal tentacles. *Amphicteis, Ampharete.*

Family Terebellidae. Prostomium bears numerous long, nonretractile tentacles. Tubicolous or sedentary burrowers. *Terebella, Amphitrite, Polymnia, Eupolymnia.*

Family Sabellidae. The fan worms or feather-duster worms. Noncalcareous tubes. *Branchiomma (Dasychone), Sabella, Myxicola, Oriopsis, Potamilla, Manyunkia, Megalomma, Spirographis.*

Family Serpulidae. The fan worms or feather-duster worms. Tubes calcareous. The coiled spirorbids are placed within a separate family (Spirorbidae) by some authorities. *Serpula, Apotomus, Spirorbis, Pomatoceros, Filograna (Salmacina), Hydroides (Eupomatus), Spirobranchus.*

Archiannelida. Most modern authorities on the annelids agree that the old phylum or class Archiannelida represents an assortment of modified annelids, such as *Polygordius* (Fig. 11–28C), *Protodrilus*, *Nerilla*, and *Dinophilus*, that properly belong to the Polychaeta. Most are members of the interstitial fauna (Fauchald, 1975).

SYSTEMATIC RÉSUMÉ OF CLASS OLIGOCHAETA

The following classification is based on that of Brinkhurst and Jamieson (1972), modified by Jamieson (1978).

Order Lumbriculida. Four pairs of setae per segment. At least one pair of male funnels in the same segment as male gonopores. Clitellum is one cell thick and includes male and female gonopores. A single family of freshwater oligochaetes, the Lumbriculidae. *Lumbriculus.*

Order Tubificida. Setal bundles usually with two or more setae, rarely absent. Setae often hairlike or otherwise modified. One pair of testes followed by one pair of ovaries, in adjacent segments. Male gonopores in segment immediately in front of or behind testicular segment. Clitellum is one cell thick including male and female gonopores.

  Suborder Tubificina. Spermathecal pores in the segment immediately in front of or behind that bearing the male pores, rarely in the same segment. Setae diverse.

    Family Tubificidae. Marine and freshwater. Some species widely distributed in poorly oxygenated and polluted waters. *(Tubifex, Branchiura, Limnodrilus).* Many genera.

    Family Naididae. Very small, aquatic, predominantly freshwater oligochaetes. Some with elongated proboscides. Reproduction often asexual. *Nais, Ripistes, Dero* (including *Aulophorus*), *Slavina, Stylaria,* the carnivorous and parasitic *Chaetogaster.*

    Family Phreodrilidae. Small family of freshwater, marine, and commensal species.

    Family Opistocystidae and Family Dorydrilidae.

  Suborder Enchytraeina. Spermathecae anterior to the testes, separated by a gap of five segments. Setae simple batons, rarely forked or absent.

    Family Enchytraeidae. Marine, freshwater, and terrestrial. *Enchytraeus.*

Order Haplotaxida. Basically having two segments with testes followed by two segments with ovaries. One pair of testes or ovaries or both often absent. If one pair of testes, these are separated from the ovaries by one or two segments. Male pores one or more segments behind the corresponding funnels.

  Suborder Haplotaxina. Four simple or forked setae per segment. Testes in segments 10 and 11. Ovaries in segments 12 and 13 or sometimes absent from 13. Male gonopores in the segment immediately behind the two testicular segments. Clitellum is one cell thick. A single family of freshwater and semiterrestrial species, the Haplotaxidae.

  Suborder Alluroidina. Four pairs of simple setae per segment. Testes in segments 10 or 10 and 11. Ovaries in 13. Clitellum is one cell thick. Families Alluroididae and Syngenodrilidae in Africa and South America. Aquatic to semiterrestrial. *Alluroides, Syngenodrilus.*

  Suborder Moniligastrina. Four pairs of simple setae per segment. One or two pairs of testes in testis sacs, each of which is suspended in the posterior septum of its segment. One or two pairs of male gonopores at the posterior border of the segment behind the corresponding testis sac. The anterior pair of ovaries is lost. Clitellum is one cell thick. A single widespread family of sometimes giant earthworms, mostly in India or Burma. *Moniligaster, Drawidia.*

  Suborder Lumbricina. Setae simple, eight or sometimes multiplied in a ring, in each segment. One or two pairs of testes, usually in segments 10 and 11. One pair (exceptionally two pairs) of male pores two or more segments behind the posterior testes. Ovaries in segment 13, rarely in 12 also. Clitellum is more than one cell thick, eggs therefore have little yolk. Mostly earthworms, some swamp worms and aquatic species.

    The suborder contains a number of small aquatic or semiaquatic families: Biwadrilidae (Japan), Spargonophilidae (North America), Almidae (Africa, South America), Criodrilidae (Europe), Lutodrilidae (North America). It also comprises small families of earthworms: Komarekionidae (North America), Kynotidae (Madagascar), Hormogastridae (Europe and North Africa), and Microchaetidae (South Africa, containing giant forms). The Lumbricina also include the following five large families of earthworms.

Family Glossoscolecidae. With esophageal gizzard. Male gonopores on or (rarely) behind the clitellum. *Glossoscolex* South America. *Pontoscolex* throughout warmer regions.

Family Lumbricidae. Male gonopores on segment 15, anterior to the clitellum. With intestinal but never esophageal gizzard. Temperate Old and New Worlds. *Lumbriculus, Eisenia, Allolobophora.*

Family Ocnerodrilidae. Male gonopores usually in segment 17 with a single pair of tubular prostate glands, less commonly in segment 18 with prostates in segments 17 and 19. Usually with esophageal diverticula in segment 9. Circumtropical. *Ocnerodrilus, Eukerria.*

Family Megascolecidae. Male gonopores usually in segment 18, fused with or near a pair of prostate glands or with prostates in segments 17 and 19. Esophageal diverticula, if present, not restricted to segment 9. Worldwide excepting western Europe. *Megascolides* includes the giant Australian worm. *Megascolex* and *Pheretima* with a ring of setae.

Family Eudrilidae. Near the Ocnerodrilidae but with distinctive prostates and, often, internal fertilization. Africa. *Eudrilus,* with one species widespread elsewhere.

Members of the family Aeolosomatidae, which are minute freshwater species, have in the past been considered primitive oligochaetes. On the basis of their reproductive systems (see p. 566), and other features, Brinkhurst and Jamieson (1972) believe that they are unrelated to other oligochaetes and should be removed from the class.

SYSTEMATIC RÉSUMÉ OF CLASS HIRUDINEA (OR HIRUDINOIDEA)

Order Acanthobdellida. A primitive order, contains a single north European species parasitic on salmonid fish. Setae and a compartmented coelom are present in the five anterior segments. No anterior sucker.

Order Rhynchobdellida. Strictly aquatic leeches with an eversible proboscis and a circulatory system that is separate from the coelomic sinuses.

Family Glossiphoniidae. Flattened leeches, typically with three annulations per segment in the midregion of the body. Includes many ectoparasites of both invertebrates and vertebrates. *Marsupiobdella, Glossiphonia, Helobdella, Placobdella, Theromyzon, Haementeria, Hemiclepsis.*

Family Piscicolidae. The fish leeches. Subcylindrical body that often bears lateral gills. Usually more than three annulations per segment. Most marine leeches belong to this family. Parasites of marine and freshwater fish and, rarely, crustaceans. *Piscicola, Pontobdella, Trachelobdella, Branchellion, Ozobranchus, Illinobdella, Myzobdella, Cystobranchus.*

Order Arnchynchobdellida. Aquatic or terrestrial leeches having a noneversible pharynx and three pairs of jaws (Hirudinidae and Haemadipsidae) or no jaws (Erpobdellidae and four other families). Five annulations per segment.

Family Hirudinidae. Chiefly amphibious or aquatic blood-sucking leeches. *Haemopsis, Hirudo* (*Hirudo medicinalis* is the European medicinal leech), *Macrobdella* (*Macrobdella decora* is the American medicinal leech), *Philobdella.*

Family Haemadipsidae. Terrestrial, tropical leeches of Australasian region attacking chiefly warm-blooded vertebrates. *Haemadipsa, Phytobdella.*

Family Erpobdellidae. Predaceous leeches. *Erpobdella, Dina.*

REFERENCES

Segmentation

Chaudonneret, J. 1979. Table ronde sur la metamerie. I. La notion de metamerie chez les invertebrés. [Round table on metamerism. I. The notion of metamerism in the invertebrates.] Bull. Soc. Zool. France. *104:*241–270.

Clark, R.B. 1964. Dynamics in Metazoan Evolution. The Origin of the Coelom and Segments. Clarendon University Press, Oxford. 313 pp.

Gilbert, S. F. 1988. Developmental Biology. Sinauer Associates, Inc., Sunderland, MA. pp. 630–661.

Meier, S. 1984. Somite formation and its relationship to metameric patterning of the mesoderm. Cell Differ. *14:*235–243.

Potswald, H. E. 1981. Abdominal segment formation in *Spirorbis moerchi* (Polychaeta). Zoomorphology. *97:*225–245.

Raff, R. A., and Kaufman, T. C. 1983. Embryos, Genes and Evolution. MacMillan Publishing Co., Inc., New York. pp. 262–286.

Sedgwick, A. On the origin of metameric segmentation and some other morphological questions. Quart. J. Microsc. Sci. *24:*43–82.

Weisblat, D. A., and Shankland, M. 1985. Cell lineage and segmentation in the leech. Phil. Trans. R. Soc. Lond. B *312:*39–56.

Coelomates and Acoelomates

Fransen, M. E. 1980. Ultrastructure of coelomic organization in annelids. I. Archiannelids and other small polychaetes. Zoomorphologie. *95:*235–249.

Hermans, C. O. (Eds.): The Ultrastructure of Polychaeta. Microfauna Marin. *4:*373–382.

Rieger, R. M. 1980. A new group of interstitial worms, Lobatocerebridae nov. fam. (Annelida) and its significance for metazoan phylogeny. Zoomorphologie. *95:*41–84.

Rieger, R. M. 1985. The phylogenetic status of the acoelomate organization within the Bilateria: A histological perspective. *In* Conway-Morris, S., George, J. D., Gibson, R. et al. Platt, H. M. (Eds.): The Origins and Relationships of Lower Invertebrates. Clarendon Press, Oxford. pp. 101–122.

Rieger, R. M. 1988. Comparative ultrastructure and the Lobatocerebridae: Keys to understanding the phylogenetic relationship of Annelida and acoelomates. *In* Westheide, W., and Rieger, R. M. 1991a. *Jennaria pulchra,* nov. gen. nov. spec., eine den psammobionten Annelidan nahestehe Gattung aus dem Kustengrundwasser von North Carolina. Ber. nat.-med. Verein Innsbruck. *78:*203–215.

Rieger, R. M. 1991b. Neue Organisationstypen aus der Sandluckenraumfauna: Die Lobatocerebriden und *Jennaria pulchra.* Verh. Dtsch. Zool. Ges. *84:*247–259.

Smith, P. R., Lombardi, J., and Rieger, R. M. 1986. Ultra-

structure of the body cavity lining in a secondary acoelomate, *Microphthalmus* cf. *listensis* Westheide (Polychaeta, Hesionidae). J. Morphol. *188:*257–271.

Cuticle Evolution

Rieger, R. M. 1984. Evolution of the cuticle in the lower Eumetazoa. *In* Bereiter-Hahn, J., Matoltsky, A. G., and Richards, K. S. (Eds.): Biology of the Integument. Vol. 1. Invertebrates. Springer-Verlag, Berlin. pp. 389–399.

Ruppert, E. E. 1991. Introduction to the aschelminth phyla: A consideration of mesoderm, body cavities, and cuticle. *In* Harrison, F. W., and Ruppert, E. E. (Eds.): Microscopic Anatomy of Invertebrates. Vol. 4. Aschelminthes. Wiley-Liss, Inc., New York. pp. 1–17.

The introductory references on page 6 include many general works and field guides that contain sections on annelids.

Annelida

Anderson, D. T. 1973. Embryology and Phylogeny in Annelids and Arthropods. Pergamon Press, Oxford. 495 pp.

Brinkhurst, R. O. 1982. Evolution in the Annelida. Can. J. Zool. *60:*1043–1059.

Cather, J. 1971. Cellular interactions in the regulation of development in annelids and molluscs. Adv. Morphogen. *9:*67–124.

Clark, R. B. 1964. Dynamics in Metazoan Evolution: The Origin of the Coelom and Segments. Clarendon Press, Oxford. 313 pp.

Dales, R. P. 1963. Annelids. Hutchinson University Library, London.

Fauvel, P., Avel, M., Harant, H. et al. 1959. Embranchement des Annelides. *In* Grassé, P. (Ed.): Traité de Zoologie. Vol. 5. pt. 1. Masson et Cie, Paris. pp. 3–686.

Harrison, F. W., and Gardiner, S. L. 1992. Microscopic Anatomy of Invertebrates. Vol. 7. Annelida. Wiley-Liss, New York. 418 pp.

Highnam, K. C., and Hill, L. 1977. The Comparative Endocrinology of the Invertebrates. 2nd Edition. University Park Press, Baltimore. 357 pp.

Klemm, D. J. 1985. A Guide to the Freshwater Annelida (Polychaeta, Naidid and Tubificid Oligochaeta, and Hirudinea) of North America. Kendall/Hunt Publishing Co., Dubuque, IA. 226 pp.

Mangum, C. P. 1970. Respiratory physiology in annelids. Amer. Sci. *58:*641–647.

Mangum, C. P. 1976. Primitive respiratory adaptations. *In* Newell, R. C. (Ed.): Adaptation to Environment. Butterworth Group Publishing, MA. pp. 191–278. (Excellent review of gas exchange and internal transport in annelids, especially polychaetes.)

Mangum, C. P. 1977. Annelid hemoglobins: A dichotomy in structure and function. *In* Reish, D. J., and Fauchald, K. (Eds.): Essays in memory of Dr. Olga Hartman. Allan

Hancock Foundation, University of Southern California, Los Angeles.

Mangum, C. P. 1982. The function of gills in several groups of invertebrate animals. *In* Houlihan, D. F., Rankin, J. C., and Shuttleworth, T. J. (Eds.): Gills. Soc. Exp. Biol. Sem. Ser. No. 16. Cambridge University Press, Cambridge.

Mangum, C. P. 1985. Oxygen transport in invertebrates. Am. J. Physiol. *248:*505–514.

Mill, P. J. (Ed.): 1978. Physiology of Annelids. Academic Press, London, 684 pp. (Chapters on various topics by different contributors.)

Olive, P. J. W., and Clark, R. B. 1978. Physiology of reproduction. *In* Mill, P. J. (Ed.): Physiology of Annelids. Academic Press, London. pp. 271–368.

Parker, S. P. (Ed.): 1982. Synopsis and Classification of Living Organisms. Vol. 2. McGraw-Hill Book Co., New York. 1236 pp.

Rieger, R. M. 1985. The phylogenetic status of the acoelomate organization within the Bilateria: A histological perspective. *In* Conway Morris, S. et al. (Eds.): The Origins and Relationships of the Lower Invertebrates. Systematics Association. Spec. Vol. No. 28. Clarendon Press, Oxford. pp. 101–122.

Trueman, E. R. 1975. The locomotion of soft-bodied animals. Edward Arnold, London. 200 pp.

Polychaeta

Baskin, D. G. 1976. Neurosecretion and the endocrinology of nereid polychaetes. Am. Zool. *16:*107–124.

Boilly, B., and Wissocq, J. C. 1977. Occurrence of striated muscle fibers in a contractile vessel of a polychaete: The dorsal heart of *Magelona papillicornis.* Biol. Cell. *28:*131–136.

Brenchley, G. A. 1976. Predator detection and avoidance: Ornamentation of tube-caps of *Diopatra* spp. Mar. Biol. *38:*179–188.

Caspers, H. 1984. Spawning periodicity and habitat of the palolo worm *Eunice viridis* in the Samoan Islands. Mar. Biol. *79:*229–236.

Chughtai, I., and Knight-Jones, E. W. 1988. Burrowing into limestone by sabellid polychaetes. Zoolog. Scrip. *17:*231–238.

Clark, L. B., and Hess, W. N. 1940. Swarming of the Atlantic palolo worm, *Leodice fucato.* Tortugas Lab. Papers. *33:*21–70.

Dales, R. P. 1957. The feeding mechanism and structure of the gut of *Owenia fusiformis.* J. Mar. Biol. Assoc. U.K. *36:*81–89.

Dales, R. P., and Peter, G. 1972. A synopsis of the pelagic Polychaeta. J. Nat. Hist. *6:*55–92.

Dauer, D. M. 1985. Functional morphology and feeding behavior of *Paraprionospio pinnata* (Polychaeta: Spionidiae). Mar. Biol. *85:*143–151.

Day, J. 1967. Polychaeta of Southern Africa. Pt. I, Errantia; Pt. II, Sedentaria. British Museum of Natural History.

Dykens, J. A., and Mangum, C. P. 1984. The regulation of body fluid volume in the estuarine annelid *Nereis succinea.* J. Comp. Physiol. B *154:*607–617.

Eckelbarger, K. J. 1983. Evolutionary radiation in polychaete ovaries and vitellogenic mechanisms: Their possible role in life history patterns. Can J. Zool. *61:*487–504.

Fauchald, K. 1975. Polychaete phylogeny: A problem in protostome evolution. Syst. Zool. *23:*493–506.

Fauchald, K. 1977. The Polychaete worms. Definitions and keys to the orders, families and genera. Nat. Hist. Mus. of Los Angeles Co. Sci. Ser. *28:*1–190.

Fauchald, K. 1983. Life diagram patterns in benthic polychaetes. Proc. Biol. Soc. Wash. *96:*160–177.

Fauchald, K., and Jumars, P. A. 1979. The diet of worms: A study of polychaete feeding guilds. Oceanogr. Mar. Biol. Ann. Rev. *17:*193–284.

Flood, P. R., and Fiala-Medioni, A. 1982. Structure of the mucous feeding filter of *Chaetopterus variopedatus.* Mar. Biol. *72:*27–34.

Fransen, M. E. 1980. Ultrastructure of coelomic organization in annelids. I. Archiannelids and other small polychaetes. Zoomorphologie. *95:*235–249.

Gardiner, S. L. 1975. Errant polychaete annelids from North Carolina. J. Elisha Mitchell Sci. Soc. *91:*77–220. (Keys, descriptions, and figures for the North Carolina fauna.)

Goerke, H. 1971. Die Ernährungweise der *Nereis*-Arten der deutschen Küsten. Veröff. Inst. Meeresforsch. Bremerh. *13:*1–50.

Goodnight, C. J. 1973. The use of aquatic macroinvertebrates as indicators of stream pollution. Trans. Am. Microsc. Soc. *92:*1–13.

Gray, J. 1939. Studies in animal locomotion. VIII. The kinetics of locomotion of *Nereis diversicolor.* J. Exp. Biol. *16:*9–17.

Hartman, O. 1959 and 1965. Catalog of the polychaetous annelids of the world. Allan Hancock Foundation Occas. Papers. Vol. 23. 628 pp. Supplement and Index (1965). 197 pp.

Hermans, C. O. 1969. The systematic position of the Archiannelida. Syst. Zool. *18:*85–102.

Kay, D. G. 1974. The distribution of the digestive enzymes in the gut of the polychaete *Neanthes virens.* Comp. Biochem. Physiol. *47*(A):573–582.

Knight-Jones, P., and Thorp, C. H. 1984. The opercular brood chambers of Spirorbidae. Zool. J. Linnean Soc. *80:*121–133.

Kudenov, J. D. 1977. The functional morphology of feeding in three species of maldanid polychaetes. Zool. J. Linn. Soc. *60:*95–109.

MacGinitie, G. E. 1939. The method of feeding of *Chaetopterus.* Biol. Bull. *77:*115–118.

Mangum, C. P. 1976. The oxygenation of hemoglobin in lugworms. Physiol. Zool. *49:*85–99.

Mangum, C. P., Woodin, B. R., Bonaventura, C. et al. 1975. The role of coelomic and vascular hemoglobin in the annelid family Terebellidae. Comp. Biochem. Physiol. *51*A:281–294.

Martin, N., and Anctil, M. 1984. Luminescence control in the tubeworm *Chaetopterus variopedatus:* Role of nerve cord and photogenic gland. Biol. Bull. *166:*583–593.

McConnaughey, B., and Fox, D. L. 1949. The anatomy and biology of the marine polychaete *Thoracophelia mucronata.* Univ. Calif. Publ. Zool. *47*(12):319–339.

Mileikovskii, S. A. 1968. Morphology of larvae systematics of Polychaeta. Zool. Zh. *47:*49–50.

Morin, J. G. 1983. Coastal bioluminescence: Patterns and functions. Bull. Marine Sci. *33:*787–817.

Nakao, T. 1974. An electron microscopic study of the circulatory system in *Nereis japonica.* J. Morphol. *144:*217–236.

Nicol, E. A. T. 1931. The feeding mechanism formation of the tube, and physiology of digestion in *Sabella pavonina.* Trans. R. Soc. Edinburgh. *56:*537–598.

Nott, J. A., and Parkes, K. R. 1975. Calcium accumulation and secretion in the serpulid polychaete *Spirorbis spirorbis* at settlement. J. Mar. Biol. Assoc. U.K. *55:*911–923.

Ockelmann, K. W., and Vahl, O. 1970. On the biology of the polychaete *Glycera alba,* especially its burrowing and feeding. Ophelia. *8:*275–294.

O'Clair, R. M., and Cloney, R. A. 1974. Patterns of morphogenesis mediated by dynamic microvilli: Chaetogenesis in *Nereis vexillosa.* Cell Tissue Res. *151:*141–157.

Pawlik, J. R. 1983. A sponge-eating worm from Bermuda: *Branchiosyllis oculata* (Polychaeta, Syllidae). P.S.Z.N.I: Mar. Ecol. *4:*65–79.

Pietsch, A., and Westheide, W. 1987. Protonephridial organs in *Myzostoma cirriferum* (Myzostomida). Acta Zool. (Stockh.) *68:*195–203.

Roe, P. 1975. Aspects of life history and of territorial behavior in young individuals of *Platynereis bicanaliculata* and *Nereis vexillosa.* Pac. Sci. *29:*341–348.

Rose, S. M. 1970. Regeneration: Key to Understanding Normal and Abnormal Growth and Development. Appleton-Century-Crofts, New York.

Ruby, E. G., and Fox, D. L. 1976. Anerobic respiration in the polychaete *Euzonus (Thoracophelia) mucronata.* Mar. Biol. *35:*149–153.

Santos, S. L., and Simon, J. L. 1974. Distribution and abundance of the polychaetous annelids in a south Florida estuary. Bull. Mar. Sci. *24:*669–689.

Schroeder, P. C. 1984. Chaetae. *In* Bereiter-Hahn, J., Matoltsy, A. G., and Richards, K. D. (Eds.): Biology of the Integument. I. Invertebrates. Springer-Verlag, Berlin. pp. 297–309.

Schroeder, P. C., and Hermans, C. O. 1975. Annelida: Polychaeta. *In* Giese, A. C., and Pearse, J. S. (Eds.): Reproduction of Marine Invertebrates. Vol. III. Academic Press, New York. pp. 1–205.

Rice, S. A. 1980. Ultrastructure of the male nephridium and its role in spermatophore formation in spionid polychaetes. Zoomorphologie. *95:*181–194.

Rice, S. A. 1987. Reproductive biology, systematics, and evolution in the polychaete family Alciopidae. Biol. Soc. Wash. Bull. *7:*114–127.

Storch, U., and Alberti, G. 1978. Ultrastructural observations on the gills of polychaetes. Helgol. Wiss. Meeresunters. *31:*169–179.

Uebelackei, J. M., and Johnson, P. G. (Eds.): 1984. Taxonomic guide to the polychaetes of the Northern Gulf of Mexico. 7 vols. NOAA Tech. Report NMFS CIRC-375.

Virnstein, R. W. 1977. The importance of predation by crabs and fishes on benthic infauna in Chesapeake Bay. Ecology. *58:*1199–1217.

Warren, L. M. 1976. A population study of the polychaete *Capitella capitata* at Plymouth. Mar. Biol. *38:*209–216.

Waxman, L. 1971. The hemoglobin of *Arenicola cristata.* J. Biol. Chem. *246:*7318–7327.

Weber, R. E. 1978. Respiration. *In* Mill, P. J. (Ed.): Physiology of Annelids. Academic Press, London. pp. 369–446.

Wells, G. P. 1950. Spontaneous activity cycles in Polychaeta worms. Symp. Soc. Exp. Biol. *4:*127–142.

Wells, G. P. 1959. Worm autobiographies. Sci. Amer. *200:*132–141.

Westheide, W. 1984. The concept of reproduction in polychaetes with small body size: Adaptations in interstitial species. Fortschr. Zool. *29:*265–287.

Westheide, W., and Hermans, C. O. (Eds.): 1988. The ultrastructure of Polychaeta. Microfauna Marina 4. Gustav Fischer Verlag, Stuttgart, New York. 494 pp.

Whitlatch, R. B. 1974. Food-resource partitioning in the deposit feeding polychaete *Pectinaria gouldii.* Biol. Bull. *147:*227–235.

Wilson, D. P. 1974. *Sabellaria* colonies at Duckpool, North Cornwall, 1971–1972, with a note for May 1973. J. Mar. Biol. Assoc. U. K. *54:*393–436.

Woodin, S. A., and Merz, R. A. 1987. Holding on by their hooks: Anchors for worms. Evolution. *41:*427–432.

Zottoli, R. A., and Carriker, M. R. 1974. Burrow morphology, tube formation, and microarchitecture of shell dissolution by the spionid polychaete *Polydora websteri.* Mar. Biol. *27:*307–316.

Oligochaeta

Bolton, P. J., and Phillipson, J. 1976. Burrowing, feeding, egestion and energy budgets of *Allolobophora rosea.* Oecologia. *23:*225–245.

Brinkhurst, R. O., and Cook, D. G. 1979. Aquatic Oligochaete Biology. Plenum Press, New York. 529 pp. (A collection of papers from a symposium, largely on ecological and taxonomic aspects of oligochaete biology.)

Brinkhurst, R. O., and Jamieson, B. G. M. 1972. Aquatic Oligochaeta of the World. Toronto University Press, Toronto. 860 pp.

Darwin, C. R. 1881. The formation of vegetable mould through the action of worms with observations on their habits. John Murray & Co., London.

Edwards, C. A., and Lofty, J. R. 1977. Biology of Earthworms. 2nd Edition. Chapman and Hall, London. (A general account of the earthworms, with special emphasis on ecology. Includes a key to common genera and methods for study.)

Friedlander, B. 1888. Über das Kriechen der Regenwürmer. Biol. Zbl. *8:*363–366.

Giere, O., and Pfannkuche, O. 1982. Biology and ecology of marine Oligochaeta, a review. Oceanog. Mar. Biol., Ann. Rev. *20:*173–308.

Gray, J., and Lissmann, H. W. 1938. Studies in animal locomotion. VII. Locomotory reflexes in the earthworm. J. Exp. Biol. *15:*506–517.

Jamieson, B. G. M. 1978. Phylogenetic and phenetic systematics of the opisthoporous Oligochaeta. Evolut. Theory. *3:*195–233.

Jamieson, B. G. M. 1981. The Ultrastructure of the Oligochaeta. Academic Press, New York. 462 pp.

Lasserre, P. 1975. Clitellata. *In* Giese, A. C., and Pearse, J. S. (Eds.): Reproduction of Marine Invertebrates. Vol. III. Academic Press, New York. pp. 215–275.

Laverack, M. S. 1963. The physiology of earthworms. Pergamon Press, MacMillan Co., New York. 205 pp.

Learner, M. A. 1972. Laboratory studies on the life histories of four enchytraeid worms which inhabit sewage percolating filters. Ann. Appl. Biol. *70*(3):251–266.

Martin, N. A. 1982. The interaction between organic matter in soil and the burrowing activity of three species of earthworms. Pedobiologia. *24:*185–190.

Meinhardt, U. 1974. Comparative observations on the laboratory biology of endemic earthworm species: II. Biology of bred species. Z. Angew. Zool. *61*(2):137–182.

Piearce, T. G. 1983. Functional morphology of lumbricid earthworms, with special reference to locomotion. J. Nat. Hist. *17:*95–111.

Rieger, R. M. 1980. A new group of interstitial worms, Lobatocerebridae nov. fam. (Annelida) and its significance for metazoan phylogeny. Zoomorphologie. *95:*41–84.

Satchell, J. E. (Ed.): 1983. Earthworm Ecology. Chapman and Hall, New York. 512 pp. (A collection of papers.)

Seymour, M. K. 1969. Locomotion and coelomic pressure in *Lumbricus*. J. Exp. Biol. *51:*47.

Singer, R. 1978. Suction-feeding in *Aeolosoma*. Trans. Am. Microsc. Soc. *97:*105–111.

Tynen, M. J. 1970. The geographical distribution of ice worms (Oligochaeta: Enchytraeidae). Can. J. Zool. *48:*1363–1367.

Hirudinea

Cross, W. H. 1976. A study of predation rates of leeches on tubificid worms under laboratory conditions. Ohio J. Sci. *76:*164–166.

Dickinson, M. H., and Lent, C. M. 1984. Feeding behavior of the medicinal leech, *Hirudo medicinalis*. J. Comp. Physiol. Sens. Neural Behav. Physiol. *154:*449–456.

Gray, J., Lissmann, H. W., and Pumphrey, R. J. 1938. The mechanism of locomotion in the leech. J. Exp. Biol. *15:*408–430.

Haupt, J. 1974. Function and ultrastructure of the nephridium of *Hirudo medicinalis* L. II. Fine structure of the central canal and the urinary bladder. Cell Tiss. Res. *152:*385–401.

Hildebrandt, J. -P. 1988. Circulation in the leech, *Hirudo medicinalis* L. J. Exp. Biol. *134:*235–246.

Klemm, D. J. 1972. Freshwater Leeches of North America. Biota of Freshwater Ecosystems Identification Manual No. 8. Environmental Protection Agency, U.S. Government Printing Office.

Lent, C. M., Fliegner, K. H., Freedman, E. et al. 1988. Ingestive behavior and physiology of the medicinal leech. J. Exp. Biol. *137:*513–527.

Mann, K. H. 1962. Leeches (Hirudinea), Their Structure, Physiology, Ecology, and Embryology. Pergamon Press, New York.

Nicholls, J. G., and Van Essen, D. 1974. The nervous system of the leech. Sci. Amer. *230:*38–48.

Sawyer, R. T. 1972. North American freshwater leeches, exclusive of the Piscicolidae, with a key to all species. Illinois Biol. Monogr. *46:*1–154.

Sawyer, R. T. 1984. Leech Biology and Behavior. 3 Vols. Oxford University Press, New York. 500 pp. (An excellent account of all aspects of the biology of leeches.)

Stent, G. S., Kristan, W. B., Friesen, W. O. et al. 1978. Neuronal generation of the leech swimming movement. Science. *200:*1348–1356.

Stent, G. S., and Weisblat, D. A. 1982. The development of a simple nervous system. Sci. Amer. *246:*136–146.

Wilde, V. 1975. Investigations on the symbiotic relationship between *Hirudo officinalis* and bacteria. Zool. Anz. *195*(5/6):289–306.

Young, S. R., Dedwylder, R. D., and Friesen, W. O. 1981. Responses of the medicinal leech *(Hirudo medicinalis)* to water waves. J. Comp. Physiol. and Sens. Neural Behav. Physiol. *144:*111–116.

Branchiobdellida

Brinkhurst, R. O., and Gelder, S. R. 1991. Branchiobdellida. *In* Thorpe, G. H., and Covich, A. P. (Eds.): Ecology and Classification of North American Freshwater Invertebrates. Academic Press, New York. pp. 434–441.

Holt, T. C. 1968. The Branchiobdellida: Epizootic annelids. The Biologist. Vol. L. Nos. 3-4. pp. 79–94.

Gelder, S. R., and Brinkhurst, R. O. 1991. An assessment of the phylogeny of the Branchiobdellida (Annelida: Clitellata) using PAUP. Can. J. Zool. *68:*1318–1326.

Jennings, J. B., and Gelder, S. R. 1979. Gut structure, feeding and digestion in the branchiobdellid oligochaete *Cambarinicola macrodonta* Ellis, 1912, an ectosymbiote of the freshwater crayfish *Procambarus clarkii.* Biol. Bull. *156:*300–314.

Pogonophora

Anderson, D. T. 1973. Embryology and Phylogeny in Annelids and Arthropods. Pergamon Press, Oxford. 495 pp.

Arp, A. J., and Childress, J. J. 1983. Sulfide binding by the blood of the hydrothermal vent tube worm *Riftia pachyptila.* Science. *219:*295–297.

Arp, A. J., Childress, J. J., and Fisher, C. R., Jr. 1985. Blood gas transport in *Riftia pachyptila. In* Jones, M. L. (Ed.): Hydrothermal Vents of the Eastern Pacific: An Overview. Bull. Biol. Soc. Wash. *6:*289–300.

Bakke, T. 1976. The early embryos of *Siboglinum fiordicum* Webb (Pogonophora) reared in the laboratory. Sarsia. *60:*1–11.

Cavanaugh, C. M., Gardiner, S. L., Jones, M. L. et al. 1981. Prokaryotic cells in the hydrothermal vent tube worm *Riftia pachyptila* Jones: Possible chemoautotropic symbionts. Science. *213:*340–342.

Cutler, E. B. 1975. The phylogeny and systematic position of the Pogonophora. Syst. Zool. *24:*512–513. (A short review of a symposium held in Copenhagen in 1973.)

Felbeck, H., and Childress, J. J. 1988. *Riftia pachyptila:* A highly integrated symbiosis. *In* Laubier, L. (Ed.): Actes du Colloque, "Les Sources Hydrothermales de la Ride du Pacifique Oriental. Biologie et Ecologie." Institut Oceanographique, Paris, 4–7 Novembre 1985. Oceanolog. Acta. Special Volume No. *8.* Montrouge: Gauthiers-Villars. pp. 131–138.

Gaill, F., and Hunt, S. 1986. Tubes of deep sea hydrothermal vent worms *Riftia pachyptila* (Vestimentifera) and *Alvinella pompejana* (Annelida). Mar. Ecol. Prog. Ser. *34:*267–274.

Gardiner, S. L., and Jones, M. L. 1993. Vestimentifera. *In* Harrison, F. W., and Rice, M. E. (Eds.): Microscopic Anatomy of Invertebrates. Vol. 12. Onychophora, Chilopoda, and Lesser Protostomata. Wiley-Liss, New York. pp. 371–460.

George, J. D., and Southward, E. C. 1973. A comparative study of the setae of Pogonophora and polychaetous Annelida. J. Mar. Biol. Assoc. U.K. *53:*403–424.

Grassle, J. F. 1985. Hydrothermal vent animals: Distribution and biology. Science. *229:*713–717.

Gupta, B. L., and Little, C. 1975. Ultrastructure, phylogeny and Pogonophora. Z. zool. Syst. Evolut.-forsch. *1975:* 43–63.

Ivanov, A. V. 1963. Pogonophora. Academic Press, New York. 479 pp.

Ivanov, A. V. 1975. Embryonalentwicklung der Pogonophora und ihre systematische Stellung. pp. 10–44.

Ivanov, A. V. 1988. Analysis of the embryonic development of Pogonophora in connection with the problems of phylogenetics. Z. Zool. Syst. Evolut.-forsch. *26:*161–185.

Jones, M. L. 1981. *Riftia pachyptila* Jones: Observations on the vestimentiferan worm from the Galapagos rift. Science. *213:*333–336.

Jones, M. L. 1985. On the Vestimentifera, a new phylum: Six new species, and other taxa, from hydrothermal vents and elsewhere. *In* Jones, M. L. (Ed.): The hydrothermal vents of the eastern Pacific: An overview. Bull. Biol. Soc. Wash. *6:*117–158.

Jones, M. L., and Gardiner, S. L. 1988. Evidence for a transient digestive tract in Vestimentifera. Proc. Biol. Soc. Wash. *101:*423–433.

Jones, M. L., and Gardiner, S. L. 1989. On the early development of the vestimentiferan tube worm *Ridgeia* sp. and observations on the nervous system and trophosome of *Ridgeia* sp. and *Riftia pachyptila.* Biol. Bull. *177:*254–276.

Norrevang, A. 1970a. The position of Pogonophora in the phylogenetic system. Z. Zool. Syst. Evolut.-forsch. 8, H. *3:*161–172.

Norrevang, A. 1970b. On the embryology of *Siboglinum* and its implications for the systematic position of the Pogonophora. Sarsia *42:*7–16.

Norrevang, A. (Ed.): 1975. The Phylogeny and Systematic Position of Pogonophora. Z. Zool. Syst. Evolut.-forsch. Sonderheft. 1975. (Proceedings of a symposium held at the University of Copenhagen in 1973.)

Southward, A. J. 1975. On the evolutionary significance of the mode of feeding of Pogonophora. Z. zool. Syst. Evolut.-forsch. *1975:*77-85.

Southward, A. J., and Southward, E. C. 1980. The significance of dissolved organic compounds in the nutrition of *Siboglinum ekmani* and other small species of Pogonophora. J. Mar. Biol. Assoc. U.K. *60:*1005–1034.

Southward, A. J., Southward, E. C., Dando, P. R. et al. 1981. Bacterial symbionts and low C/C ratios in tissues of Pogonophora indicate unusual nutrition and metabolism. Nature. *293:*616–620.

Southward, E. C. 1971a. Pogonophora of the Northwest Atlantic: Nova Scotia to Florida. Smithson. Contrib. Zool. *88:29.*

Southward, E. C. 1971b. Recent researches on the Pogonophora. Oceangr. Mar. Biol. Ann. Rev. *9:193–220.*

Southward, E. C. 1982. Bacterial symbionts in Pogonophora. J. Mar. Biol. Assoc. U.K. *62:889–906.*

Southward, E. C. 1988. Development of the gut and segmentation of newly settled stages of *Ridgeia* (Vestimentifera): Implications for relationship between Vestimentifera and Pogonophora. J. Mar. Biol. Assoc. U.K. *68:465–487.*

Southward, E. C. 1993. Pogonophora. *In* Harrison, F. W., and Rice, M. E. (Eds.): Microscopic Anatomy of Invertebrates. Vol. 12. Onychophora, Chilopoda, and Lesser Protostomata. Wiley-Liss, New York. pp. 327–369.

INTRODUCTION TO THE ARTHROPODS

596

PRINCIPLES AND EMERGING PATTERNS

THE FOSSIL RECORD

Since its systematic exploration began over 200 years ago, a rich fossil record of metazoan animals has been uncovered. Taxonomic and anatomical descriptions of fossils have occupied much of this investigation, and more recent techniques using thin sections and X-rays have yielded considerable information. Contemporary paleontologists are also searching for clues about the living habits of fossil species and the environments in which they lived.

The fossil record provides a fairly good account of the animals that have inhabited our planet and the sequence of their appearance and disappearance in geological history. It also provides some information about the evolution of lower taxa. However, the fossil record tells us almost nothing about the evolutionary origins and relationships of phyla and classes. *Knightoconus,* an apparent intermediate stage between high cone monoplacophorans and cephalopods, is one of the very few animal "missing links" at the level of class and phylum.

A number of factors account for the lack of such links. The fossil record is uneven because the preservation of a particular animal as a fossil depends on the age of the entrapping sediments, the kind of environment in which the animal lived, and most important, the type of skeleton the animal possessed. Thus, specimens of soft-bodied animals covered by very ancient sediments are rare in the fossil record. Yet such were most of the ancestral forms that gave rise to the presently known phyla.

We might have expected that the animal phyla would have appeared at different times over the course of the fossil record. In fact at least some representatives of almost all of the animal phyla and many of the classes living today appeared by the end of the great Cambrian radiation (550 million years ago) at the beginning of the Paleozoic. This is the starting point for most of the animal fossil record. One of the best of the Cambrian records is in the Burgess shale of British Columbia. This remarkable preservation contains over 120 fossil species, including cnidarians, annelids, priapulids, molluscs, arthropods, echinoderms, and chordates, plus many species that do not fit into any existing phylum (Fig. 12–1). A somewhat similar deposit has recently been found in Yunnan province of China. Some paleontologists believe that as many as 100 animal phyla existed during the Cambrian compared with the approximately 32 existing today. Without the Burgess shale our knowledge of Cambrian fossils would be very limited because fossil-bearing Cambrian strata are limited and often covered by younger sedimentary rocks.

Still older pre-Cambrian rocks have an even more restricted distribution. The few fossil-bearing outcrops that have been found in various parts of the world, such as the rocks of the Ediacara hills of South Australia, contain the oldest known animals. These Ediacaran faunas date from 565 million years ago to the beginning of the Cambrian and include fossils that look like jellyfish and worms, as well as what are probably invertebrate tracks and burrows.

There is nothing in the fossil record that reveals the evolutionary origins of major animal groups. Our speculations about those origins and the relationships of animal phyla and classes, therefore, continue to be based largely on comparative morphology and development. However, a new source of information is beginning to influence our investigations. This is the comparison of nucleotide, or amino acid, sequences in nucleic acids and proteins (p. 1042).

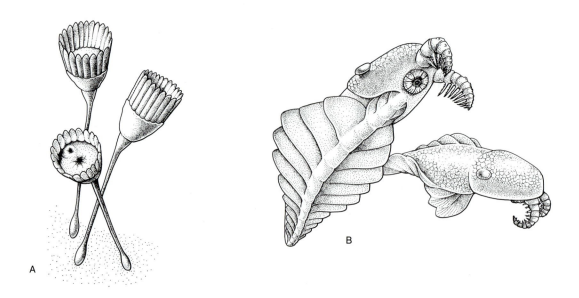

FIGURE 12–1 Reconstructions of two marine animals from the Burgess shale that may not belong to any existing phylum. **A,** The stalked flower-like *Dinomischus* had mouth and anus on the exposed upper surface (compare with Entoprocta). **B,** *Anomalocaris* had a ventral circular mouth and a pair of antennae. *(From Gould, S.J. 1989. Wonderful Life: The Burgess Shale and the Nature of History. W.W. Norton, New York.)*

THE ARTHROPODS

Arthropods are a vast assemblage of animals. At least three quarters of a million species have been described—more than three times the number of all other animal species combined (see the figure on the inside cover). The tremendous adaptive diversity of arthropods has enabled them to survive in virtually every habitat; they are perhaps the most successful of all the invaders of the terrestrial habitat.

Arthropods are protostomes and are clearly related to annelids, but whether arthropods arose from annelids or both from some common ancestor is uncertain. Nevertheless, the relationship is displayed in several ways.

    1. Arthropods, like annelids, are **segmented** (Fig. 12–2A). Segmentation is evident in the embryonic development of all arthropods and is a conspicuous feature of many adults, especially the more primitive species. As in annelids, growth in arthropods results from the addition of new segments to the region immediately anterior to the terminal section of the body (telson). The annelid prostomium and pygidium correspond, respectively, to the **acron** and **telson** of arthropods. Within many arthropod groups there has been a tendency for segmentation to become reduced. In such forms as mites, for example, it has almost disappeared. Loss of segmentation has occurred in three ways: (1) segments have disappeared; (2) segments have fused together; and (3) segmental structures, such as appendages, have become structurally and functionally differentiated from their counterparts on other segments. Different structures in two or more individuals having the same segmental origin are said to be **homologous.** Thus, the second antennae of a crab are homologous to the chelipeds (claws of scorpions), because both evolved from the original second pair of head appendages.

    The term **serial homology** is used to make comparisons of appendages from different segments within one individual. Thus, a crab's second antennae, mandibles, and chelipeds are serially homologous appendages. Serial homology is based on the similar pattern of morphogenesis of all appendages in

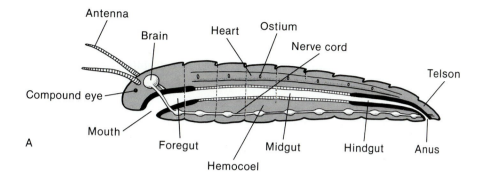

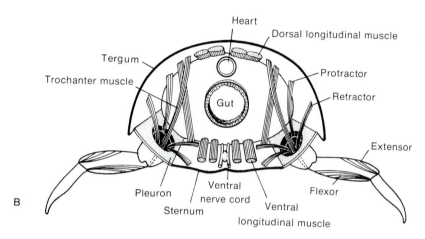

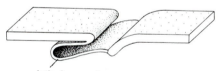

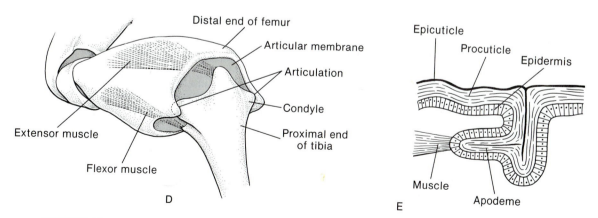

FIGURE 12–2 A and **B,** Structure of a generalized arthropod. **A,** Sagittal section. **B,** Cross section. **C,** Intersegmental articulation. Note articular membrane folded beneath segmental plate. **D,** Leg joint of an insect showing condyles and muscle attachments. **E,** An apodeme. *(C, After Weber from Vandel. D, Modified from Snodgrass. E, After Janet from Vandel.)*

the body, similarities in the component parts of appendages, and the observation that the appendages of primitive arthropods are similar in form and function.

2.   The nervous systems of annelids and arthropods are constructed on the same basic plan. In both, a **dorsal anterior brain** is followed by a **ventral nerve cord** containing ganglionic swellings in each segment (Fig. 12–2A).

3.   The embryonic development of a few arthropods still displays some degree of **spiral determinate cleavage,** with the mesoderm in these forms arising from the 4d blastomere.

4.   In the primitive condition, each arthropod segment bears a pair of **appendages** (Fig. 12–2B). This same condition is displayed by the polychaetes, in which each segment has a pair of parapodia. However, the homology between parapodia and arthropod appendages is uncertain.

The appendages of many arthropods, such as the horseshoe crabs, crustaceans, and the fossil trilobites, have two branches (See Fig. 14–5). Centipedes, millipedes, and insects have unbranched appendages. This branching or lack of branching of appendages has been considered a fundamental difference between these arthropods and is not easily reconciled by having one group derived from the other. Thus, some zoologists have argued that centipedes, millipedes, and insects had a separate evolutionary origin from that of other arthropods. Arguing against a separate evolution, a recent study proposes that the branched condition evolved from a fusion of pairs of segments in some arthropod ancestor. Two pairs of unbranched legs, one located behind the other, on originally unfused segments, may have become fused at the base, giving rise to an appendage with two branches.

The great success of arthropods is certainly correlated, in part at least, with the evolution of their appendages, which are levers that provide mechanical advantage. Coupled with rigid appendages are **cross-striated muscles** that move the appendages much faster than those of annelids, which have a different kind of muscle (obliquely striated). Moreover, in a different arthropod taxa the appendages became adapted for many different functions other than locomotion. In many ways the arthropod story is one of functional specialization of appendages.

EXOSKELETON

Although arthropods display these annelidan characteristics, they have undergone a great many profound and distinctive changes in the course of their divergence from annelids. The distinguishing feature of arthropods, and one to which many other changes are related, is the **chitinous exoskeleton,** or **cuticle,** that covers their entire body (Fig. 12–3). The cuticle can be thin and flexible, as in insect larvae, but it is usually relatively thick and rigid. Movement is made possible by the division of the cuticle into separate plates. Primitively, these plates are confined to single segments, and the plate of one segment is connected to the plate of the adjoining segment by means of an articular membrane, a region in which the cuticle is very thin and flexible (Fig. 12–2C). Basically, the cuticle of each segment is divided into four primary plates: a dorsal **tergum,** two lateral **pleura** (sing. **pleuron**), and a ventral **sternum** (Fig. 12–2B). This pattern has frequently disappeared because of either secondary fusion or subdivision. In all arthropods some degree of fusion and grouping of segmental skeletal plates has occurred, giving rise to body regions, or **tagmata** (sing. **tagma**). For example, in bees, the body is divided into three tagmata: head, thorax, and abdomen.

The cuticular skeleton of the appendages, like that of the body, has been divided into tubelike articles, or sections, connected to one another by articular membranes, thus creating a joint at each junction. Such joints enable the articles of the appendages, as well as those of the body, to move (hence the name of the phylum, Arthropoda, meaning "jointed feet"). In most arthropods the articular membrane between body segments is folded beneath the anterior segment (Fig. 12–2C). The additional development of articular condyles and sockets provides more precise control of movement and directed application of force through the attached tendon and muscle (Fig. 12–2D).

In addition to the external skeleton, what is sometimes called the **endoskeleton** has also developed. This endoskeleton may be an infolding of the cuticle (procuticle) that produces inner projections, or **apodemes,** to which the muscles are attached (Fig. 12–2E), or it may involve the sclerotization of internal connective tissue, forming free plates for muscle attachment within the body.

The arthropod skeleton is secreted by the underlying epidermis (or hypodermis). It is composed of a thin, outer **epicuticle** and a much thicker **procuticle** (Fig. 12–3). The epicuticle is composed of proteins and, in many terrestrial arthropods, wax (hydrocarbons). The fully developed procuticle consists of an outer exocuticle and an inner endocuticle. The latter

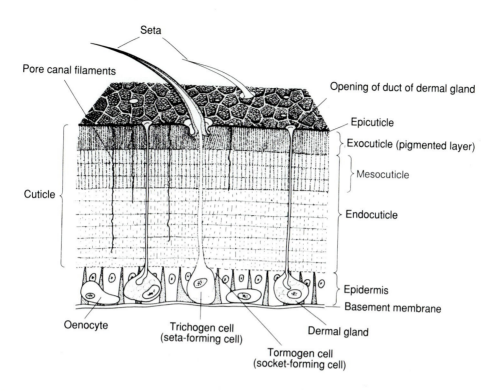

Seta

Pore canal filaments

Opening of duct of dermal gland

Epicuticle

Exocuticle (pigmented layer)

Mesocuticle

Cuticle

Endocuticle

Epidermis

Basement membrane

Oenocyte

Trichogen cell
(seta-forming cell)

Dermal gland

Tormogen cell
(socket-forming cell)

FIGURE 12–3 Diagrammatic section through the arthropod integument. *(From Hackman, R. H. 1971. In Florkin, M., and Scheer, B. T. (Eds.): Chemical Zoology. Vol. VI. Academic Press, New York.)*

two layers are composed of chitin and protein bound together to form a complex glycoprotein, but the exocuticle in addition has been **tanned;** meaning that its molecular structure has been further stabilized by the formation of additional cross linkages created using phenols. Exocuticle is absent at joints and along lines where the skeleton ruptures during molting. In many arthropods the procuticle is also impregnated with mineral salts. This is particularly true for the Crustacea, in which calcium carbonate and calcium phosphate deposition takes place in the procuticle.

Where the exoskeleton lacks a waxy epicuticle and is thin, it is relatively permeable and allows the passage of gases and water. Thus, the thin cuticle covering the gills of aquatic arthropods, such as crustaceans, is no barrier to gas exchange. In most terrestrial arthropods, the cuticle plays a very important role in reducing water loss, the reduction largely proportional to the amount of wax in the epicuticle. In many insects wax can be added to the epicuticle surface between molts by way of fine pore canals that penetrate the cuticle. A waxy epicuticle can also act as a water repellent, preventing a terrestrial arthropod

from being completely submerged in drops of rain or dew.

The arthropod cuticle is not restricted entirely to the exterior of the body. The epidermis develops from the embryonic surface ectoderm, and all infoldings of this embryonic layer, such as the fore- and hindgut, which develop from the stomodeum and the proctodeum, are lined with cuticle. Other such ectodermal derivatives include the tracheal (gas exchange) tubules of insects, chilopods, diplopods, and some arachnids; the book lungs of scorpions and spiders; and parts of the reproductive systems of some groups. All of these internal cuticular linings are also shed at the time of molting.

The color of arthropods commonly results from the deposition of brown, yellow, orange, and red melanin pigments within the cuticle. However, iridescent greens, purples, and other colors result from fine striations of the epicuticle, which cause light refraction and give the appearance of color. Often, body coloration does not originate directly in the cuticle but instead is produced by subcuticular pigment cells (chromatophores) or is caused by blood and tissue

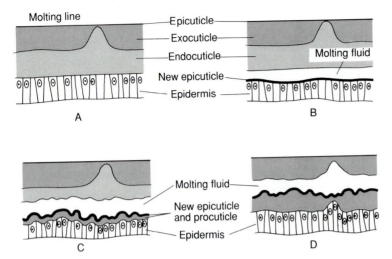

FIGURE 12–4 Molting in an arthropod. **A,** The fully formed exoskeleton and underlying epidermis between molts. **B,** Separation of the epidermis and secretion of molting fluid and the new epicuticle. **C,** Digestion of the old endocuticle and secretion of the new procuticle. **D,** The animal just before molting, encased within both new and old skeleton.

pigments, which are visible through a thin, transparent cuticle.

Despite its locomotor and supporting advantages, an external skeleton poses problems for a growing animal. The solution evolved by the arthropods has been the periodic shedding of the skeleton, a process called **molting,** or **ecdysis.**

Before the old skeleton is shed, the epidermis secretes proenzymes (inactive enzyme precursors) at the base of the skeleton. The epidermis now detaches from the skeleton and secretes a new epicuticle or at least its outermost (cuticulin) layer (Fig. 12–4B). The proenzymes secreted earlier become activated to form chitinase and protease, which digest the untanned endocuticle (Fig. 12–4C). The products of digestion are reabsorbed through the new epicuticle. With the erosion of the old endocuticle, the epidermis secretes new procuticle beneath the new epicuticle.

At this point the animal is encased within both an old and a new skeleton (Fig. 12–4D). The old skeleton now splits along certain predetermined lines, and the animal pulls out of the old encasement. The new skeleton is soft and commonly wrinkled and is stretched to accommodate the increased size of the animal. Stretching is brought about by blood pressure, facilitated by the uptake of water or air by the animal. Hardening of the new cuticle results from tanning of the protein and from stretching. Additional procuticle may be added following ecdysis, and in some arthropods, such as insects, additions are made to the epicuticle by secretions through the pore canals.

Sensory structures and muscle connections pose special problems for the molting process. Sensory structures, such as hairs, are laid down beneath the old skeleton, usually horizontally against the new skeleton. The dendrite may retain connection with the old hair until broken at ecdysis.

Muscles are attached to the exoskeleton by tonofilaments in specialized epidermal cells. The tonofilaments are anchored to an internal fold of the exoskeleton containing a fiber that runs all the way to the epicuticle (Fig. 12–5). The fibers are not digested during the molting process and maintain a connection between the old and new skeletons until severed at ecdysis.

The stages between molts are known as **instars,** and the length of the instars becomes longer as the animal becomes older. Some arthropods, such as lobsters and most crabs, continue to molt throughout their life. Other arthropods, such as insects and spiders, have more or less fixed numbers of instars, the last being attained with sexual maturity.

Although an arthropod is measurably larger and heavier following ecdysis, soft-tissue growth is actually continuous, as in most other animals. Proteins and other organic compounds are synthesized during the intermolt period, as tissue replaces water taken up following ecdysis. The uptake of water following ecdysis swells the body and unhardened cuticle. The new cuticle then hardens to a size larger than needed by the tissues within, providing room for growth.

Molting is under hormonal control. **Ecdysone,** secreted by certain endocrine glands (for example, the prothoracic glands in insects), is circulated by the bloodstream and acts directly on the epidermal cells. The production of ecdysone is in turn regulated by other hormones. Although most studied and best understood in insects and crustaceans, ecdysone controls molting in all arthropods. Molting physiology will be described in more detail for crustaceans (p. 781).

MOVEMENT AND MUSCULATURE

As movement in arthropods has become restricted to flexion between plates and cylinders of the cuticle, a related change has taken place in the nature of the body musculature. In annelids the muscles take the form of longitudinal and circular sheathlike layers of fibers lying beneath the epidermis. Contraction of the two layers exerts force on the coelomic fluid, which then functions as a hydrostatic skeleton. In arthropods, which have a different mode of locomotion involving an external skeleton, these muscular cylinders have become broken up into striated muscle bundles, which are attached to the inner surface of the skeletal system (Figs. 12–2B and 12–3).

The muscles are attached to the inner side of the exoskeleton by specialized epidermal cells (Fig. 12–5). Flexion and extension between plates are effected by the contraction of these muscles, with muscles and cuticle acting together as a lever system. This cofunctioning of the muscular system and skeletal system to bring about locomotion is similar to that in vertebrates. Extension, particularly of the appendages, is accomplished, in part or entirely (depending on the arthropod group), by an increase in blood pressure.

Arthropods employ jointed appendages as their chief means of locomotion; these act either as paddles in aquatic species or as legs in terrestrial groups. In contrast to the parapodia of polychaetes, the locomotor appendages of arthropods tend to be more slender, longer, and located more ventrally. Despite the more ventral position of the legs, the body usually hangs between the limbs. In the cycle of movement of one leg, the effective step, or stroke, during which the end of the leg is in contact with the substratum, is closer to the body than the recovery stroke, when the leg is lifted and swung outward and forward. Among the several factors determining speed of movement (e.g., cross-striated muscles), the length of stride is of obvious importance, and stride length increases with the length of the leg. Any problem of mechanical interference between legs is decreased by a reduction in the number of legs to five, four, or three pairs and by differences in leg length and the relative placement of leg tips. In arthropods that have retained a large number of legs, such as centipedes, the fields of movement of individual legs overlap those of other legs. For these animals mechanical interference is avoided by positioning a leg in effective stroke and a leg in recovery stroke at different distances from the body.

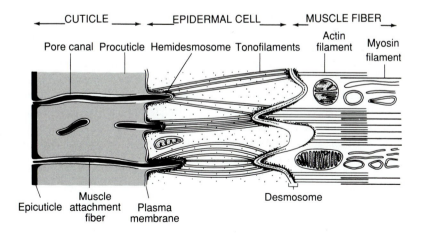

FIGURE 12–5 Diagram of the attachment of an insect muscle to the skeleton. Skeleton, epidermal cells, and muscle fibers are not drawn to scale. *(After Caveney, 1969, from Chapman, R. F. 1982. The Insects: Structure and Function. 3rd Edition. Harvard University Press, Cambridge, MA. p. 249.)*

The arthropodan gait involves a wave of leg movement, in which a posterior leg is put down just before or a little after the anterior leg is lifted. The movements of legs on opposite sides of the body alternate with one another; that is, one limb of a pair is moving through its effective stroke while its mate is making a recovery stroke. Alternate leg movement tends to induce body undulation, meaning that some of the force provided by the appendages goes into lateral, rather than forward, motion. This tendency is counteracted by increased body rigidity, such as the fused leg-bearing segments that form the thorax of insects or the cephalothorax of some crustaceans and arachnids.

An exoskeleton makes a highly efficient locomotor/skeletal system for animals that are only a few centimeters long. It provides protection in addition to support, and there is a large surface area for the attachment of muscles. The tubular construction resists bending. However, the wall buckles on impact if there is insufficient skeletal material, just as you cannot bend a heavy cylindrical can, but you can buckle one wall by kicking it. Thus, an exoskeleton imposes limits on the maximum size of arthropods. The weight of a large animal and the resulting stress produced when moving would require heavy skeletal walls. But when the arthropod molted, the soft, new skeleton would collapse under the animal's weight before hardening could occur. Significantly, the largest arthropods live in the sea, where the aquatic medium provides much more support than air.

The many muscle fibers (cells) composing vertebrate skeletal muscles are organized into motor units—one motor unit composed of a group of muscle fibers innervated by one motor neuron. When the neuron fires, all of the cells composing the motor unit contract together. The degree of contraction of the entire muscle, that is, a graded response, depends on the number of motor units involved. Such a gearing system is very efficient in animals, such as vertebrates, which have large muscles. Some large arthropod muscles operate on the motor unit principle, but most are small, and contain relatively few fibers. Because there are few fibers (cells), a graded response cannot be achieved by the recruitment of motor units. An arthropod skeletal muscle is innervated by only a small number of neurons, but each neuron may provide many axon terminals to one muscle fiber or terminals to more than one fiber in the muscle. The muscle fiber does not contract completely in

response to a given nerve impulse, as do most vertebrate skeletal motor units. Rather, depolarization of the muscle cell spreads out from the axon terminals, depending on the rapidity (frequency) of nerve impulses received. The degree of contraction results from the extent of muscle depolarization, that is, the number of sarcomeres affected. In addition, a single arthropod muscle cell may be innervated by several types of motor neurons: fast (phasic), slow (tonic), and inhibitory. The terms *fast* and *slow* refer to the rapidity of the muscle response. The burst of impulses of fast motor neurons produce rapid but brief contractions, which are often involved in rapid movements. The prolonged low-frequency impulses of slow motor neurons produce slow, powerful, prolonged contractions, which are involved in postural activities and slow movements. The impulses of inhibitory neurons block contractions. For example, the opener and closer muscles of the crayfish claw are innervated by the same motor neuron, but the two muscles function independently because each is innervated by separate inhibitory neurons.

The neuromuscular system may be further complicated, as in crustaceans, by the differentiation of the muscle fibers into phasic and tonic types, each having a distinctive ultrastructure and physiology. Some muscles are entirely phasic, some are entirely tonic, and some are mixed. Phasic motor neurons innervate only phasic muscle fibers; tonic motor neurons innervate both phasic and tonic fibers or, in some instances, only tonic fibers.

In summary, the degree of contraction of arthropod muscles depends on the extent of the spread of depolarization along the muscle cell, the type of muscle fibers contracting, the type of neuron fired, and the interaction of different types of neurons.

The organization of arthropod ganglia is like that of annelids and molluscs. Giant fiber systems are frequently well developed, and "command" systems have been identified. Arthropod neural networks and neuromuscular systems have been best studied in crustaceans.

COELOM AND BLOOD-VASCULAR SYSTEM

The well-developed, segmented coelom characteristic of the annelids has undergone drastic reduction in the arthropods and is represented by only the cavity of the gonads and in certain arthropods by the excretory organs. The change is probably related to the shift from a fluid internal skeleton to a solid external skeleton. This change is well illustrated in the development of onychophorans. Early in development, segmental blocks of mesoderm (future coelomic cavities) are conspicuous, whereas the connective tissue compartment is relatively inconspicuous. Later, the connective tissue compartment enlarges greatly to form a **hemocoel** (blood-filled space), and the mesodermal somites become restricted to the cavities of the gonads and excretory organs.

The arthropod blood-vascular system is composed of a heart, vessels, and the hemocoel. The dorsal vessel of annelids, which is contractile and the chief center for blood propulsion, may be homologous to the arthropod **heart.** The heart varies in position and length in different arthropodan groups, but in all of them the heart is a muscular tube perforated by pairs of lateral openings called **ostia** (sing. **ostium**) (Fig. 12–2A). Systole (contraction) results from the contraction of heart wall muscles, and diastole (expansion and filling) from suspensory elastic fibers and, in some species, from the contraction of suspensory muscles. The ostia enable the blood to flow into the heart during diastole from the large, surrounding sinus known as the **pericardium.** In arthropods however, the pericardium does not derive from the coelom, as in molluscs and vertebrates, but is a part of the hemocoel. After leaving the heart, blood is pumped out to the body tissues through arteries and is eventually dumped into sinuses (collectively the hemocoel) that bring it into contact with the metabolizing tissues. The blood then returns by various routes to the pericardial sinus.

The blood of arthropods contains several types of cells and, in some species, the respiratory pigment hemocyanin or, less commonly, hemoglobin. As in molluscs, arthropod **hemocyanin** is a large molecule dissolved in the plasma; however, the structure of arthropod hemocyanin indicates that it evolved independently from that of molluscs.

The blood of a number of species is known to clot, and clotting mechanisms are probably widespread. The loss of an appendage is a common accident, and the ability to plug the severed joint is avantageous. Clotting is initiated by certain blood cells (hemocytes) when subjected to altered conditions. In crustaceans, for example, hemocytes release an enzyme that converts the blood protein fibrinogen to fibrin.

Arthropods possess two types of excretory organs: malpighian tubules and saccules. **Malpighian tubules** are blind tubular evaginations of the gut that lie within the hemocoel. Wastes are secreted from the blood into the tubules and then into the gut, where they are eliminated through the anus along with fecal material. Malpighian tubules are found in centipedes, millipedes, insects, and arachnids and represent an organ system that evolved independently within these groups or their arthropod ancestors.

The other type of arthropod excretory organ are paired blind **saccules (end sacs)** that open by ducts to the outside of the body adjacent to an appendage (Fig. 12–6). The excretory organ takes the name of the appendage with which it is associated—coxal glands,

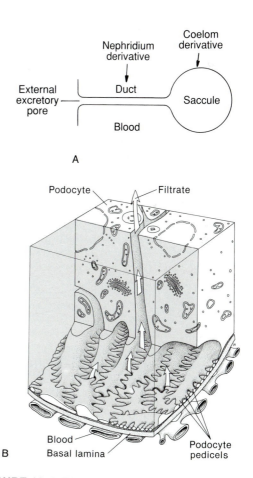

FIGURE 12–6 Diagram of an arthropod excretory end sac derived from a nephridium and coelom remnant. Podocytes form part of the saccule lining. *(From Kümmel, G. 1967. Die Podocyten. Zoolog. Beitr. N. F. 13:245–263.)*

maxillary glands, and so forth. Because the saccule is derived embryonically from the coelom, the tubule may represent an ancestral metanephridium that originally drained primary urine (filtrate) from the coelom. Typically, the saccule wall is composed of podocytes and is the site of filtration from the surrounding blood. Parts of the tubule may be modified for selective reabsorption and secretion. Although such paired excretory organs may be derived from nephridia, no living arthropod has more than a few such saccules, that is, they are not repeated in each segment.

DIGESTIVE TRACT

The arthropod gut differs from that of most other animals in having large stomodeal and proctodeal regions (Fig. 12–2A). The derivatives of these ectodermal portions are lined with chitin and constitute the foregut and hindgut. The intervening region, derived from endoderm, forms the midgut. The foregut is chiefly concerned with ingestion, trituration, and storage of food; its parts are variously modified for these functions, depending on the diet and mode of feeding. The midgut is the site of enzyme production, digestion, and absorption; however, in some arthropods, enzymes are passed forward and digestion begins in the foregut. Very commonly the surface area of the midgut is increased by outpocketings, forming pouches or large digestive glands. The hindgut functions in the absorption of water and the formation of feces.

BRAIN

There is a high degree of cephalization in arthropods. The increase in brain size is correlated with well-developed sense organs, such as eyes and antennae, and many arthropod groups display complex behavioral patterns. The arthropod brain consists of three major regions: an anterior **protocerebrum,** a median **deutocerebrum,** and a posterior **tritocerebrum** (Fig. 12–7A). The nerves from the eyes enter the protocerebrum, which contains one to three pairs of optic centers. The optic and other centers (neuropiles) of the protocerebrum function in integrating photoreception and movement and probably for initiating complex behavior.

The deutocerebrum receives the antennal nerves (first antennae in crustaceans) and contains their association centers. Antennae are lacking in the chelicerates (scorpions, spiders, mites), and in these arthropods, there is a corresponding absence of the deutocerebrum (Fig. 12–7B).

The third brain region, the tritocerebrum, gives rise to nerves that innervate the labium (lower lip), the digestive tract (stomatogastric nerves), the che-

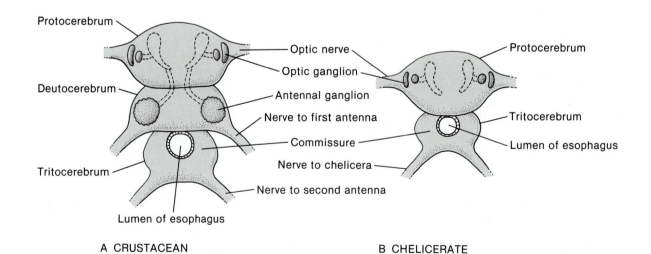

FIGURE 12–7 Arthropod brains. **A,** A mandibulate arthropod. **B,** A chelicerate arthropod. *(Both after Hanstrom from Vandel.)*

licerae (claws) of chelicerates, and the second antennae of crustaceans. The commissure of the tritocerebrum is postoral, that is, behind the foregut.

Most zoologists agree that the heads of all arthropods contain two or three preoral segments, and the antennae are segmental appendages. The tritocerebrum is a segmental ganglion that has shifted anteriorly. Its postoral commissure alone is good evidence of such an origin.

SENSE ORGANS

The sensory receptors of arthropods are usually associated with some modification of the chitinous exoskeleton, which otherwise would act as a barrier to the detection of external stimuli. A modification of the exoskeleton for the reception of environmental information other than light is called a **sensillum.** Sensilla have various shapes, depending on the type of signals they are designed to monitor. The most common form is a hair, bristle, or seta, but there are also sensilla in the form of pegs, pits, and slits. Each sensillum is composed of one or more sensory neurons plus a number of cells that produce the cuticular housing of the apparatus. Some sensilla contain a single type of receptor neuron; others encompass a number of types. For example, most insect "taste" hairs contain both chemoreceptors and mechanoreceptors. Thus, sensilla cannot always be classified by function. In general, those containing chemoreceptors have perforated walls. Mechanoreceptors are stimulated by movement of the sensilla, as in the case of hairs (Fig. 13–7C), or by changes in tension on the exoskeleton, as in the case of slit sensilla (Fig. 13–7E). The concentrations of sensilla over the arthropod body simply reflect points of most likely contact with signals to be monitored. Arthropods also possess proprioreceptors attached to the inside of the integument or to muscles.

Most arthropods have eyes, but the eyes vary greatly in complexity. Some are simple and have only a few photoreceptors. Others are large, with thousands of retinal cells, and can form a crude image. In all arthropods the skeleton contributes the transparent lens/cornea to the eye.

Insects and many crustaceans, such as crabs and shrimp, have **compound eyes** composed of many long, cylindrical units, each possessing all the elements for light reception. Each unit, or **ommatidium,** is covered at its outer end by a translucent **cornea** derived from the skeletal cuticle (Fig. 12–8). The cornea functions as a lens. The external surface of the cornea, called a **facet,** is usually hexagonal or sometimes square (as in crayfish and lobsters). Behind the cornea, the ommatidium contains a long, cylindrical or tapered element called the **crystalline cone,** which functions as a second lens. Because all of the lens elements are derived from the skeleton, they are fixed in position and cannot accommodate differences in distance.

The basal end of the ommatidium is formed by the receptor element (the retinula). The center of the **retinula** is occupied by a translucent cylinder (the **rhabdome**), around which are arranged elongated photoreceptor, or **retinular,** cells, commonly seven or eight (Fig. 12–8D). The inner photosensitive surfaces (**rhabdomeres**) of the retinular cells bear numerous microvilli that are oriented at right angles to the axis of the ommatidium (Fig. 12–8E). Some of the microvilli may be stacked so that not all are visible at a particular level in the ommatidium. The rhabdome is formed by all the rhabdomeres and the enclosed central space, when a space is present. Each ommatidium functions as a single photoreceptor unit and transmits a signal that represents a single light point. One axon extends from each retinula cell, and thus a bundle of seven or eight axons leaves each ommatidium. The axons make connections with second-order neurons in an optic ganglion within the brain or, in some crustaceans, within an eye stalk.

The retinular cells contain black or brown pigment granules, which constitute the **proximal retinal pigment.** Distally, the ommatidium is surrounded by a number of special pigment cells, forming the **distal retinal pigment.** The proximal or the distal pigment or both can migrate centrally or distally, depending on the intensity of light (Fig. 12–8D).

In bright light most compound eyes produce an **apposition image** (Fig. 12–8C). Light enters an ommatidium either at an angle or perpendicular to the facet. The proximal and distal retinal pigments are extended and act as a screen to prevent light from passing from one ommatidium to another, and thus light rays are restricted to the axial region of the crystalline cone and rhabdome.

The crystalline cone is an optic fiber through which light is channeled to the rhabdome. Studies have demonstrated that in the apposition eyes of honeybees and locusts, only 0.1 to 1% of the light reaching a rhabdome comes from facets other than its own.

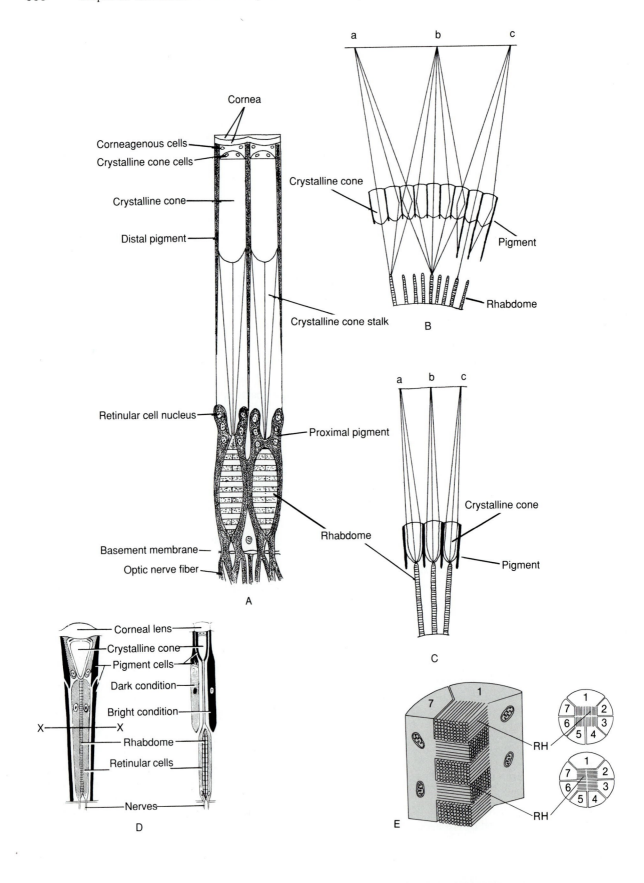

Cornea

Corneagenous cells

Crystalline cone cells

Crystalline cone

Distal pigment

Crystalline cone stalk

Crystalline cone

Pigment

Rhabdome

B

Retinular cell nucleus

Proximal pigment

Rhabdome

Basement membrane

Optic nerve fiber

A

a   b   c

Crystalline cone

Pigment

C

Corneal lens

Crystalline cone

Pigment cells

Dark condition

Bright condition

X — X

Rhabdome

Retinular cells

Nerves

D

RH

RH

E

Each ommatidium responds to a patch of light reflected from one part of the visual field with a little overlap between adjacent ommatidia. The retinula cells function more or less as a unit and register the intensity of the light received. Thus, the total image formed by the compound eye results from the mapping of "light spots" from all the stimulated ommatidia. The differences in light intensity they register provides the contrast from one part of the image to another. The image is therefore analogous to the image produced on a television screen, on which the "picture" is composed of dots of light. The apposition image formed by a compound eye is often called a **mosaic image.** Unfortunately, the term *mosaic image* implies that each ommatidium forms a fractional image, like a single tile from a mosaic in a Roman bath. This is not the case, it only forms a light point. Actually, the human eye receives and transmits light stimuli in a somewhat similar manner, a single point of light stimulating a functional unit of approximately seven cones; the principal difference is that the mosaic in the human eye is of a finer grain (composed of smaller and more numerous spots).

The image produced by a compound eye is crude at best, for the small size of the eye limits the number of photoreceptor units (ommatidia). The smallest arc of the visual field subtended by an ommatidium in the honeybee is 1 degree, compared with 1 minute of arc (1 degree = 60 minutes) in the much larger human eye. Thus, a row of closely spaced vertical bars would appear as a continuous horizontal bar to a bee. The compound eye is therefore not very good for distance vision compared with the large eyes of birds and mammals, but what matters to most arthropods is the world at close range—within 20 cm around them.

Many active, diurnal arthropods, such as bees and wasps, have high flicker fusion rates. *Flicker fusion* refers to the fusing of a rapid sequence of separate images into one, as in a motion picture. At approximately 16 frames per second, humans fuse separate images into a continuous moving image, but for honeybees the flicker fusion rate is three times greater. Thus, what we see as one continuous image a bee sees as a succession of separate images. An animal with a high flicker fusion rate can detect short increments of motion more readily than one for which the fusion rate is low. A short motion in space or time might be imperceptible to us, but to a bee it might be captured in several separate images and thereby detected.

Another advantage of the compound eye is that the total corneal surface is greatly convex, resulting in a wide visual field. This is particularly true for the stalked, compound eyes of crustaceans, in which the cornea may cover an arc of 180 degrees or more.

In weak light some compound eyes function differently and form what has been called a **superposition image** (Fig. 12–8B and D). The pigment is retracted, so no screening effect is present. Thus, light can pass from one ommatidium to another, and one rhabdome responds to light rays that originally entered several adjacent facets. In the superposition eyes of the crayfish, at least 50% of the light striking a rhabdome originally entered facets other than its own. This condition appears to be an adaptation for gathering light in semidarkness, making it more likely that the rhabdome will be activated than if it depended only on light received from its own facet.

The compound eyes of most arthropods are able to adapt, to at least some extent, to both bright and

◀**FIGURE 12–8 A,** Two ommatidia from the compound eye of the crayfish, *Astacus* (longitudinal view). **B,** Compound eye specially adapted for superposition image formation. Light rays from points a and b are being received as a superposition image. Pigment is retracted, and light rays, initially received by a number of ommatidia, are concentrated on a single ommatidium. Point of light c is being received as an apposition image. Pigment is extended, preventing light rays from crossing from one ommatidium to another. **C,** Compound eye especially adapted for apposition image formation. Each rhabdome is stimulated only by light received by its own ommatidium. **D,** Insect ommatidia, showing a diurnal (apposition) type (left) and a nocturnal (superposition) type (right). In the nocturnal type, pigment is shown in two positions, adapted for very dark conditions on the left side and for relatively bright conditions on the right. **E,** Rhabdome of compound eye of the crayfish *Procambarus clarkii.* Stereo diagram showing perpendicular orientation of the microvilli projecting from two retinular cells. Transverse sections at two levels showing the organization of the rhabdomeres (RH) of the seven retinular cells of one ommatidium of the compound eye. *(A, After Bernhards from Waterman. B and C, After Kuhn from Prosser and others. E, After Eguchi from Burr, A.H. 1982. Evolution of eyes and photoreceptor organelles in the lower phyla. In Ali, M.A. (Ed.): Photoreception and Vision in Invertebrates, Plenum Press, New York, pp. 131–178.)*

weak light, but in general they tend to be specially modified for functioning under one of these two conditions. Thus, compound eyes can be classified as either **apposition** or **superposition eyes,** although there are many gradations between the two types.

Arthropods that are diurnal and live in well-lighted habitats, such as terrestrial and littoral species, usually possess apposition eyes (Fig. 12–8C and D). The screening pigment is well developed. The length of the crystalline cone is approximately equal to its focal length, and the lower end of the cone and the upper end of the rhabdome are contiguous, or nearly so. The retinular cells are quite long, extending from the crystalline cone to the basal membrane of the retina. All of these modifications tend to confine light entering a single ommatidium and to funnel the light down the axis of the ommatidium to the rhabdome.

Superposition eyes are found in nocturnal species or those that live in poorly lighted habitats. However, there are many exceptions, and it is now recognized that not all superposition eyes can be considered nocturnal eyes. The superposition eye is especially modified for collecting and concentrating light originally striking a large patch of facets onto one ommatidium (Fig. 12–8B and D). Screening pigment is usually present but may be reduced or absent in cave-dwelling and bathypelagic species. The crystalline cone tends to be twice as long as its focal length, and there is considerable space between the end of the crystalline cone and the rhabdome, permitting the bent light rays to cross from the crystalline cone of one ommatidium to the rhabdome of another. The retinular cells are much shorter than those in apposition eyes, and they are restricted to the base of the ommatidium. In some insects a reflecting pigment is present around the retinula and may be movable.

Crayfish and lobsters have reflecting superposition eyes, in which "mirrors" at the sides of the ommatidia reflect incoming light across the space between the crystalline cone and rhabdome.

In numerous arthropods with compound eyes, as well as in many animals with other types of eyes, the membrane containing the light-sensitive pigment undergoes rhythmic degeneration and regeneration. Twenty-four-hour cycles are common, and light is a primary controlling factor. The rhythms appear to be a mechanism to bring the eye into optimum condition when the animal is most active.

Color vision has been demonstrated in a number of arthropods and has been extensively studied in some insects. Here, the photolabile pigment responds to certain wavelengths of the light spectrum, with different retinula cells exhibiting different sensitivities. For example, within the ommatidium of a bee are two retinula cells that respond to green wavelengths and three that respond to ultraviolet, and some ommatidia contain four retinula cells that respond to blue wavelengths. Variously colored flowers visited by bees elicit responses from different combinations of retinula cells.

Very important to all the visual responses described here—apposition and superposition imaging, color vision, and so on—is the way incoming signals from the receptor cells are transmitted by the subsequent levels of neurons. Just as in vertebrates, cross connections by ganglion cells make possible the pooling of signals from a number of retinula cells in very precise ways. There is also the eventual projection of signals to specific areas of the brain. Complex behavioral responses to visual information depend not only on the complexity of the eyes but also on the complexity of the associated neural circuitry.

REPRODUCTION AND DEVELOPMENT

With few exceptions, arthropods are dioecious. Copulation is common and many employ modified appendages for sperm transfer. Fertilization is always internal in terrestrial forms but may be external in aquatic species. Aquatic species do not broadcast large numbers of sperm and eggs, however, as is true of many other invertebrates. The eggs of most arthropods are rich in yolk and are **centrolecithal.** In the centrolecithal egg the nucleus is surrounded by a small island of nonyolky cytoplasm in the middle of a large mass of yolk (Fig. 12–9A). There is also a peripheral sphere of nonyolky cytoplasm. Most arthropods have a modified type of cleavage that is associated with centrolecithal eggs, called **intralecithal,** or **superficial, cleavage** (Fig. 12–9B to E).

After fertilization the centrally located nucleus undergoes mitotic divisions but without the formation of cell membranes and without any cleavage of the yolk. The result, after several such divisions, is an uncleavaged egg containing a large number of nuclei in the center. As division continues, the nuclei gradually migrate to the periphery, where cell membranes form but do not extend into the yolky interior. This stage of development represents a stereoblastula. Development continues with the formation of a primordial

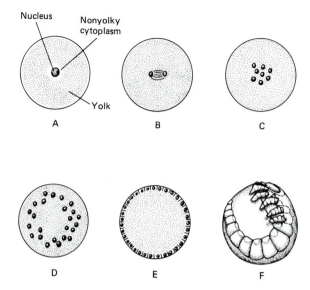

FIGURE 12–9 Superficial cleavage. **A,** Centrolecithal egg. **B** and **C,** Nuclear division. **D,** Migration of nuclei toward periphery of egg. **E,** Blastula. Cell membranes have developed, separating adjacent nuclei. Embryo of the arachnid *Thelyphonus*. *(After Kaestner, A. 1968. Invertebrate Zoology. Vol. II. John Wiley and Sons, New York.)*

germinal disc located on one side of the embryo. This germinal center proliferates endoderm and mesoderm and eventually forms the embryonic body, which appears to be wrapped around the egg (Fig. 12–9F).

ARTHROPOD CLASSIFICATION

Most zoologists today agree that there are probably four main lines of arthropod evolution, each of which we will treat as a subphylum. These lines are believed to be represented by the extinct **Trilobita;** the **Chelicerata,** containing the horseshoe crabs, scorpions, spiders, and mites; the **Crustacea,** containing the copepods, barnacles, shrimps, lobsters, and crabs; and the **Uniramia,** containing the centipedes, millipedes, and insects. In contrast to the other three arthropod lines, which have marine origins, the uniramians appear to have evolved on land. The name *uniramia* refers to the fact that the appendages are basically unbranched and have been thought to be derived from an unbranched condition.

There is some evidence from comparative morphology and embryology that at least the uniramians and perhaps even all four of these arthropod groups had a separate origin from different annelidan or near annelidan ancestors and that arthropodization (i.e., the arthropod features of a chitinous exoskeleton and jointed appendages) evolved independently at least twice or maybe four times. If this is true, the Arthropoda should be considered a superphylum, and the Trilobita, Chelicerata, Crustacea, and Uniramia should each be raised to phylum rank. This polyphyletic view of arthropod evolution is still not accepted by all specialists, particularly by many entomologists. The traditional rank of phylum has been retained for the Arthropoda in this edition, but the four lines of arthropod evolution have been recognized as subphyla.

SUMMARY

1. The arthropods are the largest assemblage of species within the Animal Kingdom. They most probably evolved from annelids or at least from some common ancestral form. The annelidan ancestry of arthropods is reflected in their segmentation, the plan of their nervous system, and their determinate cleavage.

2. In the evolution of the arthropod condition, the chitin-protein exoskeleton was a central development to which many other changes can be related. The division of the skeleton into plates and cylinders makes movement possible, and periodic molting of the exoskeleton permits growth.

3. Arthropods move by jointed segmental appendages rather than by body deformation. The coelom has become vestigial in the adult, and the musculature is organized as bundles attached to the inside of the skeleton. The skeleton and muscles operate together as a lever system.

4. There has been a general tendency among arthropods for segmentation to become reduced through fusion, loss, and differentiation of segments. The number of locomotor appendages has, in general, been reduced as a consequence of differentiation of appendages for other functions and because the small number permits greater maneuverability and speed.

5. Muscle contraction is governed by a system of multiple motor innervations: fast, slow, and inhibitory.

6. The circulatory system is a hemocoel, with a dorsal, primitively tubular heart. Paired lateral ostia permit the passage of blood into the heart from the surrounding pericardial sinus. Hemocyanin is the usual respiratory pigment.

7. Nephridia are probably represented by the paired saccular excretory organs of many arthropods. Malpighian tubules, a second type of excretory organ, are associated with the gut and are a new development in many terrestrial arthropods.

8. The sense organs usually involve some specialization of the exoskeletal barrier, which permits monitoring of environmental stimuli. Hairs or bristles are the most common type of arthropod sensilla. Many crustaceans and most insects have compound eyes, in which each of the units (ommatidia) composing the eye contains all the visual elements.

9. Most arthropods are dioecious. Copulation and internal fertilization are characteristic of the majority of species, with various appendages involved in courtship and sperm transfer.

10. The eggs of most arthropods are centrolecithal, and cleavage is commonly superficial.

FOSSIL ARTHROPODS

Arthropods make their appearance in the fossil record during the Cambrian, along with many other invertebrate groups. We would expect primitive arthropods to display a high degree of segmentation, that is, to have a body composed of many segments bearing similar appendages. One of the most common groups of fossil arthropods does indeed show such a body plan. This is the subphylum Trilobita. Trilobites were once abundant and widely distributed in Paleozoic seas. They reached their height of distribution and abundance during the Cambrian and Ordovician periods and disappeared at the end of the Paleozoic era. Over 3900 species have been described from fossil specimens.

Most trilobites ranged from 3 to 10 cm in length, although some planktonic species were only 0.5 mm long. The largest trilobites were a little less than a meter in length. The somewhat oval and flattened trilobite body was divided into three more or less equal sections: a solid, anterior cephalon; an intermediate thorax or trunk region, consisting of a varying number of separate segments; and a posterior pygidium (Fig. 12–10). Each of these body divisions was in turn divided into three regions by a pair of furrows running from anterior to posterior and forming a median axial lobe flanked on each side by a lateral lobe. The name *Trilobita* refers to this transverse trilobation of the dorsal body surface. In most specimens only this heavier dorsal surface has been preserved in the fossil record.

The anterior body section, the cephalon, was covered by a shieldlike carapace bearing a pair of lateral compound eyes. The posteriorly directed mouth was located in the middle of the underside of the cephalon just behind and beneath a liplike prominence called the labrum. On each side of the labrum was a long sensory antenna, believed to be homologous to the first pair of antennae of the crustaceans and the antennae of insects.

Behind the mouth were located four pairs of appendages similar to the appendages of the trunk and pygidium. Each appendage was biramous and consisted of an inner walking leg and an outer filament-bearing branch (Fig. 12–10). These filaments have been called gills, but in the small number of species in which they have been preserved, the filaments have the form of heavy spines, rakelike teeth, or featherlike barbs, suggesting such functions as digging, filtering, or swimming, depending on the species. The appendages of the cephalon show varying degrees of reduction in different trilobite groups, and in some, such as *Phacops,* certain pairs of coxae (basal article of the appendage) carried apposing toothed processes that must have functioned as jaws. Moreover, in all trilobites the medial bases of the appendages appear to function in moving food forward to the mouth (Fig. 12–10). The favorable ratio of surface area to volume resulting from the flattened body and the thin exoskeleton over the ventral surface may have reduced the need for specialized organs of gas exchange, at least in smaller species.

The trilobite trunk (thorax) consisted of a varying number of separately articulating segments, each of which bore ventrally a pair of appendages similar to the appendages of the cephalon. The pygidium was constructed on the same plan as the thorax except that the segments were fused and formed a solid shield.

Differences in size, shape, spination, eye size, and position indicate that trilobites displayed diverse living habits within a variety of habitats. The majority were presumably bottom dwellers and crawled over sand and mud using the walking legs. The flattened body and dorsal eyes were adaptations for this type of existence. Many could roll up (Fig. 12–11C and D), and in some species, projecting spines could have had a defensive function in the rolled position. Some trilobite groups had a shovel- or plow-shaped cephalon seemingly adapted for burrowing. Other trilobites appear to have lived in burrows, keeping their head above or near the surface of the mud or sand and seizing passing prey (Fig. 12–11E).

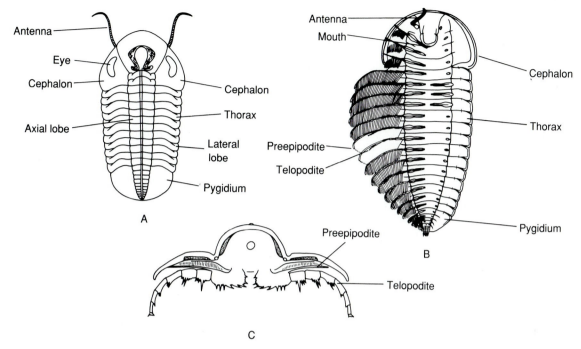

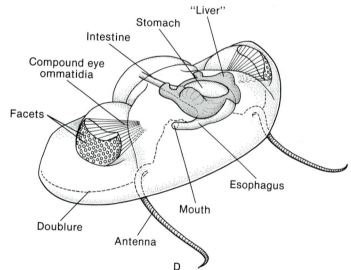

FIGURE 12–10 A, Dorsal view of *Phacops,* believed to have been a crawling-swimming, epibenthic, predacious trilobite with jaws. **B,** Ventral view of *Triarthrus eatoni.* **C,** Cross section of *Olenoides serratus.* **D,** Reconstruction of the head of *Phacops.* *(A, From Stürmer, W., and Bergström, J. 1973. New discoveries on trilobites by X-rays. Paläont. Z. 471/2:104–141. B, From Cisne, J. L. 1975. Anatomy of Triarthrus and the relationships of the Trilobita. Fossils and Strata. 4:45–63. C, From Whittington, H. B. 1975. Trilobites with appendages from the middle Cambrian, Burgess Shale, British Columbia. Fossils and Strata. 4:97–136. D, From Stürmer, W., and Bergström, J. 1973. New discoveries on trilobites by X-rays. Paläont. Z. 47$^1/_2$:104–141.)*

Some groups of trilobites were apparently not confined to the bottom but took up a swimming existence. In these forms the body was narrower, and the eyes were located on the sides of the head. Nothing is known about their appendages. The smallest species of trilobites (0.5 mm long) were planktonic (Fig. 12–11B).

Early fossil arthropods are not limited to trilobites. The Burgess shale of British Columbia, which displays a remarkable record of Cambrian inverte-

brates (p. 597), contains 19 species of arthropods, only three of which are trilobites. One is a chelicerate and two are crustaceans, the Chelicerata and Crustacea being two of the three existing subphyla of arthropods. The remaining thirteen Burgess arthropods appear to be unique, that is, they fit into no other know arthropod groups. The body plans of some are bizarre (Fig. 12–12) and some, such as *Hallucigenia,* are now assigned to extant taxa. In contrast to most modern arthropods, all of these Cambrian species

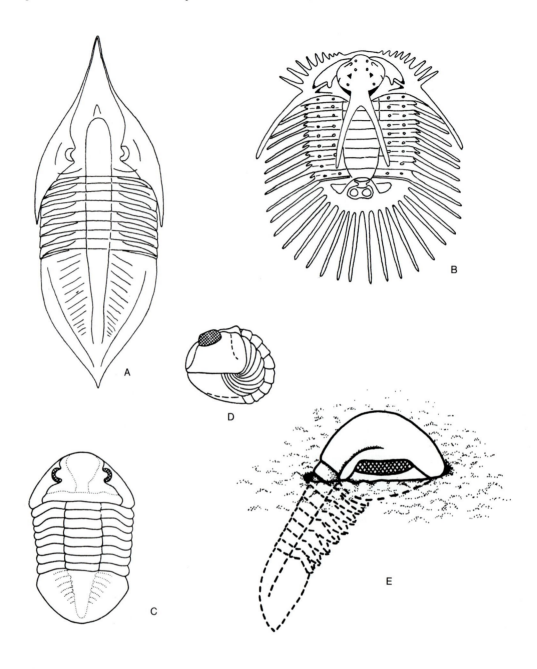

FIGURE 12–11 A, A burrowing trilobite, *Megalaspis acuticauda,* with a plow-shaped cephalon from the Ordovician period. **B,** A planktonic trilobite, *Radiaspis radiata,* from the Devonian period. The long spines may have been flotation devices. **C** and **D,** *Asaphus.* Dorsal view of extended animal **(C)** and side view of animal in enrolled condition **(D). E,** Postulated position of trilobite *Panderia* within burrow. *(A and B, After Stormer, L. 1949. Classe des Trilobites. In Grassé, P. Traité de Zoologie. Vol. VI. Masson et Cie, Paris. E, Adapted from Bergström, J. 1973. Organization, life, and systematics of trilobites. Fossils and Strata. 2 Universitetsforlaget, Oslo. 69 pp.)*

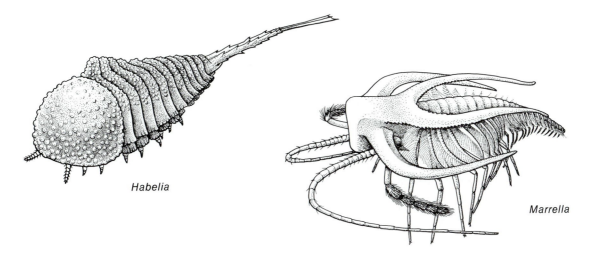

Habelia

Marrella

FIGURE 12–12 Reconstructions of two arthropods from the Burgess shale that belong to no other known arthropod groups. *(From Gould, S.J. 1989. Wonderful Life: The Burgess Shale and the Nature of History. W.W. Norton, New York. pp. 114 and 179.)*

possessed trunks of relatively numerous segments bearing similar appendages.

Assuming that all members of the phylum had a common ancestry, we conclude that the arthropod body plan probably evolved in the pre-Cambrian. During the Cambrian, specialization within that plan led to the many different designs revealed in the Burgess shales. Only a few survived. The post-Cambrian fossil arthropods are trilobites or members of one of the three subphyla that survive today.

SUMMARY

1. The marine Paleozoic trilobites (subphylum Trilobita) are a group of extinct primitive arthropods.

Each of the postoral segments carried a pair of similar appendages, which included a filament-bearing branch of uncertain function.

2. The body was composed of an anterior cephalon, a middle trunk region of unfused segments, and a posterior pygidium. The cephalon carried a pair of antennae and a pair of dorsolateral eyes.

3. Variations in form indicate that there was some diversity in trilobite living habits, such as burrowing, epibenthic crawling, planktonic, and swimming.

4. The Cambrian fossil record, especially that of the Burgess shale, contains a number of arthropods that are not trilobites and do not fit into any of the existing arthropod taxa.

REFERENCES

Anderson, D. T. 1973. Embryology and Phylogeny in Annelids and Arthropods. Pergamon Press, New York 495 pp. (The closing chapter of this work summarizes the evidence from embryonic development that supports the belief that arthropods are polyphyletic).

Atwood, H. L. 1973. An attempt to account for the diversity of crustacean muscles. Am Zool. *13:*357–378.

Atwood, H. L., and Sandeman, D. C. (Eds.): 1982. Biology of the Crustacea. Vol. III. Neurobiology: Structure and Function. Academic Press, New York.

Bergström, J. B. 1973. Organization, life, and systematics of trilobites. Fossils and Strata. No. *2.* Universitetsforlaget, Oslo.

Briggs, D. E. G. 1991. Extraordinary fossils. American Scientist. *79:*130–141.

Chapman, R. F. 1982. The Insects: Structure and Function. 3rd Edition. Harvard University Press, Cambridge, MA. 919 pp.

Cisne, J. L. 1974. Trilobites and the origin of arthropods. Science. *186:*13–18.

Conway-Morris, S. 1989. Burgess Shale Faunas and the Cambrian Explosion. Science. *246:*339–346.

Cronin, T. W. 1986. Optical design and evolutionary adaptation in crustacean compound eyes. J. Crustacean Biol. *6*(1):1–23.

Gould, S. J. 1989. Wonderful Life: The Burgess Shale and the Nature of History. W. W. Norton and Co., New York. 347 pp.

Grosberg, R. K. 1990. Out on a limb: Arthropod appendages. Science. *250:*632–633.

Gupta, A. P. (Ed.): 1979. Arthropod Phylogeny. Van Nostrand Reinhold, New York. 762 pp. (A collection of papers concerned with various aspects of arthropod origins and evolution).

Hadley, N. F. 1986. The arthropod cuticle. Scien. Am. *255*(1):104–112.

Herman, K. G. 1983. Rods, rhabdomes and rhythms. BioScience. *33*(7):432–438.

Land, M. F. 1978. Animal eyes with mirror optics. Sci. Am. *239*(6):126–134.

Land, M. F. 1981. Optics and vision in invertebrates. *In* Autrum, H. (Ed.): Handbook of Sensory Physiology. Vol. VII/6B. Springer-Verlag, Berlin. pp. 471–592.

Levi-Setti, R. 1975. Trilobites: A Photographic Atlas. University of Chicago Press, Chicago. 214 pp.

Linzen, B. et al, 1985. The structure of arthropod hemocyanins. Science. *229:*519–524.

Locke, M. 1984. Epidermal cells (Arthropoda). *In* Bereiter-Hahn, J., Matoltsky, A. G., and Richards, K. S. Biology of the Integument. 1. Invertebrates. Springer-Verlag, Berlin. pp. 502–522.

Mangum, C. P. 1985. Oxygen transport in invertebrates. Am. J. Physiol. *248:*505–514.

Manton, S. M. 1973. Arthropod phylogeny—a modern synthesis. J. Zool. *171:*111–130.

Manton, S. M. 1978. The Arthropoda: Habits, Functional Morphology and Evolution. Oxford University Press, London. 527 pp. (An elaboration of the author's lifetime study of the functional morphology of arthropods, especially their limbs.)

Moore, R. C. (Ed.): 1959. Treatise on Invertebrate Paleontology. Part O, Arthropoda 1. Geological Society of America and University of Kansas Press, Lawrence, Kans. (This volume covers trilobites.)

Nilsson, D.-E. 1989. Vision optics and evolution. BioScience. *39*(5):298–307.

Stürmer, W., and Bergström, J. 1973. New discoveries on trilobites by X-rays. Paläont. Z. *471/2*:104–141.

Wolken, J. J. 1971. Invertebrate Photoreceptors. Academic Press, New York. 179 pp. (A review of research in some invertebrate photoreceptor systems.)

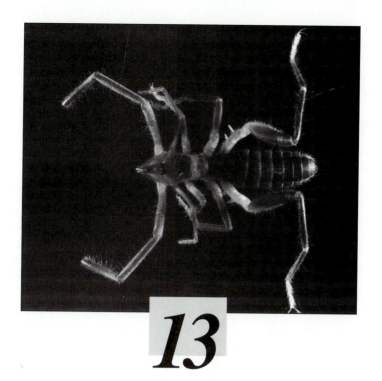

13

CHELICERATES

PRINCIPLES AND EMERGING PATTERNS

LEAF MOLD ARTHROPODS

The terrestrial counterparts to the marine interstitial fauna are the animals that live in leaf mold. Nematodes, rotifers, and members of a few other groups live between soil particles, but the spaces between fallen leaves and bits of decomposing vegetation on a forest floor is the home of an incredibly rich assemblage of mites, spiders, pseudoscorpions, various insects, and other groups of arthropods (Fig. 13–1). A few large handfuls of accumulated leaf mold can yield hundreds of individuals. Very small arthropods are subject to considerable water loss because of their large surface-area-to-volume ratio. The high humidity (low saturation deficit) of the air in the interstitial spaces of leaf mold greatly reduces the danger of desiccation to these minute inhabitants. When leaf mold is subjected to rain, some air remains trapped within spaces but much water is also absorbed. However, many arthropods can cope with such brief flooding because of the water-repelling nature (hydrofuge) of their exoskeleton, which may cause a film of air to be trapped around the body surface. The numerous hairs that clothe the bodies of some arthropods, such as many ground-dwelling spiders, have a similar function.

Bacteria and fungi bring about the gradual breakdown of the cellulose and lignin of accumulated leaves. Some of the arthropods feed directly on the decom-

FIGURE 13–1 Section through a bit of leaf mold showing a few of the many kinds of arthropod inhabitants.

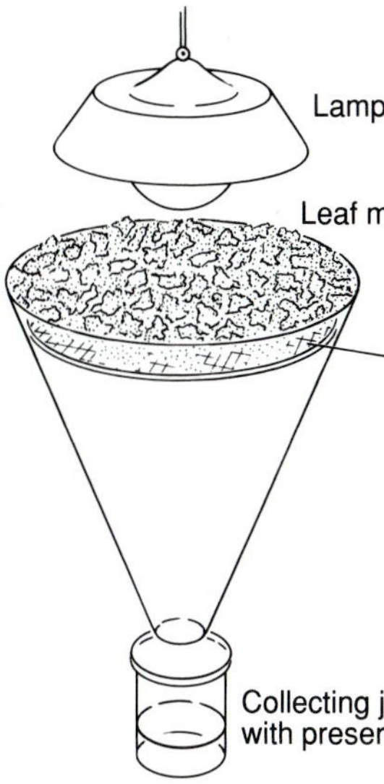

FIGURE 13–2 A Berlese funnel method for collecting small arthropods (mites, pseudoscorpions, spiders, insects, and others) from leaf mold and soil. Leaf mold is placed in the top of the funnel over an interior screen. Light, heat, naphthalene, and desiccation drive the animals downward. After one to four days many have fallen through the screen into the collecting jar. Animals can be collected alive if the collecting jar is provided with a damp substratum to prevent desiccation.

posing plant material, others on the bacteria or fungi. These detritivores are preyed on by carnivorous species; in fact, there may be several trophic levels of carnivores. Thus, there can be a relatively complex food web existing among leaf mold inhabitants.

A common method of collecting arthropods from leaf mold is with a Berlese funnel. Leaf mold (or even soil) is placed on a screen within the top of the large funnel (Fig. 13–2), and the funnel is placed beneath a lamp. Light, heat, naphthalene, and desiccation drive animals downward. After one to four days many have fallen through the screen into a collecting jar at the bottom of the funnel. Animals can be collected alive if the collecting jar is provided with a damp substratum to prevent desiccation.

INDIRECT SPERM TRANSFER

Sperm cannot be broadcast into the surrounding medium on land as is true for aquatic animals. Although most arthropods have evolved direct sperm transfer and internal fertilization, the primitive means of solving the sperm transfer problem on land was probably by indirect transfer of spermatophores. Many arthropods still possess this mode of transfer. The sperm are packaged within a spermatophore and deposited on the ground by the male. The spermatophore is commonly on a stalk. The female, on contacting the spermatophore, takes up at least the sperm globule within her genital orifice. The transfer is said to be indirect because the route of sperm transfer is from male gonopore to the ground and then to the female gonopore rather than being direct from the male gonopore to the female gonopore.

Primitively, the male abandons the spermatophore and the female locates it by chemotaxis. In most groups of arthropods, however, the male does not deposit the spermatophore until he has contacted a female, and various forms of subsequent behavior have evolved to increase the likelihood that she takes up the sperm.

CHELICERATES

All the animals described in this chapter belong to the subphylum Chelicerata, one of the three evolutionary lines of living arthropods. The body of a chelicerate is divided into a **cephalothorax** (or **prosoma**) and an **abdomen** (or **opisthosoma**). Chelicerates lack antennae and are the only subphylum of arthropods in which they are absent. This is the most distinguishing characteristic of the subphylum. The first pair of appendages are feeding structures called **chelicerae.** The second pair are called **pedipalps** and are modified to perform various functions in the different classes. The pedipalps are usually followed by four pairs of **legs.**

There are three classes of chelicerates. Two small classes, the **Merostomata** and the **Pycnogonida,** contain marine species, but most chelicerates are terrestrial and belong to the class **Arachnida.**

Class Merostomata

The Merostomata are aquatic chelicerates characterized by five or six pairs of abdominal appendages modified as **gills** and by a spikelike **telson** at the end of the body. The group can be divided into two distinct subclasses: the **Xiphosura** (horseshoe crabs) and the extinct **Eurypterida.**

Subclass Xiphosura

Although the fossil record of the Xiphosura extends back to the Cambrian period, three genera and four species compose the only living representatives of the subclass today. One of these is the horseshoe crab, *Limulus polyphemus,* common to the northwestern Atlantic coast and the Gulf of Mexico. All the other members of this group are found along Asian coasts from Japan and Korea south through the East Indies and the Philippines.

Horseshoe crabs live in shallow water on soft bottoms, plowing through the upper surface of the sand. The American *Limulus polyphemus* lives on the continental shelf but migrates to very shallow water at breeding time. They reach a length of 60 cm and are dark brown. The **carapace** is smooth, horseshoe-shaped, and convex (Fig. 13–3). Its shape not only facilitates pushing through the sand and mud but also provides a protective covering for the ventral appendages. To the outside of each of two dorsolateral

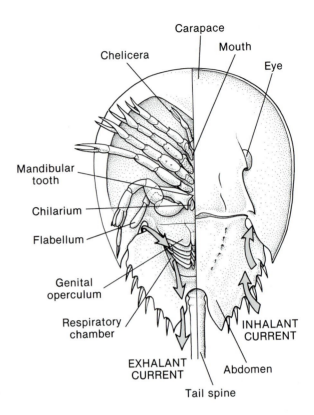

FIGURE 13–3 A horseshoe crab, *Limulus polyphemus,* in combined ventral (left) and dorsal (right) views. Arrows show water currents.

ridges is a large compound eye, and to each side of a median ridge at the anterior end is one of two small median eyes.

The anterior dorsal surface is reflected ventrally and in the front forms a large, triangular surface that tapers back toward the mouth (Fig. 13–3). A frontal organ and a pair of degenerate eyes are located on the ridge formed by this triangle. A pair of two-jointed chelicerae are attached to each side of the upper lip, or **labrum;** the last two segments form a pair of pincers, meaning that they are chelate. The mouth is located behind the labrum and is followed by a short, narrow sternum.

Five pairs of walking legs are located posteriorly to the chelicerae on the underside of the cephalothorax. The first pair is homologous to the pedipalps of other chelicerates. Except for the last pair, the walking legs are all similar and chelate, and the median side of the coxa (the most prominent segment) of each leg is heavily armed with spines. These spiny inner surfaces, called **gnathobases,** macerate and move food anteriorly. Each coxa of the fifth or last pair of legs bears a medial crushing tooth and a lateral, spatulate process known as a **flabellum,** which may be used to direct the water current entering either side of the abdomen (Fig. 13–3). Furthermore, the last pair of walking legs is not chelate. Each of these legs possesses four leaflike processes attached to the end of the first tarsal segment. This last pair of appendages is used for pushing and for clearing and sweeping away mud and silt during burrowing.

Located behind the sternum between the last pair of walking legs is a small pair of appendages known as **chilaria.** They are actually appendages of the first opisthosomal segment, which has fused with the prosoma. Each chilarium consists of a single article armed with hairs and spines, as are the gnathobases with which they function.

The abdomen (opisthosoma) is unsegmented and fits into the concavity formed by the posterior border of the cephalothorax and its lateral extensions (Figs. 13–3; 13–4). A long, triangular, spikelike tail, or caudal spine (telson), articulates with the posterior of the abdomen. This structure is not actually a true telson because it does not bear the anal opening. The tail of these animals is highly mobile and may be used for pushing and for righting the body when it is accidentally turned over. It is not used for defense, and a horseshoe crab can be picked up and carried by it.

The abdomen bears six pairs of appendages (Fig. 13–3). The fused first pair forms the genital operculum bearing the two genital pores on the underside. Posterior to the genital operculum are five pairs of flaplike, membranous appendages modified as gills. The undersurface of each flap is formed into many leaflike folds called **lamellae,** which provide the actual surface for gas exchange (Fig. 13–3). This arrangement of leaflike lamellae has caused the appendages to be called **book gills.** The movement of the gills maintains a constant circulation of water over the lamellae, and the gills also function as paddles during swimming. Water can enter the gap between cephalothorax and abdomen, the gap acting as an incurrent siphon. The water stream is then directed backward by the flabellum.

Horseshoe crabs are omnivores that feed on molluscs, worms, and other organisms, including bottom-dwelling algae. Bivalve molluscs can make up over 80% of the diet of *Limulus polyphemus.* Food material is picked up by the chelate appendages, passed to the food channel between the gnathobases (those of the fifth pair crush large food), where it is moved anteriorly to the mouth. The chilaria appear to close the back of the food channel, preventing escape of food.

The mouth, located just behind the chelicerae, opens into an esophagus; it extends anteriorly through a dilated portion, the crop, and into a grinding chamber, the gizzard (Fig. 13–4A). The longitudinal folds of cuticle in the gizzard possess denticles, and the whole structure is provided with strong muscles. After the food is ground in the gizzard, the large, undigestible particles are regurgitated through the esophagus, and the usable food material passes posteriorly through a valve into the enlarged, anterior part of the nonsclerotized midgut known as the stomach. The remainder of the midgut, called the intestine, extends posteriorly into the abdomen. Opening into each side of the stomach is a pair of ducts from one of the two large, glandular, hepatic ceca that ramify throughout the cephalothorax and abdomen. Enzyme production and digestion take place within the the midgut region. The hepatic ceca are also the principal areas for absorption of digested food materials. Wastes are egested through a short sclerotized rectum and out the anus, which is located on the ventral side of the abdomen just in front of the tail spine.

The circulatory system is well developed. A dorsal tubular heart with eight ostia is located along most of the length of the intestine (Fig. 13–4A). From the heart, blood is pumped through a well-developed arterial system. The arteries eventually terminate in tissue sinuses, and the blood collects ventrally in two large, longitudinal sinuses. From the ventral sinuses the blood flows into the book gills, where it is oxygenated. The movement of the gills not only causes the water to circulate over the outside of the lamellae but also pumps blood through these structures. From the gills the blood returns to the pericardium. The blood contains hemocyanin as well as a single type of amebocyte that functions in clotting. Because of the amount of blood that can be obtained from a large horseshoe crab, these animals have been a favorite source for physiologists and biochemists studying hemocyanin and other blood components.

Excretion takes place through four pairs of coxal glands (Fig. 13–4B), each of which opens through a common excretory pore at the base of the last pair of walking legs. The coxal glands contribute to osmoregulation by producing a dilute urine when the animal is in brackish water.

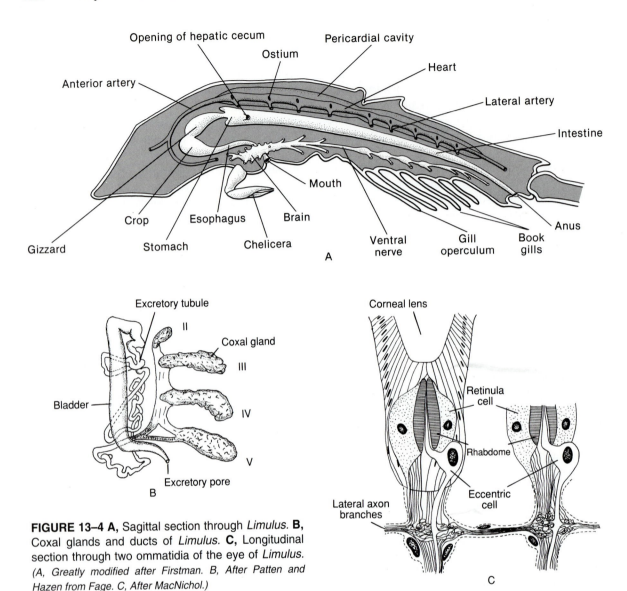

FIGURE 13–4 A, Sagittal section through *Limulus.* **B,** Coxal glands and ducts of *Limulus.* **C,** Longitudinal section through two ommatidia of the eye of *Limulus.* *(A, Greatly modified after Firstman. B, After Patten and Hazen from Fage. C, After MacNichol.)*

The nervous system displays a large degree of fusion (Fig. 13–4A). The brain forms a collar around the esophagus and includes the ganglia for the remaining first seven segments. Thus, all the appendages anterior to the genital operculum are directly innervated by the brain. A ventral nerve cord with five ganglia and lateral nerves extends through the abdomen.

The lateral eyes of horseshoe crabs are peculiar in a number of respects. They are compound eyes made up of units of 8 to 14 retinular cells grouped around a rhabdome (Fig. 13–4C). No other living chelicerate group possesses compound eyes. Each unit has a lens and a cornea and is called an ommatidium. In contrast to the ommatidia of the compound eyes of other arthropods, however, those of horseshoe crabs are not compactly arranged. Although horseshoe crabs may be able to detect movement, there are too few ommatidia in their eyes for image formation. The relationship between light stimulus and axon firing can be studied in a relatively simple state in horseshoe crab eyes, and this has resulted in their being used extensively in neurophysiological research. The median eyes are invaginated cups of retinal cells. The frontal organ is believed to be a chemoreceptor.

Horseshoe crabs are dioecious, and the gonad is located subjacent to the intestine. The sperm or eggs pass to the outside through short ducts that open onto the underside of the genital operculum.

Mating and egg laying in the American *Limulus* take place during the high tides of full and new moons in spring and summer. Males and females migrate into shallow water and congregate along the shores of sounds, bays, and estuaries. The smaller male climbs onto the abdominal carapace of the female and maintains its hold with its modified, hooklike first pair of walking legs. Meanwhile, the female, partially buried in the sand, deposits 2000 to 30,000 large eggs. The eggs are fertilized by the male during their deposition. The mating pair separate, and the eggs are covered and left in the sand.

The eggs are centrolecithal, 2 to 3 mm in diameter, and covered by a thick envelope. Cleavage is total. A **trilobite larva,** so named because of its superficial similarity to trilobites, emerges from the egg. This larva is approximately 1 cm long and actively swims about and burrows in the sand. The little caudal spine does not project beyond the abdomen. Only two of the five pairs of book gills are present, although all anterior appendages are present. As successive molts take place, the remaining book gills appear, the caudal spine increases in length, and the young animal assumes the adult form. Juvenile horseshoe crabs, which are common on intertidal sand flats, attain a carapace width of 4 cm after one year. Sexual maturity is not reached for 9 to 12 years, and the life span may be 19 years.

Subclass Eurypterida

The second group in the class Merostomata is the subclass Eurypterida (or Gigantostraca)—the extinct giant arthropods. The eurypterids were aquatic and existed from the Ordovician to the Permian period, although there are some eurypterid-like fossils known from the Cambrian. The latter are probably closer to the common ancestor of eurypterids and horseshoe crabs. The eurypterids probably attained the largest size of any of the arthropods. One species of the genus *Pterygotus* was almost 3 m long.

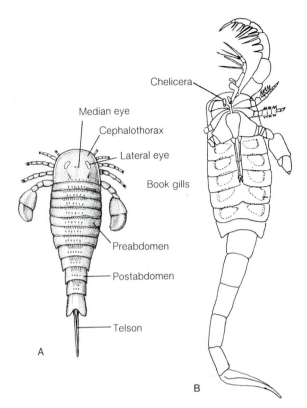

FIGURE 13–5 Eurypterida. **A,** *Eurypterus remipes* (dorsal view). **B,** *Mixopterus kiaeri* showing appendages of one side (ventral view). *(A, After Nieszkowski. B, After Stormer from Fage.)*

Eurypterids were similar to horseshoe crabs in their general body plan (Fig. 13–5), but the cephalothorax was smaller and the abdomen was composed of separate segments. Moreover, the abdomen was divided into a seven-segment preabdomen (mesosoma), bearing six pairs of gills, and a postabdomen (metasoma) of five narrower segments lacking appendages. The telson was attached to the last abdominal segment.

The chelicerae were small, and the last pair of legs was large and paddle-like. Most eurypterids, judging from the appendages, not only crawled over the bottom but also were active swimmers.

The location of fossils indicates that the eurypterids, after a marine origin, gradually invaded both brackish water and fresh water and perhaps even assumed a terrestrial existence. The aquatic merostomes certainly contain the most primitive known chelicerates, but the origin of the subphylum is still obscure. The chelicerates may have evolved from the trilobites, or they may have had a separate origin from the annelids. The account of the eurypterids, however, does not end with their extinction in the Permian period, for it is believed that they may have been the ancestors of the largest and most abundant class of chelicerates—the Arachnida.

SUMMARY

1. The subphylum Chelicerata contains the only nonantennate arthropods. The body is usually divided into an anterior cephalothorax (prosoma) and a posterior abdomen (opisthosoma). The first postoral appendages are a pair of food-handling chelicerae usually followed by a pair of pedipalps and four pairs of legs.

2. The marine origin of chelicerates is evidenced by a long fossil history beginning in the early Paleozoic.

3. Horseshoe crabs (class Merostomata) are marine chelicerates that dwell on soft-bottoms, in shallow water. The prosoma is covered by a large, horseshoe-shaped carapace, and the abdominal segments are fused together. The pedipalps are not differentiated from the posterior legs except in the mature male, which uses them to clasp the female during mating. A posterior, spikelike telson is used for pushing and righting. The ventral side of the abdomen carries five pairs of book gills.

4. Fossil evidence indicates that some species of the extinct Paleozoic eurypterids (merostomes) invaded fresh water and may have given rise to the arachnids.

CLASS ARACHNIDA

The arachnids constitute the largest and, from a human standpoint, the most important of the chelicerate classes; included are many common and familiar forms, such as **spiders, scorpions, mites,** and **ticks.** Arachnids also have the dubious honor of being the most unpopular group of arthropods as far as the layman is concerned.

The Arachnida is an old group. Fossil representatives of all the orders date to the Carboniferous period, and fossil scorpions have been found dating from the Silurian. These Silurian scorpions, however, were aquatic and contemporaries of the eurypterids, from which they may have evolved. The first terrestrial arachnid, belonging to a now extinct order

(Trigonotarbida), appears in the upper Silurian, well before the first terrestrial scorpions, which appear in the Carboniferous. Some zoologists have argued that arachnids are a polyphyletic assemblage of terrestrial species that evolved from different aquatic chelicerate ancestors.

Except for the few groups that have adopted a secondary aquatic existence, the Arachnida includes all living terrestrial chelicerates. Like other evolutionary conquests of land, this migration from an aquatic to a terrestrial environment required certain fundamental, morphological and physiological changes. The **epicuticle** became waxy, reducing water loss. The book gills became modified for use in air, resulting in the development of the arachnid **book lungs** and **tracheae.** In addition, the appendages became better adapted for terrestrial locomotion. Once a terrestrial existence was established, a great many unique innovations evolved independently along different lines. The development of silk in spiders, pseudoscorpions, and some mites and of poison glands in scorpions, spiders, and pseudoscorpions are but two examples.

The General Structure and Physiology of Arachnids

External Anatomy

Despite the diversity of forms, arachnids exhibit many features in common. The unsegmented prosoma is usually covered dorsally by a solid carapace (Fig. 13–14). The primitive abdomen is **segmented** and divided into a **preabdomen** and a **postabdomen** (Fig. 13–10). In most arachnids other than scorpions these two subdivisions are no longer conspicuous (Fig. 13–36), and segments are commonly fused.

The appendages common to all arachnids are those arising from the prosoma and consist of a pair of chelicerae, a pair of pedipalps, and four pairs of legs (Fig. 13–36). The chelicerae are used in feeding, but the pedipalps serve a number of functions and are variously modified.

Nutrition

The majority of arachnids are carnivorous, and digestion partly takes place outside the body. Prey, usually small arthropods, is captured and killed by the pedipalps and chelicerae. While the prey is held by the chelicerae, enzymes secreted by the midgut are poured out over the torn tissues of the prey. Digestion proceeds rapidly and a partially digested broth is produced. This fluid is then taken into the prebuccal cavity, located in front of the mouth.

The liquid food passes through the mouth and into the sclerotized, pumping pharynx and esophagus of the foregut (Figs. 13–21 and 13–47). The esophagus conveys the food to the midgut or mesenteron, which consists of a central tube with lateral diverticula (Fig. 13–21). The diverticula are located in both the prosoma and the abdomen and become filled with the partially digested liquid. The secretory cells of the midgut diverticula produce enzymes for external digestion and internal extracellular digestion after the food reaches the mesenteron. The absorptive cells of the diverticula are the sites of intracellular digestion and absorption. The mesenteron extends to the posterior part of the abdomen, where it is connected to the posterior anus. A short sclerotized intestine, forming the hindgut, leads to the anus.

Excretion and Water Balance

The epicuticle of arachnids has a high lipid content, which greatly reduces evaporative water loss and also acts as a water repellent (hydrofuge layer). The principal nitrogenous waste is guanine, but xanthine and uric acid are also produced; all are highly insoluble. The excretory organs are coxal glands and malpighian tubules (p. 840), the latter being derived from the terminal part of the midgut. Some groups possess both; some one or the other.

In addition to malpighian tubules and coxal glands, arachnids possess certain large phagocytic cells, called **nephrocytes,** that are localized in clusters in the hemocoel of the prosoma and abdomen.

Nervous System

The nervous system is greatly concentrated except in the relatively primitive scorpions (Fig. 13–6A). The brain, composed of the **protocerebrum** and **tritocerebrum,** is an anterior ganglionic mass lying above the esophagus. The protocerebrum contains the optic centers and optic nerves; the tritocerebrum contains the nerves supplying the chelicerae. The remainder of the central nervous system is located below the esophagus. In many orders most or all of the ganglia originally located in the thorax and abdomen have migrated anteriorly and fused with the **subesophageal ganglion**—the ganglion of the pedipalpal segment (Fig. 13–6B). Thus, the arachnid nervous system commonly resembles a collar or ring surrounding the esophagus. The posterior, ventral half of this collar gives rise on each side to nerves innervating the appendages, and a single posterior nerve bundle extends into the abdomen.

Three types of sense organs are common to most arachnids: **sensory hairs, eyes,** and **slit sense organs.** The sensory hairs can be olfactory setae, which are open at the end (Fig. 13–7D); innervated, movable setae located over the body surface; or fine hairs, called **trichobothria,** which are restricted to the appendages (Fig. 13–7C). The base of a trichobothrium is expanded to form a small ball that fits into a large, structurally complicated socket in the integument. The hair base contains a process from a sensory nerve cell of the epidermis and is stimulated by very slight vibrations or air currents. For most arachnids the sensory hairs are the primary sense organs. The eyes of all arachnids are similar. They are always composed of a combined cornea and lens, which is continuous with the cuticle but much thicker. Beneath the lens is a layer of epidermal cells known as the **vitreous body.** The retinal layer, containing the photoreceptor cells, lies behind the vitreous body.

The photoreceptors are oriented either toward the light source (a direct eye) or toward the postretinal membrane (an indirect eye). In the indirect eyes, the postretinal membrane may function as a reflector, called the **tapetum,** that reflects the light toward the receptors. Some arachnids possess only direct or only indirect eyes; many, such as spiders, have both.

A slit sense organ consists of a slitlike pit in the cuticle covered by a very thin membrane that bulges inward. The undersurface of the membrane is in contact with a hairlike process, which projects upward from a sensory cell. Slit sense organs may occur in great numbers, either singly or in groups (**lyriform organs**) on the appendages and body of most arachnids. For example, the spider, *Cupiennius salei,* has 3000 slit sense organs; about half are grouped, and most are on the appendages. The grouped, lyriform organs are near joints (Fig. 13–7E), and the simple slits are parallel to the long axis of the leg segment.

Slit sense organs respond to slight changes in the tension of the exoskeleton that result in compression (narrowing of the slit) and bowing in of the covering membrane to which the dendrite is attached. In a lyriform organ, the slits are of different lengths, which, means that varying numbers respond to a given amount of tension over that part of the skeleton. Slit sense organs may respond to load stresses in locomo-

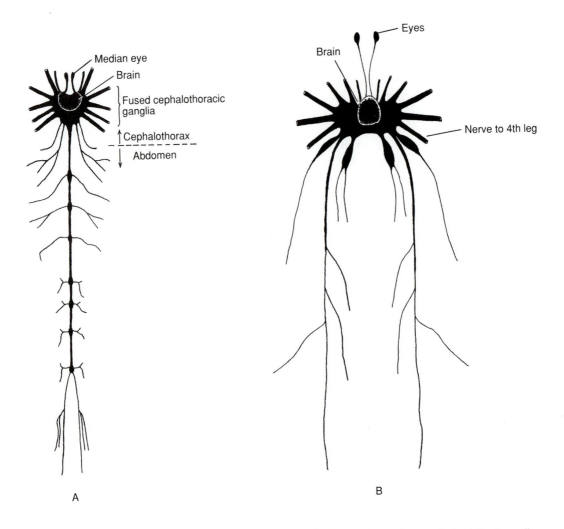

FIGURE 13–6 A, Scorpion nervous system, in which abdominal ganglia are distinct. **B,** An opilionid nervous system, in which all ventral ganglia have migrated forward and fused. *(Both after Millot, J. 1949. Traité de Zoologie. Vol. VI. Masson et Cie, Paris.)*

tion and in this respect are proprioreceptors. They may also respond to gravity, to substratum, and even to airborne vibrations.

Gas Exchange

Arachnids possess book lungs or tracheae or both. Book lungs, which are always paired, are more primitive and are probably a modification of book gills, an adaptation associated with the migration of the arachnids to a terrestrial environment. Each book lung consists of a sclerotized pocket that represents an invagination of the ventral abdominal wall (Fig. 13–8). The

wall on one side of the pocket is folded into leaflike lamellae, which are held apart by bars that enable the air to circulate freely. Diffusion of gases takes place between blood circulating within the lamellae and the air in the interlamellar spaces (Fig. 13–8). The nonfolded side of the pocket forms an air chamber (atrium) that is continuous with the interlamellar spaces and opens to the outside through a slitlike opening (**spiracle**). Some ventilation results from the contraction of a muscle attached to the dorsal side of the air chamber. This contraction dilates the chamber and opens the spiracle, but most gas movement is by diffusion.

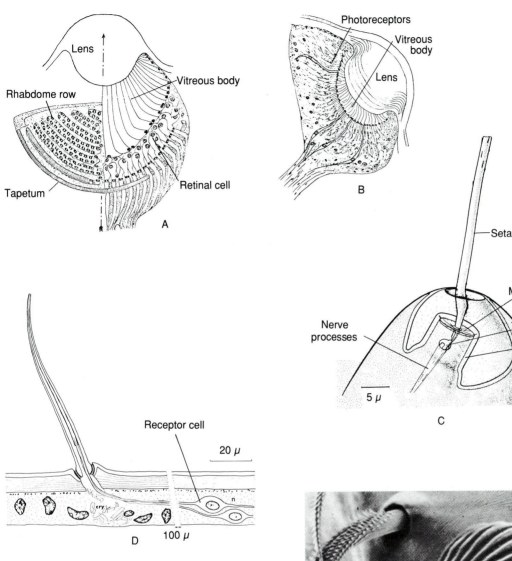

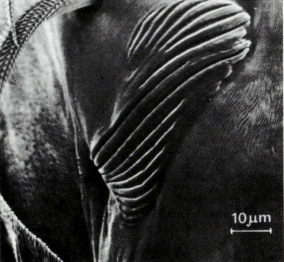

FIGURE 13–7 Arachnid sensory structures. **A,** Diagram of an indirect eye of a spider having a tapetum. Sagittal view to the right of the arrow; three-dimensional view to the left of arrow. **B,** Sagittal section of a direct eye of a spider. **C,** Trichobothrium of a spider. Only the lower part of the hair is shown. **D,** Chemosensory hair of a spider (n, nerve). **E,** Surface view of a lyriform sense organ. *(A and B, From Homann, H. 1971. Z. Morphol. Tiere. 69(3):201–273. C, From Gorner, P. 1966. Cold Spring Harbor Symposium Quant. Biol. 30:69–73. D, From Foelix, R. F. 1970. J. Morphol. 132:313–334. E, Micrograph courtesy of D. J. Harris.)*

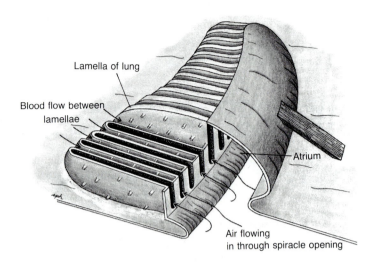

Lamella of lung

Blood flow between lamellae

Atrium

Air flowing in through spiracle opening

FIGURE 13–8 Diagrammatic section through a book lung.

The tracheal system of arachnids is similar to that of insects in being composed of internal, chitin-lined tubes through which gases diffuse, but the two systems evolved independently. Tracheae tend to be most highly developed in small arachnids, which would be subject to greater water loss with book lungs. Ricinuleids, pseudoscorpions, and some spiders possess **sieve tracheae** derived from book lungs. In tracheal systems of this type, the spiracle opens into an atrial or tubelike chamber from which arises a great bundle of tracheae. Mites, harvestmen, solifuges, and most spiders possess **tube tracheae,** in which the tracheae do not arise as a bundle but are simple, branched or unbranched tubes (Fig. 13–23B). The spiracles may open into an initial atrium or directly into a tracheal tube. At least in spiders, the trachea terminates in the hemocoel rather than in muscle or other tissues, as in insects. Blood is still the intermediate gas-transporting agent in these animals. The blood of some arachnids, such as scorpions and many spiders with book lungs, contains the respiratory pigment hemocyanin.

Internal Transport

The heart is almost always located in the anterior half of the abdomen. In its primitive condition (scorpions) the heart bears seven pairs of ostia, each corresponding to a segment. But all degrees of reduction are found in the different arachnid orders (Fig. 13–21).

A large, anterior aorta supplies the prosoma; a small, posterior aorta leads to the posterior half of the abdomen, and from each heart segment emerges a pair of small, abdominal arteries. Small arteries eventually empty the blood into tissue spaces; from these spaces blood passes into a large ventral sinus that bathes the book lungs. One or more pairs of venous channels conduct blood from the ventral sinus or the book lungs back into the pericardium.

Reproduction

The genital orifice in both sexes is usually found on the ventral side of the second abdominal or eighth body segment (Fig. 13–9). The gonads lie in the abdomen and may be either single or paired. **Indirect sperm transmission** with a **spermatophore** is characteristic of many arachnids. It appears to be the original mode of sperm transfer for the class and an adaptation for reproduction on land. There is often a complex "courtship," or precopulatory, behavior preceding mating. The sexes, especially the female, of different groups respond to chemical, tactile, or visual cues. Such cues provide recognition and elicit the receptivity and posture that indirect sperm transmission demands. This is especially important in highly predatory species, such as many arachnids.

The eggs are yolky and centrolecithal. In some groups cleavage is superficial; in others, such as spiders, pyramids of yolk are associated with the initial cleavage divisions. The blastula, however, typically consists of a uniform blastoderm surrounding a yolky mass.

Overview of the Arachnid Taxa

The class Arachnida includes the following groups:

Order Scorpiones. The scorpions.
Order Palpigradi. The palpigrades.
Order Schizomida. The schizomids.
Order Uropygi. The whip scorpions, vinegar-roons.
Order Araneae. The spiders.
Order Amblypygi. The amblypygids.
Order Ricinulei. The ricinuleids.
Order Pseudoscorpiones. The false scorpions.
Order Solifugae. The solifuges, or solpugids.
Order Opiliones. Harvestmen, or daddy longlegs.
Acari. An assemblage of seven orders containing the mites and ticks.

Order Scorpiones

Scorpions are among the oldest known terrestrial arthropods with a fossil record dating back to the Silurian period. Silurian and Devonian scorpions were aquatic, possessing gills and having no tarsal claws. Terrestrial scorpions appeared in the upper Carboniferous and are believed to have evolved from eurypterids. The 1500 to 2000 described species of scorpions are most common in tropical and subtropical areas. In North America, scorpions are most abundant in the Gulf states and the Southwest.

Scorpions are generally secretive and nocturnal, hiding by day under logs, bark, and stones and in rock crevices or in burrows in the ground. There are, however, species associated with vegetation, even in trees. A few live in the intertidal zone. Scorpions are often found near dwellings, and the desert custom of shaking out shoes in the morning is a wise precaution. Many species inhabit desert regions. In fact, Baja California hosts the greatest density of scorpion species in the world. However, scorpions are by no means restricted to arid situations. Many scorpions require a humid environment and live in tropical forests. Scorpions exhibit a spectacular fluorescence, unlike other arachnids, and can be easily observed at night with ultraviolet light.

Scorpions are large arachnids, most ranging from 3 to 9 cm in length. The smallest species is the cave-dwelling *Typhlochactas mitchelli,* which is only 9 mm long, and the largest is the African *Hadogenes troglodytes,* which reaches 21 cm. Some Carboniferous scorpions, however, attained lengths of 44 and 86 cm.

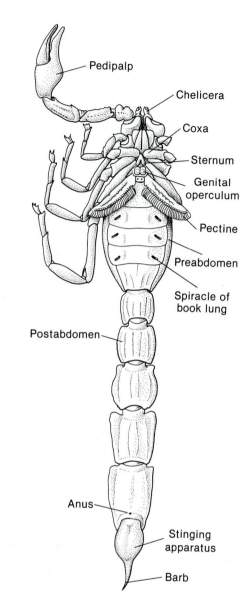

FIGURE 13–9 *Androctonus australis* (ventral view). *(After Lankester from Millot and Vachon.)*

External Anatomy

The scorpion body consists of a carapace-covered prosoma and a long abdomen, ending in a **stinging apparatus** (Figs. 13–9; 13–10). In the middle of the carapace is a pair of large, elevated, median eyes. In addition, two to five pairs of small, lateral eyes are present along the anterior, lateral margin of the carapace.

FIGURE 13–10 Successive positions of the abdomen when stinging prey in the scorpion *Hadrurus*. Total time required for the sequence is about 0.75. *(From Bub, K., and Bowerman, R. F. 1979. Prey capture by the scorpion Hadrurus arizonensis. J. Arachnol. 7:243–253.)*

The chelicerae are small and chelate, and they project anteriorly from the front of the body (Fig. 13–9). Greatly enlarged pedipalps, which are a distinguishing characteristic of scorpions, form a pair of pincers for capturing prey.

The scorpion abdomen is composed of a seven-segment preabdomen and a postabdomen of five narrow segments, so the two regions are clearly differentiated, a primitive feature. A pair of opercular plates hides the genital opening on the midventral side of the first abdominal segment. Posterior to the genital plates and attached to the second abdominal segment is a pair of sensory appendages, known as **pectines**, that is peculiar to scorpions. Each pectine is a comb-like structure that projects to each side from the point of attachment near the ventral midline. On the ventral side of each of the third through the sixth abdominal segments is a pair of transverse spiracles opening into book lungs. The segments of the postabdomen, sometimes called the tail, resemble narrow rings. The last segment bears the anal opening on the posterior ventral side and also bears the stinging apparatus characteristic of scorpions.

Stinging Apparatus and Feeding

The sting is attached to the posterior end of the last segment and consists of a bulbous base and a sharp, curved barb that injects the venom (Fig. 13–9). The venom is produced by a pair of oval glands within the base of the apparatus. By a violent contraction of the muscular envelope around the glands, the liquid venom is ejected from the lumen of the glands into a sclerotized common duct that leads to the outside through a subterminal opening of the stinging barb. The scorpion raises the postabdomen over the body, so that it is curved forward, and stabs the prey (Fig. 13–10).

The venom of most scorpions, although sufficiently toxic to kill many invertebrates, is not dangerous to humans. At most, the sting is very painful. The scorpions of the southeastern United States and Gulf Coast region fall in this category, as well as many of our midwestern and western forms.

About 25 species (all members of the family Buthidae), however, possess a highly toxic venom that can be fatal to humans. The most notorious are *Androctonus* of North Africa and various species of *Centruroides* in Mexico, Arizona, and New Mexico. Five thousand persons a year are estimated to die from scorpion stings. Humans stung by *Androctonus australis* may die in 6 to 7 h. In Mexico, species of *Centruroides* have been responsible for deaths, mostly in children. The neurotoxic venom may cause convulsions, paralysis of the respiratory muscles, or, in fatal cases, cardiac failure. Antivenoms are available for dangerous species.

Scorpions feed on invertebrates, particularly insects. Many scorpions, perhaps most, sit in an alert position and wait for prey. The pedipalp fingers are open, and the tip of the movable finger and the tip of the pectines touch the ground. Burrowing species wait near or at the burrow entrance. Prey is detected by trichobothria on the pedipalps or, at least in some scorpions, from substratum vibrations through the tarsal hairs and slit sense organs. In a few seconds, the North American desert scorpion, *Paruroctonus mesaensis,* can locate and dig out a burrowing cockroach 50 cm away. Some scorpions can catch prey in midair, detecting it with the trichobothria from a distance of 10 cm. The prey is caught and held by the large pincers while usually being killed or paralyzed by the sting (Fig. 13–10). The sting is not used, however, if the prey is readily subdued with the pedipalps. The prey is transferred to the chelicerae, which slowly crush and tear it, and digestion begins outside of the body.

Scorpions are usually most active during the first hours of darkness, but only about 10% of the population appears to be active at any one time. Many species spend 92% to 97% of their life in their burrow. When seeking prey most burrowing species stay within a meter of the burrow, but *Paruroctonus mesaensis* is known to return to its burrow from distances as great as 8 m.

Internal Structure and Physiology

Gas exchange is accomplished solely by book lungs. There are two pairs of malpighian tubules and a single pair of coxal glands, which open on the coxae of the third pair of walking legs.

Desert scorpions possess a number of adaptations for life under extremely desiccating conditions. Temperatures lethal for them are high (45–47°C); evaporative water loss through their almost impervious exoskeleton is extremely low (0.01% of body weight per hour at 25°C). They can tolerate a water loss of 40% of their body weight. In addition to being nocturnal, many are burrowers. The burrows of some species, such as *Hadrurus arizonensis* of the Sonoran Desert, may extend 90 cm. Some scorpions at times raise their bodies well off the ground. This behavior, called stilting, permits the circulation of air beneath the animal, which tends to prevent excessive elevation of body temperature and desiccation.

The scorpion nervous system, unlike that of other arachnids, retains a distinct nerve cord with seven unfused ganglia (Fig. 13–6A). Giant fibers integrate the pedipalpal grasp and the stinging thrust. The precise function of the pectines is still not clear. The ventral side of each tooth of the comb is provided with chemoreceptors and mechanoreceptors. During movement of the scorpion, the pectines are held out from the sides of the body in a horizontal position so that the teeth touch the ground. They are sensitive to vibrations and appear to determine some aspect of the substratum surface, perhaps particle size. Their removal prevents spermatophore deposition.

Reproduction and Life History

Male scorpions may have a larger abdomen than females, but the most useful character for distinguishing the sexes in scorpions is the hook present on the opercular plates of the male. The gonads are located between the midgut diverticula in the preabdomen. Prior to emptying into a single genital atrium, each

oviduct dilates to form a small, seminal receptacle. In the male, each sperm duct near the point of junction with its opposite member is modified for the formation of a spermatophore. In each sex the common genital atrium opens to the outside between the genital opercula on the first abdominal segment.

During the breeding season the male roams about and on contacting a female initiates an extended courtship. In some species the male and female face each other; each extends its abdomen high into the air and moves about in circles (Fig. 13–11B–D); in others, the male rocks. The male then seizes the female with his pedipalps, and together they walk backward and forward (promenade). This behavior may last from 10 min to hours, depending on how long it takes to locate a suitable site for spermatophore deposition. Eventually, the male deposits a spermatophore, which he attaches to the ground. A winglike lever extends from the spermatophore. The male then maneuvers the female so that her genital area is over the spermatophore. Pressure on the spermatophore lever releases the sperm mass, which is taken up into the female orifice (Fig. 13–11A).

All scorpions brood their eggs within the female reproductive tract and give birth to live, developed young. The aplacental viviparous species have large, yolky eggs; development takes place in the lumen of the ovarian tubules. The eggs of placental viviparous species possess little yolk. This type of development is found in the tropical Asian species, *Hormurus australasiae*. Its eggs develop within the diverticula of the ovary. Each diverticulum in turn develops a distal tubular appendage that contains a cluster of absorbing cells at the end (Fig. 13–11E). These cells rest against the maternal digestive ceca, from which nutritive material is absorbed. The nutritive material passes through the tubule to the embryo at the base.

Development takes several months or even a year or more, and 1 to 95 young are produced, depending on the species. At birth the young are only a few millimeters long, and they immediately crawl on the mother's back (Fig. 13–11F). The young remain there through the first molt, which occurs in one to four weeks. The young scorpions then gradually leave the mother and become independent. They reach sexual maturity in six months to six years, molting four to seven times, and some species may live as long as 25 years. Predation by such vertebrates as birds, snakes, and amphibians is a major cause of scorpion mortality.

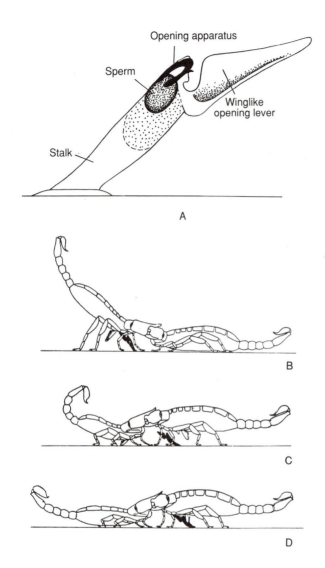

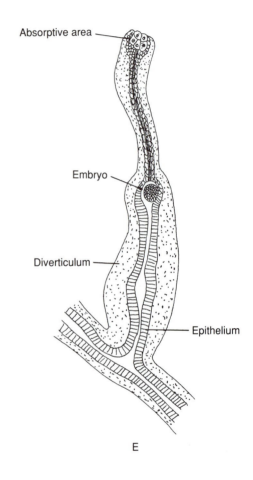

FIGURE 13–11 A, Diagram of a spermatophore of a scorpion. **B–D,** Sperm transfer in scorpions: **B,** While holding the female's pedipalps in his own, the male (on the left) deposits the spermatophore on the ground. **C,** The female is pulled over the spermatophore. **D,** The spermatophore is taken up into the female's gonopore. **E,** Diverticulum of the ovary containing a developing embryo in the tropical Asian scorpion, *Hormurus australasiae,* a placental viviparous species. **F,** Female scorpion carrying young. *(A–D, After Angermann. E, After Pflugfelder from Dawydoff. F, Photograph courtesy of H. L. Stahnke.)*

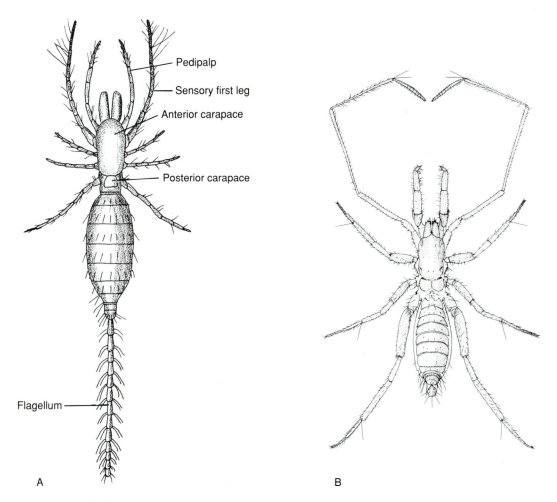

FIGURE 13–12 A, *Eukoenenia,* a palpigrade. **B,** *Schizomus sawadai,* an Asian species of the order Schizomida. The body length is only about 5 mm. *(A, After Kraepelin and Hansen from Millot. B, From Sekiguchi, K., and Yamasaki, T. 1972. A redescription of "Trithyreus sawadai" from the Bonin Islands. Acta Arachnol. 24(2):73–81.)*

Orders Palpigradi, Schizomida, and Uropygi

Three small orders of tropical and semitropical arachnids are similar in possessing a terminal, abdominal flagellum composed of numerous, small, articulating pieces (Fig. 13–12A). All live beneath wood and stones or in leaf litter and soil. The some 60 known species of Palpigradi and 80 species of Schizomida are mostly less than 3 mm in length. The palpigrades have the prosomal carapace composed of two plates (Fig. 13–12A), and that of the schizomids is composed of three plates (Fig. 13–12B).

Members of the order Uropygi, of which there are about 85 described species, are called whip scorpions and include some large arachnids. The Ameri-

can *Mastigoproctus giganteus* reaches 65 mm in length (Fig. 13–13).

The prosoma is covered by a dorsal carapace that carries a pair of anterior median eyes and three or four pairs of lateral eyes. Uropygid chelicerae have two segments, and the distal piece forms a hook, or fang, that folds against the large basal piece. The pedipalps are stout, heavy, and relatively short; the last two articles of the pedipalps are frequently modified to form a pincer used in seizing prey. The pedipalps and the long, tactile first pair of legs are held in front of the animal as it moves forward, with the tactile legs frequently touching the ground. The posterior half of the abdomen contains a pair of large anal glands, which open one on each side of the anus.

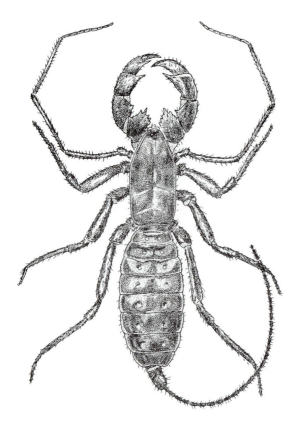

FIGURE 13–13 American whip scorpion, *Mastigo-proctus giganteus. (After Millot, J. 1949. Traité de Zoologie. Vol. VI. Masson et Cie, Paris.)*

When irritated, the animal elevates the end of the abdomen and sprays the attacker with fluid secreted from these glands. The secretion in *Mastigoproctus* is 84% acetic acid and 5% caprylic acid, the latter permitting the acetic acid to penetrate the integument of an arthropod predator. The fluid can burn human skin, and the repellent odor has given this animal the name "vinegarroon."

During feeding, the prey is seized and torn apart by the pedipalps and then passed to the chelicerae. Two pairs of book lungs are located on the ventral side of the second and third abdominal segments.

Sperm transfer is indirect in all three orders and involves the deposition of a spermatophore. During part of the complex courtship behavior, the male holds the tips of the long, modified sensory legs of the female with his chelicerae. The female picks up the sperm packages with her genital area, and in *Mastigoproctus* and *Thelyphonellus* the male then uses his pedipalps to push them into the gonopore.

The female lays her eggs in a sac attached to her body. She remains in a shelter until they have hatched and undergone several molts.

Order Araneae

Except for perhaps the Acari, which comprise the mites and ticks, spiders constitute the largest order of arachnids. Approximately 32,000 species have been described, and this probably represents only a portion of the actual number. Also, spider populations are very large. An acre of undisturbed, grassy meadow in Great Britain was estimated to contain 2,265,000 spiders.

External Structure

Spiders range from tiny species less than 0.5 mm in length to large, tropical mygalomorphs (called tarantulas, bird spiders, or monkey spiders in different parts of the world) with a body length of 9 cm; leg span can be much greater. The spider's prosoma bears a distinctive convex carapace that usually supports eight eyes anteriorly (Fig. 13–14). A large **sternum** is present on the ventral surface, and a small, median plate known as the **labium** is attached directly in front of the sternum (Fig. 13–14B).

Each chelicera consists of a **fang** and a **basal piece** containing a groove into which the fang folds (Fig. 13–15A). The female pedipalps are short and leglike, but in the male, they have become modified as copulatory organs, with the last segment greatly enlarged and knoblike. The legs vary in length and heaviness, depending on the habits of the species.

The globe-shaped or elongated abdomen is unsegmented except in a few primitive spiders, and it is connected to the prosoma by a short and narrow portion called the **pedicel** (Fig. 13–14A). Anteriorly, on the ventral side of the abdomen is a transverse groove known as the **epigastric furrow** (Fig. 13–14B). The reproductive openings are located in the middle of this furrow, and the spiracles of the book lungs are on each side of it. The end of the abdomen bears a group of modified appendages, the spinning organs, called **spinnerets** (Fig. 13–14). Primitively, there are four pairs, but in most spiders the number is reduced to three pairs, one of which is very small (Fig. 13–16A).

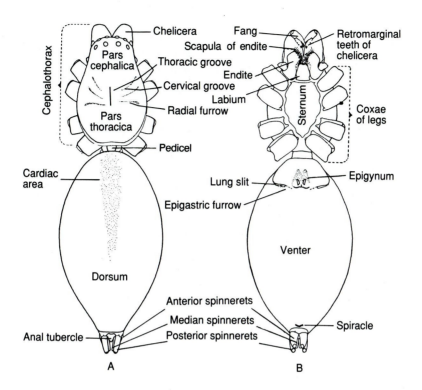

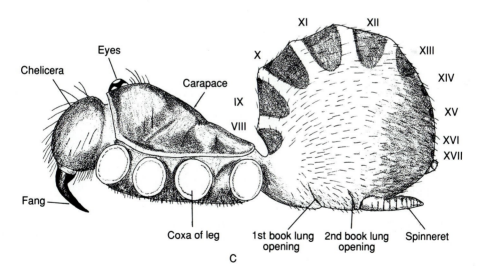

FIGURE 13–14 Dorsal **(A)** and ventral **(B)** views of a spider, showing external structure. **C,** The primitive Asian mygalomorph spider, *Liphistius malayanus,* with legs removed (side view). *(A and B, From Kaston, B. J. 1948. Spiders of Connecticut. Bull. 70. State Geol. Nat. Hist. Survey. 13. C, After Millot, J. 1949. Traité de Zoologie. Vol. VI. Masson et Cie, Paris.)*

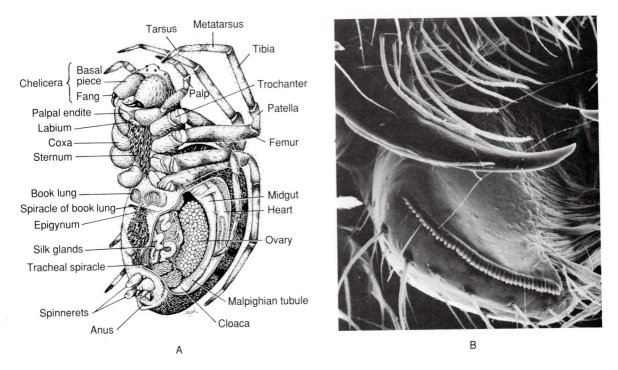

FIGURE 13–15 A, The orb weaver, *Araneus diadematus* (ventral view). **B,** Sawlike edge on the palpal endite (coxa) of *Portia,* a jumping spider. The ridge is used for cutting prey. One fang shows above the palpal endite. *(A, After Pfurtscheller from Kaestner. B, Micrograph courtesy of R. F. Foelix.)*

Silk

Silk production is a distinctive feature of spiders, and fossil species with clearly discernable spinnerets are known from as early as the Devonian. Each spinneret is a short, conical structure bearing many spigots, which are the openings from the **silk glands** (Fig. 13–16). The spinnerets are very mobile and can move independently. The silk glands themselves are large and are located within the posterior half of the abdomen.

The silk of spiders is a protein composed largely of glycine, alanine, serine, and tyrosine and is similar to the silk of caterpillars. It is emitted as a liquid, and hardening results not from exposure to air but from the drawing-out process itself, which changes its molecular configuration. Spider silk is about as strong as nylon but more elastic. A single thread is composed of several fibers, each drawn out from liquid silk supplied at a separate spigot. Drawing out occurs as the spider moves away from an anchored thread or pulls the threads with its posterior legs. Spiders produce more than one type of silk, and the various types are secreted from two to six kinds of silk glands.

Silk plays an important role in the life of the spider and is put to a variety of uses, even in the many families that do not build webs to catch prey. The original functions of silk were probably reproductive and this remains one of its important functions. The eggs of all spiders are wrapped within a silken **egg case,** and the male uses silk in the process of transferring semen to the copulatory organs. Another function of silk that is common to most spiders is as a **dragline** (Fig. 13–17). Spiders continually lay out a line of dry silk behind them as they move about. At intervals it is fastened to the substratum with adhesive silk. The dragline acts as a safety line, and the common sight of a spider suspended in midair, after being brushed off some object, results from its continual retention of its dragline. Many spiders build silken **nests** beneath bark and stones, which they may use as retreats or in which they may winter. The nest can prevent flooding by trapping an air bubble inside with

(Text continues on page 639)

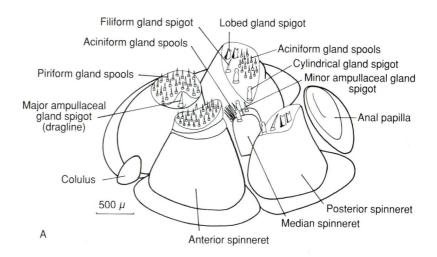

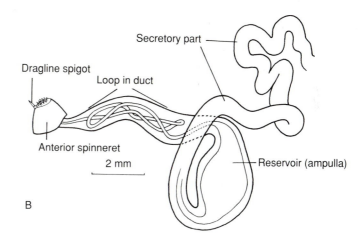

FIGURE 13–16 Spinnerets of the orb-weaving spider, *Araneus diadematus*. **A,** Distribution of spigots for different silk glands on spinnerets. **B,** Anterior spinneret and gland supplying the dragline spigot. *(From Wilson, R. S. 1969. Am. Zool. 9(1):103–111.)*

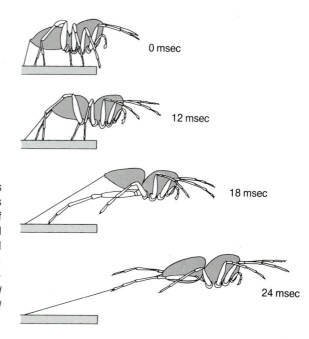

FIGURE 13–17 Tracings from motion picture frames of a jumping spider during a leap. Most of the force is provided by the rapid extension of the fourth pair of legs, which results from sudden elevation of blood pressure. Some jumping spiders jump with the third pair of legs; some with both third and fourth pairs. Note that the dragline is anchored before and retained during the jump. *(After photographs by Parry and Brown from Foelix, R. F. 1982. Biology of Spiders. Harvard University Press, Cambridge, MA. p. 155.)*

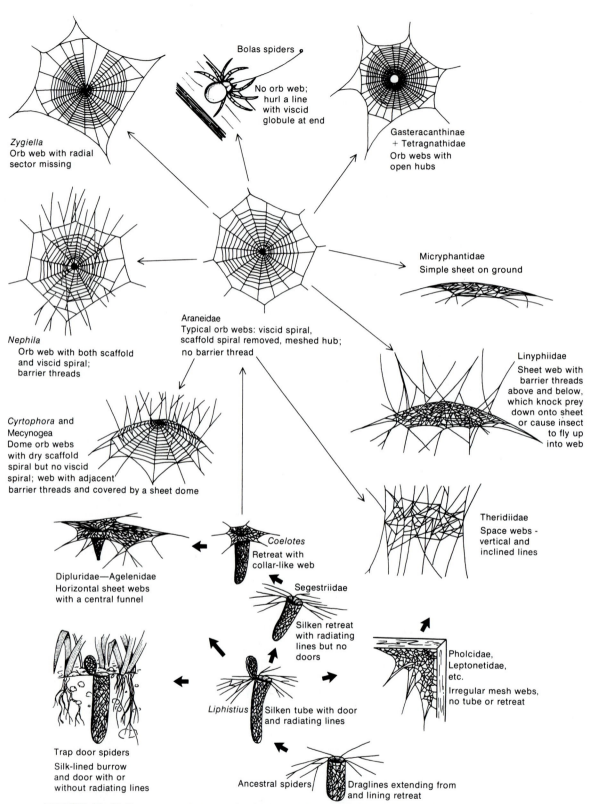

Bolas spiders
No orb web; hurl a line with viscid globule at end

Zygiella
Orb web with radial sector missing

Gasteracanthinae
+ Tetragnathidae
Orb webs with open hubs

Nephila
Orb web with both scaffold and viscid spiral; barrier threads

Araneidae
Typical orb webs: viscid spiral, scaffold spiral removed, meshed hub; no barrier thread

Micryphantidae
Simple sheet on ground

Linyphiidae
Sheet web with barrier threads above and below, which knock prey down onto sheet or cause insect to fly up into web

Cyrtophora and Mecynogea
Dome orb webs with dry scaffold spiral but no viscid spiral; web with adjacent barrier threads and covered by a sheet dome

Theridiidae
Space webs - vertical and inclined lines

Coelotes
Retreat with collar-like web

Dipluridae—Agelenidae
Horizontal sheet webs with a central funnel

Segestriidae
Silken retreat with radiating lines but no doors

Pholcidae, Leptonetidae, etc.
Irregular mesh webs, no tube or retreat

Liphistius
Silken tube with door and radiating lines

Trap door spiders
Silk-lined burrow and door with or without radiating lines

Ancestral spiders
Draglines extending from and lining retreat

FIGURE 13–18 Common web types of spiders and their possible evolution. A somewhat parallel evolution of web-building by cribellate families is not included. *(Based on Kaston, B. J. 1964. The evolution of spider webs. Am. Zool. 4:191–207; and on Levi, H. W. 1980. Orb webs: Primitive or specialized. Proc. 8th Internat. Arach. Congr., Vienna. 367–370.)*

638

the spider. After hatching, the young of some species are carried on air currents by means of a silken strand **(ballooning).** These functions of silk will be more fully described later.

Feeding

Spiders, like most other arachnids, are predatory and feed largely on insects, although small vertebrates may be captured by large species of spiders. The prey either is caught in the silken snare of members of **web-building** families or is pounced on by members of the more active **cursorial** (wandering) groups.

The dragline may have been the origin of the snare web. The first web was perhaps the radiating threads of the dragline laid down when the spider emerged from its protective retreat to catch passing prey (Fig. 13–18). Such lines might then have acquired a communicative function, informing the spider of the presence of prey when the line was struck. The retreat would also have become lined with drag line silk. From such a prototype evolved the highly developed triangles, orbs, funnels, sheets, and meshes characteristic of different spider families and species.

Many webs are composed of both dry and adhesive strands, each produced by different glands. The adhesive lines usually have an outer, unpolymerized, viscous layer of silk that greatly increases the trapping quality of the web; in some spiders (the cribellates), the adhesive is a very fine, entangling, loose "wool" attached to the line.

The construction of an orb web by a spider is a remarkable feat. Web building depends on morphological and physiological factors, such as weight, leg length, silk supply, appetite, and an instinctive behavioral pattern involving the integration of sensory information and locomotor activity. But visual information is not necessary, for a blinded spider can build a perfect web. No aspect is learned. The most complex web can be constructed by members of that species immediately after hatching. Considerable knowledge regarding web building has been provided by a pharmacologist, P. Witt, who has found the alterations in a spider's web-building behavior to reflect different drug effects. Most of the familiar orb webs are produced by members of the family Araneidae. Web construction is begun with a horizontal line, which may be quite long. To place this line, the spider may let out a strand of silk that is carried by air currents. When the free end entangles another surface, it is pulled tight and anchored. Using the original horizon-

tal line or a new one, the spider drops from the center, laying out a vertical line. The vertical line is pulled tight, converting the T to a Y frame, which provides the basic scaffolding of the web (Fig. 13–19). Additional frame lines are placed, and then the radii, or spokes, are attached from outside to the center, first on one side and then on the other, using a preexisting spoke as a guide. With the radii in place, a temporary dry spiral of thread is laid from the center to the outside. Using the dry spiral as a scaffold, the permanent adhesive spiral is placed from the outside inward, and the spider removes the scaffold in the process. Because mesh size is determined by the first legs of the spider, the size of the web is partly dependent on the size of the spider and increases with the spider's growth. The entire web, or at least the adhesive spiral, is usually replaced every day or night, for wet silk loses its stickiness within a few days. The old silk may be eaten, and very quickly much of the protein finds its way back to the silk glands. Webs are replaced in a remarkably short time. Species of *Araneus* spin a complete web in as little as 20 min. The web may require 1500 junction points and 10 to 30 m of silk.

Orb webs vary in detail according to the trapping strategies of the species. Some have a fine mesh; others have a coarse mesh. Some have a closed hub; others have an open one, which makes it possible for the spider to move quickly from one side of the web to the other (Fig. 13–18). Some species stay in the center of the web; others hide in a curled leaf retreat holding onto a signal thread from the web. There are diurnal orb weavers that add conspicuous zigzag lines of silk in the hub of the web called **stabilimenta** (Fig. 13–20D). Acting as a warning signal, the stabilimentum reduces the chance that flying birds will destroy the web.

The position and location of webs are not random. Some species spin horizontal orbs, catching insects that fly up and out of vegetation. Other species have vertical webs at characteristic elevations to trap certain species of insects. Thus, the same wooded area may support a number of species of orb weavers, each adapted for trapping a different part of the insect population.

Web-building spiders are aerialists, and they usually have legs that are more slender than those of hunting spiders (Fig. 13–20A). Climbing about the web, a spider hooks the lines by a small middle claw that lies between the two large claws found in all spi-

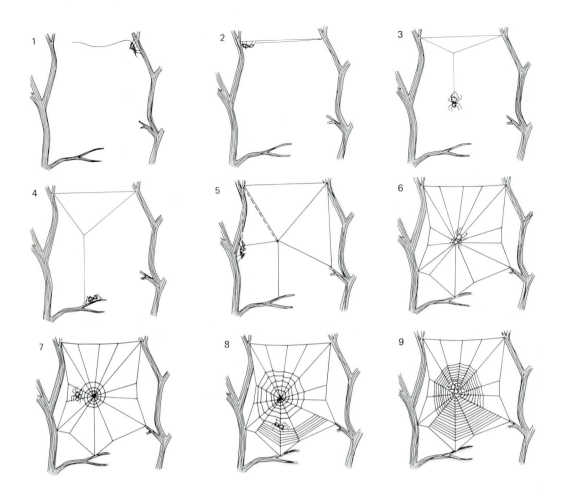

FIGURE 13–19 Stages in the construction of an orb web. See the text for an explanation. The web is usually attached to vegetation by additional frame lines, but these have been omitted to save space. *(From Levi, H. W. 1978. Orb-weaving spiders and their webs. Am. Sci. 66:734–742.)*

ders, and they are held against two outer setae called accessory claws. Why the spider does not stick to the web and how it can cut silk with its chelicerae are not known. Eyesight is not well developed in web-building spiders, but they are highly sensitive to vibrations. A web builder can determine from thread vibrations the size and location of the trapped prey, and many species respond to different stimuli with different attack patterns. Orb weavers, tangle-web spiders, and others commonly swathe prey in silk before or after biting it. Swathing aids in immobilizing prey or, when it is used after an immobilizing bite, prevents the prey from falling out of the web or, in the case of cursorial spiders, from an elevated position in vegeta-

tion. The scales on the wings of moths and butterflies are a defense against the adhesive webs of spiders, and spiders immediately give a captured moth an immobilizing bite. A few species (**kleptoparasites**) have abandoned prey trapping and steal the catch of other web builders.

Cursorial hunting forms include wolf spiders, fisher spiders, crab spiders, jumping spiders, and many mygalomorph spiders; the mygalomorphs are a more primitive group that contains the so-called tarantulas and the trap-door spiders. All of these cursorial forms are believed to be derived from web-building ancestors through loss of the web-building habit. Species that stalk their prey typically have

C

FIGURE 13–20 A, A common, web-building house spider, the comb-footed spider *(Achaearanea),* builds a web of irregularly placed threads. **B,** A crab spider, member of the cursorial family Thomisidae. **C,** The orb weaver, *Argiope aurantia,* against the stabilimentum in the hub of its web. *(A and B, Photographs courtesy of Betty M. Barnes. C, From Tolbert, W. W. 1975. Predator avoidance behaviors and web defensive structures in the orb weavers Argiope aurantia and Argiope trifasciata. Psyche. 82(1):29–52.)*

heavier legs than web builders. Also, most have a tuft of adhesive hairs **(scopula)** behind the terminal claws, which aid in adhering to surfaces and in prey capture. Some species roam about catching insects they happen to encounter; others utilize a sit-and-wait strategy. Many crab spiders (Fig. 13–20B), for example, sit in flower heads and ambush visiting insects. They can immobilize relatively large prey. Prey is detected by tactile and visual stimuli, and in some families, such as the wolf spiders and the colorful jumping spiders, the eyes are highly developed and of primary importance in prey capture. Cursorial spiders lay down a dragline, and some ground spiders tie up their prey by running around them. Trap-door spiders con-

struct silk-lined burrows that are closed by a lid covered with moss, soil, and other material (Fig. 13–32A). The spider lies in wait beneath the lid, and some species may detect passing prey by means of silk lines that radiate over the ground. If undisturbed, trap-door spiders may occupy the same burrow for many years.

Prey varies with habitat and distribution of the spider. The little wolf spider, *Pardosa amentata,* which lives on sunny, moist ground in Europe, depends on dipterans (flies and gnats) for about 70% of its diet. It catches about one insect per day, and activity is greater in the morning than in the afternoon. The North American green lynx spider, *Peucetia viri-*

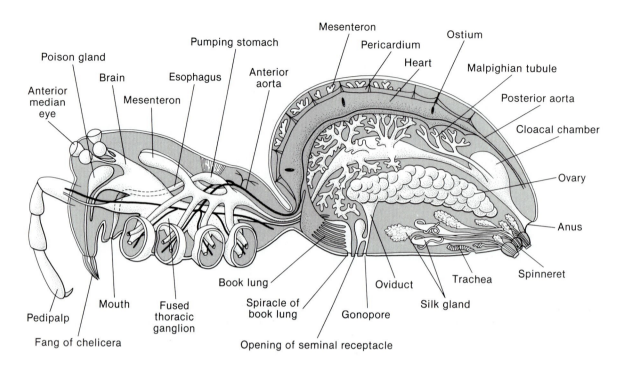

FIGURE 13–21 Internal anatomy of an araneomorph spider. *(After Comstock.)*

dans, which lives in low shrubs and herbaceous vegetation, was found to prey on 62 species of insects plus a few species of spiders.

Although most spiders are solitary, some degree of social organization has evolved in a few species within nine families. These social spiders share a communal web and cooperate in the capture of prey.

Spiders bite their prey with the chelicerae, which may also hold and macerate the tissues during digestion. The chelicerae of spiders are unique among arachnids in being provided with **poison glands** that open near the tip of the fang (Fig. 13–21). The glands themselves are located within the basal segments of the chelicerae and usually extend backward into the head. When the spider bites, the fangs are raised out of the groove in which they lie and are rammed into the prey. Simultaneously, muscles around the poison gland contract, and fluid from the gland is discharged from the fang into the body of the prey.

The venom of most spiders is not toxic to humans, but a few species have dangerous bites. Among these are species of black widows *(Latrodectus),* which are found in most parts of the world, including the United States and southern Canada (Fig. 13–22A). The venom is neurotoxic and, as in most

spiders, is composed of a mixture of protein compounds. Although the bite may be unobserved, the symptoms are severe and very painful; they include pain in the abdomen and legs, high cerebrospinal fluid pressure, nausea, muscular spasms, and respiratory paralysis. Fatal cases are rare and are more apt to result from incorrect diagnosis and mistreatment (such as removal of the appendix) than from the spider venom. Antivenom and other treatments are available. The recluse spiders, members of the genus *Loxosceles,* have a hemolytic venom and produce a local necrosis, or ulceration, of tissue that spreads from the bite. The ulcer is slow to heal. The bite of *Loxosceles reclusa,* the brown recluse spider (Fig. 13–22B), which is found in the midwestern United States and other localitites, can be dangerous, as can that of the large South American *Loxosceles laeta.* In southeastern Brazil, a ctenid, *Phoneutria,* and wolf spiders of the genus *Lycosa* have venoms that produce necrosis.

North American mygalomorph tarantulas, despite their size and reputation, are not venomous, but there are mygalomorphs, such as the Australian *Atrax* and the South American *Trechona,* both funnel-web builders, that are dangerous. Many New World tarantulas (Theraphosidae), however, have defensive, ur-

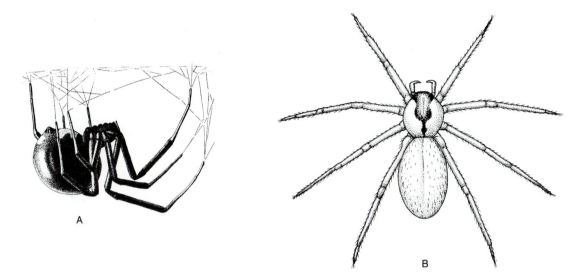

FIGURE 13–22 Venomous spiders. **A,** Female black widow, *Latrodectus mactans,* hanging in its web. Its color is shiny black with a red hourglass marking on the ventral surface of the abdomen. This species of the cosmopolitan genus is found in the southeastern and midwestern United States. **B,** The brown recluse spider, *Loxosceles reclusa,* whose bite produces a necrotic wound, difficult to heal. The brown recluse spider is found around human dwellings in the midwestern United States but has been introduced into other sections as well. *(A, From Kaston, B. J. 1970. Trans. San Diego Soc. Nat. Hist. 16(3):33–82.)*

ticating setae on the abdomen. These setae are barbed and irritating and readily penetrate the skin of a potential predator, such as small mammals, that enter the tarantula's burrow. In humans they can cause a rash.

Spiders having chelicerae with teeth (Fig. 13–14B), such as many hunting forms, chew their prey, aiding the digestion of tissues by the enzymes poured out from the mouth. The indigestible skeletal remains are discarded as a wad. Spiders with toothless chelicerae do not chew but introduce enzymes through a puncture and suck out the digested tissues. In most spiders each of the pedipalpal coxae (endites), which flank the mouth, has a sawlike edge used in cutting prey tissue and a screen of hairs that act as a filter when juices are sucked into the mouth (Fig. 13–15B). Sucking force comes from the pharynx and a posterior enlargement of the esophagus known as the **pumping stomach** (Fig. 13–21). It is located at about the middle of the prosoma. The mesenteron fills almost the entire abdomen, as well as much of the prosoma (Fig. 13–21). The posterior part of the midgut becomes enlarged in the back of the abdomen to form a cloacal chamber. Waste is collected here and then discharged from the anus through a short, sclerotized intestine.

As an adaptation for predation and an uncertain food supply, most spiders have relatively low metabolic rates, can tolerate prolonged starvation, and because of the great capacity of the extensive midgut region, can double their body weight in one feeding. For example, the wolf spider, *Lycosa lenta,* which has an estimated life span of 305 days, can survive without food for 208 days and reduce its metabolic rate by 30% to 40% during that period.

Although spiders are important insect predators, spiders themselves are preyed on by some insects and vertebrates. Two groups of wasps hunt spiders as food for their larvae. The wasp paralyzes the spider with a sting and lays an egg on its body. The spider wasps (Pompilidae) put the spider in an underground chamber; the mud daubers (Sphecidae) put a number of paralyzed spiders in a mud tube attached to elevated objects. The pressure of vertebrate predation on spiders is reflected in the fact that densities of spider populations are ten times greater on islands without vertebrate predators.

Internal Structure and Physiology

Gas exchange organs in spiders are of two forms: book lungs and tracheae. Spiders considered primitive, such as the mygalomorphs, have no tracheae but

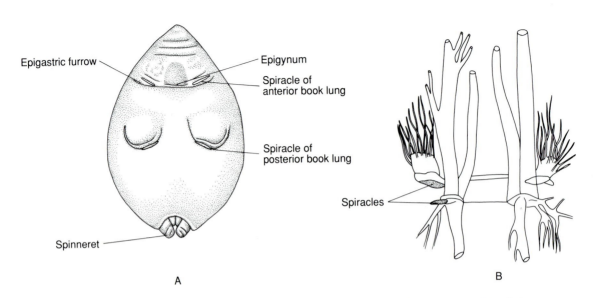

FIGURE 13–23 A, Abdomen of a mygalomorph spider, showing the two pairs of book lungs (ventral view). **B,** Central part of tracheal system of the caponiid spider, *Nops coccineus,* in which both pairs of book lungs have been replaced by tracheae. *(A, After Comstock. B, After Bertkau from Kaestner.)*

possess two pairs of book lungs derived from the second and third abdominal segments (Figs. 13–14C; 13–23A). Almost all other spiders have only a single pair of book lungs, the posterior pair having evolved into tracheae. Coinciding with this transformation in most spiders, a posterior migration and fusion of the tracheal openings have taken place, forming a single spiracle located just in front of the spinnerets (Fig. 13–14B). In several groups of small spiders the anterior book lungs have also been transformed into tracheae, so gas exchange is accomplished entirely by this means (Fig. 13–23B).

The spider's circulatory system (Fig. 13–21) is similar to that of scorpions, but as the book lungs become smaller and the tracheae increase in complexity, there is a corresponding loss in the number of heart chambers, ostia, and arteries, until in the exclusively tracheate spiders, most of which are small, there are often only two pairs of ostia.

Blood pressure in a resting spider is equal to that of humans and may double in an active spider or one ready to molt. Blood pressure causes the legs to extend in opposition to flexor muscles. Jumping spiders (Salticidae) jump by sudden elevation of blood pressure and extension of the third or fourth pair of legs. Leaps as great as 25 body lengths have been recorded.

Coxal glands are not as well developed in spiders as in other arachnids. Groups considered primitive have two pairs of coxal glands opening onto the coxae of the first and third pair of walking legs. The coxal glands are highly developed in these forms. In all others, only the anterior pair of glands remains, and it displays various stages of regression.

In most spiders the two branched malpighian tubules, which are connected to the cloacal chamber in the posterior part of the abdomen, are more important in excretion than the coxal glands (Fig. 13–21). The excretory wastes are guanine and uric acid. Certain cells located beneath the integument and over the surface of the intestinal diverticula also store guanine crystals and account for the white color of some spiders.

Sense Organs

The eyes of some spiders surpass those of all other arachnids in degree of development. Usually, there are eight eyes arranged in two rows of four each along the anterior dorsal margin of the carapace (Fig. 13–24A), but other arrangements of the eyes are characteristic of certain families. In the wolf spiders and jumping spiders, the eyes are situated over the surface of the anterior half of the carapace, resulting in a wide visual field (Fig. 13–24D). Of the two rows of four eyes usually found in spiders, the anterior median eyes are of the direct type, and all the remaining eyes are indirect. The anterior median eyes of jumping spiders are unique in having the photoreceptors arranged in four layers, each of which is believed to respond to

different wavelengths. Whether such differentiation makes possible color recognition or greater contrast is still uncertain. The few studies have been largely restricted to the jumping spiders, which have the most highly developed eyes. Because the indirect eyes are provided with a tapetum, which reflects light rays, the

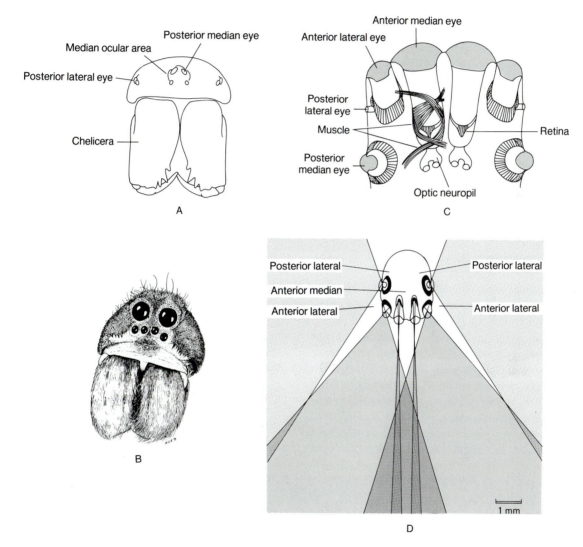

FIGURE 13–24 A, Front view of a spider, showing eight eyes arranged in anterior and posterior rows of four each, a pattern exhibited by many families. **B,** Face of a wolf spider. Posterior median eyes are very large; the posterior lateral eyes are located back over the carapace and can just be seen. **C,** Eyes of a jumping spider. Note the tubular, telephoto structure of the anterior median eyes. **D,** Visual fields of the eyes of a jumping spider *(Trite planiceps).* The posterior lateral eyes function in prey detection; the anterior median eyes function in tracking the prey after swiveling the body in the prey's direction. *(A, After Kaston. C, Modified after Land, M. F. 1969. Structure of the retinae of the principal eyes of jumping spiders in relation to visual optics. J. Exp. Biol. 51:443; and after Movements of the retinae of jumping spiders in response to visual stimuli. J. Exp. Biol. 51:771. D, Modified from Forster, L. M. 1979. Visual mechanisms of hunting behavior in Trite planiceps, a jumping spider. New Zealand J. Zool. 6:79–93.)*

anterior median eyes appear dark and the others often appear pearly white. In some families, notably the cursorial, hunting wolf spiders, the tapetum has developed to such an extent that these spiders can easily be located at night and captured by using a flashlight to look for the reflection of the eyes. Jumping spiders, on the other hand, have no tapeta at all in the indirect eyes.

The number of receptors is much greater in cursorial hunting species than in the sedentary web builders. In the cursorial spiders, eyes are important for detecting movement and locating objects. The jumping spiders are capable of perceiving a relatively sharp image of considerable size. In this family the number of receptors is large, particularly in the anterior median eyes (about 1000 photoreceptors), in which the tapered ends of the retinal cells are greatly narrowed and compact. In many species of jumping spiders there has been an increase in the depth of the anterior median eyes, resulting in a somewhat tubular telephoto structure (Fig. 13–24C). Muscles attached at the rear can rotate the tubes around the visual axis.

Chemosensitive tubular hairs on the tips of appendages, tactile hairs (most of the large body hairs), trichobothria, slit sense organs, and tarsal organs are very important sense organs in all spiders; in most spiders, especially the sedentary, web-building forms, they are more important than the eyes. The funnel-web builder, *Agelena,* for example, can determine the position of prey when it is 1 cm distant by means of the trichobothria.

Spiders have the greatest development of slit sense organs of any group of arachnids. Studies on the hunting spider, *Cupiennius,* have demonstrated the importance of grouped slit sense organs (lyriform organs) on the femur and tibia in kinesthetic orientation). Blinded individuals chased as far as 25 cm from recent captured prey are able to return to their original position, correctly determining the angle and distance of the return path. This ability is greatly reduced if the lyriform organs on the femur and tibia are destroyed.

Slit sense organs in the joint between the tarsus and the metatarsus (the last two long leg articles) of web-building spiders are especially sensitive and enable the spider to discriminate vibration frequencies transmitted through the silk strands of the web or even through the air. In species in which the spiderlings remain in the parental web after hatching, the parent can recognize its young. The spider can also discern the size of the entrapped prey or even the kind of insect if it produces buzzing vibrations. The spiders themselves may produce vibration signals. The male may tweak the strands of the female's web, or the mother may produce vibration signals that are detected by the spiderlings. **Tarsal organs,** cuplike structures situated near the tips of the legs, are probably olfactory receptors for pheromones.

Reproduction and Life History

The ovaries of the female consist of two elongated, parallel sacs located in the ventral part of the abdomen (Fig. 13–21). At maturity, the eggs rupture into the lumen of the ovary and pass into the curved oviduct leading from each ovary. Each oviduct converges to form a median tube (Fig. 13–25), which extends ventrally and backward to join with a short, chitinous vagina, opening at the middle of the epigastric furrow.

Associated with the vagina and uterus are two or more seminal receptacles and glands (Fig. 13–25). In most spiders these have separate openings to the outside and are connected internally to the vagina. These external openings are for the reception of the male copulatory organ during mating and are located just in front of the epigastric furrow on a special sclerotized plate called the **epigynum** (Figs. 13–14B; 13–25). Fertilization occurs at the time of egg laying, mating having taken place some time earlier. Sperm are stored in the seminal receptacles, or spermathecae.

The two large, tubular testes lie along each side of the male abdomen, and the two convoluted sperm ducts open through a common genital pore in the middle of the epigastric furrow. The copulatory organs of the male are not connected to the sperm duct opening but are located at the ends of the pedipalps. The tarsal segment of these appendages has become modified to form a truly remarkable structure for the transmission of sperm. Basically, each pedipalp consists of a bulblike **reservoir** from which extends an **ejaculatory duct** (Fig. 13–26B, C). This leads to a penis-like projection called the **embolus.** At rest the bulb and embolus fit into a concavity, the **alveolus,** on one side of the male's tarsal segment.

During mating (Fig. 13–27A), special regions **(hematodocha)** of the tarsal segment become engorged with blood, causing the bulb and embolus to twist and project out of the alveolus (Fig. 13–26D); the embolus is at the same time inserted into one of

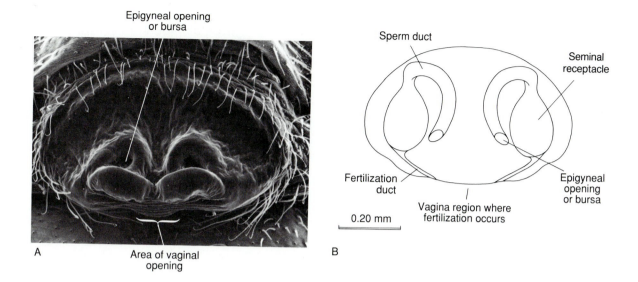

FIGURE 13–25 A, Epigynum, the chitinous plate bearing the reproductive openings of the female spider of *Philoponella republicana*. **B,** Reproductive ducts beneath the epigynum of the same species. At copulation the embolus of the male palp (see Fig. 13–26D) is inserted into the epigyneal opening to the spermatheca. The fertilization duct carries the sperm to the common oviduct (vagina), where fertilization occurs when the eggs are released through the primary genital opening (gonopore), which is hidden in these figures. *(From Opell, B. D. 1979. Revision of the genera and tropical American species of the spider family* Uloboridae. *Bull. Mus. Comp. Zool. 148(10):443–549.)*

the female reproductive openings leading to the seminal receptacles.

In primitive spiders, the pedipalpal organ consists only of the basic parts previously described (Fig. 13–26B). In the majority of families, however, the pedipalp has become much more complicated with the addition of a great many accessory parts, such as the conductor and various processes (Fig. 13–26D). These structures, in combination with the sclerotized configurations of the female epigynum, aid in orienting the pedipalp and inserting the embolus into the female openings.

The precise form of the male pedipalp and female epigynum is distinctive for each species, and therefore these organs are the primary structures used by araneologists for classifying and identifying spiders at the species level. The female and the male reproductive systems and the male pedipalpal organ are not completely formed until the last molt has taken place. The male then fills, or charges, the pedipalps with sperm in the following manner. He first spins a

tiny sperm web, or at least a strand of silk, with special silk glands that open onto the anterior ventral surface of the abdomen. On this web is ejaculated a globule of semen (Fig. 13–26A). Next the pedipalps are dipped into the globule until all of the semen is taken up into the reservoirs. With the pedipalps filled, the male spider then seeks a female with which to mate. Courtship behavior is not dependent on sperm-filled pedipalps, however.

The predatory habits of spiders, as in other arachnids, makes recognition of the sexual partner especially important, meaning that the female must identify the male as a potential mate and not as food. As a result, highly complex precopulatory behavior patterns have evolved in many species. In all spiders chemical and tactile cues are of primary importance. On encountering a dragline or web, a male spider can determine whether it was produced by a mature female of the same species, and the pheromone on the silk or body of the female may initiate trail following or the male courting response. Evidence of such a

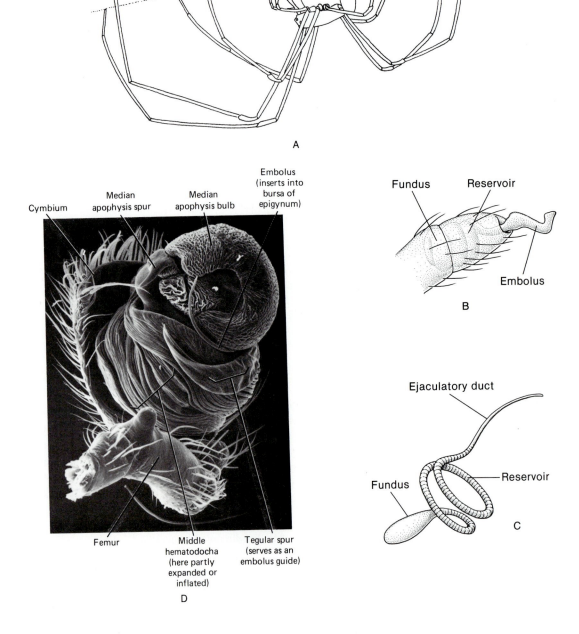

FIGURE 13–26 A, Male tetragnath spider in a sperm web, filling pedipalps from a globule of semen. B, Simple palp of *Filistata hibernalis*. C, The semen-containing part of the male pedipalp. D, Left male palp, the copulatory organ, of *Zosis geniculatus*. The severed end of the pedipalp projects at the lower left corner. *(A, After Gerhardt from Millot. B–C, After Comstock. D, From Opell, B. D. 1979. Revision of the genera and tropical American species of the spider family Uloboridae. Bull. Mus. Comp. Zool. 148(10):443–549.)*

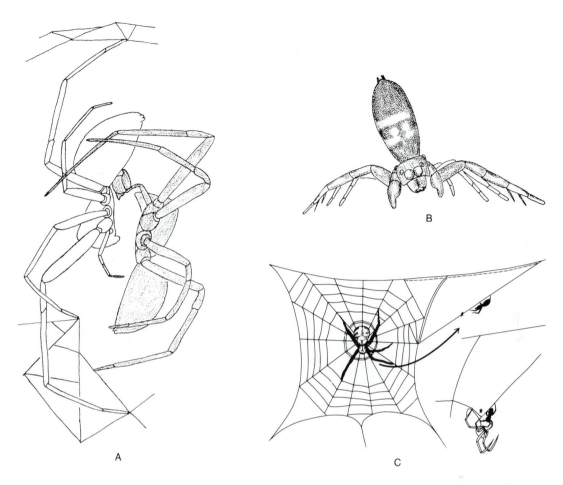

FIGURE 13–27 A, Mating position of *Chiracanthium* (male is shaded). **B,** Courting posture of the male jumping spider, *Gertschia noxiosa*. **C,** Courtship behavior of some orb weavers. The male vibrates the mating thread he has attached to the female's web. The female leaves the hub and moves onto the mating thread, where copulation occurs (star). *(A, After Gerhardt from Kaston. B, After Kaston. C, From Robinson, M. H., and Robinson, B. 1980. Comparative studies of the courtship and mating behavior of tropical araneid spiders. Pacific Insects Monograph. Bishop Museum, Honolulu, 36:1–218.)*

substance is indicated by the fact that a male will court the severed leg of a female or the evaporated washings of the female's body.

Females respond to a variety of cues produced by the male. In the sedentary web builders, the male often plucks the strands of webbing, producing vibrations that are detected and recognized by the female. The male of orb weavers plucks the radius held by the female or plucks a mating thread that he has attached to the female's web (Fig. 13–27C). The message is species-specific in the number, frequency, and inten-

sity of plucks. In some species the male is so small that he clambers ignored over the body of the female (Fig. 13–28B).

Cursorial hunting spiders display unique courtship behaviors. The approach may be direct, the male pouncing on the female, palpating her body with his pedipalps and legs and causing her to fall into an immobile state. Some male wolf spiders **stridulate** with a file and scraper located in the tibiotarsal joint of the palp. During oscillations of the joint, a group of heavy spines at the tip of the palp maintain contact

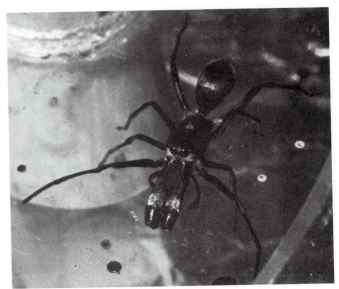

A

B

FIGURE 13–28 A, One of the courtship postures of a male jumping spider. This Australian species is one of a number in various families that mimics ants. Some feed on ants, but most mimic ants as a means of defense. The mimicry includes behavioral as well as structural features. In some species of ant mimics, like this one, the chelicerae are very long and heavy and project forward between the two small pedipalps. **B,** Male of the giant New Guinea wood spider on the dorsal surface of the female. Abdomen of the female is about 3.2 cm long. *(A, From Jackson, R. R. 1982. The biology of ant-like jumping spiders; intraspecific interactions of Myrmarachne lupata. Zool. J. Linn. Soc. 76:293–319. B, From Robinson, M. H., and Robinson, B. 1973. Ecology and behavior of the giant wood spider Nephila maculata in New Guinea. Smithson. Contrib. Zool. 149.)*

with the substratum, and the vibrations are detected by the female through the substratum rather than as airborne pressure waves.

In families with well-developed eyesight, visual cues are also important, and courtship takes the form of dancing and posturing by the male in front of the female (Figs. 13–27B; 13–28A). This involves various movements and the waving of appendages, which are often brightly colored. Such behavior is highly de-

veloped and has been studied most extensively in the colorful jumping spiders.

Although descriptions of spider courtship behavior emphasize the role of the male, the behavior of each sex depends on a sequence of reciprocal signals that release the next act in one sex or the other. Primitively, the primary releaser for courtship is probably body contact. This level is exhibited by many tarantulas, crab spiders, and certain small ground spiders.

Body pheromones or chemical or vibration signals of silk lines are more advanced primary releasers for many groups, such as wolf spiders and many web builders. The most highly evolved condition, exhibited by jumping spiders, utilizes visual cues as primary releasers.

Various copulatory positions characterize different families. On attaining the proper position, the male usually scrapes the epigynal surface rapidly with his palp until the proper orienting parts on the palp and epigynal plate connect. Then the palp becomes quickly engorged with blood, driving the embolus into the passageway of the seminal receptacle (Fig. 13–27A). Following sperm transmission with one palp, which lasts some seconds to minutes, the other palp is inserted into the opposite seminal receptacle opening, which may require the male to move to the opposite side of the female. The behavioral sequence for the wolf spider, *Lycosa rabida,* is shown in Figure 13–29. Depending on the species, there may be numerous insertions, and the entire process may last a number of hours.

Adults of some species mate a number of times during their lifetime; others mate only once. The embolus of the male palp breaks off within the female duct in some spiders. There are also many species in which a plug forms, filling the openings into the seminal receptacles and preventing a second mating. The peculiar indirect mode of sperm transmission in spiders probably evolved from an ancestral habit in which the male placed a spermatophore into the genital opening of the female with his pedipalp. As we have already seen, indirect sperm transmission by spermatophores is widespread among other arachnids, and in some groups, such as uropygids and solifuges, the male uses an appendage (the pedipalp in uropygids) to push the spermatophore or sperm mass into the female genital opening.

Some time after copulation, the female lays her eggs. Several to 3000 eggs are laid in one to several cases, depending on the species. Just before the deposition of eggs, the female spins a small basal sheet and then a cuplike wall. The eggs are fertilized with stored sperm just before leaving the female's body and laid in the silken cup. The female now covers the eggs with another sheet of silk, and the entire mass is often given an additional covering of silk so that it assumes a spherical shape—the egg sac, or cocoon (Fig.

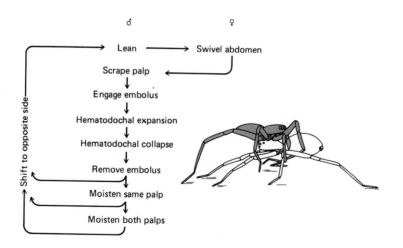

FIGURE 13–29 Sequence of events during copulation in the wolf spider, *Lycosa rabida.* "Swivel abdomen" refers to the turning of the abdomen of the female to accommodate the position of the male. The figure at the right shows a pair in copulation, with the male (black) inserting his right palp into the left seminal receptacle of the female. *(Diagram from Rovner, J. S. 1971. Mechanisms controlling copulatory behavior in wolf spiders. Psyche. 78(1):150–165. Figure from Kaston, B. J. 1948. Spiders of Connecticut. Conn. State Geol. Nat. Hist. Survey Bull. 70. p. 2006.)*

13–30). The egg sac is attached to webbing (in web-building families), hidden in the spider's retreat, or attached to the spinnerets and dragged about after the mother (as in the wolf spiders and a few others). In many spiders the female dies after completing the egg sac; in others she may produce a series of cases and may remain with the young for some time after hatching.

The spiderlings hatch inside the egg sac and remain there until they undergo the first molt. The spiderlings of many species, as well as the adults of some small forms, use a form of transportation known as ballooning, which aids in the dispersal of the species. The little spider climbs to the top of a twig or blade of grass and releases a strand of silk; as soon as air currents are sufficient to produce a tug on the strand, the spider releases its hold on the plant and is carried away by the currents. Although ballooning is important in the spread of spider populations into new habitats, desiccation and lack of food make long-distance transport unlikely.

Some spiders care for their young following hatching. Among them the female wolf spider carries

FIGURE 13–30 Black widow, *Latrodectus mactans*, with egg case. *(Photograph courtesy of W. Van Riper, Colorado Museum of Natural History.)*

A

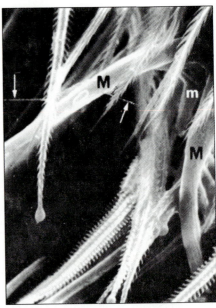

B

FIGURE 13–31 A, Female wolf spider carrying young on her abdomen. **B,** Three tarsal claws (M, m, M) at the end of a spiderling's leg as it grasps knobbed, barbed hairs on the dorsal surface of the abdomen of the wolf spider mother. *(A, Photograph courtesy of W. Van Riper, Colorado Museum of Natural History. B, From Rovner, J. S., Higashi, G. A., and Foelix, R. F. 1973. Maternal behavior in wolf spiders: The role of abdominal hairs. Science. 182:1153–1155. Copyright © 1973 by the American Association for the Advancement of Science.)*

A

B

FIGURE 13–32 Mygalomorph spiders. **A,** Trap-door spider capturing a beetle. **B,** Tarantulas from the southwestern United States. *(A, Photograph courtesy of W. Van Riper, Colorado Museum of Natural History. B, From Buchsbaum, R., and Milne, L. 1960. Lower Animals: Living Invertebrates of the World. Doubleday and Co., New York.)*

her young on her back after they hatch (Fig. 13–31A). The mother's dorsal abdominal surface is covered with special, knobbed, spiny hairs that aid in the attachment of the young (Fig. 13–31B). The spiderlings gradually disperse. The mother does not feed while carrying young.

Although some of the tarantulas (mygalomorph spiders) have been kept in captivity for as long as 25 years, most spiders live only 1 to 2 years. The number of molts required to reach sexual maturity varies, depending on the size of the species. Large species undergo up to 15 molts before the final instar; tiny species molt only a few times. Males molt fewer times than do females of the same species, and within a species there is always some variation in the number of instars before the final molt. Almost half of temperate species overwinter as immatures, usually in soil or leaf mold; others overwinter in the egg stage or as adults or in all different stages. Temperature and

photoperiod are the primary controlling factors in determining the pattern of the life cycle.

SYSTEMATIC RÉSUMÉ OF ORDER ARANEAE

Suborder Mesothelae. Original segmentation of abdomen reflected in numerous dorsal plates. Seven or eight spinnerets, the first of which is located at the level of the second pair of book lungs. The tropical Asian Liphistiidae (Fig. 13–14C).

Suborder Orthognatha (Mygalomorphae). Original segmentation of abdomen not apparent externally. Six or fewer spinnerets, located at end of abdomen. Possess two pairs of book lungs. The fang of the chelicera articulates in the same plane as the long axis of the body. This suborder includes the tarantulas (also called bird spiders or monkey spiders in various parts of the world) and trap-door spiders (Fig. 13–32). Although most stalk or lie in wait for their prey, there are some web builders. There are numerous representatives in North America. Many are large.

Suborder Labidognatha (Araneomorphae).   This suborder includes most of the spiders. Abdomen not segmented. With few exceptions, only a single pair of book lungs is present, and in all cases, the plane of articulation of the chelicerae is at right angles to the sagittal plane of the body. Descriptions of a few of the larger and more common families follows.

    Family Pholcidae.   Small and often long-legged, resembling phalangids or daddy longlegs. Members spin small webs of tangled threads in sheltered recesses. Several species commonly live in houses and, with other web-building house dwellers, are responsible for cobwebs.

    Family Theridiidae.   A large family of tangled-web (cobweb) builders known as comb–footed spiders because of the presence of a series of serrated spines on the fourth tarsus. These spines comb out a band of silk used for trussing up the prey. To this family belong the black widow, *Latrodectus* (Fig. 13–22A), and one of the most common house spiders, *Achaearanea tepidariorum* (Fig. 13–20A).

    Family Araneidae.   The orb-weaving spiders (Figs. 13–19; 13–20C). Members of this family spin circular webs. Many species are of considerable size and brightly colored, such as the black and yellow *Argiope,* or "writing" spider.

    Family Linyphiidae.   The sheet-web spiders. Webs are horizontal silken sheets or bowls (Fig. 13–18). The larger species construct their webs in vegetation. The great number of very tiny species in this family live in fallen leaves and humus.

    Family Agelenidae.   Funnel-web spiders. Although web builders, they are more closely related to certain cursorial spider groups. The web forms a funnel, the narrowed end acting as a retreat for the spider (Fig. 13–18). The web is constructed

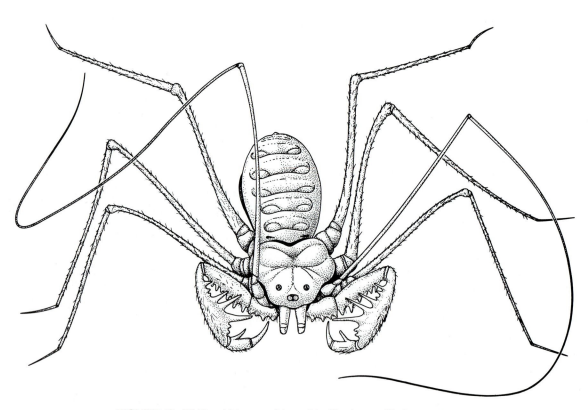

FIGURE 13–33 The African amblypygid, *Charinus milloti. (After Millot.)*

in dense vegetation or in crevices of logs or rocks, and it is easily visible, especially in grass covered with dew.

Family Lycosidae. Wolf spiders. These are rapidly moving and rather hairy spiders with dull brown and black coloration. They are most active at night and are common members of the ground fauna.

Family Pisauridae. Fisher spiders. This family is somewhat similar to the wolf spiders in appearance but with longer legs. They are common around the edges of ponds, lakes, and streams.

Family Thomisidae. Crab spiders. This and the following family commonly lie in wait on vegetation. Crab spiders get their name from their crablike movements and the position of their legs (Fig. 13–20B).

Family Salticidae. Jumping spiders. Species of this family are heavy-bodied and capable of jumping short distances (Fig. 13–17). They are often brightly colored and possess the best eyesight of all spiders. Numerous species live in the temperate and tropical regions of the world.

Order Amblypygi

The amblypygids are a tropical and semitropical group of approximately 70 species. There are many common amblypygids, but they are a nocturnal and secretive group, hiding during the day beneath logs, bark, stones, leaves, and similar objects. Amblypygids range in length from 4 to 45 mm and have a somewhat flattened body, which resembles that of spiders (Fig. 13–33). The carapace bears a pair of median eyes anteriorly and two groups of three eyes each laterally. The chelicerae are also similar to those of spiders, but the pedipalps are heavy and spiny and used to capture insect prey, which are located with the long, antenna-like first pair of legs. The gait of the amblypygid is crablike because of its flattened body and its ability to move laterally. One of the long, tactile legs is always pointed toward the direction of movement; the other may explore areas to either side of the animal. The segmented abdomen is connected to the prosoma by a narrow pedicel, like that of spiders. Two pairs of book lungs are located on the ventral side of the second and third abdominal segments.

In the species of *Charinus*, *Tarantula*, and *Admetus* in which reproductive habits have been observed, the male courts the female with trembling movements

of the antenniform legs, alternating with rocking body movements toward the female (Fig. 13–34). Eventually, a spermatophore is deposited. Using his pedipalps or first legs, the male guides the female over the spermatophore, and she takes up the sperm masses. When egg laying is at hand, the reproductive glands secrete a parchment-like membrane that holds the 6 to 60 large eggs to the underside of the female abdomen. The mother carries the eggs in this manner until hatching and the first molt of the young have taken place. The young climb onto the abdomen of the mother until the next molt.

Order Ricinulei

The ricinuleids are a small order of arachnids containing 33 uncommon species, sometimes called tick spiders. They are found in Africa *(Ricinoides)* and in the

FIGURE 13–34 Male of the amblypygid, *Heterophrynus longicornis,* tapping the body of a female with his antenniform legs. *(Photograph courtesy of Weygoldt, P. 1972. Zeitschrift des Kölner Zoo. 15(3):100.)*

FIGURE 13–35 A ricinuleid from the southwestern United States. *(Photograph courtesy of Dr. Robert Mitchell.)*

American hemisphere *(Cryptcellus)* from Brazil to the southern United States, where they have been collected from leaf mold and caves.

Ricinuleids are heavy-bodied animals that measure from 5 to 10 mm in length (Fig. 13–35). The cuticle is very thick and often sculptured. Attached to the anterior margin of the carapace is a curious hood-like structure that can be raised and lowered. When lowered, the hood covers the mouth and chelicerae. The pedipalps are leglike, and the abdomen is segmented and broadly joined to the carapace.

Sperm transfer is similar to that in spiders but involves the third legs instead of the pedipalps. The first tarsal segment bears a specialized process containing a duct that the male fills with sperm from his genital opening. At copulation the male embraces the female from the back, and the copulatory process is inserted into the female genital atrium by the encircling third legs.

Order Pseudoscorpiones

Pseudoscorpions are tiny arachnids, rarely longer than 8 mm and many only a few millimeters long. They live in leaf mold, in soil, beneath bark and stones, and in moss. Some species of several genera are common inhabitants of algae and beach drift in the intertidal zone. A cosmopolitan species, *Chelifer cancroides,* is found in houses.

Because of their small size and the nature of their habitat, these animals are rarely seen, although they are actually very common. A few handfuls of leaf mold sifted through a Berlese funnel (Fig. 13–2) usually yield at least some individuals, and densities of several hundred per square meter have been reported. Pseudoscorpions are found throughout the world, and about 2000 species have been described.

Pseudoscorpions superficially resemble the true scorpions but are much smaller and lack the long abdomen and sting (Fig. 13–36). The rectangular cara-

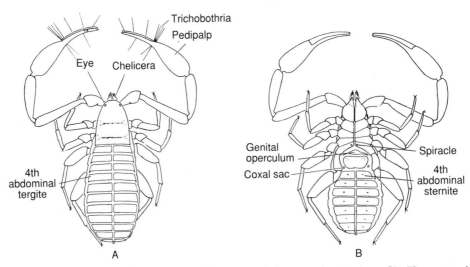

FIGURE 13–36 Dorsal **(A)** and ventral **(B)** views of the pseudoscorpion, *Chelifer cancroides* (male), showing the external structure. *(After Beier from Weygoldt, P. 1969. Biology of Pseudoscorpions. Harvard University Press, Cambridge, MA.)*

pace bears one or two eyes at each anterior lateral corner, or the eyes may be absent. The relatively wide abdomen forms a broad junction with the prosoma and is rounded posteriorly.

Pseudoscorpion chelicerae are small and chelate and bear several accessory structures, which will be described later in connection with their functions (Fig. 13–37). The chelate pedipalps are similar to those of the scorpions. These pedipalps, however, are peculiar in that each usually has a poison gland in one or both fingers or in the hand. A duct issues from the poison gland and opens at the end of a tooth at the tip of the finger. Pseudoscorpions are slow-moving and hold their pedipalps, which are supplied with tri-chobothria, to the front when they walk. If the pseudoscorpion is disturbed, it pulls the pedipalps back over the carapace and becomes immobile.

Pseudoscorpions feed on small arthropods, such as collembolans and mites. The prey is caught and paralyzed or killed by the poison glands in the pedipalps. It is then passed to the chelicerae, which tear open the exoskeleton. Digestion then takes place in typical arachnid fashion. Hairs at the front of the prebuccal chamber strain out the solid particles. When a large mass of solid particles accumulates, the chelicerae are withdrawn and short, stiff hairs ("flagella") on the chelicerae catch the mass of debris and eject it from the prebuccal chamber (Fig. 13–37). After feeding, the buccal pieces are cleaned by comblike structures (serrulae) on the fingers of the chelicerae.

Gas exchange in pseudoscorpions is accomplished by means of a tracheal system that opens through two pairs of spiracles on the ventral side of the third and fourth abdominal segments. Coxal glands that open on the coxae of the third pair of walking legs provide for excretion.

There is little secondary differentiation between sexes. The process of sperm transmission provides some clues as to possible stages in the evolution of indirect sperm transfer by spermatophores in arachnids. In some species the male deposits a stalked spermatophore on the substratum in the absence of a female. If a female encounters the spermatophore, she is attracted chemotactically. She takes a stance over the terminal sperm mass, and the sperm are taken into the female orifice. Uptake is facilitated by swelling of the sperm mass, triggered by some substance produced in the female atrium. This is probably the most primitive method of sperm transfer in pseudoscorpions. The course of subsequent evolution has led to

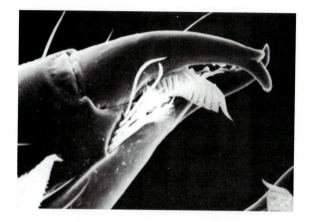

FIGURE 13–37 Scanning electron micrograph of a pseudoscorpion chelicera. The hornlike projections (galea) at the end of the movable finger contain the openings of the silk glands. The cluster of hairs ("flagella") at the base of the fixed finger is used to remove food particles that accumulate on the screening setae in the prebuccal cavity. The comblike structure (serrula) is used to clean the buccal region after feeding. *(Micrograph courtesy of Thomas Pangburn.)*

behavior that increases the chance of the female's obtaining the spermatophore. There are certain species in which the male, on encountering a female, lays down silk signal threads after depositing the spermatophore. When the female comes across the thread, she is guided to the spermatophore.

With pairing behavior, the male himself directs the female to the spermatophore in the more evolved patterns. The male grasps the female with his pedipalps and, following a promenade-like courting behavior, maneuvers the female over the spermatophore until it is in the proper position to be taken up. Finally, in the most specialized pattern, two long, tube-like organs are evaginated from the sexual region of the male abdomen. The female is attracted, and the two sexes promenade without touching each other. The male then attaches a spermatophore to the substratum and backs away. The female follows, and the male then seizes her pedipalpal femurs. When she is in the proper position, he aids in the uptake of the spermatophore by pushing with his forelegs (Fig. 13–38A–C).

The eggs are laid after sperm transmission (a month in *Chelifer cancroides*). Before laying the eggs, the female uses small bits of dead leaves and debris to build a nest and lines it with silk emitted from duct openings on hornlike processes (galeae) on

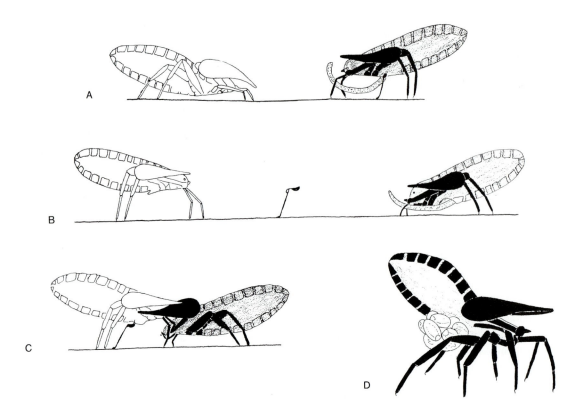

FIGURE 13–38 A–C, Courtship and sperm transmission in *Chelifer cancroides* (male is black). **A,** Male producing a spermatophore. **B,** Spermatophore attached to the substratum. **C,** Male pressing the female down onto the spermatophore. **D,** Female of *Chelifer cancroides* carrying embryos. *(All after Vachon, M. 1949. Traité de Zoologie. Vol. VI. Masson et Cie, Paris.)*

the chelicerae (Fig. 13–37). First the liquid silk is attached, and then the thread is drawn out as the chelicerae are moved to another position.

After the eggs are laid, they remain in a membranous sac attached to the genital opening on the ventral side of the female's body. Development takes place within the sac (Fig. 13–38D). In a later stage of development, the embryo is supplied with a nutritive material secreted by the mother's ovaries. The young undergo one molt before hatching and one during hatching and emerge from the brooding sac during the third instar. Depending on the species, 2 to more than 50 young may be brooded. The young pseudoscorpion molts twice again before becoming an adult. Molting takes place within a nest of silk that is constructed like the nest of the female at the time of egg production. A silk nest may be used for overwintering. Maturity is reached in a year or less, and individuals may live two to five years. Temperate species may produce several generations a year.

Order Solifugae

The solifuges are a group of about 900 tropical and semitropical arachnids, sometimes called sun spiders because of the diurnal habits displayed by many species, or wind scorpions because of the great speed with which they can run. In the United States, a few species have been found in Florida, and more than 100 have been found in the Southwest, some as far north as Colorado. Many solifuges are common in the warm desert regions of the world. They hide under stones and in crevices, and many species burrow.

Solifuges are small to large arachnids, sometimes reaching 7 cm in length. The prosoma is divided into a large, anterior carapace bearing a pair of closely placed eyes on the anterior median border, and a short posterior section (Fig. 13–39A). The segmented abdomen is large and broadly joined to the prosoma (Fig. 13–39B).

The most striking characteristic of the solifuges is the enormous size exhibited by the chelicerae, which project in front of the prosoma and can be directed upward by the flexing of the subdivided prosoma. Each chelicera is composed of two pieces forming a pair of pincers that articulate vertically. The pedipalps are leglike but terminate in a specialized adhesive organ. The first pair of legs are somewhat reduced in size and are used as tactile organs; the remaining three pairs of legs are used for running.

Solifuges are carnivorous or omnivorous, and termites are an important part of the diet of many American species. The pedipalps locate the prey, and the chelicerae kill the animal and tear apart the tis-

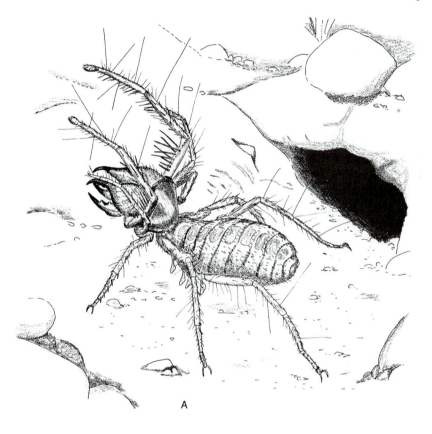

A

FIGURE 13–39 A, North African solifuge, *Galeodes arabs*. B, Cephalothorax of the solifuge *Galeodes graecus* (lateral view). *(A, After Millot, J., and Vachon, M. 1949. Traité de Zoologie. Vol. VI. Masson et Cie, Paris. B, After Kaestner from Millot and Vachon.)*

Eyes

Chelicera

Anterior carapace

1st abdominal tergite

Coxa of pedipalp

Posterior carapace

Coxa of first leg

B

sues. For gas exchange the animal uses a highly developed tracheal system; excretion is accomplished by a pair of coxal glands and a pair of malpighian tubules.

Males generally seize the females before copulation. In some solifuges there is then a brief period of stroking and palpation by the male, which throws the female into a passive state. The male turns the female over and opens her genital orifice with his chelicerae. He then emits a globule of semen on the ground, picks it up with his chelicerae, and deposits it into the genital orifice of the female. The entire act takes only a few minutes, and the male then leaps away. In American solifuges sperm transmission is direct, although there is precopulatory behavior and the male inserts the chelicerae into the female orifice before and after sperm transfer. The female deposits 50 to 200 eggs in burrows in the ground.

Order Opiliones

The order Opiliones, or Phalangida, contains the familiar long-legged arachnids known as daddy longlegs or harvestmen (Fig. 13–40). The more than 4500 described species live in both temperate and tropical climates, and most prefer humid habitats. They are abundant in vegetation, on the forest floor, on tree trunks and fallen logs, in humus, and in caves.

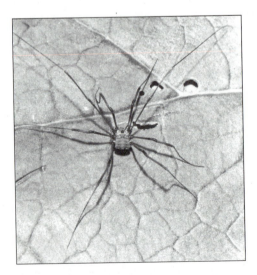

FIGURE 13–40 Dorsal view of a harvestman on a leaf. *(Photograph courtesy of Betty M. Barnes.)*

The average body length is from 5 to 10 mm, exclusive of the legs, but some of the tropical giants reach 20 mm and have a leg length of 160 mm. There also are certain tiny, short-legged, mitelike species never larger than 1 mm in length. Most species are shades of brown, but some are red, orange, yellow, green, or spotted. A number of opilionids have bizarre, spiny bodies (Fig. 13–41B).

The prosoma of opilionids is broadly joined to the short segmented abdomen with no constriction between the two divisions (Fig. 13–41A). As a result the body is rather elliptical. In the center of the carapace is a tubercle with an eye located on each side. In some groups, such as the genus *Caddo*, the tubercle extends almost the entire width of the carapace, so the eyes appear along the sides of the carapace rather than in the middle. Along the anterior, lateral margins of the carapace are the openings to a pair of **repugnatorial glands** (Fig. 13–41A), which produce secretions, often quinones and phenols, that have an acrid odor. Some harvestmen spray an intruder; certain species pick up a droplet of secretion mixed with regurgitated gut fluid and thrust it at a would-be predator.

The chelicerae are small, slender, and chelate. The pedipalps are usually short and leglike (Fig. 13–41A), but in one large suborder they are enlarged for capturing prey and possess a sickle-like claw. The legs in many opilionids are extremely long and slender and exceed the body length many times (Fig. 13–40). The tarsus is always multisegmented and flexible. When disturbed, opilionids can run very rapidly, and species living in vegetation can climb by wrapping the flexible tarsus around stems or blades of grass. The second pair of legs are usually the longest and have an important sensory function. They are moved about over the substratum, often in front of the animal. Self-amputation of a leg is an important means of defense against predators, but the legs are not regenerated.

In general, harvestmen are predatory, but scavenging is more important than in other arachnids. North American and European opilionids have been observed feeding on small invertebrates, dead animal matter, and pieces of fruits and vegetables. Predatory species feed on other small arthropods, and some species feed on snails. The prey or food is seized by the pedipalps and passed to the chelicerae, which hold and crush it. Unlike other arachnids, the ingested food is not limited to liquid material but includes small

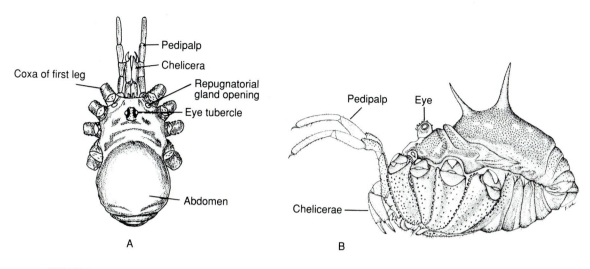

FIGURE 13–41 A, The opilionid, *Leiobunum flavum,* with only the bases of the legs shown (dorsal view). The abdomen of this species is not as conspicuously segmented as that of many other opilionids. **B,** Lateral view of an opilionid whose abdomen bears two large spines. The legs have been removed to reveal the body. *(A, After Bishop.)*

particles. Thus, a greater part of digestion must take place in the midgut.

A pair of coxal glands provides for excretion. Gas exchange takes place through tracheae, which are probably not homologous with the tracheae of other arachnids. The spiracles are located on each side of the first abdominal segment, and in many active, long-legged harvestmen, there are secondary spiracles on the tibia of the legs.

Mating is not preceded by any elaborate courtship. The male in many species faces the female and projects a tubular penis, which commonly passes between the female's chelicerae before entering the female orifice. Mites are the only other arachnids to possess a penis.

The female reproductive system includes an ovipositor, a structure that is absent in other arachnids except certain mites. In opilionids the ovipositor is a tubular, ringed organ lying in a sheath in the midventral part of the abdomen, and at the time of egg laying the ovipositor is telescoped some distance out of the genital orifice.

Shortly after mating, the female uses the long ovipositor to deposit her eggs in humus, moss, rotten wood, or snail shells. The number of eggs laid at one time ranges in the hundreds, and several batches are laid during the life of a female. In temperate regions the life span is only one year. An individual may win-

ter over in the egg or as an immature form. The name "harvestmen" refers to certain temperate species that sometimes appear in large numbers in the fall.

SYSTEMATIC RÉSUMÉ OF ORDER OPILIONES

Suborder Cyphophthalmi. Mitelike opilionids living in soil and leaf mold. Pedipalps leglike. All belong to a single family.

Suborder Laniatores. Pedipalps raptorial, with a sickle-like claw. Legs short to long. Most are inhabitants of leaf mold in the tropics and subtropics.

Suborder Palpatores. Pedipalps leglike. Legs short to long. A large and diverse group of opilionids found throughout the world. Includes many common species of temperate regions, such as members of the genera *Phalangium* and *Leiobunum.*

The Acari

The Acari are an enormously diverse arachnid assemblage containing the mites and ticks. Formerly grouped in a single order, the Acarina, they are now placed in seven orders. The Acari may be polyphyletic, perhaps having evolved from various arachnid groups, including the opilionids and ricinuleids. To simplify the coverage, we are forced to discuss the assemblage as if it were a relatively uniform group, but exceptions exist for almost every statement.

The Acari is without question the most important arachnids in terms of human economics. Numerous species are parasitic on humans, domesticated animals, and crops; others are destructive to food and other products. Mites rank as one of the most ubiquitous groups in the Animal Kingdom. Free-living, terrestrial species are extremely abundant, particularly in moss, on plants, fallen leaves, humus, soil, rotten wood, and detritus. The numbers of individuals are enormous, certainly surpassing all other arachnid orders. A small sample of leaf mold from a forest floor often contains hundreds of individuals belonging to numerous species (Fig. 13–1).

Mites even occur in fresh water and the sea. Despite their abundance, the taxonomy and biology of mites are still not so well known as the other major arachnid orders. To date, 30,000 species have been described (compared with about 32,000 described species of spiders), but some acarologists believe that this number is only a fraction of the total and that most species of mites will become extinct before they are ever known as rain forests and other habitats disappear. Because of the economic importance of mites, many zoologists direct their attention entirely to this group of arachnids; as a result, a special field has developed known as **acarology,** the study of mites.

Much of acarology belongs in the realm of parasitology, but the mites should not be considered an entirely parasitic group. Many species are free-living, and others are parasitic for only a brief period during their life cycles.

External Morphology

The majority of adult mite species are 0.25 to 0.75 mm in length, although some are larger. Their great evolutionary success is certainly correlated in part with their small size, for mites have been able to colonize many sorts of microhabitats unavailable to their large arachnid relatives. For example, there are mites small enough to live in the tracheae of insects, beneath the wings of beetles, in the quills of feathers, and in the hair follicles of mammals. Certain predatory mites live in specialized leaf pits and pouches (domatia) of a host plant from which they emerge to attack their insect prey. The ticks are the largest members of the assemblage, some species reaching 3 cm in length.

The most striking characteristic is the apparent lack of body divisions (Figs. 13–42; 13–43A, B). Abdominal segmentation has disappeared in most

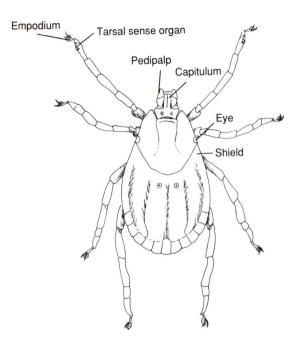

FIGURE 13–42 A tick, *Dermacentor variabilis. (After Snodgrass.)*

species, and the abdomen has fused with the prosoma. Thus, the positions of the appendages, the eyes, and the genital orifice are the only landmarks that differentiate the original body regions. Coinciding with this fusion, the entire body has become covered with a single, sclerotized shield, or carapace, in many forms (Fig. 13–44A, D).

Another general feature of the group is the change that has taken place in the head region carrying the mouth parts; this region is called the **capitulum,** or **gnathosoma** (Fig. 13–45A, B). Ventrally, the large, pedipalpal coxae extend forward to form the floor and sides of the prebuccal chamber. The roof of the chamber is formed by a labrum. These processes, which house the prebuccal chamber, together form a **buccal cone,** which fits into a sort of socket at the anterior of the mite body. The chelicerae are attached to the back wall of the socket above the buccal cone; the pedipalps are attached to both sides of the cone. The attachment of the buccal cone into the anterior socket is such that some species can extend and retract the cone.

The chelicerae and pedipalps vary in structure, depending on their function. They are usually composed of two or three articles and may be chelate

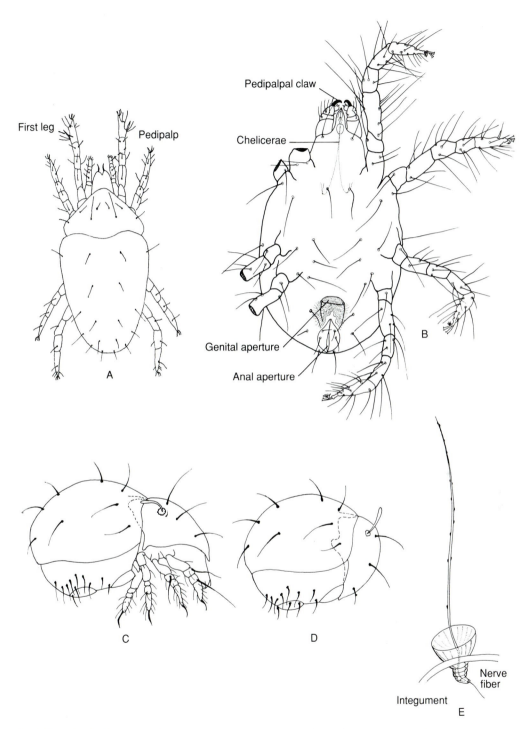

FIGURE 13–43 A, A trombidiform mite, *Tydeus starri* (dorsal view.) **B,** Ventral view of a spider mite, *Tetranychus.* **C–D,** Side view of the oribatid mite, *Mesoplophora,* **(C)** before and **(D)** after "closing up" the body. The antenna-like structure on the cephalothorax is the pseudostigmatic organ. **E,** Pseudostigmatic organ of the oribatid mite, *Belba. (A, After Baker and Wharton. B, After Krantz, G. W. 1971. Manual of Acarology. Oregon State University Press, Corvallis. C–D, After Schaller, F. 1968. Soil Animals. University of Michigan Press, Ann Arbor. E, After Beck.)*

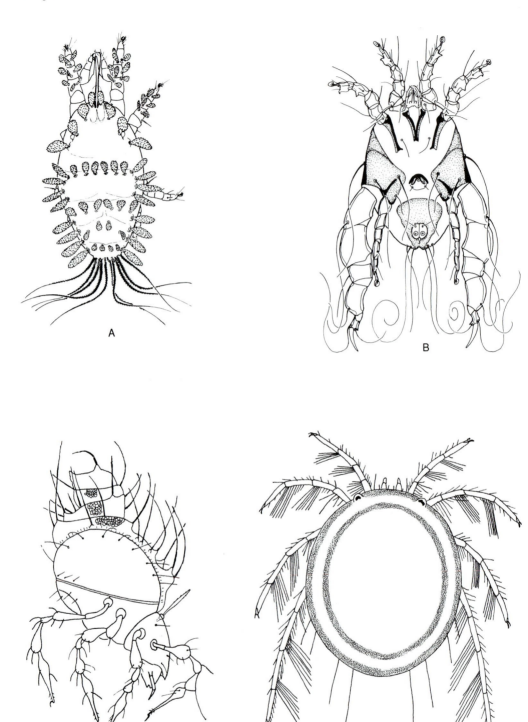

FIGURE 13–44 A, *Tuckerella,* a phytophagous mite, with symmetrically arranged, clublike setae. **B,** Ventral view of *Analges,* a feather mite. Note the greatly enlarged third pair of legs and claws. **C,** An oribatid mite, *Belba jacoti,* carrying five shed nymphal skins. **D,** The water mite, *Mideopsis orbicularis. (A–B, From Krantz, G. W. 1971. Manual of Acarology. Oregon State University Press, Corvallis. pp. 211 and 275. C, After Wilson from Baker and Wharton. D, After Soar and Williamson from Pennak.)*

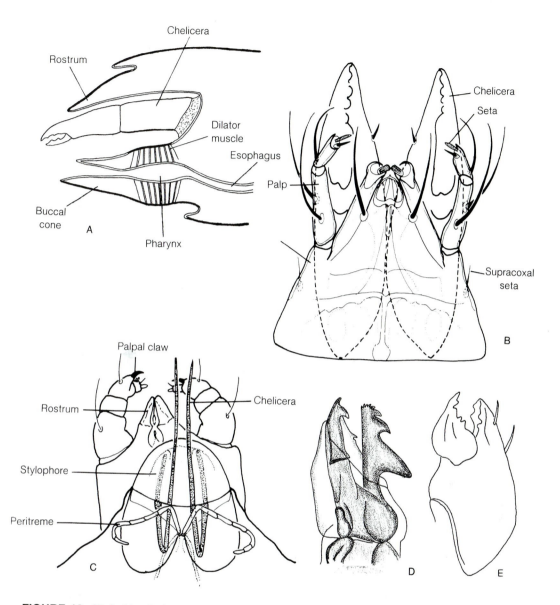

FIGURE 13–45 A, Head of a mite (sagittal section). **B,** Ventral view of the head region of *Gly-cyphagus.* **C,** Stylet-like chelicerae of the spider mite, *Tetranychus.* **D,** Hooked chelicera of the tick, *Ixodes reduvius.* **E,** Chelate chelicera of *Belba verticillipes. (A, After Snodgrass. B–C, From Krantz, G. W. 1971. Manual of Acarology. OR State University Press, Corvallis. pp. 13 and 209. D, After Neumann from Andre.)*

(Fig. 13–45B, E). The pedipalps may be relatively unmodified and leglike; they may be heavy and chelate like an additional pair of chelicerae; or in some parasitic forms, they may be vestigial. The four pairs of legs usually have six articles each, and in some groups, they have become modified for functions other than walking.

In mites and ticks the ventral side of the body is covered by plates that vary in form and number, depending on the family. The genital plate, located between the last two pairs of legs, bears the genital orifice (Figs. 13–43B; 13–44B). This location indicates a forward migration of abdominal segments, as in the harvestmen.

Hairs, or setae, cover the mite body; they vary from simple hairs to club-shaped and flattened types. Many are sensory. The nature and position of the setae in mites are extremely important characteristics in identifying and classifying species. The symmetrical arrangement of the variously shaped setae and the ornate sculpturing of their cuticle make many mites beautiful and spectacular animals (Fig. 13–44A). Most mites are varying shades of brown, but many display a wide range of hues, such as black, red, orange, green, or combinations of these colors.

Two groups of mites deserve special mention, the Oribatida (beetle mites) and several taxa of water mites. The oribatids are the largest and most studied order of free-living mites (145 families). They are very abundant in humus and moss. These little mites are usually globe-shaped, with the dorsal surface covered by a convex, highly sclerotized shield, so they resemble tiny beetles (Fig. 13–44C). Some oribatids, which are very common in leaf mold, can close up their bodies like armadillos. The cephalothorax is flexed downward and backward so that the legs and head fit into a concavity on the ventral surface of the abdomen. A plate carried on the dorsal surface of the cephalothorax covers over the withdrawn head and legs (Fig. 13–43C, D).

The water mites, containing some 2800 species, have adapted to an aquatic existence and are found in both fresh and salt water (Fig. 13–44D). The marine forms have been found from the intertidal zone to abyssal depths, and shallow-water species are frequently encountered. They do not swim but crawl about over algae, bryozoans, hydroids, and sponges. Most water mites, however, live in fresh water. Some are bright red, and many are active swimmers with long hairs on their legs.

Feeding

Mites exhibit a tremendous diversity and specialization of diets and feeding habits, although, in general, they have retained the arachnid habit of ingesting fluids, and when feeding on solid foods, there is initial external digestion and liquefaction. Carnivores that live in soil and humus feed on nematodes and small arthropods, including eggs, insect larvae, and other mites. Small crustaceans are the principal prey of water mites. The chelicerae of carnivorous mites are variously modified, depending on the prey. Some mites tear off pieces of prey; others suck out the tissues.

Many herbivorous species, such as spider mites (Tetranychidae), have chelicerae modified as needle-like stylets (Fig. 13–45C). The mites pierce plant cells and suck out the contents. A number of spider mites are serious agricultural pests of fruit trees, clover, alfalfa, cotton, and other crops. Spider mites construct protective webs from silk glands that open near the base of the chelicerae. The minute gall mites (Eriophyoidea), which also feed on plant cells, have stylet-shaped chelicerae and include some forms that are agricultural pests (Fig. 13–46D). Other herbivores include species that feed on fungi, algae, and mosses.

Many mites are carrion feeders or scavengers. Most soil-inhabiting oribatid mites feed on fungi, algae, and decomposing plant and animal material. A large number of "scavengers" have highly specialized diets. For example, different species of storage mites (Acaridae) and allied families feed on flour, dried fruit, mattress and upholstery stuffing, hay, and cheese. *Dermatophagoides* is commonly associated with house dust. The classification of scavengers should probably also include the feather mites (Fig. 13–44B) and some species that live in the fur of animals. They feed on oil, dead skin, and feather fragments and are not actually parasites.

The majority of parasitic mites are ectoparasites of animals, both vertebrates and invertebrates, but other forms of parasitism exist. Some mites have become internal parasites through an invasion of the air passageways of vertebrates and the tracheal systems of arthropods.

Many mites are parasitic only as larvae. For example, the larval stages of freshwater mites are parasitic on aquatic insects and clams. The juvenile stages of the common harvest mites (Trombiculidae) parasitize the skin of vertebrates. Larvae of species of *Trombicula* are the familiar chiggers, or red bugs. The six-legged larva emerges from an egg, which has been deposited on the soil. The larva may attack almost any group of terrestrial vertebrates, biting the host's skin and feeding on the dermal tissue, which is broken down by the external action of proteolytic enzymes (Fig. 13–46E). Feeding takes place for up to 10 days or more; then the larva drops off. After a semidormant stage, the larva undergoes a molt and becomes a free-living nymph. A later molt transforms the nymph into an adult. Both nymph and adult are predacious, feeding largely on insect eggs. Although chiggers can cause severe dermatitis, they are of much greater medical importance as vectors for

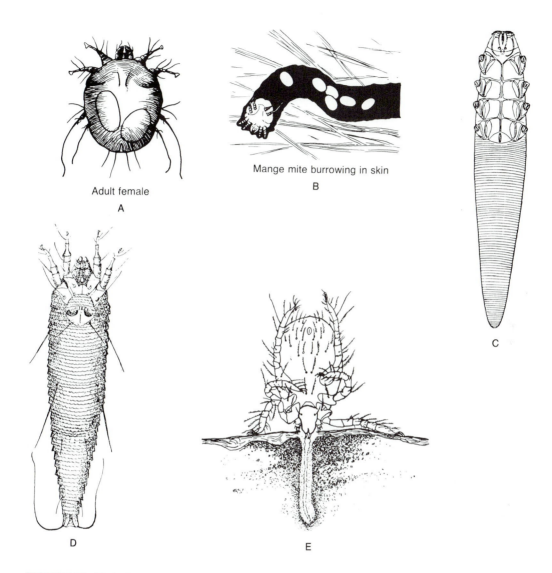

FIGURE 13–46 A–B, The mange mite, *Sarcoptes scabiei.* **A,** Adult female. **B,** Mite burrowing in the skin and depositing eggs in the tunnel. **C,** Ventral view of a sheep hair-follicle mite, *Demodex ovis* (0.23 mm). **D,** Ventral view of a gall mite, *Eriophyes,* a minute plant parasite (0.2 mm). **E,** A chigger larva of *Neotrombicula,* feeding on skin (0.25 mm). *(A–B, After Craig and Faust. C, After Hirst from Kaestner. D, After Nalepa from Kaestner. E, After Vitzthum from Kaestner. C–E, From Kaestner, A. 1968. Invertebrate Zoology. Vol. 2. Wiley-Interscience, New York.)*

pathogens, such as Asian scrub typhus. The intense itching that results from the bite of a chigger is caused by the mite's oral secretions and not, as commonly supposed, simply by the presence of the mite. Scratching quickly removes the mite, but the irritation remains for several days.

Many members of the Acari are parasitic during their entire life cycles but are attached to the host only during periods of feeding. The dermanyssid mites of birds and mammals (red chicken and other fowl mites) and the ticks illustrate this type of life cycle. Ticks penetrate the skin of the host by means of the highly specialized, hooked mouth parts and feed on blood (Fig. 13–45D). The body is not highly sclerotized and is capable of great expansion when engorged with blood. This is especially true of the female. With a few exceptions, the tick drops off the host after each feeding and undergoes a molt. Many species can live for long periods, well over a year, between feedings. Copulation occurs while the adults are feeding on the host. The female then drops to the ground and deposits an egg mass. A six-legged "seed" tick hatches from the egg.

Ticks attack all groups of terrestrial vertebrates. In humans they are responsible for the transmission of American Rocky Mountain spotted fever, tularemia, Texas cattle fever, relapsing fever, and Lyme disease. Lyme disease, which has received increasing attention in the eastern United States, is caused by a spirochete and transmitted in this part of the United States by the tick, *Ixodes dammini*. Deer are the usual host for the adult tick and deermice for the juveniles.

Finally, there are parasitic mites that spend their entire life cycles attached to the host. Included in this group are the wormlike follicle mites (Demodicidae), which live in the hair follicles of mammals (Fig. 13–46C), and the scab- and mange-producing fur mites (Psoroptidae and Sarcoptidae) of mammals. The human itch mite *(Sarcoptes scabiei)* (Fig. 13–46A, B), the cause of scabies or seven-year itch, tunnels into the epidermis. The female is less than 0.5 mm and the male less than 0.25 mm in length. Irritation is caused by the mite's secretions. The female deposits eggs in the tunnels for a period of two months, after which she dies. Up to 25 eggs are deposited every two or three days. The eggs hatch in several days, and the larvae follow the same existence as the adult. The infection can thus be endless. The mite is transmitted to another host by contact with infected areas of the skin.

Most free-living mites have a typical digestive tract (Fig. 13–47). The different feeding habits of mites make it probable that there is considerable variation in the digestive physiology of the group, however.

Internal Structure and Physiology

Excretory organs of mites consist of one to four pairs of coxal glands, or a pair of malpighian tubules, or both (Fig. 13–47). In the trombidiform mites, these typical arachnid excretory organs are lacking and have been replaced by special excretory organs modified from the hindgut.

The circulatory system is reduced and, except in a few groups, consists only of a network of sinuses. Circulation probably results from contraction of body muscles.

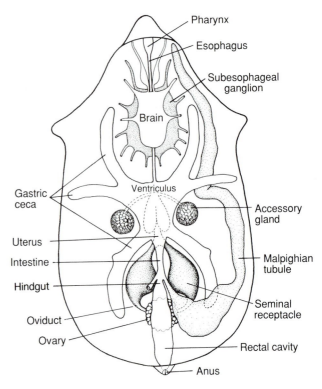

FIGURE 13–47 Internal anatomy of a mesostigmatid mite, *Caminella (After Ainscough from Krantz, G. W. 1971. Manual of Acarology. Oregon State University Press, Corvallis.)*

Although in some mites the gas exchange organs have completely disappeared, most mites have tracheae. The spiracles vary in number from one to four pairs, located on the anterior half of the body.

Sensory setae are probably the most important of the sense organs. The oribatid mites possess a peculiar form of sensory seta called a **pseudostigmatic organ,** which is probably similar to a trichobothrium. The seta itself, which has various shapes, arises out of a cupule, or pit. Two such pseudostigmatic organs are located on the cephalothorax (Fig. 13–43C–E). These setae are thought to detect air currents to which the mite responds by moving down deeper into the leaf mold. These organs thus represent an adaptation against desiccation.

Although many mites are blind, some trombidiforms and certain other groups possess simple eyes. Innervated pits and slits are common in mites and are perhaps similar to the slit sense organs of other arachnids.

Reproduction and Development

The male reproductive system consists of a pair of lobate testes, located in the midregion of the body. A vas deferens extends from each testis and may join its opposite member ventrally to open through a median gonopore or through a chitinous penis that can project through the genital orifice.

In the female there is usually a single ovary of varying size, which is connected to the genital orifice by an oviduct (Fig. 13–47). A seminal receptacle and accessory glands are also present.

As in other arachnids, sperm are transmitted indirectly in most mites. A spermatophore may be deposited on the substratum and then picked up by the female, or the spermatophore may be transferred by the male chelicerae or in some water mites by the third pair of legs. In some mites sperm are injected by the chelicerae into special openings on the female. Sperm are directly transmitted into the female by a penis in the Actinotrichida.

The number of eggs laid varies in different forms. They are deposited in soil and humus, and some mites enclose their eggs within a case. The oribatid, *Belba*, attaches its large eggs to the bodies of other individuals, which carry them about until hatching (Fig. 13–48). The oribatid mites possess an ovipositor.

After an incubation period of two to six weeks, a six-legged **"larva"** hatches from the egg. The newly hatched young lacks the fourth pair of legs and differs from the adult in certain other features. The fourth pair of legs is acquired after a molt, and the larva changes into a **protonymph.** Successive molts transform the protonymph into a **deutonymph,** a **tritonymph,** and finally an adult. During these stages adult structures are gradually attained.

The life span of mites varies greatly depending on the species, but it is generally shorter than that of other arachnids. For example, *Anblyseius brazilli*, a tropical, predatory, mesostigmatid mite, reaches adulthood in 7 days, and the female has a life span of 30 days.

SYSTEMATIC RÉSUMÉ OF THE ACARINE ORDERS

Some specialists consider the Acari to constitute a subclass. Its members are distinguished from other arachnids in having a gnathosoma and a six–legged developmental stage.

Order Opilioacarida (Notostigmata). Large, leathery, primitive mites having a segmented abdomen. They are purple to blue-green in color and resemble harvestmen. Omnivorous or predatory, they live in leaf litter and beneath stones. Subtropical and tropical.

Order Holothyrida (Tetrastigmata). Large, predatory mites of Australia, New Zealand, and the American tropics. The body is covered by a convex shield, and there are two pairs of spiracles: one pair by the third coxae and one pair behind the fourth coxae.

Order Gamasida (Mesostigmata). A large group of mites with a pair of tracheal spiracles opposite coxae II, IV, or between III and IV. Chelicerae covered by a dorsal extension of the gnathosome. Remainder of dorsal body surface may be covered with plates, which vary in number and position. Free-living and parasitic species. The Dermanyssoidea include economically important parasites of birds (red mites) and mammals.

Order Ixodida (Metastigmata). Species of large, parasitic acari known as ticks. Mouth with recurved teeth modified for piercing. A tracheal spiracle located in front or behind third or fourth pair of coxae. Hard ticks or wood ticks (Ixodidae)—*Ixodes, Dermacentor;* soft ticks (Argasidae)—*Argas.*

Order Prostigmata. (Trombidiformes). Mites with one or two pairs of spiracles located anteriorly near the mouth parts. This large, diverse order of widely distributed mites contains many plant parasites, including the so-called spider mites (Tetranychidae) and the minute gall mites (Eriophyoidea). Many are parasites on invertebrates and vertebrates. Among the latter are the harvest mites (Trombidioidea). Many of the free-

FIGURE 13–48 An oribatid mite, *Belba*, carrying eggs attached by another individual. *(From Schaller, F. 1968. Soil Animals. University of Michigan Press, Ann Arbor.)*

living species, which include the Bdellidae, or snout mites, are predacious. Also included in the suborder are the follicle mites (Demodicidae), marine water mites (Halacaridae), and freshwater mites (Hydrachnoidea, Lebertioidea, and Hygrobatoidea).

Order Astigmata (Sarcoptiformes). Small mites, most of which are parasitic at some stage in their life cycle. Chelicerae not covered by a tectum. Tracheal system absent; gas exchange occurs across the weakly sclerotized body wall. Included in this order are the storage mites (Acaridae), the scabies, mange, or itch mites (Sarcoptidae and Psoroptidae), and the feather mites.

Order Oribatida (Cryptostigmata). Oribatid or beetle mites. A large group of small, heavily sclerotized mites that are very common in leaf mold and soil. There is usually a tracheal system associated with the bases of the first and third legs, but typical spiracles are absent. A pair of specialized trichobothria (pseudostigmata) on the anterior corners of the dorsum is unique to the oribatids.

SUMMARY

1. Members of the class Arachnida are terrestrial chelicerates that lack book gills. Those species that are aquatic (some mites) represent a secondary return to fresh water or the sea.

2. Scorpions, the most primitive arachnids, have long, segmented abdomens. The highly specialized mites have lost all external evidence of metamerism, and the cephalothorax and abdomen are broadly joined together.

3. Arachnids are largely predatory chelicerates, other arthropods forming the principal prey. The pedipalps of scorpions and pseudoscorpions are used in seizing and holding prey. Poison is used to immobilize prey in several groups: spiders (with chelicerae), scorpions (with terminal abdominal barb), and pseudoscorpions (with fingers of the pedipalps). In addition, some spiders use silk in prey capture. There is some digestion outside of the body, and fluids and partially digested tissues are sucked into the gut.

4. Harvestmen and many mites are the principal exception to the arachnid predatory habit. The great diversity of feeding habits among mites is in part related to the miniaturization that characterizes the group.

5. Trichobothria and slit sense organs are important sense organs in prey capture. Although most arachnids possess eyes, a relatively small number are capable of object discrimination.

6. Large arachnids (scorpions, some spiders) possess book lungs as gas exchange organs; small forms (pseudoscorpions, some spiders, mites) possess tracheae. The heart is most highly developed in large species with book lungs, and the blood contains hemocyanin.

7. Excretory organs are coxal glands and malpighian tubules. A waxy epicuticle is an important adaptation for water conservation and has certainly contributed to the success of arachnids as terrestrial arthropods.

8. Many are secretive or nocturnal in habit. Leaf mold is the habitat for many small species, especially pseudoscorpions, mites, and spiders.

9. The primitive mode of sperm transfer is indirect by spermatophores (scorpions, pseudoscorpions, and some mites). The unique indirect sperm transfer of spiders is probably derived from handling of spermatophores by the male with the pedipalps. Sperm transfer is direct in harvestmen and many mites.

CLASS PYCNOGONIDA

The Pycnogonida, or Pantopoda, is a group of some 1000 described species of marine animals known as sea spiders. The name "sea spider" is derived from the somewhat spider-like appearance of these animals, which crawl about slowly on long legs. Although largely unknown to the layman, sea spiders are actually common animals. Careful examination of bryozoans and hydroids scraped from a wharf piling or rocks usually yields a few specimens. They live in all oceans from the Arctic and Antarctic to the tropics, and there are many littoral forms, as well as species that live at great depths.

Pycnogonids are mostly small animals with a body length ranging from 1 to 10 mm, but polar and deepwater species reach considerable size. Species of *Colossendeis* can have a leg span of 40 cm. Although most pycnogonids are drab in color, there are some that are green or purple, and some deep-sea forms are red. The body is commonly narrow and is composed of a number of distinct segments (Fig. 13–49), all of which are part of the prosoma. The opisthosoma is very much reduced. The **head,** or **cephalon,** bears on

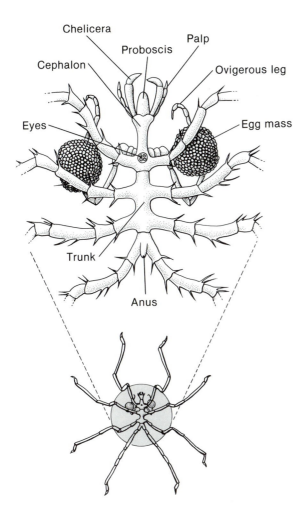

FIGURE 13–49 *Nymphon rubrum,* a sea spider. *(After Sars from Fage.)*

its dorsal surface four eyes mounted on a central tubercle and, at its anterior end, a cylindrical **proboscis.** Posterior to the head is a trunk of usually four cylindrical segments, the first of which is fused with the cephalon. The most striking feature of the trunk is the pair of large processes that projects laterally from each segment. A leg articulates with the end of each process, which often exceeds the length of the segment itself. At the rear of the trunk is a short, dorsal, conical opisthosoma, or abdomen.

The appendages consist of a pair of chelicerae (also called **chelifores**), a pair of palps, a pair of **ovigerous legs,** and usually four pairs of walking legs. The short chelicerae are attached to either side of the proboscis base. In some genera, such as *Pycno-*

gonum, the chelicerae are absent. The reduction in the chelicerae is correlated with a correspondingly greater development of the proboscis. The palps are leglike and are used not only as sensory appendages but in feeding and cleaning. The ovigerous legs, or ovigers, which are peculiar to pycnogonids, may be used for grooming and, in the male, to carry the eggs. In many species the ovigers are less well developed or are even absent in the female.

The walking legs, which in some species are very long, are attached to the lateral extensions of the trunk segments. Some nine species of pycnogonids have five or six pairs of legs instead of the usual four. This polymerous condition is not understood, for there are a number of genera that contain both eight-legged and ten-legged species.

Most pycnogonids are bottom dwellers and crawl about over hydroids and bryozoans as if on stilts (Figs. 13–50; 19–10); some are able to swim by flapping their legs alternately vertically. In fact, there are a few bathypelagic species. There are also a few pycnogonids that are interstitital and a number that are commensal or ectoparasitic on other invertebrates.

Most pycnogonids are carnivorous and feed on hydroids, soft corals, anemones, bryozoans, small polychaetes, and sponges. Molluscs are the principal hosts for parasitic forms. Pycnogonids apply the proboscis directly to the prey and suck up the tissues. The mouth at the end of the proboscis contains three liplike teeth that not only regulate the size of the opening but also act as a rasp. Not all pycnogonids are carnivores. Some feed on algae or microorganisms growing on hydroids and bryozoans or even accumulated detritus. From the mouth a triradiate pharynx extends through the proboscis (Fig. 13–51A). The pharynx not only acts as a pump but also masticates the food with bristles that project into the lumen. The pharynx leads into a short esophagus, which opens through a valve into the intestine.

The intestine constitutes the midgut of the digestive tract and is very extensive. A long, lateral cecum extends into each appendage and in the legs extends almost their entire length. Indigestible waste materials pass into a short rectum and then out through the anus at the tip of the abdomen.

The circulatory system is composed of a heart, or dorsal vessel, and a hemocoel. There are no special organs for gas exchange and excretion. The nervous system is like that of other chelicerates with a ventral nerve cord.

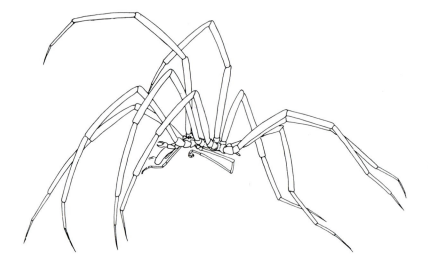

FIGURE 13–50 Side view of a ten-legged, Antarctic pycnogonid, *Decolopoda australis. (From Schram, F. R., and Hedgpeth, J. W. 1978. Locomotory mechanisms in Antarctic pycnogonids. Zool. J. Linn. Soc. 63:145–169.)*

Pycnogonids are dioecious. Females can be distinguished from males by the poorly developed condition or absence of ovigerous legs. The gonad, either testis or ovary, is single and located in the trunk above the intestine. Branches of the gonad extend far into the legs. In both males and females the reproductive openings are multiple, an unusual condition for arthropods, and are located on the ventral sides of the coxae of the second and fourth pairs of legs in males and of all legs in females. On reaching maturity, the eggs occupy lobes of the ovary in the femurs of the legs (Fig. 13–51B), and in gravid females, the femurs are conspicuously enlarged.

In those pycnogonids in which egg laying has been observed, the male hangs beneath the female so that their ventral surfaces are opposed or the male stands over the female. The eggs are fertilized as they are emitted by the female, and the male then gathers them onto his ovigerous legs, either directly or from the ovigerous legs of the female (Fig. 13–49). Glands on the femurs of the male provide cement for forming as many as 1000 eggs into an adhesive, spherical mass. The egg masses are held around the middle joints of the male's ovigerous legs and brooded until the eggs hatch.

In most pycnogonids, a larva called a **protonymphon** hatches from the egg (Fig. 13–51C). The larva has only three pairs of appendages, representing the chelicerae, palps, and ovigerous legs, and each appendage has only three segments. A short proboscis is present, but the trunk segments are still lack-

ing. Depending on the species, the larva remains on the ovigerous legs of the male, leaves to take up an independent existence, or in many shallow-water species, develops on or within hydroids and corals. In any case, the larva eventually transforms into a young pycnogonid through a sequence of molts and the addition of new appendages. The eastern Pacific *Propallene longiceps* develops from the egg to the adult in about five months.

The structure of the brain, the nature of the sense organs, the circulatory system, and the presence of chelicerae have been considered by many zoologists as justification for placing the pycnogonids among the chelicerates. Their exact relationship to arachnids and merostomes is, however, by no means clear, because pycnogonids are aberrant in many respects. The presence of multiple gonopores, ovigerous legs, the segmented trunk, and the additional pairs of walking legs in many species have no counterparts in the other chelicerate classes. S. M. Manton believes that pycnogonids are derived from an early line of marine arachnids that never became terrestrial; R.F. Schram and J.W. Hedgpeth think such a connection with arachnids is remote.

SUMMARY

1. Members of the class Pycnogonida (sea spiders) are aberrant marine arthropods that appear to be chelicerates, although their relationship to other chelicerate groups is uncertain.

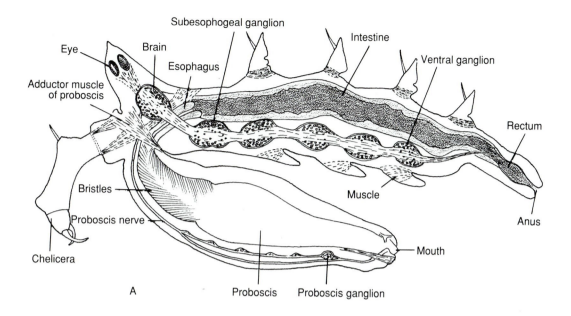

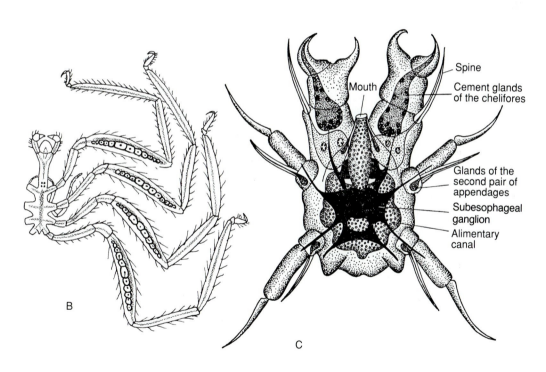

FIGURE 13–51 A, The pycnogonid, *Ascorhynchus castelli* (sagittal section). **B,** Female of *Pallene brevirostris* with eggs in the femurs. **C,** Protonymphon larva of an *Achelia echinata*. *(A, After Dohrn from Fage. B, After Sars from Fage. C, From King, P. E. 1973. Pycnogonids. St. Martin's Press, New York. 137 pp.)*

2. The often narrow body bears an anterior proboscis and usually four very long legs.

3. Pycnogonids are commonly found on sessile animals, especially hydroids and bryozoans. Some species are carnivores on sessile animals; others feed on the detritus or flora growing on such animals.

4. The absence of gas exchange and excretory organs may be correlated with their large surface and aquatic existence.

5. The eggs are carried by the male with a pair of ovigerous legs located in front of the first walking legs.

REFERENCES

The literature included here is restricted in large part to chelicerates. The introductory references on page 6 include many general works and field guides that contain sections on these animals.

General

Anderson, D. T. 1973. Embryology and Phylogeny in Annelids and Arthropods. Pergamon Press, Oxford. 495 pp. (Includes a detailed survey of chelicerate development.)

Barth, F. G. (Ed.): 1985. Neurobiology of Arachnids. Springer-Verlag, New York. 385 pp.

Buskirk, R. E. 1981. Sociality in arachnids. *In* Hermann, H. R. (Ed.): Social Insects. Academic Press, New York.

Kaestner, A. 1968. Invertebrate Zoology, Vol. 2. Wiley-Interscience, New York. 472 pp. (This volume provides a very good general account of the arachnids.)

Kovoor, J. 1977. Silk and the silk glands of Arachnida. Annee Biol. *16*(3/4):97–172.

Millot, J., Dawydoff, C., Vachon, M. et al. 1949. Classe des Arachnides. *In* Grassé, P. (Ed.): Traité de Zoologie. Vol. 6. Masson et Cie, Paris. pp. 263–905.

Schaller, F. 1968. Soil Animals. University of Michigan Press, Ann Arbor. 144 pp. (An excellent little volume on the biology of animals living in soil and in humus.)

Spiders

Anderson, J. F. 1974. Responses to starvation in the spiders *Lycosa lenta* and *Filistata hibernalis.* Ecology. *55*(3):576–585.

Anderson, J. F., and Prestwich, K. N. 1982. Respiratory gas exchange in spiders. Physiol. Zool. *55*(1):72–90.

Barth, F. G. 1978. Slit sense organs: Strain gauges in the arachnid exoskeleton. *In* Merrett, P. (Ed.): Arachnology. Symp. Zool. Soc. London. *42*:439–448. Academic Press, London.

Barth, F. G., and Pickelmann, P. 1975. Lyriform slit sense organs. J. Comp. Physiol. *103*:39–54.

Brach, V. 1977. *Anelosimus studiosus* and the evolution of quasisociality in theridiid spiders. Evolution. *31*:154–161.

Buchli, H. H. R. 1969. Hunting behavior in the Ctenizidae. Am. Zool. *9*:175–193. (An excellent account of the predatory habits of the ctenizid trap-door spiders.)

Burgess, J. W. 1976. Social spiders. Sci. Amer. *234*(3):101–106.

Cooke, J. A. L., Roth, V., and Miller, F. H. 1972. The urticating hairs of theraphosid spiders. Am. Mus. Novit. *2498*:1–43.

Dumpert, K. 1978. Spider odor receptor: Electrophysiological proof. Experimentia. *34*:754–756.

Edgar, W. D. 1970. Prey and feeding behavior of adult females of the wolf spider *Pardosa amentata.* Neth. J. Zool. *20*(4):487–491.

Eisner, T., and Nowicki, S. 1983. Spider web protection through visual advertisement: Role of the stabilimentum. Science. *219*:185–187.

Foelix, R. F. 1982. Biology of Spiders. Harvard University Press, Cambridge, MA. 306 pp. (An excellent general account of spiders, with emphasis on adaptive morphology and physiology.)

Foelix, R. F., and Chu-Wang, I-W. 1973. The morphology of spider sensilla. II. Chemoreceptors. Tissue and Cell. *5*(3):461–478.

Forster, L. 1982. Vision and prey-catching strategies in jumping spiders. Am. Sci. *70*:165–174.

Gertsch, W. J. 1979. American Spiders. 2nd Edition. Reinhold, Van Nostrand, New York. 274 pp. (The natural history of spiders.)

Harwood, R. H. 1974. Predatory behavior of *Argiope aurantia.* Am. Midl. Nat. *91*(1):130–138.

Horton, C. C. 1981. A defensive function of the stabilimenta of two orb-weaving spiders. Psyche. *87*:13–20.

Jackson, R. R. 1982. The behavior of communicating in jumping spiders. *In* Witt, P. N., and Rovner, J. S. Spider Communication: Mechanisms and Ecological Significance. Princeton University Press, Princeton. pp. 213–247.

Kaston, B. J. 1964. The evolution of spider webs. Am. Zool. *4*:191–207.

Kaston, B. J. 1970. Comparative biology of American black widow spiders. Trans. San Diego Soc. Nat. Hist. *16*(3):33–82.

Kaston, B. J. 1978. How to Know the Spiders. 3rd Edition. W. C. Brown, Dubuque, IA.

Kaston, B. J. 1981. Spiders of Connecticut. Rev. Edition. State Biol. Nat. Hist. Surv. Bull. *70:*1020. (This work is valuable for any American student interested in the taxonomy of spiders. Keys for most families and genera are provided; almost every species is illustrated. The major part of the Connecticut spider fauna is found throughout the eastern United States.)

Klaerner, D., and Barth, F. G. 1982. Vibratory signals and prey capture in orb-weaving spiders *(Zygiella X nota, Nephila clavipes).* J. Comp. Physiol. Sens. Neural Behav. Physiol. *148*(4):445–456.

Kullmann, E. J. 1972. Evolution of social behavior in spiders. Am. Zool. *12*(3):419–426.

Levi, H. W. 1967. Adaptations of respiratory systems of spiders. Evolution. *21*(3):571–583.

Levi, H. W. 1978. Orb-weaving spiders and their webs. Am. Sci. *66:*734–742.

Levi, H. W., and Levi, L. R. 1968. A Guide to Spiders and Their Kin. A Golden Nature Guide, Golden Press, New York. (An excellent small, semipopular guide for the identification of common spiders. Many colored figures and much information.)

Lubin, Y. D., and Robinson, M. H. 1982. Dispersal by swarming in a social spider. Science. *216:*319–321.

Marples, B. J., and Marples, M. J. 1972. Observations on *Cantuaria toddi* and other trapdoor spiders in central Otago, New Zealand. J. Roy. Soc. N. Z. *2*(2):179–185.

McCreve, J. D. 1969. Spider venoms: Biochemical aspects. Am. Zool. *9:*153–156.

Morse, D. H. 1979. Prey capture by the crab spider *Misumena calycina.* Oecologia. *39:*309–319.

Morse, D. H. 1983. Foraging patterns and time budgets of the crab spiders *Xysticus emertoni* and *Misumena vatia* on flowers. J. Arachnol. *11:*87–94.

Pasquet, A. 1984. Predatory site selection and adaptation of the trap in four species of orb-weaving spiders. Biol. Behav. *9*(1):3–20.

Platnick, N. I. 1971. The evolution of courtship behavior in spiders. Bull. Brit. Arach. Soc. *2*(3):40–47.

Platnick, N. I., and Gertsch, W. J. 1976. The suborders of spiders: A cladistic analysis. Am. Mus. Novit. *2607:*1–15.

Robinson, M. H. 1969. Predatory behavior of *Argiope argentata.* Am. Zool. *9:*161–173.

Robinson, M. H., and Robinson, B. 1980. Comparative Studies of the Courtship and Mating Behavior of Tropical Araneid Spiders. Pacific Insects Monographs. No. 36. Bishop Museum Press, Honolulu. 218 pp.

Roth, V. D. 1982. Handbook for Spider Identification. Published by the author, SWRS, Portal, AZ 85632. (Keys to families and genera of North American spiders.)

Rovner, J. S. 1971. Mechanisms controlling copulatory behavior in wolf spiders. Psyche. *78*(1):150–165.

Rovner, J. S. 1975. Sound production by nearctic wolf spiders: A substratum-coupled stridulatory mechanism. Science. *190:*1309–1310.

Rovner, J. S. 1978. Adhesive hairs in spiders: Behavioral functions and hydraulically mediated movement. *In* Merrett, P. (Ed.): Arachnol. Symp. Zool. Soc. Lond. *42:*99–108.

Rovner, J. S., Higashi, G. A., and Foelix, R. F. 1973. Maternal behavior in wolf spiders: The role of abdominal hairs. Science. *182:*1153–1155.

Schaefer, M. 1977. Winter ecology of spiders. Z. Angew. Entomol. *83*(2):113–134.

Schoener, T. W., and Toft, C. A. 1983. Spider populations: Extraordinarily high densities on islands without top predators. Science. *219:*1353–1355.

Seyferth, E. A., and Barth, F. G. 1972. Compound slit sense organs on the spider leg: Mechanoreceptors involved in kinesthetic orientation. J. Comp. Physiol. *78:*176–191.

Shear, W. A. (Ed.): 1986. Spiders: Webs, Behavior and Evolution. Stanford University Press, Stanford. 492 pp.

Shear, W. A., Palmer, J. M., Coddington, J. A. et al. 1989. A Devonian spinneret: Early evidence of spiders and silk use. Science. *246:*479–481.

Shultz, J. W. 1987. The origin of the spinning apparatus in spiders. Biol. Rev. *62:*89–113.

Stratton, G. E., and Uetz, G. W. 1981. Acoustic communication and reproductive isolation in two species of wolf spiders. Science. *214:*575–577.

Suter, R. B., and Renkes, G. 1982. Linyphiid spider courtship: Releaser and attractant functions of a contact sex pheromone. Anim. Behav. *30:*714–718.

Tietjen, W. J., and Rovner, J. S. 1980. Trail-following behavior in two species of wolf spiders: Sensory and etho-ecological concomitants. Anim. Behav. *28:*735–741.

Turner, M. 1979. Diet and feeding phenology of the green lynx spider, *Peucetia viridans.* J. Arachnol. *7:*149–154.

Vollrath, F. 1988. Untangling the spider's web. Trends Ecol. Evol. *3:*331–335.

Wilson, R. S. 1969. Control of drag-line spinning in certain spiders. Am. Zool. *9*(1):103–111.

Witt, P. N. 1975. The web as a means of communication. Biosci. Commun. *1:*7–23.

Witt, P. N., Reed, C. F., and Peakall, D. B. 1968. A Spider's Web. Springer-Verlag, New York. 107 pp. (An analysis of the regulatory mechanisms in web building.)

Witt, P. N., and Rovner, J. S. (Eds.): 1982. Spider Communication: Mechanisms and Ecological Significance. Princeton University Press, Princeton. (Papers from a symposium.)

Work, R. W. 1977. Mechanisms of major ampullate silk fiber formation by orb-web-spinning spiders. Trans. Am. Micros. Soc. *96:*170–189.

Other Chelicerates

Arnaud, F., and Bamber, R. N. 1987. The biology of Pycnogonida. Adv. Mar. Biol. *24:*1–96.

Balogh, J. 1972. The Oribatid Genera of the World. Akademiai Kiado, Budapest, Hungary. (Keys and descriptions of the 700 described genera of the soil-inhabiting beetle mites.)

Barlow, R. B., Powers, M. K., Howard, H. et al. 1986. Migration of *Limulus* for mating: Relation to lunar phase, tide height, and sunlight. Biol. Bull. *171:*310–329.

Barr, D. 1973. Methods for the collection, presentation and study of water mites. Royal Ontario Museum of Life Science. Misc. Publ., Ontario. 28 pp.

Binns, E. S. 1983. Phoresy as migration—some functional aspects of phoresy in mites. Biol. Rev. *57:*571–620.

Bishop, S. C. 1949. The Phalangida of New York. Proc. Rochester Acad. Sci. *9*(3):159–235. (An old but still useful taxonomic study of New York harvestmen. Keys, figures, and bibliography of taxonomic papers included.)

Bonaventura, J., Bonaventura, C., and Tesh, S. (Eds.): 1982. Physiology and Biology of Horseshoe Crabs: Studies on Normal and Environmentally Stressed Animals. A. R. Liss, New York. 334 pp.

Botton, M. L., and Ropes, J. W. 1989. Feeding ecology of horseshoe crabs on the continental shelf, New Jersey to North Carolina. Bull. Mar. Sci. *45*(3):637–647.

Brown, F. A. 1950. Selected Invertebrate Types. (Horseshoe crab) John Wiley and Sons, New York. pp. 360–381.

Brownell, P. H. 1977. Compressional and surface waves in sand: Used by desert scorpions to locate prey. Science. *197:*479–482.

Brownell, P. H. 1984. Prey detection by the sand scorpion. Sci. Am. *251*(6):86–97.

Brownell, P., and Farley, R. D. 1979. Detection of vibrations in sand by tarsal sense organs of the nocturnal scorpion, *Paruroctonus mesaenis.* J. Comp. Physiol. *131:*23–30.

Bub, K., and Bowerman, R. F. 1979. Prey capture by the scorpion *Hadrurus arizonensis.* J. Arachnol. *7:*243–253.

Cohen, J. A., and Brockmann, H. J. 1983. Breeding activity and mate selection in the horseshoe crab, *Limulud polyphemus.* Bull. Mar. Sci. *33*(2):274–281.

Edgar, A. L. 1971. Studies on the biology and ecology of Michigan Phalangida (Opiliones). Misc. Publ. Mus. Zool. Univ. Mich. *144:*1–64.

El-Banhawy, E. M. 1975. Biology and feeding behavior of the predatory mite, *Amblyseius brazilli.* Entomophaga. *20*(4):353–360.

Evans, G. O., Sheals, J. G., and MacFarlane, D. 1961. The Terrestrial Acari of the British Isles. An Introduction to Their Morphology, Biology and Classification. Vol. I. Introduction and Biology. British Museum, London.

Goddard, S. J. 1976. Population dynamics, distribution patterns and life cycles of *Neobisium muscorum* and *Chthonius orthodactylus.* J. Zool. *178:*295–304.

Hoff, C. C. 1949. The pseudoscorpions of Illinois. Bull. Illinois Nat. Surv. Vol. 24. Art. 4. (A good starting point for students interested in the taxonomy of pseudoscorpions. Keys, figures, and bibliography of taxonomic papers included.)

Jeram, A. J., Selden, P. A., and Edwards, D. 1990. Land animals in the Silurian: Arachnids and myriapods from Shropshire, England. Science. *250:*658–661.

Johnson, D. L., and Wellington, W. G. 1980. Predation of *Apochthonius minimus* on *Folsomia candida* (Collebola). I. Predation rate and size-selection. Res. Pop. Ecol. *22*(2):339–352.

King, P. E. 1973. Pycnogonids. St. Martin's Press, New York. 144 pp. (A detailed general account of the pycnogonids.)

King, P. E. 1974. British Sea Spiders. Synopses of the British Fauna (New Series) No. 5. The Linnean Society of London. Academic Press, London. 68 pp. (Guide to the British species of pycnogonids, with a brief introductory general account of the group.)

Krantz, G. W. 1970. A manual of Acarology. Oregon State University Bookstores, Inc., Convallis. 335 pp. (A very useful introduction to the techniques, the use of keys, and the literature necessary for identifying ticks and mites.)

Legg, G. 1977. Sperm transfer and mating in *Ricinoides hanseni.* J. Zool. *182*(1):51–61.

Manton, S. M. 1978. Habits, functional morphology and the evolution of pycnogonids. Zool. Linn. Soc. *63:*1–21.

McCloskey, L. R. 1973. Pycnogonida. Marine flora and fauna of the northeastern U.S. NOAA Technical Reports NMFS Circular 386. U.S. Government Printing Office.

McDaniel, B. 1979. How to Know the Mites and Ticks. W. C. Brown Co., Dubuque, IA. 335 pp.

Merrett, P. (Ed.): 1978. Arachnology. Symposia of the Zoological Soc. London. No. 42. Academic Press, London. 530 pp. (Papers from the Seventh International Congress on Arachnology.)

Muma, M. H. 1970. A synoptic review of North American, Central American, and Western Indian Solipugida. Arthropods of Florida. Vol. 5. Contribution No. 154. Bureau of Entomology, Florida Department of Agriculture and Consumer Services. 62 pp.

O'Dowd, D. J., and Willson, M. F. 1991. Associations between mites and leaf domatia. Trends Ecol. Evol. *6*(6):179–182.

Polis, G. A. (Ed.): 1990. The Biology of Scorpions. Stanford University Press, Stanford. 587 pp.

Rudloe, A. 1981. Aspects of the biology of juvenile horseshoe crabs, *Limulus polyphemus*. Bull. Mar. Sci. *31*(1):125–133.

Sankey, J. H. P., and Savory, T. H. 1974. British Harvestmen. Synopses of the British Fauna (New Series) No. 4. The Linnean Society of London. Academic Press, London. 76 pp. (Guide to the British species of Opiliones, with an introductory general account of the group.)

Sauer, J. R., and Hair, J. A. (Eds.): 1986. Morphology, Physiology, and Behavioral Biology of Ticks. Halsted Press, New York. 510 pp. (A collection of papers on tick biology.)

Schram, R. F., and Hedgpeth, J. W. 1978. Locomotor mechanisms in Antarctic pycnogonids. J. Linn. Soc. Zool. *63:*145–169.

Sharov, A. G. 1966. Basic Arthropodan Stock. Pergamon Press, New York.

Shear, W. A., and Kukalova-Peck, J. 1990. The ecology of terrestrial arthropods: The fossil evidence. Can. J. Zool. *68:*1807–1834.

Stormer, L. 1977. Arthropod invasions of land during late Silurian and Devonian times. Science. *197:*1362–1364.

Sturm, H. 1973. On the ethology of *Trithyreus sturmi* (Pedipalpi Schizopeltidia). Z. Tierpsychol. *33*(2):113–140.

Van der Hammen, L. 1977. The evolution of the coxa in mites and other groups of Chelicerata. Acarologia. *19*(1):12–19.

Weygoldt, P. 1969. The Biology of Pseudoscorpions. Harvard University Press, Cambridge, MA. 145 pp.

Weygoldt, P. 1972a. Geisselskorpione und Geisselspinnen (Uropygi und Amblypygi). Z. des Kölner Zoo. *15*(3):95–107. (A good account of the biology of these two arachnid orders.)

Weygoldt, P. 1972b. Spermatophorenbau und Samenübertragung bei Uropygen *(Mastigoproctus brasilianus)* und Amblypygen (*Charinus brasilianus* und *Admetus pumilio*). Z. Morph. Tiere. *71:*23–51.

Weygoldt, P. 1974. Indirect sperm transfer in arachnids. Verh. Dtsch. Zool. Ges. *67:*308–313.

Weygoldt, P. 1977. Agonistic and mating behaviour, spermatophore morphology and female genitalia in neotropical whip spiders (Amblypygi). Zoomorphologie. *86:*271–286.

Weygoldt, P. 1978. Mating behavior and spermatophore morphology in whip scorpions: *Thelyphonellus amazonicus* and *Typopeltis crucifer*. Zoomorphologie. *89:*145–156.

Weygoldt, P., and Paulus, H. F. 1979. Untersuchungen zür Morphologie, Taxonomie und Phylogenie der Chelicerata. Z. Zool. Syst. Evolut.-forsch. *17*(2):85–116; *17*(3):177–200.

Woolley, T. A. 1988. Acarology: Mites and Human Welfare. Wiley-Interscience, New York. 484 pp.

14

CRUSTACEANS

PRINCIPLES AND EMERGING PATTERNS
Crustacean Filter Feeding
Bioluminescence

CRUSTACEANS
External Anatomy
Locomotion and Nutrition
Internal Transport and Gas Exchange
Excretion and Osmoregulation
Nervous System
Sense Organs
Reproduction and Development
Overview of Crustacean Classification
Aspects of Crustacean Physiology
Crustacean Phylogeny

PRINCIPLES AND EMERGING PATTERNS

CRUSTACEAN FILTER FEEDING

Because crustaceans are primarily aquatic, they include most of the filter-feeding arthropods. The filter is always composed of closely placed exoskeletal setae, commonly arranged like the teeth of a comb. The setal comb is located on one or several pairs of appendages, but the location varies from one group of crustaceans to another, indicating that filter feeding has evolved numerous times within the subphylum. The filtering current is produced by the beating of certain appendages.

It has long been considered that sieving is the process by which crustaceans filter particles from a water stream. The mesh size of the filter, that is, the distance between setae or setules (side branches) (Fig. 14–1), determines the size of the diatoms or other particles collected. Studies by J. Gerritsen and K. G. Porter indicate that this is not always the case. If *Daphnia magna* is fed a mixture of three sizes of polystyrene spheres (Fig. 14–1B), two larger and one smaller (0.5 μm) than the mesh size of the filter (1 μm between setules), the water flea captures a sizable part of the smallest spheres, which should have passed through the filter. The collecting efficiency is about 60% for the smallest spheres, compared with 100% for the two larger sizes. Clearly, the collecting process is not one of simple sieving.

When a stream of water flows over or around an object, a boundary layer develops over the surface because of the adhesive nature of water molecules (viscosity). Water molecules at the surface of the object do not move at all, and those near the surface move more slowly than those farther away in the mainstream current. This boundary layer of slowly moving water can be significant when the object over which water is flowing is very small. In *Daphnia,* for example, the boundary layer that develops over one setule of the filter extends beyond the next setule. This means that little water flows between the setules unless subjected to considerable pressure. The smaller and slower a setulated appendage is, the less leaky it is. Thus, in *Daphnia,* particles are captured not by sieving but by attraction of opposite charges between the particles and the filter surface. The proportion of polystyrene spheres collected can be altered in *Daphnia* by changing their surface charges.

In copepods, the second maxillae, which are the filtering appendages, have been thought to sieve out particles passively from a forward-moving current created by other appendages. By filming dye-marked water currents around tethered copepods, investigators found that these crustaceans move their second maxillae (feeding appendages) to capture individual particles actively. When the copepod is not capturing food, the water current generated by other appendages passes over, not through the sieve of the stationary, second maxillae (Fig. 14–1C). When an algal particle in the approaching water stream is detected, however (probably by sensory setae on the antennae and other appendages), the other appendages beat asymmetrically so that the copepod orients toward the particle, and the second maxillae are spread apart. This movement causes the water stream with the particle to move to the midline (Fig. 14–1D), and the second maxillae clap over it, squeezing out much of the surrounding water (Fig. 14–1E). Variations in this behavior can be seen when the animals feed on very small or large particles. The particle is then combed off and transferred to the mandibles and mouth by the endites of the first maxillae. The property of viscosity operates here, as in *Daphnia.* The second maxillae initially act as paddles, and the captured particle is surrounded by a boundary layer of water as it is transferred. As a consequence, the particle transfer process from the second maxillae by the first maxillae has been likened to removing crumbs from one of your hands with the fingers of the other while both are immersed in molasses. The viscous forces of water to which these crustaceans are subject would be similar for many other small suspension-feeding animals. Probably most filter-feeding organisms collect particles by methods other than by sieving.

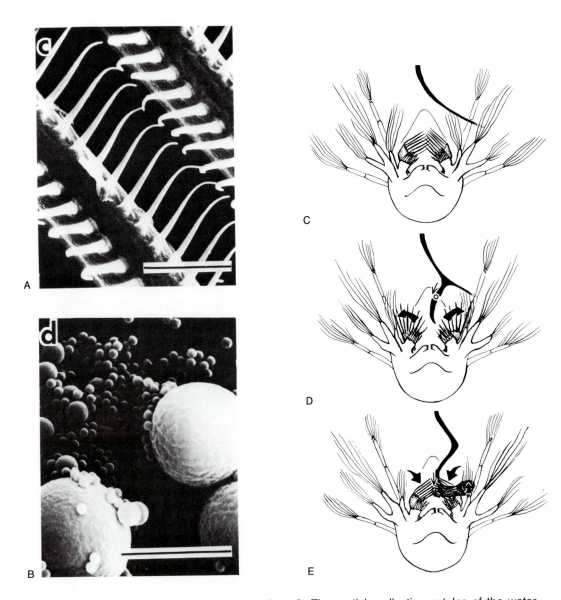

FIGURE 14–1 Crustacean suspension feeding. **A,** The particle-collecting setules of the water flea, *Daphnia*. **B,** Two sizes of polystyrene particles collected by the setules illustrated in **A.** The scale is only slightly larger than an angstrom. **C–E,** Particle collecting in a copepod. See the text for explanation. *(A and B, From Gerritsen, J., and Porter, K. G. 1982. The role of surface chemistry in filter feeding by zooplankton. Science. 216:1225–1227; C–E, From Koehl, M. A. R., and Strickler, J. R. 1981. Copepod feeding currents: Food capture at low Reynolds number. Limnol. Oceanogr. 226(6):1062–1073.)*

BIOLUMINESCENCE

The terrestrial fireflies are perhaps the most familiar example of animals capable of producing light, but this property, known as **bioluminescence,** is far more widespread among marine organisms. On a still night, light may appear to trail behind a swimmer's arms or a slow-moving boat. In shallow coastal waters, bacteria, dinoflagellates, jellyfish, ctenophores, crustaceans, brittle stars, and fishes include the majority of luminescent species, although they are but a small part of the total shallow-water fauna. On the other hand, in the mesopelagic zone of the open ocean (200–1000 m), 70% to 80% of the jellyfishes, ctenophores, shrimps, squids, and fishes are luminescent.

Bioluminescence involves the oxidation of a substrate substance, called **luciferin,** involving a molecule of oxygen and an enzyme, called **luciferase.** The luminescent product molecule is sufficiently excited to give off a photon. Luciferin and luciferase are different compounds in different luminescent groups, and the reaction pathways vary, but the reactions were understood from a study of crustaceans (ostracods). Given the variation in luminescent biochemistry and the taxonomic distribution of luminescence, the ability to produce light has clearly evolved independently many times.

The light source of luminescent animals may be symbiotic bacteria (in thaliaceans, some fishes, and squids), cellular secretions that are mixed extracellularly (in ostracod crustaceans), or specialized cells called **photocytes,** within which light is produced. The photocytes may be distributed in various parts of the body (e.g., under the comb rows in ctenophores) or within complex organs called **photophores,** which are provided with various accessory structures, such as shutters, reflectors, and lenses (in shrimps, squids, and fishes) (Fig. 14–2). Most fishes and

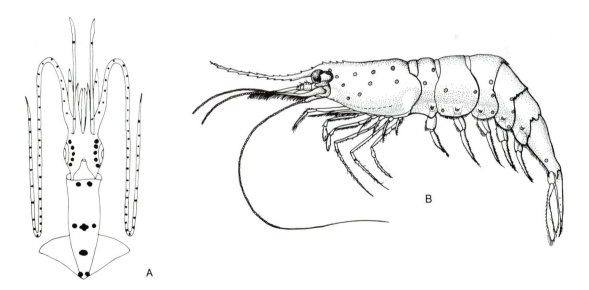

FIGURE 14–2 Light organs. **A,** Light organs (spots) in the squid, *Nematolampas.* Eye and body light organs serve for counterillumination; those on arm tips are of unknown function. **B,** Photophores (small circles) of a shrimp, *Systellapsis. (A, From Herring, P. J. 1977. Luminescence in cephalopods and fish. In Nixon, M., and Messenger, J. B. (Eds.): The Biology of Cephalopods. Symposia of the Zoological Society of London. 38. Academic Press, London. pp. 127–159. B, After Herring and Clarke, from Marshall, N. B. 1979. Deep Sea Biology, Garland STPM Press. New York, 566 pp.)*

some squids and shrimps have complex light organs, called glandular light organs, in which the light-producing system is provided by symbiotic bacteria or by glandular secretion.

The light produced by luminescent animals may be emitted as a flash or as a sustained glow, depending on the species and the specific functional role of the luminescence. Some species can flash and glow, and one species can have several kinds of photophores depending on functional requirements. The light may serve one or several functions, but the most common and widespread function is believed to be predator avoidance, although this hypothesis has rarely been tested. The predator may be startled and repelled by the light, which is usually a flash on contact, or the light may save the animal from a predator by disguising its form, by creating a confusing substitute shape (luminescent cloud), or by making it invisible. Invisibility is achieved in such oceanic animals as squids and fishes by **counterillumination.** Prey above a deeper-swimming predator stand out as a silhouette against light coming down from the surface. Glowing light organs located over the ventral surface can provide counterillumination to reduce or obliterate the silhouette. Intraspecific communication is the second most common function of bioluminescence. Flashes or patterns of glowing light over the body act as signals to other individuals and may serve to bring together members of the opposite sex. The functional significance of light production has been well established for relatively few marine animals, and in gelatinous zooplankton it is especially poorly understood.

CRUSTACEANS

The more than 38,000 known species of the subphylum Crustacea include some of the most familiar arthropods, such as crabs, shrimps, lobsters, crayfish, and wood lice. In addition, myriad tiny crustaceans living in the seas, lakes, and ponds of the world occupy an important position in aquatic food chains.

The Crustacea is the only large subphylum of arthropods whose members are primarily aquatic. Most crustaceans are marine, but there are many freshwater species. In addition, there are some semi-terrestrial and terrestrial groups, but in general, the terrestrial crustaceans have never undergone extensive adaptive evolution for life on land.

EXTERNAL ANATOMY

As you will soon appreciate, crustacean diversity is so great that a description of a "typical" form is impossible. Here we can only point out some commonly recurring features. Structurally, the head is more or less uniform among members of the subphylum. It bears five pairs of appendages (Figs. 14–3A; 14–45B). Anteriormost is a pair of **first antennae,** or **antennules,** which are generally considered homologous to the antennae of the other mandibulate classes. The first antennae are followed by the **second antennae,** often referred to simply as the **antennae.** The presence of two pairs of antennae is a distinguishing feature of crustaceans. The second antennae are postoral in embryological origin and are probably homologous to the arachnid chelicerae, both of which are innervated by the tritocerebrum. Flanking and often covering the ventral mouth is the third pair of appendages, the **mandibles** (Fig. 14–3A). These are composed of a number of articles but are usually short and heavy with opposing grinding and biting surfaces. Behind the mandibles are two pairs of accessory feeding appendages, the **first maxillae,** or **maxillules,** and the **second maxillae.**

The **trunk** is much less uniform within the subphylum than the head. Primitively, the trunk is composed of a series of many distinct and similar segments and a terminal telson bearing the anus at its base (Fig. 14–9B). In most crustaceans the trunk segments are characterized by varying degrees of regional specialization, of reduction or restriction in number, of fusion, and of other modifications. In a number of groups the trunk is divided into a thorax and abdomen (Fig. 14–11A), but the number of segments that each contains varies from group to group.

In some crustaceans the first one to three trunk segments are fused with the head, forming a **head shield,** and one or several of these trunk appendages are turned forward and modified for feeding, in which case they are called **maxillipeds.** In many common species, the thorax, or anterior trunk segments, is covered by a dorsal **carapace** (Fig. 14–11A). The carapace generally arises as a posteriorly directed fold of the body wall of the head and may be fused with a varying number of segments behind it (Fig. 14–4). Usually, the lateral margins of the carapace overhang the sides of the body at least to some extent, and in extreme cases the carapace may completely enclose the entire body like the valves of a clam (Fig. 14–74D).

Crustacean appendages are typically **biramous** (Fig. 14–5). There is a basal **protopod** composed of two pieces: a **coxa** and a **basis.** To the basis is attached an inner branch (the **endopod**) and an outer branch (the **exopod**), each of which may be composed of one to many articles. There are innumerable variations of this basic plan. Sometimes an appendage has lost one of the branches and has become secondarily uniramous. Often, different parts of the appendage bear highly developed processes or extensions that have been given special names.

In primitive crustaceans, the appendages are numerous and similar and together perform a number of functions. But as in other arthropods, the evolutionary tendency in crustaceans has been toward a reduction in the number of appendages and a regional specialization of different appendages for particular functions. Thus, in a family of swimming crabs, one pair of appendages is modified for swimming, and others are modified for crawling, prehension, sperm transmission, egg brooding, and food handling.

The cuticle of most large crustaceans, in contrast to that of most other arthropods, is usually calcified. Both the epicuticle and the procuticle contain depositions of calcium salts, and the outer layer of the procuticle is also pigmented and contains tanned proteins.

LOCOMOTION AND NUTRITION

Ancestral crustaceans were probably small, swimming, epibenthic suspension feeders, and many modern forms have retained this primitive mode of existence. Most larger crustaceans have taken up a benthic habit, and certain appendages have usually become heavier and adapted for crawling and burrowing. Crustaceans, have a great range of diets and feeding mechanisms. Commonly, certain anterior trunk appendages are adapted for suspension feeding, predation, or picking up food, and the maxillae and mandibles function in holding, biting, and directing food into the mouth.

The subphylum Crustacea embraces the largest number of arthropod filter–suspension feeders. Filter feeding in the arthropods involves setae instead of cilia. Fine setae on certain appendages function in the collection of food particles, and the spacing of the setae has been thought to determine the size of the particle collected. The process is now known, however, to be more complicated than simple sieving (p. 679). The necessary water current is produced by the beating of the filtering appendages or, more commonly, by other appendages modified for this purpose. The collected particles are removed from the filter setae by special combing or brushing setae, and these particles are transported to the mouthparts by other appendages or, sometimes, in a ventral food groove.

Filter feeding undoubtedly evolved independently a number of times within the Crustacea, and virtually every pair of appendages, even the antennae and mandibles, may be modified in one group or another for filter feeding. Filter feeding probably first arose in connection with swimming and is therefore primitively associated with the trunk, the same limbs creating both the swimming and the feeding currents. The tendency in most groups has been for the filtering apparatus to develop forward, nearer the mouth, and to involve only the anterior trunk appendages or the head appendages.

The mouth is ventral, and the digestive tract is almost always straight (Fig. 14–3B). The foregut is commonly enlarged and functions as a triturating stomach, the walls of which bear apposing chitinous ridges, denticles, and calcareous ossicles (Fig. 14–29B). The midgut varies greatly in length, and one to several pairs of ceca are almost always present. At least one pair of midgut ceca, especially in large crustaceans, has usually become modified to form large, spongy digestive glands (the **hepatopancreas**) composed of ducts and blind secretory tubules. Its secretions are the primary source of digestive enzymes. The action of the digestive fluid takes place in the midgut and in the triturating stomach of the foregut when this chamber is present. Absorption is confined

(Text continues on page 686)

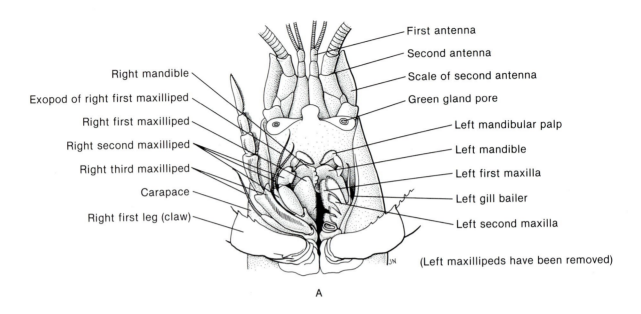

Right mandible
Exopod of right first maxilliped
Right first maxilliped
Right second maxilliped
Right third maxilliped
Carapace
Right first leg (claw)

First antenna
Second antenna
Scale of second antenna
Green gland pore
Left mandibular palp
Left mandible
Left first maxilla
Left gill bailer
Left second maxilla

(Left maxillipeds have been removed)

A

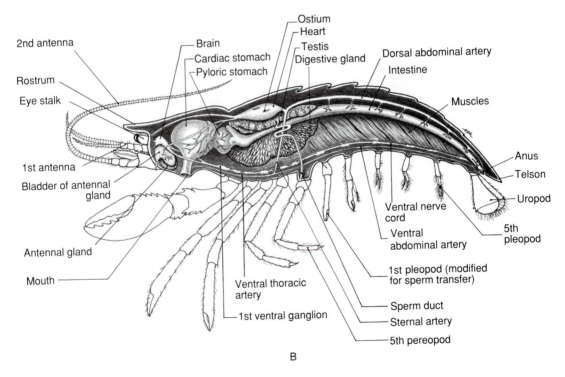

2nd antenna
Rostrum
Eye stalk
1st antenna
Bladder of antennal gland
Antennal gland
Mouth

Brain
Cardiac stomach
Pyloric stomach

Ostium
Heart
Testis
Digestive gland

Dorsal abdominal artery
Intestine
Muscles

Anus
Telson
Uropod
5th pleopod

Ventral nerve cord
Ventral abdominal artery

Ventral thoracic artery
1st ventral ganglion

1st pleopod (modified for sperm transfer)
Sperm duct
Sternal artery
5th pereopod

B

FIGURE 14–3 Crustacean structure. **A,** Head appendages of a lobster. More posterior mouth parts on the left side have been removed. **B,** Internal structure of a crayfish (lateral view). **C,** Cross section of a crayfish just behind third pair of legs. *(B and C, After Howes.)*

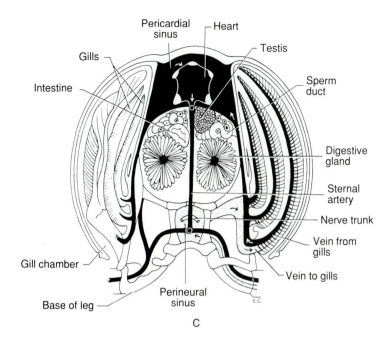

Heart
Pericardial
sinus
Testis
Gills
Sperm
duct
Intestine
Digestive
gland
Sternal
artery
Nerve trunk
Vein from
gills
Gill chamber
Vein to gills
Base of leg
Perineural
sinus

C

FIGURE 14–3 (*Continued*)

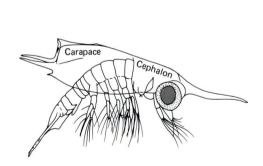

Carapace
Cephalon

FIGURE 14–4 Larva of a stomatopod showing the developing of the carapace as a fold of the head (cephalon). *(From Newman, W. A., and Knight, M. D. 1984. The carapace and crustacean evolution—A rebuttal. J. Crust. Biol. 4(4):683.)*

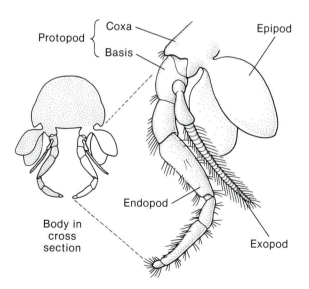

Protopod { Coxa
Basis
Epipod
Body in
cross
section
Endopod
Exopod

FIGURE 14–5 Right 5th pereopod, or leg, of the syncarid, *Anaspides tasmaniae,* showing the basic structure of a crustacean appendage. *(After Waterman and Chace.)*

to the midgut walls and the tubules of the hepatopancreas; however, the hepatopancreas also contains cells for glycogen, fat, and calcium storage.

INTERNAL TRANSPORT AND GAS EXCHANGE

The circulatory system of crustaceans is similar to that of other arthropods, but the dorsal heart varies in form from a long tube to a compact vesicle. Small crustaceans have limited blood vessels, but in large crustaceans, there is an extensive arterial system (Fig. 14–3B), even capillary beds in some tissues. Although the arterial walls lack muscle tissue, they are supported by collagen and elastic fibers, and can function to some degree as pressure reservoirs. Muscular sphincters in the arteries provide some control over the distribution of blood.

The blood contains small hyaline and larger granular amebocytes that, in addition to phagocytosis, are involved in clotting. Under certain irritating conditions, such as amputation, amebocytes called **explosive cells** disintegrate and liberate a substance that converts plasma fibrinogen to fibrin. Islands of coagulated plasma then appear, to which other islands connect and in which blood cells are trapped, thus forming a clot.

Gills are the usual gas exchange organs of crustaceans and are typically associated with the appendages, but they vary greatly in form, location, and derivation. The water current for ventilation is generally provided by the beating of certain appendages. Within the blood (hemolymph), oxygen is transported either in simple solution or bound to plasma hemoglobin or hemocyanin. Hemocyanin is found in the large species, but both pigments have a sporadic distribution.

EXCRETION AND OSMOREGULATION

The excretory organs of crustaceans are similar in structure and origin to the coxal glands of chelicerates and thus are metanephridial systems. Crustacean excretory organs are paired and composed of an end sac, an excretory canal, and a short exit duct, all located in the head (Fig. 14–6).

The end sac arises from an anterior coelomic compartment adjacent to the antennal or second maxillary segments, and the excretory organs are therefore called **antennal** or **maxillary glands.** The excretory pores of the antennal glands open onto the underside of the bases of the second antennae, and those of the maxillary glands open onto or near the bases of the second maxillae. Both antennal and maxillary glands are commonly present in crustacean larvae, but usually only one pair persists in the adult.

The end-sac walls are composed of cells similar to the podocytes of vertebrate glomeruli, and it is through the slits between podocyte extensions that filtration from the blood into the end sac occurs (Fig. 14–6C). Processing of the filtrate by selective reabsorption or secretion occurs within the excretory canal, and this region may be variously modified, depending on the extent of these processes.

The gills are a principal site for the excretion of ammonia, and therefore in most crustaceans, the antennal and maxillary glands must function in regulating other metabolites and ions and controlling internal fluid volume. Also, in most crustaceans the antennal and maxillary glands do not play an important role in osmoregulation. Most crustaceans, even many freshwater and terrestrial species, produce a urine that is isosmotic with the blood. This is not true, however, of the freshwater crayfish, in which the antennal glands do elaborate a hyposmotic urine. For most crustaceans, the gills are the chief organs for maintaining salt balance. In freshwater and brackish-water forms the gills actively absorb salts.

Crustaceans, like most other arthropods, possess **nephrocytes**—cells capable of picking up and accumulating waste particles. Nephrocytes are most commonly situated in the hemocoelic spaces of the gill axes and in the bases of the legs.

NERVOUS SYSTEM

In crustaceans, as in most other arthropods, the nervous system displays a tendency toward concentration and fusion of ganglia. In crayfish, for example, medial fusion has taken place, resulting in a single cord with single, rather than obviously paired, ganglia. In most crabs, all ventral ganglia have fused with the subesophageal ganglia to form a single mass. Giant fibers, which conduct impulses rapidly, are found in the central nervous system of many crustaceans. The giant fibers are particularly well developed in such forms as shrimps and crayfish, which can dart rapidly backward by a sudden flexion of the tail. We cannot discuss the numerous studies on crustacean reflexes in the context of this book, but see the references for elaboration on this topic.

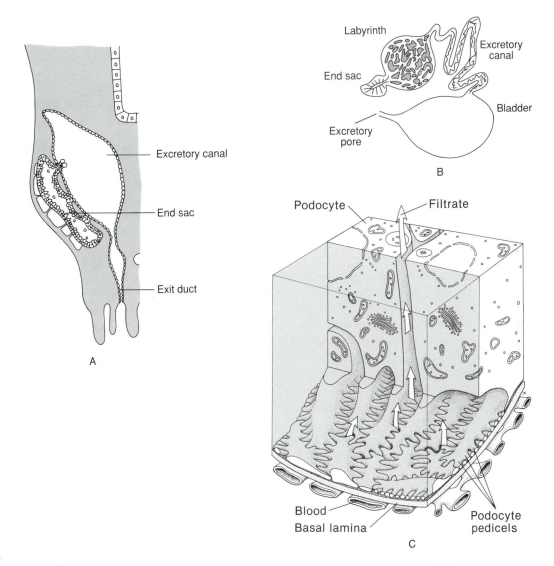

FIGURE 14–6 A, Section through the excretory organ (maxillary gland) of a barnacle. **B,** The excretory organ of a crayfish. Excretory pore opens onto the base of the second antenna. Ultrafiltration of the hemolymph occurs across the wall of the end sac (coelomosac); ion transport and protein reabsorption occur in the labyrinth; storage of urine occurs in the bladder. **C,** End sac wall of a crayfish, showing podocytes. Arrow shows the direction of filtration flow toward the lumen at the top of page. *(A, From White, K. N., and Walker, G. 1981. The barnacle excretory organ. J. Mar. Biol. Assoc. U.K. 61:529–547. Copyrighted and reprinted by permission of Cambridge University Press. C, Modified from Kümmel, G. 1967. Die Podocyten. Zoolog. Beitr. N. F. 13:245–263.)*

SENSE ORGANS

The sense organs of crustaceans include eyes, statocysts, sensory hairs, and proprioceptors. The eyes are of two types: median and compound. A median eye is a characteristic feature of the nauplius larva of crustaceans and is therefore often referred to as the **naupliar eye** (Figs. 14–8; 14–45A). It may degenerate or persist in the adult. The naupliar eye is composed of three or sometimes four small, pigment-cup ocelli each containing a few photoreceptors. The cups are located directly over the protocerebrum, where they either form a compact mass or are somewhat separated. The naupliar eye is largely an organ of orientation. As such, the eye enables the animal to determine the direction of a light source and thus to locate the upper surface of the water or, in burrowing forms, to locate the surface of the substratum.

Two **compound (lateral) eyes** are found in the adults of most species. The eyes may be at the end of a usually movable stalk (peduncle) (Fig. 14–3B), or they may be sessile (Fig. 14–62B). The total corneal surface is often greatly convex, resulting in a wide visual field. This is particularly true for stalked compound eyes, in which the cornea may cover an arc of 180 degrees or more. The number of ommatidia in crustacean compound eyes varies enormously. A single eye in the wood louse, *Armadillidium,* is composed of not more than 25 ommatidia; the eye of the lobster, *Homarus,* may possess as many as 14,000. Those with well-developed compound eyes, such as some shrimps and crabs, show some ability to discriminate form and size.

Color discrimination has been demonstrated in a number of crustaceans and may be of wide occurrence among members of the subphylum. For example, the hermit crab, *Pagurus,* can discriminate between painted yellow and blue snail shells and shells colored different shades of gray. The chromatophores of the shrimp, *Crangon,* adapt to a background of yellow, orange, or red but not to any shade of gray (the chromatophore changes are mediated through the eyes, p. 784).

Statocysts are restricted to a few groups of large crustaceans. Only a single pair is present and is located in the base of the antennules or in the base of the abdomen, uropods, or telson. Each statocyst arises as an ectodermal invagination and usually retains an opening to the exterior. The statolith may be secreted, but commonly it is composed of a mass of agglutinated sand grains. The physiology of crustacean statocysts is best known among the decapods (shrimps, crayfish, lobsters, and crabs) and is discussed with these groups.

Various types of sensory hairs are located over the body surface, especially the appendages; among these are chemoreceptors, called **aesthetascs,** which are found in the majority of crustaceans. Aesthetasc sensory hairs are usually present in rows on the first antennae (Fig. 14–7). In terrestrial crustaceans, the aesthetascs have the form of tiny plates instead of hairs.

REPRODUCTION AND DEVELOPMENT

Most crustaceans are dioecious. The gonads are typically elongated, paired organs lying in the dorsal portion of the trunk (Fig. 14–3B, C). The oviducts and

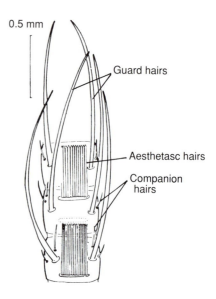

FIGURE 14–7 Rows of chemosensory hairs, called aesthetascs, on the antenna of the lobster *Panulirus.* Only one of the two rows on each segment has been shown. *(From Laverack, M. S. 1968. Oceanogr. Mar. Biol. Ann. Rev. 6:249–324.)*

sperm ducts are usually paired, simple tubules that open either at the base of a pair of trunk appendages or on a sternite. The segments that bear the gonopores, however, vary from one group to another.

Copulation is the general rule in crustaceans, the male usually having certain appendages modified for clasping the female. In many crustaceans, including shrimps, lobsters, and crabs, the sperm lacks a flagellum and is nonmotile*. In some crustaceans the sperm are transmitted as spermatophores (Fig. 14–39). The sperm ducts may open at the end of a penis, or, commonly, certain appendages may be modified for the transmission of sperm. A seminal receptacle is sometimes present in the female. The seminal receptacle may be located near the base of the oviduct, but frequently it is a separate, pouchlike, ectodermal invagination of the genital segment or a neighboring segment.

Most crustaceans brood their eggs for various lengths of time, depending on the group. The eggs may be attached to certain appendages, contained within a brood chamber in various parts of the body,

---

*Flagellated sperm are found in the Branchiopoda, Cirripedia (barnacles), Mystacocarida, and Ostracoda.

or retained within a sac secreted when the eggs are expelled.

The eggs of many higher crustaceans are centrolecithal, and cleavage is superficial; in lower groups, the eggs are small, and holoblastic cleavage is common. In some, it is determinate and shows traces of a spiral pattern (barnacles, copepods, cladocerans).

A free-swimming, planktonic larva is characteristic of most marine species and even some freshwater forms. The earliest and basic type of crustacean larva is a **nauplius** (Fig. 14–8). Only three pairs of appendages are present: the first antennae, the second antennae, and the mandibles. The second antennae and mandibles bear swimming setae. No trunk segmentation is evident, and a single median, or naupliar, eye is borne on the front of the head.

In the course of successive molts, trunk segments and additional appendages are usually gradually acquired. Development proceeds through the proliferation of trunk segments at the anterior margin of the

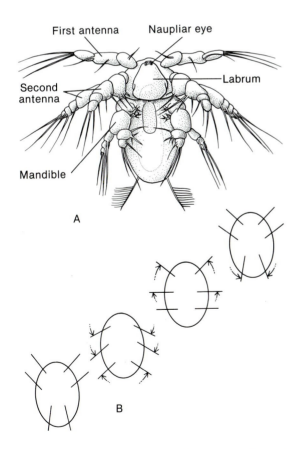

First antenna   Naupliar eye

Second antenna

Labrum

Mandible

A

B

FIGURE 14–8 A, Ventral view of a nauplius larva. **B,** Locomotion of a nauplius.

telson, which is homologous with the pygidium of annelid larvae. A distinctive larval stage, called a **zoea** (Fig. 14–46A), occurs in large crustaceans, such as crabs and shrimps, when growth results in eight appendage-bearing segments posterior to the carapace. By contrast with the naupliar larva, the new thoracic appendages provide for propulsion in swimming. Among other crustaceans, a structurally distinctive zoeal stage is absent, and as a result, the term *zoea* is not applied. With the acquisition of a full complement of functional appendages, including abdominal appendages, the young crustacean is called a **postlarva.** The postlarva may be quite similar to the adult in general appearance or may still be strikingly different in some respects. For example, the postlarva of crabs has a crablike cephalothorax, but the abdomen is still large and extended (Fig. 14–46E). After additional molts all of the adult features are attained except for size and sexual maturity.

The basic developmental pattern of nauplius, zoea (or its equivalent), and postlarva is frequently modified. In many groups there has been a tendency for some or all of the larval stages to be suppressed. In most shrimps and crabs, for example, the naupliar stage is passed in the egg, and the zoea is the hatching stage. Both the nauplius and the zoea may be suppressed, and the young hatch as postlarvae. All larval development is suppressed in some crustaceans, such as crayfish. Because the postlarval stages of different groups are usually distinctive, these stages have often received special names, such as the megalops of crabs, and the acanthosoma of sergestid prawns. Moreover, the zoea, if particularly distinctive, may be given a special designation, such as the mysis of lobsters. The intermediate stages have also received different names. For example, the later nauplius instars are called **metanauplii,** and the prezoeal instar is called a **protozoea.**

Other aspects of crustacean behavior and physiology are discussed at the end of this chapter, following the survey of crustacean groups.

SUMMARY

1. The 38,000 species of the subphylum Crustacea constitute the only major group of aquatic arthropods. Most are marine, but there are many freshwater species, and there have been a number of invasions of the terrestrial environment.

2. Crustaceans are extremely diverse in structure and habit, but they are unique among arthropods in having two pairs of antennae. Other characteristic head appendages are one pair of mandibles and two pairs of maxillae. The trunk specialization varies greatly, but a carapace that covers all or part of the body is common.

    3. Crustacean appendages are typically biramous and, depending on the group, have become adapted for many different functions.

    4. Gills, which are absent only in species of minute sizes, are typically associated with the appendages, but the location, number, and form vary greatly.

    5. Excretory organs are a pair of blind sacs in the hemocoel of the head that open onto the bases of the second pair of antennae (antennal glands) or the second pair of maxillae (maxillary glands).

    6. Crustacean sense organs include two types of eyes: a pair of compound eyes and a small, median, dorsal naupliar eye composed of three or four closely placed ocelli. Some groups lack compound eyes, and the naupliar eye, characteristic of the crustacean larva, does not persist in the adult of many groups.

    7. Copulation is typical of most crustaceans, and egg brooding is common. The earliest hatching stage is a naupliar larva, which possesses a median naupliar eye and only the first three pairs of body appendages.

OVERVIEW OF CRUSTACEAN CLASSIFICATION

The diversity of crustaceans requires a division of the classification hierarchy into more levels than is usually necessary for other animal groups. Each of the major groups in the subphylum Crustacea (38,000) is given the rank of class.[†] The parenthetical number indicates the number of species within the class.

Subphylum Crustacea (38,000)
    Class Remipedia (2)
    Class Cephalocarida (9)
    Class Branchiopoda (821)
    Class Ostracoda (5,650)
    Class Copepoda (8,405)
    Class Mystacocarida (9)
    Class Tantulocarida (4)
    Class Branchiura (150)
    Class Cirripedia (900)
    Class Malacostraca (22,651) ✓

[†]The name Entomostraca, encountered in older literature, included all of the crustacean classes except the Malacostraca.

Primitive Crustaceans: Classes Remipedia and Cephalocarida*

The first crustaceans were probably small marine animals that swam or crawled over the bottom. The trunk would have been composed of numerous unfused segments, each bearing a pair of similar biramous appendages. The two classes of crustaceans that exhibit such a primitive condition are the Remipedia and Cephalocarida. The class Remipedia, first described in 1981, is now represented by ten or eleven species, all from island caves that connect to the sea (Turks and Caicos, Bahamas, and the Canary Islands). The body of these small crustaceans is elongated, resembling that of a polychaete, with numerous biramous appendages (Fig. 14–9). They swim by a rowing action of the trunk appendages, which sweep down the body as a metachronal wave. Remipedians are carnivores, and the first pair of trunk appendages is modified as prehensile feeding appendages (maxillipeds).

    The nine species of Cephalocarida have been collected from the bottom sand and mud in many parts of the world and to depths of 1550 m. The first was found in Long Island Sound in 1953. Cephalocarids are tiny crustaceans, less than 4 mm in length (Fig. 14–10). The horseshoe-shaped head is followed by an elongated trunk of 20 segments, of which only the first 8 bear appendages. The trunk appendages are not only all similar (Fig. 14–10), but they are also not markedly different from the second pair of maxillae. Cephalocarids crawl over the sediment and are selective deposit feeders.

    The members of both groups are blind and hermaphroditic, characteristics that are unlikely to be primitive for crustaceans. Although compound eyes are present in cephalocarids, they are buried in the head. The hatching stage of cephalocarids is a metanauplius; the development of remipedians is still unknown.

Class Malacostraca

The class Malacostraca contains over half (22,651) of all the known species (38,000) of crustaceans, as well as most of the larger forms, such as crabs, lobsters, and shrimps. The trunk of a malacostracan is typically

---

*The root *carid,* so widely used in the names of crustacean taxa, means "shrimp."

A

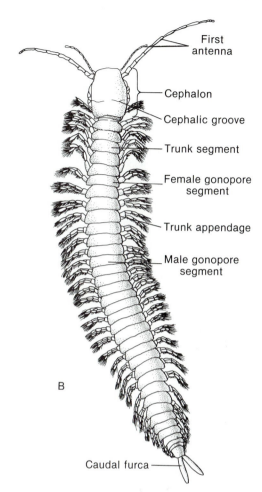

First
antenna

Cephalon

Cephalic groove

Trunk segment

Female gonopore
segment

Trunk appendage

Male gonopore
segment

B

Caudal furca

FIGURE 14–9 A, *Lasionectes entrichomas,* a member of the newly discovered, primitive class Remipedia from caves in the Turks and Caicos Islands. Note the long, wormlike body with numerous pairs of similar appendages. **B,** Dorsal view of the body of *Speleonectes tulumensis. (A, Photograph courtesy of Dennis W. Williams. B, From Felgenhauer, B. E., Abele, L. G., and Felder, D. L. 1992. Remipedia. In Harrison, F. W., and Humes, A. G. (Eds.): Microscopic Anatomy of Invertebrates Vol. 9. Crustacea. Wiley-Liss, New York. pp. 225–247.)*

composed of 14 segments, plus the telson, of which the first 8 segments form the thorax and the last 6 the abdomen (Fig. 14–11A). The thorax may or may not be covered by a carapace. All of the segments bear appendages. The first antennae are often biramous. The exopod of the second antenna is frequently in the form of a flattened scale.

Primitively, the thoracic appendages, or legs, (pereopods), are similar, and the endopod is the more highly developed of the two branches of these appendages, being used for crawling or prehension (Fig. 14–11B). In most malacostracans the first one, two, or three pairs of thoracic appendages have turned forward and become modified to form maxillipeds and are used in feeding.

The anterior (usually the first five pairs) abdominal appendages, called **pleopods,** are similar and biramous. The pleopods may be used for swimming, burrowing, ventilating, carrying eggs in the female, or sometimes for gas exchange. In the male, the first one or two pairs of pleopods are usually modified, forming copulatory organs. Usually, each ramus of the sixth abdominal appendages (**uropods**) is composed of a large flattened piece and, together with the usually flattened telson, forms a tail fan, which is most frequently used in escape swimming.

In most malacostracans, the foregut is modified as a two-chambered stomach bearing triturating teeth and comblike, filtering setae. Within the stomach, food is chewed and digestion begins. The fine particu-

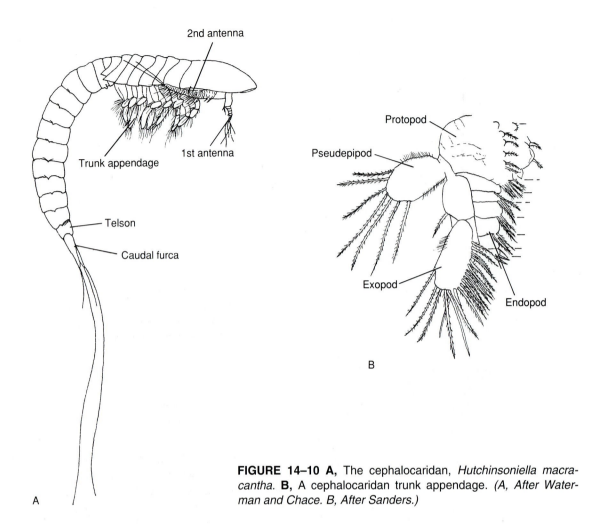

FIGURE 14–10 **A,** The cephalocaridan, *Hutchinsoniella macracantha.* **B,** A cephalocaridan trunk appendage. *(A, After Waterman and Chace. B, After Sanders.)*

late products of this action are filtered out and passed to the midgut and then to its outpocketings, called the digestive glands, or hepatopancreas.

The female gonopores are always located on the sixth thoracic segment, and the male gonopores are on the eighth. The naupliar larva is usually passed within the egg.

Because of the size and diversity of the Malacostraca, we will discuss each of the orders separately and begin with the Decapoda, which includes the most familiar crustaceans.

Order Decapoda

The order Decapoda contains the familiar shrimps, crayfish, lobsters, and crabs and is the largest order of crustaceans. The approximately 10,000 described decapods represent almost one quarter of the known species of crustaceans. Most of the decapods are marine, but the crayfish and some shrimps and crabs have invaded fresh water. There are also some terrestrial crabs. Decapods are distinguished from the euphausiaceans, as well as from other malacostracans, in that their first three pairs of thoracic appendages are modified as maxillipeds. The remaining five pairs of thoracic appendages are legs, from which the name Decapoda is derived (Fig. 14–18B). The first pair, or sometimes the second, is frequently much enlarged and chelate, and when so constructed, the limb is called a cheliped (Fig. 14–18B). The legs usually lack exopods; that is, they are not biramous. The sides of the overhanging carapace enclose the gills within well-defined, lateral branchial chambers (Fig. 14–30E).

Among the smallest decapods are species of the brachyuran crab, *Dissodactylus,* which are commen-

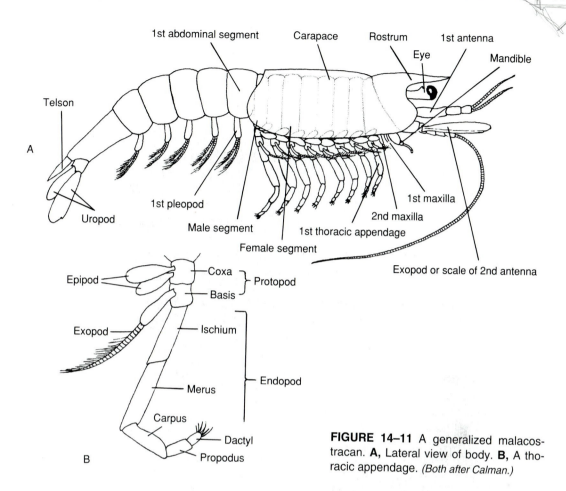

FIGURE 14-11 A generalized malacostracan. **A,** Lateral view of body. **B,** A thoracic appendage. *(Both after Calman.)*

sal on sand dollars (Fig. 14–20B). The cephalothorax of this crab is only a few millimeters wide. *Macrocheira kaempferi,* a Japanese spider crab, has the greatest leg span of any living arthropod (Fig. 14–20A). The cephalothorax may attain a length of 45 cm, and the chelipeds, a span of 4 m. Some lobsters, however, have longer bodies.

Decapod Diversity: Locomotion and Habitation

The wide range of body designs among decapods is easiest to understand in terms of adaptations for locomotion and habitation.

There are three separate, distinct groups of decapods that are called shrimps, or prawns, one of which—the Penaeidea—includes the most primitive decapods (see p. 696). The bodies of shrimps tend to be cylindrical or laterally compressed with well-developed abdomens (Fig. 14–12), and the cephalothorax often bears a keel-shaped, serrated rostrum. The

legs are usually slender, and chelipeds may be present or absent. The exoskeleton is commonly thin and flexible.

The only pelagic decapods are shrimps, mostly members of three families (Penaeidae, Sergestidae, and Oplophoridae) (Fig. 14–12B). Although pelagic shrimps occur at all depths, the majority are found in the epipelagic and mesopelagic zones (the upper 1000 m) and typically migrate from 100 to over 800 m upward at night (Fig. 14–13).

The pereopods, which are large and fringed, are the principal swimming organs, although rapid ventral flexion of the abdomen with the tail fan is used for quick backward darts. Abdominal flexion also provides for vertical steering. The legs, which may be elongated, are used for flotation and, probably, as sensory structures.

Species living in the upper 500 m (epipelagic and upper mesopelagic) are transparent or semitranspar-

(Text continues on page 696)

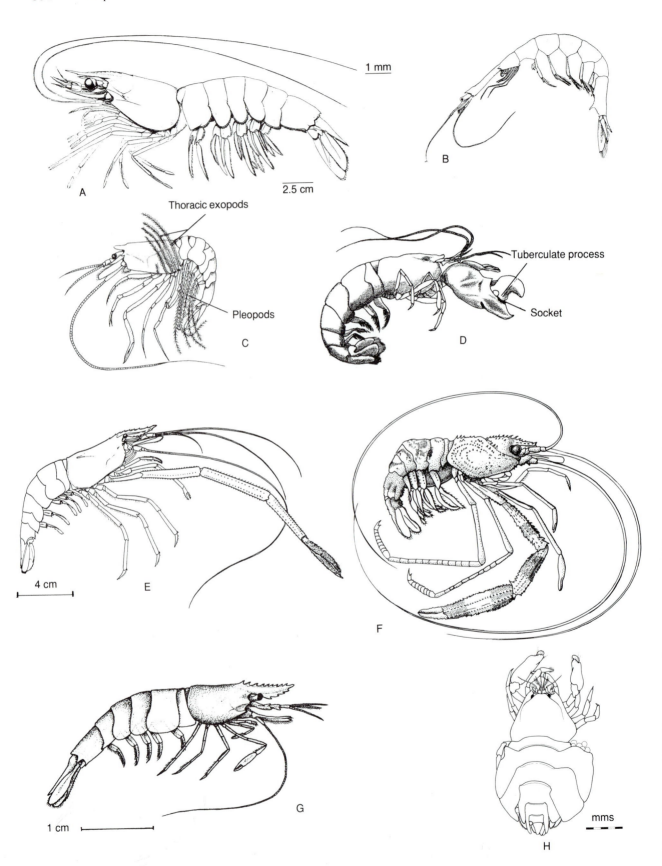

1 mm

2.5 cm

A

B

Thoracic exopods

Pleopods

C

Tuberculate process

Socket

D

4 cm

E

F

1 cm

G

mms

H

◀ **FIGURE 14–12** Shrimps. **A,** *Penaeus setiferus* (Penaeidea), an important species for the shrimping industry along the east coast of the United States. **B,** *Lucifer faxoni* (Penaeidea), a widely distributed pelagic species. **C,** *Psathyrocaris fragilis* (Caridea) with well-developed thoracic exopods. **D,** *Alpheus* (Caridea), a pistol, or snapping, shrimp. **E,** *Macrobrachium acanthurus* (Caridea), a river shrimp of eastern United States that ranges from brackish water to as much as 100 miles above the river's mouth. **F,** *Stenopus hispidus* (Stenopodidea), the banded coral shrimp, a cleaner shrimp that occurs primarily in association with coral reefs. **G,** *Palaemonetes* (Caridea), a large genus of small, transparent, estuarine or freshwater shrimps; many are common inhabitants of eel grass and other submerged vegetation. **H,** *Paratypton siebenrocki* (Caridea), an almost spherical Indo-Pacific, pontoniid shrimp that lives in cystlike chambers induced in the scleractinian coral, *Acropora*. *(A, B, and E, From Williams, A. B. 1965. Marine decapod crustaceans of the Carolinas. U. S. Fishery Bull. 65(1):1–298. C, After Alcock from Calman. D, After Schmitt from Bruce, A. J. 1976. Shrimps and prawns of coral reefs, with special reference to commensalism. In Jones, O. A., and Endean, R. (Eds.): Biology and Geology of Coral Reefs. Vol. III: Biology 2. Academic Press, New York. p. 49. F, After Limbaugh et al. from Williams, A. B. 1984. Shrimps, Lobsters, and Crabs of the Atlantic Coast of the Eastern United States, Maine to Florida. Smithsonian Institution Press, Washington, DC. 550 pp.)*

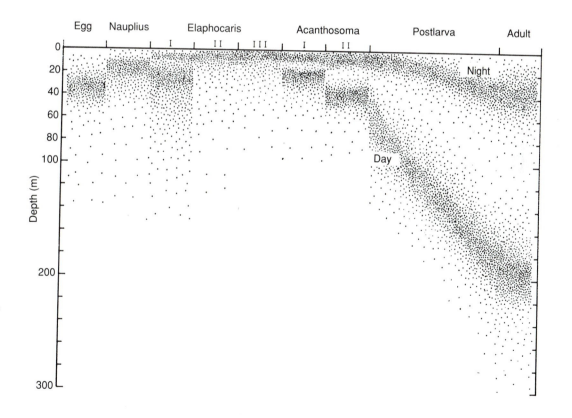

FIGURE 14–13 Vertical distribution of the pelagic shrimp, *Sergia lucens,* from egg to adult. *(After Omori, M. 1974. The biology of pelagic shrimps in the ocean. Adv. Mar. Biol. Academic Press, London. 12:233–324.)*

ent; those pelagic forms living below about 500 m during the day are red. Many of the latter group also possess luminescent organs, or photophores, located internally or anywhere on the body surface (Fig. 14–2B). Among the best known luminescent species are those belonging to the genera *Sergestes* and *Sergia.*

Most shrimps are not pelagic. They are bottom dwellers, using the legs for crawling, and swim intermittently with the pleopods. They live among algae and sea grasses, beneath stones and shells, and within holes and crevices in coral and rock. Some, including many penaeids and the sand shrimps *(Crangon),* are shallow burrowers in soft bottoms and use the beating pleopods for excavating. Their activity patterns may be controlled by light or tides. The commercially important *Penaeus duorarum* on the east coast of the United States is quiescent on the surface or buried during the day and is an active crawler at night. The European *Crangon crangon* lies buried near the low-water mark at low tides on sandy beaches but emerges at high tide to swim over the lower part of the submerged beach.

The pistol, or snapping, shrimps (Alpheidae) are a common and widely distributed family. These little shrimps, which are 3 to 6 cm long, have one of the chelipeds greatly enlarged (Fig. 14–12D). The base of the movable finger contains a large, tuberculate process that fits into a socket on the immovable finger. The movable finger is locked, or cocked, when contact is made between two specialized discs, one at the base of the elevated finger and one on the hand. The adhesive force between the discs prevents closing of the finger until the contracting muscle has generated a counteracting pull. Then the finger closes with great rapidity and force, producing a snapping or popping noise and a little water jet. Although some species have been reported to use the snapping mechanism in predation (cracking small clams and stunning fish), it also functions in threat (agonistic) displays between individuals and probably ensures spacing among members of a population.

Snapping shrimps live in holes and crevices beneath shells, rocks and coral rubble, or they construct retreats or burrows. There are a number of snapping shrimps, as well as other species of shrimps, that live in sponges, tunicates, bivalves, and corals or with molluscs, sea urchins, and sea anemones. A large

sponge may become a veritable apartment house for shrimps.

A number of unrelated groups of shrimps, called cleaning shrimps, remove ectoparasites and other unwanted materials from the surfaces of certain reef fish. The shrimp may climb over the fish and even insert its chelipeds into the fish's gill region. Some species of *Periclimenes* use the tentacles of sea anemones as their homes, and cleaning stations to which the fish must come are located nearby. The shrimp signals the fish with a "dance" of antennae waving and body rocking, and the fish responds with a distinct stationary pose. The shrimp then strokes the fish with its antennae and boards it to clean. The shrimps are protected from the nematocysts of the sea anemone in much the same manner as clown fish that live in sea anemones.

Species of *Penaeus* and related groups are the most important commercial shrimps throughout the world (Fig. 14–12A). In the United States, the shrimp fisheries are centered primarily along the southeastern Atlantic coast and the Gulf of Mexico. In the United States and other parts of the world, marine shrimps are caught by trawling, in which a long, V-shaped net is towed behind the shrimp boat (Fig. 14–14). Each arm of the V is connected by ropes to the boat, and at the apex the net forms a bag. The nets are towed along the bottom at slow speeds for half an hour to an hour, depending on the catch. Penaeid shrimp farming is now carried out in many parts of South and Central America and Asia but has been most highly developed in Japan, where shrimps are reared from eggs to marketable size.

All of the remaining decapod groups are benthic animals that have become more highly adapted for crawling than have most shrimps. The body tends to be dorsoventrally flattened, at least to some degree. The legs are usually heavier than those of shrimps, and the first pair of legs are typically powerful chelipeds. The pleopods are never adapted for swimming.

In many groups, the more primitive, long-bodied form of the shrimp has been greatly shortened through folding of the abdomen ventrally beneath the anterior part of the body (Fig. 14–15). This evolution of the "crab" form occurred independently a number of times in decapod evolution as an adaptation for locomotion or habitation.

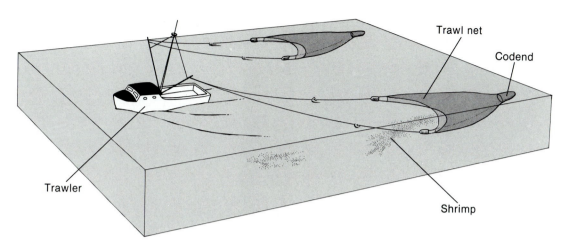

FIGURE 14–14 A shrimp trawler pulling an otter trawl.

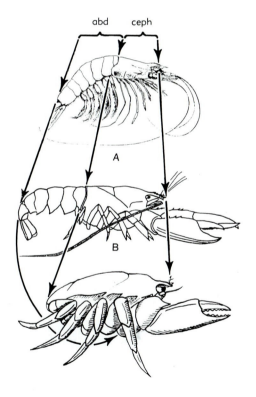

FIGURE 14–15 Comparison of cephalothorax and abdomen in a shrimp, lobster, and crab and the origin of the short body form. *(From Glaessner, M. F. 1969. Decapoda. In Moore, R. C. Treatise on Invertebrate Paleontology Part R. Arthropoda 4. Vol. 2. Geological Society of America and University of Kansas Press, Lawrence p. 401.)*

Lobsters, crayfish, and burrowing shrimps still possess a large, extended abdomen bearing the full complement of appendages, and the carapace is always longer than it is broad (Fig. 14–16). Lobsters and crayfish crawl with the thoracic legs but can move rapidly backward to escape by flexing the abdomen ventrally. The pleopods are used for ventilation. Lobsters are heavy-bodied decapods and are generally inhabitants of holes and crevices of rocky and coralline bottoms. The Nephropidae have large chelipeds and are similar in form to crayfish. The American lobster, *Homarus americanus,* may reach a length of 60 cm and a weight of 22 kg (Fig. 14–16B). They are caught commercially in pots, or traps, which the animal enters seeking shelter or bait. The lobster population along the New England and Canadian coasts has been badly overfished, and the largest catches are now taken offshore along the continental shelf and shelf edge. The European lobster, *Homarus gammarus,* is similar but smaller (Fig. 14–16C). The frozen lobster tails sold in food markets are mostly species of spiny and slipper lobsters shipped from various tropical and subtropical parts of the world. Neither of these two groups of lobsters has chelipeds (Fig. 14–16D, E).

Members of the infraorder Brachyura, the true crabs, have the most highly specialized short body form and, judging from the number of species (over 4500), are probably the most successful decapods. The abdomen is greatly reduced and fits tightly be-

FIGURE 14–16 Lobsters, crayfish, and mud shrimps. **A,** *Cambarus coosae* (bottom), a crayfish from Georgia, beside a specimen of the giant crayfish, *Astacopsis gouldi,* from Tasmania. The rule is 15 cm long. **B,** *Homarus americanus,* the lobster of commercial importance along the northeastern coast of the United States. **C,** *Nephrops norvegicus* (Norway lobster). This species and *Homarus gammarus* are small lobsters taken commercially along European coasts. **D,** *Panulirus argus* (spiny lobster) from the West Indies. Members of this genus have large antennae but lack large chelipeds. **E,** *Scyllarides aequinoctialis* (Spanish or slipper lobster) from the West Indies. Members of this family have short, flat antennae and lack large chelipeds. *(Photographs courtesy of Betty M. Barnes.)*

neath the cephalothorax (Fig. 14–18). Uropods have disappeared in all but a few primitive forms. In the female, pleopods are retained for brooding eggs (Fig. 14–18C); in the male, only the anterior two pairs of copulatory pleopods are present (Fig. 14–18D). The carapace is broad, often as wide as it is long and commonly much wider, thus increasing the flattened appearance of the body. The evolution of abdominal reduction and flexion in brachyurans was probably a locomotor adaptation, shifting the center of gravity forward to a point beneath the locomotor appendages. Crabs can crawl forward slowly, but they commonly move sideways, especially when crawling rapidly. In such a gait, the leading legs pull by flexing, and the

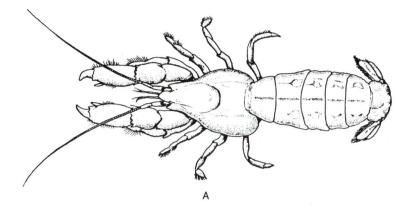

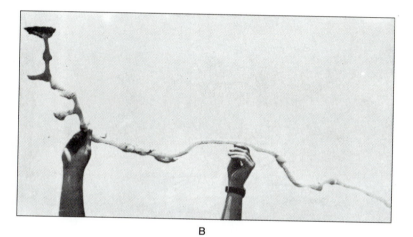

FIGURE 14–17 A, *Upogebia affinis,* a species of burrowing shrimp from the southeast coast of the United States. **B,** Mold of a burrow of *Upogebia affinis* from the coast of Georgia. *(A, From Ruppert, E. E. and Fox, R. S. 1988. Seashore Animals of the Southeast. Univ. of So. Carol. Press, Columbia. B, Photograph courtesy of Frey, R. W., and Howard, J. D. 1969. Trans. Gulf Coast Assoc. Geol. Soc. 19:427–444.)*

trailing legs push by extending. Chelipeds are not usually used in crawling.

Brachyuran crabs are found in all types of habitats and to great depths. Many deep-sea crabs have long, slender legs, which bear flattened terminal articles, for crawling about over soft bottoms. At the other extreme is the common tropical *Grapsus grapsus,* which lives just above the water on wave-washed rocks, climbing and clinging to vertical surfaces with great agility. Terrestrial and freshwater crabs will be described later (p. 716).

Most crabs cannot swim, but members of the family Portunidae, which includes the common edible blue crab *(Callinectes sapidus)* of the Atlantic coast, are the most powerful and agile swimmers of all crustaceans. The last pair of legs in members of this family terminate in broad, flattened paddles (Fig. 14–18B) and during swimming describe figure eights in their movement. The action is essentially like that of a propeller, and the counter-beating fourth pair of legs act as stabilizers. Portunids can swim sideways, backward, and sometimes forward with great rapidity. They are benthic animals, however, like other crabs, and only swim intermittently.

Although the chelipeds are important in defense, other protective devices and habits have evolved in

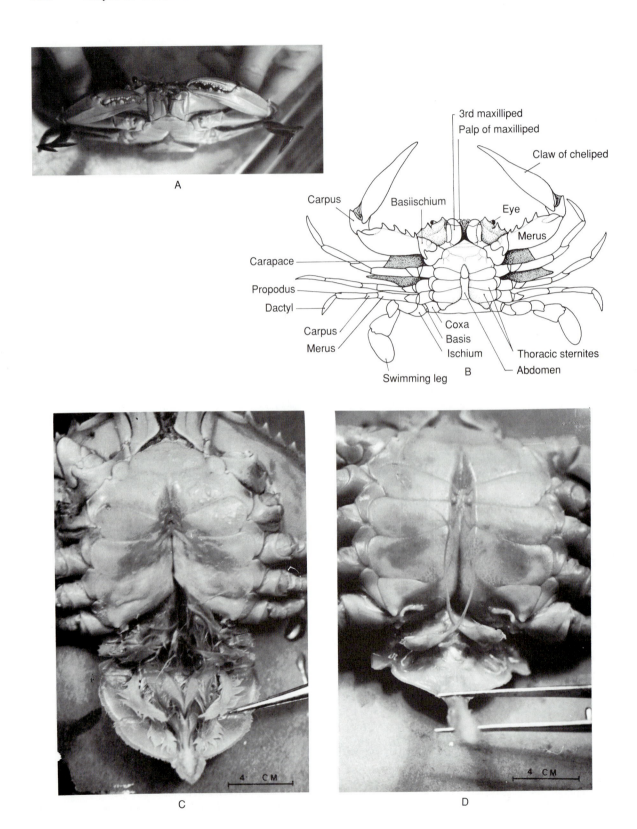

A

3rd maxilliped
Palp of maxilliped
Claw of cheliped
Carpus
Basiischium
Eye
Merus
Carapace
Propodus
Dactyl
Carpus
Merus
Coxa
Basis
Ischium
Thoracic sternites
Abdomen
Swimming leg
B

C

D

4 CM

◀ **FIGURE 14–18** A brachyuran crab of the family Portunidae. The fifth pair of legs is adapted for swimming. **A,** Frontal view, showing the dimorphic claws. The crushing claw is on the crab's right, and the cutter claw is on the left. Between the claw tips are the large, third maxillipeds covering the other mouthparts. **B,** Ventral view. **C,** Abdomen of a female. Pleopods are used for carrying eggs. **D,** Abdomen of a male. Note that it is much narrower than that of the female. Only the anterior two pairs of pleopods, which are used as copulatory organs, are present. *(B, After Schmitt from Rathbun; C and D, Photographs courtesy of Betty M. Barnes.)*

many brachyurans. Some species carry sea anemones on their chelipeds. Some spider crabs, which have triangular convex bodies and slender legs, are covered with hooked setae to which foreign objects become attached. This "decorating" habit has been highly developed in some species, and the body becomes completely overgrown with algae, sponges, and other sessile organisms (Fig. 14–19A). The decorator crab remains relatively immobile and camouflaged during the day, when predators are active.

The Dromiidae and Dorippidae use the small, dorsally directed fourth and fifth pairs of legs to hold objects over the body. *Hypoconcha* and *Dorippe* usually cover themselves with half of a bivalve shell, although *Dorippe* may use other objects, including an old fish head. Species of *Dromia* and related genera cut out a cap of sponge with the chelipeds and fit the cap on the back like an oversized beret. Members of the deep-water family Homolidae also carry objects, which seems strange given the total darkness of the environment in which they live.

Most of the commensal crabs belong to the family Pinnotheridae, called pea crabs because of the small size of many species. In addition to inhabitants

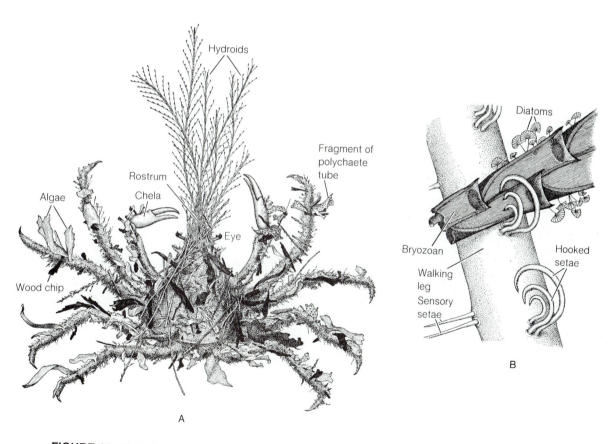

FIGURE 14–19 A, decorator crab (spider crab), *Oregonia gracilis,* camouflaged with attached organisms. **B,** Hooked setae on the leg of the decorator crab, *Podochela hemphilli. (From Wicksten, M. K. 1980. Decorator crabs. Sci. Amer. 242(2):149.)*

of polychaete tubes and burrows, there are species that live in the mantle cavities of bivalves and snails, on sand dollars (Fig. 14–20B), in tunicates, and in other animals. Often the body has become considerably modified for commensal existence. For example, the female of the oyster crab, *Pinnotheres ostreum,* has a soft exoskeleton. The male of this species, which leaves its host to find a female, has the usual chitinization. Pea crabs respond positively to substances produced by the host and can detect and move up a water stream that has passed over a host.

The family Hapalocarcinidae contains the coral gall crabs. These little tropical crabs cause certain species of corals to form gall-like chambers, within which the female lives. Small openings in the coral allow the entrance of the tiny male, as well as the plankton on which the crab feeds.

Many species of crabs are caught and eaten by humans throughout the world. The portunid swimming crab, *Callinectes sapidus,* or blue crab, is the commercially important crab occurring along the east and Gulf coasts of the United States, especially in the Chesapeake Bay. It is caught in shallow water with a trap or line, but commercial fishermen also catch them in large numbers by trawling when fishing for shrimp, or in winter by means of a crab dredge, which dislodges crabs that have buried themselves in the sea bottom. Soft-shell crabs are simply newly molted blue crabs. On the west coast of the United States and in Europe, species of *Cancer,* a genus of nonswimming crabs, are used as food and are caught by trapping. *Cancer magister,* the dungeness or market crab, is the most important species off the coast of California.

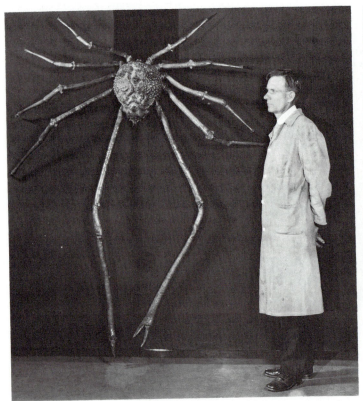

A

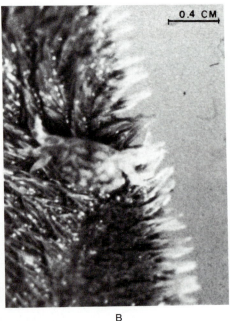

B

FIGURE 14–20 Brachyurans. **A,** The Japanese spider crab, *Macrocheira kaempferi,* the largest living arthropod. **B,** *Dissodactylus,* a commensal pea crab that lives on sand dollars. This is one of the smallest decapods. *(A, Neg. Trans. No. 312007. Courtesy of the Department Library Services, American Museum of Natural History.)*

The remaining decapods are a diverse assemblage of forms composing the infraorder Anomura. Some are crablike. The abdomen is never as reduced as in brachyurans, however, and uropods are usually present. All anomurans are similar in having the fifth pair of legs small and either located beneath the sides of the carapace or directed dorsally (Fig. 14–26A). Most also have longer second antennae than do most brachyuran crabs, and they are located lateral to the eyes rather than between the eyes, as is true of brachyuran crabs.

Burrowing shrimps or mud shrimps (*Callianassa, Upogebia,* and *Thalassina*) are shallow-water or intertidal decapods that live in long, deep burrows excavated in sand or mud (Fig. 14–27). There are many common species along both temperate and tropical coastlines. A species of *Thalassina* in southeast Asia invades rice paddies, where it builds 50-cm chimneys over its burrow openings that are a nuisance for the local farmers. The fifth pair of legs is not greatly reduced, and these decapods are not always placed within the Anomura. The exoskeleton is soft and flexible, and the coloration is typically pale.

Decapods of the anomuran superfamily Paguroidea have evolved the habit of housing the abdomen within gastropod shells and are called hermit crabs. The hermit crab condition probably had its origin in forms that utilized crevices and holes as protective retreats. Indeed, among the primitive, symmetrical Pomatochelidae and even the more specialized Paguridae there are species that still utilize such retreats; others live in bamboo or hollow mangrove roots (Figs. 14–21E, F; 14–23).

In most hermit crabs, the abdomen is not flexed beneath the cephalothorax but is modified to fit within the spiral chamber of gastropod shells (Fig. 14–21B). The curved or twisted abdomen is covered with a thin, soft, nonsegmented cuticle. On the outside of the curve, the pleopods have been reduced or lost, whereas those on the inside of the curve are retained in the females to carry eggs. The twist of the abdomen is adapted for right-handed spirals, although left-handed shells may be used.

The shell is held in several ways. The uropods are modified, and the larger left one is used to grasp the columella of the shell. Contraction of the longitudinal abdominal muscles presses the surface of the abdomen against the inner walls of the shell, and the last two pairs of thoracic legs press against the wall of the shell opening. The contact surfaces on both the legs and the uropods are covered with tiny scales that provide traction in gripping the shell (Fig. 14–21D). One or both chelipeds may be adapted for blocking the aperture of the shell when the crab is withdrawn.

Hermit crabs always use empty shells and never kill the original occupant. Most species inhabit the shells of a number of different gastropods, depending on what is available, and there can be a relatively high turnover of the shell supply.

When the crab becomes too large for its shell, it seeks another but does not leave the old shell until a suitable new one has been found. Hermit crabs locate a prospective shell with their eyes and then inspect it with their chelae, inserting one into the interior. If the shell appears suitable, the crab leaves its old shell and tries out the new one. If the fit, weight, or mobility is bad, the crab returns to the old shell (Fig. 14–22). Two hermit crabs will fight for possession of an empty shell or the shell inhabited by one of the combatants. The importance of shells to hermit crabs as a housing resource is reflected by the fact that in some areas hermit crab growth and reproduction are limited by a short shell supply.

Coral, stones, tooth shells, wood, and other structures have been adapted as houses by certain species (Fig. 14–23). There are also species of hermit crabs that live commensally with species of the sea anemone, *Calliactis.* By tapping and massaging the sea anemone, the hermit crab is able to bring about the sea anemone's release and transfer to its own shell. *Calliactis parasitica,* transfers to the shell of the hermit crab without the aid of the crab.

From hermit ancestors evolved two groups of anomurans, the coconut crabs *(Birgus)* and the lithodid crabs, which have short body forms and no longer house the abdomen within shells (Fig. 14–23). The abdomen, although folded crablike beneath the cephalothorax, has asymmetrical appendages (Fig. 14–25B). Coconut crabs are large, terrestrial crabs inhabiting tropical Pacific islands (Fig. 14–24) (p. 715). The lithodid crabs, which have heavy and sometimes sculptured carapaces, are mostly inhabitants of cold oceans (Fig. 14–25A, B). To this group belongs the large Alaskan king crab, *Paralithodes camtschatica,* which is trapped commercially in the north Pacific (Fig. 14–25C).

The anomuran Galatheoidea include the little porcelain crabs. These animals have a flexed, crablike abdomen, but the abdomen is symmetrical and not greatly reduced, as in brachyurans. The porcelain

(Text continues on page 707)

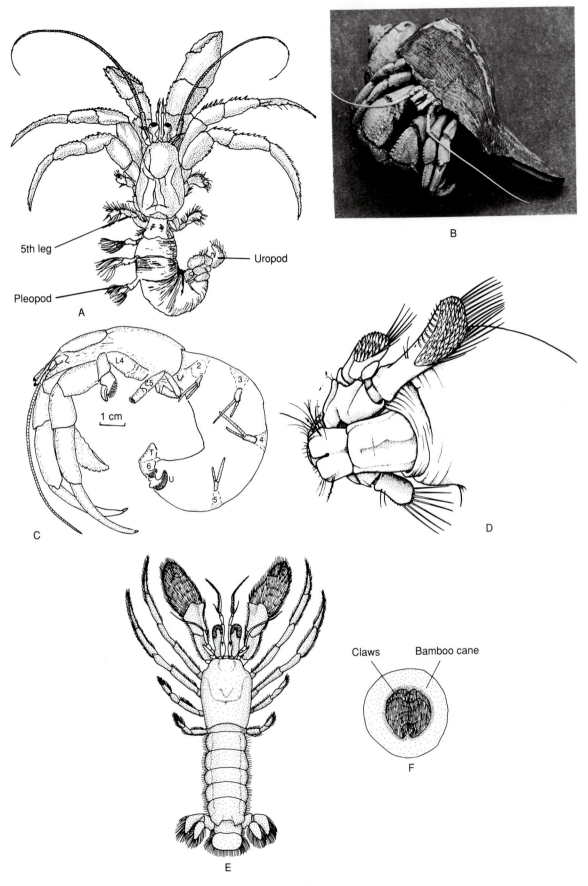

5th leg

Uropod

Pleopod

A

B

L4
L5
2
3
4
1 cm
T
6
U
5

C

D

Claws

Bamboo cane

F

E

FIGURE 14–21 Anomuran pagurid hermit crabs. **A,** Dorsal view of *Pagurus* out of its shell, showing the asymmetry of the abdomen. **B,** A pagurid in a gastropod shell. **C,** Side view of *Clibanarius*. L4 and L5 indicate fourth and fifth legs; U, uropod. Pleopods are present on the left side of the abdomen but are absent on the right. **D,** End of the abdomen of *Clibanarius,* showing the tuberculate gripping surfaces on the large left uropod. **E,** *Pylocheles miersi,* an anomuran that inhabits sections of old bamboo canes (dorsal view). Note how little the abdomen of this anomuran is modified compared with that of the pagurid illustrated in **A. F,** *Pylocheles* in a bamboo cane with the claws blocking opening. *(B, Neg. Trans. No. 39558. Courtesy of the Department Library Services, American Museum of Natural History. E–F, After Alcock from Calman.)*

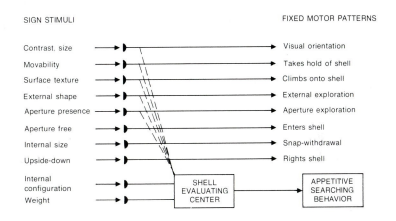

FIGURE 14–22 Sequential behavioral responses of a shell-seeking hermit crab to various shell stimuli. The searching behavior is reduced with increasingly positive responses to shell conditions. *(From Reese, E. S. 1963. The behavioral mechanisms underlying shell selection by hermit crabs. Behaviour. 21:78–126.)*

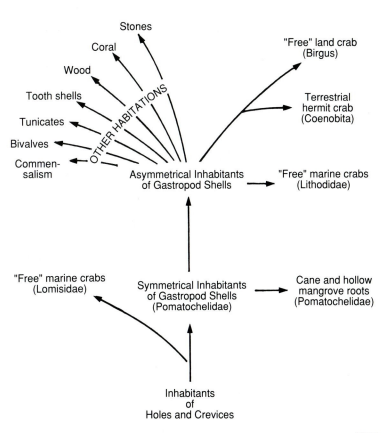

FIGURE 14–23 Adaptive evolution of the anomuran superfamily Paguroidea.

A

FIGURE 14–24 A, The south Pacific robber, or coconut, crab, *Birgus latro.* **B,** Juvenile individual of *Birgus,* in which the abdomen is housed within a gastropod shell in typical hermit crab manner. *(A, Photograph courtesy of Betty M. Barnes. B, From Reese, E. S. 1968. Science. 161:385–386. Copyright 1968 by the Association for the Advancement of Science.)*

B

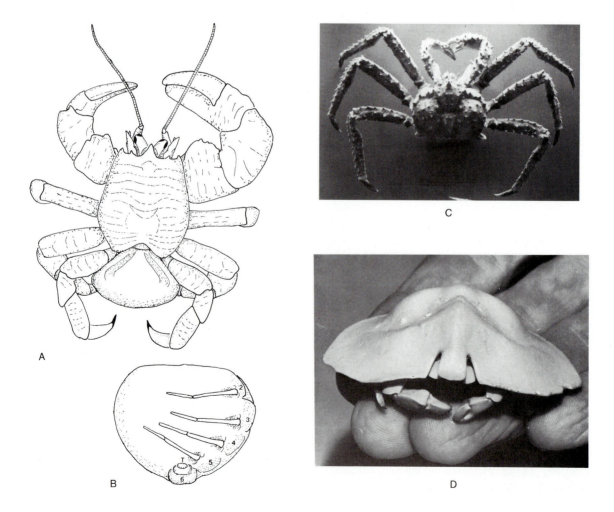

A

B

C

D

crabs are common, shallow-water decapods in many parts of the world and often occur in large numbers—hundreds per square meter. They live beneath stones and in their general appearance and crawling locomo-tion look very much like brachyuran crabs (Fig. 14–26). When disturbed, however, many species swim by flapping their abdomen.

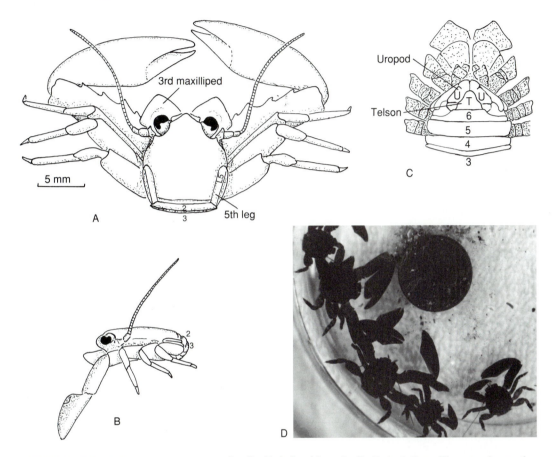

FIGURE 14–26 The anomuran superfamily Galatheoidea. **A–C,** *Petrolisthes,* The members of this genus, like others in the family Porcellanidae (porcelain crabs), are often very common beneath stones in shallow water along rocky coasts. They look remarkably like brachyuran crabs, but swim like shrimps by extension and flexion of the abdomen. Note the reduced and folded fifth pair of legs and the retention of the uropods and telson on the folded abdomen. **A,** Dorsal view. **B,** Lateral view. **C,** Ventral view of the abdomen and overlying cephalothorax. Numbers indicate abdominal segments or legs (L): U and T refer to uropods and telson, respectively. **D,** Intertidal porcelain crabs collected from beneath stones in the Gulf of California.

◀ **FIGURE 14–25** Paguroid lithodid crabs. A, Dorsal view of *Dermaturus mandteii* from Alaska. The fifth pair of legs is hidden by the carapace. **B,** Ventral view of the abdomen, showing asymmetrical pleopods. Numbers indicate abdominal segments, and T stands for telson. **C,** *Paralithodes camtschatica* (king crab). This large north Pacific species is a lithode crab, although it looks somewhat like a brachyuran spider crab. The leg span of the specimen in the photograph is over 1 m. **D,** *Cryptolithodes sitchensis,* a little lithodid crab with a helmet-like carapace. *(C–D, Photographs courtesy of Betty M. Barnes.)*

The superfamily Hippoidea contains the sand crabs, or mole crabs. Abdominal flexion in this group appears to be an adaptation for burrowing backward in sand. The mole crab body is somewhat cylindrical, and there are no chelipeds (Fig. 14–27). Species of *Emerita* live on open beaches and dig with the uropods and fourth pair of legs. In surf, mole crabs are usually washed out with each wave; while the wave is receding, they rapidly burrow backward in the soft sand until only the antennae are visible.

Nutrition

The ventral mouth of a decapod is flanked by the feeding appendages, which lie on top of one another (Fig. 14–3A). The third maxillipeds are the outermost appendages and cover the other appendages. In a brachyuran crab, the third maxillipeds are rectangular plates and completely fill the usually square buccal frame, covering the inner mouth appendages like double doors (Fig. 14–18A). Food is caught or picked up with the chelipeds and then passed to the third maxillipeds, which push it between the other mouthparts. While a portion is bitten or held by the mandibles, the remainder is torn away by the maxillae and maxillipeds. The severed piece is then directed into the mouth.

Decapods exhibit a wide range of feeding habits and diets, but most species combine predatory feeding with scavenging. The relative importance of the two habits varies with the species and also with the available food resource. Large invertebrates are common prey. For example, echinoderms and bivalves are the main food of the Alaskan king crab, and polychaetes, crustaceans, and bivalves are prey for snow crabs in the same area.

Herbivores include most freshwater and terrestrial decapods and some marine species. Herbivorous and scavenging habits grade into detritus feeding. Scavenging–detritus feeding is characteristic of most hermit crabs and many shrimps. *Palaemonetes pugio*, the abundant grass shrimp of tidal marshes along the east coast of the United States, plucks out and consumes small bits of cellulose matrix from large detrital fragments and thereby plays an important role in the breakdown of algae and grasses of tidal marsh ecosystems.

The chelipeds of crabs and other decapods often reflect feeding habits. Species that scrape algae from rocks or feed on detritus from the surface of sand and mud commonly have chelipeds with spoon-shaped fingers. The many species that include molluscs in their diet have dimorphic chelipeds. The heavier right claw bears blunt proximal teeth in the fingers and is adapted for crushing; the more slender left claw is adapted for cutting (Figs. 14–18; 14–28). The blue crab, *Callinectes sapidus*, of the east coast of the United States, preys heavily on juvenile oysters, and some members of the box crab family, Calappidae, can literally peel the shell from a snail.

Detritus feeding grades into filter feeding. For example, fiddler crabs scoop up mud and detritus with their small cheliped. The doorlike third maxillipeds open, and the collected material is dumped into the buccal frame. Using water pumped in from the branchial chamber, the material is worked over and shifted by brush and straining setae on the second and first maxillipeds. After organic material has been removed, the mineral residue is spit out as round pellets, which may eventually surround the burrows or cover the surface of the beach among species that feed on detritus left by the receding tide (Fig. 14–35A).

Examples of decapods that filter feed from a water current include many burrowing shrimps (*Callianassa, Callichirus, Upogebia*), some commensal pea crabs, most porcelain crabs, and the mole crabs. The maxillipeds are used as filtering appendages in many anomurans and in the commensal pea crabs, although some pea crabs living in oysters are known to feed on the mucous food strings collected by the host. Some burrowing shrimps filter with the first two pairs of legs, and the anomuran porcelain crabs use the third pair of maxillipeds. Porcelain crabs also scrape detritus from rocks with their chelipeds.

Among the most interesting filter feeders are the mole crabs, *Emerita* (Fig. 14–27). The long, densely fringed, second antennae project above the sand surface after the crabs are buried. In species such as *Emerita talpoida*, which inhabit open beaches, the second antennae filter plankton and detritus from the receding wave current. The crab buries itself seaward and its outstretched second antennae form a characteristic V on the sand surface during the backwash of each wave; in species that live in quieter water, the projecting antennae are moved about.

The typical decapod foregut consists of a short esophagus leading into a capacious **cardiac chamber** (Figs. 14–3B; 14–29) and a smaller, posterior, **pyloric chamber** separated from the cardiac portion by a valve.

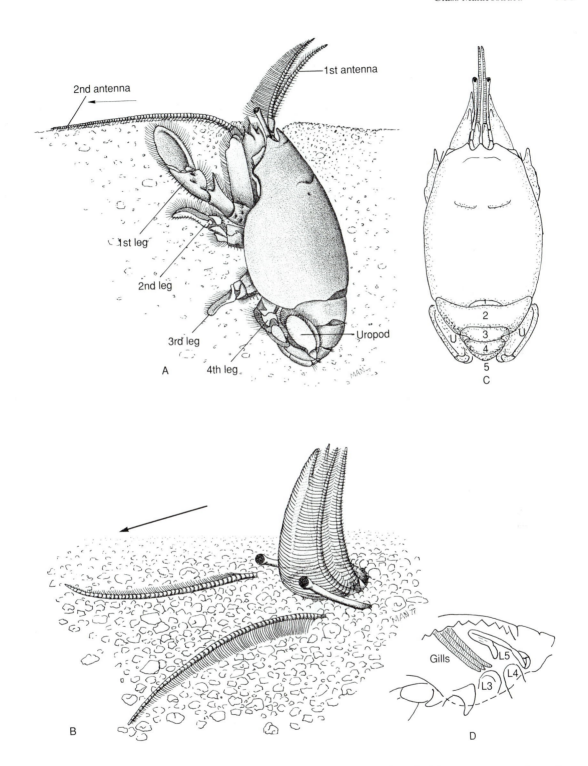

FIGURE 14–27 The anomuran superfamily Hippoidea, mole crabs. *Emerita talpoida,* a common mole crab on surf beaches along the east coast of the United States. **A,** Lateral view of the animal buried in the sand with the second antennae in filtering position. **B,** Surface view of the buried animal. First antennae form a siphon for ventilating the current. Arrow indicates the direction of the receding wave. **C,** Dorsal view. **D,** Lateral view of one side of the body showing tiny the fifth pair of legs folded beneath the carapace, which has been cut away.

FIGURE 14–28 The crab, *Calappa flammea,* opening the shell of the marine snail, *Fasciolaria.* The right crushing claw bears large tubercles at the base of the movable finger. *(From Shoup, J. B. 1968. Shell opening by crabs of the genus Calappa. Science. 160:887–888. Copyright 1968 by the American Association for the Advancement of Science.)*

The esophagus and the cardiac and pyloric chambers are lined with chitinous exoskeleton, which is variously thickened to form a number of ossicles in the walls of the cardiac and pyloric chambers. The ossicles provide support and sites for muscle attachment externally. Certain ossicles give rise internally to a median dorsal tooth and to two lateral teeth, one on either side of the median tooth. These three teeth, which sit internally at the posterior portion of the cardiac chamber, form the so-called **gastric mill,** where food is broken down mechanically (Fig. 14–29B). The triturating action of the gastric mill and the movement of the stomach walls are controlled by a complex series of muscles surrounding the stomach. The pyloric chamber is divided into a dorsal portion, which leads directly to the intestine, and a complex, ventral, bilobed gland filter (ampulla), which leads to the hepatopancreas by way of two large ducts, one from each lobe of the gland filter. The dorsal portion of the pyloric chamber is separated from the ventral by a row of paired denticles, which prevent large particles from entering the gland filter.

The hepatopancreas, or digestive gland, is a large, bilobed organ consisting of numerous blind tubules. Each tubule is composed of cells that function in enzyme secretion, endocytosis and intracellular digestion of particulate food, absorption and storage of nutrients, and vesicular packaging of indigestible wastes and their removal by exocytosis. Digestive secretions produced by the hepatopancreas move into both the cardiac and the pyloric chambers.

Material that is too large to pass through the gland filter and enter the hepatopancreas is transported from the dorsal portion of the pyloric chamber into the intestine. Here the midgut epithelium at the anterior end of the intestine secretes a clear, membranous tube, the peritrophic membrane, which encloses the material to be voided by the cuticle-lined hindgut as fecal pellets.

Gas Exchange

Primitively, there are four gills on each side of every thoracic segment. The gills arise from the body wall on or near the point of attachment of the appendage and have been given different names depending on their position (Fig. 14–30E). If all of the gill series were present on all segments, there would be a total of 32 gills on each side of the body, but no decapod has retained this maximum number. The penaeid shrimp, *Benthesicymus,* has the greatest number, 24,

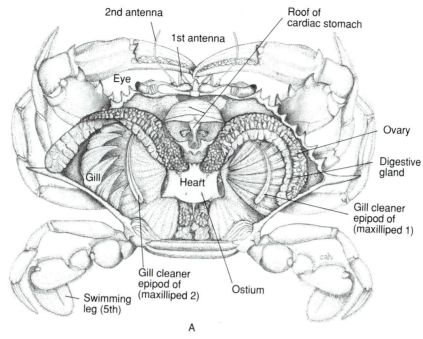

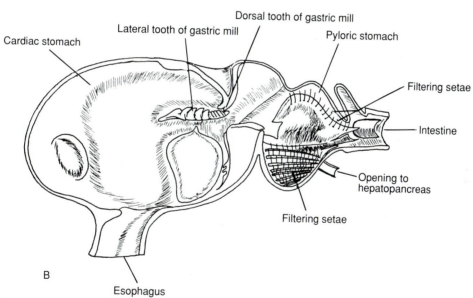

FIGURE 14–29 A, The brachyuran blue crab, *Callinectes sapidus* (dorsal dissection). **B,** Stomach of the crayfish, *Astacus* (lateral view). *(A, Drawing by Carolyn Herbert. B, After Kaestner, A. 1970. Invertebrate Zoology. Vol. 3. Wiley-Interscience, New York.)*

but reduction to far fewer than this is the general rule. For example, the lobster, *Homarus,* has a complement of 20 gills on each side of the body, distributed as follows: one gill on the second and three gills on the third maxillipeds, and four gills on each of the legs except the first, which bears three, and the last, which

bears only one. A similar pattern is found in crayfish. Nine pairs of gills are common in marine crabs, but the little pea crab, *Pinnotheres,* has only three gills to a side.

The gill is composed of a central axis along which lateral extensions, or branches are arranged.

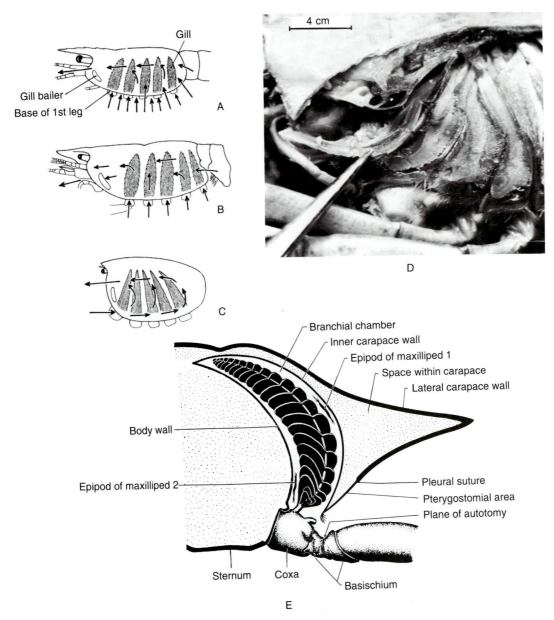

FIGURE 14–30 A–C, Paths of water circulation through the gill chamber of three decapods, showing progressive restriction of openings into the chamber. **A,** Shrimp. Water enters along entire ventral and posterior margin of the carapace. **B,** Crayfish. Water enters at the bases of the legs and at the posterior carapace margin. **C,** Crab. Water enters only at the base of the cheliped. **D,** Gill bailer (held by forceps) of a crayfish, showing its position within the branchial chamber. **E,** Cross section through the gill chamber of a crab. *(D, Photograph courtesy of Betty M. Barnes. E, From Kaestner, A. 1970. Invertebrate Zoology. Vol. 3. Wiley-Interscience, New York.)*

The structure of the gill branches varies among decapods (Fig. 14–31). In the axis of each gill runs an afferent and an efferent branchial channel (Fig. 14–31A). Blood flows from the afferent channel into each filament or lamella and then back into the efferent channel.

The blood of decapods contains hemocyanin dissolved in the blood plasma, and in large, active forms, such as the swimming crabs, the hemocyanin transports about 90% of the blood's oxygen.

The ventilating current is produced by the beating of a paddle-like **scaphognathite,** or **gill bailer,** a

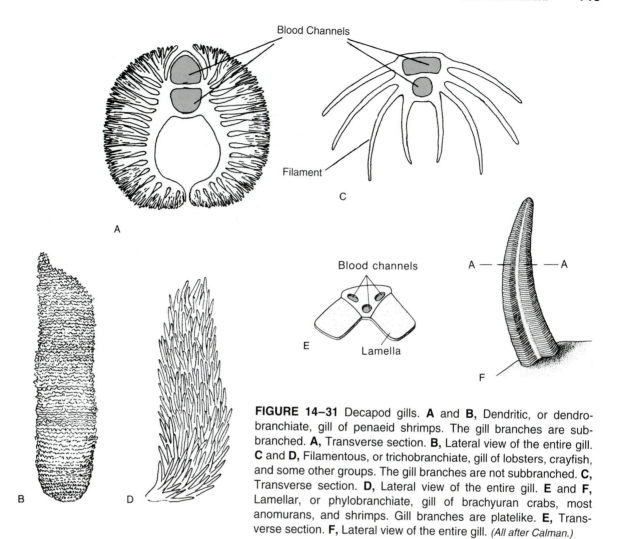

FIGURE 14–31 Decapod gills. **A** and **B,** Dendritic, or dendrobranchiate, gill of penaeid shrimps. The gill branches are subbranched. **A,** Transverse section. **B,** Lateral view of the entire gill. **C** and **D,** Filamentous, or trichobranchiate, gill of lobsters, crayfish, and some other groups. The gill branches are not subbranched. **C,** Transverse section. **D,** Lateral view of the entire gill. **E** and **F,** Lamellar, or phylobranchiate, gill of brachyuran crabs, most anomurans, and shrimps. Gill branches are platelike. **E,** Transverse section. **F,** Lateral view of the entire gill. *(All after Calman.)*

projection of the second maxilla (Fig. 14–30D). Water is pulled forward, and the exhalant current flows out anteriorly in front of the head. In shrimps, the ventral margins of the carapace fit loosely against the sides of the body, and water can enter the branchial chamber at any point along the posterior and ventral edges of the carapace (Fig. 14–30A). In other decapods, the carapace fits somewhat more tightly, and the entrance of water is limited to the posterior carapace margins and around the bases of the legs (Fig. 14–30B). The point of entrance of the ventilating stream is most restricted in the brachyuran crabs, in which the inhalant opening is located around the bases of the chelipeds (Figs. 14–30C; 14–32A). The forward position of the inhalant openings in

brachyurans results in water taking a U-shaped course through the gill chambers. On entering the inhalant opening, the water passes posteriorly into the hypobranchial part of the chamber and then moves dorsally, passing between the gill lamellae. The exhalant current flows anteriorly in the upper part of the gill chamber and issues from paired openings in the upper lateral corners of the buccal frame (Fig. 14–32A).

Because the majority of decapods are bottom dwellers and includes many burrowers, a variety of mechanisms have evolved to prevent clogging of the gills with silt and debris. The bases of the chelipeds in crabs and the coxae of the legs of crayfish and lobsters bear setae that filter the incurrent stream. The

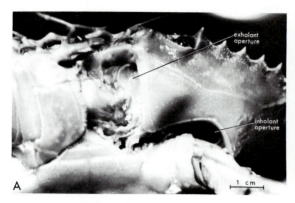
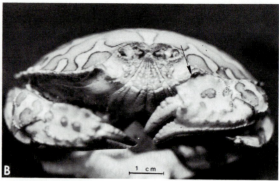

FIGURE 14–32 **A,** Anterior view of a blue crab, *Callinectes.* The left maxillipeds have been removed, and the gill bailer can be seen within the exhalant aperture. **B,** The box crab, *Hepatus.* Arrow indicates the opening of the left siphon formed by the opposing carapace and the cheliped. The siphonal groove on the carapace can be seen on the right side. *(Both photographs courtesy of Betty M. Barnes.)*

gills are cleaned in some shrimps with the first and second pairs of legs and in some anomurans with the reduced last pair of legs. The gills are cleaned in crabs by the fringed epipods of the three pairs of maxillipeds. These processes are elongated, especially those of the first maxillipeds (Fig. 14–30E), and they sweep up and down the surface of the gills, removing detritus. As a further aid to the cleaning of the gills and branchial chamber, the gill bailer of many decapods periodically reverses its beat and thus reverses the direction of flow through the chamber.

In some burrowing species, when the ventral parts of the body are covered by mud and sand, a reversed current is used for ventilating purposes; when the body is free, a forward current is used. A further ventilating adaptation in many burrowing decapods is the development of inhalant siphons. For example, in some species of brachyuran crabs, mole crabs, and shrimps, the first antennae are held together, forming an inhalant passageway between them. The large box crabs of the genera *Calappa* and *Hepatus,* which burrow just below the surface, have greatly flattened chelipeds, which when folded fit tightly against the face but leave an inhalant channel between the inner side of the cheliped and the carapace (Fig. 14–32B).

Internal Transport

The heart is a box-shaped sac located in the thorax and is provided with three pairs of ostia (Fig. 14–29A). Five arteries leave the heart anteriorly (Fig. 14–3B). A median abdominal artery leaves the heart

posteriorly, and a sternal artery arises either from the underside of the heart or from the base of the abdominal artery. Each of these major arteries branches extensively, supplying various organs and structures. After passing into the tissue sinuses, blood eventually drains into a large, median, ventral "sternal sinus" prior to passing through the gills and returning to the pericardium and heart by way of the branchiopericardial vessels. Complete circulation has been estimated to take from 40 to 60 s in large decapods.

Excretion and Salt Balance

Antennal (or green) glands are the excretory organs of decapods and reach their highest degree of development in this group. The end sac, or saccule, and the first part of the tubule, which is modified as a labyrinth, lies in front of and to both sides of the esophagus. Fluid collects by filtration within the saccule, whose walls contain podocytes. The labyrinth walls are greatly folded and glandular, forming a spongy mass, and appear to be an important site of reabsorption. The labyrinth leads into a bladder of varying complexity by way of an excretory canal (Fig. 14–6B). From the bladder a short duct extends into the basal segment of the second antenna, where it opens to the outside on the summit of a little papilla (Fig. 14–3A). In brachyurans, the excretory pore is covered by a little movable operculum.

Although the antennal glands are called excretory organs, most nitrogenous waste (NH_3) is diffused from the body surface where the exoskeleton is thin, as on the gills. The antennal glands appear to control internal fluid pressure and, to some extent, ion content, such as Mg^{++}. Uptake of water by the blood increases the filtration pressure within the antennal gland and the passage of fluid into the saccule. A copious amount of urine may be excreted, depending on internal and external conditions. In undiluted seawater (3.4%), the European crab, *Carcinus maenas*, produces a daily amount of urine equivalent to 3.6% of the body weight; in brackish water (1.4%), daily urine production is equal to one third of the body weight. The ability to maintain a constant hemocoelic volume is indicated by the fact that the size increase of a newly molted blue crab *(Callinectes)* is the same regardless of the salinity of the environment. The antennal glands apparently function to maintain this constancy.

Although the antennal glands may function to regulate specific ions, such as Mg^{++}, they do not play a major role in osmotic regulation. In most decapods, the urine is isosmotic with the blood even when the animal is in brackish water. Within the higher salinity ranges of sea water, most decapods act as osmoconformers; that is, the blood salt content is the same as that of the external medium. Those species that can tolerate lower salinities become osmoregulators, maintaining a blood salt concentration that is hyperosmotic to the surrounding medium. As in fish, the gills pick up salts from the ventilating current, replacing ions lost in the urine and elsewhere. Freshwater crabs produce an isosmotic urine but have a cuticle that is relatively impermeable to water, therefore reducing water intake. Crayfish, on the other hand, have an antennal gland with a long tubule between the labyrinth and the bladder. Reabsorption of salts by the tubule enables these crustaceans to produce a diluted urine that aids in osmoregulation.

Freshwater and Terrestrial Decapods

There have been numerous invasions of fresh water and land by decapods, but except for crayfish and the true freshwater crabs (superfamily Potamoidea), the colonization record has not been especially successful.

Freshwater shrimps belong to the families Atyidae and Palaemonidae. The Atyidae are scraping-filter feeders inhabiting streams, pools, and lakes in tropical and subtropical parts of the world. The Palaemonidae include many marine and brackish shallow-water species, such as *Palaemonetes*, in both tropical and temperate regions. Freshwater species include some river, stream, and pool inhabitants, such as the large, pantropical *Macrobrachium*, which is trapped or farmed in many parts of the world. The Palaemonidae osmoregulate by means of the gills, and this is probably also the case with Atyidae.

Crayfish are the most successful freshwater decapods. The more than 500 species are found throughout the world in streams, ponds, lakes, and caves. Some live beneath stones or among debris. Many species excavate burrows, which they use as retreats and for overwintering. Burrowing crayfish commonly inhabit bottom lands where the water table is not too far below the surface, and they may be semiterrestrial. A chimney of excavated mud typically rises above the burrow opening. Most crayfish are less than 10 cm long, but some Australian species reach the size of lobsters (Fig. 14–16A). Crayfish are omnivores. The antennal glands of crayfish, unlike those of other decapods, can excrete a hypoosmotic urine and play an important role in salt balance.

The anomurans include one group of terrestrial decapods, the tropical land hermit crabs *(Coenobita)* which are often sold in pet shops, and the closely related coconut crab *(Birgus)* (Fig. 14–24A). Both are found in Indo-Pacific regions, and *Coenobita* also occurs in the tropical western Atlantic and Caribbean. The land hermits live close to the shore and use either available water in pools or crevices or sea water to wet the body and interior of the shell. Adult coconut crabs have abandoned the hermit crab habit and have acquired a crablike form with a flexed abdomen. The adults live in burrows farther back from the sea but are still a coastal species. Coconut crabs feed on carrion and both decaying and fresh vegetation, and they can husk and open fallen coconuts. They obtain water by drinking. Both *Coenobita* and *Birgus* have reduced gills and have converted the moist branchial chamber into lungs with highly vascularized areas for gas exchange. At least in *Birgus*, most of the nitrogenous waste is excreted as water-conserving uric acid.

Many brachyuran crabs can tolerate either brackish or fresh water but must return to salt water to breed. In this category belongs the Chinese mitten crab, *Eriocheir* (Fig. 14–33A), which is found in the rivers and rice paddies of Asia and has invaded the rivers of Europe, probably by way of ship ballast, beginning about 1912. Three specimens were collected from Lake Erie in 1973. Another is the brachyuran, *Rhithropanopeus*, of the American Atlantic coast,

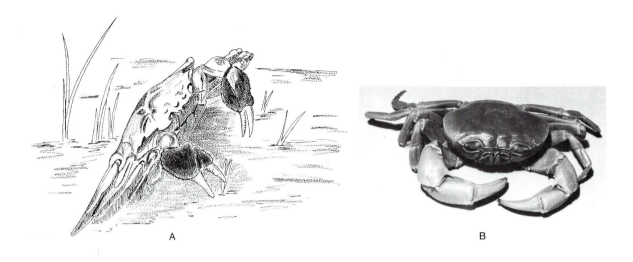

A   B

FIGURE 14–33 A, The Chinese mitten crab, *Eriocheir sinensis,* in a rice paddy. This amphibious freshwater brachyuran is native to southern Asia but has been introduced into the Rhine and Elbe rivers of Europe. **B,** A river crab, *Potamon anomalus* (Potamonidae). *(A, After Schmitt. B, Photograph courtesy of Betty M. Barnes.)*

which is found in salt water as well as in freshwater ditches. *Rhithropanopeus* was introduced into northern Europe and San Francisco Bay, possibly in the same way as *Eriocheir.* The blue crab, *Callinectes sapidus,* is quite euryhaline and ranges far up into brackish estuaries and bays. Some very successful but poorly known freshwater decapods are the so-called river crabs (superfamily Potamoidea) (Fig. 14–33). These occur in all tropical regions, from the mouths of rivers to mountain elevations of 5000 m. They complete their entire life cycle in fresh water. All of these crabs, both temporary and permanent residents of brackish and fresh water, excrete urine that is isosmotic with the blood and regulate their salts by means of ion absorption through the gills.

Except for the land hermits and coconut crabs, most of the decapods that have invaded land are brachyuran crabs. This is not surprising, considering the motility of crabs and their tightly enclosed branchial chambers. Although the living terrestrial crabs are derived from a number of different land invasions, most display similar adaptations.

Gills continue to be utilized for gas exchange, but there is also a tendency for the branchial chamber to become rather like a lung, with some surface other than the gills given over to gas exchange. The gill bailer continues to provide for ventilation, but it moves air rather than water.

Water loss by evaporation is much greater in land crabs than in other terrestrial arthropods (arachnids, insects), and most of this is through the carapace. Water is replaced, for the most part, by drinking, and many can maintain gill moisture by taking up water from soil, dew, or rain. No land crab can excrete a hyperosmotic urine. Such a capability would aid in conservation of water. Land crabs, however, save water by excreting very little urine, and we now know that the coconut crab can excrete uric acid, eliminating it by way of the digestive gland and feces.

Land crabs tend to live in burrows, or at least beneath stones, and are usually active only at night. They are primarily vegetarians and scavengers. All can run rapidly. Eyes are generally well developed.

The name *land crab* usually refers to the members of the family Gecarcinidae, all of which are terrestrial. They are found in tropical and subtropical America, West Africa, and the Indo-Pacific area. The family includes *Cardisoma* (Fig. 14–34) and *Gecarcinus,* which live in coastal fields and woods in Texas, southern Florida, tropical America, and the West Indies.

Although terrestrial as adults, the gecarcinids are nevertheless coastal in distribution, for females must return to the sea to release their spawn. The terrestrial potamonids are found the greatest distance from the coast, often hundreds of miles. These crabs, however,

FIGURE 14–34 The gecarcinid land crab, *Cardisoma guanhumi,* of the West Indies and Florida. *(Photograph courtesy of Betty M. Barnes.)*

have direct development and do not need to return to the sea to breed.

The family Ocypodidae includes some of the most familiar amphibious and terrestrial crabs, such as the fiddler crabs and ghost crabs. The known species of semiterrestrial fiddler crabs *(Uca)* number 62. They live on protected sand and mud beaches of

bays and estuaries, in brackish marshes, and in mangroves. Although most species are tropical and semitropical, fiddler crabs are found on both the east and the west coasts of North America.

The burrows of fiddler crabs are located in the intertidal zone, and at low tide the crabs come out to feed and court (Fig. 14–35A); enormous numbers may cover a beach. Tropical species tend to be active only during diurnal low tides, but temperate species emerge at night as well, perhaps compensating for the winter period, when they must remain dormant within their burrows. The simple burrows are usually L-shaped; those of North American species are generally no deeper than about 36 cm (Fig. 14–35B). The excavated sand is carried to the surface as small balls cradled in the legs of one side. As the tide comes in, the crabs return to their burrows and plug the entrance with sand or mud. The burrows of some tropical species that inhabit upper zones of mangrove are flooded only during spring tides.

Some fiddler crabs feed within 20 to 50 cm of the burrow, making spokelike movements from the burrow mouth; others may feed up to 50 m away. When disturbed, fiddler crabs quickly flee into their bur-

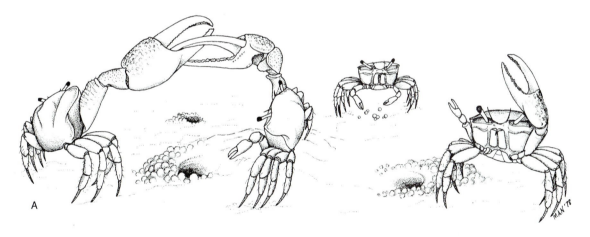

A

FIGURE 14–35 A, Fiddler crabs on a beach at low tide. Three burrow openings are shown. The large balls of sand have been excavated from one of the burrows; the small pellets are composed of sand that has been filtered of organic material and ejected by the mouth parts. The crab in the background with the two small claws is a female. The two males on the left are engaging in ritualized combat. The male on the right is waving the large claw in courtship display. **B,** Mold of the burrow of the fiddler crab, *Uca pugilator.* *(B, Photograph courtesy of Frey, R. W., and Howard, J. D. 1969. Trans. Gulf Coast Assoc. Geol. Soc. 19:427–444.)*

B

rows. Many crabs do not find their original burrows, however, and they either occupy a vacant hole or dig a new one.

Ghost crabs *(Ocypode)* are common in many parts of the world, including the east coast of the United States (Fig. 14–36). They are several times larger than fiddler crabs and never occur in large aggregations. Their widely separated burrows are found mostly above the high-tide mark on the upper beach or in dunes. Ghost crabs are largely nocturnal and many go down to the lower beach to prey on clams, mole crabs, even sea turtle hatchlings, or to scavenge for food.

The Indo-Pacific soldier crabs (Fig. 14–37) of the family Mictyridae are similar in habit to fiddler crabs

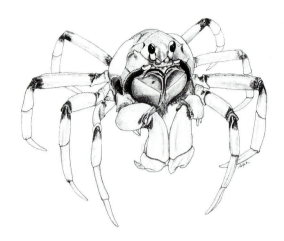

FIGURE 14–37 The Indo-Pacific soldier crab, *Mictyris longicarpus. (Based on a photograph by Healy and Yaldwyn.)*

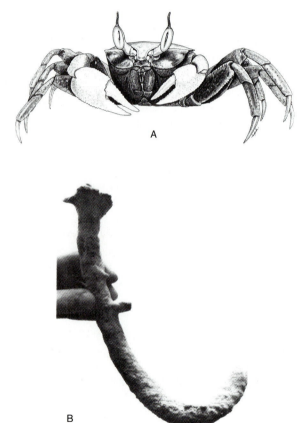

FIGURE 14–36 A, The Indo-Pacific ghost crab, *Ocypode ceratophthalma.* **B,** Cast of the burrow of the ghost crab, *Ocypode quadrata,* from the coast of Georgia. *(A, Based on a photograph by Healy and Yaldwyn. B, From Frey, R. W. 1971. Decapod burrows in holocene barrier island beaches and washover fans, Georgia. Senckenbergiana Marit. 3:53–77.)*

(Uca). When disturbed, a soldier crab quickly burrows sideways like a corkscrew.

The family Grapsidae contains perhaps the most ecologically diverse assemblage of crabs. There are marine, brackish-water, freshwater, amphibious, and terrestrial species. Amphibious grapsids include *Sesarma,* which lives beneath drift and stones; the agile, rock-inhabiting *Grapsus;* and numerous mangrove inhabitants, such as *Aratus,* which live in trees. The Jamaican *Metapaulias depressus* spends part of its life in the water-holding leaves of arboreal bromeliads.

Nervous System and Sense Organs

The ventral nerve cord of shrimps and lobsters, which have large extended abdomens, contains a series of separate ganglia for all but the first two thoracic segments. The ventral ganglia in other groups show varying degrees of additional longitudinal fusion, culminating in the brachyurans, in which all of the abdominal ganglia have migrated anteriorly to fuse with the thoracic ganglia, forming a single ventral mass.

As in other crustaceans, the legs and antennae are important sites for the reception of environmental information, and aesthetascs—the chemosensory hairs on the first pair of antennae—are commonly well developed (Fig. 14–7). The aesthetascs aid not only in locating food but also in recognizing other individuals and their sexual state. The frequent grooming of the antennae by many decapods prevents fouling of the receptor sites.

Although there are some blind decapods, particularly among deep-sea forms and cave-dwelling crayfish, compound eyes are usually highly developed, and the eye stalk is more mobile than in most other crustaceans.

A pair of statocysts is present in nearly all decapods and is located in the basal segment of the first antennae. The sac is always open to the exterior. The aperture may be occluded partially by setae, although in crabs and in some other decapods, the opening is reduced to a slit and is functionally closed. The statolith may be composed of fine sand grains bound together by secretions from the statocyst wall (Fig. 14–38). Because the statocyst is an ectodermal invagination, its lining is shed along with the statolith at each molt. The sand grain statoliths are replaced when the head is buried in sand or the animal actually inserts sand grains into the sac. Along the floor of the sac are a number of rows of sensory hairs with which the statolith is directly connected or is in intermittent contact. The hairs arise from receptor cells in the sac

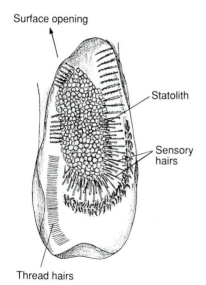

FIGURE 14–38 Statocyst of the American lobster, *Homarus americanus* (with dorsal wall removed). Note the four crescent-shaped rows of sensory hairs; the inner three are in contact with the statolith. The row of long, delicate hairs (lower left) are thread hairs. The opening of the statocyst to the exterior is at upper left beyond the edge of the illustration. *(After Cohen from Cohen and Dijkgraaf.)*

wall, and the receptor cells are innervated by a branch of the antennular nerve.

When the animal is in a normal horizontal position, the floor of the statocyst is inclined about 30 degrees, which results in a medial gravitational pull on the statolith, but one statocyst counterbalances the other. When the floors of both statocysts are tilted, the one in which the receptor cells are most stimulated dominates the other. Crabs and some other decapods have complex statocysts, which indicate not only the gravitational position of the animal but also its position in regard to movement. Such statocysts are functionally similar to the semicircular canals in vertebrates.

Reproduction and Development

The decapod sperm is tack- or star-shaped and lacks a middle piece and flagellum (Fig. 14–39D). Decapods transmit sperm in spermatophores, which in most species are delivered to the female by the anterior two pairs of copulatory pleopods of the male. The paired but connected testes lie in the thorax but may extend into the abdomen. The sperm duct is glandular and modified, depending on the degree of spermatophore formation, and the terminal end of the sperm duct is a muscular ejaculatory duct, which opens onto or near the coxae of the last pair of legs. At copulation, the spermatophores pass from the gonopore to the copulatory pleopods for transmission. For example, in some brachyuran crabs, the first pleopod is in the form of a cylinder into which the piston-like second pleopod fits. Soft spermatophores are delivered to the pleopods by a nipple-like penis associated with the gonopore opening on each side and pumped through the anterior conducting pleopod of each pair (Fig. 14–40).

The ovaries are similar in structure and location to the testes (Fig. 14–29A). The oviducts are usually unmodified and open via gonopores onto the coxae or near the coxae of the third pair of legs.

Primitively, the spermatophores are simply deposited on the female sternal surface. In many decapods, however, this surface has become specialized, commonly as spermathecae (seminal receptacles), that receive spermatophores by way of spermathecal openings. Finally, in the higher groups of crabs, the oviducts connect directly with the spermathecae and use their openings as gonopores.

Paniluran lobsters and anomuran crabs transmit the spermatophores directly from the male gonopore

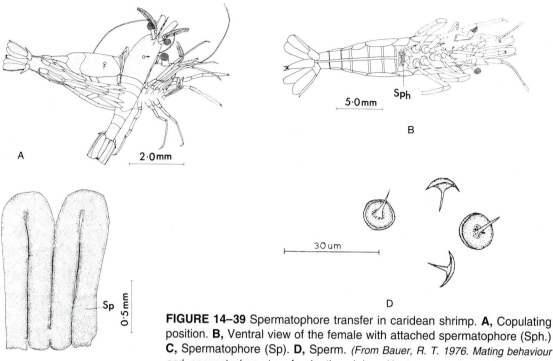

FIGURE 14–39 Spermatophore transfer in caridean shrimp. **A,** Copulating position. **B,** Ventral view of the female with attached spermatophore (Sph.) **C,** Spermatophore (Sp). **D,** Sperm. *(From Bauer, R. T. 1976. Mating behaviour and spermatophore transfer in the shrimp Heptacarpus pictus. J. Nat. Hist. 10:415–440.)*

to the female sternum. In the case of the anomurans, this transfer is aided by a long tubular penis.

Mating in most aquatic decapods occurs shortly after female molting, and the sexes are attracted to each other by pheromones before or after molting, depending on the group. Some sort of precopulatory courtship is typical. For example, the male hermit crab holds the female with one cheliped and taps and strokes her with the other, or pulls her back and forth.

In some brachyuran families, such as the Cancridae *(Cancer)* and Portunidae *(Callinectes, Portunus,* and *Carcinus),* there is a premolt attendance of the female by the male in which the male carries the female about beneath him, her carapace beneath his sternum. He releases her so that she can molt. Copulation occurs shortly afterwards. In contrast to these brachyuran crabs, some male mole crabs are minute and neotenous and attach to the body of the female (Fig. 14–41).

Visual and acoustical signals are of special importance for attraction in terrestrial forms. The semiterrestrial fiddler crabs *(Uca)* go through an elaborate courtship behavior, which has been studied in considerable detail. Sexually active males entice females into their burrows, where mating occurs, and the females remain there until their eggs hatch. Males use the greatly enlarged claw to attract females and to defend their burrow territory against other males (Fig. 14–35A). In male–male encounters there is a highly ritualized combat, in which the large claw is held like a shield. Combat movements involve variations of pushing and extension.

The male fiddler crab attracts a female by waving the large claw in a semaphore fashion. Each species has a characteristic pattern of movement (Fig. 14–42). The male also attracts the female with acoustical signals produced by rapping the elbow (propodus) of the claw against the substratum or by rapid flexion of the ambulatory legs. The number of raps (pulses) in a series and the interval between series is also characteristic of the species. The sounds are picked up largely through the substratum and have been studied with amplifying microphones. Females can detect the signals 50 to 100 cm away from the male by means of special "myochordotonal" organs in the legs.

The courting of tropical species is restricted to daylight hours and, depending on the species, reaches a peak each month whenever the low tides occur

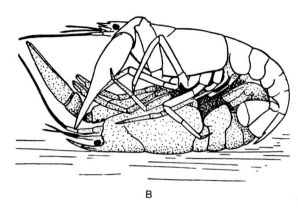

B

A

FIGURE 14–40 Copulatory structures of the brachyuran crab, *Ranina.* The nipple-like structure in front of the end of the dissecting needle is the penis. Just above the penis is the base of the second copulatory pleopod, which projects into the somewhat cylindrical first pleopod. The two pleopods are directed forward and lie in the sternal depression covered by the abdomen. **B,** Copulating crayfish. The female is stippled. *(A, Photograph courtesy of E. Ryan. B, From Andrews in Kaestner, A. 1970. Invertebrate Zoology. Vol. 3. Wiley-Interscience, New York.)*

within certain periods. Temperate species court during both diurnal and nocturnal low tides, but activity is concentrated during two periods within the tidal month (Fig. 14–43). A male of *Uca pugilator* attracts females during the day by waving the claw. If a female approaches, the male also raps his cheliped against the substratum and eventually backs into the burrow, where rapping alone entices the female to

join him (Fig. 14–42). At night, the male attracts the female by rapping only at the mouth of his burrow. These special breeding burrows have chambers that house the female and one burrow may have three chambers, each containing one female. Here mating and egg laying take place.

Male ghost crabs do not wave the cheliped to attract females but use acoustical signals like fiddler

FIGURE 14–41 Dwarf male of the mole crab, *Emerita rathbunae.* **A,** Male (2.5 mm) attached to base of the third leg of the female. **B,** Enlarged view of the male showing attachment by a pair of large spermatophores. *(From Efford, I. E. 1967. Neoteny in sand crabs of the genus Emerita. Crustaceana. 13:81–83)*

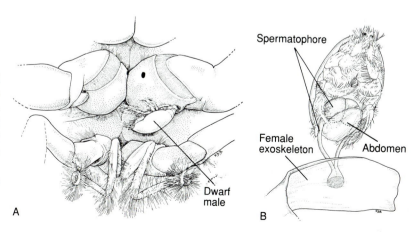

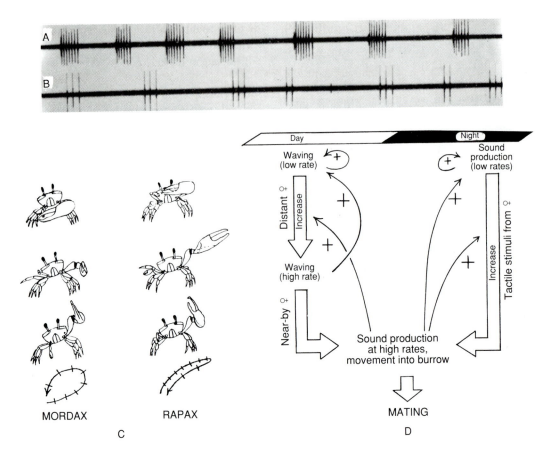

FIGURE 14–42 Courtship behavior in fiddler crab. **A** and **B,** Acoustical signals produced by cheliped rapping in *Uca pugilator* **(A)** and *Uca speciosa* **(B).** Each vertical line on the oscillograph recording is one rap (pulse). **C,** Cheliped waving in two species of *Uca*. The starting position is shown at the top of each of the figures, and the path of the cheliped is described at the bottom by an arrow. The cheliped is commonly moved in a series of jerks, which are indicated by the cross bar on the arrow path. **D,** Diagram of the sequence of behavior in courtship ending in mating. See the text for explanation. *(A–C, From Salmon, M. 1967. Coastal distribution, display and sound production by Florida fiddler crabs (genus Uca). Anim. Behav. 15:449–459. D, From Salmon, M., and Horch, K. W. 1972. Acoustic signalling and detection by semiterrestrial crabs of the family* Ocypodidae. *In Winn, H. E., and Olla, B. L. (Eds.): Behavior of Marine Animals. Vol. I. Plenum Press, New York. pp. 60–96.)*

crabs. Some male ghost crabs, such as *Ocypode ceratophthalmus* along the Red Sea, construct sand pyramids that attract females and also provide orientation.

At copulation, a pair of shrimps is commonly oriented at right angles to each other with the genital regions opposing one another (Fig. 14–39A). The somewhat modified first and second pairs of pleopods are used to transfer a spermatophore to the median receptacle between the thoracic legs of the female. In crayfish and lobsters, the male turns the female over

and pins back her chelipeds with his own (Fig. 14–40B). The first two pairs of pleopods are then lowered to a 45-degree angle and held in this position by one of the last pair of legs (folded beneath the pleopods). In crayfish possessing a seminal receptacle, the tips of the first pleopods are inserted into the chamber, and spermatophores flow along grooves in the pleopods. When the seminal receptacle is absent in lobsters, the spermatophores are attached to the body of the female, particularly at the bases of the last

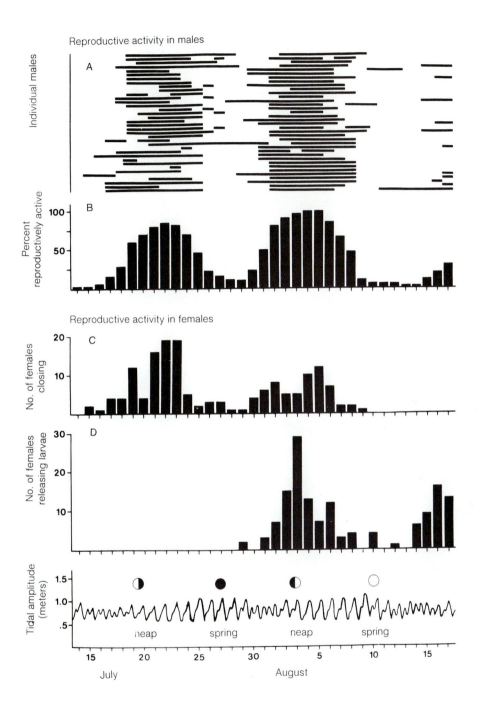

FIGURE 14–43 Relationship of the reproductive activity of the fiddler crab, *Uca pugilator,* to the tidal cycle on the west coast of Florida. Number of females closing refers to the number of females entering the male's burrow, where copulation occurs. The male closes the burrow and waits with the female until she emerges to release larvae. *(From Christy, J. H. 1978. Adaptive significance of reproductive cycles in the fiddler crab Uca pugilator: A hypothesis. Science. 199:453–455. Copyright 1978 by the American Association for the Advancement of Science.)*

two pairs of legs. These are frequently seen as "tar spots" on the females of the spiny lobsters *(Panulirus).*

Hermit crabs partially emerge from their shells to mate. The ventral surfaces are appressed, and spermatophores and eggs are released simultaneously.

During copulation in crabs, the female lies beneath the male or in the reverse position, with ventral surfaces opposing each other. The first pair of pleopods, which conduct the sperm, are inserted into the openings of the female (Fig. 14–44).

Egg laying takes place shortly after copulation in forms with no seminal receptacles, but when sperm are deposited in seminal receptacles, which are commonly sealed and plugged by the semen (as in brachyuran crabs), egg laying may not take place until some time later.

Penaeid and related shrimps shed their eggs directly into the sea water. In all other decapods, the eggs are typically attached to the pleopods with cementing material. The cementing material is associated with the egg membrane.

Fertilization is internal in brachyurans, but in most decapods the eggs are probably fertilized at the moment of egg laying. In crayfish the female lies on her back and curls the abdomen far forward, creating a chamber into which the eggs are driven by a water current produced with the beating pleopods. In crabs, the usually tightly flexed abdomen is lowered to a considerable degree to permit brooding, and the egg mass, which often becomes orange in color, is sometimes called a "sponge."

The hatching stage varies greatly. In all dendrobranchiate shrimps (penaeids and sergestids), where the eggs are shed directly into the water, the hatching stage is a naupliar or metanaupliar larva (Fig. 14–45). In all other decapods, the eggs are carried on the pleopods of the female. In marine species, hatching

FIGURE 14–44 Copulating land crabs *(Gecarcinus lateralis).* The female is uppermost. *(From Bliss, D. E., van Montfrans, J., van Montfrans, M. et al. 1978. Behavior and growth of the land crab Gecarcinus lateralis (Freminville) in southern Florida. Bull. Am. Mus. Nat. Hist. 160(2):137.)*

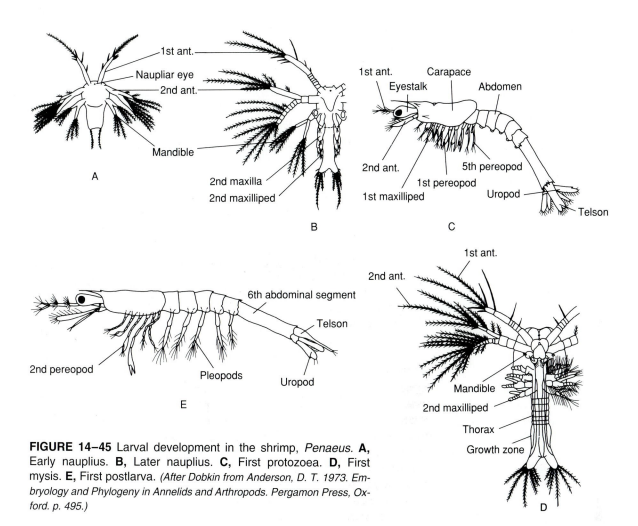

FIGURE 14–45 Larval development in the shrimp, *Penaeus*. **A,** Early nauplius. **B,** Later nauplius. **C,** First protozoea. **D,** First mysis. **E,** First postlarva. *(After Dobkin from Anderson, D. T. 1973. Embryology and Phylogeny in Annelids and Arthropods. Pergamon Press, Oxford. p. 495.)*

takes place at the protozoea and zoea stage. The special names given the zoea and postlarval stage of the different decapod groups are listed in Table 14–1.

The zoea larval stages of most crabs are easily recognized by the very long, rostral spine and sometimes by a pair of lateral spines from the posterior margin of the carapace (Fig. 14–46A). At least in some species, the spines appear to reduce predation by small fish. The postlarval stage, called a **megalops,** has a large or flexed abdomen and the full complement of appendages (Fig. 14–46E).

Many decapods with planktonic larvae inhabit shallow coastal waters, estuaries, or salt marshes as adults or during juvenile stages. For example, estuaries and salt marshes are the nursing grounds for many species of penaeid shrimps. The larvae of such decapods display various locomotor responses to different environmental cues, such as salinity and tidal changes, which help to keep them from being dispersed out to sea.

As is true of many other invertebrates, there is a tendency for larval life to be shortened in decapods that inhabit cold oceans or abyssal depths. Larval stages are usually absent in the strictly freshwater decapods (see p. 175). Brackish-water forms, such as blue crabs *(Callinectes),* and river immigrants, such as *Eriocheir,* return to more saline water for breeding.

Likewise, development in terrestrial anomurans and brachyurans takes place in the sea. The female migrates to the shore and into the water, at which time the zoeae hatch and are liberated. The reproductive events in the common American land crab, *Gecarcinus lateralis,* are facilitated by zonation of the population from the shore landward. During the reproductive period, which usually coincides with the rainy season, the zone nearest the shore is occupied

TABLE 14–1 Types of Postembryonic Development and Larvae in Decapods

Group	Postembryonic Development	Larvae
Suborder Dendrobranchiata		
Family Penaeidae	Slightly metamorphic	Nauplius→protozoea→mysis→mastigopus (zoea) (postlarva)
Family Sergestidae	Metamorphic	Nauplius→elaphocaris→acanthosoma→mastigopus (protozoea) (zoea) (postlarva)
Suborder Pleocyemata		
Infraorder Caridea	Metamorphic	Protozoea→zoea→parva (postlarva)
Infraorder Stenopodidea	Metamorphic	Protozoea→zoea→parva
Infraorder Palinura	Metamorphic	Phyllosoma→puerulus, nisto, or pseudibaccus (zoea) (postlarva)
Infraorder Astacidea	Slightly metamorphic	Mysis→postlarva (zoea)
Infraorder Anomura	Metamorphic	Zoea→glaucothoë in pagurids, grimothea (postlarva)
Infraorder Brachyura	Metamorphic	Zoea→megalopa (postlarva)

Modified from Waterman and Chace, *In* Waterman, T. H. (Ed.): 1960. Physiology of Crustacea. Vol. 1. Academic Press, New York.

by courting males, females near ovulation, and females carrying eggs, and it is here that reproduction occurs. The female migrates into the water to liberate the larvae, vigorously shaking herself as she is washed by waves. A female may produce three broods during the reproductive period. The composition of the various population zones changes over the years, and individuals migrate shoreward to establish themselves within the reproductive zone.

The potamonid crabs, which are terrestrial or amphibious, return to fresh water to breed, although there are no free larval stages.

Order Euphausiacea and Superorder Eucarida

The closest relatives of decapods are members of the small planktonic order Euphausiacea. The Decapoda and Euphausiacea are placed together within the superorder Eucarida because both have a shrimplike body form with a highly developed carapace that is fused with all the thoracic segments.

Euphausiaceans, known as krill, are pelagic, shrimplike crustaceans approximately 3 cm in length. All of the approximately 85 described species are marine. The sides of the carapace do not extend down to tightly enclose the gills, as is true of many decapods (Fig. 14–47A). The thoracic appendages are biramous, and none are specialized as maxillipeds. Euphausiaceans swim with the large, setose pleopods, and most are filter feeders.

Each of the six to eight leglike thoracic appendages bears a long fringe of setae on one side, and together with the other limbs of that side they form one half of a funnel-shaped net or basket (Fig. 14–47A). In at least those species, such as members of the genus, *Euphausia,* that feed on phytoplankton, the filter-feeding process occurs in bouts in response to chemical cues, at which time water is pumped out of the basket. Many krill consume zooplankton, and a few are predacious. Of the 28 species in the central Pacific, 22 depend largely on animal food and only 4 on phytoplankton.

Large, exposed, filamentous gills are present on the thoracic appendages (Fig. 14–47B), and the ventilating current is produced by the thoracic exopods.

Many euphausiaceans are bioluminescent. The light-producing material is not secreted, as in ostracods and other entomostracans, but is intracellular, located within special light-producing organs (photophores) (Fig. 14–47B) (p. 681). The luminescence is probably an adaptation for swarming and reproduction.

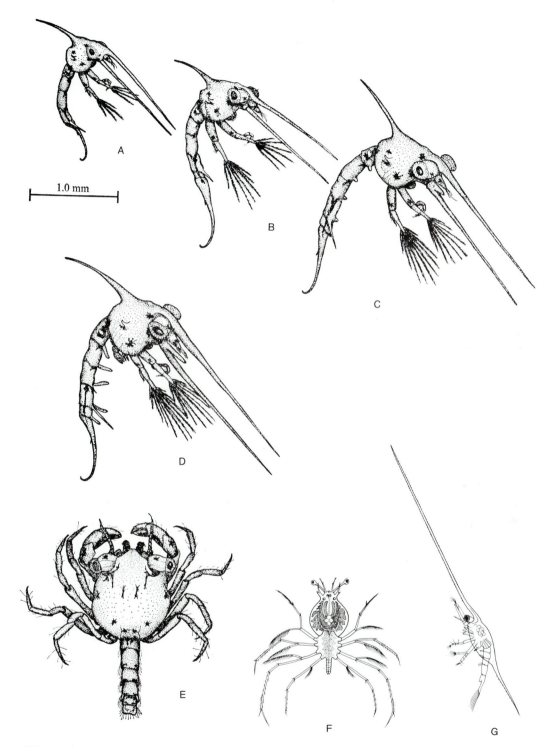

FIGURE 14–46 Decapod larvae. Four zoeal stages (**A, B, C,** and **D**) and megalops **(E)** of the brachyuran crab, *Rhithropanopeus harrisii*. **F,** Phyllosoma (zoea) larva of the spiny lobster, *Palinurus.* **G,** Metazoea of the anomuran crab, *Porcellana. (A–E, From Costlow, J. D., and Bookhout, C. G. 1971. Fourth European Marine Biology Symposium. Cambridge University Press, London. p. 214. Copyrighted and reprinted by permission of the publisher. F, Modified after Claus.)*

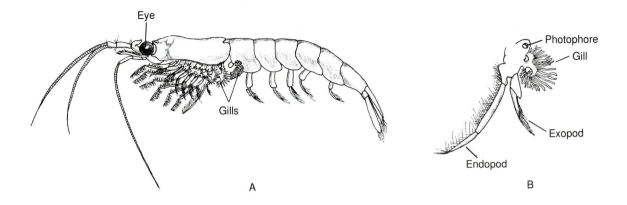

FIGURE 14–47 A, *Meganyctiphanes,* a euphausiacean. **B,** Seventh thoracic appendage. *(Both after Calman.)*

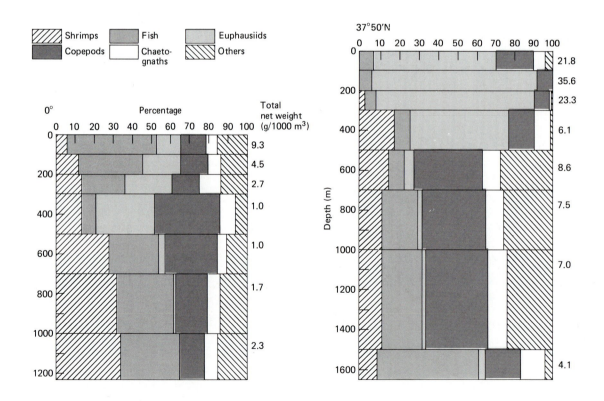

FIGURE 14–48 Percentage composition of the major groups of planktonic and small nectonic animals at various depths at night at tropical and temperate stations in the Pacific along longitude 150° E. *(From Omori, M. 1974. The biology of pelagic shrimps in the ocean. Adv. Mar. Biol. (Academic Press, London) 12:233–324).*

The sperm are transferred to the female in spermatophores, and the eggs may be liberated into the sea water or brooded briefly. A nonfeeding nauplius is the hatching stage. Euphausiaceans are unusual among crustaceans in continuing to molt frequently even after reaching adulthood, and in the absence of an adequate food supply can even become smaller with successive molts.

Many species of euphausiaceans are important components of the pelagic fauna (Fig. 14–48). Such species as the Antarctic *Euphausia superba,* which is about 5 cm in length and transparent, are surface forms; others live at deeper levels or undergo vertical migrations. Many Antarctic species, such as *Euphausia superba,* live in great swarms and constitute the chief food of many other animals. Blue whales may consume a ton of euphausiaceans at one feeding and make up to four such feedings a day. A euphausiacean swarm may cover an area equivalent to several city blocks, and one seen from the air looks like a giant ameba slowly moving and changing shape. Although the swarm may occupy a layer 5 or more meters thick, the several meters of surface water contain the greatest concentration and may reach densities of 60,000 individuals per cubic meter, all of which are adults or near adults. Swarming *Euphausia superba* molt very rapidly—within a second—and if alarmed, many will literally jump out of their skins. The shed molt remains behind and perhaps functions as a decoy.

Fishing for Antarctic krill *(Euphausia superba),* has now become a major industry. Over 395,000 tons were harvested between 1989 and 1990 for human consumption, Russian and Japanese trawlers taking about 95% of the catch.

Subclass Hoplocarida and Order Stomatopoda

The 300 marine crustaceans called mantis shrimp constitute the Stomatopoda, which is the only order of hoplocaridans. Mantis shrimp are highly specialized predators of fishes, crabs, shrimp, and molluscs, and many of their distinctive features are related to their predatory behavior. The body is dorsoventrally flattened with a small, shieldlike carapace and a large, broad, distinctly segmented abdomen (Fig. 14–49).

The well-developed compound eyes are large and stalked, and between them is a naupliar eye. A distinctive feature of stomatopods is the structure of

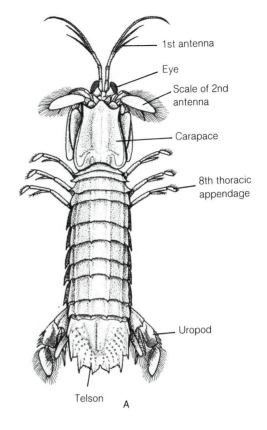

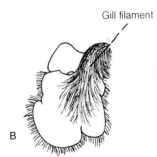

FIGURE 14–49 *Squilla empusa,* a stomatopod. **A,** Dorsal view. Pleopods hidden (see Fig. 14–50). **B,** Second pleopod and gill. *(B, After Calman.)*

the thoracic appendages. The first five pairs are uniramous and subchelate, and the second pair is enormously developed for raptorial feeding. The inner edge of the movable finger is provided with long spines or is shaped like the blade of a knife. Hoplocarida means "armed shrimp" in reference to these appendages, which can be extended rapidly in prey capture or defense. The pleopods are well developed and bear filamentous gills (Fig. 14–49B). The uropods and telson are large.

Mantis shrimp range in size from small species approximately 5 cm long to giant forms greater than 36 cm in length. Most mantis shrimp are tropical, but *Squilla empusa,* which is about 18 cm in length, is a common species inhabiting the North American Atlantic coast and is frequently caught in shrimp trawls. Many stomatopods are brilliantly colored. Green, blue, and red with deep mottling are common, and some species are striped or display other patterns.

Most stomatopods live in rock or coral crevices or in burrows excavated in the bottom. *Squilla empusa* lives in U-shaped burrows but may build vertical burrows with a single opening that extend to 4 m for winter quarters. Some stomatopods use burrows excavated by other animals. One coral-inhabiting species, *Echinosquilla guerini,* has radiating spines that ornament the entire surface of its telson. This armored telson is used to plug the entrance to the burrow when the mantis shrimp is inside. From the exterior, the telson mimics a small sea urchin attached to the coral surface. *Gonodactylus bredini* closes its burrow entrance with debris at night and blocks it during the day with the raptorial appendages. The short carapace probably contributes to the body flexibility needed to turn around and manuever within their burrow.

Many mantis shrimp leave the burrow to feed. They crawl about over the bottom or swim by the powerful oarlike beating of the pleopods. The large, antennal scales and uropods serve as rudders.

Mantis shrimp either spear or smash their prey with the large, second, thoracic raptorial appendage. The distal finger of this appendage lies folded within a groove of the penultant articles much like the blade of a pocket knife within the handle. During a strike the finger unfolds with great force and speed (no more than 4 msec). Some mantis shrimp forage for prey; others lie in wait at the mouths of their burrows. Spearers feed on soft-bodied invertebrates, such as shrimp and fish. The prey is speared by an extremely rapid extension and retraction of the movable finger of the second thoracic appendage, which is provided with barbed spines (Fig. 14–50).

Smashers, especially species of the Gonodactylidae, which live in holes and crevices on rocky or coralline bottoms, stalk their prey: mostly snails, clams, and crabs. Their hard-bodied victims are smashed with the heavy heel of the unfolded raptorial appendage (Fig. 14–51). A crab is disabled by a blow from behind or by having its claws broken immediately. Its legs and carapace are then smashed, and the carcass is dragged back to the hole, where it is consumed. Snails and clams are broken within the burrow, and the shells are deposited outside. The blows of the raptorial appendage are so powerful that captured specimens have cracked the glass walls of aquaria.

Many stomatopods defend their burrow territories against other stomatopods, and those species that live in holes and crevices within coralline rock and rubble have especially complex social behavior. The raptorial appendages are the offensive weapons and

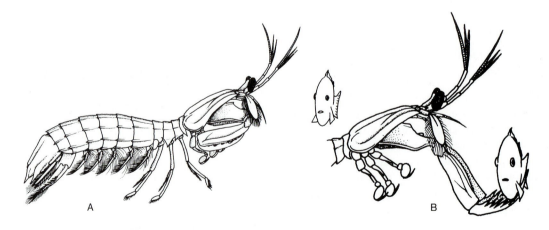

FIGURE 14–50 Prey capture by a mantis shrimp that spears its prey. The second thoracic appendage rapidly unfolds, and the barbs on the terminal finger are driven into the body of the prey.

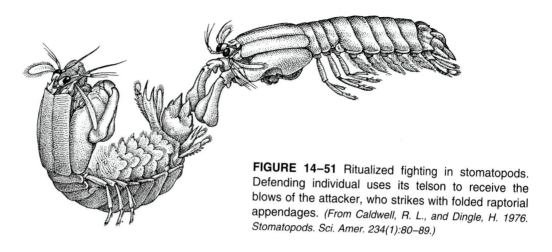

FIGURE 14–51 Ritualized fighting in stomatopods. Defending individual uses its telson to receive the blows of the attacker, who strikes with folded raptorial appendages. *(From Caldwell, R. L., and Dingle, H. 1976. Stomatopods. Sci. Amer. 234(1):80–89.)*

are used as clubs even in those species that spear their prey. A gonodactylid defends itself against opponents' clubs with its large telson, which is heavily armored and thrust forward by the abdomen while the animal lies on its back (Fig. 14–51). Some gonodactylids have large, colored depressions on the basal piece (merus) of the raptorial appendage, which functions in threat display.

Mantis shrimp have the most highly developed compound eyes among crustaceans. Not only do the eyes detect moving objects but, judging from the proximity and position in which the eyes can be held and the accuracy with which the animal can swim to prey, depth perception must be possible. The antennae are important sites of chemoreception and are also used in prey detection when the range is short enough.

Some mantis shrimp pair for life, sharing the same burrow or retreat. Others come together only at the time of mating. The eggs are agglutinated to form a globular mass by means of an adhesive secretion. The egg mass, which may be the size of a walnut with as many as 50,000 eggs *(Squilla),* is carried by the smaller, subchelate appendages and is constantly turned and cleaned (Fig. 14–52A). The female does not feed while she is brooding. Some species keep the egg mass inside the burrow. For example, a Bahamian species of *Gonodactylus,* which lies curled up in a coral crevice, holds the egg mass over the back of the carapace (Fig. 14–52B).

A zoea is the hatching stage and bears a relatively larger carapace than the adult (Fig. 14–53). Planktonic larval life may last for three months.

A

B

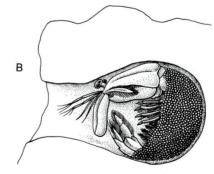

FIGURE 14–52 Two species of stomatopods caring for eggs. **A,** *Squilla mantis* carrying an egg mass. **B,** *Gonodactylus oerstedi* with egg mass in a burrow. *(Modified after Schmitt.)*

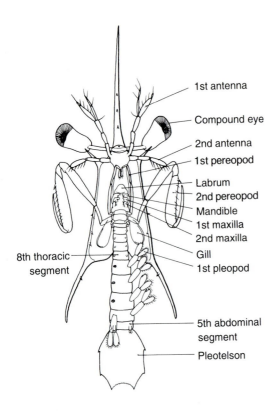

1st antenna

Compound eye

2nd antenna

1st pereopod

Labrum

2nd pereopod

Mandible

1st maxilla

2nd maxilla

Gill

1st pleopod

8th thoracic segment

5th abdominal segment

Pleotelson

FIGURE 14–53 First pelagic, antizoea larva of *Squilla*, a mantis shrimp. (*After Brooks, W. K. 1878. The larval stages of Squilla empusa. Chesapeake Zoological Laboratory: Scientific Reports. pp. 143–170.*)

SUMMARY (MALACOSTRACA THROUGH STOMATOPODA)

1. The class Malacostraca contains the largest number of crustacean species and most of the larger forms. The trunk is composed of a thoracic region of eight segments and an abdominal region of six segments. All trunk segments bear appendages. Usually, one or more thoracic appendages have turned forward and function in food handling. The other thoracic appendages (legs, or pereopods) are used in prehension or crawling. The abdominal appendages are pleopods and one or more pairs of terminal uropods.

2. The superorder Eucarida is distinguished by stalked, compound eyes and a carapace that is fused with all the thoracic tergites.

3. The eucarid order Decapoda contains almost one quarter of the species of crustaceans. The first three pairs of thoracic appendages are modified as maxillipeds, leaving five pairs of legs, hence the name of the order. The first pair of legs is commonly modified as a large claw, or cheliped. Most of the larger species of crustaceans are decapods.

4. Shrimp, lobsters, and crayfish have well-developed abdomens (long-bodied forms), but many decapods have the abdomen reduced and folded beneath the thorax. The short body form evolved independently a number of times, and all such species, called crabs, are not necessarily related.

5. The three groups of shrimp are adapted for swimming. The body is laterally compressed, and the pleopods, which are the swimming appendages, are large and fringed. The legs are long and slender, chelipeds may or may not be present, and the exoskeleton is relatively flexible. Most shrimp are bottom dwellers and swim intermittently.

6. Most other decapod groups (crayfish, lobsters, and crabs) are benthic decapods and adapted for crawling. The legs are heavier than those of shrimp, and the pleopods are never used in swimming. The pleopods are retained only for reproductive functions. The body is somewhat dorsoventrally flattened, and the exoskeleton is relatively rigid. Crayfish and lobsters (infraorders Astacidea and Palinura) have retained a well-developed abdomen.

7. Members of the infraorder Brachyura, or "true" crabs, are the most successful of the decapods with short bodies, for this group includes more than 4500 species (one half of the decapods) and they are diverse and widely distributed. In brachyurans, folding of the abdomen beneath the thorax, which throws the center of gravity forward beneath the legs, appears to have enhanced motility.

8. The infraorder Anomura is characterized by reduction or dorsal position of the fifth pair of legs, but the modification appears to have evolved independently in the different anomuran groups.

9. Mole crabs (anomuran superfamily Hippoidea) are adapted for suspension feeding and burrowing backward in sand. Some inhabit surf beaches. Porcelain crabs (anomuran superfamily Galatheoidea) live beneath stones and move rapidly. The short body form of this group appears to be an adaptation for motility.

10. The central group of the anomuran superfamily Paguroidea is that of the hermit crabs, which are adapted for housing the abdomen within gastropod shells. The ancestors of this group probably were decapods that backed into holes or other retreats, and the primitive modern forms have symmetrical abdomens. Various species have become secondarily

adapted for utilizing other objects for housing, and the coconut crabs *(Birgus)* and the lithodid crabs (Lithodidae) have abandoned the hermit habit and evolved the short body form. Their hermit ancestry, however, is reflected in the unpaired pleopods of the female.

11. The greatest number of decapods couple predation with scavenging, but there are species that are filter feeders, detritus feeders, and herbivores. Enzymes produced by the midgut hepatopancreas pass forward into the large foregut cardiac stomach, where digestion begins. The cardiac stomach is adapted for trituration and the pyloric stomach for filtering material into the hepatopancreas, where absorption occurs.

12. The gills of decapods are dorsal evaginations of the body wall near the junction of the thoracic appendages and the trunk. They are protected by the overhanging carapace, which forms a gill chamber. The ventilation current is produced by the gill bailer, a process of the second maxilla. It generally pulls water through the gill chamber and therefore is at the point of exit of the ventilation current. The intake point is determined by how closely the carapace fits against the trunk.

13. The majority of decapods are marine osmoconformers, but numerous species have invaded fresh water and land. The gills are a major site for excretion of ammonia, however, and except for crayfish and a few shrimp, the urine of the antennal glands is isosmotic with the blood. For all freshwater decapods, the gills are an important site of ion absorption to balance loss through the urine.

14. All terrestrial decapods are burrowers and nocturnal. None can produce a hyperosmotic urine. Except for the Potamoidea, land crabs must return to the sea to breed, and development is indirect.

15. Most male decapods have the anterior pleopods adapted as copulatory organs. Most female decapods brood their eggs attached to the pleopods. A zoea is the typical larval hatching stage. Development in many freshwater species is direct.

16. The small eucarid order Euphausiacea includes marine planktonic species. They are shrimplike in appearance, but none of the anterior thoracic appendages are modified as maxillipeds. Some of the filter-feeding species occur in enormous numbers in waters of high primary productivity and are important in marine food chains.

17. The superorder Hoplocarida, order Stomatopoda (mantis shrimp), is a group of relatively large, raptorial malacostracans in which the large, subchelate, second thoracic appendage is adapted for spearing or smashing prey. They live in burrows or in natural retreats and tend to be territorial.

Superorder Peracarida

The superorder Peracarida is a large assemblage of seven orders, which together with the decapods make up the majority of malacostracans, indeed the majority of crustaceans. Approximately 30% of crustaceans are peracaridans. Although most are less than 2 cm in length and therefore not as conspicuous as decapods, peracaridans, especially isopods and amphipods, are abundant and widespread in the sea, in fresh water, and on land. The distinctive characteristic of the group is the presence of a ventral brood pouch, or **marsupium,** in the female. The marsupium is formed by large, platelike processes (**oostegites**) on certain thoracic coxae. The oostegites project inward horizontally and overlap with one another to form the floor of the marsupium (Fig. 14–54). The thoracic sternites form the roof. The marsupial oostegites may make their appearance as small projections during juvenile instars, but their development is under hor-

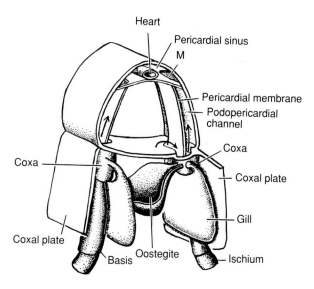

FIGURE 14–54 Diagrammatic cross section through the body of an amphipod. Only one marsupial plate, or oostegite, is shown. Arrows indicate blood flow. *(After Klovekorn and others, from Kaestner, A. 1970. Invertebrate Zoology. Vol. 3. Wiley-Interscience, New York.)*

monal control, and they are not completely formed until the reproductive instar. Development is direct, and on release from the brood chamber the young (postlarvae) have most of the adult features.

A carapace may be present or absent. The naupliar eye never persists in the adult. Of the eight pairs of thoracic appendages, the first is usually a pair of maxillipeds and the remaining seven are legs.

Primitively, peracaridans are maxillary suspension feeders, as are many mysids, cumaceans, and tanaidaceans, but the tendency in most higher peracaridans has been toward other modes of feeding.

We begin our survey of peracaridans with the two largest orders: the Amphipoda and Isopoda.

Order Amphipoda

The more than 6000 species of amphipods constitute the largest group of peracaridans. The order includes a great diversity of species, which are placed within over 100 families. Most are marine, but there are many freshwater species and a family of terrestrial forms.

The body of amphipods tends to be laterally compressed, giving the animal a somewhat shrimplike appearance (Fig. 14–55). The compound eyes are sessile. A carapace is absent, although the first thoracic segment and sometimes the second are fused with the head. Also, the abdomen is usually not distinctly demarcated from the thorax in either size or shape. The first and second antennae are usually well developed, and the first pair of thoracic appendages are modified to form maxillipeds. The coxae of the seven pairs of legs are usually long, flattened plates that increase the appearance of lateral body compression (Fig. 14–55). The second and third thoracic appendages (pereopods, or legs) are usually enlarged and subchelate for prehension and are called **gnathopods.** Gills are borne on the bases of the thoracic appendages. The anterior three pairs of abdominal appendages are pleopods and are used in swimming and for ventilation. Unlike other malacostracans, the posterior three pairs are uropods, are directed backward, and are used for jumping, burrowing, or swimming, depending on the species.

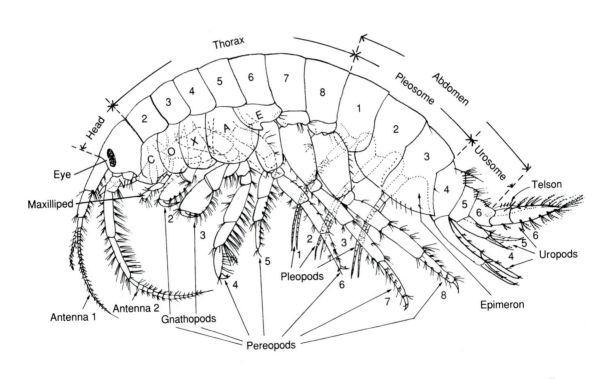

FIGURE 14–55 External structure of a gammaridean amphipod in lateral view. *(From Bousfield, E. L. 1973. Shallow-Water Gammaridean Amphipoda of New England. Cornell University Press, Ithaca, NY. 344 pp.)*

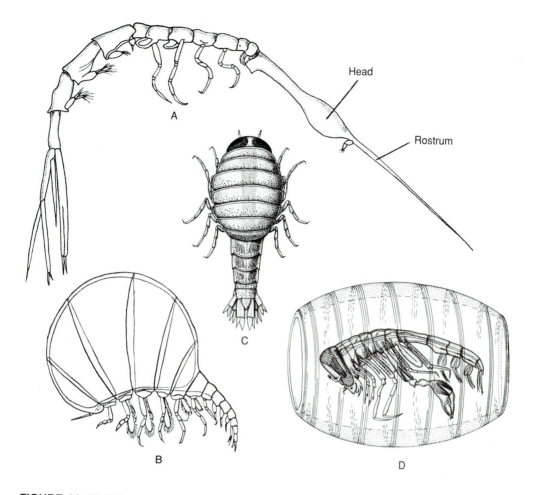

FIGURE 14–56 Pelagic hyperiidean amphipods. **A,** *Rhabdosoma,* an amphipod with elongated head and needle-like rostrum. **B,** *Mimonectes,* 2.5 cm long. **C,** *Hyperia,* 2 cm long. Species of this genus are commonly found attached to pelagic coelenterates and ctenophores. **D,** *Phronima sedentaria,* 3 cm long, which lives within salps. *(A, After Stebbing from Calman. B, After Schellenberg from Kaestner, A. 1970. Invertebrate Zoology. Vol. 3. Wiley-Interscience, New York. C, After Sars. D, After Claus.)*

Exceptions to all of these typical amphipod features exist, and some groups have diverged widely from the general plan.

Most amphipods are between 5 and 15 mm long. The giant of the order was thought to be the marine *Alicella gigantea,* which may reach 14 cm in length, but in 1968, an undescribed 28 cm benthic lysianassid amphipod was photographed from 5300 m in the Pacific. The smallest forms, mostly blind interstitial species, are less than 1 mm long. Most amphipods are translucent, brown, or gray, but some species are red, green, or blue-green.

Locomotion and Habitation

Hyperiidean amphipods are pelagic, and for at least part of their life cycle probably all live commensally with such gelatinous planktonic animals as salps, siphonophores, jellyfish, and ctenophores. Members of the family Phronimidae live in "barrels" constructed from salps (Fig. 14–56D), which they propel through the water. The strange swollen shape of the head and thorax of some species (*Hyperia, Mimonectes,* and *Rhabdosoma*) may be related to their commensal habits or contribute to flotation. Many have very large eyes (Fig. 14–56C, D). The Antarctic

Hyperiella dilatata, which is preyed on by fish, has evolved a remarkable defense. It carries on its back a sea butterfly (an opisthobranch mollusc—*Clione limacina*) that fish find distasteful.

Gammarids include many of the common amphipods. They are essentially bottom dwellers, but most can swim, even if infrequently. Propulsion for swimming is provided by the pleopods and in many groups, by the uropods. Usually, they swim intermittently between crawling and burrowing; in leaving the substratum, initial thrust is commonly gained by a backward flip of the abdomen. Walking is effected by the legs, but in rapid movements over the bottom, both legs and pleopods are used, and the animal often leans far over to one side.

Some amphipods, especially the caprellids, are adapted for climbing. Caprellids, called skeleton shrimp, have long slender bodies with greatly reduced abdomens (Fig. 14–57B). The tips of the legs are provided with grasping claws for clinging to hydroids, bryzoans, and algae over which many species crawl like inchworms. Some caprellids live on sea stars or even on spider crabs.

Many amphipods are accomplished burrowers, and some construct tubes of mud or of secreted material (Fig. 14–58). The burrows may be horizontal, vertical, or U-shaped with two openings. Sometimes the walls are simply packed, but often they are reinforced with secreted material.

Tube construction is peculiar to corophioidean and ampeliscoidean amphipods, and a variety of materials, such as mud, clay, sand grains, and shell and plant fragments may be used. The material is usually bound together with a cementing secretion produced from glands in the fourth and fifth thoracic appendages. *Haploops* mixes the secretion with clay, and the resulting stiff mass is then drawn out between the end of the abdomen and the gnathopods and applied to the walls.

Several tube dwellers, such as species of *Siphonoecetes* and *Cerapus,* build unattached tubes of shell fragments and sand grains and carry the tubes about with them (Fig. 14–58C). *Cyrtophium* makes its tubes out of a section of a hollow plant stem, which it lines with secreted materials; the animal can even swim with the tube, beating the antennae in an oarlike fashion. A species of *Pseudamphithoides* builds its mobile home selectively from a certain species of brown alga. The chemical defenses of the alga are thus conferred upon the amphipod, protecting it against reef fish.

A few amphipods have rather unusual retreats. Some species of *Siphonoecetes* live in old tooth shells, and *Photis conchicola* builds its tubes in an empty gastropod shell. Like the isopod, *Limnoria,* the amphipod, *Chelura terebrans,* bores in wood. The Stenothoidae are commensals of sponges and ascidians, and *Dulichia rhabdoplastis* lives on sea urchin spines (Fig. 14–59E). Members of the suborder

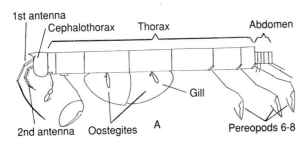

1st antenna
Cephalothorax   *Thorax*   *Abdomen*
Gill
2nd antenna   *Oostegites*   *A*   *Pereopods 6-8*

B

FIGURE 14–57 Skeleton shrimp (Caprellidea). **A,** Diagram of a typical caprellidean skeleton shrimp. The abdomen is stippled; marsupial plates and bat-shaped gills appear in the center of the thorax. **B,** *Caprella equilibra* clinging to seaweed. *(A, From Laubitz, D. R. 1970. Studies on the Caprellidea of the American North Pacific. Nat. Mus. Can. Publ. Biol. Oceanogr. 1:1–89. B, Photograph courtesy of D. P. Wilson.)*

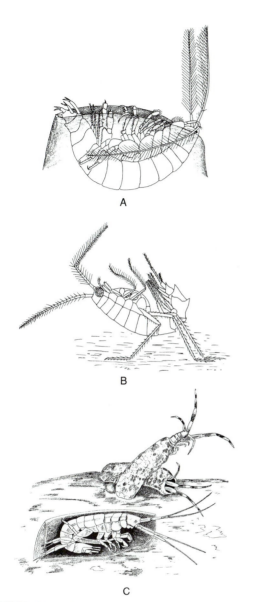

FIGURE 14–58 A, *Haploops tubicola,* hanging on the edge of its tube in a feeding position. The margin of the tube is shaded but shown as transparent. **B,** *Melphidippella macra* in feeding stance. **C,** *Cerapus,* a tubicolous amphipod. The tube, which the animal can carry about, is composed of an inner, secreted layer covered with fragments of foreign material. *(A–B, After Enequist. C, Modified after Schmitt.)*

Ingolfiellidea, some of which are less than 1 mm long, are adapted for an interstitial existence in both marine and freshwater habitats (Fig. 14–59F).

Freshwater amphipods, mostly gammarids, are common benthic animals in algae and other vegeta-

tion of streams, ponds, and lakes and in subterranean waters and are sometimes found in great numbers. Population densities for *Gammarus lacustris* of 10,000 per square meter have been reported from certain springs. Lake Baikal in Siberia contains a remarkable endemic gammarid fauna of nearly 300 species, some brightly colored red and blue and some reaching a size of over 6 cm. These are both pelagic and benthic species.

In contrast to isopods, amphipods have been less successful in adapting to life on land. The terrestrial species are members of the family Talitridae. The beach fleas live beneath drift or stones or burrow in the sand near the high-tide mark, but the leaf litter hoppers of the Southern Hemisphere are found in moist humus and soil away from the shore. They continue to use gills as gas exchange surfaces and most are not very resistant to desiccation.

Beach fleas scull rapidly over the sand, gaining additional power with pushing strokes from the abdomen. They can jump, using a sudden backward extension of the abdomen and telson. *Talorchestia,* 2 cm in length, can leap forward 1 m. Most beach fleas also burrow. In *Talorchestia,* the body is braced with the second and third pairs of legs while the first pair of gnathopods sweep the substratum material back to the uropods and telson, which flip it away. Like the isopod, *Tylos,* the beach flea, *Talitrus,* has been shown to use its eyes to obtain astronomical clues for locating the high-tide zone, in which these amphipods normally live. If displaced either above or below the high-tide mark, the animals migrate accurately back to their normal zone. The angle of the sun is used as a compass in conjunction with a map sense of the east–west orientation of the particular beach they inhabit.

The angle of the sun is a primary clue for orientation, because the direction of movement of the animal can be controlled in experiments that reflect the rays of the sun from different angles. An internal clock mechanism provides interpolation for the changing angle of the sun during the course of the day. This aspect of the mechanism was supported by transporting *Talitrus* in the dark to a beach at different longitude. On liberation, the immigrants operated on the same time as that of the original location. In the absence of direct sunlight, these amphipods are reported to use sky polarization in the same manner. Other factors besides the celestial clues are utilized in orientation—horizon level, beach slope, sand moisture, and grain size—but visual clues are most important.

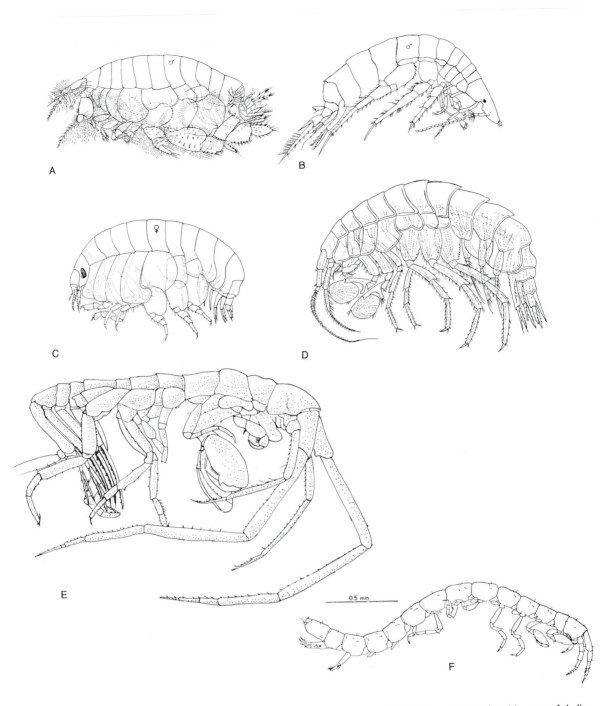

FIGURE 14–59 Diversity in amphipods. **A,** Burrowing *Haustorius canadensis* (Haustoriidae), with powerful digging legs and antennae and with filter-feeding mouthparts. **B,** Burrowing *Platyischnopus herdmani* (Platyischnopidae), with slender body and "shark snout." **C,** *Orchomenella minuta* (Lysianassidae), a flesh-scavenger. **D,** *Eusirus cuspidatus* (Eusiridae), a predatory and free-swimming species. **E,** *Dulichia rhabdoplastis,* commensal on sea urchins along the north Pacific coast of the United States. The amphipod lives on detritus strands stretched between the sea urchin spine tips and constructed from its own fecal pellets. The amphipod feeds on diatoms growing on the strands or filters detritus and plankton from the surrounding water with outstretched antennae. **F,** *Ingolfiella putealis* (Ingofiellidea), an interstitial amphipod, with vermiform body and no eyes. *(B, From Barnard, J. L. 1969. The families and genera of marine gammaridean Amphipoda. U. S. Nat. Mus. Bull. 27:1–535. D, After Sars, from Bousfield, E. L. 1973. Shallow-Water Gammaridean Amphipoda of New England. Cornell University Press, Ithaca, NY. E, From McCloskey, L. R. 1970. A new species of Dulichia commensal with a sea urchin. Pac. Sci. 24:90–98. By permission of the University of Hawaii Press. F, From Stock, J. H. 1976. A new member of the crustacean suborder. Ingolfiellidea from Bonaire, with a review of the entire suborder. Vitg. Naturwet. Studiekring Suriname Ned. Antillen. 86:57–75.)*

Nutrition

There are some herbivorous amphipods, but most are detritus feeders or scavengers. Mud or animal and plant remains are picked up with the gnathopods, or detritus is raked from the bottom with the antennae, particularly the second pair. Some burrowing forms scrape detritus and diatoms from sand grains. Daily consumption of detritus in gammarids may be as great as 100% of the body weight for juveniles and 60% for adults.

A number of amphipods are filter feeders, but different appendages are adapted for food collection. *Aora, Corophium,* and other tube builders strain fine detritus through filter setae on the gnathopods, the feeding current being provided by the pleopods. Many gammarids, including the tube-dwelling *Haploops,* extend the antennae into the natural water current as a net. *Haploops* hangs upside down in the mouth of the tube, clinging to the rim with its specialized legs (Fig. 14–58A). *Melphidippella,* which sits upside down on the bottom, uses not only the second antennae but also the third and fourth pairs of thoracic appendages as a filter. All three limbs are held outstretched and catch falling detritus (Fig. 14–58B). At intervals, the limbs are scraped by the gnathopods. The beach burrower, *Haustorius arenarius,* uses a maxillary filter.

Although many amphipods supplement their diet by catching small animals, strictly predacious feeding is not common. The most notable examples of predacious feeders are the pelagic hyperiids, two families of gammarid (eusirids and pardaliscids), and some caprellids (skeleton shrimp). A raptorial caprellid attaches itself to a hydroid or bryozoan stem with the last pair of thoracic appendages and projects itself motionless and outstretched, waiting to seize passing prey, such as copepods, with its gnathopods. Many caprellids, however, feed on diatoms and detritus that collect on the surfaces of the organisms on which they live, or filter feed with their antennae. The young of *Caprella scaura,* a common skeleton shrimp of tropical waters, scrape epiphytes from the exoskeleton of their mother, on which they live for a short time after hatching and leaving the marsupium.

Parasitism is much less prevalent among amphipods than among pericaridan isopods. There are a few ectoparasites of fish, with suctorial mouth parts. The cyamids, called whale lice, have legs adapted for clinging to the host, but they probably feed on diatoms and debris that accumulates on the whale's skin (Fig. 14–60).

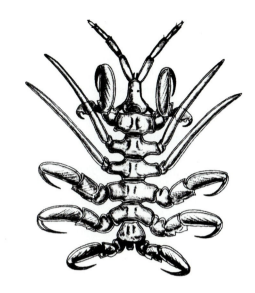

FIGURE 14–60 *Cyamus boopis,* an amphipod commensal on whales. *(After Sars.)*

Gas Exchange, Internal Transport, and Excretion

The gills, which are thoracic, are usually lamellae or sacs attached to the inner face of the coxae of the legs (pereopods 2 to 7) (Fig. 14–54). Typically, the posteriorly flowing ventilating current is produced by the pleopods. A ventilating current is particularly important in those amphipods that dwell in burrows or tubes. Oxygen is transported in the blood by hemocyanin.

Although they live out of water, the Talitridae have retained the gills. The air must be humid, and talitrids are thus restricted to living in moist sand beneath drift or in damp forest leaf litter away from the sea. They feed at night, when there is less danger of desiccation.

The amphipod heart is tubular, with one to three pairs of ostia, and lies in the thorax above the coxal gills. The arterial system is not greatly developed. The excretory organs are antennal glands.

Reproduction

The gonads are paired and tubular. The male gonopores open at the end of a pair of long penis papillae on the sternum of the last thoracic segment, and the female oviducts open on the sixth thoracic coxae. In most species, pleopods do not function as copulatory organs. By means of sensory aesthetascs on the first antennae, males are attracted by female pheromones. In species of *Gammarus* and some other genera, the male carries the female around beneath him for days, clasping the thoracic region on the coxal plates of the female with his gnathopods. The animals separate briefly to permit the final, preadult molt of the female. Actual sperm transfer is accomplished quickly. The male twists his abdomen around so that his uropods touch the female marsupium; when sperm are emitted, they are swept into the marsupium by the ventilating current of the female. The pair separate, and the eggs are soon released into the brood chamber, where fertilization takes place. Development is direct. The ventilating current also provides for the ventilation of the eggs in the marsupium. The marsupia of most gammarideans bear interlocking marginal setae that aid in preventing the eggs from falling out. One annual brood of 15 to 50 eggs is common in most temperate freshwater species. In marine forms, there may be 2 to 750 eggs in a clutch, and more than one brood per year is common, but the life span is generally about one year. For example, the intertidal estuarine *Gammarus palustris,* which lives along the east coast of the United States, overwinters as an adult; the first eggs appear in February (Maryland). Overwintering populations die out during the spring and early summer, and the new generation produces broods during late summer and early fall.

Order Isopoda

The order Isopoda is the second largest order of crustaceans. Most of the 4000 described species live in the sea, where they are widely distributed and occupy all types of habitats. One group (suborder Paraselloidea) forms one of the most abundant components of the deep-sea benthic fauna. There are a considerable number of freshwater isopods, and the pill bugs, or wood lice, are the largest group of truly terrestrial crustaceans. The order also includes many parasitic forms.

The most striking characteristic of isopods is the dorsoventrally flattened body (Figs. 14–61; 14–62B). The head is usually shield-shaped, and the terga of the thoracic and the abdominal segments tend to project laterally. As in amphipods, a carapace is absent, although the first one or two thoracic segments are fused with the head. The abdominal segments may be distinct or fused to varying degrees. The last abdominal segment is almost always fused with the telson; in the Asellota, which includes the greatest number of freshwater species, all but the first or second abdominal segments are fused with the telson to form a large abdominal plate (Fig. 14–61A). The abdomen is usually the same width as the thorax, so that the two regions may not be clearly demarcated dorsally.

The first antennae are short and uniramous, and in terrestrial isopods they are vestigial. The compound eyes are always sessile. The first pair of thoracic appendages is modified to form maxillipeds; the remaining seven pairs of thoracic appendages are legs usually adapted for crawling. In some groups, the more anterior pairs are modified as prehensile gnathopods (Fig. 14–61A). Unlike those in most other crustaceans, some of the isopod pleopods are used for gas exchange.

Most isopods are 5 to 15 mm in length. The giant of the group is the deep-sea *Bathynomus giganteus,* which reaches a length of 42 cm and a width of 15 cm. The coloration in isopods is usually drab, shades of gray being the most common. Chromatophores adapt the body coloration to the background in many species.

Locomotion

Isopods are benthic animals, and most are adapted for crawling. *Ligia,* one of the most common isopods along coastlines, can run very rapidly over exposed

---

FIGURE 14–61 A, The freshwater isopod, *Asellus* (dorsal view). **B,** Female *Asellus* (ventral view). Appendages ▶ complete only on one side. **C–D,** *Sphaeroma quadridentatum,* a common, shallow-water marine isopod found on algae and pilings. All but one abdominal tergite are fused with the telson. The members of this genus are capable of rolling into a ball, as are many terrestrial isopods. **C,** Dorsal view. **D,** Ventral view. **E,** *Cyathura,* a genus of marine isopods adapted for living in burrows in mud and sand bottoms. The many other species of this large family have a similar habit. **F,** *Idotea pelagica* (dorsal view). This and other members of the large marine family Idoteidae are common inhabitants of algae in shallow water. *(A, After Pennak, R. W. 1978. Freshwater Invertebrates of the United States. 2nd Edition. John Wiley and Sons, New York. B, After Van Name, W. G. 1936. The American land and freshwater isopod Crustacea. Bull. Amer. Mus. Nat. Hist. 71:7. E, After Gruner from Kaestner. F, After Sars.)*

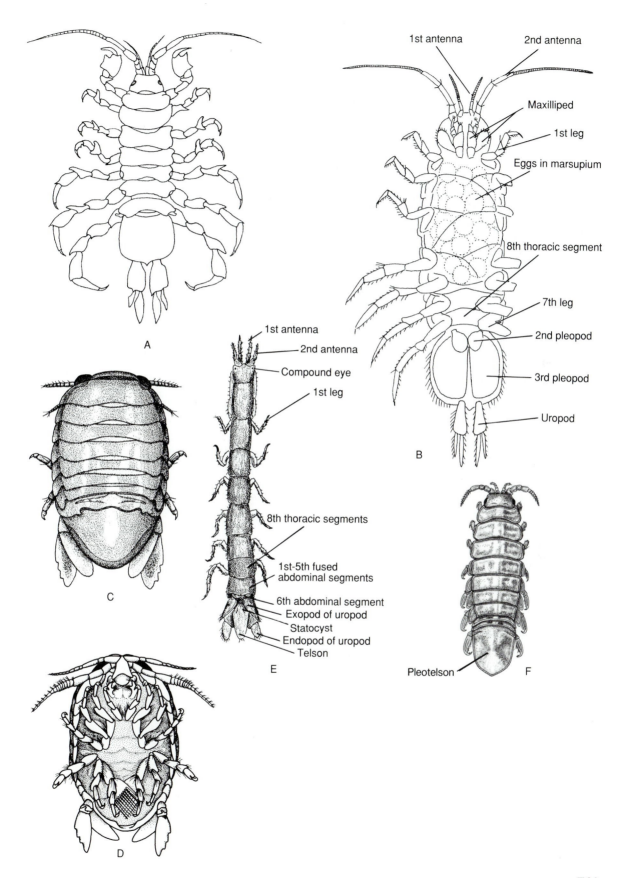

A

B
- 1st antenna
- 2nd antenna
- Maxilliped
- 1st leg
- Eggs in marsupium
- 8th thoracic segment
- 7th leg
- 2nd pleopod
- 3rd pleopod
- Uropod

C

D

E
- 1st antenna
- 2nd antenna
- Compound eye
- 1st leg
- 8th thoracic segments
- 1st-5th fused abdominal segments
- 6th abdominal segment
- Exopod of uropod
- Statocyst
- Endopod of uropod
- Telson

F
- Pleotelson

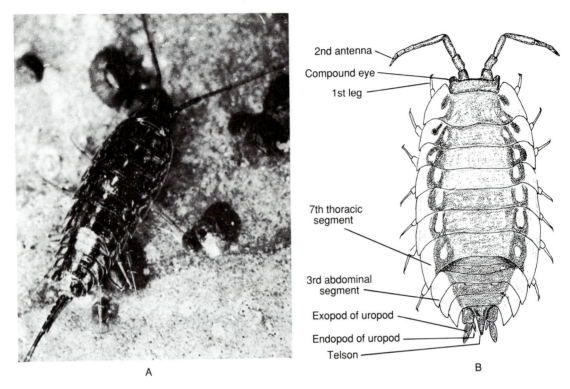

2nd antenna
Compound eye
1st leg
7th thoracic
segment
3rd abdominal
segment
Exopod of uropod
Endopod of uropod
Telson

A   B

FIGURE 14–62 Terrestrial oniscoidean isopods. **A,** *Ligia* on a sea wall at low tide. The barnacles are little less than 1 cm in diameter. The members of this widely distributed genus are fast-moving species that live at the edge of the sea just above the water's edge on rocks, pilings, and sea walls. **B,** *Oniscus asellus,* a European pill bug common in gardens, hot houses, and human habitations. It has been introduced into North America. *(A, Photograph courtesy of Betty M. Barnes. B, After Paulmier from Van Name.)*

wharf pilings and rocks, even upside down. Many aquatic isopods burrow, and some construct tunnels through the substratum, packing excavated material against the walls. Some species of *Limnoria* tunnel through wood and can cause extensive damage to docks and pilings (Fig. 14–63); others bore into the holdfasts of kelps. *Sphaeroma tenebrans* bores into the prop roots of mangroves.

Aquatic isopods can usually swim as well as crawl. Most commonly the pleopods are used for swimming, and in the families Sphaeromatidae and Serolidae the first three pairs of pleopods are especially adapted for swimming; gas exchange is restricted to the more posterior pleopods.

The ability to roll up in a ball has evolved in many terrestrial Oniscoidea. Many marine sphaeromids also roll up, with the sharp and spiny tips of the uropods and telson exposed.

Nutrition

Most isopods are scavengers and omnivores, although some tend toward a herbivorous diet. Deposit feeding is common. Wood lice feed on algae, fungi, moss, bark, and any decaying vegetable or animal matter. A few wood lice are carnivorous, as are some marine species, such as the intertidal *Cirolana* and the large, pelagic *Bathynomus.*

Wood-boring marine isopods feed on wood, and their hepatopancreatic secretions include cellulase. At settling, the wood-boring species of *Limnoria* are attracted to fungi in the wood. The fungi add nitrogen to their largely cellulose diet. In terrestrial wood lice cellulose digestion results from bacteria, and the hindgut plays a major role in the digestive process.

There are several groups of parasitic isopods. The Gnathiidae in the larval stages and the adult Cymothoidae are ectoparasitic on the skin of fish (Fig.

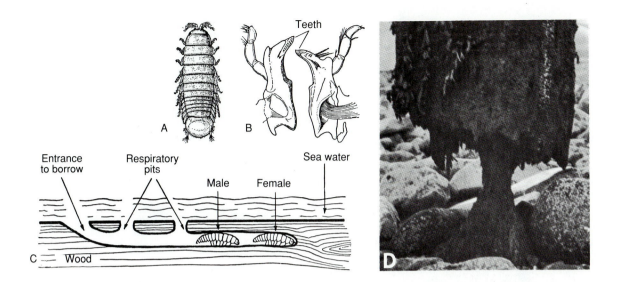

FIGURE 14–63 A, The wood-boring isopod, *Limnoria lignorum*. **B,** Mandibles of *Limnoria lignorum*. **C,** Diagrammatic section of a burrow of *Limnoria lignorum*. **D,** Jetty piling nearly eaten through at the base by *Limnoria lignorum*. *(A, After Sars from Yonge. B, After Hoek from Yonge. C, From Yonge, C. M. 1949. The Seashore. Collins, London.)*

14–64C) and have mandibles adapted for piercing. Piercing mouth parts are also present in the parasites composing the suborder Epicaridea, all of which are bloodsuckers, parasitic on many crustacean groups. Many epicaridans are highly modified and show little resemblance to free-living forms (Fig. 14–64D–I).

Gas Exchange and Excretion

The pleopods of isopods provide for gas exchange. In primitive forms, each pleopod ramus is modified as a large, flat lamella, and both rami of each pleopod function in both gas exchange and swimming (Fig. 14–61B). There is, however, usually some modification of this arrangement. In some isopods, gas exchange and swimming are divided among the pleopods, the anterior ones being fringed and coupled for swimming and the posterior pleopods being for gas exchange.

The pleopods typically lie flat against the underside of the abdomen and are often protected by a covering (the operculum) formed by the first pair of pleopods or by the exopods of one or more pairs. In the marine Valvifera, the uropods are greatly elongated and meet at the midline ventrally to form a gill covering resembling two doors. If gas exchange

through the gill is blocked in the terrestrial *Ligia* and *Oniscus,* oxygen consumption is reduced by only about 50%, indicating that the general integumentary surface is equally important as a site of gas exchange. The blood of isopods contains hemocyanin. The excretory organs are maxillary glands.

Adaptations for Life on Land

The terrestrial isopods—the wood lice, or pill bugs—are members of the suborder Oniscoidea. They are believed to have invaded land directly from the sea, rather than by way of fresh water, and have come to occupy a wide range of habitats and to display varying degrees of toleration to desiccating conditions (Fig. 14–65). Some live at the edge of the sea and some in marshes. Many species live beneath stones, in bark, and in leaf mold in both temperate and tropical regions, but there are also species that live in deserts.

Shore-inhabiting forms include the widespread *Ligia,* which lives on pilings, jetties, and rocks at the water's edge, and *Tylos,* which lives beneath beach drift or sand at the high-tide mark. *Tylos* emerges at low tide to feed on algae and other debris and orients to the high-tide mark by means of beach slope, sand

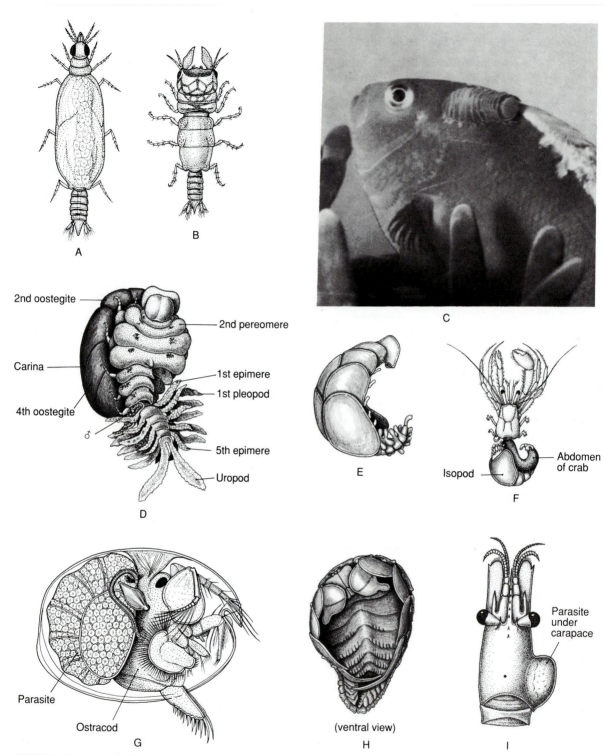

FIGURE 14–64 A–C, Parasitic isopods. **A** and **B,** An aberrant gnathidean isopod, *Gnathia maxillaris,* that looks like an insect. The larval stage **(B)** is parasitic on fish and possesses crowded sucking mouthparts. Both larva and adult **(C)** are less than 3 mm long. **C,** An isopod fish louse, *Rocinela,* on a fish living with a sea anemone. **D–I,** Epicaridean parasitic isopods. **D,** The bopyrid, *Cancricepon elegans,* parasitic in the gill chambers of certain crabs. **E** and **F,** *Athelges tenuicaudis* on the abdomen of a hermit crab. **G,** Isopod, *Cyproniscus,* in ostracod, *Cypridina.* **H** and **I,** Ventral view of *Bopyrus squillarum* and location in branchial cavity of a shrimp. *(A and B, Based on living specimens and figures by Sars. C, Photograph courtesy of Mariscal, R. N. 1969. Crustaceana. 14:7–104. D–I, After Sars.)*

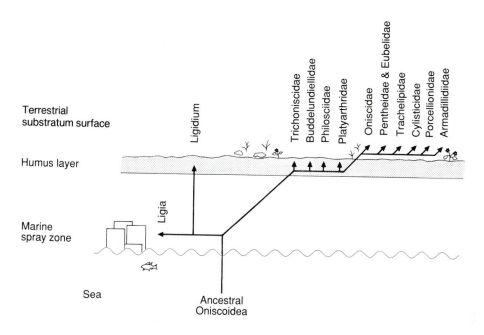

FIGURE 14–65 Evolutionary stages in the isopod invasion of land based on comparisons of extant species. Adaptations to reduce water loss, such as pseudotracheae and rolling into a ball, are best developed in the surface-dwelling families. *(Adapted from Schmalfuss, H. 1975. Morphologie, Funktion und Evolution der Tergithöcker bei Landisopoden. Z. Morph. Tiere. 80:287–316.)*

moisture, or horizon height. Some species can also use the sun's angle for orientation.

Most terrestrial isopods possess some adaptations to reduce water loss, but the group as a whole is considerably less well adapted in this regard than other terrestrial arthropods. Wood lice tend to be nocturnal and live beneath stones and in other places where the environment is humid. They have never evolved a waxy epicuticle of the type responsible for reducing integumental evaporation in insects and spiders. The thin, ventral exoskeleton is a primary site of evaporation, and the ability to roll up into a ball is probably, for some species, an adaptation to cut down on water loss. In general, wood lice are photonegative and strongly thigmotactic and can discriminate between relatively slight differences in humidity, all of which tends to keep them beneath protective retreats during the day. The commonly observed aggregation of wood lice under certain stones or wood may result in part from their being attracted by the body odor of other individuals of their own species and from a common response to the same environmental conditions.

The eyes of wood lice are poorly developed, probably related to their nocturnal, secretive behavior and a diet of decaying vegetation that does not require vision to locate. Repugnatorial glands are used in defense against such predators as spiders and ants. Tubercles and tergal plates, which in some species may be large and even spinelike, serve as protection, especially in forms that roll up into a ball (Fig. 14–66). In addition to protection, the tergal tubercles may function in digging, strengthening of the tergal plates, and reducing evaporative water loss.

Wood lice continue to use gills for gas exchange, but in the more terrestrial species the operculum contains a lunglike cavity (in the Oniscidae), tubelike invaginations, or pseudotracheae (as in in the Porcellionidae and Armadillidiidae) (Fig. 14–67B). Wood lice with pseudotracheae can tolerate much drier air.

The replacement of water lost from integumental evaporation usually comes from moist food and drinking, but some desert species replace lost water by ingesting moist burrow sand and by cutaneous absorption of water from moist burrow air. The gills must retain a covering film of moisture. The fact that

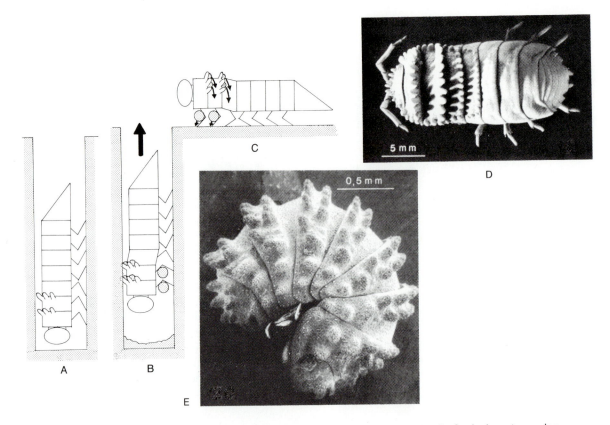

FIGURE 14–66 Adaptations of tergal tubercles in terrestrial isopods. **A–C,** A desert species, *Hemilepistus reaumuri*, that uses the anterior tergal tubercles for digging. The soil is transported with the anterior legs. **D,** *Hemilepistus aphganicus*, a related species in which the anterior tergal tubercles are used for both digging and transporting soil. **E,** *Armadillo tuberculatus* with large projecting protective tubercles. *(From Schmalfuss, H. 1975. Morphologie, Funktion und Evolution der Tergithöcker bei Landisopoden. Z. Morph. Tiere. 80:287–316.)*

they lie within a depression of the covering exopod facilitates water retention, but water must be replaced periodically (Fig. 14–67A). In some wood lice, the two endopods of the uropods are held together like a tube and dipped into a droplet of dew or rain. Water taken up is distributed to the gills. Other wood lice possess a system of surface channels that carry any water that comes in contact with the animal's back to the ventral surface and then back to the gills.

The maxillary glands of wood lice are poorly developed, and these isopods release nitrogenous wastes as gaseous ammonia.

Reproduction and Development

The gonads are paired and separate. The male sperm ducts open onto the sternum of the genital segment by way of either separate or united papillae. The first two pairs of pleopods are copulatory organs in most terrestrial isopods. The first pleopod of the male bends the papilla back to the endopods of the second pleopods, and a cavity in the second pleopods is filled with sperm. The female gonopores are also median, paired, sternal openings, and each oviduct is enlarged basally to form a seminal receptacle. During copulation the male presses its ventral side against one side of the female and injects sperm into one of her gonopores with the second, copulatory pleopod. The male then moves to the other side of the female's body, where the process is repeated. The eggs are fertilized in the oviduct. In many species, copulation occurs during or just after the female's molt, and there may be a long precopulatory attendance by the male. In the freshwater asellids mating can occur only during

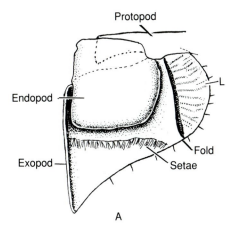

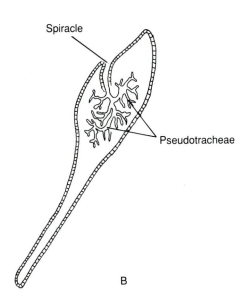

FIGURE 14–67 A, Fifth pleopod of the wood louse, *Oniscus asellus,* showing depression in the expod in which the respiratory endopod lies. Setae keep out foreign material in water entering the depression. **B,** Section through the pleopod exopod of the wood louse, *Porcellio scaber,* showing spiracle and pseudotracheae. *(Both after Unwin from Kaestner, A. 1970. Invertebrate Zoology. Vol. 3. Wiley-Interscience, New York.)*

the short period between the molting of the posterior half and the anterior half of the female's body.

In most of the terrestrial oniscoids (wood lice), copulation occurs during the intermolt, and the blind, saclike seminal receptacles do not make connection with the oviducts until the next molt takes place.

The eggs are brooded in the marsupium (Fig. 14–61B). The marsupium of a wood louse is kept filled with fluid, so development of the young is essentially aquatic despite the terrestrial habit of the adults.

Few to several hundred eggs are usually brooded, and the hatching stage is a postlarva (manca stage), with the last pair of legs incompletely developed. The young usually do not remain with the female after they leave the marsupium, but in *Arcturus* the female carries the young about attached to her long antennae. Most isopods in temperate regions—marine, freshwater, or terrestrial—produce one to two broods each summer and live for two to three years.

Order Cumacea

Cumaceans are marine peracaridans that live buried in sand and mud. The body is several centimeters long and distinctively shaped (Fig. 14–68). The head and thorax are somewhat enlarged; the abdomen is narrow and terminates in slender, elongated uropods. When eyes are present, they are located on a common median prominence situated above the base of the rostrum. The antennae are vestigial in the female; in the male, they are extremely long and are borne folded back along the sides of the body, sometimes in a groove. The animal burrows backward, using the more posterior legs, and ends up in an inclined position with its head projecting above the surface (Fig. 14–68).

A series of filamentous gills is located on each first maxilliped, which also functions as a ventilating pump. In *Diastylis,* the inhalant ventilating current, while passing over the mouthparts, is filtered for food particles by setae on the second maxillae. Not all cumaceans are filter feeders. Some scrape organic matter from sand grains.

Swarming at the time of mating is characteristic of many cumaceans. Large numbers leave their burrows and swim to the surface. In the European cumacean, *Diastylis rathkei,* swarming, which occurs at night, is associated with molting, and successive pelagic phases over the late summer involve successively older juveniles, with the first adults appearing in the fall.

More than 800 species of cumaceans have been described, most of which live at depths of less than 200 m, where bottom conditions are favorable. Some species may attain population densities of hundreds per square meter.

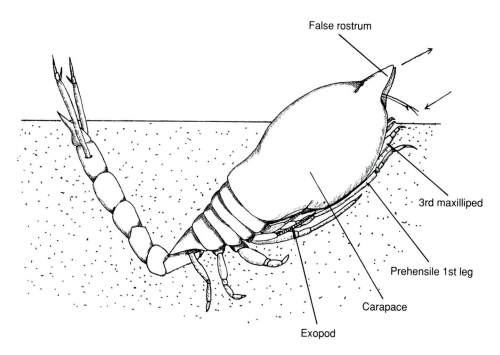

False rostrum

3rd maxilliped

Prehensile 1st leg

Carapace

Exopod

FIGURE 14–68 *Diastylis,* a cumacean, buried in sand. Arrows indicate the direction of the feeding/ventilating current.

Order Tanaidacea

The Tanaidacea display similarities to both the Cumacea and the Isopoda, with which they were formerly classified. These peracaridans, of which some 550 species have been described, are almost exclusively marine and generally small, most of them being only 2 to 7 mm in length (Fig. 14–69). The majority are bottom inhabitants of the littoral zone, where they live buried in mud, construct tubes, or live in small holes and crevices in rocks, but there a number of species that inhabit the deep sea. *Leptochelia dubia* is a widespread, tubicolous inhabitant of shallow water in many parts of the world. In the soft bottoms of Tomales Bay, California, this tanaidacean reaches densities as great as 30,000/m².

A small carapace covers the anterior part of the body and is fused to the first and second thoracic segments. The inner surface of the carapace functions as a gill. Many species lack eyes, but when eyes are present, they are located laterally on immovable processes. The first pair of thoracic appendages are maxillipeds, and the second pair (gnathopods) are large and chelate, a distinctive feature of the

tanaidaceans. The third pair of thoracic appendages are adapted for burrowing.

Some tanaidaceans are suspension feeders using the second maxillae or maxillipeds, but *Leptochelia dubia* and *Tanais cavolinii* collect diatoms, algae, and other material from around the mouths of their burrows with their chelipeds. There are also raptorial species, and deep-sea tanaidaceans are believed to feed on organic detritus.

Some tanaidaceans are hermaphrodites, but the eggs are brooded as in other peracaridans.

Order Mysidacea

Mysidaceans look much like little shrimp, and because they possess a ventral marsupium, they are sometimes called opossum shrimp. The majority are from 2 to 30 mm in length, but some, such as the bathypelagic *Gnathophausia,* may attain a length of 35 cm (Fig. 14–70A). Some species live in fresh water, but most are marine and are found at all depths.

The thorax is covered by a carapace, but unlike true shrimp, the carapace is not united with the last four thoracic segments. Anteriorly, the carapace often

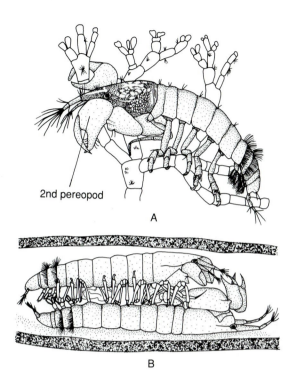

2nd pereopod

A

B

FIGURE 14-69 *Tanais cavolinii,* an intertidal tanaidacean from the Norwegian coast. This species lives within tubes constructed on calcareous algae. **A,** The male leaves its tube to find a female. **B,** Copulating pair of *Tanais cavolinii* within the tube of the female. The male genital cones penetrate slits in the ovisac to release sperm. *(A, From Johnson, S. B., and Attramadal, Y. G. 1982. A functional morphological model of Tanais cavolinii adapted to a tubicolous life-strategy. Sarsia. 67:29–42. B, From Johnson, S. B., and Attramdal, Y. G. 1982. Reproductive behaviour and larval development of Tanais cavolinii. Mar. Biol. 71:11–16.)*

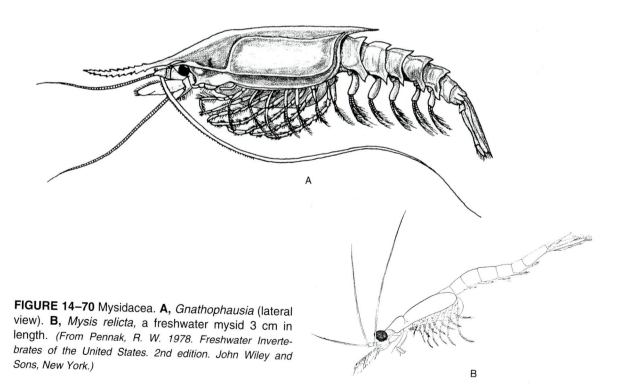

A

B

FIGURE 14-70 Mysidacea. **A,** *Gnathophausia* (lateral view). **B,** *Mysis relicta,* a freshwater mysid 3 cm in length. *(From Pennak, R. W. 1978. Freshwater Invertebrates of the United States. 2nd edition. John Wiley and Sons, New York.)*

extends forward as a rostrum, below which project stalked, compound eyes. The first thoracic appendages and sometimes the second pair as well (in the Mysidae) are modified as maxillipeds. The remaining six or seven thoracic appendages are more or less similar, and the exopods are filamentous and sometimes bear swimming setae. The pleopods are commonly reduced. Many mysidaceans (suborder Mysida) have a statocyst in the uropod endopod. Its visibility through the translucent exoskeleton is a convenient way to distinguish mysidaceans from other, similar planktonic crustaceans.

The thoracic exopods and pleopods are used for swimming, or the thoracic exopods alone when the pleopods are reduced, as in *Mysis*. Benthic forms crawl over the bottom or plough through the surface sand or mud. Thoracic gills are present in some species, but in the Mysidae, the inner surface of the carapace acts as a gill.

Most mysidaceans are omnivores and are capable of feeding on small particles groomed from the body surface and seizing of relatively large planktonic organisms within a basket formed by the thoracic endopods. Many bathypelagic forms are scavengers.

About 780 species of mysidaceans have been described. Marine species often live in large swarms and form an important part of the diet of such fish as shad and flounder, but many marine forms are found in algae and tidal grass. There are some 42 freshwater species, including those that live in ground water. *Mysis relicta* is confined to cold, deep-water lakes of the northern United States, Canada, and Europe (Fig. 14–70B). Lake trout feed extensively on this species.

The stalked, compound eyes; well-developed carapace; thoracic gills; long, tubular heart; and the presence of both antennal and maxillary glands are believed by some specialists to be primitive features and indicate that the mysidaceans stem from near the base of the peracaridan line.

SUMMARY (PERACARIDA)

1. The superorder Peracarida contains approximately 30% of the described species of crustaceans. Although most are smaller in size than the average decapod, peracaridans are diverse and widespread. They have invaded all of the major habitats, including land. Peracaridans are distinguished by the presence of a brood chamber, or marsupium, beneath the thorax formed by shelflike coxal plates.

2. The Amphipoda and Isopoda, the two largest orders, comprise most of the peracaridans. In addition to a somewhat similar size range (most are between 0.5 and 1.5 cm), these two groups share a number of characteristics: no carapace, sessile compound eyes, one pair of maxillipeds and seven pairs of legs, and no sharp demarcation between thorax and abdomen.

3. Amphipods are usually laterally compressed; most isopods are dorsoventrally flattened. Amphipods, moreover, possess thoracic gills, and isopods have abdominal pleopods modified as gills.

4. Most amphipods and isopods are benthic crustaceans, although intermittent swimming is common, especially among amphipods. There are few truly pelagic species, but hyperiidean amphipods live with planktonic gelatinous animals, such as jellyfish. There are many burrowing members of both groups, and there are numerous tube-dwelling amphipod species.

5. Many isopods and amphipods are scavengers and detritus feeders, but there are some carnivorous members of both groups. The amphipods also include some suspension feeders. There are many parasitic isopods, which, like parasitic copepods, are called fish lice.

6. Although most amphipods and isopods are marine, there are species of both groups that live in fresh water and on land. The terrestrial amphipods are members of a single family, the Talitridae (beach fleas and leaf litter hoppers), and are common inhabitants of drift along the shore throughout the world and of forest litter in New Zealand. Terrestrial isopods (wood lice) are widely distributed throughout the world, including deserts, and are the largest and most successful group of terrestrial crustaceans.

7. The shrimplike Mysidacea have stalked compound eyes and a thorax covered by a carapace. Species that occur in large swarms are an important food source for certain fish and other animals.

8. Cumaceans are marine burrowers having a carapace that encloses a branchial chamber around the gills on the maxillipeds.

9. Tanaidaceans are minute burrowing or tube-dwelling peracaridans. A small carapace covers only the first two thoracic segments and the second pair of pereopods are large chelipeds.

10. The eggs of peracaridans are deposited, fertilized, and brooded within the marsupium, and development is direct.

Subclass Phyllocarida; Order Leptostraca

This cosmopolitan order is composed of about 20 species of small marine crustaceans that differ from the basic malacostracan plan in that they have eight abdominal segments instead of six. Morphologically, the phyllocaridans are believed to represent the most primitive existing malacostracans. The fossil record bears this out, for the earliest known malacostracans were phyllocaridans and appeared in the Cambrian period.

Most are shallow-water suspension feeders. *Nebalia bipes,* which reaches about 12 mm in length, lives in bottom mud and in seaweed along the Atlantic coast of the United States, as well as in many other parts of the world (Fig. 14–71).

Subclass Eumalacostraca; Superorder Syncarida; Orders Anaspidacea and Bathynellacea

The Syncarida are represented by two small orders of primitive freshwater crustaceans, the Anaspidacea and the Bathynellacea, in which the thoracic appendages are biramous and similar. A carapace is absent, and either one thoracic segment is fused with the head (anaspidaceans) or the first thoracic segment is free (bathynellaceans). Although living syncarids are found only in fresh water, there are many fossil marine species. Members of the Bathynellacea are long, blind, interstitial or groundwater species and include some of the smallest malacostracans. *Brasilibathynella florianopolis* from Brazil is just over 0.5 mm long, and *Parabathynella neotropica* is only 1.2 mm in length (Fig. 14–72). Bathynellaceans have been reported from most parts of the world, including North America.

The anaspidaceans, which may reach a length of 5 cm, inhabit lakes, pools, and groundwater in Australia, New Zealand, and South America (Fig. 14–73). The larger species swim and crawl over the bottom or on aquatic vegetation. They are suspension feeders or omnivores.

Class Branchiopoda

Branchiopods are small crustaceans that are almost entirely restricted to fresh water. Although several structurally diverse groups compose the class, all are characterized by trunk appendages that have a flattened, leaflike structure (Fig. 14–77B). The exopod and endopod each consists of a single, flattened lobe bearing dense setae along the margin. The coxa is provided with a flattened epipod that serves as a gill,

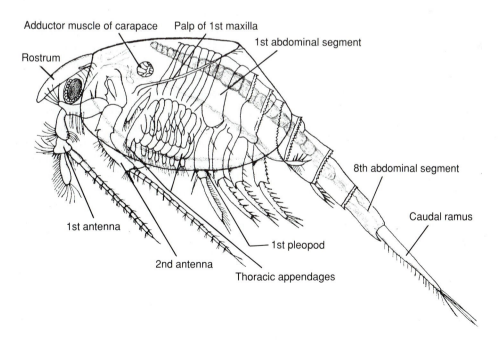

FIGURE 14–71 Female *Nebalia bipes,* a leptostracan. *(After Claus from Calman.)*

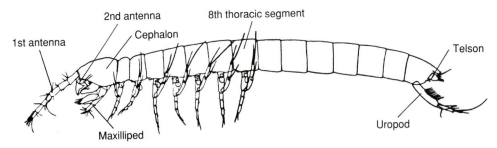

FIGURE 14–72 Lateral view of *Parabathynella neotropica,* a bathynellacean. Total length is only 1.2 mm. *(From Noodt, W. 1965. Natürliches System und Biogeographie der Syncarida. Gewässer und Abwässer. 37–38:77–186.)*

hence the name Branchiopoda, meaning "gill feet." In addition to gas exchange, the trunk appendages are usually adapted for suspension feeding and commonly for locomotion. The first antennae and second maxillae are vestigial. The last abdominal, or anal, segment bears a pair of large, terminal processes called **cercopods.**

The some 800 described species of Branchiopoda belong to four distinct groups. Fairy shrimp (order Anostraca) are characterized by an elongated trunk containing 20 or more segments (Fig. 14–74A). A carapace is absent, and the compound eyes are

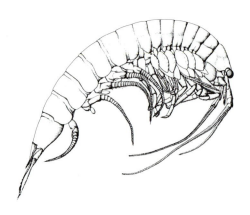

FIGURE 14–73 *Anaspides tasmaniae,* a syncaridan. (From Schminke, Von H. K. 1978. Die phylogenetische Stellung der Stygocarididae (Crustacea, Syncarida)—unter besonderer Berücksichtigung morphologischer Ähnlichkeiten mit Larvenformen der Eucarida. Z. Zool. Syst. Evolut.-forsch. 16:225–229.)

stalked. In tadpole shrimp (order Notostraca), the head and anterior half of the trunk are covered by a large, shieldlike carapace (Fig. 14–74C). Only vestiges of the second antennae remain. The compound eyes are sessile and are located close together beneath the carapace. There is a long, flexible, ringed abdomen bearing up to 70 pairs of appendages. The telson at the end of the abdomen carries two long, cercopods (furcal processes).

Members of two groups make up the order Diplostraca and include clam shrimp and water fleas; these are laterally compressed, and the body is at least partially enclosed within a bivalve carapace. In clam shrimp (suborder Conchostraca), the entire body is nearly or completely enclosed within the carapace and the animal looks strikingly like a little clam (Fig. 14–74D, E). Ten to 32 trunk segments are present, each bearing a pair of appendages. The second antennae are well developed, biramous, and setose, and the compound eyes are sessile.

The water fleas (suborder Cladocera) constitute half of the branchiopods and include many widespread and common species, such as those belonging to the genus *Daphnia* (Fig. 14–75A). The carapace encloses the trunk but not the head and often terminates posteriorly in an apical spine. The head has a ventrally and somewhat posteriorly directed projection, so the body has the appearance of a plump bird. The number of trunk appendages is reduced to five or six pairs. The tip end of the trunk, commonly called the **postabdomen,** is turned ventrally and forward and bears special claws and spines for cleaning the carapace.

The majority of branchiopods are only a few millimeters in length, and some are as small as 0.25 mm. The largest are the fairy shrimp, which are usually

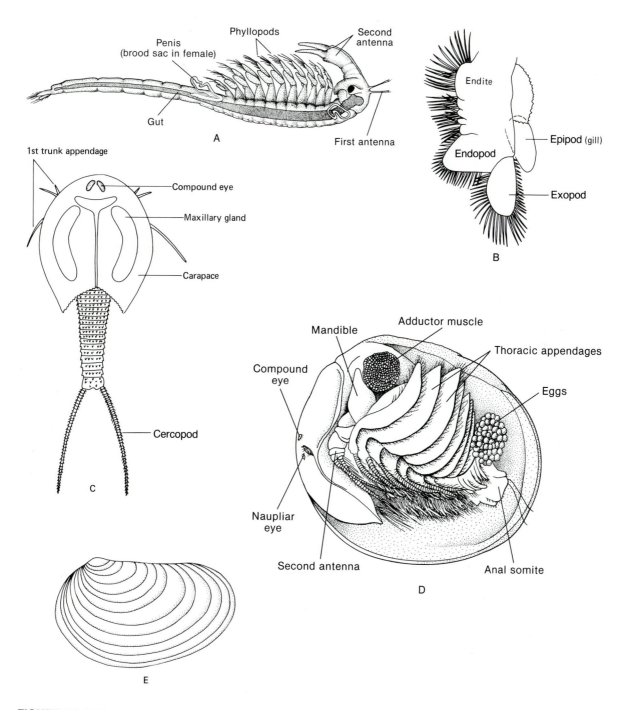

FIGURE 14–74 Branchiopoda. **A,** Fairy shrimp, *Branchinecta,* an anostracan (lateral view). **B,** Trunk appendage of *Branchinecta paludosa,* showing foliaceous structure. **C,** Tadpole shrimp, *Triops,* a notostracan (dorsal view). **D,** Lateral view of the conchostracan, *Lynceus gracilicornis,* with the left valve removed. **E,** Left shell valve of the conchostracan, *Cyzicus.* (*A, After Sars from Martin, J. W. 1992. Branchiopoda. In Harrison, F. W., and Humes, A. G. (Eds.): Microscopic Anatomy of Invertebrates. Vol. 9. Crustacea. Wiley-Liss, New York. pp. 25–224. B, After Sars from Pennak. C, After Pennak, R. W. 1978. Freshwater Invertebrates of the United States. 2nd Edition. John Wiley and Sons, New York. D, From Martin, J. W., Felgenhauer, B. E., and Abele, L. G. 1986. Redescription of the clam shrimp Lynceus gracilicornis from Florida, with notes on its biology. Zool. Scripta. 15(3):221–232. E, After Sars from Calman.*)

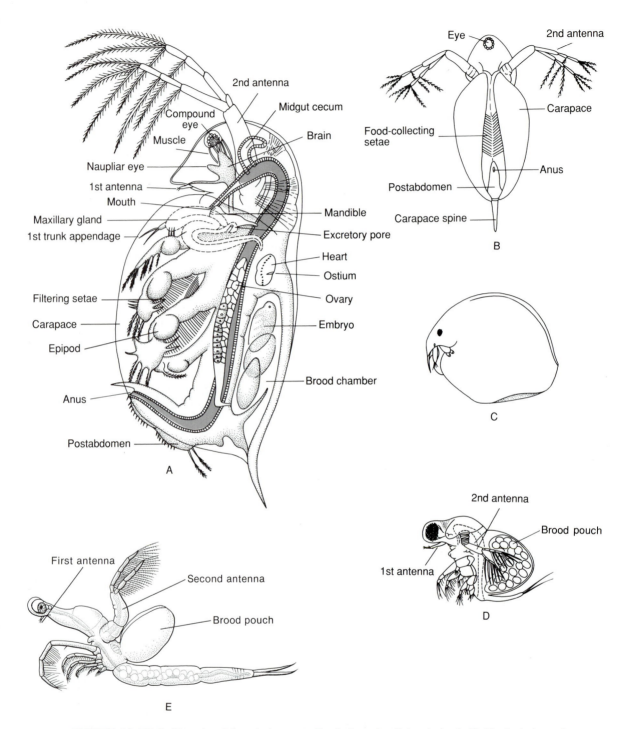

FIGURE 14–75 A, Female of the cladoceran, *Daphnia pulex* (lateral view). **B,** Ventral view of *Daphnia*. **C,** *Chydorus gibbus,* a cladoceran with a more rounded body than *Daphnia pulex.* Appendages are not shown. **D** and **E,** Two aberrant cladocerans. **D,** *Polyphemus pediculus.* **E,** The predatory *Leptodora kindtii. (A, After Matthes from Kaestner. C, From Pennak, R. W. 1978. Freshwater Invertebrates of the United States. 2nd Edition. John Wiley and Sons, New York. D, After Lilljeborg from Pennak. E, From Martin, J. W. 1992. Branchiopoda. In Harrison, F. W., and Humes, A. G. (Eds.): Microscopic Anatomy of Invertebrates. Vol. 9. Crustacea. Wiley-Liss, New York. pp. 25–224.)*

more than 1 cm long and may attain a length of 10 cm. Most branchiopods are pale and transparent, but rose or red colors, caused by the presence of hemoglobin within the body, are sometimes found.

Branchiopods live almost exclusively in fresh water; only cladocerans inhabit streams, large ponds, and lakes. Most branchiopods are confined to temporary pools, springs, and small ponds. Moreover, most species are highly ephemeral, appearing only briefly during the short existence of pools formed by melting snows or spring rains. The restriction of the larger branchiopods to small bodies of fresh water may be correlated with the absence of predatory fish in such habitats. Some fairy shrimp can tolerate saline pools, and species of *Artemia* (brine shrimp) are found in salt lakes and ponds throughout the world.

Locomotion

The water fleas swim by means of powerful second antennae (Fig. 14–75A). Movement is largely vertical and usually jerky. The downstroke of the antennae propels the animal upward; then it slowly sinks, using the antennae in the manner of a parachute. The little plumose setae at the end of the abdomen act as stabilizers, for if they are removed, the water flea rotates ventral side up.

Other branchiopods use the trunk appendages in swimming, although clam shrimp use the second antennae as well. Fairy shrimp usually swim upside down; however, when lighted from below in the aquarium, they roll over and swim right side up. Many clam shrimp and tadpole shrimp swim and crawl over the bottom, and some clam shrimp plow through the bottom sediment. Many cladocerans have also taken up a bottom or near-bottom existence.

The naupliar eye is persistent in almost all branchiopods (Fig. 14–75A). The sessile, cladoceran compound eyes are unusual not only in being fused into a single median eye but also in that this single median eye can be rotated by special muscles (Fig. 14–75A). The compound eye, at least in many cladocerans, is used to orient the animal in swimming.

Nutrition

Most branchiopods are suspension feeders and collect food particles with fine setae on the trunk appendages. Particle collection does not involve sieving, however (see p. 679). In the fairy shrimp, the space between limbs increases as the limbs move forward. Water is sucked into this space from the mid-

ventral line, and the lateral setae of the appendages collect particles from the incoming stream. On the backstroke, water is forced out of the interlimb space posteriolaterally and distally. By several complex mechanisms, the collected food particles are transferred to a midventral food groove. Here they become entangled in mucus and then moved forward to the maxillae, which push the food into the mouth.

The feeding mechanism of other branchiopods works on the same principle but with various modifications. In clam shrimp, only the anterior trunk appendages are used for collecting food, and the posterior appendages are modified as jaws for grinding large particles. In the cladocerans, only certain of the four to six pairs of trunk appendages are adapted for particle collection, and the filter setae are usually arranged on the appendage to form a distinct comb (Fig. 14–75). The water current passes from anterior to posterior, and collected particles are transferred into the food groove by special setae (gnathobases) at the basal part of the appendages. The efficiency of the collecting mechanism of *Daphnia* is indicated by the ability of some members of this genus to grow on a diet of bacteria.

Some branchiopods scrape up food material from plant and other surfaces, and some are predacious. The planktonic cladocerans, *Leptodora, Bythotrephes,* and *Polyphemus,* have anterior appendages modified for grasping.

The branchiopod foregut forms a short esophagus, and the midgut is often enlarged to form a stomach. In cladocerans, however, the midgut is more or less tubular and not easily distinguished from other parts of the digestive tract. There are often two small, digestive ceca (Fig. 14–75A). The intestine in some cladocerans is coiled one to several times.

Gas Exchange, Internal Transport, and Osmoregulation

The thin, vesicular or lamellar epipod on each of the trunk appendages of branchiopods is called a gill, but it is probable that the entire appendage and even the general integumentary surface are also important in gas exchange (Fig. 14–74B). Hemoglobin has been found in the blood, muscles, nervous tissue, and eggs of many branchiopods. The presence of hemoglobin frequently depends on the amount of oxygen in the water, so the animals are colorless in well-aerated water and pink in stagnant water.

The heart of cladocerans is a small, globular sac with only one pair of ostia, and it lies dorsally at the anterior of the trunk (Fig. 14–75A). In all other branchiopods, the heart is tubular with many pairs of ostia. The arterial system is restricted to a short, unbranched anterior aorta.

The excretory organs are maxillary glands, usually called shell glands when the duct can be seen coiled within the carapace wall. Little is known regarding water balance mechanisms except in the brine shrimp, *Artemia,* which has been studied extensively. This crustacean can tolerate salinities ranging from 10% sea water to the saturation point for sodium chloride. The internal osmotic pressure varies only slightly with external conditions. Ionic regulation is maintained by the absorption or excretion of salts through the gills. Also, *Artemia* is known to be capable of excreting a urine hyperosmotic to its blood. In brine, the osmotic pressure of the urine is four times that of the blood.

Reproduction and Development

The united or separate male gonopores of cladocerans open near the anus or the postabdomen, which may be modified in the form of a copulatory organ, and the oviducts open into a dorsal brood chamber located beneath the carapace. In other branchiopod species, the gonopores open ventrally on varying segments located more toward the middle of the trunk.

During copulation, the male fairy shrimp clasps the dorsal side of the female abdomen with the second antennae, which have become modified for this purpose, and then, twisting the abdomen around, inserts the paired, reversible copulatory processes containing the openings of the vas deferens into the single, median female gonopore. Similar copulatory behavior is displayed by the water fleas, although the male clasps with the first rather than the second antennae.

Branchiopods brood their eggs for varying lengths of time, depending on the species. In fairy shrimp, a special sac (Fig. 14–74A) is formed on extrusion of the eggs from the glandular uterine chamber. Both cladocerans and clam shrimp brood their eggs dorsally beneath the carapace (Fig. 14–75A).

The eggs are produced in clutches of two to several hundred, and a single female may produce several clutches. Development in most cladocerans is direct, and the young are released from the brood chamber by the ventral flexion of the postabdomen of the female. In the other branchiopods, embryonated eggs are released by the female and fall to the bottom after only a brief brooding period, or they may remain attached to the female and reach the bottom when she dies. In any case, the eggs typically hatch as nauplii. Parthenogenesis is common in branchiopods, and in some species males are uncommon or unknown.

In cladocerans, the reproductive pattern is strikingly like that of many rotifers. Parthenogenetic diploid eggs hatch into females for several generations, and one female can produce a succession of broods. Development is direct, and when the young leave the brood chamber beneath the carapace, the skeleton is molted and a new batch of eggs is released into the new brood chamber. The changeover from one brood to the next can take place in 5 min. At some point certain factors, such as change in water temperature or the decrease of the food supply as a result of population increase, induce the appearance of males, and fertilized eggs are produced. The fertilized eggs are large, and only two are produced in a single clutch, one from each ovary. The walls of the brood chamber are now transformed into a protective, saddle-like capsule (**ephippium**). This is cast off at the next molt, either separating from or remaining with the rest of the detached exoskeleton (Fig. 14–76A). The ephippia float, sink to the bottom, or adhere to objects and can withstand drying and freezing and even passage through the gut of fish and of fish-eating birds and mammals. By means of such protected resting eggs, cladocerans may be dispersed by wind or animals for some distances and can overwinter and survive summer droughts.

Thin-shelled summer eggs and thick-shelled resting (dormant) eggs are also produced by many species of the other branchiopod groups, but both types of eggs may be either parthenogenetic or fertilized. Development in thin-shelled eggs is rapid, and hatching may take place while the eggs are attached to the female. Production of dormant eggs may be stimulated by a variety of external factors, such as population density, temperature, and photoperiod. Dormancy (or diapause) may occur after an initial period of development. Thus, the eggs are prepared for rapid hatching when the precise necessary environmental conditions are present. Controlling factors that break dormancy, including oxygen, salinity, temperature, and illumination, vary and are related to the species habitat. Population pulses, or cycles, are common, particularly in many cladoceran species. Some

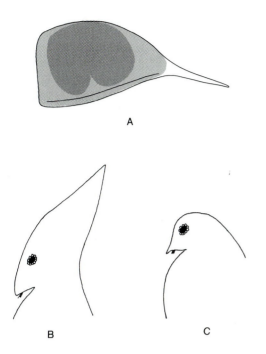

A

B   C

FIGURE 14–76 A, Ephippium of *Daphnia pulex.* **B** and **C,** Cyclomorphosis in *Daphnia.* **B,** Summer long-head form. **C,** Spring and fall round-head form. *(A, From a photograph by Pennak. B–C, After Pennak, R. W. 1978. Freshwater Invertebrates of the United States. 2nd Edition. John Wiley and Sons, New York.)*

cladocerans, such as species of *Diaphanosoma, Moina,* and *Chydorus,* may exhibit a single population rise and fall during the warmer months, whereas others are dicyclic and exhibit both spring and fall population peaks. The phenomenon is by no means always predictable. For example, some species of *Daphnia* may be monocyclic, dicyclic, or even acyclic, depending on the lake or pond in which they live.

In some cladoceran species, such as the lake-inhabiting *Daphnia dubia* and *Daphnia retrocurva,* the head progressively changes from a round to a helmet-like shape between spring and midsummer (Fig. 14–76B, C). From midsummer to fall, the head changes back to the normal, round shape. Such **cyclomorphosis** is still poorly understood. It may, as in rotifers, result from internal factors, or it may result from an interaction between external conditions (temperature) and internal factors (perhaps genetic). In at least some species, it appears to reduce predation by

producing a size less acceptable to certain predators.

A summary of the branchiopods appears on page 779.

Class Ostracoda

Ostracods or ostracodes, sometimes called mussel or seed shrimp, are small crustaceans that are widely distributed in the sea and in all types of freshwater habitats. Some 5650 living species have been described. They superficially resemble clam shrimp in having a body completely enclosed in a bivalve carapace. Perhaps these animals, rather than the Conchostraca, should have the name clam shrimp, for the ostracod carapace has evolved along lines that are even more strikingly like those of bivalves. In ostracods the elliptical valves are impregnated with calcium carbonate; there is a distinct, dorsal hinge line formed by a noncalcified strip of cuticle; and the valves are closed by a cluster of transverse adductor muscle fibers that are inserted near the center of each valve (Fig. 14–77). In some ostracods the valves may be locked by hinge teeth and ridges at the hinge line. The surface of the valves may be covered with setae and sculptured with pits, tubercles, or irregular projections (Fig. 14–77B). In the order Myodocopa there is a notch in the anterior margin of the valves permitting the protrusion of the antennae when the valves are closed (Fig. 14–77A).

The ostracod fossil record is continuous from the Cambrian period and is the most extensive of any group of crustaceans. The small size and the calcification of the valves have undoubtedly been primary factors in preservation. The valves are virtually the sole remains of the more than 10,000 fossil species, and the classification of fossil forms is based entirely on valve morphology.

The head region constitutes much of the ostracod body, for the trunk is very much reduced in size (Fig. 14–77C). The head appendages, particularly the antennules and antennae, are well developed. All external trunk segmentation has disappeared, and the trunk appendages are reduced to no more than two pairs. The trunk appendages, as well as the maxillae, may be more or less leglike or modified for swimming, feeding, clasping, or cleaning debris from within the valves.

Most ostracods are minute, several millimeters or less in length. The giant of the group is the pelagic deep-sea *Gigantocypris mulleri,* which can reach a length of 3 cm. Hues of gray, brown, or green are the most common.

Locomotion and Nutrition

Although there are some planktonic ostracods, such as the neutrally buoyant *Gigantocypris,* the majority live near the bottom, where they swim intermittently or crawl over or plough through the upper layer of mud and detritus. There are burrowing and interstitial species and species that live on the surface of algae, water plants, or other submerged objects. Some are commensal on other animals, living, for example, among the leg setae of crayfish. One or both pairs of antennae are the principal locomotor appendages and

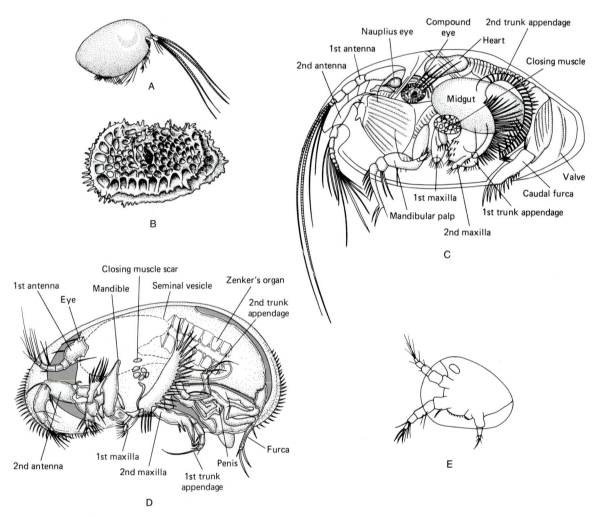

FIGURE 14–77 A, A myodocopid ostracod with antennal notches in the valves (lateral view). **B,** Lateral view of the sculptured valve of *Agrenocythere spinosa,* a benthic marine ostracod. **C,** Female marine myodocopid ostracod, *Skogsbergia,* with left valve and left appendages removed (lateral view). This is a marine scavenging carnivore that is able to swim over the bottom. **D,** *Candona suburbana,* a very common freshwater podocopid ostracod, left valve removed. This nonswimming bottom dweller feeds on algae and decomposing vegetation. Zenker's organ, which is part of the male reproductive system, ejects sperm to the penis. **E,** Lateral view of an ostracod naupliar larva. Only one member of each of the three pairs of appendages is shown. *(C, After Claus from Calman. C–D, From Cohen, A. C. 1982. Synopsis and Classification of Living Organisms. Vol. 2. McGraw-Hill Book Co., New York. pp. 191 and 192. E, After Schreiber from Pennak, R. W. 1978. Freshwater Invertebrates of the United States. 2nd Edition. John Wiley and Sons, New York. p. 427.)*

are variously modified, depending on the habits of the animals. Kicking forward with the caudal furca is a common aid to locomotion in many benthic species. Perhaps the most remarkable ostracods are the New Zealand and South African terrestrial species of *Mesocypris,* which can plough through forest humus.

Ostracods display diverse feeding habits; some are carnivores and others herbivores, scavengers, or filter feeders. Algae are a common plant food, and the prey of carnivorous species includes other crustaceans, small snails, and annelids. The large *Gigantocypris* is predacious and is reported to capture other crustaceans and even small fish with its antennae. Detritus particles, often stirred up by the antennae or mandibles, are also a common source of food. In the suspension-feeding species, the collecting setae are located on one of the pairs of maxillae.

Internal Structure and Physiology

Gills are lacking, and gas exchange is integumentary, the locomotor and feeding currents providing ventilation. A heart and blood vessels are present only in the marine order Myodocopa. Blood circulates between the valve walls in all ostracods. Some ostracods possess antennal glands, some have maxillary glands, and some are among the few crustaceans that possess both types of excretory organs in the adult. The maxillary glands are large and coiled and lie between the inner and outer walls of the valves.

A naupliar eye is present in all ostracods, but sessile compound eyes appear only in the Myodocopa (Fig. 14–77C). The most important sense organs are probably the sensory hairs found not only on the appendages but also on the valves.

Reproduction and Development

The female gonopores are located ventrally between the last pair of appendages at the caudal end of the trunk. The paired sperm ducts open between or through one of the two large, sclerotized penes projecting ventrally in front of the caudal furca. The sperm are motile and in some cyprids are of a remarkably large size. *Pontocypris monstrosa,* which is less than 1 mm in length, has sperm as long as 6 mm.

Some ostracods use bioluminescence as a sexual attractant, much like fireflies. Ostracods were the first crustaceans in which bioluminescence was observed. Small puffs of bluish light are produced externally by secretions from a gland in the labrum and are often patterned as a string of distinct flashes lasting only 1

to 2 s. On a reef, males of the genus *Vargula* may synchronize their flashes, producing a spectacular display.

During copulation in some ostracods, the female is clasped dorsally and posteriorly by the second antennae or the first pair of legs of the male, and the penes are inserted between the valves of the female into the gonopores.

Most commonly the eggs are shed freely in the water or are attached singly or in groups to vegetation and other objects on the bottom, but the eggs are brooded in the dorsal part of the shell cavity in some ostracods. Some species, especially those in temporary bodies of fresh water, produce eggs very resistant to dessication.

On hatching the juveniles are enclosed in a bivalve carapace like that of the adult and take up the adult living pattern. The valve skeleton is shed at each molt in the ostracod life history, along with the rest of the exoskeleton, but molting ceases at adulthood.

Parthenogenesis occurs in some freshwater species, especially those that live in ephemeral habitats.

A summary of the ostracods is provided on p. 779.

Class Copepoda

The Copepoda is the largest class of small crustaceans, over 8500 species having been described. Most copepods are marine, but there are many freshwater species and a few that live in moss, soil-water films, and leaf litter. Also, there are many that are parasitic on various marine and freshwater animals, particularly fish. Marine copepods exist in enormous numbers and are usually the most abundant and conspicuous component of a plankton sample. Because most planktonic species feed on phytoplankton, they are the principal link between phytoplankton and higher trophic levels in many marine food chains. A major part of the diet of many marine animals is composed of copepods.

Most copepods range in length from less than 1 mm to more than 5 mm, although there are larger (17 mm) free-living species. Some parasitic forms are very large, over 32 cm in length (Fig. 14–82). Although most copepods are rather pale and transparent, some species may be brilliant red, orange, purple, blue, or black. Many luminescent species have been reported.

The body of a free-living copepod is commonly tapered from anterior to posterior and is somewhat cylindrical (Fig. 14–78), but there are many exceptions to this generalization. The trunk is composed of a thorax and abdomen. The anterior end is either rounded or pointed. Compound eyes are absent, but the median naupliar eye is a typical and conspicuous feature of most copepods. Also conspicuous are the uniramous first antennae, which are generally long and held outstretched at right angles to the long axis of the body.

The head is fused with the first of the six thoracic segments and sometimes with the second thoracic segment as well. The first pair of thoracic appendages have become modified to form maxillipeds for feeding. The remaining five thoracic appendages, except the last one or two pairs, are all more or less similar and rather symmetrically biramous. Left and right members of each pair of thoracic appendages are rigidly linked together by an exoskeletal plate between their coxal articles. The link ensures that both legs of a pair beat simultaneously.

The abdomen is composed of five segments, which are commonly narrower than those of the thorax. There are no abdominal appendages except for a single pair of caudal rami in the telson. In some planktonic marine species these caudal rami are spectacularly developed. For example, in *Calocalanus pavo,* each ramus is turned laterally and bears four long, feather-like setae.

Of the five free-living orders of copepods, the calanoids are largely planktonic; the harpacticoids, which include over 50% of copepod species, are largely benthic; and the cyclopoids include both planktonic and epibenthic species. Variations in body shape are related to the habitat of the species. Planktonic forms have rather cylindrical bodies with a narrow abdomen. Those that live high in the water column tend to be more slender and fusiform than those that swim several meters over the bottom. Epibenthic species, which crawl and swim just over the bottom, have somewhat broader bodies. Benthic species that live on algae and sea grasses may be broad and flattened; interstitial species are narrow and wormlike (Fig. 14–79).

Nutrition

Copepods display a range of feeding habits, depending in part on where they live. Planktonic copepods are chiefly suspension feeders, and the second maxil-

lae are modified to capture food. Recent studies indicate, however, that these animals do not really sieve particles from suspension (p. 679). Phytoplankton constitutes the principal part of the diet of most suspension-feeding species, but some rely heavily on detritus particles as well. Using radioactive diatom cultures, *Calanus finmarchicus,* which is about 5 mm long, was found to collect and ingest from 11,000 to 373,000 diatoms every 24 h, depending on their size. When feeding on particles of varying size, larger particles are selected because they can be handled more efficiently, but under natural conditions planktonic copepods probably accept whatever food predominates.

Not all planktonic copepods are herbivorous suspension feeders; some are omnivorous, and some are strictly predacious. Species of *Anomalocera* and *Pareuchaeta* even capture young fish. Species of *Tisbe,* planktonic epibenthic harpacticoids, are known to swarm over a small fish and eat its fins, thus immobilizing it. They then devour the body as it falls to the bottom. Some freshwater Cyclopidae are herbivorous; others are carnivorous. Members of the common freshwater genus, *Cyclops,* are predatory, as are some of the other cyclopoid genera. Most bottom-dwelling harpacticoids feed on microorganisms and detritus attached to sand grains, algae, or sea grasses.

Calanoids store nutrients in a fat body or in a midgut oil sac, which often gives the body a red or blue color. The oil in many species contributes to buoyancy.

Locomotion

Both the thoracic appendages and the second antennae are used in rapid swimming. The second antennae, the two branches of which beat in a rotary manner, appear to be of primary importance in calanoids, and the thoracic appendages are most important in cyclopoids. The first antennae, which are long and setose in primarily slow sinking planktonic forms, act like parachutes. They may lie back against the body when movement is rapid and then extend laterally again when movement is slowed.

Carnivorous species cruise continually as they seek potential prey. Herbivorous species alternate cruising with feeding. During a feeding bout, which lasts 10 to 30 s *(Eucalanus crassus),* the first antennae act both as parachutes and as sensors for detecting algae; the other anterior appendages set up a flow field that brings water to the animal. The slight ten-

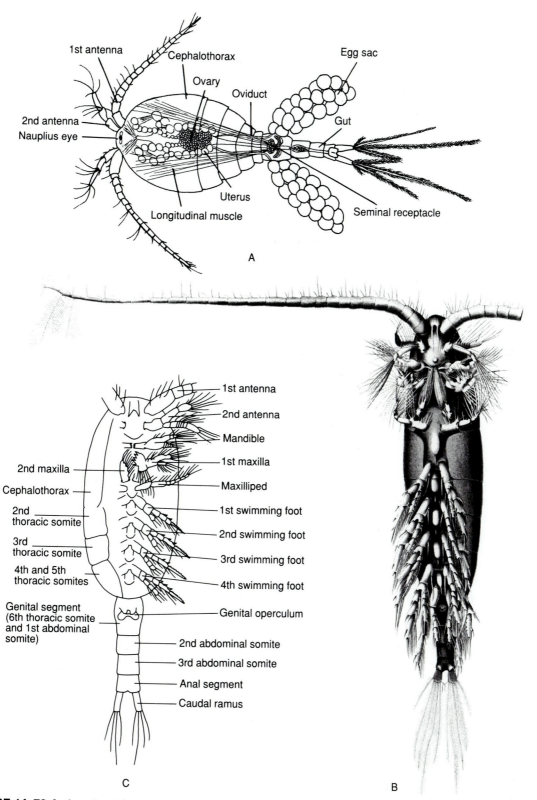

FIGURE 14–78 A, A cyclopoid copepod, *Macrocyclops albidus* (dorsal view). **B,** Ventral view of *Calanus*, a typical calanoid copepod, with appendages shown. **C,** Diagrammatic ventral view of *Pseudocalanus*, showing appendages. *(A, After Matthes from Kaestner. B, After Giesbrecht, W. 1892, Fauna and Flora Golfes Neapel. Monogr. 19:1–831. C, From Corkett, J., and McLaren, I. A. 1978. The biology of Pseudocalanus. Adv. Mar. Biol. 15:2–231.)*

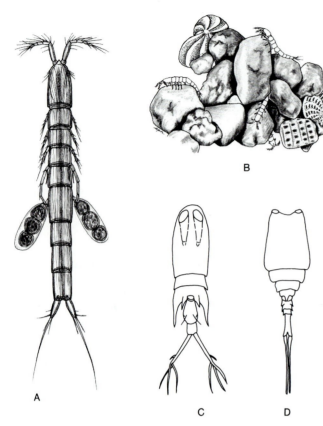

A

B

C   D

FIGURE 14–79 **A,** A cylindropsyllid harpacticoid copepod (dorsal view). **B,** Interstitial harpacticoid copepods crawling among sand grains. Two foram shells are among the sand grains. **C–D,** Two marine cyclopoid copepods, *Corycaeus* **(C)** and *Copilia* **(D),** in which the naupliar eye is very large and divided. Copepods of these and related genera are often brilliantly colored. *(A, After Sars. C–D, From Smith, D. L. 1977. A Guide to Marine Coastal Plankton and Marine Invertebrate Larvae. Kendall/Hunt Publishing Co., Dubuque, IA.)*

dency to sink is important in enabling the copepod to maintain a proper orientation for its flow field.

After a feeding bout, the copepod cruises or sinks to a new position for another feeding bout. Swimming positions vary greatly among species (upside down, vertical, etc.), and the caudal rami can be held in varying positions or act as a rudder. For example, *Eucalanus crassus* swims backwards in a vertical position.

Although most planktonic copepods live in the upper 50 m of the sea, many species are found at greater depths, even in the deep sea (Fig. 14–80). Vertical movement is oriented by light, and many species exhibit daily vertical migrations.

The bottom-dwelling harpacticoids and some cyclopoids crawl over or burrow through the substratum, and many harpacticoids live between sand grains (Fig. 14–79B). The thoracic limbs are used in crawling, and this is accompanied in harpacticoids by lateral undulations of the wormlike body.

Internal Structure

There are no gills in free-living copepods, and except in the calanoids and some parasitic species, there is neither heart nor blood vessels. The excretory organs are maxillary glands.

Reproduction and Development

Male copepods are commonly smaller than females and are usually outnumbered by females. Copepods are among the few small crustaceans that form spermatophores, and the lower end of the sperm duct is modified for this purpose. The male opening is located on the first abdominal segment of most copepods, as are the female gonopores and openings to the seminal receptacles. During spermatophore transfer, the male clasps the female with one or both first antennae and, in most calanoids, with the last pair or modified thoracic appendages as well. The spermatophores are transferred to the female by the thoracic appendages of the male and adhere to the receptacle openings by means of a special cement.

Most calanoids shed their eggs singly into the water. The eggs of other copepods, however, are usually enclosed within an ovisac. The ovisac is produced by oviduct secretions when the eggs are emitted from the oviduct and remains attached to the female genital segment, where it functions as a brood

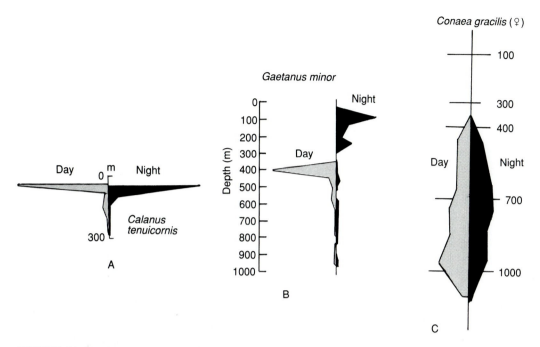

FIGURE 14–80 Vertical distribution of three species of mid-Atlantic marine planktonic copepods. The black side indicates the distribution during night, and the gray indicates distribution during the day; width reflects population density. **A** and **B,** Calanoids, *Calanus tenuicornis* and *Gaetanus minor.* **C,** The cyclopoid, *Conaea gracilis.* Note the restricted vertical distribution of *Calanus tenuicornis* as compared with *Conaea gracilis* and the extensive day/night vertical migration of *Gaetanus minor.* (A–B, From Roe, H. S. J. 1972. The vertical distributions and diurnal migrations of calanoid copepods collected on the Sond Cruise, 1965, pt. II. J. Mar. Biol. Assoc. U.K. 52:315–343. C, From Boxshall, G. A. 1977. The depth distributions and community organization of the planktonic cyclopods of the Cape Verde Islands region. J. Mar. Biol. Assoc. U.K. 57:543–562. Copyrighted and reprinted by permission of Cambridge University Press.)

chamber (Fig. 14–78A). One or two sacs are formed, depending on the number of oviducts. Each sac contains a few to 50 or more eggs, and clutches may be produced at frequent intervals. For example, among freshwater copepods, species of diaptomids shift back and forth between a gravid and nongravid condition every four days, and mating is required for each clutch of eggs.

The eggs typically hatch as nauplii. After five or six naupliar instars, the larva passes into the first copepodid stage. The first copepodid larva displays the general adult features, but the abdomen is usually still unsegmented, and there may be only three pairs of thoracic limbs. The adult structure is attained typically after six naupliar and five copepodid stages (Fig. 14–81); molting then ceases. The entire course of development may take as little as one week or nearly as

long as one year. Six months to a little over a year is the maximum life span of most free-living species. Studies on the copepodid stages of 20 planktonic species in the Adriatic indicate that there were from three to six generations a year.

Some freshwater calanoids and harpacticoids produce both thin-shelled eggs and thick-shelled dormant eggs, and overwintering resting eggs have now been reported for a number of marine calanoids. In many freshwater copepods, the copepodid stages (or even adults) secrete an organic, cystlike covering and become inactive under unfavorable conditions. Such cysts, buried in mud, are well adapted to withstand desiccation and enable the copepod to estivate when its pool or pond dries up. They also provide a means of dispersal when carried on the muddy feet or bodies of birds and other animals.

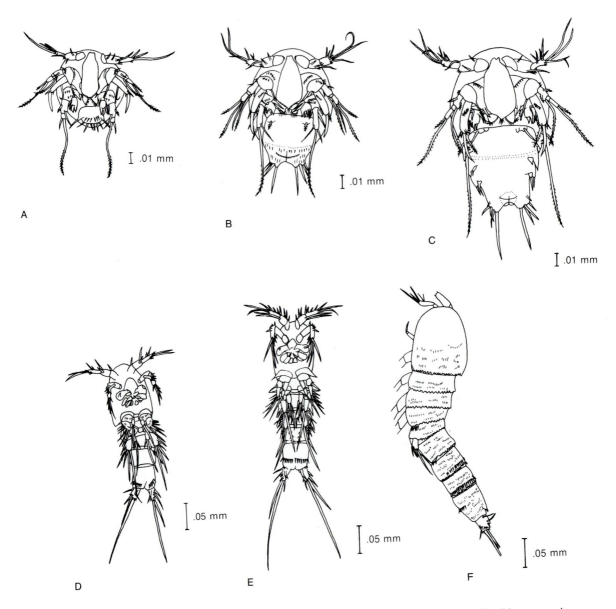

FIGURE 14–81 Ventral views of the developmental stages of the harpacticoid copepod, *Elaphoidella bidens coronata.* The six naupliar and five copepodid stages are each separated by a molt. **A,** First nauplius. **B,** Fourth nauplius. **C,** Sixth nauplius. **D,** First copepodid. **E,** Second copepodid. **F,** Fifth copepodid (dorsal view). *(Adapted from Carter, M. E., and Bradford, J. M. 1972. Postembryonic development of three species of harpacticoid Copepoda. Smithson. Contrib. Zool. 119:1–26.)*

Parasitic Copepods

The Copepoda includes over 1000 species of parasitic crustaceans. Some copepods are ectoparasites on fish and attach to the gill filaments, the fins, or the general integument (Fig. 14–82). Other copepods are commensal or endoparasitic within polychaete worms, the intestine of echinoderms (particularly crinoids), and in tunicates and bivalves. Cnidarians, especially anthozoans, are the hosts for many species of copepods. All degrees of modification from the free-living copepod form are exhibited by these parasites. Ancestral forms are usually ectoparasites and resemble free-living species (Fig. 14–83A). On the other hand, some ectoparasitic and endoparasitic copepods are so

FIGURE 14–82 Parasitic copepods, *Penella exocoeti,* on a flying fish. The copepods are in turn carrying the barnacle, *Conchoderma virgatum* (striped body). *(Modified after Schmitt.)*

highly modified and bizarre that they no longer have any resemblance to the free-living species (Fig. 14–83C–E).

Among ectoparasitic copepods certain appendages have usually become specialized as holdfast organs, and the mouthparts are adapted for piercing and sucking. In some, a frontal gland produces a button (the **bulla**), which is attached to the gill filament of the fish. The second maxillae are inserted into the button for permanent attachment (Fig. 14–83B).

In most parasitic copepods, the adults are adapted for parasitism, and the swimming larval stages are usually similar to those of free-living copepods. Contact with the host occurs at various times during the life cycle of the copepod, and modifications appear with each molt. The salmon gill maggot, *Salmincola salmoneus,* which is parasitic on the gills of the European salmon, has a typical life cycle. When the salmon enters the coastal estuaries on its migration to fresh water, the copepod, in the form of a first copepodid larva, attaches to the gills. The larva attaches by a structure resembling a button and thread (bulla) held by the second maxillae; the larva then undergoes a series of molts. The male matures first, and copula-tion takes place before the female is mature. The male then dies. The female undergoes a final molt and becomes permanently attached to the host by the second maxillae fused with the bulla, which is embedded in the tissue of the host (Fig. 14–83B). Egg sacs are then formed and may be as long as 11 mm. Several clutches are produced by a single female.

A summary of the copepods is provided on p. 779.

Class Mystacocarida

The Mystacocarida is a small class of interstitial crustaceans that are probably related to the copepods. The Mystacocarida were first described in 1943 from specimens collected off Massachusetts, but nine species have since been reported from many other coasts, principally around the North and South Atlantic oceans. The majority of these little crustaceans are approximately 0.5 mm in length and are adapted for living between intertidal sand grains (Fig. 14–84). The body is long and cylindrical but divided like that of copepods. Both pairs of antennae and the mandibles are long and prominent and rather similar to the antennae of naupliar larvae. The maxillae and

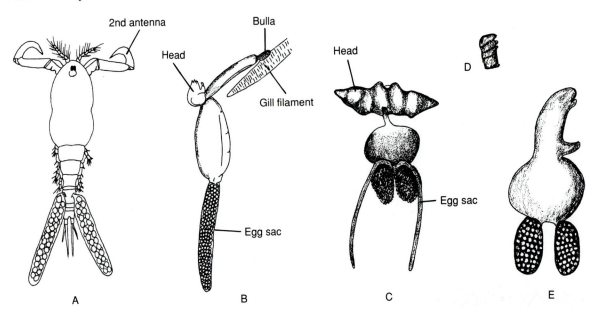

FIGURE 14–83 Parasitic copepods. **A,** *Ergasilus versicolor,* a cyclopoid parasite that lives on gills of freshwater fish. Only the adult female is parasitic, hooking to fish with clasping antennae. **B,** *Salmincola salmonea,* mature female attached to the gill of European salmon. **C,** Spyrion. The head is embedded in the skin of a fish; the remainder of the body hangs free. **D** and **E,** Male **(D)** and female **(E)** of *Brachiella obesa* live on the gills of red gurnard. *(A, After Wilson from Pennak. B, After Friend. C, From Parker and Haswell. D–E, After Green, J. 1961. A Biology of Crustacea. Quadrangle Books, Chicago. p. 113.)*

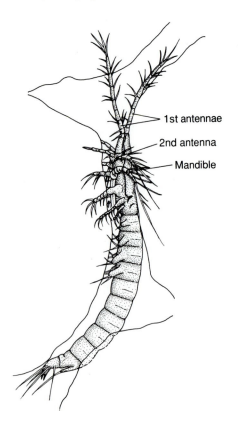

maxillipeds are provided with setae, probably for collecting detritus and other particulate matter. Only the naupliar eye is present. The sexes are separate, and a nauplius is the hatching stage.

Classes Branchiura, Tantulocarida, and Pentastomida

There are three classes of parasitic crustaceans believed to be related to copepods. The class Branchiura includes approximately 130 species of ectoparasites on the skin or in the gill cavities of freshwater and marine fish. They feed on the host's mucus and blood. The most striking attributes of branchiurans that dis-

◄ **FIGURE 14–84** The interstitial *Derocheilocaris typica* crawling among sand grains. The pushing force for movement is produced by the second antennae and mandibles, the branches of each appendage operating against the substratum both above and below the head of the animal. *(From Lombardi, J. and Ruppert, E. E. 1982. Functional morphology of locomotion in Derocheilocaris typica. Zoomorphology. 100:1–10.)*

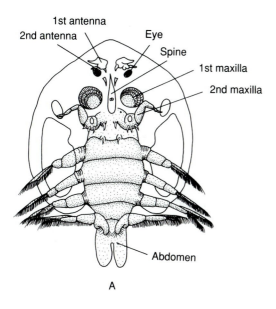

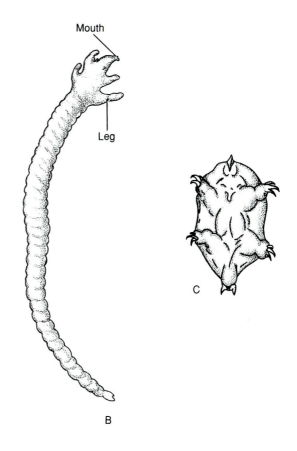

FIGURE 14–85 A, *Argulus foliaceus,* a branchiuran fish parasite. **B,** *Cephalobaena tetrapoda* from the lung of a snake. **C,** Larva of *Porocephalus crotali. (A, After Wagler from Kaestner. B, After Heymons from Cuénot. C, After Penn.)*

tinguish them from copepods are the presence of a pair of sessile compound eyes and a large, shieldlike carapace covering the head and thorax (Fig. 14–85A).

The abdomen is small, bilobed, and unsegmented. Both pairs of antennae are minute, and the first pair is provided with a large claw for attachment to the host. Also important in attachment (except in *Dolops*) are two large suckers modified from the bases of the first maxillae, the rest of the appendage being vestigial. The mouthparts are adapted for feeding on the mucus and blood of the host. The four thoracic appendages are well developed; branchiurans can detach and swim or crawl from one host to another.

Copulation occurs while the parasites are on the host, but the eggs are deposited on the bottom. A postnaupliar stage hatches from the egg and is parasitic like the adult.

The recently described class Tantulocarida includes a small number of ectoparasites that live on other deep-water crustaceans. They are somewhat similar to copepods but lack all posterior appendages.

Pentastomids constitute a class of approximately 90 parasitic species that live within the lungs or nasal passageways of vertebrates. The hosts of 90% of pentastomids are reptiles, such as snakes and crocodiles, but some species parasitize mammals and birds. Although largely tropical, pentastomids have been reported from North America, Europe, Australia, and even from Arctic birds.

The wormlike body is 2 to 13 cm long and bears five short, anterior protuberances, from which the inappropriate name *pentastomida* ("five mouths") is derived (Fig. 14–85B). Four of these projections, two on each side of the body, are leglike, bearing claws. The fifth projection is an anterior, median, snoutlike process bearing the mouth. Not infrequently the legs are reduced to nothing more than the claws, which are used for clinging to the tissues of the host.

The body is covered by a nonchitinous cuticle that is ringed over the abdomen and molted periodically during larval development. The digestive tract is a relatively simple, straight tube with the anterior end modified to pump in the blood of the host, on which these parasites subsist. There are no gas exchange, circulatory, or excretory organs.

The sexes are separate, with a well-developed genital system. Fertilization is internal; the embryonated eggs move from the lungs into the digestive system of the host and then to the outside in the feces of the host. In most pentastomids, the life cycle requires an intermediate host, which may be fish (pentastomids in crocodiles) or herbivorous or omnivorous mammals, such as rodents, rabbits (intermediate host for *Linguatula* in dogs), or small ungulates. Actually, larval pentastomids have been reported from almost every class of vertebrates. Larval development takes place within the intermediate host and involves a number of molts. The larvae possess four to six leg-like appendages (Fig. 14–85C). When the intermediate host is eaten, the parasite is transferred to the stomach of the primary host and reaches the lungs and nasal passageways through the esophagus. Pentastomid larvae have been reported in humans but are killed by calcareous encapsulation.

The taxonomic status of the pentastomids has been uncertain, and they commonly have been placed in a separate phylum. There is a growing consensus, however, that they are highly aberrant crustaceans related to branchiurans and copepods. Evidence for this occurs in unique similarities of pentastomid and crustacean sperm ultrastructure and from the analysis and comparison of ribsomal RNA gene sequences.

Class Cirripedia

The Cirripedia include the familiar marine animals known as barnacles. Cirripedes are the only sessile group of crustaceans, aside from the parasitic forms, and as a result they are one of the most aberrant groups of Crustacea. In fact, it was not until 1830, when the larval stages were first discovered, that the relationship between barnacles and other crustaceans was fully recognized, and barnacles were removed from the phylum Mollusca.

Barnacles are exclusively marine. Approximately two thirds of the nearly 900 described species are free-living, attaching to rocks, shells, coral, floating timber, and other objects. Some barnacles are commensal on whales, turtles, fish, and other animals, and a large number of barnacles are parasites. The entirely parasitic Rhizocephala are so highly specialized that all traces of arthropod structure have disappeared in the adult.

External Structure

Louis Agassiz described a barnacle as "nothing more than a little shrimplike animal, standing on its head in a limestone house and kicking food into its mouth." An additional analogy might be drawn if one can imagine an ostracod turned upside down and attached to the substratum by the anterior end.

The larval **cypris** of a barnacle, which indeed looks very much like the ostracod, *Cypris,* for which it is named, settles to the bottom and attaches to the substratum by means of **cement glands** located in the base of the first antenna (Fig. 14–93B). The larval carapace, which encloses the entire body, as in ostracods, persists as the enveloping carapace or mantle of adult barnacles. It becomes covered externally with calcareous plates in the ordinary barnacles (Thoracica). The carapace opening is therefore directed upward and enables the animal to project its long, thoracic appendages for scooping plankton.

The free-living, or thoracican, barnacles can be divided into stalked and sessile (meaning stalkless in this case) types. In stalked barnacles, sometimes called goose neck barnacles, there is a muscular, flexible stalk (**peduncle**) that is attached to the substratum at one end and bears the major part of the body (**capitulum**) at the other (Fig. 14–86A). The peduncle represents the preoral end of the animal and contains the vestiges of the larval first antennae and the cement glands that opened on them. The capitulum contains all but the preoral part of the body and includes the surrounding carapace (mantle). The mantle surface is covered by two pairs of calcareous plates (scuta and terga) (Fig. 14–87). The carapace margin can be pulled together for protection or opened for the extension of the appendages. A large adductor muscle runs transversely between two of the plates (scuta) (Fig. 14–86A).

There is no peduncle in sessile barnacles (Fig. 14–88). The attached undersurface of the barnacle is called the basis and is either membranous or calcareous. This is the preoral region of the animal and contains the cement glands. A vertical **wall** of plates completely rings the animal (Fig. 14–88B, C), and within the wall the animal is covered by an operculum, formed by the paired movable terga and scuta. The plates composing the wall overlap one another and may be held together by living tissue only or by interlocking teeth or may actually be fused to some extent. They are anchored to the bases by muscle fibers (Fig. 14–94).

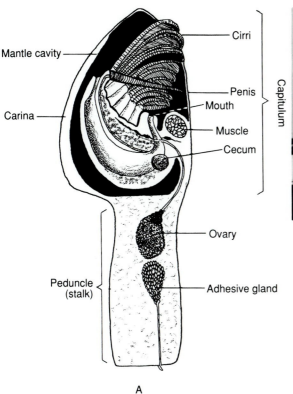

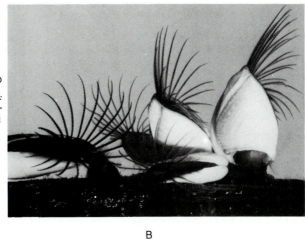

B

FIGURE 14–86 A, *Lepas,* a stalked barnacle. **B,** A species of *Lepas* with extended cirri. *(A, After Broch from Kaestner. B, © Jen and Des Bartlett, The National Audobon Society Collection.)*

Within the mantle the body of either a stalked or a sessile barnacle is flexed backward so that the appendages are directed upward toward the mantle aperture rather than toward the side (Figs. 14–86A; 14–88A). The major part of the body consists of a cephalic region and an anterior trunk (thoracic) region. External segmentation is indistinct.

The first antennae are vestigial except for the cement glands, and the second antennae are present only in the larva. The oral appendages are variously modified. Typically, six pairs of long, biramous thoracic appendages called **cirri** (from which the name Cirripedia is derived) are used in suspension feeding, and each branch is provided with many long setae.

When not encrusted with other sessile organisms, barnacles are usually white, pink, or purple. Stalked barnacles, including the peduncle, range from a few millimeters to 75 cm in length. The majority of sessile species are a few centimeters in diameter, but some are considerably larger. *Balanus psittacus* from the west coast of South America reaches a height of 23 cm and a diameter of 8 cm and is a popular local seafood.

The calcareous plates of barnacles have provided an extensive fossil record that dates back to the Silurian.

Evolution and Ecological Distribution of Barnacles

Of the two types of common free-living (thoracican) barnacles, the stalked barnacles are considered more primitive. On the basis of fossil forms, the cypris larva common to all cirripedes (Fig. 14–93B), and other evidence, the ancestral barnacle was probably a cypris-like, bivalved crustacean attached to the substratum by its first antennae (Fig. 14–87). Initially, the valves were not covered by plates. Subsequently, the ventral aperture became guarded by the paired terga and scuta, and these were supported by a posterior dorsal plate, the **carina.** From such an early stalked barnacle, two principal lines are believed to have evolved: one leading to the existing stalked barnacles known as the lepadids, and another leading to the living stalked barnacles known as the scalpellids. An ancestral scalpellid gave rise to the sessile barnacles. Not surprisingly, a commensal habit—attachment on other animals—has evolved in all three lines. Wher-

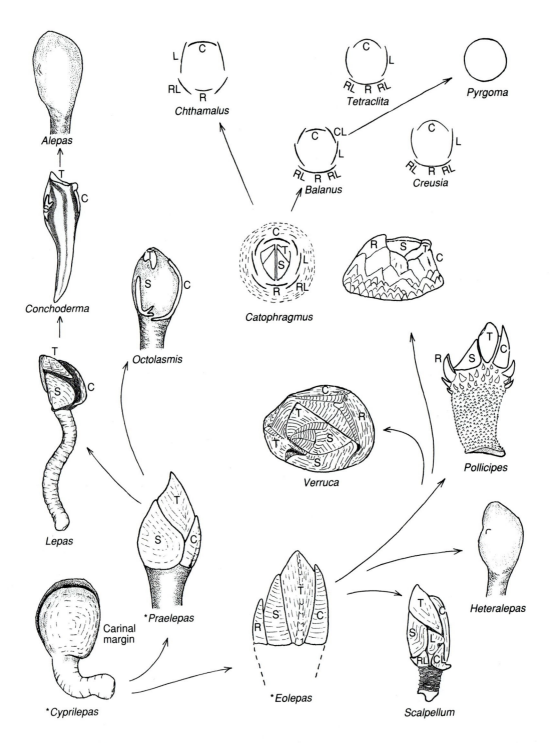

FIGURE 14–87 The probable phylogeny of the barnacles, showing the tendency toward reduction of shell plates. The genera illustrated represent types within the principal phylogenetic lines. *Cyprilepas, Praelepas,* and *Eolepas* are extinct genera(*). The tergum and scutum are omitted from plans of the sessile barnacles except for *Verruca* (Verrucomorpha) and *Catophragmus.* Plates designated include carina (C), carinolateral (CL), lateral (L), rostrum (R), rostrolateral (RL), scutum (S), and tergum (T).

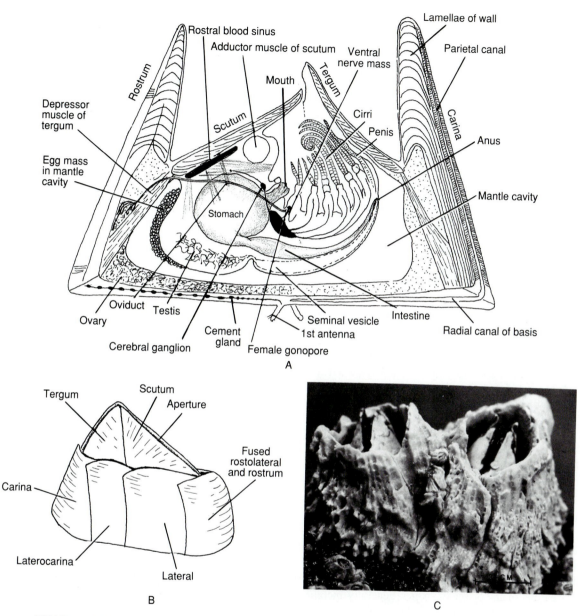

FIGURE 14–88 A, *Balanus,* a sessile barnacle (vertical section). **B,** *Balanus* showing number and position of shell plates (diagrammatic lateral view). **C,** *Tetraclita,* a sessile barnacle, showing the circular wall and the projecting movable tergal plates. *(A, After Gruvel from Calman. B, After Broch from Kaestner. C, Photograph courtesy of Betty M. Barnes.)*

ever this has occurred, there has been a tendency for the calcareous plates to become reduced or even lost, since the host offers some protection to the barnacle.

In the lepadids, which attach to floating objects, wood, coconuts, bottles, tar, or to other animals, the peduncle, or stalk, has remained naked, and the capitulum is covered with no more than the five original,

basic plates: one carina, two terga, and two scuta (Fig. 14–87). This form is well illustrated by the many common and widely distributed species of *Lepas.* Commensals include *Conchoderma,* which lives on whales, turtles, and ships's bottoms, and *Alepas,* which occurs only on jellyfish (Figs. 14–87; 14–89B).

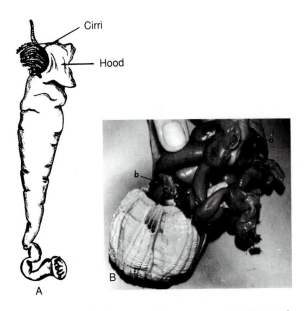

FIGURE 14–89 A, *Xenobalanus,* a sessile barnacle that lives commensally on the fins of cetaceans. Note the loss of opercular plates. **B,** A cluster of the stalked barnacle, *Conchoderma,* attached to *Coronula,* a sessile whale barnacle. Opening into the mantle cavity of *Conchoderma* (a) and *Coronula* (b). *(A, After Gruvel from Calman. B, Photograph courtesy of V. B. Scheffer, U. S. Fish and Wildlife Service.)*

A related family of small, stalked barnacles (Poecilasmatidae) includes many common commensals of crustaceans, such as a species of *Octolasmis* that lives on the gills of lobsters and crabs.

The scalpellids are bottom dwellers, and although some, such as the common Californian *Pollicipes polymerus,* live on intertidal rocks, most are found in deep water. The peduncle is covered with calcearous plates or scales, which generally increase in size toward the capitulum. In many genera, the base of the capitulum is surrounded by several whorls of accessory plates (Fig. 14–87). The extreme in plate reduction is seen in the naked *Heterolepas,* which is commensal on other crustaceans, such as lobsters. One group of scalpellids has become adapted for boring in coralline rock (Fig. 14–90C).

The sessile barnacles (balanomorphs) are thought to have arisen during the Jurassic from the stalked scalpellids by a shortening and disappearance of the peduncle. The large attachment surface and the low, heavy, circular wall make sessile barnacles highly adapted for life on current-swept and wave-pounded intertidal rocks. The wall (**mural plates**) is believed to have formed from the lateral plates covering the capitulum of stalked barnacles. The most primitive living sessile barnacle, *Neoverruca,* recently discovered living near abyssal hydrothermal vents in the western Pacific, passes through several juvenile pedunculate stages before metamorphosing into the sessile form. In the primitive genus *Catophragmus,* the wall is composed of many whorls of imbricated plates, with the interior and largest whorl consisting of eight plates: three paired mural plates, the carina, and the rostrum (Fig. 14–87). In most other sessile barnacles, only this inner whorl of eight plates remains and composes the wall; these plates are usually reduced in number through loss or fusion (Fig. 14–87), perhaps as an evolutionary response to predation by snails, which attack sessile barnacles at plate junctions.

Although there are some deep-water balanomorphs, the group as a whole is intertidal or just subtidal, and species are typically restricted to particular zones. Some are limited to the low-tide mark, and some live in the midintertidal zone. A few are adapted for life in the spray zone at the high-tide mark on wave-splashed rocks. Intertidal barnacles commonly occur in enormous numbers, as indicated by the figures in Table 14–2, which gives the location and aggregation of two species of barnacles on the English coast. Barnacles are less conspicuous on tropical rocky shores, perhaps because of prolonged high temperatures that must be tolerated at the upper intertidal levels.

Many sessile barnacles have become adapted for life on other surfaces besides rocks; a number of species have colonized intertidal grasses and mangroves. They are commensal with a wide range of hosts—sponges, hydrozoans, octocorals, scleractinian corals, crabs (Fig. 14–90B), sea snakes, sea turtles, manatees, porpoises, and whales.

Barnacles are among the most serious fouling problems on ship bottoms, buoys, and pilings, and many species have been transported all over the world by shipping. Sessile barnacles are acquired by ships in coastal waters; larvae from floating lepadids settle on ships at sea. The speed of a badly fouled ship may be reduced by 30%. Much effort and money have been expended toward the development of special paints and other antifouling measures.

Most peculiar are the boxlike, mostly deep-water species of the suborder Verrucomorpha, in which one

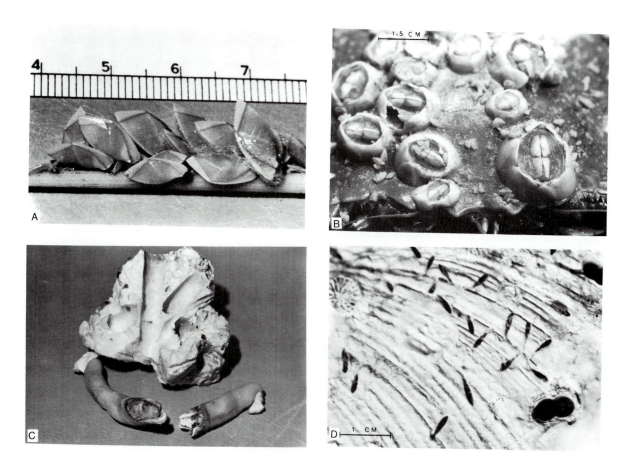

FIGURE 14–90 A, A species of *Poecilasma* on a sea urchin spine. The members of this family of stalked barnacles (Poecilasmatidae) are related to the lepadids but are attached to bottom objects, including sea urchins and crustaceans. *Octolasmis* (shown in Fig. 14–87), which is found on crab gills, is also a member of this family. **B,** *Chelonibia* on the carapace of a blue crab. **C,** *Lithotrya,* a scalpellid barnacle that bores into coralline rock. **D,** Slitlike openings of the burrows of the boring acrothoracican barnacle, *Kochlorine,* in an old clam shell. The dumbbell-shaped openings are the burrows of a boring clam. *(All photographs courtesy of Betty M. Barnes.)*

tergum and scutum form the lid, and the other tergum and scutum are part of the wall (Fig. 14–87). They are believed to have diverged early from the sessile line, because the recently discovered *Neoverruca* mentioned earlier displays both verrucomorph and balanomorph features.

The barnacles described thus far and illustrated in Figure 14–87 are all members of the order Thoracica. In the early evolution of barnacles there appeared another line, represented today by some 30 species of the order Acrothoracica, which are believed to have stemmed from the Scalpellids. Acrothoracicans are the smallest cirripedes, usually only a few millimeters in length. They bore into coral or old mollusc shells using chitinous teeth on the mantle as well as chemical dissolution (Fig. 14–90D). When feeding, the naked, sac-shaped animal projects its cirri through the slitlike opening of its burrow, sometimes holding the appendages open like a fan and turning them several times before retracting them into the burrow.

Parasitic Barnacles

Considering the common occurrence of commensal barnacles, it is not surprising that parasitism has evolved within the class. There are a few parasitic

TABLE 14–2 Vertical Distribution of Two Species of Sessile Barnacles on the Plymouth Coast of England

| | Numbers of Individuals per Square Meter | | | | | |
| | Balanus balanoides | | | Chthamalus stellatus | | |
Height above Mean Tide	Adults	Young	Total	Adults	Young	Total
+3.4 m.	0	0	0	0	0	0
+2.7 m.	0	0	0	15,200	9,200	24,400
+1.8 m.	0	400	400	54,000	38,000	92,000
+0.8 m.	4,000	12,400	16,400	55,600	35,200	90,800
−0.2 m.	40,400	20,400	60,800	400	4,800	5,200
−2.1 m.	0	0	0	0	0	0

Moore, 1936.

Thoracica, and two other orders are exclusively parasitic. The most important of these is the order Rhizocephala, which is largely parasitic on decapod crustaceans. The body is saccular, and the mantle is devoid of calcareous plates. There is also a complete absence of appendages and segmentation.

A cypris larva of *Lernaeodiscus porcellanae,* one destined to become female, enters the gill chamber of the host crab and attaches to the gill. Following metamorphosis of the cypris, a perforation in the host's integument is made, permitting the entrance of a mass of dedifferentiated cells (**kentrogon**) of the parasite (Fig. 14–91A). Within the host, growth of the parasite takes place through the ramification of a nutrient-absorbing rootlike system (called the **interna**). Sexual development involves the formation of an external brood chamber (called the **externa,** which resembles the female crab's egg sponge), which projects mushroom-like through a new opening produced in the crab's integument near the base of the abdomen. A male cypris can then attach itself to the opening of the brood chamber and extrude dedifferentiated cells, which migrate into a special chamber in the female, in which they redifferentiate into a testis. The female now is functioning as a hermaphrodite, and fertilization and development to the naupliar stage occur within the brood chamber. The effects of this parasitism on the crab are numerous and severe. Two of the most striking effects are the inhibition of molting and parasitic sterilization. The development of the gonads is retarded, or the gonads atrophy.

Members of the order Ascothoracica are ectoparasites on sea lilies and serpent stars, or endoparasites in octocorallian corals and in the coelom of sea stars and echinoids (Fig. 14–91B). The ascothoracicans are now believed to be the most primitive of the Cirripedia, despite their adaptations for parasitism. A chitinous, basically bivalved carapace encloses the body, which includes a limbless abdomen. There are no second antennae, but all species possess prehensile first antennae. The six pairs of thoracic appendages are not cirriform but are like the swimming appendages of the cypris.

Internal Anatomy and Physiology

During feeding, the paired scuta and terga open, and the cirri unroll and extend through the aperture (Fig. 14–86B). When outstretched, a number of cirri on each side form one side of a basket. The two sides sweep toward each other and downward, each half acting as a scoop net. The action is somewhat similar to the opening and closing of your two fists simultaneously when the bases of the palms are placed together. In currents some barnacles may hold the cirri outstretched for a period, and some, such as *Pollicipes polymerus,* the intertidal scalpellid common on the west coast of the United States, do not rhythmically beat the cirri but extend them into the flow through surge channels.

On the closing stroke, food particles suspended in the water are trapped by the setae, and the first one to three pairs of cirri are used to scrape these particles off and transfer them to the mouthparts. Both the mandibles and maxillae are well developed and are used to macerate food. The size of plankton used for food varies. Some species of *Lepas, Pollicipes,* and *Tetraclita* capture copepods, isopods, amphipods, and other relatively large organisms and could therefore

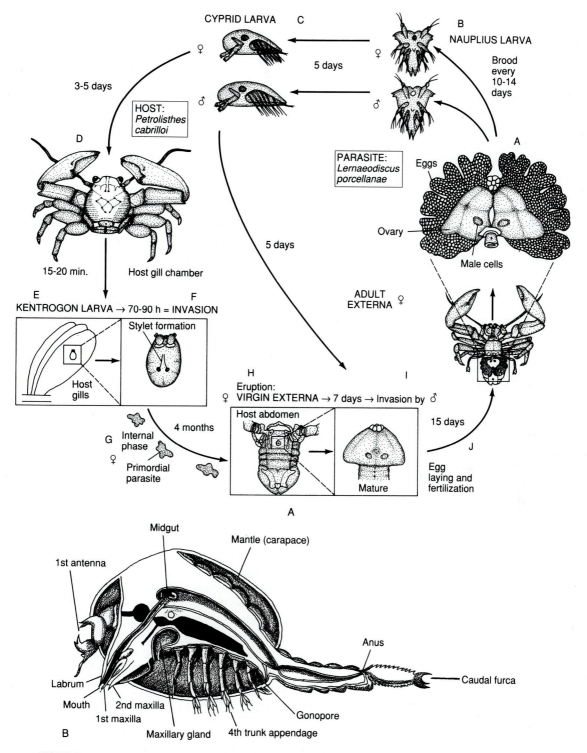

FIGURE 14–91 A, Life cycle of a rhizocephalan barnacle, *Lernaeodiscus porcellanae,* parasitic on the porcelain crab, *Petrolisthes cabrilloi,* on the California coast. See the text for additional explanation. **B,** *Ascothorax ophioctenis,* a parasite in the bursae of brittle stars. *(A, From Ritchie, L. E., and Hoeg, J. T. 1981. The life history of Lernaeodiscus porcellanae and coevolution with its porcellanid host. J. Crust. Biol. 1(3):334–347. B, After Wagin.)*

Labels in figure:

CYPRID LARVA — C

NAUPLIUS LARVA — B

5 days

Brood every 10-14 days

♀

♂

3-5 days

HOST: *Petrolisthes cabrilloi*

D

PARASITE: *Lernaeodiscus porcellanae*

A

Eggs

Ovary

Male cells

15-20 min.   Host gill chamber

5 days

ADULT EXTERNA ♀

E   KENTROGON LARVA → 70-90 h = INVASION   F

Stylet formation

Host gills

H   Eruption: ♀ VIRGIN EXTERNA → 7 days → Invasion by ♂   I

Host abdomen

Mature

15 days

J

Egg laying and fertilization

G ♀   Internal phase

Primordial parasite

4 months

A

Midgut

Mantle (carapace)

1st antenna

Anus

Caudal furca

Labrum

Mouth

1st maxilla

2nd maxilla

Maxillary gland

4th trunk appendage

Gonopore

B

be classified as predaceous rather than suspension feeders.

Many barnacles feed on small plankton, but in at least some of these barnacles, such as *Balanus improvisus* and *Balanus balanoides,* large and small plankton are filtered. The last four pairs of cirri perform the usual scooping motion and capture particles over several micrometers in diameter. The water current produced by scooping is directed over the first two pairs of cirri, which remain within the mantle cavity close to the mouth. These anterior cirri have the setae closely placed, and the cirri collect finer particles, such as phytoplankton, from the water current.

The gut of barnacles is divided into a cuticle-lined foregut and hindgut, as well as a digestive midgut. The foregut of thoracican barnacles has a plate of thickened cuticle against which the mandibles grind the food. The midgut bears seven digestive ceca and a pair of pancreatic glands. The glands secrete enzymes for extracellular digestion in the midgut. The hindgut forms fecal pellets.

A heart and arteries are absent, but circulation of the blood is facilitated by a blood pump, consisting of a large sinus located between the esophagus and the adductor muscle (Fig. 14–88A). Blood circulates through the body mantle walls and also into the peduncle.

Gills are lacking, and the mantle and cirri are probably the principal sites of gas exchange. The excretory organs are maxillary glands. The naupliar eye divides and is retained in the adult as two lateral components and one median component.

Reproduction and Development

Most parasitic barnacles and the boring acrothoracicans are dioecious. Thoracican barnacles are mostly hermaphroditic and are the only large group of hermaphroditic crustaceans. Cross fertilization is generally the rule, however, because a suitable substratum almost always contains a large number of adjacent individuals. The ovaries of sessile barnacles (Fig. 14–88A) lie in the basis and in the walls of the mantle and they are located in the peduncle in stalked forms, (Fig. 14–86A). The paired oviducts open at or near the basis of the first pair of cirri. Just before reaching the gonopore, each oviduct dilates as an oviductal gland that secretes a thin, elastic ovisac at the time of egg deposition. As the ovisac receives eggs, it swells and stretches, emerging from the gonopore and com-

ing to lie within the mantle cavity. The testes are located in the cephalic region, and the two sperm ducts unite within a long penis, which lies in front of the anus (Fig. 14–86A). The penis is quite extensible and can be protruded out of the body and into the mantle cavity of another individual for the deposition of sperm (Fig. 14–92A). Functional males can recognize a functional female, which may be inseminated by more than one male. The sperm are deposited as a mass near the first cirri and must penetrate the ovisac to reach the eggs.

Dwarf males that attach themselves to female or hermaphroditic individuals are found in some of the pedunculate genera, such as *Scalpellum* and *Ibla,* in a few species of *Balanus,* and in the boring Acrothoracica (Fig. 14–92B–D). In addition to being minute, these males show all degrees of modification through degeneracy or loss of structures. When attached to a hermaphrodite, the males are called **complemental males.** When the males are greatly modified (Fig. 14–92D), the species is usually dioecious, and the larger, "host" individuals are females.

The eggs are brooded within the ovisac in the mantle cavity in all barnacles, but the ovisac gradually deteriorates during the incubation period. In *Balanus balanoides,* for example, eggs are fertilized in the autumn and brooded until March. A nauplius represents the hatching stage in most species and may be easily recognized by the triangular, shield-shaped carapace (Fig. 14–93A). Six naupliar instars are succeeded by a nonfeeding larval cypris. The entire body is enclosed within a bivalve carapace and possesses a pair of sessile compound eyes and six pairs of thoracic appendages. The cypris is the settling stage and attaches to a suitable substratum, at first temporarily, with the secretions of discs located on the first antennae, and then permanently, with the antennal cement glands (Fig. 14–93B).

A number of factors appear to increase the likelihood of dense settling, on which reproduction in the adult depends. Simultaneous shedding of the nauplius by aggregations of individuals has been observed in some species. A protein in the exoskeleton of older attached individuals has been demonstrated to attract settling larvae. Illumination, roughness, position, depth, and bacterial film of the substratum may also be important in selection, and attachment may not occur at the first surface contact. Following attachment, metamorphosis takes place: the cirri elongate, the body undergoes flexion, and the primordial plates

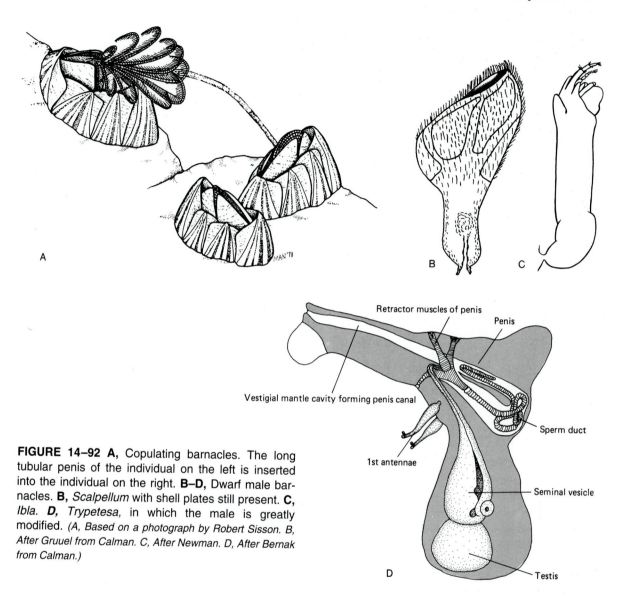

FIGURE 14–92 A, Copulating barnacles. The long tubular penis of the individual on the left is inserted into the individual on the right. **B–D,** Dwarf male barnacles. **B,** *Scalpellum* with shell plates still present. **C,** *Ibla.* **D,** *Trypetesa,* in which the male is greatly modified. *(A, Based on a photograph by Robert Sisson. B, After Gruuel from Calman. C, After Newman. D, After Bernak from Calman.)*

appear on the next exoskeleton lying beneath the old valves of the cypris.

Shell growth is more or less continuous and independent of body growth and ecdysis. In *Balanus improvisus,* the first 20 molts take place two to three days apart, on average. The cuticle, or exoskeleton, lining the interior of the mantle cavity and covering the appendages is molted periodically, as in other arthropods. The calcareous plates are secreted by the underlying mantle and are not shed at ecdysis. Growth of the plates takes place by the continual addition of material to their margins and interior surfaces, thus increasing their thickness and diameter.

Microscopic growth bands are produced, as in the calcareous shells and skeletons of other animals. In sessile barnacles, a wedge of mantle tissue lying between the junction of the basis and the mural plates adds material to the periphery of the basis and to the lower margin of the mural plates (Fig. 14–94). This mantle tissue thus permits a continual outward and upward growth of the mural plates to accommodate the increase in diameter and height. The young barnacle is about 3 mm in diameter at the end of a month.

Cement is elaborated throughout the life of an individual, and repair of partial detachment is possible. Detachment, however, is most likely rare, because the

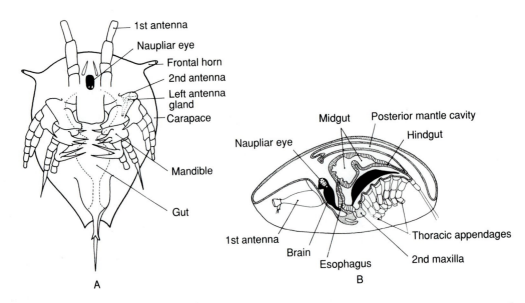

1st antenna
Naupliar eye
Frontal horn
2nd antenna
Left antenna gland
Carapace

Mandible

Gut

A

Midgut   Posterior mantle cavity
Naupliar eye   Hindgut
1st antenna
Brain   Thoracic appendages
Esophagus   2nd maxilla
B

FIGURE 14–93 A, Nauplius of *Balanus.* The setae are omitted. **B,** Free-swimming cypris larva of *Balanus. (Both from Walley, L. J. 1970. Phil. Trans. Roy. Soc. London. B. Biol. Sci. 256(807):237–280.)*

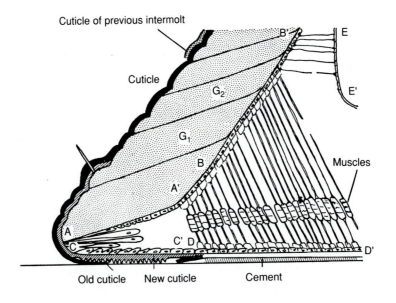

Cuticle of previous intermolt
Cuticle
G_2
G_1
B
B'
E
E'
A'
Muscles
A
C
C' D
D'
Old cuticle   New cuticle   Cement

FIGURE 14–94 Diagrammatic section through the junction of wall and basis of *Balanus,* showing growth zones. A–A' is the region of maximum calcium carbonate secretion. B–B' region of minimum secretion. C–C' cuticle-secreting region. D–D' region in the basis of cement ducts and no secretion of calcium carbonate. E–E' inner wall of mantle, in which exoskeleton is shed at each molt. G_1 and G_2 are growth bands of calcium carbonate. *(From Bourget, E., and Crisp, D. J. 1975. Analysis of the growth bands and ridges of barnacle shell plates. J. Mar. Biol. Assoc. U.K. 55(2):439–462. Copyrighted and reprinted by permission of Cambridge University Press.)*

protein cement provides a very strong bond that can withstand enormous pressures.

Most barnacles that survive the heavy mortality in the period immediately following settling probably live for one to ten years. In *Balanus glandula* on the Pacific coast of the United States, where their reproductive biology is well known, two to six broods of 1000 to 30,000 nauplii each are produced in the winter and spring. The maximum basal diameter of the adult is 22 mm, of which 7 to 12 mm is reached by the end of the first year and 14 to 17 mm at the end of the third. The life span is eight to ten years. Longevity and growth rates, however, are greatly influenced by environmental conditions. Smothering by other individuals and other sessile organisms is a common cause of death.

SUMMARY (CLASSES BRANCHIOPODA THROUGH CIRRIPEDIA)

1. The three major classes of small crustaceans are the Branchiopoda, the Ostracoda, and the Copepoda. Branchiopods are distinguished by their foliaceous appendages, which in many species are adapted for suspension feeding. In other respects, branchiopods are diverse. Water fleas (cladocerans) have a carapace that encloses the trunk but not the head; clam shrimp (conchostracans) have a bivalve carapace that encloses the entire body; fairy shrimp and brine shrimp (anostracans) lack a carapace.

2. Branchiopods are largely inhabitants of fresh water, especially temporary pools and ditches. Only cladocerans are also represented in permanent fresh, and to a lesser extent, marine waters.

3. Ostracods have a bivalve carapace impregnated with calcium carbonate that encloses the entire body. Most ostracods are less than 2 mm long. They are mostly marine benthic animals, but some species live in fresh water. Ostracods have an extensive fossil record.

4. Copepods possess more or less cylindrical, tapered bodies. Long, laterally projecting first antennae and a persistent naupliar eye are distinctive features of many species. The trunk appendages are markedly biramous. Most copepods are less than 2 mm long.

5. Most copepods are marine, but some species are common in freshwater lakes and pools. There are planktonic, epibenthic, and interstitial species. There are also many parasitic forms.

6. Feeding habits among these classes of small crustaceans are diverse. Suspension feeding is especially characteristic of most branchiopods and most planktonic copepods. Suspension-feeding, planktonic copepods are of great importance in marine food chains.

7. Many freshwater forms, especially branchiopods, undergo parthenogenesis, produce both dormant and rapidly hatching eggs, and encyst.

8. Barnacles, members of the class Cirripedia, are unique among crustaceans, indeed most arthropods, in being sessile. A number of peculiarities of barnacles, such as a carapace covered with calcareous plates, suspension-feeding cirri, hermaphroditism, long tubular penis, and dwarf males, may be correlated with their sessility.

9. Free-living barnacles are attached by a peduncle (stalk) or are sessile (stalkless). The peduncle represents the preoral part of the body and contains the cement glands. The oldest known fossil barnacles are pedunculate. Lepadids, which are extant pedunculate barnacles, attach to floating, inanimate objects, such as wood, or to pelagic animals. Scalpellids attach to rocks and, in addition to the five large principal plates, have many small plates covering the peduncle and capitulum.

10. The sessile barnacles, which are believed to have evolved from the scalpellids, are stalkless, the preoral region being represented by the basis, which contains the cement glands. Only the paired terga and scuta are movable. The other capitular plates form a circular wall around the barnacle. Sessile barnacles are especially adapted for intertidal life on hard substrates that are subjected to waves, surge, and currents.

11. Commensalism, which has evolved in all three major lines of barnacles, has resulted in reduction of the protective calcareous plates. Commensalism has undoubtedly been the avenue to parasitism, which is characteristic of one third of the species of cirripedes.

ASPECTS OF CRUSTACEAN PHYSIOLOGY

Hormones

The hormones of crustaceans and insects are the best known of any invertebrates. Crustacean hormones are either neurosecretions or secretions of one of three endocrine tissues: the Y-organ, the androgenic gland, or the ovary. Located between the two basal optic ganglia is a body called the sinus "gland," which is a center for hormone release (Fig. 14–95). The **sinus gland** is composed of the swollen endings of nerve fibers that have their origin in neurosecretory cell bodies located in several different places within the eye stalk ganglia. The cell body clusters are called **X-organs** with various designations to indicate the specific site, such as sensory pore X-organ. The neurosecretory material synthesized within the cell bodies, probably a different hormone in each cluster, migrates along the axons to the swollen endings composing the sinus gland. Here it is released into the hemolymph. In addition to the sinus glands, neurosecretions are released into the blood in the tritocerebrum (postcommissural organs) and in the region of the heart (pericardial organs).

Hormones are known to regulate many functions in crustaceans, including various aspects of general metabolism.

Regulation of Reproduction

Although sex in crustaceans is determined genetically, the development and function of the gonads and the development of secondary sexual characteristics is under hormonal control (Fig. 14–96). If the Y-organ, located on either side in the anterior part of the cephalothorax, is removed prior to sexual maturity, gonadal development is seriously impaired; if the animal is adult, the gonads are unaffected. In blue crabs, a female pheromone initiates a courtship response in the male but not if the male's eyestalks have been removed. Either the eyestalk contains pheromone-responding neural pathways or the eyestalk is controlling the androgenic gland, which in turn regulates courtship behavior.

In female malacostracans, a hormonal interrelationship exists between the ovary and the X-organ/sinus gland complex that is somewhat similar to the pituitary/ovary relationship of vertebrates. The sinus gland produces a hormone that inhibits the development of eggs during the nonbreeding periods of the year. During the breeding season, a gonad-stimulating hormone is secreted, probably by the central nervous system (Fig. 14–96). As a result of the presence of this stimulatory hormone, the blood level of gonad-inhibiting hormone declines, egg development begins, and the ovary elaborates a hormone, initiating structural changes preparatory for egg brooding, such

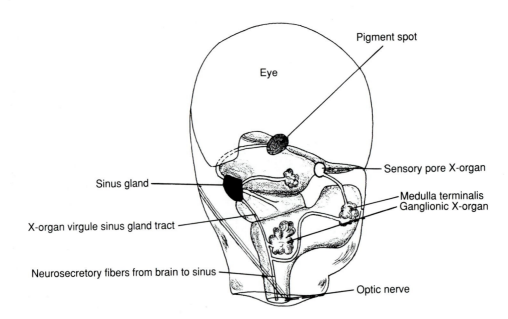

FIGURE 14–95 Eye and eyestalk of the shrimp, *Palaemon,* showing neurosecretory centers and tracts. *(After Carlisle, D. B., and Knowles, F. 1959. Endocrine Control in Crustaceans. Cambridge University Press, London. Copyrighted and reprinted by permission of the publisher.)*

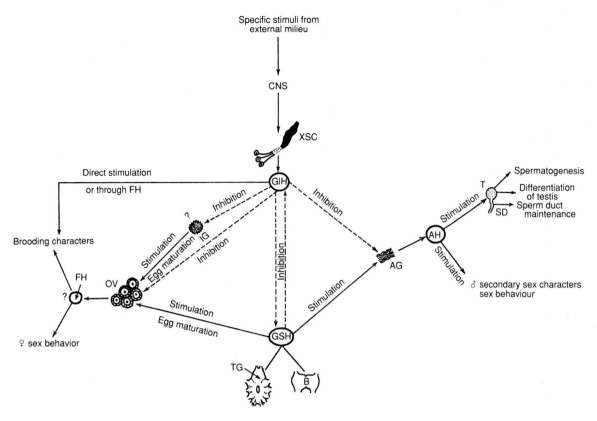

FIGURE 14–96 Diagram illustrating possible hormonal control of reproduction in a decapod crustacean. Key to lettering: AG, androgenic gland; AH, androgenic hormone; B, brain; CNS, central nervous system; FH, female hormone; GIH, gonad-inhibiting hormone; GSH, gonad-stimulating hormone; IG, hypothetical intermediate gland; OV, ovaries; SD, sperm duct; T, testes; TG, thoracic ganglion; XSC, X-organ/sinus-gland complex. *(From Adiyodi, K. G., and Adiyodi, R. G. 1970. Biol. Rev. 45:121–165. Copyrighted and reprinted by permission of Cambridge University Press.)*

as the development of ovigerous setae on the pleopods or the development of oostegites (marsupium) in peracaridans. These characteristics appear at the next molt.

The development of the testes and male sexual characteristics in malacostracans is controlled by hormones produced in a small mass of secretory tissue, the androgenic gland. This gland is located at the end of the vas deferens (except in isopods, in which it appears to be located in the testis itself). Removal of the androgenic gland is followed by a loss of male characteristics and conversion of the testes into ovarian tissue. If an androgenic gland is transplanted into a female, the ovaries become testes and male characteristics appear.

Molting and Growth

In many crustaceans, including barnacles, crayfish, and the lobster, *Homarus,* molting and growth continue throughout the life of the individual, although molts become spaced further and further apart. Such crustaceans may live to be quite old, and some may become very large. In others, such as some crabs, molting and growth cease with the attainment of sexual maturity or of a certain size or after a certain number of instars. The process of molting has probably been investigated in more detail in crustaceans, especially decapods, than in any other arthropods.

Molting is virtually a continuous process in the life of a crustacean; 90% or more of the period between actual molts may be involved with concluding and preparatory processes associated with the preceding and the future molts. This is especially true in species that molt year round. In species that molt seasonally, such as species of the crayfish, *Cambarus,* and the fiddler crab, *Uca,* there is a rather definite intervening rest period during the intermolt. But even during the rest period food reserves are accumulated for the next molt.

Physiologists generally recognize four stages in the molt cycle: **proecdysis, ecdysis, postecdysis,** and **intermolt** (see Fig. 14–97). The preparatory phase, or proecdysis (or premolt), is marked by a continuing accumulation of food reserves and a rise in blood calcium probably resulting from the release of stored calcium from the hepatopancreas and resorption of calcium from the cuticle. In some crustaceans, such as crayfish and land crabs, the stomach epithelium se-

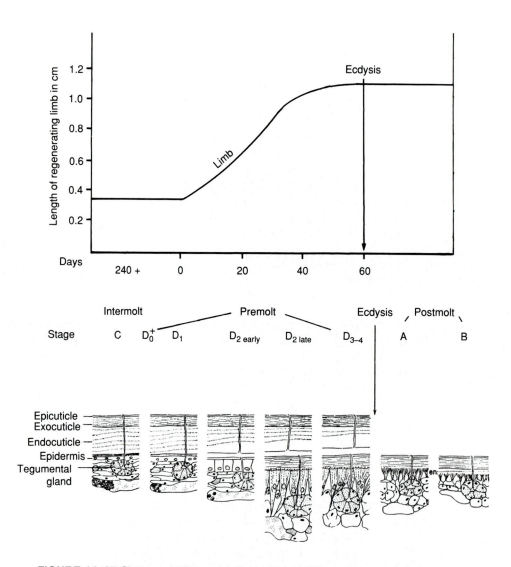

FIGURE 14–97 Changes in the exoskeleton and associated tissues during the molt cycle of the land crab, *Gecarcinus.* Levels of the molting hormone, ecdysone, and growth of regenerating limbs at corresponding states are plotted above. *(Adapted from Skinner, D. M. 1962. The structure and metabolism of a crustacean integumentary tissue during a molt cycle. Biol. Bull. 123(3):635–647.)*

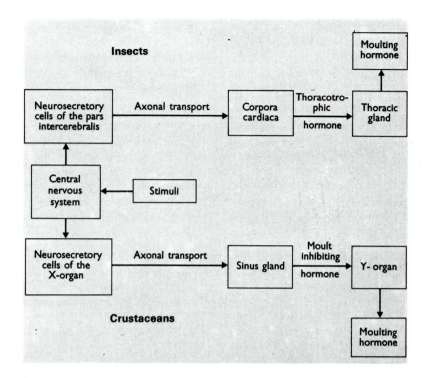

FIGURE 14–98 Comparison of the hormonal control of molting in crustaceans with that in insects. The insect thoracic glands and the crustacean Y-organs are analogous, producing ecdysone in both. In insects, however, the hormone from the corpora cardiaca stimulates secretion of ecdysone, whereas in crustaceans the sinus gland hormone inhibits secretion. *(From Highnam, K. C., and Hill, L. 1977. The Comparative Endocrinology of the Invertebrates. 2nd Edition. Edward Arnold [Publishers], Ltd., London. p. 215.)*

cretes calcareous concretions, called **gastroliths,** that function as calcium storage centers. Eventually, the membranous layers and part of the calcified layers of the old exoskeleton are digested away by the epidermis (Fig. 14–97). Resorption of calcium and digestion of the calcified layer are especially great where splitting later occurs or where the old skeleton must be stretched or broken to permit extraction of a large terminal part of an appendage, such as a claw. After the separation of the old cuticle from the epidermis and the secretion of the new epicuticle and exocuticle, the animal is prepared for the actual brief process of ecdysis and usually seeks some protected retreat or remains in its burrow. The body swells from the uptake of water through the gills or midgut and quickly emerges from the old skeleton, which is commonly eaten later for its calcium salts. The precise mechanism of water absorption is uncertain, but the elevated salt level of the hemolymph probably establishes an osmotic gradient along which water enters. The amount of water absorbed may equal almost half of the premolt body weight. The gastroliths, if present, are dissolved by digestive secretions, and the calcium salts are absorbed back into the blood.

During postecdysis, or metecdysis, the endocuticle is secreted, and calcification and hardening of the skeleton take place around the water-swollen body (Fig. 14–97). As the new exoskeleton hardens excess body water is eliminated, and the soft tissues shrink away from the now slightly oversized exoskeleton to allow for tissue growth during the intermolt period. The animal remains in its retreat and does not feed during the first part of this phase.

The intermolt may be long or short, depending on whether or not the animal molts seasonally. Although the exoskeleton is completely formed, food reserves are accumulated for the next molt.

The physiological processes involved in molting are regulated by hormonal interactions that are essentially like those of insects (Fig. 14–98). The Y-organs, which are analogous to the prothoracic glands of insects, produce ecdysone, which acts on the epidermal cells and the hepatopancreas to initiate proecdysis. The production of ecdysone by the Y-organs is inhibited by a hormone released from the sinus gland. Thus, the removal of the Y-organs prevents molting; removal of the eyestalks initiates premature proecdysis. The inhibitory action of a sinus gland hormone on the production of ecdysone by the Y-organs is an important difference from control in insects, where the corpora cardiaca stimulate rather than inhibit the prothoracic glands to produce ecdysone (Fig. 14–98).

The regulation of molting hormones, and thereby the regulation of the actual molt cycle, depends on different stimuli operating on the central nervous system. In crayfish, which molt seasonally, day length is the controlling factor; in the crab, *Carcinus,* tissue growth is the controlling factor.

Limb loss is a common event in the life of many crustaceans, and in decapods limbs may even be self-amputated. Severance often takes place at a preformed breakage plane, which runs across the basiischium, a proximal leg joint. Internally, there is a corresponding double membranous fold, which is perforated by a nerve and blood vessels. When the limb is cast off, the plane of severance passes between the two membranes, leaving one membrane attached to the basal stub. The membrane constricts around the perforations, so there is very little bleeding. In some species, severance can take place only if the limb is pulled either by the animal itself or by an outside force; in its most highly developed state, however, as in most decapods (except shrimp), autotomy is a unisegmental reflex. If a leg is caught or damaged by a predator, a reflex is set up, and an autotomizer muscle (one of the locomotor muscles), is stimulated to undergo extreme contraction, fracturing the limb along the breakage plane.

Following severance and scab formation, a small papilla representing the new limb bud grows out from the stub. Growth then halts until the premolt period, when rapid growth and regeneration are completed. The new limb unfolds from a sac at the time of molting, but at this molt is not quite as large as the original. Multiple amputation of limbs induces molting, but the premolt regeneration phase must first be accomplished. If partially regenerated limbs are removed, the molting cycle is delayed until new limb buds are formed.

Chromatophores

A characteristic and striking feature of the integument of many malacostracans is the presence of **chromatophores.** The crustacean chromatophore is a pigment-bearing cell with branched, noncontractile processes. Each cell is tightly bound with a number of similar chromatophores to form a multicellular organ with radiating processes (Fig. 14–99A). Pigment granules flow into the processes in the dispersed, or stellate, state and are confined to the center of the cell in the concentrated, or punctate, state. The chromatophores are located in the subepidermal connective tissue, and where present, the overlying exoskeleton is sufficiently thin or transparent to make them visible.

White, red, yellow, blue, brown, and black pigments may be present. The red, yellow, and blue pigments are carotene derivatives obtained from the diet. The red compound so conspicuous in boiled crabs, lobsters, and shrimps is **astaxanthin** (carotenalbumin). In the exoskeleton of the living animal, this pigment is conjugated with a protein and is blue or some other color characteristic of the conjugated state. Curiously, a single chromatophore may possess one, two, three, or even four color pigments, any one of which can move independently of another. In general, polychromatic chromatophores are found only in shrimp. The most common type of rapid (physiological) color change is a simple blanching (or lightening) and darkening. This response is typical of many crabs, such as the fiddler crab, *Uca.* Many crustaceans, however, especially shrimp, can adapt to a wide range of colors. The little shrimp, *Palaemonetes,* for example, possesses trichromatic chromatophores with red, yellow, and blue pigments; through the independent movement of these three primary colors it can adapt to any background color, even black. Other species have similar abilities.

The movement of chromatophore pigments is controlled by hormones elaborated by the neurosecretory system in the eyestalk (or below the eye, when the eyes are sessile) and in other parts of the central nervous system. In shrimp (aside from *Crangon*) that have red, yellow, blue, and white pigments, removal of the eyestalks results in a darkening of the body through dispersion of the red and yellow pigments. If these animals are then injected with sinus gland extract, the white pigment disperses, and the body color

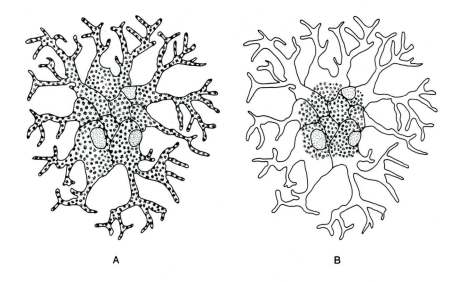

A   B

C

FIGURE 14–99 A–B, Crustacean chromatophores in which pigment is dispersed **(A)** and concentrated **(B)**. Each body is actually a cluster of cells, each a chromatophore. Processes radiate in all directions, not just in the plane shown above. **C,** Fiddler crabs in pale (nighttime) and dark (daytime) phases. *(A and B, Based on a photograph by McNamara. 1981. C, From Prosser, A. L. 1973. Comparative Animal Physiology. W. B. Saunders Co., Philadelphia. p. 924.)*

rapidly blanches. The opposite occurs in some crabs that have black, red, yellow, and white pigments. Removal of the eyestalks causes blanching of the body color, resulting from a concentration of the black pigments and dispersal of the white. Injections of sinus gland extract cause a darkening.

The current understanding of hormonal control of chromatophores, although still incomplete, provides an explanation for the opposite effects of eyestalk removal in shrimp and crabs. There are separate, antagonistic pairs of **chromatophorotropins** (hormones) for each pigment; one brings about dispersion and one concentrates. Moreover, some at least appear

not to be species-specific but are found in a large number of decapods. These hormones are released not only by the sinus gland in the eyestalk but also by the commissural gland in the tritocerebrum. Whether blanching of the body occurs when the eyestalks are removed depends on whether the sinus gland is the releasing site for a number of those chromatophorotropins that have a dispersing effect on the pigment granules of the chromatophores.

Eyestalk hormones control not only the functioning of the chromatophores but other pigment changes as well. In different crustaceans, any one or a combination of the three retinal pigments—distal, proximal,

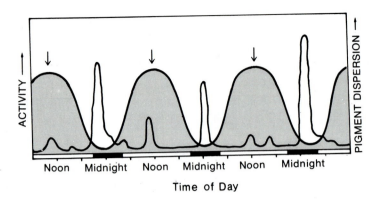

FIGURE 14–100 Rhythmic motor activity and chromatophore changes (shaded) in the crab, *Carcinus*. Motor activity exhibits a tidal rhythm and is greatest at high tides. The chromatophores exhibit a diurnal rhythm, with pigment dispersed (animals dark) during the day (indicated by arrows). *(Modified from Palmer, J. D. 1976. An Introduction to Biological Rhythms. Academic Press, New York. p. 99; and 1975. Biological clocks in the tidal zone. Sci. Am. 232(2):74.)*

and reflecting pigments—may migrate distally or proximally in adapting the eye to bright or weak light. Experimental evidence indicates that at least the movement of the distal retinal pigment and the reflecting pigment is under the control of sinus gland hormones.

Physiological Rhythms

Pigment movement and other physiological processes often display a rhythmic activity in crustaceans. Such physiological rhythms, or physiological "clock mechanisms," have been studied extensively in a number of species, especially the green crab, *Carcinus*, and the fiddler crab, *Uca*. Both crabs live on sand flats and in the intertidal zone of protected beaches, but *Carcinus* is active at high tide and *Uca* at low tide. Through dispersion and concentration of chromatophore pigments, both species are pale at night and darken during the day (Fig. 14–100). The rhythm persists when the crabs are kept in constant light or darkness, and fiddler crabs flown from Woods Hole, Massachusetts, to California continued to blanch and darken on Woods Hole time. Removal of the eyestalks, which secrete the pigment-controlling hormone, disrupts the rhythm in *Carcinus* and reduces its amplitude in *Uca*.

Locomotor activity in both crabs follows the tide and is governed by a lunar rhythm or a biological clock set on lunar time (Fig. 14–100). After about a week under constant conditions, the rhythm is lost. It can be restored by simply immersing *Uca* in sea water and cooling *Carcinus*. The clocks of both species can be set to a new tidal rhythm by subjecting the crabs to periods of cooling or high pressure that correspond to the desired tidal rhythm.

Similar rhythms have been recorded for other crustaceans, such as the blue crab, *Callinectes sapidus,* and the intertidal isopod, *Ligia exotica.* Species of crayfish display a diurnal rhythm of locomotor activity, but such a rhythm is never present in cave species.

SUMMARY

1. Hormones are known to control a number of functions in crustaceans, of which reproduction, molting and growth, and chromatophore changes have been most studied (in decapods). There are several centers of hormone secretion, and the sinus gland in the eyestalk of decapods is a principal center of hormone release.

2. Some crustaceans molt throughout their lives; others cease on reaching sexual maturity. Many important aspects of molting physiology take place during proecdysis (e.g., calcium resorption), during postecdysis (e.g., calcium deposition), and during the intermolt (accumulation of food reserves).

3. The integument of many malacostracans contains branched chromatophores, within which pigment granules of one or more colors may become dispersed or concentrated, changing the coloration of the animal. Adaptation to background is a common function of chromatophores.

4. Chromatophore changes and other functions of crustaceans may display rhythmic activity that coincides with tidal or diurnal rhythms.

5. Many malacostracans are capable of self-amputation of limbs (autotomy), aiding in escape from predators. The limbs are regenerated in connection with the molting cycle.

CRUSTACEAN PHYLOGENY

Nearly all taxa of crustaceans are known to have existed since the Cambrian period, but their origin and their relationship to the other arthropod groups are obscure. The patchy fossil record is not very informative.

Given the comparative anatomy of living forms, the ancestral crustacean was probably a small, swimming, epibenthic animal possessing a head and a trunk of numerous similar segments. The head bore two pairs of antennae, a pair of mandibles, two pairs of maxillae, a pair of stalked compound eyes, and a naupliar eye (Fig. 14–101). The mouth was directed backward. The trunk appendages were numerous and similar, not only to each other but probably also to the maxillae, and probably served in locomotion, gas exchange, and feeding. Certainly among living crustaceans, the cephalocarids and the remipedians most closely resemble such a hypothetical ancestral crustacean, despite the fact that each possesses a number of specialized features.

During the evolution of the existing crustacean groups, the ancestral stock probably divided early into three principal lines: one leading to the Branchiopoda; one (Maxillopoda) to the Branchiura, Cirripedia, and Copepoda; and one to the malacostracans. The Cephalocarida and Remipedia are considered by many to be sufficiently generalized to stand near the stem line of all three lines (Fig. 14–101). A number of alternative schemes have been recently proposed, however. The malacostracan line culminates in two major groups, the peracaridans and eucaridans.

SYSTEMATIC RÉSUMÉ OF THE CRUSTACEA

Class Remipedia. Crustaceans with long, segmented, wormlike bodies, each segment bearing a pair of biramous appendages. Inhabitants of marine caves. *Lasionectes.*

Class Cephalocarida. Minute crustaceans having elongated segmented trunks, the first eight segments bearing similar biramous appendages. Living in marine surface sediments. *Hutchinsoniella.*

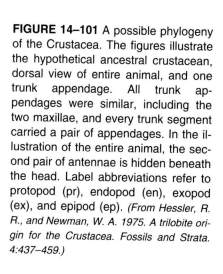

FIGURE 14–101 A possible phylogeny of the Crustacea. The figures illustrate the hypothetical ancestral crustacean, dorsal view of entire animal, and one trunk appendage. All trunk appendages were similar, including the two maxillae, and every trunk segment carried a pair of appendages. In the illustration of the entire animal, the second pair of antennae is hidden beneath the head. Label abbreviations refer to protopod (pr), endopod (en), exopod (ex), and epipod (ep). *(From Hessler, R. R., and Newman, W. A. 1975. A trilobite origin for the Crustacea. Fossils and Strata. 4:437–459.)*

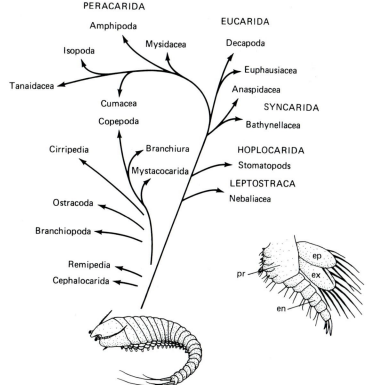

Class Branchiopoda.   Small crustaceans with foliaceous trunk appendages. Mostly inhabitants of
fresh water.
Subclass Calamanostraca.
Order Notostraca.   Tadpole shrimp. Body composed of a thorax with appendages and an abdomen without appendages. Thorax covered
by a carapace. *Triops.*
Subclass Diplostraca.   Body enclosed within a laterally compressed carapace.
Order Conchostraca.   Clam shrimp. Body completely enclosed within the carapace. Trunk
with numerous appendages. *Lynceus.*
Order Cladocera.   Water fleas. Trunk but not
head enclosed within bivalve carapace. Head
bears single median compound eye. Antennae
commonly large and used in swimming. Four
to six pairs of trunk appendages. Although
most species occur in fresh water, there are a
small number of marine forms. *Daphnia, Leptodora.*
Subclass Sarsostraca.
Order Anostraca.   Fairy shrimp. Trunk composed of 11 to 18 segments with appendages.
Carapace absent. Eyes compound and stalked.
Branchinecta, Artemia (inhabits salt lakes and
ponds).
Class Ostracoda. Ostracods.   Minute crustaceans
having the body enclosed within a hinged, bivalved
carapace, which is commonly impregnated with
calcium carbonate. Reduced trunk having no more
than two pairs of appendages.
Subclass Myodocopa.   Marine ostracods having
the valves with an antennal notch. The second
antennae usually adapted for swimming. Two
pairs of trunk appendages.
Order Myodocopida.   The long, wormlike second pair of trunk appendages is adapted for
cleaning interior of valves. *Cypridina, Gigantocypris, Skogsbergia.*
Order Halocyprida.   Marine ostracods, with first
pair of trunk appendages present or absent and
second pair absent or short and leglike. Includes most of the planktonic species.
Subclass Podocopa.   Valves without an antennal
notch. One or two pairs of trunk appendages.
Order Podocopida.   First pair of trunk appendages leglike. This large order includes
marine as well as freshwater species. *Cypris,
Pontocypris, Candona, Cypridopsis, Mesocypris, Darwinula, Cythere.*

Order Platycopida.   Marine, benthic species
having the first pair of trunk appendages not
leglike.
Class Copepoda§.   Copepods. Mostly small crustaceans with cylindrical bodies. Trunk composed
of a thorax bearing five pairs of biramous appendages and an abdomen without appendages.
First antennae longer than second pair and held
outstretched. Compound eyes absent; naupliar eye
present.
Order Calanoida.   Free-living, largely planktonic
copepods with long first antennae (usually with
22 or more articles). *Calanus, Calocalanus, Diaptomus, Metridia, Pleuromamma, Centropages,
Lucicutia, Acartia.*
Order Misophrioida.   A small group of pelagic
copepods living on or above the bottom surface.
First antennae shorter than those of calanoids (13
to 17 articles). *Misophria.*
Order Harpacticoida.   Mostly free-living, marine
and freshwater, bottom-dwelling copepods.
Some planktonic; many interstitial. First antennae short (fewer than ten articles). *Harpacticus,
Canthocamptus.*
Order Monstrilloida.   Marine copepods with larval
stages parasitic in polychaetes and gastropods.
Nonfeeding planktonic adults lack second antennae and mouthparts. *Monstrilla, Xenocoeloma.*
Order Siphonostomatoida.   Freshwater and marine
copepods parasitic as adults on fish and invertebrates. *Nemesis, Clevella, Caligus, Lepeoph-
theirus, Salmincola, Penella.*
Order Cyclopoida.   Marine and freshwater, planktonic and benthic copepods. Includes some commensals and parasites. First antennae short (10 to
16 articles) and second antennae uniramous. *Cyclops, Sapphirina, Oncaea, Lernaea, Doropy-
gus.*
Order Poecilostomatoida.   Marine copepods parasitic as adults on invertbrates and fish. *Ergasilus,
Chondracanthus.*
Class Mystacocarida.   Marine interstitial crustaceans
with long cyclindrical bodies. Only the first five
trunk appendages bear appendages. *Derocheilocaris.*
Class Branchiura.   Crustaceans ectoparasitic on marine and freshwater fish. Head and thorax covered

---

§The classes Copepoda, Branchiura, Mystacocarida, Tantulocarida,
and Cirripedia are grouped together within the class Maxillopoda
in some current classifications. All have no more than 11 trunk
segments and the first pair of trunk appendages are maxillipeds.

by a shieldlike carapace; abdomen small. Both pairs of antennae reduced and modified for attachment. Compound eyes present. *Argulus, Dolops.*

Class Pentastomida. Wormlike parasites of the respiratory system of reptiles. Anterior end with snoutlike process bearing the mouth and four projections bearing claws. *Linguatula.*

Class Tantulocarida. Small marine crustaceans ectoparasitic on deepwater crustaceans. Body somewhat similar to that of copepods but lacking trunk appendages.

Class Cirripedia. Barnacles. Sessile free-living and parasitic crustaceans having the body enclosed within a bivalved carapace. Both pairs of antennae reduced or absent.

Order Ascothoracica. Naked barnacles that parasitize echinoderms and corals. Bivalve or saccular mantle. Prehensile first antennae and abdomen are usually present. *Dendrogaster, Ascothorax.*

Order Acrothoracica. Naked, boring barnacles with a chitinous attachment disc and three to five pairs of cirri (one at mouth, remainder terminal). They live in any calcareous substratum, especially shells and corals. *Trypetesa, Kochlorine, Berndtia.*

Order Thoracica. Free-living and commensal barnacles with six pairs of well-developed cirri. Mantle usually covered with calcareous plates.

Suborder Lepadomorpha. Pedunculate barnacles. *Lepas, Scalpellum, Pollicipes, Conchoderma, Heterolepas.*

Suborder Verrucomorpha. Asymmetrical, sessile barnacles. Wall composed of carina and rostrum, one tergum, and one scutum; the other tergum and scutum form the operculum. *Verruca.*

Suborder Balanomorpha. Symmetrical, sessile barnacles with a wall surmounted by the paired movable terga and scuta. *Balanus, Chthamalus, Catophragmus, Octomeris, Tetraclita, Pyrgoma, Coronula, Xenobalanus.*

Order Rhizocephala. Naked barnacles. Parasitic primarily on decapod crustaceans; a few parasitic on tunicates. Appendages and digestive tract are absent; peduncle forms footlike absorptive processes. *Peltogasterella, Lernaeodiscus, Sacculina.*

Class Malacostraca. Crustaceans having a trunk composed of 14 segments all of which bear appendages. The first eight segments compose the

thorax and the last six, the abdomen. Female gonopores open onto the ventral side of the fifth thoracic segment and the male onto the eighth. Compound eyes present in most species.

Subclass Phyllocarida.

Order Nebaliacea. Malacostracans having a seventh abdominal segment lacking appendages. Thorax with foliaceous appendages and enclosed within a bivalved carapace. *Nebalia, Paranebolia.*

Subclass Hoplocarida.

Order Stomatopoda. Mantis shrimp. Malacostracans having the second pair of thoracic appendages large and subchelate, adapted for prey capture. First antennae with three flagella. Small shieldlike carapace does not cover the last two thoracic segments. Compound eyes stalked. *Squilla, Gonodactylus, Lysiosquilla.*

Subclass Eumalacostraca. Antennae without three flagella, and abdomen without seventh segment.

Superorder Syncarida. Malacostracans without a carapace and the eight thoracic segments free or the first fused with the head. Thoracic appendages biramous and similar.

Order Anaspidacea. First thoracic segment fused with head. Inhabitants of freshwater in Australia, New Zealand, and South America. *Anaspides.*

Order Bathynellacea. First thoracic segment not fused with the head. Minute syncarids inhabiting groundwater or fresh-water sediments in various parts of the world.

Superorder Pancarida.

Order Thermosbaenacea. Minute malacostracans in which a reduced carapace is fused only with the first thoracic segment. Blind. Inhabitants of hot springs and ground water; reported from Texas, West Indies, and Mediterranean region.

Superorder Eucarida. Malacostracans having a carapace fused with all of the thoracic segments. Gills located near base of thoracic appendages. Eyes stalked.

Order Euphausiacea. Krill. Thoracic appendages biramous and anterior ones not modified as maxillipeds. Gills not tightly enclosed by carapace. Marine, pelagic. *Euphausia.*

Order Decapoda. Decapods. First three pairs

of thoracic appendages are maxillipeds. Gills tightly enclosed by sides of carapace.

Suborder Dendrobranchiata. Shrimp (in part). Gills dendrobranchiate; body laterally compressed; first three pairs of legs chelate but not with enlarged chelipeds. Eggs planktonic (not carried by female on pleopods), and a nauplius is the first larval stage. *Penaeus, Sicyonia,* and the pelagic shrimp *Sergestes, Acetes,* and *Lucifer.*

Suborder Pleocyemata. Decapods with phyllobranchiate and trichobranchiate gills; eggs carried by female on pleopods and hatch as zoeae.

Infraorder Stenopodidea. Shrimp. Cephalothorax more or less cylindrical; first three pairs of legs chelate, and one member of the third pair is enlarged; pleura of second abdominal segment not overlapping those of the first. Gills trichobranchiate. *Stenopus.*

Infraorder Caridea. Shrimp. Cephalothorax more or less cylindrical; first two pairs of legs chelate or subchelate, and either first or second pair commonly heavier or longer than the others, third pair chelate; pleura of second abdominal segment overlapping those of the first and third. Gills phyllobranchiate. The Caridea include the greatest number of species of shrimp. Sand shrimp, *Crangon;* the snapping shrimp *Alpheus* and *Synalpheus;* the marine, brackish-water, and freshwater Palaemonidae—*Palaemonetes, Macrobrachium, Leander,* and *Periclimenes* (cleaning shrimp); the freshwater Atyidae—*Atya;* the pelagic Olophoridae; *Hippolysmata* (cleaning shrimp), *Hippolyte,* and *Tozeuma.*

Infraorder Astacidea. Crayfish and lobsters with large claws. Cephalothorax more or less cylindrical; abdomen well developed and somewhat flattened dorsoventrally. First three pairs of legs chelate and first pair greatly enlarged. The lobsters *Nephrops* and *Homarus;* the freshwater crayfish *Procambarus, Astacus, Parastacus,* and *Cambarus.*

Infraorder Thalassinidea. Compressed carapace. First pair of legs in form of chelipeds but asymmetrical; third pair not chelate. Abdomen well developed, extended, and flattened. Marine burrowing shrimp. *Thalassina, Axius, Callianassa, Callichirus,* and *Upogebia.*

Infraorder Palinura. Cephalothorax more or less cylindrical; abdomen well developed and somewhat flattened dorsoventrally. Legs may be chelate or subchelate, but first pair in most species not enlarged. The relic *Neoglyphea;* spiny lobsters, *Panulirus,* with long, heavy antennae; Spanish, slipper lobster, or shovel-nosed lobsters, *Scyllarus,* with flattened carapace and large, short, flattened antennae; the deep-water *Polycheles,* with broad, depressed carapace and first four pairs of legs chelate.

Infraorder Anomura. Carapace usually depressed; third legs never chelate; fifth legs reduced or turned upward. Abdomen variable. In crablike forms, eyes medial to antennae.

Superfamily Paguroidea. Carapace oval. Abdomen usually asymmetrical and housed within a gastropod shell or other objects (hermit crabs) or folded beneath the carapace. First legs in form of chelipeds. The hermit crabs, *Paguristes, Pomatocheles, Petrochirus, Clibanarius, Coenobita* (land hermit crab), *Pagurus,* and *Pylopagurus.* This group also contains a number of crablike forms that are probably secondarily derived from the hermit crabs. The abdomen is flexed beneath the thorax as in the true crabs and well chitinized, but like that in hermit crabs, the abdomen is not symmetrical, and females have pleopods on left side. Includes the coconut crab, *Birgus,* and the lithodid (stone) crabs, *Lithodes* and *Paralithodes* (the commercial king crab of the North Pacific).

Superfamily Galatheoidea. Mostly crablike forms with symmetrical abdomen curled or flexed beneath thorax. Well-developed tail fan. Rostrum often well developed; carapace not fused with epistome. First legs in form of chelipeds; fifth legs minute and folded along side of carapace. The lobster-like *Galathea* and *Munida* with abdomen only curled beneath cephalothorax; the porcelain crabs, *Petrolisthes, Pachycheles, Porcellana,* and *Polyonyx;* the South American freshwater *Aegla.*

Superfamily Hippoidea. Mole crabs. Abdomen flexed symmetrical; telson beneath thorax. Rostrum reduced or absent. Cephalothorax flattened or more or less cylindrical. First legs chelate or subchelate but never in form of chelipeds; fifth legs greatly reduced and folded, often located beneath carapace. Posterior end of abdomen folded ventrally and forward. Commonly in sand in surf zone. *Hippa, Emerita, Blepharipoda, Lepidopa,* and *Albunea.*

Infraorder Brachyura. Carapace broad and fused with epistome. First legs in form of heavy chelipeds; third legs never chelate. Eyes usually lateral to second antennae. Symmetrical abdomen reduced and tightly flexed beneath cephalothorax.

Section Dromiacea. Primitive brachyurans; marine. Last pair of legs often dorsal in position and modified for holding objects over the crab. Carapace never broader than long. Uropods may be present but much reduced. Genital apertures on coxae. *Homolodromia, Dromia, Dromidia, Hypoconcha,* and *Homola.*

Section Archeobrachyura. Carapace may be longer than wide and anterior abdominal segments may be visible from above. Some deepwater species carry objects with the dorsally located fifth pair of legs. The Raninidae are burrowers. *Ranilia.*

Section Oxystomata. Last pair of legs normal or modified as in Dromiacea. Mouth frame triangular rather than quadrate. Uropods absent. Female reproductive openings on sternum. Marine. *Dorippe;* the box crabs, *Calappa* and *Hepatus; Persephona* and *Ebalia.*

Section Oxyrhyncha. Carapace narrowed anteriorly into a rostrum. Body shape roughly triangular. Mouth frame quadrate. Marine. The decorator and spider crabs: *Maja, Inachus, Macrocheira, Hyas, Libinia, Pelia, Parthenope, Pugettia, Loxorhynchus,* and *Stenorhynchus.*

Section Cancridea. Carapace elongated or transversely oval or hexagonal; interocular distance narrow. Orbits of eyes with two fissures above. First antennae folded longitudinally or obliquely. Mouth frame quadrate. Female openings on sternum. Corystidae—*Corystes.* The cancer crabs, Cancridae—*Cancer.*

Section Brachyrhyncha. Carapace not narrowed anteriorly; interocular distance wide. Body shape round, transversely oval, or square. Orbits well developed and complete. Mouth frame quadrate. Female openings on sternum. This section contains the majority of crabs. The swimming crabs, Portunidae—*Portunus, Callinectes, Carcinus, Arenaeus,* and *Ovalipes.* The fresh water crabs, Superfamily Potamoidea—*Potamon* and *Pseudothelphusa.* The mud crabs, Xanthidae—*Xantho, Menippe* (stone crab), *Pilumnus, Rhithropanopeus, Panopeus, Neopanopeus,* and *Eurypanopeus.* The commensal pea crabs, Pinnotheridae—*Pinnotheres, Pinnixa, Dissodactylus.* The amphibious crabs, Ocypodidae—*Ocypode* (ghost crabs), *Uca* (fiddler crabs), *Dotilla,* and

Ucides; Mictyridae (soldier crabs). Marine, freshwater, amphibious, and terrestrial Grapsidae —*Geograpsus, Grapsus, Pachygrapsus, Planes, Sesarma, Eriocheir,* (Chinese mitten crab), *Aratus, Metaplax,* and *Metopaulias.* The land crabs, Gecarcinidae—*Cardisoma, Epigrapsus, Gecarcoidea,* and *Gecarcinus.* The coral gall crabs, Hapalocarcinidae—*Hapalocarcinus, Cryptochirus,* and *Troglocarcinus.*

Superorder Peracarida. Malacostracans in which the eggs are brooded within a brood chamber, or marsupium, beneath the thorax, formed by shelflike plates projecting inward from the thoracic coxae. First thoracic segment fused with the head and at least the first pair of thoracic appendages modified as maxillipeds. Development direct.

Order Mysidacea. Mysids. Thorax covered by a carapace which is unfused with at least the four posterior thoracic segments. Eyes stalked. Pelagic and benthic species, mostly marine but a few found in lakes and caves. *Gnathophausia, Mysis, Neomysis.*

Order Cumacea. Cumaceans. Marine burrowing peracaridans in which a carapace encloses the anterior thoracic segments to form a gill chamber. First three pairs of thoracic appendages are modified as maxillipeds, and the first pair bear gills. When present, eyes usually fused and located on an anterior prominence. *Diastylis.*

Order Amphipoda. Amphipods. Laterally compressed peracaridans, having sessile compound eyes and lacking a carapace. Thorax with one pair of maxillipeds and seven pairs of pereopods; last three pairs of abdominal appendages modified as uropods.

Suborder Gammaridea. Head not fused with second thoracic segment. Maxilliped with palp. Thoracic legs with well-developed coxal plates. Eyes normally present, pigmented, lateral. Abdomen strong; pleopods and uropods well developed. This is the largest suborder of amphipods with more than 4700 described species. Marine forms include the free-burrowing Phoxocephalidae, Haustoriidae, Platyischnopidae, and Oedicerotidae; the flesh-scavenging Lysianassidae; the free-swimming and predatory Eusiridae and Pardaliscidae; the tube-building Ampeliscidae and Corophioidea; the sponge and ascidian inhabiting Leucothoidae and Colomastigidae; the parasitic and predatory Stenothoidae and Acanthonotozomatidae; the fish parasites Lafystiidae and Laphystiopsidae; and a host of free-swimming, clinging, crawling, and nestling species of Melitidae, Pontogeneiidae, Pleustidae, and Atylidae, among others.

The principal freshwater families include (in the Northern Hemisphere) the mainly epigean Gammaridae and Pontoporeiidae; the mainly hypogean Crangonyctidae, Niphargidae, and Hadziidae; (mainly in the Southern Hemisphere) the essentially epigean Hyalellidae, Paramelitidae, Neoniphargidae, and Calliopiidae; and in tropic and warm-temperate regions, the hypogean Bogidiellidae. Terrestrial and semiterrestrial species are in the single family Talitridae.

Suborder Hyperiidea. Head not fused with second thoracic segment. Maxilliped lacks a palp. Thoracic coxae often small or fused with the body. Abdomen and pleopods powerfully developed, last two abdominal segments fused. Eyes often very large, covering greater part of head. Body generally transparent. The more than 350 known species are entirely marine and may all be commensal on gelatinous pelagic animals. The infraorder Physosomata contains species with inflated bodies—*Scina, Mimonectes, Lanceola.* The infraorder Physocephalata contains species with inflated heads—*Vibilia, Parathemisto, Hyperia, Primno, Phronima, Rhabdosoma, Platyscelus.*

Suborder Caprellidea. Head partly fused with second thoracic segment. Maxilliped with palp. Thoracic coxae vestigal or lacking. Abdominal segments much reduced, with vestigial appendages. Eyes small. Body elongated, cylindrical, or short and flattened. The 250 known species are entirely marine and include the skeleton shrimp, Caprellidae, and others—*Cercops, Aeginella, Deutella,* and *Caprella;* the whale lice, Cyamidae—*Cyamus;* and the monotypic family Caprogammaridae, intermediate between gammarids and caprellids.

Suborder Ingolfiellidea. Head fused or not with second thoracic segment. Maxillipeds with palp. Thoracic coxae minute. Body elongated, cylindrical. Abdominal segments not fused; uropods independent, pleopods much reduced. Pigmented sessile eyes lacking, but ocular lobes present. The some 30 species of this primitive group are mostly hypogean or interstitial in fresh and brackish water. There are a few deep-sea forms.

Order Isopoda. Isopods. Dorsoventrally flattened peracaridans, having sessile compound eyes and lacking a carapace. Thorax with one pair of maxillipeds and seven pairs of pereopods. Some of the abdominal segments may be fused with the telson. At least some of the pleopods modified as gills.

Suborder Gnathiidea. Thoracic segments 1 and 7 reduced, so that only five large segments are visible dorsally. Eighth thoracic appendages are absent. Abdomen small and much narrower than thorax. Manca stage is ectoparasitic on marine fish; adult nonfeeding. *Gnathia.*

Suborder Anthuridea. Body long and cylindrical. First pair of legs are heavy and subchelate; first pair of pleopods form an operculum covering other pairs. Most are marine burrowers except for a few freshwater species. *Anthura, Cyathura, Cruregens.*

Suborder Microcerberidea. A small group of minute, blind, interstitial and freshwater cave species.

Suborder Flabellifera. Body is more or less flattened; some abdominal segments may be fused together; the last segment is fused with the telson. The uropods are fan-shaped and form a tail fan together with the telson. This suborder contains the most common shallow-water marine species. A few are found in fresh water. The wood borer, *Limnoria; Cirolana; Bathynomus; Serolis; Sphaeroma;* the ectoparasites of fish, the Cymothoidae.

Suborder Valvifera. Abdominal segments 3 to 6 always fused with telson, and in some species segments 2 and 1 as well. Uropods form an operculum over gills. Mostly marine, and many species are common in seaweed. *Astacilla, Arcturus, Idotea.*

Suborder Asellota. Abdominal segments 3 to 6 always fused with telson, and in some species segments 2 and 1 as well. Uropods are commonly styliform. Pleopods 3 to 5 form gills and covered by anterior two opercular pleopods. A large suborder that includes many common marine species as well as some freshwater forms. *Asellus, Lirceus, Munnopsis, Jaera.*

Suborder Phreatoicoidea. Abdominal segments are not fused with each other. Elongated body is laterally compressed. Uropods are styliform. Includes certain freshwater isopods of Australia, New Zealand, and South Africa. *Phreatoicus.*

Suborder Epicaridea. Parasites on crustaceans. Females are often greatly modified, some without segmentation or appendages. *Bopyrus, Entoniscus, Portunion, Liriopsis.*

Suborder Oniscoidea. Five abdominal segments are usually distinct dorsally. First antennae are vestigial. Amphibious and terrestrial members (wood lice). *Ligia, Oniscus, Porcellio, Armadillidium, Tylos.*

Order Tanaidacea. Small, benthic marine peracaridans having the anterior two thoracic segments covered by and fused with a carapace. Second thoracic appendages are gnathopods. *Leptochelia, Tanais.*

REFERENCES

The literature included here is restricted in large part to crustaceans. The introductory references on page 6 include many general works and field guides that contain sections on these animals.

A. Kaestner's (1970) and F. R. Schram's (1986) volumes provide detailed general accounts of the Crustacea. The journal *Crustaceana* and the *Journal of Crustacean Biology* are devoted to publication of research on crustaceans.

General

Anderson, D. T. 1973. Embryology and Phylogeny in Annelids and Arthropods. Pergamon Press, Oxford. 495 pp. (Includes a detailed survey of crustacean embryology.)

Bliss, D. E. (Ed.): 1982–1985. Biology of the Crustacea. 10 vols. Academic Press, New York. (This series covers many aspects of the ecology, behavior, and physiology of crustaceans.)

Chang, E. S., and O'Connor, J. D. 1988. Crustacea: Molting. *In* Laufer, H., and Downer, R. G. H. (Eds.): Endocrinology of Selected Invertebrate Types. Alan R. Liss, New York. pp. 259–278.

Cronin, T. W. 1986. Optical design and evolutionary adaptation in crustacean compound eyes. J. Crust. Biol. *6*(1):1–23.

Felgenhauer, B. E., Watling, L., and Thistle, A. B. (Eds.): 1989. Functional Morphology of Feeding and Grooming in Crustacea. A. A. Balkema, Rotterdam.

Fingerman, M. 1987. The endocrine mechanisms of crustaceans. J. Crust. Biol. *7*(1):1–24.

Fitzpatrick, J. F. 1983. How to Know the Freshwater Crustacea. W. C. Brown Co., Dubuque, IA.

Gibson, R., and Barker, P. L. 1979. The decapod hepatopancreas. Oceanogr. Mar. Biol. Ann. Rev. *17*:285–346.

Gregoire, C. 1971. Hemolymph coagulation in arthropods. *In* Florkin, M., and Scheer, B. T. (Eds.): Chemical Zoology. Vol. 6. Academic Press, New York. pp. 145–189.

Hastings, J. W. 1983. Chemistry and control of luminescence in marine organisms. Bull. Mar. Sci. *33*(4):818–828.

Herring, P. J. 1987. Systematic distribution of bioluminescence in living organisms. J. Biolumin. Chemolumin. *1*:147–163.

Kaestner, A. 1970. Invertebrate Zoology. Vol. 3. Crustacea. Wiley-Interscience, New York. 523 pp. (An old but excellent general account of the crustaceans; systematically arranged and covering all aspects of the biology.)

LaBarbera, M. 1984. Feeding currents and particle capture mechanisms in suspension feeding animals. Am. Zool. *24*:71–84.

McLaughlin, P. 1980. Comparative Morphology of Recent Crustacea. W. H. Freeman and Co., San Francisco. 159 pp. (Detailed anatomical descriptions of representatives of the higher taxa of crustaceans.)

McMahon, B. R., and Burnett, L. E. 1990. The crustacean open circulatory system: A reexamination. Physiol. Zool. *63*(1):35–71.

McNamara, J. C. 1981. Morphological organization of crustacean pigmentary effectors. Biol. Bull. *161*:270–280.

Moore, R. C. (Ed.): 1979. Treatise on Invertebrate Paleontology. Pt. R. Arthropoda 4. Vols. 1 and 2. Geological Society of America and University of Kansas Press, Lawrence. (These two volumes cover the Crustacea except for ostracods, which are treated in Part Q. Arthropoda 3. 1961.)

Morin, J. G. 1983. Coastal bioluminescence: Patterns and functions. Bull. Mar. Sci. *33*(4):787–817.

Palmer, J. D. 1976. An Introduction to Biological Rhythms. Academic Press, New York.

Rebach, S., and Dunham, D. W. (Eds.): 1983. Studies in Adaptation: The Behavior of Higher Crustacea. John Wiley and Sons, New York. 282 pp.

Schmitt, W. L. 1975. Crustaceans. University of Michigan Press, Ann Arbor. (An interesting popular treatment of crustaceans.)

Schram, F. R. (Ed.): 1983. Crustacean Phylogeny. A. A. Balkema, Rotterdam. 372 pp. (A collection of papers from a symposium on crustacean phylogeny.)

Schram, F. R. 1986. Crustacea. Oxford University Press, New York. 700 pp. (A detailed account of the crustacean groups.)

Young, R. E. 1983. Oceanic bioluminescence: An overview of general functions. Bull. Mar. Sci. *33*(4):829–847.

Branchiopods, Ostracods, Copepods, Barnacles

Barnes, H., Barnes, M., and Klepal, W. 1977. Studies on the reproduction of cirripedes: I. Introduction: Copulation, release of oocytes, and formation of the egg lamellae. J. Exp. Mar. Biol. Ecol. *27*(3):195–218.

Bate, R. H., Robinson, E., and Sheppard, L. M. (Eds.): 1982. Fossil and Recent Ostracods. Horwood, Chichester, England. 494 pp. (A collection of papers on ostracods.)

Bourget, E., and Crisp, D. J. 1975. Factors affecting deposition of the shell in *Balanus balanoides*. J. Mar. Biol. Ass. U.K. *55*:231–249.

Boxshall, G. A., and Lincoln, R. J. 1983. Tantulocarida, a new class of Crustacea ectoparasitic on other crustaceans. J. Crust. Biol. *3*(1):1–16.

Cohen, A. C., and Morin, J. C. 1990. Patterns of reproduction in ostracodes: A review. J. Crust. Biol. *10*(2):184–211.

Corkett, C. J., and McLaren, I. A. 1978. The biology of *Pseudocalanus.* Adv. Mar. Biol. *15:*1–231.

Coull, B. C. 1977. Copepoda: Harpacticoida. Marine flora and fauna of the northeastern U.S. NOAA Technical Report NMFS circular 399. U.S. Government Printing Office, Washington, DC.

Crisp, D. J., and Bourget, E. 1985. Growth in Barnacles. Adv. Mar. Biol. *22:*199–244.

Darwin, C. 1851–1854. A Monograph on the Subclass Cirripedia. 2 Vols. Ray Society, London. (A classic and still valuable account of the barnacles.)

Foster, B. A. 1971. Desiccation as a factor in the intertidal zonation of barnacles. Mar. Biol. *8:*29.

Foster, B. A. 1974. The barnacles of Fiji, with observations on the ecology of barnacles on tropical shores. Pac. Sci. *28*(1):35–56.

Frey, D. G. 1982. Contrasting strategies of gametogenesis in northern and southern populations of Cladocera. Ecology. *63*(1):223–241.

Frost, B. W. 1977. Feeding behavior of *Calanus pacificus* in mixtures of food particles. Limnol. Oceanogr. *22*(3):472–491.

Gerritsen, J., and Porter, K. G. 1982. The role of surface chemistry in filter feeding by zooplankton. Science. *216:*1225–1227.

Gotto, R. V. 1979. The association of copepods with marine invertebrates. Adv. Mar. Biol. *16:*1–109.

Grice, G. D., and Marcus, N. H. 1981. Dormant eggs of marine copepods. Ann. Rev. Oceanogr. Mar. Biol. *19:*125–140.

Griffiths, A. M., and Frost, B. W. 1976. Chemical communication in the marine planktonic copepods *Calanus pacificus* and *Pseudocalanus* sp. Crustaceana. *30*(1):1–8.

Henderson, P. A. 1990. Freshwater Ostracods. Synopses British Fauna. Linnean Soc. London and the Estuarine and Coastal Sciences Association. Universal Book Service/Dr. W. Backhuys, Oegstgeest, The Netherlands. 228 pp.

Hicks, G. F., and Coull, B. C. 1983. The ecology of marine meiobenthic harpacticoid copepods. Oceanogr. Mar. Biol., Ann. Rev. *21:*67–175.

Ho, J. S. 1977. Copepoda: Lernaeopodidae and Sphyriidae. Marine flora and fauna of the northeastern U. S. NOAA Technical Reports NMFS Circular 406. U. S. Gov. Printing Office, Washington, DC. 13 pp.

Ho, J. S. 1978. Copepoda: Cyclopoids parasitic on fishes. Marine flora and fauna of the northeastern U. S. NOAA Technical Reports NMFS Circular 409. U. S. Gov. Printing Office, Washington, DC. 11 pp.

Koehl, M. A. R., and Strickler, J. R. 1981. Copepod feeding currents: Food capture at low Reynolds number. Limnol. Oceanogr. *26*(6):1062–1073.

Lewis, C. A. 1978. A review of substratum selection in free-living and symbiotic cirripeds. *In* Chia, F. -S., and Rice, M. E. (Eds.): Settlement and Metamorphosis of Marine Invertebrate Larvae. Elsevier, North Holland, New York. pp. 207–218.

Marcotte, B. M. 1977. An introduction to the architecture and kinematics of harpacticoid feeding: *Tisbe furcata.* Mikrofauna Meeresboden. *61:*183–196.

Marcotte, B. M. 1983. The imperatives of copepod diversity: Perception, cognition, competion and predation. *In* Schram, F. R. (Ed.): Crustacean Phylogeny. A. A. Balkema, Rotterdam. pp. 47–72.

Marcus, N. H. 1982. Photoperiodic and temperature regulation of diapause in *Labidocera aestiva.* Biol. Bull. *162:*45–52.

Marshall, S. M. 1973. Respiration and feeding in copepods. Adv. Mar. Biol. *11:*57–120.

Mellors, W. K. 1975. Selective predation of ephippial *Daphnia* and the resistance of ephippial eggs to digestion. Ecology. *56:*975–980.

Morin, J. G. 1986. "Fireflies" of the sea: Luminescent signaling in marine ostracode crustaceans. Florida Entomolog. *69*(1):105–121.

Moyse, J., and Hui, E. 1981. Avoidance by *Balanus balanoides* cyprids of settlement on conspecific adults. J. Mar. Biol. Assoc. U. K. *61:*449–469.

Newman, W. A. 1989. Juvenile ontogeny and metamorphosis in the most primitive living sessile barnacle, *Neoverruca,* from abyssal hydrothermal springs. Bull. Mar. Sci. *45*(2):467–477.

Newman, W. A., and Ross, A. 1976. Revision of the balanomorph barnacles; including a catalog of the species. San Diego Soc. Nat. Hist. Memoir. *9:*1–108.

Palmer, A. R. 1982. Predation and parallel evolution: Recurrent parietal plate reduction in balanomorph barnacles. Paleobiology. *8*(1):31–44.

Pennak, R. W., and Zinn, D. J. 1943. Mystacocarida, a new order of Crustacea from intertidal beaches in Massachusetts and Connecticut. Smithson. Misc. Collect. *103:*1–11.

Poulet, S. A. 1976. Feeding of *Pseudocalanus minutus* on living and nonliving particles. Mar. Biol. *34*(2):117–125.

Sanders, H. L. 1963. The Cephalocarida. Memoirs Conn. Acad. Arts Sci. *15:*1–180.

Sarvala, J. 1979. Benthic resting periods of pelagic cyclopoids in an oligotrophic lake. Holarctic Ecol. 2:88–100.

Shmeleva, A. A., and Kovalev, A. V. 1974. Biologic cycles of copepods of the Adriatic Sea. Boll. Pesca Piscic. Hydrobiol. 29(1):49–70.

Strickler, J. R. 1982. Calanoid copeods, feeding currents, and the role of gravity. Science. 218:158–160.

Stubbings, H. G. 1975. *Balanus balanoides.* Liverpool Mar. Biol. Comm. Mem. No. 37 (A detailed account of one of the most common sessile intertidal barnacles of north temperate Atlantic and Pacific waters.)

Tomlinson, J. T. 1969. The burrowing barnacles (Cirripedia: Order Acrothoracica). Bull. U. S. Nat. Mus. 269:1–162.

Uye, S., and Kasahara, S. 1978. Life history of marine planktonic copepods in neritic region with special reference to the role of resting eggs. Bull. Plankton Soc. Japan. 25(2):109–122.

Walker, G., and Yule, A. B. 1984. Temporary adhesion of the barnacle cyprid: The existence of an antennular adhesive secretion. J. Mar. Biol. Assoc. U. K. 64:679–686.

Watras, C. J. 1983. Mate location by diaptomid copepods. J. Plankton Res. 5(3):417–423.

Watras, C. J., and Haney, J. F. 1980. Oscillations in the reproductive condition of *Diaptomus leptopus* and their relation to rates of egg-clutch production. Oecologia. 45:94–103.

Wiman, F. H. 1981. Mating behavior in the *Streptocephalus* fairy shrimps. Southwest. Nat. 25(4):541–546.

Wong, C. K. 1981. Cyclomorphosis in *Bosmina* and copepod predation. Can. J. Zool. 59:2049–2052.

Yager, J. 1981. Remipedia, a new class of Crustacea from a marine cave in the Bahamas. J. Crust. Biol. 1(3):328–333.

Decapods

Barker, P. L., and Gibson, R. 1977. Observations on the feeding mechanism, structure of the gut, and digestive physiology of the European lobster *Homarus gammarus.* J. Exp. Mar. Biol. Ecol. 26(3):297–324.

Bauer, R. T. 1977. Antifouling adaptations of marine shrimps: Functional morphology and adaptive significance of antennular preening by the third maxillipeds. Mar. Biol. 40:261–276.

Bauer, R. T. 1981. Grooming behavior and morphology in the decapod Crustacea. J. Crust. Biol. 1(2):153–173.

Bauer, R. T. 1986. Phylogenetic trends in sperm transfer and storage complexity in decapod crustaceans. J. Crust. Biol. 6(3):313–325.

Bertness, M. D. 1981. Pattern and plasticity in tropical hermit crab growth and reproduction. Am. Nat. 117(5):754–773.

Bliss, D. E. 1982. Shrimps, Lobsters and Crabs. New Century Publishers, Piscataway, NJ. 242 pp. (A semipopular account of the decapods.)

Bliss, D. E., van Montfrans, J., van Montfrans, M. et al. 1978. Behavior and growth of the land crab *Cecarcinus lateralis* (Freminville) in southern Florida. Bull. Am. Mus. Nat. Hist. 160(2):113–151.

Bruce, A. J. 1976. Shrimps and prawns of coral reefs, with special reference to commensalism. *In* Jones, O. A., and Endean, R. (Eds.): Biology and Geology of Coral Reefs. Vol. III: Biol. 2. Academic Press, New York. pp. 37–94.

Caine, E. A. 1975. Feeding and masticatory structures of selected Anomura. J. Exp. Mar. Biol. Ecol. 18:277–301.

Cameron, J. N. 1985. Molting in the blue crab. Sci. Amer. 252(5):105–109.

Christy, J. H. 1978. Adaptive significance of reproductive cycles in the fiddler crab *Uca pugilator:* A hypothesis. Science. 199:453–455.

Christy, J. H. 1982. Burrow structure and use in the sand fiddler crab, *Uca pugilator.* Anim. Behav. 30:687–694.

Christy, J. H. 1986. Timing of larval release by intertidal crabs on an exposed shore. Bull. Mar. Sci. 39(2):176–191.

Christy, J. H. 1987. Competitive mating, mate choice and mating associations of brachyuran crabs. Bull. Mar. Sci. 41(2):177–191.

Cobb, J. S., and Phillips, B. F. (Eds.): 1980. The Biology and Management of Lobsters. Vol. 1. Physiology and Behavior. 462 pp. Vol. 2. Ecology and Management. 390 pp. Academic Press, New York.

Cornell, J. C. 1979. Salt and water balance in two marine spider crabs, *Libinia emarginata* and *Pugettia producta.* I. Urine production and magnesium regulation. Biol. Bull. 157(2):221–233.

Crane, J. 1975. Fiddler Crabs of the World. Princeton University Press, Princeton, 660 pp. (An exhaustive treatment of these familiar intertidal crabs. First part covers taxonomy; second part, biology.)

Cronin, T. W., and Forward, R. B. 1979. Tidal vertical migration: An endogenous rhythm in estuarine crab larvae. Science. 205:1020–1022.

Derby, C. D. 1982. Structure and function of cuticular sensilla of the lobster *Homarus americanus.* J. Crust. Biol. 2(1):1–21.

Diaz, H., and Rodriguez, G. 1977. The branchial chamber in terrestrial crabs: A comparative study. Biol. Bull. 153:485–504.

Eggleton, D. B. 1990. Foraging behavior of the blue crab, *Callinectes sapidus,* on juvenile oysters, *Crassostrea virginica:* Effects of prey density and size. Bull. Mar. Sci. 46(1):62–82.

Fontaine, M. T., Passelecq-gerin, E., Bauchau, A. G. 1982. Structures chemoreceptrices des antennules du crabe *Carcinus maenas*. Crustaceana. *43*(3):271–283.

Forward, R. B. 1987: Larval release rhythms of decapod crustaceans: An overview. Bull. Mar. Sci. *41*(2):165–176.

Gibson, R., and Barker, P. L. 1979. The decapod hepatopancreas. Oceanogr. Mar. Biol., Ann. Rev. *1979:*285–346.

Gleeson, R. A., Adams, M. A., and Smith, A. B. 1987. Hormonal modulation of pheromone-mediated behavior in a crustacean. Biol. Bull. *172:*1–9.

Greenaway, P., and Morris, S. 1989. Adaptations to a terrestrial existence by the robber crab, *Birgus latro*. J. Exp. Biol. *143:*333–346.

Greenspan, B. N. 1982. Semi-monthly reproductive cycles in male and female fiddler crabs, *Uca pugnax*. Sci. Am. *30:*1084–1092.

Hart, J. F. L. 1982. Crabs and Their Relatives of British Columbia. British Columbia Provincial Museum, Victoria. 267 pp.

Hartnoll, R. G. 1979. Mating in Brachyura. Crustaceana. *16:*161–181.

Hazlett, B. A. 1970. The effect of shell size and weight on the agonistic behavior of a hermit crab. Z. Tierpsychol. *27*(3):369–374.

Hazlett, B. A. 1982. Chemical induction of visual orientation in the hermit crab *Clibanarius vittatus*. Anim. Behav. *30*(4):1259–1260.

Hazlett, B. A. 1987. Hermit crab shell exchange as a model system. Bull. Mar. Sci. *41*(2):99–107.

Herring, P. J. 1976. Bioluminescence in decapod Crustacea. J. Mar. Biol. Assoc. U.K. *56:*1029–1047.

Hobbs, H. H. 1972. Crayfishes (Astacidae) of North and Middle America. Biota of Freshwater Ecosystems. Identification Manual No. 9. EPA. U.S. Government Printing Office, Washington, DC. 173 pp.

Hopkins, S. P., and Nott, J. A. 1980. Studies on the digestive cycle of the shore crab *Carcinus maenas* with special reference to the B cells in the hepatopancreas. J. Mar. Biol. Assoc. U.K. *60:*891–907.

Ingle, R. W. 1983. Shallow-Water Crabs. Synopses of the British Fauna No. 25. Cambridge University Press, London. 206 pp.

Ivanov, B. G. 1970. On the biology of the Antarctic krill *Euphausia superba*. Mar. Biol. *7:*340.

Jonasson, M. 1887. Fish cleaning behaviour of shrimp. J. Zool. London. *213:*117–131.

Johnson, P. T. 1980. The Histology of the Blue Crab, *Callinectes sapidus*. Praeger Publishers, New York. 440 pp.

Klaassen, F. 1975. Ecological and ethological studies on the reproductive biology in *Gecarcinus lateralis*. Form. Funct. *8*(2):101–174.

Levine, D. M., and Blanchard, O. J. 1980. Acclimation of two shrimps of the genus *Periclimenes* to sea anemones. Bull. Mar. Sci. *30*(2):460–466.

Mangum, C. P., and Weiland, A. L. 1975. The function of hemocyanin in respiration of the blue crab *Callinectes sapidus*. J. Exp. Zool. *193*(3):257–263.

McLaughlin, P. A. 1974. The hermit crabs of northwestern North America. Zool. Verh., Leiden. *130:*1–396.

McLay, C. L. 1983. Dispersal and use of sponges and ascidians as camouflage by *Cryptodromia hilgendorfi*. Mar. Biol. *76:*17–32.

Mesce, K. A. 1982. Calcium-bearing objects elicit shell selection behavior in a hermit crab. Science. *215:*993–995.

Morgan, S. G. 1989. Adaptive significance of spination in estuarine crab zoeae. Ecology. *70*(2):464–482.

Nolan, B. A., and Salmon, M. 1970. The behavior and ecology of snapping shrimp. Form. Funct. *4:*289–335.

Omori, M. 1974. The biology of pelagic shrimps in the ocean. Adv. Mar. Biol. *12:*233–324.

Palmer, J. D. 1975. Biological clocks of the tidal zone. Sci. Am. *232*(2):70–79.

Peterson, D. R., and Loizzi, R. F. 1974. Ultrastructure of the crayfish kidney—coelomosac, labyrinth, nephridial canal. J. Morph. *142*(3):241–263.

Reese, E. S. 1963. The behavior mechanisms underlying shell selection by hermit crabs. Behaviour. *21:*78–126.

Salmon, M. 1967. Coastal distribution, display and sound production by Florida fiddler crabs. Anim. Behav. *15:*449–459.

Salmon, M., and Horch, K. W. 1972. Acoustical signalling and detection by semiterrestrial crabs of the family Ocypodidae. *In* Winn, H. E., and Olla, B. L. (Eds.): Behavior of Marine Animals. Vol. 1. Invertebrates. Plenum Press, New York. pp. 60–96.

Sargent, R. C., and Wagenbach, G. E. 1975. Cleaning behavior of the shrimp, *Periclimenes anthophilus*. Bull. Mar. Sci. *25*(4):466–472.

Schembri, P. J. 1982. Feeding behaviour of fifteen species of hermit crabs from the Otago region, southeastern New Zealand. J. Nat. Hist. *16:*859–878.

Skinner, D. M. 1985. Molting and regeneration. *In* Bliss, D. E. (Ed.): Biology of the Crustacea, vol. 9. Academic Press, New York. pp. 43–146

Smith, L. D. 1990. Patterns of limb loss in the blue crab, *Callinectes sapidus*, and the effects of autotomy on growth. Bull. Mar. Sci. *46*(1):23–36.

Spight, T. M. 1977. Availability and use of shells by intertidal hermit crabs. Biol. Bull. *152:*120–133.

Warner, G. F. 1977. The Biology of Crabs. Van Nostrand Reinhold Co., New York. 202 pp.

Warner, J. A., Latz, M. I., and Case, J. F. 1979. Cryptic bioluminescence in a midwater shrimp. Science. *203:*1109–1110.

Welsh, B. L. 1975. The role of grass shrimp, *Palaemonetes pugio,* in a tidal marsh ecosystem. Ecology. *56*(3):513–530.

Wickins, J. F. 1976. Prawn biology and culture. Oceanogr. Mar. Biol. Ann. Rev. *14*:435–507.

Wicksten, M. K. 1980. Decorator crabs. Sci. Am. *242*(2):146–154.

Wicksten, M. K. 1985. Carrying behavior in the family Homolidae, J. Crust. Biol. *5*(3):476–479.

Wilber, T. P., and Herrnkind, W. 1982. Rate of new shell acquisition by hermit crabs in a salt marsh habitat. J. Crust. Biol. *2*(4):588–592.

Williams, A. B. 1974. Crustacea: Decapoda. Marine flora and fauna of the northeastern U.S. NOAA Technical Reports NMFS Circular 389. U.S. Government Printing Office, Washington, DC. 49 pp.

Williams, A. B. 1984. Shrimps, Lobsters, and Crabs of the Atlantic Coast of the Eastern United States, Maine to Florida. Smithsonian Institution Press, Washington, DC.

Wolcott, T. G. 1984. Uptake of interstitial water from soil: mechanisms and ecological significance in the ghost crab *Ocypode quadrata* and two gecarcinid land crabs. Physiol. Zool. *57*(1):161–184.

Wolcott, T. G., and Wolcott, D. L. 1982. Larval loss and spawning behavior in the land crab *Gecarcinus lateralis.* J. Crust. Biol. *2*(4):477–485.

Wolcott, T. C., and Wolcott, D. L. 1990. Wet behind the gills. Nat. Hist. *1990* (Oct.):47–54. (Water uptake and gas exchange in land crabs).

Other Malacostracans

Barnard, J. L. 1969. The families and genera of marine gammaridean Amphipoda. U. S. Nat. Mus. Bull. *271*:1–535.

Barnard, J. L., and Barnard, C. M. 1983. Freshwater Amphipoda of the World: I. Evolutionary patterns. II. Handbook and bibiliography. Hayfield Associates, Alexandria, VA. 830 pp.

Bek, T. A. 1972. Feeding of intertidal gammarids. Vestn. Mosk, Univ. Ser. G.. Biol. Pochvoved. *27*(1):106–107.

Berkes, F. 1975. Some aspects of feeding mechanisms of euphausiid. Crustaceana. *29*(3):266–270.

Bousfield, E. L. 1973. Shallow-water Gammaridean Amphipoda of New England. Cornell University Press, Ithaca, NY. 344 pp.

Bousfield, E. L. 1977. A new look at the systematics of Gammaroidean amphipods of the world. Crustaceana (Supplement). *4*:282–316.

Bousfield, E. L. 1978. A revised classification and phylogeny of amphipod crustaceans. Trans. R. Soc. Can. (Ser. IV) *16*:343–390.

Brooks, H. K. 1962. On the fossil Anaspidacea, with a revision of the classification of the Syncarida. Crustaceana. *4*(3):229–242.

Burrows, M. 1969. The mechanics and neural control of the prey capture and strike in the mantid shrimps Squilla and Hemisquilla. Z. Vergl. Physiol. *62*(4):361–381.

Caine, E. A. 1974. Comparative functional morphology of feeding in three species of caprellids from the northwestern Florida Gulf Coast. J. Exp. Mar. Biol. Ecol. *15*:81–96.

Caine, E. A. 1978. Habitat adaptations of North American caprellid Amphipoda. Biol. Bull. *155*:288–296.

Caldwell, R. L. 1979. Cavity occupation and defensive behaviour in the stomatopod *Gonodactylus festai:* Evidence for chemically mediated individual recognition. Anim. Behav. *27*:194–201.

Caldwell, R. L., and Dingle, H. 1976. Stomatopods. Sci. Am. *234*(1):80–89.

Coenen-Strasse, D. 1981. Some aspects of water balance of two desert woodlice, *Hemilepistus aphganicus* and *Hemilepistus reaumuri.* Comp. Biochem. Physiol. Comp. Physiol. *70*(3):405–420.

Diebel, C. E. 1988. Observations on the anatomy and behavior of *Phronima sedentaria.* J. Crust. Biol. *8*(1):79–90.

Dingle, H., and Caldwell, R. L. 1978. Ecology and morphology of feeding and agonistic behavior in mudflat stomatopods. Biol. Bull. *155*(1):134–149.

Enequist, P. 1949. Studies on the soft-bottom amphipods of the Skagerrak. Zool. Bidrag. Fran Uppsala. *18*:297–492.

Farr, J. A. 1978. Orientation and social behavior in the supralittoral isopod *Ligia exotica.* Bull. Mar. Sci. *28*(4):659–666.

Fox, R. S., and Bynum, D. H. 1975. The amphipod crustaceans of North Carolina estuarine waters. Chesapeake Sci. *16*(4):223–237.

Friend, J. A., and Richardson, A. M. M. 1986. Biology of terrestrial amphipods. Ann. Rev. Entomol. *31*:25–48.

Geyer, H., and Becker, G. 1980. Attractive effects of several marine fungi on *Limnoria tripunctata.* Mater Org. *15*(1):53–78.

Greenway, P., and Morris, S. 1989. Adaptations to a terrestrial existence by the robber crab, *Birgus latro.* J. Exp. Biol. *143*:333–346.

Grossnickle, N. E. 1982. Feeding habits of *Mysis relicta:* An overview. Hydrobiologia. *93*(1/2):101–108.

Hamner, W. M. 1984. Aspects of schooling in *Euphausia superba.* J. Crust. Biol. *4* (spec. no. 1):67–74.

Hamner, W. M., Hamner, P. P., Strand, S. W. et al. 1983. Behavior of Antarctic krill, *Euphausia superba:* Chemoreception, feeding, schooling, and molting. Science. *220*:433–435.

Hamner, W. M., Smith M., and Mulford, E. K. 1969. The behavior and life history of a sand-beach isopod, *Tylos punctatus*. Ecology. *50*(3):442–453.

Hartwick, R. F. 1976. Beach orientation in talitrid amphipods: Capacities and strategies. Behav. Ecol. Socibiol. *1*(4):447–458.

Hassall, M., and Jennings, J. B. 1975. Adaptive features of gut structure and digestive physiology in the terrestrial isopod *Philoscia muscorum*. Biol. Bull. *149:*348–364.

Hay, M. E., and Duffy, J. E. 1990. Host-Plant specialization decreases predation on a marine amphipod: A herbivore in plant's clothing. Ecology. *71*(2):733–743.

Herring, P. J., and Locket, N. A. 1978. The luminescence and photophores of euphausiid crustaceans. J. Zool. *186:*431–462.

Holdich, D. M., and Jones, J. A. 1983. Tanaids. Synopses of the British Fauna No. 27. Cambridge University Press, London. 98 pp.

Holsinger, J. R. 1972. The freshwater amphipod crustaceans (Gammaridae) of North America. Biota of Freshwater Ecosystems, Identification Manual 5. U. S. Environmental Protection Agency, Washington, DC.

Johnson, S. B., and Attramadal, Y. G. 1982a. Reproductive behaviour and larval development of *Tanais cavolinii*. Mar. Biol. *71:*11016.

Johnson, S. B., and Attramadal, Y. G. 1982b. A functional-morphological model of *Tanais cavolinii* adapted to a tubicolous life-strategy. Sarsia. *67:*29–42.

Jones, N. S. 1976. British Cumaceans. Synopses of the British Fauna No. 7. Linn. Soc. London. Academic Press, London. 62 pp.

Land, M. F. 1992. Locomotion and visual behaviour of mid-water crustaceans. Mar. Biol. Ass. U.K. *72:*41–60.

Laval, P. 1980. Hyperiid amphipods as crustacean parasitoids associated with gelatinous zooplankton. Oceanogr. Mar. Biol. Ann. Rev. *18:*11–56.

Lim, S. T. A., and Alexander, C. G. 1986. Reproductive behaviour of the caprellid amphipod, *Caprella scaura*. Mar. Behav. Physiol. *12:*217–230.

Manning, R. B. 1974. Crustacea: Stomatopoda. Marine flora and fauna of the northeastern United States. NOAA Technical Report NMFS Circular 386. U. S. Government Printing Office, Washington, DC. 6 pp.

Mauchline, J. 1980. The biology of mysids and euphausids. Adv. Mar. Biol. *18:*3–677.

McCain, J. C. 1968. The Caprellidae (Crustacea: Amphipoda) of the western North Atlantic. Bull. U. S. Nat. Mus. *278:*1–147.

McClintock, J. B., and Janssen, J. 1990. Pteropod abduction as a chemical defence in a pelagic Antarctic amphipod. Nature. *346*(6283):462–464.

Mendoza, J. A. 1982. Some aspects of the autecology of *Leptochelia dubia*. Crustaceana. *43*(3):225–240.

Naylor, E. 1972. British Marine Isopods. Synopses of the British Fauna No. 3. Linn. Soc. London. Academic Press, London. 86 pp.

Nicol, S. 1990. The age-old problem of krill longevity. BioScience. *40*(11):833–836.

Rees, C. P. 1975. Life cycle of the amphipod *Gammarus palustris*. Estuar. Coast. Mar. Sci. *3*(4):413–419.

Ross, R. M., and Quetin, L. B. 1986. How productive are Antarctic krill? BioScience. *36*(4):264–269.

Schabes, M., and Hamner, W. 1992. Mysid locomotion and feeding: Kinematics and water-flow patterns of *Antarctomysis* sp., *Acanthomysis sculpta,* and *Neomysis rayii.* J. Crust. Biol. *12*(1):1–10.

Schmalfuss, H. 1975. Morphologie, Funktion und Evolution der Tergithocker bei Landisopoden Z. Morph. Tiere. *80:*287–316.

Schminke, H. K. 1981. Adaptation of Bathynellacea to life in the interstitial ("Zoea Theory"). Int. Revue ges. Hydrobiol. *66*(4):575–637.

Schmitz, E. H., and Schultz, T. W. 1969. Digestive anatomy of terrestrial Isopoda: *Armadillidium vulgare* and *Armadillidium nasutum*. Am. Midl. Nat. *82*(1):163–181.

Schultz, G. A. 1969. How to Know the Marine Isopod Crustaceans. W. C. Brown Co., Dubuque, IA. 359 pp.

Spicer, J. I., Moore, P. G., and Taylor, A. C. 1987. The physiological ecology of land invasion by the Talitridae. Proc. Roy. Soc. Lond. B. *232:*95–124.

Storch, V. 1987. Microscopic anatomy and ultrastructure of the stomach of *Porcellio scaber*. Zoomorphology. *106:*301–311.

Sutton, S. L. 1972. Woodlice. Ginn & Co., London. 144 pp. (An informative general account of terrestrial isopods plus a guide to British species.)

Sutton, S. L., and Holdich, D. M. (Eds.): 1984. The Biology of Terrestrial Isopods. Oxford Univ. Press, Oxford. 518 pp. (Papers from a symposium.)

Vader, W. 1978. Associations between amphipods and echinoderms. Astarte. *11:*123–134.

Valentin, C., and Anger, K. 1977. *In situ* studies on the life cycle of *Diastylis rathkei*. Mar. Biol. *39*(1):71–76.

Van Name, W. G. 1936. The American land and freshwater isopod crustaceans. Am. Mus. Bull. *71:*1–535. (A very complete taxonomic treatment of the American wood lice and freshwater isopods.)

Williams, W. D. 1972. Freshwater Isopods (Asellidae) of North America. Biota of Freshwater Ecosystems. Identification Manual No. 7. EPA U. S. Government Printing Office, Washington, DC. 45 pp.

MYRIAPODS

15

PRINCIPLES AND EMERGING PATTERNS

Chemical Defenses

Chemical defenses against predators have evolved in numerous groups of animals. The compounds are commonly repellent but may also be toxic, and the predators in most cases appear to be vertebrates, which may reflect to some degree a bias in observation and testing. In some aquatic groups the compounds are dispersed throughout the body (as in sponges) or throughout the integument (as in sea slugs) and are deterrents against fish predators. A toxic substance (holothurin) is found in the body wall of certain sea cucumbers or is liberated from their everted defensive tubules. An irritant on the calcareous setae of polychaete fire worms is an effective fish deterrent. Certain amphipods gain protection from fish by using noxious algae in their tube construction or by carrying around a noxious sea butterfly.

Terrestrial arthropods, especially insects, display a wide range of chemical defenses against terrestrial vertebrate predators, particularly birds. The body fluids of some insects contain repellent or toxic compounds, and lady bugs and blister beetles release these compounds by bleeding through the leg joints when disturbed. The toxic glycosides found in the blood of monarch butterflies are derived from the milkweed on which they feed.

Repellent and toxic compounds of many arthropods are produced in specialized **repugnatorial glands** located on specific parts of the body. A mixture of volatile compounds are produced that usually include quinones. The actual defensive substances may be secreted as precursors that react with other compounds liberated into a chamber within the exoskeleton, thereby protecting the tissues of the arthropod from its own secretions. Such repugnatorial glands are found in harvestmen (chelicerates), millipedes (myriapods), and various groups of insects, both adults and larvae. Some insects can spray their defensive secretions. Most notable is the bombardier beetle, which ejects a mixture of hot quinones from the tip of its abdomen at the predator.

MYRIAPODS

Five groups of related arthropods, including centipedes, millipedes, and insects, are believed to have had a common terrestrial origin. They are called **uniramians** because of the apparent unbranched nature of their appendages. The appendages of crustaceans and those of primitive chelicerates, on the other hand, are composed of several branches (e.g., the biramous abdominal appendages of shrimps and crabs). Moreover, in crustaceans the mandibles are branched, jointed appendages in which the basal piece (gnathobase) performs the grinding, biting, or cutting function. Grinding appears to have been the primitive function of the crustacean mandibles, with food moving forward into a posteriorly directed mouth. Forward-moving food macerated by coxal gnathobases is also the primitive condition in chelicerates, such as

horseshoe crabs and presumably, trilobites. In uniramians, on the other hand, the mandibles are nonjointed, unbranched limbs without palps and are used to handle food from below, not from behind.

The most distinctive characteristic of uniramians is the presence of a single pair of antennae, which are believed to be appendages of the second head somite, thus corresponding to the first antennae of crustaceans. The gas exchange organs are tracheae, accounting for the fact that these animals are sometimes called **tracheates.** The name is not entirely appropriate, however, because, as we have seen, many arachnids also possess tracheae. The excretory organs are malpighian tubules, but, like tracheae, the malpighian tubules originated independently from those of arachnids.

Although all contemporary zoologists agree that the uniramians constitute a major evolutionary line of arthropods, they do not agree on the origin and relationship of uniramians to other members of the phylum. Are the arthropods monophyletic, meaning that they all derived from a common arthropod ancestor, or are they polyphyletic, meaning that at least the tracheates had a separate origin from a nonarthropod ancestor? S. M. Manton, formerly of the British Museum, was the principal proponent of the uniramian concept, which she derived from her extensive work on the functional anatomy of centipedes and millipedes. Tracheates are terrestrial animals, and Manton has argued that the arthropods evolved on land independent of the marine chelicerates and crustaceans.

Many zoologists, especially entomologists, have been unconvinced by Manton's arguments, and some recent evidence has raised further doubts. Fossils of some giant Paleozoic insects have been described as having a branch (exite) at the base of their legs; they therefore were not uniramians, that is, they did not have unbranched appendages. The molecular structure of the hemocyanin of certain centipedes can, moreover, be most easily explained as having a common origin with that of chelicerates and crustaceans, and a recent comparative study of rRNA sequences in various arthropods supports a monophyletic origin of arthropods. Mandibles and compound eyes are found in both crustaceans and tracheates. These would have to be convergent structures; that is, they evolved independently in tracheates and crustaceans if the uniramian concept is correct.

The oldest fossil tracheates are terrestrial Devonian forms similar to centipedes and millipedes. The small number of marine millipedes and insects and the more numerous freshwater insects are all clearly secondary invaders of the aquatic environment. The freshwater larvae of some insects also represent a specialization, because primitive insects lack larvae. If uniramian arthropods had a separate origin from other arthropods, the most likely ancestor is believed by many zoologists to have been an organism resembling an onychophoran, a taxon discussed at the end of this chapter (p. 817). These small, terrestrial, caterpillar-like animals are known from the beginning of the Paleozoic and are represented today by a small number of tropical and south temperate species.

MYRIAPODUS ARTHROPODS

Four groups of uniramians comprising some 10,500 species—the centipedes, the millipedes, the pauropods, and the symphylans—have a body composed of a **head** and an enlongated **trunk** with many leg-bearing segments. This common feature was formerly considered sufficient reason for uniting all four groups within a single class, the Myriapoda. Although these arthropods are probably more closely related to each other than to the Insecta, they exhibit fundamental differences. Most zoologists have therefore abandoned the Myriapoda except as a convenient collective name, and each of the four groups is now considered a separate class.

Most myriapods require a humid environment, because their relatively permeable epicuticle usually lacks the high lipid content found in spiders and insects. The lipids present are probably more important in repelling water (functioning as a hydrofuge) than in reducing water loss. Myriapods live beneath stones and wood and in soil and humus and are widely distributed in both temperate and tropical regions.

The head bears a pair of **antennae** and usually ocelli, but except in certain centipedes, true compound eyes are never present. The mouthparts lie on the ventral side of the head and are directed forward. An **epistome** and **labrum** form the upper lip and the roof of a preoral cavity. The lower lip is formed by either a first or second pair of **maxillae,** and enclosed within the preoral cavity are the pair of **mandibles** and a **hypopharynx.** Gas exchange is typically by a **tracheal system** in which the spiracles cannot be closed (another path of water loss). Excretion takes place through **malpighian tubules.** The heart is a dorsal tube extending through the length of the trunk, with a pair of ostia in each segment. A branched system of arteries is rarely present. The nervous system is not concentrated, and the ventral nerve cord contains a ganglion in each segment. Indirect sperm transfer by spermatophore is highly developed, and the myriapods parallel the arachnids in many aspects of this process.

Class Chilopoda

The members of the class Chilopoda, known as centipedes, are perhaps the most familiar of the myriapodous arthropods. They are distributed throughout the world in both temperate and tropical regions, where they live in soil and humus and beneath stones, bark, and logs. The approximately 2500 described species are distributed within four principal orders (Fig. 15–1). The order Geophilomorpha is composed

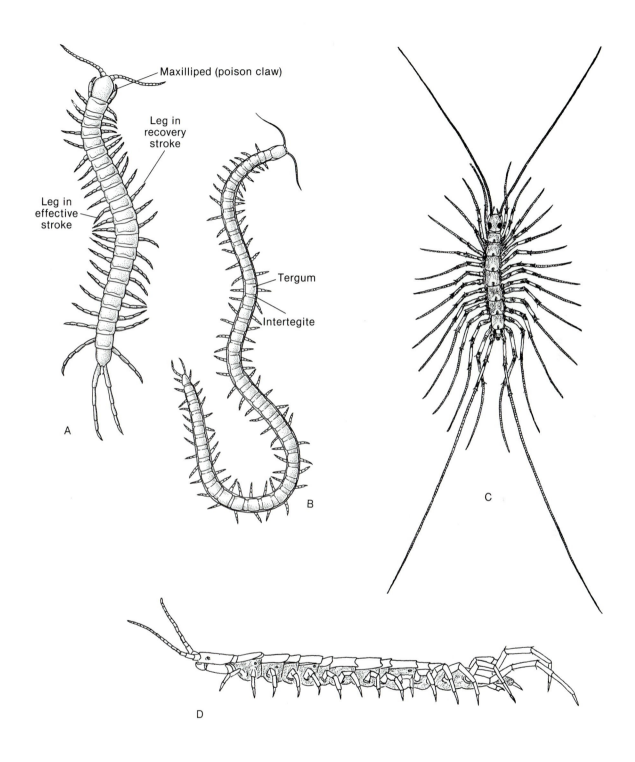

Maxilliped (poison claw)

Leg in
recovery
stroke

Leg in
effective
stroke

Tergum

Intertegite

A

B

C

D

FIGURE 15–1 Chilopoda. **A,** *Otocryptops sexspinnosa,* a scolopendromorph centipede. **B,** A geophilomorph centipede. **C,** *Scutigera coleoptrata,* the common house centipede, a scutigeromorph. **D,** *Lithobius,* a lithobiomorph centipede. *(All after Snodgrass.)*

of long, threadlike centipedes that are adapted for living in soil. The orders Scolopendromorpha and Lithobiomorpha both contain heavy-bodied, flat centipedes that live in crevices beneath stones, bark, and logs and in soil. The Scutigeromorpha are long-legged forms, some of which live in and around human habitations. *Scutigera coleoptrata,* which is found in both Europe and North America, is frequently found trapped in bathtubs and wash basins.

The largest centipede is the tropical American *Scolopendra gigantea,* which may reach almost 30 cm in length. Many other tropical forms, particularly

scolopendrids, are over 20 cm in length, but most North American and European species are only 3 to 6 cm long. Temperate zone centipedes are most commonly reddish brown, but many tropical forms, especially the scolopendromorphs, are red, green, yellow, blue, or combinations of colors.

The head is convex in scutigeromorphs but flattened in other centipedes, with the antennae on the front margin (Figs. 15–1A; 15–2A). The mandible, which bears teeth and a thick fringe of setae, lies beneath the ventrolateral surface of the head. Beneath the mandibles are two pairs of maxillae. Covering the

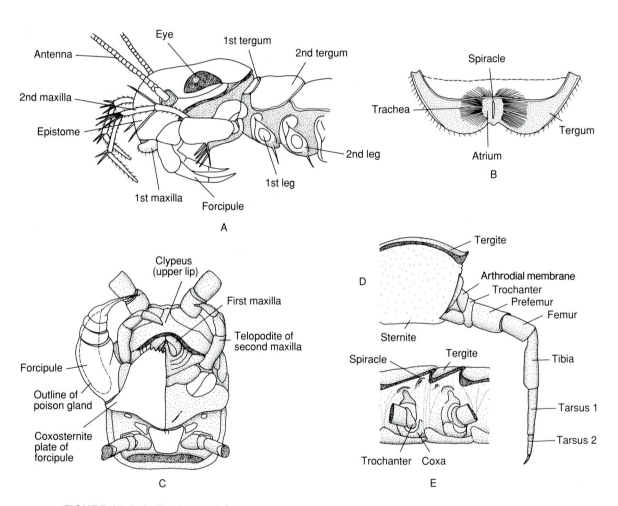

FIGURE 15–2 A, The head of *Scutigera coleoptrata* (lateral view). **B,** Posterior of *Scutigera tergum* showing a pair of tracheal lungs. **C,** Ventral view of head of *Lithobius forficatus.* Only one forcipule (poison claw) is shown. **D,** Cross section of *Lithobius forficatus.* **E,** Lateral view of two segments of *Lithobius forficatus.* Only the basal leg segments are shown. *(A and B, After Snodgrass. C, After Rilling, 1968. D and E, After Manton, 1965. C–E, From Lewis, J. G. E. 1981. The Biology of Centipedes. Cambridge University Press, London. pp. 9, 11. Copyrighted and reprinted by permission of the publisher.)*

mouth appendages are a large pair of **forcipules,** commonly called poison claws, which are actually the appendages of the first trunk segment (Fig. 15–2A, C). Each appendage is curved toward the midventral line and bears a terminal, pointed fang, which is the outlet for the duct of a poison gland. The large coxae of the forcipules and the associated sternite of that segment form a large plate that covers the underside of the head.

Posterior to the first trunk segment, which carries the forcipules, are 15 or more leg-bearing segments. The last pair of legs is sensory or defensive and not locomotor, and the last two trunk segments are legless. The tergal plates vary considerably in size and number, and the differences are correlated with locomotor habits.

Protection

The hiding places of centipedes afford some protection not only from possible predators but also from desiccation. At night they emerge to hunt for food or new living quarters. Scolopendromorphs construct a burrow system in soil or beneath stones and logs, with a chamber into which the animal retreats.

Although centipedes are equipped with poison claws, they have other adaptations for defense. The last pair of legs in centipedes is the longest, and in some scolopendromorphs they can be used in defense by pinching. Many scolopendromorphs and geophilomorphs possess repugnatorial glands on the ventral side of each segment. Male lithobiomorph centipedes possess glands on the coxae of the last four pairs of legs that were thought to be involved in water uptake. They are now believed to produce a pheromone.

Locomotion

Two orders of centipedes are adapted for running. The scolopendromorphs have long legs, all approximately the same length, and correspondingly long strides. The scutigeromorphs can run three times faster, however, and many of their structural peculiarities are associated with the evolution of a rapid gait. In scutigeromorphs the effective leg stroke is faster than the recovery stroke, unlike the scolopendromorphs, whose strokes are equal in duration. Moreover, scutigeromorphs have a marked progressive increase in leg length from anterior to posterior, which enables the posterior legs to move to the outside of the anterior legs, thus reducing interference. For example, in *Scutigera* the posterior legs are twice as long as the first pair (Fig. 15–1C).

To overcome the tendency to undulate, the trunk is strengthened by more or less alternately long and short tergal plates in the lithobiomorphs, and by a reduced number of large, overlapping, tergal plates in the scutigeromorphs. Finally, the annulated distal leg segments of the scutigeromorphs enable the animal to place a considerable section of the end of the leg against the substratum, very much like a foot, to decrease slippage.

In contrast to the other centipedes, the wormlike geophilomorphs are adapted for burrowing through loose soil or humus. The pushing force is provided not by the legs, as in millipedes, but by extension and contraction of the trunk, as in earthworms. A British species of *Stigmogaster,* for example, can increase its body length by as much as 68%. Powerfully developed longitudinal muscles of the body wall, an elastic pleural wall, an increased number of segments, and a small intersternite and intertergite between each of the larger, main sternal and tergal plates all facilitate great extension and contraction of the trunk in burrowing (Fig. 15–1B). The legs are short and in burrows anchor the body like the setae of an earthworm. Geophilomorphs walk with the legs, but there is little overlap of leg movement.

Nutrition

Most centipedes are believed to be predacious. Small arthropods form the major part of their diet, but there are numerous reports of large scolopendrids feeding on frogs, toads, snakes, birds, and even mice. Some centipedes, especially geophilomorphs, feed on earthworms, snails, and nematodes. Prey is detected and located with the antennae, or with the legs in *Scutigera,* and then is captured and killed or stunned with the forcipules. *Scolopendra* attacks glass beads (smeared with a food extract) when it first touches them with its antennae; *Lithobius* does not feed if deprived of its antennae. Large tropical centipedes are often held in dread, but the neurotoxic venom of most species, although painful, is not sufficiently toxic to be lethal to people, or even to small children. Reports of fatalities from older literature are difficult to authenticate. The effect of the bite of such species is generally similar to that of a severe sting from a yellow jacket or hornet.

Following capture, the prey is held by the second maxillae and the forcipules while the mandibles and first maxillae perform the manipulative action required for ingestion. Certain geophilomorphs, which possess weakly armed and less mobile mandibles, may partially digest their prey before ingestion. The digestive tract is a straight tube, with the foregut (pharynx and esophagus) occupying from one seventh to two thirds of the length, depending on the species (Fig. 15–3). The hindgut is short. Salivary secretions are provided by glands associated with the feeding appendages and buccal region.

Gas Exchange and Excretion

Except in the scutigeromorphs, the spiracles of the tracheal system lie in the membranous pleural region above and just behind the coxae (Fig. 15–2E). There is basically one pair of spiracles per segment, but some segments lack them, and the pattern of distribution varies in different groups. The unclosable spiracle opens into an atrium lined with cuticular hairs (tri-

chomes), which may reduce desiccation or prevent the entrance of dust particles (Fig. 15–4). The tracheal tubes open at the base of the atrium. Depending on the order, the tracheal system may contain longitudinal trunks, a network of tubes, or unconnected tubes.

Perhaps associated with their more active habits, and thus with a higher metabolic rate, the tracheal system of the scutigeromorphs is lunglike and probably evolved independently from that of the other centipedes. The spiracles are located middorsally near the posterior margin of the tergal plates covering the leg-bearing segments (Fig. 15–2B). Each spiracle opens into an atrium from which extend two large fans of short tracheal tubes. The tracheae are bathed in the blood of the pericardial cavity. The blood of

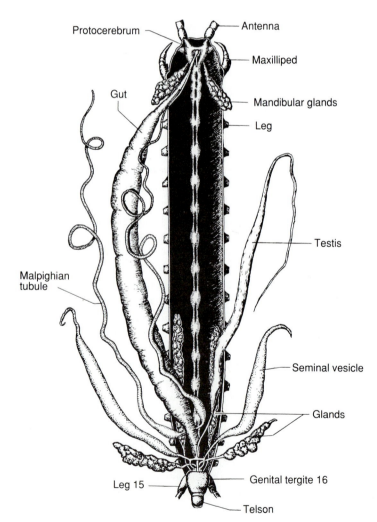

FIGURE 15–3 Internal structure of the centipede, *Lithobius*. *(From Kaestner, A. 1968. Invertebrate Zoology. Vol. II. Wiley-Interscience, New York.)*

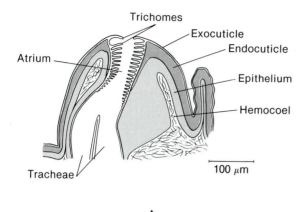

A

B

FIGURE 15–4 A, Longitudinal section through the spiracle of *Lithobius forficatus.* **B,** Pore plate of the coxal glands of a posterior leg of a male of *Lithobius forficatus.* *(A, From Curry, A. 1974. The spiracle structure and resistance to desiccation of centipedes. In Blower, J. G. (Ed.): Myriapoda. Academic Press, London. p. 368. B, From Littlewood, P. M. H., and Blower, J. G. 1987. The chemosensory behaviour of Lithobius forficatus. 1. Evidence for a pheromone released by the coxal organs. J. Zool. 211:65–32.)*

scutigeromorphs contains the respiratory pigment hemocyanin, which is absent from other uniramians.

There is usually a single pair of malpighian tubules (Fig. 15–3), but much of the nitrogenous waste is excreted as ammonia rather than as uric acid.

Sense Organs

All geophilomorphs and some scolopendromorphs lack eyes. Other centipedes possess few to many ocelli. In the scutigeromorphs, the ocelli are so clustered and organized that they form compound eyes. The optic units, which number up to 200, form a compact group on each side of the head and tend to be elongated with converging optic rods. In *Scutigera,* the combined corneal surface is greatly convex, as are the compound eyes of insects and crustaceans, and each unit is remarkably similar to an ommatidium (Fig. 15–2A). There is no evidence, however, that the compound eyes of *Scutigera,* nor the eyes of any other centipedes, function in more than the simple detection of light and dark. Many centipedes are negatively phototactic.

A pair of **organs of Tömösvary** is present on the head at the base of the antennae in lithobiomorphs and scutigeromorphs. Each sense organ consists of a disc with a central pore into which the endings of subcuticular sensory cells converge. The few studies on the function of organs of Tömösvary are conflicting. There is evidence for vibration detection and monitoring of humidity.

Centipedes possess various types of sensory hairs and other cuticular sensory structures on the antennae and other parts of the body. The long, last pair of legs of many centipedes have a sensory function, especially in lithobiomorphs and scutigeromorphs; they are modified to form a pair of posteriorly directed, antennae-like appendages.

Reproduction and Development

The ovary is a single, tubular organ located above the gut, and the oviduct opens through a median atrium and aperture onto the ventral surface of the posterior, legless genital segment. A pair of seminal receptacles also opens into the genital atrium. In the male, one to many testes are located above the midgut. The testes are connected to a single pair of sperm ducts, which open through a median gonopore on the ventral side of the posterior genital segment (Fig. 15–3). The genital segment of both sexes carries small appendages, called **gonopods,** which function in reproduction.

Sperm transmission is indirect in centipedes, as in other myriapods. Except in scutigeromorphs, the male constructs a little web of silk strands secreted by a spinneret located in the genital atrium. A spermatophore as large as several millimeters is emitted and placed on the webbing. The female picks up the spermatophore and takes it into her reproductive opening. The gonopods of each sex aid in handling the spermatophore. In centipedes the male usually does not produce a spermatophore until a female is encountered. Moreover, there is often initial courtship behavior that varies in detail from species to species. The two sexes may palpate each other's posterior end with their antennae while moving in a circle. This behavior may last as long as an hour before the male spins a spermatophore web and deposits a spermatophore. Following spermatophore deposition, the male "signals" the female in various ways. For example, in species of *Lithobius,* the male keeps his posterior pair of legs to either side of the spermatophore and webbing while turning the anterior part of his body and stroking the antennae of the female. She responds by crawling across the posterior end of his body and picking up the spermatophore.

Both the scolopendromorphs and the geophilomorphs brood their eggs in clusters of 15 or more. These centipedes locate themselves in cavities hollowed out in a piece of decayed wood or soil and then wind themselves about the egg mass. The female guards the eggs in this manner through the hatching period and the dispersal of the young (Fig. 15–5C). Female lithobiomorphs and scutigeromorphs carry the eggs about for a short time between the gonopods and then deposit them singly in the soil. In the brooding orders Scolopendromorpha and Geophilomorpha, development is **epimorphic;** that is, the young display the full complement of segments when they hatch. Development in the other two orders is **anamorphic,** meaning that on hatching, the young have only a part of the adult complement of segments. For example, newly hatched young in *Scutigera* have 4 pairs of legs, and in the subsequent six molts, pass through stages with 5, 7, 9, 11, and 13 pairs of legs. The life span of many centipedes is from four to six years or more.

SYSTEMATIC RÉSUMÉ OF CLASS CHILOPODA

Subclass Epimorpha. Eggs brooded; young possess all segments on hatching. Adult has 21 or more pairs of legs.

Order Geophilomorpha. Slender, burrowing centipedes, with 31 to 170 pairs of legs. Intercalary tergal plates located between tergal plates of more or less equal length. Eyes absent. Widely distributed in both temperate and tropical regions throughout the world. *Geophilus, Strigamia, Mecistocephalus.*

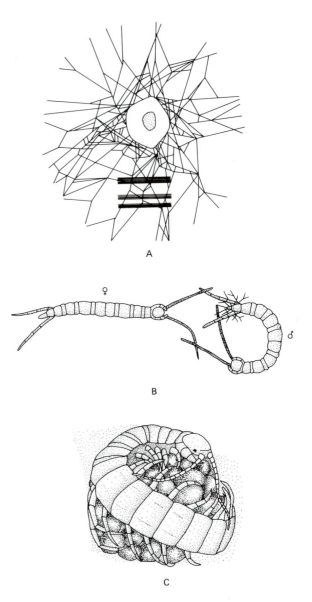

FIGURE 15–5 A, Web and spermatophore of *Lithobius forficatus.* **B,** Male *Lithobius forficatus* with posterior over the web signaling the female to pick up the spermatophore. **C,** A female of *Scolopendra* brooding her eggs. *(A and B, After Klingel, 1960, from Lewis, J. G. E. 1981. The Biology of Centipedes. Cambridge University Press, London. p. 281. C, After Brehm, from Lewis, J. G. E. 1981. The Biology of Centipedes. Cambridge University Press, London. p. 272.)*

Order Scolopendromorpha. Most species have 21 pairs of legs, but some possess 23 pairs. Tergal plates not alternating in size. With or without eyes. Many species distributed throughout the world, especially in the tropics. *Scolopendra, Theatops, Otocryptops.*
Subclass Anamorpha. Brooding absent; young do not possess full complement of segments on hatching. Adult has 15 pairs of legs.
Order Lithobiomorpha. Alternating large and small tergal plates; spiracles paired and lateral. Worldwide in distribution, but most genera and species are found in temperate and subtropical zones. *Lithobius, Bothropolys.*
Order Scutigeromorpha. Legs and antennae very long. Eyes large and compound. Spiracles unpaired and located middorsally on tergal plates. A single family distributed throughout the world, especially in the tropics. *Scutigera.*

Class Symphyla

The Symphyla is a small class of approximately 160 described species that live in soil and leaf mold in most parts of the world. They have evoked considerable interest among some zoologists as being myriapods that display a number of characteristics similar to those of insects.

Symphylans are between 1 and 8 mm in length and superficially resemble lithobiomorph centipedes (Fig. 15–6A). The trunk contains 12 leg-bearing segments, which are covered by 15 to 24 tergal plates. The last (14th) segment bears a pair of spinnerets, or cerci, and a pair of long, sensory hairs (trichobothria). The trunk terminates in a tiny oval telson.

The head projects in front of the laterally placed antennae (Fig. 15–6B). The mandibles are covered ventrally by a pair of long, first maxillae. The second pair of maxillae are fused, forming a labium (Fig. 15–6C). The apparent similarity of symphylan mouthparts to those of insects has often been cited as evidence for the supposed affinity of the two groups. The mouthparts may be only superficially similar, however, and are functionally very different.

The trunk structure, especially the presence of the additional tergal plates, which increases dorsoventral flexibility, is undoubtedly correlated with the locomotor habits of these animals. Many

symphylans can run very rapidly and can twist, turn, and loop their bodies when crawling through the crevices within humus. This ability is probably an adaptation for escaping predators in the network of living and decayed vegetation in which they live and feed. Scutigerellids attack plant roots and can be a serious pest to vegetable and flower crops, especially in greenhouses.

A single pair of spiracles opens onto the sides of the head, and the tracheae supply only the first three trunk segments. Attached to the body wall beneath the base of each leg are an eversible coxal sac and a small appendage (the stylus), structures also present in primitive insects. The coxal sacs take up moisture. The function of the stylus is unknown, although it is probably of a sensory nature. There are no eyes, but two organs of Tömösvary are well developed (Fig. 15–6B).

The genital openings are located on the ventral side of the fourth trunk segment. The copulatory behavior of *Scutigerella* is known and is most unusual. The male deposits 150 to 450 spermatophores, each at the end of a stalk. The female, on encountering a spermatophore, eats it, but instead of swallowing the sperm, stores the contents of the spermatophore in special buccal pouches. After removing the eggs with her mouth from the single gonopore, she attaches them to the substratum and then works them over with her mouth, smearing each egg with sperm and fertilizing it (Fig. 15–6D, E).

The eggs are laid in clusters of about 8 to 12 and are attached to the walls of crevices or to moss or lichen. Parthenogenesis is common. The role of the spinning organs in reproduction is unknown. Development is anamorphic; on hatching, the young have six or seven pairs of legs. *Scutigerella immaculata* lives as long as four years and molts throughout its lifetime.

Class Diplopoda

Members of the Diplopoda are commonly known as millipedes or thousand-leggers. Millipedes are secretive and largely shun light, living beneath leaves, stones, bark, and logs and in soil. Quite a number of millipedes are cave inhabitants. The some 10,000 described species constitute the greatest number of myriapodous arthropods. They live throughout the world,

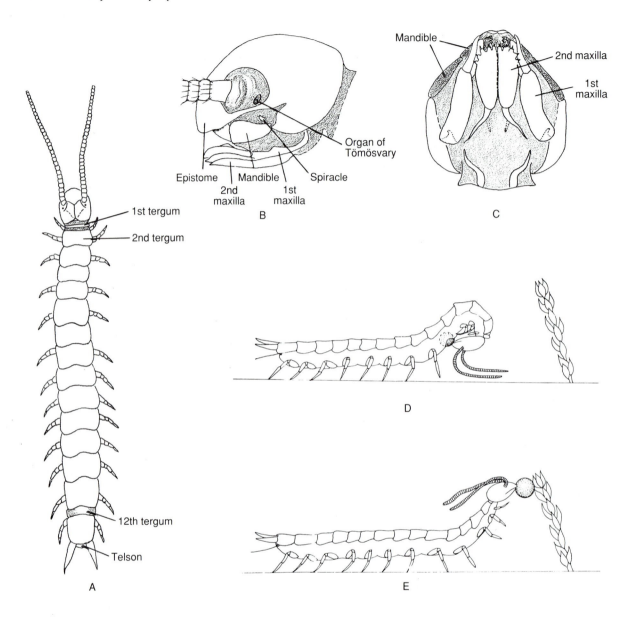

FIGURE 15–6 Symphyla. **A,** *Scutigerella immaculata* (dorsal view). **B,** The head of *Hanseniella* (lateral view). **C,** The head of *Scutigerella immaculata* (ventral view). **D–E,** Female of *Scutigerella* removing egg from gonopore with her mouthparts and attaching it to moss. When carried by the mouthparts, the egg is smeared with semen stored in buccal pouches. *(A–C, After Snodgrass. D and E, After Juberthie-Jupeau.)*

especially in the tropics, but the best known faunas are those of North America and Europe.

A distinguishing feature of the class is the presence of doubled trunk segments, or **diplosegments,** derived from the fusion of two originally separate somites. Each diplosegment bears two pairs of legs, from which the name of the class is derived (Figs. 15–7; 15–8B, C). The diplosegmented condition is

also evident internally, for there are two pairs of ventral ganglia and two pairs of heart ostia within each segment. It is from such a diplosegment that one current speculation derives the biramous appendages of crustaceans (p. 600).

The diplopod head tends to be convex dorsally and flattened ventrally (Fig. 15–9). The sides of the head are covered by the convex bases of the very

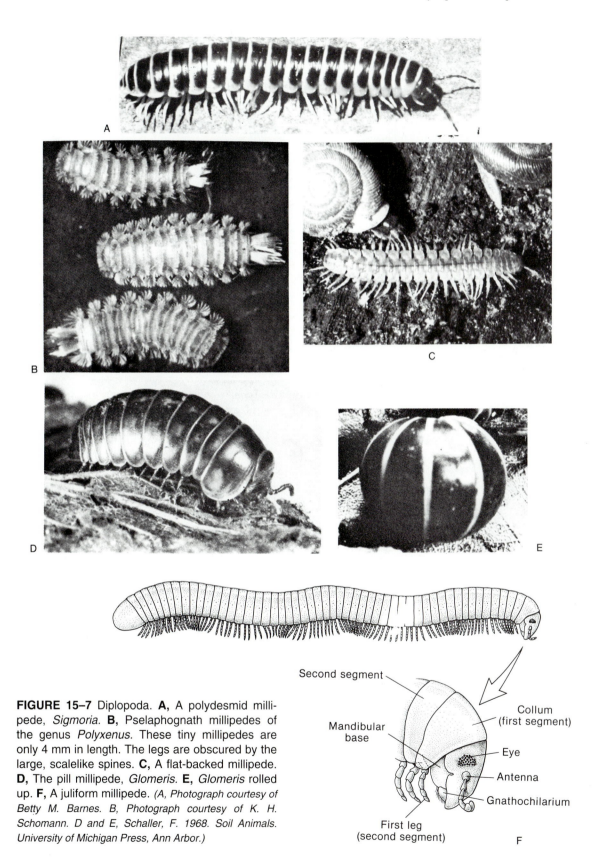

FIGURE 15–7 Diplopoda. **A,** A polydesmid milli-pede, *Sigmoria*. **B,** Pselaphognath millipedes of the genus *Polyxenus*. These tiny millipedes are only 4 mm in length. The legs are obscured by the large, scalelike spines. **C,** A flat-backed millipede. **D,** The pill millipede, *Glomeris*. **E,** *Glomeris* rolled up. **F,** A juliform millipede. *(A, Photograph courtesy of Betty M. Barnes. B, Photograph courtesy of K. H. Schomann. D and E, Schaller, F. 1968. Soil Animals. University of Michigan Press, Ann Arbor.)*

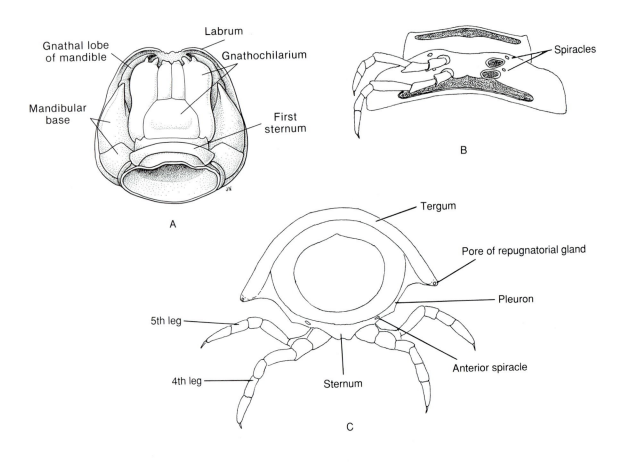

FIGURE 15–8 A, Head of *Habrostrepus,* a juliform millipede (ventral view). **B,** A diplosegment of *Apheloria,* a flat-backed millipede (ventral view). **C,** A diplosegment of *Apheloria* (transverse section). *(All after Snodgrass.)*

large mandibles. The biting edge of the mandible bears teeth and a rasping surface. The floor of the pre-oral chamber is formed by a fused pair of maxillae, often called the **gnathochilarium** (Figs. 15–8A; 15–9). It is a broad, flattened plate attached to the posterior, ventral surface of the head and bearing four sensory palps distally. The head of a diplopod does not contain a second maxillary segment.

The trunk may appear to be dorsoventrally flattened in the so-called flat-backed millipedes because of lateral, shelflike projections of the terga (Fig. 15–7C). In the familiar millipedes of the order Julida and a number of large tropical orders, the trunk is essentially cylindrical (Fig. 15–7F). Such species are said to be juliform. In primitive diplopods the tergal, sternal, and pleural sclerites composing a segment may be separate and distinct, but coalescence of vary-

ing degrees has usually taken place. In flat-backed and juliform millipedes all the sclerites are fused, and in the latter group, they form a nearly cylindrical ring.

The extreme anterior segments differ from the others in that the first (the **collum**) is legless and forms a large collar behind the head (Fig. 15–7F). The collum is not a diplosegment and is probably homologous to the segment that carries the second pair of maxillae in other uniramians. The second, third, and fourth segments carry only a single pair of legs (anterior pair suppressed). In some millipedes, such as the flat-backed species, the last one to five segments are also legless. The body terminates in the telson, on which the anus opens ventrally.

The integument is hard, particularly the tergites, and similar to the crustacean integument, it is impregnated with calcium salts. The surface is often smooth,

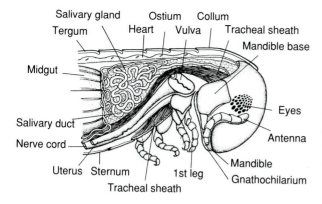

Salivary gland · Ostium · Collum
Tergum · Heart · Vulva · Tracheal sheath
Mandible base
Midgut
Salivary duct
Nerve cord · Eyes
Antenna
Uterus · Sternum · 1st leg · Mandible
Tracheal sheath · Gnathochilarium

FIGURE 15–9 Head and anterior trunk segments of the juliform millipede, *Narceus* (lateral view). *(After Buck and Keister.)*

but in some groups, the terga bear ridges, tubercles, spines, or isolated bristles and in the little, soft-bodied pselaphognaths, the body is covered with tufts and rows of scalelike spines (Fig. 15–7B).

Diplopods vary greatly in size. The Penicillata contains minute forms, some species of *Polyxenus* being only 2 mm long. The largest millipedes are tropical species of the family Spirostreptidae, which may be 30 cm long and have a trunk with as many as 90 segments.

Most diplopods are black and shades of brown; some species are red or orange, and mottled or spotted patterns are not uncommon. Some southern California millipedes are bioluminescent.

Locomotioïn

In general, most species of diplopods crawl slowly about over the ground. Unlike the alternating pattern of stepping that characterizes centipedes and most other arthropods, the effective stroke of the legs on one side of the body of millipedes coincides with that on the opposite side. The gaits of diplopods, although slow, are adapted for exerting a powerful pushing force, enabling the animal to push its way through humus, leaves, and loose soil. The force is exerted entirely by the legs, and the diplosegmented structure is probably associated with the evolution of such a gait. The backward, pushing stroke is activated in waves along the length of the body and is of longer duration than the forestroke. Thus, at any moment, more legs are in contact with the substratum than are raised. The number of legs involved in a single wave is propor-

tional to the amount of force required for pushing. Thus, while the animal is running, 12 legs or fewer may compose a wave, but when pushing, a single wave may involve as many as 52 legs in some juliform millipedes.

The head-on pushing habit has been most highly developed in the juliform species, which burrow into relatively compact leaf mold and soil. This habit is reflected in the smooth, fused, rigid cylindrical segments, the rounded head, and the placement of the legs close to the midline of the body. The flat-backed millipedes, which are the most powerful, open up cracks and crevices by pushing with the whole dorsal surface of their bodies. The lateral keels in these millipedes provide a protected working space for the more laterally placed legs. Ability to climb is particularly striking in some colobognaths and lysiopetalids, which inhabit rocky situations. These millipedes can climb up smooth surfaces by gripping with opposite legs. These rock dwellers also include the swiftest of the millipedes. Speed is correlated with their predatory and scavenging feeding habits and the need to cover great distances to find food.

Protection

To compensate for the lack of speed in fleeing from predators, a number of protective mechanisms have evolved in millipedes. The calcareous exoskeleton offers some protection to the upper and lateral sides of the body. The long, many-segmented millipedes, such as the colobognaths and the juliform groups, protect the more vulnerable ventral surface by coiling the trunk into a spiral when at rest or disturbed. Members of the order Glomerida (superorder Oniscomorpha), called pill millipedes, as well as some others, can roll up into a ball (Fig. 15–7D, E). When rolled up, some tropical species are larger than golf balls.

Repugnatorial glands are present in many millipedes, including the flat-backed and juliform groups. There is usually only one pair of glands per segment, and the openings are located on the sides of the tergal plates or (in the flat-backed millipedes) on the margins of the tergal lobes (Fig. 15–8C). Each gland consists of a large secretory sac, which empties into a duct and out through an external pore. The principal component of the secretion varies in different species. Aldehydes, quinones, phenols, and hydrogen cyanide have been identified. The hydrogen cyanide is liberated when a precursor and an enzyme are mixed from a double-chambered gland. The secretion is toxic or repellent to other small animals; the secretion of some large tropical species is reportedly caustic to human skin. The fluid is usually exuded slowly, but large, tropical, juliform spirobolids can discharge it as a spray or jet for 20 to 30 cm.

Nutrition

Most millipedes are herbivorous, feeding mostly on decomposing vegetation. Food is usually moistened by secretions and chewed or scraped by the mandibles. In the tropical Siphonophoridae, however, the labrum and gnathochilarium are modified to form a long, piercing beak for feeding on plant juices. A carnivorous or omnivorous diet has been adopted by the rock-inhabiting lysiopetalids and some other millipedes. It has been reported that prey includes phalangids, insects, centipedes, and earthworms. Like earthworms, some millipedes ingest soil from which they digest the organic matter.

The digestive tract is typically a straight tube with a long midgut. Salivary glands open into the preoral cavity. The midgut produces a peritrophic membrane that surrounds the food, as in insects and crustaceans (see p. 836).

Gas Exchange, Internal Transport, and Excretion

There are four spiracles per diplosegment, located on the sterna, an unusual position among arthropods (Fig. 15–8B, C). Each spiracle opens into an internal tracheal pouch from which arise numerous tracheae.

The heart ends blindly at the posterior end of the trunk, but anteriorly a short aorta continues into the head (Fig. 15–9). There are two pairs of lateral ostia for each segment. Two malpighian tubules arise from each side of the midgut/hindgut junction and are often long and looped. Like centipedes, millipedes excrete more ammonia than uric acid. Also like centipedes, millipedes possess nephridia-like maxillary glands, but their function is still unknown.

Although most millipedes cannot tolerate desiccating conditions, some colobognaths and lysiopetalids live in arid habitats. These species possess coxal sacs, which supposedly take up water, such as dew drops. The ability of many millipedes to coil the body or roll up into a ball may contribute to reduction of water loss when they are inactive. The appearance of large numbers of ground millipedes on tree trunks, rocks, or walls is probably related to humidity, the animals tending to move upward when the air is more saturated with water.

Sense Organs

Eyes may be totally lacking, as in the flat-backed millipedes, or there may be 2 to 80 ocelli arranged about the antennae (Fig. 15–9). Most millipedes are negatively phototactic, and even those without eyes have integumental photoreceptors. The antennae contain tactile hairs and peglike and conelike projections richly supplied with what are probably chemoreceptors. The animal continually taps the substratum with the antennae as it moves along.

As in centipedes, organs of Tömösvary are present in many millipedes and may have an olfactory function or monitor water vapor.

Reproduction and Development

A pair of long, fused, tubular ovaries lies between the midgut and the ventral nerve cord. Two oviducts extend anteriorly to the third, or genital, segment, where each opens into a protractable, pouchlike atrium, or vulva behind the coxae of the second pair of legs (Figs. 15–9; 15–10C, D). When retracted, a vulva is covered externally by a sclerotized, hoodlike piece. At the bottom of the vulva, a groove leads into a seminal receptable.

The testes occupy positions corresponding to those of the ovary but are paired tubes with transverse connections. Anteriorly, near the region of the genital segment (the third), each testis passes into a sperm duct, which either opens through a pair of papillae on or near the coxae of the second pair of legs or opens through a single median papilla into a median groovelike depression between the coxal bases. Note that in contrast to posteriorly located gonopores of centipedes, those of millipedes are located at the anterior end of the trunk.

Sperm transfer in millipedes is indirect. The actual copulatory organs are usually modified trunk appendages (gonopods) (Fig. 15–10A, B), and these are critical structures in identification of species. In most millipedes, one or both pairs of legs of the seventh diplosegment serve as gonopods. In the flat-backed millipedes, *Apheloria*, for example, in which the first pair of legs on the seventh segment are gonopods, the male charges the gonopods with sperm by bending the anterior part of the body ventrally and posteriorly. In this position sperm are deposited from the two coxal papillae on the third segment into the reservoir at the base of the gonopods. In the juliform order Spirobolida, both pairs of legs of the seventh segment are gonopods, but the first pair forms a pro-

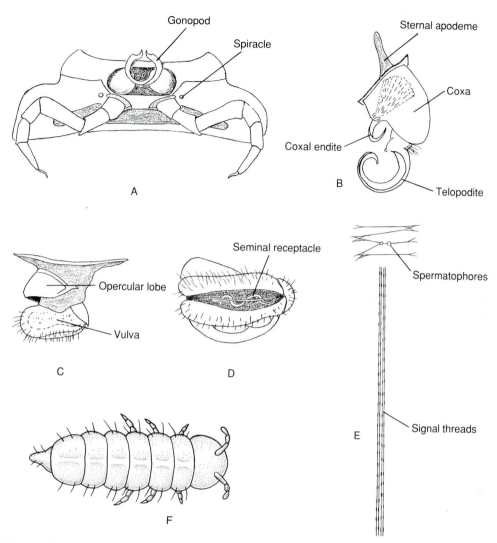

FIGURE 15–10 A, Seventh diplosegment of a male *Apheloria,* showing gonopods and legs (ventral view). **B,** Left gonopod of *Apheloria.* **C,** Right vulva of third segment of *Apheloria* (lateral view). **D,** Vulva (ventral view). **E,** Signal threads leading to spermatophore web of *Polyxenus.* **F,** A newly hatched millipede. *(A–D, After Snodgrass. E, After Schomann. F, After Cloudsley-Thompson, J. L. 1958. Spiders, Scorpions, Centipedes and Mites. Pergamon Press, New York.)*

tective shield over the second pair, which contains sperm reservoirs and canals.

Males communicate their identity and intent to females in a variety of ways. The signal is tactile in most juliform millipedes, when the male climbs onto the back of the female, clinging by special leg pads. Antennal tapping, head drumming, and stridulating are utilized by various other millipedes. Many millipedes produce pheromones, which either initiate mating behavior or continue the sequence of steps initiated by other signals.

During copulation the body of the male is twisted about or stretched out against that of the female so that the gonopods are opposite the vulvae, and the body of the female is held by the legs of the male. The gonopods are protracted, and sperm are transferred through the ringlike tip of the gonopod (telopodite) into the vulva. In the pentazonian millipedes there are no gonopods, and the male uses his mouthparts to transfer sperm.

Parthenogenesis is common in the pselaphognaths, and males are rare.

The diplopod eggs are fertilized at the moment of laying, and anywhere from 10 to 300 eggs are produced at one time, depending on the species. Some deposit their eggs in clusters in soil or humus; others, such as *Narceus*, regurgitate a material that is molded into a cup with the head and anterior legs. A single egg is laid in the cup, which is then sealed and polished. The capsule is deposited in humus and crevices, and it is eaten by the young millipede on hatching. The European pill millipede, *Glomeris*, has similar habits but forms the capsule with excrement.

Many millipedes construct a nest for the deposition of the eggs. Some flat-backed species and colobognaths construct the nest from excrement, building a thin-walled, domed chamber topped by a chimney. The vulvae are applied against the chimney opening, and the eggs fall into the chamber as they are laid. The opening is then sealed, and the chamber is covered with grass and other debris. The female, and in some species the male, may remain coiled about the nest for several weeks.

Some flat-backed and juliform millipedes construct the nest in soil, reinforcing the walls from the inside with excrement. The flat-backed *Strongylosoma pallipes* builds several such nests, each containing 40 to 50 eggs, and closes them from the outside. The Chordeumatida enclose their eggs within silk cocoons, and *Polyxenus* covers its egg clusters with shed tail setae.

Development is anamorphic. The eggs of most species hatch in several weeks, and the newly hatched young usually have only the first three pairs of legs and not more than seven trunk rings (Fig. 15–10F). With each molt, additional segments and legs are added to the trunk. Many millipedes undergo ecdysis within specially constructed molting chambers similar to the egg nests, and it is within the molting chamber that many tropical species survive the dry season. The shed exoskeleton is generally eaten, perhaps to aid in calcium replacement. Millipedes live from one to ten or more years, depending on the species.

SYSTEMATIC RÉSUMÉ OF CLASS DIPLOPODA

Subclass Pencillata (Pselaphognatha). Minute millipedes with soft integument bearing tufts and rows of serrated scalelike setae. Trunk bears 13 to 17 pairs of legs. No gonopods. No repugnatorial glands. Fewer than 100 species, but the group is represented throughout the world. *Polyxenus, Lophoproctus.*

Subclass Pentazonia. Tergal plates arched. Last two pairs of legs modified for clasping.

Order Glomeridesmida (Limacomorpha). Small, eyeless, tropical millipedes. Trunk is composed of 22 arched segments. Cannot roll into a ball. No repugnatorial glands. *Glomeridesmus.*

Order Sphaerotheriida. Giant pill millipedes. Trunk with 13 tergites. Mostly tropical; South Africa, Asia, Australia, and New Zealand.

Order Glomerida (Oniscomorpha). Pill millipedes. Trunk is covered with 11 to 12 arched tergites; the second and last are enlarged, enabling body to roll into a tight ball that conceals head and legs. No repugnatorial glands. Largely Palearctic. *Glomeris* is common in Europe.

Subclass Helminthomorpha. Segments more or less cylindrical or somewhat flattened in cross section. At least one pair of legs (gonopods) of the seventh segment in the male is modified for sperm transfer. This subclass contains the greatest number of millipede species. Only half of the 11 orders are mentioned here.

Order Polyzoniida. Gonopods leglike and preceded by eight pairs of ordinary legs. *Polyzonium.*

Order Spirobolida. Cylindrical body composed of 35 to 60 ringlike segments. Gonopods modified. Chiefly tropical millipedes, somewhat similar to the temperate juliform species. *Narceus, Rhinocricus.*

Order Spirostreptida. Segments cylindrical. First pair of legs modified in male. Anterior pair of gonopods of male modified for sperm transfer; posterior pair commonly reduced. This tropical order contains the largest species of millipedes. *Orthoporus.*

Order Julida. Trunk composed of 30 to 90 cylindrical segments. Sternites fused with pleurotergal arch. In general, smaller than millipedes of the previous two orders. Both pairs of legs of seventh segment modified as gonopods in male. Mostly temperate species. Widespread. *Julus, Blaniulus, Nemasoma, Cylindroiulus.*

Order Polydesmida. Flat-backed millipedes. No eyes. Trunk usually composed of 20 rings with prominent, lateral, tergal keels. Sternites fused with pleurotergal arch. Widely distributed. Many species. *Polydesmus, Oxidus, Apheloria.*

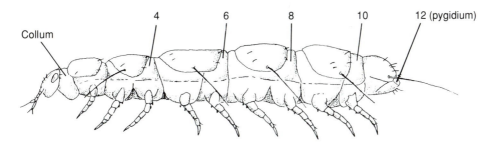

FIGURE 15–11 The pauropod, *Pauropus silvaticus* (lateral view). *(After Tiegs from Snodgrass.)*

Order Chordeumatida. Usually with eyes. Trunk composed of 26 to 32 rings, which are either cylindrical or with lateral tergal keels. End of trunk bears no spinnerets. Sternites not fused with pleurotergal arch. No repugnatorial glands. These species have longer legs and move more rapidly than most other millipedes. Widely distributed. *Cleidogona, Chordeuma.*

Class Pauropoda

The pauropods constitute a small class of soft-bodied, rather grublike animals that inhabit leaf mold and soil (Fig. 15–11). All are minute, ranging from 0.5 to 1.5 mm in length. Although once considered rare, pauropods have now been found to be frequently abundant in forest litter. There are approximately 500 described species, which are widespread in both temperate and tropical regions.

Pauropods are similar to millipedes in a number of ways. The trunk usually contains 11 segments, 9 of which bear a pair of legs. The first segment (collum) and the eleventh segment and telson are legless. Certain of the dorsal tergal plates are very large and overlap adjacent segments. Five of the terga carry a pair of long, laterally placed setae. Unlike the collum of a diplopod, that of the pauropod is very inconspicuous dorsally and expanded ventrally.

On each side of the head there is a peculiar, disclike sensory organ that is perhaps homologous to the organ of Tömösvary of other myriapods. The antennae are biramous. One division terminates in a single flagellum; the other in two flagella and a peculiar club-shaped sensory structure. The mandibles are adapted for grinding or piercing. The lower lip is probably homologous to the gnathochilarium of diplopods, because it apparently represents the first maxillae.

Most pauropods feed on fungi or decomposing plant tissue, but some are predatory. There is neither heart nor (except in some primitive species) tracheae, their absence probably being associated with the small size of these animals.

As in diplopods, the third trunk segment is the genital segment. Sperm are transferred via a spermatophore, which is deposited by the male along with two signal threads in the female's absence. The eggs are laid in humus, either singly or in clusters. Development is anamorphic, and as in diplopods, the young hatch with only three pairs of legs. In *Pauropus sylvaticus,* development to sexual maturity takes about 14 weeks.

PHYLUM ONYCHOPHORA

Onychophorans are not usually considered arthropods, but rather a phylum of animals closely related to arthropods. We include them here only for purposes of comparison, because an onychophoran-like organism is believed by some zoologists to be the ancestor of the uniramians.

There are only about 70 living species of onychophorans, but the phylum is an ancient one and does not appear to have changed greatly since the Cambrian period, from which the only possible fossil specimen has been taken.

The geographical distribution of living species is relatively restricted. All onychophorans live in tropical regions (the East Indies, the Himalayas, the Congo, the West Indies, and northern South America) or southern temperate regions (Australia, New Zealand, South Africa, and the Andes). No species have been found north of the Tropic of Cancer.

Most onychophorans are confined to humid terrestrial habitats, such as tropical rain forests, beneath

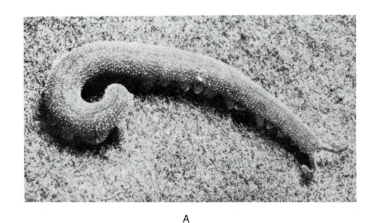

A

Antenna

Peribuccal lobes

Oral papilla

Mandible

First leg

B

Ventral organ

Opening of coxal vesicle

Nephridiopore

Crural papilla

Gonopore

Anus

Anal gland opening

C

FIGURE 15–12 Onychophora. **A,** *Peripatus.* **B,** Anterior of *Peripatopsis capensis* (ventral view). **C,** Posterior of a male *Peripatus corradoi* (ventral view). *(A, Photograph courtesy of H. Sturm. B, After Cuénot, L. 19: Les Onychophores. In Grassé, P. (Ed.): Traité de Zoologie. Vol. 6. Masson et Cie, Paris. pp. 3–75. C, After Bouvier from Cuénot.)*

logs, stones, and leaves, or along stream banks. During winter snows and low temperatures or during dry periods, they become inactive and remain in protective burrows or other retreats.

External Structure

Onychophorans have been described as a missing link between annelids and arthropods because of their many similarities to both groups. Onychophorans look very much like slugs with legs (Fig. 15–12A); in fact, they were thought to be molluscs when first discovered by L. Guilding in 1826. The body is more or less cylindrical and ranges from 1.4 cm to 15 cm in length. The anterior end bears a pair of large, annulated antennae and a ventral mouth, which is flanked by a pair of clawlike mandibles and by a pair of short, conical, oral papillae. The mandibles represent modified segmental appendages, as in arthropods. The legs vary in number from 14 to 43 pairs, depending on the species and the sex. Each leg is a large, conical, non-jointed protuberance bearing a pair of terminal claws (Fig. 15–12B). The entire surface of the body is covered by large and small tubercles, which are arranged in rings or bands encircling the legs and trunk. The tubercles are covered by minute scales. Onychophorans are blue, green, orange, or black, and the papillae and scales give the body surface a velvety and iridescent appearance.

Internal Structure and Physiology

The body surface is covered by a chitinous cuticle that is composed of the same layers as in arthropods and is molted, but unlike the exoskeleton of arthropods, the cuticle of onychophorans is thin, flexible, and very permeable, and it is not divided into articulating plates (Fig. 15–13A). The absence of a rigid exoskeleton enables onychophorans to squeeze their bodies into tight places. In fact, a number of their peculiarities are probably related to this habit. Beneath the cuticle is a single layer of epidermis and three layers of smooth muscle fibers: circular, diagonal, and longitudinal. The body wall is thus constructed similarly to that of most soft-bodied worms, including annelids. The coelom, however, is reduced to the gonadal cavities and to small sacs associated with the nephridia, whereas the blood-vascular system is expanded to form a hemocoel.

Onychophorans crawl slowly by means of the legs and by extension and contraction of the body, which is held off the ground. The body fluids act as a hydrostatic skeleton. When a segment is extended, the legs are lifted from the ground and moved forward. The legs are located more ventrally than are the parapodia of annelids, but the effective stroke of a leg

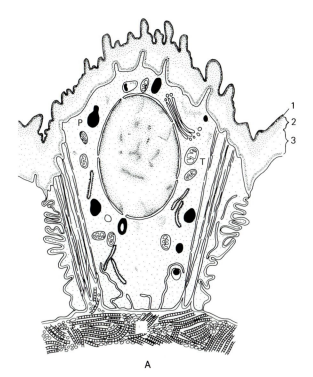

A

B

FIGURE 15–13 A, An epidermal cell of *Peripatus acacioi,* showing the three layers of the surface cuticle: epicuticle (1), exocuticle (2), and endocuticle (3). Tonofibrils anchoring the muscle layer to the cuticle are labeled T. **B,** Scanning electromicrograph of the body surface of *Peripatopsis moseleyi,* showing a papilla with terminal sensory bristle. *(A, After Lavallard from Storch, V. 1984. Onychophora. In Bereiter-Hahn, J., Matoltsy, A. G., and Richards, K. S. Biology of the Integument. 1. Invertebrates. Springer-Verlag, Berlin. p. 704. B, SEM courtesy of Storch, V., and Ruhberg, H. 1977. Fine structure of the sensilla of Peripatopsis moseleyi. Cell Tiss. Res. 177:539–553.)*

does not alternate with that on the opposite side of the body as in most arthropods.

Most species are predacious and feed on small invertebrates, such as snails, insects, and worms. For prey catching and defense, onychophorans secrete an adhesive material from slime glands opening at the ends of the oral papillae (Fig. 15–12B). The secretion is discharged as two streams from a distance as great as 15 cm; it hardens almost immediately, entangling the prey in a net of adhesive threads.

The mouth is located at the base of the prebuccal depression. Within the depression lie the lateral claw-like mandibles, which are used for grasping and cutting prey. Salivary secretions are passed into the body of the prey, and the partially digested tissues are then sucked into the mouth. The prebuccal depression opens into the chitin-lined foregut, composed of a pharynx and an esophagus (Fig. 15–14B). A large,

straight intestine is immediately posterior to the esophagus and is the site of the remaining digestion and of absorption. As in insects and crustaceans, certain of the intestinal cells secrete a **peritrophic membrane,** which confines the food and creates a space between the membrane and the intestinal wall (see p. 836). The tubular hindgut (rectum) opens through the anus on the ventral side at the end of the body.

The circulatory system is similar to that of the arthropods. A tubular heart, open at each end and provided with a pair of lateral ostia in each segment, lies within the pericardial sinus and propels blood forward into the general hemocoel. The partitions between sinuses are perforated by openings that facilitate blood circulation. The blood is colorless and contains phagocytic amebocytes.

Each segment contains a single pair of nephridia located in the ventrolateral sinuses. The ciliated fun-

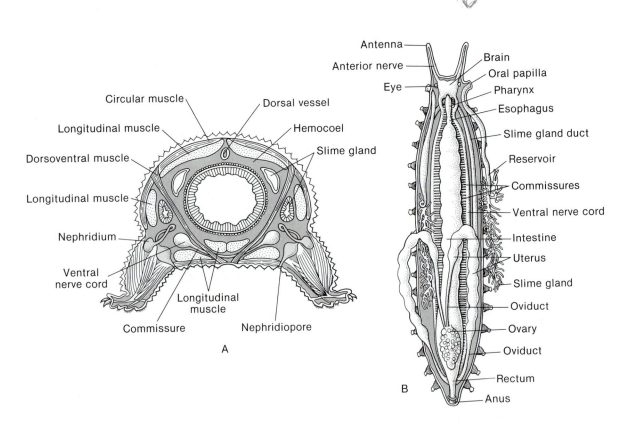

FIGURE 15–14 A, Diagrammatic cross section through the body of an onychophoran. **B,** Internal anatomy of an onychophoran, dorsal view. *(Both after Snodgrass, R. E. 1938. Smithson. Misc. Coll. 97(6).)*

nel and nephrostome lie within an end sac, which is a vestige of the coelom. The saccule wall contains podocytes. The nephridiopore is located on the inner base of each leg (Fig. 15–12C).

Many onychophorans possess eversible sacs, one on the inner side of each leg (Fig. 15–12C). These are believed to function in the uptake of moisture, similar to the coxal sacs of some myriapods.

The gas exchange organs are tracheae. The spiracles are minute openings and are present in large numbers all over the surface of the body between bands of tubercles. Each spiracle opens into a very short atrium, at the end of which arises a tuft of minute tracheae. Each trachea is a simple, straight tube and extends directly to the tissue that it is supplying.

The nervous system is composed of a large, bilobed brain lying over the pharynx and a pair of ventral nerve cords connected by commissures. The brain supplies nerves to the tentacles, the eyes, and the mouth region. In each segment the ventral nerve cords contain a ganglionic swelling and give rise to a number of paired nerves supplying the legs and body wall (Fig. 15–14).

There is a small eye at the base of each antenna. Onychophorans avoid light and are largely nocturnal. Sensory bristles are located on the antennae and on the tubercles scattered over the body surface (Fig. 15–13B).

Reproduction

The sexes are always separate. The ovaries are a pair of fused, elongated organs, located in the posterior part of the female's body. Each ovary is connected to an elaborate genital tract that in some species contains a seminal receptacle and a uterus. The ends of each uterus join together and open to the exterior through a common genital pore, situated ventrally near the posterior of the body (Fig. 15–14B).

The male system contains two elongated, nonfused testes and relatively complex, paired genital tracts. Prior to reaching the exterior, the two tracts join to form a single tube in which the sperm are formed into spermatophores. The male gonopore is ventral and posterior, like that of the female (Fig. 15–12C). Males also possess one or two crural glands of unknown function on the inner side of each leg (Fig. 15–12C).

In the mating of the South African *Peripatopsis,* which lacks seminal receptacles, the male crawls over the body of the female and deposits a spermatophore at random on her side or back, and over time a female may accumulate many spermatophores. The spermatophore stimulates blood amebocytes to bring about dissolution of the underlying integument. Sperm then pass from the spermatophore into the female's hemolymph. They eventually reach the ovaries, where fertilization of the eggs takes place. Sperm transfer in onychophorans with seminal receptacles is not understood.

Onychophorans are oviparous, ovoviviparous, or viviparous. Oviparous forms lay large, yolky eggs in moist situations, each egg enclosed in a chitinous shell. As in most arthropods, cleavage is superficial. All other onychophorans give birth to their young, with either yolk or the mother providing for embryonic nutrition, and the eggs develop within the uterus. The eggs of viviparous species, which includes the largest number of onychophorans, are small and have little yolk, and cleavage has become holoblastic. Uterine secretions provide for the nutrition of the embryo, and the nutritive material is obtained by the embryo through a special embryonic membrane or through a "placental" connection to the uterine wall.

During the life span of some six years *(Peripatopsis),* molting occurs frequently, as often as every two weeks.

Geographical Distribution

The geographical distribution of onychophorans is peculiar in a number of respects. The phylum consists of two families. Each has a wide, discontinuous distribution around the world, but neither is found in the same area with species of the other family. The Peripatidae are more or less equatorial in distribution, extending no further north than the Himalayas, Caribbean Sea, and central Mexico; the Peripatopsidae are limited to the Southern Hemisphere. Thus, the South African species are more closely related to those in Chile than to those in equatorial Africa.

Considering the antiquity of the onychophorans and the improbability of their being spread by other animals, the distribution of the phylum can be accounted for by the past geological connection of the American and African land masses and subsequent continental drift.

Phylogenetic Relationships

The structure of the body wall, the nephridia, the thin, flexible cuticle, and the nonjointed appendages are certainly annelidan in character. But in other respects, onychophorans are more similar to arthropods. The coelom is reduced, and the cuticle is chitinous and is molted. A pair of appendages are modified for feeding, the tracheae are gas exchange organs, and there is a hemocoel with a dorsal, tubular heart containing ostia. Although onychophorans reflect an annelidan ancestry, their origins are obscure. Their development is somewhat similar to that of clitellate annelids (oligochaetes/leeches), and recent studies on the ultrastructure of onychophoran sperm have also revealed striking similarities to the sperm of clitellates. Some zoologists postulate that clitellates and onychophorans are a monophyletic group and the sister group of the myriapod/insect assemblage.

SUMMARY

1. Members of the subphylum Uniramia include the centipedes, millipedes, and insects: terrestrial arthropods with appendages that are primitively nonbranched. They possess one pair of antennae, and the mouthparts include a pair of mandibles. The mandible is a nonjointed unbranched limb, and in the primitive condition, food is not brought forward from behind, as in other arthropods, but is picked up directly beneath the mouth.

2. The uniramians are believed to have evolved from terrestrial ancestors, which may have resembled members of the phylum Onychophora.

3. The myriapodous arthropods include the centipedes (class Chilopoda) and millipedes (class Diplopoda), plus two other small classes (Symphyla and Pauropoda). All have long trunks with many segments and appendages. Tracheae provide for gas exchange and malpighian tubules for excretion.

4. Myriapods live in leaf litter and beneath stones, logs, and bark. Many of their structural features are adaptations for locomotion.

5. Centipedes possess one pair of legs per segment. In many groups the trunk has been strengthened for a running gait by overlapping tergites or tergites of unequal size, the larger extending onto adjacent segments.

6. Centipedes are largely predacious, and prey (mostly other small arthropods) are caught and killed with a pair of anterior forcipules.

7. Millipedes possess two pairs of legs per segment, a condition derived from the fusion of two original segments. The millipede diplosegments appear to be an adaptation for a pushing gait. The trunk is strengthened to withstand the pushing force generated by a large number of legs.

8. Most millipedes feed on decomposing vegetation. Depending on the group, protection is gained from repugnatorial glands, coiling, and rolling up.

9. Both centipedes and millipedes transfer sperm indirectly by spermatophores. The gonopores are located at the posterior end of the trunk in centipedes and at the anterior end of the trunk (third trunk segment) in millipedes.

10. The 70 species of onychophorans are terrestrial, caterpillar-like animals of the tropics and Southern Hemisphere. The soft body, which is covered by a thin, flexible, chitinous cuticle, is adapted for squeezing beneath stones, logs, and other objects.

11. Onychophorans possess a pair of antennae, a pair of clawlike mandibles, and many pairs of nonjointed, peglike legs. Internally, there is a combination of arthropod and annelidan features: body wall of circular and longitudinal muscles, segmental nephridia, reduced coelom, hemocoel and tubular heart, and tracheae. The chitinous cuticle is periodically molted.

12. In some species, sperm are transferred as spermatophores. Some onychophorans are oviparous, but many brood their eggs internally and give birth to their young.

REFERENCES

The literature included here is restricted in large part to myriapods. The introductory references on page 6 include many general works and field guides that contain sections on these animals.

Albert, A. M. 1983. Life cycle of Lithobiidae, with a discussion of the r- and K-selection theory. Oceologia. *56*:272–279.

Anderson, D. T. 1973. Embryology and Phylogeny in Annelids and Arthropods. Pergamon Press, New York.

495 pp. (Detailed accounts of the embryology of ony-chophorans and uniramian arthropods and the phylo-genetic implications of the embryonic patterns.)

Attems, G. 1926–1940. *In* Kukenthal, W. and Krumbach, T. (Eds.): Handbuch der Zoologie. Vol. 4. Progoneata, Chilopoda. 1926; Vol. 52. Myriapodia, Geophilomor-pha. 1929; Vol. 54. Chilopoda, Scolopendromorpha, 1930; Vols. 68–70. Diplopoda, Polydesmoidea. 1937–1940. W. de Gruyter, Berlin and Leipzig. (This and the works of Verhöff (see below) contain the most extensive and detailed accounts of the myriapodous classes.)

Blower, J. G. (Ed.): 1974. Myriapoda. Academic Press, London. (Papers presented at the Second International Congress of Myriapodology at the University of Man-chester in 1972.)

Buck, J. B., and Keister, M. L. 1950. *Spirobolus margina-tus*. *In* Brown, F. A. (Ed.): Selected Invertebrate Types. John Wiley and Sons, New York, pp. 462–475.

Camatini, M. (Ed.): 1979. Myriapod Biology. Academic Press, London. 456 pp. (A collection of papers from an international symposium on myriapods.)

Cloudsley-Thompson, J. L. 1948. *Hydroschendyla subma-rina* in Yorkshire: With an historical review of the ma-rine Myriapoda. Naturalist. *827*:149–152.

Cloudsley-Thompson, J. L. 1958. Spiders, Scorpions, Cen-tipedes, and Mites. Pergamon Press, New York. (A discussion of the natural history and ecology of the myriapodous arthropods is presented in Chapters 2, 3, and 4.)

Cuénot, L. 1949. Les Onychophores, Les Tardigrades, et Les Pentastomides. *In* Grassé, P. (Ed.): Traité de Zo-ologie. Vol. 6. Masson et Cie, Paris. pp. 3–75.

Haacker, U. 1974. Patterns of communication in courtship and mating behavior of millipedes. *In* Blower, J. G. (Ed.): 1974. Myriapoda. Academic Press, London. pp. 317–328. (Papers presented at the Second International Congress of Myriapodology at the University of Man-chester in 1972.)

Hoffman, R. L. 1979. Classification of the Diplopoda. Mu-seum d'Histoire Naturelle, Geneve. 237 pp.

Hoffman, R. L., and Payne, J. A. 1969. Diplopods as carni-vores. Ecology. *50*(6):1096–1098.

Jamieson, B. G. 1986. Onychophoran-euclitellate relation-ships: Evidence from spermatozoal ultrastructure. Zool. Scrip. *15*(2):141–155.

Kaestner, A. 1968. Invertebrate Zoology. Vol. 2. Wiley-In-terscience, New York. 472 pp.

Klingel, H. 1960. Die Paarung des *Lithobius forficatus*. Verh. Dtsch. zool. Ges. *23*:326–332.

Kukalova-Peck, J. 1983. Origin of the insect wing and wing articulation from the insect leg. Can. J. Zool. *61*(7):1618–1669.

Kukalova-Peck, J. 1992. The "Uniramia" do not exist: The ground plan of the Pterygota as revealed by Permian

Diaphanopterodea from Russia. Can. J. Zool. *70*:236–255.

Lewis, J. G. E. 1981. The Biology of Centipedes. Cam-bridge University Press, London. 476 pp.

Littlewood, H. 1991. The water relations of *Lithobius forfi-catus* and the role of the coxal organs. J. Zool. *223*:653–665.

Littlewood, H., and Blower, J. G. 1987. The chemosensory behavior of *Lithobius forficatus*. 1. Evidence for a pheromone released by the coxal organs. J. Zool. *211*:65–82.

Mangum, C. P. et al. 1985. Centipedal hemocyanin: Its structure and its implications for arthropod phylogeny. Proc. Natl. Acad. Sci. *82*:3721–3725.

Manton, S. M. 1952–1961. The evolution of arthropodan locomotory mechanisms. J. Linn. Soc. (Zool.), 1952. Pt. 3. The locomotion of the Chilopoda and Pauropo-da. *42*:118–166; 1954. Pt. 4. The structure, habits, and evolution of the Diplopoda. *42*:229–368; 1956. Pt. 5. The structure, habits, and evolution of the Pselaphog-natha (Diplopoda). *43*:153–187; 1958. Pt. 6. Habits and evolution of the Lysiopetaloidea (Diplopoda), some principles of the leg design in Diplopoda and Chilopoda, and limb structure in Diplopoda. *43*:487–556; 1961. Pt. 7. Functional requirements and body design in Colobognatha (Diplopoda), together with a comparative account of diplopod burrowing techniques, trunk musculature, and segmentation. *44*:383–461.

Manton, S. M. 1964. Mandibular mechanisms and the evo-lution of Arthropods. Phil. Trans. R. Soc., London, B. *247*:1–183.

Manton, S. M. 1965. The evolution of arthropod locomoto-ry mechanisms. Pt. 8. Functional requirements and body design in Chilopoda, together with a comparative account of their skeletomuscular systems and an ap-pendix on the comparison between burrowing forces of annelids and chilopods and its bearing upon the evolution of the arthropodan haemocoel. J. Linn. Soc. (Zool.). *46*:251–483.

Manton, S. M. 1973a. Arthropod phylogeny—a modern synthesis. J. Zool. *171*:111–130.

Manton, S. M., 1973b. The evolution of arthropodan loco-motory mechanisms. Pt. 2. Habits, morphology and evolution of the Uniramia (Onychophora, Myriapoda, Hexapoda) and comparisons with the Arachnida, to-gether with a functional review of uniramian muscula-ture. Zool. J. Linn. *53*:257–375.

Manton, S. M. 1977. The Arthropoda: Habits, Functional Morphology, and Evolution. Clarendon Press, Oxford. 527 pp. (A synthesis of the author's lifelong study of the functional morphology of arthropod limbs and its implications for arthropod evolution.)

Moore, R. C. (Ed.): 1969. Treatise on Invertebrate Paleon-tology. Pt. R. Arthropoda 4. Vol. 2. Geological Society of America and University of Kansas Press, Lawrence. (This volume covers the myriapods.)

Rilling, G. 1968. *Lithobius forficatus. In* Grosses Zoologisches Praktikum. Pt. 13b. Fischer, Stuttgart.

Sakwa, W. N. 1974. A consideration of the chemical basis of food preference in millipedes *In* Blower, J. G. (Ed.): 1974. Myriapoda. Academic Press, London. pp. 329–346. (Papers presented at the Second International Congress of Myriapodology at the University of Manchester in 1972.)

Schaller, F. 1968. Soil Animals. University of Michigan Press, Ann Arbor. 144 pp.

Sharov, A. G. 1966. Basic Arthropodan Stock. Pergamon Press, New York.

Turbeville, J. M., Pfeifer, D. M., Field, K. G. and Raff, R. A. 1991. The phylogenetic status of arthropods, as inferred from 18S rRNA sequences, Mol. Biol. Evol. *8*:669–686.

Verhöff, K. W., 1926–1934. *In* Bronn, H. G., (Ed.): Klassen und Ordnungen des Tierreichs. Chilopoda. Bd. 5. II (1); Diplopoda. Bd. 5. II (2); Symphyla and Pauropoda. Bd. 5. II (3).

Weygoldt, P. 1986. Arthropod interrelationships—the phylogenetic-systematic approach. Zeitschr. Zool. Systematik Evolut.-forschung. *24*(1):19–35.

INSECTS

PRINCIPLES AND EMERGING PATTERNS

SPECIES NUMBERS AND EVOLUTIONARY SUCCESS

Throughout this text we have cited the numbers of described species contained within various groups of invertebrates. How are these numbers obtained? A described species is one for which there is a published description, including a newly constructed name. In small groups, such as cephalocarid and remipedian crustaceans, the number of such descriptions is small enough and recent enough to be counted. In large groups, however, this is much more difficult because the literature is widely scattered in many journals published throughout the world and over a long period of time. Compounding the problem is the fact that many species are described more than once, the second author unaware that the species is already known. Some orb-weaving spiders, for example, have been named three and four times. Thus, for large groups of animals the number of described species is an estimate, usually based on revisions (taxonomic studies) of their subgroups. A reasonably good estimate of the number of described species of metazoan animals is about 1,035,250, of which three quarters are insects.

Because new species are continually being described, we obviously know only a part of the world's fauna. Estimates of the numbers of undescribed species are based on the annual rate of new descriptions. But some groups are much better known than others. There are very few descriptions of new birds and mammals, which indicates that the number of described species is close to the actual number of existing species. On the other hand, the rate of new descriptions of insects is very high. Terry Erwin, an entomologist working on the small beetles living in the canopy of Brazilian rainforests, found that each of the forest types he surveyed harbored large numbers of endemic species (species found nowhere else), and most are undescribed. Extrapolating from his work on beetles, Erwin speculates that there may be as many as 30 million species of insects! Mites are another largely undescribed group of animals; described species may represent only about 20% of the actual fauna. Polychaete annelids are about 60% known, but echinoderms (sea stars and sea urchins) are about 90% known. Most of the undescribed species of echinoderms are believed to live in the deep sea.

Zoologists often use the term *successful* to describe certain groups of animals. Evolutionary success can be measured in a number of ways, but certainly the number of species the group contains and the extent of their geographical distribution are significant and easy criteria for measuring success. By these two standards alone, insects would be a very successful group of animals.

Success is usually attributed to certain adaptive features that evolved in the ancestral members of the group. These features permitted an adaptive radiation of species colonizing many different habitats and filling new niches. The great success of insects might, therefore, be attributed to the evolution of five significant features: a waxy epicuticle, reducing desiccation; wings, enhancing access to food and other resources, as well as helping in evading predators; wing folding at rest, permitting utilization of confined microhabitats; a resistant egg shell, permitting exposure to more extreme environmental conditions; and a development that includes a larva, permitting the juvenile insect to utilize different resources than the adult.

INSECTS

The class Insecta, or Hexapoda, containing more than 750,000 described species, is the largest group of animals; in fact, it is three times larger than all the other animal groups combined. Only a brief, rather superficial treatment of insects is possible here. For more extensive accounts, especially those dealing with the details of morphology of the insect orders, you must refer to textbooks of entomology.*

Insects are distinguished from other arthropods by having **three pairs of legs** and usually **two pairs of wings** carried on the middle, or thoracic, region of the body. In addition, the head typically bears a single **pair of antennae** and a **pair of compound eyes** (Fig. 16–1C). A tracheal system provides for gas exchange, and the gonoducts open at the posterior end of the abdomen.

The success of insects is evidenced by the tremendous number of species and individuals and by their great adaptive radiation. Although they are essentially terrestrial animals and have occupied virtually every environmental niche on land, insects have also invaded the aquatic habitats and are absent only from the subtidal waters of the sea. This success of insects can be attributed to a number of factors, but certainly the evolution of flight endowed these animals with a distinct advantage over other terrestrial invertebrates. Dispersal, escape from predators, and access to food or optimum environmental conditions were all greatly enhanced. The powers of flight also evolved in reptiles, birds, and mammals, but the first flying animals were insects.

Insects are of great ecological significance in the terrestrial environment. Two thirds of all flowering plants depend on insects for pollination. Insects are also of enormous importance for humans. Mosquitoes, lice, fleas, bedbugs, and a host of flies can contribute directly to human misery. More importantly, these and others affect us indirectly as vectors of human diseases or of diseases of domesticated animals: mosquitoes (malaria, elephantiasis, and yellow fever); tsetse fly (sleeping sickness); lice (typhus and relapsing fever); fleas (bubonic plague); and the

housefly (typhoid fever and dysentery). Our domesticated plants are dependent on some insects for pollination but are destroyed by others. Vast sums are expended to control insect pests, which can greatly reduce the high agricultural yields necessary to support large human populations. The overzealous use of pesticides, however, can in turn be hazardous to the environment and to human health.

EXTERNAL MORPHOLOGY

Although the exoskeleton of an arthropod segment is composed of a tergum, a sternum, and two pleura, some parts are more highly sclerotized or are more conspicuous than others. Such thickened areas of cuticle, called **sclerites,** are prominent features of the insect body surface. They are often separated by **sutures,** which reflect ridges on the inner side of the skeleton or lines of thinner exoskeleton. The sclerites and sutures have received detailed attention from insect morphologists, and an extensive anatomical nomenclature has been developed. Only those general features of insect anatomy that provide a basis for comparison with other arthropods will be discussed in the following paragraphs.

Body Regions

The heads of most insects are oriented so that the mouthparts are directed downward (**hypognathus;** Fig. 16–1B). A more specialized, anteriorly directed position (**prognathus**) is found in some predaceous species, such as the carabid and tiger beetles; a posteriorly directed position (**opisthognathus**) is present in the hemipterans and homopterans, which have sucking beaks.

The head skeleton of most insects forms a complete, external capsule surrounding the soft, inner tissues. The more lateral and dorsal surfaces of the head bear one pair of **compound eyes** and one pair of **antennae.** Between the eyes and the antennae are usually three **ocelli** (Fig. 16–1A). Three pairs of appendages contribute to the mouthparts (Fig. 16–1A, B). One pair of **mandibles** is located anteriorly, followed by a pair of **maxillae** and then by the **labium.** Although single, the labium actually represents a fused pair of second maxillae. Anteriorly, the mandible is covered by a shelflike extension of the head, forming an upper lip, or **labrum.** From the floor

---

*This chapter has been designed to meet the requirements of those invertebrate zoology courses that include a brief coverage of insects. This account is not intended to be equivalent to the extensive treatment accorded to the other invertebrate groups.

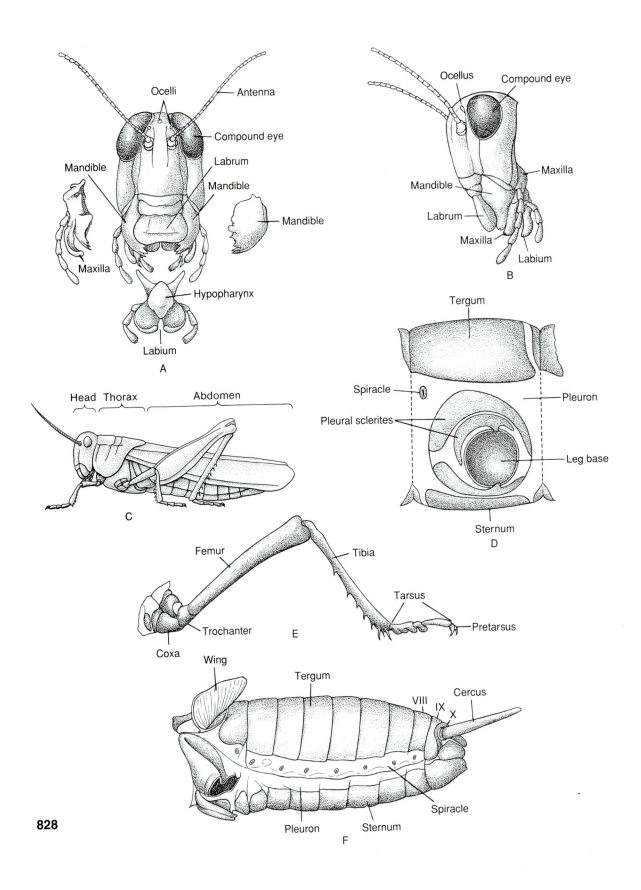

Ocelli — Antenna

Compound eye

Mandible

Labrum

Mandible

Mandible

Maxilla

Hypopharynx

Labium

A

Ocellus — Compound eye

Maxilla

Mandible

Labrum

Maxilla

Labium

B

Head Thorax Abdomen

C

Tergum

Spiracle — Pleuron

Pleural sclerites — Leg base

Sternum

D

Femur — Tibia

Tarsus

Pretarsus

Trochanter

Coxa E

Wing — Tergum

VIII IX X — Cercus

Spiracle

Pleuron Sternum

F

828

of the prebuccal cavity projects a median lobelike process, called the **hypopharynx.** The hypopharynx arises behind the maxillae near the base of the labium. The modifications of the mouthparts associated with different feeding habits will be discussed later.

The **thorax,** which forms the middle region of the insect body, is composed of three segments: a **prothorax,** a **mesothorax,** and a **metathorax.** A pair of legs articulates with the pleura on each of the three segments. (Fig. 16–1D). The thoracic terga of insects are called **nota,** and it is with the notal and pleural processes of the mesothorax and the metathorax that the two pairs of wings articulate. The basal section of the leg articulating with sclerites in the pleural area is the **coxa,** which is followed by a short **trochanter** (Fig. 16–1E). The remaining sections consist of a **femur,** a **tibia,** a **tarsus,** and a **pretarsus.** The tarsus is composed of one to five articles. The pretarsus is represented chiefly by a pair of claws. In other arthropods, such as arachnids, the pretarsus is not usually considered separate from the tarsus. The legs of insects are generally adapted for walking or running, and during their effective strokes, the forelegs pull while the middle and hind legs push. The middle legs step outside of the other two pairs, which reduces interference. One or more pairs may be modified for such functions as grasping prey, jumping, swimming, and digging.

The **abdomen** is composed of 9 to 11 segments plus a telson, but the telson is complete only in the primitive proturans and in embryos (Fig. 16–1F). The only abdominal appendages in the adult are a terminal pair of sensory cerci borne on the eleventh segment. The reproductive structures are thought by some entomologists to represent segmental appendages, but except for the female ovipositors, this is by no means certain. A variety of abdominal appendages serving different functions are present in many insect larvae.

INSECTS WINGS AND FLIGHT

Wings are characteristic features of insects, but a wingless condition occurs in a number of groups. In some the absence of wings is obviously secondary. For example, ants and termites have wings only at certain periods of the life cycle; workers always lack wings. Some parasitic insect orders, such as the lice and fleas, have lost the wings completely. On the other hand, it is fairly certain that the insects in the orders Protura, Thysanura, Collembola, and a few others arose from ancestral wingless insects. Those groups in which the wingless condition is considered primary are called apterygotes and are believed to represent the most primitive members of the class. Winged insects and those that are secondarily wingless are referred to as pterygotes. Some entomologists consider the wingless apterygotes to be related to insects but not insects themselves.

The sequence of events in the evolutionary development of insect wings is unknown. The earliest known fossil insect is a bristletail, a wingless insect, from the early Devonian period. Winged insects appear later. No intermediate types have yet been discovered. The most widely accepted theory of wing origin is that wings were originally flat, lateral flanges of the notum, which enabled the insect to alight right side up when jumping. These flanges then gradually enlarged into winglike structures, making gliding possible. The last step was the development of hinges, enabling wings to move. The earliest functional wing was thought to be a fan-shaped membrane with trusslike supporting veins.

Because wings are evaginations, or folds, of the integument, they are composed of two sheets of cuticle. The two cuticular membranes are separated by tubular thickenings called veins, which form effective supporting skeletal rods for the wing. Wing veins open into the body and contain circulating blood and commonly tracheae and sensory nerve branches. The blood is important in maintaining the proper water content of the cuticle.

The wing venation of the more primitive insects is netlike, but there has been a general tendency in the evolution of wings toward reduction to a few longitudinal veins and cross veins, thus giving a stronger support system to the wing. The arrangement of veins in a wing is very specific in certain genera and fami-

◄ **FIGURE 16–1** External morphology. **A,** Anterior surface of the head of a grasshopper. **B,** Lateral view of the head of a grasshopper. **C,** Lateral view of the body. **D,** Lateral view of a wingless thoracic segment. **E,** Leg of a grasshopper. **F,** Lateral view of the abdomen of a male cricket. *(A and B, After Snodgrass from Ross. D–E, From Snodgrass.)*

lies of insects and provides a useful tool for systematists; the principal veins and their branches are all named and numbered.

Primitively, wings are held outstretched, as in dragonflies. The evolution of sclerites in the wing base, which permits many insects to fold the wings over the abdomen and thus keep them out of the way when at rest, was an important event in the evolution of the class. The ancestors of many modern orders were then able to radiate into microhabitats, beneath bark and stones and in soil, dung, and wood, where outstretched wings would have been a serious impediment. Wing folding was probably accompanied by reduction in the body size of many groups.

There is great variation in the wings of insects. Many of these variations represent modifications that accommodate the demands of flight characteristic to the particular group of insects. Primitively, as in damselflies, roaches, and termites, the two pairs of wings beat independently of each other, but this requires that the hind wings operate in the air turbulence created by the forewings. Thus, in many insects the two wings on each side are coupled by interlocking devices or by simple overlapping so that the wings operate together. Only the second pair of wings is used for flight in beetles; the front pair has been adapted as hard protective plates, called elytra (Fig. 16–14).

Each wing articulates with the edge of the tergum, but its inner end rests on a dorsal pleural process, which acts as a fulcrum (Fig. 16–2). The wing is thus analogous to a seesaw off center. Upward movement of the wing results indirectly from the contraction of vertical muscles within the thorax, depressing the tergum. Downward movement of the wings is produced directly, by contraction of muscles attached to the wing base (dragonflies and roaches), indirectly by the contraction of longitudinal muscles raising the tergum (bees, wasps, and flies), or by both direct and indirect muscles (grasshoppers and beetles).

Up and down movement alone is not sufficient for flight. The wings must at the same time be moved forward and backward. A complete cycle of a single wing beat describes an ellipse (grasshoppers) or a figure eight (bees and flies), during which the wings are held at different angles to provide both lift and forward thrust (Fig. 16–2C, E). The wings of many insects describe a "clap and peel" motion in the beat cycle. At the top of the upstroke the wings touch, or clap, and then peel away from each other beginning

with the front margins (Fig. 16–2D). Air rushes between them and forms a vortex around each wing.

In addition to the angle at which the wing is held, insects can also achieve lift by raising and lowering wing veins, thereby changing the wing shape, or contour. Some insects can reduce the wing surface area by changing the degree of fore- and hindwing overlap. All of these changes—angle, contour, and surface area—can take place in the course of one wing beat cycle and increase net upward lift. They reflect, in part, the ability of the wing to deform, rather like a sail.

The raising or lowering of the wings resulting from the contraction of one set of flight muscles stretches the antagonistic muscles, which then also contract. Insect wing beat thus involves the alternate contraction of these antagonistic elastic systems. The beat frequency varies greatly—4 to 20 bps (beats per second) in butterflies and grasshoppers; 190 bps in the honeybee and housefly; and 1000 bps in certain midges. At low frequencies (30 bps or fewer) there is usually one nerve impulse to one muscle contraction. At higher frequencies, however, the contraction is myogenic, originating from the stretching caused by the contraction of the antagonistic muscles, and there are a number of beats, or oscillations, associated with each nerve impulse.

Rapid contraction is facilitated by the nature of the muscle insertion. A very slight decrease in muscle length during contraction can bring about a large movement of the wing (as a seesaw with the fulcrum near one end). The elastic nature of the thoracic skeleton and the joints of wing articulation also contribute to the beat motion. If, for example, the stable position is down, as in grasshoppers, elastic forces to return to the down position are set up when the wings are raised, much like stretching a rubber band. Some flies and beetles have both stable down and stable up positions, and at some point in the movement of the wing from one position to the other, the opposite elastic forces take over. This arrangement is called a click mechanism.

Flying ability varies greatly. Many butterflies and damselflies have a relatively slow wing beat and limited maneuverability. At the other extreme, some flies, bees, and moths can hover and dart. From the standpoint of maneuverability, a housefly can outperform any bird. Not only can a housefly fly a rapid straight course and hover, it can fly upside down and turn in the distance of one body length. The fastest

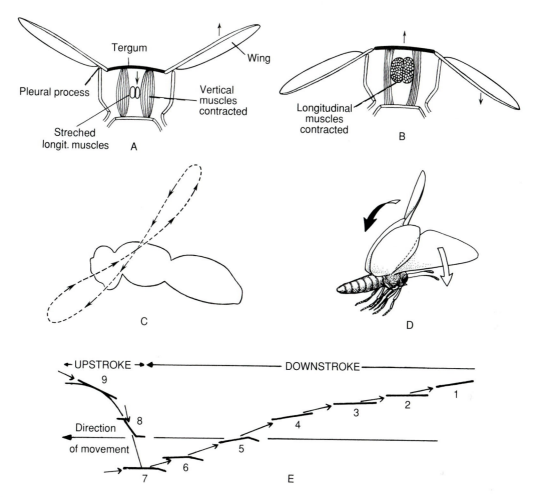

FIGURE 16–2 Diagrams showing relationship of wings to tergum and pleura and the mechanism of the basic wing strokes in an insect. **A,** Upstroke resulting from the depression of the tergum through the contraction of vertical muscles. **B,** Downstroke resulting from the arching of the tergum through the contraction of longitudinal muscles. **C,** An insect in flight, showing the figure eight described by the wing during an upstroke and a downstroke. **D,** Wing peeling following a "clap" at the top of the upstroke. **E,** Changes in the position of the forewing of a grasshopper during the course of a single beat. Short arrows indicate direction of wind flowing over wing, and numbers indicate consecutive wing positions. *(A and B, After Ross, H. H. 1965. A Textbook of Entomology. 3rd Edition. John Wiley and Sons, New York. C, After Magnan from Fox and Fox. E, After Jensen from Chapman. 1971. The Insects: Structure and Function. University of London Press Ltd., London. Courtesy of the American Elsevier Publishing Co., Inc.)*

flying insects are hummingbird moths and botflies, which have been clocked at 25 mph. Honeybees can cruise at 15 mph. Gliding, an important form of flight in birds, occurs in only a few large insects.

There is no single flight control center in the insect nervous system, but the eyes and sensory receptors on the antennae, head, wings and other parts of the body provide continuous feedback information for flight control. Horizontal stability is maintained in part by a dorsal light reaction: the insect keeps the dorsal ommatidia of the eyes under maximum illumination from above. Deviation because of rolling is corrected by slight changes in wing position to bring the dorsal part of the eyes back to maximum illumination.

Members of the order Diptera (flies, gnats, and mosquitoes) have the second pair of wings reduced to knobs, called **halteres** (Fig. 16–20R). The halteres beat with the same frequency as the forewings and function as gyroscopes to offset flight instability. Receptors on the haltere base detect deviating forces, such as tendencies to pitch, roll, and yaw, and from this information corrections in wing position are made.

Flight speed is probably determined by air flow over receptors on the antennae and movement of objects from front to back across the eyes. Flight is inhibited by contact of the tarsi with a solid surface.

Insect flight muscles are very powerful. The fibrils are relatively large, and the mitochondria are huge (about half the size of a human red blood cell), reflecting the high respiratory rate of these cells. Insects are the only ectothermic fliers, and a low body temperature and a correspondingly low metabolic rate impose limitations on mobility. On a cool morning many flying insects literally warm up before flight. They remain stationary on a tree trunk or some other location and move the wings up and down until sufficient internal heat is generated to permit the stroke rate necessary for flight, or more commonly, contract the flight muscles while in a decoupled state or "neutral gear."

INTERNAL ANATOMY AND PHYSIOLOGY

Nutrition

Insects have adapted to all types of diets. The mouthparts may be highly modified, but the modifications are associated less with diet than with the method by which food is obtained. Primitive mouthparts are adapted for chewing, and it is mouthparts of this type that are described and illustrated in the beginning section on external structure (Fig. 16–1A, B). The mandibles are heavy and capable of cutting, tearing, and crushing, and the maxillae and labium function in food handling. The hypopharynx aids in swallowing. Insects with chewing mouthparts include the primitive apterygotes, dragonflies, crickets, grasshoppers, beetles, and many others. The larvae of such insects as moths and butterflies also have chewing mouthparts, although the adult's mouthparts are highly modified. The diets of chewing insects may be herbivorous or carnivorous, but some diets are highly specific.

The specialization of insect mouthparts has been primarily in modifications for piercing and sucking. Adaptations for the same feeding habits are not uniform, however, because a sucking or piercing feeding habit evolved independently in different insect orders.

Moreover, the mouthparts may be adapted for more than one function: chewing and sucking, cutting and sucking, or piercing and sucking.

The mouthparts of moths and butterflies are adapted for sucking liquid food, such as nectar, from flowers. A part (the galea) of each of the two greatly modified maxillae forms a long tube through which food is sucked (Fig. 16–3A). When the insect is not feeding, the tube is coiled. The other mouthparts are absent or are vestigial.

Piercing mouthparts are characteristic of herbivorous insects, such as aphids and leafhoppers, which feed on plant juices. Some predaceous insects, such as assassin bugs and mosquitoes, which utilize the body fluids of other animals as food, also have piercing mouthparts. In all these insects the mouthparts are elongated and are organized in various ways to form a beak. For example, the beak of the plant-feeding and predaceous bugs (Hemiptera and Homoptera) consists of a stylet composed of the mandibles and the maxillae that lie in a groove on the heavier labium. The stylet contains one lumen for the outward passage of salivary secretions and another for sucking in fluids (Fig. 16–3B–F). Other parts of the beak do not penetrate.

Bees and wasps have mouthparts adapted for both chewing and sucking. A bee, for example, gathers nectar by the elongated maxillae and the labium. Pollen and wax are handled by the labrum and mandibles, which have retained the chewing form.

In biting flies, such as horseflies, the knifelike mandibles produce a wound. Blood is collected from the wound by a spongelike labium and is conveyed to the mouth by a tube formed from the hypopharynx and epipharynx (the inner side of the labrum) (Fig. 16–4A). Some predatory flies and hemipterans inject salivary secretions into the prey and suck up the digested tissues. Certain nonbiting flies, such as houseflies, use the spongelike labium alone for obtaining food, the mandibles and maxillae being reduced. Such insects are not restricted to liquid foods. Saliva can be exuded through the labium onto the solid material, and the resulting fluid then can be sucked back into the mouth (Fig. 16–4B).

Food taken into the mouth passes into the foregut, which is commonly subdivided into an anterior **pharynx,** an **esophagus,** a **crop,** and a narrower **proventriculus** (Fig. 16–5A). The pharynx is highly modified as a pump in sucking insects. The crop, when present, is a storage chamber. The proventricu-

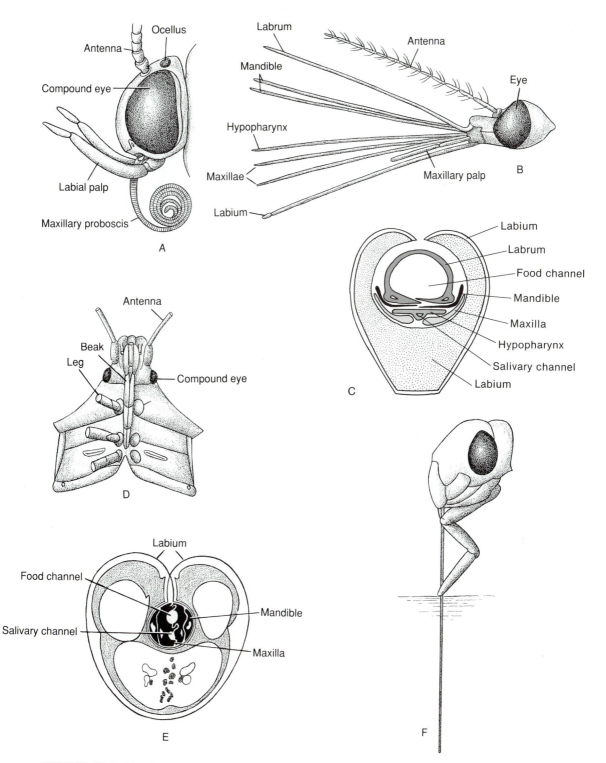

FIGURE 16–3 Mouthparts of sucking and piercing insects. **A,** Lateral view of the head of a moth, a sucking insect **B,** Lateral view of the head of a mosquito, showing separated mouthparts. **C,** Cross section of mouthparts of a mosquito in their normal functional position. **D,** Ventral view of anterior half of a hemipteran, showing beak. **E,** Cross section through a hemipteran beak, showing the food and the salivary channels enclosed within the stylet-like maxillae and the mandibles. **F,** A hemipteran penetrating plant tissue with its stylets. *(A, After Snodgrass, B and C, After Waldbauer from Ross. D, After Hickmann. E, After Poisson. F, After Kullenberg.)*

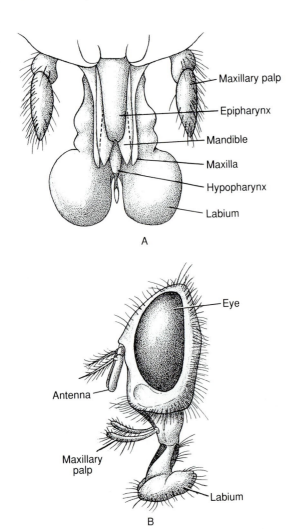

FIGURE 16–4 A, Mouthparts of false blackfly adapted for cutting and sponging. **B,** Lateral view of the head and sponging mouthparts of a housefly. **C,** Scanning electronmicrograph of the ventral surface of the spongelike labium of a face fly. The letter P is in the mouth. *(A, After Ross, H. H. 1965. A Textbook of Entomology. 3rd Edition. John Wiley and Sons, New York. B, After Snodgrass. C, From Elzinga, R. J. 1987. Fundamentals of Entomology. 3rd Edition. Prentice-Hall, Englewood Cliffs, NJ.)*

lus is variable in structure and function. In insects that eat solid food, the proventriculus is usually modified as a gizzard and bears teeth or hard protuberances for triturating food (Fig. 16–5C). In sucking insects, on the other hand, the proventriculus consists only of a simple valve opening into the midgut (Fig. 16–5D). Between these two extremes are some beetles and honeybees, in which the proventriculus acts as a regulatory valve permitting fluids but not solid food to enter the midgut. This function is particularly important in the separation of pollen from nectar in bees.

Most insects possess a pair of salivary, or labial, glands, that lie below the midgut and have a common duct opening into the buccal cavity (Fig. 16–5B). Mandibular glands, in addition to salivary glands, are functional in the apterygotes and in a few groups of pterygotes. The function of salivary glands varies and has not been determined in all insects. The glands usually secrete saliva, which moistens the mouthparts and may be a solvent for the food. The salivary glands may also produce digestive enzymes, which are mixed into the food mass before it is swallowed. In some lepidopterans (moths) and hymenopterans (bees and wasps), the glands secrete silk used to make the pupal cells. Other special secretions of the salivary glands in various insects include mucoid materials, a pectinase that hydrolyzes the pectin of cell walls, venomous spreading agents, anticoagulants and agglutinins, and an antigen that produces the typical mosquito-bite reaction in humans.

A stomodeal valve separates the foregut from the midgut. The insect midgut, which is also called the

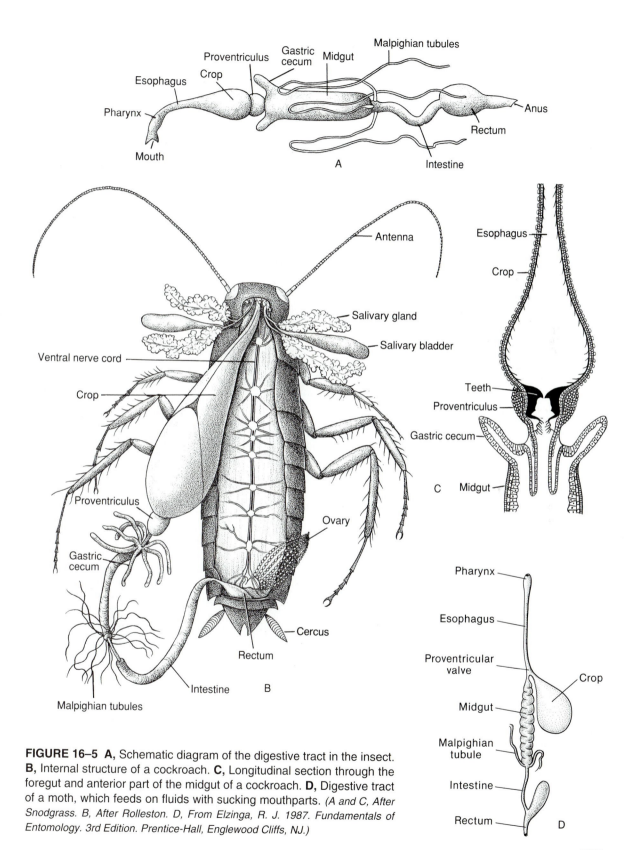

FIGURE 16–5 **A,** Schematic diagram of the digestive tract in the insect. **B,** Internal structure of a cockroach. **C,** Longitudinal section through the foregut and anterior part of the midgut of a cockroach. **D,** Digestive tract of a moth, which feeds on fluids with sucking mouthparts. *(A and C, After Snodgrass. B, After Rolleston. D, From Elzinga, R. J. 1987. Fundamentals of Entomology. 3rd Edition. Prentice-Hall, Englewood Cliffs, NJ.)*

ventriculus, or stomach, is usually tubular and, as in other arthropods, is the principal site of enzyme production, digestion, and absorption. A characteristic feature of the midgut of most insects, other than hemipterans, is the presence of a **peritrophic membrane.** This membrane, composed of a very thin layer of protein and chitin, is periodically delaminated by the midgut lining (grasshoppers) or is continuously secreted by the epithelial cells near the valve at the end of the foregut (flies). The membrane forms a covering around the food mass moving through the midgut (Fig. 16–6). The peritrophic membrane protects the delicate midgut walls from abrasion by the food mass and more importantly conserves enzymes by dividing the gut lumen into two compartments. The membrane is permeable to some enzymes and the products of digestion. The initial products of digestion pass through the membrane, where they are attacked by a second order of enzymes that are restricted to the space between the gut wall and peritrophic membrane.

Most insects possess outpocketings of the midgut called **gastric ceca,** commonly located at the anterior end of the midgut (Fig. 16–5). The gastric ceca are the final site of digestion and, along with the anterior end of the gut, are the principal sites of food absorption. Water is also absorbed here, although some water enters at the posterior end of the midgut (Fig. 16–6). Fluid commonly circulates through the midgut via the following pathway: posterior within the peritrophic membrane, anterior outside of the membrane, and then into and out of the gastric ceca.

There are many exceptions to the generalizations described above. In grasshoppers, roaches, and beetles the crop is an important site of digestion for enzymes that pass forward from the midgut.

The hindgut, or proctodeum, consists of an anterior intestine and a posterior rectum, both of which are lined by cuticle (Fig. 16–6). The hindgut functions in the egestion of waste and in water and salt balance. In most insects, **rectal pads,** or glands, occur in the epithelium. These organs are the principal sites

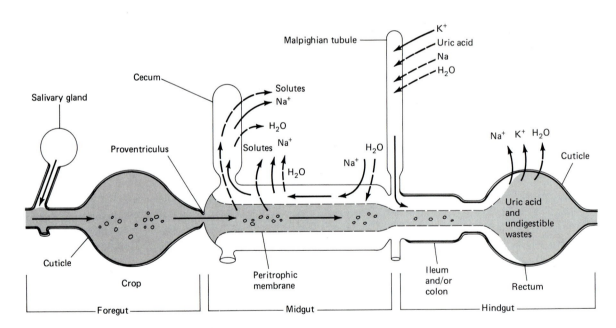

FIGURE 16–6 Diagram of the digestive tract of an insect showing the passage of food (small circles) through the gut, the absorption of food products in the ceca, and the secretion of wastes in the malpighian tubule. Active transport of salts (solid arrows) leads to passive diffusion of water and other substances (dashed arrows). *(Modified from Berridge, M. J. 1970. A structural analysis of intestinal absorption. In Neville, A. C. (Ed.): Insect Ultrastructure. Sympos. Roy. Ent. Soc. 5:135–151; and from Evans, H. E. 1984. Insect Biology: A Textbook of Entomology. Addison-Wesley Publishing Co., Reading, MA. p. 85.)*

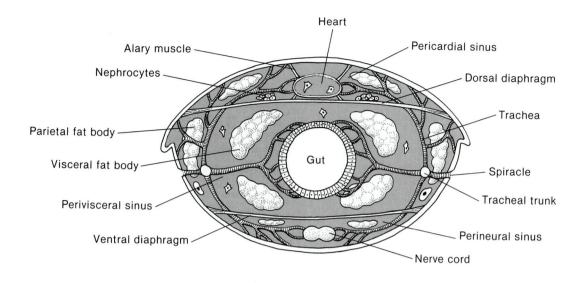

FIGURE 16–7 Cross section of the abdomen of an insect. *(Modified from Davies, R. G. 1988. Outlines of Entomology. 7th Edition. Chapman and Hall, London.)*

of water reabsorption. Digestion of cellulose by some termites, certain wood-eating roaches, and cockroaches is made possible by the action of enzymes produced by protozoa or bacteria (cockroaches) that inhabit the hindgut of these insects (p. 34). Acetic acid formed by the breakdown of wood is actively absorbed by the hindgut epithelium in these insects. Higher termites secrete cellulase but possess nitrogen-fixing gut bacteria.

Fat bodies are present in various places within the hemocoel, depending on the species, and function somewhat like the chlorogogen tissue in annelids and the liver of vertebrates. This tissue is a site of synthesis and long- and short-term storage. Glycogen reserves in the fat body can be rapidly mobilized by hormones to release sugar into the blood. Many insects that do not feed as adults rely on the fats, proteins, and glycogen stored in the fat body during immaturity (Fig. 16–7).

Internal Transport

The heart of the insect lies within a dorsal pericardial sinus that is separated by a perforated, dorsal diaphragm from the perivisceral sinus surrounding the gut (Fig. 16–7). There may also be a ventral diaphragm separating the nerve cord from the perivisceral sinus. The heart is tubular and in most species,

extends through the first nine abdominal segments. In each segment, a pair of alary muscles extends laterally from the heart to the body wall within the double-layered dorsal diaphragm. Contractions of the alary muscles cause the heart to expand and blood to pass through the heart ostia. This filling phase is followed by a wave of contractions of the myoepithelial cells of the heart wall, and blood is driven forward. The heart is closed posteriorly but anteriorly is continuous with an aorta that runs into the head. Blood normally flows from posterior to anterior in the heart and from anterior to posterior within the perivisceral and perineural sinus. Perforations in the dorsal diaphragm permit return of the blood into the pericardial sinus. Blood flow may be augmented by accessory pulsating structures in the head, thorax, legs, or wings and by contractions of the dorsal diaphragm. In many rapid-flying insects there is an additional thoracic "heart," which draws blood through the wings and discharges it into the aorta. Blood flow is also facilitated by various body movements, such as the ventilating abdominal contractions.

In addition to effecting blood transport, localized elevations of blood pressure may serve a variety of functions, such as the jettisoning of wings from termites, the unrolling of the proboscis in Lepidoptera, the eversion of various organs, the egestion of fecal pellets, and the swelling of the body during molting and hatching.

The blood of insects is usually colorless or green with a number of types of hemocytes, some of which are phagocytic. Some insects possess clotting agents in the blood, but most species close wounds with a plug of cells. Because tissue gas exchange is handled directly by the tracheal system, the blood plays a very minor role in gas transport. Most animals rely on inorganic ions, such as sodium and chloride ions, as osmotic regulators of the body fluid. In insects, organic molecules, especially free amino acids, are more important in this function. Hemolymph also contains high concentrations of dissolved uric acid, organic phosphates, and a nonreducing sugar—trehalose.

Many insects of temperate regions can survive freezing temperatures by accummulating compounds, such as glycerol, sorbitol, and trehalose, that act as antifreeze agents. Some insects are able to supercool the blood and cellular fluids to $-30°C$ without freezing; others exhibit controlled freezing, permitting ice crystals to form only in extracellular spaces. Special proteins may be produced that act as nuclei in the formation of the ice crystals.

Gas Exchange

Gas exchange in insects occurs through a system of tracheae, which has been more extensively studied than that of any other arthropods. A pair of spiracles is usually located above the second and third pairs of legs or only above the last pair. The first seven or eight abdominal segments possess a spiracle on each lateral surface (Figs. 16–1D, F; 16–8B). Thus, there is a maximum of ten pairs of spiracles. Tracheal spiracles in their simplest form are merely holes in the integument, as in some Apterygota. In most insects, however, the spiracles open into a pit, or atrium, from which the tracheae arise. The spiracle is generally provided with a closing mechanism, and in many terrestrial insects, the atrium contains filtering devices (Fig. 16–8A). The closing mechanism of the spiracle reduces water loss, and the filtering structures prevent the entrance of dust and parasites as well as reducing water loss.

The pattern of the internal tracheal system is variable, but a pair of longitudinal trunks with cross connections form the ground plan of most species (Fig. 16–8). The tracheae are supported by thickened spiral rings of cuticle, the **taenidia** (Fig. 16–8A). The rings resist compression (i.e., prevent collapse) but permit stretching of the tube. The epicuticle of the tracheae lacks the waxy component typical of the external skeleton. The tracheae themselves are seldom uniform in size but widen in various places, forming internal air sacs (Fig. 16–8D), especially in insects capable of rapid flight. The air sacs provide for both oxygen storage and ventilation.

The smallest subdivisions of the tracheae, the **tracheoles,** are generally less than 1 μm in diameter. These fine, cuticular tubes are often given off in clusters from the tracheae and then further branch into a fine network over the tissue cells (Fig. 16–8C). A number of tracheoles may be formed by a single tracheole cell. In the flight muscles of some insects, the tracheoles even push into the fibrils. The tracheole cuticle is not shed during molting, as is that of the tracheae, and after molting, new tracheae are joined to old tracheoles.

Exchange through the tracheae has been thought to occur primarily by diffusion; however, the spiracles are closed most of the time, and exchange is probably a result of both diffusion and ventilation. Recent studies have demonstrated that the spiracles open very briefly (200 ms) and not all at once in response to a localized reduction in hemocoelic pressure. The spiracle is literally sucked open, and a "gulp" of air is taken in. The pressure drop results from intersegmental muscle contraction and is under the control of the nervous system, which in turn may be regulated by the oxygen/carbon dioxide tension of the blood. More spiracles are therefore open during flight than when the insect is at rest. Because an insect must balance oxygen need against the danger of water loss, the number and duration of open spiracles are generally held to the lowest possible level.

Ventilating pressure gradients result from body movements, largely abdominal, which bring about compression of the air sacs and the longitudinal extension and contraction of trachea. Ventilation is facilitated by the sequence in which certain spiracles are opened and closed.

At the tissue/tracheole level gases are exchanged by diffusion down a concentration gradient. Tracheoles are permeable to liquids, and in most insects their tips are filled with fluid. This fluid is believed to be involved in the final transport of oxygen. The volume of fluid has been shown to increase and decrease, depending on the osmotic pressure in the surrounding tissue.

Some very small insects that live in moist surroundings, such as collembolans and proturans, lack

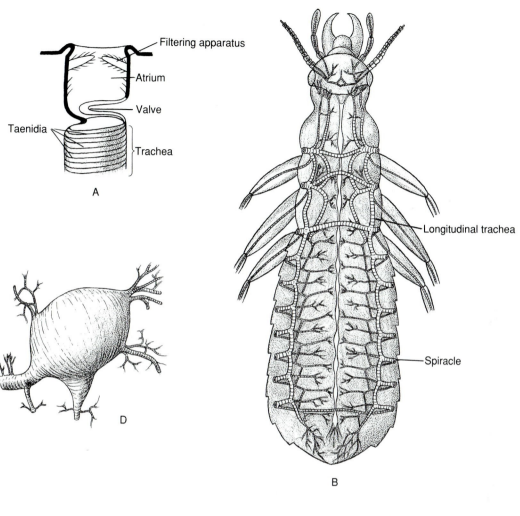

A

B

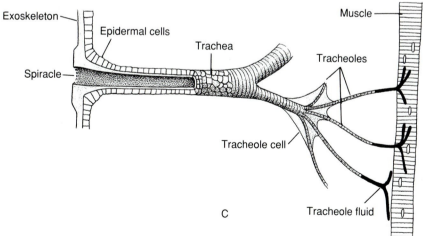

C

FIGURE 16–8 A, A spiracle with atrium, filtering apparatus, and valve. **B,** A tracheal system of an insect. **C,** Diagram showing relationship of spiracle and tracheoles to tracheae. **D,** An air sac. *(B and C, After Ross, H. H. 1965. A Textbook of Entomology. 3rd Edition. John Wiley and Sons, New York. A and D, After Snodgrass.)*

tracheae, and gas exchange occurs over the general body surface. Some aquatic immature insects also lack tracheae, especially during the early stages. Tracheae are usually present, however, in aquatic immatures and always in adult insects that live in water. The adults merely utilize air from air bubbles or films held against the body surface by special "unwettable" (hydrofuge) hairs, but the nymphs and larvae of certain groups may possess special adaptations for gas exchange in water. Damselfly and mayfly nymphs possess abdominal gills. The gills are provided with closed tracheae, and gas exchange occurs across the gill surface between the water and tracheae. Dragonfly nymphs pump water in and out of the rectum, which contains gills supplied with tracheae. Usually, gas exchange in aquatic immature insects occurs across the general integument between the tracheae and the water. Some larvae, such as those of mosquitoes, have a few functional spiracles associated with one or more breathing tubes. The larva rises to the surface periodically and obtains air through the tube.

Excretion and Water Balance

The chief organs of excretion in insects are the malpighian tubules, although they are absent in some forms (collembolans and aphids). The malpighian tubules lie more or less free in the hemocoel, with the proximal end attached usually at the junction of the midgut and the hindgut (Figs. 16–5; 16–6). The number of tubules varies from 2 (coccids) to about 250 (grasshopper *Schistocerca*). The tubule lumen is lined by large, cuboidal, epithelial cells. The outer layer of the tubule wall, which is in contact with the hemolymph, is composed of elastic connective tissue and muscle fibers. The tubules are capable of peristalsis and can undergo movement within the hemocoel.

The contents of the malpighian tubules are derived from secretion rather than from filtration. Water enters passively as a consequence of the active transport of ions, particularly potassium, into the lumen (Fig. 16–6). Uric acid, formed in the tissues and passed into the hemolymph, is then taken up by the malpighian tubule cells along with amino acids and salts. Together these substances are secreted into the lumen of the tubule. In many insects the lower pH that develops in the contents toward the junction with the hindgut causes the uric acid to precipitate. Reabsorption of water, some of the salts, and other nutritive substances occurs in the course of elimina-

tion. Some reabsorption of water and inorganic ions may take place in the proximal parts of the tubules themselves, these substances being returned to the hemolymph, but in most insects, the rectal epithelium transfers these substances back into the hemolymph. The hindgut may also be the site of uric acid precipitation.

Not all the waste products are removed by the malpighian tubules; some are stored in specialized cells. Pericardial cells, or **nephrocytes,** typically located on or near the heart, pick up particulate or complex waste for intracellular degradation. Cells of the fat bodies can also be the site of uric acid deposition in some insects, such as collembolans.

Of the terrestrial arthropods, insects are among the best adapted for the prevention of water loss. The epicuticle is impregnated with waxy compounds, which reduce surface evaporation, and spiracle closure reduces evaporation from the tracheal system. The excretion of uric acid reduces loss of water due to protein metabolism. And the reabsorption of water by the rectum further conserves water that would be lost through excretion and egestion. Insects are one of the few groups of invertebrates that can produce a hyperosmotic urine.

As might be expected, aquatic insects excrete ammonia rather than uric acid, and salts are conserved by the hindgut.

Nervous System and Sense Organs

The insect nervous system is basically like that of other arthropods. The brain is composed of a protocerebrum with eyes, deutocerebrum with antennae, and a tritocerebrum. The ventral nerve cord forms a chain of median segmental ganglia (Fig. 16–5B). As in other arthropods the ventral segmental ganglia, both thoracic and abdominal, are often fused. The greatest number of free ganglia is three in the thorax and eight in the abdomen (apterygotes). The subesophageal ganglion is always composed of three pairs of fused ganglia, which control the mouthparts, the salivary glands, and some of the cervical muscles. Giant fibers generally occur throughout the class and have been well studied in the cockroach.

Associated with a hypocerebral ganglion lying over the foregut and just beneath the brain are two pairs of glandular bodies: the **corpora cardiaca** and the **corpora allata.** These two bodies, together with the **prothoracic glands** and certain neurosecretory cells in the protocerebrum, are the principal endocrine centers of insects. Their role in controlling growth and metamorphosis will be described in the section on development. Other endocrine functions include regulation of water reabsorption, heartbeat, and certain metabolic processes.

Sensilla, sense organs other than eyes and ocelli, are scattered over the body but are especially numerous on the appendages. Most are believed to be derived from simple setae, but many have become modified as bristles, pegs, scales, domes, plates, and so on (Figs. 16–9; 16–10). They exhibit a range of receptor functions, as described on page 607.

Auditory receptors that resemble drumheads, called **tympanic organs,** are a specialized type of chordotonal organ. Chordotonal organs have receptors that are stretched between two points on the inner side of the integument and respond to tension changes. Tympanic organs are found in grasshoppers, crickets, and cicadas, which also have sound-producing organs. Tympanic organs develop from the fusion of parts of a tracheal dilation and the body wall. The sensory receptors are attached to the tympanum, and an air sac beneath the tympanum permits vibrations that excite the attached receptors.

The photoreceptors are the ocelli and the compound eyes. Ocelli are absent in many adult insects, but when present, there are usually three, found on the anterior dorsal surface of the head (Fig. 16–1A). The photoreceptor cells on each ocellus are organized somewhat like those of a single ommatidium of a compound eye. Ocelli can detect changes in light intensity and may be very sensitive to low intensities. They function in orientation and in some way appear to have a general stimulatory effect on sensitivity, enhancing the reception of stimuli by other sensory structures. The number of facets in the compound eyes is greatest in flying insects that depend on vision for feeding, and the facets are larger in nocturnal than in diurnal insects.

Reproduction and Development

Reproduction

The typical female reproductive system consists of two ovaries, one on each side, and two lateral oviducts (Fig. 16–11). Each ovary consists of a group of tubules, the ovarioles. The paired oviducts usually unite to form a common oviduct, which leads into a vagina. The vagina in turn opens onto the ventral sur-

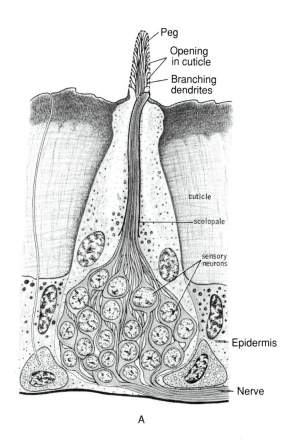

A

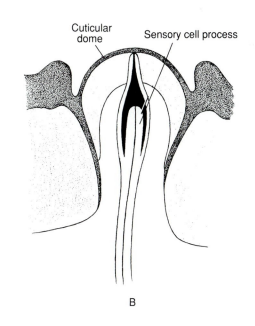

B

FIGURE 16–9 A, A chemosensory peg organ from the antenna of a grasshopper. **B,** A campaniform sense organ. *(A, After Slifer et al. B, After Snodgrass.)*

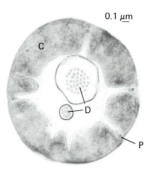

FIGURE 16–10 Cross section of an olfactory hair from the antenna of a male moth. The thick cuticular wall (C) is perforated (P) and contains two dendrites (D). *(From Steinbrecht, R. A. 1984. Arthropoda: Chemo-, hygro-, and thermoreceptors. In Bereiter-Hahn, J., Matoltsy, A. G., and Richards, K. S. Biology of the Integument. 1. Invertebrates. Springer-Verlag, Berlin. p. 528.)*

face behind the seventh, eighth, or ninth segment. Diverticula of the common oviduct or vagina include a spermatheca (seminal receptacle) and paired accessory glands. Some Lepidoptera have a separate copulatory gonopore located behind the vaginal opening.

The copulatory gonopore leads to a bursa copulatrix that receives the spermatophores. Sperm reach the spermatheca by a special duct from the bursa to the vagina.

The male reproductive system includes a pair of testes, a pair of lateral ducts, and a median duct opening through a ventral penis, called the aedeagus, associated with the eighth segment (Fig. 16–12A, B). Each testis consists of a group of sperm tubes containing spermatozoa in various stages of development. These tubes empty into a lateral duct, the vas deferens, which joins the duct from the other side to form a common ejaculatory duct. A section of each vas deferens is usually enlarged into a seminal vesicle, where sperm are stored. Accessory glands, which secrete seminal fluid, are commonly present as pouches from the upper end of the ejaculatory duct.

In copulation the often extensible, or eversible, penis of the male is inserted into the female vagina or bursa copulatrix. The males of many orders—dragonflies, true flies, butterflies, and moths, to name a few—possess clasping organs to hold the female abdomen (Fig. 16–12C). The clasping organs are derived from parts of the terminal segments and vary greatly in structure.

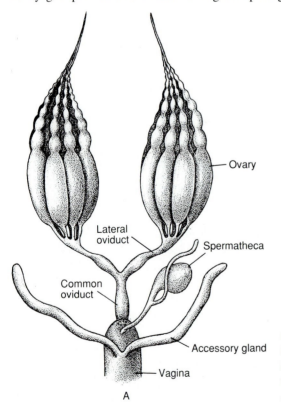

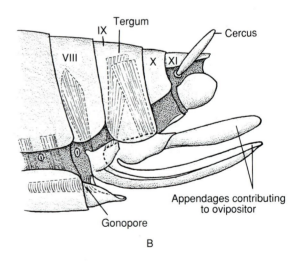

FIGURE 16–11 A, Reproductive system in a female insect. **B,** Lateral view of the posterior end of the abdomen, showing reproductive opening and appendages, forming ovipositor. (The roman numerals refer to segment numbers) *(Both after Snodgrass.)*

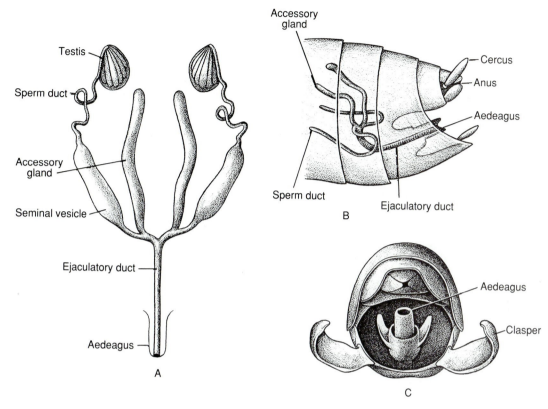

FIGURE 16–12 Reproductive system in a male insect. **A,** General plan of the system. **B,** Lateral view of the posterior end of the abdomen, showing reproductive opening and other structures. **C,** Posterior view of the abdomen, showing aedeagus and claspers. *(All after Snodgrass.)*

Although some insects, such as honeybees, do not have spermatophores, most insects use spermatophores to transfer sperm. Among a few primitive, wingless insects, such as thysanurans and collembolans, the spermatophore is deposited on the ground and then taken up by the female. In most insects, however, the spermatophores are deposited directly into the female vagina or bursa copulatrix at copulation. Following deposition, sperm are released from the spermatophores and are soon found within the spermatheca, where they are stored until the eggs are laid. Fertilization occurs as the eggs pass through the oviduct at the time of egg deposition. At each mating a large number of sperm are transferred to the spermatheca of the female. This number is sufficient for fertilization of more than one batch of eggs. Many insects mate only once in their lifetime, and none mate more than a few times.

When the eggs reach the oviduct, they are already surrounded by a shell-like membrane (**chorion**) secreted by ovarian follicle cells. This shell may be up to seven layers thick and is perforated by one or more **micropyles.** The sperm enter through these minute openings. The evolution of a protective encasement for the development of the egg in a desiccating terrestrial environment is another factor that has contributed to the success of insects. Egg deposition in the majority of insects is through an ovipositor, derived from parts of the eighth or ninth segment (Fig. 16–11B). The site of egg laying varies tremendously and in large part depends on the habitat and life-style of the adult. Adhesive materials for attaching the eggs to the substratum or to each other are produced by the accessory glands.

An interesting type of egg deposition is that associated with gall formation. The females of gall wasps (hymenopterans) and gall gnats (dipterans) deposit their eggs in plant tissues. The plant tissue surrounding the eggs is induced to undergo abnormal growth and forms a gall, which has a shape characteristic of the insect producing it. The gall forms a protective chamber for the developing eggs, larvae, and often even the pupae. The larvae feed on the gall tissues. The gall-inducing agent is a substance secreted by the female when she deposits the egg and by another substance produced later by the larva. In North America alone there are about 2000 species of gall-inducing insects.

Development

Superficial cleavage is characteristic of most insects. Insect young vary in the degree of development at hatching. Young apterygotes are like the adults except in size and sexual maturity. Newly hatched grasshoppers, cockroaches, stoneflies, leaf hoppers, and bugs (hemipterans) resemble the adults, except

that the wings and reproductive organs are undeveloped. The wings of early nymphs are merely external pads, which begin to look like wings only at the preadult molt (Fig. 16–13). The adult form is reached gradually with successive molts. This type of development is called **gradual,** or **incomplete, metamorphosis (hemimetabolous development);** all the immature stages from hatching to the adult are termed **nymphs** or, when aquatic, **naiads.**

In many insects, including bees, wasps, flies, and beetles, the wing rudiments develop internally; the wings seem to appear suddenly in the adults. This type of development is a **complete metamorphosis (holometabolous development)** and consists of three distinct stages (Fig. 16–14). The newly hatched **larval stage,** which has no wings, is the caterpillar of

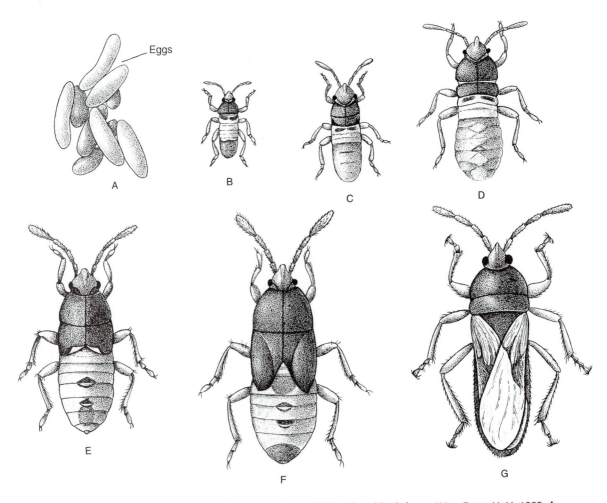

FIGURE 16–13 Stages in the gradual metamorphosis of a chinch bug. *(After Ross, H. H. 1965. A Textbook of Entomology. 3rd Edition. John Wiley and Sons, New York.)*

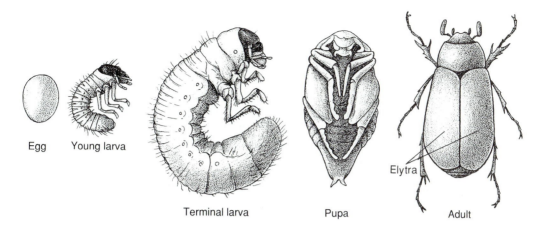

FIGURE 16–14 Stages in the complete metamorphosis of a beetle. *(After Ross, H. H. 1965. A Textbook of Entomology. 3rd Edition. John Wiley and Sons, New York.)*

butterflies, the maggot of flies, and the grub of beetles. This is an active feeding stage, although the food is usually quite different from that of the adult. In some species the larvae and the adults have different kinds of mouthparts. For example, caterpillar larvae have chewing mouthparts, whereas the adults have sucking mouthparts. Some parasitic groups may have two or more different larval habits and structures (**hypermetamorphosis).**

It is important to keep in mind that insect larvae represent a specialized, not a primitive, condition and evolved through suppression and delay of the development of adult features.

At the end of the larval period the young become nonfeeding and quiescent. This stage is called a **pupa** and is usually passed in protective locations, such as the ground, a cocoon, or plant tissues. During pupation adult structures are developed from embryonic rudiments (**imaginal discs and buds),** and except in neuropterans and beetles, few larval structures are carried over to the adult stage. The number of molts required to reach the adult stage ranges from as few as 3 to more than 30 and depends in large part on the type of development.

Holometabolous development was of great adaptive significance in the evolution of the higher orders of insects because the larvae can utilize different food sources, habitats, and life-styles than those of adults. Of the 26 orders of living insects, 9 have holometabolous development, but these 9 include about 80% percent of the species.

The transformation of immature insects into reproducing adults is under endocrine control. A hormonal secretion from the brain stimulates a gland in the prothorax, the prothoracic gland, which produces **ecdysone,** a hormone that stimulates growth and molting. During the larval stages another hormone, the **juvenile hormone,** is secreted by the corpora allata of the brain. This hormone is responsible for the maintenance of larval structures and inhibits metamorphosis. The juvenile hormone can exert its effect only after the molting process has been initiated. It thus must act in conjunction with the ecdysone. When a relatively high level of juvenile hormone is present in the blood, the result is a larva-to-larva molt. When the level of the juvenile hormone is lower, the molt is larva-to-pupa, and in the absence of the juvenile hormone, there is a pupa-to-adult molt.

There are many insects whose life cycle includes a period of arrested development, called **diapause.** Diapause, which is under hormonal control, may be passed in any stage of development, depending on the species, and enables the insect to survive adverse environmental conditions, such as long dry or cold periods. Diapause is typically passed in the ground, a cocoon, or some other protected place. Day length and temperature are important factors for many insects in inducing or breaking diapause.

There are several groups of insects in which multiple generations live in different habitats and have different methods of reproduction. A life cycle of this type is exemplified by the aphids, which have remarkable powers of reproduction. Eggs laid in the autumn hatch in the spring and develop into wingless, parthenogenetic females. The females, sometimes called stem mothers, are viviparous and may give birth to any number of broods of wingless, parthenogenetic females like themselves. One of these generations eventually gives birth to winged forms, which are still parthenogenetic and viviparous females. This generation usually moves to other plants but continues to produce either winged or wingless parthenogenetic females. As fall approaches, a sexual generation of both males and females is produced. These mate, eggs are laid, and the cycle is repeated. The reproductive abilities and life cycles of aphids appear to be adaptations for rapid population increase from low initial levels and are similar to those of water fleas, rotifers, and other inhabitants of temporary bodies of fresh water.

Polyembryony, in which the initial mass of embryonic cells gives rise to more than one embryo, is highly developed in some parasitic hymenopterans and is a classic example of this developmental phenomenon. Two to many larvae are formed from a simple egg deposited in the body of another insect host. The extreme occurs in the tiny chalcid, *Litomastix*. The female deposits a few eggs into the body of a large caterpillar. From these eggs several thousand chalcid larvae may develop, completely devouring the caterpillar host.

INSECT/PLANT INTERACTIONS

Plants are a major resource for thousands of species of insects. Virtually every part of a plant is a food source for some adult or juvenile insect. Not surprisingly, the evolution of plants and insects has been closely connected: insects have been a major factor in the selection of certain features in the evolution of plants, and plants have determined various adaptations in insects. This coevolution is particularly striking in pollination, where the interaction of plants and insects is mutually beneficial. About 67% of flowering plants are pollinated by insects, and the great diversity of floral structure reflects, in large part, adaptations for pollination. Bees, wasps, butterflies, moths, and flies are the principal pollinators. Colors, odors, and nectaries are insect attractants, and floral form often determines landing sites and orientation. Concealment of nectaries deep in the flower forces the pollinator to touch reproductive structures. There

has been a general tendency in the evolution of flowers for the floral design to restrict possible pollinators to a relatively small number of insect species, thus increasing the likelihood that the visiting insects brings pollen from the same plant species.

Insects can also damage plants, and in recent years there has been much interest in plant defenses against insects and insect counteroffenses. The principal plant defense against being eaten is chemical. Plants have evolved certain toxic compounds deposited in leaves or other parts, or the tissues contain high concentrations of indigestible compounds, such as tannins, resins, and silica. The plant defenses have led to various evolutionary strategies among insect herbivores. Some feed on a wide range of plant species (polyphagous insects) but must be selective or utilize less desirable parts because of toxic compounds. Other insects are restricted to one or a few plant species (monophagous insects) and have evolved enzyme systems that can detoxify young, desirable tissues or utilize other countermeasures. For example, many that feed on milkweeds cut the vein delivering the toxic lactose. They then feed downstream of the blocked system. The life cycles of many insects may also be adapted to coincide with the availability of the plant food source.

PARASITISM

There are many parasitic insects, and the condition has evolved many times within the class. Insect parasitism often is an adaptation of a stage in the life cycle to exploit a habitat and nutritional source that is unexploited by other stages. A relatively small number of insects, however, are parasitic for the entire life cycle and are always in contact with the host. This is true of lice, most of which are blood-sucking parasites of birds and mammals (Fig. 16–15) (there is one group of chewing lice). The hemipteran bedbugs are bloodsuckers throughout their life cycle, but they do not live continually on the host.

For many insects parasitism provides a new food source and habitat for the adults only. For example, adult fleas (Fig. 16–20T) are blood-sucking ectoparasites on the skin of birds and mammals. The eggs and immature stages of fleas develop off the host in its nest or den, but even the adult may not always be located on the host. The order Diptera contains a large number of species that are bloodsuckers of birds and mammals as adults. Included here are mosquitoes,

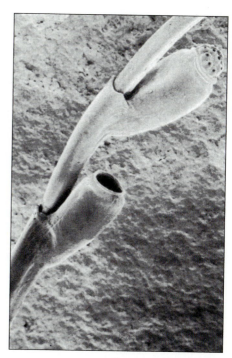

FIGURE 16–15 The crab, or pubic, louse of humans. **A,** An adult gripping two pubic hairs. **B,** Two egg cases attached to a pubic hair. The bottom case is empty; the upper contains an unhatched nymph. *(From Elzinga, R. J. 1987. Fundamentals of Entomology. 3rd Edition. Prentice-Hall, Englewood Cliffs, NJ.)*

black flies, no-see-ums (punkies), horse flies, deer flies, and tsetse flies.

For some insects parasitism offers a new food source for the larva, and the adult is free-living. Many species of wasps and flies illustrate larval parasitism. The screwworm fly, a species of blowfly and a pest of domestic animals, lays its eggs in the wounds of mammals, and the larvae feed on living tissue. The evolution of this parasitic condition was probably preceded by the deposition of eggs in carrion, for this is the habit of many nonparasitic species of blowflies. Both flies are remarkable in invading the digestive tract of horses and other mammals. The adult fly lays its eggs on the hairs of the host, such as a horse. When the host licks egg-bearing hairs, the eggs hatch and the larvae gain entry to the digestive tract. They develop attached to the wall of the stomach or intestine and eventually exit with feces, where they pupate.

Many insect parasites utilize other hosts besides vertebrates. Insects are common hosts, especially of wasps. The larval braconid and ichneumonid wasps parasitize beetles, butterflies, moths, flies, other wasps, and spiders, and depending on the species of parasite, may attack egg, larva, pupa, or adult of the host. Parasitic wasps include some of the best examples of **hyperparasitism,** in which one parasitic wasp is in turn parasitized by another wasp. For example, the ichneumonid wasp, *Aphidius smithi,* lays eggs in certain aphids. Over an eight-day period the larva consumes the aphid and then pupates within a silken cocoon inside the aphid skin. The adult wasp cuts a hole in the aphid skin following pupation. Another parasitic wasp, the hyperparasite, lays an egg in the *Aphidius* larva within the aphid host. Its egg, however, does not hatch until *Aphidius* has reached the pupa stage, then the hyperparasite larva consumes the pupa. There can even be a tertiary parasitic wasp that attacks the secondary one.

COMMUNICATION

Both social and nonsocial insects use chemical, tactile, visual, and auditory signals to communicate. Chemical communication by pheromones has been studied more extensively in insects than in any other group of animals, and many examples are now known. Many species use pheromones to attract one sex to another, and the much-studied moth, *Bombyx mori,* is a classic example. Pheromones also mark trails or territories in some species. For example, substances deposited on the ground by ants returning from a foraging trip serve as a trail marker for other ants. This type of communication is especially important in the complex movements of tropical army ants. Substances produced by the death and decomposition of the body of an ant within the colony stimulate other workers to remove the body. If a live ant is painted with an extract from a decomposing body, the painted ant is carried live and struggling from the nest. There are also pheromones that produce alarm responses in other individuals of an aggregation or colony.

Among the more unusual visual signals are the bioluminescent flashings of fireflies, which function in sexual attraction. In species of *Photinus*, for example, flying males flash at definite intervals. Females located on vegetation flash in response if the male is sufficiently close. The male then redirects his flight toward her, and further flashing occurs.

Sound production is especially notable in grasshoppers, crickets, and cicadas. The chirping sounds of the first two are produced by rasping. The front margin of the forewing (crickets) or the hind leg (katydids) acts as a scraper and is rubbed over a file formed by a vein of the forewing. Where scraper and file are both located on the forewings, as in crickets, the wings cross over, and one forewing functions as a scraper and the other as a file. Each species of cricket produces a number of songs that differ from the songs of other species. Cricket songs function in sexual attraction and aggression. The static-like sounds of cicadas, which serve to aggregate individuals, are produced by vibrations of special chitinous, abdominal membranes.

The remarkable mechanisms of communication in bees are described in the next section.

SOCIAL INSECTS

Colonial organization has evolved in a number of animal phyla, but only among a few spiders and some insects and vertebrates are individuals functionally interdependent yet morphologically separate. The condition is therefore usually described as a social organization.

Social organizations have evolved in two orders of insects, the Isoptera, which comprise the termites, and the Hymenoptera, which include the ants, bees, and wasps. In all social insects no individual can exist outside of the colony nor can it be a member of any colony but the one in which it developed. There is cooperative brood care and an overlap of generations. All social insects exhibit some degree of polymorphism, and the different types of individuals in a colony are termed **castes.** The principal castes are male, female (or queen), and worker. Males function for the insemination of the queen, which produces new individuals for the colony. The workers provide for the support and maintenance of the colony. Caste determination is a developmental phenomenon regulated by the presence or absence of certain substances provided in the immature stages by other members of the colony.

Termites live in a nest usually constructed in soil, and in many species the nest may be huge and structurally complex. Termites differ from social hymenopterans in that workers are sterile individuals of both sexes and because termites are hemimetabolous, workers may be juveniles or adults. The reproductive male is a permanent member of the colony. The colony is built and maintained by workers and may include a soldier caste (Fig. 16–16). The soldiers have large heads and mandibles and defend the colony. Workers and soldiers are wingless; wings are present in the males and queens only during a brief nuptial flight, during which pairing and dispersal occur.

Except for the fungus-growing species, most termites depend on cellulose as a food source and on symbiotic flagellates for cellulose digestion. Because the symbionts are obtained by anal secretions passed from one termite to another, the symbiosis was probably an important factor in the evolution of social behavior in termites, which are generally agreed to be closely related to cockroaches.

Ant colonies resemble those of termites and are usually housed within a gallery system in soil or wood or beneath stones. There may be a soldier caste in addition to workers. Some species of ants raid the nests of other species and carry away their larvae and pupae to be raised as "slaves." Ants exhibit a wide range of diet. Many are scavengers, but some, like the harvester ants, feed on seeds, which they also store. The leaf cutter ants take bits of fresh leaves to their subterranean gardens, where they grow the fungus on which they feed. The cultivation of fungal gardens evolved independently in ants and termites. In contrast to most ants, tropical army ants do not live in permanent nests but "bivouac" together in the open. They are predatory, and during their nomadic mass

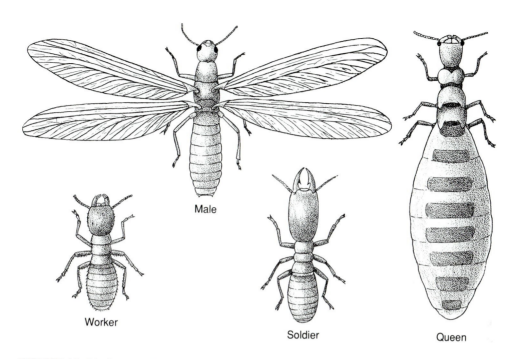

Male

Worker

Soldier

Queen

FIGURE 16–16 Castes of the common North American termite, *Termes flavipes. (After Lutz.)*

movements they consume virtually all invertebrates they encounter. Many fascinating aspects of ant biology are described by Hölldobler and Wilson (1990).

As in bees and wasps, ant soldiers and workers are always sterile females. Wings are present only during the nuptial period of reproductive males and females. Copulation occurs at this time, and the male never becomes a functional part of the colony.

Polymorphism is less highly developed in wasps and bees. There is no soldier caste and workers are winged, but many of these insects exhibit remarkable adaptations for a social organization.

The honeybee, *Apis mellifera,* is the best known social insect. This species is believed to have originated in Africa and to be a recent invader of temperate regions, for unlike other social bees and wasps of temperate regions, the honeybee colony survives the winter. Multiplication occurs by the division of the colony, a process called swarming. Stimulated at least in part by the crowding of workers (20,000 to 80,000 in a single colony), the mother queen leaves the hive along with some of the workers (a swarm) to found a new colony. The old colony is left with developing queens. On hatching, a new queen takes several nuptial flights during which copulation with males

(drones) occurs, and she accumulates enough sperm to last her lifetime. The male dies following copulation, when his reproductive organs are literally exploded into the female. A new queen may also depart with some of the workers as an afterswarm, leaving the remaining workers to yet another developing queen. Eventually, the old colony consists of about one third of the original number of workers and their new queen.

Honeybee colonies are large. The workers' life spans are not long, and a queen may lay 1000 eggs per day. The diet provided these larvae by the nursing workers results in their developing into sterile females, meaning additional workers. The nursing behavior of the workers is a response to a pheromone ("queen substance") produced by the queen's mandibular glands. At the advent of swarming, or when the vitality of the queen diminishes, the production of this pheromone declines. In the absence of the inhibiting effect of the pheromone, the nursing workers construct royal cells, into which eggs, royal jelly, and a greater amount of food are placed. Royal jelly is a secretion from the hypopharyngeal glands of the worker nurses, and those larvae that feed on it develop into queens in about 16 days. At the same time that queens are being produced, unfertilized eggs are deposited into cells similar to those for workers. These haploid eggs develop into drones.

A remarkable feature of honeybee social organization is the temporal division of the workers. As can be seen from the graph in Figure 16–17A, the first activities of the worker are maintenance tasks within the hive. During this period there is secretion by wax, mandibular, and other glands involved in comb construction, food storage, and larval care. After about three weeks, such glandular activity declines, and the bee begins a period of foraging outside of the hive, its final service to the colony. The many functions performed in the lifetime of a worker are not strictly sequential; rather, a worker shifts from one task to another (Fig. 16–17A). A large amount of time is spent by the older worker bees in resting and patrolling. Patrolling, or "determination" of hive needs, plus the ability of workers to change tasks, enables a colony to adjust to changing environmental conditions and is believed to have contributed to the success of the species.

Communication between members of a honeybee colony is highly evolved; some aspects, such as the tail-wagging dance, set the honeybees apart from all other social insects. A successful foraging scout returns to the hive and communicates to other workers the nature, direction, and distance of a food source. The nectar and pollen on the scout's body provide the information about the kind of food that has been found. The scout bee also executes an excited dance that is a ritualization of the flight path. The dancing bee circles to the right and to the left, with a straight-line run between the two semicircles (Fig. 16–17B). During the straight-line run, the bee wags her abdomen and emits audible pulsations. Karl von Frisch, a pioneer in the study of communication in bees, discovered that the orientation of the circular movements shows the direction of the food and that the frequency of the tail-wagging runs indicates the distance. The closer the food source is to the hive, the greater the frequency of tailwagging runs. The sound pulsations of the dancer apparently also indicate the distance of the food source from the hive. The average number of vibrations is proportional to the distance of the food from the hive.

Bees use the angle of the sun and light polarization as a means of orientation, and the dance of the scout bee indicates the location of the food in reference to the sun's position. If the tail-wagging run is directed upward, the food is located toward the sun; if the tail-wagging run is directed downward, the food is located away from the sun. The inclination of the run to the right or to the left of vertical indicates the angle of the sun to the right or to the left of the food source. An internal "clock" compensates for the passage of time between discovery of the food and the start of the dance, so the information is correct even though the sun has moved during the interval. On cloudy days the polarization of the light rays and ultraviolet light act as indirect references in the absence of the sun. If the food source is closer than 80 m, the clues provided by chemoreception are sufficient for finding the food, and the tail-wagging dance is not performed by the scout bee.

The tail-wagging dance cannot be communicated to other bees visually because the hive is dark. The dancing bee, however, continually vibrates its wings and the resulting air-particle oscillations are detected by the surrounding bees.

Although all entomologists agree that the honeybee dance contains information about location and distance of food sources, some argue that there is no good evidence that other bees actually use the information. They maintain that experimental evidence supports odor as being the primary means by which the bees communicate information about food sources and locate them. The debate is detailed in a recent book by Adrian Wenner and Patrick Wells of the University of California at Santa Barbara.

SYSTEMATIC RÉSUMÉ OF CLASS INSECTA, OR HEXAPODA

Living insects may be placed in 26 orders, comprising nearly 1000 families and many thousands of genera. More than 84,000 species of insects from all 26 orders have been described from North America, but it is estimated that nearly 25,000 species remain to be discovered on that continent alone. And there may be as many as 1,000,000 to 30,000,000 more undescribed insect species in the world.

Subclass Entognatha.   Mouthparts sunk into a pouch. Compound eyes and Malpighian tubules absent or reduced. Wings are absent and the wingless condition is primary. The three entognath orders along with two wingless ectognath orders are commonly grouped together as the Apterygota, the wingless insects, but the evidence indicates that the entognath and ectognath condition is the more natural division.

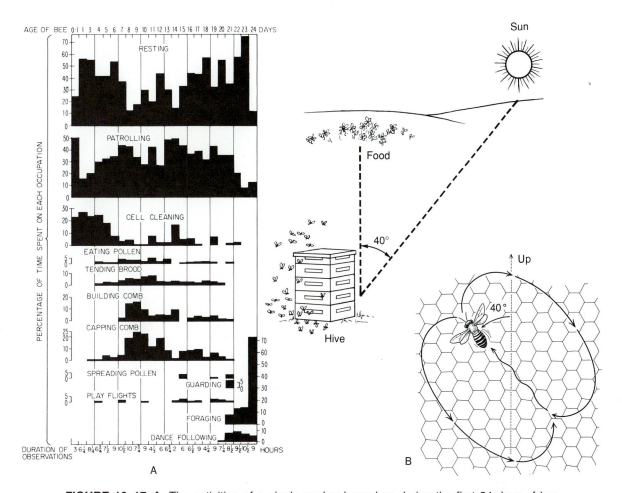

FIGURE 16–17 A, The activities of a single worker honeybee during the first 24 days of her adult life. **B,** Diagram illustrating the inclination of the straight tail-wagging run by a scout bee to indicate the location of a food source by reference to the sun. The food source is located at an angle 40 degrees to the left of the sun. The tail-wagging run of the scout bee is therefore upward (indicating that food is toward the sun) and inclined 40 degrees to the left (indicating the angle of the food source to the sun). *(A, Redrawn from Ribbands, 1953; based on data of Lindauer, 1952. B, After von Frisch)*

Order Protura. Small eyeless insects with a cone-shaped head lacking antennae. First pair of legs used as antennae. First three pairs of abdominal appendages bear sensory styli, vestiges of legs. Proturans live in damp humus and soil and feed on decayed organic matter (Fig. 16–18A).

Order Collembola. Springtails. Small wingless insects with an abdominal jumping organ, well-developed legs, and antennae; eyes either absent or represented by isolated ommatidia. Abundant in moist leaf mold, soil, and rotten wood. Most feed on decaying plant material and fungi (Fig. 16–18B).

Order Diplura. Small, blind, wingless insects. End of abdomen bears two caudal filaments. They live in damp humus, decomposing wood, and beneath stones and bark (Fig. 16–18C).

Subclass Ectognatha. Mouthparts not sunk into a head pouch. Among other characters, there is an ovipositor derived from the eighth and ninth abdominal segments. Except for the orders Microcoryphia and Thysanura, both of which exhibit a pri-

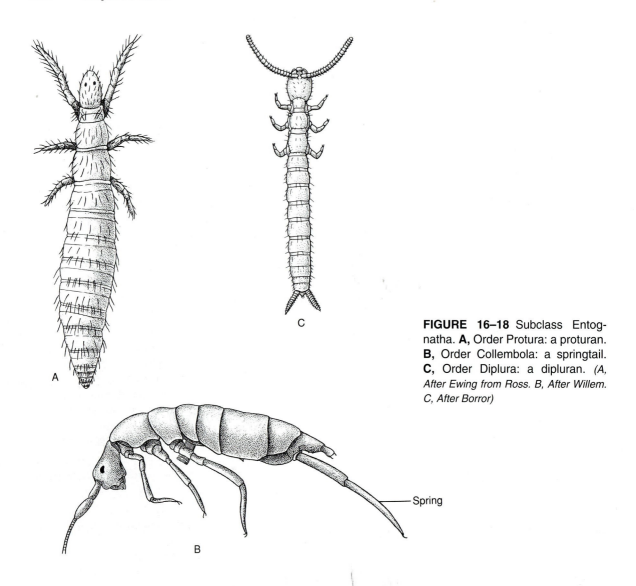

FIGURE 16–18 Subclass Entognatha. **A,** Order Protura: a proturan. **B,** Order Collembola: a springtail. **C,** Order Diplura: a dipluran. *(A, After Ewing from Ross. B, After Willem. C, After Borror)*

mary wingless condition, all of the remaining insects are winged or at least derived from winged forms. They are commonly grouped together as the Pterygota.

Order Microcoryphia, or Archeognatha. Jumping bristletails. Wingless insects with rather cylindrical bodies. Abdomen bears three terminal styliform appendages. Large compound eyes and ocelli present. Found beneath bark, stones, and in leaf mold and can jump when disturbed. They feed on algae, lichens, and moss (Fig. 16–19A).

Order Thysanura. Silverfish. Fast-running insects with three styliform appendages on the abdomen. Body rather depressed. Compound eyes small, and ocelli may be absent. They live in dead leaves and wood and around stones. Some species are found in houses where they eat books and clothing (Fig. 16–19B).

Order Ephemeroptera. Mayflies. Elongated insects with net-veined wings, of which the first is larger than the second; wings held vertically at rest. Two or three caudal, filiform appendages. Antennae small and mouthparts of short-lived adults vestigial. Gradual metamorphosis with aquatic nymphs (Fig. 16–20A).

Order Odonata. Dragonflies and damselflies. Predacious insects with long, narrow, net-veined wings; large eyes; and chewing mouthparts. Gradual metamorphosis. Nymphs aquatic. Dragonflies are stout-bodied; damselflies are slender

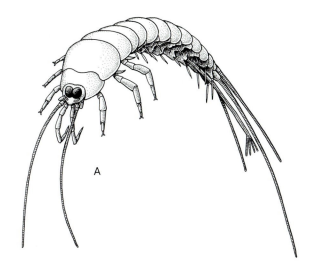

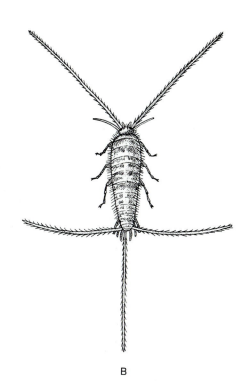

FIGURE 16–19 Subclass Ectognatha, apterygotes. **A,** Order Microcoryphia: a jumping bristletail. **B,** Order Thysanura: a silverfish. *(A, After Borror. B, After Lutz)*

and delicate, weak fliers. Wings are held vertically at rest in damselflies and horizontally in dragonflies (Fig. 16–20B).

Order Orthoptera. Grasshoppers, katydids, crickets, roaches, mantids, and walkingsticks. Large-headed insects with strong chewing mouthparts and compound eyes. Femur of hind leg enlarged for jumping in many species. Winged and wingless species. Most winged forms have membranous hindwings folded fanlike beneath leathery forewings. Largely herbivorous, sometimes causing vast crop damage. Gradual metamorphosis. Some classifications place the roaches, mantids, and walkingsticks in separate orders (Fig. 16–20C).

Order Isoptera. Termites. Social insects. Winged and wingless individuals composing the colony. Soft-bodied, pale, with abdomen broadly joined to thorax. Forewings and hindwings of equal size and held horizontally over abdomen (unlike flying ants, which are darker, with elbowed antennae, narrow waists, hindwings smaller than forewings, and wings held vertically over abdomen). Gradual metamorphosis (Fig. 16–16.)

Order Plecoptera. Stone flies. Adults have long antennae; chewing mouthparts; and two pairs of well-developed, membranous wings, or vestigial wings. Abdomen with two multisegmented, caudal cerci of varying length. Gradual metamorphosis. Nymphs aquatic. Adults emerge during winter in certain groups (Fig. 16–20D).

Order Dermaptera. Earwigs. Elongated insects resembling beetles, with chewing mouthparts, compound eyes, and large, forceps-like cerci. Most species have fan-shaped wings and elytra. Nocturnal, with omnivorous food habits. Gradual metamorphosis (Fig. 16–20E).

Order Embioptera. Web spinners. Small, slender, soft-bodied insects with large heads and eyes. They feed on plants and live in silken tunnels, which they weave ahead of themselves to create routes. Silk glands and spinning hairs located on front tarsi. They are gregarious, and many individuals may live together. Mostly tropical. Gradual metamorphosis (Fig. 16–20F).

Order Psocoptera, or Corrodentia. Book lice, bark lice, and psocids. Small, fragile, pale insects with chewing mouthparts. Wings membranous

(Text continued on page 859)

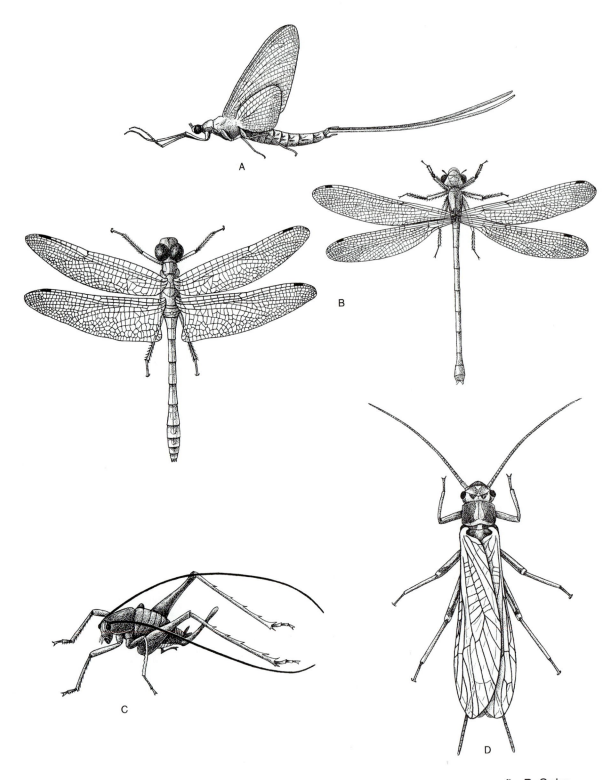

FIGURE 16–20 Subclass Ectognatha: pterygotes. **A,** Order Ephemeroptera: a mayfly. **B,** Order Odonata: a dragonfly (left) and a damselfly (right). **C,** Order Orthoptera: a camel cricket. **D,** Order Plecoptera: a stone fly.

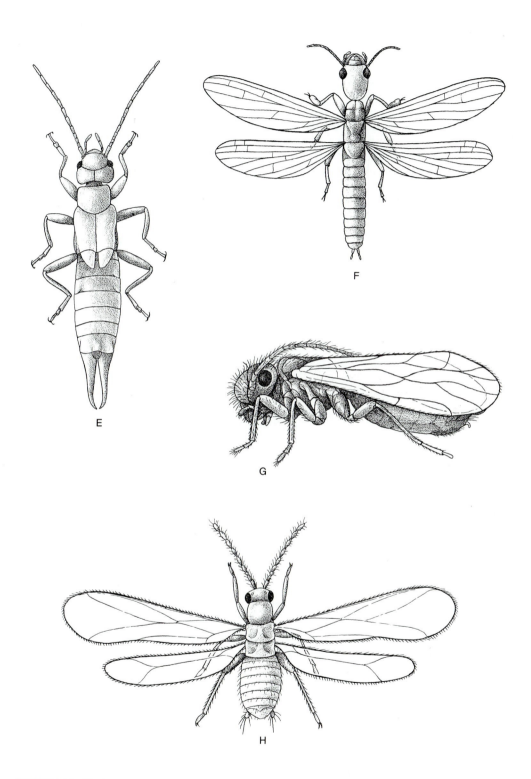

FIGURE 16–20 (continued) **E,** Order Dermaptera: an earwig. **F,** Order Embioptera: a web spinner. **G,** Order Psocoptera: a psocid. **H,** Order Zoraptera: *Zorotypus.*

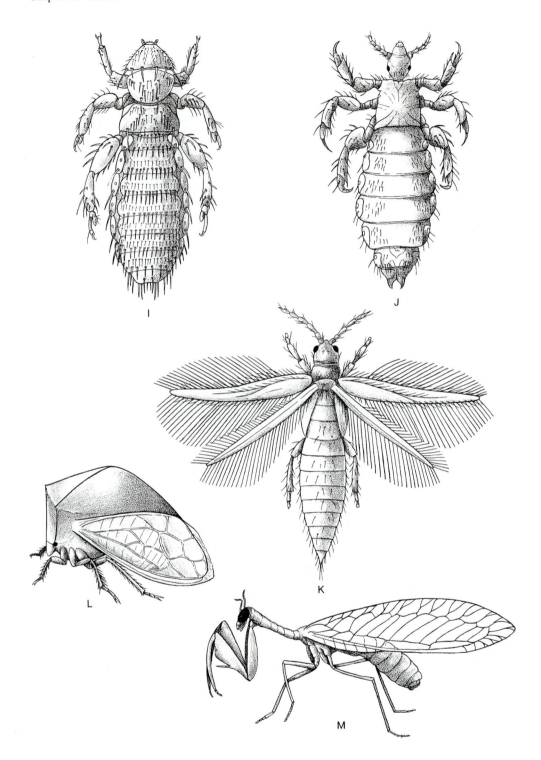

FIGURE 16–20 (continued) **I**, Order Mallophaga: a guinea pig louse. **J**, Order Anoplura: a body louse of humans. **K**, Order Thysanoptera: a thrip. **L**, Order Homoptera: buffalo treehopper. **M**, Order Neuroptera: mantispid.

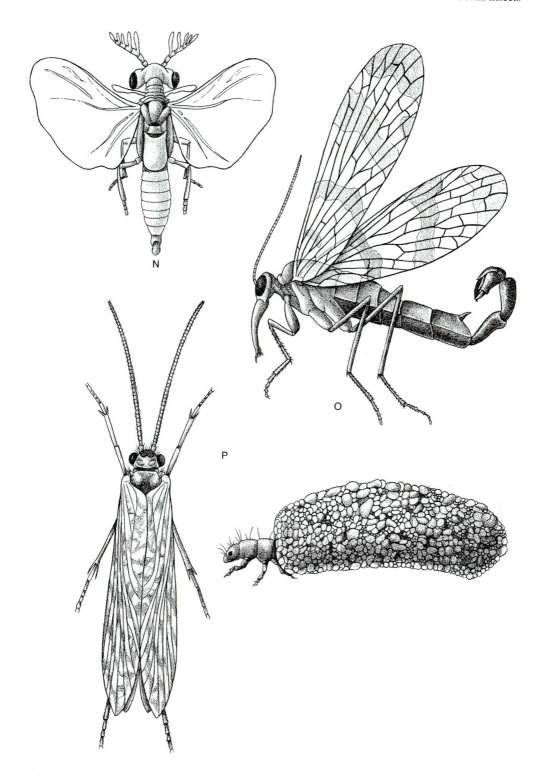

FIGURE 16–20 (continued) **N**, Order Strepsiptera: twisted-wing parasite. **O**, Order Mecoptera: a scorpion fly. **P**, Order Trichoptera: an adult caddis fly and larva in case.

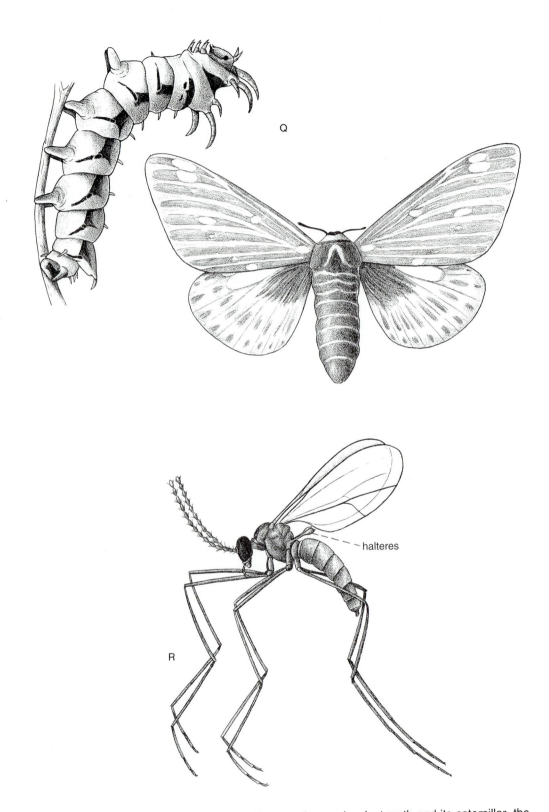

FIGURE 16–20 (continued) **Q,** Order Lepidoptera: the royal walnut moth and its caterpillar, the hickory-horned devil. **R,** Order Diptera: a gall gnat.

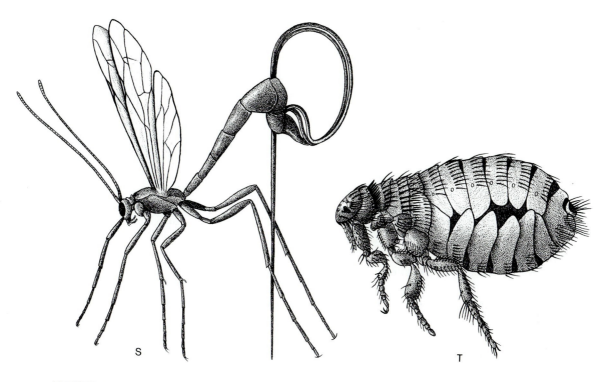

FIGURE 16–20 (continued) **S,** Order Hymenoptera: a parasitic female ichneumon fly. **T,** Order Siphonaptera: a flea. *(A, After Ross. B, After Kennedy from Ross. C, After Lutz. D, After Ross from Illinois Nat. Hist. Survey. E, After Fulton from Borror and DeLong. F, After Enderlein from Comstock. G, After Sommerman from Ross. H, After Caudell from Comstock. I and J, From Grassé, P. (Ed.): Traité de Zoologie. K, After Moulton from Grassé, P. (Ed.): Traité de Zoologie. L, After Irving from H. Curran. 1954. Golden Playbook of Insect Stamps. Simon & Schuster, New York. M, After Banks from Borror & DeLong. N, After Bohart; Entomological Society of America from Borror & DeLong. O, After Taft from Borror & DeLong. P, After Mohr; Illinois Nat. Hist. Survey, from Ross, Q, After Lutz. R, After Usda from Borror & DeLong. S, After Lutz. T, After Bouche from Borror & DeLong.)*

and front pair a little larger than hind pair, or wingless. Gradual metamorphosis. They live in a wide variety of habitats—bark, foliage, under stones. Some species infest buildings and are found in books (Fig. 16–20G).

Order Zoraptera.   Small, pale, soft-bodied insects resembling tiny termites with chewing mouthparts. Both winged and wingless forms. Gradual metamorphosis. These are rare insects, living in colonies under dead wood in warm climates. Only some 20 known species and one genus (Fig. 16–20H).

Order Mallophaga.   Chewing lice and bird lice. Wingless, flattened insects that live as ectoparasites on birds; two families parasitic on mammals. The eyes are reduced or absent; legs are short; and thorax is small. Gradual metamorphosis. Feed on scales, feathers, hair, skin, and sometimes dried blood around wounds. Many species infest domestic birds and livestock and cause considerable damage by skin irritation (Fig. 16–20I).

Order Anoplura. Sucking lice.   Similar to chewing lice but mouthparts adapted for sucking. Ectoparasites of birds and mammals. A number of species are parasitic on domestic animals, and the head louse and crab louse are parasites of humans. More serious than the irritation produced by these parasites is their role as vectors of disease, such as typhus fever (Fig. 16–20J).

Order Thysanoptera. Thrips.   Small, slender insects with mouthparts adapted for rasping and sucking. Winged and wingless. Wings are narrow, with few veins, and fringed with hairs. Development is peculiar in that there are nymphlike early instars but a preadult pupa stage. A large number of species feed on sap in flowers, but some thysanopterans are predaceous and feed on mites and smaller insects (Fig. 16–20K).

Order Hemiptera. True bugs. Piercing and sucking mouthparts, in which the beak arises from the front of the head and extends ventrally and posteriorly. Forewings with a thickened basal and distal membranous section. The membranous portions of the forewings overlap when at rest over the abdomen. Hindwings entirely membranous. Gradual metamorphosis. Herbivorous and predaceous (Fig. 16–13).

Order Homoptera. Cicadas, leafhoppers, and aphids. Herbivorous insects related to the hemipterans, but beak arises from back of head. Forewings typically membranous. Wings are commonly held in a tentlike position over the body. Gradual metamorphosis (Fig. 16–20L).

Order Neuroptera. Lacewings, ant lions, mantispids, snake flies, and dobsonflies. Adults have chewing mouthparts and long antennae. The two pairs of similar membranous wings with many veins are held tentlike over abdomen. Complete metamorphosis. Larvae are predaceous and usually terrestrial (Fig. 16–20M).

Order Coleoptera. Beetles, weevils. Largest order of insects (over 300,000 species), with hard bodies and chewing mouthparts. Adults usually have two pairs of wings, of which the front pair are modified as heavy protective covers (elytra); the hind pair are membranous. Complete metamorphosis. The majority are plant feeders, but there are predatory families. Some aquatic species (Fig. 16–14).

Order Strepsiptera. Minute, beetle-like insects, mostly parasitic on other insects. Only males possess wings; in females, forewings reduced to club-shaped appendages. Complete metamorphosis (Fig. 16–20N).

Order Mecoptera. Scorpion flies. Slender-bodied insects, often vividly colored. Biting mouthparts prolonged as a beak. Most species with long narrow wings, which have many veins. Complete metamorphosis. Adults are omnivorous, and grublike larvae feed on organic matter (Fig. 16–20O).

Order Trichoptera. Caddis flies and water moths. Soft-bodied insects, with two pairs of hairy, membranous wings and poorly developed, chewing mouthparts. Complete metamorphosis. Larvae are aquatic and build portable cases of various materials (Fig. 16–20P).

Order Lepidoptera. Butterflies and moths. Soft-bodied insects with wings, body, and appendages covered with pigmented scales. Mouthparts are modified as coiled proboscis used for sucking flower nectar. Compound eyes large. Complete metamorphosis. Larvae are caterpillars and usually are plant feeders; adults feed little or not at all (Fig. 16–20Q).

Order Diptera. True flies. Large order, all of which have functional front wings and reduced, knoblike hind wings (halteres). Mouthparts are variable, as is body form. Complete metamorphosis. Group includes mosquitoes, horseflies, midges, and gnats. Adults often vectors of diseases; larvae frequently damaging to vegetables and domestic animals (Fig. 16–20R).

Order Hymenoptera. Ants, bees, wasps, and sawflies. A large and varied order, all with chewing mouthparts but also modified for lapping or sucking in some forms. Winged and wingless species. Wings are transparent with only a few veins. Complete metamorphosis. Larvae are caterpillars or are grublike with chewing mouthparts (Fig. 16–20S).

Order Siphonaptera. Fleas. Small, wingless insects with laterally flattened bodies. Legs are long with large coxae and are adapted for jumping. These insects have piercing and sucking mouthparts and feed on the blood of mammals and birds. They are vectors of bubonic plague. Complete metamorphosis (Fig. 16–20T).

SUMMARY

1. The class Insecta contains the largest number of species of any group of animals. At least three quarters of a million species have been described. Insects rank with vertebrates as being the most successful inhabitants of the terrestrial environment.

2. Insects are distinguished from other uniramian arthropods in having the body divided into a head, thorax, and abdomen. The head bears one pair

of antennae and the feeding appendages; the thorax carries three pairs of legs; the abdomen lacks appendages.

3. The ability of most insects to fly has contributed greatly to their success. Flight has enhanced distribution, exploitation of food sources and habitats, escape from predators, and reproductive processes. Most insects have two pairs of thoracic wings, although one pair is reduced, modified, or lost in various groups. Flight evolved early in the evolutionary history of insects, but some groups (apterygotes), such as collembolans and thysanurans, are primitively wingless.

4. The mouthparts consist of a pair of mandibles, a pair of maxillae, and a labium (fused second maxillae). Primitively, the mouthparts are adapted for chewing plant material, but they have become modified for a wide range of diets and feeding modes, including piercing and sucking.

5. A tracheal system provides for gas exchange. Spiracles are located along the sides of the thorax and abdomen but vary in number depending on the species.

6. The nitrogenous waste of insects is uric acid, which is excreted through malpighian tubules. Insects are capable of producing a hyperosmotic urine, which together with the waxy epicuticle, is an important adaptation for reducing water loss and has contributed to the success of insects as terrestrial animals.

7. The tubular heart is located in the dorsal part of the abdomen and propels blood anteriorly through a short aorta. The remainder of the blood-vascular system is a hemocoel.

8. Most insects possess a pair of large, lateral, compound eyes, three ocelli on the top or front of the head, and a great variety of types of sensilla located over the body surface, especially on the antennae and legs.

9. Most insects transfer sperm in spermatophores. Primitively, transfer is indirect, as in many other terrestrial arthropods, but in the majority of insects, the male deposits the spermatophores directly within the female reproductive system. The female deposits eggs encased in protective coverings. Cleavage is typically superficial.

10. In primitive insects, the juvenile stages are similar to the adult. In higher orders, the juvenile gradually acquires certain structures, such as the wings, during the course of development. Development with larval stages and complete metamorphosis is a specialization of the orders containing beetles, flies, bees, and wasps. Development of this type enables juveniles and adults to exploit different habitats and food sources.

11. Parasitism has evolved a number of times in the evolution of insects. Juveniles, adults, or both may be parasitic.

12. Highly developed social (colonial) organization has evolved within two orders, the Isoptera (termites) and the Hymenoptera (ants, bees, and wasps). Only some hymenopterans are social, and there is a great range in the complexity of social organization.

REFERENCES

The literature here is restricted to insects. The introductory references on page 6 include many general works and field guides that contain sections on these animals. The literature on insects is voluminous. The following introductory entomology texts provide good general accounts of insects. They vary in their approach and emphasis, but all contain references to more specialized topics. An introductory entomology text is also a useful starting point for the student interested in taxonomy.

Borror, D. J., Triplehorn, C. A., and Johnson, N. F. 1981. An Introduction to the Study of Insects. 6th Edition. Saunders College Publishing, Philadelphia. 875 pp. (This text emphasizes insect groups.)

Davies, R. G. 1988. Outlines of Entomology. 7th Edition. Chapman and Hall, London, 408 pp.

Elzinga, R. J. 1987. Fundamentals of Entomology. 3rd Edition. Prentice-Hall, Inc., Englewood Cliffs, NJ. 456 pp.

Evans, H. E. 1984. Insect Biology: A Textbook of Entomology. Addison-Wesley Publishing Co., Reading, MA. 436 pp. (This text emphasizes the relationships of insects to the world around them.)

Gillot, C. 1980. Entomology. Plenum Publishing, New York. 747 pp.

Little, V. A. 1972. General and Applied Entomology. 3rd Edition. Harper and Row, New York. 527 pp.

Romoser, W. S. 1981. The Science of Entomology. 2nd Edition. Macmillan Publishing Co., New York. 544 pp.

Ross, H. H., Ross, J. R. P., and Ross, C. A. 1982. A Textbook of Entomology. 4th Edition. John Wiley and Sons, New York. 696 pp.

Other General References

Arnett, R. H. 1985. American Insects: A Handbook of the Insects of North America. Van Nostrand Reinhold, New York. 850 pp.

Arnett, R. H., Downie, N. M., and Jaques, H. E. 1980. How to Know the Beetles. 2nd Edition. W. C. Brown, Dubuque, IA. 416 pp.

Askew, R. R. 1971. Parasitic Insects. Elsevier-North Holland Publishing Co., New York. (See also the general parasitology texts listed on p. 204.)

Bland, R. G., and Jaques, H. E. 1978. How to Know the Insects. 3rd Edition. W. C. Brown, Dubuque, IA. 409 pp.

Blum, M. S. (Ed.): 1985. Fundamentals of Insect Physiology. Wiley-Interscience, New York. 598 pp.

Borror, D. J., and White, R. E. 1974. A Field Guide to the Insects of America North of Mexico. Peterson Field Guide Series. Houghton Mifflin Co., Boston.

Brian, M. V. 1983. Social Insects: Ecology and Behavioural Biology. Chapman and Hall, London. 377 pp.

Chapman, R. F. 1982. The Insects: Structure and Function. 3rd Edition. Harvard University Press, Cambridge, MA. 919 pp.

Chu, H. F. 1949. How to Know the Immature Insects. W. C. Brown, Dubuque, IA. 234 pp.

Covell, C. V. 1984. A Field Guide to the Moths of Eastern North America. Peterson Field Guide Series. Houghton Mifflin Co., Boston. 496 pp.

Dussourd, D. E., and Eisner, T. 1987. Vein-cutting behavior: Insect counterploy to the latex defense of plants. Science. 237:898–901.

Ehrlich, P. R., and Ehrlich, A. H. 1961. How to Know the Butterflies. W. C. Brown, Dubuque, IA. 262 pp.

Evans, D. L., and Schmidt, J. O. (Eds.): 1990. Insect Defenses. State University of New York Press, Albany. 482 pp.

Frisch, K. von 1971. Bees: Their Vision, Chemical Senses and Language. 2nd Edition. Cornell University Press, Ithaca, NY. 176 pp.

Funk, D. H. 1989. The mating of tree crickets. Sci. Am. 261(2):50–57

Goldsworthy, G. J., and Wheeler, C. H. 1989. Insect Flight. CRC Press, Boca Raton, FL. 352 pp.

Hölldobler, B., and Wilson, E. O. 1990. The Ants. Harvard University Press, Cambridge, MA. 732 pp.

Lee, R. E. 1989. Insect cold-hardiness: To freeze or not to freeze. BioScience. 39(5):308–311.

Merritt, R. W., and Cummins, K. W. 1984. An Introduction to the Aquatic Insects of North America, 2nd Edition. Kendall Hunt Publishing Co., Dubuque, IA. 722 pp.

Metcalf, C. L., Flint, W. P., and Metcalf, R. L. 1962. Destructive and Useful Insects. 4th Edition. McGraw-Hill Book Co., New York. 1087 pp.

Moran, N. A. 1992. The evolution of aphid life cycles. Ann. Rev. Entomol. 37:321–348.

Peters, T. M. 1988. Insects and Human Society. Van Nostrand Reinhold Co., New York. 450 pp.

Price, P. W. 1984. Insect Ecology. 2nd Edition. John Wiley and Sons, New York. 607 pp.

Rockstein, M. (Ed.): 1973–1974. The Physiology of Insecta. 2nd Edition. Vols I–VI. Academic Press, New York. (An extensive treatment of the physiology and behavior of insects. Well organized by topics contributed by many physiologists and entomologists.)

Slama, K. 1988. A new look at insect respiration. Biol. Bull. 175:289–300.

Sullivan, D. J. 1987. Insect hyperparasitism. Ann. Rev. Entomol. 32:49–70.

Terra, W. R. 1990. The evolution of digestive systems of insects. Ann. Rev. Entomol. 35:181–200.

Topoff, H. 1990. Slave-making in ants. Am. Sci. 78:520–528.

Wenner, A., and Wells, P. H. 1990. Anatomy of a Controversy: The Question of a Language Among Bees. Columbia University Press, New York. 399 pp.

Wilson, E. O. 1971. The Insect Societies. Harvard University Press, Cambridge, MA. 548 pp. (An excellent detailed account of social insects.)

Wooton, R. J. 1990. The mechanical design of insect wings. Sci. Am. 263(5):114–120.

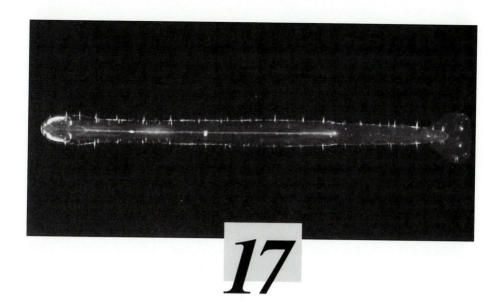

17

PROTOCHORDATES AND CHAETOGNATHS

PRINCIPLES AND EMERGING PATTERNS

HOMOLOGY AND ANALOGY

All biological similarities, whether among molecules, organelles, cells, tissues, or organs, are either the result of a close genealogical relationship (family members resemble each other more closely than nonfamily members), or they are not closely related (insects, bats, birds), but have independently evolved similar adaptations (wings) to solve a common functional challenge (flight in air). Similarity that results from close evolutionary relationship is called **homology,** whereas similarity that derives from parallel functional adaptations among distant relatives is known as **analogy** and is the result of evolutionary **convergence.**

If homologies are the result of evolutionary relationship, then the discovery of a homology in two or more species means that those species are related and share a recent common ancestor. Thus, it is the recognition of homology that permits biologists to reconstruct the evolutionary history, or **phylogeny,** of organisms. Scientists engaged in the search for homologies are **systematists,** and their science is called **systematics,** or simply, comparative biology.

Accurate reconstructions of evolutionary history hinge on the ability of systematists to discriminate between homology and analogy, a distinction that is not always easy to make. It would be unacceptable, for example, for bats and birds to be interpreted as close relatives simply because both have wings and fly. On the other hand, birds and certain ground-dwelling dinosaurs (coelurosaurs) are likely to be sister groups because both are not only small, lightweight, and birdlike in appearance, but also are alike in the configuration of their limb bones. Thus, the key to discriminating between homology and analogy lies in the study of structural detail and not in superficial similarity.

Two sources of details are available to comparative biologists for use in discriminating between homology and analogy: (1) If a structure is homologous in two or more species, then it should occupy a *similar anatomical position* in those species. For example, the compound eyes of both crustaceans and insects occur on the acron and receive innervation from the protocerebrum—the first part of the brain. They therefore satisfy the first criteron of homology, a **positional correspondence.** (2) If two or more structures are homologous with each other, then they should be *similar in composition.* For example, insect and crustacean compound eyes both consist of many ommatidia, each composed of a lens, crystalline cone, pigment cells, and rhabdomeres. Thus, in addition to fulfilling the first criterion, the eyes also satisfy the second criterion of homology, a **compositional correspondence.** Because the eyes correspond in position and composition, it can be hypothesized that these compound eyes are homologous organs, and on the basis of this homology, that crustaceans and insects shared a common ancestor that had compound eyes.

In practice, a systematist, like other scientists, is often confronted with contradictory data; homology may be suggested by a comparison of some organs but not others. For example, it might be suspected that there is a close relationship between birds and reptiles because of the apparent homology of their clawed feet and scaly legs, but birds have feathered wings and toothless beaks; reptiles, on the other hand, have scaly, clawed forelegs and teeth set in jaws. In such cases, where some clues match and other do not, uncertainty compels the systematist to search for and evaluate additional details. The details are usually sought for in other species, living or extinct, and in developmental stages of the species in question. In the example of birds and reptiles, the fossil species *Archaeopteryx lithographica* was an organism with scaly, clawed hindlegs; feathered, clawed wings; and a toothy jaw. Thus, the organization of *Archaeopteryx* bridges the differences in the compositional criterion between reptiles and birds, that is, a winged forelimb can have claws, and a feathered animal can have jaws beset with teeth.

Intermediates in development are equally useful in fulfilling homology criteria. Juvenile hoatzins (a bird found in South American wetlands) have reptilian claws on their wings that are used for grasping the dense foliage, but are lost as the animal matures. The transient presence of claws in hoatzins again indicates that reptile and bird forelimbs are homologues and that birds shared a common ancestor with reptiles. Thus, developmental intermediates, like structural intermediates provided by other species, are a rich source of information in homology recognition and evolutionary reconstruction. Such intermediates are also useful among invertebrates. For example, the close evolutionary relationship between hemichordates and echinoderms is based largely on the structural similarities of their larvae (tornaria, bipinnaria), and it would be impossible to know that acorn and rhizocephalan cirripedes were both barnacles were it not for the common occurrence of a cypris larva in both groups.

Figure 17–1 illustrates three similar animals that, when alive, closely resemble each other in fishlike body form and swimming behavior. Of the three, however, only two—amphioxus and ammocoetes—are closely related to each other. The third is a polychaete, a member of the family Opheliidae, that evolved convergently on the two chordates, but which can be distinguished from them on close examination.

PROTOCHORDATES AND CHAETOGNATHS

Historically, three groups of animals—the Hemichordata, Urochordata, and Cephalochordata—were classified as chordates (along with vertebrates) and were seen to embody clues to the early evolutionary history of the phylum. When the homology of the chordate notochord and a similar structure in hemichordates was thrown into doubt, the hemichordates were removed from the Chordata and isolated in their own phylum. Despite this discrepancy, the biology of hemichordates continues to be a source of information used in recent models of chordate evolution. For this reason, the hemichordates, cephalochordates, and urochordates are lumped together here under the informal heading of protochordates, or "first chordates."

PHYLUM HEMICHORDATA

The hemichordates are large solitary worms (Enteropneusta) or hydroid-like colonies of tiny zooids (Pterobranchia) inhabiting the sea. The bodies of hemichordates are divided into three regions—a locomotory preoral lobe, a short collar, and a trunk—reflecting an underlying tricoelomate organization. A complex heart/kidney occurs in the preoral lobe and is supported by a notochord-like structure. The principal nervous concentration occurs dorsally in the collar region and, in the Enteropneusta, is hollow. A pharynx, which bears one or more pairs of gill slits, is found in the trunk.

The occurrence, or rudimentary expression, of three chordate features—notochord, dorsal hollow nerve cord, and gill slits—was formerly the basis for including the hemichordates within the chordates.

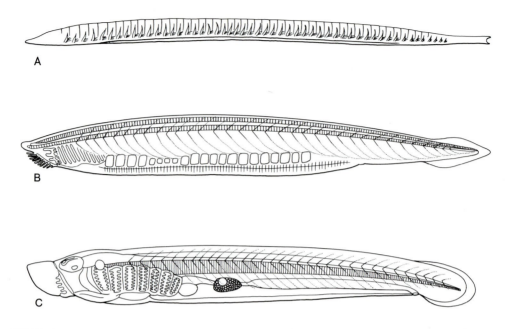

FIGURE 17–1. Homology and analogy. These three animals, the polychaete, *Armandia brevis* **(A)**, the lancelet, *Branchiostoma caribaeum* **(B)**, and the ammocoetes larva of the brook lamprey *Lampetra* spp. **(C)** all swim in water and burrow in sand using rapid lateral undulations of the body. All have laterally compressed bodies, a tail fin, segmental muscles, and a streamlined head, but these similarities are mutually homologous only in the lancelet **(B)** and the ammocoetes **(C)**, which are both members of the phylum Chordata. The analogous structures in *Armandia* are believed to be the result of an evolutionary convergence on a similar lifestyle.

This is no longer an accepted classification, but the hemichordates nevertheless are close relatives of the chordates and provide many clues to their functional organization and ancestry.

Class Enteropneusta

The 70 species of Enteropneusta, or acorn worms, are primarily benthic inhabitants of shallow water, but some occur in the deep sea and in association with hydrothermal vents. Some live under stones and shells, and many common species, including those of *Saccoglossus* and *Balanoglossus,* construct mucus-lined burrows in mud and sand. Exposed tidal flats are frequently dotted with the coiled, ropelike castings of these animals.

External Organization

Acorn worms are relatively large animals, the majority ranging between 9 and 45 cm in length. *Saccoglossus (Dolichoglossus) kowalevskii* is typically 15 cm long, and the common *Balanoglossus aurantiacus* of the southeastern coast of the United States reaches approximately 1 m in length (Fig. 17–2). The largest described acorn worm, *Balanoglossus gigas,* which ranges from Brazil to Cape Hatteras, North Carolina, exceeds 2.5 m in length and constructs 3 m-long burrows.

The cylindrical and rather flaccid body is composed of an anterior **proboscis,** a middle **collar,** and a long posterior **trunk** (Fig. 17–2). These regions correspond to the typical deuterostome body divisions: protosome, mesosome, and metasome.

The proboscis is usually short and more or less conical and is connected to the collar by a slender but sturdy **stalk.** The collar is a short cylinder that anteriorly overlaps the proboscis stalk, embraces the posterior end of the proboscis, and contains the mouth on the ventral side. In some species, the combined proboscis and collar resemble an acorn and its cap, and the common name—acorn worm—was derived from this similarity.

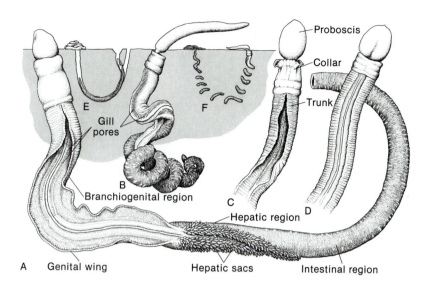

FIGURE 17–2. Enteropneust diversity and burrow structure. **A,** *Balanoglossus auriantiacus.* **B,** *Saccoglossus (Dolichoglossus) kowalevskii.* **C,** *Ptychodera jamaicensis.* **D,** *Schizocardium brasiliense.* The intestinal region of *Balanoglossus aurantiacus* has been shortened. **E,** Burrow arrangement and feeding position of *Balanoglossus aurantiacus* showing fecal cast, feeding funnel, and ventilation shaft. **F,** Burrow arrangement and feeding position of *Saccoglossus kowalevskii.* The burrow is helical, and the trunk is thrown into a permanent twist. *(Adapted from Ruppert, E. E. and Fox, R. S. 1988. Seashore Animals of the Southeast. University of South Carolina Press, Columbia. 429 pp.)*

The trunk constitutes the major part of the body and is divisible from anterior to posterior into three regions: the branchiogenital region, the hepatic region, and the intestinal region. The **branchiogenital region** bears a longitudinal row of **gill pores** on each side of the dorsal midline (Fig. 17–2A). Lateral to the two rows of gill pores, the trunk bears either two low **genital ridges** *(Saccoglossus, Schizocardium)* or two large flaplike **genital wings** *(Balanoglossus, Ptychodera),* which arch over the gill region and enclose a water channel (Fig. 17–2A). A specialized **hepatic region** follows the branchiogenital region posteriorly and is characterized by numerous fingerlike **hepatic sacs** (except *Saccoglossus*) that project dorsally from the surface of the body (Fig. 17–2A). The long cylindrical **intestinal region** completes the trunk and extends rearward to the terminal anus.

Body Wall and Internal Organization

The enteropneust epidermis lacks a cuticle but is densely ciliated and richly glandular, and bears numerous short microvilli (Fig. 17–3). The gland cells,

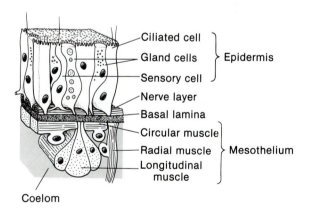

FIGURE 17–3. Hemichordate body walls lack a cuticle and are either monociliated (pterobranchs) or multiciliated (enteropneusts). The coelom is lined by a mesothelium that forms the musculature; a peritoneum is absent.

which are especially abundant at the tip of the proboscis and on the collar, are responsible for secreting the mucus covering of the body. Some of the gland cells produce bromine-containing compounds that impart a strong medicinal odor to acorn worms and may protect them from bacterial infection or animal predation. A basal lamina and a connective tissue dermis lie beneath the epidermis, but the dermal fibers are not well developed or well organized, perhaps accounting for the fragility of the body.

Acorn worms are tricoelomate, like most lophophorates and echinoderms. A single unpaired **protocoel** occupies the proboscis, a pair of **mesocoels** is found in the collar, and a pair of **metacoels** is found in the trunk (Fig. 17–4A). The protocoel opens to the exterior on the proboscis stalk via one (sometimes two) small opening, the **proboscis pore** (Fig. 17–4A). Each of the mesocoels has a duct that leads to the exterior by way of the first gill pores; the metacoel lacks coelomic ducts. As in echinoderms and most chordates, there is a trend in hemichordates for specialized

diverticula to develop from one coelomic region and project into another. In enteropneusts, the clearest example of this phenomenon is the extension of two tubular anterior outgrowths of the metacoels, the **perihemal coeloms,** into the collar region of the body, one on each side of the dorsal blood vessel. The linings of the perihemal coeloms are muscular, and their action may help to circulate blood in the dorsal blood vessel.

The general coelomic lining is a simple nonstratified mesothelium, and most of the cells are differentiated as smooth muscle cells, each bearing a single cilium, which may help to circulate the coelomic fluid. The muscle cells are pseudostratified into a poorly developed outer circular musculature and an inner, well-developed, longitudinal musculature (Fig. 17–3). In addition, radial muscle fibers span the coelom in all body regions and may functionally replace the weak circular muscles. The musculature is best developed in the proboscis, the locomotory organ, and in the trunk (Figs. 17–4A; 17–5A), where

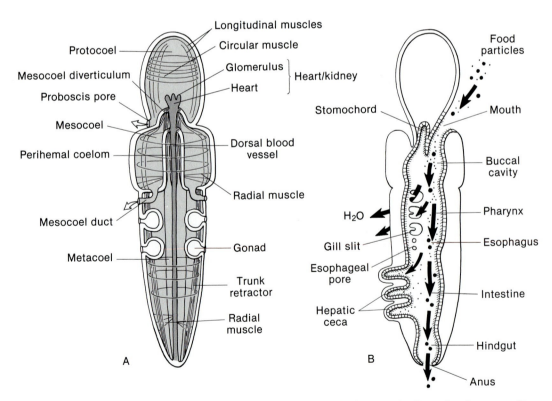

FIGURE 17–4. A, Musculature, blood-vascular, and coelomic organization of enteropneusts, dorsal view. **B,** Digestive system, showing the path of food and removal of water, via the gill slits, in the pharynx.

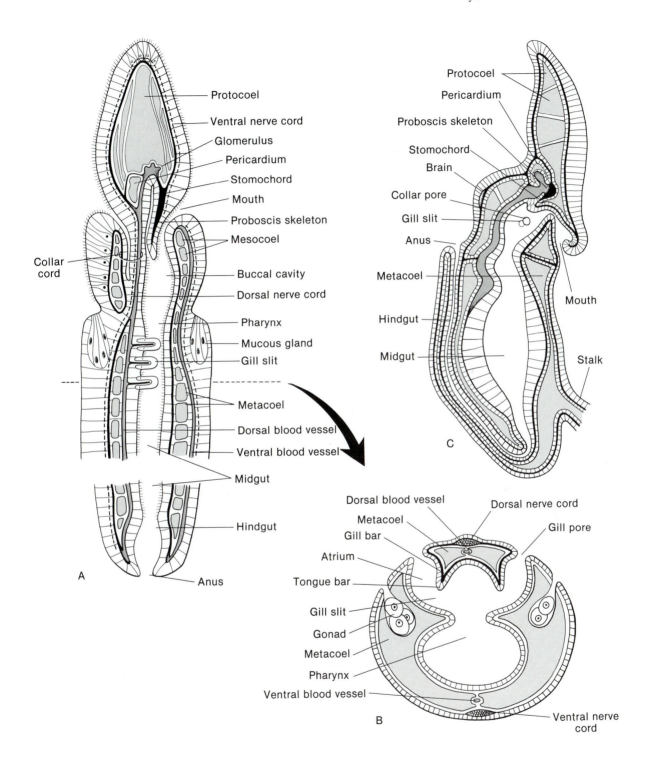

FIGURE 17–5. A, Diagrammatic sagittal section of an enteropneust. **B,** Transverse section through the pharynx. **C,** Diagrammatic sagittal section of a pterobranch.

two ventrolateral bands form the retractor muscles that shorten the trunk. Generally, acorn worms are sluggish animals that move slowly using retrograde peristaltic contractions of the proboscis. After the proboscis advances and anchors, the trunk and collar are pulled forward and the cycle is repeated (Fig. 17–6A–D).

Acorn worms may be deposit feeders, suspension feeders, or both. Burrowing species are predominantly deposit feeders, consuming sand and mud and digesting out the organic matter, but a few species protrude the proboscis from the burrow opening and collect suspended material from the water. Nonburrowing species are primarily suspension feeders but may also ingest deposited material. Food in the form of detritus and plankton, or sediment in burrowing species, is trapped in mucus on the proboscis and transported posteriorly over its surface toward the mouth. Some of the material enters the mouth ventrally after passing over a shallow ciliary groove, the **preoral ciliary organ,** located on the posterior face

of the proboscis. Particles that are not ingested are transported rearward by cilia over the collar and trunk (Fig. 17–6E).

The digestive tract is a straight tube that is histologically differentiated into a number of regions (Fig. 17–4B). The large mouth, which may be closed by withdrawing the proboscis into it like a plug, is the opening between the ventral anterior margin of the collar and the proboscis stalk. The mouth leads into a **buccal cavity** within the collar. The slender proboscis stalk attaches to the inner dorsal wall of the buccal tube and is supported internally by a cartilagenous, wishbone-shaped **proboscis skeleton.** A middorsal diverticulum, the **stomochord,** extends from the buccal cavity through the proboscis stalk and into the proboscis, where it lies directly below the heart and nephridium (discussed below) (Figs. 17–4B; 17–5A; 17–7).

The buccal cavity passes posteriorly into the pharynx, which occupies the branchial region of the trunk, and is laterally perforated by the gill slits. In

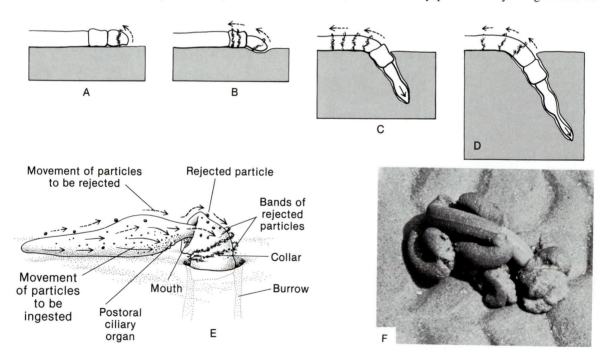

FIGURE 17–6. A–D, Burrowing in acorn worms is largely the result of retrograde peristaltic waves along the muscular proboscis but is assisted by surface cilia, which transport sand rearward along the body. **E,** While feeding, particles are trapped in surface mucus and transported by cilia to the mouth. Rejected particles are carried over the rim of the collar and moved posteriorly in rings. **F,** The posterior end and ropy fecal cast of the giant acorn worm, *Balanoglossus gigas,* on the surface of an exposed sand flat at low tide. The cast is approximately 1 cm in diameter. *(E, Modified and redrawn after Burdon-Jones, C. 1956. Observations on the enteropneust, Protoglossus kohleri. Proc. Zool. Soc. Lond. 127:35.)*

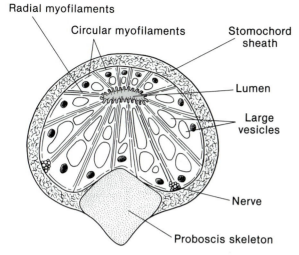

Radial myofilaments
Circular myofilaments
Stomochord sheath
Lumen
Large vesicles
Nerve
Proboscis skeleton

FIGURE 17–7. Cross section of the stomochord of an acorn worm. The stomochord owes some of its stiffness to turgid intracellular vesicles and muscle filaments, but a highly thickened and cross-linked part of the extracellular sheath, the proboscis skeleton, is the principal skeletal element. It secures the proboscis to the collar and, with the stomochord, forms a rigid platform to support the heart.

many acorn worms, such as species of *Balanoglossus* and *Ptychodera,* the pharynx is divided into an upper respiratory chamber that bears the gill slits and through which water flows, and a ventral food chamber that transports the food (Fig. 17–5B).

Behind the pharynx the gut continues as a short **esophagus,** where food particles and mucus are molded into a cord. Simple, oval, **esophageal pores** open from the esophagus directly to the exterior in some families and may help to remove excess water (Fig. 17–4B). The esophageal pores may be an early stage in the development of gill slits, which are added continuously as the animal grows.

The **intestine** constitutes the remainder of the gut. The anterior hepatic region is the site of extracellular and intracellular digestion and nutrient storage (hepatic sacs). The posterior intestine compacts the fecal material and conveys it to the terminal anus, which releases the feces in a long ropy casting (Fig. 17–6F).

Two common burrowing enteropneusts, *Saccoglossus kowalevskii* (Fig. 17–2B) and *Balanoglossus aurantiacus* (Fig. 17–2A), are both surface deposit feeders that occupy U-shaped burrows but have markedly different feeding styles. *Saccoglossus*

emerges from its burrow and extends its long proboscis radially over the surface from the burrow opening (Fig. 17–2F). While extended, the proboscis conveys the surface detritus back to the mouth. After feeding for a time, the proboscis is retracted and then reextended in a new direction. After several cycles of proboscis movements, *Saccoglossus* creates a spoke-like series of radiating furrows, a **feeding rosette,** around the mouth of its burrow. The burrow of *Balanoglossus* typically has two surface openings and a third crater-like depression, the feeding funnel, which lies above a branch from the head end of the U-shaped burrow (Fig. 17–2E). While feeding, *Balanoglossus* occupies the branch and positions its proboscis immediately below the bottom of the surface crater. As it removes sand from the base of the funnel, surface detritus tumbles into the funnel, falls to the bottom, and is ingested by the worm. Thus, *Balanoglossus* is able to feed on the organically rich surface detritus without exposing any part of its body above its protective burrow.

The blood-vascular system of hemichordates consists of an anterior heart and two main contractile vessels, one middorsal carrying the colorless blood forward, and one midventral, carrying blood posteriorly. Smaller vessels supply the body wall and major organs.

The **heart/kidney** is located in the posterior of the protocoel (Fig. 17–4A). It is supported beneath by the stomochord and rigid proboscis skeleton and pressurized above by a small contractile coelom, the pericardium (Fig. 17–5A). Blood enters the heart in a connective tissue channel between the stomochord and pericardium. As the blood is pressurized by the slow contractions of the pericardium, approximately 6/min, it flows into efferent vessels for transport elsewhere in the body and into numerous blind-ending folds of the heart wall. These folds greatly increase the area of the heart wall in contact with the protocoel, forming the **glomerulus** (Figs. 17–4A; 17–5A). The wall of the heart is composed of podocytes. As the heart contracts, blood is ultrafiltered across the wall of the heart into the protocoel, forming primary urine, which may be modified by the muscle cells within the protocoel. Urine then passes from the protocoel to the exterior through the proboscis duct and pore on the left side of the proboscis stalk (Fig. 17–4A).

A pair of ciliated collar ducts, from the mesocoels, transports coelomic fluid to the exterior by way of the first gill pores, but no vascular filtration site has, as yet, been identified in association with these ducts (Fig. 17–4A). On the other hand, podocyte-covered blood vessels associated with the gill bars occur in the trunk coelom and may be sites of nutrient release from the blood into the metacoels; nephridial ducts are absent from the metacoels. The physiology of excretion in hemichordates has not been studied.

The pharyngeal gill slits are assumed to be gas exchange structures. The number of slits can range from a few to as many as 200 or more pairs. Each slit opens through the side of the pharyngeal wall as a narrow U-shaped cleft (Figs. 17–5A; 17–8). The pharyngeal wall between the clefts (the **gill arches**) and that part of the wall projecting downward between the arms of the U (the **tongue bar**) are supported by skeletal thickenings of the dermis of the pharyngeal epithelium. In many species, but not in *Saccoglossus,* narrow horizontal braces, called **synapticles,** bridge the gill slit and join the gill arches and tongue bars. The synapticles help to support the gill slits and are vascularized, like the gill arches and tongue bars, thus supplying blood to the gills.

Each gill slit opens into a saclike **atrium,** which in turn, opens to the exterior via a dorsolateral gill pore (Fig. 17–5B). All of the pores on each side of the body are often located in a longitudinal groove. The walls around the pharyngeal clefts are ciliated and contain a plexus of blood vessels, which are involved in gas exchange. The coelomic surface of some of the vessels also bear podocytes, as mentioned above. The beating cilia produce a stream of water passing into the mouth and out through the gill slits and pores.

The gill slits of both hemichordates and chordates probably originated as an adaptation to eliminate excess water from ingested food and only later adopted a gas exchange function.

The main features of the enteropneust nervous system are an intraepidermal nerve plexus and nerve cords (dorsal and ventral), several circumferential nerve rings, especially in the proboscis, and dorsally, a more or less hollow, subepidermal **collar cord.**

The morphogenesis of the hollow collar cord is essentially like neurulation in the chordates (Fig. 17–46). Neurulation results in an internal hollow tube lined by a ciliated epithelium and opening to the exterior via an anterior **neuropore.** The neuropore persists throughout life in acorn worms and is located immediately inside the dorsal margin of the collar. Sensory cells of unknown function project into the cavity of the cord and perhaps monitor the water flowing into the mouth. The nervous layer at the base of the epithelial cells is continuous with the basiepidermal dorsal nerve anterior and posterior to the collar. Other sensory structures in acorn worms are ciliated neurosensory cells scattered throughout the epidermis and, perhaps, the postoral ciliary organ (Fig. 17–6E).

Acorn worms are exceptionally fragile animals and invariably tear or break posterior to the branchiogenital region when handled. It is likely that they also

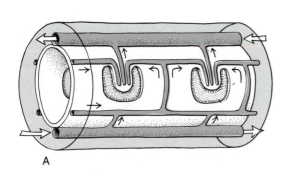

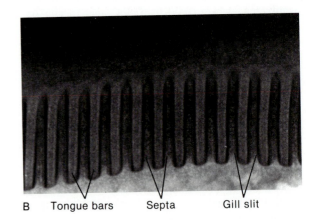

B Tongue bars   Septa   Gill slit

FIGURE 17–8. Enteropneust gill circulation and gill slits. **A,** Diagrammatic section of pharyngeal circulation; anterior is to the left. Arrows indicate direction of blood flow. The distance between the U-shaped gill slits is greatly exaggerated. **B,** Photograph of gill slits from the interior of the pharynx of a living acorn worm.

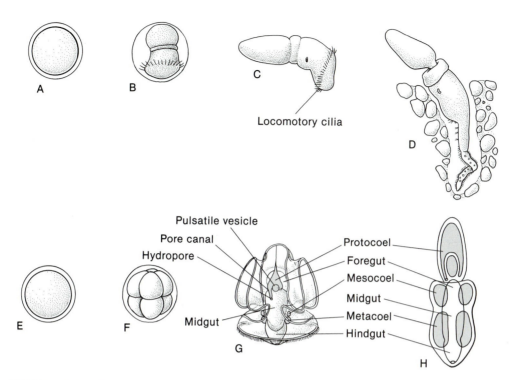

FIGURE 17–9. Enteropneust development. **A–D,** Indirect development in *Saccoglossus kowalevskii.* The large yolky eggs of this species develop into a short-lived, lecithotrophic, swimming larva **(C).** After settlement **(D),** the juvenile develops a postanal tail. **E–F,** Indirect development in *Schizocardium brasiliense.* The small, nonyolky eggs develop into a long-lived planktotrophic larva called a tornaria **(G)** before undergoing a rapid metamorphosis to form a juvenile worm **(H).** *(D, Redrawn after Burdon-Jones, C. 1952. Development and biology of the larva of Saccoglossus horsti. Phil. Trans. Roy. Soc. Lond. B. 236:553–590.)*

suffer damage under natural conditions and then heal the wounds and regenerate missing parts. The Atlantic auger snail, *Terebra dislocata,* profits from the regenerative capacity of acorn worms by feeding repeatedly on a renewable part of the body. *Terebra* lingers on the sandy fecal mounds of *Balanoglossus aurantiacus* and nips off the worm's posterior end as it exposes itself at the surface to defecate. Asexual reproduction by fragmentation has been reported for several species, including members of the genera *Glossobalanus* and *Balanoglossus.*

Acorn worms are dioecious animals that release eggs and sperm into the surrounding seawater where fertilization occurs. The numerous follicular gonads are situated in the connective tissue space of the genital ridges (Harrimaniidae and Spengelidae) or genital wings (Ptychoderidae). Each follicle has its own surface pore. Male gonads are often orange and female gonads, lavender or grey. The zygotes undergo holoblastic, equal, radial cleavage (Fig. 17–9E, F). Gastrulation is by invagination, and the blastopore becomes the anus. The mesoderm and coeloms originate from the archenteron, but all morphogenetic modes—enterocoely, schizocoely, and other patterns—are represented within the class, one or another associated with a given species.

A few groups, such as species of *Saccoglossus,* have large yolky eggs that develop into short-lived lecithotrophic larvae before settling and undergoing metamorphosis (Fig. 17–9A–D). The recently settled juvenile of *Saccoglossus* has a long postanal tail, reminiscent of the pterobranch stalk, that helps to anchor them in the sand. Most enteropneusts have small nonyolky eggs that develop into a transparent, long-lived, planktotrophic larva, called a **tornaria** (meaning "turner", (Fig. 17–9E–H) referring to the rotating metachronal wave of locomotory cilia in the telotroch. The organization of the tornaria is strikingly like that of some echinoderm larvae and provides the main evidence for assuming a close evolutionary relationship between the hemichordates and echinoderms. Like many echinoderm larvae, the tornaria has an anterior protocoel from which a short pore canal extends to a hydropore on the left side of the dorsal midline (Fig. 17–9G). These structures together with a small pulsating vesicle form the larval nephridium, as in echinoderms.

Class Pterobranchia

The pterobranchs are colonial, tube-dwelling, epibenthic hemichordates that superficially resemble bryozoans and hydroids. As in these other two groups, pterobranch zooids are small, not exceeding 5 mm in body length *(Cephalodiscus),* and many species are less than 1 mm long. But despite their small body size, pterobranch organ systems are similar to those of their larger relatives, the enteropneusts. Because of these similarities, only the distinctive features of pterobranchs will be discussed in the following section.

Pterobranchs are uncommon marine animals that rarely occur in shallow water where they grow on the surfaces of shells and rocks. More often, they are dredged from deep water and are best known from the seas around Antarctica. There is growing evidence, however, that pterobranchs are widespread, but easily overlooked, in shallow tropical and temperate seas, including the coasts of Florida and Bermuda. There are 21 species of pterobranchs in three genera: *Cephalodiscus, Atubaria,* and *Rhabdopleura. Atubaria,* however, closely resembles a *Cephalodiscus* that has abandoned its tube, which it often does, and is known only from preserved collections dredged from a depth of 200 to 300 m off the coast of Japan.

Although pterobranchs are rarely seen by most biologists, they have nevertheless figured prominently in theories of deuterostome evolution. Most models of the evolution of echinoderms and chordates place a pterobranch-like ancestor at the base of the evolutionary tree, and within the hemichordates, the pterobranchs are primitive to the enteropneusts in most of their shared structural traits.

External Organization

The tripartite body consists of an anterior oral shield, a short collar, and a saccate trunk (Figs. 17–5C; 17–10C). The broad, flattened **oral shield** is a glandular, ciliated disc on which the zooids glide over the inner surface of their tubes. The short collar region arches dorsally over the oral shield and bears the **arms** and their pinnately arranged **tentacles.** In *Cephalodiscus,* two to eighteen arms, depending on body size and the species in question, radiate from the collar to form a spherical basket above the zooid (Fig. 17–11C), whereas in *Rhabdopleura* (Fig. 17–10C), two long arms are held in a V in front of the zooid and above the tube opening. A pair of ciliated folds of the body wall, the **oral lamellae,** occurs at the base of the arms, one lamella on each side of the collar. A pair of large ciliated pores, the **mesocoel pores,** one from each of the paired collar coeloms, opens laterally at the collar/trunk boundary (Fig. 17–11G). In addition, a single pair of gill pores opens immediately behind the mesocoel pores in *Cephalodiscus,* but they are absent in *Rhabdopleura.* The trunk bears a conspicuous fleshy stalk, which unites zooids together into a colony (Figs. 17–10; 17–11).

Colonial Organization and Locomotion

Pterobranch colonies inhabit a branching tubular network, called a **coenecium,** that is secreted from glands in the oral shields of the zooids. In *Rhabdopleura,* the coenecium consists of prostrate tubes that give rise to a series of unbranched vertical tubes, each of which has a simple circular opening and is occupied by a single zooid (Fig. 17–10A). The stalks of all zooids in the colony join a common threadlike stolon that runs through the prostrate tubes (Fig. 17–10B). Most colonies of *Rhabdopleura* are between 1 and 25 cm in diameter and 5 to 10 mm in height. The coenecium of some species of *Cephalodiscus* resembles that of *Rhabdopleura* but is typically larger (Fig. 17–11A). In others, however, the coenecium is a bulky cakelike mass that reaches 30 cm in diameter and 10 cm in height. The tube openings may have smooth circular rims, as in *Rhabdopleura* (Fig. 17–10D), but in many species the rim bears one or more long spines on which the zooids perch while feeding (Fig. 17–11B, C). A *Cephalodiscus* coenecium is occupied by many unjoined zooids, each of which produces buds from an adhesive disc at the base of the stalk to form a small colony (Fig. 17–11B). Thus, each coenecium of *Cephalodiscus* contains an aggregation of colonies.

Individual zooids move to and from their feeding apertures and within the coenecium by gliding on the ciliated ventral surface of the oral shield. As they

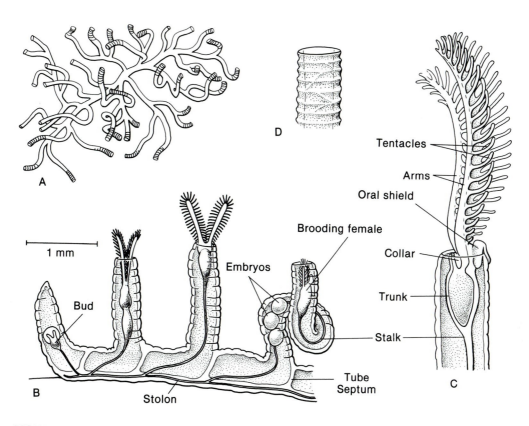

D

Tentacles

Arms

Oral shield

Brooding female

Collar

Embryos

Trunk

1 mm

Bud

Stalk

Tube
Septum

C

B

Stolon

FIGURE 17–10. The pterobranch, *Rhabdopleura*. **A, B,** Colonial organization. **C,** Feeding position of zooid with oral shield folded over the tube rim. While in this position, the oral shield secretes a new tube section, several of which are shown in **D.** *(A, Redrawn after van der Horst, C. J. 1939. Hemichordata. In Bronn, H. C. (Ed.): Kl. Ord. Tierreichs. 4:1–735.)*

move about, they remain tethered to the stolon *(Rhabdopleura)* or adhesive disc *(Cephalodiscus)* by their contractile stalks, which can be stretched to ten times their contracted lengths. The adhesive disc of each *Cephalodiscus* colony is also motile, gliding on surfaces like the oral shield. Under certain conditions, the adhesive disc and its attached zooids abandon the coenecium, creep over the substratum, and secrete a new coenecium.

Body Wall and Internal Organization

The pterobranch body wall is similar to that of enteropneusts, but in pterobranchs, the cells are always monociliated, a primitive feature (Fig. 17–3). The musculature consists of longitudinal and radial fibers only; circular body wall muscles are absent (Fig. 17–5C). The radial muscles are common in the oral shield, collar, and arms. The principal longitudinal

muscles are a pair of smooth stalk retractors that draw the zooid into the coenecium, cross-striated tentacular muscles that flick the tentacles toward the food groove, and striated dilator muscles of the mesocoel ducts that open the mesocoel pores.

All of the coelomic lining cells, including the myocytes, are monociliated and help to circulate the coelomic fluid. Some of the mesothelial cells on the gut wall in the trunk store nutrients. A heart/kidney occurs in the unpaired protocoel, which opens to the exterior via two tiny protocoel pores, one on each side of the dorsal midline. Cilia in the large mesocoel ducts and pores transport coelomic fluid to the exterior and may have some excretory function.

The organization of the pterobranch blood-vascular system is imperfectly known, but the major vessels include the heart; dorsal blood vessel; an enlargement of the dorsal vessel in the trunk to form the

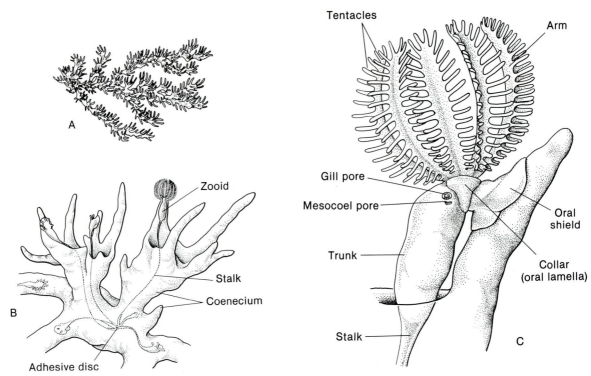

FIGURE 17–11. The pterobranch, *Cephalodiscus.* **A, B,** Colonial organization. Zooids emerge from openings in the coenecium and perch on branch tips while feeding **(B).** The leftmost zooid in **B** occupies the common coenecium but belongs to a colony separate from that of the other zooids. **C,** Feeding posture. *(A, Modified and redrawn from van der Horst, C. J. 1939. Hemichordata. In Bronn, H. C. (Ed.): Kl. Ord. Tierreichs. 4:1–735. B, C, Modified and redrawn after Lester, S. M. 1985. Cephalodiscus sp. (Hemichordata: Pterobranchia): Observations of functional morphology, behavior and occurrence in shallow water around Bermuda. Mar. Biol. 85:263–268.)*

genital blood sinus, which surrounds the gonad; and a ventral vessel (Fig. 17–5C). Smaller blood vessels are known to occur (for example, one extends into each tentacle), but details of the blood circuit and direction of flow of the colorless blood are unknown. All blood vessels and sinuses are simple, unlined channels in the connective tissue between epithelia. Gas exchange most likely occurs across the arms, tentacles, and general body surface, all of which are ciliated and generate water flow over the surface of the body.

The nervous system is entirely within the epidermis, and a hollow, submerged collar cord is absent, although a concentration of neurons ("brain") occurs dorsally in the collar region of the body (Fig. 17–5C). A dorsal nerve extends anteriorly and posteriorly from the collar ganglion. The anterior extension branches into the arms, and one nerve, which probably carries both motor and sensory information, innervates each tentacle. A nerve ring (or rings) passes around the pharynx from the ganglion and gives rise anteriorly to a nerve plexus in the ventral oral shield, and posteriorly to a ventral nerve cord. The ventral cord passes into the stalk.

Innervation of the muscles apparently occurs by diffusion of neurotransmitters across the epidermal basal lamina to the coelomic myocytes. Individual ciliated sensory cells are scattered on the abfrontal surface of the tentacles and perhaps elsewhere on the body.

Feeding and Digestive System

Pterobranchs filter feed using their hollow ciliated arms and tentacles. Lateral cilia on the tentacles create a flow of water and particles over the frontal surface of the tentacles and arms. Small particles are intercepted by these frontal surfaces or are batted onto them by the flicking tentacles. Once captured, cilia transport the particles along the length of the arms to a ciliated groove that passes under the undulating oral lamellae before joining the mouth. The oral lamellae, like lophophorate epistomes, may help to direct food into the mouth and may also be areas where excess water is eliminated and food concentrated. The frontal surface of the arms transports large inedible particles away from the mouth to the arm tips, where they are discharged downstream in the exhaust flow.

Feeding zooids of *Rhabdopleura* grip the rim of the tube opening with the oral shield and extend the arms and tentacles, which resemble two tiny feathers, into the surrounding water (Fig. 17–10C). As feeding continues, each zooid rotates around the tube rim, thus exposing the frontal surface of its feeding apparatus in all compass directions. Like *Rhabdopleura, Cephalodiscus* also filters suspended particles from 360 degrees around the zooid body but does not rotate around its tube aperture. Instead, *Cephalodiscus* has a radially symmetrical feeding apparatus, which gathers particles from all directions.

Pterobranchs have a U-shaped gut. The mouth is located ventrally under the posterior margin of the oral disc, and the anus is middorsal on the trunk, immediately posterior to the collar. Movement of food through the gut is by ciliary transport. Food entering the mouth passes into a short **pharynx,** which in *Cephalodiscus* has one pair of gill clefts, which eliminate excess water from the food. Food passes from the pharynx into the saccate **stomach** located posteriorly and ventrally in the body. From the posterior end of the stomach, the slender intestine curves dorsally and anteriorly and passes into a short hindgut before opening to the exterior via the anus (Fig. 17–5C). Digestion probably occurs in the stomach, and fecal pellets are formed in the intestine. Absorption may occur in the stomach and part of the intestine.

Reproduction

Pterobranch zooids arise asexually, by budding from the stolon (*Rhabdopleura;* Fig. 17–10B) or adhesive disc (*Cephalodiscus;* Fig. 17–11B) during colony growth. Moreover, an individual adhesive disc of *Cephalodiscus* may be capable of fragmentation or fission to form new colonies.

Pterobranch colonies are believed to be dioecious, and the male and female zooids, which have one (*Rhabdopleura)* or two (*Cephalodiscus)* gonads, often show sexual dimorphism. In *Cephalodiscus sibogae,* for example, male zooids have a rudimentary lophophore, and much of the trunk is occupied by the large testes. Male and neuter zooids of *Rhabdopleura normani,* however, retain a fully developed lophophore, and that of female zooids is reduced (Fig. 17–10B). The details of fertilization are not fully known, but the ripe males probably release sperm or spermatophores into the water. Fertilization is internal, and development begins in the body of the female. Later developmental stages are brooded within a modified part of the coenecium. In *Rhabdopleura normani,* embryos are brooded in the lower part of a coiled vertical tube that is occupied by the nonfeeding female zooid (Fig. 17–10B). Cleavage is initially radial but soon becomes bilateral and results in a coeloblastula (Fig. 17–12A–E). Gastrulation is by ingression or delamination, and the coeloms form schizocoelously before the gut differentiates.

The embryos develop into short-lived, uniformly ciliated, lecithotrophic larvae, which are released from the coenecium into the sea (Fig. 17–12F). After a brief swimming existence of a day or two, the larvae settle and secrete a cocoon (coenecium), which encloses and glues them to the substratum. A gradual metamorphosis occurs within the cocoon, resulting in the morphogenesis of an ancestrula, which breaks through the upper wall of the cocoon to feed and later buds to form a colony (Fig. 17–12G–J).

PHYLUM CHORDATA: SUBPHYLUM UROCHORDATA

The chordates are the largest phylum of deuterostomes, but most chordates are vertebrates and fall outside the scope of this book. Two subphyla, the Urochordata and Cephalochordata, lack a backbone but possess the four distinguishing chordate characteristics: at some time in the life cycle, there can be found a **notochord,** a **dorsal hollow nerve cord, gill slits** (pharyngeal clefts), and a **postanal tail.**

Adult urochordates, commonly known as **tunicates,** are common and conspicuous marine animals that little resemble other chordates. Most are sessile, and the body is covered by a complex exoskeletal tunic. There is a highly developed, perforated pharynx, but the notochord and nerve cord are absent in the adult. Only the larval stage, which resembles a microscopic tadpole, possesses distinct chordate char-

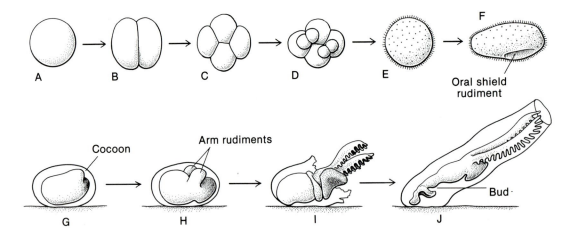

FIGURE 17–12. The yolky eggs of *Rhabdopleura* **(A)** show a form of biradial cleavage **(D)** leading to a planula-like, ciliated, lecithotrophic larva **(F)**. The ventral furrow of the larva is not the blastopore but rather the invaginated oral shield rudiment **(G–J)**. After a brief swimming existence, the larva settles, secretes a cocoon (tube rudiment), and gradually differentiates into a feeding zooid that produces other colony members by budding. *(Recombined, modified, and redrawn after Lester, S. M. 1988. Settlement and metamorphosis of Rhabdopleura normani (Hemichordata: Pterobranchia). Acta Zool. 69:111–120.)*

acteristics. The tunicates include three classes: the Ascidiacea, the Thaliacea, and the Larvacea. The ascidians contain the majority of species and the most common tunicates. The other two classes are adapted for planktonic existence. Approximately 1250 species of urochordates have been described.

Class Ascidiacea

External Organization

Ascidians, often called sea squirts, are common sessile marine invertebrates throughout the world. Although some species are solitary and large, the many colonial species have colonies comprising few to hundreds of tiny zooids (Fig. 17–13). The majority of ascidians are found in shallow waters, where they attach to rocks, shells, pilings, and ship bottoms or are sometimes fixed in mud and sand by filaments or a stalk. A great diversity of species inhabit shallow tropical seas, and many minute colonial forms inhabit crevices in old coral heads and the underside of coralline rock. Others form large, conspicuous clusters on gorgonian corals and mangrove roots, or massive rubbery lobes on rocks and pilings (Fig. 17–14). Over 119 species have been taken from depths greater

than 200 m (Fig. 17–15), and there are even a few interstitial ascidians.

The bodies of solitary species range from spherical or cylindrical to irregular in shape. One surface is attached to the substratum, and the opposite bears two openings, the **buccal** and **atrial siphons** (Fig. 17–16A). All colors are found in ascidians, and some colonial species are strikingly beautiful. The body of solitary species ranges in size from that of a seed (1 mm or so in diameter) to that of a large potato, which some species closely resemble (*Cnemidocarpa verrucosa* from the Antarctic reaches 18 cm). *Halocynthia pyriformis,* which is found on the Atlantic coast north of Cape Cod, is called the sea peach, which it resembles in size, shape, and color. Colonial species typically range in size from centimeters to a meter or more and may be several centimeters in thickness.

In contrast to shallow water ascidians, 95% of which live attached to rigid surfaces, most deep-sea species inhabit soft bottoms. The greatest number are tiny, spherical zooids anchored by fibrils, but there are also some large, stalked forms and some strange, transparent, irregular, raglike species that float while tethered over a small attachment point (Fig. 17–15).

Although they were originally thought to be molluscs because of their tough exterior tunics, soft fleshy bodies, and large gills (pharynx), ascidians were shown to be chordates by the Russian zoologist, Alexander Kowalevsky, in the latter half of the nineteenth century. Ascidians, as sessile filter feeders, nevertheless show convergent similarities to other filter-feeding groups, such as sponges and bivalve molluscs.

Colonial Organization

Most of the larger ascidians, such as *Styela, Ascidia, Microcosmos,* and *Molgula* are solitary and often are called **simple ascidians** (Fig. 17–13D, F, G). There are, however, many colonial, or **compound ascidians** (Fig. 17–13A–C, E). Colonial organization has arisen independently a number of times within the class, and a number of types of colonies occur. In general, the zooids composing a colony are very small, although the colony itself may reach considerable size, often a meter or more (Fig. 17–14).

In the simplest colonies, the zooids are spaced apart but are united by **stolons.** For example, in *Perophora,* the colony is like a vine to which the globular zooids are attached (Fig. 17–13C). In others of these compound forms, such as *Ecteinascidia turbinata,* the stolons are short and the zooids form tuftlike clusters.

A more intimate association is seen in some ascidians, such as species of *Clavelina,* in which not only the stolons but also the basal parts of the body tunics are joined to other zooids (Fig. 17–13B).

In the most specialized colonial families, all the zooids composing the colony are completely embedded in a common tunic, and the colonies may be

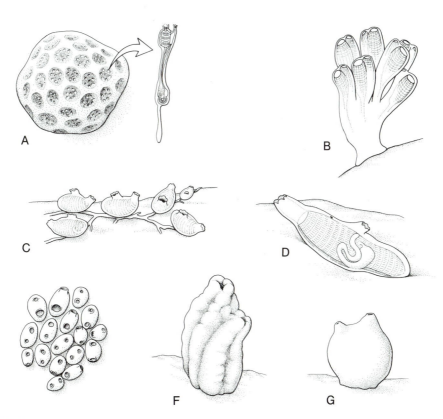

FIGURE 17–13. Ascidian diversity. **(A)** Aplousobranch colonies of *Aplidium stellatum* (sea pork), and **(B)** the light bulb tunicate, *Clavelina oblonga.* **(C)** The colonial phlebobranch, *Perophora viridis,* and **(D)** the solitary phlebobranch, *Ascidia curvata.* The colonial stolidobranch, *Symplegma rubra,* and **(F)** solitary stolidobranchs, *Styela plicata* and **(G)** the sea grape, *Molgula manhattensis.*

FIGURE 17–14. Sea liver, *Eudistoma hepaticum,* grows into tough, massive colonies on floats, jetties, and pilings along the southeastern coast of the United States.

small or large, encrusting and thin, or massive and thick (Figs. 17–13A; 17–14). In some species, each microscopic zooid is more or less independent of its neighbors, and both of its siphons open onto the surface of the common tunic. In others, however, the colonial zooids are integrated into organized **systems,** and all the zooids in each system share a common atrial aperture, usually surrounded by the individual openings of their buccal siphons. *Botryllus schlosseri* and similar species form flat, encrusting colonies in which such systems are each in the form of a star (Fig. 17–17A). The buccal siphons of each zooid open separately to the exterior in a ring, but the atria open into a common cloacal chamber, which has one aperture in the center of the ring (Fig. 17–17B). The zooids of *Botryllus* are only a few millimeters in diameter, but because a single colony may contain many star-shaped systems, the entire colony may be 12 to 15 cm in diameter.

Body Regionation

Within the tunic, the ascidian zooid can be conveniently divided into three regions: an anterior, or distal, **thoracic region** containing the pharynx; an **ab-**

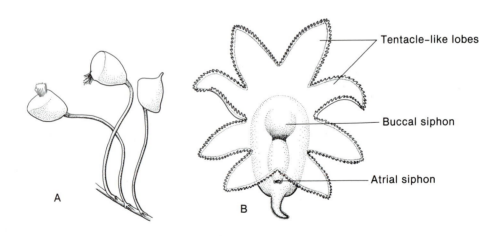

FIGURE 17–15. Deep-sea tunicates. Most live on soft bottoms in contrast to predominately hard-bottom forms of shallow water. Note the long slender stalks or rootlike process by which fixation in soft bottoms is made possible. **A,** *Coleolus suhmi.* **B,** *Octonemus ingolfi.* This last species, unlike most other tunicates, is believed to be predaceous. Prey are trapped by means of the large siphon lobes. *(Drawn from photographs by Monniot and Monniot, 1975.)*

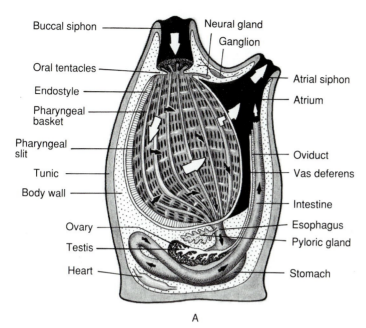

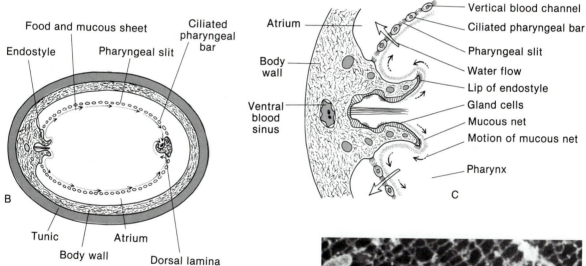

FIGURE 17–16. A, Diagrammatic lateral view of a solitary ascidian showing internal organs. Large arrows show the path of water flow, small arrows that of the mucous net. **B,** Cross section of the pharynx in **A. C,** Enlargement of the endostylar region of **B** to show secretion of the mucous net and the water flow through gill slits. **D,** Scanning electron micrograph of the mucous net of *Ciona intestinalis.* (D, From Flood, P. R., and Fiala-Medioni, A. 1979. Filter characteristics of ascidian food trapping mucous films. Acta Zool. 60:271.)

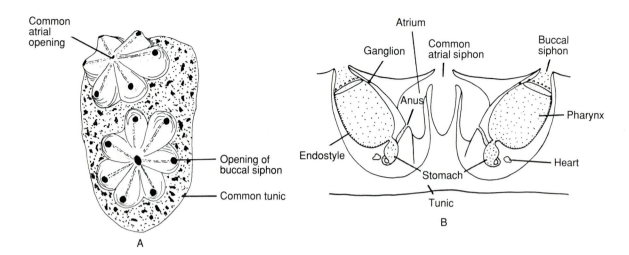

FIGURE 17–17. A, A compound ascidian, *Botryllus schlosseri* **B,** Vertical section through *Botryllus* showing arrangement of zooids in a system. *(A, After Milne-Edwards from Yonge; B, After Delage and Herouard.)*

dominal region containing the digestive tract and associated structures; and the **postabdominal region** (Fig. 17–18). The postabdomen, which may be long and threadlike, is the most basal (posterior) part of the body and contains the heart and reproductive organs. All three regions are distinct only in the microscopic zooids of some colonial species (aplousobranchs). Most ascidians lack a postabdomen, and many of the species with large zooids (*Ascidia, Styela, Molgula, Microcosmos*) even lack an abdomen, the visceral organs having shifted anteriorly to one side of the pharynx.

Larval Origins of Adult Body Form

The attached condition of adult ascidians has led to such specialization that it is necessary to examine the larva and its metamorphosis to appreciate fully the chordate affinities and the origins of the adult form. The tadpole larva swims with a long, posterior tail, which contains a notochord and a neural tube (Fig. 17–19). Dorsally, the anterior half of the larva* contains an ocellus and statocyst both within the dilated end of the neural tube, which later gives rise to the adult cerebral ganglion and neural gland. The mouth, which becomes the buccal siphon, is located anteriorly but is not open during larval life; the larva is a lecithotroph. The mouth leads into the pharynx, which in turn is followed by a twisted digestive loop with a dorsally directed intestine. The pharynx is perforated by a few slits that open into a surrounding pocket, the atrium. Although the atrium does not open to the exterior in the larva, it later becomes exposed to the surface and forms the atrial siphon (Fig. 17–19A).

At the end of the free-swimming stage, the larva settles to the bottom, attaching by three anterior adhesive papillae. A radical metamorphosis now ensues, during which the tail is retracted and the notochord and nerve cord absorbed. As a result of rapid growth of the region between the adhesive papillae and mouth, the entire body is rotated 180 degrees. The mouth, or buccal siphon, is shifted backward to the end opposite that of attachment. The atrium expands to capture the anus and to form a jacket around the pharynx. The number of pharyngeal slits rapidly increases. The siphons open to the exterior as the larval

---

*This description is based on aplousobranch larvae. The larvae of many ascidians are less well developed, and differentiation of structures occurs only during metamorphosis.

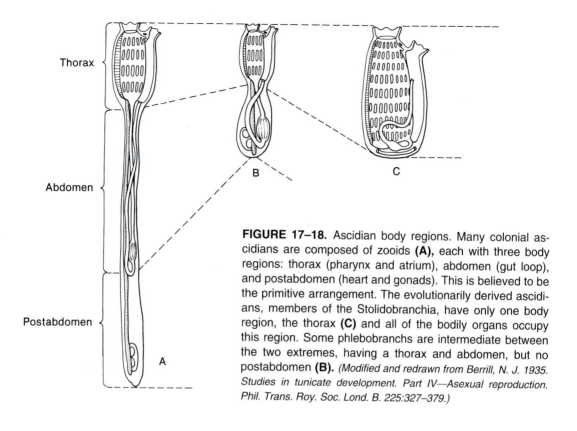

FIGURE 17–18. Ascidian body regions. Many colonial ascidians are composed of zooids **(A),** each with three body regions: thorax (pharynx and atrium), abdomen (gut loop), and postabdomen (heart and gonads). This is believed to be the primitive arrangement. The evolutionarily derived ascidians, members of the Stolidobranchia, have only one body region, the thorax **(C)** and all of the bodily organs occupy this region. Some phlebobranchs are intermediate between the two extremes, having a thorax and abdomen, but no postabdomen **(B).** *(Modified and redrawn from Berrill, N. J. 1935. Studies in tunicate development. Part IV—Asexual reproduction. Phil. Trans. Roy. Soc. Lond. B. 225:327–379.)*

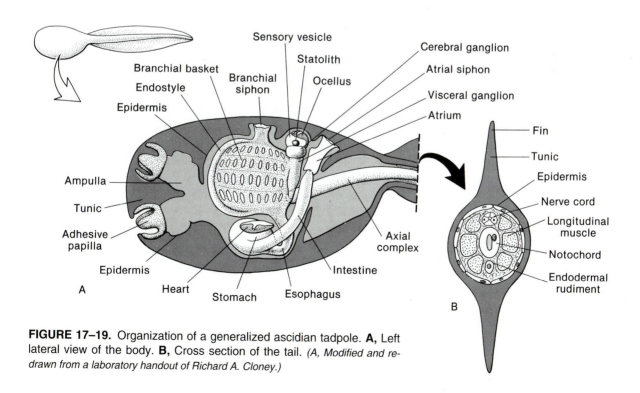

FIGURE 17–19. Organization of a generalized ascidian tadpole. **A,** Left lateral view of the body. **B,** Cross section of the tail. *(A, Modified and redrawn from a laboratory handout of Richard A. Cloney.)*

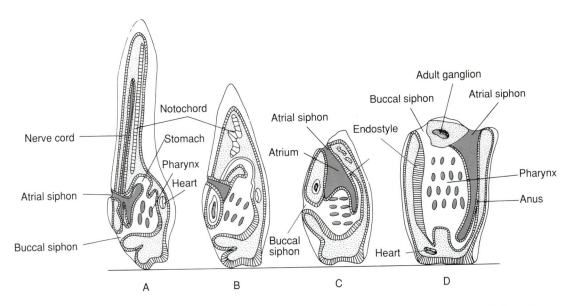

FIGURE 17–20. Metamorphosis of the ascidian tadpole larva. **A,** Diagrammatic lateral view of an ascidian tadpole larva that has just attached to the substratum by its anterior end. **B** and **C,** Metamorphosis. **D,** Young individual immediately after metamorphosis. *(Modified and redrawn from Seeliger.)*

tunic is molted and the adult tunic established, and the juvenile (ascidiozooid, or blastozooid) begins to feed (Fig. 17–20).

Body Wall

The body of an ascidian is covered by a single layer of epithelial cells, but this epidermal covering does not form the external surface. Instead, the entire body is invested with a special cuticular covering, the **tunic,** which is the characteristic exoskeleton of most members of the subphylum and from which the name *tunicate* is derived (Fig. 17–21). The tunic is usually quite thick but varies from a soft delicate consistency to one that is tough and similar to cartilage. The tunic of *Aplidium (Amaroucium) stellatum,* called sea pork, has both the appearance and texture of pork fat. On the other hand, the transparent gelatinous tunic of *Diplosoma* species encloses numerous colonial zooids that resemble frog eggs in their jelly mass.

Like most cuticles, the tunic is composed of varying amounts of proteins, carbohydrates, and water but it also has several features that are unique to urochordates (Fig. 17–21). One is the presence of structural fibers composed of a kind of cellulose, called **tunicin.** These parallel fibers usually occur in thin and successive sheets, each of which has its fibers oriented at a different angle from that preceding it, stacked within the tunic like successive plys of plywood. This fiber arrangement, which is similar to that of some arthropod cuticles, gives the tunic toughness and strength. Another striking feature is the presence of blood cells in the tunic, and in some species, **tunic blood vessels** invade the tunic from the body, circulating blood through the exoskeleton, which can thus be viewed as a living tissue akin to a connective tissue within the body. The third unique characteristic is that ascidians are the only animals that *do not molt* as they grow *inside* a thick exoskeleton. This means that the tunic must grow along with the zooid itself, and the tunic blood cells, along with the epidermis, are responsible for the enlargement and maintenance of the tunic.

The tunic clearly has supportive and protective functions that are enhanced in some species by the addition of formed elements. Species of didemnids and some pyurids synthesize calcareous spicules or plates and embed them in the tunic. Some species, such as *Eudistoma carolinense,* incorporate sand grains in the tunic to such an extent that their bodies are rock hard and are nearly unrecognizable as living organisms. Others, for example *Didemnum psammathodes,* whose colonies spread over rocks like

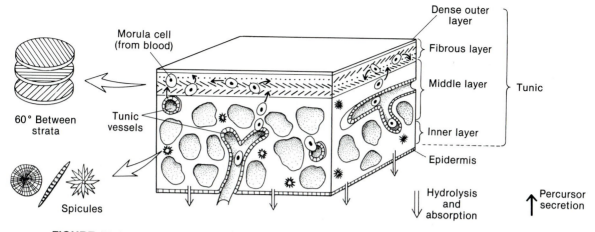

FIGURE 17–21. Generalized ascidian tunic. The extracellular tunic of ascidians is a dynamic exoskeleton that enlarges, without being molted, as the body grows. Although the process of tunic growth is not fully understood, tunic precursors are transported to the tunic by the blood in species having a vascularized tunic and released into the outer surface of the tunic. Some removal of the inner layer of the tunic by the epidermis may also occur.

melted chocolate, incorporate fecal material in the tunic, perhaps for camouflage and to increase its stiffness.

It is by means of the tunic that ascidians adhere to the substratum, and the tunic is often roughened and papillose in this region. Often, rootlike stolons ramify from the base of the body, and these too are covered by the tunic.

A basal lamina and thick gelatinous connective tissue layer underlie the epidermis. The muscles, nerves, blood channels, and amoeboid cells occur in the connective tissue layer. The musculature consists of distinct outer circular and inner longitudinal bands of smooth muscle. The longitudinal fibers extend from the body to the siphons and on contraction withdraw the siphons (Fig. 17–22). The circular muscles predominate on the siphons as sphincters that regulate the openings. Contraction of the body wall musculature occurs periodically, compressing the body, forcing water out through the siphons in jets, and eventually closing the siphons. The **periodic squirting** of water from the siphons, characteristic of all ascidians, and often in response to unwanted matter in the water or other disturbance, is the reason for their common name—"sea squirt."

Pharynx and Atrium

The anterior **buccal siphon** opens internally into a large pharynx, and a circlet of **buccal tentacles** pro-

jecting from within the siphon prevents large objects from entering with the water current (Fig. 17–16A). The walls of the pharynx are perforated with small gill slits (stigmata, each a stigma) that permit water to pass from the pharyngeal cavity into the surrounding atrium and then out by way of the atrial siphon. The pharynx is completely surrounded by the atrium except where the pharynx is attached to the body wall. Cordlike mesenteries cross the atrium between the atrial and pharyngeal walls, suspending the unattached part of the pharynx, which lacks a visceral skeleton, and prevents it from collapse. Dorsally, the atrium opens to the exterior through the atrial siphon. The atrial region just inside the siphon is sometimes called the **cloaca** because its receives both wastes

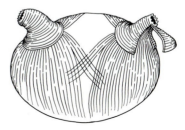

FIGURE 17–22. Musculature of *Pyura ostreophila*. *(Modified and redrawn from Kott, P. 1989. Form and function in the Ascidiacea. Bull. Mar. Sci. 45:253–276.)*

from the anus and gametes from the gonoducts. The entire atrial lining is derived from ectoderm and is continuous through the atrial siphon with the ectodermally derived epidermis. Traditionally, the pharynx is believed to be derived from embryonic endoderm, but a recent developmental study of the edible Japanese ascidian, *Halocynthia roretzi*, by Hiroki Nishida (1987) indicates that the pharynx, like the atrium, originates from ectoderm.

Nutrition

Ascidians are filter feeders and remove plankton from the water current that passes through the pharynx. The water current is produced by the beating of lateral cilia on the margins of the gill slits, and an enormous quantity of water is strained for food. A specimen of *Ascidia nigra*, itself only a few centimeters in length can pass 173 l of water through its body in 24 h (Fig. 17–23A).

On the ventral side of the pharynx—the side opposite the atrial siphon—a deep gutter, the **endostyle,** extends the length of the pharyngeal wall (Fig. 17–16B, C). The endostyle, which is the evolutionary forerunner of the vertebrate thyroid gland, is the principal center for the elaboration of a specialized mucus—a complex mucoprotein containing iodine bound to the amino acid tyrosine. The long flagella at the bottom of the gutter deflect mucus to the side, and the keyhole shape of the organ perhaps ensures proper mixing from different zones of secretion to form the final product. Endostylar flagella cast mucus onto the lining of the pharynx in two sheets, one on each side of the endostyle. Each sheet, which microscopically resembles mosquito netting, is slowly transported from the endostyle (ventral) to the opposite side of the pharynx (dorsal) by the action of frontal cilia on the pharyngeal lining (Fig. 17–16D). As the nets move over the slits, water passes through

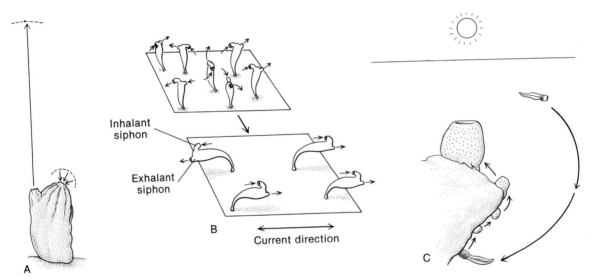

FIGURE 17–23. A, Water movement around siphons of the tropical *Ascidia nigra.* **B,** Recently settled juveniles of *Styela montereyensis* may attach with siphons oriented randomly in a bidirectional water current (arrows), but through differential mortality, those remaining have siphons oriented in the plane of the current, which enhances water flow through the animal. **C,** Movement and light adaptation in the tropical compound ascidian, *Didemnum molle.* This species harbors the photosynthetic symbiont, *Prochloron.* Larvae settle, and transparent juvenile colonies develop on the shaded undersides of coral rubble, but as the colony grows it migrates slowly over the substratum into the fully lighted waters on the surface of the rock. This migration, which may take a month or more to complete, coincides with the development in the ascidian of masking pigments and calcareous spicules. These presumably filter the sunlight and control the intensity falling on the symbiotic algae. *(A, Redrawn from Berrill, N. J. 1950. The Tunicata. Quaritch, Ltd. London. 354 pp. B, Combined from Young, C. M., and Braithwaite, L. F. 1980. Orientation and current-induced flow in the stalked ascidian Styela montereyensis. Biol. Bull. 159:428–440. C, Redrawn after Olson, R. R. 1983. Ascidian–Prochloron symbiosis: The role of larval photadaptations in midday larval release and settlement. Biol. Bull. 165:221–240.)*

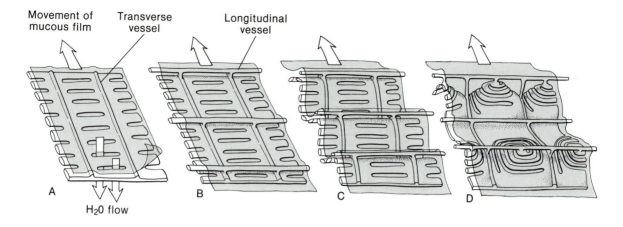

Movement of mucous film Transverse vessel Longitudinal vessel

A B C D

H₂0 flow

FIGURE 17–24. The ascidian pharynx shows an evolutionary trend to increase the surface area of its lining, gill slits, and vascular supply. The planar linings of aplousobranchs, such as *Aplidium* and *Eudistoma* **(A),** have only transverse vessels, whereas those of some phlebobranchs, for example *Ciona* **(B),** also have longitudinal vessels. The new longitudinal vessels not only increase the vascular supply to the pharynx but also support and transport the mucous net across the lining. In stolidobranchs, the lining is pleated **(C, D)** to increase its area, and in molgulids **(D)** the gill slits are coiled. *(Modified and redrawn from Kott, P. 1989. Form and function in the Ascidiacea. Bull. Mar. Sci. 45:253–276.)*

them by the action of lateral cilia on the gill slits, and food particles suspended in the water are trapped on the nets. When the food-laden nets reach the dorsal midline of the pharyngeal lining, they are rolled together into a single threadlike cord by a curved ciliated ridge, the **dorsal lamina** (Fig. 17–16B), or a row of tonguelike **languets,** that extends posteriorly to the opening of the esophagus. Cilia on the dorsal lamina convey the food cord into the esophagus.

When examined with an electron microscope, the pharyngeal mucous net can be observed to consist of a regular meshwork with openings of approximately 0.5 μm (Fig. 17–16D). Using such a filter, ascidians readily remove the smallest plankton from the water as well as bacteria. Ascidians can halt feeding by closing the buccal siphon, by arresting ciliary beat, or by stopping the flow of mucus from the endostyle.

In general, the number and form of gill slits, and the degree of folding and vascularization of the pharyngeal wall are adaptations to increase surface area for food capture and gas exchange, and are thus roughly correlated with body size. They are also used by taxonomists to classify ascidians. The pharynx lining is more or less smooth in the tiny colonial zooids of aplousobranch ascidians, smooth or undulating in the larger phlebobranchs, such as *Ciona* or *Ascidia,* and highly pleated in the stolidobranchs, for example

the large-bodied *Styela,* but also in the relatively small zooids of *Botryllus* or *Symplegma* (Fig. 17–24).

The tiny aplousobranch zooids generally have few oval gill slits arranged in transverse rows, whereas those of larger ascidians are numerous, often reaching several hundred in the largest species. In the molgulids, which include the familiar sea grape, *Molgula,* the large zooids not only have many gill slits but each slit is spiralled, further increasing its surface area (Fig. 17–24D).

The specialized bands that bear the frontal cilia and transport the mucous net typically surmount blood vessels that bulge into the pharynx like varicose veins. They occur in simplest forms as **transverse vessels,** which extend between the rows of gill slits in aplousobranch ascidians (Fig. 17–24A). Because these vessels bulge above the level of the pharyngeal lining, they elevate the mucous net slightly, thus improving water flow through it. In some phlebobranchs, such as *Ascidia,* numerous blind-ending papillae grow upward (inward) from the transverse vessels and further elevate the mucous net off the lining of the pharynx. Finally, in the stolidobranchs and some phlebobranchs, the papillae from adjacent transverse rows have fused together to form **longitudinal vessels** that intersect the transverse vessels at right angles (Fig. 17–24B–D). Like the unmodified transverse vessels from which they originate, the papillae and longitudinal vessels bear the frontal cilia and each contains a blood channel.

There are a few soft-bottom ascidians that feed on deposited material from the surface of surrounding sediment. There are also some deep-water species that feed on small animals, such as nematodes and small, epibenthic crustaceans, caught with lobes around the buccal siphon. The pharynx is small and glandular with only a few gill slits.

The base of the pharynx on the dorsal side contains the esophageal opening, toward which the dorsal lamina is directed. The postpharyngeal part of the digestive tract is located in the abdomen and is arranged in a U-shaped loop (Fig. 17–16A).

The esophagus forms the descending arm. The stomach is an enlargement of the digestive loop at the turn of the U. It is lined with secretory cells and is the site of extracellular digestion. The ascending arm of the digestive tract is formed by the intestine, the terminal end of which is modified as a rectum and opens through the anus into the atrium. The intestine is the site of fecal pellet formation and is probably the site of absorption (Fig. 17–16A).

In all tunicates, there is a network of vesicles and tubules (**pyloric glands**) on the outer walls of the anterior intestine (Fig. 17–16A). By way of one or many collecting canals, this network opens into the intestine near its attachment to the stomach. The products of the pyloric glands are secretory in nature, but whether they are enzymatic, are involved in pH regulation, or have some other function is uncertain.

Some members of the colonial family Didemnidae contain symbiotic algae within the tunic and cloacal lining. One of these symbionts, *Prochloron,* is found only in ascidians and has a prokaryotic structure but the photosynthetic pigments typical of eukaryotic green algae. These same ascidians also may have blue-green algal (cyanobacteria) symbionts embedded in the tunic. Excess photosynthate produced by the symbionts is assumed to be used by the tunicate as an auxilliary food source. At least one *Didemnum* species may migrate over the substratum as it grows to optimize the light intensity for photosynthesis by its *Prochloron* symbionts (Fig. 17–23C).

Internal Transport and Gas Exchange

The tunicates have a well developed blood-vascular system, including a heart, that is rich in blood cells of various kinds. The usually well-defined vessels lack endothelia and are simple channels in connective tissue. Although the blood is frequently pigmented, often pale green, none of the pigments are respiratory in nature, and gases are transported in physical solution in the plasma. The large surface area of the gill (pharynx), large volumes of water pumped, and low activity of the animals all probably account for the absence of a respiratory protein. The blood corpuscles have a variety of other functions, however, including tunic synthesis, internal defense, and even defense against external predators and competitors.

The heart is a short, curved, or U-shaped cylinder at the base of the digestive loop (Fig. 17–25A). On close examination, it consists of a circular cylinder with its outer wall folded inward along its length to form a tube within a tube (Fig. 17–25B). The outer tube is the fluid-filled **pericardium,** and the inner tube encloses the blood-containing **heart sinus.** The wall of the heart sinus, formed by the muscular invaginated wall of the pericardium, is the **myocardium.** The contractile filaments in the myocardial wall are arranged in two crossed helixes with an angle of 60 degrees between them. Their action causes the heart to contract in a twisting, wringing motion that passes from one end of the heart to the other, pushing the blood forward and preventing backflow.

The blood circuit is not clearly established for any ascidian, but the placement of major vessels and the flow pattern through the pharynx are well understood. From the ventral (anterior) end of the heart, blood passes beneath (outside) the endostyle by way of a large, subendostylar vessel that runs along the ventral side of the pharynx supplying the pharyngeal vessels. From the pharynx, blood is collected in a middorsal vessel and delivered to channels supplying the digestive loop and other viscera. These vessels eventually drain into the dorsal end of the heart. In those ascidians, such as *Ciona* and *Ascidia,* in which the tunic is provided with blood vessels, a vessel to the tunic usually arises from each of the main channels at opposite ends of the heart.

There is growing evidence in tunicates (based largely on thaliacean anatomy, p. 893) that several of the consuming tissues—gut, gonads, pharynx, and parts of the nervous system—are arranged in series, rather than parallel, along the blood circuit (Fig. 17–25D, E). This unusual arrangement, if confirmed, may explain a remarkable and unique feature of the tunicate circulatory system, the periodic reversal of

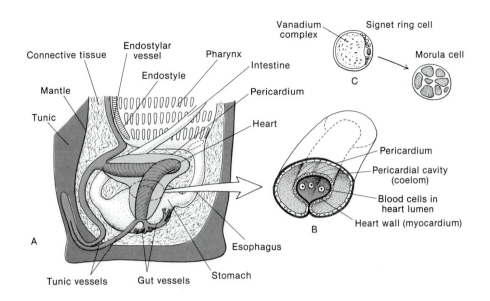

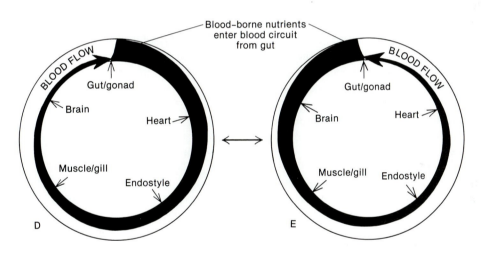

FIGURE 17–25. The ascidian blood-vascular system. **A,** The heart and major vessels in *Ciona*. **B,** Section of the heart wall from **A. C,** Immature and mature vanadocytes (morula cells) from *Ascidia*. **C, D,** Diagram showing presumed series arrangement of organs in the blood circuit in forward **(D)** and reverse **(E)** blood flow. Shaded arrows show depletion of blood-borne substances as blood flows through the circuit. Reversal of blood flow may be a means of averaging the distribution of substances, such as oxygen and nutrients, to the organs. *(A, Modified and redrawn after Berrill, N. J. 1936. Studies in tunicate development. Part VI—The evolution and classification of ascidians. Phil. Trans. Roy. Soc. Lond. B. 226:43–70. B, Modeled after Oliphant, L. W., and Cloney, R. A. 1972. The ascidian myocardium: Sarcoplasmic reticulum and excitation–contraction coupling. Z. Zellforsch. 129:395–412. C, Drawn after electronmicrographs from Pirie, B. J. S., and Bell, M. V. 1984. The localization of inorganic elements, particularly vanadium and sulfur, in haemolymph from the ascidians Ascidia mentula (Muller) and Ascidiella aspersa (Muller). J. Exp. Mar. Biol. Ecol. 74:187–194; D–E, Based on discussion of thaliacean circulatory system in Heron, A. C. 1975. Advantages of heart reversal in pelagic tunicates. J. Mar. Biol. Ass. U.K. 55:959–963.)*

blood flow. **Heartbeat reversal** occurs every 2 or 3 min as heartbeat is momentarily arrested, then resumed in the opposite direction. The heart is myogenic, and there is an excitation center at each end of the heart that is responsible for initiating contractile waves over the heart from that point, each alternating in dominance over the other. Pacemaker fatigue, or rising back pressure may generate the stimulus for beat reversal. If the consuming tissues in the blood circuit are arranged in series, one after the other, then those first in the series will have access to the highest concentrations of blood-borne nutrients and oxygen, whereas those at the end of the series will receive reduced supplies of these substances. After heartbeat reversal, however, blood flow is reversed, and tissues formerly at the end of the series are now at the beginning, where they can receive substances in high concentration. Thus, periodic heartbeat reversal in tunicates may be an adaptation for averaging the distribution of metabolites to tissues (Fig. 17–25D, E).

The periodic reversal of blood flow through the pharyngeal gill precludes the possibility of countercurrent gas exchange, as typically occurs, for example, across molluscan gill surfaces. In tunicate gills, blood flow and water flow are always perpendicular to each other, ensuring that the rate of gas exchange remains constant regardless of the direction of blood flow.

The blood of tunicates contains several kinds of cells that can be grouped into four categories: **lymphocytes,** which are formed in the connective tissue and give rise to all other types of cells; **nutritive phagocytic amebocytes; morula cells;** and **nephrocytes.**

Morula cells concentrate heavy metals in berry-like intracellular clusters of vesicles (Fig. 17–25C). Members of the Pyuridae concentrate iron, others concentrate niobium or tantalum, and species of the Ascidiidae and Perophoridae accumulate and store high concentrations of vanadium in their yellowish green morula cells **(vanadocytes).** The concentration of vanadium in seawater is approximately 5×10^{-8} M, whereas in vanadocytes, the concentration reaches 1 M (a 100 million-fold increase in vanadium concentration). The only other organism known to concentrate vanadium above the very low levels found in animals in general is the death cup mushroom, *Amanita muscaria.*

Vanadium itself is a potent inhibitor of many enzymes, including those responsible for the sodium/potassium pump, ciliary motility, and muscular contraction. It is also a powerful reducing agent. In ascidians, vanadium occurs in the +3 oxidation state, which is stable only at a pH of less than 2; concentrated sulfuric acid is found in the cellular vesicles containing vanadium. Typically in ascidians, the vanadocytes enter the tunic from the blood and may align themselves immediately below the outer tunic surface, where they are sometimes called bladder cells. The vanadocytes eventually disintegrate in the tunic, releasing their contents, or if the tunic is damaged by some external agency, vanadium and sulfuric acid are discharged to the exterior.

Vanadium in tunicates appears to have at least two functions. The reducing power of vanadium (and iron) is used in the polymerization of tunicin filaments, hence in **tunic synthesis.** Second, it has been suggested that the toxicity of vanadium and the unpalatability of sulfuric acid may be used as **antibiotics** to discourage predation on tunicates and fouling of the tunic by the attachment and growth of other organisms.

Excretion

Ascidians, like other tunicates, are primarily ammonotelic animals that release protein nitrogen in the form of ammonia across the pharynx and into the exhaust flow. Other metabolic byproducts, however, such as uric acid and calcium oxalate (urates), are not excreted but are stored internally and released only on the death of the zooid. This phenomenon is known as **storage excretion.** Urates may be stored in nephrocytes, which then accumulate in various tissues of the body, including the mantle, digestive loop, or gonads. Urates may also be secreted by epithelial tissues into extracellular compartments, such as the pericardium *(Ascidia)* or two specialized organs (Fig. 17–26A).

One, called the **epicardium,** occurs in many aplousobranchs and a few phlebobranchs *(Ciona).* It consists of one (sometimes two) long, blind-ending diverticulum from the lower wall of the pharynx that extends posteriorly to overlay the heart. Urate crystals, presumably secreted by the epicardial walls, occur within the epicardium, which remains open at its origin to the pharyngeal cavity (Fig. 17–26B).

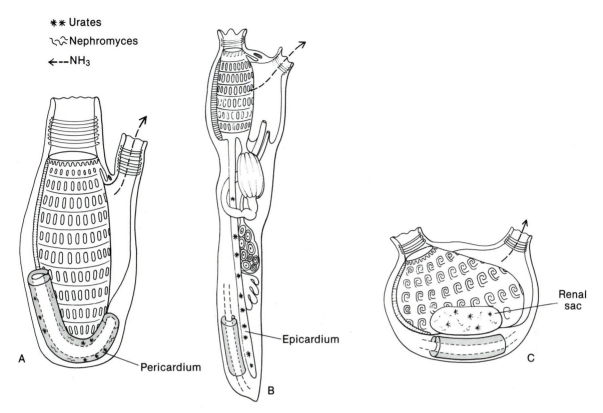

※※ Urates

〰️ Nephromyces

←--NH₃

FIGURE 17–26. Storage excretion in ascidians. **A,** In *Ascidia,* crystallized excretory wastes (urates) are stored in the pericardium in association with a symbiotic fungus, called *Nephromyces.* **B,** In some aplousobranchs, this storage function is adopted by one or two diverticula of the pharynx called epicardia, the lining of which secretes the urates into the epicardial lumen. The epicardium remains open to the pharynx in aplousobranchs, but in the stolidobranch *Molgula* **(C)** and its relatives, the epicardium pinches free of the pharynx during development and becomes a closed renal sac. The renal sac contains urates and *Nephromyces,* which may help to break down the stored urate crystals.

In *Molgula* and its relatives, a specialization of the epicardium, called the **renal sac,** is located on the anterior (upper) surface of the heart (Fig. 17–26C). It is a closed, bean-shaped sac that lacks any opening to the exterior or interior. Urates are secreted into the sac, where they form a conspicuous whitish or yellowish mass.

A unicellular fungus, called **nephromyces,** occurs in the renal sacs of molgulids and in the pericardium of *Ascidia.* The nephromyces cells, in turn, harbor bacterial symbionts. The functional role of nephromyces is unknown, but it has been suggested that the fungi, perhaps in collaboration with their bacteria, may break down the stored urates.

Nervous System

The nervous system is relatively simple and consists of a cylindrical to spherical **cerebral ganglion,** or "brain," located in the connective tissue between the two siphons (Figs. 17–16A; 17–27). The nerves arising from the anterior end of the ganglion supply the buccal siphon and mantle musculature; those issuing from the posterior end innervate the greater part of the body—the atrial siphon, the mantle musculature, the pharynx (control ciliary arrest), and the visceral organs.

In some and perhaps all colonial ascidians, the nonnervous epidermally derived walls of the blood vessels propagate nerve impulses and are responsible

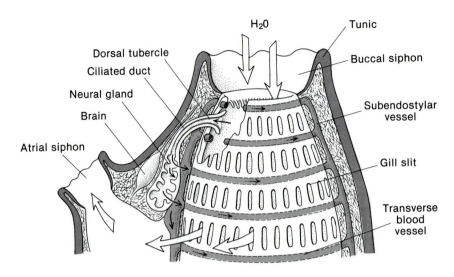

FIGURE 17–27. In ascidians, as in other urochordates, the neural gland, and its associated dorsal tubercle and ciliated duct, pump seawater into the blood and thus help to regulate the fluid volume of the blood. The dorsal tubercle excludes large particles from the incurrent, and the gland removes small particles by phagocytosis. *(Modified after Ruppert, E. E. 1990. Structure, ultrastructure and function of the neural gland complex of* Ascidia interrupta *(Chordata, Ascidiacea): Clarification of hypotheses regarding the evolution of the vertebrate anterior pituitary. Acta Zool. 71:135–149.)*

for vessel contraction and coordinated retraction of the zooids. It has been suggested that tunicate epithelial tissue in general, like the heart in particular, may have rhythmic electrical activity and conductive ability.

Beneath the cerebral ganglion lies a hollow spongy organ called the **neural gland** (Fig. 17–27), which, like the ganglion, originates from the embryonic neural tube. Contrary to its name, nerves are absent from the neural gland, and the presence of gland cells is questionable. The lumen of the gland extends anteriorly as a ciliated duct, opening into the pharynx by way of a large, ciliated funnel (**dorsal tubercle**). Cilia in the duct create an incurrent of water from the pharynx into the gland lumen and then across its wall into the branchial blood vessels. Dense cilia on the dorsal tubercle prevent large particles, such as sand grains, from entering the duct, while phagocytes lining the gland remove bacteria and other small particulates from the water. The dorsal tubercle, ciliated duct, and neural gland together function to restore and maintain the fluid volume of the blood, and are thus analogous to the madreporite, stone canal, and Tiedemann's bodies of some echinoderms (p. 931). It has long been speculated that the neural gland is an evolutionary precursor of the vertebrate anterior pituitary gland, and there are unconfirmed reports of pituitary-like hormones in neural gland tissue.

There are no special sensory organs in ascidians, but sensory cells are abundant on the internal and external surfaces of the siphons, on the buccal tentacles, and in the atrium, and very likely play a role in controlling the current of water passing through the pharynx.

Asexual Reproduction

All ascidians regenerate well, but asexual reproduction is characteristic only of colonial ascidians. Asexual reproduction takes place by means of budding but is complex and exceedingly variable, perhaps more so than in any other group of metazoans.

A tunicate bud is called a **blastozooid,** and it arises from the body of the **oozooid** (the zooid developing from the fertilized egg). There is great variation, depending on the species, in the site of bud formation, and there is a corresponding variation in the germinal tissues included within the bud. The primitive type of budding appears in *Perophora* and similar species, in which the bud arises from the stolon. In other families, the buds may arise in the abdomen, postabdomen, or even precociously in the larval stage, as in species of *Diplosoma* ("two bodies").

Sexual Reproduction

With few exceptions, tunicates are **hermaphrodites,** but only a few species are self-fertilized; cross fertilization is typical of the class. There are usually a single testis and a single ovary that lie in close association with the digestive loop (Fig. 17–16A). The oviduct and sperm duct run parallel to the intestine and open into the cloaca in front of the anus. In the order Stolidobranchia (e.g., *Styela*), one to many gonads are usually located in the body wall.

Solitary ascidians generally have small eggs with little yolk. The eggs are shed from the atrial siphon, and fertilization takes place in the sea. The eggs of such oviparous species are frequently surrounded by special membranes that act as flotation devices. The eggs of colonial species are typically richer in yolk and are usually brooded in the oviduct or atrium, which sometimes contains special incubating pockets. Hatching usually takes place at the larval stage, and the larva then leaves the parent, but the entire course of development may occur within the atrial cavity. In general, development in brooding species with considerable yolk is more rapid and condensed than in nonbrooding forms.

Cleavage is complete and leads to a coeloblastula. Gastrulation is by epiboly and invagination, and the large archenteron obliterates the old blastocoel. The blastopore marks the posterior end of the embryo and closes while the embryo elongates along the anteroposterior axis. Along the middorsal line, the archenteron gives rise to a supporting rod—the notochord (Figs. 17–19B; 17–20A). Laterally, the archenteron proliferates mesodermal cells that form a cord of cells along each side of the body. In this respect, development departs from that shown in amphioxus and other deuterostomes, because there is no pouching of the archenteron. A coelomic cavity never appears, nor is there any segmentation. The ectoderm along the middorsal line differentiates as a neural plate, sinks inward, and rolls up as an internal neural tube.

Continued development leads to the tadpole larva described at the outset of the section (Figs. 17–19; 17–20). The only additional features that should be mentioned here are that the larva is covered by a tunic secreted by the surface ectoderm and that in the tail region, the tunic is extended dorsally and ventrally to form a fin. In some colonial ascidians, however, the larval tail rotates 90 degrees during development, and as a result the fins are horizontal rather than vertical in position. The pharynx contains a ventral endostyle. The tadpole larva has a free-swimming period of 36 h or less, and in *Botryllus,* the larva may settle in a few minutes.

There has been a tendency toward a shorter free-swimming larval stage among tunicates that inhabit sand and mud bottoms. This reduction seems to be associated with the fact that in such species, reaching a suitable substratum is no problem. For species requiring certain kinds of firm substrata for attachment, a free-swimming larva is indispensible.

Most ascidians have a life span of one to three years, although colonies may have a longer life.

SYSTEMATIC RÉSUMÉ OF CLASS ASCIDIACEA

Order Aplousobranchia. Colonial tunicates with simple pharyngeal linings. Epicardia present. Gonads within gut loop. Some species with postabdomen. *Aplidium, Clavelina, Didemnum.*

Order Phlebobranchia. Solitary or colonial tunicates with longitudinal vessels within pharynx. Gonads within gut loop. No postabdomen. Gut loop behind or beside pharynx. *Ascidia, Ciona, Corella, Diazona, Ecteinascidia, Perophora, Phallusia.*

Order Stolidobranchia. Solitary or colonial tunicates with pleated pharyngeal linings bearing longitudinal vessels. Gut loop located to one side of pharynx. Epicardia absent. *Botryllus, Halocynthia, Microcosmos, Molgula, Polycarpa, Symplegma, Styela:* the macrophagous, deep-sea *Hexacrobylus* and *Gasterascidia.*

Class Thaliacea

The other two classes of tunicates, Thaliacea and Larvacea, are both specialized for a free-swimming, **planktonic** existence. Thaliacean zooids, which include the luminescent pyrosomes, cask-shaped doliolids, and chainlike salps, differ from those of ascidians in having the buccal and atrial siphons at opposite ends of the body. The water current is thus used not only for feeding and gas exchange but also for **jet propulsion.** Like many other planktonic animals, thaliaceans are **transparent.** The tunic, which may be thin or thick, and the connective tissue are gelatinous and buoyant. Like ascidians, thaliaceans are filter feeders that remove large quantities of minute suspended material from seawater. They may occur in enormous numbers and be distributed over large areas of the sea surface. Thaliaceans have **complex life cycles,** often involving more than one stage between the larva and adult, and at least one of the stages is **colonial** in organization. The class contains only six genera, but species of thaliaceans are widespread in open oceanic waters; most species live in tropical and semitropical seas.

The tropical pyrosomes, such as species of *Pyrosoma,* are brilliantly luminescent colonial thaliaceans, having the form of a cylinder that is closed at one end (Fig. 17–28A, B). The length ranges from a few centimeters *(Pyrosoma atlanticum)* to over 3 m *(Pyrosoma spinosum).* The zooids are oriented in the wall

(common tunic) of the colony so that the buccal siphons open to the outside and the atrial siphons are directed into the central cloacal cavity (Fig. 17–28C). Thus, the inflows from numerous zooids are combined in the cloaca to produce a forceful thrust of water from the single exhaust aperture (Fig. 17–28B). Pharyngeal cilia produce both the feeding and locomotory currents. Two **luminescent organs,** one on each side of the pharynx, contain bioluminescent bacteria and are responsible for producing the intense light when the colony is disturbed (Fig. 17–28C).

Pyrosome colonies grow by the addition of new zooids, which originate as buds from the heart region of the parent zooids (Fig. 17–28C). Sexual reproduction begins when the single egg within each zooid is fertilized internally. Development is internal, direct, and condensed, and the rudimentary oozooid (cyatho-

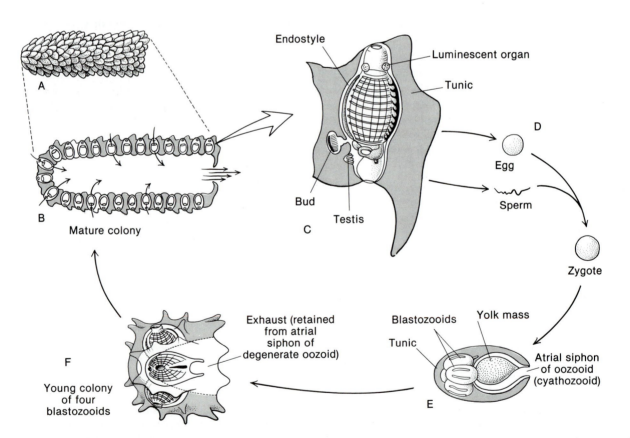

FIGURE 17–28. A, Adult colony of *Pyrosoma atlanticum.* **B,** Longitudinal section of **A,** showing zooids, common cloaca, and exhaust aperture. Arrows indicate path of water flow through the colony. **C,** Enlargement of one of the zooids in **B. D,** Gametes. **E,** Lecithotrophic oozooid (cyathozooid) and its four precocious buds (blastozooids). **F,** Young colony composed of four zooids.

zooid) undergoes precocious budding to form four blastozooids—the first four colonial zooids (Fig. 17–28E). The oozooid soon degenerates, except for its atrial siphon, which is retained by the growing colony as its exhaust aperture (Fig. 17–28F).

Salps and doliolids are thaliaceans in which an asexual polymorphic colony periodically releases aggregates of sexual zooids. Some of the salp aggregates, called **chains,** reach a meter or more in length (Fig. 17–29). They are found in the upper levels of all oceans but are more common in warmer seas. The body of the **solitary zooid** in both is somewhat barrel-shaped, particularly in doliolids (Figs. 17–30A–C; 17–31A). Hooplike circular muscle bands, complete in doliolids and incomplete in salps, produce contractions of the body wall that drive water from the atrium out through the exhaust aperture. Doliolids have retained the ciliary feeding current, but salps use the flow generated by muscular contraction for feeding, gas exchange, and locomotion. Salps are peculiar in having only two gill clefts, which are so enormous that there are virtually no side walls of the pharynx remaining (Figs. 17–31A; 17–32). They feed using a fine mucous net, as do ascidians; the net, secreted by the endostyle, projects rearward into the pharynx, similar to a plankton net. As the net moves posteriorly, it is rolled up and passed into the esophagus by esophageal cilia and cilia on the gill bar.

Asexual reproduction in salps and doliolids is the most unusual and complex among animals, although parallel adaptations occur in some siphonophores. The oozooids reproduce asexually by budding from the heart region to produce a stolon, short in doliolids and long in salps (17–30A; 17–31A). In both, buds arising from the stolon give rise to colonies of blastozooids but, beyond these generalizations, their patterns of reproduction differ. In doliolids, rudimentary buds migrate from the stolon, through the body of the oozooid, and embed themselves on a long, slender dorsal appendage, the **spur,** which trails behind the oozooid (nurse) (Fig. 17–30A). Once attached to the spur, the buds differentiate into one of three different kinds of zooids. **Trophozooids,** which each have an enormous buccal opening and pharynx, are responsible for feeding the colony (the digestive system of the oozooid degenerates) (Fig. 17–30E); **phorozooids** are locomotory zooids that have a short posterior spur on which buds differentiate into **gonozooids,** the sexually-reproducing members of the colony (Fig. 17–30B, C). When fully developed, the phorozooids

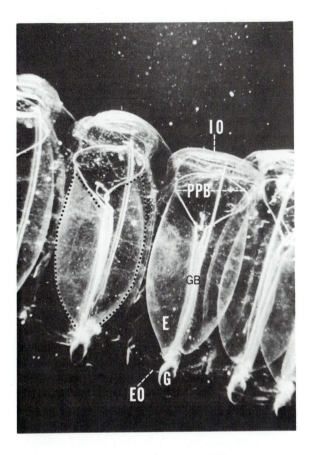

FIGURE 17–29. Part of a chain of aggregated salps *(Pegea confoederata)* filter feeding. Dotted line outlines the mucous net extending back to the esophagus (E) from peripharyngeal band (PPB). Rows of cilia on gill bar (GB) appear as bars. Water enters the buccal siphon (IO) and leaves through the atrial siphon (EO) in the region of the gut (G). The atrial siphon is very faint in the photograph. Each zooid is 6 cm in length. *(From Madin, L. P. 1974. Field observations on the feeding behavior of salps. Mar. Biol. 25:143–147.)*

and their attached gonozooids break free of the colony and take up an independent existence. Fertilization and early development are internal in the gonozooid, and a single tadpole larva is released into the plankton from each gonozooid (Fig. 17–30C, D). Metamorphosis of the tadpole gives rise to the oozooid and completes the cycle.

The stolon produced by each salp oozooid is long (up to several meters) and trails ventrally from the body (Fig. 17–31A). Within the stolon, blastozooids differentiate in groups, which are separated from each

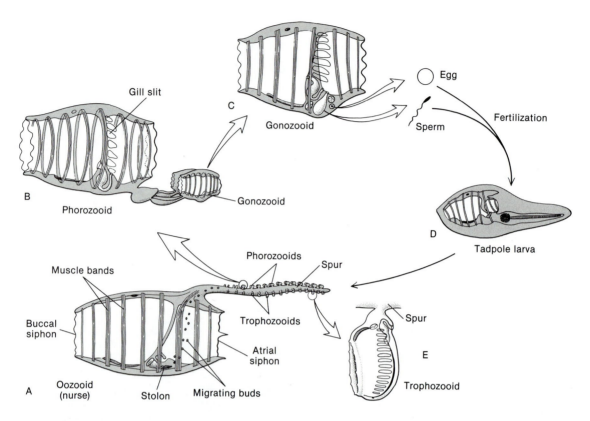

FIGURE 17–30. Doliolid organization and life cycle. **A,** The nurse is essentially a colony of polymorphic zooids. The nurse herself developed from the fertilized egg and subsequent tadpole larva **(D),** but the other members of the colony arose by budding from the nurse's stolon. The undifferentiated buds migrate from the stolon of the nurse, through the connective tissue, and then lodge in her trailing spur, which may reach 50 cm or more in length. Once attached to the spur, the buds may differentiate into trophozooids **(E),** which are specialized for feeding the colony (the nurse's digestive system degenerates), or phorozooids **(B),** which eventually break free of the spur and jet away under their own power. Buds attached to the phorozooids differentiate into the sexually reproductive gonozooids **(C).** Fertilization is probably internal in the gonozooids but a free-swimming tadpole **(D)** is released that undergoes metamorphosis in the plankton to form a young nurse, thus completing the life cycle.

other along the length of the stolon by **deploying points** (Fig. 17–31A). At intervals, clusters of blastozooids break free of the stolon at these deploying points and take up an independent existence as aggregates of sexually-reproducing members. Sexual reproduction occurs within the blastozooid. A single egg is fertilized internally and develops directly within a special pouch provided with a placental attachment to the parent's blood (Fig. 17–31B). The juvenile grows within the parent (nurse) and eventually occupies its entire body before casting off the nurse's hull and becoming a independent oozooid (Fig. 17–31C).

Class Larvacea

The Larvacea, or Appendicularia, contains some 70 species of small, transparent animals that in most respects, are the most specialized of all tunicates. They are found in the surface marine plankton throughout the world, often in astronomical numbers. A population density of 25,620 individuals of *Oikopleura dioica* per cubic meter of water was reported by H. Seki from British Columbia. The Larvacea are so named because the adults, most of which are approximately 5 mm in length, resemble ascidian tadpole larvae (Fig. 17–33). A tail is present, and the body looks

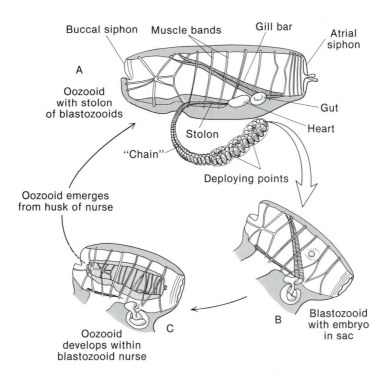

FIGURE 17–31. Salp organization and life cycle. **A,** Oozooid of *Cyclosalpa* trailing its stolon of differentiating buds and blastozooids. The stolon breaks at predetermined points and deploys clusters, or chains, of blastozooids, which swim away from the oozooid and other members of the parent colony. Each blastzooid **(B)** bears a single egg that is fertilized internally and develops in a special brood sac, complete with a placental connection to the circulatory system of the blastozooid. The growing embryo eventually occupies the entire volume of the blastozooid-nurse's body **(C)** and then breaks free as a young oozooid, thus completing the life cycle.

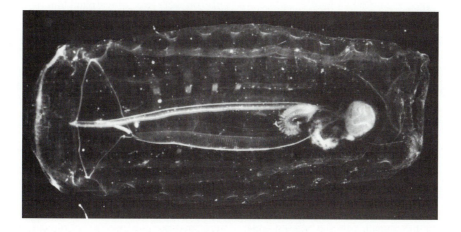

FIGURE 17–32. A living oozooid of the salp, *Thalia democratica,* collected from the Gulf Stream off southern Florida. Zooid is approximately 2 cm in length. Its buccal siphon (left), muscle bands, slender whitish endostyle, and comblike gill bar can be seen. The developing stolon is the annulated J-shaped structure at the posterior end of the endostyle, and the gut, which is whitish and circular, lies behind the stolon.

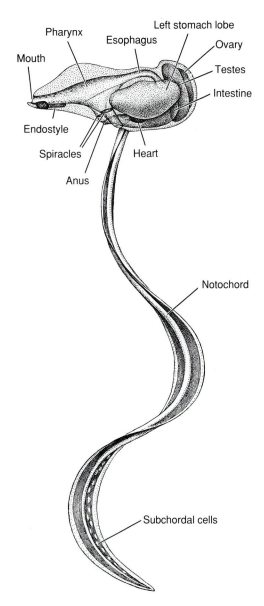

Pharynx

Left stomach lobe

Esophagus

Mouth

Ovary

Testes

Intestine

Endostyle

Spiracles

Heart

Anus

Notochord

Subchordal cells

FIGURE 17–33. Lateral view of the larvacean, *Oikopleura albicans. (From Alldredge, A. 1976. Sci. Am. 235:95–102.)*

somewhat like a typical ascidian tadpole bent at a right angle or in the shape of a U. The mouth is located at the anterior of the body, and the intestine opens directly to the outside on the ventral side. There are only *two* gill slits (spiracles), one on each side, and each opens directly to the exterior.

A remarkable feature of the Larvacea is the **"house"** in which the body is enclosed or to which it is attached (Fig. 17–34). There is no cellulose tunic in larvaceans, the surface epidermis (oikoplast epithelium) secretes a delicate gelatinous material that, when expanded by an influx of water, may completely enclose the body. In *Fritillaria,* however, the animal lies outside of the house but is attached beneath it. In species of the common *Oikopleura,* on the other hand, the house is somewhat spherical and ranges in size from that of a pea to a walnut. Some giant deep-water species with body sizes of 75 mm have houses that reach a meter in diameter.

Of the three families of appendicularians, the common Oikopleuridae is the best known. The interior of the oikopleurid house in which the animal is suspended contains a number of interconnecting passageways and both an incurrent and an excurrent orifice (Fig. 17–34B). By beating its tail, the animal creates a water current that passes through the house. The orifice through which the water enters is covered by a grid, or screen, of fine fibers that exclude all but the finest plankton (0.1–8.0 μm, depending on the species). During its passage through the house the water is strained a second time through two very fine filters. The collected plankton is then transported to the mouth in a continuous stream, apparently resulting from the actions of the tail and the ciliated gill slits. Food entering the mouth is trapped on mucus secreted by the endostyle. A pair of peripharyngeal bands, which coils around the pharyngeal lining, winds the food-laden mucus into a string and conveys it into the esophagus. Appendicularians feed primarily on very fine phytoplankton (nanoplankton), including bacteria, that is unavailable to many other planktonic filter feeders. They exhibit efficient clearing rates; according to Paffenhöfer (1973), one animal can remove 250,000 phytoplankton cells from the 300 ml of water filtered through the house each day.

The house is shed repeatedly and then replaced, and a single house is kept no longer than 4 h in *Oikopleura.* House construction involves two stages. The house is first prefabricated by the **oikoplast epithelium,** which secretes it as a collapsed surface layer. In *Oikopleura dioica,* the unexpanded house fits over the oral end of the trunk like a muzzle. When the old house is discarded, the new one is expanded in a few seconds by movements of the animal's body and filling with water. One animal can produce 4 to 16 houses a day, depending on temperature and food availability, and up to 46 houses can be secreted during the life span of one individual. The shed houses

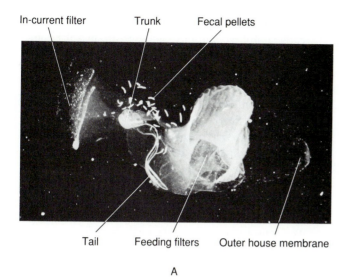

In-current filter   Trunk   Fecal pellets

Tail   Feeding filters   Outer house membrane

A

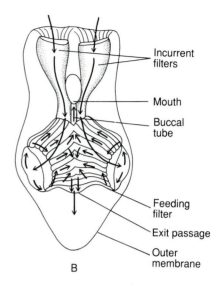

Incurrent filters

Mouth

Buccal tube

Feeding filter

Exit passage

Outer membrane

B

FIGURE 17–34. A, The larvacean, *Megalocercus,* within its mucous house. **B,** Diagram of the house of *Megalocercus.* Large arrows indicate the path of the water current; small arrows show the path of filtered particles. The oval represents the trunk of the animal. *(A, By James M. King. B, After Alldredge, A. 1976. Appendicularians. Sci. Am. 235:95–102.)*

are a significant contribution to "marine snow," (suspended organic material that resembles atmospheric snow) and is an important substrate for microbial decomposition and nutrient recycling in the sea. House production ceases at the time of spawning.

Only sexual reproduction occurs in the Larvacea. Development leads to a free-swimming tadpole larva that undergoes metamorphosis without settling.

SUMMARY

1. The great majority (class Ascidiacea) of the some 1300 species of the chordate subphylum Urochordata are adapted for a sessile existence. The tunic, filtering pharynx, and hermaphroditism are all correlated with sessility.

2. The external tunic of urochordates is unique in containing cellulose and in housing ameboid corpuscles and blood vessels (in many) (although it lies external to the epidermis), and in not being molted with growth.

3. The pharynx has become highly specialized for filter feeding. A mucous film (the filter) is produced in the endostyle and carried across the inner surface of the pharynx by frontal cilia.

4. Ascidians possess a blood-vascular system that supplies not only the internal organs and the pharynx but in some species, the tunic as well. The

system is unique in the periodic reversal of flow through the circuit. Some ascidians possess a type of blood cell (vanadocyte) within which vanadium has been concentrated from the trace amounts in sea water. The cell passes into the tunic where the vanadium compound functions as an antibiotic and in the deposition of cellulose.

5. Ascidians are ammonotelic animals but have unusual excretory organs that fulfill other excretory functions. Pericardia, epicardia, and renal sacs can function in storage excretion, and the neural gland helps to regulate blood volume.

6. Most ascidians are simultaneous hermaphrodites. Fertilization occurs externally or within the atrium. Brooding within the atrium is common in aplousobranchs.

7. Development leads to a tadpole larva, which possesses all of the chordate characteristics. Following a free-swimming existence of varying duration, the larva settles and attaches by the anterior end. Metamorphosis involves degeneration of the tail, containing the notochord and dorsal nerve cord. Differential growth results in rotation of the siphons to the end opposite the point of attachment.

8. There are two small classes of pelagic urochordates, the Thaliacea and the Larvacea. Thaliaceans have the buccal and atrial siphons at opposite ends of the body. They swim by using the water currents passing through the pharynx and atrium. The current is generated by muscular contraction of the body wall. The larvaceans may be progenetic urochordates that live within unique, secreted mucous houses. Plankton is filtered from the water stream that flows through the house.

PHYLUM CHORDATA: SUBPHYLUM CEPHALOCHORDATA

The cephalochordates, or Acrania, are fishlike chordates that are collectively known as lancelets or individually called amphioxus. Cephalochordates were first described, as molluscs, in 1744 and it was not until 60 years later that the Italian naturalist, Gabriel Costa, redescribed them as a kind of fish, closely related to lampreys and hagfishes. In the decades that followed Costa's redescription, the importance of amphioxus as an evolutionary intermediate between invertebrates and vertebrates became clear and resulted in the publication of several books on vertebrate evolution. Because of their prevertebrate organization, active research on these animals continues to the present.

Although all cephalochordates are similar in body form, approximately 25 species have been found in tropical and temperate oceans worldwide. Lancelets are benthic animals that burrow in coarse, shelly, current-swept sands, usually in shallow nearshore areas. Their heads poke up above the sand into the water, from which they filter out suspended food particles.

Cephalochordates are important as food in parts of Asia. Despite their modest size, lancelets are clean muscular animals which, unlike fish, lack bones. Fresh animals may be fried for immediate consumption or dried for later use. One amphioxus fishery in southern China recorded an annual catch of 35 tons, or approximately 1 billion lancelets (*Branchiostoma belcheri*), from a fishing ground 1 mi wide and 6 mi long. Five thousand individuals of *Branchiostoma caribaeum* per square meter have been reported from Discovery Bay in Jamaica. In parts of Brazil, chickens are herded onto beaches to feed on amphioxus.

Body Organization and Locomotion

Cephalochordates are slender, laterally compressed, and eellike in appearance and behavior. The body ta-

pers acutely to a point at both anterior and posterior ends, and it is from this characteristic that the common name, *amphioxus,* meaning "opposite points," was coined (Fig. 17–35A, B). Adult lancelets range in size from 4 to 8 cm. In life, most of the internal organs and tissues can be seen through the pink, translucent, and iridescent skin.

The body can be divided into a poorly developed **head,** a trunk, and a **tail** (Fig. 17–35B). The head terminates in a short blunt snout, or **rostrum,** which helps the animal push aside the sand while burrowing. Immediately posterior to the rostrum, the head bears a ventral **mouth** that is surrounded by a veil of finger-like sensory projections, the **oral cirri.** The common generic name, *Branchiostoma,* meaning "gill mouth," was given in reference to the oral cirri, which were erroneously thought to be gills. The bulk of the trunk is occupied by the pharynx and gonads, comprising the branchiogenital region. This region of the body is triangular in cross section, and two low longitudinal ridges, the **metapleural folds,** are located ventrolaterally at the lower corners of the triangle (Fig. 17–35C). The branchiogenital region ends at the large midventral **atriopore,** which corresponds to the tunicate atrial siphon. The **intestinal region** extends from the atriopore to the tiny midventral **anus,** located immediately in front of the tail. A tapered postanal tail bears a well developed fin. A continuous **dorsal fin** extends from the head to the tail, and a short, low **ventral fin** is located between the atriopore and tail.

Cephalochordates lack a cuticle or tunic, and the epidermis is a simple, nonciliated glandular epithelium, never stratified as in the vertebrates (Fig. 17–35D). A basal lamina lies below the epidermis, and beneath the basal lamina is a thick connective tissue **dermis.** The dermis is laced with cross-helically wound fibers, which support and toughen the body and are responsible for its iridescence.

Although they are primarily sedentary infaunal animals (Fig. 17–36B), cephalochordates can vacate their burrows and swim using short bursts of rapid, eellike, lateral undulations of the body (Fig. 17–37). They swim equally well in either head-first or tail-first directions. The same swimming movements are used for burrowing in sand (Fig. 17–36C, D). Living specimens confined to an aquarium are nearly impossible to catch with fingers or forceps because of their streamlined shape, rapidity of movement, and slippery surface, and must instead be netted, like small fish.

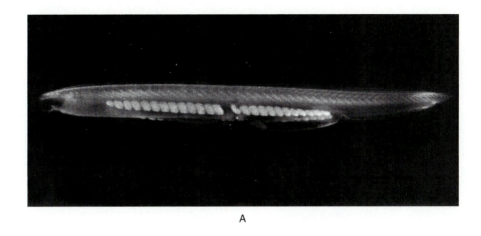

A

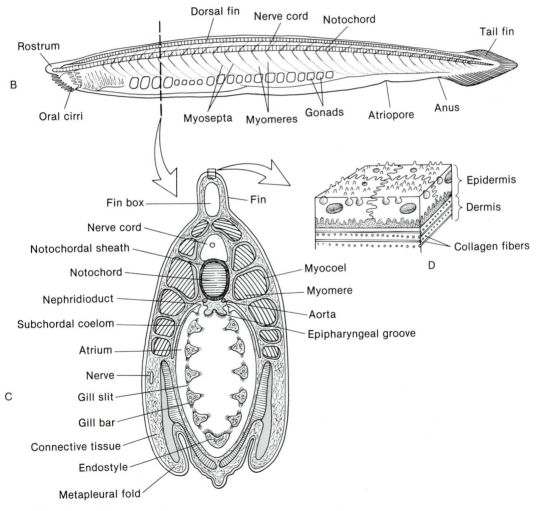

FIGURE 17–35. A, Photograph of a living cephalochordate, *Branchiostoma caribaeum,* in left lateral view. **B,** Left lateral view of a generalized lancelet. **C,** Cross section through the pharynx of a generalized lancelet, gonads omitted. **D,** Enlargement of epidermis, showing the absence of cuticle but presence of well-developed, fibrous dermis.

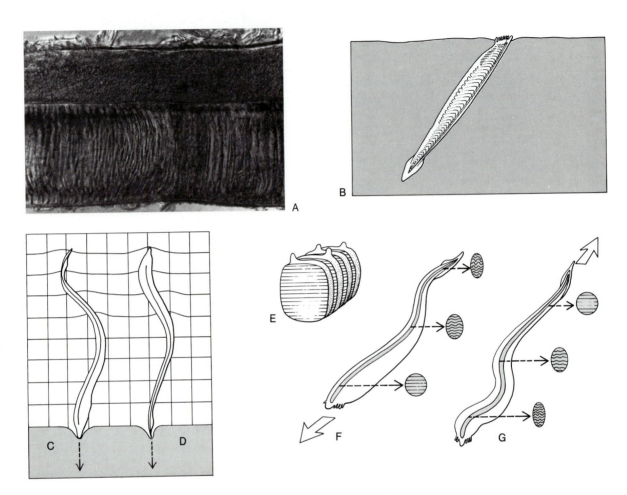

FIGURE 17–36. A, Photograph of the nerve cord (above) and notochord (below) dissected from a living amphioxus; lateral view. Several of the thin, platelike notochordal muscle cells can be seen in edge-on view in the photograph. **B,** Feeding position of amphioxus. Head-first **(C)** and tail-first **(D)** burrowing in amphioxus. **E,** Three of the many muscle plates composing the notochord. The myofilaments in each plate run transversely from left to right. **F, G,** Hypothetical changes in the cross-sectional profile of the notochord during forward **(F)** and rearward **(G)** swimming. While swimming forward, the posterior notochordal muscle may be contracted as shown to lower the bending resistance of the tail and increase its amplitude, thus creating greater forward thrust. Conversely, the anterior notochord may be contracted while the animal swims rearward

Internal Organization

The nervous system of amphioxus consists of a dorsal, hollow, longitudinal cord and segmental sensory (and some motor) nerves, which leave the cord dorsally and innervate a variety of structures, especially the rostrum, oral cirri, velum, and tail (Fig. 17–38B). The cord extends from the base of the rostrum to nearly the tip of the tail. As in tunicates, the central canal opens to the exterior via a permanent ciliated neuropore (Kölliker's pit) which is situated on the left side of the anterior extremity of the cord (Fig. 17–38B). A short distance behind the neuropore, the cord enlarges slightly to form a small hollow brain. A pair of pigment cup ocelli is located in the wall of the brain. Additional ocelli, up to 1500 or more in *Branchiostoma lanceolatum,* are located in the cord posterior to the brain. Curiously, the ocelli on the left side of the cord face dorsally, and those on the right and ventral sides face ventrally. Amphioxus is negatively

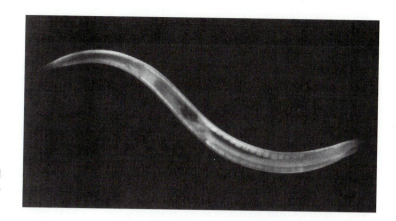

FIGURE 17–37. Instantaneous ventral view of an amphioxus swimming in the forward direction (right).

phototactic, and continuous illumination seems to arrest locomotion, causing the animals to remain in their burrows. A sudden pulse of bright light in dark-adapted animals, however, will cause them to leave the sand and swim.

The notochord is located directly below the nerve cord and extends from the tip of the tail forward into the rostrum (Figs. 17–36A; 17–38). The name cephalochordate, meaning "head chordate," refers to the unique continuation of the notochord anterior to the brain in amphioxus. The notochord consists of a longitudinal series of disc-shaped cells, arranged like a stack of coins (Fig. 17–36A, E). The entire structure is enclosed in a fibrous extracellular matrix, the **notochordal sheath,** which resembles the fibrous mesh below the epidermis and probably helps to stiffen the notochord and control its shape (Fig.

17–35C). Each notochordal cell is a specialized muscle cell in which the contractile filaments run transversely, from left to right. Thus, the amphioxus notochord is a muscle that is designed for a skeletal, rather than a motile function. Presumably, a contraction of all or some of the notochordal muscle cells would alter the stiffness of the notochord. Using its muscular notochord, amphioxus may be able to adjust its skeletal system to achieve optimal performance while swimming forward, rearward (Fig. 17–36F, G), through water, or through sand. Each notochordal cell is innervated via a short, slender outgrowth of the muscle cell itself, which makes contact with the overlying nerve cord (Fig. 17–36E).

Cephalochordates are segmented, coelomate animals, like annelids (and vertebrates). Lancelet segmentation is obvious in the arrangement of the

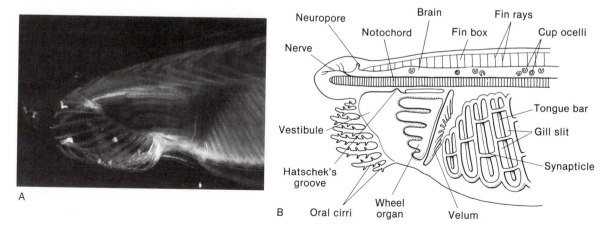

FIGURE 17–38. The head and oral region of amphioxus. **A,** Left lateral view of the head of *Branchiostoma caribaeum.* **B,** Left lateral cross section of the head of a generalized cephalochordate.

swimming muscles into a longitudinal series of 50 to 75 V-shaped units, called **myomeres** ("muscle segments"), on each side of the body. The myomeres, which are out of register on opposite sides of the body, are separated from each other by **myosepta** (Fig. 17–35B). The myomeres form the longitudinal cross-striated musculature of the body wall and are responsible for the undulatory swimming and burrowing movements of the body. Circular muscles are absent.

Segmentation in lancelets is more similar to that of annelids than is obvious from superficial examination. Each amphioxus myomere is actually a segmental coelomic cavity lined by a simple (nonstratified) mesothelium (p. 183). Where the wall of the coelom abuts the nerve cord (Fig. 5–9F, G), the mesothelial cells are greatly enlarged myocytes, which bulge into the coelomic cavity and together form the swimming muscles. Because all of the muscle cells differentiate in contact with the outer wall of the nerve cord, they are innervated by motorneurons within the nerve cord, in contrast to the vertebrates in which motorneurons leave the cord by way of the ventral roots to innervate distant muscles. The enlarged myocytes nearly fill the coelomic space, but a small cavity remains, called the **myocoel** (Fig. 17–35C). In addition to the myocytes, the myocoel is lined by low, noncontractile, ciliated cells that complete the coelomic lining and may help to circulate the coelomic fluid within.

The dorsal and ventral fins, but not the tail fin, are supported by a longitudinal series of small unpaired coeloms, called **fin boxes** (Fig. 17–38). The number of fin boxes in the dorsal fin, for example, often exceeds 200, and thus the number of fin "segments" does not correspond to the number of myomeres. The septa between the fin boxes form the **fin rays.**

A paired nonsegmented **subchordal coelom** is located in the branchial region between the atrium and the notochord (Fig. 17–35C). This coelom is the urinary space in which the excretory organs are located, discussed below. A paired unsegmented **perivisceral coelom** surrounds the ventral half of the pharynx and the entire intestine. It extends into the metapleural folds in the branchiogenital region, and its lining provides muscles that compress the atrium, forcibly squirting water from the atriopore. In the intestinal region, the coelomic lining contributes some poorly developed gut muscles and circulates coelomic fluid.

The digestive system of amphioxus is strikingly similar to that of hemichordates, urochordates, and primitive vertebrates. As in these other groups, lancelets ingest large quantities of particles from which they extract organic material while eliminating water through the gill slits and mineral particles through the anus (Fig. 17–39). Water entering the pharynx passes through the gill slits into a circumpharyngeal water jacket, the atrium, which opens to the exterior posteriorly at a midventral atriopore (Fig. 17–35A). Ciliary action accounts for the transport of particles and water through the entire gut.

The large mouth is surrounded by oral cirri that form a coarse screen over the mouth opening and prevent entry of large particles. Sensory cells on each

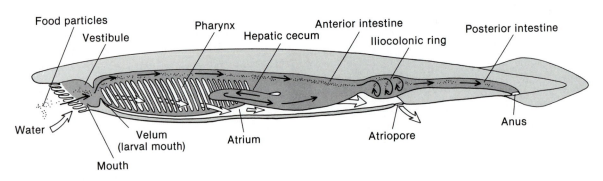

FIGURE 17–39. Functional organization of the digestive system of amphioxus. Open arrows show passage of water, small closed arrows show the path of food particles and digestive enzymes.

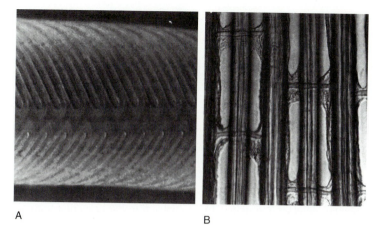

A   B

FIGURE 17–40. **A,** Ventral view of the exterior surface of the pharynx from a living amphioxus showing the elongated U-shaped gill slits and slender tongue bars. The dark zig-zag band along the ventral midline is probably a blood vessel. **B,** High-magnification view of part of two gill slits showing the transverse synapticles that support the tongue bars.

cirrus probably provide the animal with information on water quality. The **vestibule** (buccal cavity), which is enclosed by the **oral hood,** is a short section of the foregut that receives particles and water from the mouth, and its inner, ciliated surface bears two specialized structures. Longitudinal ridges project forward from the rear wall of the vestibule and bear dense cilia whose action creates a powerful water flow from the mouth into the pharynx. The coordinated action of these cilia generates a metachronal wave around the circumference of the vestibule, and for this reason the ciliated ridges are collectively known as the **wheel organ** (Fig. 17–38B). **Hatschek's groove** is a shallow ciliated outpocket of the dorsal wall of the vestibule. Recent research, using immunocytochemical techniques, indicates that Hatschek's groove is lined partly by endocrine cells, which are believed to secrete pituitary-like hormones into the blood.

A ring of fingerlike tentacles, the **velum,** separates the vestibule from the pharynx. The velum bears a sphincter muscle and is well supplied with sensory cells (Figs. 17–38B; 17–39).

The pharynx of amphioxus is a long, spacious tube perforated by 180 or more pairs of slender gill slits and surrounded by the atrium. Early in development, the gill slits are segmental, corresponding in number to the myomeres, but later the number of gill slits exceeds that of the myomeres. Each gill slit de-

velops as a circular or oval opening in the pharyngeal wall but soon elongates vertically and is nearly divided into two halves by the downgrowth of a tongue bar (Figs. 17–38; 17–40). Lateral cilia in the slits move water from the pharynx into the atrium while frontal cilia on the gill and tongue bars transport food and other particulates trapped in the mucous net, secreted by the endostyle, to the dorsal **epipharyngeal groove** (Fig. 17–35C). The ciliated epipharyngeal groove, like the tunicate dorsal lamina (or languets), gathers the food-laden mucous net and transports it rearward into the intestine. Like the proboscis skeleton and gill bars of acorn worms, the visceral skeleton of cephalochordates is composed of a cartilage-like material in the basement membrane of the gill bars, tongue bars, and oral cirri.

At its junction with the pharynx the anterior intestine gives rise to a single unbranched diverticulum, the **hepatic cecum** (Fig. 17–39). The cecum extends anteriorly from its origin and bulges into the atrium along the right side of the pharynx. The hepatic cecum is a digestive gland that secretes enzymes into the intestine and is the site of intracellular digestion and nutrient storage.

Posterior to the hepatic cecum, the anterior intestine narrows before opening into a short, wider section, called the **iliocolonic ring** (Fig. 17–39). Here, mucus-entangled particles delivered to the intestine from the epipharyngeal groove are rotated by cilia on the wall of the iliocolon. As the food mass rotates, it is subjected to digestive enzymes from the hepatic cecum. Small, partially digested organic material is released from the rotating mass and transported by specialized ciliary tracts into the cecum for the intracellular phase of digestion. Indigestible mineral particles move from the iliocolon into the **posterior intestine** where they are compacted into a slender fecal thread and released through the anus.

The blood-vascular system is well developed, consisting of arteries, veins, and smaller vessels but no specialized heart (Fig. 17–41). The blood, which is colorless and contains few cells, is circulated by muscles in the vessel walls. Blood flows from capillaries in the gills, where gas exchange occurs, into paired dorsal aortae, which unite in the intestinal region of the body to form an unpaired dorsal aorta. The aortae transport blood posteriorly. The dorsal aortae give rise to three primary branches: segmental arteries leading to capillaries in the myosepta, segmental arteries leading to capillaries in the atrium and gonads, and intestinal arteries to capillaries in the gut wall. Blood returning from these capillary beds may pass into one of the two cardinal veins, which carries blood anteriorly to the gills, or first into the hepatic portal vein. Blood in the hepatic portal vein enters a capillary bed in the hepatic cecum before being collected by the hepatic vein. As blood leaves the hepatic vein, it enters the strongly contractile sinus venosus ("heart") and endostylar artery (ventral aorta), which pump the blood through either the car-

dinal veins for circulation through the gills or directly back to the dorsal aortae.

The blood vessels of lancelets lack the endothelium typical of vertebrates and are lined instead by the basal lamina of overlying cell layers or are simply open channels in the connective tissue layer (p. 185). Thus, although the anatomy of the cephalochordate circulatory system is like that of vertebrates, it is histologically similar to the circulatory systems of invertebrates.

Cephalochordates are probably ammonotelic animals, but paired, segmental, filtration nephridia are present in the branchial region of the body in the subchordal coelom adjacent to the dorsal aortae (Fig. 17–42). As segmental organs, one pair of nephridia corresponds to each pair of primary gill slits (and myomeres). The anteriormost nephridium, **Hatschek's nephridium,** however, is unpaired and is associated with the velum (the embryonic mouth), which itself is believed to be a modified gill slit. Each nephridium consists of a cluster of podocytes on the wall of a short branch from the dorsal aorta. The apical end of each podocyte bears a flagellum encircled by a collar of microvilli. The microvillar collar and the flagellum enter a short ciliated tubule that leads from the subchordal coelom into the atrium (Fig. 17–42). Although the podocytes, by virtue of their collars, resemble protonephridial terminal cells, it is now believed that cephalochordates have a metanephridial system. Blood is thought to be ultrafiltered across the blood vessel wall to form primary urine in the sub-

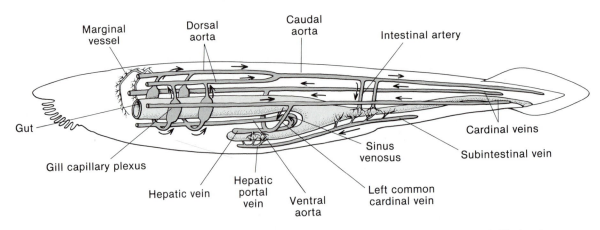

FIGURE 17–41. Simplified anatomy of the cephalochordate circulatory system. *(Modified and redrawn from Rähr, H. 1979. The circulatory system of amphioxus (Branchiostoma lanceolatum Pallas). A light-microscopic investigation based on intravascular injection technique. Acta Zool. 60:1–18.)*

chordal coelom. The primary urine may then be directed into the tubule, for modification to final urine, by the action of the podocyte flagella. Urine leaves the body through the atriopore, but uric acid, an end-product of purine metabolism, is reported to be stored in the tissues associated with the gonads. The physiological significance of the nephridia is unknown.

Cephalochordates are dioecious animals with external fertilization and a complex life cycle that includes a benthic adult and a planktotrophic swimming larva. In *Branchiostoma,* there are usually 26 pairs of gonads (Fig. 17–35A, B), often corresponding to myomeres 10 through 35, but in *Asymmetron* and

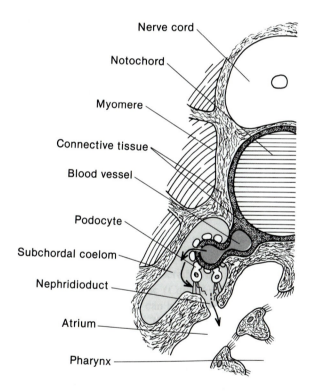

Nerve cord

Notochord

Myomere

Connective tissue

Blood vessel

Podocyte

Subchordal coelom

Nephridioduct

Atrium

Pharynx

FIGURE 17–42. Amphioxus has a bilateral series of segmental filtration nephridia, and one nephridium is shown here in cross-sectional view. Presumably, primary urine is formed by ultrafiltration of the blood into the subchordal coelom. Apparently, the urine is modified in the subchordal coelom before it is swept by cilia into the nephridioduct and atrium by cilia, where it eventually leaves the body via the atriopore. The cells at the filtration site are intermediate in form between podocytes and protonephridial terminal cells.

Epigonichthys, gonads occur only on the right side of the body. Spawning occurs at sunset in the European *Branchiostoma lanceolatum* as ripe gametes rupture from the gonads and are carried to the exterior in the exhaust from the atriopore. The zygotes undergo equal, holoblastic, radial cleavage to form a coeloblastula. Gastrulation is by invagination, and the anterior coeloms, myomeres, and notochord all form by enterocoely (Fig. 17–43).

The larva is a tiny, fishlike sliver (Fig. 17–44). Oral cirri, oral hood, atrium, and dorsal and ventral fins are absent, but there is a well-developed tail fin. Anteriorly, the body is strikingly asymmetric: a large mouth is located on the left side of the head, and the first gill openings are found on the right side (later, they migrate to form the left gill slits). The precursor of the wheel organ and Hatschek's groove, the **preoral pit,** opens on the left side of the head (Fig. 17–44A). During the gradual metamorphosis, the larval mouth becomes the velum, and the oral hood develops to enclose the vestibule and form the adult mouth and oral cirri. The metapleural folds enclose the pharynx to form the atrium, and the developing gills slits become bilaterally distributed along the pharynx (Fig. 17–44C).

SUMMARY

1. Cephalochordates, or lancelets, are fishlike chordates that live in coarse marine sands.

2. The body is segmented into V-shaped myomeres, which produce the characteristic lateral swimming undulations. A muscular notochord prevents longitudinal shortening and controls the lateral flexibility of the body. The gill slits, nephridia, and gonads are also segmental organs.

3. Lancelets filter feed in a manner similar to that of ascidians, using a pharyngeal mucous net. The net is produced by an endostyle and transported into the gut by an epipharyngeal groove.

4. The dorsal nerve cord is hollow and remains open to the exterior at an anterior neuropore.

5. A blood-vascular system is present, but the blood is colorless. The coeloms are numerous and highly specialized to form the fins, myomeres, notochord, and other structures.

6. Cephalochordates are dioecious animals with external fertilization. Radial cleavage leads to an asymmetric planktotrophic larva.

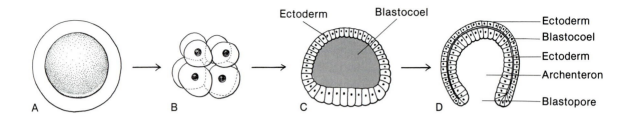

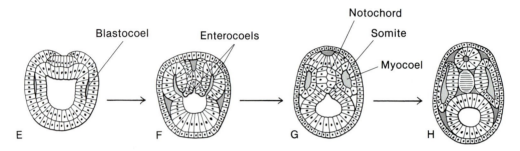

FIGURE 17–43. Amphioxus development. **A,** Zygote. **B,** Eight-cell stage, showing radial cleavage pattern. **C,** Blastula. **D,** Gastrula. **E,** Onset of neurulation and somitogenesis (segment formation). **F,** Onset of notochordogenesis and later stages of neurulation and somitogenesis. **G, H,** Completion of organogenesis. *(Modified and redrawn from Wiley, A. 1894. Amphioxus and the Ancestry of the Vertebrates. MacMillan and Co., New York. 316 pp. Conklin, E. G. 1932. Embryology of amphioxus. J. Morphol. 54:69–151.)*

PHYLUM CHORDATA: EVOLUTION OF THE VERTEBRATES

The evolution of vertebrates from the invertebrate chordates has long attracted the curiosity and stimulated the imagination of biologists. While reflecting on the evolution of vertebrates, it is important to emphasize that they did not evolve from any one of the existing phyla or subphyla of protochordates. Instead, groups, such as acorn worms, sea squirts, and lancelets, are cousins of the vertebrates, and each provides clues to the organization of the common ancestor. The clues reveal themselves in homologues—common patterns and trends (p. 864). When such common patterns are identified, it can usually be assumed that they were present in the common ancestor. The organization of the ancestral vertebrate can be reconstructed, piece by piece, from the characteristics shared with the protochordates. It is beyond the scope of this book to attempt a comprehensive reconstruction of the hypothetical ancestral vertebrate, but the evolution of a few of its characteristics can be traced as an example of how such a reconstruction is accomplished.

It seems probable that the ancestral vertebrate had the typical chordate features of a notochord, dorsal hollow nerve cord, and gill slits because each of these occurs not only in the vertebrates, but also in the protochordates, at least in rudimentary form. A knowledge of anatomical, embryological, and functional details of each of the structures allows us to hypothesize the ancestral design and function of each structure and to trace its evolution.

The notochord probably evolved from a fluid-filled coelomic cavity, lined by a typical mesothelium, as is found in the notochord of larvaceans and the coeloms of many nonchordate invertebrates (Fig. 17–45). Invertebrate coeloms commonly have a skeletal function, and it is not surprising that the chordates, like other invertebrates, have developed skeletal specializations from a simple hydrostat. The morphogenesis of the notochord in amphioxus, a cephalochordate, provides further evidence for the evolution of the notochord from a hydrostatic coelom. In amphioxus, the notochord arises enterocoelously from the archenteron as do the anterior somites, which are also coeloms. As the notochord differenti-

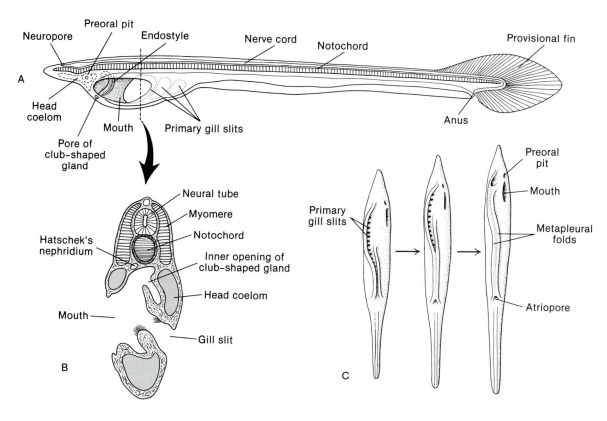

FIGURE 17–44. Amphioxus larva and metamorphosis. **A,** Left lateral cross-sectional view of the larva of *Branchiostoma caribaeum.* **B,** Cross section through the mouth of the larva in **A,** showing asymmetry. **C,** Ventral view of stages in metamorphosis showing the enclosure of the larval gill slits by the metapleural folds to form the atrium. The figures also show the beginning of the ventral migration of the mouth and preoral pit from their original positions on the left side of the head. The larval mouth corresponds to the adult velum, and the preoral pit becomes Hatschek's groove in the roof of the vestibule. *(C, Modified and redrawn from Willey, A. 1894. Amphioxus and the Ancestry of the Vertebrates. MacMillan and Co., New York. 316 pp.)*

ates, its mesothelium becomes a thickened muscular layer in a manner similar to the morphogenesis of body wall muscles from the coelomic lining in groups such as echinoderms, hemichordates, and some annelids (p. 183; Fig. 5–9). In amphioxus, moreover, the fibrous notochordal sheath is prominent and also plays an important skeletal role. The notochord of primitive vertebrates, such as lampreys and hagfish, which lack vertebrae, owes its stiffness to noncontractile intracellular fibers and, as in amphioxus, to a well-developed extracellular sheath. Among the higher vertebrates, the extracellular matrix around the notochord calcifies during development to form the vertebrae of the vertebral column. With the advent of the bony vertebral column, the skeletal importance of the notochord diminished and was therefore freed to adopt another function. The notochord persists in higher vertebrates as a contributor to the intervertebral discs that cushion the vertebrae.

The evolutionary history of the pharynx, gill slits, and gill skeleton can also be traced with reasonable certainty, although many points in this scenario, like that of the notochord, are subjects for further clarification. The universal occurrence in chordates of a ciliated pharynx provided with clefts for the elimination of water while feeding is evidence that the pharynx evolved as a food-concentrating structure. The notion that pharyngeal clefts evolved as outlets for excess water ingested with food is supported by the widespread occurrence of similar analogous adap-

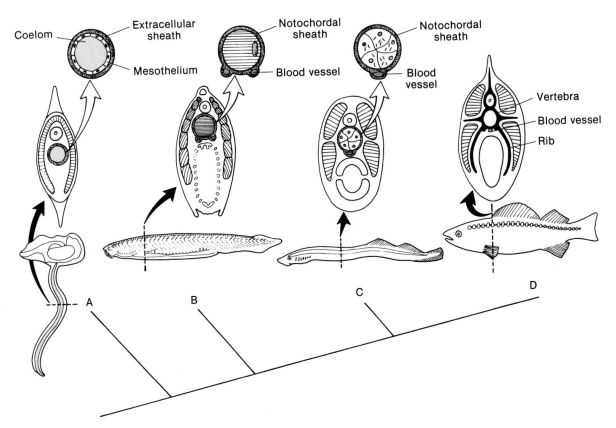

FIGURE 17–45. Evolution of the notochord and axial skeleton in chordates. **A,** The notochord of larvaceans is a fluid-filled coelomic cavity surrounded by a simple mesothelium and its extracellular sheath. **B,** In cephalochordates, the notochord develops enterocoelously, but the lining cells obliterate the cavity and are muscular. The notochordal sheath is thick and fibrous. **C,** The notochord of primitive vertebrates, such as the agnathan fishes (lampreys and hagfish), is well developed and cellular, each cell containing large vesicles and intracellular supportive fibers. The notochordal sheath is very thick and fibrous. **D,** Finally, in the vertebrates, the notochord is nearly eliminated by calcification of the surrounding connective tissue to form the vertebrae and other bones. The notochord of higher vertebrates persists only as the intervertebral discs that cushion adjacent vertebrae. *(A, After Olsson, R. 1965. Comparative morphology and physiology of the Oikopleura notochord. Israel J. of Zool. 14:213–220. C, After Schwarz, W. 1961. Elekronmikroskopische Untersuchungen an der Chordazellen von Petromyzon. Z. Zellforsch. 55:597–609.)*

tations outside of the chordates: for example, the pharyngeal pores of macrodasyid gastrotrichs, the intestinal siphons of echiurans and echinoids, and the intestinal groove of sipunculans, all of which concentrate food by removing water. As body size increased beyond that of the microscopic pterobranch hemichordates and larvaceans, the pharyngeal surface area and number of clefts increased accordingly, and a visceral skeleton evolved to support the large surface. The enlarged surface area of the pharynx and the increased water flow through it probably created two other functional opportunities for the use of the pharynx: (1) as a gill and (2) as a filtering surface for the entrapment of suspended food. (This latter function is analogous to the adoption of a suspension-feeding function by the enlarged gill of lamellibranch bivalves.) Primitively, food particles were probably trapped in mucus secreted by the pharyngeal lining as a whole. Later, a specialized region, called the endostyle, which secreted iodinated proteins in the mucus, evolved to produce an organized net of appropriate mesh size.

The pharyngeal apparatus, as it appears for example in amphioxus, carried over essentially unchanged into the vertebrates but lost its food-collecting function. Today, it can be found in the ammocoetes larva of lampreys, but during larval metamorphosis, the pharynx undergoes a profound reorganization. This metamorphosis, in addition to the comparative anatomy, development, and paleontology of vertebrates provides a rather clear understanding of the evolution of the vertebrate pharynx. Comparative vertebrate anatomy courses routinely trace the fates of the gill slits and gill arches during vertebrate evolution, but the derivation of one structure, the endostyle, is noted here. It is virtually certain that the protochordate endostyle evolved into the vertebrate thyroid gland. During the evolutionary transformation, the endostyle abandoned the production of iodinated mucus for use in filter feeding, but the thyroid still requires iodine in the synthesis of thyroxin, the principal hormone secreted by the gland.

The chordate dorsal, hollow, neural tube probably evolved from a simple intraepidermal nerve cord,

such as that found in the hemichordates. As in the hemichordates, it consisted of a concentration of nerve cells sandwiched between the epidermal cells and the underlying basal lamina of the ciliated dorsal epidermis. (Cilia on the epidermal cells produced a posterior flow of oxygenated surface water over the body to provide a countercurrent for the forward flow of blood in the dorsal blood vessel.)

During chordate development, a strip of dorsal epithelium is internalized, in a process called neurulation, to form the neural tube. This strip, the neural plate, submerges below the surface and is covered over by the midline fusion of left and right neural folds, which arch over the sunken neural plate and enclose a cavity, the neurocoel (Fig. 17–46B–D). The ectodermal cells lining the neurocoel (ependymal cells) grow cilia, which extend into the neurocoel and beat with posteriorly directed effective strokes. The nerve cells differentiate as a thick layer between the ependymal cells and the basal lamina. The neural folds seal the neural tube from posterior to anterior but the anteriormost end of the neural tube remains

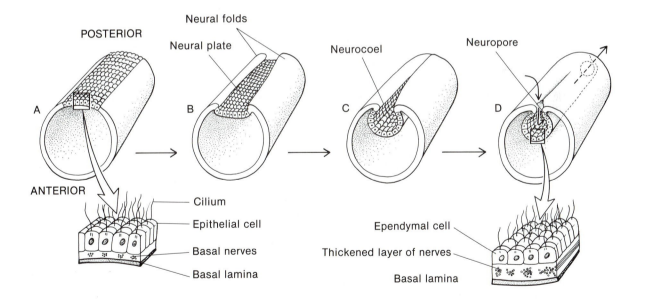

FIGURE 17–46. A, During chordate development, the neural plate cells submerge below the surface of the dorsal midline **(B),** roll into a tube and are covered over by the neural folds **(B–D).** The original ciliated surface epithelium in **A** (the protochordate ancestor) persists as the ciliated lining (ependymal cells) of the neural tube, whereas the intraepidermal nerves persist as the nervous layer of the nerve cord (enlargement, **D**). In the vertebrates, the neuropore **(D)** is a transient structure present only during development, but in the protochordates, it persists throughout the life of the animal.

unsealed in all protochordates (Fig. 17–46D). Through this persistent opening, called the neuropore, seawater flows into the neurocoel by the action of cilia on the ependymal cells. The incoming seawater eventually passes across the wall of the neural tube and enters the circulatory system. In vertebrates, the neuropore closes, sealing off the neural canal from the exterior. Its fluid becomes the cerebrospinal fluid.

According to this analysis of the three principal chordate features, the ancestral chordate probably had a fluid-filled hydrostatic notochord, a filter-feeding pharynx bearing gills slits and a mucus-producing endostyle, and a dorsal hollow nerve cord that opened anteriorly to the exterior via a neuropore. Although the analysis is limited to only three characteristics, the presence of a notochord in the reconstruction suggests that the hypothetical ancestor was motile rather than sessile in its life pattern.

There are many theories to explain the evolutionary origin of the chordates. One of the most common proposes that the chordate ancestor was a sessile, pterobranch-like organism that gave rise to the echinoderms on the one hand, and the ascidians on the other. From the larva of a sessile ascidian, the first motile chordate evolved by progenesis (or neoteny). According to this view, the larvaceans, cephalochordates, and vertebrates all evolved from this progenetic ancestor. Recently published research by L. Z. Holland (1989), however, casts some doubt on this interpretation. In a detailed study of tunicate gametes, Holland found that larvacean gametes were more primitive than those of ascidians or thaliaceans. Her

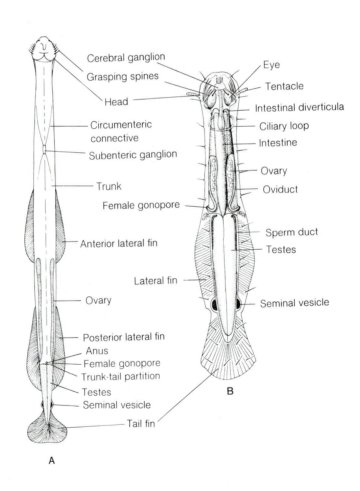

FIGURE 17–47. Phylum Chaetognatha. **A,** *Sagitta elegans* (ventral view). **B,** *Spadella* (dorsal view). *(A, After Ritter-Zahony. B, After Hertwig.)*

results imply that the motile larvacean body design is primitive to that of sessile ascidians, and thus the first chordates were motile organisms.

PHYLUM CHAETOGNATHA

The chaetognaths, known as arrowworms, are common animals found in marine plankton. The entire phylum of some 70 species is marine, and except for the benthic genus *Spadella,* all arrowworms are adapted for a planktonic existence. Most are found in tropical waters and in the upper 900 m, but the phylum is represented in the plankton of all oceans. The adults possess none of the features common to the other deuterostome phyla, and they are like aschelminths in many respects. Only the development of arrowworms would suggest a deuterostome position for these animals.

External Structure

The body of an arrowworm is shaped like a torpedo or feathered dart and ranges in length from 2 to 120 mm (Fig. 17–47). The body is divided into a head, a trunk, and a postanal tail region; a narrowed neck separates the head and trunk. On the underside of the rounded head is a large chamber (the vestibule) that leads into the mouth (Fig. 17–48).

Hanging down from each side of the head and flanking the vestibule are 4 to 14 large, curved, chitinous hooks, the **grasping spines,** that are used in seizing prey (Fig. 17–48). Several rows of much shorter spines (anterior and posterior teeth) that are curved around the front of the head may also assist in capturing prey or in the injection of venom. A pair of **eyes** is located posteriorly on the dorsal surface. In the neck region is a peculiar fold of the body wall, the **hood,** that can be pulled forward to enclose the entire head. The hood is thought to perhaps protect the spines when they are not in use and reduce water resistance during swimming.

The remainder of the body is composed of the elongated trunk and the tail. A characteristic feature of the chaetognaths is the lateral **fins** that border these regions of the body. In some arrowworms, such as *Sagitta,* there are two pairs of lateral fins, but in most species, there is a single pair (Fig. 17–47). Posteriorly, a large, spatula-like caudal fin encompasses the end of the tail.

Internal Structure and Physiology

The epidermis, which is covered on the outer surface with a thin surface coat, is **multilayered** (stratified) except on the ventral surface of the head and inside of the hood and contains large, vacuolated cells around

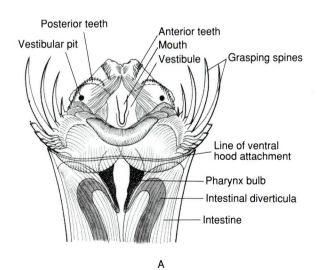

A

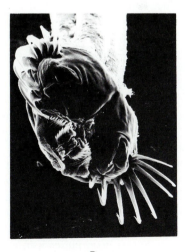

B

FIGURE 17–48. A, Head of *Sagitta elegans* (ventral view). **B,** Anterioventral scanning electronmicrograph view of head of *Sagitta. (A, After Ritter-Zahony. B, Scanning Electronmicrograph courtesy of H. Spero, D. Hagan, and A. Vastano.)*

the neck. The ventral surface of the head bears a thick cuticle. A basement membrane lies beneath the epidermis and is thickened to form the supporting rays between the two epithelial layers of the fins. The cross-striated longitudinal muscles of the body wall are arranged in two dorsolateral and two ventrolateral bands. In the head are special muscles for operating the hood, the teeth, the grasping spines, and other structures.

The compartmented coelom is lined by a mesothelium that is stratified on the body wall side into the thick musculature, noted previously, and a thin noncontractile **peritoneum.** On the outer surface of the gut, the mesothelium is a simple myoepithelium. The head contains a single coelomic space that is separated by a septum from the paired trunk coelomic spaces. Another septum separates the trunk from the one or two coelomic compartments that occupy the tail. Thus, chaetognaths are tricoelomate in organization.

Chaetognaths alternately swim and float, with the fins acting as flotation devices. When the body begins to sink, the longitudinal trunk muscles contract rapidly producing a dorsoventral undulation, and the animal darts swiftly forward. This forward motion is then followed by an interval of gliding and floating. The benthic *Spadella* adheres to bottom objects by means of special posterior adhesive papillae, but it can swim short distances.

Arrowworms are all carnivorous and feed on other planktonic animals, particularly copepods, which they detect from vibrations. They are thus of great importance as a trophic link between the primary and higher consumers in the sea. *Sagitta* consumes young fish and other arrowworms, including those of its own species, as large as itself. Arrowworms are voracious feeders. *Sagitta nagae,* for example, consumes 37% of its own weight each day in prey. In capturing prey, chaetognaths dart forward or laterally, withdrawing the hood and spreading the grasping spines. The prey is seized with the grasping spines, and its body may be pierced with the teeth. The chaetognath then injects a bacteria-derived toxin (tetrodotoxin), a potent sodium ion channel blocker, that immobilizes the prey. Tetrodotoxin is well known in certain toxic puffer fish, made famous in Japanese cuisine, but is used as a venom to capture prey in only one other animal, the blue-ringed octopus, *Hapalochlaena maculosa,* of the western Pacific (p. 474), whose bite can be fatal to humans. The sites

of toxin production and release in chaetognaths are unknown.

The digestive tract is simple (Figs. 17–47). The mouth leads into the bulbous, muscular pharynx that penetrates the head/trunk septum to join a straight intestine. The intestine extends through the length of the trunk, and at its anterior gives rise to a pair of lateral diverticula.

After capture the prey is pushed into the mouth, where it is lubricated by pharyngeal secretions and then passed to the posterior of the intestine. Here the food is rotated and often moved back and forth until it is broken down. Digestion is probably extracellular.

There are no gas exchange or excretory organs, and the coelomic fluid acts as a circulatory medium. Unless there are blood vessels present (none yet reported), gut-derived nutrients in the trunk coeloms must diffuse across the trunk/tail, and trunk/head septa to supply the tail and head regions.

The nervous system consists of six ganglia in the head and a large ventral ganglion in the trunk. The prominent cerebral ganglion is located above the esophagus and from it a nerve ring encircles the gut. The ring bears a pair of large **vestibular ganglia** laterally and a pair of **esophageal ganglia** ventrally, from which esophageal nerves running dorsally and ventrally innervate the gut myoepithelium. From each side of the cerebral ganglion, a large lateral nerve extends posteriorly to the very large ventral ganglion in the trunk. Twelve pairs of nerves issue from the ganglion and join an extensive intraepidermal nerve plexus, which innervates the somatic musculature and sensory organs. Sense organs include the eyes (Fig. 17–47) and sensory hairs. The eyes, at least in *Sagitta,* are of the pigment-cup type, and the photoreceptoral membranes are derived from cilia. The sensory hairs are arranged in fanlike arrays (ciliary fences) in longitudinal rows along the length of the trunk and function in detecting water-borne vibrations, as does the lateral line system of fish.

Reproduction

Chaetognaths are hermaphrodites; a pair of elongated ovaries is located in the trunk coelom in front of the trunk/tail septum, and a pair of testes are behind the septum (Fig. 17–47). From each testis a sperm duct passes posteriorly and laterally to terminate in a seminal vesicle embedded in the lateral body wall. Sperm leave the testis as spermatogonia, and spermatogenesis is completed in the tail coelom. When mature, the sperm pass into the ciliated funnel of the sperm duct and from there into a seminal vesicle in which the sperm are formed into a single spermatophore. The seminal vesicle ruptures, enabling the spermatophore to escape.

An oviduct runs along the lateral side of each ovary and opens to the exterior through two gonopores, one on each side of the body just in front of the trunk/tail septum. The eggs do not begin to mature until after spermatogenesis has commenced in the tail coelom.

The process of spermatophore transfer is known in detail only for species of *Spadella*. In *Spadella schizoptera*, one individual acts as a donor and one as a receiver, and the spermatophore is transferred very rapidly to the recipient's female gonopore after a brief, preliminary signaling behavior.

In *Sagitta*, the eggs are fertilized when they pass into the oviduct and are then emitted through the gonopore. The eggs are planktonic and are surrounded by a coat of jelly. In other arrowworms, the eggs may be attached to the body surface of the parent and carried about for some time. *Spadella* deposits its eggs in small clusters on algae or other objects.

Cleavage is radial, complete, and equal and leads to a coeloblastula. Gastrulation is accomplished by invagination, obliterating the blastocoel. The anterior end wall of the archenteron invaginates, folding backward on each side and cutting off two pairs of lateral, coelomic sacs (Fig. 17–49). The coelom thus originates by an unusual form of enterocoely. Further development is direct, and although the young are called larvae when they hatch, they are similar to the adult, and no metamorphosis occurs.

Despite historical claims of structural similarities between chaetognaths and some protostomes (aschelminths, molluscs, arthropods), the development of the phylum appears to be deuterostome in nature, especially in regard to the tricoelomate arrangement of body cavities. There is no larval stage comparable to that of the echinoderms and the hemichordates, however. Thus, the chaetognaths cannot be allied with any specific deuterostome phylum. If chaetognaths are really deuterostomes, the phylum must have departed very early from the base of the deuterostome line and is only remotely related to the other deuterostome groups. Nielsen (1985), who places chaetognaths with the aschelminths, and Bone and colleagues (1991) review embryonic and adult features of the group.

SUMMARY

1. Members of the phylum Chaetognatha, called arrowworms, are common marine planktonic animals. The small, translucent, torpedo-shaped body bears lateral and caudal fins, which enable the animal to glide and float following rapid, dartlike swimming.

2. Chaetognaths are predacious on other planktonic animals, which are seized using grasping spines located on either side of the head. There are no special organs for gas exchange, excretion, or internal transport.

3. Chaetognaths are hermaphrodites with internal fertilization via spermatophores. Development is planktonic and direct.

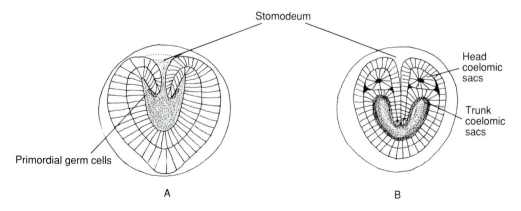

FIGURE 17–49. Coelom formation in *Sagitta*. **A,** Initial folding of archenteron walls. **B,** Separation of head coelomic sacs. *(Both after Burfield from Hyman.)*

REFERENCES

Homology and Analogy

Haszprunar, G. 1992. The types of homology and their significance for evolutionary biology and phylogenetics. J. Evol. Biol. *5:*13–24.

Hyman, L. H. 1942. Comparative Vertebrate Anatomy. The University of Chicago Press, Chicago. pp. 3–4. (Concise definitions and criteria for homology and analogy.)

Remane, A. 1956. Die Gründlagen des natürlichen Systems, der vergleichenden Anatomie und der Phylogenetik. [The foundation of the natural system, comparative anatomy, and phylogenetics]. Akademische Verlagsges. Geest & Portig K.-G., Leipzig. Reprinted by Otto Koeltz. Königstein-Taunus 1971. (A thorough and erudite account, unfortunately not available in English translation.)

Rieger, R. M., and Tyler, S. 1979. The homology theorem in ultrastructure research. Am. Zool. *19:*655–664. (This paper offers a modern discussion of the traditional concept of homology.)

Tyler, S. 1988. The role of function in determination of homology and convergence—Examples from invertebrate adhesive organs. Fortschr. Zool. *36:*331–347. (An advanced but lucid treatment of the use of functional criteria in homology recognition.)

Wagner, G. P. 1989. The biological homology concept. Ann. Rev. Ecol. Syst. *20:*51–69.

The literature included below is restricted to books and papers on urochordates, hemichordates, and chaetognaths. The introductory references on page 6 list many general works and field guides that contain sections on these groups.

Hemichordates

Armstrong, W. G., Dilly, P. N., and Urbanek, A. 1984. Collagen in the pterobranch coenecium and the problem of graptolite affinities. Lethaia. *17*(2):145–152.

Balser, E. J., and Ruppert, E. E. 1990. Structure, ultrastructure and function of the preoral heart–kidney in *Saccoglossus kowalevskii* (Hemichordata, Enteropneusta) including new data on the stomochord. Acta Zool. *71:*235–249.

Barnes, R. D. 1977. New record of a pterobranch hemichordate from the Western Hemisphere. Bull. Mar. Sci. *27*(2):340–343.

Barrington, E. 1965a. Observations of feeding and digestion in *Glossobalanus.* Q. J. Micr. Sci. *82:*227–260.

Barrington, E. 1965b. The Biology of Hemichordata and Protochordata. W. H. Freeman, San Francisco.

Dawydoff, C. 1948. Embranchement des Stomocordes. *In* Grassé, P. (Ed.): Traité de Zoologie. Vol. 11. Echinodermes, Stomocordes, Procordes. Masson et Cie, Paris. pp. 367–551.

Dilly, P. N., Welsch, U., and Rehkämper, G. 1986a. Fine structure of heart, pericardium and glomerular vessel in *Cephalodiscus gracilis* M'Intosh, 1882 (Pterobranchia, Hemichordata). Acta Zool. *67:*173–179.

Dilly, P. N., Welsch, U., and Rehkämper, G. 1986b. Fine structure of the tentacles, arms and associated coelomic structures of *Cephalodiscus gracilis* (Pterobranchia, Hemichordata). Acta Zool. *67:*181–191.

Dilly, P. N., Welsch, U., and Rehkämper, G. 1986c. On the fine structure of the alimentary tract of *Cephalodiscus gracilis* (Pterobranchia, Hemichordata). Acta Zool. *67:*87–95.

Duncan, P. B. 1987. Burrow structure and burrowing activity of the funnel-feeding enteropneust *Balanoglossus aurantiacus* in Bogue Sound, North Carolina, USA. P.S.Z.N.I.: Mar. Ecol. *8:*75–95.

Hadfield, M. G. 1975. Hemichordata. *In* Giese, A. C., and Pearse, J. S. (Eds.): Reproduction of Marine Invertebrates. Vol. II. Academic Press, New York. pp. 185–240.

Knight-Jones, E. W. 1952. On the nervous system of *Saccoglossus cambrensis.* Phil. Trans. R. Soc. Lond. B. *236:*315–354.

Lester, S. M. 1985. *Cephalodiscus* sp. *[gracilis]* (Hemichordata:Pterobranchia): Observations of functional morphology, behavior and occurrence in shallow water around Bermuda. Mar. Biol. *85:*263–268.

Lester, S. M. 1988a. Ultrastructure of adult gonads and development and structure of the larva of *Rhabdopleura normani* (Hemichordata: Pterobranchia). Acta Zool. *69:*95–109.

Lester, S. M. 1988b. Settlement and metamorphosis of *Rhabdopleura normani* (Hemichordata: Pterobranchia). Acta Zool. *69:*111–120.

Pardos, F. 1988. Fine structure and function of pharynx cilia in *Glossobalanus minutus* Kowalewsky (Enteropneusta). Acta Zool. *69:*1–12.

Pardos, F., and Benito, J. 1988. Blood vessels and related structures in the gill bars of *Glossobalanus minutus* (Enteropneusta). Acta Zool. *69:*87–94.

Petersen, J. A., and Ditadi, A. S. F. 1971. Asexual reproduction in *Glossobalanus crozieri* (Ptychoderidae, Enteropneusta, Hemichordata). Mar. Biol. *9:*78–85.

Romero-Wetzel, M. B. 1989. Branched burrow-systems of the enteropneust *Stereobalanus canadensis* (Spengel) in deep-sea sediments of the Voring-Plateau, Norwegian Sea. Sarsia. *74:*85–89.

Welsch, U., Dilly, P. N., and Rehkämper, G. 1987. Fine

structure of the stomochord in *Cephalodiscus gracilis* M'Intosh 1882 (Hemichordata, Pterobranchia). Zool. Anz. *218*:209–218.

Urochordates

Alldredge, A. L. 1976. Appendicularians. Sci. Am. *235*(1):94–102. (An excellent brief account of the Larvacea.)

Alldredge, A. L. 1977. House morphology and mechanisms of feeding in the Oikopleuridae (Tunicata, Appendicularia). J. Zool. London. *181*:175–188.

Alldredge, A. L., and Madin, L. P. 1982. Pelagic tunicates: Unique herbivores in the marine plankton. BioSci. *32*(8):655–663.

Arkett, S. A., Mackie, G. O., and Singla, C. L. 1989. Neuronal organization of the ascidian (Urochordata) branchial basket revealed by cholinesterase activity. Cell Tissue Res. *27*:285–294.

Barham, E. G. 1979. Giant larvacean houses: Observations from deep submersibles. Science. *205*:1129–1131.

Berrill, N. J. 1936. Studies in tunicate development. Part V. The evolution and classification of ascidians. Phil. Trans. R. Soc. Lond. B. *226*:43–70.

Berrill, N. J. 1950. The Tunicata. Quaritch Ltd., London. 354 pp.

Berrill, N. J. 1975. Chordata: Tunicata. *In* Giese, A. C., and Pearse, J. S. (Eds.): Reproduction of Marine Invertebrates. Vol. II. Academic Press, New York. pp. 241–282.

Brien, P. 1948. Embranchement des Tuniciers. *In* Grassé, P. (Ed.): Traité de Zoologie. Vol. 11. Echinodermes, Stomocordes, Procordes. Masson et Cie, Paris. pp. 553–930.

Bullough, W. S. 1958. Practical Invertebrate Anatomy. Macmillan, New York. pp. 446–464. (Descriptions of representative tunicates, including Thaliacea and the Larvacea.)

Burighel, P., and Milanesi, C. 1975. Fine structure of the gastric epithelium of the ascidian *Botryllus schlosseri*, mucous, endocrine and plicated cells. Cell Tissue Res. *158*:481–496.

Burighel, P., and Milanesi, C. 1977. Fine structure of the intestinal epithelium of the colonial ascidian *Botryllus schlosseri*. Cell Tissue Res. *182*:357–369.

Cloney, R. A. 1982. Ascidian larvae and the events of metamorphosis. Am. Zool. *22*:817–826.

Deibel, D. 1986. Feeding mechanism and house of the appendicularian *Oikopleura vanhoeffeni*. Mar. Biol. *93*:429–436.

Fenaux, R. 1985. Rhythm of secretion of oikopleurid's houses. Bull. Mar. Sci. *37*(2):498–503.

Fenaux, R. 1986. The house of *Oikopleura dioica* (Tunicata, Appendicularia): Structure and function. Zoomorphology. *106*:224–231.

Flood, P. R. 1982. Transport speed of the mucous feeding filter in *Clavelina lepadiformis* (Aplousobranchia, Tunicata). Acta Zool. *63*:17–23.

Flood, P. R. 1990. Visualization of the transparent, gelatinous house of the pelagic tunicate *Oikopleura vanhoeffeni* using *Sepia* ink. Biol. Bull. *178*:118–125.

Flood, P. R. 1991. A simple technique for preservation and staining of the delicate houses of oikopleurid tunicates. Mar. Biol. *108*:105–110.

Godeaux, J. E. A. 1989. Functions of the endostyle in the tunicates. Bull. Mar. Sci. *45*:228–242. (Godeaux is one of the foremost students of the Thaliacea, and references to his earlier publications on this group, which are cited in this paper, should be consulted for further information.)

Goodbody, I. 1974. The physiology of ascidians. Adv. Mar. Biol. *12*:1–149. (An excellent review.)

Harbison, G. R., and McAlister, V. L. 1979. The filter-feeding rates and particle retention efficiencies of three species of *Cyclosalpa*. Limnol. Oceanogr. *24*(5):875–892.

Heron, A. C. 1975. Advantages of heart reversal in pelagic tunicates. J. Mar. Biol. Ass. U.K. *55*:959–963.

Heron, A. C. 1976. A new type of excretory mechanism in the tunicates. Mar. Biol. *36*:191–197. (Heron describes the release of excretory crystals from the salp. *Thalia*, which presumably would compromise the transparency of the animal if stored throughout life as in some sessile ascidians.)

Holland, L. Z. 1989. Fine structure of spermatids and sperm of *Dolioletta gegenbauri* and *Doliolum nationalis* (Tunicata: Thaliacea): Implications for tunicate phylogeny. Mar. Biol. *101*:83–95.

Jones, J. C. 1971. On the heart of the orange tunicate, *Ecteinascidia turbinata*. Biol. Bull. *141*:130–145.

Katz, M. J. 1983. Comparative anatomy of the tunicate tadpole, *Ciona intestinalis*. Biol. Bull. *164*:1–27.

Kott, P. 1984. Prokaryotic symbionts with a range of ascidian hosts. Bull. Mar. Sci. *34*:308–312.

Kott, P. 1989. Form and function in the Ascidiacea. Bull. Mar. Sci. *45*:253–276. (A good review and synthesis.)

Lane, D. J. W., and Wilkes, S. L. 1988. Localisation of vanadium, sulfur and bromine within the vandocytes of *Ascidia mentula* Muller. A quantitative electron probe X-ray microanalytical study. Acta Zool. *69*:135–145.

Macara, I. G. 1980. Vanadium—An element in search of a role. Trends Biochem. Sci. *5*:92–94.

Mackie, G. O. 1978. Luminescence and associated effector

activity in *Pyrosoma* (Tunicata: Pyrosomida). Proc. R. Soc. Lond. B. *202*:483–495.

Mackie, G. O. 1983. Coordination of compound ascidians by epithelial conduction in the colonial blood vessels. Biol. Bull. *165*:209–220.

Mackie, G. O. 1986. From aggregates to integrates: Physiological aspects of modularity in colonial animals. Phil. Trans. R. Soc. Lond. B. *313*:175–196. (In addition to provocative ideas on the evolution of coloniality, Mackie discusses colonial coordination in ascidians and thaliaceans.)

Madin, L. P. 1974. Field observations on the feeding behavior of salps. Mar. Biol. *25*:143–147.

Markus, J. A., and Lambert, C. C. 1983. Urea and ammonia excretion by solitary ascidians. J. Exp. Mar. Biol. Ecol. *66*:1–10.

Millar, R. H. 1970. British Ascidians. Synopses of the British Fauna No. 1. Academic Press, London. (Keys and notes for the identification of British species.)

Millar, R. H. 1971. The biology of ascidians. Adv. Mar. Biol. *9*:1–100 (A good review of ascidian biology.)

Mirre, C., and Thouveny, Y. 1977. Étude ultrastructurale de la glande pylorique de l'ascidié *Botryllus schlosseri* P. Bull. Soc. Zool. France *102*:439–443.

Monniot, C., and Monniot, F. 1975. Abyssal tunicates: An ecological paradox. Ann. Inst. Oceanogr. *51*(1):99–129.

Monniot, C., and Monniot, F. 1978. Recent work on the deep-sea tunicates. Oceanogr. Mar. Biol. Ann. Rev. *16*:181–228.

Monniot, F. 1979. Microfiltres et ciliatures branchiales des ascidiés littorales en microscopie électronique. Bull. Mus. Hist. Nat. Paris. Ser. 4. Vol. 1. Sec. A. (4):843–859.

Nishida, H. 1987. Cell lineage in ascidian embryos by intracellular injection of a tracer enzyme. III. Up to the tissue restricted stage. Devel. Biol. *121*:526–541.

Olsson, R. 1965. Comparative morphology and physiology of the *Oikopleura* notochord. Israel J. Zool. *14*:213–220. (The original description of the coelomic organization of the larvacean notochord.)

Olson, R. R. 1983. Ascidian–*Prochloron* symbiosis: The role of larval photoadaptations in midday larval release and settlement. Biol. Bull. *165*:221–240.

Paffenhöfer, G. A. 1973. The conservation of an appendicularian through numerous generations. Mar. Biol. *22*:183–185.

Parry, D. L., and Kott, P. 1988. Co-symbiosis in the Ascidiacea. Bull. Mar. Sci. *42*:149–153.

Pennachetti, C. A. 1984. Functional morphology of the branchial basket of *Ascidia paratropa* (Tunicata, Ascidiacea). Zoomorphology. *104*:216–222.

Ruppert, E. E. 1990. Structure, ultrastructure and function of the neural gland complex of *Ascidia interrupta* (Chordata, Ascidiacea): Clarification of hypotheses regarding the evolution of the vertebrate anterior pituitary. Acta Zool. *71*:135–149.

Saffo, M. B. 1981. The enigmatic protist *Nephromyces*. BioSystems. *14*:487–490.

Saffo, M. B., and Nelson, R. 1983. The cells of *Nephromyces*: Developmental stages of a single life cycle. Can. J. Bot. *61*:3230–3239.

Stoecker, D. 1978. Resistance of a tunicate to fouling. Biol. Bull. *155*:615–626. (Discusses the possibility that vanadium discourages fouling of the tunic of *Phallusia nigra*.)

Stoecker, D. 1980. Chemical defenses of ascidians against predators. Ecology. *61*:1327–1334.

Young, C. M., and Bingham, B. L. 1987. Chemical defense and aposematic coloration in larvae of the ascidian *Ecteinascidia turbinata*. Mar. Biol. *96*:539–544.

Young, C. M., and Braithwaite, L. F. 1980. Orientation and current-induced flow in the stalked ascidian *Styela montereyensis*. Biol. Bull. *159*:428–440.

Cephalochordates

Conklin, E. G. 1932. The embryology of amphioxus. J. Morphol. *54*:69–151.

Eakin, R. M., and Westfall, J. A. 1962. Fine structure of the notochord of amphioxus. J. Cell Biol. *12*:646–651.

Flood, P. 1970. The connection between spinal cord and notochord in amphioxus (*Branchiostoma lanceolatum*). Z. Zellforsch. *103*:115–128.

Flood, P. 1975. Fine structure of the notochord of amphioxus. Symp. Zool. Soc. Lond. *36*:81–104.

Holland, N. D., and Holland, L. Z. 1991. The histochemistry and fine structure of the nutritional reserves in the fin rays of a lancelet, *Branchiostoma lanceolatum* (Cephalochordata = Acrania). Acta Zool. *72*:203–207.

Light, S. F. 1923. Amphioxus fisheries near the University of Amoy, China. Science. *58*:57–60.

Rähr, H. 1979. The circulatory system of amphioxus (*Branchiostoma lanceolatum* (Pallas). Acta Zool. *60*:1–18.

Webb, J. E. 1973. The role of the notochord in forward and reverse swimming in the amphioxus, *Branchiostoma lanceolatum*. J. Zool. Lond. *170*:325–338.

Chordate Phylogeny

Bateson, W. 1886. The ancestry of the Chordata. Q. J. Microsc. Sci. *26*:535–571.

Berrill, N. J. 1955. The Origin of Vertebrates. Oxford University Press, London.

Berrill, N. J. 1987. Early chordate evolution. Part 1. Amphioxus, the riddle of the sands. Int. J. Invert. Reprod. Dev. *11:*1–14.

Bone, Q. 1960. The origin of the chordates. J. Linn. Soc. Lond. *44:*252–269.

Bone, Q. 1972. The origin of chordates. Oxford Biology Readers. Number 18. Oxford University Press, London. pp. 1–16.

Bone, Q. 1981. The neotenic origin of chordates. Atti Conv. Lincei. *49:*465–486.

Gans, C., and Northcutt, R. G. 1983. Neural crest and the origin of vertebrates: A new head. Science. *220:*268–274.

Garstang, W. 1928. The morphology of the Tunicata and its bearing on the phylogeny of the Chordata. Q. J. Microsc. Sci. *72:*51–187.

Gislen, T. 1930. Affinities between the Echinodermata, Enteropneusta, and Chordonia. Zool. Bidrag. Uppsala. *12:*199–304.

Holland, L. Z., Gorsky, G., and Fenaux, R. 1988. Fertilization in *Oikopleura dioica* (Tunicata, Appendicularia): Acrosome reaction, cortical reaction and sperm-egg fusion. Zoomorphology. *108:*229–243. (The new observations of reproductive biology that are reported in this paper are used to test alternative theories for the evolution of urochordates. In a thorough discussion, the authors favor an appendicularian-like ancestor for the tunicates.)

Jefferies, R. P. S. 1986. The Ancestry of Vertebrates. Cambridge University Press, Melbourne.

Jollie, M. 1982. What are the "Calcichordata?" and the larger question of the origin of the chordates. Zool. J. Linn. Soc. *75:*167–188.

Nielsen, C. 1985. Animal phylogeny in the light of the trochaea theory. Biol. J. Linn. Soc. *25:*243–299.

Randall, D. J., and Davie, P. S. 1980. The hearts of urochordates and cephalochordates. *In* Hearts and Heart-Like Organs. Vol. 1. Academic Press. pp. 41–59.

Schaeffer, B. 1987. Deuterostome monophyly and phylogeny. Evol. Biol. *21:*179–235.

Sedgwick, A. 1886. The original function of the canal of the central nervous system of Vertebrata. Stud. Morphol. Lab. Univ. Camb. *2:*160–164.

Willey, A. 1894. Amphioxus and the Ancestry of the Vertebrates. Columbia University Biological Series. II. MacMillan and Co., New York and London. 314 pp. (An excellent account of cephalochordates as well as a discussion of chordate phylogeny.)

Chaetognaths

Ahnelt, P. 1984. Chaetognatha. *In* Bereiter-Hahn, J., Matoltsky, A. G., and Richards, K. S. (Eds.): Biology of the Integument. 1. Invertebrates. Springer-Verlag, Berlin. pp. 746–755.

Alvarino, A. 1965. Chaetognaths. Oceanogr. Mar. Biol. Ann. Rev. *3:*115–194.

Bone, Q., Kapp, H., and Pierrot-Bults, A. C. 1991. The Biology of Chaetognaths. Oxford University Press. 173 pp.

Bieri, R., and Thuesen, E. V. 1990. The strange worm *Bathybelos.* Am. Sci. *78:*542–549.

Feigenbaum, D. L. 1978. Hair-fan patterns in the Chaetognatha. Can. J. Zool. *56(4):*536–546.

Ghirardelli, E. 1968. Some aspects of the biology of the chaetognaths. Adv. Mar. Biol. *6:*271–375.

Goto, T., and Yoshida, M. 1984. Photoreception in Chaetognatha. *In* Ali, M. A. Photoreception and Vision in Invertebrates. Plenum Press, New York. pp. 727–742.

Goto, T., and Yoshida, M. 1985. The mating sequence of the benthic arrowworm *Spadella schizoptera.* Biol. Bull. *169:*328–333.

Hyman, L. H. 1959. The Invertebrates. Vol. 5. Smaller Coelomate Groups. McGraw-Hill, New York. pp. 1–71. (A general account of the chaetognaths.)

Nagasawa, S., and Marumo, R. 1972. Feeding of a pelagic chaetognath, *Sagitta nagae,* in Suruga Bay Central Japan. J. Oceanogr. Soc. Japan. *32:*209–218.

Reeve, M. R., and Cosper, T. C. 1975. Chaetognatha. *In* Giese, A. C., and Pearse, J. S. (Eds.): Reproduction of Marine Invertebrates. Vol. 2. Academic Press, New York.

Shinn, G. L. 1992. Ultrastructure of somatic tissues in the ovaries of a chaetognath *(Ferosagitta hispida).* J. Morphol. *211:*221–241.

Sullivan, B. K. 1980. *In situ* feeding behavior of *Sagitta elegans* and *Eukrohnia hamata* (Chaetognatha) in relation to the vertical distribution and abundance of prey at Ocean Station "P." Limnol. Oceanogr. *25:*317–326.

Thuesen, E. V., and Bieri, R. 1987. Tooth structure and buccal pores in the chaetognath *Flaccisagitta hexaptera* and their relation to the capture of fish larvae and copepods. Can. J. Zool. *65:*181–187.

Thuesen, E. V., Kogure, K., Hashimoto, K. et al. 1988. Poison arrowworms: A tetrodotoxin venom in the marine phylum Chaetognatha. J. Exp. Mar. Biol. Ecol. *116:*249–256.

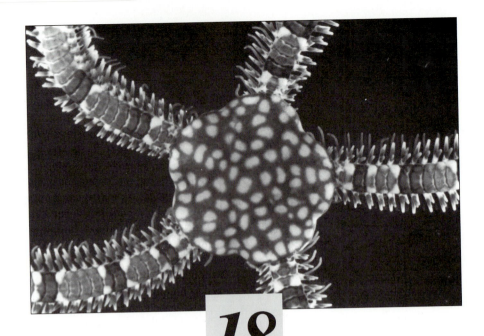

18

ECHINODERMS

PRINCIPLES AND EMERGING PATTERNS

MUTABLE CONNECTIVE TISSUE

The echinoderms have an endoskeleton located beneath the epidermis and composed of two parts: a layer of collagenous connective tissue (**dermis**) and small calcareous ossicles, which are embedded in the dermis. The presence of ossicles in the dermis increases the rigidity of the overall skeleton as has been noted already in discussions of sponge and cnidarian skeletons (Chapters 3 and 4). Among echinoderms, the stiffness of the skeleton ranges from the rigid test of most sea urchins, in which the ossicles are fused together, to the pliable, wormlike body wall of many sea cucumbers, which have microscopic ossicles widely dispersed in the dermis. Thus, the stiffness of an echinoderm skeleton would seem to be determined by the proportions of connective tissue and ossicles, but nothing could be farther from the truth. Echinoderms, unlike any other animals, can voluntarily and rapidly change the stiffness of the connective tissue itself. This unique property is called **mutable (or catch) connective tissue.**

Mutable connective tissue can be made rigid or flexible, depending on the voluntary response to circumstances faced by a particular animal. The degree of change in the rigidity of mutable connective tissue is almost as great as that of ice to water. Indeed, people who have firsthand experience of echinoderm tissues passing from rigid to flexible states, liken the transition to liquification. A softening of the connective tissue is made use of by brittle stars, many of which will readily cast off an arm in response to being seized by a potential predator. In this case, a rapid depolymerization occurs at a break point, and the arm literally falls off. A similar, local softening of connective tissue occurs in sea cucumbers, enabling them to forcibly eject parts of their viscera (evisceration) or parts of their body wall (Fig. 18–1) in response to predators and other factors. Hardening of the connective tissue also has a functional importance. For example, sea urchins normally wave their spines about, but when some species enter rocky crevices for protection, they elevate and lock their spines, wedging themselves in the crevice. The spines lock in place because the flexible connective tissue ligaments at the base of each spine polymerize and become rigid. Although relatively flexible, starfish also can stiffen their body by a similar mechanism. For example, while attempting to feed on a clam, the starfish dermis stiffens and the body becomes a rigid scaffold from which the tube feet extend and pull on the clam's valves.

The physiological control of mutable connective tissue and the molecular mechanisms of hardening and softening are active areas of modern research. Although the dermis contains some muscle, nerve, and other types of cells, it is the extracellular matrix that changes in stiffness. In most careful observations of the dermis, nerves have been found to terminate in the extracellular matrix. Apparently, there are two types of nerves present: one whose action may harden the matrix, another whose action softens it. The stiffness of the matrix is affected by changes in calcium ion concentrations and by experimental manipulations of other cations. In general, an increase in the calcium ion concentration stiffens the matrix; a decrease softens it. Such a result suggests that calcium ions may form cross bridges between macromolecules in the matrix.

Whatever the underlying mechanism of mutable connective tissue, the echinoderms have made use of this unique attribute in defense, feeding, entering and an-

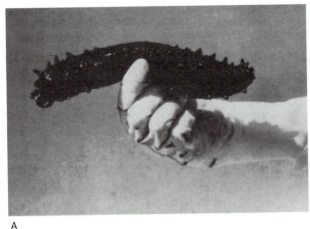

B

FIGURE 18–1 Mutable connective tissue of echinoderms is graphically illustrated in these two photographs of the sea cucumber, *Stichopus chloronotus*. When touched **(A)**, the animal stiffens the connective tissue in its body wall, but after being rubbed vigorously, the body wall becomes so soft that it flows between the fingers of the experimenter **(B).** *(From Motokawa, T. 1988. Catch connective tissue: A key character for echinoderms' success. In Burke, R. D., Mladenov, P. V., Lambert, P. et al. (Eds.): Echinoderm Biology. A. A. Balkema, Rotterdam. pp. 39–54.)*

choring in refuges, and asexual reproduction. One of the first students of mutable connective tissue, John Eylers, suggested that its sensitivity to environmental ionic concentrations might even account for the absence of echinoderms in freshwater habitats.

RESOURCE PARTITIONING

Every species occupies some niche within its habitat. The term **niche** refers to the mode of existence of the animal, that is, how it is oriented to the substratum, how it feeds, where it reproduces, and so forth. Theoretically, no two species can occupy the same niche within the same habitat, because one would eliminate the other through competition. Selection to avoid competition has led to partitioning, or dividing up, of such resources as food and nesting sites, so that two or more species with somewhat similar living patterns can be supported without competition. Some of their niche dimensions (characteristics) may overlap but not all. Clearly, resource partitioning leads to species specialization.

Canadian zoologist, Norman Sloan, observed five species of sea cucumbers living beneath boulders in the intertidal zone on Aldabra Atoll in the Indian Ocean. His study asks the question: What are these five species doing differently that makes it possible for them to live together within the same habitat, that is, how are they partitioning the available resources?

Two species were found to live only beneath rugose boulders. One of these attached to the undersurface of the boulder, and the other attached to slab rock beneath the boulder. The former was a deposit feeder; the latter fed on material deposited on the rock surface. The other three species of sea cucumbers lived in sand and gravel located beneath any type of boulder, but one of these was restricted to sites where there was standing water at low tide. All three were deposit feeders but each fed on different size particles. Moreover, the one living in standing water fed continuously, whereas the other four sea cucumbers only fed at high tide.

Notice that the five species of sea cucumbers were separated in three ways: by where they lived (spatial separation); by their food source (trophic separation); and by the time at which they fed (temporal separation). Each species was separated from the others by at least two of the three niche dimensions.

Although contemporary ecologists no longer believe that the one-species/one-niche rule is invariable, resource partitioning has certainly been a major factor in the evolution of animal species.

ECHINODERMS

Members of the phylum Echinodermata are among the most familiar marine invertebrates, and representatives, such as the sea stars, have become virtually a symbol of sea life. The phylum contains some 6000 known species and constitutes the only major group of deuterostome invertebrates.

Echinoderms are exclusively marine and are largely bottom dwellers. All are relatively large animals, most being at least several centimeters in diameter. The most striking characteristic of the group is their **pentamerous radial symmetry**—that is, the body can usually be divided into five parts arranged around a central axis. This radial symmetry, however, has been secondarily derived from a bilateral ancestor, and the echinoderms, are not closely related to the other radiate phyla.

Characteristic of all echinoderms is the presence of an internal skeleton. The skeleton is composed of **calcareous ossicles** that may articulate with one another, as in sea stars and brittle stars, or may be sutured together to form a rigid skeletal test, as in sea urchins and sand dollars. Commonly, the skeleton bears projecting spines or tubercles that give the body surface a warty or spiny appearance, hence the name echinoderm, meaning "spiny skin."

The most distinctive feature of echinoderms is the presence of a unique system of coelomic canals and surface appendages composing the **water-vascular system.** Primitively, the water-vascular system probably functioned in collecting and transporting food, but in many echinoderms, it has assumed a locomotor function.

Echinoderms possess a spacious coelom in which a well-developed digestive tract is suspended. There are no excretory organs. Gas exchange structures vary from one group to another and appear to have arisen independently within the different classes. Most members of the phylum are dioecious. The reproductive tracts are very simple, for there is no copulation and fertilization is usually external in seawater.

Echinoderm Development

A brief general description of echinoderm development will aid in understanding the secondary nature of their symmetry as well as some of the structural features that will be described in the following section. The eggs are typically homolecithal, and the early development is relatively uniform throughout the phylum and displays radial and indeterminate cleavage, the basic features of deuterostome development (p. 196).

The blastula contains a large blastocoel, and gastrulation takes place primarily by invagination, forming a narrow, tubular archenteron (primitive gut). The archenteron grows forward and eventually connects with the anterior stomodeum, which will form the mouth. The blastopore remains as the larval anus.

Before the mouth forms, the advancing distal end of the archenteron, primitively at least, gives rise to a pair of lateral pockets or pouches that eventually separate from the archenteron (Figs. 18–42; 5–19). The cavities of the pouches represent the future coelomic cavity, and the cells composing the pouch wall become the mesoderm. The two original pouches, one on each side, each give rise by evagination or subdivision to coelomic vesicles arranged one behind the other and called, respectively, the **axocoel,** the **hydrocoel,** and the **somatocoel** (Fig. 18–2A). These coelomic vesicles correspond to the paired protocoel, mesocoel, and metacoel of other deuterostome coelomates (hemichordates) with a tripartite (oligomerous)

body. The two somatocoels meet above and below the gut to form the gut mesenteries. The left axocoel opens dorsally through a pore called the **hydropore.** This somewhat generalized plan of the coelomic vesicles is considerably modified in the development of existing echinoderms.

The gastrula rapidly develops into a free-swimming larva (Fig. 18–3). The most striking feature of the echinoderm larva is its bilateral symmetry, which is in marked contrast to the radial symmetry of the

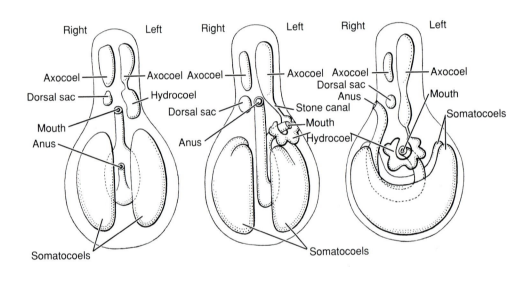

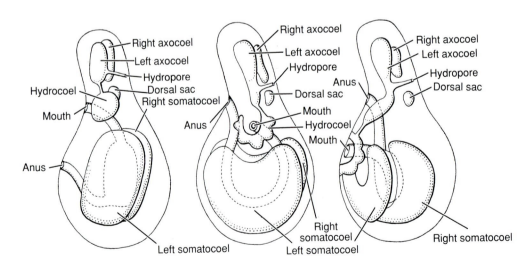

FIGURE 18–2 Generalized echinoderm metamorphosis. Top row, ventral view; bottom row, side view. *(From Ubaghs, G. 1967. In Moore, R. C. (Ed.): Treatise on Invertebrate Paleontology. Pt. S. Vol. 1. Courtesy of the Geological Society of America and the University of Kansas, Lawrence.)*

adult. Wound over the surface of the larva are varied numbers of ciliated **locomotor bands.** There is a complete, functional digestive tract, and food particles are obtained by the ciliated bands and stomodeal cilia. Later larval development in many echinoderms involves the formation of short or long slender projections (**arms**) from the body wall, and the nature and position of these arms, or the lack of them, distinguishes the larvae of the different echinoderm classes. The arms disappear in later development and are not

equivalent to the arms of certain adult echinoderms, such as sea stars and brittle stars.

After a free-swimming planktonic existence, the bilateral larva undergoes a metamorphosis in which the larval arms and, in some echinoderms, the mouth and parts of the gut degenerate. The left side of the body becomes the oral surface and the right side the aboral surface. Only the left axohydrocoel forms the water-vascular system (Fig. 18–2C). Such a planktotrophic larval development is characteristic of only

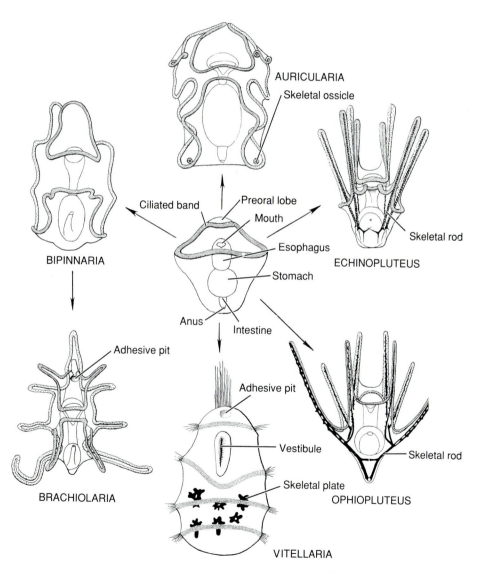

FIGURE 18–3 Comparison of different types of echinoderm larvae showing relationship to a hypothetical dipleurula-type ancestral larva. *(From Ubaghs, G. 1967. In Moore, R. C. (Ed.): Treatise on Invertebrate Paleontology. Pt. S. Vol. 1. Courtesy of the Geological Society of America and the University of Kansas, Lawrence. pp. S3–S60.)*

40% to 50% of the phylum. The remainder have lecithotrophic larvae or direct development.

Although the crinoids (sea lilies) are often regarded as the most primitive class of living echinoderms, the more familiar asteroids (sea stars) are treated first in the discussion of the echinoderm classes as an introduction to the basic features of echinoderm structure.

CLASS ASTEROIDEA

The class Asteroidea contains the sea stars, or starfish. These are star-shaped, free-moving echinoderms in which the body is composed of **rays,** or **arms,** projecting from a **central disc.** The 1500 described species are common, familiar animals that crawl about over rocks and shells or live on sandy or muddy bottoms. They are found throughout the world, largely in coastal waters, but the northeast Pacific, particularly from Puget Sound to the Aleutian Islands, possesses the greatest concentration of asteroid species. Seventy species are endemic to the Vancouver Island area alone.

Sea stars are commonly red, orange, blue, purple, or green or exhibit combinations of colors.

External Structure

Sea stars are typically pentamerous, with most species possessing five arms. The sun stars, however, possess 7 to 40 or more arms (Fig. 18–4B). Most asteroids range from 12 to 24 cm in diameter, but some are less than 2 cm in diameter, and the many-rayed star, *Pycnopodia,* of the northwest coast of the United States may measure almost 1 m across.

The arms of asteroids are not sharply set off from the central disc; that is, the arm usually grades into the disc, and some species have very short arms. In the cushion stars, *Plinthaster* and *Goniaster,* each arm has the shape of an isosceles triangle, and in *Culcita* (Fig. 18–4C), the arms are so short that the body appears pentagonal.

The mouth is located in the center of the underside of the disc, and the entire undersurface of the disc and arms is called the **oral surface.** From the mouth a wide furrow extends radially into each arm (Fig. 18–10B). Each furrow **(ambulacral groove)** contains two or four rows of small, tubular projections, called **tube feet,** or **podia.**

The margins of the ambulacral grooves are guarded by movable spines that are capable of closing over the groove. The tip of each arm bears one or more small, tentacle-like sensory tube feet and a red pigment spot (Fig. 18–10A). The **aboral surface** (upper) bears both the inconspicuous anus, when present, in the center of the disc, and a large, button-like structure (the **madreporite**) toward one side of the disc between two of the arms. The general body surface may appear smooth or be covered with spines, tubercles, or ridges. In some species the arms and disc are bordered by large, conspicuous plates (Fig. 18–4A).

Body Wall

The outer surface of the body is covered by an epidermis composed of monociliated and nonciliated epithelial cells, mucous cells, and ciliated sensory cells (Fig. 18–5). Detritus that falls on the body is trapped in the mucus and then swept away by the epidermal cilia. Within the basal part of the epidermis is a layer of nerve cells forming an intraepidermal plexus.

Below the integument a thick layer of body wall connective tissue (the dermis) houses the hard skeletal pieces (ossicles). Although ossicles are the conspicuous part of the skeletal system, especially in sea urchins and sand dollars, the organic extracellular matrix is also an important skeletal element. Unlike any other phylum of animals, echinoderms can vary, under nervous control, the rigidity of their connective tissue. This unique phenomenon is termed mutable (or catch) connective tissue (p. 921).

Asteroid ossicles are in the shape of rods, crosses, or plates arranged in a lattice network and bound together by connective tissue (Fig. 18–6A, B). The ossicles are irregularly perforated (fenestrated), perhaps representing an adaptation for reduction of weight and increase in strength. The most remarkable characteristic is that each ossicle of the echinoderm skeleton represents a single crystal of magnesium-rich calcite, $6(Ca,Mg)CO_3$. The crystal is formed within a cell of the dermis, but as the crystal increases in size it becomes surrounded by a large number of cells, all of which are daughter cells of the original cell that initiated the formation of the crystal.

FIGURE 18–4 A, *Astropecten duplicatus,* a burrowing sea star (aboral view). **B,** The sun star *Crossaster papposus* (aboral view). **C,** Oral view of *Culcita,* a sea star with very short arms. **D,** The Pacific coral-eating *Acanthaster planci. (B, Photograph courtesy of D. P. Wilson. C, Photograph courtesy of the National Museum of Natural History. D, Photograph courtesy of Betty M. Barnes.)*

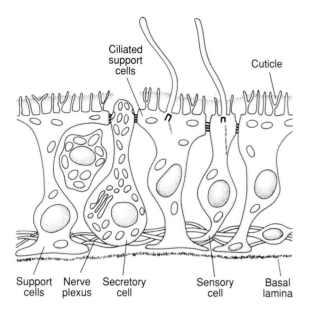

Ciliated support cells

Cuticle

Support cells

Nerve plexus

Secretory cell

Sensory cell

Basal lamina

FIGURE 18–5 Diagram of the echinoderm integument. *(From Holland, N. D. 1984. Epidermal cells (Echinodermata). In Bereiter-Hahn, J., Matoltsy, A. G., and Richards, K. S. (Eds.): Biology of the Integument. 1. Invertebrates. Springer-Verlag, Berlin. p. 757.)*

Spines and tubercles are part of the endoskeleton and, as such, are also covered by the epidermis. In the paxillosid and valvatid sea stars, the aboral surface bears special ossicles, and the central portion of each is raised above the body surface and even extended out like a parasol. The raised part of the ossicle is crowned with small, movable spines. Such an ossicle and its associated spines is called a **paxilla;** it is an adaptation for the burrowing existence of many sea stars belonging to this group. Adjacent paxillae create a protective space above the aboral integument. Through this space flow respiratory and feeding currents. The surrounding sediment is held back by the

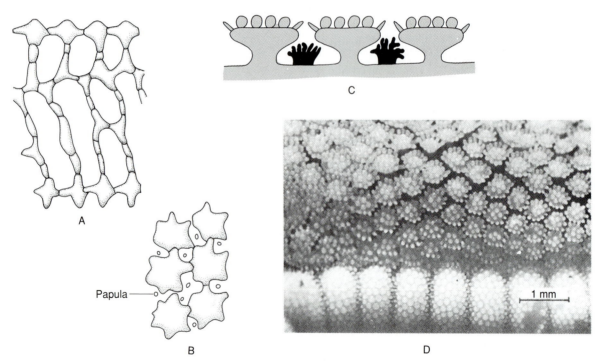

Papula

A

B

C

D

1 mm

FIGURE 18–6 A, Lattice-like arrangement of skeletal ossicles in the arm of an asteriid sea star. **B,** Small section of endoskeletal system of a paxillosid sea star. **C,** Diagrammatic cross section through several paxillae of *Luidia*. The raised, table-shaped ossicles bear small rounded spines on the surface and flat movable spines along the edge. Dendritic papulae (black) are located in the spaces between the projecting edges of the paxillae and associated spines. **D,** Surface of paxillae of *Astropecten* (compare with Fig. 18–4A). *(A, After Fisher from Hyman. B, After Hyman, L. H. 1955. The Invertebrates. Vol. IV. McGraw-Hill Book Co., New York. D, Photograph courtesy of Betty M. Barnes.)*

paxillae (Fig. 18–6C, D). Paxillae account for the smooth appearance of the aboral surface of many sea stars.

Beneath the dermis is a muscle layer composed of outer circular and inner longitudinal smooth fibers, which are involved in the bending of the arms. The inner surface of the muscle layer borders the coelom and is covered with an often ciliated peritoneum.

In two orders of sea stars, the body surface bears small, specialized jawlike appendages **(pedicellariae)** that are used for protection, especially against small animals or larvae that might settle on the body surface of the sea star. The pedicellariae are of two types: stalked and sessile. The stalked pedicellariae are characteristic of the order Forcipulata, which includes *Asterias, Pycnopodia,* and *Pisaster.* Each pedicellaria consists of a short, fleshy stalk surmounted by a jaw-like apparatus composed of three small, movable ossicles that are arranged to form forceps or scissors (Fig. 18–7). Stalked pedicellariae may be scattered over the body surface, situated on the spines, or commonly form a wreath around the base of the spines (Figs. 18–8A, B; 18–11B).

Sessile pedicellariae are largely limited to the order Valvatida and are composed of two or more short, movable spines on the same or adjacent ossicles. The spines, which in some species are shaped like the valves of a clam, oppose each other and articulate against each other, acting as pincers.

The **papulae** are numerous, small evaginations of the body wall scattered over the aboral body surface. The papulae and the podia are discussed in connection with gas exchange and the water-vascular system.

Water-Vascular System

The water-vascular system, unique to echinoderms, consists of canals and appendages of the body wall. Because the entire system is derived from the coelom, the canals are lined with a ciliated epithelium and filled with fluid. The water-vascular system is well developed in asteroids and functions as a means of locomotion.

The internal canals of the water-vascular system connect to the outside through the button-shaped madreporite on the aboral surface (Fig. 18–9A). The surface of the madreporite is creased with many furrows covered by the ciliated epithelium of the body surface. The bottom of each furrow contains many pores that open into pore canals passing downward through the madreporite. The pore canals eventually lead into a vertical **stone canal** that descends to the oral side of the disc (Fig. 18–9A). The stone canal is so named because of the calcareous deposits in its walls. On reaching the oral side of the disc, the stone canal joins a circular canal (the **ring canal**) just to the inner side of the ossicles that ring the mouth.

The inner side of the ring canal (water ring) gives rise to four or, more usually, five pairs of greatly folded pouches called **Tiedemanns's bodies** (Fig. 18–9A). Each pair of these pouches has an interradial position. Also attached interradially to the inner side of the ring canal in many asteroids, although not in *Asterias,* are one to five elongated, muscular sacs, which are suspended in the coelom. These sacs are known as **polian vesicles.**

From the ring canal, a long, ciliated, **radial canal** extends into each arm (Fig. 18–9A). The radial canal runs on the oral side of the ossicles that form the center of the ambulacral groove **(ambulacral ossicles)** (Fig. 18–20), and it ends in a small, external tentacle at the tip of the arm. **Lateral canals** arise alternately from each side of the radial canal along its entire length and pass between the ambulacral ossicles on each side of the groove (Fig. 18–9B).

Each lateral canal is provided with a valve and terminates in a bulb and a tube foot (Fig. 18–9). The bulb, or **ampulla,** is a small, muscular sac that bulges into the aboral side of the perivisceral coelom. The

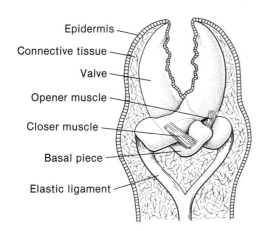

Epidermis

Connective tissue

Valve

Opener muscle

Closer muscle

Basal piece

Elastic ligament

FIGURE 18–7   Distal ends of a scissors-type pedicellaria from *Asterias. (After Hyman, L. H. 1955. The Invertebrates. Vol. IV. McGraw-Hill Book Co., New York.)*

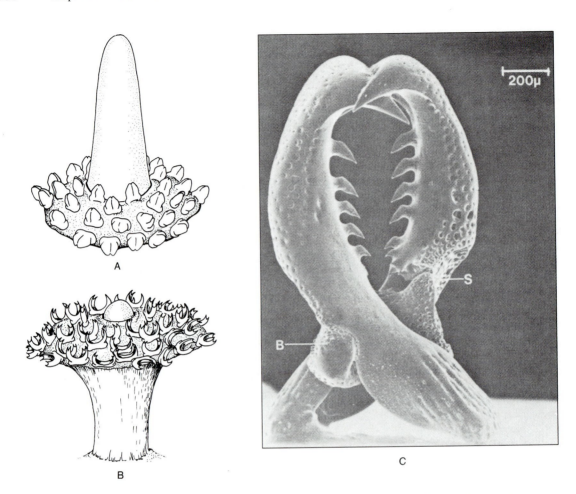

FIGURE 18–8 Fish-catching pedicellariae of *Stylasterias.* **A,** Wreath of pedicellariae at rest around the spine. **B,** Wreath raised when stimulated by potential prey. **C,** Scanning electronmicrograph of ossicle. (B, basal piece; S, muscle attachment scar.) *(From Chia, F. S., and Amerongen, H. 1975. On the prey-catching pedicellariae of a starfish, Stylasterias forreri. Can. J. Zool. 53:748–755. Reproduced by permission of the National Research Council of Canada.)*

ampulla opens directly into a canal that passes downward between the ambulacral ossicles and leads into the tube foot, or podium.

The podium is a short, tubular, external projection of the body wall located in the ambulacral groove (Fig. 18–10A). Commonly, the tip of the podium is flattened, forming a sucker. Like the body wall, the podium is covered on the outside with a ciliated epithelium and internally with peritoneum. Between these two layers lie connective tissue and longitudinal muscle fibers. Contraction of the muscles on one side of the podium brings about bending of the appendage. The podia are arranged in two rows where the lateral canals are all of the same length or four rows where they are alternately long and short.

The entire water-vascular system is filled with fluid that is similar to sea water except that it contains coelomocytes, a little protein, and a high potassium ion content. The system operates during locomotion as a hydraulic system. When the ampulla contracts, the valve in the lateral canal closes, and water is forced into the podium, which elongates. When the podium comes in contact with the substratum, the sucker adheres. Adhesion is largely chemical, the podium secreting a substance that bonds with surface films. Another secretion breaks the bonds and brings about release. Note the similarity to the duogland systems of flatworms and gastrotrichs. When attachment is prolonged or a large force is generated, adhesion by suction probably also takes place.

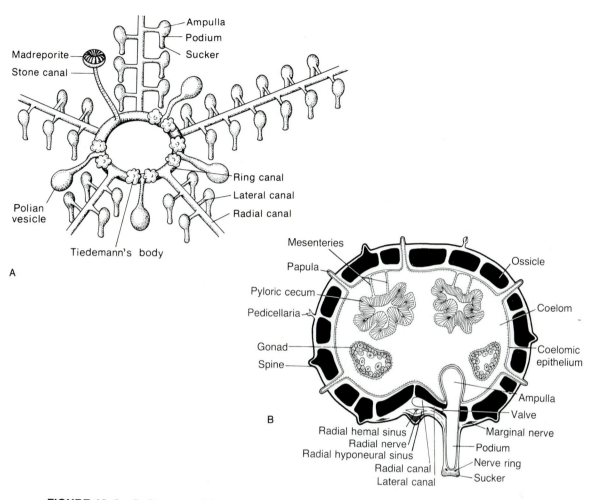

A

B

FIGURE 18–9 A, Diagram of the asteroid water-vascular system. **B,** Diagrammatic cross section through the arm of a sea star.

After the sucker adheres to the substratum, the longitudinal muscles of the podium contract, shortening the podium and forcing fluid back into the ampulla. It has been generally thought that other parts of the water-vascular system—the madreporite, the stone canal, the ring canal, the muscular polian vesicles, and the radial canal—perhaps function in maintaining the proper water volume necessary for the operation of the ampullae and podia, because there is some leakage across the podial wall during fluid pressure elevation.

The porous madreporite bears surface cilia, which prevent large particles, such as sand and other debris, from clogging the pores. Cilia in the pores beat outward but are at least periodically overcome by the inward beat of powerful cilia in the stone canal. Studies using dyes and radioactive and flourescent tracers all confirm that there is some inflow of water across the madreporite and into other parts of the water-vascular system and, perhaps, into the perivisceral coelom. The Tiedemann's bodies contain phagocytes, which remove foreign matter, such as bacteria, from the incoming water. The unique occurrence in echinoderms of a seawater inlet to replenish coelomic fluid (which occurs by osmosis or drinking in other coelomates) may be related to an exceptional rate of fluid loss across the numerous active tube feet.

During movement each podium performs a sort of stepping motion. The podium swings forward, grips the substratum, and then moves backward. In a particular section of an arm most of the tube feet are performing the same step, and the animal moves forward. The action of the podia is highly coordinated.

During progression one or two arms act as leading arms, and the podia in all the arms move in the same direction, but not necessarily in unison (Fig. 18–10B). The combined action of the podial suckers exerts a powerful force for adhesion and enables sea stars to climb vertically over rocks or up the side of an aquarium.

If a sea star is turned over, it can right itself by folding. The distal end of one or two arms twists, bringing the tube feet in contact with the substratum. Once the substratum has been gripped, these arms move back beneath the animal so that the rest of the body is folded over. The sea star may also right itself by arching its body and rising on the tips of its arms. It then rolls over onto its oral surface. In general, sea stars move rather slowly and tend to remain within a more or less restricted area.

Podia of many but not all sea stars that live on soft bottoms, such as *Astropecten* and *Luidia,* lack suckers. Rather, the tip is pointed to facilitate thrusting of the podium into the sand. Associated with this adaptation of the podia is the presence of bilobed am-

pullae, which provide increased force for driving the podia into the substratum. In addition to enabling a sea star to move over a soft bottom, podia of this type may also be used to burrow and even to plaster the walls of the burrow with mucus.

Nutrition

The digestive system is radial, extending between the oral and aboral sides of the disc (Fig. 18–11). The mouth is located in the middle of a tough, circular, peristomial membrane, which is muscular and provided with a sphincter. The mouth leads into a short esophagus, which opens in turn into a large stomach that fills most of the interior of the disc and is divided by a horizontal constriction into a large, oral chamber (the **cardiac stomach**) and a smaller, flattened, aboral chamber (the **pyloric stomach**). The walls of the cardiac stomach are pouched and connected to the ambulacral ossicles of each arm by ten pairs of triangular mesenteries called gastric ligaments.

A

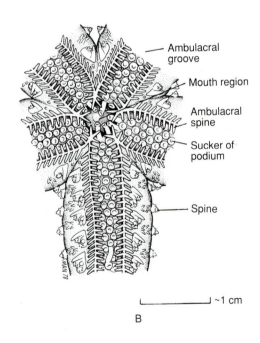

— Ambulacral groove

— Mouth region

— Ambulacral spine

— Sucker of podium

— Spine

⌞_____⌟ ~1 cm

B

FIGURE 18–10   **A,** Upturned tip of one arm of *Asterias forbesi,* showing tube feet, some of which bear white suckers. **B,** Oral surface of *Asterias* (disc and part of one arm).

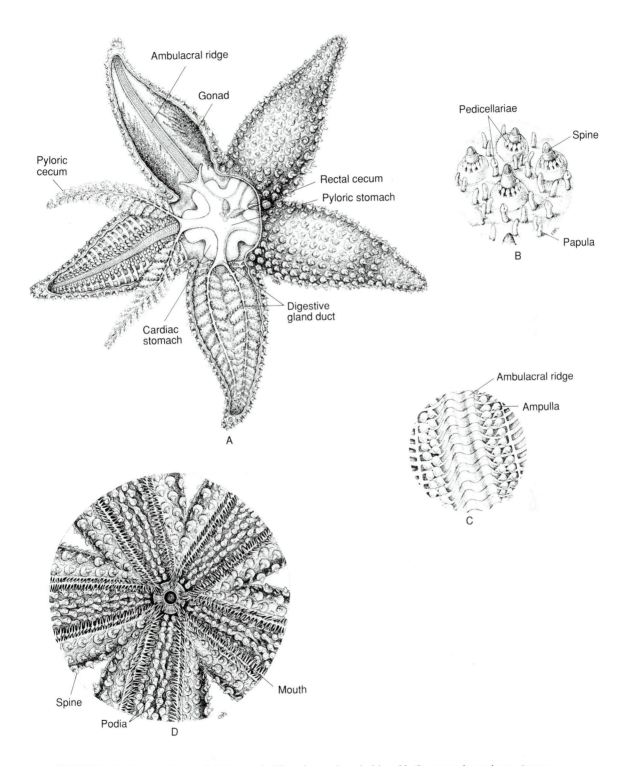

FIGURE 18–11 Anatomy of *Asterias*. **A,** View from aboral side with the arms in various stages of dissection. **B,** A small area of the aboral surface enlarged. **C,** Interior view of a part of one of the ambulacral ridges, showing ampullae to either side. **D,** Oral surface of the disc.

The smaller, aboral, pyloric stomach, which is often star-shaped, is simply the reception chamber of the ducts from the **pyloric ceca** (Fig. 18–11). There are two pyloric ceca, or digestive glands, in each arm, each of which is composed of an elongated mass of glandular cells suspended in the coelom of the arm by a dorsal mesentery (Fig. 18–9B).

A short, tubular intestine extends from the aboral side of the pyloric stomach to open through a minute anus in the middle of the aboral surface of the disc. The intestine commonly bears a number of small out-pocketings called **rectal ceca** (Fig. 18–11). The entire digestive tract is lined with a ciliated epithelium, and the cilia are in the ducts of the pyloric ceca arranged to create fluid currents, both incoming and outgoing. Gland cells are particularly abundant in the cardiac stomach lining.

Most asteroids are scavengers and carnivores and feed on all sorts of invertebrates, especially snails, bivalves, crustaceans, polychaetes, other echinoderms, and even fish. Some have very restricted diets; others utilize a wide range of prey but may exhibit preferences, depending on availability. For example, the Chilean *Meyenaster* feeds on 40 types of echinoderm and molluscan prey found within its habitat. Most asteroids detect and locate prey by substances the prey releases into the water, and many prey species have evolved escape responses to the slow-moving asteroids. Some soft-bottom sea stars, including species of *Luidia* and *Astropecten,* can locate buried prey and then dig down into the substratum to reach it. *Stylasterias forreri* and *Astrometis sertulifera* along the west coast and *Leptasterias tenera* of the east coast of the United States catch small fish, amphipods, and crabs with the pedicellariae when the prey comes to rest against the aboral surface of the sea star.

There are some asteroids that feed on sponges, sea anemones, and the polyps of hydroids and corals. The tropical Pacific *Acanthaster planci* (crown-of-thorns sea star) has attained considerable notoriety as a result of its consumption of coral polyps. High population levels of this sea star, as great as 15 adults per square meter, have temporarily devastated large numbers of reef corals in some areas. Branching and plate corals are preferred over massive and encrusting types, and with its everted stomach one sea star can consume an area as great as its own disc in one day. Whether or not the high population levels of this sea star have resulted from human modification of the environment has been debated. Current evidence suggests that the larval periods of high population levels coincided with plankton blooms. The blooms appear to have resulted from nutrient runoff following heavy rains, which in turn followed a dry period of several years.

Some sea stars are suspension feeders. Plankton and detritus *(Porania, Henricia)* or mud *(Ctenodiscus)* that comes in contact with the body surface is trapped in mucus and then swept toward the oral surface by the epidermal cilia. On reaching the ambulacral grooves, the food-laden mucous strands are carried by ciliary currents to the mouth. Some sea stars, such as *Astropecten* and *Luidia,* which are largely carnivorous, utilize ciliary feeding as an auxiliary method of obtaining food.

In primitive groups of sea stars, including *Astropecten* and *Luidia,* which cannot evert their stomachs and have suckerless tube feet, the prey is swallowed whole and digested within the stomach, although the stomach wall must be in contact with the tissues being digested. Shells and other indigestible material are then cast out of the mouth. Other asteroids (Valvatida, Spinulosida, Forcipulata) feed extraorally. Through the contraction of the body wall muscles, the coelomic fluid exerts pressure on the cardiac stomach, causing it to be everted through the mouth. The everted stomach, which is anchored by the gastric ligaments, engulfs the prey. The prey then may be brought into the stomach by retraction, or digestion may begin outside the body. The soft parts of the victim are reduced to a thick broth, which is then passed into the body in ciliated gutters. When digestion is completed, the stomach muscles contract, retracting the stomach into the interior of the disc.

Many sea stars feed almost exclusively on bivalves and are notorious predators of oyster beds. During feeding, such a sea star extends itself over a clam, holding the gape of the clam upward against its mouth and applying its arms against the sides of the clam valves. The sea star inserts the everted stomach through minute openings between the imperfectly sealed edges of the valves, or the pull exerted by the sea star produces a very slight gape in the clam. The gape is produced quite rapidly and not by causing the clam to fatigue over a long period. The everted cardiac stomach of some sea stars can squeeze through a space as slight as 0.1 mm. The gape increases as digestion ensues, and the clam's adductor muscles are attacked. Japanese species of *Asterias* require 2.5 to 8 h to consume a bivalve, depending on the species of bivalve being attacked.

Asteroids are of considerable economic importance as predators of oysters, and they are sometimes removed from commercial oyster beds by a large, moplike apparatus dragged over the bottom. The sea stars grasp or become entangled in the mop threads with their pedicellariae and are brought to the surface and destroyed.

The everted stomach is also an effective feeding organ for many omnivores and nonpredacious sea stars. The American west coast bat star, *Patiria miniata,* spreads its stomach over the bottom, digesting all types of organic matter it encounters. In the same manner the tropical cushion sea star, *Culcita,* and the oreasterids, which inhabit reef flats, feed on sponges, algal felt, and epibenthic film.

Digestion in asteroids appears to be primarily extracellular, and a complex of enzymes is produced by the stomach wall and pyloric ceca. Products of stomach digestion are carried by ciliary tracts up the stomach wall and through the pyloric ducts into the pyloric ceca, where digestion, both intracellular and extracellular, and absorption occur. Products of digestion may be stored in the cells of the pyloric ceca or passed through the ceca into the coelom for distribution. The pyloric ducts also convey wastes out to the rectum, where the rectal ceca, when present, act as pumps for expulsion through the anus.

Internal Transport, Gas Exchange, and Excretion

Asteroids, like other echinoderms, rely primarily on coelomic circulation for internal transport of gases and some nutrients. The blood-vascular system, called the **hemal system** in echinoderms, is rudimentary in asteroids, although it plays a role in nutrient transport.

Asteroids have four coelomic circulatory systems: the **perivisceral coelom,** which occupies the disc and arms and supplies the viscera; the water-vascular system, already described, which supplies the locomotory muscles of the tube feet; the **hyponeural sinus system,** which supplies the nervous system; and the **genital coelom,** which supplies the gonads. Cilia on the peritoneal lining of these coeloms create a continuous circulation of the coelomic fluid. The body

fluids of all asteroids, as well as those of other echinoderms, are isosmotic with seawater. Their inability to osmoregulate prevents most species from inhabiting estuarine waters. The coelomic fluid contains phagocytic coelomocytes that are produced by the coelomic peritoneum and can form a clot in response to tissue damage.

The hyponeural sinuses parallel the canals of the water-vascular system, but ampullae and tube feet are absent (Fig. 18–12). Each sinus in the hyponeural system is double, the left being separated from the right by a mesentery. The hyponeural system has a ring sinus (hyponeural ring sinus) that accompanies the nerve ring, and a radial sinus (radial hyponeural sinus) along the radial nerve in each arm. A vertical sinus called the **axial sinus,** lies beside the stone canal and joins the hyponeural ring to the madreporite and the madreporic end of the stone canal, thus linking the water-vascular and hyponeural systems.

Throughout the hyponeural system, a blood vessel is suspended in the mesentery between the two halves of each sinus. Other blood vessels occur in mesenteries of the perivisceral coelom, especially in association with the pyloric ceca and the genital coelom. Eventually, all blood vessels unite with the **heart** (dorsal sac and hemal sinus) situated just internal and to one side of the madreporite. Immediately below the heart in the axial sinus, one blood vessel becomes enlarged and folded, and is called the **axial gland.** Two similar vascular structures, called **gastric hemal tufts,** bulge into the perivisceral coelom near the junction of pyloric ceca vessels with the heart. Both the axial gland and gastric hemal tufts are covered with podocytes.

The heart beats rhythmically (approximately 6 beats per minute in *Asterias forbesi*), but the pattern of circulation of the colorless blood is unknown. There is some evidence that the blood has a role in nutrient transport, but the functional role of the glomerulus-like structures (axial gland, gastric hemal tufts) is uncertain. The asteroid heart, axial sinus, and axial gland are similar in structure to, and derive from the same embryonic rudiments as the heart, proboscis coelom, and glomerulus of hemichordates (p. 871).

Removal of nitrogenous wastes (NH_3) is accomplished by general diffusion through thin areas of the body surface, such as the tube feet and the papulae. Studies involving the injection of dyes into the coelom indicate that coelomocytes engulf waste, and

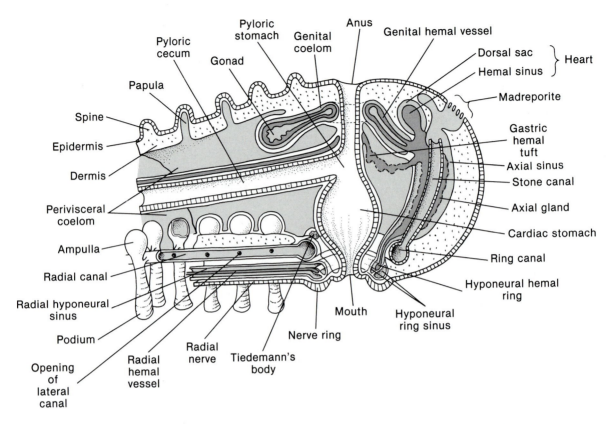

FIGURE 18–12 Diagrammatic view of coelomic and blood-vascular (hemal) systems of an asteroid. (*Combined and modified from various sources.*)

when laden, some migrate to the papulae and podia, where they collect at the distal end. The tip of the papulae then constricts and pinches off, discharging the coelomocytes to the outside. Other coelomocytes may pass to the outside through the epithelium of the suckers of the podia or at other sites.

The papulae and tube feet provide the principal gas exchange surfaces for asteroids. The ciliated peritoneum that forms the internal lining of the papulae and tube feet produces an internal current of coelomic fluid; the outer, ciliated epidermal investment produces a current of seawater flowing over the papulae (Fig. 18–13). In burrowing species the branched papulae are protected by the paxillae, and the ventilating current flows through the channel-like spaces beneath these spines (Fig. 18–6C).

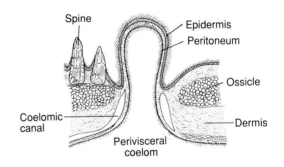

FIGURE 18–13 Section through an asteroid papula. (After Cuénot, L. 1948. Échinodermes. *In* Grassé, P. (Ed.): Traité de Zoologie, Vol. XI. Masson et Cie, Paris.)

Nervous System

The asteroid nervous system, like that of some other echinoderms, is not conspicuously ganglionated, and the greater part of the system is intimately associated with the epidermis. The nervous center is a somewhat pentagonal, **circumoral nerve ring,** which lies within the base of the peristomial epidermis. From each angle of the ring a large **radial nerve** extends into each of the arms, forming a large, intraepidermal, V-shaped mass along the midline of the oral surface of the ambulacral groove (Fig. 18–9B). The radial nerve supplies fibers to the podia and ampullae and is continuous with the general intraepidermal nerve plexus. At the margins of the ambulacral groove, the epidermal nervous layer is thickened to form a pair of marginal nerve cords that extends the length of the arm (Fig. 18–9B), and there are also motor centers located in the vicinity of the podia and the podia/ampullae junctions.

There are many experimental studies on the role of the nervous system in the movement of sea stars. The integrity of both the radial nerves and the circumoral nerve ring is essential in the coordination of the podia. Although the podia may not all "step" in unison, they are coordinated to the extent of their involvement in stepping—that is, to step or not to step—and they are coordinated in that they step in the proper direction, depending on which arm is leading.

Each arm has a nerve center, probably at the junction of the radial nerves and nerve ring. A leading arm exerts a temporary dominance over the nerve centers of the other arms. In the majority of sea stars, including *Asterias,* any arm can act as a dominant arm, and such dominance is determined by reaction to external stimuli. In a few species one arm is permanently dominant. Of all the reactions to external stimuli, contact of the podia with the substratum appears to be dominant and probably accounts for the righting reaction.

With the exception of the eye spots at the tips of the arms, there are no specialized sense organs in the asteroids. The dispersed sensory cells contained within the epidermis are the primary sensory receptors and probably function for the reception of light, contact, and chemical stimuli. This is true of other echinoderms as well. These epidermal sensory cells are particularly prevalent on the suckers of the tube feet, on the terminal, tentacle-like sensory tube feet, and along the margins of the ambulacral groove, where 70,000 sensory cells per square millimeter have been reported.

The eye spot at the end of each arm lies beneath the tentacle on the oral side of the arm tip and is composed of a mass of 80 to 200 pigment-cup ocelli that form an optic cushion. The importance of the optic cushions (Fig. 18–10A) in reactions to light stimuli varies in different species, but most asteroids are positively phototactic.

Regeneration and Reproduction

Asteroids exhibit considerable powers of regeneration. Any part of the arm can be regenerated, and destroyed sections of the central disc are replaced. Studies on *Asterias vulgaris* have shown that if there is at least one fifth of the central disc attached to an arm, an entire starfish will be regenerated. If the remaining section of the disc contains the madreporite, even less of the disc is required. Regeneration is typically slow and may require as long as one year for complete reformation to take place.

A number of asteroids normally reproduce asexually. Commonly, this involves a division of the central disc so that the animal breaks into two parts. Each half then regenerates the missing portion of the disc and arms, although extra arms are commonly produced (Fig. 18–14A). Species of *Linckia,* a genus of common sea stars in the Pacific and other parts of the world, are remarkable in being able to cast off their arms near the base of the disc. Unlike those of other asteroids, the severed arm regenerates a new disc and rays (Fig. 18–14B).

With few exceptions asteroids are dioecious, and there are ten gonads, two in each arm (Fig. 18–11A). The gonads are double-walled and tuftlike or resemble a cluster of grapes. Normally, they occupy only a small area at the base of the arm. When filled with eggs or sperm, however, the gonads almost completely fill each of the arms. There is a gonopore or gonopore cluster for each gonad, usually located between the bases of the arms. In a number of astropectinids, as well as in some other groups, each arm contains many gonads, which are arranged in rows along the length of the arm. In such species the gonopores open on the oral surface. There are a few hermaphro-

A

B

FIGURE 18–14 A, Regenerating arms in a specimen of *Coscinasterias,* which reproduces asexually by division of the disc. **B,** Comet of *Linckia.* Regeneration of body at base of detached arm. *(A, Photograph courtesy of Betty M. Barnes. B, After Richters from Hyman.)*

ditic asteroid species, such as the common European sea star, *Asterina gibbosa,* which is protandric.

In the majority of asteroids the eggs and sperm are shed freely into the seawater, where fertilization takes place. There is usually only one breeding season per year, and a single female may shed as many as 2,500,000 eggs.

In most asteroids the liberated eggs and individuals in the later developmental stages are planktonic. Some sea stars, especially Arctic and Antarctic species, however, brood large, yolky eggs beneath the disc in depressions on the aboral surface of the disc, in brooding baskets formed by spines between the bases of the arms, or even in the cardiac stomach. In all the brooding species, development is direct. Although not a brooding species, *Asterina gibbosa* is unusual in that it attaches its eggs to stones and other objects.

Development

The early stages of development conform to the pattern described in the introduction of this chapter. In most species the coelom arises from the tip of the ad-

vancing archenteron as two lateral pouches. The tricoelomate condition is attained by a complex subdivision, but the axocoel and hydrocoel never completely separate (Figs. 18–15A; 18–16). The left axohydrocoel connects with the dorsal surface to form the hydropore.

The asteroid embryo becomes free-swimming at some point between the blastula and gastrula stages. Later the surface ciliation becomes confined to a definite locomotor band (Fig. 18–15A). The preoral loop later separates or in some cases arises separately. After the formation of the locomotor bands, projections, called arms, arise from the body surface. The locomotor bands extend along the arms. This larval stage is then known as a **bipinnaria larva.**

The ciliated bands function in both locomotion and feeding, and the larval arms increase the surface area of the bands (Fig. 18–15B). Phytoplankton and other fine suspended particles that constitute the food of echinoderm larvae are kept on the upstream side of the ciliary bands by localized reversal of ciliary beat. In this manner they are bounced down the ciliated bands to the stomodeum. The hydropore opening from the ciliated canal of the left hydrocoel can be in-

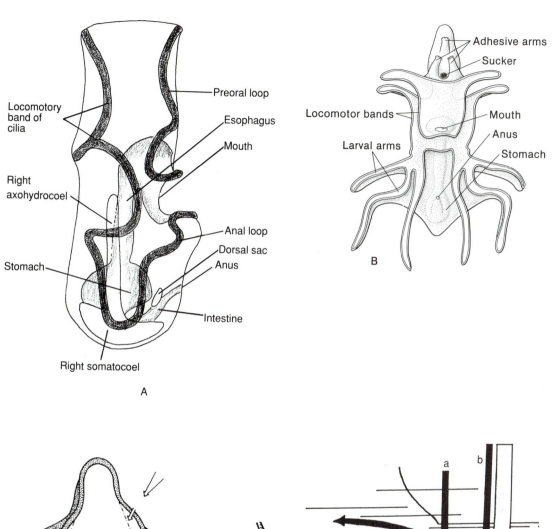

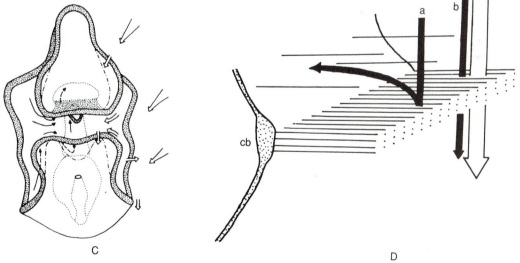

FIGURE 18–15 **A,** Early (14 days) bipinnaria larva of *Astropecten auranciacus,* lateral view. **B,** Brachiolaria larva of *Asterias,* ventral view. **C,** Water currents (white arrows) and paths of food particles (black arrows) produced by the ciliated band of a bipinnaria larva. **D,** Diagrammatic view of a section of the ciliated band (cb) of an echinoderm larva, showing passage of water current (white arrow), uncollected particle (black arrow b), and collected particle (black arrow a) resulting from a localized reversal of ciliary beat. *(A, After Hörstadius from Hyman. B, After Agassiz from Cuénot. D, From Strathmann, R. R., 1972 and 1975. Biol. Bull. 142:505–519, and Am. Zool. 15:717–730.)*

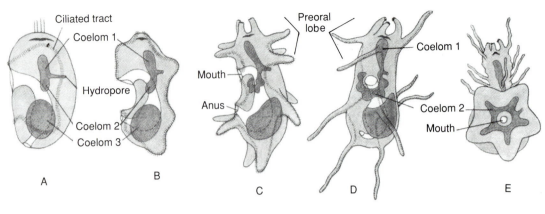

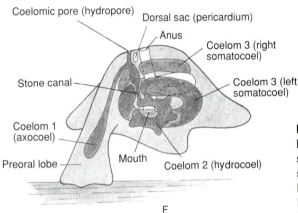

FIGURE 18–16 **A-F,** Diagrammatic lateral views of larval development and metamorphosis of a sea star, showing development of coelom and water-vascular system. **A** and **B,** Early bipinnaria larva. **C,** Brachiolaria larva. **D,** Attached metamorphosing larva. **E** and **F,** Young starfish developing from posterior part of old larva.

terpreted as a nephridial outlet. The wall of the axocoel (coelom) contains podocytes through which fluid filters from the blastocoel. The fluid is flushed to the outside through the canal and hydropore. The system regulates body fluid volume, and perhaps density, and up to 14% of the body fluid is cleared per hour.

The bipinnaria larva becomes a **brachiolaria larva** with the appearance of three additional arms at the anterior end (Fig. 18–15B). These arms are short, ventral in position, and covered with adhesive cells at the tip. Between the bases of the three arms is a glandular, adhesive area that forms a sucker. The three arms and the sucker represent an attachment device, and the brachiolaria then settles to the bottom. There are some species, such as *Luidia* and *Astropecten* and the common European sea star *Asterina,* in which a brachiolaria never forms.

Metamorphosis then takes place. The anterior end of the larva degenerates and forms only an attachment stalk, and the adult body develops from the rounded, posterior end of the larva (Fig. 18–16E, F).

The left side becomes the oral surface, and the right side becomes the aboral surface. The adult arms appear as extensions of the body.

Internally, the mouth, the esophagus, part of the intestine, and the anus degenerate. All these parts are formed anew and in a position coinciding with the adult radial symmetry. The somatocoel forms the major part of the coelom. The left axohydrocoel forms the water-vascular system, and from the hydrocoel develop five pairs of projections, two in each of the developing arms. These projections represent the cavity and coelomic lining of the first pair of podia in each arm. As soon as additional podia are formed, they begin to grip the substratum and soon free the body from the substratum. At about this time the skeletal system appears. As in other echinoderms, the first adult ossicles to be formed are those around the aboral pole, which contains the anus. New ossicles are then added peripherally to the initial skeleton. The detached baby starfish is less than 1 mm in diameter, with very short, stubby arms.

Growth rates and life spans are variable, as illustrated by two intertidal species from the Pacific Coast of the United States. *Leptasterias hexactis* broods a small number of yolky eggs during the winter, and the young mature sexually in two years, when they weigh about 2 g. The average life span is ten years. *Pisaster ochraceus* releases a large number of eggs each spring, and development is planktonic. Sexual maturity is reached in five years, at which time the animal weighs 70 to 90 g. Individuals may live 34 years, reproducing annually.

SUMMARY

1. The phylum Echinodermata is composed of marine animals that are distinguished by a pentamerous radial symmetry, an endoskeleton of calcareous ossicles, spiny ossicles on the body surface, mutable connective tissue, and a water-vascular system of coelomic canals and body appendages, or podia, that is used in feeding or locomotion.

2. The ciliated coelom is well developed and important in gas and nutrient transport; the blood-vascular (hemal) system is less well developed and has little role in gas transport.

3. In general, the sexes are separate, fertilization is external, and development is planktonic. There is commonly a bilateral larva that swims and feeds by means of ciliated bands wound over the body.

4. The class Asteroidea contains echinoderms in which the body is composed of a central disc and radiating arms. The arms are not sharply set off from the central disc.

5. Asteroids move by means of podia, which are located within ambulacral grooves. Podia are extended by hydraulic pressure generated by the contraction of a bulblike ampulla. In many species suckers at the ends of the podia permit attachment to the substratum.

6. The arms can bend and twist, permitting the sea star to move over irregular surfaces, grasp prey, and right itself. Arm movement is made possible by a flexible, lattice-like arrangement of ossicles within the dermis and by circular and longitudinal muscle layers in the body wall.

7. The large coelom provides for internal transport, and evaginations of the body wall (papulae) are the sites of gas exchange and excretion. The thin walls of the podia are a significant additional exchange surface, however.

8. Feeding behavior is related not only to diet but also to arm length. Predatory species with short arms swallow the entire prey. Those with long arms evert the stomach and partially digest the prey outside the body. Those sea stars that prey on bivalve molluscs slide the stomach between the valves of the mollusc. Some species use the everted stomach like a mop to remove organic material from various surfaces.

9. Sea stars inhabiting soft bottoms generally possess pointed tube feet and double ampullae; paxillae keep the papulae clear of sediment.

10. Pedicellariae, which are restricted to certain groups of sea stars, probably function to clear the body surface of settling organisms.

11. There are usually two gonads in each arm, and the gametes exit by interradial gonopores. Development leads to a bipinnaria larva, in which ciliated bands are located on long larval arms. With the formation of attachment structures, the larva is called a brachiolaria and is prepared for settling. Following settlement and attachment, the larva undergoes metamorphosis, in which the larval arms degenerate, the left side becomes the oral surface, and the adult body is derived from the posterior part of the larval body.

CLASS OPHIUROIDEA

The class Ophiuroidea contains those echinoderms known as basket stars and serpent stars, or brittle stars (Fig. 18–17). The 2000 described species make this the largest of the classes of echinoderms. They are found in all types of marine habitats, and they are often abundant on soft bottoms in shallow water and the deep sea.

Ophiuroids, like asteroids, have arms, but, in other respects the two classes are quite different. The extremely long arms of ophiuroids are more sharply set off from the central disc. There is no ambulacral groove, and the podia play little role in locomotion except in a few rare species. Moreover, the arms have a relatively solid construction compared with those of the asteroids.

External Structure

Ophiuroids are relatively small echinoderms. The disc in most species ranges from 1 to 3 cm in diameter, although the arms may be quite long. The basket stars are the largest members of the class, and the disc

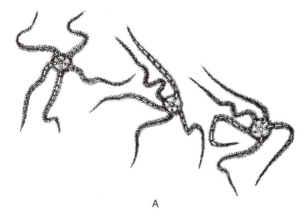

A

B

FIGURE 18–17 **A,** Caribbean brittle star, shown in a drawing based on repetitive flash photographs, pulling itself along with its two anterior arms and shoving with the other three. Ophiuroids are far more agile and flexible than sea stars. **B,** A basket star. *(A, Based on a photograph by Fitz Goro. B, Photography courtesy of Photo Researchers, New York.)*

in some species of this group may attain a diameter of almost 12 cm. A great variety of colors is found in the ophiuroids, and mottled and banded patterns are common. The central disc is flattened, displaying a rounded or somewhat pentagonal circumference (Fig. 18–18). The aboral surface varies from smooth to granular and may bear small calcareous plates, called **shields,** and small tubercles or spines (Fig. 18–18B).

There are typically only five arms; however, in basket stars the arms branch at either the base or more distally, and the subdivisions repeatedly branch to produce a great mass of coils that resemble tentacles (Fig. 18–17B).

The arms of ophiuroids appear jointed because of the presence of four longitudinal rows of shields (Figs. 18–18A, B; 18–19). There are two rows of lateral shields, one row of aboral shields, and one row of oral shields. A single set—that is, one aboral, one oral, and two lateral shields—completely surrounds the arm and corresponds in position to an internal skeletal ossicle, called a **vertebra,** which is described later. Not infrequently, the oral and aboral shields are reduced by the large size of the lateral shields, which may even meet on the oral and aboral surfaces (Fig. 18–18A). Each lateral shield usually bears 2 to 15 large spines arranged in a vertical row (Fig. 18–19). These spines vary considerably in size and shape, depending on the species.

In contrast to the asteroids, an ambulacral groove (open ambulacrum) is absent on the oral surface of the arms. The ambulacral ossicles have sunk inward and enlarged to form the vertebrae, and the radial water canal lies sandwiched between the vertebral ossicles and the oral shields (Fig. 18–20). The ambulacrum is said to be closed. The podia are small, tentacle-like, papillate appendages that extend between the oral and the lateral shields, and there is typically one pair of podia per joint (Fig. 18–19). Neither papulae nor pedicellariae are present in ophiuroids.

The center of the oral surface of the disc is occupied by a complex series of large plates (oral shields) that frame the mouth area and also form a chewing apparatus with five triangular, interradial **jaws** (Fig. 18–18A). In most ophiuroids, one oral shield is modified, forming a madreporite. Thus, the madreporite is located on the oral surface, in contrast to its aboral position in asteroids.

Body Wall and Skeleton

The epidermis lacks conspicuous surface cilia except in certain areas. The dermis contains the more superficial skeletal shields as well as the large, deeper ossicles of the arms. Each ossicle is a large, bilaterally symmetrical, skeletal piece (a vertebra) that almost fills the entire interior of the arm and greatly reduces the coelom (Figs. 18–19; 18–20; 18–21). The

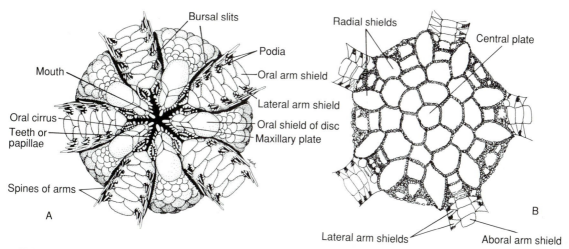

FIGURE 18–18 **A,** The disc of *Ophiura sarsi* (oral view). **B,** The disc of *Ophiolepis* (aboral view). *(A, After Strelkov. B, After Hyman, L. H. 1955. The Invertebrates. Vol. IV. McGraw-Hill Book Co., New York.)*

vertebral ossicles are arranged linearly from one end of the arm to the other, and each ossicle is covered by the four superficial arm shields.

The two end surfaces of a vertebral ossicle bear nodes and sockets, which articulate with corresponding surfaces on adjacent vertebrae, and pits for the insertion of large intervertebral muscles, which move

the arm (Fig. 18–19). In many brittle stars this articulation allows great lateral mobility of the arm but little vertical movement. In the basket stars and some brittle stars, however, the arms can bend and coil in any direction. As in other animals with calcareous skeletons, the vertebral ossicles of brittle stars have been found to show microscopic growth bands.

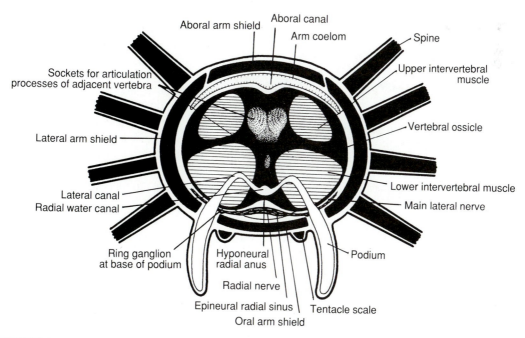

FIGURE 18–19 Diagrammatic section through the arm of a brittle star. Spine, shield, and vertebral muscles attached to the face of the vertebra are shown with horizontal lines.

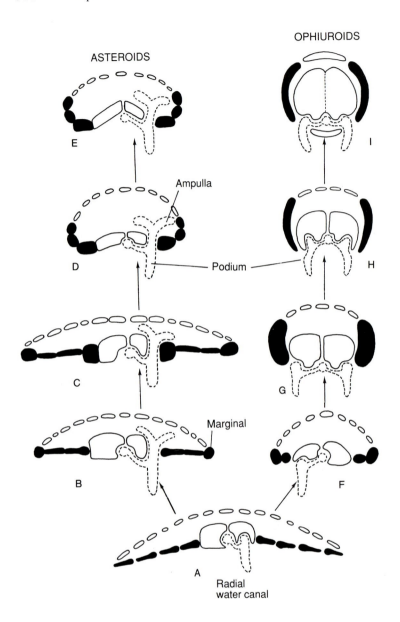

ASTEROIDS

OPHIUROIDS

FIGURE 18–20 Evolution of the arm skeleton in asteroids and ophiuroids. Arms are shown in cross section with ossicles white or black, and the water-vascular system is dotted. Bottom diagram is of an ancestral somasteroid. *(Adapted from Nichols, 1969.)*

Locomotion and Habitation

The ophiuroids are the most mobile echinoderms. During movement the disc is held above the substratum, with one or two arms extended forward and one or two arms trailing behind. The remaining two lateral arms perform a rapid rowing movement against the substratum that propels the animal forward in leaps or jerks. The spines provide traction.

Brittle stars show no arm preference and can move in any direction. In clambering over rocks or in seaweed and hydroid colonies, the ends of the supple arms often coil about objects (Fig. 18–22C). Although most ophiuroids use the arms to move, a few, such as *Ophionereis annulata* along the American west coast, creep about on their podia.

There are burrowing ophiuroids, notably members of the family Amphiuridae (*Amphiura, Amphiodia,* and *Amphipholis*). The animal excavates a mucus-lined burrow with tubelike channels to the surface in mud or sand (Fig. 18–22D), using undulatory waves of the arms and digging movements of the tube feet. The animal never leaves the burrows unless dislodged, although some of the arms are projected

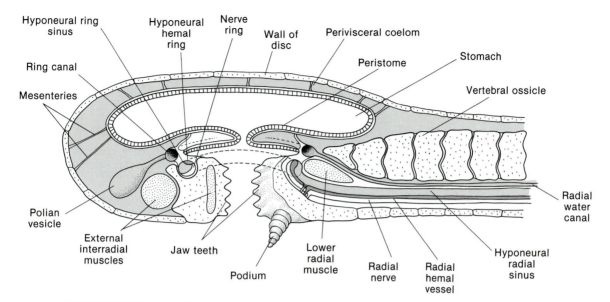

FIGURE 18–21 The disc and base of an arm of a brittle star (vertical section). *(After Ludwig.)*

above the burrow into the water. Arm undulation provides for burrow ventilation.

Underwater photographs from the deep sea as well as shallow water reveal high population densities of ophiuroids where conditions are favorable. Some species live in dense aggregations. The European *Ophiothrix fragilis* responds to the presence of other individuals and in favorable currents on muddy gravel bottoms attains densities as great as 1000 to 2000 per square meter.

Some echinologists consider the ophiuroids the most successful group of living echinoderms; this success in part is attributed to their motility, small size, and ability to utilize the protective cover of crevices, holes, spaces beneath stones, and other natural retreats (Fig. 18–22A).

The only commensal echinoderms are ophiuroids. Large sponges may contain great numbers of ophiuroids living in their water canals (Fig. 18–22B). Other ophiuroids inhabit arborescent octocorallians and scleractinian corals. Species of *Ophiomaza*, an Indo-Pacific brittle star, live on the oral surfaces of feather stars, clutching the feather star's calyx with its arms. There are also several tiny Indo-Pacific brittle stars that live on the undersurfaces of sand dollars.

Water-Vascular System

The oral shield that forms the madreporite in ophiuroids usually bears but a single pore and canal, and

the stone canal ascends to the ring canal, which is located in a groove on the aboral surface of the jaws. The water ring bears four polian vesicles and also gives rise to the radial canals, which penetrate the lower side of the vertebral ossicles of the arms (Fig. 18–19). In each ossicle the radial canal gives rise to a pair of lateral canals, which lead to the podia. The paired lateral canals of ophiuroids contrast with the staggered arrangement of other echinoderms. Ampullae are absent, probably correlated with the reduction of the arm coelom, but a valve is present between the podium and the lateral canals. Fluid pressure for protraction is generated by a dilated, ampulla-like section of the podial canal and in some forms by localized contraction of the radial water canal.

Nutrition

Ophiuroids are carnivores, scavengers, deposit feeders, or filter feeders. Most use several feeding modes, but one is generally predominant. The Atlantic brittle star, *Ophiocomina nigra,* which is primarily a suspension feeder, can be used to illustrate all four feeding habits.

In filter feeding, the arms of *Ophiocomina* are lifted from the bottom and waved about in the water. Plankton and detritus adhere to mucous strands strung between the adjacent arm spines. The trapped particles may be swept downward toward the tentacular scale by ciliary currents or collected from the spines

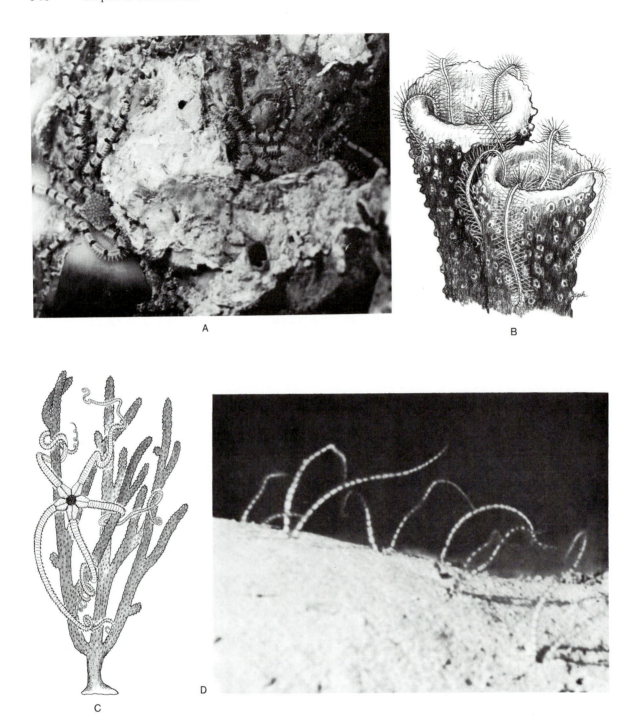

FIGURE 18–22 **A,** Two specimens of a West Indian brittle star *(Ophionereis)* lodged in crevices on the underside of a coral head. **B,** Two brittle stars in a sponge. **C,** A euryalous brittle star climbing on a gorgonian coral. These brittle stars are related to the basket stars and are capable of coiling their arms vertically. **D,** Specimens of *Amphioplus* projecting two arms from the tubelike burrows and trapping suspended particles from the passing water current. *(A, Photography courtesy of Betty M. Barnes. C, Modified from Hyman, L. H. 1955. The Invertebrates. Vol. IV. McGraw Hill Book Co., New York. D, From Fricke, H. W. 1970. Helgo. wiss. Meeresunters. 21:124–133.)*

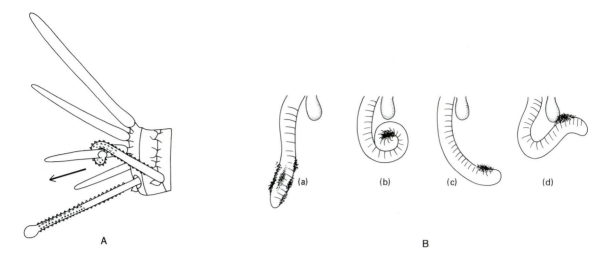

FIGURE 18–23 Feeding activity of podia in brittle stars. **A,** Spine wiping in *Ophiocoma wendtii*, a West Indian brittle star. Note that the podium of one side wipes the spines on the opposite side of the arm. **B,** Particle consolidation and transfer in the suspension-feeding brittle star, *Ophionereis fasciata*. (a), Particles collected by podium from spines. (b), Particles consolidated by podium into one mass. (c–d), Mass transferred from podium to tentacle scale. (A, From Sides, E. M., and Woodley, J. D. 1985. Niche separation in three species of *Ophiocoma* in Jamaica, West Indies. Bull. Mar. Sci. *36*(3):701–715. B, From Pentreath, R. J. 1970. J. Zool. *161*:395–429.)

by the tube feet, which extend upward for this purpose (Fig. 18–23). A tentacular scale is a reduced spine. The tube feet are then scraped across the tentacular scales, depositing collected particles in front of the scale (Fig. 18–23). This is also where the ciliary tracts deposit their material. On each side the food particles are picked up by adjacent podia, compacted into a bolus, and passed along the midoral line of the arm toward the mouth. The food balls are moved by the podia until they reach the proximal parts of the arm, where movement toward the mouth is facilitated by cilia.

Deposit feeding on intermediate particles is performed by the podia. The podia collect the particles from the substratum, compact them into food balls, and move them toward the mouth.

Large food material, such as dead animal matter, is swept into the mouth by the looping motion of an arm. Browsing over algae or carrion, the animal utilizes its teeth or oral tube feet.

Ophiothrix fragilis uses its papillate podia for filter feeding. The feeding arms are elevated and twisted so that the oral surface is directed toward the current. The podia are extended well beyond the spines, forming comblike filtering series on either side of the arm (Fig. 18–24A). Collected particles are periodically removed and transported as a growing bolus by a wave action of the podia that travels down the arm toward the mouth (Fig. 18–24B, C).

Such mechanisms of deposit and filter feeding have the advantage of permitting the animal to extend only two or three feeding arms from its protective retreat as well as to utilize a variety of food sources (Fig. 18–22).

Brittle stars that are predominantly carnivores feed largely on polychaetes, molluscs, and small crustaceans. Food is usually captured and brought to the mouth by arm looping.

Basket stars, of which there are only about 100 species, are suspension feeders but capture zooplankton of relatively large size (10–30 mm: crustaceans, polychaetes, and others). Perched above the bottom, the basket star extends its arms in a parabolic filtration fan with the concave and aboral sides directed toward the prevailing water current (Fig. 18–24D). Prey is seized with the ends of the many arm branches, which coil about the catch, and minute hooks on the arm surface prevent escape. Periodically, the basket star removes the collected plankton from the arms by passing them through comblike oral papillae. The

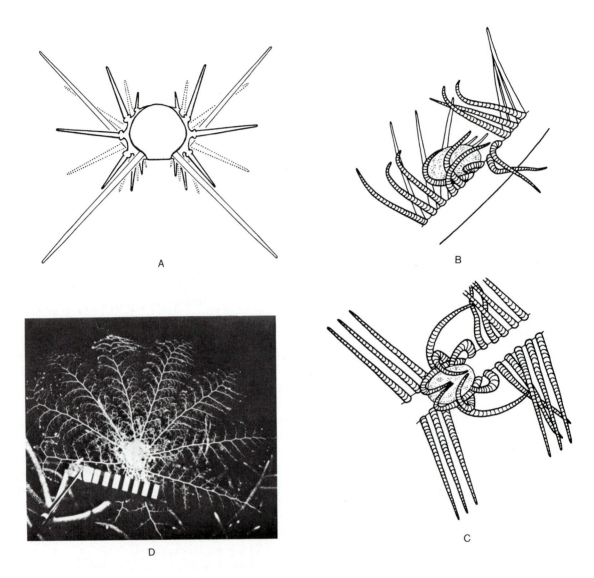

A

B

C

D

FIGURE 18–24 Filter feeding in ophiuroids. **A,** End view of arm section of the brittle star, *Ophiothrix fragilis,* showing spines (s) and position of tube feet (p). Alternate tube feet are directed orally and aborally. The tube feet may also be extended laterally, forming a single filtering series on either side of the arm. **B–C,** Particle collection and transport in *Ophiothrix fragilis* seen in side view **(B)** and orally **(C).** Particles are added to a bolus, which is about 1 mm in diameter on reaching the mouth. **D,** Feeding position of the basket star, *Astrophyton.* The arms form a parabolic fan with the tips directed toward the current, which in this figure is moving away from the viewer. *(B and C, From Warner, G. F., and Woodley, J. D. 1975. Suspension-feeding in the brittle star Ophiothrix fragilis. J. Mar. Biol. Assoc. U. K. 55:199–210. Copyrighted and reprinted by permission of Cambridge University Press. D, Photograph courtesy of Meyer, D. L., and Lane, N. G. 1976. The feeding behavior of some Paleozoic crinoids and recent basket stars. J. Paleontol. 50(3):472–480.)*

tube feet do not play the role of food transport, as in brittle stars.

The digestive tract is simple (Fig. 18–21). The jaws frame a shallow, prebuccal cavity, which is roofed aborally by the peristomial membrane containing the mouth. The esophagus connects the mouth with a large, saclike stomach. The stomach fills most of the interior of the disc, and in most ophiuroids, the margins are infolded to form ten pouches. There is no intestine or anus, and the digestive tract does not extend into the arms. Extracellular and intracellular digestion and absorption occur largely within the stomach.

Gas Exchange, Excretion, and Internal Transport

Gas exchange in ophiuroids takes place across the tube feet and by means of ten internal sacs (**bursae**) that represent invaginations of the oral surface of the disc. The bursae are connected to the outside by slits that run along the margins of the arms on the oral surface of the disc (Fig. 18–18A). The bursae may be lined with ciliated epithelium, especially the slits. The beating cilia create a current of water that enters the peripheral end of the slit, passes through the bursae, and flows out the oral end of the slit. Many species also pump water into and out of the bursae by raising and lowering the oral or aboral disc wall or by contracting certain disc muscles associated with the bursae.

A few species of ophiuroids, such as *Hemipholis elongata* and some species of *Ophiactis*, have hemoglobin in coelomocytes of the water-vascular system. *Hemipholis* wraps around the tube of the polychaete, *Diopatra cuprea*, which resides in anoxic muds.

The thin-walled respiratory bursae may well be the principal center for removing wastes, including waste-laden coelomocytes. The coelom in ophiuroids is much reduced compared with that of other echinoderms. The vertebral ossicles restrict the coelom to the aboral part of the arms (Fig. 18–19); the stomach, bursae, and gonads leave only small coelomic spaces in the disc. The hemal system is essentially like that of asteroids.

Nervous System

The nervous system is composed of a circumoral nerve ring and radial nerves, as in asteroids (Figs. 18–19; 18–21). Specialized sense organs are absent; dispersed epithelial sensory cells compose the sensory system. Most ophiuroids are negatively phototropic, and are also able to detect food without contact.

Regeneration and Reproduction

Many ophiuroids can cast off, or autotomize, one or more arms if disturbed or seized by a predator. A break can occur at any point beyond the disc; the lost portion is then regenerated. There are some ophiuroids, notably six-armed species of *Ophiactis*, in which asexual reproduction takes place by division of the disc into two pieces, each piece with three arms.

The majority of ophiuroids are dioecious. The gonads are small sacs attached to the coelomic side of the bursae near the bursal slit. There may be one, two, or numerous gonads per bursa with various positions of attachment. Hermaphroditic species are not uncommon. Some bear separate testes and ovaries; others are protandric.

When the gonads are ripe, they discharge into the bursae, probably by rupture, and the sex cells are carried out of the body in the ventilating water current. Fertilization and development take place in the seawater in many species, but brooding is common. The bursae are commonly used as brood chambers, but the female of some species broods her eggs in the ovary or coelom. Development takes place within the mother until the juvenile stage is reached. In most species only a few young are brooded in each bursa.

Development

In nonbrooding, oviparous ophiuroids early development is similar to that in the asteroids. The larva of many species, called an **ophiopluteus,** displays four pairs of elongated arms supported by calcareous rods and bearing ciliated bands. The shape of the larva is distinctive and different from that of the brachiolaria of asteroids (Fig. 18–25). Metamorphosis takes place while the larva is still free-swimming, and there is no attachment stage. The tiny brittle star sinks to the bottom and takes up an adult existence. For the some six species for which information is available, development to this point takes 14 to 40 days. To reach the same stage in the single brooding species studied takes three to seven months. Of the 2000 described species of ophiuroids, larvae were known for 71, and 55 were known to brood by 1975. The numbers are

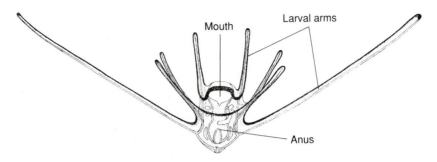

FIGURE 18–25 Ophiopluteus larva of *Ophiomaza* (oral view). *(After Mortensen.)*

probably a little higher today, but the relative proportion is probably the same. The developmental patterns of the remainder are still unknown.

SUMMARY

1. In the class Ophiuroidea the long, slender, snaky arms are sharply set off from the central disc.

2. Ophiuroids are considered the most successful group of echinoderms. Their success is probably correlated with their motility, diversity of feeding habits, and small size, which have enabled them to exploit habitats unavailable to most other echinoderms.

3. Ophiuroids move rapidly by pushing and pulling with their flexible arms. Lateral arm spines provide traction. The arm is occupied by large ossicles (vertebrae) that articulate with each other in a horizontal column. Intervertebral muscles provide movement. Podia generally are not used in locomotion.

4. The vertebrae are covered by flat, superficial ossicles, called shields, with which the spines are associated. The vertebral ossicles restrict the coelom to a small, dorsal chamber.

5. Perhaps because of coelom reduction, the water-vascular system lacks ampullae. The lateral canal or radial canals assume the function of ampullae. The madreporite is located on one of the oral shields.

6. The reduced arm coelom restricts much of gas exchange to five pairs of pouchlike invaginations (bursae) on the oral side of the disc.

7. The feeding of ophiuroids includes one or all of the following mechanisms for a given species: scavenging by arm raking, deposit feeding by the podia, and suspension feeding by podia and mucous strands slung between spines. These methods enable many species to feed without leaving protective retreats. The principal functions of ophiuroid podia are food collection and transport. Basket stars use their arms to form a parabolic fan at right angles to the water current and catch zooplankton with the tips of the arm branches.

8. The gonads of ophiuroids are associated with the coelomic side of the bursae, which provide the exit for the gametes and the site of development in brooding species. In nonbrooding species, development leads to an ophiopluteus larva, which undergoes metamorphosis prior to settling.

CLASS ECHINOIDEA

The echinoids are free-moving echinoderms commonly known as sea urchins, heart urchins, and sand dollars. About 950 species have been described. The name Echinoidea, meaning "like a hedgehog (porcupine)," refers to the movable spines that cover the bodies of these animals. The echinoid body does not possess arms. Rather, the shape is circular or oval, and the body is spherical or greatly flattened along the oral/aboral axis. The class is particularly interesting from the standpoint of symmetry, because although the sea urchins are radially symmetrical, many members of the class that live in soft bottoms display various stages in the attainment of a secondary bilateral symmetry. A third distinctive feature of echinoid structure is the flattening and suturing together of the skeletal ossicles into a solid case (the **test**).

External Structure

Regular Echinoids

The radial, or regular, members of the class are known as sea urchins. In these forms the body is more or less spherical and armed with relatively long, mov-

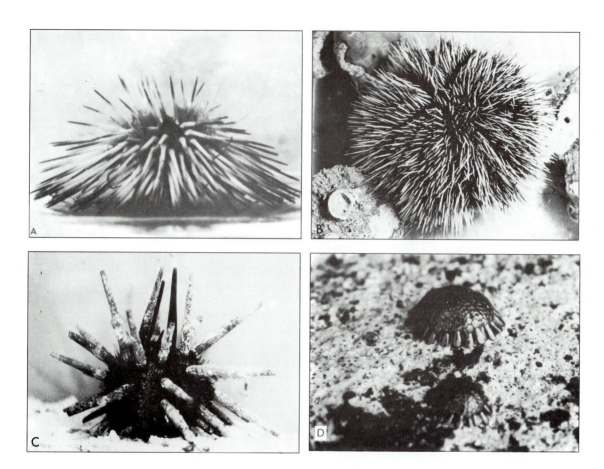

FIGURE 18–26 Regular urchins. **A,** Side view of the common Atlantic sea urchin, *Arbacia punctulata,* showing long spines and podia. **B,** A West Indian species of *Tripneustes* viewed from above. **C,** A species of *Eucidaris* with very small secondary spines around the base of the heavy primary spines. **D,** *Colobocentrotus,* a Pacific sea urchin with blunt aboral spines that fit together to form a smooth surface. Such spines are perhaps an adaptation for living on intertidal rocks. *(All photographs courtesy of Betty M. Barnes.)*

able spines (Fig. 18–26). Sea urchins are brown, black, purple, green, white, and red, and some are multicolored. Most are 6 to 12 cm in diameter, but some Indo-Pacific species may attain a diameter of nearly 36 cm. The sea urchin body can be divided into an aboral and an oral hemisphere, with the parts arranged radially around the polar axis. The oral pole bears the mouth and is directed against the substratum. The mouth is surrounded by a peristomial membrane that bears a number of different structures arranged in a radial manner. There are five pairs of short, heavy, modified podia, called **buccal podia,** and five pairs of bushy projections, called **gills** (Fig. 18–27A). In addition to the buccal podia and the gills, the area around the peristome bears small spines and pedicellariae.

The aboral pole contains the anal region, known as the **periproct** (Figs. 18–27B; 18–28A). The periproct is a small, circular membrane containing the anus, usually in the center, and a variable number of embedded plates, depending on the species (Figs. 18–27B; 18–28A). The globose body surface can be divided into ten radial sections, which converge at the oral and aboral poles (Fig. 18–27A). Five sections contain tube feet and are called **ambulacral areas.** The ambulacral areas alternate with sections devoid of podia, known as **interambulacral areas.**

The skeletal plates are arranged in rows running from the oral pole to the aboral pole. Each ambulacral area is composed of two rows of ambulacral plates, and each interambulacral area is composed of two rows of interambulacral plates. There are thus 20

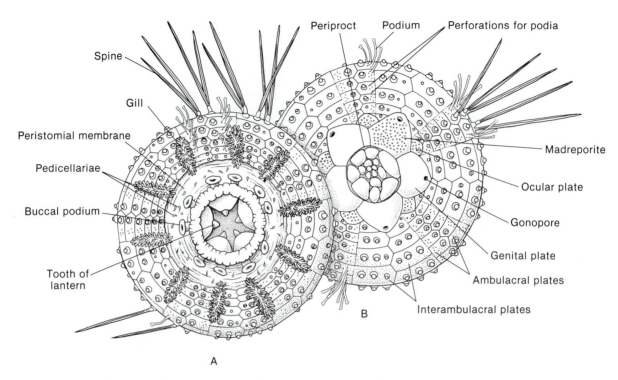

FIGURE 18–27   The regular urchin, *Arbacia punctulata*. **A,** Oral view. **B,** Aboral view. *(After Reid, W. M. In Brown, F. A. Selected Invertebrate Types. John Wiley and Sons, New York.)*

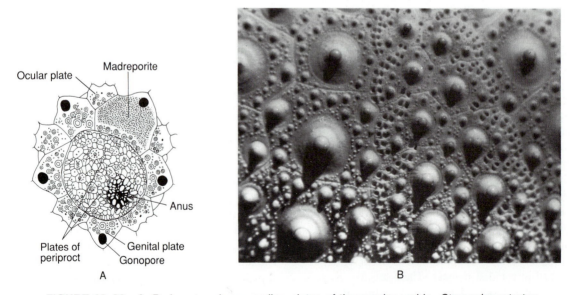

FIGURE 18–28   A, Periproct and surrounding plates of the regular urchin, *Strongylocentrotus*. **B,** Surface view of a section of a sea urchin test, showing tubercles on which spines are located, paired perforations for tube feet, and junction line (groove) of sutured ossicles. Compare with Figure 18–36. *(A, After Loven from Ludwig. B, Photograph courtesy of Betty M. Barnes.)*

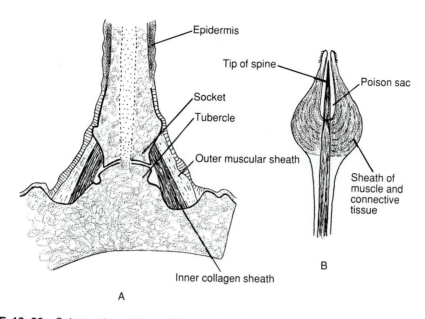

FIGURE 18–29 Spines of regular urchins. **A,** Section through the base of a *Cidaris* spine, showing muscular and collagen sheaths. **B,** Poison spine of *Asthenosoma varium,* an Indo-Pacific species. (A, After Cuénot, L. 1948. Échinodermes. *In* Grassé, P. (Ed.): Traité de Zoologie. Vol. XI. Masson et Cie, Paris. B, After Sarasins from Hyman.)

rows of plates: 10 ambulacral and 10 interambulacral (Fig. 18–27B). The ambulacral plates are pierced by holes forming canals that connect the internal ampullae and external podia (Fig. 18–27B). In sea stars and brittle stars, the canals pass between ossicles.

Around the periproct is a series of plates. These consist of five large, **genital plates,** one of which is porous and serves as the madreporite, and five smaller ocular plates (Figs. 18–27B; 18–28A). The genital plates, each of which bears a gonopore, line up with the interambulacral areas and alternate with the ocular plates, which coincide with the ambulacral areas.

The movable spines, which are so characteristic of sea urchins, are arranged more or less symmetrically in the ambulacral and interambulacral areas. The spines are longest around the equator and shortest at the poles. Most sea urchins possess long (**primary**) and short (**secondary**) **spines,** the two types being more or less equally distributed over the body surface. *Arbacia punctulata,* the common sea urchin along the Atlantic coast of North America, however, possesses only the long type (Fig. 18–26A).

Each spine contains a concave socket at the base that fits over a corresponding tubercle on the test (Fig. 18–29A). Two sheaths of fibers that extend between the spine base and the test encircle the ball-and-socket joint. Contractions of the outer muscular sheath incline the spines in one direction or another. The inner sheath of collagen fibers (catch fibers) can rapidly shift from a soft to a hard condition on stimulation, thereby causing the spine to be rigidly erect (p. 921).

The spines are usually cylindrical and taper to a point, but many species depart from this generalization. Species of *Diadema,* which are common on tropical reefs, have very long, needle-like spines, which can be rapidly tilted and waved in the direction of intruders that cast a shadow on the urchin (Fig. 18–30A). The spines, which can be regenerated, are hollow, brittle, and provided with an irritant, and the outer surface is covered with circlets of small barbs directed toward the spine tip (Fig. 18–30B). This urchin can inflict serious, painful wounds if stepped on. The primary spines of the slate-pencil urchins, species of *Heterocentrotus,* and some species of cidaroids are heavy and blunt (Fig. 18–26C). The aboral spines of the intertidal Indo-Pacific genus *Colobocentrotus* are short and heavy, and the blunt tips are polygonal in cross section (Fig. 18–26D). These spines fit together like tiles, providing an effec-

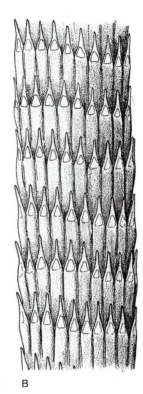

FIGURE 18–30   A, *Diadema.* Species of this genus, which occurs in both the Caribbean and the Indo-Pacific, possess long, hollow, needle-like spines, which can inflict painful punctures when handled or stepped on. This West Indian species is common on reefs, where it lives in sheltered or protected recesses on exposed or sand bottoms. **B,** Section of a spine of *Diadema.* Circlets of barbs are directed toward the spine tip. *(A, Photograph courtesy of C. Gebelein.)*

tive wave-resistant surface and protection against desiccation. The deep-water sea urchins of the family Echinothuridae bears special poison spines on the aboral surface (Fig. 18–29B).

Pedicellariae, which are characteristic of all echinoids, are located over the general body surface as well as on the peristome. The echinoid pedicellaria is composed of a long stalk surmounted by jaws. The stalk may contain a supporting skeletal rod, and there are usually three opposing jaws (Fig. 18–36). Muscles at the base of the stalk provide for elevation and direction of the pedicellariae in response to certain stimuli.

A species of urchins usually possesses several types of pedicellariae, one of which may contain poison glands. Poisonous pedicellariae reach their greatest degree of development in members of the widespread subtropical and tropical family Toxop-

neustidae *(Lytechinus, Tripneustes)* (Fig. 18–36). The outer side of each jaw is surrounded by one or two large poison sacs that open by one or two ducts just below the terminal tooth of the jaw. The poison has a rapid paralyzing effect on small animals and drives larger enemies away. The spines frequently incline away from the poison pedicellariae so that these pedicellariae are more exposed. The poison pedicellariae of the Indo-Pacific *Toxoptneustes* look like little parasols when open and can produce a painful reaction in humans.

Other types of pedicellariae are used for defense or for cleaning the body surface, biting, and breaking up small particles of debris, which are then removed by the surface cilia. When the pedicellariae are touched on the outside, they snap open; when touched on the inside, they snap shut. Pedicellariae also respond to chemical stimuli.

Irregular Echinoids

The bilateral, or irregular, echinoids include the heart urchins, cake urchins, and sand dollars. Most of their peculiarities are adaptations for burrowing in sand. In contrast to sea urchins, the test is clothed with many small spines, which are used in locomotion and keeping sediment off the body surface. The heart urchins (spatangoids) are more or less oval, the long axis representing the anteroposterior axis of the body (Fig. 18–31). The oral surface is flattened, and the aboral surface is convex. The entire center of the oral surface, containing the mouth and peristome, has mi-grated anteriorly. The center of the aboral surface usually remains in the center of the upper or dorsal surface, but the periproct and anus have migrated to the posterior end in what now becomes the posterior interambulacrum. Podia are degenerate or absent around the circumference of the body, so functional podia are confined to the oral and aboral surfaces. The conspicuous aboral ambulacral areas are each shaped like a petal radiating from the center and are known as **petaloids.** The podia of the petaloids are modified for gas exchange. The oral ambulacral areas (**phyllodes**) contain specialized podia for obtaining food

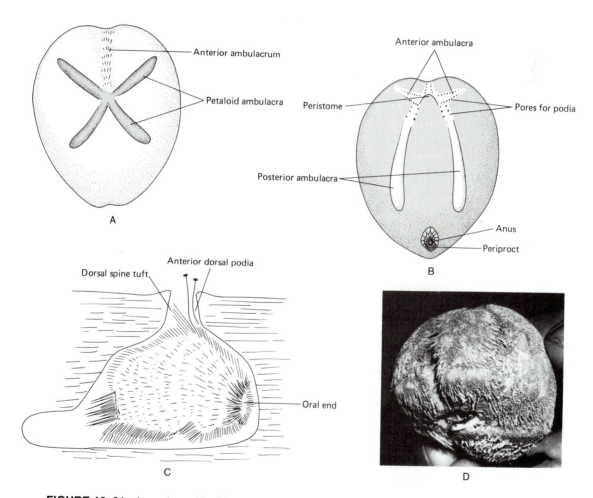

FIGURE 18–31 Irregular echinoids, spatangoids (heart urchins). **A,** *Meoma ventricosa* from the West Indies (aboral view). **B,** *Meoma* (oral view). **C,** *Echinocardium flavescens* in its sand burrow (lateral view). **D,** Anterior end of *Moira atropos,* a heart urchin from the Atlantic coast of the southeastern United States. *(A and B, From Hyman, L. H. 1955. The Invertebrates. Vol. IV. McGraw-Hill Book Co., New York. C, Modified after Gandolfi-Hornyold. D, Photograph courtesy of Betty M. Barnes.)*

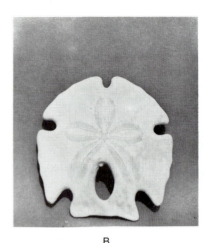

A   B

FIGURE 18–32   Irregular echinoids, clypeasteroids. **A,** Side view of the test of sea biscuit, *Clypeaster.* **B,** Aboral surface of the test of the arrowhead sand dollar, a species of *Encope.* *(Both photographs courtesy of Betty M. Barnes.)*

particles. The small spines form a dense covering over the body surface but have the same basic structure as those of sea urchins.

The cake urchins or sand dollars (clypeasteroids) differ from the heart urchins in a number of respects. A few species, such as the sea biscuits, *Clypeaster* (Fig. 18–32A), are shaped somewhat like heart urchins, but the typical sand dollar has a greatly flattened body displaying a circular circumference (Fig. 18–33). Burrowing and covering are thus facilitated. The aboral center and the oral center, which contains the mouth, are both centrally located. The periproct, however, is ventral and, like that of the heart urchins, is located in the posterior interambulacrum. The aboral surface bears conspicuous **petaloids,** and the oral surface contains distinct radiating grooves.

The bodies of some common sand dollars, called keyhole sand dollars *(Mellita),* contain large, elongated notches or openings known as **lunules** (Fig. 18–33A–C). Lunules vary in number from two to many and are symmetrically arranged. In most cases the lunules arise from indentations that form along the circumference of the animal and then become enclosed in the process of growth.

Body Wall

The body wall of echinoids is composed of the same layers as in asteroids. A ciliated epidermis covers the outer surface, including the spines. Beneath the epidermis lie a nervous layer and then a connective tissue dermis that contains the flattened and sutured skeletal plates. A muscle layer is absent, because the ossicles are immovable, and the inner surface of the test is covered by the peritoneum, composed of ciliated columnar epithelium.

Locomotion

Sea urchins are adapted for life on both hard and soft bottoms, and spines and podia are used in movement. The tube feet function in the same manner as those of the sea stars, and spines may be used for pushing and raising the oral surface off the substratum. Sea urchins can move in any direction, and any one of the ambulacral areas can act as the leading section. If overturned, these animals right themselves by attaching the more aboral podia of one of the ambulacral areas. Attachment of the podia progresses in an oral direction, gradually turning the animal over onto the oral side. Righting may also involve specialized movements of the spines.

Movement of sea urchins is closely related to feeding activity. For example, *Strongylocentrotus franciscanus* in kelp beds off the California coast exhibits mean movements of 7.5 cm per day, but where food supplies are lower, movement may be as great as 50 cm per day.

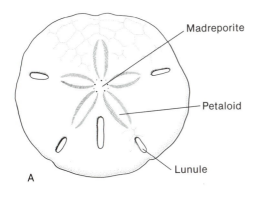

A

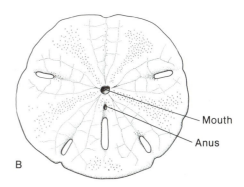

B

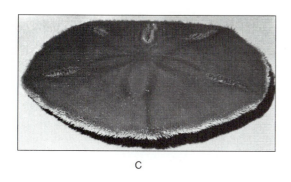

C

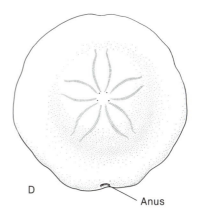

D

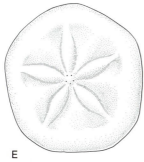

E

F

FIGURE 18–33 Irregular echinoids, clypeasteroids. **A–C,** *Mellita quinquiesperforata,* the five-slotted sand dollar from the Atlantic coast of the United States. **A,** Oral view of the test. **B,** Aboral view of test. **C,** Anterior view of a specimen with spines intact. **D,** Aboral view of the test of *Dendraster excentricus* from the Pacific coast of the United States. Aboral center is closest to the posterior edge of the test. **E,** Aboral view of test of species of *Laganum* from the Indo-Pacific. **F,** Scanning electronmicrograph of the aboral spines of *Mellita. (F, Scanning Electronmicrograph courtesy of J. Ghiold.)*

Some sea urchins tend to seek rocky depressions, and some species are actually capable of increasing the depth of such depressions or even of excavating burrows in rock and other firm material. Boring is performed largely by the scraping action of the chewing apparatus.

Boring behavior appears to be an adaptation to counteract excessive wave action, and these species are largely found in habitats that are exposed to rough water. One of the most notable boring sea urchins is *Paracentrotus lividus,* which lives along the coast of Europe. This sea urchin riddles rock walls with burrows. When the burrows are shallow, the animal leaves to feed, but it remains permanently within deeper burrows, which often have entrances too small

to permit exit. Echinometrids are common boring species on tropical reefs. The urchins can usually be seen within their shallow, irregular excavations but are very difficult to remove without breaking the surrounding rock. The West Indian *Echinometra* honeycombs coralline rock in surge areas (Fig. 18–34), but nonburrowing populations of this species are sometimes encountered. *Strongylocentrotus purpuratus* is a surge-loving sea urchin found along the Pacific coast of North America that commonly burrows in soft rock.

The irregular echinoids are adapted for a life of burrowing in sand. The animal burrows with its anterior end forward, and movement results from action of the spines, the podia being modified for other func-

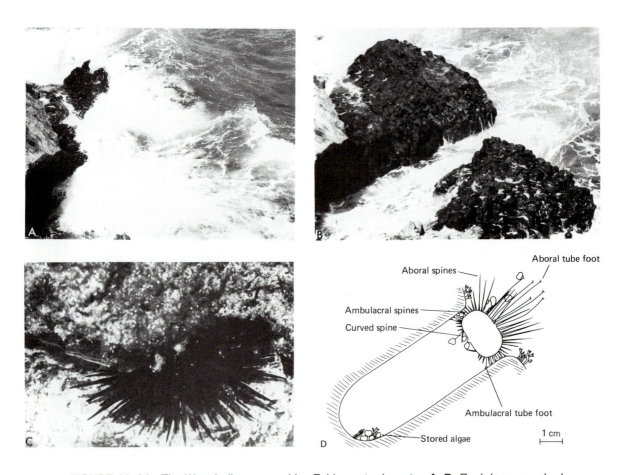

FIGURE 18–34 The West Indian sea urchin, *Echinometra lacunter.* **A–B,** Rock honeycombed with *Echinometra* burrows, shown covered **(A)** and uncovered **(B)** during wave surges. **C,** Urchin in burrow. **D,** The Indo-Pacific *Echinostrephus molaris* feeding on pieces of algae collected at the entrance to its burrow. *(A–C, Photographs courtesy of Betty M. Barnes. D, From De Ridder, C., and Lawrence, J. M. 1982. Food and feeding mechanisms: Echinoidea. In Jangoux, M., and Lawrence, J. M. (Eds.): 1982. Echinoderm Nutrition. A. A. Balkema, Rotterdam. p. 90.)*

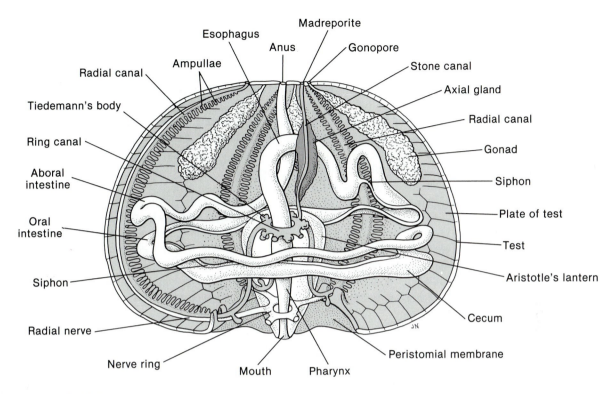

FIGURE 18–35 Internal structure of the regular urchin, *Arbacia* (side view). *(Modified after Petrunkevitch from Reid.)*

tions. A heart urchin burrows into the sand by inclining the anterior end downward and moving sand with specially modified, paddle-shaped spines on the anterior sides of the body. Some heart urchins tend to remain buried in one spot below the surface of the sand. Contact with the surface in those species that bury themselves more than several centimeters is maintained by a funnel-like opening in the sand over the aboral side of the animal. Maintenance of the opening and of the subterranean chamber wall is carried out by specialized podia and apical spines (Fig. 18–31C). A short, blind tunnel extends posteriorly behind the anus. Tracts of ciliated spines (**fascioles**) pump water into the surface opening and out the posterior tunnel. Some intertidal species, such as the Indo-Pacific *Lovenia*, come to the surface at low tide and rebury themselves when the tide comes in. *Lovenia* can bury itself in 1 min, but other species may take as long as 50 min.

Sand dollars burrow just beneath the surface of the sand. Some species, such as the common sand dollars of the east coast of the United States, *Mellita quinquiesperforata*, cover themselves completely

with sand; in others, such as the Pacific coast *Dendraster excentricus*, the posterior end projects obliquely above the sand surface in quiet water. The sea biscuit, *Clypeaster rosaceus*, of Florida and the West Indies does not burrow but sits on the surface of sandy bottoms.

Some sand dollars can right themselves if turned over. In righting itself, the animal burrows its anterior end into the sand, gradually elevates its posterior end, and eventually flips its body over. The Atlantic five-lunuled species partially elevates its body and then apparently depends on water currents to turn it back over onto its oral surface. The flattened shape of the sand dollars lying just beneath the surface subjects them to lift and dislodgment by water currents. The slots or lunules characteristic of many species appear to be an adaptation to reduce lift.

Water-Vascular System

The water-vascular system of echinoids is essentially like that of the sea stars (Fig. 18–35). One of the genital plates around the periproct contains pores and pore canals and functions as the madreporite (Fig. 18–28A). A stone canal descends orally to the ring canal, which lies above the peristome in heart urchins or just above the chewing apparatus in regular urchins and sand dollars. The radial canals extend from the ring canal and run along the underside of the ambulacral areas of the test. Each radial canal terminates in a small protrusion called a terminal tentacle, which penetrates the most apical ambulacral plate. The lateral canals of one side of the radial canal alternate with those of the other side. The canals connecting the ampullae and podia, unlike those in other echinoderms, penetrate the ambulacral ossicles rather than pass between them. These canals are also peculiar in being doubled; that is, from each ampulla two canals pierce the ambulacral plate and become confluent on the outer surface to enter a single podium (Fig. 18–36). The suckers of the podia of sea urchins are

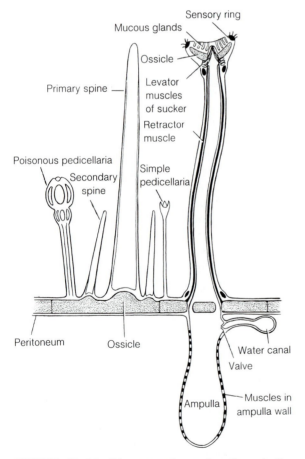

FIGURE 18–36 Diagrammatic section through the body wall of a sea urchin, showing one ambulacral and one interambulacral ossicle and associated structures. *(After Nichols, in part.)*

highly developed and have a system of muscles and supporting ossicles.

In irregular urchins, which use the spines to move, the podia are modified for a number of functions that will be described in the following sections. Those on the oral surfaces of sand dollars are widely dispersed, penetrate the test via a single canal, and are served by elaborately branching canals.

Nutrition

Sea urchins feed with a highly developed scraping apparatus called **Aristotle's lantern.** The apparatus is composed of five large, calcareous plates called **pyramids,** each of which is shaped somewhat like an arrowhead with the point projected toward the mouth (Fig. 18–37). The pyramids are arranged radially, with each side connected to that of the adjacent pyramid by means of transverse muscle fibers. Passing down the midline along the inner side of each pyramid is a long, calcareous band. The oral end of the band projects beyond the tip of the pyramid as an extremely hard, pointed tooth. Because there is one such tooth band for each pyramid, there are five teeth projecting from the oral end of the lantern. The curled, upper end of the band is enclosed within a dental sac and is the area of new tooth formation. In *Paracentrotus lividus,* new tooth material is produced at a rate of about 1 to 1.5 mm per week.

In addition to the teeth and pyramids, Aristotle's lantern is composed of a number of smaller, rodlike pieces at the aboral end. By means of special muscles the lantern can be partially protruded and retracted through the mouth. Other muscles control the teeth, which can be opened and closed. In the more primitive cidaroids, the scraping action of the lantern is largely restricted to opening and closing the teeth, but in higher groups, the lantern can also be swung laterally. The ability of the lantern to be protruded and retracted makes possible pulling and tearing in addition to scraping. Aristotle's lantern is absent from heart urchins but functional in sand dollars.

The majority of sea urchins are grazers, scraping the substratum surface on which they live with their teeth. Although algae are usually the most important food, most sea urchins are generalists and include a wide range of plant and animal material in their diets. Moreover, the diet of a particular species varies from area to area, depending on what is available. *Lytechinus variegatus,* an inhabitant of turtle grass beds, consumes about 1 g of grass per day. The grazing effect

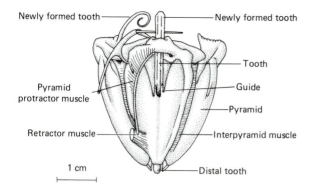

Newly formed tooth
Newly formed tooth
Tooth
Guide
Pyramid protractor muscle
Pyramid
Retractor muscle
Interpyramid muscle
1 cm
Distal tooth

FIGURE 18–37 Lateral view of Aristotle's lantern of *Sphaerechinus granularis*. (From De Ridder, C., and Lawrence, J. M. 1982. Food and feeding mechanisms: Echinoidea. In Jangoux, M., and Lawrence, J. M. (Eds.): 1982. Echinoderm Nutrition. A. A. Balkema, Rotterdam. p. 81.)

of some sea urchins was dramatically revealed following the 1983 crash in Caribbean populations of the long-spined *Diadema antillarum*, where in some areas the algal mat increased from 1–2 mm to 20–30 mm.

Boring sea urchins feed on encrusting and endolithic algae on the walls of their burrows, as well as algal fragments and other organic debris that is washed in. *Echinostrephus molaris* of the Indo-Pacific sits at the mouth of its burrow, and when debris touches the spines or long podia, the spine tips close and grasps it.

The interior of Aristotle's lantern contains both a buccal cavity and a pharynx that ascends through the apparatus and passes into an esophagus (Fig. 18–35). The esophagus descends along the outer side of the lantern and joins a tubular stomach (Fig. 18–35). At the junction of the esophagus and stomach, a blind pouch, or cecum, is usually present. The stomach makes a complete turn around the inner side of the test wall, from which it is suspended. It then passes into the thinner-walled aboral intestine, which makes a complete turn in the opposite direction. The intestine then ascends to join the rectum, which empties through the anus within the periproct.

In most echinoids a narrow tube, called a **siphon,** parallels the stomach for about half its length. The ends of the siphon open into the lumen of the intestine. Extracellular digestion begins in the stomach and is completed in the intestine, where absorption also occurs. The siphon functions to remove excess water before the food reaches the region of extracellular digestion.

Irregular urchins are selective deposit feeders. All feed on organic material in the sand in which they burrow. The heart urchins obtain food by means of modified podia on the oral surface. During feeding these podia grope about the sand surface of the chamber, picking up food particles.

Sand dollars feed on particles picked up by podia from the substratum beneath the oral surface of the animal. These podia are irregularly scattered across the oral surface (the radial water canal has lateral lobes). Particles are passed from podium to podium to the food grooves and then down the grooves to the mouth (Fig. 18–38). For *Dendraster excentricus* on the west coast of the United States, suspended particles are a major source of food because in quiet water this sand dollar lives with the posterior half of the body projecting above the sand surface. Food includes not only particles that pass between the spines but also diatoms and algal fragments collected by the tube feet and small crustaceans caught by the pedicellariae.

The alimentary canal of irregular echinoids is more or less like that described for sea urchins, although the rectum extends posteriorly to the anus. Most spatangoids, including *Echinocardium*, build one or two drains at the back of the burrow to collect

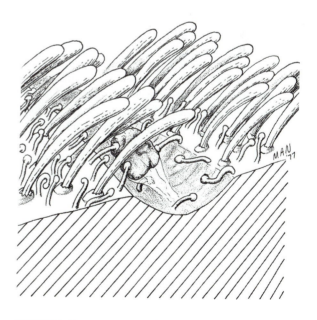

FIGURE 18–38 Section through a food groove on the oral surface of a sand dollar, showing food bolus being moved by podia. Many club-shaped spines rise above the podia.

feces (Fig. 18–31C). The drain is constructed by specialized posterior podia around the anus.

Predators of sea urchins include sea otters, fish, certain gastropods, and sea stars. Sea urchin eggs are eaten by humans in various parts of the world, especially in Japan. The Japanese support a large sea urchin fishery and import large quantities of sea urchin roe from other parts of the world.

Internal Transport, Gas Exchange, and Excretion

Coelomic fluid is the principal circulatory medium, and coelomocytes are abundant. The hemal system has the same basic plan of structure as in the asteroids but is better developed.

In regular echinoids the five pairs of peristomial gills are important centers of gas exchange (Fig. 18–27A). Each gill is a highly branched outpocketing of the body wall and is therefore lined within and without by a ciliated epithelium. Coelomic fluid from the lantern coelom is pumped into and out of the gills by a system of muscles and ossicles associated with Aristotle's lantern.

As in other echinoderms, all the podia contribute to gas exchange. In most sea urchins the more aboral podia are modified in various ways for this function. The podia and ampullae are commonly septate with a two-way circulation of fluid through these structures (Fig. 18–39). The aboral podia may lack suckers, and on *Arbacia* they are flattened (Fig. 18–39B).

There are no peristomial gills in heart urchins and sand dollars. In these animals the modified podia of the petaloids, which are short and flattened, act as gas exchange structures (Fig. 18–40). The water current produced by external cilia flows in the opposite direction of the current within the podium, a countercurrent system ensuring a gradient favoring uptake of oxygen by the coelomic fluid. A similar countercurrent exists between the fluid within the ampulla and the surrounding fluid of the perivisceral coelom.

The coelomocytes are active in the removal of particulate waste and carry these accumulations to the gills, podia, and axial organ for disposal. The axial organ complex is similar to that of asteroids.

Nervous System

The nervous system is basically like that of the asteroids (Fig. 18–35). The circumoral ring encircles the pharynx inside the lantern, and the radial nerves pass

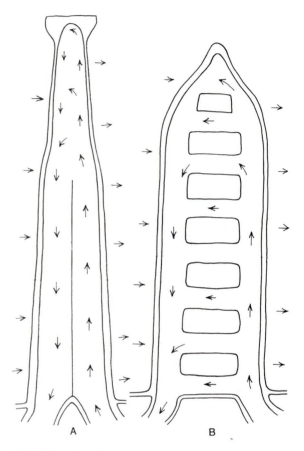

FIGURE 18–39 Respiratory adaptations of sea urchin podia. **A,** A podium of *Strongylocentrotus*. All podia are suckered and partially septate, dividing the interior into two channels with a circulatory flow of fluid through podium, test, and ampulla. **B,** Flattened, suckerless, respiratory podium of *Arbacia*. Rippled wall divides the interior into two channels with circulating fluid. Arrows indicate internal and external fluid flow. *(From Fenner, D. H. 1973. The respiratory adaptations of the podia and ampullae of echinoids. Biol. Bull. 145:323–339.)*

between the pyramids of the lantern and run along the underside of the test, lying just beneath the radial canals of the water-vascular system.

The numerous sensory cells in the epithelium, particularly on the spines, pedicellariae, and podia, compose the major part of the echinoid sensory system. The buccal podia of sea urchins, the podia around the circumference of heart urchins, and the podia of the oral surface of sand dollars are all important in sensory reception.

Echinoids possess statocysts, which are located with hard spherical bodies called **spheridia.** In sea

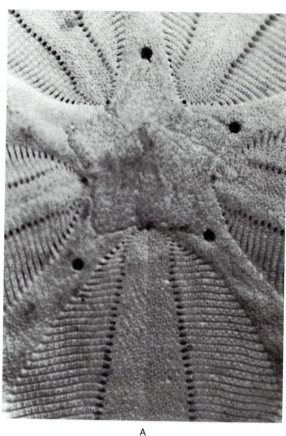

FIGURE 18–40 **A,** Surface of the aboral center of the test of the sand dollar, *Mellita.* Holes opposite the points of the central star (madreporite) are the gonopores. **A)** gonad is absent in the posterior interambulacrum. In the petaloid included in the lower half of the photograph, the grooves in which the modified branchial podia are located are conspicuous (compare with diagram in **B**). At the ends of the grooves can be seen the perforations for the podial canals penetrating the test. The inner canal is very conspicuous; in this photograph, the outer canals can be seen only in the areas opposite the gonopores. **B,** Diagram of branchial podia across one petaloid of a sand dollar. Arrows show countercurrent flow of seawater (external) and fluid of water-vascular system. *(A, Photograph courtesy of Betty M. Barnes. B, Adapted from Fenner.)*

A

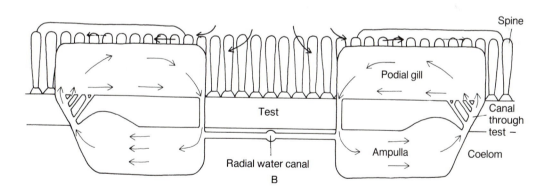

B

urchins they are stalked, few to many, and located at various places along the ambulacra. In sand dollars spheridia are located near the peristome and are buried in the test. The spheridia function in orienting the animal to gravitational pull.

Echinoids are, in general, negatively phototropic and many tend to seek the shade of crevices in rocks and shells. Some species of sea urchins, such as *Tripneustes, Lytechinus, Strongylocentrotus,* and the sea biscuit, *Clypeaster,* cover themselves with shell fragments and other objects, using the tube feet (Fig. 18–41A). The significance is uncertain but clearly seems to be a light response in some, for *Tripneustes* covers more in the summer than in the winter, and the

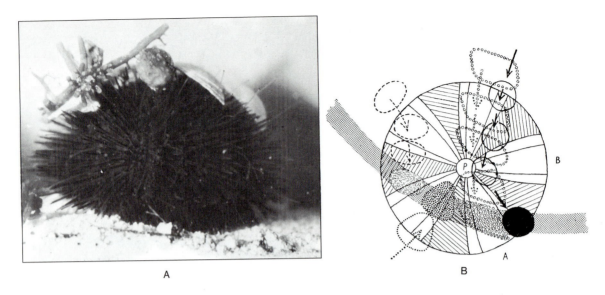

A

B

FIGURE 18–41 **A,** *Lytechinus variegatus,* a sea urchin that covers its aboral surface with shells, stones, and algae. **B,** Specimen of *Lytechinus* is subject to a narrow band of light, which cuts across one side of the test. In response, the sea urchin picks up small stones and passes them from three different directions onto the illuminated region of the test. *(A, Photograph courtesy of Betty M. Barnes. B, From Millott, N. 1956. The covering reaction of sea-urchins. J. Exp. Biol. 33:508–523.)*

related *Lytechinus* drops its cover at night and will cover only an experimental band of light crossing the body (Fig. 18–41B).

Reproduction

All echinoids are dioecious. A regular echinoid has five gonads suspended along the interambulacra on the inner side of the test (Fig. 18–35), but in most irregular echinoids, the gonad of the posterior interambulacrum has disappeared. A short gonoduct extends aborally from each gonad and opens through a gonopore located on one of the five genital plates (Figs. 18–28A; 18–35; and 18–40A). Some burrowing sand dollars have long genital papillae, permitting release of eggs or sperm above the sand's surface.

Sperm and eggs are shed into the seawater, where fertilization takes place. Brooding is displayed by some cold-water sea urchins and heart urchins, and there is a brooding species of sand dollar. Brooding sea urchins retain their eggs on the peristome or around the periproct and use the spines to hold the eggs in position. The irregular forms brood their eggs in deep concavities on the petaloids.

Development

Cleavage is equal, up to the eight-cell stage, after which the blastomeres at the vegetal pole proliferate a number of small micromeres. A typical blastula ensues and becomes ciliated and free-swimming within 12 h after fertilization.

Gastrulation is typical but is preceded by an interior proliferation of cells by the micromeres, which form the mesenchyme. The coelom is formed by the separation of the free end of the archenteron. This separated portion then divides into right and left pouches, or lateral divisions may appear before the end separates from the main portion of the archenteron (Fig. 18–42A). The gastrula becomes somewhat cone-shaped and gradually develops into a planktonic larva, the **echinopluteus,** which bears six pairs of long larval arms and is very similar to the ophiopluteus of ophiuroids (Fig. 18–42B). The echinopluteus swims and feeds as long as several months. During later larval life the adult skeleton begins to form, first the five genital plates, then the ocular. The echinopluteus gradually sinks to the bottom; however, unlike asteroids, an attachment is not made, and metamor-

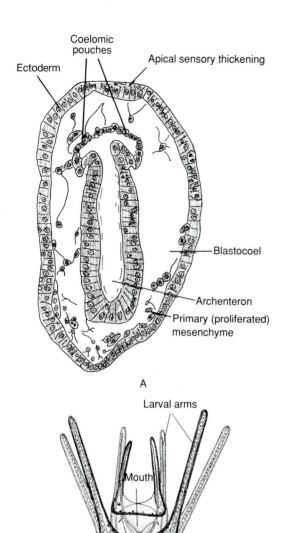

A

B

FIGURE 18–42 **A,** Gastrula of the sea urchin, *Echinus esculentus.* **B,** Echinopluteus larva of the sand dollar, *Fibularia craniola. (A, After McBride from Hyman. B, After Mortensen.)*

phosis is extremely rapid, taking place in about an hour. Young urchins are no larger than 1 mm.

The larvae of *Dendraster excentricus,* the common sand dollar along the Pacific coast of the United States, has been found to settle and metamorphose in response to a substance released by adults located in the same site. Such preferential settlement would explain why this species, like many other sand dollars, occurs in sand beds with high population densities. The life span of *Dendraster* is about eight years, and sand dollar beds persist for decades.

Growth rates are known for only a few echinoids. Two sand dollars from the Gulf of California, *Encope grandis* and *Mellita grantii,* which reach diameters of 74 mm and 38 mm, respectively, require five years to attain 95% of their maximum size. The annual mortality rate is 18% for *Encope* and 58% for *Mellita. Strongylocentrotus purpuratus* off the California coast, one of the best known sea urchins, reaches sexual maturity during its second year when only 25 mm in diameter but may live 30 years or more.

SUMMARY

1. In the class Echinoidea the spherical or flattened body is not drawn out into arms. The surface is covered with movable spines, which articulate on a test of sutured ossicles. Ambulacral areas containing the podia alternate with interambulacral areas arranged in meridians around the body. The plates of the test are perforated for the exit of gametes and for the canals connecting podia and ampullae. One genital plate functions as the madreporite. Correlated with the presence of a rigid skeletal test, the body wall lacks an internal muscle layer. Stalked, tridentate pedicellariae provide protection against settling organisms.

2. The regular echinoids, or sea urchins, are in general, adapted for living on firm substrata. The radial globose body with long spines is believed to be primitive for the class. Sea urchins move by using the podia and pushing with the spines.

3. Most sea urchins feed by scraping algae, encrusting organisms, and detritus from hard surfaces. The scraping apparatus is a complex organ composed of numerous ossicles, five of which function as teeth.

4. Five pairs of oral evaginations (peristomial gills) function in gas exchange.

5. The irregular echinoids are adapted for burrowing in soft bottoms. The body is covered with a great number of minute spines. The spines serve not only for locomotion and burrowing but also to keep sediment off the body surface. The greatly flattened form of sand dollars is probably an adaptation for shallow burrowing.

6. Because of the burrowing habit, the same ambulacrum is always directed anteriorly, and varying degrees of secondary bilaterality have developed. In all irregular echinoids the anus has moved out of the aboral center to the posterior margin or posterior lunule. In sea biscuits and sand dollars the mouth remains in the center of the oral surface; in heart urchins the entire oral center has shifted forward.

7.   Irregular echinoids are largely deposit feeders. Podia are used in food collection (heart urchins) or food transport (sand dollars).

8.   In irregular echinoids modified aboral podia (petaloids) function in gas exchange.

9.   The larva of echinoids is an echinopluteus. Metamorphosis occurs toward the end of planktonic life and at settling, but there is no attached stage.

CLASS HOLOTHUROIDEA

The holothuroids are a class of some 900 echinoderms known as sea cucumbers. Like echinoids, the body of the holothuroid is not drawn out into arms, and the mouth and anus are located at opposite poles. Also, there are ambulacral and interambulacral areas arranged meridionally around the polar axis. Holothuroids, however, are distinguished from other echinoderms in having the polar axis greatly lengthened, which results in the elongated cucumber shape (Fig. 18–43). This shape forces the animal to lie with the side of the body, rather than the oral pole, against the substratum. The class is further distinguished from other echinoderms by the reduction of the skeleton to **microscopic ossicles** and by the modification of the buccal podia into a circle of **tentacles** around the mouth.

External Anatomy

Most sea cucumbers are black, brown, or olive green, but other colors and patterns are encountered. There is considerable range in size. The smallest species are less than 3 cm in length (oral to aboral end), whereas *Stichopus* from the Philippines may attain a length of 1 m and a diameter of 24 cm. Most of the common North American and European species, such as *Cucumaria, Holothuria, Thyone (Sclerodactyla),* and *Leptosynapta,* range from 10 to 30 cm in length.

The body shape varies from almost spherical to long and wormlike, as in the synaptid sea cucumbers (Fig. 18–44). Not infrequently, the mouth and anus are turned dorsally at the ends of the long axis of the body, and some sea cucumbers are even U-shaped.

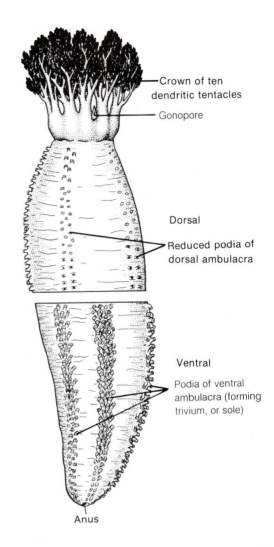

Crown of ten dendritic tentacles

Gonopore

Dorsal

Reduced podia of dorsal ambulacra

Ventral

Podia of ventral ambulacra (forming trivium, or sole)

Anus

FIGURE 18–43 The North Atlantic sea cucumber, *Cucumaria frondosa.*

Although there are a few genera, such as *Psolus* and *Ceto,* that have a protective armor of calcareous plates (modified surface ossicles), the body surface of the majority of sea cucumbers is leathery.

Holothuroids lie with one side of the body against the substratum, and this ventral surface is composed of three ambulacral areas (the **trivium**), commonly called the **sole** (Fig. 18–43). The dorsal surface consists of two ambulacral areas. As might be expected, the dorsal podia are usually reduced to warts or tubercles or are absent altogether, thus producing a secondary bilateral symmetry. Note that the bilateral symmetry of holothuroids has evolved in an

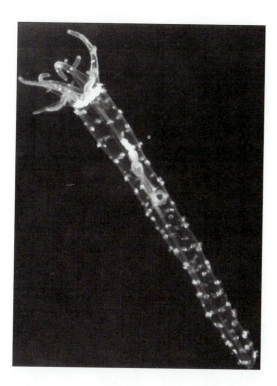

FIGURE 18–44 *Synaptula hydriformis,* a member of the order Apodida, a wormlike holothuroid that lacks podia but clings to algae using anchor-shaped ossicles seen as white patches in the photograph.

entirely different manner from that of the irregular echinoids.

The podia of some sea cucumbers, whether reduced or not, have lost their radial distribution and become more or less randomly scattered over the body surface. The primitive radial configuration is seen in *Cucumaria*, where the podia are more or less restricted to the five ambulacral areas; in *Thyone (Sclerodactyla)*, on the other hand, podia are scattered over the entire body surface. Members of orders Apodida (*Synapta, Leptosynapta, Synaptula,* and *Euapta*), which are elongated and wormlike, and Molpadiida completely lack podia (Fig. 18–44). Holothuroid podia may terminate in suckers, which are used for adhesion, or they may be suckerless; sometimes both types occur on one individual. Slender suckerless podia are called papillae.

The mouth is always surrounded by 10 to 30 tentacles, which represent modified buccal podia and are thus part of the water-vascular system. The tentacles are highly retractile, and the animal can completely retract both mouth and tentacles by pulling the adjacent body wall over them. The form and branching of the tentacles vary considerably.

Body Wall

The epidermis is nonciliated and covered externally by a thin cuticle. The thick dermal layer contains microscopic ossicles (called sclerites), which display a great variety of shapes (Fig. 18–45). These different shapes are important in the taxonomy of holothuroids. Beneath the dermis is a layer of circular muscle that overlies five single or double bands of longitudinal fibers located in the ambulacral areas. Recent studies have disclosed that the mechanical properties of the dermis can change under chemical stimulation. Thus, the body wall can be sufficiently flexible to permit the sea cucumber to squeeze through restricted passages or become so rigid that the animal cannot be dislodged (p. 921).

The body wall of sea cucumbers is a culinary delicacy in the Orient. Large species of sea cucumbers are collected and boiled, which causes the bodies to contract and thicken and also brings about evisceration of the internal organs. The body wall is then dried and sold, mostly to Chinese, as *trepang* or *bêche-de-mer*. *Trepang* imparts a distinctive texture and flavor to food.

Locomotion

Sea cucumbers are relatively sluggish animals and live on the bottom surface or burrow in sand and

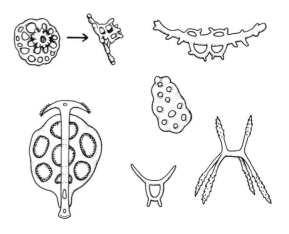

FIGURE 18–45 Microscopic ossicles of sea cucumbers. *(After Bell.)*

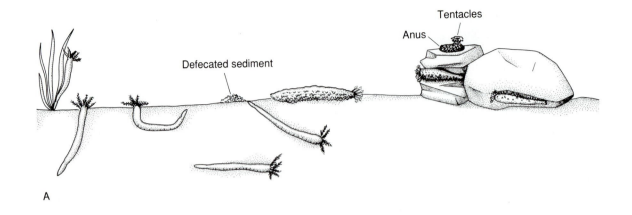

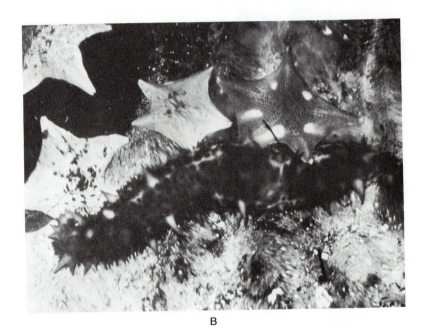

FIGURE 18–46 **A,** Modes of existence of shallow-water sea cucumbers. Animals are not drawn to scale. The genera named contain common species that illustrate a particular life pattern. Inhabitants of rock crevices *(Cucumaria, Holothuria)*; inhabitants of the undersurfaces of large stones *(Holothuria)*; rock and sediment surface dwellers *(Stichopus, Holothuria, some synaptids)*; inhabitants of marine plant surfaces (synaptids); burrowers with two openings to sediment surface *(Thyone, Cucumaria, Echinocucumis)*; burrowers with tentacles projecting to surface *(Synapta)*; burrowers with anus projecting to surface *(Leptopentacta, Caudina)*. **B,** The north Pacific *Stichopus californicus* in a tank with bat stars, *Patiria miniata*, both from the west coast of the United States. Species of the widely distributed genus *Stichopus* are large sea cucumbers (over 36 cm) with tough body walls. *(B, Photograph courtesy of Betty M. Barnes.)*

mud. Species with podia may creep along on the sole, with the podia functioning as in asteroids. Righting is accomplished by twisting the oral end around until the podia touch the substratum. Many hard-bottom species live beneath stones, in rock and coral crevices, and among large algal holdfasts. Some, such as species of *Holothuria, Cucumaria,* and *Psolus,* are so sedentary that the podia are used more for attachment than for locomotion. Others, such as species of the large, tough *Stichopus,* crawl exposed on the surface (Fig. 18–46A). There are also a few species that live on algae.

Burrowing species include the Apodida and the order Molpadiida, which lack podia, as well as some holothuroids that have podia, such as the common *Thyone (Sclerodactyla)* (Fig. 18–47). Burrowing is accomplished by alternate contraction of longitudinal and circular muscle layers of the body wall in the manner employed by earthworms. The tentacles aid by pushing away the sand. Burrowing sea cucumbers are oriented in various ways to the substratum. Some construct U-shaped burrows; others are oriented with either the anus or oral end at the surface (Fig. 18–46). The tentacles extend from one opening of the burrow, and a pulsating, anal ventilating current is maintained through the opposite opening if the burrow is U-shaped. Sedentary burrowers move very little once they have attained the proper position.

The Elasipodida are a curious group of entirely deep-sea holothuroids, some of which are benthic and some pelagic. The podia may be greatly enlarged and used for walking, because most deep-sea holothuroids live on the surface of the sea floor. The sole is quite flattened, and the tentacular crown is turned ventrally,

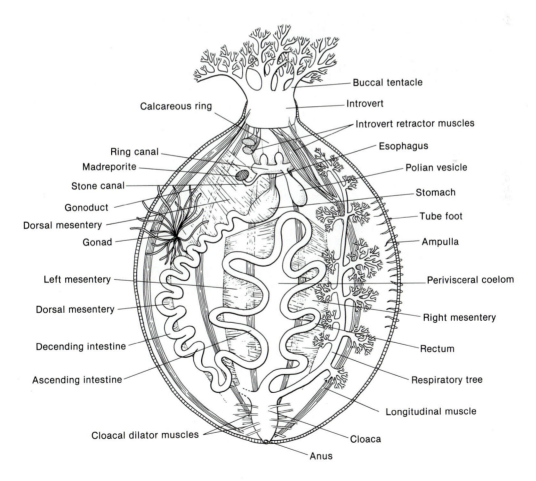

FIGURE 18–47 Internal structure of *Thyone (Sclerodactyla) briaereus,* a common sea cucumber that inhabits North Atlantic coastal waters. *(After Coe from Hyman.)*

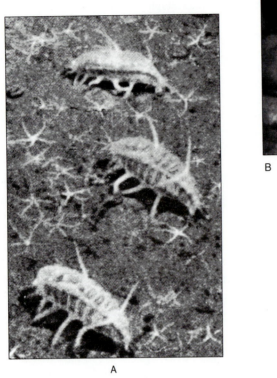

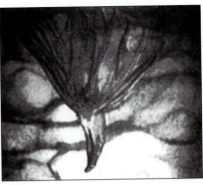

B

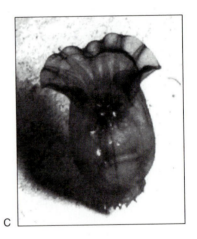

C

FIGURE 18–48 Holothuroids of the order Elasipodida. **A,** Three specimens of *Scotoplanes* crawling over the bottom in the San Diego trough (1060 m deep). The leglike structures are podia. **B,** *Pelagothuria natatrix,* a pelagic species having a circlet of long, webbed papillae behind the mouth and buccal podia. This species alternately hangs suspended in the water and swims using a rearward thrust of the web. **C,** Benthic species, *Enypniastes eximia,* leaving the bottom on a brief swimming excursion. This species and *Pelagothuria* are medusa-like in behavior and in the gelatinous nature of their bodies. *(A, Photograph courtesy of R. F. Dill through the courtesy of Barham, E. G. In Hansen, B. 1972. Deep Sea Res. 19:461–462. B, C, From Miller, J. E., and Pawson, D. L. 1990. Swimming sea cucumbers (Echinodermata: Holothuroidea): A survey, with analysis of swimming behavior in four bathyal species. Smithson. Contr. Mar. Sci. 35:1–18.)*

giving them a markedly bilateral symmetry (Fig. 18–48). Pelagic species have papillae webbed together in various ways to form fins, sails, or medusa-like bells. The transparent bathypelagic *Peniagone diaphana* lives up to 70 m off the bottom, holding its body in a vertical position, tentacles upward. Approximately half of the species of the Elasipodida are able to swim (Fig. 18–48B, C).

Nutrition

Sea cucumbers are chiefly deposit or suspension feeders. They stretch out their branched tentacles and either sweep them over the bottom or hold them out in the seawater. In either case, particulate material is trapped on adhesive papillae on the tentacular surfaces; the structure of the surface is related to the particle sizes normally selected for food. One at a time the tentacles are then stuffed into the pharynx, and the adhering food particles are wiped off as the tentacles are pulled out of the mouth.

Many sedentary species that live on hard surfaces beneath stones, such as members of the genus *Cucumaria,* are suspension feeders. More mobile epibenthic forms, such as the large *Stichopus* and the deep-sea elasipods, are deposit feeders, grazing on the bottom with their tentacles. They may be selective or not. Nonselective forms may literally shovel sand into the mouth with their tentacles, and the sand or mud castings of such species are conspicuous. *Parastichopus parvimensis,* a shallow-water, epibenthic deposit feeder from the west coast of North America, has been found to be about 22% efficient in its utilization of organic deposits, of which the plant components are largely undigested.

The epibenthic and infaunal Apodida and Molpadiida are also deposit feeders. The column of ingested sand in some of these wormlike species is easily visible through their transparent body wall. The European *Leptosynapta tenuis,* which lives in U-shaped burrows, ingests sand from the bottom of a funnel-like opening to the surface. Species of *Molpadia* live upside down in the sediment and produce a pile of castings where the anus comes to the surface.

The mouth is located in the middle of a buccal membrane at the base of the tentacular crown. The mouth opens into a pharynx, usually muscular, which is surrounded anteriorly by a **calcareous ring** of ossicles (Fig. 18–47). The calcareous ring provides support for the pharynx and the ring canal, and it serves as the site for the anterior insertion of the longitudinal muscles of the body wall and the retractor muscles of the tentacles and mouth region. The tentacles and mouth can be pulled completely within the anterior end of the body when the animal is disturbed. Protraction is brought about through elevation of the coelomic fluid pressure.

The pharynx opens into an esophagus and then into a stomach (Fig. 18–47). The stomach is absent in many holothuroids, but when present, it functions as a gizzard. A long, looped intestine composes most of the digestive tract and is the site of digestion and absorption. The digestive process is still poorly understood. In many holothuroids, the intestine terminates in a cloaca prior to opening to the outside through the anus.

Gas Exchange, Excretion, and Internal Transport

The coelom is large, and the peritoneal cilia produce a current of coelomic fluid that contributes to the general circulation of materials within the body. Of the several types of coelomocytes, one flattened discoidal type, called a hemocyte, contains hemoglobin. When present in large numbers, hemocytes give a red color

to the coelomic fluid, especially the water-vascular system, [*Thyone (Sclerodactyla), Cucumaria*], and the hemal system (some molpadiids).

Gas exchange in most holothuroids is accomplished by a remarkable system of tubules called **respiratory trees.** The respiratory trees are located in the coelom on the right and left sides of the digestive tract (Fig. 18–47). Each tree consists of a main trunk with many branches, each of which ends in a tiny vesicle. The trunks of the two trees emerge from the upper end of the cloaca either separately or by way of a common trunk.

Water circulates through the tubules by means of the pumping action of the cloaca and the respiratory trees. The cloaca dilates, filling with seawater. The anal sphincter then closes, the cloaca contracts, and water is forced into the respiratory trees. Water leaves the system because of the contraction of the tubules and the reverse action of the cloaca. Pumping is slow; *Holothuria* requires six to ten cloacal dilations and contractions to fill the trees, each contraction taking a minute or more. All the water is expelled in one action.

The slender tropical pearlfish, which is about 15 cm long, makes its home in the trunk of a respiratory tree of certain sea cucumbers. The fish leaves the host at night to search for food; after such excursions the fish forces its way into the anus and cloaca and back to the shelter of the respiratory tree.

Pelagic and benthic deep water Elasipodida and burrowing Apodida lack respiratory trees. They obtain oxygen through the general body surface. The podia of elasipods are especially important as sites of gas exchange.

Most ammonia probably exits by diffusion through the respiratory trees. Particulate waste, as well as nitrogenous material in crystalline form, is carried by coelomocytes from various parts of the body to the gonadal tubules, the respiratory trees, and the intestine. Waste then leaves the body through these organs.

Holothuroids, especially the Aspidochirotida (*Holothuria* and *Stichopus*), also possess the most highly developed hemal system of any of the echinoderm classes (Fig. 18–49). The general organization of the system is similar to that of other echinoderms but a systematic heart is absent. A hemal ring and radial hemal sinuses parallel the water ring and the radial canals of the water-vascular system. The most conspicuous features of the system, at least in larger species, are a dorsal and a ventral vessel that accom-

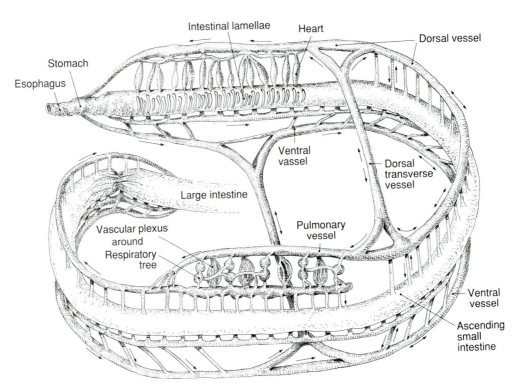

FIGURE 18–49 Hemal system of *Isostichopus badionotus*. Arrows indicate direction of blood flow. *(After Herreid, C. F., LaRussa, V. F., and DeFesi, C. R. 1976. Blood vascular system of the sea cucumber Stichopus moebii. J. Morph. 150(2):423–451.)*

pany the intestine. These main intestinal sinuses supply the intestinal wall with a large number of smaller channels. In the region of the descending small intestine, at least in *Isostichopus badionotus,* 120 to 150 single-chambered hearts pump blood from the dorsal vessel through a system of intestinal lamellae that project into the intestinal lumen. The left respiratory tree is intimately associated with the hemal system of the ascending small intestine.

Most of the hemal vessels are well developed, containing a muscle and a connective tissue layer, and covered on the outside with ciliated peritoneum and on the inside with endothelium. The blood is essentially like the coelomic fluid; in fact, coelomocytes are formed in the walls of certain vessels. Peristaltic contractions of the dorsal vessel are of primary importance in propelling blood, but although circulation is largely unidirectional, it is not rapid, and there is some ebb and flow. The holothuroid hemal system is clearly involved in some gas transport and appears to play some role in absorption or nutrient transport.

Water-Vascular System

Although the water-vascular system of holothuroids is basically like that of other echinoderms, the madreporite in most species is peculiar in having lost its connection with the body surface and in being unattached in the coelom (Fig. 18–47). Perivisceral coelomic fluid, rather than seawater, enters the system. The madreporite hangs just beneath the base of the pharynx and is connected to the water ring by a short stone canal.

Although a surface madreporite is generally absent, the wall of the cloaca is perforated by short ciliated ducts in many if not most holothuroids, and these connect the coelom with the exterior. Coelomic fluid is lost through these ducts when the body contracts strongly, and seawater may enter via ciliary action when the animal relaxes.

The ring canal encircles the base of the pharynx and gives rise to polian vesicles, which hang into the coelom (Fig. 18–47). The vesicles may function as expansion chambers in maintaining pressure within the water-vascular system or as pumps to aid in circulation of fluid in the water-vascular system. The five radial canals give off canals to the tentacles prior to reaching the podia. Ampullae are present for both podia and tentacles, although when the podia are reduced, there is a corresponding reduction in the ampullae. In the Apodida, which lack tube feet, the water-vascular system is limited to the oral ring canal, the polian vesicles, and the buccal podia (tentacles). Tiedemann's bodies are absent but some sea cucumbers have specialized blind funnels attached to the wall of the perivisceral coelom that remove unwanted particles from the coelomic fluid.

Nervous System

The circumoral nerve ring lies in the buccal membrane near the base of the tentacles (Fig. 18–50). The ring supplies nerves to the tentacles and also to the pharynx. The five radial nerves, on leaving the ring, pass through the notch in the radial plates of the calcareous ring and run the length of the ambulacra in the coelomic side of the dermis.

The burrowing Apodida, which tend to keep the oral end directed downward, possess one statocyst adjacent to each radial nerve, located near the point at which the nerve leaves the calcareous ring.

Evisceration and Regeneration

The expulsion of sticky tubules from the anal region is commonly associated with sea cucumbers, but this defensive phenomenon is actually limited to some species of the genera *Holothuria* and *Actinopyga*. Such sea cucumbers possess from a few to a large mass of white, pink, or red blind tubules (**tubules of Cuvier**) attached to the base of one (frequently the left) or both respiratory trees or to the common trunk of the two trees (Fig. 18–51). When these sea cucumbers are irritated or attacked by some predator, the anus is directed toward the intruder, the body wall contracts, and by rupture of the cloaca the tubules are shot out of the anus.

The Cuvierian tubules of some species are not adhesive but liberate a toxic substance, **holothurin** (a saponin). Holothurin is also found in the body wall of some species. South Pacific islanders have long used the macerated bodies of certain sea cucumbers to catch tide pool fish.

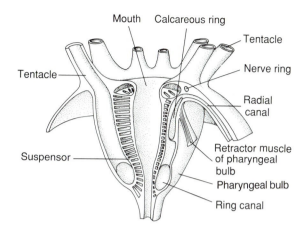

FIGURE 18–50 Section through anterior end of *Ocnus planci*. (After Herouard from Feral, J.-P., and Massin, C. 1982. Digestive systems: Holothuroids. In Jangoux, M., and Lawrence, J. M. (Eds.): 1982. Echinoderm Nutrition. A. A. Balkema, Rotterdam. p. 199.)

During the process of expulsion each tubule is greatly elongated by water forced into its lumen, and the tubules break free from their attachment to the respiratory tree. In *Holothuria* the detached tubules are sticky and entangle the intruder in a mesh of adhesive threads. Small crabs and lobsters may be rendered completely helpless and left to die slowly, while the sea cucumber crawls away. After discharge the tubules of Cuvier are regenerated.

Sometimes confused with the discharge of the tubules of Cuvier is a more common phenomenon (called **evisceration**) that occurs in many holothuroids. Depending on the species, the anterior or posterior end ruptures and parts of the gut and associated organs are expelled.

Eviscerated specimens or individuals in the process of regeneration have been reported from natural habitats during certain times of the year, and it is a normal seasonal phenomenon in some species, perhaps initiating a period of inactivity when food supply is low or eliminating wastes stored in internal tissues. Evisceration is later followed by regeneration of the lost parts.

Reproduction

Holothuroids differ from all other living echinoderms in possessing a single gonad. Most cucumbers are dioecious, and the gonad is located anteriorly in the

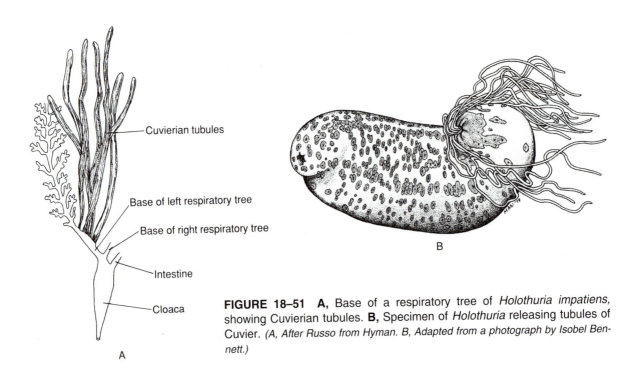

FIGURE 18–51 **A,** Base of a respiratory tree of *Holothuria impatiens,* showing Cuvierian tubules. **B,** Specimen of *Holothuria* releasing tubules of Cuvier. *(A, After Russo from Hyman. B, Adapted from a photograph by Isobel Bennett.)*

coelom beneath the middorsal interambulacrum (Fig. 18–47). The gonopore is located middorsally between the bases of two tentacles or just behind the tentacular collar (Fig. 18–43).

Some 30 brooding species are known, over half of which are cold-water forms, largely Antarctic. During spawning the eggs are caught by the tentacles and transferred to the sole or to the dorsal body surface for incubation. Even more remarkable is coelomic incubation, which takes place in the Californian *Thyone rubra,* in *Leptosynapta* from the North Sea, in *Synaptula hydriformis* from Florida and the Caribbean, and in a few species from other parts of the world. The eggs pass from the gonads into the coelom and are fertilized in an undiscovered manner. Development takes place within the coelom, and the young leave the body of the mother through rupture in the anal region.

Development

Except in brooding species, development takes place externally in the seawater, and the embryo is planktonic. Development through gastrulation is like that of asteroids. The anterior half of the archenteron separates to develop as the coelom, leaving a shorter, posterior portion to become the gut. The right axohydrocoel never forms.

By the third day of development a larval stage called an **auricularia** has been reached (Fig. 18–52A). The auricularia is very similar to the bipinnaria of the asteroids and possesses a ciliated locomotor band that conforms to the same development as the locomotor band of the bipinnaria. Further development leads to a barrel-shaped larva, called a **doliolaria,** in which the original ciliated band has become broken up into three to five ciliated girdles (Fig. 18–52B).

Gradual metamorphosis during the latter part of planktonic existence results in a young sea cucumber. The tentacles, which are equivalent to buccal podia, appear prior to the appearance of the functional podia. At this stage the metamorphosing animal is sometimes called a **pentactula.** Eventually, the young sea cucumber settles to the bottom and assumes the adult mode of existence. The life span of many sea cucumbers is between five and ten years. There are many species of holothuroids (Dendrochirotida) that possess a nonfeeding, barrel-shaped **vitellaria** (Fig. 18–3G). This type of larva, which is found in crinoids and a few ophiuroids, possesses cilated bands but no arms and is probably a specialized condition.

SUMMARY

1. Members of the class Holothuroidea are distinguished by cylindrical bodies, in which the oral/aboral axis is greatly elongated, by a skeleton of microscopic ossicles, and by tentacular buccal podia.

2. As a consequence of the elongated oral/aboral axis, holothuroids lie on their side. Because most species lie on the same three ambulacra (sole), this posture has led to some secondary bilateral symmetry. The ventral ambulacra in bilateral forms have well-developed podia; the dorsal ambulacra have reduced podia.

3. Some holothuroids are bottom surface dwellers, some live beneath stones or lodge in crevices, some burrow, and a few (mostly deep-sea forms) are pelagic. The podia are used for crawling and gripping the substratum. Two groups of wormlike burrowers have lost the locomotor podia and move by peristaltic contractions.

4. Holothuroids are deposit feeders and suspension feeders. The mucus-covered tentacular surface traps particles when swept across the bottom or held in the water. Collected material is removed by the sucking action of the pharynx when the tentacles are stuffed in the mouth.

5. The water-vascular system is peculiar only in having the madreporite in the perivisceral coelom and coelomic pores through the wall of the cloaca. Branched internal evaginations of the posterior gut wall (respiratory trees) are gas exchange organs. The coelomic fluid contributes to internal transport, but many holothuroids have in addition a well-developed hemal system.

6. Gametes from the single gonad exit through an intertentacular gonopore. Development leads to an auricularia larva. Metamorphosis occurs prior to settling.

CLASS CRINOIDEA

The crinoids are the most ancient and in some respects the most primitive of the living classes of echinoderms. Attached, stalked crinoids, called sea lilies, flourished during the Paleozoic era, and some 80 species still exist today. Modern sea lilies, however, live at depths of 100 m or more and are therefore not commonly encountered. The majority of living crinoids belong to a more modern branch of the class, the order Comatulida. The comatulids, or feather stars, are free-living crinoids that live from the

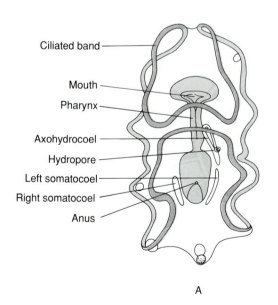

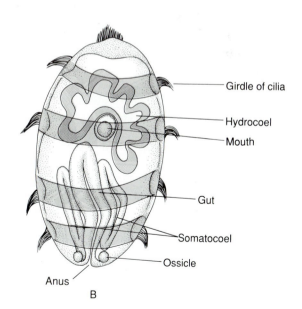

A

B

FIGURE 18–52   A, An auricularia larva (oral view). **B,** The doliolaria larva of *Leptosynapta inhaerens;* a common North Atlantic holothuroid (oral view). *(A, After Mortensen from Hyman. B, After Runnström from Cuénot.)*

intertidal zone to great depths, and some occur in large numbers on coral reefs. There are approximately 550 species found primarily in Indo-Pacific and polar waters.

External Structure

The body of existing crinoids is composed of a basal attachment **stalk** and a pentamerous body proper, called the **crown** (Fig. 18–53A). A well-developed stalk is present in sea lilies but is lost during the post-larval development of the free-moving feather stars

(Fig. 18–53B). In the sessile sea lilies the stalk may reach almost 1 m in length but is usually much shorter. There are fossil species, however, with 20-m stalks. The basal end bears a flattened disc or rootlike extensions by which the animal is fixed to hard or soft substrata.

The internal skeletal ossicles give the stalk a characteristic jointed appearance. The stalk of many crinoids bears small, slender, jointed appendages **(cirri)** that are displayed in whorls around the stalk (Fig. 18–54A). Although the stalk is lost in comatulids, the most proximal cirri of the stalk remain and

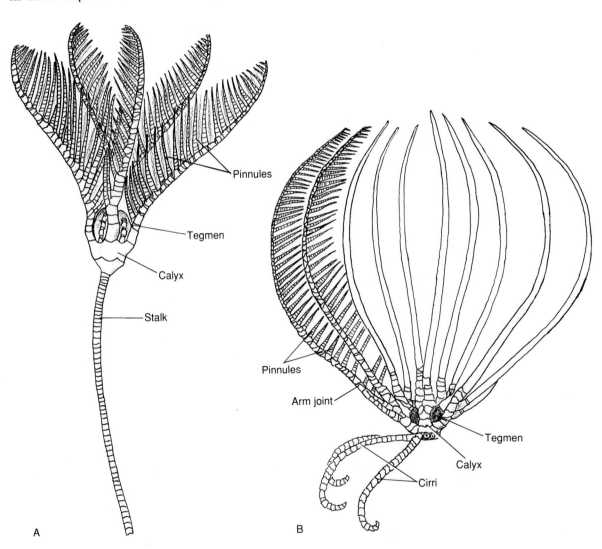

FIGURE 18–53 **A,** *Ptilocrinus pinnatus,* a stalked crinoid (or sea lily) with five arms. **B,** A Philippine 30-armed comatulid (or feather star), *Neometra acanthaster.* Not all arms shown. *(Both after Clark from Hyman.)*

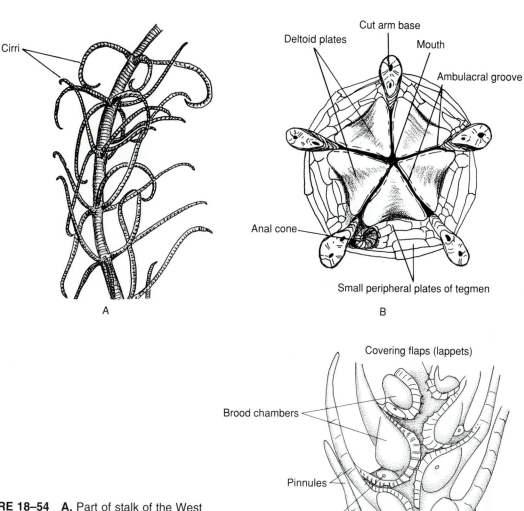

FIGURE 18–54   A, Part of stalk of the West Indian crinoid, *Cenocrinus asteria,* showing whorls of cirri. **B,** Tegmen of *Hyocrinus,* a stalked crinoid (oral view). **C,** An arm section from *Notocrinus virile,* a comatulid (oral view). Podia not shown. *(A and B, After Carpenter from Hyman. C, Modified after Hyman, L. H. 1955. The Invertebrates. Vol. IV. McGraw Hill Book Co., New York.)*

spring as one or more circles from around the base of the crown (Fig. 18–53B). The cirri of comatulids are used for grasping the substratum when the animal comes to rest. They are long and slender in forms that rest on soft bottoms and stout and curved in species that grasp rocks, seaweed, and other objects. The pentamerous body (the crown) is equivalent to the body of other echinoderms and, like those of the asteroids and the ophiuroids, is drawn out into arms. The crown is attached to the stalk by its aboral side; thus, in con-

trast to other living echinoderms, the oral surface is directed upward. The skeletal ossicles are best developed in the aboral body wall, usually called the **calyx,** which is thus somewhat cuplike. The oral wall **(tegmen)** forms a more or less membranous covering for the calyx cup (Figs. 18–53A; 18–54B). The mouth is located in or near the center of the oral surface; five ambulacral grooves extend peripherally from the mouth to the arms. The anus opens onto the oral surface and is usually located in one of the interambulacral areas at the top of a prominence called the **anal cone** (Fig. 18–54B).

The arms issue from the periphery of the crown and have a jointed appearance like the stalk. Although there are some primitive species that possess five arms (Fig. 18–53A), in most crinoids each arm forks immediately on leaving the crown, forming a total of ten arms. Further branching results in additional arms, and some comatulids possess 80 to 200 arms. The arms are usually less than 10 cm in length but may reach almost 35 cm in some species.

On each side of the arm is a row of jointed appendages called **pinnules,** from which the name *feather star* is derived (Figs. 18–53; 18–54C). The ambulacral grooves on the oral surface extend along the length of both the arms and the pinnules. The margins of the grooves are bordered by movable flaps, called **lappets,** which can expose or cover the groove. On the inner side of each lappet are three podia united at the base (Figs. 18–54C; 18–55B).

Both podia and lappets also extend onto the pinnules.

Cold-water crinoids and those of the eastern Pacific are usually brown, but littoral species from tropical waters, especially comatulids, display a variety of colors in brilliant solid and variegated patterns.

Body Wall

The epidermis is nonciliated except in the ambulacral grooves, and most of the dermis is occupied by the skeletal ossicles. The stalk, cirri, arm, and pinnules are of a solid construction, being composed almost entirely of a series of thick, disc-shaped ossicles; this accounts for the jointed appearance of these appendages (Figs. 18–53; 18–56). The surfaces of the arm ossicles articulate to permit at least some movement, similar to the ossicles composing the ophiuroid arms. The ossicles of the stalk are more securely interlocked than those of the arms, but even here some bending is possible. The ossicles of the stalk, cirri,

A

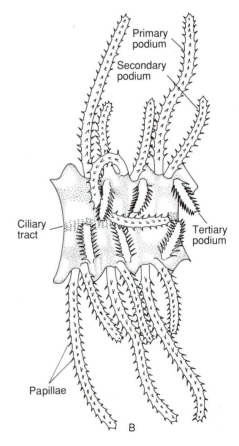

B

FIGURE 18–55 Function of crinoid podia in feeding. **A,** Extended position of the primary podia along the pinnules of the feather star, *Florometra serratissima.* **B,** View of the ambulacral groove of a section of pinnule of *Antedon bifida.* One podium is wiping against ciliary current of ambulacral groove and against tertiary podium. *(A, From Byrne, M., and Fontaine, A. R. 1980. The feeding behavior of Florometra serratissima. Can. J. Zool. 59:11–18. B, From Lahaye, M. C., and Jangoux, M. 1985. Functional morphology of the podia and ambulacral grooves of the comatulid crinoid Antedon bifida. Mar. Biol. 86:307–318.)*

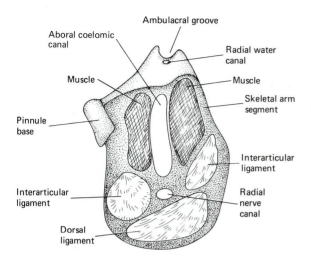

FIGURE 18–56 Diagrammatic cross section of the arm of the feather star, *Florometra serratissima*. The muscles and ligaments are attached to the face of the large arm ossicle, which fills most of the section. *(Based on a photograph by Meyer, D. L. 1971. The collagenous nature of problematical ligaments in crinoids. Mar. Biol. 9(3):238.)*

Labels: Aboral coelomic canal; Ambulacral groove; Radial water canal; Muscle; Muscle; Skeletal arm segment; Pinnule base; Interarticular ligament; Interarticular ligament; Radial nerve canal; Dorsal ligament

and arms are bound together by collagen fiber bands, called **ligaments,** which penetrate into the porous skeletal material. As in other echinoderms, these fibers can rapidly change from a soft to a rigid state, permitting the animal to roll up the arms by the flexor muscles or lock them in an extended position.

Locomotion

The sessile sea lilies are limited to bending movements of the stalk and flexion and extension of the arms. The stalkless comatulids, however, are free-moving and are capable of both swimming and crawling. The oral surface is always directed upward, and the cirri are strongly thigmotactic and appear to control the righting reflex.

A sea lily swims by raising and lowering one set of arms alternately with certain others. In the ten-arm species every other arm sweeps downward while the alternate set moves upward. In species with more than ten arms, the arms still move in sets of five, but sequentially. To crawl, the animal lifts its body from the substratum and moves about on its arms. The arms and pinnules, which have minute terminal hooks, are often used to grasp and pull the animal over irregular and vertical surfaces.

Feather stars swim and crawl only for short distances, and swimming is largely an escape response. They cling to the bottom for long periods by means of the grasping cirri. Many shallow-water species are nocturnal. Three Red Sea species of *Lamprometra, Capillaster,* and *Comissia* inhabiting coral reefs hide during the day in crevices and deep within branching corals, keeping their arms tightly rolled (Fig. 18–57A). Stimulated by the lowered light intensities at sunset, they crawl upward out of their hiding places to exposed positions, where the arms are extended to feed. Species inhabiting deeper water may be stationary. Many specimens of two species of *Decametra* and *Oligometra* were reported clinging to a gorgonian coral at about 30 m for several months in the same position. They roll their arms in response to daytime illumination.

Nutrition

Crinoids are suspension feeders. During feeding the arms and pinnules are held outstretched and the podia are erect. The podia are shaped like small tentacles and bear mucus-secreting papillae along their length (Fig. 18–55). The three podia forming the triplets along the pinnules have different functions in the feeding process. The primary podium is the long, conspicuous, outstretched member. On touching a food particle, the primary and secondary podia whip into the ambulacral groove with the particle adhered to its surface (Fig. 18–55B). Depending on the species, the animal removes the particle by wiping the podium against the ciliary current of the ambulacral groove or against the tertiary podium (Fig. 18–55B) or by scraping it between adjacent lappets. Transport of particles down the groove is by ciliary action. The lappets function largely as side walls for the ambulacral groove.

There is still uncertainty about the precise nature of crinoid food. The gut contents of Red Sea crinoids were largely zooplankton, but *Antedon bifida* appears to feed largely on resuspended detritus.

The feeding position of the arms varies with the environment. Many crinoids form a planar, vertical filtration fan that is oriented more or less at a right angle to the prevailing current or a circular fan that is tilted into the current. Sea lilies achieve the tilt by bending the stalk. The arm tips may be turned back toward the current so that the animal has the shape of an umbrella (Fig. 18–57C). In the fan position the ad-

FIGURE 18–57 Arm positions of crinoids. **A,** Diurnal, nonfeeding, rolled arm position of the feather star, *Heterometra.* **B,** The Red Sea feather star, *Commissia,* which inhabits crevices in coral reefs, projecting its arms to feed. **C,** Feeding crinoids in the Straits of Florida at 600 to 700 m. A feather star atop a sponge is holding its arms in the form of a vertical fan. Three inclined sea lilies *(Diplocrinus)* form parabolic fans with the arm tips directed toward the current. *(A and B, From Fishelson, L. 1974. Ecology of the northern Red Sea crinoids and their epi- and endozoic fauna. Mar. Biol. 26:183–192. C, Photograph courtesy of C. Neumann, from Macurda, D. B., and Meyer, D. L. 1976. The identification and interpretation of stalked crinoids from deep-water photographs. Bull. Mar. Sci. 26(2):205–215.)*

jacent pinnules and podia form a relatively tight mesh and an efficient filter. Shallow-water comatulids may employ a vertical fan held at a right angle to currents, or they may remain in more protective positions within crevices or coral beds and extend the arms in several directions (Fig. 18–57B).

There is a correlation between podial spacing and habitat. Those species that live on reefs in spaces between corals have longer and more widely spaced podia than those that live exposed and form more regular filtration fans. There is also some correlation between the number and length of the arms and the food supply of the habitat. Crinoids living at great depths or in cold water, where detritus or plankton is rich, usually have a small number of arms (ten or fewer); the reverse is true of littoral, warm-water species. The total length of the food-trapping ambulacral surface may be enormous. The Japanese stalked crinoid, *Metacrinus rotundus,* with 56 arms 24 cm in length, possesses a total ambulacral groove length of 80 m.

Crinoids are believed to display the primitive method of feeding used by the echinoderms and also to illustrate the original function of the water-vascular system—that is, the water-vascular system originally evolved as a means of capturing food and secondarily assumed a locomotor function in those groups that have become free-moving and inverted.

The mouth leads into the short esophagus, which then opens into an intestine (Fig. 18–58). The intestine descends and makes a complete turn around the inner side of the calyx wall. The terminal portion then passes upward into the short rectum, which opens through the anus at the tip of the anal cone.

The details of digestion are still unknown. Wastes are egested as large, compact, mucus-cemented balls that fall from the anal cone onto the surface of the disc and then drop off the body.

Internal Transport, Gas Exchange, and Excretion

Numerous, calcified mesenteries partition the coelom into a network of communicating spaces. In the oral side of the arms, the coelom extends as five parallel canals, and in the stalk, five coelomic canals pass through a central perforation in the ossicles and give rise to one canal into each cirrus. The hemal system consists of an oral hemal ring, which joins a network

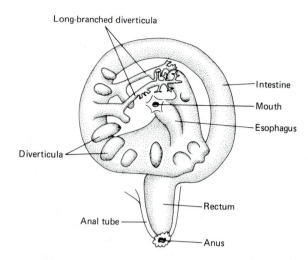

FIGURE 18–58 Digestive tract of the comatulid *Antedon. (After Chadwick from Hyman.)*

of vessels and sinuses in some of the coelomic mesenteries, and gives rise to one or two radial vessels in each arm. The podia are undoubtedly the principal sites of gas exchange, and the great surface area presented by the branching arms makes unnecessary any special respiratory surfaces, such as those found in most other echinoderms.

Wastes gathered by coelomocytes are believed to be deposited in little saccules located in rows along the sides of the ambulacral grooves. Supposedly, the saccules periodically discharge to the exterior.

Water-Vascular System

A single madreporite is absent in crinoids. Instead, numerous (often hundreds) of separate surface pores and pore canals perforate the tegmen and open into the coelom near the stone canals. The ring canal encircles the mouth and at each interradius gives off a large number of short stone canals, which open into the perivisceral coelom (the feather star, *Antedon,* possesses about 50 canals at each interradius). At each radius of the ring canal a radial canal extends into each arm just beneath the ambulacral groove and forks into all the branches and into the pinnules (Fig. 18–56). From the radial canals extend lateral canals supplying the podia. There are no ampullae, and one lateral canal supplies the cluster of three podia except in the buccal region. Hydraulic pressure for extension of the podia is generated by contraction of the radial water canal, which is provided with muscle fibers that span the canal.

Nervous System

The crinoid nervous system is composed of three interconnecting divisions. The chief motor system is an aboral (or entoneural) system located as a cup-shaped mass in the apex of the calyx. The aboral system provides nerves to the cirri and five brachial nerves to the arms and pinnules (Fig. 18–56). The oral (ectoneural) nerve system, which is sensory, is homologous to the principal system of other echinoderms and consists of a subepidermal radial nerve that runs just beneath the ambulacral groove in the arms and a nerve ring around the mouth. Just below the oral system is a deeper, hyponeural sensory system that has a central ring from which arises a pair of lateral brachial nerves to each arm. These nerves innervate the pinnules and podia.

Regeneration and Reproduction

Crinoids possess considerable powers of regeneration and, in this respect, are similar to the asteroids and the ophiuroids. Part or all of an arm can be cast off if seized or if subjected to unfavorable environmental conditions. The lost arm is then regenerated. The visceral mass within the calyx can be regenerated in several weeks; such regeneration may be important in surviving fish predation.

Crinoids are all dioecious, and there are no distinct gonads. The gametes develop from germinal epithelium within an expanded extension of the coelom (the **genital canal**) located within the pinnules, as in *Antedon,* or within the arms (Fig. 18–54C). Not all the pinnules are involved in the formation of sex cells, but only those along the proximal half of the arm length. When the eggs or sperm are mature, spawning takes place by rupture of the pinnule walls, and the eggs and sperm are shed into the seawater. In *Antedon* and others, the eggs are cemented to the outer surface of the pinnules by the secretion of epidermal gland cells. Hatching takes place at the larval stage. Brooding by cold-water crinoids (many Antarctic forms) is displayed, as in other echinoderms. The brood chambers are saclike invaginations of the arm or the pinnule walls adjacent to the genital canals, and

the eggs probably enter the brood chamber by rupture (Fig. 18–54C).

Development

Development through the early gastrula stage is essentially like that in asteroids and holothuroids. During the formation of the coelomic sacs the embryo elongates, and development proceeds toward a free-swimming larval stage. The crinoid larva, a nonfeeding vitellaria, is essentially like the vitellaria of holothuroids, somewhat barrel-shaped with an anterior apical tuft and a number of transverse, ciliated bands (Fig. 18–59A).

After a free-swimming existence the vitellaria settles to the bottom and attaches, employing a glandular midventral depression (the adhesive pit) located near the apical tuft. There ensues an extended metamorphosis resulting in the formation of a minute, stalked, sessile crinoid. In the comatulids, metamorphosis also results in a stalked sessile stage (the **pentacrinoid**) that resembles a minute sea lily (Fig. 18–59B). The pentacrinoid of *Antedon* is a little over 3 mm long when the arms appear, and it requires about six weeks from the time of attachment of the vitellaria to attain this stage. After up to several months as a pentacrinoid, during which time the cirri are formed, the crown breaks free from the stalk, and the young animal assumes the adult, free-swimming existence.

SUMMARY

1. Members of the class Crinoidea, which include the stalked and attached sea lilies and the stalkless and free feather stars, are the only living echinoderms in which the oral surface is directed upward. This condition is also true of most Paleozoic echinoderms.

2. The crown of both stalked and stalkless crinoids is composed of multiple arms around a heavy, central calyx, which is covered by a membranous oral wall, or tegmen. The tegmen contains the mouth in the center and the anus to one side.

3. The multiplicity of arms results from basal branching of an originally pentamerous arrangement. The arms bear numerous small, lateral branches (pinnules), and the oral surface of all branches, including the pinnules, contains a ciliated, ambulacral groove.

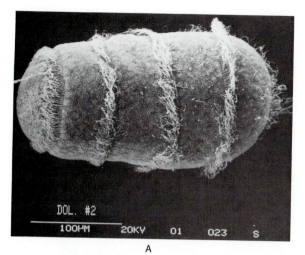

A   B

FIGURE 18–59   A, Scanning electron micrograph of the doliolaria larva of the feather star, *Florometra serratissima.* The cilated bands are numbered, and the arrow points to the apical tuft of cila. **B,** A cluster of pentacrinoids attached to a cirrus of an adult *Comactinia. (B, Photograph courtesy of E. J. Balser.)*

4. Heavy ossicles compose much of the relatively solid stalk, cirri, arms, and pinnules. Heavy ossicles are also located within the calyx wall.

5. The attached sea lilies can bend the stalk and unroll the arms when feeding. Feather stars perch with their cirri and crawl and swim with the arms.

6. Crinoids are suspension feeders, and the podia, on touching zooplankton or other suspended particles, undergo flicking action, driving the particles into the ambulacral groove. Ambulacral cilia carry the mucus-entrapped particles down the arms into the mouth. The arms are held as a funnel or, when in a current, as a planar or circular fan. The multiple arms and pinnules provide the necessary surface area for this mode of feeding.

7. Gametes are produced in the arms, which are also the site of brooding, when it occurs. Development leads to a barrel-shaped vitellaria larva. Metamorphosis occurs following settling and attachment. Feather stars pass through a stalked stage (pentacrinoid) before the crown becomes free.

Class Concentricycloidea

In 1983 and 1984, specimens of a minute disc-shaped echinoderm were collected attached to sunken wood in a thousand meters of water off the coast of New Zealand. They were quickly recognized as representatives of a new class of echinoderms, which was named Concentricycloidea and the species *Xyloplax medusiformis.* A second species, *Xyloplax turnerae,* was subsequently discovered in wood recovered from great depths in the Caribbean. The little animals are only 2 to 9 mm across and are covered aborally with platelike ossicles (Fig. 18–60A). Marginal spines are located around the periphery. The water-vascular system is peculiar in apparently having two ring canals, with the podia arising from the outer one. A gut is absent in *Xyloplax medusiformis* but one is present in *Xyloplax turnerae.* A pair of gonads is associated with each of five bursae. Despite the lack of arms, the passage of podia between ossicles and the growth pattern are more like those of stellaroids than those of any of the other echinoderm classes. The mode of feeding and other aspects of the biology of these strange little echinoderms are still unknown.

Fossil Echinoderms and Echinoderm Phylogeny

The echinoderms rank with molluscs, brachiopods, and arthropods in having one of the richest and oldest fossil records of any group in the Animal Kingdom. Echinoderms first appeared in the early Cambrian period and were extremely abundant during the later periods of the Paleozoic era, when a number of fossil classes reached the peak of their evolutionary development.

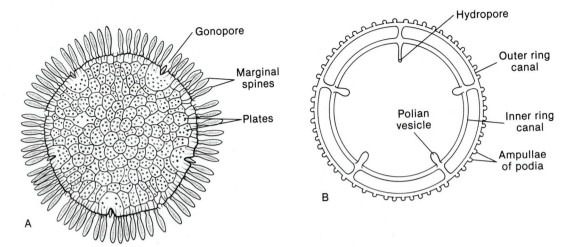

FIGURE 18–60 *Xyloplax medusiformis* of the new class Concentricycloidea. **A,** Dorsal view. **B,** Diagram of the water-vascular system. *(From Baker, A. N. et al. 1986. A new class of Echinodermata from New Zealand. Nature. 321:862–864.)*

The crinoids are the only living echinoderms that are attached, but an attached condition was characteristic of the majority of Paleozoic forms, including a number of extinct classes. These groups are believed to display many primitive features of the phylum.

Like living crinoids, most fossil forms were probably suspension feeders in which the podia and upward-directed ambulacral grooves were food-catching and food-conducting structures. In most attached fossil species, the grooves branched onto a slender, pinnule-like projection (the **brachiole**) (Fig. 18–63B). The term *brachiole* is used to distinguish these body extensions from the heavier arms of crinoids, asteroids, and ophiuroids. Brachioles varied in number from a few to hundreds and, like the arms and pinnules of crinoids, represented an adaptation that increased the surface area for food collection. Because these fossil forms were sessile, and because the oral surface was always directed upwards, the original function of the podia could not have been locomotive.

The skeletal system of the extinct attached echinoderms was somewhat like that of the crinoids, with the exception that the ossicles (plates) of the crown were not limited to the aboral surface (calyx) but extended orally to the mouth, thus enclosing the internal organs within a test (or **theca**). Protective cover plates, which could fold over the ambulacral grooves and mouth, were generally present. The anus was usually located eccentrically in one of the interradii, as in the crinoids.

The oldest group of extinct echinoderms is the Eocrinoidea, known from the early Cambrian to the Ordovician. Eocrinoids were stalked or stalkless echinoderms with an enclosed theca (Fig. 18–61C). At the upper, or oral, end were five ambulacra and five to many brachioles. The pentamerous and oral position of the ambulacra and brachioles gives these animals a superficial resemblance to crinoids (Fig. 18–62A), although they are actually more similar to cystoids, the following group.

The cystoids (classes Rhombifera and Diplorita) were attached species that ranged from the middle Ordovician period to the Permian. The theca of cystoids was more or less oval, with the oral end directed upward and the aboral end attached to the substratum directly or sometimes by a stem (Fig. 18–63A). A characteristic feature of cystoids was a system of pores that perforated the theca, probably part of a system for gas exchange. The three or, more commonly, five ambulacra were radially arranged and extended outward and downward to varying degrees over the sides of the theca, in some species to the aboral pole. Small brachioles were either located around the

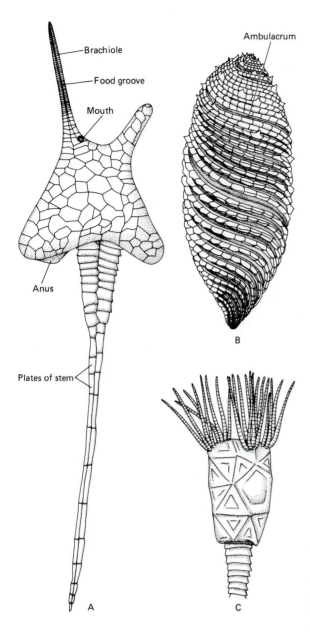

mouth or mounted on the plates along each side of the ambulacral grooves (Fig. 18–63B). Branches from the groove extended up each brachiole.

The class Blastoidea is another group of extinct, attached echinoderms that flourished in the Paleozoic seas (Fig. 18–64). The blastoids first appeared in the Ordovician period, reached their peak in the Mississippian, and became extinct during the Permian.

Like cystoids, blastoids possessed an oval theca that was attached aborally by a short stalk or attached directly to the substratum. Five ambulacra extended from the mouth at the oral pole down the sides of the theca and frequently were elevated as broad ridges that alternated with the depressed interambulacral areas. The margins of the ambulacra were bordered by a single row of slender, closely placed brachioles, into which extended branches from the ambulacral groove. In oral view, the brachioles gave the appearance of a thick fringe along the border of a five-pointed star.

Peculiar to blastoids was a system of folds, or folds and pores, called **hydrospires,** located at each side of the ambulacra and thought to represent a gas exchange system.

The Edrioasteroidea is a class of extinct attached echinoderms that first appeared in the early Cambrian period and ranged into the Pennsylvanian period. The theca was oval or discoid and composed of imbricated, abutting, or fused plates. The presence of tubercles on the plates indicates that some species possessed spines. Some edrioasteroids had stalks (Fig. 18–65A), but the discoid species were stalkless and lived attached to shells or other objects (Fig. 18–65B). There were no brachioles, and the five ambulacra, which extended over the theca, were straight or curved. Some of these animals thus looked like a brittle star wrapped around a ball. Of particular interest was the presence of pores between the ambulacral plates through which the podia extended. Movable cover plates protected the podia and food grooves.

The crinoids possess the richest fossil record of all the echinoderm classes and are the only attached species that still exist today. Typical crinoids first appeared in the Ordovician, and they left a fossil record in every succeeding period. The class reached its climax in the Mississippian, although there was a second somewhat lesser evolutionary development during the Permian. During these periods certain shallow

FIGURE 18–61 A, *Dendrocystites,* a later carpoid possessing one brachiole. **B,** A hypothetical restoration of a partially contracted helicoplacoid, *Helicoplacus.* **C,** *Mimocystites,* an eocrinoid. *(A, Modified from Bather. B, After Durham and Caster. C, After Ubaghs, G. 1967. Treatise on Invertebrate Paleontology. Pt. S. Vol. 1. Courtesy of Geological Society of America and the University of Kansas, Lawrence.)*

FIGURE 18–62 Reconstructions of the marine fauna of Paleozoic seas. **A,** Crinoids. A coiled cephalopod is on right. **B,** A long stalked crinoid in front of a cluster of rugose corals. To the left of the rugose corals is a brachiopod, and in the front left is a gastropod. *(Photographs courtesy of Betty M. Barnes. Models are in the National Museum of Natural History, Smithsonian Institution.)*

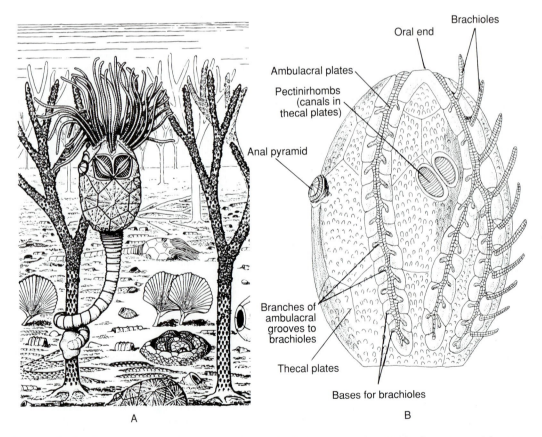

FIGURE 18–63 **A,** Reconstruction of an Ordovician cystoid, *Lepadocystis,* attached to an erect bryozoan. An edrioasteroid is attached to the bottom in front of two brachiopods. **B,** Lateral view of the cystoid, *Callocystites.* (A, From Kesling, R. V. 1967. In Moore, R. C. (Ed.): Treatise on Invertebrate Paleontology. Pt. S. Vol. 1. Courtesy of the Geological Society of America and the University of Kansas, Lawrence. pp. S85–S286. B, After Hyman, L. H. 1955. The Invertebrates. Vol. IV. McGraw-Hill Book Co., New York.)

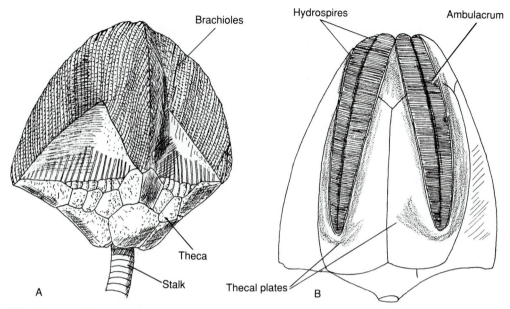

FIGURE 18–64 Class Blastoidea. **A,** *Blastoidocrinus,* showing brachioles (lateral view). **B,** *Pentremites* (lateral view, brachioles not shown). *(A, After Jaekel from Hyman. B, After Hyman, L. H. 1955. The Invertebrates. Vol. IV. McGraw-Hill Book Co., New York.)*

seas supported enormous faunas. The Burlington (Mississippian) limestone of Iowa, Illinois, and Missouri is composed almost entirely of crinoids, with 15,000 crinoids per cubic meter of limestone. Paleozoic crinoids were all stalked, and the earlier species lacked pinnules. Modern crinoids, which contain the stalkless comatulids and belong to different orders from the Paleozoic species, did not appear until the Mesozoic era.

An early, small group of extinct echinoderms was the carpoids, known from the Cambrian to the Devonian. Formerly placed in the class Carpoidea, they are now separated into four classes (p. 989). These animals are of especial interest because they were asymmetrical (Fig. 18–61A). Some were stalkless and rested directly on the bottom; others were stalked, but the body was apparently bent over so that the crown was oriented horizontally to the substratum—that is, one side of the oral/aboral axis faced toward the substratum and the other side faced away from the substratum. The oral surface carried openings that have been interpreted as anus and mouth. Two of the classes possessed a single, armlike projection bearing an ambulacral groove. Another nonradial class of fossil echinoderms is the Helicoplacoidea. The helicoplacoids were spindle-shaped animals with the mouth at one end of the body (Fig. 18–61B). There was a single, branched ambulacrum. The body

wall was covered with pleated plates, which allowed expansion and contraction. This flexibility perhaps permitted the animal to retract its body, which may have been situated vertically in the sand. The expanded anterior end projected above the surface of the sand when the animal was feeding.

As would be expected from the nature of the skeleton, the holothuroids have the poorest fossil record and the echinoids have the best fossil record of the living classes of unattached echinoderms, but compared with attached forms, the fossil record of

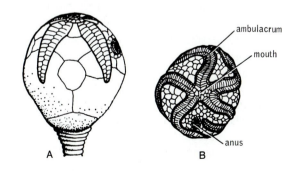

FIGURE 18–65 Edrioasteroids. **A,** *Steganoblastus,* a stalked species with straight ambulacra. **B,** *Edrioaster,* an unattached species with curved ambulacra. *(A and B, After Nichols, D. 1969. Echinoderms. Hutchinson and Co., London.)*

unattached species is sparse. Although fossil asteroids and echinoids first appeared in the Ordovician period and ophiuroids in the Mississippian, Paleozoic species are relatively rare; these classes are much better represented in Mesozoic and Cenozoic rocks. The earlier echinoids, including a number of extinct orders, were all regular forms, and the first fossil heart urchins and sand dollars appeared in the Triassic period.

The origin of the echinoderms and the phylogenetic relationships of the subgroups continue to be unsolved questions and subjects of much speculation. Despite the extensive fossil record, paleontological evidence is still insufficient in many important areas. Before reviewing some of the older ideas that are still current, as well as some more recent views, it may be helpful to enumerate a few reasonable assumptions accepted by many zoologists.

1. Echinoderms evolved from motile, bilaterally symmetrical ancestors that possessed a tripartite coelom.

2. As is true of other animals, the echinoderm skeleton and radial symmetry probably represent adaptations to a sessile existence, at least initially.

3. The original function of the water-vascular system was feeding and not locomotion.

A classical theory, and one that still finds many supporters, holds that from a bilateral, tricoelomate, free-moving ancestor, some group became attached to the bottom and assumed a sessile mode of life. As in many other animal groups, such a sessile existence resulted in a shift to a more adaptive radial symmetry. Attachment apparently took place at the anterior end of the animal. Based on the metamorphosis of living echinoderms, the change in symmetry involved a clockwise, 90-degree rotation of the animal so that the left side became the upper (oral) side and the right side became the lower (aboral) side. Simultaneously, the mouth moved around to the original left side, now the upper side. The two right, anterior, coelomic sacs (axohydrocoel) were reduced, and the two left sacs became the water-vascular system, which functioned in suspension feeding.

During the course of these changes, the echinoderm skeleton probably evolved as a supportive and protective structure for the sessile animal. The pentamerous form of radial symmetry arose in conjunction with the skeleton. The suture planes—the junction between two skeletal plates—represent a weak point in the body wall from a structural standpoint, and it would have been advantageous to the animal not to have had two such suture planes opposite each other. This advantage could be attained only by an odd number of ossicles forming the circumference of the body wall. The smallest number would be five if the animal were to be truly radial. At this point we have a suspension-feeding echinoderm that was both sessile and radial. This stage in the evolution of echinoderm symmetry is illustrated by the extinct and living Crinozoa.

After attaining a radial symmetry, some of these sessile echinoderms became detached and reassumed a free-moving existence. The radial symmetry was retained, but the oral surface, which was directed upward in sessile forms, was placed against the substratum; the aboral surface became the functional upper side of the animal, and the water-vascular system was utilized in locomotion. Sea stars, brittle stars, and sea urchins all illustrate such a free-moving, radial existence.

This theory is supported by the extinct and living crinozoan fauna and by embryological evidence, such as the attached metamorphosis of asteroids. There are also, however, some objections to this theory. The oldest echinoderms, all from the early Cambrian, are three very diverse groups: the eocrinoids, the edrioasteroids, and the carpoids. The upward-directed oral surface of the eocrinoids and the edrioasteroids is compatible with the idea that the first echinoderms were radially symmetrical and attached animals, but the irregular carpoids are not. Either the carpoid asymmetry is secondary, in which case the earliest echinoderms are Precambrian, or the first echinoderms were not pentamerous, radially symmetrical animals.

A second principal problem with this classical theory of echinoderm evolution is the lack of any forms that bridge the great gap between those echinoderms in which the oral surface is directed upward and those, like sea stars and sea urchins, in which it is directed downward. The echinozoan edrioasteroids have frequently been suggested as a bridge group, for they were directed upward and some were unattached. It is difficult, however, to visualize how the oral surface could have come to be directed downward with sufficient intermediate stages to permit the necessary functional adjustments.

Some zoologists postulate that the free habit, not the attached one, was the most primitive mode of existence in echinoderms, and that the sessile condition arose secondarily. The living echinoids and most of the holothuroids are believed to be descended from forms that were always free-moving. The helicoplacoids are regarded as being close to the ancestral echinoderms, and the earliest echinoderms were perhaps benthic deposit feeders. This view has the advantage of reconciling the two conflicting types of body orientation within the phylum, but they do not convincingly explain the original adaptive significance of the echinoderm skeleton, which is a basic feature of the phylum, nor the pentamerous radial symmetry of such forms as the echinoids.

There has also been much speculation about the nature of the ancestral preechinoderms. Was the water-vascular system a new development of the coelom in the evolution of the phylum or was there a functional precursor? A number of zoologists have noted the considerable similarity between the coelom associated with a lophophore and that which is part of the echinoderm water-vascular system. Both are derived from the mesocoel, and both have a water ring (base of lophophore) from which extend radial canals (lophophore tentacles). Could the ancestors of echinoderms have possessed a lophophore-like structure from which the water-vascular system was derived?

SUMMARY

1.   The Paleozoic echinoderms, which include a number of extinct classes in addition to the crinoids, were largely attached and had the oral surface directed upward.

2.   Assuming a feeding mechanism like that of living crinoids, the food-collecting surface area in Paleozoic forms was provided either by brachioles (eocrinoids, blastoids, and cystoids) or by arms and pinnules (crinoids).

3.   A bilateral larva that bears ciliated bands is a characteristic developmental feature of some members of all echinoderm classes and indicates that the pentamerous radial symmetry of echinoderms is secondary. The ancestors of echinoderms were probably bilateral, motile tricoelomates.

4.   The evolution of a pentamerous radial symmetry and an endoskeleton of calcareous ossicles was

perhaps correlated with an early assumption of a sessile existence.

5.   The evolution of the water-vascular system may have originally been an adaptation for suspension feeding.

6.   The relationship between most modern classes of echinoderms, in which the oral surface is directed downward, and the Paleozoic forms, in which the oral surface is directed upward, is obscure.

SYSTEMATIC RÉSUMÉ OF THE ECHINODERMATA

Subphylum Homalozoa.   Carpoids. Paleozoic echinoderms lacking any evidence of radial symmetry.

Subphylum Crinozoa.   Radially symmetrical echinoderms having a globoid or cup-shaped theca and brachioles or arms. Mostly attached, with oral surface directed upward. This subphylum contains the fossil eocrinoids, cystoids, and the fossil and living crinoids. Only the classification of the crinoids is covered here.

Class Crinoidea.   Sea lilies and feather stars. Stalked and free-moving echinoderms having the oral side of the body directed upward. Arms branched and bearing pinnules.

Subclasses Inadunata, Flexibilia, and Camerata.   Stalked Paleozoic crinoids with or without cirri, some without pinnules. Organization of calyx ossicles important in distinguishing these fossil groups.

Subclass Articulata.   Extinct as well as all living crinoids belong to this subclass. The latter are placed within five orders.

Order Millericrinida.   Sea lilies without cirri. *Hyocrinus, Calamocrinus.*

Order Cyrtocrinida.   This order contains two living species from the Caribbean and mid-Atlantic in which the aboral end of the crown is attached directly to the substratum. *Holopus.*

Order Bourgueticrinida.   Mostly small sea lilies. The slender stalk lacks cirri. *Rhizocrinus, Bathycrinus.*

Order Isocrinida.   Sea lilies with cirri. *Metacrinus, Cenocrinus.*

Order Comatulida.   Feather stars. Stalkless, unattached crinoids. *Antedon, Florometra.*

Subphylum Asterozoa.   Unattached, radially symmetrical echinoderms having a body composed of a flattened central disc and radially arranged arms, or rays.

Class Asteroidea. Sea stars. Arms not sharply set off from the central disc. Ambulacral grooves are open, and a large coelomic cavity is present in the relatively wide arms.

Order Platyasterida. Feet without suckers. Primitive, mostly extinct sea stars; the two living genera are *Platyasterias* and the common soft-bottom *Luidia*.

Orders Paxillosida and Valvatida. These two orders were formerly united within the order Phanerozonia. The Paxillosida lack suckers on the tube feet; the Valvatida possess suckers. Sea stars with marginal plates and usually with paxillae on the aboral surface. Pedicellariae of the sessile type. *Astropecten, Ctenodiscus, Culcita, Goniaster, Oreaster, Linckia, Porania.*

Order Spinulosida. The members of this order are not always distinct from those of the two orders above, but in general, conspicuous marginal plates are absent, and the tube feet are suckered. The aboral surface is covered with low spines, which give the order its name. There are no pedicellariae. *Asterina, Patiria, Echinaster, Henricia, Acanthaster, Crossaster, Pteraster.*

Order Forcipulata. Sea stars with pedicellariae composed of a short stalk and three skeletal ossicles. Tube feet with suckers. *Heliaster, Pycnopodia, Asterias, Leptasterias, Pisaster, Brisinga, Zoroaster.*

Class Ophiuroidea. Brittle stars. Arms sharply set off from the central disc. Ambulacral groove absent. Arms largely filled by vertebral ossicles.

Order Oegophiurida. A largely fossil group with a single living species (from Indonesia). No dorsal and ventral arm shields or bursae. Madreporites at edge of disc.

Order Phrynophiurida. Ophiuroids in which dorsal arm shields are absent.

Suborder Ophiomyxina. Primitive brittle stars in which disc and arm plates are covered by a thick soft skin. *Ophiocanops, Ophiomyxa.*

Suborder Euryalina. Arms simple or branched (basket stars) but capable of coiling vertically. *Asteronyx, Gorgonocephalus.*

Order Ophiurida. Mostly small ophiuroids, usually with five arms. Arms capable of transverse movement only. Dorsal arm shields present. This order contains most of the brittle stars, or serpent stars. *Amphiura, Amphipholis, Ophiopholis, Ophiactis, Ophiothrix, Ophioderma, Ophiocoma, Ophiolepis, Ophiomusium, Ophiomaza, Ophionereis.*

Class Concentricycloidea. Minute deep-water echinoderms with disc-shaped bodies. Tentatively placed in the Asterozoa because of certain similarities to the asteroids (p. 983)

Subphylum Echinozoa. Radially symmetrical globoid or discoid echinoderms without arms or brachioles. Mostly unattached. Contains the fossil helicoplacoids and edrioasteroids as well as the echinoids and holothuroids. Only the last two are covered here.

Class Echinoidea. Sea urchins, heart urchins, sand dollars. Echinoderms having more or less spherical or orally/aborally flattened bodies without arms. Ossicles fused to form an internal test on which movable spines are mounted.

Subclass Perischoechinoidea. Largely primitive fossil urchins of the Paleozoic seas, which made their first appearance in the Ordovician period with *Bothriocidaris.*

Order Cidaroida. Of the four orders of the subclass Perischoechinoidea, this is the only one with two rows of plates for each ambulacrum and interambulacrum. It is also the only one that survived the Paleozoic era and became the ancestor of the remaining echinoids, most of which belong to the next subclass. The existing members of the Cidaroida are characterized by widely separated primary spines and small secondary spines. Gills are absent. *Eucidaris, Cidaris, Notocidaris.*

Subclass Euechinoidea. This subclass contains the majority of living species of echinoids.

Superorder Diadematacea. Sea urchins with perforated tubercles. Gills usually present.

Order Pedinoida. Rigid test with solid spines. Ten buccal plates on peristomial membrane. *Caenopedina* is the only living genus.

Order Diadematoida.   Rigid or flexible test with hollow spines. Ten buccal plates on peristomial membrane. *Diadema, Plesiodiadema.*

Order Echinothuroida.   Flexible test with poisonous secondary spines. Simple ambulacral plates on peristomial membrane. Gills inconspicuous or lost. *Asthenosoma.*

Superorder Echinacea.   Sea urchins with rigid test and solid spines. Gills present. Peristomial membrane with ten buccal plates.

Order Arbacioida.   Periproct with four or five plates. *Arbacia.*

Order Salenoida.   Anus located eccentrically within periproct because of the presence of a large plate (suranal plate). *Acrosalenia.*

Order Temnopleuroida.   Test sculptured in some. Camarodont lantern (large epiphyses are fused across top of each pyramid). *Tripneustes, Toxopneustes, Lytechinus.*

Order Phymosomatoida.   Like the previous order but primary tubercles imperforate. *Glyptocidaris.*

Order Echinoida.   Camarodont lantern and nonsculptured test with imperforate tubercles. *Echinus, Psammechinus, Paracentrotus, Echinometra, Echinostrephus, Colobocentrotus, Heterocentrotus, Strongylocentrotus.*

Superorder Gnathostomata.   Irregular urchins. Mouth is in center of oral surface but anus has shifted out of apical center. Lantern present.

Order Holectypoida.   No petaloids. Many fossil members were essentially regular in shape. The two living genera, *Echinoneus* and *Micropetalon,* are oval.

Order Clypeasteroida.   True sand dollars. Petaloids present. Test greatly flattened. No phyllodes. *Clypeaster, Fibularia, Mellita, Encope, Rotula.*

Superorder Atelostomata.   Irregular urchins. No lantern.

Order Holasteroida.   Oval or bottle-shaped echinoids with thin, delicate test.

Petaloids and phyllodes not developed. Deep-water species. *Pourtalesia.*

Order Spatangoida.   Heart urchins. Oval and elongated echinoids. Oral center shifted anteriorly, and anus shifted out of aboral apical center. Petaloids present but may be sunk into grooves. Phyllodes present. *Spatangus, Echinocardium, Moira, Meoma, Lovenia.*

Order Cassiduloida.   Mostly extinct echinoids with round to oval test and a central or slightly anterior apical center. Phyllodes with intervening smaller areas (bourrelets). Poorly developed petaloids. The few existing species are tropical burrowers and somewhat similar to heart urchins. *Echinolampas.*

Class Holothuroidea.   Sea cucumbers. Echinoderms having the body elongated along the oral/aboral axis. Oral podia modified as tentacles. Skeleton reduced to microscopic ossicles.

Order Dactylochirotida.   Primitive sea cucumbers. Tentacles are simple, and the body is enclosed within a flexible test. Body U-shaped. *Sphaerothuria, Echinocucumis.*

Order Dendrochirotida.   Buccal podia, or tentacles, are dendritic and not provided with ampullae. Podia occurring on the sole, on all the ambulacra, or over the entire surface. *Cucumaria, Thyone, Psolus.*

Order Aspidochirotida.   Tentacles peltate, or shieldlike. Podia present, sometimes forming a well-developed sole. *Holothuria, Actinopyga, Stichopus.*

Order Elasipodida.   Aberrant sea cucumbers with large, conical papillae and other appendages. Tentacles peltate. Almost all are deep-sea species. *Pelagothuria, Peniagone.*

Order Molpadiida.   Posterior end of body narrowed to a tail. Fifteen digitate tentacles, but regular podia absent. *Molpadia, Caudina.*

Order Apodida.   Wormlike sea cucumbers with only buccal podia, or tentacles, present. Tentacles digitate or pinnate. *Leptosynapta, Synapta, Euapta.*

REFERENCES

Mutable Connective Tissue

Eylers, J. P. 1976. Aspects of skeletal mechanics of the starfish *Asterias forbesi*. J. Morphol. *149:*353–368.

Eylers, J. P. 1982. Ion-dependent viscosity of holothurian body wall and its implications for the functional morphology of echinoderms. J. Exp. Biol. *99:*1–8.

Florey, E., and Cahill, M. A. 1977. Ultrastructure of sea urchin tube feet. Evidence for connective tissue involvement in motor control. Cell Tiss. Res. *177:*195–214.

Hidaka, M., and Takahashi, K. 1983. Fine structure and mechanical properties of the catch apparatus of the sea-urchin spine, a collagenous connective tissue with muscle-like holding capacity. J. Exp. Biol. *103:*1–14.

Motokawa, T. 1984a. Catch connective tissue: The connective tissue with adjustable mechanical properties. *In* Keegan, B. F., and O'Connor, B. D. S. Echiondermata. Proceedings of the Fifth International Echinoderm Conference, Galway. A. A. Balkema, Rotterdam, Boston. pp. 69–73.

Motokawa, T. 1984b. Connective tissue catch in echinoderms. Biol. Rev. *59:*255–270.

Smith, D. S., Wainwright, S. A., Baker, J. et al. 1981. Structural features associated with movement and "catch" of sea-urchin spines. Tiss. Cell. *13:*299–320.

Smith, G. N., Jr., and Greenberg, M. J. 1973. Chemical control of the evisceration process in *Thyone briareus*. Biol. Bull. *144:*421–436.

Wilkie, I. C. 1984. Variable tensility in echinoderm collagenous tissues: A review. Mar. Behav. Physiol. *11:*1–34.

Wilkie, I. C., Emson, R. H., and Mladenov, P. V. 1984. Morphological and mechanical aspects of fission in *Ophiocomella ophiactoides* (Echinodermata, Ophiuroidea). Zoomorphology. *104:*310–322.

The literature below is restricted to echinoderms. The introductory references on page 6 include many general works and field guides that contain sections on these animals.

General

Binyon, J. 1972. Physiology of Echinoderms. Pergamon Press, Oxford.

Cuénot, L. 1948. Anatomie, Ethologie, et Systematique des Échinodermes. *In* Grassé, P. (Ed.): Traité de Zoologie. Vol. II. Échinodermes, Stomocordes, Procordes. Masson et Cie, Paris. pp. 1–363.

Donnay, G., and Pawson, D. L. 1969. X-ray diffraction studies of echinoderm plates. Science. *166:*1147–1150.

Emson, R. H., and Wilkie, I. C. 1980. Fission and autotomy in echinoderms. Oceanogr. Mar. Biol. Ann. Rev. *18:*155–250.

Hamann, A. 1885. Beiträge zur Histologie der Echinodermen. Heft 2. Die Asteriden.

Jangoux, M. (Ed.): 1980. Echinoderms: Present and Past. Proceedings of the European Colloquium on Echinoderms, Brussels, 1979. A. A. Balkema, Rotterdam.

Jangoux, M., and Lawrence, J. M. (Eds.): 1982. Echinoderm Nutrition. A. A. Balkema, Rotterdam. 654 pp.

Jangoux, M., and Lawrence, J. M. (Eds.): 1983. Echinoderms Studies. Vol. I. A. A. Balkema, Rotterdam. 203 pp. (A collection of reviews on various aspects of echinoderm biology.)

Lawrence, J. M. 1987. A Functional Biology of Echinoderms. Johns Hopkins University Press, Baltimore. 340 pp.

Moore, R. C. (Ed.): 1966 to 1978. Treatise on Invertebrate Paleontology, Echinodermata. Pts. S to U. Geological Society of America and University of Kansas Press, Lawrence.

Nichols, D. 1969. Echinoderms. 4th Edition. Hutchinson University Library, London.

Ruppert, E. E., and Balser, E. J. 1986. Nephridia in the larvae of hemichordates and echinoderms. Biol. Bull. *171:*188–196.

Sloan, N. A., and Campbell, A. C. 1982. Perception of food. *In* Jangoux, M., and Lawrence, J. M., (Eds.): 1982. Echinoderm Nutrition. A. A. Balkema, Rotterdam.

Strathmann, R. R. 1975. Larval feeding in echinoderms. Am. Zool. *15:*717–730.

Welsch, U., and Rehkämper, G. 1987. Podocytes in the axial organ of echinoderms. J. Zool. *213:*45–50.

Yoshida, M., Takasu, N., and Tamotsu, S. 1984. Photoreception in echinoderms. *In* Ali, M. A. Photoreception and Vision in Invertebrates. Plenum Press, New York. pp. 743–772.

Asteroids

Anderson, J. M. 1978. Studies on functional morphology in the digestive system of *Oreaster reticulatus*. Biol. Bull. *154:*1–14.

Binyon, J. 1964. On the mode of functioning of the water vascular system of *Asterias rubens*. J. Mar. Biol. Assoc. U.K. *44:*577–588.

Birkeland, C. 1982. Terrestrial runoff as a cause of outbreaks of *Acanthaster planci*. Mar. Biol. *69:*175–185.

Blake, D. B. 1990. Adaptive zones of the class Asteroidea. Bull. Marine. Sci. *46*(3):701–718.

Burnett, A. L. 1960. The mechanism employed by the starfish *Asterias forbesi* to gain access to the interior of the bivalve *Venus mercenaria*. Ecology. *41:*583–584.

Chia, F. S., and Amerongen, H. 1975. On the prey-catching pedicellariae of a starfish, *Stylasterias forreri*. Can. J. Zool. *53*:748–755.

Christensen, A. M. 1970. Feeding biology of the sea star *Astropecten irregularis*. Ophelia. *8*:1–134.

Downey, M. E. 1973. Starfishes from the Caribbean and the Gulf of Mexico. Smithson. Contrib. Zool. *126*. 158 pp.

Feder, H. M. 1955. On the methods used by the starfish *Pisaster ochraceus* in opening three types of bivalved mollusks. Ecology. *36*:764–767.

Ferguson, J. C. 1984. Translocative functions of the enigmatic organs of starfish—the axial organ, hemal vessels, Tiedemann's bodies, and rectal caeca: An autoradiographic study. Biol. Bull. *166*:140–155.

Ferguson, J. C. 1989. Rate of water admission through the modreporite of a starfish. J. Exp. Biol. *145*:147–156.

Ferguson, J. C. 1990. Seawater inflow through the modreporite and internal body regions of a starfish *(Leptasterias hexactis)* as demonstrated with flourescent microbeads. J. Exp. Zool. *255*:262–271.

Jangoux, M. 1982. Food and feeding mechanisms: Asteroidea; Digestive systems: Asteroidea and Ophiuroidea. *In* Jangoux, M., and Lawrence, J. M. (Eds.): 1982. Echinoderm Nutrition. A. A. Balkema, Rotterdam.

Laxton, J. H. 1974. Aspects of the ecology of the coral-eating starfish, *Acanthaster planci*. Biol. J. Linn. Soc. *6*(1):19–45.

Mauzey, K. P., Birkeland, C., and Dayton, P. K. 1968. Feeding behavior of asteroids and escape responses of their prey in the Puget Sound region. Ecology. *49*(4):603–619.

Menge, B. 1975. Brood or broadcast? The adaptive significance of different reproductive strategies in the two intertidal sea stars *Leptasterias hexactis* and *Pisaster Ochraceus*. Mar. Biol. *31*(1):87–100.

Moran, P. J. 1990. *Acanthaster planci* (L.): Biographical data. Coral Reefs. *9*:95–96.

Scheibling, R. E. 1980. The microphagous feeding behavior of *Oreaster reticulatus*. Mar. Behav. Physiol. *7*(3):225–232.

Shick, J. M., Edwards, K. C., and Dearborn, J. H. 1981. Physiological ecology of the deposit-feeding sea star *Ctenodiscus crispatus:* Ciliated surfaces and animal-sediment interactions. Mar. Ecol. Prog. Ser. *5*:165–184.

Sloan, N. A. 1980. Aspects of the feeding biology of asteroids. Oceanogr. Mar. Biol. Ann. Rev. *18*:57–124.

Sloan, N. A., and Northway, S. M. 1982. Chemoreception by the asteroid *Crossaster papposus*. J. Exp. Mar. Biol. Ecol. *61*:85–98.

Thomas, L. A., and Hermans, C. O. 1985. Adhesive interactions between the tube feet of a starfish, *Leptasterias hexactis,* and substrata. Biol. Bull. *169*:675–688.

Thomassin, B. A. 1976. Feeding behavior of the felt-, sponge-, and coral-feeding sea stars, mainly *Culcita*

schmideliana. Helgol. Wiss. Meeresunters. *28*(1):51–65.

Wilkinson, C. R., and Macintyre, I. G. (Eds.): 1992. The *Acanthaster* debate. Coral Reefs. *22*(2). (A special issue devoted to the crown-of-thorns seastar.)

Ophiuroids

Brehm, P., and Morin, J. G. 1977. Localization and characterization of luminescent cells in *Ophiopsila californica* and *Amphipholis squamata*. Biol. Bull. *152*:12–25.

Broom, D. M. 1975. Aggregation behavior of the brittlestar *Ophiothrix fragilis*. J. Mar. Biol. Assoc. U.K. *55*:191–197.

Emson, R. H., and Woodley, J. D. 1987. Submersible and laboratory observations on *Asteroschema tenue,* a long-armed euryaline brittle star epizoic on gorgonians. Mar. Biol. *96*:31–45.

Fontaine, A. R. 1965. Feeding mechanisms of the ophiuroid *Ophiocomina nigra*. J. Mar. Biol. Assoc. U.K. *45*:373–385.

Fricke, H. W. 1968. Beiträge zur Biologie der Gorgonenhaupter *Astrophyton muricatum* (Lamarck) und *Astroboa nuda* (Lyman). Ernst-Reuter Gesellschaft, Berlin. 197 pp.

Gage, J. D. 1990. Skeletal growth bands in brittle stars: Microstructure and significance as age markers. J. Mar. Biol. Assn. U.K. *70*:209–224.

Hendler, G. 1975. Adaptational significance of the patterns of ophiuroid development. Am. Zool. *15*:691–715.

Hendler, G. 1982. Slow flicks show star tricks: Elapsed-time analysis of basketstar *(Astrophyton muricatum)* feeding behavior. Bull. Mar. Sci. *32*:909–918.

Thomas, L., and Hermans, C. O. 1985. Adhesive interactions between the tube feet of a starfish, *Leptasterias hexactis,* and substrata. Biol. Bull. *169*:675–688.

Tyler, P. A. 1980. Deep-sea ophiuroids. Oceanogr. Mar. Biol. Ann. Rev. *18*:125–153.

Warner, G. F. 1982. Food and feeding mechanisms: Ophiuroidea. *In* Jangoux M., and Lawrence, J. M. (Eds.): 1982. Echinoderm Nutriton. A. A. Balkema, Rotterdam. pp. 161–181.

Warner, G. F., and Woodley, J. D. 1975. Suspension-feeding in the brittle star *Ophiothrix fragilis*. J. Mar. Biol. Assoc. U.K. *55*:199–210.

Woodley, J. D. 1975. The behavior of some amphiurid brittle stars. J. Exp. Mar. Biol. Ecol. *18*:29–46.

Echinoids

Burke, R. D. 1984. Pheromonal control of metamorphosis in the Pacific sand dollar, *Dendraster excentricus*. Science. *225*:442–443.

Chia, F. S. 1969. Some observations of the locomotion and feeding of the sand dollar, *Dendraster excentricus.* J. Exp. Mar. Biol. Ecol. *3*(2):162–170.

De Ridder, C., and Lawrence, J. M. 1982. Food and feeding mechanisms: Echinoidea. *In* Jangoux, M., and Lawrence, J. M. (Eds.): 1982. Echinoderm Nutrition. A. A. Balkema, Rotterdam. 57–115.

Ebert, T. A., and Dexter, D. M. 1975. A natural history study of *Encope grandis* and *Mellita grantii,* two sand dollars in the northern Gulf of California. Mar. Biol. *32*(4):397–407.

Ellers, O., and Telford, M. 1984. Collection of food by oral surface podia in the sand dollar, *Echinarachnius parma.* Biol. Bull. *166:*574–582.

Fenner, D. H. 1973. The respiratory adaptations of the podia and ampullae of echinoids. Biol. Bull. *145:*323–339.

Ferber, I., and Lawrence, J. M. 1976. Distribution, substratum preference, and burrowing behavior of *Lovenia elongata* in the Gulf of Elat and Red Sea. J. Exp. Mar. Biol. Ecol. *22:*207–225.

Ghiold, J. 1979. Spine morphology and its significance in feeding and burrowing in the sand dollar, *Mellita quinquiesperforata.* Bull. Mar. Sci. *29*(4):481–490.

Greenway, M. 1976. The grazing of *Thalassia testudinum* in Kingston Harbor, Jamaica. Aquat. Bot. *2*(2):117–126.

Highsmith, R. C. 1982. Induced settlement and metamorphosis of sand dollar *(Dendraster excentricus)* larvae in predator-free sites: Adult sand dollar beds. Ecology. *63*(2):329–337.

Hyman, L. H. 1955. The Invertebrates. Vol. 4. Echinodermata. McGraw-Hill, New York.

Lawrence, J. M. 1975. On the relationship between marine plants and sea urchins. Oceanogr. Mar. Biol. Ann. Rev. *13:*213–286.

Lewin, R. 1988. Sea urchin massacre is a natural experiment. Science. *239:*867.

Lewis, J. B. 1968. The function of the sphaeridia of sea urchins. Can. J. Zool. *46:*1135–1138.

Millott, N. 1975. The photosensitivity of echinoids. Adv. Mar. Biol. *13:*1–52.

Ogden, J. C., Brown, R. A., and Salesky, N. 1973. Grazing by the echinoid *Diadema antillarum* Philippi: Formation of halos around West Indian patch reefs. Science. *182:*715–717.

Serafy, D. K., and Fell. F. J., 1985. Marine Flora and Fauna of the Northeastern United States: Echinoidea. U.S. Department of Commerce, Scientific Publications Office. 27 pp.

Smith A. 1984. Echinoid Paleobiology. Allen and Unwin, London. 190 pp.

Telford, M. 1983. An experimental analysis of lunule function in the sand dollar *Mellita quinquesperforata.* Mar. Biol. *76:*125–134.

Telford, M., Mooi, R., and Ellers, O. 1985. A new model of podial deposit feeding in the sand dollar, *Mellita quinquiesperforata:* The sieve hypothesis challenged. Biol. Bull. *69:*431–448.

Timko, P. 1976. Sand dollars as suspension feeders: A new description of feeding in *Dendraster excentricus.* Biol. Bull. *151*(1):247–259.

Holothuroids

Bakus, G. J. 1973. The biology and ecology of tropical holothurians. *In* Jones, O. A., and Endean, R. (Eds.): Biology and Geology of Coral Reefs. Vol. II. Biology 1. Academic Press, New York. pp. 326–368.

Byrne, M. 1985. Evisceration behaviour and the seasonal incidence of evisceration in the holothurian *Eupentacta quinquesemita.* Ophelia. *24*(2):75–90.

Fankboner, P. V. 1981. A re-examination of mucus feeding by the sea cucumber *Leptopentacta (Cucumaria) elongata.* J. Mar. Biol. Assoc. U.K. *61:*679–683.

Fankboner, P. V., and Cameron, J. L. 1985: Seasonal atrophy of the visceral organs in a sea cucumber. Can. J. Zool. *63:*2888–2892.

Fish, J. D. 1967. Biology of *Cucumaria elongata.* J. Mar. Biol. Assoc. U.K. *47:*129–143.

Herreid, C. F., La Russa, V. F., and DeFesi, C. R. 1976. Blood vascular system of the sea cucumber, *Stichopus moebii.* J. Morphol. *150*(2):423–451.

Pawson, D. L. 1977. Echinodermata: Holothuroidea. Marine flora and fauna of the northeastern U.S. NOAA Technical Reports NMFS Circular 405. U.S. Government Printing Office, Washington, DC. 13 pp.

Smith, T. B. 1983. Tentacular ultrastructure and feeding behavior of *Neopentadactyla mixta.* J. Mar. Biol. Assoc. U.K. *63:*301–311.

VandenSpiegel, D., and Jangoux, M. 1987. Cuvierian tubules of the holothuroid *Holothuria forskali* (Echinodermata): A morphofunctional study. Mar. Biol. *96:*263–275.

Yingst, J. Y. 1976. The utilization of organic matter in shallow marine sediments by an epibenthic deposit-feeding holothurian. J. Exp. Mar. Biol. Ecol. *23:*55–69.

Crinoids

Baker, A. N., Rowe, W. E., and Clark, H. E. S. 1986. A new class of Echinodermata from New Zealand. Nature. *321:*862–864.

Balser, E. J., and Ruppert, E. E. 1993. Ultrastructure of axial vascular and coelomic organs in comasterid featherstars (Echinodermata: Crinoidea). Acta Zool. (Stockh.) *74:*87–101.

Byrne, M., and Fontaine, A. R. 1981. The feeding behaviour of *Florometra serratissima.* Can. J. Zool. *59:*11–18.

Fishelson, L. 1974. Ecology of the northern Red Sea crinoids and their epi- and endozoic fauna. Mar. Biol. *26:*183–192.

LaHaye, M. C., and Jangoux, M. 1985. Functional morphology of the podia and ambulacral grooves of the comatulid crinoid *Antedon bifida.* Mar. Biol. *86:*307–318.

La Touche, R. W. 1978. The feeding behavior of the featherstar *Antedon bifida.* J. Mar. Biol. Assoc. U.K. *58:*877–890.

La Touche, R. W., and West, A. B. 1980. Observations on the food of *Antedon bifida.* Mar. Biol. *61*(1):39–46.

Macurda, D. B., and Meyer, D. L. 1974. Feeding posture of modern stalked crinoids. Nature. *247:*394–396.

Macurda, D. B., and Meyer, D. L. 1976. The identification and interpretation of stalked crinoids from deep-water photographs. Bull. Mar. Sci. *26*(2):205–215.

Macurda, D. B., and Meyer, D. L. 1983. Sea lilies and feather stars. Am. Scientist. *71:*354–364.

Meyer, D. L., 1982. Food and feeding mechanisms: Crinozoa. *In* Jangoux, M., and Lawrence, J. M. (Eds.): 1982. Echinoderm Nutrition. A. A. Balkema, Rotterdam. pp. 25–45.

Meyer, D. L., and Lane, N. G. 1976. The feeding behavior of some Paleozoic crinoids and recent basket stars. J. Paleontol. *50*(3):472–480.

Meyer, D. L., LaHaye, C. A., Holland, N. D. et al. 1984. Time-lapse cinematography of feather stars on the Great Barrier Reef, Australia: Demonstrations of posture changes, locomotion, spawning and possible predation by fish. Mar. Biol. *78:*179–184.

Mladenov, P. V., and Chia, F. S. 1983. Development, settling behaviour, metamorphosis and pentacrinoid feeding and growth of the feather star *Florometra serratissima.* Mar. Biol. *73:*309–323.

Rutman, J., and Fishelson, L. 1969. Food composition and feeding behavior of shallow water crinoids at Eilat (Red Sea). Mar. Biol. *3*(1):46–57.

Fossil Echinoderms and Evolution

Durham, J. W., and Caster, K. E. 1963. Helicoplacoidea, a new class of echinoderms. Science. *140:*820–822.

Fell, H. B. 1965. The early evolution of the Echinozoa. Breviora. *219:*1–19.

Fell, H. B. 1966. Ancient echinoderms in modern seas. Oceanogr. Mar. Biol. Ann. Rev. *4:*233–245.

Nichols, D. 1967. The origin of echinoderms. *In* Millott, N. (Ed.): Echinoderm Biology. Academic Press, New York. pp. 209–229.

Nichols, D. 1986. A new class of echinoderms. Nature. *321:*808. (A commentary on the newly discovered class Concentricycloidea.)

Paul, C. R. C., and Smith, A. B. (Eds.): 1988. Echinoderm Phylogeny and Evolutionary Biology. Clarendon Press, Oxford.

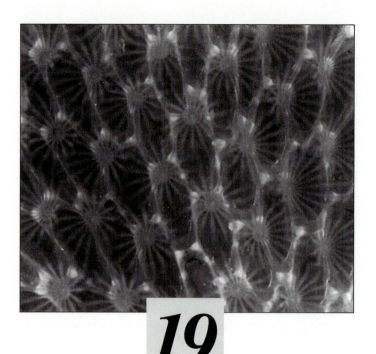

19

LOPHOPHORATES AND ENTOPROCTS

PRINCIPLES AND EMERGING PATTERNS

COLONIAL ORGANIZATION 2

Species of invertebrates that form colonies are widespread among cnidarians, chordates, and lophophorates, and a few occur in several other groups. Colonies are typically sessile, attached, and plantlike but some, such as siphonophores and salps, are motile and pelagic. A **colony** is composed of member **zooids,** which are physically interjoined, equipped to share resources, and often specialized for particular functions. Thus, colonies are, by definition, physiologically integrated entities and not mere aggregates of individuals. Resource sharing can also occur among members of societies, for example among honeybees and humans, and such societies gain many of the same advantages that result from colonial organization.

Colonies probably evolved when the rate of asexual budding exceeded the rate of separation of the daughter zooids. The first colonies were probably monomorphic, composed of functionally similar zooids, each providing for its own needs and each retaining, more or less, its individual character. As colonial intergration advanced, the zooids became specialized for particular functions, providing for a division of labor within the colony as a whole. At this point, the polymorphic colony behaved as an individual, and the zooids were merely functionally dedicated parts of it, like specialized tissues and organs within a body. For this reason, zoologists do not usually speak of zooids as *individuals* but instead use terms, such as *zooid* or *member* or even *module,* that do not imply individuality.

Several features of colonial organization are adaptive for sessile invertebrates. Most are suspension feeders, and the colonial surface area is large in relation to its volume. Moreover, because sessile colonies usually grow over the substratum in sheets or above it on slender branches supported by the water, the surface-area-to-volume ratio of the colony does not decrease with growth as it would in solitary animals.

Growth is typically indeterminate, and there is an exponential rate of increase of zooids. These properties enable the colony to adopt the configuration of the substratum and to expand over it rapidly. Given the limited availability of substrata for attachment, colonial invertebrates may gain a competitive edge over solitary species that rely primarily on sexual reproduction and larval settlement for colonization. These same properties also allow for the attainment of large body size and, potentially, great longevity. D. C. Potts has estimated, for example, that some living corals may be 1000 years old. Large body size may reduce the effect of predation because predators are unlikely to consume the entire colony, and consumed parts regenerate quickly. Indeterminate and rapid growth may also encourage dispersal by fragmentation. Colonial nervous systems enable an entire colony to respond defensively to an attack on one local part. For example, local stimulation of a bryozoan colony with a hair causes the retraction of the feeding members and, in pyrosomes, a localized stimulus causes a spread of bioluminescence over the surface of the colony.

Pelagic colonies of siphonophores and salps benefit by combining many locomotory zooids into one unit. By uniting several swimming bells or zooids together in linear chains or in a streamlined colony, drag is reduced and each propulsive zooid works less to achieve a given velocity than it would if it were a solitary individual.

THE GULFWEED COMMUNITY

With the exception of the plankton and the interstitial fauna, where microscopic organisms predominate, communities of large animals and plants develop in the sea wherever there is a stable substratum and water movement. The substratum provides a refuge or site for attachment, and the water flow transports materials to and from the organism. Familiar examples of this phenomenon include intertidal communities along rocky coasts, as well as communities on pilings, mangrove roots, oyster reefs, and coral reefs. The principle is also illustrated by an odd community of animals in the Sargasso Sea that lives permanently on floating Gulfweed *(Sargassum),* represented by two species of brown algae.

The Gulfweed itself is unusual because it is one of the few large marine algae that floats unattached at the surface of the sea, rather than being anchored to the bottom in nearshore areas. It now seems certain that floating Gulfweed evolved from an ancestral attached species, and the animals that accompany it also evolved from benthic ancestors.

The Gulfweed community is confined to an area of nearly two million square miles and is centered in the western Atlantic, roughly between the Azores in the east and Florida and the West Indies in the west. This vast area is actually an eddy in the center of four great oceanic currents, including the Gulf Stream, that circulate clockwise around the Atlantic basin. Because the Sargasso Sea is hemmed in by currents at its periphery and because there is no upwelling of bottom material into its waters, it receives few nutrients and, as a result, is the marine equivalent of a terrestrial desert. Without the Gulfweed community at its surface, the Sargasso Sea would be biologically impoverished. The Sargasso Sea is estimated to support approximately 10 million tons of Gulfweed at any given time. Occasionally, storms blow some of the Gulfweed ashore where it can accumulate in windrows on beaches.

For the animals in the Gulfweed community (Fig. 19–1), the algae provide food and a surface for attachment and concealment. Most animals, including bryozoans, goose barnacles, and especially hydroids, are premanently attached. These, in turn, are grazed by such predators as sea slugs, pycnogonids, polychaetes, and fishes. One predator, the polychaete fireworm, *Amphinome rostrata,* resembles a stinging caterpillar, and feeds on attached goose barnacles *(Lepas),* swallowing them whole. It later egests cleaned and unbroken barnacle plates. Proximity to sunlight at the surface of the sea is exploited by *Sargassum* itself and also by the small anemone, *Anemonia sargassensis,* and the acoel turbellarian, *Amphiscolops sargassi,* both of which harbor zooxanthellae in their tissues. *Sargassum* provides immediate concealment for tiny animals, but larger organisms, such as crabs and molluscs often adopt forms of camouflage. For example, the grazing nudibranch, *Scyllaea pelagica,* mimics the Gulfweed in form and color. It even nips off and swallows the plant's air bladders, presumably to improve its own buoyancy (and probably to digest the attached hydroids). Like *Scyllaea,* the small Gulfweed fish, *Histrio histrio,* also mimics the Gulfweed and fishes using a lure derived from its first dorsal spine. The Gulfweed crab, *Planes minutus,* ironically belongs to the family of shore crabs (Grapsidae) but has evolved into the open sea with the floating weeds and avoids detection by its small size and protective coloration.

LOPHOPHORATES AND ENTOPROCTS

Three phyla—the Bryozoa, the Phoronida, and the Brachiopoda—are similar in possessing a food-catching tentacular organ called a lophophore and are often grouped together as the lophophorate coelomates. The arms and tentacles of pterobranch hemichordates (p. 874) also qualify as a lophophore. Hemichordates, however, are not traditionally classified among the lophophorates, although the evolutionary ties between lophophorates and deuterostomes have been strengthened in recent years. The entoprocts historically were included in the Bryozoa, but they are acoelomate protostomes. As a result, most zoologists now believe that the entoprocts are only convergently similar to bryozoans. Although entoprocts are clearly spiralians, their relationship to any other spiralian phylum is uncertain, and for that reason and as an example of evolutionary convergence, they are covered in this chapter.

Lophophorates have a mixture of protostome and deuterostome characteristics. Similar to primitive deuterostomes, the body is usually tricoelomic, cleavage is radial, the blastopore may become the anus (some brachiopods), and coeloms may arise by enterocoely (brachiopods). Protostome characteristics, on the other hand, include the presence of larval protonephridia in phoronids and the derivation of the mouth from the blastopore in bryozoans and phoronids.

The **lophophore** is a circular or horseshoe-shaped fold of the body wall that encircles the mouth and bears numerous ciliated tentacles (Fig. 19–2A). The lophophore contains the mesocoel in the form of a ring from which a hollow outgrowth extends into each tentacle. The ciliary tracts on the tentacles drive a current of water through the lophophore, and suspended food is collected in the process.

In addition to possessing a lophophore, nearly every member of these phyla is sessile, has a poorly-developed head, secretes a protective covering, and (except for some of the brachiopods) possess a U-shaped digestive tract (Fig. 19–2A). All these characteristics are, however, correlated with a sessile existence and, in part at least, may represent evolutionary convergences.

PHYLUM BRYOZOA

The phylum Bryozoa (Polyzoa or Ectoprocta) is the largest and the most common of the lophophorate phyla and contains approximately 5000 living species. The group constitutes a major animal phylum, but because of their small body size, bryozoans are less familiar to most people than are most other large groups of invertebrates.

Bryozoans are sessile colonies composed of zooids, each of which is approximately 0.5 mm in length. Zooids are often polymorphic, but typically the body of each consists of a stationary trunk and an eversible introvert, which bears the lophophore. In most species the trunk of each zooid secretes a protective covering and bears a specialized opening for the protrusion of the lophophore. The interior of the body is occupied largely by a spacious coelom (metacoel) and the U-shaped digestive tract. Organs specialized for gas exchange and excretion are absent. Most of the peculiarities of bryozoans are associated with their small zooid size, colonial organization, polymorphism, and rigid skeletal covering.

The phylum is divided into three classes: the Phylactolaemata, the Gymnolaemata, and the Stenolaemata. The class Stenolaemata contains some living marine species and over 500 fossil genera. The class Gymnolaemata is almost entirely marine and includes the great majority of living bryozoans, as well as many fossil species. The class Phylactolaemata is restricted to fresh water and, although it is widely distributed, contains only about 50 species.

Structure of the Zooid

Each bryozoan zooid is boxlike, oval, or tubular (Figs. 19–2; 19–3). In gymnolaemates, on which the following description is based, the cuticle (**zooecium**) is composed of protein-chitin or calcium carbonate (Fig. 19–20). Thus, many bryozoans possess a rigid exoskeleton, although in others, such as the common *Bugula neritina*, calcification is slight.

In extended zooids, the **introvert (tentacle sheath),** which bears the lophophore, is sometimes surrounded at its base by a fold of the body wall, called the **collar.** When the zooid retracts its feeding apparatus, the tentacles bunch together into a bundle, and the bundle is pulled directly inward as the introvert inverts. A **sphincter muscle** constricts the tentacular sheath above the tips of the retracted tentacles. During the final phase of retraction, the collar is pulled inward, thereby inverting a short section of the trunk. The opening through which the collar, tentacles, and introvert are extended is called the **orifice** (Fig. 19–4). In many marine bryozoans, the orifice is

(Text continues on p. 1002)

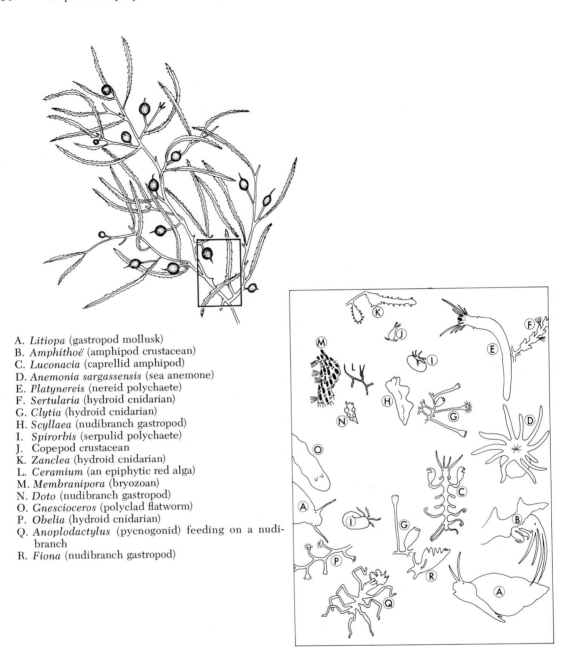

A. *Litiopa* (gastropod mollusk)
B. *Amphithoë* (amphipod crustacean)
C. *Luconacia* (caprellid amphipod)
D. *Anemonia sargassensis* (sea anemone)
E. *Platynereis* (nereid polychaete)
F. *Sertularia* (hydroid cnidarian)
G. *Clytia* (hydroid cnidarian)
H. *Scyllaea* (nudibranch gastropod)
I. *Spirorbis* (serpulid polychaete)
J. Copepod crustacean
K. *Zanclea* (hydroid cnidarian)
L. *Ceramium* (an epiphytic red alga)
M. *Membranipora* (bryozoan)
N. *Doto* (nudibranch gastropod)
O. *Gnesioceros* (polyclad flatworm)
P. *Obelia* (hydroid cnidarian)
Q. *Anoplodactylus* (pycnogonid) feeding on a nudibranch
R. *Fiona* (nudibranch gastropod)

FIGURE 19–1 Invertebrate animals that live on floating *Sargassum* (Gulfweed), brown algae in the Sargasso Sea. Although probably derived from West Indian ancestors, many of these species have become especially adapted for an epiphytic life on floating *Sargassum*. The figure in part **I** shows a large piece of *Sargassum* from which part **II** has been taken. **A,** *Litiopa* (gastropod mollusc). **B,** *Amphithoë* (amphipod crustacean). **C,** *Luconacia* (caprellid amphipod). **D,** *Anemonia sargassensis* (sea anemone). **E,** *Platynereis* (nereid polychaete). **F,** *Sertularia* (hydroid cnidarian). **G,** *Clytia* (hydroid cnidarian). **H,** *Scyllaea* (nudibranch gastropod). **I,** *Spirorbis* (spirorbid polychaete). **J,** Copepod crustacean. **K,** *Zanclea* (hydroid cnidarian). **L,** *Ceramium* (an epiphytic red alga). **M,** *Membranipora* (bryozoan). **N,** *Doto* (nudibranch gastropod). **O,** *Gnesioceros* (polyclad flatworm). **P,** *Obelia* (hydroid cnidarian). **Q,** *Anoplodactylus* (pycnogonid) feeding on a nudibranch. **R,** *Fiona* (nudibranch gastropod).

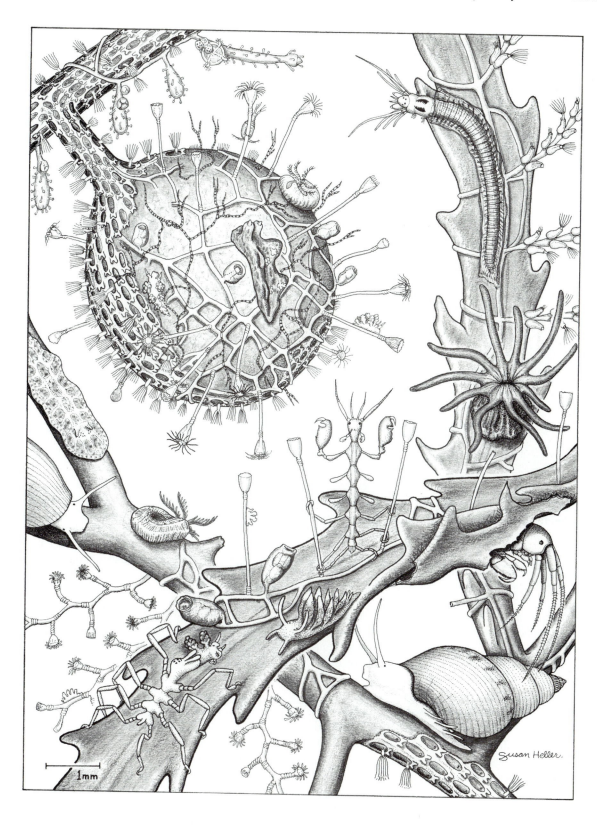

Susan Heller.

1mm

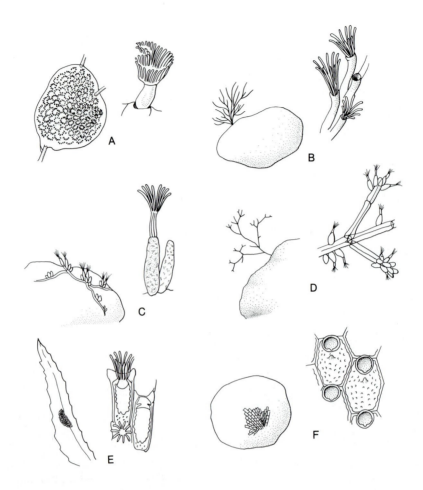

FIGURE 19–2 Bryozoan diversity. **A,** The freshwater bryozoan, *Pectinatella magnifica* (Class Phylactolaemata). **B,** The marine bryozoan, *Crisia eburnea* (Class Stenolaemata). **C–F,** Marine bryozoans (Class Gymnolaemata). **C,** The stolonate ctenostome, *Bowerbankia maxima.* **D,** The erect ctenostome, *Zoobotryon verticillatum,* showing jointed stolon composed of tubular kenozooids. **E,** The anascan cheilostome, *Membranipora tuberculata,* encrusting a leaflet of *Sargassum.* **F,** The ascophoran cheilostome, *Schizoporella unicornis,* encrusting the surface of a rock.

provided with a lid **(operculum),** which closes when the lophophore is withdrawn (Fig. 19–10D–G). In species in which a collar is present, the space occupied by the retracted collar and inverted section of the trunk is called the **atrium** (Fig. 19–10A–C).

The outer cuticular covering is secreted by the epidermis of the body wall (Figs. 19–4; 19–20A). As would be expected, complete muscle layers are absent in the body wall that underlies the rigid exoskeleton of gymnolaemates, and the epidermis covers a thin, delicate mesothelium.

In gymnolaemates, the lophophore is circular and consists of a simple ridge bearing 8 to 30 or more **tentacles** (Figs. 19–2B–E; 19–3; 19–9; 19–14C).

When retracted, the tentacles are bunched together; when protruded, the tentacles fan out, forming a bell-shaped funnel with the mouth at its base.

The lateral surfaces and the inner frontal surface of each tentacle bear a longitudinal tract of multiciliated epidermal cells. Below the epidermis are a longitudinal nerve, a layer of collagenous connective tissue, and a mesothelium enclosing a coelom. Some of the cells of the coelomic lining are enlarged to form the longitudinal muscle tract of the tentacle.

The **mouth** at the center of the lophophore opens into a **U-shaped digestive tract** (Fig. 19–8A). The **anus** opens through the dorsal side of the introvert and is situated *outside* the lophophore, hence the

FIGURE 19–3 Living bryozoan colonies. **A,** Biserially arranged autozooids of the erect cheilostome species, *Bugula neritina.* **B,** Autozooids on the surface of a branch (kenozooid) of the stolonate ctenostome, *Zoobotryon verticillatum.* **C,** Erect calcified colony of *Scrupocellaria regularis,* an ascophoran cheilostome. **D,** Surface view of the encrusting ascophoran, *Watersipora subovoidea,* showing one fully expanded and one partly expanded polypide.

name *Ectoprocta* (meaning "outside anus"), which is sometimes used for the phylum. The coelom is partially divided by a septum into (a) an anterior ring that occupies the base of the lophophore and extends into each tentacle as part of the **mesocoel,** and (b) a larger posterior or trunk coelom **(metacoel).** The two divisions are connected by one or two pores. The trunk coelom is crossed by muscle fibers and by a single or branching tube of mesothelial tissue, which constitutes the **funiculus** (Fig. 19–4), and plays a role in nutrient transport between zooids.

The nervous system is composed of a nerve ring around the pharynx with a ganglionic mass on its dorsal side (Fig. 19–4). The ganglion and ring give rise to nerves that extend into each of the tentacles and into other parts of the body. There are no specialized

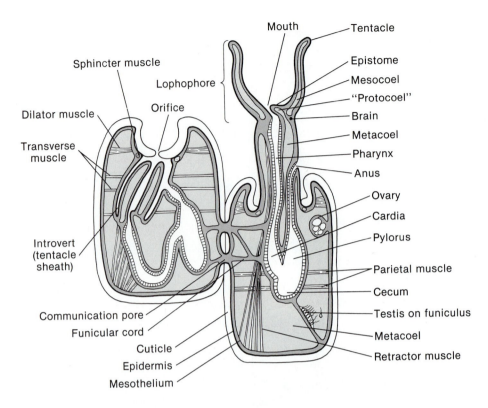

FIGURE 19–4 Organization of two generalized bryozoan zooids. *(Modified and redrawn after Marcus from Hyman, L. H. 1959. The Invertebrates. Vol. V. The lesser coelomates. McGraw-Hill Book Co., New York.)*

sense organs in bryozoans, but individual sensory cilia occur on the tentacles.

The terms *cystid* and *polypide* will be encountered by students who deal with the literature on bryozoans. The term **cystid** refers to the exoskeleton and body wall of the stationary trunk. The term **polypide** refers to the introvert, lophophore, and movable contents within the trunk, that is, the gut, muscles, and other structures.

Zooids of the freshwater Phylactolaemata differ from those of gymnolaemates in many ways. The body wall contains circular and longitudinal muscle layers, and the epidermis is covered by a cuticle which, in *Pectinatella* for example, is a thick gelatinous layer. The lophophore of freshwater phylactolaemates, with the exception of the circular lophophore of *Fredericella*, is horseshoe-shaped and is composed of two ridges bearing a total of 16 to 106 tentacles (Figs. 19–2A; 19–5). Such a shape provides a greater food-collecting surface for these zooids, which tend to be larger than the marine gymnolae-

mates. As in phoronids, one ridge passes above and one below the mouth at the bend of the horseshoe. A dorsal hollow lip (**epistome**) overhangs the mouth and is responsible for the name *Phylactolaemata*, which means "covered throat." (*Gymnolaemate* means "naked throat" in reference to the absence of an epistome. *Stenolaemate* means "narrow throat.") The epistome is lined by a coelom that may be homologous to the protocoel in the preoral lobe of hemichordates, phoronids, and brachiopods. The phylactolaemate protocoel, however, is incompletely separated from the metacoel (Figs. 19–4; 19–5B).

Colony Organization

The form of a bryozoan colony, or **zoarium,** depends on the pattern of asexual budding of the zooids, the degree of polymorphism, the arrangement of the polymorphs within the colony, and the composition and extent of secretion of the skeletal material. In addition to the support provided by the substratum on which the colony grows, there are three sources of skeletal support in bryozoan colonies. These are turgor pressure of the coelomic fluid (e.g., *Zoobotryon,* approximately 2 atmospheres of pressure), calcification (most bryozoans), and the production of a gelatinous (e.g., *Pectinatella*) or rubbery *(Alcyonidium)* extracellular material.

Gymnolaemates are very common and abundant marine animals that exhibit a wide range of colonial forms. Although some species have been recorded from depths as great as 8200 m, most species are found in coastal waters attached to rocks, pilings, shells, algae, and other animals. Members of such genera as *Bowerbankia, Amathia,* and *Zoobotryon* form **stoloniferous** colonies that resemble the hydroid colonies. (Figs. 19–2C, D; 19–3B). The erect or creeping **stolons**—stemlike sections that link feeding members together—are composed of modified zooids

that give the stolon a jointed appearance. Unmodified feeding zooids are often completely separate from one another but are attached by their posterior ends to the common stolon. The exoskeleton of stoloniferous bryozoans usually lacks calcium carbonate.

The vast majority of marine bryozoans are not stoloniferous, and the colony is formed by the more direct attachment and fusion of adjacent zooids. Moreover, the orientation of the body to the substratum is different. The dorsal surface is attached to the substratum, or to other zooids, and the ventral surface, now called the **frontal surface,** is exposed to the surrounding water (Fig. 19–2E, F).

The growth patterns of nonstoloniferous bryozoan colonies vary greatly (Fig. 19–2). Many slightly calcified species, such as the common Atlantic *Bugula neritina,* form erect, bushy colonies that look superficially similar to seaweed. In *Bugula,* for exam-

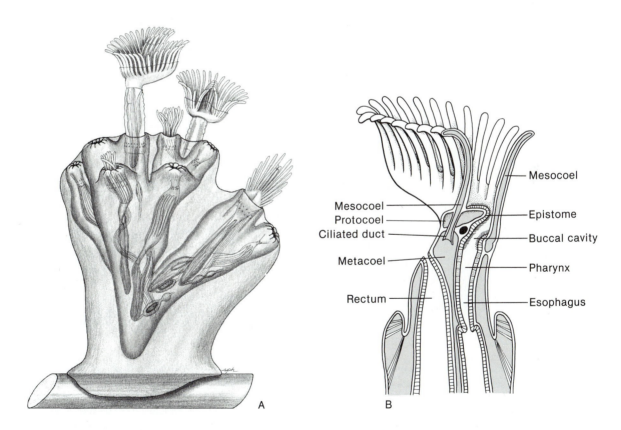

FIGURE 19–5 Phylactolaemate (freshwater) bryozoans. **A,** Small colony of *Lophopus crystallinus* attached to a water plant. **B,** Organization of *Plumatella fungosa,* shown in parasagittal section. One of the two ducts linking the metacoel and mesocoel is shown. *(A, After Allman. B, Modified and redrawn from Brien, P. 1960. Sous-Classe des Phylactolèmes ou Phylactolémates. In Grassé, P.-P. (Ed.): Traité de Zoologie. Vol 5. No. 2. Masson et Cie, Paris.)*

ple, such a plantlike growth form is characteristically attained through a biserial attachment of zooids (Fig. 19–3). The most common type of colony is the **encrusting** form, in which the zooids are organized as a lichen-like sheet attached to rocks and shells. The exoskeleton is usually calcareous, and because the lateral and end walls of the zooids are fused, the orifice has typically migrated toward the exposed frontal surface. *Membranipora*, *Micro-* *porella*, and *Schizoporella* are very common encrusting genera (Figs. 19–2E, F; 19–3D).

Erect **foliaceous** colonies are composed of one sheet of zooids or two sheets attached back to back. The foliaceous cheilostome, *Thalamoporella*, resembles an open head of lettuce, each leaf of which is as rigid and fragile as a thin potato chip. Other colonies are tuftlike, and in some, the zooids are radially arranged.

The colonies of freshwater phylactolaemates are of two types. In forms such as *Lophopus*, *Cristatella*, and *Pectinatella*, the zooids project from one side of the soft colony sac, resembling the fingers of a glove (Fig. 19–5A). Colonies of *Pectinatella* secrete a gelatinous base, sometimes several feet in diameter,

to which the zooids adhere (Fig. 19–2A). The other type of colony, of which *Plumatella*, *Fredericella*, and *Stolella* are examples, has a more or less plantlike growth form, in which there are either erect or creeping branches composed of a succession of zooids. The colonies of freshwater bryozoans are attached to vegetation, submerged wood, rocks, and other objects. *Cristatella*, in which the colony is a flattened gelatinous ribbon, is not fixed and creeps over the substratum at a rate of up to 10 cm a day (Fig. 19–6C). Small colonies of *Pectinatella magnifica* and *Lophopus crystallinus* can also move, but only about 2 cm per day. The mechanism of locomotion is unknown.

Although the zooids are microscopic, the colonies themselves are one to several centimeters in diameter or height and may attain a much greater size. Some stoloniferous species reach a meter or more in size, and some encrusting colonies may attain a diameter of more than 50 cm and contain as many as 2,000,000 zooids. White or pale tints are typical of most colonies, but darker colors, especially orange, may be found. The taxonomy of marine bryozoans is based almost exclusively on the structure of the exoskeleton and the colonial organization.

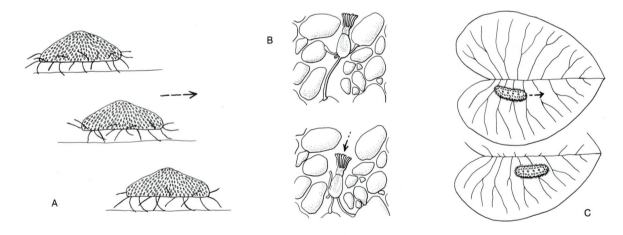

FIGURE 19–6 Bryozoan locomotion. Although of rare occurrence, a few species of bryozoans are able to move over or through the substratum. *Selenaria maculata* is a small discoid colony that moves over sand using its marginal vibracula as stepping appendages. As the leglike vibracula move, the colony lurches ahead in 3 mm-long increments. **B,** *Monobryozoon ambulans* is a tiny interstitial zooid that uses contractile processes bearing sticky tips to anchor and pull itself through the spaces between sand grains. New buds typically form on the processes, which are sometimes called "pseudostolons." **C,** A colony of the phylactolaemate, *Cristatella mucedo*, creeps on its flattened lower surface. The creeping mechanism is unknown, although muscles are thought to be involved. (*A, Drawn from photographs and description of Cook, P. L., and Chimonides, P. J. 1978. Observations on living colonies of* Selenaria *(Bryozoa, Cheilostomata). I. Cah. Biol. Mar. 19:147–158. B, Modified from Swedmark, B. 1964. The interstitial fauna of marine sand. Biol. Rev. 39:1–42. C, After Wesenburg-Lund from Hyman, L. H. 1959. The Invertebrates. Vol. 5. Lesser Coelomates. McGraw-Hill Book Co., New York.*)

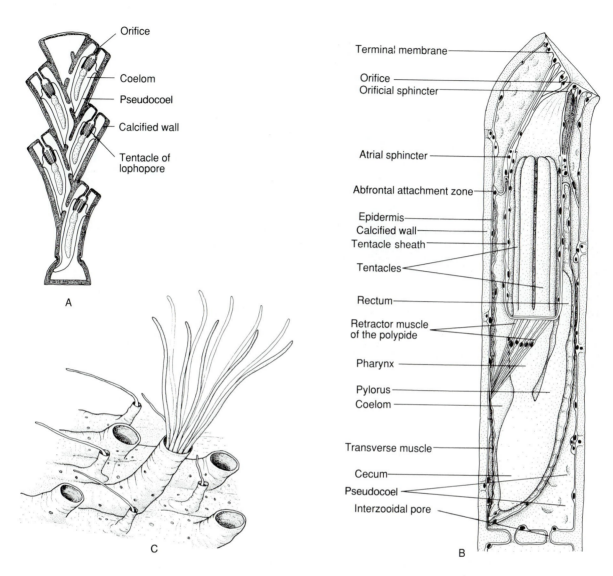

FIGURE 19–7 Stenolaemate bryozoans. **A,** Diagram of a colony of *Crisia*. **B,** Diagram of one zooid of *Crisia*. **C,** Five autozooids and four heterozooids of *Diplosolen*. The sweeping motion of the single tentacle of the heterozooid is believed to keep the surface of the colony clean. *(A, B, From Nielsen, C., and Pedersen, K. J. 1979. Cystid structure and protrusion of the polypide in* Crisia *(Bryozoa, Cyclostomata). Acta Zool. 60:65–88. C, From Silen, L., and Harmelin, J.-G. 1974. Observations on living Diastoporidae, with special regard to polymorphism. Acta Zool. 55:81–96.)*

Colony Integration

The zooids of a colony are connected by **pores** that are located in the transverse end walls, the lateral walls, or both, depending on the growth pattern of the colony (Figs. 19–4; 19–7B; 19–8). In two classes (Phylactolaemata and Stenolaemata), at least some of the pores are open, and fluid may flow between the interconnected zooids of the colony. In phylactolae-mates, coelomic fluid passes through the pores but, in stenolaemates, connective tissue fluid (hemolymph) is exchanged between the zooids (Fig. 19–9C, D). The pores of gymnolaemates differ in being closed by a plug of cells, which are arranged in an ordered and polarized manner. Some of these cells are associated with branches of the funiculus (Fig. 19–9B, E).

Electrophysiological studies have confirmed that conduction of nerve impulses occurs in colonies of

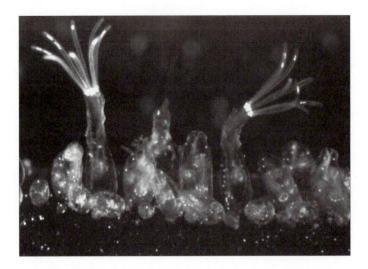

FIGURE 19–8 Living zooids of the ctenostome, *Zoobotryon verticillatum*. The two fully extended zooids are actively orienting their lophophores while feeding. The right zooid is using a tentacle flick to bat a food particle (not visible) toward the mouth. Several retracted autozooids as well as new autozooid buds are visible in the photograph. All of the autozooids have arisen from and are attached to a large kenozooid (stolon segment) at the bottom of the photograph.

some nonstoloniferous bryozoans, such as *Electra* and *Membranipora.* Certain nerves that encircle the zooid wall enter the pores and make connections with similar nerves in adjacent zooids. This colonial nervous system coordinates lophophore feeding activity and orientation within the colony.

Polymorphism

Although the colonies of phylactolaemates are strictly monomorphic, those of most gymnolaemates are polymorphic. In such polymorphic colonies, a typical feeding zooid, called an **autozooid,** makes up the bulk of the colony. Reduced or modified zooids that serve other functions are known as **heterozooids.** A common type of heterozooid is one modified to form stolons (**kenozooid**), attachment discs, rootlike structures, and other such vegetative parts of the colony. These zooids are so modified that they consist of little more than the body wall and strands of funicular tissue passing through their interior (Fig. 19–2D; 19–3B; 19–9B).

Two other types of heterozooids, called avicularia and vibracula, are found in many cheilostomes. An **avicularium** is usually smaller than an autozooid, and its internal structure is greatly reduced (Fig. 19–10A, B). The operculum and its muscles are typi-

cally highly developed and modified, however, converting the lid into a powerful movable jaw. Avicularia may be sessile or stalked. When stalked, they make repetitive nodding motions, and the movable jaw (operculum) can be snapped closed like the wire bail on a mouse trap. Stalked avicularia are found in *Bugula* (but not in the common *Bugula neritina,* which may use chemical defenses instead) and resemble little bird heads attached to the colony. Avicularia typically defend the colony against small organisms, including the settling larvae of other animals. The avicularia of *Bugula* however, appear to be more important in defending against larger crawling animals (0.5–4 mm), such as tube-building amphipods and polychaetes, whose appendages are seized by the jaws of the avicularium. In a case study of *Reptadeonella costulata,* it was shown that a sessile avicularium, like a mousetrap, could immobilize and strangle a syllid polychaete.

In a **vibraculum** the operculum is modified to form a long bristle, sometimes called a seta, which can be moved in more than one plane. The vibraculum is used in species of the sand-dwelling *Discoporella* and *Cupuladria* to sweep away fouling material as well as to move these bryozoans vertically through, and to a limited extent, over the sand (Figs. 19–10C, D; 19–11B). Species of the discoid *Sele-*

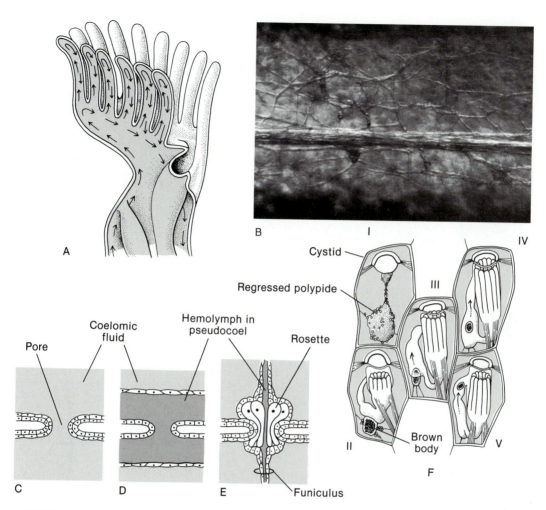

FIGURE 19–9 Internal transport and polypide regression in bryozoans. **A,** Circulation of coelomic fluid using cilia in the phylactolaemate, *Pectinatella*. **B,** Photograph of the main funiculus and its branches in a kenozooid (stolon zooid) of the ctenostome, *Zoobotryon verticillatum*. The bases of a few attached autozooids can be seen at the top of the photograph. **C–E,** A communication pore between two zooids of a phylactolaemate **(C),** stenolaemate **(D),** and gymnolaemate **(E). F,** Polypide regression (I) and brown body formation and expulsion (II–V) in the anascan cheilostome, *Electra pilosa*. (A, After Oka from Hyman, L. H. 1955. The Invertebrates. Vol. 5. The Lesser Coelomates. McGraw-Hill Book Co., New York. C–E, Modified and redrawn from Mackie, G. O. 1986. From aggregates to integrates: Physiological aspects of modularity in colonial animals. Phil. Trans. R. Soc. Lond. B 313:175–196. F, Redrawn and rearranged after Marcus from Brien, P. 1960. Classe des Bryozoaires. In Grassé, P.-P. (Ed.): Traité de Zoologie. Vol. 5. Part 2. Masson et Cie, Paris.)

naria move efficiently over sand using coordinated movements of marginal vibracula (Fig. 19–6A). Modification of zooids for reproductive purposes is very common and will be discussed later.

Feeding and Digestion

During feeding the lophophore is pushed outward through the atrium and orifice, causing the tentacular sheath to evert. The tentacles then expand, forming a bell-shaped funnel. Protraction of the lophophore is affected in all cases by a muscular elevation of the coelomic fluid pressure, although the mechanism by which this is accomplished varies. In the freshwater phylactolaemates the coelomic fluid pressure is elevated by contraction of the body wall musculature, which compresses the trunk and everts the polypide.

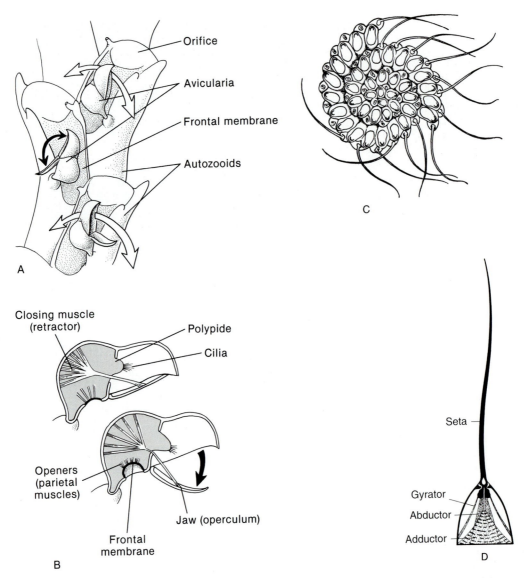

FIGURE 19–10 Bryozoan heterozooids. **A,** Bird's head avicularia of the anascan cheilostome, *Bugula fulva.* Open arrows show avicularium movement; solid arrows, movement of jaw (operculum). **B,** Simplified anatomy of a bird's head avicularium. Compare with Figure 19–12D, E. **C,** Part of a colony of *Heliodoma,* showing marginal vibracula. **D,** Anatomy of a vibraculum. *(A, Modified and redrawn from Maturo, F. J. S., Jr. 1966. Bryozoa of the southeast coast of the United States: Bugulidae and Beaniidae (Cheilostomata: Anasca). Bull. Mar. Sci. 16:556–583. B, Modified and re-drawn after Calvet from Brien, P. 1960. Classe des Bryozoaires. In Grassé, P.-P. (Ed.): Traité de Zoologie. Vol 5. Part 2. Masson et Cie, Paris. C, From Moore, R. C. 1953. Treatise on Invertebrate Paleontology. Geological Society of America and University of Kansas Press Lawrence. D, Based on Marcus and Ryland.)*

The mechanism of polypide protrusion in steno-laemates, such as *Crisia,* depends on the contraction of circular body wall muscles, as in the phylactolae-mates. Unlike the freshwater bryozoans, however, the cuticle of stenolaemates is calcified and rigid, and the body does not compress during lophophore protru-sion. Instead, the body wall muscles and coelom are widely separated from the epidermis and cuticle by an expanded fluid-filled connective tissue (a pseudo-coel), thus forming a chamber into which the coelom and its contents can expand (Fig. 19–7A, B).

In gymnolaemates that have a flexible, chitinous cuticle *(Bowerbankia, Amathia),* the contraction of transverse **parietal muscle bands,** which are attached

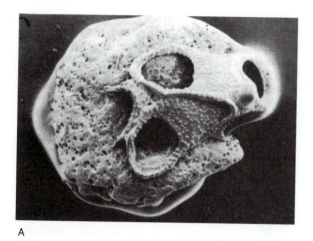

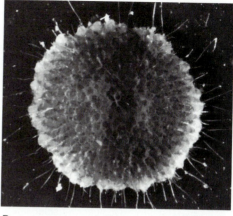

A   B

FIGURE 19–11 Bryozoans living on or in sand. **A,** A small colony of *Membranipora triangularis* encrusts a single grain of sand. Such colonies can spread from grain to grain by asexual division. Although not a common habitat for bryozoans, the surface of sand grains is exploited by at least 35 species. **B,** A domed colony of *Cupuladria doma* from an offshore sand bottom near Ft. Pierce, Florida. The slender vibracula enable the species to remove sand from its surface and to move over and through the sand. *(A, After Winston from Edwards, D. D. 1987. Home on the grain. Science News. 131:156–157.)*

to opposite walls (Figs. 19–4; 19–12A–C), causes the flexible wall to bow inward, thereby elevating the coelomic fluid pressure.

In some calcareous gymnolaemates (anascan cheilostomes), such as *Membranipora,* and even some forms with rigid chitinous exoskeletons, the exposed frontal surface contains a window covered by a thin chitinous membrane (the **frontal membrane**) (Figs. 19–2E; 19–11A; 19–12D, E). The parietal muscles are inserted on the inner side of the membrane. On contraction, the flexible frontal membrane bows inward, increasing the coelomic pressure. This arrangement, however, leaves the zooid vulnerable to attack on its exposed, unprotected, soft, frontal surface. Probably because of this vulnerability, many calcareous gymnolaemates (ascophoran cheilostomes and others) protect the frontal surface by calcifying it. In these forms having a calcified frontal wall, the pressure-regulating membrane is internalized in the form of a sac, called an **ascus,** which opens to the exterior via one or two tiny pores. The parietal muscles attach to the undersurface of the ascus and on contraction increase the volume of the sac, causing water to enter through the surface pores. As the volume of the ascus increases, the pressure of the coelomic fluid is raised, and the polypide everts (Fig. 19–12F, G).

Retraction of the lophophore and introvert through the orifice is effected by the contraction of the striated **lophophoral retractor muscles** (Figs.

19–4; 19–12). Retraction is rapid, often requiring less than 60 ms to complete. Reextension is slower and often tentative. Before complete reextension, many zooids partly extend the still tightly bundled tentacles, exposing the sensory cilia at their tips to the surrounding water.

When the lophophore is protruded, the lateral ciliated tracts on the tentacles create a current that sweeps downward into the funnel and passes outward between the tentacles (Fig. 19–13E). Small phytoplanktonic organisms, which are probably the principal food of bryozoans, are driven into the funnel with the water current, trapped on the tentacles, and delivered to the mouth by cilia on the inner frontal surface of each tentacle. The mechanism by which food particles are filtered from the water is not fully understood, but two ideas have been advanced. The ciliary reversal theory suggests that when suspended particles touch the lateral cilia, they cause a local reversal of beat, which kicks the particle back onto the upstream, frontal side of the tentacle for transport to the mouth. The impingement theory, on the other hand, proposes that particles are removed from the water stream as they strike the frontal surface of each tentacle. However the particles are cleared from the water, the general mode of feeding may be classified as an upstream ciliary collecting system.

Tentacle flicking is a common accessory feeding mechanism in many species (even primary in a few).

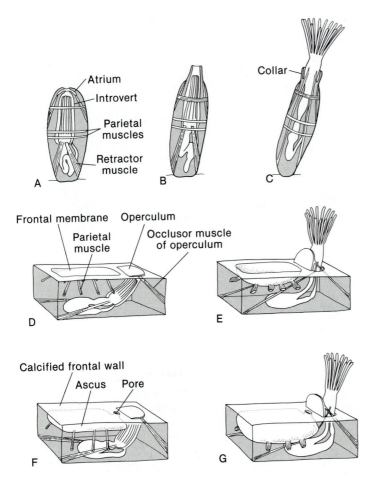

FIGURE 19–12 Polypide protrusion in gymnolaemate bryozoans. **A–C,** Contraction of parietal muscles in ctenostomes compresses the body to elevate the coelomic pressure, protruding the polypide. The collar in many ctenostomes **(C)** is pleated longitudinally, and each pleat resembles a tooth on a comb. The name *ctenostome,* meaning "comb mouth," is a reference to this feature. **D, E,** Although the cystid wall is largely calcified in the anascan cheilostomes, such as *Membranipora,* one surface, the frontal membrane, remains uncalcified and flexible. Contraction of the parietal muscles bows the frontal membrane inward **(E)** which elevates the coelomic pressure and protrudes the polypide. **F, G,** In the anascan and cribrimorph cheilostomes, the frontal wall is calcified, and the flexible surface is internalized as a sac, or ascus. The action of the parietal muscles is to bow one wall of the ascus inward **(G),** which elevates the coelomic pressure and extends the polypide. Water enters the ascus via the surface pore during polypide protrusion.

A particle is batted toward the mouth by a rapid inward flick of one tentacle (Fig. 19–13B). *Bugula neritina* captures zooplankton by closing the tips of the tentacles to form a cage around the prey. Many species scan for particles by rotating or bending the lophophore (Fig. 19–8). Adjacent expanded funnels may form an extensive filtering surface. In encrusting forms, the large volume of water passing below the filters exits through "chimneys," which are blank areas formed by lophophores tilted away from each other or by the space occupied by one or more modified zooids.

Food particles accumulate beneath the epistome in phylactolaemates and within the expanded mouth of gymnolaemates. When the bolus reaches a certain size, the muscular sucking **pharynx** dilates rapidly and, together with the **esophagus** (when present), pumps the food into the stomach (Fig. 19–13B, C). Particles may be rejected by mouth closure, tentacle flicking, funnel closure, or simply being passed between the tentacles (Fig. 19–13E).

From the mouth and foregut, food particles pass into the large stomach, which composes much of the U-shaped gut (Fig. 19–4). The dilated anterior tubular part of the stomach, called the **cardia,** is separated from the esophagus by a valve. A valvelike constriction separates the posterior stomach region, or **pylorus,** from the **intestine** and **rectum.** A large **cecum** projects backward from the central part of the stomach. In some bryozoans, such as *Bowerbankia* and *Amathia,* the cardia is modified as a **gizzard,** which crushes diatoms, thus making their contents available for digestion (Fig. 19–13D). The gizzard has a well-developed, circular muscle layer, and the lining epithelial cells bear chitinous teeth, which are ultrastructurally identical to annelid setae.

Digestion is both extracellular and intracellular within the stomach, with the cecum being the princi-

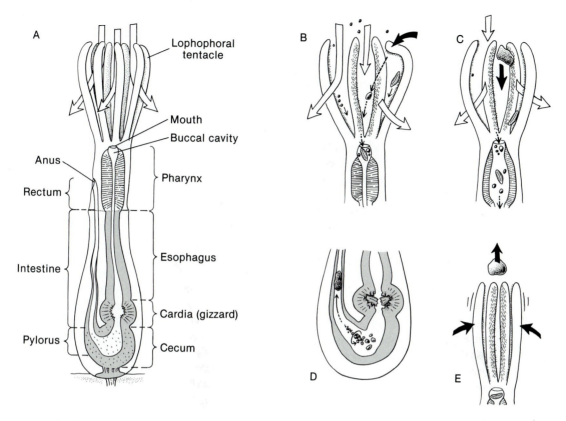

FIGURE 19–13 Digestive system and feeding in gymnolaemates (based on *Zoobotryon*). **A,** Water currents (open arrows) and digestive system. **B–D,** Water currents (open arrows) bring in food particles **(B),** which are either batted toward the mouth by a tentacular flick (large solid arrow, dashed arrow) or caught on the tentacle surface and moved toward the mouth by frontal cilia (small arrows). The collected particles accumulate in the ciliated buccal cavity **(B)** and are engulfed by a rapid dilation of the muscular pharynx **(C).** The food particles move rapidly down the esophagus, pass through the muscular gizzard, which crushes diatoms, before entering the cecum **(D).** Once in the cecum, the food is rotated by cilia and digested; indigestible material is compacted into fecal pellets in the intestine **(D).** Rejected particles may pass between tentacles and be carried away in the exhaust, but sometimes the tentacles bunch together rapidly and particles are rejected as shown in **E.**

pal site of intracellular digestion. Food passes through the stomach by peristaltic contractions, but the pylorus rotates and compacts waste materials, which then pass into the intestine (Fig. 19–13D).

Gas Exchange, Internal Transport, and Excretion

Gas exchange occurs across the exposed body surface, including the lophophore. Internal transport of gases, some food, and wastes is provided by the coelomic fluid. Circulation of coelomic fluid using cilia is known only in the relatively large (2–4 mm) zooids of some phylactolaemates (*Paludicella, Pectinatella*) (Fig. 19–9A). The coelomic fluid contains coelomocytes (but no respiratory pigments), which engulf and store waste materials. In stenolaemates, both coelomic and pseudocoelic fluids are probably important in transport (Fig. 19–9D). The funicular system provides for at least some nutrient transport and is the main system for the colony-wide dispersal of metabolites in gymnolaemates (Fig. 19–9B, E).

The absence of nephridia in bryozoans is unusual and cannot be attributed to small body size, because most small invertebrates, including larval stages, have nephridia. It is likely that ammonia diffuses across the general body surface, and other wastes, such as uric acid, may be stored in body tissues. A kind of storage excretion may be associated with the phenomenon of **polypide regression** (Fig. 19–9F).

Regardless of the age of a colony, the lophophore and gut of a zooid degenerate after a few weeks leaving the cystid unaffected. Some components are phagocytized, but a large residual mass of necrotic cells containing stored waste products remains lodged in the coelom as a conspicuous dark ball, called a **brown body.** Polypide regression is followed by the regeneration of a new lophophore and gut from the wall of the cystid. In some species, the old brown body is a permanent resident of the coelom; in others, it is incorporated within the stomach of the regenerating polypide and expelled during the first defecation (Fig. 19–9F).

Alternating phases of degeneration and regeneration are especially common in bryozoan colonies that live a number of years, and even in some annual species. A large colony may exhibit zones of individuals in various stages of development. For example, at the outer perimeter of an encrusting colony, or at the tips of erect branches in a form such as *Bugula,* budding and development of new individuals occur. Further inward is a zone of fully developed individuals, which are feeding and reproducing. Still further inward is a zone of degenerating members containing brown bodies. To the inner side of the degenerate zone may be a zone of regenerated feeding zooids.

Evolution and Distribution of Bryozoans

Of the three classes of living bryozoans, the freshwater phylactolaemates are considered to be the most primitive. The cylindrical zooids, the anterior orifice, the horseshoe-shaped lophophore, the presence of an epistome, and the nonpolymorphic colonies are all considered primitive features. In many other ways, phylactolaemates are specialized, especially in their reproduction, which is evolutionarily derived.

Although not usually common, freshwater bryozoans are widely distributed in lakes and streams that do not contain excessive mud or silt. Many species, such as *Fredericella sultana* and *Plumatella repens,* are cosmopolitan.

Unfortunately, there are no fossil phylactolaemates, and thus paleontology contributes no information regarding their relationship with the marine classes. The first known marine bryozoan is a questionable fossil species from the late Cambrian. But beginning with the Ordovician there is a rich fossil record, and thousands of fossil species have been described. Stenolaemates, of which there are three distinct orders, dominate the Paleozoic fauna, although there were Paleozoic ctenostome gymnolaemates. Cheilostomes, the dominant marine forms today, made their appearance in the late Jurassic.

Marine bryozoans successfully exploit all types of hard surfaces, including rock, shells, coral, and wood (e.g., mangrove roots), and even the surface of sand grains (Fig. 19–11A). A few species of *Monobryozoon* live and move about in the interstitial spaces of marine sand (Fig. 19–6B; p. 206). Some species bore in calcareous substrata. The stolons of boring species are located in tunnels within the substratum, and the autozooids open to the surface. A few bryozoans also live on soft bottoms. The circular, shield-shaped colony of *Cupuladria doma* reaches the size of a small coin. It rests free on the bottom so that the convex surface bearing the frontal surface of the zooids is directed upward (Fig. 19–11B).

From an economic standpoint, marine bryozoans are one of the most important groups of organisms that foul ships' bottoms. About 120 species, of which different species of *Bugula* are among the most abundant, have been taken from ships' bottoms.

Along with hydroids, bryozoans rank among the most abundant marine epiphytic animals. Large brown algae are colonized by many species, which display distinct preferences for certain types of algae. Evidence indicates that at the time of settling, the larvae of epiphytic species are attracted to the algal substratum, perhaps by some substance produced by the algae (p. 364). The widely distributed, encrusting *Membranipora,* a species of which is abundant on floating *Sargassum* (Fig. 19–1M), displays a number of adaptations for an epiphytic life. The frontal surface is noncalcified, and the lateral walls are jointed (i.e., broken), permitting some flexibility when the algal thallus bends. Although colonies encrust over large portions of kelp, growth of the colonies is predominantly in the direction of the stalk of the kelp thallus and is controlled by the movement of water over the algal surface.

Reproduction

All freshwater bryozoans and most marine species are **hermaphrodites.** Simultaneous production of eggs and sperm may take place, but a tendency toward protandry is more common. A colony may thus be composed of male and female zooids. The one or two ovaries and the one to many testes are bulging masses of developing gametes covered by mesothelium. The ovaries are located in the distal end of the animal, whereas the testes occur in the basal end (Fig. 19–4). Genital ducts are absent, and the eggs and sperm rupture into the coelom.

Some marine species (*Electra* and *Membranipora*) shed small eggs directly into the seawater. The vast majority of bryozoans brood their eggs, which are almost always large, yolky, and few in number. A variety of brooding mechanisms are employed. A few species brood their eggs within the coelom, but most brood them externally. The digestive tract and lophophore often degenerate to provide space for the egg. The cavity of the introvert or invaginations of the atrial wall are common sites for brooding (Fig. 19–4). Many cheilostomes, including the common *Bugula*, brood their eggs in a special external chamber called an **ovicell** (Fig. 19–14C, D). The body wall

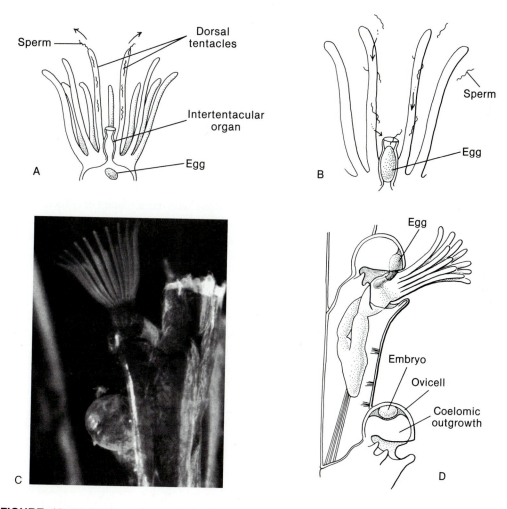

FIGURE 19–14 Sperm release and egg brooding in bryozoans. **A, B,** Dorsal view of the lophophore of *Electra posidoniae.* **A,** Sperm release and entry of egg into the intertentacular organ. **B,** Sperm entry and fertilization. **C,** Autozooid and nearby ovicell containing a developing embryo in *Bugula neritina.* **D,** *Bugula avicularia,* showing extrusion of egg from the supraneural pore of an autozooid into the ovicell (above) and egg positioned within an ovicell (below). *(A, B, Modified and redrawn from Silen, L. 1966. On the fertilization problem in the gymnolaematous bryozoa. Ophelia. 3:113–140. D, Modified and redrawn from Brien, P. 1960. Classe des Bryozoaires. In Grassé, P.-P (Ed.): Traité de Zoologie. Vol. 5. Part 2. Masson et Cie, Paris.)*

at the distal end of the zooid grows outward, forming a large hood (the ovicell). A second, smaller evagination, which is directly connected to the coelom, bulges into the space formed by the ovicell. A single egg is brooded in the space between the two evaginations (Fig. 19–14D). The developing embryo may derive its nutrition entirely from contained yolk, but in many species, including *Bugula*, placenta-like connections from the funiculus to the ovicell provide food material from the maternal zooid. An ovicell is thought to represent a modified zooid and thus contributes to the polymorphic nature of the colony.

When the eggs are shed into the seawater or have been brooded internally, they move through the coelom along a ciliated gutter, and escape to the exterior by way of a special opening between the bases of the two dorsalmost tentacles. This **coelomopore** may be a simple opening (supraneural pore) or may be mounted at the end of a projection called an **intertentacular organ** (Fig. 19–14A, B). In those species with ovicells, the deformable egg is extruded through the coelomopore as a "stream" and then rounds up within the ovicell cavity (Fig. 19–14D).

In all species that have been studied (*Electra, Membranipora, Schizoporella, Bugula,* and others), the sperm are shed though terminal pores of two or more tentacles of the lophophore and are disseminated into the seawater (Fig. 19–14A). The liberated sperm, when caught in the feeding currents of other individuals, adhere to the tentacular surfaces of the lophophore or enter the intertentacular organ. In the former case, the eggs are fertilized as they leave the coelomopore. In those species in which the sperm enter the intertentacular organ *(Electra, Membranipora),* this structure is the site of fertilization (Fig. 19–14B). The coelomopore may be the means by which sperm enter the coelom in species that brood their eggs internally. Fertilization between zooids of the same colony is probably common, but sufficient cross-fertilization between colonies occurs to ensure outbreeding.

Cleavage in marine species is biradial (Fig. 19–15) and leads to a larva, which escapes from the brood chamber. The stenolaemates are remarkable in exhibiting a **polyembryony,** in which the primary embryo asexually produces numerous secondary embryos, which may then produce tertiary embryos all of which are genetically identical (clones). The larvae of bryozoans vary considerably in form (Fig. 19–16), but all possess a locomotor ciliated girdle (corona), an anterior tuft of long cilia, and a posterior adhesive sac (**internal sac;** Fig. 19–29D). The larvae of some species of nonbrooding gymnolaemates, such as *Electra* and *Membranipora,* are called **cyphonautes larvae.** Cyphonautes larvae are triangular and greatly compressed, and are encased by a bivalved chitinous shell (Figs. 19–16A; 19–29C, D). Only the larvae of nonbrooding bryozoans possess a functional digestive tract and feed during larval existence. Feeding larvae may have a larval life of several months, but the larvae of brooding species, which are nonfeeding, have a brief larval existence prior to settling.

During settling the internal sac everts and becomes fastened to the substratum by means of adhesive secretions. The larval structures of the attached larva undergo retraction and histolysis, followed by

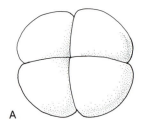
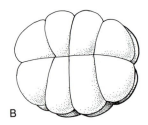
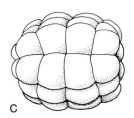

A   B   C

FIGURE 19–15 Biradial cleavage in the gymnolaemate, *Schizoporella erata.* Animal pole views of 8- **(A),** 16- **(B),** and 32-cell **(C)** stages. *(Redrawn after Zimmer from Reed, C. G. 1991. Bryozoa. In Giese, A. C., Pearse, J. S., and Pearse, V. B. (Eds.): Reproduction of Marine Invertebrates. Vol. VI. Echinoderms and Lophophorates. The Boxwood Press, Pacific Grove, CA. pp. 85–245.)*

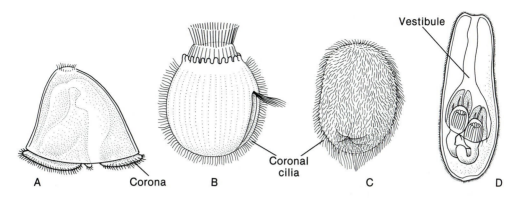

FIGURE 19–16 Bryozoan larvae. **A,** A planktotrophic cyphonautes larva, as found in the genera *Membranipora* and *Electra*. The larva is laterally compressed and bears a shell (see also Fig. 19–29). **B–D,** Lecithotrophic larvae of the gymnolaemate, *Bugula neritina* **(B),** the stenolaemate, *Crisia eburnea* **(C),** and a phylactolaemate **(D).** In phylactolaemates, autozooids differentiate precociously within the larval body. *(B, Redrawn from Nielsen, C. 1971. Entoproct life-cycles and the entoproct/ectoproct relationship. Ophelia. 9:209–341. C, Redrawn from Nielsen, C. 1970. On metamorphosis and ancestrula formation in cyclostomatous bryozoans. Ophelia. 7:217–256. D, Redrawn from Brien, P. 1960. Les Bryozoaires. In Grassé, P.-P. (Ed.): Traité de Zoologie. Vol. 5. Part 2. Masson et Cie, Paris.)*

development into an adult. The first zooid is called an **ancestrula** (Fig. 19–17E). By means of asexual budding, the ancestrula gives rise to a series of other zooids that often show changes in size and shape. These zooids, in turn, bud off new members, and thus by subsequent asexual reproduction the colony gradually increases in size (Fig. 19–18). Budding involves the cutting off of a part of the parent zooid by the formation of a body wall partition. The new chamber evaginates by the mitotic activity of the distal cells, and the skeleton is "stretched" by the addition of new material (Fig. 19–19B). New internal structures develop from the ectoderm and mesothelium of the body wall. The exact pattern of budding (i.e., the number of location of buds) determines the growth pattern of the colony. In the erect, dendritic *Bugula,* growth occurs at the tips of each branch. Encrusting species have a peripheral growing edge (Fig. 19–19B), and in stoloniferous bryozoans, new buds arise from the surface and tips of the stolons (Fig. 19–8).

The life span of bryozoan colonies varies greatly. Some live only a single year, and in forms living in temperate regions, growth takes place when the water temperature rises. Liberation of larvae typically occurs in the summer, and in turn, marks the end of the life of the colony. Annual life spans are especially characteristic of bryozoans that live on algae. Some epiphytic species may pass through several generations in one season. Many species live for two or more years (up to 12 years in the European *Flustra foliacea*) with growth slowing down or halting during the winter. The colonies of some species, such as *Bugula,* may die back to the stolons or holdfast and re-form the following season. Sexual reproduction may occur during a restricted period, or, commonly in perennial species, it may occur throughout the growing period.

Development in freshwater phylactolaemates takes place within an embryo sac that originates as an external invagination of the cystid wall and bulges into the coelom. A placenta develops between the maternal zooid and the embryo. Development leads to the formation of a larva whose ciliated surface may correspond to the coronal cilia of gymnolaemate larvae (Fig. 19–16D). One end of the larva is invaginated to form a vestibule from which one to several polypides have been precociously budded. The larva thus contains a prefabricated young colony. The larva swims about for a short time prior to settling. After larval attachment, the once-ciliated outer epithelium of the larva begins to be retracted and subsequently degenerates. The young colony continues to bud successive zooids until an adult colony is formed. A parent zooid dies after producing a number of daughter zooids. Thus, in branching colonies, only the tips of branches contain living zooids; in the flattened gelatinous colonies, living zooids are restricted to the periphery.

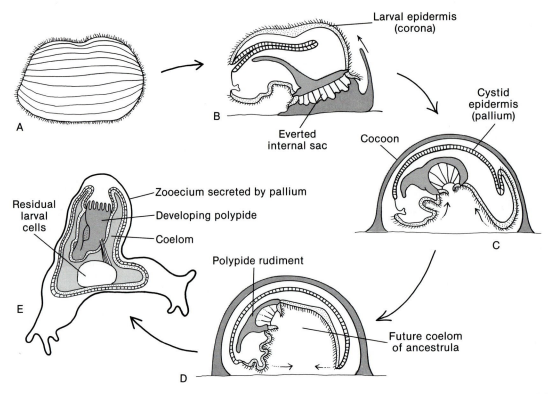

FIGURE 19–17 Larval metamorphosis of the ctenostome gymnolaemate, *Bowerbankia.* The settling larva **(A)** attaches to a substratum using its everted internal sac **(B),** which also secretes a substance that forms a cocoon around the settled larva **(C).** Once enclosed in the cocoon, the larva detaches the internal sac from the substratum and rolls its ciliated epidermis inward **(C, D).** The inrolled larval epidermis eventually encloses a cavity that gives rise to the ancestrula's coelom **(D, E).** Residual larval cells provide nutrition for the developing ancestrula. **B–D** requires approximately 2 min to complete. The ancestrula begins to feed five days after the onset of metamorphosis. *(A–D, Simplified from Reed, C. G. 1991. Bryozoa. In Giese, A. C., Pearse, J. S., and Pearse, V. B. (Eds.): Reproduction of Marine Invertebrates. Vol. VI. Echinoderms and Lophophorates. The Boxwood Press, Pacific Grove, CA. pp. 85–245. E, Modified according to the description of Reed from d'Hondt, J.-L. 1977. Structure larvaire et histogenèse post-larvaire chez* Bowerbankia imbricata *(Adams, 1798) bryozoaire cténostome (vesicularines). Arch. Zool. Exp. Gen. 118:211–243.)*

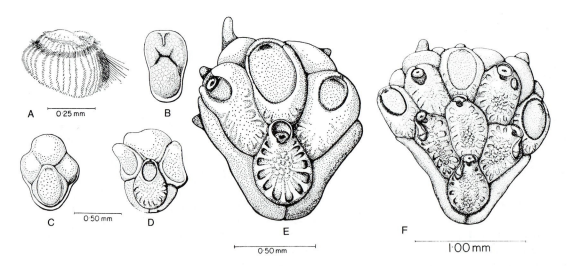

FIGURE 19–18 Ancestrula and early colony formation in *Metarabdotos unquiculatum,* an encrusting, shallow-water bryozoan found on stones on the west coast of Africa. **A,** Larva. **B–C,** Formation of ancestrula, which is one of an initial tetrad of zooids. **B** is 2 h after settlement; **C** is 28 h after settlement. **F,** Formation of additional members of colony at 140 h after settlement. *(From Cook, P. 1973. Settlement and early colony development in some Cheilostomata. In Larwood, G. P. (Ed.): Living and Fossil Bryozoa. Academic Press, London. pp. 65–71.)*

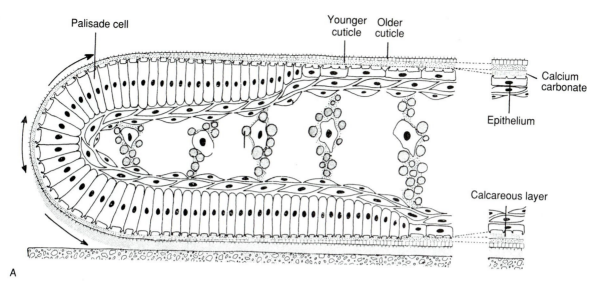

A

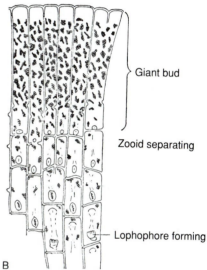

Giant bud

Zooid separating

Lophophore forming

B

FIGURE 19–19 A, Wall formation and development at the growing margin of a cheilostome colony. Completely developed wall is shown at the far right. **B,** Growing edge of the encrusting *Membranipora.* New zooids are cut off at the rear of the giant buds. *(A, From Tavener-Smith, R., and Williams, A. 1972. The secretion and structure of the skeleton of living and fossil Bryozoa. Phil. Trans. R. Soc. Lon. B 264:101. B, From Lutaud, G. 1961. Ann. Soc. R. Zool. Belg. 91:157–300.)*

In addition to sexual reproduction and to budding, freshwater bryozoans also reproduce asexually by means of special resistant bodies called **statoblasts** (Fig. 19–20), which are similar to the gemmules of freshwater sponges. One to several statoblasts develop on the funiculus and bulge into the coelom to form masses of mesothelial cells that contain stored food material. The statoblast masses eventually become covered by epidermal cells that migrate to the site of statoblast formation. After organizing cellularly, each mass secretes both an upper and a lower chitinous valve that form a protective covering for the internal cells. Because the rims of the valves often project peripherally to a considerable extent, the statoblasts are usually somewhat disc-shaped. Statoblasts are continuously formed during the summer and fall. Some types of statoblasts adhere to the par-

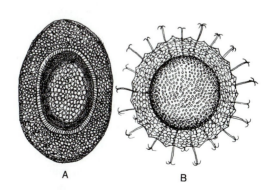

A   B

FIGURE 19–20 Statoblasts of freshwater bryozoans. **A,** A floating statoblast of *Hyalinella punctata.* **B,** Statoblast with hooks from *Cristatella mucedo.* *(A, After Rogick from Pennak. B, After Allman.)*

ent colony or fall to the bottom; others contain air spaces and float. These floating statoblasts are sometimes armed with attachment hooks around the margins.

Statoblasts remain dormant for a variable length of time. During this period they may attach to and be spread considerable distances by animals, floating vegetation, or other agents and are able to withstand desiccation and freezing. When environmental conditions become favorable, as in the spring, germination takes place, the two valves separate, and a zooid develops from the internal mass of cells. The number of statoblasts produced by a freshwater bryozoan is enormous. Drifts of statoblast valves 1 to 4 ft wide have been reported along the shores of Douglas Lake in Michigan, and *Plumatella repens* colonies in a 1-m² patch of littoral lake vegetation have been estimated to produce 800,000 statoblasts.

SUMMARY

1. Lophophorates are a group of three phyla that possess an anterior, ciliated, tentacular structure, called a lophophore. Each tentacle contains an extension of the coelom. The lophophore is used in suspension feeding.

2. The phylum Bryozoa contains minute sessile animals that form colonies of a centimeter or more in length or diameter. The majority of the more than 5000 species are marine, but there are some freshwater species.

3. Many of the peculiarities of the bryozoans are related to their miniaturization, colonial organization, and sessile life habit.

4. The body of most species (marine forms) is covered by a chitinous cuticle or a cuticle overlying calcium carbonate. The exoskeleton accounts for the great fossil record of bryozoans.

5. The lophophore can be retracted within the body encasement. It is protruded by elevation of coelomic fluid pressure, usually produced by compression of some area of the body. The skeletal covering, as well as the colonial organization, has restricted the area of possible compression to the frontal surface of most bryozoans.

6. Particles collected by the lophophore are passed into the large stomach, which makes up the greater part of the U-shaped gut. Digestion is both extracellular and intracellular.

7. As might be expected from the minute size of individual zooids, bryozoans lack typical systems for gas exchange and circulation.

8. Some bryozoan colonies are stoloniferous, but most species have erect or encrusting colonies of contiguous zooids.

9. Polymorphism is common, and physiological exchange occurs among zooids of a colony via pores in their walls through which extends a nutrient-carrying mesenchymal cord, called the funiculus.

10. Bryozoans are hermaphrodites. Their gonads lack gonoducts, and gametes break into the coelom and exit by way of tentacular pores (sperm) or by an elevated intertentacular organ (eggs).

11. Some bryozoans are oviparous, but most brood their eggs, which undergo biradial cleavage. A lecithotrophic larva is typically present.

SYSTEMATIC RÉSUMÉ OF PHYLUM BRYOZOA

Class Phylactolaemata. Freshwater bryozoans in which the cylindrical zooid possesses a horseshoe-shaped lophophore (except in *Fredericella*), an epistome, a body wall musculature, and a noncalcified covering. The coelom is continuous between individuals. Colonies are nonpolymorphic. *Fredericella, Plumatella, Pectinatella, Lophopus, Cristatella.*

Class Stenolaemata. Marine bryozoans. Zooids tubular, with calcified walls that are fused with adjacent zooids. Orifices circular and terminal. Lophophore protrusion not dependent on deformation of body wall.

Order Cyclostomata. Contains some living and many fossil species: *Crisia, Lichenopora, Stomatopora, Tubulipora.*

Orders Cystoporata, Trepostomata, and Cryptostomata all became extinct at the end of the Paleozoic.

Class Gymnolaemata. Primarily marine bryozoans with polymorphic colonies. Zooid cylindrical or flattened; lophophore circular; an epistome and body wall intrinsic musculature are lacking. Protrusion of the circular lophophore depends on body wall deformation.

Order Ctenostomata. Stoloniferous or compact colonies in which the noncalcified exoskeleton is membranous, chitinous, or gelatinous. The usually terminal orifice lacks an operculum. *Amathia, Alcyonidium, Aeverrillia, Bowerbankia, Monobryozoon, Zoobotryon*, the freshwater *Paludicella.*

Order Cheilostomata. Colonies composed of box-like zooids that are adjacent but have separate calcareous walls. Orifice is provided with an

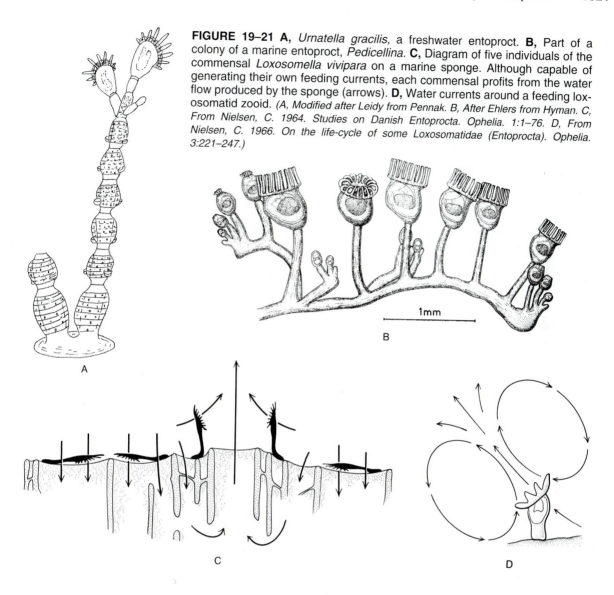

FIGURE 19–21 A, *Urnatella gracilis,* a freshwater entoproct. **B,** Part of a colony of a marine entoproct, *Pedicellina.* **C,** Diagram of five individuals of the commensal *Loxosomella vivipara* on a marine sponge. Although capable of generating their own feeding currents, each commensal profits from the water flow produced by the sponge (arrows). **D,** Water currents around a feeding loxosomatid zooid. *(A, Modified after Leidy from Pennak. B, After Ehlers from Hyman. C, From Nielsen, C. 1964. Studies on Danish Entoprocta. Ophelia. 1:1–76. D, From Nielsen, C. 1966. On the life-cycle of some Loxosomatidae (Entoprocta). Ophelia. 3:221–247.)*

operculum (except in *Bugula*). Avicularia, vibracula, or both may be present. Eggs are commonly brooded in ovicells.

Suborder Anasca.   Frontal membrane membranous. *Aetea, Callopora, Electra, Flustra, Membranipora, Tendra, Bugula, Scrupocellaria, Cupuladria, Discoporella, Thalamoporella, Cellaria.*

Suborder Cribrimorpha.   Frontal membrane covered by a vault of overarching spines, which may be partially fused. *Cribrilina.*

Suborder Ascophora.   Frontal wall calcified. Lophophore protrusion involves dilation of an invaginated sac, or ascus. *Microporella, Schizoporella, Smittina, Watersipora.*

PHYLUM ENTOPROCTA

The Entoprocta is a small phylum of approximately 150 species that are mostly sessile and superficially similar to bryozoans. They were formerly included in the phylum Bryozoa but were removed because of a number of differing features, including a spiral cleavage pattern and the lack of a coelom.

Except for the single freshwater genus, *Urnatella* (Fig. 19–21A), all entoprocts are marine and live at-

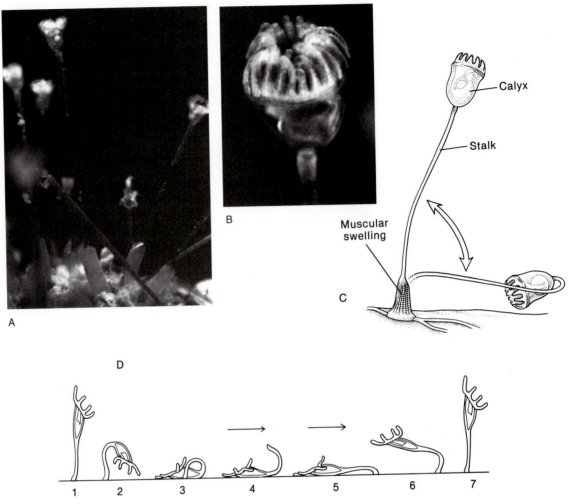

FIGURE 19–22 A, B, Living colony of *Barentsia laxa.* **A,** Zooids have muscular stalks and swollen bases, which enable them to nod and twist as shown in **C. B,** View of the calyx and tentacles. **D,** Somersaulting in *Loxosoma agile. (D, From Nielsen, C. 1964. Studies on Danish Entoprocta. Ophelia. 1:1–76.)*

tached to rocks, shells, or pilings or are commensal on sponges, polychaetes, bryozoans, and other marine animals. Members of the commensal family Loxosomatidae are solitary; the other two families are colonial. All are less than 5 mm in length. The body (Figs. 19–21; 19–22B) consists of a somewhat ovoid or boat-shaped structure called the **calyx,** which contains the internal organs, and a **stalk** by which the calyx is attached to the substratum (Figs. 19–21; 19–22A). The attached underside of the calyx was the original embryonic dorsal surface. The upper margin of the calyx bears an encircling crown of 8 to 30 solid tentacles, which represent extensions of the body wall.

The area enclosed by the tentacles, which is called the **vestibule** or **atrium,** contains the mouth at one end and the anus at the other. Both mouth and anus, however, are located within the tentacular crown, hence the name *Entoprocta,* meaning "inside anus." The mouth and anus mark the anterior and posterior ends of the animal, respectively. The bases of the tentacles are united by a membrane that pulls partly over the crown when the tentacles contract and folds inward over the vestibule (Fig. 19–23).

There may be a single stalk, as in the solitary *Loxosoma* (Fig. 19–21C, D), or several stalks may originate from a common attachment disc. Alternatively, as in *Pedicellina,* numerous stalks arise from a horizontal, creeping stolon or upright, branching stems (Fig. 19–21B). The stalk is separated from the calyx by a septum-like fold of the body wall and is commonly partitioned into short cylinders, or segments. In many species *(Barentsia),* certain segments are swollen with longitudinal muscle fibers that, on sudden contraction, produce a curious nodding motion in members of the colony (Fig. 19–22A, C). Many solitary species of *Loxosoma* can move about on their stalk, which is provided with a sucker (Fig. 19–21C), whereas others, such as *Loxosoma agile,* move by a somersaulting action (Fig. 19–22D).

The body wall consists of a fibrous cuticle and underlying epidermis. The muscle layer is limited to longitudinal fibers that occur along the inner wall of the tentacles, in the tentacular membrane, and in certain areas of the calyx. The connective tissue is well developed, loose, and at least partly fluid. Some of this fluid is reported to flow between the calyx and stalk as a result of movements of groups of contractile mesenchymal cells. The connective tissue contains both free and fixed cells.

Entoprocts are filter feeders, consuming organic particles and small plankton. The beating of cilia on the sides of the tentacles causes a water current to pass into the vestibule between the tentacles. Subsequently, the current passes upward and out, unlike the downward flow in bryozoans (Figs. 19–21D; 19–23). When suspended food particles pass between the tentacles, they become trapped by the lateral cilia (a downstream collecting system, instead of the upstream system of bryozoans). They are then transported by the frontal cilia, which are located on the inner tentacular face and beat downward, carrying the food particles to the base of the tentacle. Here the

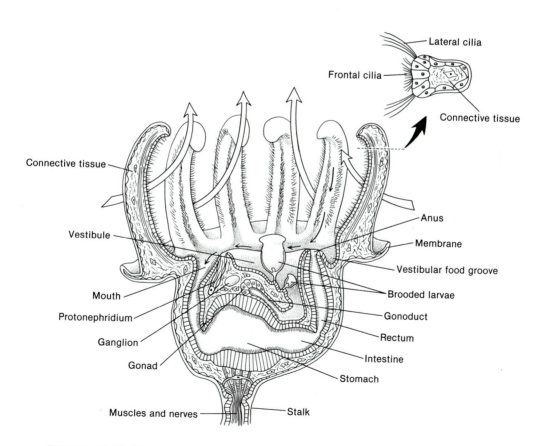

FIGURE 19–23 Entoproct anatomy. Open arrows show water flow generated by lateral cilia; solid arrows show path of food.

food particles are carried in ciliated **vestibular grooves** that run along the inner base of the tentacular crown on both sides toward the mouth (Fig. 19–23).

The digestive tract is U-shaped with a large, bulbous stomach making up the major part (Fig. 19–23). Two protonephridia open through a common nephridiopore just behind the mouth. The nervous system lies subepidermally in the connective tissue. It consists of a single, large, median ganglion situated between the stomach and the vestibule, with nerves to the tentacles, stalk, and calyx (Fig. 19–23).

Asexual reproduction by budding is common in all entoprocts, and it is by this means that extensive colonies are formed. In most species, the buds arise from segments of the stolon (Fig. 19–21B) or from the upright branches. In the solitary entoprocts, the buds develop from the calyx (Fig. 19–23A), become separated from the parent, and then attach as new individuals.

The phylum is probably entirely hermaphroditic, with some protandric species. The one or two pairs of gonads are located between the vestibule and the stomach (Fig. 19–23). The simple gonoducts join together and empty through a single, median gonopore located just posterior to the nephridiopore. The eggs are believed to be fertilized in the ovaries and are brooded externally in the vestibule (Fig. 19–23). Cleavage is spiral, and a ciliated, free-swimming larva hatches from the egg. The larva, which is a trochophore more or less like that of annelids and molluscs, has an apical tuft of cilia at the anterior end, a frontal organ, a ciliated girdle around the ventral margin of the body, and a ciliated foot (Fig. 19–24A). After a short free-swimming existence, the larva settles to the bottom where it creeps over the surface with the ciliated foot and eventually attaches with the frontal organ. In some species, the larva undergoes a complex metamorphosis, in which the future calyx

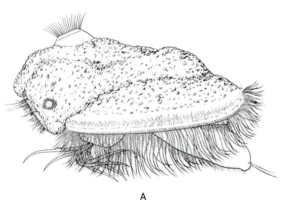

A

FIGURE 19–24 Larva of *Loxosomella harmeri* **(A).** Generalized larval metamorphosis **(B–E).** *(A, From Nielsen, C. 1971. Entoproct life-cycles and the entoproct/ectoproct relationship. Ophelia 9:209–341. B, Highly modified after Cori from Hyman, L. H. 1951. The Invertebrates. Vol. 3. McGraw-Hill Book Co., New York. 572 pp.)*

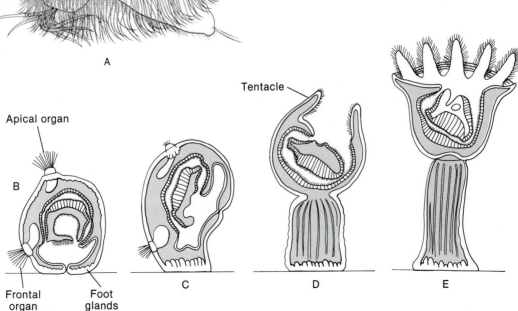

Apical organ

B

Tentacle

Frontal organ   Foot glands

C

D

E

rotates 180 degrees to attain the inverted condition of the adult (Fig. 19–24B–E). Occasionally, the larva does not develop into an adult but produces buds from which the adults are derived.

On the basis of their development, some zoologists believe that entoprocts are related to other groups that have trochophore larvae and probably had a common ancestry with the coelomate bryozoans. Others believe that the similarities to bryozoans are convergences based on the adoption of a sessile filter-feeding habit. The differences between upstream and downstream collecting systems would seem to underscore the independent evolution of filter feeding in the two phyla.

SUMMARY

1. The phylum Entoprocta contains approximately 150 species of minute, tentaculate, sessile animals that were formerly included in the Bryozoa.

2. In contrast to bryozoans, however, the anus of entoprocts is located within the circle of tentacles, a coelom is absent and an exoskeleton is lacking.

3. A pair of protonephridia is present.

4. Entoprocts are almost entirely marine and may be solitary or colonial.

5. Entoprocts are hermaphrodites and many brood their eggs.

6. Spiral cleavage typically leads to a planktotrophic larva.

PHYLUM PHORONIDA

The phoronids are a small group consisting of two genera and 14 species of wormlike animals. They are externally bilateral but internally they show a left-side dominance. All members are marine and live within a chitinous tube that is either buried in sand or attached to rocks, shells, and other objects in shallow water (Figs. 19–25; 19–26A). A few species, such as the primitive *Phoronis ovalis,* bore into mollusc shells or calcareous rock. *Phoronis vancouverensis* along the Pacific coast of the United States often forms masses of intertwined individuals.

The cylindrical body, which in most species is less than 20 cm long, lacks appendages and regional differentiation except the conspicuous lophophore anteriorly and a bulbous **ampulla** at the posterior end (Fig. 19–26). The lophophore is primitively a circular ring of tentacles *(Phoronis ovalis)* around the central mouth; the anus is located dorsally, outside the ring

FIGURE 19–25 *Phoronis hippocrepia* from the coasts of Europe, South Africa, and Brazil. Aggregations of tubes encrust rocks, shells, and coral. *(From Emig, C. C. 1974. The systematics and evolution of the phylum Phoronida. Z. Zool. Syst. Evolut.-forsch. 12:128–151.)*

of tentacles. Frequently, however, the ring folds inward dorsally to form a crescent or horseshoe of two parallel, tentacle-bearing ridges. The arc of the crescent is located ventrally, with one ridge passing above and the other ridge passing below the mouth (Fig. 19–27). The horns of the ridges are directed dorsally and may be rolled up as a spiral. Lophophore folding and coiling increase the area for food collection and gas exchange and are correlated with species having large body sizes (Fig. 19–28).

The body wall is composed of a naked glandular epidermis composed of tall cells. Ciliated cells on the lophophore, in contrast to those of bryozoans (and entoprocts), bear only one cilium per cell. The nervous system lies within the base of the epidermis and above the basal lamina. A mesothelium lines the spacious body cavity. On the body wall side of the coelom, the mesothelium is differentiated into pseudostratified, circular, and well-developed longitudinal muscles. Elsewhere on the gut, mesenteries, and blood vessels, the mesothelium is low and consists of noncontractile cells, myoepithelial cells, and cells in-

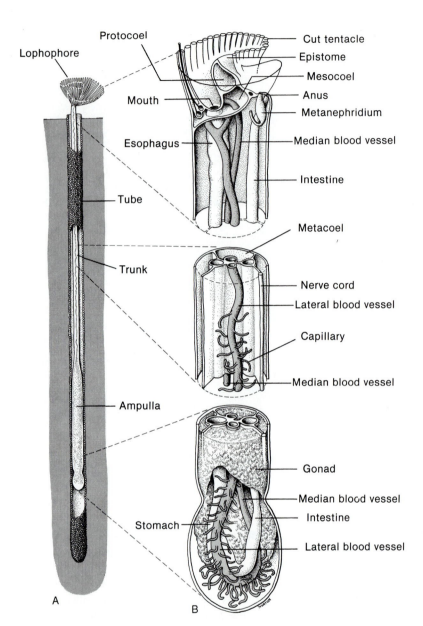

FIGURE 19–26 Phoronid body organization **(A)** and internal anatomy **(B)**. *(From Vandergon, T. L., and Colacino, J. M. 1991. Hemoglobin function in the lophophorate* Phoronis architecta *(Phoronida). Physiol. Zool. 64:1561–1577.)*

volved in nutrient storage. The coelom is divided into three parts: a posterior coelom (metacoel), which occupies the trunk and ampulla; a lophophoral coelom (mesocoel), which lies in a ring at the base of the lophophore and sends a branch into each of the tentacles; and an anterior coelom (**protocoel**), which occupies the epistome (Fig. 19–26B). The trunk coelom is divided by four mesenteries into left, right, dorsal, and ventral chambers. The mesocoel and protocoel are unpaired.

Except for retraction into their tubes, phoronid movements are slow and limited to emergence from the tube, bending movements of the extended trunk, and tentacular movements, including flicking. When disturbed, phoronids commonly jettison the lophophore and, later, regenerate it. Freshly collected specimens often show lophophores in various stages of regeneration.

Like all lophophorates, phoronids are filter feeders. The tentacular cilia beat downward, creating a water current from which plankton and suspended detritus are collected and entangled in mucus on contact with the tentacles. Cilia in the groove between the two ridges of the lophophore convey the food parti-

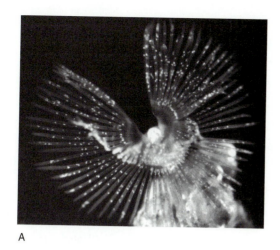

A

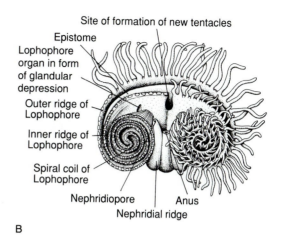

Site of formation of new tentacles
Epistome
Lophophore
organ in form
of glandular
depression
Outer ridge of
Lophophore
Inner ridge of
Lophophore
Spiral coil of
Lophophore
Nephridiopore   Anus
Nephridial ridge

B

FIGURE 19–27 The lophophores of *Phoronis architecta* **(A)** and *Phoronis australis*. **(B,** anterior view). *(B, After Shipley.)*

cles under the epistome and into the mouth. The digestive tract is U-shaped, and digestion occurs extracellularly within the esophagus and stomach (Fig. 19–26B).

Phoronids possess a blood-vascular system with corpuscles that contain hemoglobin. The presence of hemoglobin in such small animals is apparently an adaptation to life in anoxic or microoxic habitats, such as fine marine sediments. Unlike many other tubicolous animals, such as some polychaetes, phoronids do not irrigate their tubes with oxygenated surface water. As a result, oxygen to support aerobic metabolism enters through the exposed lophophore, and most is transported to the body bound to the hemoglobin in the vascular system. The blood of *Phoronis architecta* has an oxygen-carrying capacity equal to that of most vertebrates, and its volume in relation to the body volume is twice that of humans.

In *Phoronis ovalis,* the circulatory system consists of a dorsal (median) vessel and two lateral vessels that join afferent and efferent lophophoral ring vessels anteriorly and produce blind-ending capillaries in the ampulla that supply the stomach and gonads. Each tentacle receives a single vessel into which the blood flows from the afferent lophophoral vessel and ebbs into the efferent vessel. Blood flows anteriorly in the dorsal vessel and posteriorly in the lateral vessels. In all other phoronids, except *Phoronis ovalis,* the right lateral vessel is absent.

A pair of metanephridia leads from the metacoel internally to a nephridiopore on each side of the anus

(Figs. 19–26B; 19–27B). Often, the left nephridium is larger than the right. Ultrafiltration apparently occurs on the contractile dorsal vessel, on which podocytes have been found.

The intraepidermal nervous system consists of a nerve ring at the base of the lophophore from which nerves arise to supply the tentacles and body wall muscles (Fig. 19–26B). In *Phoronis ovalis,* there are two lateral, nonsegmented, nerve cords, but in all

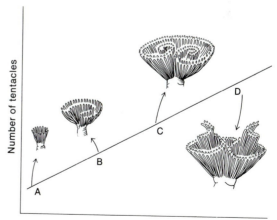

Number of tentacles

Body length (mm)

FIGURE 19–28 Relationship between body size and lophophore shape. **A,** *Phoronis ovalis*. **B,** *Phoronis architecta*. **C,** *Phoronis australis*. **D,** *Phoronopsis californica*. *(Modified from Abele, L. G., Gilmour, T., and Gilchrist, S. 1983. Size and shape in the phylum Phoronida. J. Zool. Lond. 200:317–323.)*

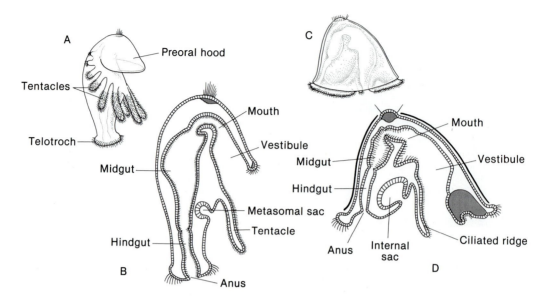

FIGURE 19–29 The actinotrocha larva of phoronids in external view **(A)** and in section **(B)** compared with a bryozoan cyphonautes larva **(C, D).** *(D, Simplified from Stricker, S. A., Reed, C. G. and Zimmer, R. L. 1988. The cyphonautes larva of the marine bryozoan* Membranipora membranacea. *I. General morphology, body wall, and gut. Can. J. Zool. 66:368-383.)*

other phoronids, only the left cord is present. Giant axons occur within the cord.

Phoronids reproduce sexually and asexually, by budding and by transverse fission. Approximately half of the species are hermaphrodites. The gonad is situated around the stomach in the ampulla, and it receives numerous capillaries from the lateral blood vessel (Fig. 19–26B). The gametes are shed into the coelom of the trunk and escape to the outside by way of the nephridia. The sperm are transferred in spermatophores produced by a pair of **lophophoral organs** (Fig. 19–27B). The eggs appear to be fertilized internally and either are planktonic or are brooded in the concavity formed by the two arms of the lophophore.

Cleavage is radial, the mesoderm originates as mesenchyme but soon becomes epithelial and forms the larval coeloms (protocoel and metacoel), and the blastopore becomes the mouth. The origin of the mesocoel is undetermined. The gastrula develops into an elongated, tentacular, ciliated larva called an **actinotrocha** (Fig. 19–29A). A telotroch, which is probably the principal locomotor organ, rings the posterior of the trunk. After a long or short, free-swimming and feeding planktonic existence, the actinotrocha undergoes a rapid metamorphosis and

sinks to the bottom, where it secretes a tube and takes up an adult existence (Fig. 19–30).

Among the 14 species of phoronids, *Phoronis ovalis* is believed to embody many primitive phoronid traits. It has a simple oval lophophore, and its blood-vascular, nervous, and excretory systems are all bilaterally symmetrical. Other phoronids show a greater or lesser reduction of these organs on the right side of the body. Such left-side dominance is reminiscent of several groups of deuterostomes, including enteropneust hemichordates, echinoderms, and cephalochordates, all of which have larval and adult asymmetries favoring the left side of the body.

SUMMARY

1. The phylum Phoronida contains 14 species of elongated marine lophophorates that live in chitinous tubes fixed in sediment or attached to rocks.

2. When the animal feeds, the lophophore is projected from the end of the tube.

3. The blood-vascular system is well developed and contains red blood cells.

4. Species are either dioecious or hermaphroditic, cleavage is radial, the blastopore becomes the mouth.

5. The life cycle typically includes a planktotrophic actinotrocha larva.

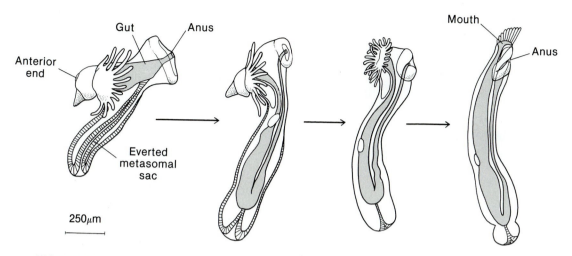

FIGURE 19-30 Metamorphosis from planktonic actinotrocha larva to benthic juvenile can be accomplished in 15 to 20 min. *(From Hermann, K. 1975. Einfluss von Bacterien auf die Metamorphose-Auslösung und deren Verlauf bei Actinotrocha branchiata (Phoronis muelleri). Verh. Dtsch. Zool. Ges. 67:112–115.)*

PHYLUM BRACHIOPODA

The last of the three lophophorate phyla is the phylum Brachiopoda, commonly known as lamp shells. These animals resemble bivalve molluscs in possessing a mantle and a calcareous shell of two valves that approximates that of small molluscs in size. In fact, the phylum was not separated from the molluscs until the middle of the nineteenth century. The resemblance to molluscs is superficial, however, because in brachiopods the two valves enclose the body dorsally and ventrally instead of laterally as in bivalve molluscs, and the ventral valve is typically larger than the dorsal (Fig. 19–31A, C).

All brachiopods are marine, and are found from the intertidal zone to the deep sea. Most species live attached to rocks or other firm substrata, but some forms, such as *Lingula,* live in vertical burrows in sand and mud bottoms (Fig. 19–32A). Although fossil species were widely distributed and abundant on reefs, modern species are largely inhabitants of cold waters and have a scattered distribution.

The approximately 325 species of living brachiopods are but a fraction of the 12,000 described fossil species that flourished in the seas of the Paleozoic and Mesozoic eras. The phylum made its appearance in the Cambrian period and reached its peak of evolutionary development during the Ordovician period. Brachiopods were especially hard hit by the great Permian extinction, and the limited numbers and restricted distribution of extant species are a result of that extinction event. The genus, *Lingula* (but none of the present living species), dates back to the Ordovician period.

External Anatomy

Each of the two valves is bilaterally symmetrical and is usually convex. The smaller dorsal shell fits over the larger ventral shell, the apex of which in some groups contains a hole like that in a Roman lamp—hence the name "lamp shell" (Fig. 19–31A, C). In the burrowing lingulids, the valves are flattened and more equal in size. The valves may be ornamented with concentric growth lines and a fluted, ridged, or even spiny surface. The shells of most living brachiopods are dull yellow or gray, but some species have orange or red shells.

The two valves articulate with one another along the posterior line of contact, called the hinge line (Fig. 19–31A, C), and the nature of the articulation is the basis for the division of the phylum into two classes, the Inarticulata and the Articulata. In inarticulate brachiopods, such as *Lingula* and *Glottidia,* the two valves are held together only by muscles. Valve opening occurs by backward retraction of the soft body, which produces outward pressure through the

(Text continues on p. 1032)

A

2 cm

B

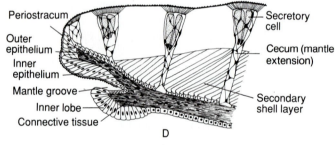

'Catch' adductor

'Quick' adductor

Socket wall

Hinge axis

Pedicle

Tooth

Diductor

Dorsal valve

Ventral valve

5 mm

C

Periostracum

Outer epithelium

Inner epithelium

Mantle groove

Inner lobe

Connective tissue

Secretory cell

Cecum (mantle extension)

Secondary shell layer

D

FIGURE 19–31 A, The articulate brachiopod, *Tere-bratella*. Ventral valve with pedicle formamen at top; dorsal valve at bottom; two valves in apposing position at right. **B,** *Paraspirifer,* a fossil articulate brachiopod. **C,** Sagittal section through the terebratulid, *Waltonia,* showing relationship of valves, muscles, and pedicle. **D,** Section through mantle edge of an articulate brachiopod. **E,** Internal surface of dorsal valve of an articulate brachiopod. *(A, B, Photographs courtesy of Betty M. Barnes. C, From Rudwick, M. J. S. 1970. Living and Fossil Brachiopods. Hutchinson and Co., London. D, From Williams and Rowell. E, After Davidson from Hyman.)*

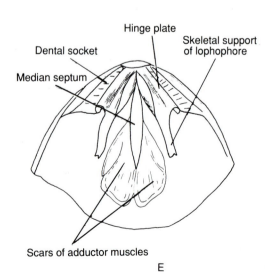

Hinge plate

Dental socket

Skeletal support of lophophore

Median septum

Scars of adductor muscles

E

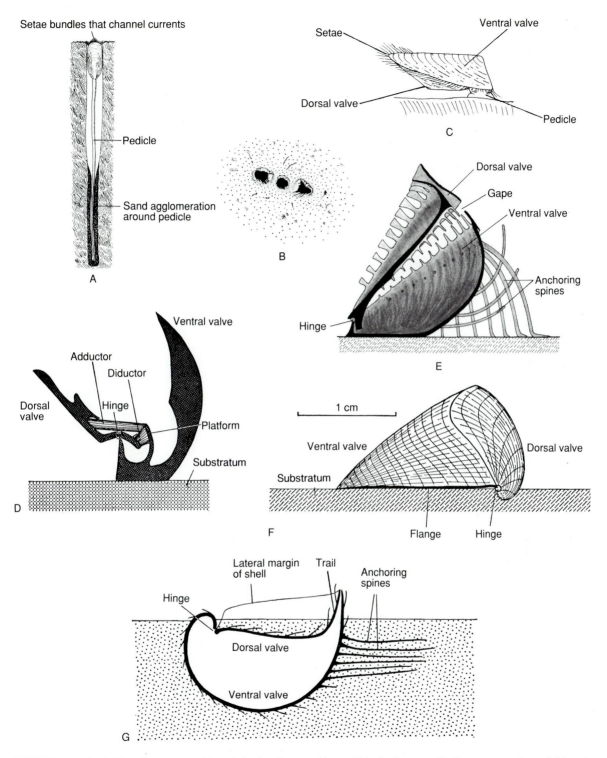

FIGURE 19–32 A, The inarticulate *Lingula* in feeding position within its burrow. **B,** Burrow opening of *Lingula* when feeding. Setae around the middle exhalant and lateral inhalant apertures. **C,** Living epibenthic inarticulate *Discinisca,* attached by pedicle. **D,** Diagrammatic lateral view of living articulate *Lacazella,* which lives attached directly to the substratum by its ventral valve. **E,** Sagittal section through Permian articulate *Chonosteges,* which has cemented its ventral valve directly to the substratum. **F,** Devonian articulate *Spyringospira,* which is believed to have lived unattached on soft bottoms. **G,** Permian articulate *Waagenoconchia,* which is believed to have lived partially buried in soft bottoms. *(A, Modified from Francois. B, Modified from Rudwick. C, After Morse from Hyman. D, F, G, From Rudwick, M. J. S. 1970. Living and Fossil Brachiopods. Hutchinson and Co., London. E, Modified from Rudwick.)*

coelomic fluid on the dorsal and ventral valves. Lingulids burrow using scissor-like movements of the valves to slice through the sediment (Fig. 19–36A, B). In articulate brachiopods, on the other hand, the ventral valve bears a pair of hinge teeth that fit into opposing sockets on the underside of the hinge line of the dorsal valve (Fig. 19–31E). This articulating mechanism locks the valves securely together and allows only a slight anterior gape of approximately 10 degrees. A pair of adductor muscles closes the valves. Another pair of muscles (diductors) opens the valves (Fig. 19–31C).

Although inarticulates are considered more primitive than articulate forms based on comparative anatomy, both groups are present from the beginning of the fossil record.

The shell is secreted by the underlying dorsal and ventral mantle lobes (Fig. 19–31D). The shell of inarticulates is usually composed of a mixture of calcium phosphate and chitin (chitinophosphate), and the shell of articulates is composed of calcium carbonate in the form of calcite. The chitinophosphate shell appears to be the more primitive, because the Cambrian shells were of this type; calcite shells did not appear until the Ordovician. The outer surface of the shell is covered by a thin organic periostracum. As in most molluscs, the periostracum and the outer layers of mineral deposition are secreted by the mantle edge, and the inner layer of shell is secreted by the entire outer mantle surface.

In addition to secreting the shell, the mantle edge in most species bears long, chitinous setae that may have a protective and perhaps a sensory function (Figs. 19–32C; 19–33).

The body proper of brachiopods occupies only the posterior part of the chamber formed by the two valves (Fig. 19–33). Anteriorly, the space between the mantle lobes (the mantle cavity) is filled by the lophophore.

Most brachiopods are attached to the substratum by a cylindrical extension of the ventral body wall, called the **pedicle.** The pedicle of the inarticulate lingulids (*Lingula* and *Glottidia*) is long, fluid-filled, and provided with muscles; it emerges at the posterior of the animal between the two valves (Figs. 19–32A; 19–33). The lingulids live in U-shaped burrows exca-

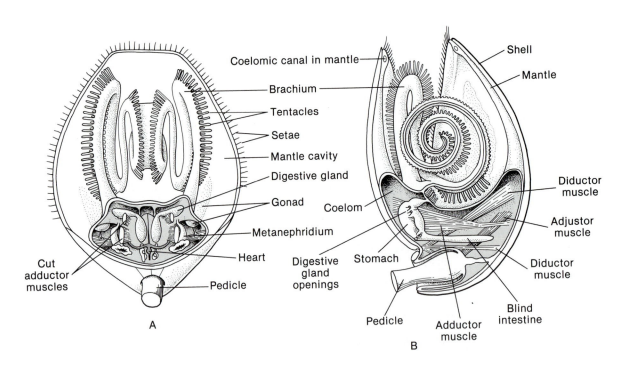

FIGURE 19–33 Brachiopod anatomy in dorsal view **(A)** and sagittal section **(B)**. (*Based on   after Delage and Herouard modified from de Beauchamp, P. 1960. Classe des Brachiopodes. In Grassé P.-P. (Ed.): Traité de Zoologie. Vol. 5. Part 2. Masson et Cie, Paris.*)

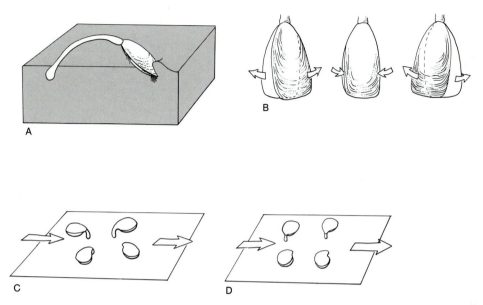

FIGURE 19–34 Brachiopod movements. Burrowing *Lingula* and *Glottidia* **(A)** arch the body using the pedicle and slice into the sediment using scissor-like movements of the dorsal and ventral valves **(B).** Individuals of the articulate, *Terebratalia transversa,* placed randomly in a unidirectional current **(C)** rotate on their pedicles, positioning themselves with the longitudinal body axis perpendicular to the water flow **(D).** In this orientation, the water currents aid, rather than oppose, the ciliary current produced by the lophophore. *(A, Modified from Trueman, E. R., and Wong, T. M. 1987. The role of the coelom as a hydrostatic skeleton in lingulid brachiopods. J. Zool. Lond. 213:221–232 and personal observations of E.E.R. B, Based on data and discussion by LaBarbera, M. 1977. Brachiopod orientation to water movement. I. Theory, laboratory behavior, and field orientations. Paleobiology. 3:270–287.)*

vated in sand and mud. The anterior ends of the valves are directed toward the burrow opening, and the pedicle extends downward toward the bottom of the burrow and is encased in sand. When the animal is feeding, the gaping valves are near the opening of the burrow, and the long mantle setae are bunched together in three groups to form inhalant (two lateral) and exhalant (one central) siphons that project above the surface of the sand and prevent entry of sediment particles into the mantle cavity (Fig. 19–32B). When the animal is disturbed, the pedicle contracts and pulls the animal downward into the burrow.

The pedicle of an articulate brachiopod may be short and lacking muscles, or it may be a flexible, muscular tether (Figs. 19–32C; 19–33). Moreover, the pedicle emerges either from a notch at the hinge line of the ventral valve or through a hole at the up-turned apex (Fig. 19–31A). This means, of course, that the pedicle emerges from the dorsal side of the ventral valve, which extends posteriorly considerably beyond the dorsal valve. An articulate brachiopod is

attached to the substratum upside-down with the valves held in a horizontal position (Fig. 19–32C) or with the hinge end down and the gape directed upward. Muscles within the valves, which are inserted on the pedicle base, permit erection, flexion, and even rotation of the animal on the pedicle (Fig. 19–34C, D). The end of the pedicle adheres to the substratum by means of rootlike extensions or short papillae.

The pedicle has been lost completely in a few brachiopods of both classes, such as *Crania* (Inarticulata) and *Lacazella* (Articulata). Such species cement their ventral valve directly to the substratum and are thus oriented in a normal manner with the dorsal side up (Fig. 19–32D). The more posterior part of the ventral valve being the point of attachment, the anterior margin is directed somewhat upward and clear of the substratum. Some species of fossil brachiopods that attached by cementation were important contributors to Paleozoic reefs.

The shell form of a number of fossil groups suggests that they were adapted for living free on the surface of soft bottoms. Spines, long shell "wings," and flattened ventral surfaces appear to have been devices to prevent sinking (Fig. 19–32F, G). Among living species, the New Zealand species *Neothyris lenticularis* lives free on gravel or coarse sand bottoms, and *Terebratella sanguinea* lives either attached to rock or free on sand and mud (Fig. 19–35).

FIGURE 19–35 A rock wall on the coast of New Zealand supporting three species of brachiopods. The white, branching organism is a stylasterine hydrocoral. Photograph covers approximately 1 m². *(Photograph courtesy of P. J. Hill, DSIR, New Zealand.)*

Lophophore and Feeding

As in other lophophorates, the brachiopod lophophore is basically a crown of hollow tentacles surrounding the mouth. In order to increase surface area, however, the lophophore projects anteriorly as two arms, or **brachia,** from which the name *brachiopod* is derived. In its simplest form the lophophore is horseshoe-shaped, and each arm, or **brachium,** projects anteriorly. The arms may be further looped or spiraled in complex ways, greatly increasing the collecting surface area of the lophophore (Figs. 19–33; 19–36). Each brachium bears a row of tentacles, and the tentacle-bearing ridge is flanked by a **brachial groove** at its base. The lophophore is supported by a cartilaginous axis and a fluid-filled canal within each brachium (Fig. 19–36B). Also, in many brachiopods, the dorsal valve bears complex supporting processes, and the processes and inner valve surface may be grooved and ridged for the reception of the lophophore (Fig. 19–31E).

When a brachiopod feeds, water enters and leaves the valve gape through distinct inhalant and exhalant apertures and chambers created by the lophophore (Fig. 19–36A). In its circuit over the lophophore, water is driven between the tentacles by the lateral tentacular cilia. Particles, especially fine

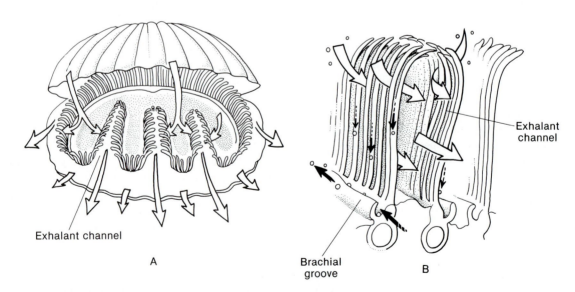

FIGURE 19–36 A, View into the anterior gape of *Megathyris,* showing lophophore and filtering (solid arrows) and exhalant (open arrows) water currents. **B,** Enlargement of a part of the lophophore of *Megathyris* showing water currents (open arrows) and the direction of ciliary tracts (small arrows) carrying food particles to the brachial grooves. *(A, Modified from Rudwick. B, After Atkins.)*

phytoplankton, are screened by the lateral cilia of two adjacent filaments and then transported down the tentacles to the brachial groove in the same manner as in bryozoans. The brachial groove conducts food to the mouth (Fig. 19–36). Rejected particles are carried away in the median, outward-flowing current. Brachiopod ciliated cells, like those of phoronids, have only one cilium per cell.

The mouth leads into an esophagus that extends dorsally and a short distance forward prior to turning posteriorly and joining a dilated stomach (Fig. 19–33). The stomach is surrounded by a digestive gland that opens through the stomach wall by means of one to three ducts on each side. At least in *Lingula* digestion is largely intracellular within the digestive gland. In inarticulates, the intestine extends from the stomach to the rectum, which opens to the outside via a posterior anus located between the valves. In articulates, the intestine is blind.

Internal Anatomy and Physiology

The coelom of a brachiopod extends into the mantle, where it is partitioned into channels. The coelomic fluid contains coelomocytes of several sorts, one of which contains hemerythrin in the burrowing inarticulates. There are no specialized gas exchange organs other than the lophophore and mantle lobes. Oxygen transport is probably provided by the coelomic fluid, for there is a definite circulation of coelomic fluid through the mantle channels, and oxygen is carried, at least in part, by hemerythrin in the coelomocytes.

In addition to the coelomic spaces and channels, brachiopods also have a blood-vascular system, but the blood is colorless and contains few cells. There is a contractile vesicle (heart) located over the stomach in the dorsal mesentery, and from the heart extends an anterior and a posterior channel supplying various parts of the body (Fig. 19–33A). The function of the blood-vascular system of brachiopods is uncertain. The circulation of nutrients is perhaps its primary function.

One or two pairs of metanephridia are present in brachiopods. The nephrostomes open into the metacoel on each side of the posterior end of the stomach, and the tubules then extend anteriorly to open into the mantle cavity through a nephridiopore situated posteriorly and to each side of the mouth (Fig. 19–33).

An esophageal nerve ring with a small ganglion on its dorsal side and a larger ganglion on the ventral side forms the nerve center of brachiopods. From the ganglia and their commissures, nerves extend anteriorly and posteriorly to innervate the lophophore, the mantle lobes, and the valve muscles.

As in bivalve molluscs, the mantle margin of brachiopods is probably the most important site of sensory reception. The mantle setae, although not directly associated with sensory neurons, probably do transmit tactile stimuli to receptors in adjacent mantle epidermis. In some brachiopods, the setae are long and form a protective "sensory grille" over the gaping valves.

Reproduction and Development

With a few exceptions, brachiopods are dioecious, and the gonads, usually four in number, are masses of developing gametes beneath the mesothelium of the mantle coelom (articulates) or visceral mesenteries (inarticulates) (Fig. 19–33). When ripe, the gametes pass into the coelom and are discharged to the exterior by way of the nephridia.

Except for a few brooding species, the eggs are shed into the seawater and fertilized at the time of spawning. Cleavage is radial and nearly equal and leads to a coeloblastula that usually undergoes gastrulation by invagination. The blastopore becomes the anus and the mouth forms secondarily, as in deuterostomes. In contrast to the typical method of mesoderm formation in protostomes, the mesoderm in brachiopods appears to be enterocoelic, as in deuterostomes—that is, it arises as an outpocket of the archenteron.

The embryo eventually develops into a free-swimming larva. The inarticulate larva is planktotrophic and resembles a minute brachiopod (Fig. 19–37E). The pair of mantle lobes and the larval valves enclose the body, and the ciliated lophophore, which can be extended outward from the mantle cavity, is the larval swimming organ. The pedicle, which in this group is derived from the mantle, is coiled in the back of the mantle cavity. As additional shell is laid down, the larva becomes heavier and sinks to the bottom. There is no metamorphosis in *Lingula;* the pedicle attaches to the substratum, and the young brachiopod takes up an adult existence. The articulate larva differs in having a ciliated anterior lobe representing the body and lophophore, a posterior lobe that forms the pedicle, and a mantle lobe that is directed backward (Fig. 19–37A). In *Terebratulina,* the larva

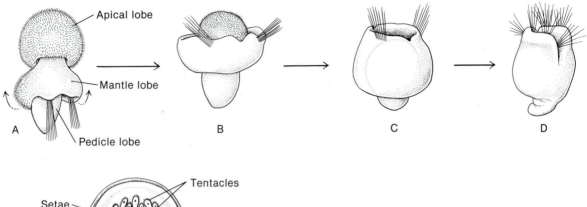

FIGURE 19–37 Brachiopod larvae and larval metamorphosis. **A–D,** Metamorphosis of an articulate larva **(A),** showing reversal of the mantle. **E,** Larva of *Lingula* and *Glottidia.* The ciliated tentacles project from the shell as the animal swims. *(E, After Yatsu.)*

settles after a short free-swimming existence of approximately 24 to 30 h, and then it undergoes metamorphosis. The mantle lobe reverses position and begins the secretion of the valves, and the adult structures develop from their larval precursors (Fig. 19–37B–D).

LOPHOPHORATE EVOLUTION

Most recent ideas regarding the evolution of the lophophorates argue that the common ancestor to brachiopods, phoronids, and bryozoans had the following characteristics: trimeric body, tentacles from the mesocoelic body region, a U-shaped gut, a secreted encasement, and a sessile life habit. It must be noted, however, that on the basis of these characteristics, the pterobranch hemichordates cannot be excluded from consideration with the lophophorates. This fact suggests a close relationship between deuterostomes and lophophorates, and most recent evidence favors such a view. Although an evolutionary relationship between brachiopods and deuterostomes has long been suspected, Claus Nielsen's study (1991) of the development of the *Crania* confirms the deuterostome nature of at least one species. Embryological studies of phoronids have also made clear that the phoronid cleavage pattern is radial and not spiral, as was previously thought. Bryozoan cleavage is also radial, as is the case in most deuterostomes. On the other hand, phoronid developmental features indicate not only an

alliance with deuterostomes, but also with the protostomes. For example, the embryonic blastopore becomes the mouth, and the actinotrocha has a pair of protonephridia, which are reminiscent of those found in some annelid larvae. Thus, the phoronids have traits that overlap both deuterostomes and protostomes. For this reason, a thorough knowledge of the biology of phoronids and other lophophorates may someday provide important clues to solve the mystery of the relationship between the two major lines of animal evolution.

SUMMARY

1.   The phylum Brachiopoda contains about 335 marine lophophorates with two calcareous valves and a shell enclosing the body. The calcareous valves account for the well-known fossil history of brachiopods. In contrast to other lophophorates, most brachiopods are relatively large (1–6 cm).

2.   Although brachiopod shells superficially resemble those of bivalve molluscs, the shell valves of brachiopods are oriented dorsoventrally, and in most species, the ventral valve is larger than the dorsal one.

3.   Similar to other lophophorates, brachiopods are sessile, and the body is usually anchored to the substratum by a flexible pedicle, which emerges through the articulating end of the ventral valve.

4.   The lophophore is horseshoe-shaped, with the arms often looped or spiraled.

5. Brachiopods possess a poorly developed blood-vascular system, and the coelom, which contains hemerythrocytes in lingulid inarticulates, is responsible for gas transport. There are one or two pairs of metanephridia. The sexes are separate. Gametes mature in the coelom and exit through the nephridia.

6. Fertilization is external, cleavage is radial, coeloms form schizocoelously or enterocoelously, the blastopore becomes the anus, and there is a free-swimming larva.

SYSTEMATIC RÉSUMÉ OF THE PHYLUM BRACHIOPODA

Class Inarticulata. The two shell valves are held together by muscles and other tissues; shell teeth and sockets are absent. The shell contains chitin and calcium phosphate (except for the craniids, whose shells are of calcium carbonate). Direct muscular action closes the valves, but valve opening, as well as pedicle and brachial turgor, result from hydrostatic coelomic pressure. Larva has two body re-gions: a mantle and a body lobe. Members of the order Lingulida *(Lingula, Glottidia)* are sediment burrowers (having hemerythrin as a respiratory pigment), whereas those in the order Acrotretida attach to the surface of hard objects with a short pedicle *(Discinisca)* or directly with the ventral valve *(Crania)*.

Class Articulata. The two shell valves are articulated by means of teeth (ventral) and sockets (dorsal). The shell is composed of calcium carbonate. Shell opening and closing results from the direct action of specialized muscles. The pedicle, when present, is supported by connective tissue. The lophophore is suspended on specialized calcareous supports. The gut ends blindly, and an anus is absent. The larva has three body regions, consisting of mantle, body, and pedicle lobes. Most articulates attach to the surfaces of hard objects, but a few deep sea forms *Chlidonophora)* anchor into sediment with a branched rootlike pedicle. *Hemithiris, Argyrotheca, Terebratalia.*

REFERENCES

The literature included here is restricted chiefly to books and papers on lophophorates alone. The introductory references on page 6 list many general works and field guides that contain sections on these animals.

Colonial Organization

Buss, L. W. 1979. Habitat selection, directional growth and spatial refuges: Why colonial animals have more hiding places. *In* Larwood, G., and Rosen, B. R. (Eds.): Biology and Systematics of Colonial Organisms. The Systematics Association Special Vol. No. 11. Academic Press, London. pp. 459–497.

Jackson, J. B. C. 1977. Competition on marine hard substrata: The adaptive significance of solitary and colonial strategies. Am. Nat. *111:*743–769.

Mackie, G. O. 1986. From aggregates to integrates: Physiological aspects of modularity in colonial animals. Phil. Trans. R. Soc. Lond. B *313:*175–196.

Gulfweed Community

Ryther, J. H. 1956. The Sargasso Sea. Scientific American.

General

Hyman, L. H. 1959. The Invertebrates. Vol. 5. Smaller Coelomate Groups. McGraw-Hill, New York. pp. 228–609.

Mackie, G. O. 1986. From aggregates to integrates: Physiological aspects of modularity in colonial animals. Phil. Trans. R. Soc. Lond. B *313:*175–196.

Nielsen, C. 1977. The relationship of Entoprocta, Ectoprocta, and Phoronida. Am. Zool. *17:*149–150.

Nielsen, C. 1985. Animal phylogeny in the light of the trochaea theory. Biol. J. Linn. Soc. *25:*243–299.

Strathmann, R. 1973. Function of lateral cilia in suspension feeding of lophophorates. Mar. Biol. *23*(2):129–136.

Zimmer, R. L. 1973. Morphological and developmental affinities of the lophophorates. *In* Larwood, G. P. (Ed.): Living and Fossil Bryozoa: Recent Advances in Research. Academic Press, New York. pp. 593–599.

Bryozoa

Banta, W. C., McKinney, F. K., and Zimmer, R. L. 1974. Bryozoan monticules: Excurrent water outlets? Science. *185:*783–784.

Best, M. A., and Thorpe, J. P. 1985. Autoradiographic study of feeding and the colonial transport of metabolites in the marine bryozoan *Membranipora membranacea.* Mar. Biol. *84:*295–300.

Bobin, G. 1977. Interzooecial communications and the funicular system. *In* Woollacott, R. M., and Zimmer, R. L. (Eds.): Biology of Bryozoans. Academic Press, New York.

Brown, C. J. D. 1933. A limnological study of certain freshwater Polyzoa with special reference to their statoblasts. Trans. Am. Microsc. Soc. *52:*271–316.

Carle, K. J., and Ruppert, E. E. 1983. Comparative ultrastructure of the bryozoan funiculus: A blood vessel homologue. Z. Zool. Syst. Evolut.-forsch. *21:*181–193.

Cheetham, A. H. 1986. Branching, biomechanics and bryozoan evolution. Proc. R. Soc. Lond, B *228:*151–171.

Cook, P. L. 1977. Colony water currents in living Bryozoa. Cah. Biol. Mar. *18:*31–47.

Cook, P. L., and Chimonides, P. J. 1978. Observations on living colonies of *Selenaria* (Bryozoa, Cheilostomata). I. Cah. Biol. Mar. *19:*147–158.

Cook, P. L., and Chimonides, P. J. 1980. Further observations on water current patterns in living Bryozoa. Cah. Biol. Mar. *21:*393–402.

Farmer, J. D., Valentine, J. W., and Cowen, R. 1973. Adaptive strategies leading to the ectoproct groundplan. Syst. Zool. 22(3):233–239.

Gordon, D. P. 1975. Ultrastructure and function of the gut of a marine bryozoan. Cah. Biol. Mar. *16:*367–382.

Hayward, P. J., and Ryland, J. S. 1979. British Ascophoran Bryozoans. Synopsis of the British Fauna No. 10. Academic Press, London. 312 pp.

Lacourt, A. W. 1968. A monograph of the freshwater Bryozoa—Phylactolaemata. Zool. Verhand. *93:*1–159, plus 18 plates.

Larwood, G. P. (Ed.): 1973. Living and Fossil Bryozoa: Recent Advances in Research. Academic Press, London. 652 pp.

Larwood, G. P., and Abbott, M. B. (Eds.): 1979. Advances in Bryozoology. Syst. Assoc. Spec. Vol. 13. Academic Press, London. 638 pp.

Larwood, G. P., and Nielsen, C. (Eds.): 1981. Recent and Fossil Bryozoa. Olsen and Olsen, Fredensborg, Denmark. 334 pp. (Papers from a symposium.)

Lidgard, S. 1985. Zooid and colony growth in encrusting cheilostome bryozoans. Palaeontology. *28:*255–291.

Lidgard, S., and Jackson J. B. C. 1989. Growth in encrusting cheilostome bryozoans: I. Evolutionary trends. Paleobiology. *15:*255–282.

Markham, J. B., and Ryland, J. S. 1987. Function of the gizzard in Bryozoa. J. Exp. Mar. Biol. Ecol. *107:*21–37.

Moore, R. C. (Ed.): 1953. Treatise on Invertebrate Paleontology. Bryozoa, Pt. G. 253 pp. 1965. Brachiopoda. Pt. H. Vols. 1 and 2. 926 pp. Geological Society of America and University of Kansas Press, Lawrence.

Mukai, H., and Oda, S. 1980. Comparative studies on the statoblasts of higher phylactolaemate bryozoans. J. Morphol. *165:*131–155.

Mundy, S. P. 1980. Stereoscan studies of phylactolaemate bryozoan statoblasts including a key to the statoblasts of the British and European Phylactolaemata. J. Zool. *192:*511–530.

Nielsen, C., and Larwood, G. P. (Eds.): 1985. Bryozoa: Ordovician to Recent. Proceedings of the 6th International Conference on Bryozoa, Vienna, 1983. Olsen and Olsen, Fredensborg, Denmark.

Nielsen, C., and Pedersen, K. J. 1979. Cystid structure and protrusion of the polypide in *Crisia.* Acta Zool. *60:*65–88.

Reed, C. G. 1991. Bryozoa. *In* Giese, A. C., Pearse, J. S., and Pearse, V. B. (Eds.): Reproduction of Marine Invertebrates. Vol. VI. Echinoderms and Lophophorates. The Boxwood Press, Pacific Grove, CA. pp. 85–245.

Reed, C. G., and Cloney, R. A. 1982. The settlement and metamorphosis of the marine bryozoan *Bowerbankia gracilis* (Ctenostomata: Vesicularioidea). Zoomorphology. *101:*103–132.

Rider, J., and Cowen, R. 1977. Adaptive architectural trends in encrusting ectoprocts. Lethaia. *10:*29–41.

Ross, J. R. P. (Ed.): 1987. Bryozoa: Present and Past. Proceedings of the Seventh Conference of the International Bryozoology Association, Bellingham, WA, 1986. Western Washington University.

Ryland, J. S. 1970. Bryozoans. Hutchinson University Library, Hutchinson and Co., London. (An excellent general account of living and fossil bryozoans.)

Ryland, J. S. 1976. Physiology and ecology and marine bryozoans. Adv. Mar. Biol. *14:*285–443.

Ryland, J. S., and Hayward, P. J. 1977. British Anascan Bryozoans. Synopses of the British Fauna No. 10. Academic Press, London. 188 pp.

Silen, L. 1972. Fertilization in the Bryozoa. Ophelia. 10(1):27–34.

Steele-Petrovic, H. M. 1976. Brachiopod food and feeding processes. Paleontology (Lond.). *19:*417–436.

Strathmann, R. R. 1982. Cinefilms of particle capture by an induced local change of beat of lateral cilia of a bryozoan. J. Mar. Biol. Ecol. *62:*225–236.

Stricker, S. A. 1988. Metamorphosis of the marine bryozoan *Membranipora membranacea:* An ultrastructural study of rapid morphogenetic movements. J. Morphol. *196:*53–72.

Stricker, S. A. 1989. Settlement and metamorphosis of the marine bryozoan *Membranipora membranacea.* Bull. Mar. Sci. *45:*387–405.

Stricker, S. A., Reed, C. G., and Zimmer, R. L. 1988. The cyphonautes larva of the marine bryozoan *Membranipora membranacea.* II. Internal sac, musculature, and pyriform organ. Can J. Zool. *66:*384–398.

Thorpe, J. P., Shelton, G. A., and Laverack, M. S. 1975. Colonial nervous control of lophophore retraction in cheilostome Bryozoa. Science. *189:*60–61.

Winston, J. E. 1978. Polypide morphology and feeding behavior in marine ectoprocts. Bull. Mar. Sci. 28(1):1–31.

Winston, J. E. 1984. Why bryozoans have avicularia—a review of the evidence. Am. Mus. Novitates No. 2789. 1–26.

Winston, J. E. 1986. Victims of avicularia. P.S.Z.N.I.: Mar. Ecol. *7:*193–199.

Winston, J. E., and Jackson, J. B. C. 1984. Ecology of cryptic coral reef communities. IV. Community development and life histories of encrusting cheilostome Bryozoa. J. Exp. Mar. Biol. Ecol. *76:*1–21.

Woollacott, R. M., and Zimmer, R. L. (Eds.): 1977. Biology of Bryozoans. Academic Press, New York. 566 pp. (A collection of reviews of various aspects of bryozoan biology.)

Zimmer, R. L., and Woollacott, R. M. 1989. Larval morphology of the bryozoan *Watersipora arcuata* (Cheilostomata: Ascophora). J. Morphol. *199:* 125–150.

Entoprocts

Emschermann, P. 1982. Les Kamptozoaires. États actuel de nos connaissances sur leur anatomie, leur développement, leur biologie et leur position phylogénétique. [The kamptozoans. Present state of our knowledge of their anatomy, development, biology and phylogenetic position.) Soc. Zool. France. *107:*316–344.

Hyman, L. H. 1951. The Invertebrates. Vol. 3. Acanthocephala, Aschelminthes and Entoprocta. McGraw-Hill, New York.

Mariscal, R. 1965. The adult and larval morphology of and life history of the entoproct *Barentsia gracilis* (M. Sars, 1835). J. Morphol. *116:*311–338.

Mariscal, R. 1975. Entoprocta. *In* Giese, A. C. and Pearse, J. S. (Eds.): Reproduction of Marine Invertebrates. Vol. II. Entoprocts and Lesser Coelomates. Academic Press, New York. pp. 1–41.

Mukai, H., and Makioka, T. 1980. Some observations on the sex differentiation of an entoproct, *Barentsia discreta* (Busk). J. Exp. Zool. *213:*45–59.

Nielsen, C. 1964. Studies on Danish Entoprocta. Ophelia. *1*(1):1–76.

Nielsen, C. 1971. Entoproct life cycles and the entoproct/ectoproct relationship. Ophelia. *9*(2):209–341.

Nielsen, C., and Rostgaard, J. 1976. Structure and function of an entoproct tentacle with a discussion of ciliary feeding types. Ophelia. *15:*115–140.

Phoronida

Abele, L. G., Gilmour, T., and Gilchrist, S. 1983. Size and shape in the phylum Phoronida. J. Zool. Lond. *200:*317–323.

Bartolomaeus, T. 1989. Ultrastructure and relationship between protonephridia and metanephridia in *Phoronis muelleri* (Phoronida). Zoomorphology. *109:*113–122.

Emig, C. C. 1974. The systematics and evolution of the phylum Phoronida. Z. Zool. Syst. Evolutions.-forsch. *12:*(2):128–151.

Emig, C. C. 1977. Embryology of Phoronida. Am. Zool. *17:*21–37.

Emig, C. C. 1979. British and other Phoronids. Synopsis of the British Fauna No. 13. Academic Press, London. 58 pp.

Emig, C. C. 1982. The biology of Phoronida. Adv. Mar. Biol. *19:*1–89.

Hay-Schmidt, A. 1990. Distribution of catecholamine-containing, serotonin-like and neuropeptide FMRFamide-like immunoreactive neurons and processes in the nervous system of the actinotroch larva of *Phoronis muelleri* (Phoronida). Cell Tissue Res. *259:*105–118.

Lacalli, T. C. 1990. Structure and function of the nervous system in the actinotroch larva of *Phoronis vancouverensis*. Proc. R. Soc. Lond. B *327:*655–683.

Vandergon, T. L., and Colacino, J. M. 1991. Hemoglobin function in the lophophorate *Phoronis architecta* (Phoronida). Physiol. Zool. *64:*1561–1577.

Vandermeulen, J. H. 1970. Functional morphology of the digestive tract epithelium in *Phoronis vancouverensis:* An ultrastructural and histochemical study. J. Morphol. *130:*271–286.

Zimmer, R. L. 1991. Phoronida. *In* Giese, A. C., Pearse, J. S., and Pearse, V. B. (Eds.): Reproduction of Marine Invertebrates. Vol. VI. Echinoderms and Lophophorates. The Boxwood Press. Pacific Grove, CA. pp. 1–45.

Brachiopoda

Atkins, D. 1959. The growth stages of the lophophore and loop of the brachiopod *Terebratalia transversa* (Sowerby). J. Morphol. *105:*401–426.

Chuang, S. H. 1959. Structure and function of the alimentary canal in *Lingula unguis*. Proc. Zool. Soc. Lon. *132:*293–311.

Cooper, G. A. 1977. Brachiopods from the Caribbean Sea and adjacent waters. Studies in Tropical Oceanography No. 14. University of Miami Press, Coral Gables. 211 pp.

Emig, C. C. 1981. Observations on the ecology of *Lingula reevei*. J. Exp. Mar. Biol. Ecol. *52*(1):47–62.

Eshelman, W. P., Wilkens, J. L., and Cavey, M. J. 1982. Electrophoretic and electron microscopic examination of the adductor and diductor muscles of an articulate brachiopod, *Terebratalia transversa*. Can. J. Zool. *60:*550–559.

Gould, S. J., and Calloway, C. B. 1980. Clams and brachiopods—ships that pass in the night. Paleobiology. *6*(4):383–396.

Gustus, R. M., and Cloney, R. A. 1972. Ultrastructural similarities between the setae of brachiopods and polychaetes. Acta Zool. *53:*229–233.

Gutmann, W. F., Vogel, K., and Zorn, H. 1978. Brachiopods: Biochemical interdependences governing their origin and phylogeny. Science. *199:*890–893.

LaBarbera, M. 1977. Brachiopod orientation to water movement. 1. Theory, laboratory behavior, and field observations. Paleobiology. *3:*270–287.

LaBarbera, M. 1981. Water flow pattern in and around three species of articulate brachiopods. J. Exp. Mar. Biol. Ecol. *55:*185–206.

Long, J. A., and Stricker, S. A. 1991. Brachiopoda. *In* Giese, A. C., Pearse, J. S., and Pearse, V. B. (Eds.): Reproduction of Marine Invertebrates. VI. Echinoderms and Lophophorates. The Boxwood Press. Pacific Grove, CA. pp. 47–84.

Nielsen, C. 1991. The development of the brachiopod *Crania (Neocrania) anomala* (O. F. Mueller) and its phylogenetic significance. Acta Zool. *72:*7–28.

Paine, R. T. 1963. Ecology of the brachiopod *Glottidia pyramidata.* Ecol. Monogr. *33:*187–213.

Reed, C. G., and Cloney, R. A. 1977. Brachiopod tentacles: Ultrastructure and functional significance of the connective tissue and myoepithelial cells in *Terebratalia.* Cell Tissue Res. *185:*17–42.

Richardson, J. R. 1981. Brachiopods in mud: Resolution of a dilemma. Science. *211:*1161–1162.

Rudwick. M. J. S. 1970. Living and Fossil Brachiopods. Hutchinson University Library, Hutchinson and Co., London. 199 pp. (An excellent general account of living and fossil brachiopods.)

Stricker, S. A., and Reed, C. G. 1985a. The ontogeny of shell secretion in *Terebratalia transversa* (Brachiopoda: Articulata). I. Development of the mantle. J. Morphol. *183:*223–250.

Stricker, S. A., and Reed, C. G. 1985b. The ontogeny of shell secretion in *Terebratalia transversa* (Brachiopoda: Articulata). II. Formation of the protegulum and juvenile shell. J. Morphol. *183:*251–171.

Stricker, S. A., and Reed, C. G. 1985c. The protegulum and juvenile shell of a recent articulate brachiopod: Patterns of growth and chemical composition. Lethaia. *18:*295–303.

Stricker, S. A., and Reed, C. G. 1985d. Development of the pedicle in the articulate brachiopod *Terebratalia transversa* (Brachiopoda, Terebratulida). Zoomorphology. *105:*253–264.

Suchanek, T. H., and Levinton, J. 1974. Articulate brachiopod food. J. Paleontol. *48*(1):1–5.

Trueman, E. R., and Wong, T. M. 1987. The role of the coelom as a hydrostatic skeleton in lingulid brachiopods. J. Zool. Lond. *213:*221–232.

Valentine, J. W., and Jablonski, D. 1983. Larval adaptations and patterns of brachiopod diversity in space and time. Evolution. *37:*1052–1061.

Wilkens, J. L. 1978a. Adductor muscles of brachiopods: Activation and contraction. Can. J. Zool. *56:*315–323.

Wilkens, J. L. 1978b. Diductor muscles of brachiopods: Activation and very slow contraction. Can. J. Zool. *56:*324–332.

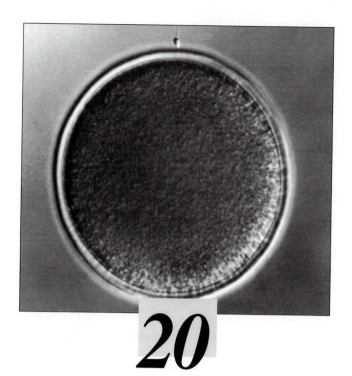

PATTERNS OF INVERTEBRATE EVOLUTION

PRINCIPLES AND EMERGING PATTERNS

The phylogeny of metazoans, of which the invertebrates comprise the vast majority, is understood only in broad brushstrokes. Even the relationships of major evolutionary lines, such as acoelomates and coelomates or annelids and arthropods, are not established with certainty. Intuition is a common source of phylogenetic ideas, and because many of these have been difficult to test scientifically, invertebrate phylogenies are often creative and controversial. Sometimes, the controversies have reached emotional levels, especially among authorities who represent different schools of thought. For the pragmatic Libbie Hyman, who compiled the first American treatise on invertebrates and who was an expert on turbellarians, the phylogenetic speculation by the prominent German biologist, Adolf Remane, that acoelomate flatworms evolved from coelomate ancestors, was incomprehensible. She reacted to his proposal by writing that the theory was ". . . fantastic nonsense, for which there does not exist a single scrap of evidence."[1] To this criticism, Remane later replied, "It is better to avoid terms like . . . fantastic nonsense. These . . . [questions] . . . are very complex, and only after a thorough study of much material would it be possible to give a definite answer."[2]

Controversies abound where there are few facts. The problem underlying any attempt to reconstruct the evolution of metazoans is that there are few homologues with which to determine the proper relationships. The challenge for those aspiring to clarify metazoan phylogeny is to find new homologues, as discussed at the beginning of Chapter 17. This may be accomplished by intensive comparative studies using traditional techniques of observation or by application of a modern technology that reveals previously unseen structures. In the past 25 years, two such technologies have become available: transmission electron microscopy and molecular biological techniques.

Use of the electron microscope opened a window on ultrastructures—cellular and subcellular details—either that were previously unseen, such as the 9+2 axoneme of cilia and flagella, or were unresolved by light microscopy, such as the structural details of cuticles and exoskeletons. Electron microscopy has also clarified the anatomy of many small organisms and developmental stages.

The tools of molecular biology are opening windows on the structure of the genome and especially the nucleotide sequences of genes. Once obtained, these sequences can be compared and degrees of similarity between them ranked to provide an estimate of evolutionary relationship. Although various genes have been sequenced partially or completely, sequences of ribosomal RNA genes are primarily used to infer relationships among phyla of animals. This is because the variability of ribosomal RNA gene sequences is large enough to distinguish phyla as separate entities but small enough to disclose similarities (sequence homology) resulting from evolutionary relationship.

---

[1]Hyman, L. H. 1959. The Invertebrates. Smaller Coelomate Groups. McGraw-Hill Book Co., New York. p. 750

[2]Remane, A. 1963. The enterocelic origin of the coelom. *In* Dougherty, E. C. (Ed.): The Lower Metazoa. University of California Press, Berkeley. p. 79.

A METHOD OF PHYLOGENETIC CLASSIFICATION (CLADISTICS)

Cladistics is a rational method of inferring evolutionary relationships and portraying them in a logical classification, typically a dichotomously branching tree, or **cladogram.** For a classification to meaningfully reflect evolutionary history, the organisms comprising each group (e.g., species, genus, family, phylum) should be defined by a unique characteristic, or set of characteristics, called a homologue. For example, gill slits, notochord, dorsal hollow nerve cord, and postanal tail are considered by most to be homologues that uniquely define a group of animals called chordates (Fig. 20–1). All chordates express these homologues, at least at some point in their life history, and in all other animals this particular set of homologues is absent. The urochordates, cephalochordates, and vertebrates are assumed, therefore, to have diverged from a common ancestor and are more closely related to, each other than any one is to another group of animals, that is, they are **monophyletic.** If, instead, each subphylum had evolved independently from a separate ancestor, the chordates would be **polyphyletic** in origin.

Variations in the structure of homologues between the subphyla provide additional information useful for inferring the evolutionary relationships of the chordates. Focusing on the notochord, recall that it is expressed as a coelom in the urochordates (larvaceans), as a stack of muscle cells in the cephalochordates, and as a cord of noncontractile, fibrous cells in the vertebrates (Fig. 17–45), all variations on a common theme. Because a coelom occurs in phyla other than the chordates and because the cephalochordate notochord develops enterocoelously, it therefore seems likely that the coelomic layout of the urochordate notochord is the common theme. It is a carryover from the coelomate ancestor of the chordates, and as such is regarded as the **primitive,** or **plesiomorphic,** form of the notochord. The muscular notochord of the cephalochordates and the fibrous notochord of the vertebrates (both with well-developed sheaths) are variations on the theme and are considered

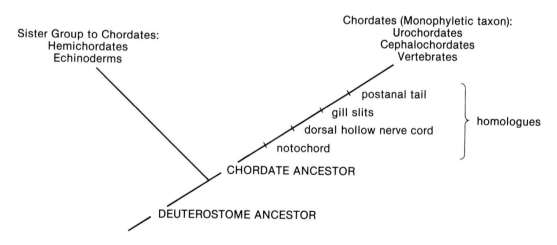

FIGURE 20–1 An example of how homologues are used to determine evolutionarily related groups of organisms. The set of four homologues is unique to the urochordates, cephalochordates, and vertebrates. Thus, these three taxa are more closely related to each other than to any other group of animals. They shared a recent common ancestor, from which they evolved as a monophyletic taxon, the Chordata.

to be evolutionarily **derived,** or **apomorphic,** forms of the notochord (Fig. 20–2). Just as a common theme is general and its variations specific, so too a primitive expression of a homologue characterizes a broad, more inclusive group, whereas derived expressions of a homologue characterize narrow, more exclusive, groups. As a result, a classification based on homology identification is multileveled, or **hierarchic,** in structure. Because homologous similarities are believed to be the result of evolutionary descent with modification, such a classification is viewed as an evolutionary (phylogenetic) classification.

After considering such an obvious example, it may be surprising to learn that few of the major evolutionary lines of metazoans are as well defined as that of the chordates, and many groups of phyla, for example the lophophorates or the aschelminths, lack even one uniquely distinguishing homologue. For this reason, the interrelationships of most of these phyla and their specific placement within the general framework of metazoan phylogeny is uncertain.

Principal Results of Phylogenetic Research

Observations from classical anatomy and embryology, made primarily during the nineteenth century, resulted in our general understanding of the evolutionary relationships of animals, but many questions remain unanswered: are the metazoans monophyletic, or diphyletic, with the sponges constituting a separate evolutionary line? Were the ancestral bilaterians acoelomate or coelomate? Where do the aschelminth, lophophorate, pogonophoran, chaetognath, nemertean, sipunculan, and "mesozoan" branches attach to the evolutionary tree? Are the coelomates as well as the arthropods monophyletic, diphyletic, or polyphyletic taxa?

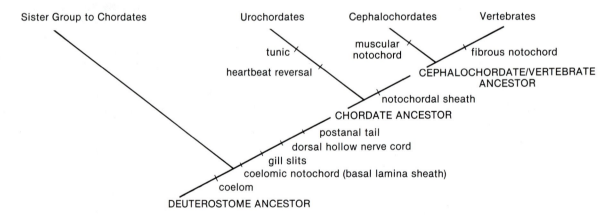

FIGURE 20–2 Example of how the composition, or states, of a homologue are used to determine levels of relatedness in an evolutionary reconstruction. Although the histology of the chordate notochord varies widely, the coelomic notochord of larvacean urochordates is probably the primitive expression because a coelom occurs in all non-chordate deuterostomes. A shared-derived change in the notochord, the evolution of a thick extracellular sheath, defines the cephalochordates and vertebrates as being more closely related to each other than either is to the urochordates. The muscular notochord of cephalochordates and a notochord composed of cells bearing intracellular fibers in the vertebrates are homologues that define each of these groups.

Provisional answers to these and other questions of metazoan phylogeny awaited the means to observe and compare detailed information at other levels of organization. As data from microanatomy (electron microscopy) and molecular biology continue to be assembled and analyzed, several of these issues have been clarified but not finally answered.

Phylum-Level Relationships Suggested by the Application of Electron Microscopy

1. Metazoans are probably a monophyletic kingdom, based on the widespread occurrence of choanocyte-like cells and uniflagellated sperm.

2. Metazoans probably evolved from colonies of flagellates that resembled the extant choanoflagellates, based on similarities of choanoflagellate and metazoan flagella.

3. Coelomates probably constitute a monophyletic taxon, because of similarities of their mesothelia, blood vessels, and nephridia.

4. Protostomes and deuterostomes are each likely to be monophyletic taxa, based on the nature of their photoreceptoral cells.

5. Nemerteans are probably coelomate animals, as indicated by the peritoneal lining of the rhynchocoel and "blood vessels," as well as the schizocoelous mode of development of the "blood vessels."

6. Developmental and structural evidence supports a close relationship between sipunculans and molluscs.

7. Gastrotrichs and nematodes (possibly nematomorphs) constitute a monophyletic group, based on similarities of their pharynx and cuticle.

8. Rotifers and acanthocephalans probably constitute a monophyletic group because of the similarity of their integument.

9. Pentastomids and crustaceans may be a monophyletic group based on similarities of their sperm.

10. Pogonophorans, echiurans, and annelids may constitute a monophyletic group because of the identical structure of their setae (but similar setae also occur in brachiopods, bryozoans, and octopods).

General Evolutionary Trends and Reinterpretations Documented by Electron Microscopy

1. Cells bearing one flagellum (or cilium) are more primitive than those bearing more than one, a conclusion supported by comparative and developmental analyses.

2. Ciliary photoreceptors are probably more primitive than microvillar (rhabdomeric) photoreceptors, based on comparative and devolopmental studies. Combined with item 4 in the previous list, this inference suggests that the deuterostomes are more primitive than the protostomes.

3. An epithelial musculature (myoepithelial mesothelium) is probably more primitive than a subperitoneal musculature in coelomates, based on comparative and developmental studies.

4. Cross-striated muscle and a type of smooth muscle, in which the thin actin filaments attach and anchor to the cell membrane only (e.g., in cnidarians and ctenophores), are more primitive than other types of striated and smooth muscle arrangements.

5. Neurons having multiple functions (sensory-motor interneurons, as in some cnidarians) are more primitive than those having a specialized function.

6. A sperm composed of an acrosome, spherical nucleus, short mitochondrial midpiece and a single flagellum is more primitive than other designs of sperm.

7. Filtration nephridia are the primitive excretory organs of bilaterians.

8. Coelomates having blood vessels and metanephridial systems can evolve into acoelomates having protonephridia and lacking blood vessels, as evidenced in many species of annelids. Such "regressive" evolution is actually a positive adaptation to small body size.

9. Many small nematodes, all gastrotrichs, and kinorhynchs are acoelomate and not pseudocoelomate in body organization.

10. There is evidence of a true coelom, lined by peritoneum, in a species of the pseudocoelomate Priapulida.

11. Species that have a cuticle can evolve into species that lack one, as evidenced in the Annelida. As with the loss of a coelom, the loss of cuticle is correlated with small body size or the adoption of an endoparasitic habit (e.g., some nematodes).

Phylum-Level Evolutionary Indications from Molecular Analyses

1. Although the ciliates constitute a monophyletic group, the flagellates are almost certainly polyphyletic in origin (rRNA).

2. The Metazoa is probably a monophyletic taxon (rRNA).

3. Rhombozoan "mesozoans" are probably rooted in the diploblastic animals and not in the bilaterians (rRNA).

4. Coelomates are probably a monophyletic group (rRNA).

5. Protostomes and deuterostomes are each likely to be monophyletic taxa (rRNA).

6. Platyhelminths are an early, perhaps the earliest, lineage within the Bilateria (rRNA).

7. Nemerteans and coelomates are likely to form a monophyletic taxon (rRNA).

8. Sipunculans and molluscs are probably a monophyletic group (rRNA).

9. Annelids and pogonophorans are likely to be a monophyletic taxon (rRNA).

10. Arthropods are probably a monophyletic group (rRNA, hemocyanin).

11. Within the coelomate protostomes, the annelids and molluscs may form one monophyletic taxon, and the arthropods another. If this is true, the traditional closer relationship between annelids and arthropods would be rejected.

12. Pentastomids and crustaceans may constitute a monophyletic group (rRNA).

13. Onychophorans and arthropods may be a monophyletic taxon (rRNA).

Despite these advances, many of the evolutionary questions posed at the beginning of this section remain unanswered, including the nature of the earliest bilaterians—acoelomate or coelomate, small or large? The ultrastructural data have more or less supported the possibility that an acoelomate body can evolve from a coelomate design, thus supporting Remane's evolutionary viewpoint. Hyman's position, however, is supported by preliminary molecular data. They suggest that the acoelomate platyhelminths evolved before the lineage, or lineages, leading to the coelomate animals. At this point in our discussion, it may be useful to construct a series of phylogenies of metazoans to illustrate better the provisional nature of evolutionary classifications in general and to highlight hypotheses of general significance, in particular the evolution of bilaterally symmetrical animals.

PHYLOGENETIC CLASSIFICATION OF METAZOANS

A first approach to metazoan evolutionary classification can be based on three homologues: eukaryotic cells, tissues, and organs, which are often regarded as "levels of organization." The assumption here is that the eukaryotic cell evolved in the protists (including protozoa) and was retained, not independently evolved, in the lines to the fungi, metaphytes, and metazoans. The eukaryotic cell, therefore, is a homologue of protists, fungi, metaphytes, and metazoans and is one indication that they form a monophyletic taxon. Similarly, once tissues, such as epithelial and connective tissues, evolved in lower metazoans, they were retained in higher metazoans. Thus, animal tissues are unique to metazoans, they define metazoans as monophyletic, and they distinguish metazoans from protozoans. The same is assumed to be true for the capacity to organize tissues into organs. Although sponges and placozoans (and "mesozoans") have tissues, they lack the organs found in all other metazoans and can be distinguished from them on that basis. The capacity to form organs, therefore, is a homologue of "higher" organ-bearing animals setting them apart as a monophyletic taxon within the Metazoa as a whole. Thus, these three homologues provide a first approximation to the phylogeny of metazoans (Fig. 20–3A).

The body symmetry of animals can be used to refine the evolutionary classification shown in Figure 20–3A. Sponges and placozoans are largely asymmetric in body form; cnidarians, ctenophores, and "mesozoans" are radial or biradial, as are the early developmental stages of most other animals; but bilateral symmetry distinguishes adult higher animals from cnidarians and ctenophores. Addition of the homologue bilateral symmetry to the classification in Figure 20–3A, therefore, resolves the organ-level animals into two groups: the more primitive asymmetric and radially symmetric phyla and the evolutionarily derived, bilaterally symmetrical animals (Bilateria) (Fig. 20–3B).

At the next level of consideration, degrees of compartmentation of the body by epithelia can be used as characteristics with which to further specify the evolutionary tree. Although a protoepithelium occurs already in sponges, a true epithelial epidermis is expressed in placozoans, an epidermis and gastrodermis (with connective tissue in between) occur in cnidarians and all higher animals. A subset of the higher animals, called coelomates (e.g., annelids, arthropods, molluscs, echinoderms) have a third epithelium (mesothelium) that encloses a coelom, a homologue unique to the coelomate assemblage of phyla. The occurrence of a mesothelium splits the Bilateria (Fig. 20–3B) into the primitive acoelomates and pseudocoelomates, and the evolutionarily derived coelomates (Fig. 20–3C). Although this general classification receives wide support among zoologists and is the outline followed in this book, several researchers regard the acoelomate and pseudocoelomate animals as secondarily simplified coelomates. While completing an evolutionary tree for the Metazoa, we will examine the factual basis for the persistence of this controversy.

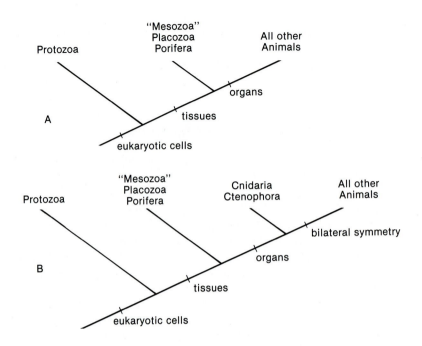

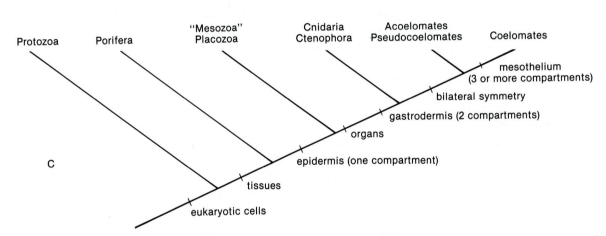

FIGURE 20–3 Progressive models of metazoan phylogeny based on **A,** levels of organization; **B,** levels of organization and body symmetry; **C,** levels of organization, body symmetry, and number of epithelium-lined body compartments.

Phylogeny of the Bilateria

Although bilateral symmetry unambiguously defines the bilaterally symmetrical animals, alternative classifications result from assumptions regarding the primitive or derived states of other characteristics. In the earliest bilaterians, did the blastopore give rise to the mouth or anus, or to the mouth and anus? Were they coelomate or acoelomate? Was the nephridium a protonephridium or a metanephridial system? Was a blood-vascular system present or absent? Were the animals large or small?

Depending on how these questions are answered, different evolutionary classifications result.

Three alternative classifications for the phylogeny of the Bilateria are shown in Figure 20–4, and each embodies only one unlikely assumption, making a choice between them difficult. The unlikely assumptions are that some presumed homologue evolved more than once (and thus the two structures, because of their independent appearances, are actually non-homologous). The classification shown in Figure 20–4A shows the acoelomates as primitive bilaterians, but assumes that radial cleavage evolved twice, once in the cnidarians and again in the line to the deuterostomes. Figure 20–4B shows the coelomates as primitive bilaterians and radial cleavage as monophyletic but suggests a secondary origin for acoelomate and pseudocoelomate body designs. In Figure 20–4C, the acoelomates are again shown as primitive bilaterians, but radial cleavage is monophyletic. Spiral cleavage, however, is diphyletic, having evolved once in the acoelomates and again in the coelomates. Ultimately, a choice between these alternatives, or the establishment of another more accurate tree, lies in the discovery of new facts and a careful analysis of homology.

Phylogeny of Acoelomate Spiralians

Acoelomate organization implies small body size, a characteristic that may be primitive in Bilateria, but one that also has evolved independently in several groups of animals. The advent of small body size in aquatic animals is correlated with the use of cilia for locomotion, the absence of a coelom and blood-vascular system, and the occurrence of protonephridia, all of which are homologues of acoelomates. Because all of these can be reduced to a single trait, small body size, the assemblage may be polyphyletic rather than monophyletic in origin. With this possibility in mind, the spiralian acoelomates may be said to include the gnathostomulids, platyhelminths, and entoprocts.

The gnathostomulids are regarded as the primitive taxon in the assemblage because their epidermal cells and protonephridial terminal cells bear only one cilium each. In the entoprocts and platyhelminths, on the other hand, the epidermal and protonephridial cells are multiciliated, a shared-derived homologue (Fig. 20–5).

Phylogeny of Coelomate Protostomes

The evolutionary tree shown in Figure 20–6 divides the coelomate protostomes into two distinct lineages. One lineage, set apart by the advent of segmentation, embraces the annelids, pogonophorans, echiurans, onychophorans, and arthropods. Although the echiurans lack segmentation, they are grouped with the annelids because of the shared occurrence of an annelidan cross. The absence of segmentation in echiurans, therefore, is viewed as secondary feature. The second lineage includes the sipunculans and molluscs, which are shown as having evolved from a nonsegmented ancestor with few coelomic cavities (oligomery). An alternative to this phylogeny is one in which segmentation appears earlier, in the ancestor to both annelids and molluscs. In that case, the absence of segmentation in sipunculans and molluscs, except perhaps in *Neopilina,* would have to be interpreted as a secondary loss.

(Text continues on p. 1052)

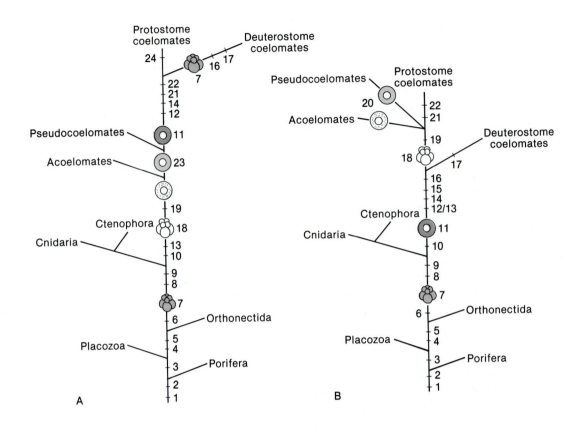

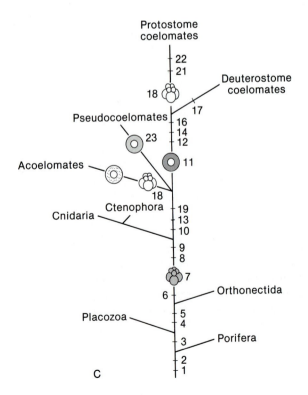

24 ventral nervous system
23 pseudocoel
22 trochophore larva
21 schizocoely
20 coelom loss (modification of mesothelium)
19 blastopore becomes mouth
18 spiral cleavage
17 blasopore becomes anus
16 enterocoely
15 blastopore becomes mouth and anus
14 blood-vascular system
13 protonephridium
12 metanephridial system (protonephridium in larva)
11 coelom (mesothelium)
10 bilateral symmetry
 9 organs
 8 gastrodermal epithelium
 7 radial cleavage
 6 radial symmetry
 5 nerve cells
 4 muscle cells
 3 epidermal epithelium
 2 embryos
 1 connective tissue

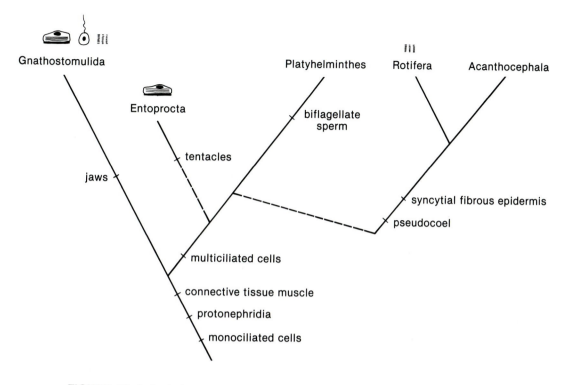

FIGURE 20–5 A phylogeny of spiralian acoelomates. Symbols represent myoepithelial cells, monociliated cells, and cross-striated muscles, all presumed primitive traits.

◀ **FIGURE 20–4** Three alternatives for the evolution of bilaterally symmetrical animals. Homologues are numbered; diphyletic or polyphyletic characteristics are boxed. **A,** Acoelomates are the most primitive bilaterians and spiral cleavage (18) evolved once, but radial cleavage (7) is diphyletic. This evolutionary tree largely reflects Hyman's viewpoint, is widely accepted among zoologists, and receives support from the preliminary results of molecular systematics. It indicates that body cavities evolved in the order: acoelomate; pseudocoelomate, coelomate. **B,** Coelomates are the most primitive bilaterians, but the loss of a coelom in the acoelomates and pseudocoelomates is secondary and difficult to demonstrate. This reconstruction approximates Remane's viewpoint and has some supporters in the United States and Europe but is not as popular as the scheme shown in **A**. Nevertheless it gains support from comparative ultrastructural data. **C,** This reconstruction is a variation on **A** that assumes acoelomates are the most primitive bilaterians and that radial cleavage (7) is monophyletic. Under these conditions, however, spiral cleavage (18) is diphyletic. Although logically possible, this tree is not widely discussed among systematists.

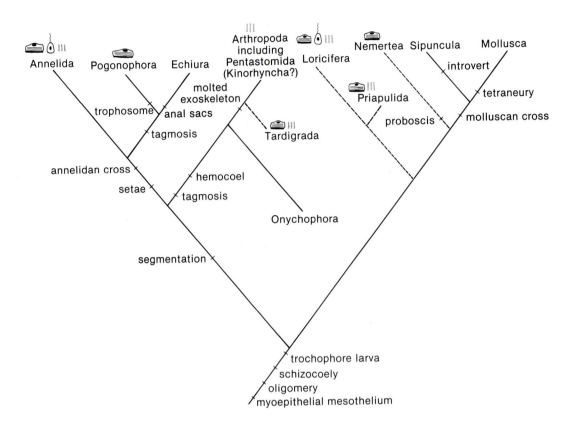

FIGURE 20–6 A phylogeny of coelomate protostomes. Symbols as in Figure 20–5.

Phylogeny of Deuterostomes

The deuterostomes, including the chaetognaths and lophophorates, may constitute a monophyletic taxon based on the common occurrence of enterocoely and a tricoelomate body, but the specific placement in the tree of phyla in relation to each other is problematic (Fig. 20–7A). Placement of the lophophorates in the deuterostomes suggests that the absence of enterocoely in bryozoans and phoronids is a derived characteristic, as is the derivation of the mouth from the blastopore, a typical protostome trait, in the phoronids. If it is assumed that hemichordate and chordate gill slits are homologous structures and that the complex protocoelic nephridia of hemichordate and echinoderm larvae are also homologues, then the absence of gill slits in echinoderms is probably secondary (There is an indication of gill pores in some extinct fossil echinoderms, called carpoids.) (Fig. 20–7A). This scenario is supported by preliminary molecular data from the analysis of rRNA sequences.

An alternative is to consider the hemichordates and chordates together as a monophyletic group and the echinoderms as a separate, more primitive taxon within the deuterostomes (Fig. 20–7B). In that case, the homologue, gill slits, would uniquely characterize the hemichordates and chordates. The protocoelic nephridium,

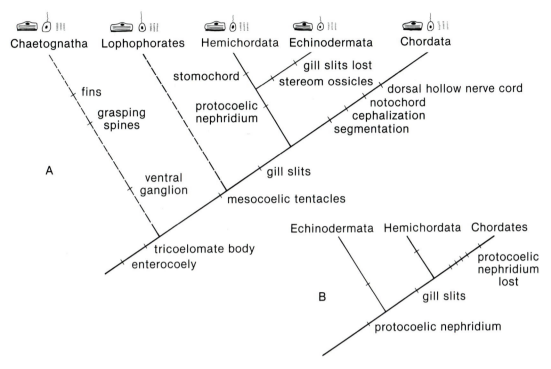

FIGURE 20–7 A, A phylogeny of deuterostomes. Symbols as in Figure 20–5.

however, would either have had to evolve twice independently, once in the echinoderms and again in the hemichordates, or it evolved in the common ancestor to all three phyla. If so, the chordates have secondarily lost the protocoelic nephridium. (A remnant may persist in cephalochordates as the larval preoral pit.) The systematic position of the chaetognaths relative to other deuterostomes, a perennial question in deuterostome phylogenetics, remains unanswered for lack of a shared-derived homologue that links them with any other phylum.

Phylogeny of the Aschelminth Phyla

The aschelminth assemblage of phyla, because of its pseudocoelomate grade of organization, has long been placed between acoelomates and coelomates as a group of evolutionarily intermediate animals. Although this may be a correct interpretation, a significant amount of new information speaks against a close relationship (monophyly) of these taxa and even rejects the idea that all are pseudocoelomate. As a result, it seems likely that some aschelminth phyla, or groups of them, could be more closely related to some non-aschelminth phylum, then they are to each other. The question is: to which non-aschelminth phylum or phyla is any aschelminth group linked? Rotifers and acanthocephalans provide evidence of spiral cleavage, but does this indicate an alliance with the acoelomate protostomes or the coelomate protostomes? The probable monophyly of gastrotrichs, nematodes, and nematomorphs has already been mentioned. All have bilateral cleavage patterns, and in many gastrotrichs, the blastopore forms both the mouth and anus. Should these three phyla be aligned with protostomes or deuterostomes? The unfortunate fact is that on the basis of current information, it is impossible to find a characteristic that uniquely links any aschelminth phylum with any non-aschelminth phylum.

CONCLUDING REMARKS ON ANIMAL PHYLOGENY

A thorough phylogenetic analysis of a group of organisms as large and diverse as the Metazoa involves such complex issues and such a magnitude of information as to be beyond the comprehension of any one individual. Many of the accomplished practitioners have a catholic knowledge of structure and function, a talent for synthesis, and sound logical ability. Such individuals, many of whose names are mentioned in this book, have been largely responsible for our current knowledge of animal relationships and for new ideas about the evolution of organs and other structures. That the ideas of some are in disagreement with those of others is simply a reflection of the partial knowledge on which each is based, although sometimes errors in logic or interpretation are made. In this respect, phylogenetic speculations are hypotheses similar to others in science, that is, they are provisional ideas that are subject to scrutiny and reevaluation. As we have tried to show in this chapter, real progress has been made, and is being made, in advancing our knowledge of animal relationships, although much remains to be accomplished. The progress may not be rapid, but with the growing interest in evolutionary studies, future advances are assured. As Remane noted 30 years ago, ". . . even after critical study of homologies the facts may still not convincingly favor a given theory, but may suggest two or three hypotheses. In such instances we can only treat the problem as a continuing challenge and constantly seek to refine our knowledge and render our discrimination more acute."[3]

REFERENCES

Phylogenetic Methods

Crowson, R. A. 1970. Classification and Biology. Atherton Press. New York.

Eldredge, N., and Cracraft, J. 1980. Phylogenetic Patterns and the Evolutionary Process. Columbia University Press. New York.

Hennig, W. 1966. Phylogenetic Systematics. University of Illinois Press, Urbana.

Mayr, E. 1969. Principles of Sytematic Zoology. McGraw-Hill Book Co., New York.

Wiley, E. O. 1981. Phylogenetics: The Theory and Practice of Phylogenetic Systematics. John Wiley and Sons, New York.

Molecular Systematics

Abele, L. G., Kim, W., and Felgenhaur, B. E. 1989. Molecular evidence for inclusion of the phylum Pentastomida in the Crustacea. Mol. Biol. Evol. *6:*685–691.

Ballard, J. W. O., G. J. Olsen, D. P. Faith, W. A. Odgers, D. M. Rowell and P. W. Atkinson. 1992. Evidence from 12S ribosomal RNA sequences that onychophorans are modified arthropods. Science. *258:*1345–1348.

Bergstrom, J. 1986. Metazoan evolution—a new model. Zool. Scripta. *15:*189–200.

Christen, R., Ratto, A., Baroin, A. et al. 1991. An analysis of the origin of metazoans, using comparisons of partial sequences of the 28S RNA, reveals an early emergence of the triploblasts. EMBO J. *10:*499–503.

Field, K. G., Olsen, G. J., Lane, D. J. et al. 1988. Molecular

---

[3]Remane, A. 1963. The evolution of the Metazoa from colonial flagellates vs. plasmodial ciliates. *In* Dougherty, E. C. (Ed.): The Lower Metazoa. University of California Press, Berkeley. p. 30.

phylogeny of the animal kingdom. Science. *239:*748–753.

Ghiselin, M. T. 1988. The origin of molluscs in the light of molecular evidence. Oxford Sur. Evol. Biol. *5:*66–95.

Hori, H., Muto, A., Osawa, S. et al. 1988. Evolution of Turbellaria as deduced from 5S ribosomal RNA sequences. Fortschr. Zool. *36:*163–167.

Hori, H., and Osawa, S. 1987. Origin and evolution of organisms as deduced from 5S ribosomal RNA sequences. Mol. Biol. Evol. *4:*445–472.

Lake, J. A. 1990. Origin of the metazoa. Proc. Natl. Acad. Sci., USA. *87:*763–766.

Mangum, C. P., Scott, J. L., Black, R. E. L., et al. 1985. Centipedal hemocyanin: Its structure and its implications for arthropod phylogeny. Proc. Natl. Acad. Sci., USA. *82:*3721–3725.

Ohama, T., Kumazaki, T., Hori, H. et al. 1984. Evolution of multicellular animals as deduced from 5S rRNA sequences: A possible early emergence of the Mesozoa. Nucl. Acids Res. *12:*5101–5108.

Turbeville, J. M., Field, K. G., and Raff, R. A. 1992. Phylogenetic position of phylum Nemertini, inferred from 18S rRNA sequences: Molecular data as a test of morphological character homology. Mol. Biol. Evol. *9:*235–249.

Turbeville, J. M., Pfeiffer, D. M., Field, K. G. et al. 1991. The phylogenetic status of arthropods, as inferred from 18S rRNA sequences. Mol. Biol. Evol. *8:*669–686.

Morphological Systematics

Anderson, D. T. 1981. Origins and relationships among the animal phyla. Proc. Linn. Soc. N. S. Wales. *106:*151–166.

Ax, P. 1985. The position of the Gnathostomulida and Plathelminthes in the phylogenetic system of the Bilateria. *In* Conway-Morris, S., George, J. D., Gibson, R. et al. (Eds.): The Origins and Relationships of Lower Invertebrates. Clarendon Press, Oxford. pp. 168–180.

Ax, P. 1987. The Phylogenetic System. The Systematization of Organisms on the Basis of Their Phylogenesis. Jefferies, R. P. S., Transl. John Wiley and Sons, New York.

Ax, P. 1989. Basic phylogenetic systematization of the Metazoa. *In* Fernholm, B., Bremer, K., and Jornvall, H. (Eds.): The Hierarchy of Life. Elsevier, Amsterdam. pp. 229–245.

Barnes, R. D. 1985. Current perspectives on the origins and relationships of lower invertebrates. *In* Conway-Morris, S., and Bone, Q. 1960. The Origin of the Chordates. Oxford University Press, London.

Boudreaux, H. B. 1979. *In* Gupta, A. P. (Ed.): Arthropod Phylogeny. Van Nostrand-Reinhold, New York. pp. 551–586.

George, J. D., Gibson, R., and Platt, H. M. (Eds.): The Origins and Relationships of Lower Invertebrates. Clarendon Press, Oxford.

Clark, R. B. 1964. Dynamics in Metazoan Evolution. The Origin of the Coelom and Segments. Clarendon Press, Oxford.

Clark, R. B. 1979. Radiation of the Metazoa. *In* House, M. R. (Ed.): The Origin and Evolution of Major Invertebrate Groups. Vol. 12. Academic Press, London. pp. 55–102.

Eernisse, D. J., Albert, J. S., and Anderson, F. E. 1992. Annelida and Arthropoda are not sister taxa: A phylogenetic analysis of spiralian metazoan morphology. Syst. Zool. *41:*305–330.

Grell, K. G. 1981. *Trichoplax adhaerens* and the origin of Metazoa. Atti Conv. Lincei. *49:*107–121.

Gutmann, W. F. 1981. Relationships between invertebrate phyla based on functional-mechanical analysis of the hydrostatic skeleton. Am. Zool. *21:*63–81.

Hadzi, J. 1963. The Evolution of the Metazoa. Macmillan, New York.

Hanson, E. D. 1977. The Origin and Early Evolution of Animals. Pitman, London.

Haszprunar, G., Rieger, R. M., and Schuchert, P. 1991. Extant "problematica" within or near the Metazoa. *In* Conway-Morris, S., and Simonetta, A. (Eds.): The Early Evolution of the Metazoa and the Significance of Problematic Taxa. Cambridge University Press. pp. 99–105.

Hyman, L. H. 1940. The Invertebrates. Vol. 1. Protozoa through Ctenophora. McGraw-Hill Book Co., New York. pp.

Hyman, L. H. 1951. The Invertebrates. Vol. 2. Platyhelminthes and Rhynchocoela. The Acoelomate Bilateria. McGraw-Hill Book Co., New York. pp.

Hyman, L. H. 1959. The Invertebrates. Vol 5. Smaller Coelomate Groups. McGraw-Hill Book Co., New York. pp. 783.

Inglis, W. G. 1985. Evolutionary waves: Patterns in the origins of animal phyla. Austral. J. Zool. *33:*153–178.

Ivanov, A. V. 1988. On the early evolution of the Bilateria. Fortschr. Zool. *36:*349–352.

Jägersten, G. 1955. On the early phylogeny of the metazoa. Zool. Bidrag Uppsala. *30:*321–354.

Jägersten, G. 1959. Further remarks on the early phylogeny of the metazoa. Zool. Bidrag Uppsala. *33:*79–108.

Lankester, E. R. 1873. On the primitive cell layers of the embryo as the basis of geneological classification of animals, and on the origin of vascular and lymph systems. Ann. Nat. Hist. *11:*321–337.

Lankester, E. R. 1877. Notes on the embryology and classification of the animal kingdom. Q. J. Microsc. Sci. *17:*399–454.

Lorenzen, S. 1985. Phylogenetic aspects of pseudocoelomate evolution. *In* Conway-Morris, S., George, J. D., Gibson, R. et al. (Eds.): The Origins and Relationships of Lower Invertebrates. Clarendon Press, Oxford. pp. 210–223.

Manton, S. M. 1977. The Arthropoda. Clarendon Press, Oxford.

Marcus, E. 1958. On the evolution of the animal phyla. Q. Rev. Biol. *33:*24–58.

Nielsen, C. 1985. Animal phylogeny in the light of the trochaea theory. Biol. J. Linn. Soc. *25:*243–299.

Nielsen, C. 1987. Structure and function of metazoan ciliary bands and their phylogenetic significance. Acta Zool. (Stockh.) *68:*205–262.

Pantin, C. 1960. Diploblastic animals. Proc. Linn. Soc. Lond. *171:*1–14.

Remane, A. 1963a. The evolution of the metazoa from colonial flagellates vs. plasmodial ciliates. *In* Dougherty, E. C. (Ed.): The Lower Metazoa. University of California Press, Berkeley. pp. 23–32.

Remane, A. 1963b. The evolutionary origin of the coelom. *In* Dougherty, E. C. (Ed.): The Lower Metazoa. University of California Press, Berkeley. pp. 78–90.

Remane, A. 1963c. The systematic position and phylogeny of the pseudocoelomates. *In* Dougherty, E. C. (Ed.): The Lower Metazoa. University of California Press, Berkeley. pp. 247–255.

Remane, A. 1967. Die Geschichte der Tiere. *In* Heberer, G. (Ed.): Die Evolution der Organismen. Vol. 1. Gustav Fischer Verlag, Stuttgart. pp. 589–677.

Rieger, R. M. 1976. Monociliated epidermal cells in Gastrotricha: Significance for concepts of early metazoan evolution. Z. Zool. Syst. Evolut.-forsch. *14:*198–226.

Rieger, R. M. 1985. The phylogenetic status of the acoelomate organization within the Bilateria: A histological perspective. *In* Conway-Morris, S., Platt, H. M., Gibson, R. et al. (Eds.): The Origins and Relationships of Lower Invertebrates. Clarendon Press, Oxford. pp. 101–122.

Rieger, R. M., Haszprunar, G., and Schuchert, P. 1991. On the origin of the Bilateria: Traditional views and recent alternative concepts. Proceedings of the International Symposium in Camerino. 1989. *In* Simonetta, A. M.,

and Conway-Morris, S. (Eds.): The Early Evolution of Metazoa and the Significance of Problematic Taxa. pp. 107–112.

Rieger, R. M., and Lombardi, J. L. 1987. Ultrastructure of coelomic lining in echinoderm podia: Significance for concepts in the evolution of muscle and peritoneal cells. Zoomorphology. *107:*191–208.

Ruppert, E. E. 1982. Comparative ultrastructure of the gastrotrich pharynx and the evolution of myoepithelial foreguts in Aschelminthes. Zoomorphology. *99:*181–200.

Salvini-Plawen, L. v. 1978. On the origin and evolution of the lower Metazoa. Z. Zool. Syst. Evolut.-forsch. *16:*40–88.

Salvini-Plawen, L. v. 1980. Phylogenetischer Status und Bedeutung der mesenchymaten Bilateria. Zool. Jahrb. Ab. Anat. *103:*354–373.

Salvini-Plawen, L. v. 1982. A paedomorphic orgin for the oligomerous animals? Zool. Scripta. *11:*77–81.

Schaeffer, B. 1987. Deuterostome monophyly and phylogeny. Evol. Biol. *21:*179–235.

Scheltema, A. 1993. Aplacophora as progenetic aculiferans and the coelomate origin of mollusks as the sister taxon of Sipuncula. Biol. Bull. *184:*57–78.

Sedgwick, A. 1884. On the origin of metameric segmentation and some other morphological questions. Q. J. Microsc. Sci. *24:*43–82.

Siewing, R. 1979. Homology of cleavage types. Fortschr. Zool. Syst. Evolut. *1:*7–18.

Siewing, R. 1980. Das Archicoelomatenkonzept. Zool. Jahrb. Ab. Anat. *103:*439–482.

Weygoldt, P. 1986. Arthropod interrelationships—the phylogenetic-systematic approach. Z. Zool. Syst. Evolut.-forsch. *24:*19–35.

Willmer, P. 1990. Invertebrate Relationships. Cambridge University Press. 400 pp.

Willmer, P. G., Holland, P. W. H. 1991. Modern approaches to metazoan relationships. J. Zool. Lond. *224:*689–694.

GLOSSARY

An excellent glossary of invertebrate terms has been published by M. Stachowitsch (1992, *The Invertebrates. An Illustrated Glossary.* Wiley-Liss & Sons, Inc., New York. 676 pp.) and may be consulted as a supplement to this glossary.

Abyssal   Referring to the bottom region of ocean basins.

Abyssal zone   Zone comprised of the abyssal plain.

Acanthella   Developmental stage following the acanthor that develops in the intermediate host.

Acanthor   First larval stage of acanthocephalans; the larva has a rostellum with hooks that are used in penetrating the host's tissues.

Aciculum (pl. Acicula)   Chitinous rod that internally supports the divisions of the parapodium.

Acoelomate   Body organization of bilaterally symmetrical animals in which a fluid-filled cavity is absent between the epidermis and gastrodermis.

Acontium (pl. Acontia)   A thread originating from the middle lobe of an anthozoan septal filament that projects freely into the gastrovascular cavity.

Acron   The anterior nonsegmental part of the body of a segmented animal.

Acrorhagus (pl. Acrorhagi)   Cnidocyte-covered elevations on specialized sweeper tentacles or on the column of certain anthozoans.

Actinotrocha   Tentaculate ciliated larva of phoronids.

Actinula   A polyp-like larva of certain hydrozoans that resembles a short stemless hydranth.

Adductor   Typically, one of a pair of muscles that close the valves of a bivalve shell.

Adoral zone   Region within the buccal cavity of certain ciliates.

Aesthetasc   An elongated chemoreceptive sensillum situated on or near the antennae of most crustaceans.

Aestivation   A dormant state in which some animals pass hot, dry seasons.

Agamete   Nucleus within the plasmodium of an orthonectid "mesozoan" that divides mitotically and eventually gives rise to a sexual adult.

Alveolus (pl. Alveoli)   One of many flattened vesicles that form a more or less continuous layer beneath the cell membrane of ciliates and a few other protozoans.

Ambulacrum (pl. Ambulacra)   Groove, ridge, or band of tube feet, radial canal, and associated body wall of echinoderms.

Amictic egg   Type of rotifer egg that is thin shelled,

cannot be fertilized, and develops by parthenogenesis into amictic females.

Amphiblastula   Sponge larva which is hollow. One hemisphere is composed of small flagellated cells and the other is composed of large nonflagellated macromeres.

Amphid   Paired, anterior chemosensory organs of many nematodes.

Ampulla (pl. Ampullae)   Small, muscular sac attached to an echinoderm tube foot that bulges into the perivisceral coelom. The posterior, usually expanded, end of the phoronid body.

Analogy   Structural similarity resulting from evolutionary convergence rather than common ancestry.

Anamorphic   Refers to development in which the young, when hatched, have only a part of the adult complement of segments.

Ancestrula (pl. Ancestrulae)   Zooid that develops from the egg in bryozoans.

Anisomyarian   A reduction of the anterior adductor muscle in clams.

Aphotic zone   Region from the transition zone down to the ocean floor where total darkness prevails.

Apical field   The area devoid of cilia inside the circumapical band of rotifers.

Apodeme   Inner projection of the arthropod endoskeleton to which the muscles are attached.

Apomorphic   Refers to an evolutionarily derived state of a homologue.

Apopyle   Outlet from a flagellated chamber to an excurrent canal in leuconoid sponges.

Archenteron   The embryonic gut formed during gastrulation.

Architomy   Form of fission in which some planarians simultaneously fragment the body into several pieces.

Aristotle's lantern   Highly developed chewing apparatus used for feeding by sea urchins.

Article   Any section, between joints, of an arthropod appendage.

Asconoid   A sponge body that is a simple cylinder and always small.

Ascus   Internal pressure-regulating sac of some cheilostome bryozoans.

Astaxanthin   The red compound in some crustaceans.

Athecate   Refers to those hydroids that lack a hydrotheca.

Atoke   In polychaetes showing epitoky, the non-reproductive, benthic individual.

Atoll   Reef that rests on the summit of a submerged volcano.

Atrium (pl. Atria)   Internal cavity through which water flows in asconoid sponges (spongocoel). The internal cavity that receives the outflow of water from the pharynx in hemichordates and chordates.

Auricularia   Primary larval stage in holothuroid development.

Autogamy   Nuclear reorganization without conjugation or exchange of micronuclear material between two protozoans.

Autotomy   Self amputation. Deliberate loss of appendages, typically at specialized breakage points.

Autotrophic   Refers to type of nutrition in which organic compounds used in metabolism are obtained by synthesis from inorganic compounds.

Autozooid   Typical feeding zooids of bryozoans and some colonial anthozoans.

Avicularium (pl. Avicularia)   Jawed heterozooid found in many cheilostome bryozoans.

Axial rod   Tough, collagenous endoskeleton of gorgonians.

Axopodium (pl. Axopodia)   Fine, needle-like pseudopodium that contains a central bundle of microtubules.

Axoneme   Microtubules and other proteins that compose the inner core of flagella and cilia.

Barrier reef   Reef whose platform is separated from the adjacent land mass by a lagoon.

Basal body   An organelle equivalent to a centriole at the base of flagellum or a cilium.

Basal lamina   Thin, collagenous, fibrous sheet secreted by epithelial cells and on which they rest.

Basis   A bulbous, secreted structure that supports the hoplonemertean proboscis stylet. The attached calcified floor of a sessile barnacle. The second of two basal articles of the crustacean limb.

Bathyal zone   Zone formed by the continental slopes.

Benthic   Bottom-dwelling.

Benthos   Community of organisms that lives on or in the bottom.

Bilateral symmetry   The upper side of the body (dorsal side) is different from the lower (ventral side) so that only one plane of symmetry divides the body into mirror-image halves.

Binary fission   Asexual division that produces two similar individuals.

Bipectinate   Refers to a gill in which the filaments occur on both sides of the gill axis.

Bipinnaria   Primary free-swimming larval stage of asteroids.

Biradial symmetry   Having similar parts arranged around a central axis, but each of the four arcs or sides of the body is identical to the opposite side but not to the adjacent sides.

Biramous   Refers to the division of annelid and some arthropod appendages into two branches.

Blastaea   Hypothetical ancestor that is suggested by the blastula stage which occurs in the development of all animals.

Blastema   Dome-shaped mass of unspecialized cells that forms beneath the epidermis prior to healing and regeneration and is the source of new cells.

Blastocoel   The fluid or gel-filled embryonic cavity beneath the germ layers. The embryonic connective-tissue compartment.

Blastomere   A cell resulting from the cleavage divisions of the zygote.

Blastopore   Primary opening of the archenteron to the exterior of the embryo.

Blastostyle   A reduced, finger-like gonozooid that bears gonophores.

Blastozooid   A tunicate bud that arises from the body of the oozooid.

Blastula (pl. Blastulae)   A sphere of blastomeres created by repeated cleavage divisions of the zygote.

Blood-vascular system   Circulatory system of bilaterally symmetrical animals that develops within the connective tissue.

Body ciliature   Cilia distributed over the general body surface of ciliates.

Body whorl   The last and largest whorl of the gastropod shell that opens at the aperture from which the head and foot of the living animal protrude.

Bonellin   Echiuran dermal pigment that may have antibiotic properties.

Brachiolaria   Second asteroid larva, following the bipinnaria, marked by the appearance of three adhesive arms at the anterior end.

Brachiole   Slender, pinnule-like projection of fossil echinoderms.

Brood   To care for the eggs during at least the early part of development either inside or outside of the female body. The male may be involved in brooding in some animals.

Brown body   A large residual mass of necrotic cells containing stored waste products that remains lodged in the coelom as a conspicuous dark ball.

Buccal cavity   Cavity within the mouth opening.

Buccal field   A large ventral ciliated area which surrounds the mouth of some rotifers.

Bud   Protozoans: The smaller of two progeny cells resulting from fission. Metazoans: Asexually-produced progeny that either remains attached to the parent as a colonial zooid or undergoes differentiation before being released as a separate individual.

Bulbous pharynx   Platyhelminth pharynx characterized by a sucking muscular bulb.

Bursa (pl. Bursae)   A pouchlike structure. Commonly refers to a female reproductive chamber for the reception and temporary storage of sperm received at copulation. The ten respiratory invaginations are at the bases of the arms of many ophiuroids.

Byssus   A tough protein secretion produced by a gland in the bivalve foot and commonly in the form of threads used for attachment.

Calcareous   Composed of calcium carbonate.

Calcium carbonate   Limestone. Laid down within a secreted organic framework in animals.

Calymma   A broad vacuolated cortex formed by extracapsular cytoplasm that surrounds the central capsule of radiolarians.

Calyx   Skeletal cup of a crinoid disc. The body and tentacles of an entoproct.

Capitulum   The major part of the body of stalked barnacles.

Captaculum (pl. Captacula)   A threadlike tentacle of scaphopod molluscs.

Carapace   The shieldlike exoskeletal plate that covers at least part of the anterior dorsal surface of many arthropods.

Carina   Posterior dorsal plate of the barnacle exoskeleton. One of the five primary plates.

Casting   Continuous pile of defecated organic and mineral matter.

Caudal gland   Spinneret typical of many free-living nematodes.

Central capsule   The membrane-enclosed, innermost cytoplasm of a radiolarian cell.

Centriole   Microscopic cylindrical structure, composed of microtubules, which is situated at each pole of the mitotic spindle and is distributed to daughter cells during mitosis. There it may function as a basal body and give rise to a flagellum or cilium.

Centrolecithal   Refers to an arthropod egg in which the yolk is centralized, surrounded by nonyolky cytoplasm that eventually forms the embryo.

Centrosome   Structure from which bundles of microtubules radiate outwards.

Cephalic gland   Slime secreting gland of nemerteans.

Cephalization   Head development.

Cephalothorax   The combined head and thorax.

Ceras (pl. Cerata)   Projection from the dorsal body surface of many nudibranchs that may be club-shaped, branched, or look like a cluster of grapes.

Cercaria (pl. Cercariae)   Free-swimming developmental stage of digeneans that has a digestive tract, suckers and a tail. Develops from a redia and differentiates into a metacercaria.

Cerebral organ   One of a pair of ciliated sensory canals associated with the nemertean brain.

Chain   A free-swimming aggregate of sexual zooids in salps.

Chelate   Refers to appendages that are pincer-like or clawlike; usually applied to an arthropod appendage where

the terminal piece forms a movable finger that moves against an immovable finger on the subterminal piece.

Chelicera (pl. Chelicerae)   The first pair of appendages of chelicerates that are used as feeding structures.

Cheliped   A clawed or chelate thoracic appendage of decapod crustaceans.

Chilarium (pl. Chilaria)   One of a pair of appendages of the first opisthosomal segment that has fused with the prosoma of horsehoe crabs; consists of a single article armed with hairs and spines.

Chitin   A complex polysaccharide found in the exoskeleton of some invertebrates.

Chlorocruorin   Type of polychaete hemoglobin that is green in color.

Chondrophore   A ventrally directed flange in the hinge area of certain bivalves that provides a large surface area for the attachment of the inner hinge ligament.

Chorion   The shell-like membrane secreted by ovarian follicle cells that surrounds the eggs when they reach the oviduct.

Chromatophore   A pigment cell in the body wall that can expand or contract to expose or conceal its pigment.

Chromatophorotropin   Hormone involved in the functioning of chromatophores.

Cilium (pl. Cilia)   Characteristic of many protozoan and metazoan cells, a motile outgrowth of the cell surface that is typically short and its effective stroke is stiff and oarlike.

Cingulum   Dinoflagellates: Horizontal or transverse groove that bears the transverse flagellum. Rotifers: Posterior band of cilia of the divided corona.

Circumapical band   A crownlike ring of cilia that extends around the anterior margin of the rotifer head.

Cirrus (pl. Cirri)   Name given to various appendages, usually tentacle-like and curved, in different animal groups. Literally, a curl.

Cirrus sac   Contains the internal seminal vesicle, prostate glands, and cirrus of some platyhelminths.

Cnida (pl. Cnidae)   An eversible cnidarian organelle that occurs in a cnidocyte.

Cnidocil   A short, stiff, bristle-like cilium that is borne on a cnidocyte.

Cnidocyte   A cnidarian cell that contains an eversible cnida.

Cnidosac   Distal tip of a ceras of cnidarian-eating nudibranchs. The sac is an extension of the gut and contains undischarged nematocysts acquired from the prey.

Coenecium   A branching tubular network inhabited by pterobranch colonies that is secreted from glands in the oral shields of the zooids.

Coelenteron   The body cavity and gut of cnidarians and ctenophores. Gastrovascular cavity.

Coeloblastula   Blastula having a well developed blastocoel.

Coelom   Body cavity lined by a mesodermally derived epithelium.

Coelomate   An animal having a coelom.

Coelomocyte   A circulating coelomic cell which may or may not contain a respiratory protein.

Coelomoduct   A mesodermally derived duct leading from a coelom to the exterior. Usually a gonoduct.

Coenenchyme   All of the tissue situated between polyps in anthozoan colonies.

Coenosarc   Ther living tissue underlying the cuticular perisarc of hydroids.

Collagen   Common animal fibrous protein that forms extracellular skeletal materials.

Collar   Anthozoans: Circular fold at the junction of the column and the oral disc. Enteropneusts: The second of three body divisions.

Collencyte   A fixed cell of sponges that is anchored by long, cytoplasmic strands and secretes dispersed collagen fibers (not spongin).

Colloblast   An adhesive cell situated on the tentacles of ctenophores.

Collum   The first anterior, legless segment of millipedes that forms a collar behind the head.

Colony   Body composed of structurally joined zooids that share resources.

Columella   Central axis of asymmetrical shells around which whorls are coiled.

Columnar epithelium   Epithelium in which the cells are elongated and fit together like the logs of a pallisade.

Comb   A flat paddle of fused cilia in ctenophores.

Comb row   One of eight ciliary bands of ctenophores, each composed of a series of combs.

Commensalism   A type of symbiotic relationship in which one species benefits from the relationship and the other species (host) is neither benefited nor harmed.

Commissure   A nerve that transversely joins two ganglia or some other part of the nervous system.

Complemental male   Dwarf male barnacle that attaches to a hermaphrodite.

Complete cleavage   Cleavage divisions completely cut through the egg mass. Also referred to as holoblastic cleavage.

Compound eye   An image-forming eye of many arthropods composed of multiple lenses and photoreceptors called ommatidia.

Conchiferan   Meaning "shell-bearers," the term refers collectively to monoplacophoran, gastropod, bivalve, scaphopod, and cephalopod molluscs.

Conjugant   One of a pair of fused ciliates in the process of exchanging genetic material.

Connective   A nerve that longitudinally joins two ganglia.

Connective tissue   Body layer between epithelia, composed of a fluid or gel extracellular matrix with or without cells.

Connective tissue compartment   Body layer occupied by connective tissue.

Continental rise   That part of the continental margin that is between the continental slope and the abyssal plain.

Continental slope   That part of the continental margin that is between the continental shelf and the continental rise.

Contractile vacuole   Large spherical vesicle responsible for osmoregulation in protozoans and some sponge cells.

Contractile vacuole complex   Protozoan system of water and ion pumping organelles.

Convergence   Independent evolution of similar structures.

Coracidium   A ciliated free-swimming developmental stage of cestodes.

Corona   Ciliated organ at anterior end of rotifers used for feeding or swimming.

Cortex   An outer ectoplasmic layer.

Coxa (pl. Coxae)   The basal article of an arthropod appendage.

Ctenidium (pl. Ctenidia)   A molluscan gill in which the filaments alternate on opposite sides of the axis.

Cuboidal epithelium   Epithelium in which the cells are roughly cubical in shape.

Cuticle   Protective or supportive, external, nonliving, extracellular layer.

Cyclomorphosis   Seasonal changes in body shape or proportions.

Cydippid   A free-swimming ctenophore larva having an ovoid or spherical body.

Cyphonautes   Planktotrophic larva of some species of nonbrooding gymnolaemate bryozoans. The larval body is triangular, compressed, and enclosed in a bivalved shell.

Cypris   A settling larval stage of barnacles that follows the naupliar stage. It has six pairs of appendages and resembles an ostracod.

Cysticercus   Developmental stage of certain tapeworms, following the oncosphere, and characterized by a fluid-filled oval body with an invaginated scolex.

Cystid   Exoskeleton and body wall of the stationary trunk of bryozoans.

Cytopharynx   Permanent oral canal, or passageway, of ciliates that is separated from the cytoplasm by the cell membrane.

Cytoproct   Permanent cellular anus of some ciliates.

Cytostome   Cell mouth.

Dactylozooid   A finger-shaped, defensive, hydrozoan polyp.

Dedifferentiation   Loss of specialized cellular features returning to a more generalized condition. Characteristic of certain aspects of development, especially regeneration.

Deploying point   A site of separation of an asexually-produced group of salp blastozooids from other such groups.

Deposit feeding   Feeding upon detritus that has settled to the bottom of marine and freshwater environments.

Desiccation   Drying out.

Determinate cleavage   Developmental process during which the fates of the blastomeres are fixed early in cleavage; mosaic development.

Detritus   Fragments of plant or animal remains.

Deuterostome   Member of a major branch of the animal kingdom in which the site of the blastopore is posterior—far from the mouth, which forms as a new opening at the anterior end.

Diapause   A period of arrested development which can occur in any stage of insect development and which enables the insect to survive adverse environmental conditions.

Dioecious   Having separate sexes; i.e., some individuals contain the male reproductive system and other individuals contain the female system.

Diploblastic   Refers to having only two embryonic germ layers.

Diplosegment   Doubled trunk segments derived from the fusion of two originally separate somites; it is the distinguishing feature of millipedes.

Direct development   Lacking a larval stage in the course of development. On hatching the young have the adult body form.

Directive   Either of two pairs of septa at each edge of the compressed anthozoan pharynx.

Distal tip cell   The innermost cell of each nematode gonad that secretes a mitosis-promoting substance.

Doliolaria   Barrel-shaped larval stage, following the auricularia, of holothuroids.

Dormant egg   Fertilized egg that is encased in a heavy resistant shell. It is capable of withstanding desiccation and other adverse conditions and may not hatch for several months or even years.

Dorsal lamina   Longitudinal tissue fold along the inner dorsal pharyngeal wall of some ascidians. Gathers mucous net and food and conveys them into the esophagus.

Dwarf male   A minute male that shows all degrees of modification through degeneracy or loss of structures and attaches itself to a female or a hermaphroditic individual.

Ecdysis   Molting or the periodic shedding of the skeleton.

Ecdysone   Hormone responsible for molting process.

Echinopluteus   Planktotrophic larva of echinoid echinoderms that bears six pairs of long larval arms.

Ectoderm   Embryonic germ layer composing the outer wall of the gastrula.

Ectoparasite   Parasite that lives on the outside of its host.

Elytrum (pl. Elytra)   Platelike scale, modified from a dorsal cirrus, that is borne on a short stalk on the dorsal side of the body of scaleworm polychaetes.

Encystment   Forming resistant cysts in response to unfavorable conditions such as lack of food or desiccation.

Endemic   Refers to species having a restricted distribution, i.e., species endemic to the Hawaiian Islands are found nowhere else.

Endocytosis   Process in which some extracellular materials enter a cell in minute pits on the cell's membrane that later pinch off internally.

Endoderm   Embryonic germ layer composing the archenteron wall.

Endoparasite   Parasite that lives inside of its host.

Endoral membrane   Ciliate undulating membrane that runs transversely along the right wall and marks the junction of the vestibule and buccal cavity.

Enterocoel   Coelomic cavity formed from an outpocketing of the embryonic archenteron.

Enteronephric   Refers to either typical or modified nephridia that open into various parts of the digestive tract of earthworms.

Enzymatic-gland cell   Cell responsible for the secretion of digestive enzymes into the cnidarian coelenteron.

Ephippium (pl. Ephippia)   A protective, saddle-like capsule formed from the walls of the brood chamber in cladocerans.

Ephyra (pl. Ephyrae)   An immature scyphomedusa.

Epibenthic   Refers to the bottom surface of a body of water.

Epiboly   Type of morphogenetic movement in gastrulation in which ectodermal cells overgrow the inner germ layers.

Epicuticle   Thin, outer layer of the arthropod skeleton composed of proteins and sometimes wax.

Epidermal replacement cell   A platyhelminth parenchymal cell that migrates from the parenchyma to the body surface and replaces a damaged or destroyed epidermal cell.

Epidermis   Outer epithelial layer of the body.

Epifauna   The animals that live on the surface of ocean, lake, and stream bottoms.

Epigastric furrow   A transverse groove on the ventral side of the arachnid abdomen that contains the reproductive openings.

Epimorphic   Development in which the young display the full complement of segments when they hatch.

Epiplasm   Dense supportive mesh formed by filamentous proteins in the cortical cytoplasm.

Epitheliomuscular cell   A cnidarian contractile cell that has characteristics of both epithelial and muscular cells.

Epitoky   Reproductive phenomenon in some polychaetes: The production, either by transformation or budding, of a reproductive individual (epitoke) adapted for a pelagic existence from a nonreproductive individual adapted for a benthic existence.

Epizoic   A plant or animal living on the surface of another animal.

Esthete   A sensory organ lodged in a minute vertical canal in the upper layer of the shell plate of chitons.

Estuary   Embayment at the junction of a river with the sea, typically containing water of low salinity (brackish water).

Eukaryotic   Refers to a cell characterized by the presence of membrane-bound organelles.

Eulamellibranch gill   Bivalve gill in which the filaments are joined together by continuous sheets of tissue.

Euphotic zone   Zone in which light sufficient to allow photosynthesis to exceed respiration penetrates to only a short distance below the surface or to depths as great as 200 m, depending on the turbidity of the water.

Eutely   Having an invariant and genetically-fixed number of cells.

Euryhaline   Tolerant to a relatively wide range of environmental salinities, usually salinities lower than those of the open ocean.

Evert   Protrusion by turning inside out.

Evisceration   When the anterior or posterior end of a species ruptures and parts of the gut and associated organs are expelled.

Exconjugant   Ciliates that have separated after sexual reproduction.

Exocytosis   Process in which indigestible material is released from a cell to the exterior by fusion of the residual vesicle with the cell membrane.

Exumbrella   Aboral, upper surface of the bell of a medusa.

Fasciole   One of several ciliated spines of certain echinoids that together form a siphon.

Filibranch gill   Bivalve gill in which individual filaments are more or less separate and are held together only by tufts of specialized cilia.

Filopodium (pl. Filopodia)   Pseudopodium that is slender, clear, and sometimes branched.

Filter feeding   A type of suspension feeding in which particles (plankton and detritus) are removed from a water current by a filter.

Fin box   One in a longitudinal series of small, median, unpaired coelomic cavities that form and help to support the dorsal and ventral fins of cephalochordates.

Fin ray   Any of several stiff, slender structures that support a fin.

Fission   Asexual division of an organism into two or more progeny.

Fixed parenchymal cell   A large, branched, mesodermal cell of platyhelminths that makes contact with and interjoins other cells and tissues.

Flabellum (pl. Flabella)   Lateral, spatulate process on each of the fifth appendages of horseshoe crabs which is used to direct the dorsal incurrent of water over the gills.

Flagellum (pl. Flagella)   A characteristic of many protozoan and metazoan cells; it is typically long and its motion is a complex whip-like undulation.

Flame bulb   A protonephridial terminal cell that has many flagella which beat synchronously and resemble a minute flickering flame; its nucleus is offset to one side of the flame.

Flame cell   A protonephridial terminal cell that has many flagella, which beat synchronously and resemble a minute flickering flame; its nucleus is at the base of the flame.

Foliaceous   Composed of one sheet of zooids or of two sheets, attached back to back.

Food vacuole   Cellular vesicle containing ingested food.

Foot   Ventral surface of a mollusc that is muscular and flattened, forming a creeping sole.

Forcipule   One of a pair of appendages of the first trunk segment in centipedes that covers the mouth appendages; commonly called a poison claw.

Fringing reef   Reef that extends seaward directly from the shore.

Frontal gland   Anterior aggregation of secretory cells in platyhelminths.

Funiculus (pl. Funiculi)   A mesenchymal cord that extends across the coelom from the underside of the stomach and plays a role in nutrient transport between bryozoan zooids.

Gamogony   Multiple fission that forms gametes that fuse to form a zygote.

Ganglion (pl. Ganglia)   An aggregation of neuronal cell bodies.

Gap junction   Intercellular junction that allows for intercellular communication, such as electrical coupling of muscle cells.

Gastric filament   One of several cnidocyte-bearing threads that extend into the scyphozoan stomach from the septa between gastric pockets.

Gastric mill   Posterior portion of the malacostracan cardiac stomach where food is broken down mechanically by internal teeth.

Gastric pouch or pocket   One of four pockets in the wall of the scyphozoan stomach.

Gastrodermis   Cellular epithelial lining of the gastrovascular cavity of cnidarians and ctenophores and the midgut lining of bilaterally symmetrical animals.

Gastrolith   A calcareous concretion secreted by the stomach epithelium of some crustaceans which functions as a calcium storage center.

Gastrovascular cavity   Internal extracellular cavity of cnidarians and ctenophores lined by gastrodermis.

Gastrozooid   Nutritive or feeding polyp of cnidarians which is similar to a short hydra.

Gastrula   A two-layered embryo.

Gastrulation   The developmental establishment of germ layers.

Genital atrium   A small chamber in parasitic platyhelminths that receives the openings of both the male and female reproductive systems.

Girdle   The peripheral area of the polyplacophoran mantle that is thick and stiff, and extends beyond the lateral margins of the shell plates.

Glycocalyx (pl. Glycocalyces)   The carbohydrate and protein surface coat of eukaryotic cells.

Gnathobase   Spiny medial surface of the basal articles of many crustacean limbs.

Gnathochilarium (pl. Gnathochilaria)   The floor of the preoral chamber formed by a fused pair of maxillae; broad flattened plate attached to the posterior ventral surface of the head of millipedes.

Gnathopod   Each of the second and third thoracic appendages of amphipods which are usually enlarged and subchelate for prehension.

Gonangium (pl. Gonangia)   Type of gonozooid that consists of a central blastostyle bearing gonophores and is surrounded by an extension of the perisarc (gonotheca).

Gonoduct   Principal duct providing for the transport of sperm or eggs in any reproductive system.

Gonophore   A hydroid reproductive bud that bears the germ cells and may become a free-swimming medusa or a variously modified sessile medusa. Medusoid.

Gonopore   External opening of any reproductive system.

Gonotheca (pl. Gonothecae)   An extension of the perisarc around a gonozooid.

Gonozooid   A hydrozoan reproductive polyp which is often reduced, lacking mouth and tentacles, and bears gonophores. A sexually reproductive zooid of thaliaceans.

Gorgonin   A tanned collagen.

Growth zone   Region that includes all of the larva between the mouth and telotroch on the fully developed trochophore larva.

Hadal   Referring to oceanic trenches; great depths.

Halteres   Second pair of wings of dipteran insects which is reduced to knobs and functions as a gyroscope to offset flight instability.

Haptocyst   Special adhesive organelle borne on the tentacles of suctorians.

Haptor   Attachment organ that bears hooks and suckers.

Hatschek's groove   A shallow ciliated outpocket from the dorsal wall of the vestibule of cephalochordates.

Hemal system   Blood-vascular system.

Hemocoel   A voluminous, blood-filled cavity, occupying all or most of the body.

Hepatopancreas   Large, spongy digestive diverticula from the arthropod midgut that are composed of ducts and blind secretory tubules.

Hermaphroditic   Having both male and female reproductive systems in the same individual. When both systems are present at the same time, the hermaphroditism is said to be simultaneous; when the male system appears and functions first and is followed by the female system, the hermaphroditism is said to be protandric.

Heterotrophic   Refers to the type of nutrition in which organic compounds used in metabolism are obtained by consuming the bodies or products of other organisms.

Heterozooid   Reduced or modified zooids that have functions other than feeding.

Higgins larva   Larval stage of loriciferans.

Hinge ligament   A noncalcified, elastic, proteinaceous band that attaches the two valves of a bivalve.

Holoblastic cleavage   Cleavage divisions completely cut through the egg mass. Also referred to as complete cleavage.

Holonephridium (pl. Holonephridia)   A typical, segmental metanephridial duct of an oligochaete.

Holoplankton   Marine zooplankton that spend their entire lives in the plankton.

Holothurin   Toxic substance released in the Cuvierian tubules of certain holothuroids.

Homolecithal egg   Egg in which the yolk is uniformly distributed. Isolecithal.

Homology   Similarity of structures attributable to common ancestry in two or more species.

Homologue   An attribute, or characteristic, of one species that is similar in anatomical position and structural composition to that of another species. Differences between the characteristics may be bridged by attributes of intermediate form in other species or developmental stages.

Hydranth   The oral end of a hydroid polyp bearing the mouth and the tentacles.

Hydrocaulus   The stalk of a hydroid polyp.

Hydrocoral   Colonial, calcified polypoid hydrozoan with either an encrusting or an upright growth form.

Hydroid colony   A collection of polyps in which each polyp is connected to the others.

Hydromedusa (pl. Hydromedusae)   Hydrozoan medusa.

Hydrorhiza (pl. Hydrorhizae)   Horizontal rootlike stolon of a hydroid colony that grows over the substratum.

Hydrotheca (pl. Hydrothecae)   A cuticle that encloses the hydranth. Theca.

Hyperparasitism   When one parasitic species is parasitized by another.

Hypobranchial gland   One of two patches of mucus-secreting epithelium on the mantle roof that lie downstream to each gill and trap sediment in the excurrent of water.

Hypognathus   Insect head orientation that causes the mouthparts to be directed downward.

Hypostome   A mound or cone that bears the mouth of hydropolyps. Manubrium.

Incomplete cleavage   Cleavage divisions that do not completely cut through the egg mass.

Incurrent canal   Tubular invagination of the sponge pinacoderm that leads into the flagellated chambers.

Incurrent pore   Small opening on the surface of sponges that leads into an incurrent canal. Ostium.

Indeterminate cleavage   Fate of the blastomeres is fixed relatively late in development. Regulative development.

Indirect development   Having a larval stage in the course of development.

Infauna   Animals that live within bottom sediments.

Infraciliary system   The entire assemblage of ciliary basal bodies, or kinetosomes, and the fibers that link them together in the cell cortex of ciliates.

Infusoriform larva The final free-swimming larval stage of rhombozoans.

Ingression Mode of gastrulation in which cells of the blastula wall proliferate cells into the blastocoel.

Instar Each of the several stages between molts in arthropods.

Intercellular junction Membrane specialization that binds cells together, promotes communication between cells, or helps to regulate transport across an epithelium.

Intermediate host The host for the larval or developmental stage of a parasite.

Interstitial cell A small, rounded totipotent cnidarian cell, sandwiched between cells of the epidermis and gastrodermis.

Interstitial fauna Animals that live in the spaces between sediment particles (sand grains).

Intertidal Lying between the low and high tide levels.

Introvert Eversible part of the bryozoan, priapulid, or sipunculan body.

Invagination Infolding. In gastrulation, this refers to a type of morphogenetic movement in which the cells of the vegetal hemisphere fold into the interior to form the archenteron.

Isolecithal egg Egg in which the yolk is uniformly distributed. Homolecithal.

Isomyarian Refers to more or less equal development of clam adductor muscles.

Jellyfish A cnidarian medusa.

Kenozooid A bryozoan heterozooid modified to form a stolon.

Kinetodesma (pl. Kinetodesmata) A fine striated fiber that connects kinetosomes (basal bodies) of ciliates.

Kinetoplast Conspicuous mass of DNA that is situated within the single, large, elongated mitochondrion of kinetoplastid (trypanosome) protozoans.

Kinetosome A ciliary or flagellar basal body.

Kinety (pl. Kineties) One row of cilia, kinetosomes, and kinetodesmata of ciliates.

Lacunar canal system Unique circulatory system located within the syncytial epidermis of acanthocephalans.

Lamella (pl. Lamellae) A sheet of tissue. In lamellibranch bivalves, each of the four gill flaps.

Languet One of several folds of tissue along the dorsal pharyngeal wall of some ascidians which together convey food to the esophagus. A discontinuous dorsal lamina.

Lappet Lobe formed by the margin of the scalloped scyphozoan bell. Movable flaps that can expose or cover the ambulacral groove of crinoids.

Larva (pl. Larvae) An independent, motile, developmental stage that typically is just barely visible to the naked eye and looks different from the adult.

Larviparous Eggs brooded internally within the female that are later released as larvae.

Lateral canal Part of the echinoderm water-vascular system that joins the radial canal and tube feet.

Laurer's canal Short, inconspicuous canal that extends from the seminal receptacle of trematodes to the dorsal surface, where it may open at a minute pore.

Lecithotrophic larva A nonfeeding larva that utilizes yolk as a source of nutrition.

Lemniscus (pl. Lemnisci) One of a pair of tegumental invaginations that is filled with fluid and functions as part of the hydraulic system in proboscis eversion or in fluid transport to and from the proboscis of acanthocephalans.

Leuconoid Refers to a type of sponge body organization built around flagellated chambers and an extensive system of canals.

Ligament sac A hollow, elongated, connective-tissue sac suspended in the pseudocoel of acanthocephalans which houses the gonads.

Littoral Intertidal; lying between the low and high tide levels.

Lobopodium (pl. Lobopodia) A pseudopodium that is rather wide with rounded or blunt tips, is commonly tubular, and is composed of both ectoplasm and endoplasm.

Longitudinal cord Ridge that entends the length of the body created by the inward expansion of the epidermis in nematodes and some gastrotrichs.

Lophophoral organ One of pair of glandular areas near the nephridial openings of the lophophore of phoronids. Used in the formation of spermatophores.

Lophophore A circular or horseshoe-shaped fold of the mesosomal body wall that encircles the mouth and bears numerous hollow ciliated tentacles.

Lorica A girdle-like skeleton.

Luciferase Enzyme that catalyzes the bioluminescence reaction.

Luciferin A substrate that is capable of bioluminescence.

Lunule One of the large, elongated notches or openings in the bodies of some clypeasteroids (sand dollars).

Lyriform organ Group of slit sense organs found on the appendages and bodies of most arachnids.

Macromere One of several large blastomeres located in the yolky vegetal hemisphere of early embryos.

Macronucleus (pl. Macronuclei) Large, usually polyploid, ciliate nucleus concerned with the synthesis of RNA, as well as DNA, and therefore directly responsible for the phenotype of the cell.

Macrophagous Collecting and ingesting large food particles.

Madreporite Pore or sieve plate of the echinoderm water-vascular system that connects the stone canal to the exterior seawater (most echinoderms) or to the perivisceral coelomic fluid (crinoids and holothuroids).

Malpighian tubule Arthropod excretory organ; one of several blind, tubular, contractile evaginations of the midgut that lie within the hemocoel.

Mandible One member of the third pair of chewing or grinding appendages of crustaceans. Jaw.

Mangrove   A small tropical tree or shrub adapted for living in the marine intertidal zone. Most highly developed in estuaries.

Mantle   A body wall fold that secretes a shell, as in molluscs and brachiopods; or the body wall beneath the ascidian tunic.

Mantle cavity   Protective chamber created by the overhang of a mantle. Pallial cavity.

Manubrium (pl. Manubria)   Tubelike extension, bearing the mouth, that hangs down from the center of the subumbrella of cnidarian medusae. Hypostome of hydroid polyps.

Marsupium (pl. Marsupia)   Brood pouch found outside the reproductive system.

Mastax   Pharynx of a rotifer.

Mastigoneme   One of the many fine, lateral branches of some flagella.

Mastigont system   Complex formed by groups of flagella and several microtubular and fibrillar organelles.

Medulla   Central part of the heliozoan cell that is composed of dense endoplasm, containing one to many nuclei and the bases of the axial rods.

Medusa (pl. Medusae)   Form of cnidarian that has a well developed, gelatinous mesoglea and is generally free-swimming.

Megalops (Megalopa, pl. Megalopae)   Postlarval stage of crabs that has a large or flexed abdomen and the full complement of appendages.

Megasclere   A large spicule forming one of the chief supporting elements in the skeleton of sponges.

Mehlis's gland   Conspicuous unicellular gland cells associated with the reproductive system of trematodes which play a role in egg capsule formation.

Meiofauna   Very small multicellular animals; usually referring to those living in the interstitial spaces of sand.

Membranelle   Type of ciliary organelle derived from two or three short rows of cilia, all of which adhere to form a more or less triangular or fan-shaped plate.

Meroplankton   Marine zooplankton whose larvae enter and leave the plankton at different points in the course of their development.

Merozoites   Individuals produced by multiple fission of sporozoan trophozoites.

Mesenchyme   A network of loosely associated, often motile, embryonic cells, that are usually, but not always, of mesodermal origin. The term is still applied to adult connective tissues of some groups of animals.

Mesentery (pl. Mesenteries)   A longitudinal sheet of tissue that divides the body cavity of bilaterally-symmetrical animals.

Mesentoblast   Blastomere associated with spirally cleaving zygotes that contains an unidentified cytoplasmic factor which causes the cell and its progeny to form mesoderm.

Mesoderm   Embryonic germ layer that forms the tissues situated between ectoderm and endoderm.

Mesoglea   Connective-tissue layer between the epidermis and gastrodermis of cnidarians and ctenophores.

Mesohyl   Sponge connective tissue. Lies beneath the pinacoderm and consists of a gelatinous proteinaceous matrix containing skeletal materal and ameboid cells.

Mesothelium (pl. Mesothelia)   Simple, nonstratified coelomic lining epithelium.

Metacercaria (pl. Metacercariae)   Encysted final stage of digenean development.

Metachrony   Wave pattern that results from the sequential coordinated action of cilia or flagella over the surface of a cell or organism.

Metamerism   The division of an animal's body into a linear series of similar parts, or segments. Segmentation.

Metamorphosis (pl. Metamorphoses)   Transformation from a larva into an adult.

Metanauplius (pl. Metanauplii)   One of several instars following the nauplius, e.g., in copepods, mystacocarids.

Metanephridial system   Excretory system composed of a vascular ultrafiltration site, a coelomic cavity, and a metanephridium (metanephridial tubule).

Metanephridium (pl. Metanephridia)   An excretory tubule that opens into the coelom by a ciliated funnel and to the exterior by a nephridiopore.

Metatroch   A second girdle of cilia that develops posterior to the prototroch of a trochophore.

Micromere   One of many small blastomeres located in the animal hemisphere of the cleaving zygote.

Micronucleus (pl. Micronuclei)   Small, usually diploid, ciliate nucleus concerned primarily with the synthesis of DNA. It undergoes meiosis before functioning in sexual reproduction.

Microphagous   Collecting and ingesting small food particles.

Micropyle   An opening in a shell of an egg or resting stage from which the primordium eventually emerges.

Microsclere   A tiny sponge spicule.

Microtrich   Type of microvillus found on the tegument of tapeworms.

Microtubule organizing center (MTOC)   A region around basal bodies and centrioles that controls the organized assembly of microtubules.

Mictic egg   Type of fertilized rotifer egg that is thin-shelled and haploid.

Miracidium (pl. Miracidia)   Ciliated, free-swimming, first larva of digenean trematodes.

Molt   To shed the old cuticle as a new cuticle is being secreted. Ecdysis.

Monomyarian   Refers to the absence, in bivalves, of the anterior adductor muscle.

Monopectinate   Refers to a gill in which the filaments occur on only one side of the axis, like teeth on a comb.

Monophyletic group All of the species descended from a common ancestor.

Mosaic development Embryonic fate determination in which cell fate is determined early in development and is the result of the action of specific factors that are unevenly distributed, like pieces of a mosaic, in the cytoplasm of the uncleaved egg.

Mucocysts Mucigenic bodies that are arranged in rows, similar to ciliate trichocysts, and discharge a mucoid material.

Mucus Animal secretion utilized in a variety of ways as an adhesive, protective cover, or lubricant.

Mutualism A type of symbiotic relationship in which both species benefit from the relationship.

Myocyte Type of sponge mesohyl cell which displays some similarities to a smooth muscle cell in shape and contractility. A muscle cell.

Myoepithelial cell A muscle cell that is part of an epithelium.

Myoneme A bundle of contractile filaments that lies in the pellicle of some protozoans.

Nacre The innermost, lustrous, shell layer of molluscs.

Naiad Aquatic immature stage of development between hatching and adult insect.

Naupliar eye Median, simple eye of the naupliar larva of crustaceans.

Nauplius Earliest and basic type of crustacean larva; has three pairs of appendages.

Nectophore Mouthless, pulsating swimming bell of siphonophores.

Nematocyst Stinging cnida of cnidarians.

Nematodesma (pl. Nematodesmata) One of several microtubular rods that line and support the wall of the ciliate cytopharynx and assist in the inward transport of food vacuoles.

Nematogen Adult dicyemid.

Neoblast A totipotent cell that is important in wound healing and regeneration.

Nephridium (pl. Nephridia) An excretory tubule that usually opens to the exterior via a nephridiopore. The inner end of the tubule may be blind (protonephridium), ending in terminal cells, or may open into the coelom (metanephridium) through a ciliated funnel.

Nephrocyte A large phagocytic cell, alone or in clusters, in the hemocoel of many arthropods.

Nephromyces A unicellular fungus that occurs in the renal sacs or the pericardium of some ascidians.

Nephrostome An open ciliated funnel at the inner coelomic end of a metanephridium.

Neritic zone Zone comprised of the waters over the continental shelves.

Nerve net Plexus.

Neuropodium (pl. Neuropodia) The ventral branch of a polychaete parapodium.

Niche Refers to the mode of existence or life habit of an animal.

Nonselective deposit feeding Feeding in which animals ingest both organic detritus particles and the surrounding mineral particles (sand grains).

Notopodium (pl. Notopodia) The dorsal branch of a polychaete parapodium.

Nuchal organ One of a pair of ciliated chemosensory pits or slits that are often eversible and are situated in the head region of most polychaetes.

Nutritive-muscle cell A muscle cell in the cnidarian gastrodermis that usually bears a cilium and is responsible for intracellular digestion.

Obturaculum (pl. Obturacula) One of two elongated, medially fused structures which arise anteriorly from the head of vestimentiferan pogonophores and bear and support the gills.

Occluding junction Sealing junction between cells.

Oceanic zone Zone comprised of the waters beyond the continental shelf.

Ocellus (pl. Ocelli) A small cluster of photoreceptors, i.e., a simple eye.

Odontophore An elongated, muscular, cartilagenous mass in the floor of the buccal cavity of many molluscs.

Oligomery Division of the body into three linear regions, characteristic of many deuterostome animals. Tricoelomate.

Oncomiracidium (pl. Oncomiracidia) The ciliated larva of monogeneans.

Oncosphere Encapsulated first stage in the life cycle of certain tapeworms that bears six hooks and cilia. Typically referred to as a coracidium when released into the water.

Oostegite A large platelike process on several thoracic coxae that together form a marsupium.

Ootype Small, centrally-positioned sac within the female reproductive system of most parasitic platyhelminths.

Oozooid The zooid developing from the fertilized egg of urochordates.

Operculum (pl. Opercula) A lid or covering to a chamber-like structure.

Ophiopluteus (pl. Ophioplutei) Planktotrophic larva of many species of ophiuroids.

Opisthognathus Posteriorly directed position of insect head.

Opisthosoma The posterior end of the pogonophore body which is composed of numerous (up to 95) segments; the posterior tagma of a chelicerate, also called the abdomen.

Oral arm One of the four, often frilly extensions of the scyphozoan manubrium.

Oral ciliature Cilia that are associated with the mouth region of ciliates.

Oral disc Area around the mouth of an anthozoan polyp which bears eight to several hundred hollow tentacles.

Oral shield One of a series of large plates that frame the ophiuroid mouth and also form a chewing apparatus with five triangular, interradial jaws at the center.

Oral sucker   Organ that surrounds the trematode mouth, prevents dislodgement and aids in feeding.

Organ of Tömösvary   One of a pair of sensory organs situated at the bases of the antennae on the head of some centipedes.

Osculum (pl. Oscula)   The excurrent opening of the water circulation system of the sponge.

Osmoconformity   The maintenance of internal body fluids in osmotic equilibrium with the aqueous environment; i.e., the salt concentrations of the two remain about the same.

Osmoregulation   The maintenance of internal body fluids at a different osmotic pressure (usually higher) than that of the external aqueous environment; i.e., the salt concentration of internal body fluids is maintained at a different level from that of the environment.

Ossicle   An internal skeletal piece, commonly calcareous as in echinoderms.

Ostium (pl. Ostia)   A small incurrent opening or pore on the surface of a body or gill.

Ovigerous leg   One of the third pair of appendages of pycnogonids. Used by the male to brood the fertilized eggs.

Oviparous   Egg-laying.

Pallial line   The line of mantle attachment which is impressed on the inner surface of the shell as a scar.

Palmella   Nonflagellated stage of flagellated protozoans.

Papula (pl. Papulae)   Finger-like, respiratory evagination of the aboral body wall of some asteroids.

Paramylon   Photosynthetic storage product of euglenoids.

Parapodium (pl. Parapodia)   Lateral, fleshy, paddle-like appendage on polychaete annelids.

Parasitism   A type of symbiotic relationship in which one species (parasite) benefits from the relationship and the other species (host) is harmed.

Paratomy   The phenomenon of linear budding in some turbellarians and annelids.

Parenchyma   Connective tissue compartment between the body wall musculature and gut of platyhelminths.

Parenchymula   A sponge larva that lacks an internal cavity and bears flagellated cells over all of its outer surface except, often, the posterior pole. Parenchymella.

Parthenogenesis   The embryonic development of unfertilized, usually diploid, eggs.

Patch reef   A small circular or irregular reef that rises from the floor of a lagoon behind a barrier reef or within an atoll.

Paxilla (pl. Paxillae)   An echinoderm ossicle crowned with small, movable spines.

Pectinate   Having teeth or side branches arranged like a comb.

Pectine   One of a pair of sensory appendages that is peculiar to scorpions.

Pedal disc   In some sea anemones, a flattened disc at the aboral end of the column for attachment.

Pedal laceration   Method of asexual reproduction in some anemones in which parts of the pedal disc are left behind as the animal moves.

Pedicellaria (pl. Pedicellariae)   A small, specialized jawlike appendage of asteroids and echinoids which is used for protection and feeding.

Pedicle   Muscular, flexible stalk that attaches articulate brachiopods to the substratum.

Pedipalp   One of the second pair of appendages of chelicerates which is modified to perform various functions in the different classes.

Peduncle   Muscular, flexible stalk of barnacles that attaches to the substratum at one end and bears the major part of the body at the other.

Pelagic   Living, floating, or swimming above the water column.

Pelagosphera   Secondary planktotrophic larva of sipunculans.

Pellicle   Protozoan "body wall" composed of cell membrane, cytoskeleton, and other organelles.

Penetration anchor   Kind of anchor that holds one part of a burrowing animal's body in place as another part penetrates and advances into the sediment.

Peniculus (pl. Peniculi)   A modified membranelle that is greatly lengthened and thus tends to be similar to an undulating membrane in function.

Pentactula   Metamorphosing stage of holothuroid development that bears five primary tentacles.

Pentamerous   Divided into five parts, characteristic of the body of echinoderms.

Periostracum   The outer layer of a molluscan shell that is composed of a quinone-tanned, horny protein material called conchiolin or conchin.

Periproct   The membranous area, often bearing ossicles, around the anus of echinoids.

Perisarc   A supporting, nonliving chitinous cuticle secreted by the epidermis surrounding most hydroids.

Peristalsis   A wave of muscular contraction moving over a body or along an internal tube or vessel. The motion moves material along the tube or over the body, or moves the body through the external medium.

Peristome   Buccal cavity of ciliates. The membranous area around the mouth of some echinoderms, i.e., sea urchins.

Peristomium   The first true segment, immediately posterior to the prostomium, of an annelid. Usually lacks locomotory appendages.

Peritoneum   The innermost, noncontractile layer of a stratified coelomic epithelium; separates the coelomic fluid from the musculature.

Petaloid   One of five petal-shaped areas on the aboral surface of irregular urchins that bear specialized respiratory podia.

Phagocytosis   The engulfment of large particles, such as bacteria and protozoans, by active outgrowth of the cell surface.

Pharynx (pl. Pharynges)   The anterior part of the gut. Often used interchangeably with esophagus, foregut, and even proboscis.

Phorozooid   A locomotory zooid of doliolids that has a short posterior spur upon which buds differentiate into gonozooids.

Photocyte   Specialized cell within which light is produced.

Photophore   A light-producing organ.

Photosynthate   The organic carbon fixed by photosynthetic organisms.

Phyllode   Each of five oral ambulacral areas of irregular echinoids that contains specialized podia for obtaining food particles.

Phytoflagellate   A flagellate that usually bears one or two flagella and typically possesses chloroplasts.

Phytoplankton   Microscopic algae suspended in that part of the water column of lakes and seas that is penetrated by light.

Pilidium   A free-swimming and planktotrophic larva of many heteronemerteans which is characterized by an apical tuft of cilia and is somewhat helmet-shaped.

Pinacocyte   One of the epithelial-like flattened cells which together make up the sponge pinacoderm.

Pinnate   Having side branches, like a feather.

Pinnule   Side branch of an appendage, i.e., on octocoral tentacles, crinoid arms.

Pinocytosis   A nonspecific form of endocytosis in which the rate of uptake is in simple proportion to the external concentration of the material being absorbed.

Plankton   Organisms that live suspended in the water column but are unable to counter water current because of small size or insufficient motility.

Planktotrophic larva   A larva that feeds on other planktonic organisms.

Planula (pl. Planulae)   A cnidarian larva that is elongated and radially symmetrical but with anterior and posterior ends.

Plasmodium (pl. Plasmodia)   Amoeboid syncytial mass.

Pleopod   Anterior abdominal appendage of malacostracans.

Plerocercoid   The final stage in the life cycle of certain tapeworms.

Plesiomorphic   Refers to an evolutionarily-primitive state of a homologue.

Pleuron (pl. Pleura)   Either of the two primary, lateral, exoskeletal plates of each segment of an anthropod.

Plicate   Folded.

Podocyst   A contractile, vesicle-like extension of the embryonic foot of some pulmonate snails which functions in absorption of albumen as well as in gas exchange and excretion.

Podocyte   Specialized cell with branching toelike or finger-like processes that interlock with each other, usually over the surface of a blood vessel. An adaptation for ultrafiltration.

Polyembryony   Developmental phenomenon in which the initial mass of embryonic cells gives rise to more than one embryo.

Polymorphism   When two or more members of a species or zooids of a colony are structurally modified for different functions.

Polyp   Form of cnidarian that has a thin layer of mesoglea and is generally sessile.

Polyphyletic   A group of species that are similar but derived from two or more different ancestral stocks.

Polypide   The introvert, lophophore, and moveable contents within the trunk of bryozoans.

Polypide regression   The phenomenon by which bryozoan polypides periodically degenerate and are replaced by regeneration from the wall of the cystid.

Porocyte   A sponge cell that surrounds an opening which extends from the external surface to the spongocoel.

Preoral pit   The developmental precursor of the wheel organ and Hatschek's groove that opens on the left side of the head of larval cephalochordates.

Pressure drag   The difference in pressure at the front end (higher pressure) of a forward-moving organism as compared to the rear end (lower pressure).

Pretrochal region   Region consisting of the apical plate, prototroch, and the area about the mouth on the fully developed trochophore larva.

Primary host   The host for the adult stage of a parasite. Definitive host.

Proboscis (pl. Proboscides)   Any tubular process of the head or anterior part of the gut, usually used in feeding and often extensible.

Proboscis apparatus   The complex, eversible, prey-capturing organ of nemerteans.

Proboscis pore   The opening of the proboscis apparatus at or near the anterior tip of a nemertean.

Procercoid   Developmental stage of certain tapeworms between oncosphere and plerocercoid.

Proctodeum   The embryonic area of infolded ectoderm that connects the anus with the endodermal tube (midgut) and forms the hindgut.

Procuticle   Thick, inner layer of the anthropod skeleton.

Proglottid   One of the linearly arranged segment-like sections that make up the strobila of a tapeworm.

Prognathus   Anteriorly directed position of insect head; more specialized than hypognathus.

Propodium (pl. Propodia)   The front of a gastropod foot which acts like a plough and anchor.

Prosopyle   Internal opening of a sponge through which water flows from the incurrent canal into a radial canal or flagellated chambers.

Protandry   Type of hermaphroditism in which the individual is first a male and then changes to a female.

Protoconch   The shell that develops spirally in the veliger and may remain at the apex of the adult shell for some time until it disappears through differential growth and erosion.

Protonephridium (pl. Protonephridia)   A ciliated excretory tubule that is capped internally by one or more terminal cells, which are specialized for ultrafiltration.

Protopod   The basal part of the crustacean appendage, corresponding to the combined coxa and basis.

Protostome   Member of a major branch of the Animal Kingdom, in which the blastopore contributes to the formation of the mouth.

Prototroch   Preoral ring of cilia of a trochophore larva.

Protozoea   Third larval stage of a natant decapod (shrimp); follows the metanauplius and precedes the zoea.

Pseudocoel   Fluid-filled body cavity that occupies the connective tissue compartment. Differs from the hemocoel only in the lack of a heart.

Pseudofeces   In filter feeders such as bivalves, material removed from the water flow, aggregated, and rejected before it enters the gut.

Pseudolamellibranch gill   Bivalve gill in which the filaments are bound together with some (though not extensive) interfilamentous tissue junctions.

Pseudopodium (pl. Pseudopodia)   A flowing extension of a cell.

Ptychocyst   A cnida that discharges a thread used to weave a tube.

Pygidium (pl. Pygidia)   The terminal, nonsegmental part of the body of a segmented animal. Typically bears the anus. Telson.

Pupa (pl. Pupae)   Developmental stage of insects at the end of the larval period that is nonfeeding, quiescent and usually passed in protected locations such as under ground, in a cocoon, or in plants.

Pyramid   Large calcareous plate that composes Aristotle's lantern; shaped somewhat like an arrowhead with the point projected toward the mouth.

Radial canal   One of five fluid-filled channels of the echinoderm water-vascular system that join the ring canal to the lateral canals.

Radial cleavage   Type of cleavage pattern in which the cleavage spindles or cleavage planes are at right angles or parallel to the polar axis of the egg.

Radial symmetry   The arrangement of similar parts around a central axis.

Radiole   Each of the several pinnate tentacles on the head of a sabellid, serpulid, or spirorbid polychaete.

Radula (pl. Radulae)   A membranous belt, found in most molluscs, that bears transverse rows of teeth and extends medially over the odontophore and around its anterior end.

Radula sac   Deep outpocket from the posterior wall of the buccal cavity from which the molluscan radula arises.

Raptorial   Referring to animals that capture prey.

Redia (pl. Rediae)   Stage in the digenean life cycle between the sporocyst and cercaria.

Regulative development   Embryonic fate determination in which cell fates are determined by a network of cellular communication in the embryo.

Repugnatorial gland   Specialized gland in many arthropods that produces repellent and toxic compounds for chemical defense.

Reserve stylet   One of several accessory reserve stylets present on each side of the nemertean central proboscis stylet.

Respiratory tree   One of two respiratory organs of most holothuroid echinoderms. Consists of a network of thin-walled tubules in the perivisceral coelom that originates from the cloacal wall.

Reticulopodium (pl. Reticulopodia)   A pseudopodium that forms a threadlike branched mesh and contains axial microtubules.

Retractor muscle   Muscle that withdraws an eversible or protrusible body part.

Rhabdite   Platyhelminth epidermal secretion droplets which are characterized microscopically by a specific, layered ultrastructure.

Rhagon   Developmental stage immediately following the metamorphosis of a demosponge larva. Typically, it is asconoid or syconoid in structure.

Rhinophore   One of the second pair of nudibranch tentacles located behind the first pair.

Rhombogen   A dicyemid rhombozoan similar to a nematogen but whose axial cell is in the process of forming an infusoriform larva. A sexually reproductive nematogen.

Rhopalium (pl. Rhopalia)   A club-shaped, marginal sensory organ of scyphozoans.

Rhopalial lappet   One of two small, specialized flaps on a rhopalium.

Rhynchocoel   A fluid-filled coelomic cavity that houses the retracted nemertean proboscis.

Rhynchodeum   In nemerteans, the short anterior canal that joins the proboscis pore to the proboscis.

Ring canal   Part of the echinoderm water-vascular system that joins the stone canal to the radial canals. The marginal canal of the gastrovascular system of some medusae.

Rostroconchida   An extinct class of molluscs that may have been ancestral to modern bivalves.

Rostrum   Middorsal projection in some rotifers that bears cilia and sensory bristles at its tip and is also adhesive. Median, anteriorly-directed spine from the carapace and head of some decapod crustaceans.

Scaphognathite   In Crustacea, a paddle-like projection of the second maxilla that produces a ventilating current; also called a gill bailer.

Schizocoel   Coelomic cavity derived from the separation, or splitting apart, of a solid mass of mesodermal cells.

Schizogamy   Apicomplexa. Multiple fission that produces merozoites.

Sclerite   Thickened area of cuticle in the exoskeleton of arthropods.

Scleroseptum (pl. Sclerosepta)   One of the many radiating calcareous partitions in the skeletal cup of stony corals.

Sclerotized   Refers to highly tanned (hardened) and often thickened arthropod exoskeleton.

Scolex (pl. Scoleces)   Anterior head region of tapeworms that is adapted for adhering to the host.

Scutum (pl. Scuta)   One of a pair of calcareous plates that cover the mantle surface of barnacles. The unpaired carina and the paired scuta and terga comprise the five primary plates of a barnacle.

Scyphistoma (pl. Scyphistomae)   A scyphozoan polyp.

Segmentation   Characteristic of animals whose bodies are divided into a longitudinal series of similar, repeated units (segments or metameres). Metamerism.

Selective deposit feeding   Feeding in which animals selectively remove organic detritus particles from the surrounding sand particles.

Seminal receptacle   Chamber in the female reproductive system for the reception and storage of sperm.

Seminal vesicle   Part of the male reproductive system that functions in the storage of sperm.

Sensillum (pl. Sensilla)   Sense organ, other than eyes and ocelli, that involves a specialized part of the exoskeleton.

Septum (pl. Septa)   A double-walled tissue partition in the cross-sectional plane of a bilaterian or a radial plane of a cnidarian.

Septal filament   The free edge of an anthozoan septum that is trilobed.

Seta (pl. Setae)   Chitinous bristle. Chaeta.

Shell   The calcified covering of molluscs and brachiopods.

Shield   A small calcareous plate in certain echinoderms, especially ophiuroids.

Sieve tracheae   Arthropod tracheal system in which the spiracle opens into an atrial or tubelike chamber from which a great bundle of tracheae arises.

Siliceous   Composed of silica.

Sinus   Large saclike space.

Siphon   An accessory gut of echiurans and some echinoids. A tubular fold of the molluscan mantle used to direct water to and from the mantle cavity. Inhalant and exhalant apertures of urochordates.

Siphonoglyph   Ciliated groove in the pharyngeal wall of some anthozoans that moves water into the coelenteron.

Siphonozooid   A highly-modified pennatulacean polyp that pumps water into, or allows it to escape from, the interconnected gastrovascular cavities of the colony.

Siphuncle   In shelled cephalopods, a slender outgrowth of the body wall and its thin calcareous enclosure. Secretes gas into the chambered shell for buoyancy.

Slug   Shell-less opisthobranch or pulmonate in which the shell is absent, or reduced and buried within the mantle.

Solenocyte   A protonephridial terminal cell that bears one flagellum and has a long tubular collar of microvilli.

Spasmin   Ciliate contractile protein which requires ATP for extension.

Spermatheca (pl. Spermathecae)   Another term for a seminal receptacle.

Spheridium (pl. Spheridia)   An echinoid statocyst.

Spicule   A small needle-like or rodlike skeletal piece.

Spinneret   Spinning organ of spiders.

Spiracle   Slitlike opening to the outside of the arthropod tracheal system.

Spiral cleavage   Type of cleavage pattern in which the cleavage spindles or cleavage planes are oblique to the polar axis of the egg.

Spire   The whorls of a gastropod shell above the body whorl.

Spirocyst   Cnida with a long adhesive thread that functions in capture of prey and in attachment to a substratum.

Spongin   A large, collagenous, connective tissue fiber of sponges.

Spongiome   System of small vesicles or tubules that surrounds the contractile vacuole in the contractile vacuole complex of ciliates.

Spongocoel   Interior cavity of asconoid sponges. Atrium.

Sporocyst   Nonciliated second stage in the life cycle of digeneans. Arises from a miracidium and gives rise to rediae.

Sporosac   Incomplete gonophore (made up of only the gonadal tissue) that remains attached to the polypoid colony.

Sporozoite   Apicomplexa. Infective sporelike stage that results from meiosis of the zygote.

Spur   A long, slender dorsal appendage of doliolids that trails behind the oozooid and bears buds. Cadophore.

Squamous epithelium   Epithelium in which the cells have a flattened shape and fit together like tiles.

Stabilimentum (pl. Stabilimenta)   Conspicuous zigzag lines of silk in the hub of an orb spider's web that are a warning signal to flying birds and can be camouflage for the spider.

Statocyst   A sense organ that can provide orientation to the pull of gravity. Typically composed of a chamber containing concretions (statoliths) in contact with receptor cells.

Stenohaline   Tolerant to only a relatively narrow range of environmental salinities, usually those of the open ocean.

Stereoblastula   A solid blastula, lacking an internal cavity or blastocoel.

Stereogastrula   A solid gastrula, lacking an archenteron cavity.

Sternum (pl. Sterna)   The primary ventral plate of the cuticle of each segment of an arthropod.

Stolon Rootlike extension of the body that interconnects colonial zooids.

Stomodeum The embryonic area of infolded ectoderm that joins the mouth with the endodermal tube (midgut). Mouth cavity and associated structures are typically derived from the stomodeum.

Stone canal Part of the echinoderm water-vascular system that joins the madreporite with the ring canal. Usually, but not always calcified.

Storage excretion When some excretory products, such as uric acid, are stored internally and released only upon the death of the organism.

Strobila (pl. Strobilae) A scyphozoan polyp that buds medusae; or the posterior part of a tapeworm that consists of proglottids.

Strobilation Process by which scyphomedusae arise as buds that are released by transverse fission of the oral end of the scyphistoma.

Stylet A dagger-like structure associated with various systems of different animal groups.

Subchelate Referring to an arthropod appendage in which the terminal piece folds against the subterminal piece, the latter lacking a movable finger.

Subradula organ A cushion-shaped chemosensory structure used by chitons in their nutrition process.

Subumbrella Lower oral surface of a medusa.

Sulcus A longitudinal groove of dinoflagellates that bears the posteriorly directed flagellum.

Suspension feeding Feeding upon particles (plankton and detritus) suspended in water.

Suture The junction between the septum and the wall of a shell.

Syconoid sponge A radially symmetrical sponge that has a body wall folded into radially oriented canals.

Symbiosis An intimate physical relationship between two species, in which at least one of the species is dependent, to various degrees, upon the other. Parasitism, commensalism, and mutualism are the three types of symbiotic relationships.

Symmetrogenic Producing mirror-image daughter cells as a result of fission.

Syncytium Tissue in which nuclei are not separated by cell membranes.

Synkaryon Zygotic nucleus of ciliates.

Tagma (pl. Tagmata) A body region of arthropods (i.e., head, thorax, abdomen).

Tanned When the exocuticle of an arthropod has been further stabilized by the formation of additional cross linkages.

Tapetum A reflective layer within an image-forming eye.

Tarsal organ Cuplike structure situated near the tips of spiders' legs that is probably an olfactory receptor for pheromones.

Tegmen Membranous oral wall of the crinoid disc.

Tegument The nonciliated outer syncytial layer of the body wall of parasitic platyhelminths and acanthocephalans.

Telolecithal Type of egg in which the yolk material is concentrated to one side (vegetal) of the egg.

Telotroch A ring of cilia that encircles the anus at the posterior end of a trochophore larva.

Tentacle Evagination of the body wall surrounding the mouth which aids in the capture and ingestion of food.

Tergum (pl. Terga) The primary, dorsal, exoskeletal plate of each arthropod segment. One of a pair of primary plates covering a barnacle.

Terminal anchor Anchor at the leading end of a burrowing animal.

Terminal cell Tubular flagellated cell that is attached to the inner end of the protonephridial tubule.

Test An encasing or shell-like skeleton, typically covered externally by cytoplasm or living tissue.

Theca (pl. Thecae) The nonliving cuticle around the hydranths of thecate hydroids. Hydrotheca.

Thecate Refers to hydroids with a hydrotheca surrounding the polyp proper.

Thigmotactic Responding to touch or surface contact.

Tiedemann's body One of the interradial outpockets of the ring canal of many echinoderms. Removes unwanted particulates from the water-vascular system.

Tongue bar A downgrowth of pharyngeal tissue that divides a developing gill opening into two slits.

Tornaria Transparent, long-lived, planktotrophic larva of enteropneusts.

Torsion The 90- or 180-degree counterclockwise twist of the gastropod visceral mass with respect to the head and foot.

Toxicyst A vesicular organelle in the pellicle of gymnostome ciliates which discharges long threads with bulbous bases; used for defense or capturing prey.

Trichocyst A bottle-shaped extrusible organelle of the ciliate pellicle.

Trilobite (or **Trilobitid**) **larva** Horseshoe crab larva that superficially resembles trilobites; it is approximately 1 cm long and actively swims about and burrows in sand.

Triploblastic Refers to embryos that possess all three germ layers: Ectoderm, endoderm, and mesoderm.

Trochophore Type of larva found in molluscs, annelids, and other groups in which the larval body is ringed by a girdle of cilia, the prototroch.

Trochus The anterior band of cilia of the divided corona of some rotifers.

Trophi Teeth and other hard parts of the rotifer mastax.

Trophosome Central mass of tissue in the trunk of the pogonophore that is packed with symbiotic bacteria.

Trophozoite Apicomplexa. Feeding stage that occurs when the sporozoite invades the host.

Trophozooid Nutritive or feeding zooid of doliolid urochordates.

Tube tracheae Tracheal system in which the tracheae do not arise as a bundle but are simple branched or unbranched tubes.

Tubicolous Tube-dwelling.

Tubules of Cuvier Eversible toxic or sticky tubules associated with the bases of the respiratory trees of some holothuoid echinoderms.

Tubulus (pl. Tubuli) Sensory papilla on the trunk of some aschelminths.

Tunic Special cuticular covering of the body of ascidians.

Tunicate A urochordate.

Tunicin A kind of cellulose that forms structural fibers in ascidian tunics.

Typhlosole Oligochaeta. A ridge or fold that projects internally from the middorsal wall of the intestine and increases its surface area.

Ultrafiltration The passage of body fluid across a filter that retains proteins and larger particles but allows smaller substances to pass through.

Umbo (pl. Umbos, Umbones) A dorsal protuberance of a bivalve valve that rises above the line of articulation on the valve and is the oldest part of the shell.

Uncinus (pl. Uncini) A minute seta modified into a hook.

Undulating membrane Type of ciliary organelle that is a row of adhering cilia forming a sheet.

Uniramians Five groups of related arthropods that are believed to have had a common terrestrial origin and are so named because of the apparent unbranched nature of the appendages.

Uropods Sixth abdominal appendages of malacostracans.

Vanadocytes Yellowish-green ascidian blood cells that contain high concentrations of vanadium.

Vascular plug Specialized nemertean exchange site across which an ultrafiltrate passes from the blood to the rhynchocoel.

Vegetative nucleus Macronucleus.

Velarium Velum-like structure of cubozoans.

Veliger Molluscan planktotrophic larva that supersedes the trochophore, and in which the shell, foot, velum and other structures make their appearance.

Velum Shelf formed by the margin of the umbrella projected inward which is characteristic of most

hydromedusae; or one of the two ciliated flaps with which a veliger larva swims and feeds.

Vermiform embryo Asexually-produced young of dicyemids that has the same form as the parent; formed within the axial cell of the parent.

Vessel A small tubular blood channel.

Vestibule Preoral chamber.

Vestigial Representing an evolutionary remnant.

Vestimentum The collar-like body region of a vestimentiferan that helps to secrete the animal's tube.

Vibraculum (pl. Vibracula) Bristle-like heterozooid found in some cheilostome bryozoans.

Visceral mass One of three primary parts of the molluscan body; bears the internal organs. The other two parts are the foot and the head.

Viscous drag Friction that results from the tendency of the polar water molecules to stick to each other and to surfaces.

Vitellarium (pl. Vitellaria) Specialized part of the ovary for the production of yolk-filled nurse cells; or nonfeeding barrel-shaped larval stage of some echinoderms.

Viviparous Embryos brooded internally within the female where supplemental nutrition is supplied.

Whorl Any one complete turn (360 degrees) of a coiled molluscan shell.

Zoarium (pl. Zoaria) The form of a bryozoan colony.

Zoea (pl. Zoeae) Penultimate larval stage of many decapod crustaceans. Precedes the postlarval stage.

Zoochlorella (pl. Zoochlorellae) A green algal symbiont of certain animals, especially freshwater sponges and freshwater and marine cnidarians and turbellarians.

Zooecium (pl. Zooecia) The cuticle, or exoskeleton, of a bryozoan zooid.

Zooflagellate A flagellate that has one to many flagella, lacks chloroplasts, and is heterotrophic.

Zooplankton Microscopic animals that are free-swimming or suspended in the water of both oceans and freshwater lakes.

Zooxanthella (pl. Zooxanthellae) A golden-brown alga, usually a dinoflagellate, that is symbiotic with various marine animals, especially cnidarians.

INDEX

Italicized page numbers indicate pages containing figures.